Active Control of Noise and Vibration

SECOND EDITION

VOLUME II

Active Control of Noise and Vibration

SECOND EDITION

VOLUME II

Colin Hansen ▪ Scott Snyder ▪ Xiaojun Qiu
Laura Brooks ▪ Danielle Moreau

CRC Press
Taylor & Francis Group
Boca Raton London New York

CRC Press is an imprint of the
Taylor & Francis Group, an **informa** business

CRC Press
Taylor & Francis Group
6000 Broken Sound Parkway NW, Suite 300
Boca Raton, FL 33487-2742

Printed in the United States of America on acid-free paper
Version Date: 2012924

International Standard Book Number: 978-1-4665-6336-0 (Hardback)

Library of Congress Cataloging-in-Publication Data

Active control of noise and vibration / Colin Hansen ... [et al.]. -- 2nd ed.
p. cm.
Includes bibliographical references and index.
ISBN 978-1-4665-6336-0 (v. 1 : hardcover : alk. paper) -- ISBN 978-1-4665-6339-1 (v. 2 : hardcover : alk. paper)
1. Active noise and vibration control. 2. Damping (Mechanics) 3. Signal processing. I. Hansen, Colin H., 1951-

TK5981.5.A36 2013
620.2'3--dc23 2012029903

Visit the Taylor & Francis Web site at
http://www.taylorandfrancis.com

and the CRC Press Web site at
http://www.crcpress.com

CONTENTS

CHAPTER 8

Active Control of Free-Field Sound Radiation

8.1 INTRODUCTION

In this chapter, the second of the general classes of active noise control problems, the active control of sound radiation into free space, will be examined. As with the previous chapter on controlling sound propagation in ducts, principal consideration will be given to the 'physical' part of the active noise control system; models will be derived that enable prediction of the effects of applying active noise control, both to the acoustic field and to the acoustic radiation characteristics of the noise sources. These models can then be used in analysis of the physical mechanisms of active control, and in system design exercises, where the aim is to find an arrangement of sources and sensors that is capable of providing the maximum level of disturbance attenuation. In practice, this arrangement would be driven by the feedforward control systems described in Chapter 6, or the feedback arrangements described in Chapter 5.

Some of the first attempts to control sound radiation into free space can be found in the work directed at controlling tonal noise radiated by large electrical transformers (Conover, 1956; Ross, 1978; Hesselman, 1978; Angevine, 1981, 1992, 1993; Berge et al., 1987, 1988). This work involved the use of loudspeakers placed near the transformer tank to control the noise radiated to one or more community locations. It was found that if global control was to be achieved in all directions, then it was necessary to use a large array of speakers, almost as large as the transformer tank itself. Use of one or two speakers resulted in reduced noise levels in some directions at the expense of increased noise levels in others. In addition, the angular spread of the directions of reduced levels was generally quite narrow. Although not stated by the authors, the physical reason for this behaviour is that to achieve global noise reduction using acoustic control sources, it is necessary for the sources to change the radiation impedance 'seen' by the transformer tank. This can only be done by using a large array of control sources. If only one or two small sources are used, areas of reduced noise level are achieved solely by local destructive interference effects at the expense of other areas of increased level where constructive interference takes place. Thus, in this latter case, the radiation impedance 'seen' by the transformer (and hence its radiated sound power) is barely changed, and as a result the overall radiated sound power of the transformer plus control sources is generally larger than the sound power radiated by the transformer itself, even though there will be some locations (particularly error sensor locations) where the sound level will be reduced.

In the past, a number of researchers have investigated active control systems to control the sound power radiated by vibrating surfaces. Knyazev and Tartakovskii (1967) were the first to investigate the control of sound radiation by using control forces on the vibrating structure that was responsible for the original sound. In 1988, Deffayet and Nelson analysed the active control of sound radiated by a finite rectangular plate using acoustic control sources. Hansen et al. (1989) compared the relative effectiveness of control forces and acoustic sources for controlling sound radiated by a rectangular plate. Analytical modelling of the active control of sound radiated by a rectangular plates has been undertaken by Walker (1976), Fuller (1990), Pan et al. (1992) and Wang and Fuller (1992).

More recently, research efforts have focused upon the control of orthotropic plates (Meirovitch and Thangjitham, 1990a), the control of fluid loaded plates (Meirovitch and Thangjitham, 1990b; Gu and Fuller, 1993), combined active and passive control (Koshigoe and Murdock, 1993), broadband disturbances (Baumann et al., 1992), the use of piezoceramic crystals to provide the control forces (Hansen et al., 1989; Wang et al., 1990; Fuller et al., 1991a, 1991b; Hansen et al., 1991; Metcalf et al., 1992) and shaped PVDF (poly vinyl diflouride) sensors instead of microphones to provide the required controller error signal (Lee and Moon, 1990; Clark and Fuller, 1992a,b,c; Elliott and Johnson, 1993; Naghshineh and Koopmann, 1993; Snyder and Tanaka, 1993b, 1994a,b). Use of vibration error sensors to minimise sound radiation is much more convenient than the use of microphones mounted in the far-field of a noise radiating surface, and the problem of shaping vibration sensors to provide such a signal is an interesting and complex problem, which is why so much effort has been devoted to research in this area.

Investigation of the physical mechanisms (Hansen et al., 1989; Snyder and Hansen, 1991; Snyder, 1991; Elliott et al., 1991a; Elliott and Johnson, 1992) involved in controlling sound radiation from a simple vibrating surface has provided an understanding of the complexity of the problem and has also resulted in the determination of the influence of geometric and structural/acoustic variables on the maximum achievable reduction in sound power. This work has led to the formulation of strategies for the optimum design of multi-channel systems for the simultaneous control of a number of sources and error sensors (Hansen et al., 1990; Thi et al., 1991; Snyder and Hansen, 1991).

This chapter will begin with an examination of the active control of harmonic sound radiation from a set of monopole sources in free space by the introduction of a second set of monopole sources. While this problem is somewhat idealised, it will provide a variety of physical and methodological insights, which will prove valuable in examining ‘more realistic’ problems. Problems of controlling sound radiation from vibrating structures will then be discussed, including the use of vibration sources to modify the velocity distribution of the source for the purpose of attenuating acoustic radiation from the surface. It will be shown that while the control of sound radiation using acoustic and vibration control sources may appear very different on the surface, the physical mechanisms behind the attenuation are actually closely related.

Once control of harmonic sound radiation from idealised monopole sources, and more realistic structure models, has been studied, attention will be given to the control of non-harmonic acoustic radiation, random noise and impulse excitation. It will be shown that these problems are related to the harmonic excitation cases studied earlier in the chapter, both in terms of the control mechanism and optimal performance considerations. The chapter will conclude with a brief discussion of the relationship between feedback control of a vibrating structure, discussed in more detail in Chapter 11, and feedback control of acoustic radiation from that structure.

8.2 CONTROL OF HARMONIC SOUND PRESSURE AT A POINT

Possibly the most basic active noise control problem that could be envisaged is the minimisation, or ‘cancellation’, at some point of harmonic sound, radiated by one acoustic monopole (primary) source, through the introduction of a second (control) monopole source, the geometry of which is shown in Figure 8.1. As such, an examination of this problem provides an ideal introduction to the problem of free-field active noise control.

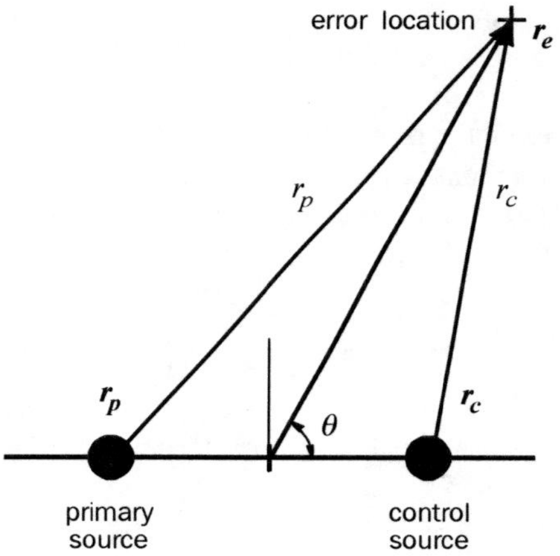

Figure 8.1 Single monopole primary source/single monopole control source system arrangement.

It was shown in Chapter 2 that the sound pressure at some location $\boldsymbol{r}_e$ in space due to the harmonic operation of an acoustic monopole source located at a position $\boldsymbol{r}_q$ is:

$$p(\boldsymbol{r}_e) = \frac{j\omega\rho_0 q \,\mathrm{e}^{-\mathrm{j}kr}}{4\pi r} \tag{8.2.1}$$

where $r = |\boldsymbol{r}_e - \boldsymbol{r}_q|$; q is the volume velocity, or strength, of the monopole source; k is the acoustic wavenumber at the frequency ω of interest; and positive harmonic time dependence of the form, $\mathrm{e}^{\mathrm{j}\omega t}$ is implicit in the equation (this is the case throughout this book). It will be assumed throughout this chapter that the acoustic sources are all constant volume velocity sources (the volume velocity of the source does not change, regardless of any changes or additions to the environment in which it is placed). Physically, this translates into the ideal that the internal impedance of the source is infinite. However, many practical sources and environments, such as a loudspeaker (with a small airtight backing enclosure) operating in air, exhibit characteristics that are sufficiently close to this. That is, for these sources, their mechanical impedance is several orders of magnitude greater than the radiation impedance that they 'see' looking into the acoustic environment, so that the results obtained theoretically using this idealisation accurately represent real-world phenomena.

As the system being considered is linear, the concept of superposition is valid, so that the sound pressure at some location $\boldsymbol{r}_e$ during the operation of both the primary and control sources is simply the sum of the individual sound pressures:

$$p(\boldsymbol{r}_e) = p_p(\boldsymbol{r}_e) + p_c(\boldsymbol{r}_e) \tag{8.2.2}$$

where the subscripts p and c denote primary and control sources respectively. Using Equation (8.2.1), this can be rewritten as:

$$p(\boldsymbol{r_e}) = \frac{j\omega\rho_0 q_p \mathrm{e}^{-jkr_p}}{4\pi r_p} + \frac{j\omega\rho_0 q_c \mathrm{e}^{-jkr_c}}{4\pi r_c} \tag{8.2.3}$$

where r_p and r_c are the distances between the primary and control sources and the point in space of interest, $r_p = |\boldsymbol{r}_e - \boldsymbol{r}_p|$, $r_c = |\boldsymbol{r}_e - \boldsymbol{r}_c|$.

The object of the exercise here is, given a primary source volume velocity q_p, to find the volume velocity q_c that will minimise the acoustic pressure at some location, $\boldsymbol{r}_e$. For a single control source and single 'error' location, the minimised acoustic pressure will, in fact, be zero. This objective is easily accomplished using Equation (8.2.3). For $p(\boldsymbol{r}_e) = 0$:

$$\frac{j\omega\rho_0 q_p \mathrm{e}^{-jkr_p}}{4\pi r_p} = -\frac{j\omega\rho_0 q_c \mathrm{e}^{-jkr_c}}{4\pi r_c} \tag{8.2.4}$$

or

$$q_{c,p} = -\frac{r_c}{r_p} q_p \mathrm{e}^{-jk(r_p - r_c)} \tag{8.2.5}$$

where the subscripts c,p are used to denote a control source volume velocity that has been derived with the aim of minimising the acoustic pressure at some discrete location. Equation (8.2.5) describes the relationship between the primary and control source volume velocities that will result in the minimisation of the harmonic sound pressure at some point in space. It states that the control source volume velocity, relative to the primary source volume velocity, must be proportional in amplitude to the relative distances to the error location of interest (r_c/r_p), and produce a pressure signal that is 180° out of phase with the primary source generated pressure when it arrives at the error location $(-\mathrm{e}^{-jk(r_p - r_c)})$.

For this extremely simple system, it is quite straightforward to derive the relationship between the primary and control sources that will result in the minimisation of acoustic pressure at some point in space. It will not, however, always be so easy to simply write down the solution to the problem, and so it is useful to re-derive the result of Equation (8.2.5) in a way that will be more easily generalised later in this chapter. Therefore, consider again the problem of minimising the acoustic pressure amplitude at some point in space. This is equivalent to minimising the squared pressure amplitude:

$$|p(\boldsymbol{r})|^2 = p^*(\boldsymbol{r})p(\boldsymbol{r}) \tag{8.2.6}$$

where * denotes the complex conjugate. During the operation of both the primary and control sources, the pressure at any point in space is as defined in Equation (8.2.3). Substituting this into Equation (8.2.6) gives:

$$|p(\boldsymbol{r_e})|^2 = \left(\frac{j\omega\rho_0 q_p \mathrm{e}^{-jkr_p}}{4\pi r_p} + \frac{j\omega\rho_0 q_c \mathrm{e}^{-jkr_c}}{4\pi r_c}\right)^* \left(\frac{j\omega\rho_0 q_p \mathrm{e}^{-jkr_p}}{4\pi r_p} + \frac{j\omega\rho_0 q_c \mathrm{e}^{-jkr_c}}{4\pi r_c}\right)$$
$$= q_c^* a q_c + q_c^* b + b^* q_c + c \tag{8.2.7a,b}$$

where

$$a = \left(\frac{\omega\rho_0}{4\pi r_c}\right)^2; \; b = \left(\frac{\omega\rho_0}{4\pi}\right)^2 \frac{1}{r_c r_p} q_p \mathrm{e}^{-\mathrm{j}k(r_p - r_c)}; \; c = \left(\frac{\omega\rho_0}{4\pi r_p}\right)^2 \tag{8.2.8a–c}$$

Equation (8.2.7) shows that the squared acoustic pressure amplitude at any location in space is a real quadratic function of the complex control source volume velocity q_c. This is perhaps more easily seen by rewriting the equation in terms of its real and imaginary parts as follows:

$$|p(\boldsymbol{r})|^2 = aq_{c\mathrm{R}}^2 + 2b_R q_{c\mathrm{R}} + aq_{c\mathrm{I}}^2 + 2b_\mathrm{I} q_{c\mathrm{I}} + c \tag{8.2.9}$$

where the subscripts I and R denote the real and imaginary components respectively. Thus, a plot of the squared acoustic pressure amplitude as a function of the real and imaginary components of the control source volume velocity forms a 'bowl', as shown in Figure 8.2. The bottom of the bowl defines the optimum control source volume velocity that will minimise the error criterion, the squared acoustic pressure amplitude. This value of volume velocity can be found by differentiating Equation (8.2.9) with respect to its real and imaginary components, and setting the result equal to zero. Doing this gives:

$$\frac{\partial |p(\boldsymbol{r})|^2}{\partial q_{c\mathrm{R}}} = 2aq_{c\mathrm{R}} + 2b_\mathrm{R} = 0 \tag{8.2.10a,b}$$

and

$$\frac{\partial |p(\boldsymbol{r})|^2}{\partial q_{c\mathrm{I}}} = 2aq_{c\mathrm{I}} + 2b_\mathrm{I} = 0 \tag{8.2.11a,b}$$

Multiplying the result of Equation (8.2.11) by the imaginary number j and putting it together with the result of Equation (8.2.10), produces the following expression for the optimum control source volume velocity (Nelson et al., 1985, 1987):

$$q_{c,\mathrm{p}} = -a^{-1}b \tag{8.2.12}$$

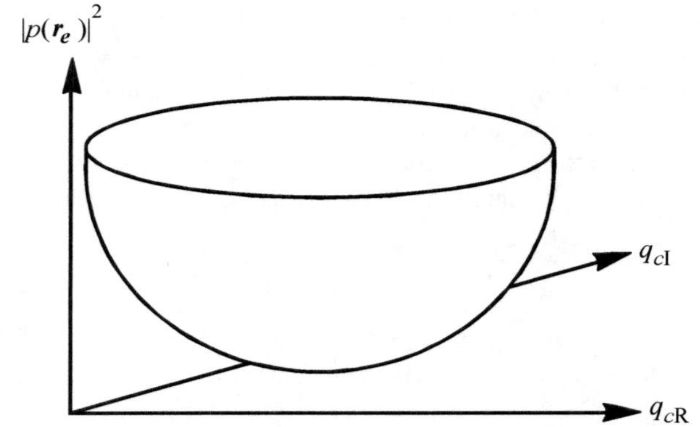

Figure 8.2 Typical plot of squared acoustic pressure amplitude as a function of the real and imaginary parts of control source volume velocity.

For the particular case of interest here, if the definitions of Equation (8.2.8) are substituted into Equation (8.2.12), the optimum control source volume velocity is found to be:

$$q_{c,p} = -\frac{r_c}{r_p} q_p \mathrm{e}^{-jk(r_p - r_c)} \tag{8.2.13}$$

This result is, of course, identical to that derived in Equation (8.2.5). The difference is that in this instance it was derived using quadratic optimisation techniques, which will be the methodology adopted throughout this chapter when considering the minimisation of acoustic pressure and power for more complex problems.

There are two points that should be briefly noted here while on the topic of quadratic optimisation. First, the result of Equation (8.2.12) is unique only if the value of a is non-zero, and will define a minimum only if the value of a is positive. In other words, for quadratic optimisation to produce the desired result for an optimum control source volume velocity, a must be positive definite. From the definition of a given in Equation (8.2.8), it is obvious that this property is true in this instance. Second, if the result of Equation (8.2.12) is substituted back into Equation (8.2.7), the minimum squared acoustic pressure amplitude at the error sensing location is found to be:

$$|p(\boldsymbol{r})|^2_{\min} = c - b^* a^{-1} b = c + b^* q_{c,p} \tag{8.2.14a,b}$$

If the terms in Equation (8.2.14) are expanded using the definitions of a, b and c given in Equation (8.2.7) for this problem, the minimum acoustic pressure will be found to be equal to zero.

It does not automatically follow that minimising the sound pressure at a single point in space will minimise, or even reduce, the overall levels of sound pressure in the combined radiated field. It is well known that two coherent sound sources will interfere to produce (in a linear system) a pattern of constructive and destructive interference 'fringes', as shown in Figure 8.3. The question arises: how close must the control source be to the primary source if, when minimising the acoustic pressure at some point in space, no fringes are to appear? In other words, how close must the control source be to the primary source before minimising the sound pressure at some particular point in space will guarantee a pressure reduction at every point in the far-field? The answer to this question can be found by making some far-field approximations to the result of Equation (8.2.5).

Consider the primary/control source arrangement shown in Figure 8.4, where $\boldsymbol{r}_f$ is some location in the far-field such that $r_c/r_p \approx 1.0$. If the separation distance between the primary and control sources is d, then the optimum relationship between the control source and primary source volume velocities for minimising the acoustic pressure amplitude at this point, stated in its general form in Equation (8.2.5), can be approximated as:

$$q_{c,p} \approx -q_p \mathrm{e}^{-jk(d \cos\theta_f)} \tag{8.2.15}$$

Substituting this relationship into Equation (8.2.3) gives the resultant sound pressure at any point $\boldsymbol{r}$ in space during the operation of both the primary and control sources as:

$$p(\boldsymbol{r}) = \frac{j\omega\rho_0}{4\pi} q_p \left(\frac{\mathrm{e}^{-jkr_p}}{r_p} - \frac{\mathrm{e}^{-jk(r_c + d\cos\theta_f)}}{r_c} \right) \tag{8.2.16}$$

or

$$p(\boldsymbol{r}) = \frac{j\omega\rho_0}{4\pi} q_p \mathrm{e}^{-\mathrm{j}kr_p} \left(\frac{1}{r_p} - \frac{\mathrm{e}^{-\mathrm{j}k(r_c - r_p + d\cos\theta_f)}}{r_c} \right) \quad (8.2.17)$$

Figure 8.3 Interference pattern of two coherent sources.

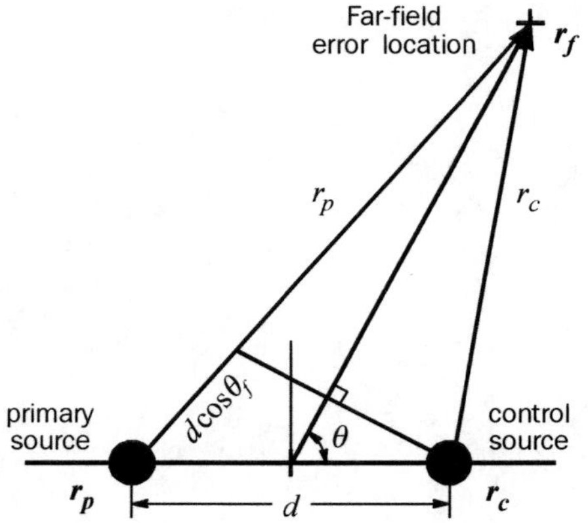

Figure 8.4 Single monopole primary source/single monopole control source system arrangement with far-field approximations.

Confining consideration to any point in the far-field, and again invoking the far-field assumptions already outlined, this can be written as:

$$p(\boldsymbol{r}) \approx \frac{j\omega\rho_0}{4\pi r_f} q_p \mathrm{e}^{-\mathrm{j}kr_p} \left(1 - \mathrm{e}^{-\mathrm{j}kd(\cos\theta_f - \cos\theta)}\right) \quad (8.2.18)$$

If the acoustic pressure amplitude is reduced as a result of the operation of the control source, then,

$$\frac{|p(\boldsymbol{r})|^2}{|p_p(\boldsymbol{r})|^2} < 1 \tag{8.2.19}$$

Substituting Equations (8.2.1) and (8.2.18) into Equation (8.2.19) yields the criterion:

$$2\left(1 - \cos\left[kd\,\left\{\cos\theta_f - \cos\theta\right\}\right]\right) < 1 \tag{8.2.20}$$

As the maximum value of the expression in the curly brackets is equal to 2.0, this criterion reduces to (Nelson and Elliott, 1992):

$$1 - \cos 2kd < 1/2 \tag{8.2.21}$$

Therefore, for the sound pressure to be reduced at all locations in the far-field as a result of minimising the sound pressure at some arbitrary error location, $kd < \pi/6$, or, in terms of wavelength of the frequency of sound of interest, $d < \lambda/12$.

The extremely strict separation criterion of Equation (8.2.21) corresponds to a worst case situation, where $(\cos\theta_f - \cos\theta) = 2$, or $\theta_f = 0$ and $\theta = \pi$. The criterion can be relaxed a bit with some *a priori* knowledge of a more judicious positioning of the error location. Consider, for example, the case where $\theta_f = \pi/2$, reducing the maximum value of the expression in the curly brackets in Equation (8.2.20) to 1.0, at $\theta = \pi$. For this case, the maximum allowable separation distance for attenuation of the sound field in all directions is increased to $d = \lambda/6$. This fact is borne out in the plots of the residual sound field for the minimisation of acoustic pressure at $\theta_f = 0$ and π, shown in Figure 8.5.

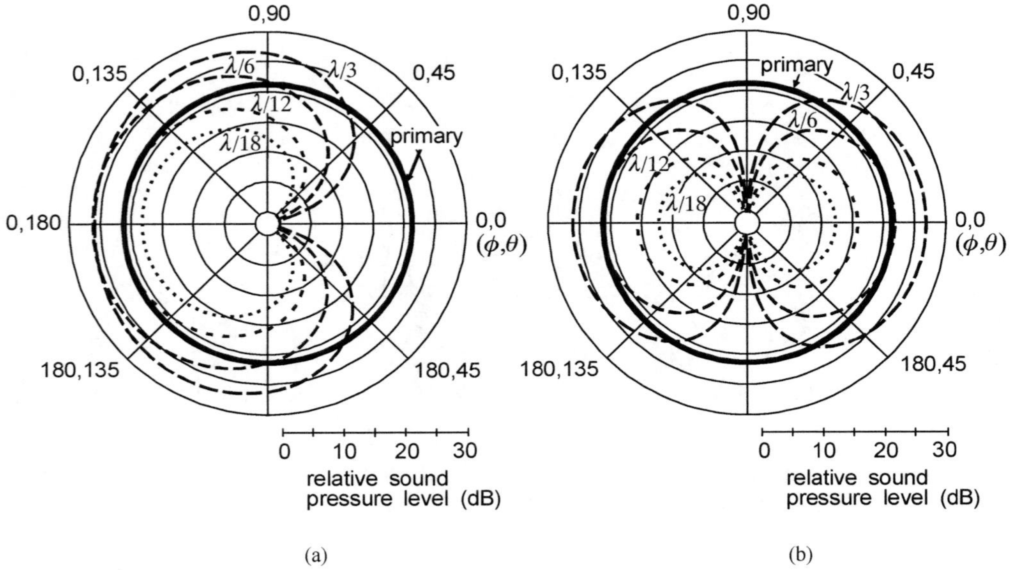

Figure 8.5 Residual sound field for control of one monopole source by another. (a) $\theta_f = 0$. (b) $\theta_f = \pi$.

The lessons from this exercise are: the control source must be located in close proximity to the primary source for overall sound attenuation to be achieved; and the error sensor location has a significant effect upon the overall levels of sound attenuation that are achieved by an active noise control system. The problem of optimising the error sensor location will be considered three sections from now, following consideration of the problem of establishing a basis for evaluating how much attenuation of the total radiated acoustic power is physically possible.

8.3 MINIMUM ACOUSTIC POWER OUTPUT OF TWO FREE-FIELD MONOPOLE SOURCES

In the previous section, the problem of minimising at some point in space, the acoustic pressure radiated from one harmonically-oscillating monopole (primary) source, by the introduction of a second (control) source, radiating at the same frequency, was considered. It was seen, however, that the result of such an exercise is not necessarily a reduction in the amplitude of the total radiated sound field, or total radiated acoustic power. To be able to assess the quality of the control effect provided by minimising the sound pressure at a discrete error location, two questions arise. How much attenuation in the total radiated acoustic power is actually possible, and what is the relationship between the primary and control sources that will achieve this result?

To answer these questions, the problem of using one harmonically radiating point source to control the output of another is again considered, but this time using as an error criterion, minimisation of the total radiated acoustic power. It was seen in Chapter 2 that the total radiated acoustic power is equal to the integration of the real acoustic intensity radiated out of a sphere encompassing the sound sources. Using the geometry of Figure 8.6, the acoustic power output W of the sources is:

$$W = \int_0^{2\pi}\int_0^{\pi/2} \frac{|p(\boldsymbol{r})|^2}{2\rho_0 c_0}\, r^2 \sin\theta \; d\theta \; d\varphi = 2\pi r^2 \int_0^{\pi} \frac{|p(\boldsymbol{r})|^2}{2\rho_0 c_0} \sin\theta \; d\theta \tag{8.3.1a,b}$$

where r is the radius of the enclosing sphere, which is in the far-field of the sources ($R >> d$). The total acoustic pressure amplitude squared at any point in the space during the operation of both the primary and control sources was stated in the previous section, in Equation (8.2.7).

Substituting Equation (8.7) into Equation (8.3.1), the total acoustic power output of the primary source/control source combination can be written as:

$$W_{tot} = q_c^* a_w q_c + q_c^* b_w + b_w^* q_c + c_w \tag{8.3.2}$$

where

$$a_w = \frac{\pi r^2}{\rho_0 c_0} \int_0^{\pi} \left(\frac{\omega \rho_0}{4\pi r_c} \right)^2 \sin\theta d\theta \tag{8.3.3}$$

$$b_w = \frac{\pi r^2}{\rho_0 c_0} \int_0^{\pi} \left(\frac{\omega \rho_0}{4\pi} \right)^2 \frac{1}{r_c r_p} q_p e^{-jk(r_p - r_c)} \sin\theta d\theta \tag{8.3.4}$$

$$c_w = \frac{\pi r^2}{\rho_0 c_0}\int_0^{\pi}\left(\frac{\omega\rho_0}{4\pi r_p}\right)^2 \sin\theta\,\mathrm{d}\theta \tag{8.3.5}$$

Thus, as with the squared acoustic pressure amplitude at a point, the total radiated acoustic power is a real quadratic function of the control source volume velocity. Using the result of Equation (8.2.12), the optimum control source volume velocity can be written directly as:

$$q_{c,w} = -a_w^{-1} b_w \tag{8.3.6}$$

where the subscript *c,w* is used to denote a control source volume velocity that has been derived with the aim of minimising the total radiated acoustic power.

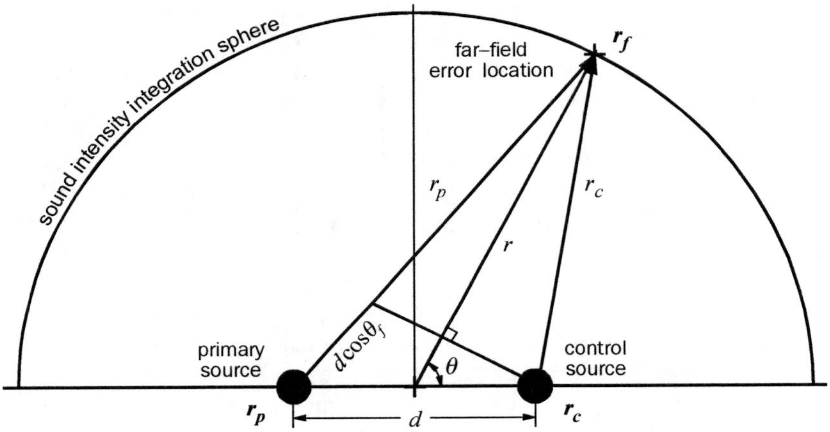

Figure 8.6 Geometry used to determine sound power by far-field integration of acoustic intensity.

Thus, to evaluate the optimum control source volume velocity with regard to minimising the total acoustic power output of the system, it is necessary to solve two integrals, and a further third integral, if the minimum power output is required (see Equation (8.3.15)). For the case of two monopole acoustic sources being considered here, the integrals of Equation (8.3.3) can be solved analytically, thereby greatly simplifying the final result. To do this, note that b_w can be rewritten using the far-field approximation of Equation (8.2.15) as:

$$b_w \approx \frac{\pi r^2}{\rho_0 c_0}\int_0^{\pi}\left(\frac{\omega\rho_0}{4\pi}\right)^2 \frac{1}{r^2} q_p \mathrm{e}^{-\mathrm{j}k(d\cos\theta_f)} \sin\theta\,\mathrm{d}\theta \tag{8.3.7}$$

and that:

$$\int_0^{\pi}\cos(kd\cos\theta)\sin\theta\,\mathrm{d}\theta = \frac{2\sin(kd)}{kd} = 2\ \mathrm{sinc}\, kd$$

$$\int_0^{\pi}\sin(kd\cos\theta)\sin\theta\, d\theta = 0 \tag{8.3.8a–c}$$

Using these results, the terms in Equations (8.3.3) to (8.3.5) can be evaluated as:

$$a_w = \frac{\omega k \rho_0}{8\pi} = \frac{\omega^2 \rho_0}{8\pi c_0} \tag{8.3.9a,b}$$

$$b_w = \frac{\omega k \rho_0}{8\pi} q_p \operatorname{sinc} kd \tag{8.3.10}$$

$$c_w = \frac{\omega k \rho_0}{8\pi} \tag{8.3.11}$$

resulting in the optimum control source volume velocity of Equation (8.3.6) being expressed as (Nelson and Elliott, 1986):

$$q_{c,w} = -q_p \operatorname{sinc} kd \tag{8.3.12}$$

The minimum acoustic power output of the primary source/control source pair can be found by substituting the result of Equation (8.3.12) back into Equation (8.3.2), producing:

$$W_{\min} = c_w - b_w^* a_w^{-1} b_w = c_w + b_w^* q_{c,w} = \frac{\omega^2 \rho_0}{8\pi c_0} |q_p|^2 (1 - \operatorname{sinc}^2 kd) \tag{8.3.13a–c}$$

It will be shown shortly (in Equation (8.3.24)) that the acoustic power radiated by the primary source operating alone is:

$$W_{p,unc} = \frac{\omega^2 \rho_0}{8\pi c_0} |q_p|^2 \tag{8.3.14}$$

Therefore, the reduction in total radiated acoustic power, which is possible by introducing the control source, is (Nelson and Elliott, 1986):

$$\frac{W_{\min}}{W_p} = 1 - \operatorname{sinc}^2 kd \tag{8.3.15}$$

Before discussing the significance of the results obtained so far, it will be worthwhile to note that the control source volume velocity definition derived in Equation (8.3.12) using far-field intensity integration can be arrived at in a slightly different manner using a ‘near-field’ analysis technique (Levine, 1980a, 1980b, 1980c; Nelson et al., 1987). It will be worthwhile re-deriving the previously obtained results using this methodology, as it will provide a basis for the work on minimising the acoustic power output from more than one primary source by using more than one control source, which is considered later on in this chapter. This methodology is more akin to that used in the previous chapter to formulate the optimum control source volume velocity characteristics for minimising the acoustic power propagation in air handling ducts, in that the acoustic power output at the source location is used as the basis of examination, rather than integrating to obtain the acoustic power in the far-field.

It was stated in Chapter 2 that the real part of the acoustic power output of a source can be calculated by integrating the real part of the acoustic intensity travelling out of a surface enclosing the source:

$$W = \int_S \boldsymbol{I}\cdot\boldsymbol{n}\,\mathrm{d}S \tag{8.3.16}$$

where $\boldsymbol{n}$ is the outward normal vector from the surface S enclosing the source. Performing this integration over a sphere, which is enclosing the monopole source gives:

$$W = \int_S \frac{1}{2}\mathrm{Re}\left\{\bar{p}^*(\boldsymbol{r})\,\bar{u}(\boldsymbol{r})\right\}\cdot n\,\mathrm{d}S \tag{8.3.17}$$

or

$$W = \frac{1}{2}\mathrm{Re}\left\{\bar{p}^*(\boldsymbol{r})\bar{q}\right\} \tag{8.3.18}$$

It should be noted that $\bar{p}$ and $\bar{q}$ in Equation (8.3.18) are the acoustic pressure and volume velocity complex amplitudes respectively.

From Equation (8.3.18) it can be deduced that, with a knowledge of the source volume velocity, calculation of the acoustic pressure at the monopole source location will enable the calculation of the acoustic power radiated by the source. The acoustic pressure due to the operation of the source at some location $\boldsymbol{r}$ relative to it, defined in Equation (8.2.1), can be expressed in the form of an impedance as follows:

$$p(\boldsymbol{r}) = q\,z(\boldsymbol{r}) \tag{8.3.19}$$

where z is a complex impedance given by:

$$z(\boldsymbol{r}) = j\omega\rho_0\frac{\mathrm{e}^{-jkr}}{4\pi r} \tag{8.3.20}$$

where r is the distance between the source location $\boldsymbol{r}_q$ and the point of interest $\boldsymbol{r}$ given by:

$$r = |\boldsymbol{r}-\boldsymbol{r}_q| \tag{8.3.21}$$

The impedance can be expressed in terms of real and imaginary parts as:

$$z(\boldsymbol{r}) = \frac{\omega k\rho_0}{4\pi}\left(\frac{\sin kr}{kr} + \mathrm{j}\frac{\cos kr}{kr}\right) \tag{8.3.22}$$

Note that as $r\to 0$, the impedance $z(\boldsymbol{r})$ becomes infinite. However, on closer inspection it can be seen that this is due only to the imaginary part becoming infinite, as:

$$\lim_{r\to 0}\frac{\sin kr}{kr} = 1 \tag{8.3.23}$$

Therefore, substituting Equation (8.3.23) into Equation (8.3.22) and then into Equation (8.3.18) produces the expression for the acoustic power radiated by the source:

$$W = \frac{1}{2}|\bar{q}|^2 \,\mathrm{Re}\{z(\boldsymbol{r})\} = \frac{1}{2}|\bar{q}|^2 z_0 \tag{8.3.24a,b}$$

where

$$z_0 = \frac{\omega k \rho_0}{4\pi} \tag{8.3.25}$$

Note that this is the result that was 'anticipated' prior to Equation (8.3.14).

Having now obtained a set of expressions for calculating the power output of a monopole source based upon the characteristics of the acoustic environment at its location, the power output of the monopole primary and control sources when they are operating together can now be calculated. Consider first the primary monopole source. From Equation (8.3.18), the sound power radiated by the primary monopole source in the presence of the sound field produced by the control monopole source will be:

$$W_p = \frac{1}{2}\mathrm{Re}\left\{\left(\bar{p}_p(\boldsymbol{r}_p) + \bar{p}_c(\boldsymbol{r}_p)\right)^* \bar{q}_p\right\} \tag{8.3.26}$$

where $\bar{p}_p(\boldsymbol{r}_p)$ and $\bar{p}_c(\boldsymbol{r}_p)$ are the acoustic pressure amplitudes at the location of the primary monopole source $\boldsymbol{r}_p$ due to the operation of the primary and control sources respectively. Using Equation (8.3.19) this can be re-expressed in terms of impedance as:

$$W_p = \frac{1}{2}\mathrm{Re}\left\{\left(\bar{q}_p z_{p/p} + \bar{q}_p z_{p/c}\right)^* \bar{q}_p\right\} \tag{8.3.27}$$

where $z_{p/p}$ is the component of the primary source impedance due to self-operation, given by:

$$z_{p/p} = \frac{p_p(\boldsymbol{r}_p)}{q_p} \tag{8.3.28}$$

and $z_{p/c}$ is the component of the primary source impedance due to the operation of the control source, given by:

$$z_{p/c} = \frac{p_c(\boldsymbol{r}_p)}{q_p} \tag{8.3.29}$$

Using the expression for the acoustic pressure at some location $\boldsymbol{r}$ relative to a monopole source, Equation (8.2.1), and the definition of z_0 given in Equation (8.3.25), Equation (8.3.27) can be simplified to:

$$\begin{aligned} W_p &= \frac{1}{2}|\bar{q}_p|^2 \,\mathrm{Re}\left\{\left(z_{p/p} + z_{p/c}\right)^*\right\} \\ &= \frac{1}{2}|\bar{q}_p|^2 z_0 \mathrm{Re}\left\{1 + \frac{\bar{q}_c^*}{\bar{q}_p^*}\frac{\sin kd}{kd}\right\} = \frac{1}{2}|\bar{q}_p|^2 z_0 \mathrm{Re}\left\{1 + \frac{\bar{q}_c^*}{\bar{q}_p^*}\,\mathrm{sinc}\, kd\right\} \end{aligned} \tag{8.3.30a–c}$$

Using a similar analysis, the acoustic power output of the control source operating in the presence of the primary source sound field can be evaluated as:

$$W_c = \frac{1}{2}|\bar{q}_c|^2 \mathrm{Re}\left\{(z_{c/c} + z_{c/p})^*\right\} = \frac{1}{2}|\bar{q}_c|^2 z_0 \mathrm{Re}\left\{1 + \frac{\bar{q}_p^*}{\bar{q}_c^*} \operatorname{sinc} kd\right\} \tag{8.3.31a,b}$$

Combining Equations (8.3.30) and (8.3.31), the total sound power output of the actively controlled system is:

$$W_{tot} = W_p + W_c = \frac{1}{2}\Big(|\bar{q}_c|^2 \mathrm{Re}\left\{z_{c/c}^*\right\} + \bar{q}_c^* \mathrm{Re}\left\{z_{c/p}^*\right\}\bar{q}_p + \bar{q}_p^* \mathrm{Re}\left\{z_{p/c}^*\right\}\bar{q}_c + |\bar{q}_p|^2 \mathrm{Re}\left\{z_{p/p}^*\right\} \Big) \tag{8.3.32a,b}$$

Noting that from reciprocity $\mathrm{Re}\{z_{p/c}\} = \mathrm{Re}\{z_{c/p}\}$, Equation (8.3.32) can be re-expressed in a form more amenable to solution by the quadratic optimisation techniques used earlier:

$$W_{tot} = \bar{q}_c^* a_w \bar{q}_c + \bar{q}_c^* b_w + b_w^* \bar{q}_c + c_w \tag{8.3.33}$$

where

$$a_w = \frac{1}{2}\mathrm{Re}\left\{z_{c/c}^*\right\} = \frac{\omega k \rho_0}{8\pi} \tag{8.3.34a,b}$$

$$b_w = \frac{1}{2}\mathrm{Re}\left\{z_{c/p}^*\right\}\bar{q}_p = \frac{\omega k \rho_0}{8\pi}\bar{q}_p \operatorname{sinc} kd \tag{8.3.35a,b}$$

$$c_w = \frac{1}{2}|\bar{q}_p|^2 \mathrm{Re}\left\{z_{p/p}^*\right\} = |\bar{q}_p|^2 \frac{\omega k \rho_0}{8\pi} \tag{8.3.36a,b}$$

These terms, however, show that the expression for the total radiated acoustic power given in Equation (8.3.33), formulated using the near-field technique, is exactly the same as that obtained using far-field integration, stated in Equation (8.3.2) using the terms defined in Equations (8.3.9) to (8.3.11) (as well it should be!). Using quadratic optimisation techniques on Equation (8.3.33) to determine the optimum control source volume velocity amplitude that will minimise the total radiated acoustic power will produce a result identical to that stated in Equation (8.3.12):

$$\bar{q}_{c,w} = -a_w^{-1} b_w \tag{8.3.37}$$

Therefore, whether considering the minimisation of the real part of the source power output (near-field analysis) or total radiated acoustic power (far-field intensity integration), the optimum control source volume velocity is the same.

There are two ways to derive the necessary quadratic equation for acoustic power output: by considering some far-field measure of acoustic power output (integration of acoustic intensity in the far-field); or by directly considering the real acoustic power output at the source location. The merits of using either method of formulation will vary from case to case; however, the end result should be the same no matter what method is used.

There are several points of interest to note concerning the optimum control source volume velocity, stated in Equation (8.3.12), and the resultant minimum acoustic power output, stated in Equation (8.3.13). The first is that the amplitude of the optimum control source volume velocity is always *less* than the primary source volume velocity. This can be contrasted to the problem of nulling the acoustic pressure at a single point, considered in the previous section, where it was seen that the control source volume velocity amplitude was greater than the primary source volume velocity amplitude for points closer to the primary source than the control source, and less than the primary source volume velocity amplitude for points closer to the control source. Also, note that the control source volume velocity is constrained to be exactly 180° out of phase (for separation distances of odd intervals of half wavelengths) or in-phase (for separation distances of even intervals of half wavelengths). This is in contrast to the problem of nulling the acoustic pressure at a single point, where the only constraint on phase is that the primary and control source pressures must be out of phase at the error sensing location. (Both of these facts have obvious implications for optimum error sensor placement, discussed later in this chapter.) These characteristics of the control source volume velocity can be seen in Figure 8.7, which illustrates the optimum control source volume velocity as a function of separation distance *d*.

The next point to note is that the maximum achievable levels of acoustic power attenuation drop off rapidly as the separation distance between the control source and primary source increases, as shown in Figure 8.8. These levels represent the absolute best result that can be achieved, no matter how good the electronics or how large the number of error sensors. The appearance of the integrations in Equation (8.3.3) to (8.3.5) used in formulating the result is equivalent to using an infinite number of error sensors, although the maximum levels of acoustic power attenuation can be achieved using a number which is greatly reduced from this in practice, in fact only one for this problem, with some judicious placement. This point will be considered later in the chapter. So, for example, if 10 dB of acoustic power attenuation is desired, the control source must be placed within approximately 1/10 wavelength from the primary source as a prerequisite, before considering the practical problems of error sensor placement and electronic control system design.

One question that may arise out of the previously derived results is whether a reduction in the total radiated sound power via the control source volume velocity of Equation (8.3.12) necessarily corresponds to a reduction in the acoustic pressure amplitude at all locations in the far-field of the control source/primary source location. To answer this question, the control source volume velocity defined in Equation (8.3.12) can be substituted into the expression for far-field acoustic pressure derived in the previous section, Equation (8.2.18), producing:

$$\bar{p}(\boldsymbol{r}) = \bar{p}_p(\boldsymbol{r})\left(1 - \operatorname{sinc} kd\, \mathrm{e}^{\mathrm{j}k(d\cos\theta)}\right) \tag{8.3.38}$$

where $\bar{p}_p(\boldsymbol{r})$ is the complex acoustic pressure amplitude at a point $\boldsymbol{r}$ in the far-field due only to the primary source.

Therefore, the ratio of the squared acoustic pressure amplitudes at any point in the far-field before and after the application of active control is:

$$\frac{|\bar{p}(\boldsymbol{r})|^2}{|\bar{p}_p(\boldsymbol{r})|^2} = 1 + \operatorname{sinc}^2 kd - 2\operatorname{sinc} kd \cos(kd\cos\theta) \tag{8.3.39}$$

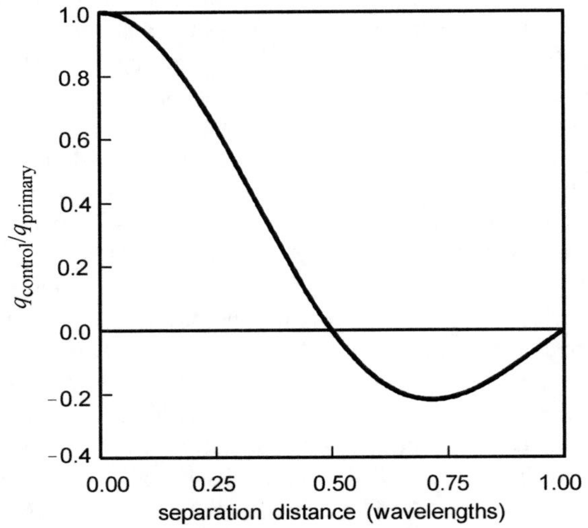

Figure 8.7 Optimum control source volume velocity as a function of primary monopole source/control monopole source separation distance.

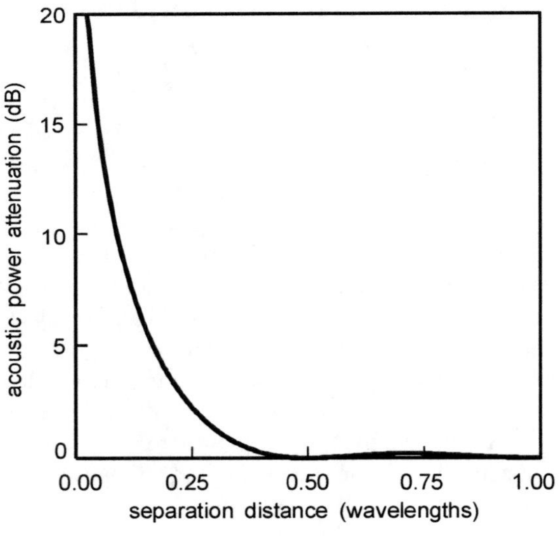

Figure 8.8 Maximum acoustic power attenuation as a function of primary monopole source/control monopole source separation distance.

For this to be less than one for all θ:

$$\text{sinc}^2 kd - 2\,\text{sinc}\,kd\cos kd < 0 \tag{8.3.40}$$

or

$$\text{sinc}\,kd < 2\cos kd \tag{8.3.41}$$

This criterion is valid for $\approx d < \lambda/6$, which is more stringent than the criterion for achieving a reduction in the total radiated acoustic power. Therefore, a reduction in the total radiated acoustic power is not necessarily accompanied by a reduction in the acoustic pressure at all points in the far-field. This fact is borne out in Figure 8.9, which illustrates the residual far-field sound pressure levels when minimising total radiated acoustic power at various separation distances.

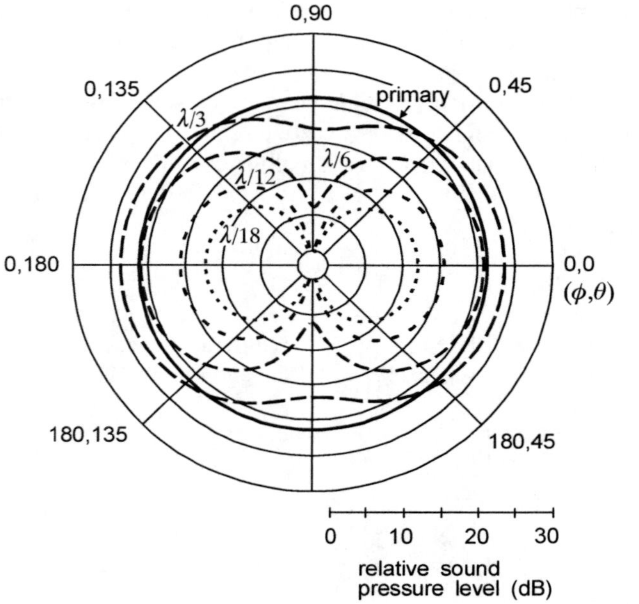

Figure 8.9 Residual acoustic pressure distribution for various separation distances (labelled as a fraction of the wavelength of the acoustic disturbance) during maximum power attenuation.

In viewing the far-field pressure distributions of Figure 8.9, it is readily apparent that when the separation distance is small, the residual sound field looks virtually identical to that of a dipole. However, as the separation distance increases, this characteristic becomes less exact. It is interesting to compare the optimal control source volume velocity with that of a dipole. For a dipole, the control source volume velocity $q_c = -q_p$, which is the limiting case of the optimum control source volume velocity of Equation (8.3.12) as the separation distance tends towards zero. If this control source volume velocity is substituted into Equation (8.3.32), the total acoustic power output of the primary source/control source pair is found to be:

$$W_d = \frac{1}{2}|\bar{q}_p|^2 z_0 \{2(1 - \operatorname{sinc} kd)\} \tag{8.3.42}$$

resulting in a change over the sole operation of the primary source of (Ffowcs Williams, 1984):

$$\frac{W_d}{W_p} = 2(1 - \operatorname{sinc} kd) \tag{8.3.43}$$

This result is directly comparable to that given in Equation (8.3.15) for the change obtained when the optimum control source volume velocity is used. Observe that for the dipole arrangement, as the separation distance of the primary source and control source increases, the acoustic power output of the dipole becomes greater than that of the primary source operating alone, tending towards double the original power output, as the two sources effectively become completely independent. The trend is different than what was found when using the optimum control source volume velocity, where as the separation distance increases, the acoustic power reduction simply tends towards zero, as the optimum volume velocity 'law' dictates that the control source actually turns itself off when there are large separation distances involved.

It will be interesting to consider the time domain equivalent of the optimum control source volume velocity relationship stated in Equation (8.3.12). To transform this frequency domain result into the time domain, the derivation of the control source volume velocity can be viewed as a filtering operation, as illustrated in Figure 8.10. The frequency response of the filter, from Equation (8.3.12), is:

$$G(\omega) = -\operatorname{sinc} kd = -\frac{\sin kd}{kd} \tag{8.3.44a,b}$$

The impulse response of this filter, found by taking the inverse Fourier transform of Equation (8.3.44), is (Elliott and Nelson, 1986; Nelson and Elliott, 1986):

$$g(t) = -\frac{\left\{H\left(t+\frac{d}{c_0}\right) - H\left(t-\frac{d}{c_0}\right)\right\}}{\frac{2d}{c_0}} \tag{8.3.45}$$

where $H(\alpha)$ is the Heaviside unit step function defined as:

$$H(\alpha) = \begin{cases} 0 & \alpha < 0 \\ 1 & \alpha > 0 \end{cases} \tag{8.3.46}$$

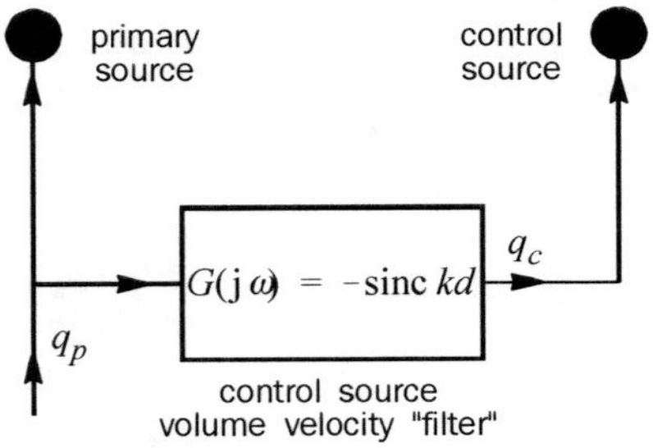

Figure 8.10 Control source volume velocity derivation viewed as a filtering operation.

The impulse response of Equation (8.3.45) is shown on Figure 8.11. Evaluation of this impulse response function requires a knowledge of what happened a time period (d/c_0) ago (the time of propagation between the two sources), as well as a knowledge of what will happen at a time (d/c_0) in the future. Hence, the use of the control source volume velocity of

Equation (8.3.12) results in a non-causal control system. This situation is quite satisfactory for harmonic sound problems, which are what are being considered here (and are possibly what constitutes the majority of noise problems targeted for active control). However, for non-periodic excitation, this control strategy is not appropriate. This point will be considered in more depth later in this chapter.

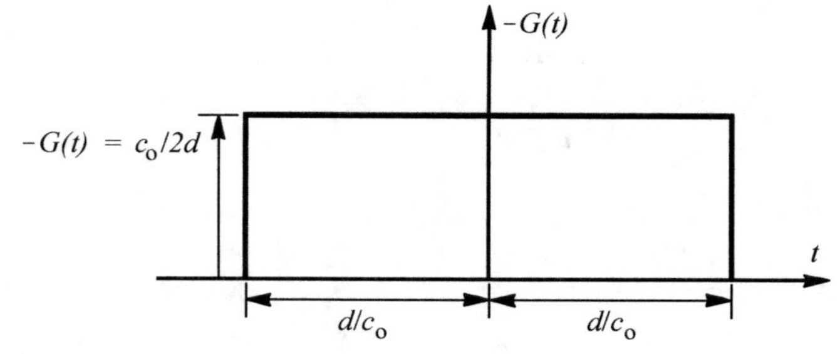

Figure 8.11 Impulse response of the optimum control source volume velocity relationship.

The final point to discuss is perhaps the most fundamental. What is the physical mechanism by which a reduction in the total radiated acoustic power is achieved? To answer this question, it is informative to calculate the acoustic power output of the primary and control sources during the operation of the active control system, where the control source volume velocity is as defined by Equation (8.3.12). Consider first the control source. If the optimum volume velocity of Equation (8.3.12) is substituted into the equation defining the control source power output derived using the near-field technique, Equation (8.3.31), the result is:

$$W_c = \frac{1}{2}|\bar{q}_c|^2 z_0 \mathrm{Re}\left\{1 + \frac{\bar{q}_p^*}{-\bar{q}_p^* \mathrm{sinc}\, kd}\ \mathrm{sinc}\, kd\right\} = 0 \tag{8.3.47a,b}$$

In other words, when the control source is driven to produce the optimal volume velocity, its power output is zero. If a similar substitution is done into Equation (8.3.30) for the primary source, the actively controlled primary source acoustic output is found to be:

$$W_p = \frac{1}{2}|\bar{q}_p|^2 z_0 \mathrm{Re}\left\{1 - \frac{-\bar{q}_p^* \mathrm{sinc}\, kd}{\bar{q}_p^*} \mathrm{sinc}\, kd\right\} = \frac{1}{2}|\bar{q}_p|^2 z_0 (1 - \mathrm{sinc}^2 kd) \tag{8.3.48a,b}$$

Comparing this to the original (uncontrolled) primary source power output, using Equation (8.3.24):

$$W_{p,\mathrm{unc}} = \frac{1}{2}|\bar{q}_p|^2 z_0 \tag{8.3.49}$$

it can be deduced that the introduction of the control source has reduced the power output of the primary source by a factor of $(1 - \mathrm{sinc}^2 kd)$. But how has this reduction been accomplished? It was stated in the beginning of Section 8.2 that the acoustic sources being considered here are all assumed to have a constant volume velocity, so that a change in this quantity can be ruled out as a mechanism for acoustic power reduction. Referring back to the defining equation for the primary source acoustic power, Equation (8.3.18), it can therefore be deduced that a reduction in acoustic power output could only have come about by a reduction in the component of pressure at the source in-phase with the source velocity. This is perhaps better stated in terms of the real, or active, radiation impedance presented to the source. For the primary monopole source, the real part of the radiation impedance 'seen' under optimally controlled conditions is, from Equation (8.3.30):

$$\mathrm{Re}\{z_{p/p} + z_{p/c}\} = z_0 \mathrm{Re}\left\{1 + \frac{\bar{q}_c^*}{\bar{q}_p^*}\,\mathrm{sinc}\,kd\right\} = z_0 \mathrm{Re}\{1 - \mathrm{sinc}^2 kd\} \tag{8.3.50a,b}$$

Equation (8.3.50) states that during the action of active control, the real part of the radiation impedance seen by the primary source is reduced. In other words, the source is unloaded by the introduction of the control source into the system. For the control source, under optimum controlled conditions, the impedance is, from Equation (8.3.31):

$$\mathrm{Re}\{z_{c/c} + z_{c/p}\} = z_0 \mathrm{Re}\left\{1 + \frac{\bar{q}_p^*}{\bar{q}_{cw}^*}\,\mathrm{sinc}\,kd\right\} = z_0 \mathrm{Re}\{1 - 1\} = 0 \tag{8.3.51a–c}$$

This shows that while in the action of applying active control, the control source is itself unloaded, in fact perfectly so (the real radiation impedance is equal to zero). It can therefore be concluded that the physical mechanism by which attenuation of the total radiated acoustic power is obtained is a mutual unloading, or a mutual reduction in the real radiation impedance of the sources. The sources continue to displace the fluid medium (they are constant volume velocity sources), but this action is simply no longer efficiently converted into propagating acoustic power. This mechanism can also be referred to as one of acoustic power *suppression*. In fact, it has been known for many years that sound sources operating in close proximity to each other will mutually alter each other's radiation impedance, and hence acoustic power output; active control simply aims to optimise this phenomenon to minimise the total acoustic power output.

The reader may be wondering at this stage why the optimal control mechanism for free-field radiation is exclusively one of suppression, and not a combination of suppression and absorption as was the case for two acoustic sources interacting in an air handling duct, considered in the previous chapter. The difference in mechanism is a direct result of the validity of acoustic reciprocity in free space, a concept which may not hold true in a non-rigidly terminated duct. This will be considered in greater depth later in this chapter.

It should be pointed out that the result of Equation (8.3.51) does not imply that the control source cannot be *forced* to absorb sound; rather, in this instance it is simply not the optimal way of minimising the total radiated sound power. To demonstrate this, consider the case where there are two monopole sources (control and primary) in close proximity, and constrained to have identical volume velocities, while allowed a variable phase relationship, such that:

$$\bar{q}_c = \bar{q}_p \mathrm{e}^{\mathrm{j}\varphi} \tag{8.3.52}$$

With this scenario, the primary source, control source, and total system acoustic power output are, from Equation (8.3.30) to (8.3.32):

$$W_p = \frac{1}{2}|\bar{q}_p|^2 z_0 \mathrm{Re}\left\{1 + \frac{\bar{q}_c^*}{\bar{q}_p^*}\operatorname{sinc} kd\right\} = \frac{1}{2}|\bar{q}_p|^2 z_0\left(1 + \frac{\sin(kd - \varphi)}{kd}\right) \tag{8.3.53a,b}$$

$$W_c = \frac{1}{2}|\bar{q}_c|^2 z_0 \mathrm{Re}\left\{1 + \frac{\bar{q}_p^*}{\bar{q}_c^*}\operatorname{sinc} kd\right\} = \frac{1}{2}|\bar{q}_p|^2 z_0\left(1 + \frac{\sin(kd + \varphi)}{kd}\right) \tag{8.3.54a,b}$$

$$W_{tot} = W_p + W_c = \frac{1}{2}|\bar{q}_p|^2 z_0\left\{\left(1 + \frac{\sin(kd - \varphi)}{kd}\right) + \left(1 + \frac{\sin(kd + \varphi)}{kd}\right)\right\} \tag{8.3.55a,b}$$

Using these three equations, Figure 8.12 illustrates the source and total acoustic power output for this scenario. It can be seen that at some source phase differences, especially 90° and 270°, one of the sources is absorbing acoustic power, which 'induces' the other source to substantially increase its acoustic power output. The absorption does not completely compensate for the increased output from the other source, however, and the result is a non-optimal change in the total acoustic power flow.

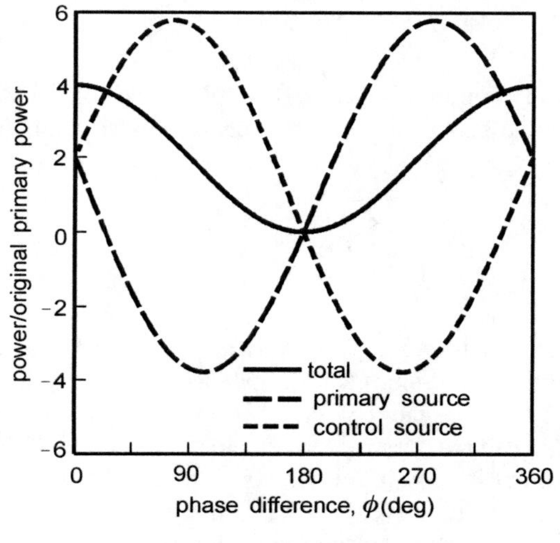

Figure 8.12 Typical plot of primary source, control source, and total acoustic power output as a function of phase difference between the sources, where the sources are of equal volume velocity.

To continue with this line of thought, it will be interesting to consider what would happen if, rather than trying to minimise the total radiated acoustic power, the error criterion was changed to one of maximising the power absorption of the control source. Such an arrangement could be viewed as equivalent to having a 'passive' control element being acted upon by the sound field (Elliott et al., 1991). To derive the optimum control source volume velocity with the aim of producing such a system, the control source power output as defined in Equation (8.3.31) can be re-expressed as:

$$W_c = \bar{q}_c^* a_w \bar{q}_c + \frac{1}{2}\left(\bar{q}_c^* z_{c/p} \bar{q}_p + \bar{q}_p^* z_{c/p}^* \bar{q}_p\right) \tag{8.3.56}$$

where a_w is as defined in Equation (8.3.34). Differentiating this equation with respect to the control source volume velocity, and setting the resultant gradient expression equal to zero, produces an expression for the control source volume velocity amplitude that will minimise the control source power output (that is, maximise absorption):

$$\bar{q}_{ca} = -\frac{1}{2} a_w^{-1} z_{c/p} \bar{q}_p = -\frac{1}{2} \bar{q}_p \left(\frac{\sin kd}{kd} + \mathrm{j} \frac{\cos kd}{kd} \right) \tag{8.3.57a,b}$$

where the subscript *c,a* denotes that the control source volume velocity has been derived with respect to maximising absorption. The control source volume velocity is now based upon the total acoustic pressure from the primary source presented to it, rather than just the real component, as was used when minimising the total acoustic power output. If this control source volume velocity is substituted into the defining equation for the primary source power output, Equation (8.3.30), the controlled power output, written in terms of the original (uncontrolled) primary source power, is found to be:

$$W_p = W_{p,\mathrm{unc}} \left\{ 1 + \frac{1}{2} \left[\left(\frac{\cos kd}{kd} \right)^2 - \left(\frac{\sin kd}{kd} \right)^2 \right] \right\} = W_{p,\mathrm{unc}} \left\{ 1 + \frac{\cos 2kd}{2(kd)^2} \right\} \tag{8.3.58a,b}$$

Making a similar substitution into Equation (8.3.30) for the control source, its power output while being configured to maximally absorb, expressed in terms of the uncontrolled primary source power, is:

$$W_p = -W_{p,\mathrm{unc}} \left\{ \frac{1}{4} \left[\left(\frac{\cos kd}{kd} \right)^2 - \left(\frac{\sin kd}{kd} \right)^2 \right] \right\} = -W_{p,\mathrm{unc}} \left\{ \frac{1}{4(kd)^2} \right\} \tag{8.3.59a,b}$$

where the negative sign indicates that the power flow is into the control source (power absorption). Using these expressions, the primary source, control source and total power flow are shown in Figure 8.13, plotted as a function of source separation distance. It can be seen that within a small band, between approximately $\lambda/12$ and $5\lambda/12$, some attenuation of the total system power output is achieved. However, as the separation distance becomes small, such that $kd \rightarrow 0$, the power output of the primary source increases dramatically. The absorption of the control source also increases, but not sufficiently to compensate for it; hence the total system power output also increases dramatically. This behaviour is predicted by Equations (8.3.58) and (8.3.59), which show the limiting total power flow as the separation distance approaches zero to be (Elliott et al., 1991):

$$\lim_{kd \rightarrow 0} W_{tot} = W_{p,\mathrm{unc}} \frac{1}{4(kd)^2} \tag{8.3.60}$$

Thus, the net effect of an active control system configured to have the control source maximally absorb sound is to produce an enormous *increase* in the total system acoustic power output as the control source comes within close proximity of the primary source (Elliott et al., 1991).

In concluding this discussion of the minimum power output of two free-field monopole sources, it is useful to summarise a few important results. First, it is possible to formulate analytically an expression for a control source volume velocity that will minimise the total acoustic power output of the system, using quadratic optimisation techniques. This volume velocity was shown in Equation (8.3.12) to be $\bar{q}_c = -\bar{q}_p \text{sinc}\, kd$, which will reduce the total acoustic power output by a factor $\Delta W = 1 - \text{sinc}^2 kd$. From these expressions it is clear that the control source and primary source must be in close proximity to each other (say, within $\lambda/10$) for any appreciable reduction in radiated acoustic power to be obtained. The optimum control source volume velocity is less than that of the primary source, decreasing as the separation distance increases. In the limiting case, as the separation distance $kd \to 0$, the control source and primary source form a dipole, but *only* in the limiting case. If the control source is constrained to be a dipole, the results become significantly degraded as the separation distance increases. Also, the optimal control source volume velocity produces a system that acts non-causally with respect to the primary source. Finally, the mechanism of control for maximum acoustic power attenuation is exclusively one of power suppression, owing to the fact that the impedances seen by the primary and control sources when operating alone are equal. Any forcing of the control source to absorb the acoustic power output of the primary source will degrade the control effect; this effect will become more significant as the separation distance decreases and the rate of absorption increases.

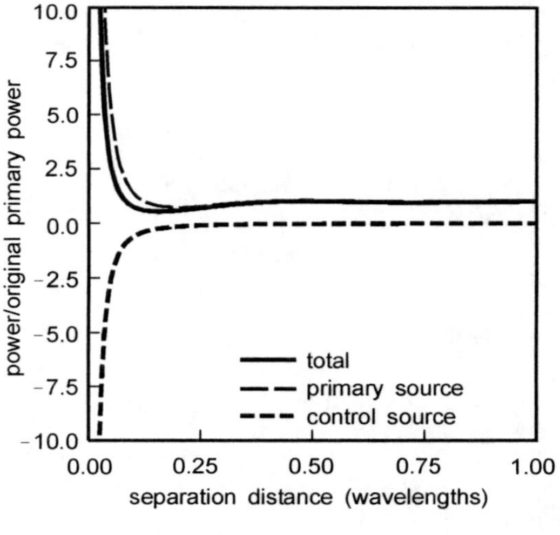

Figure 8.13 Primary source, control source and total acoustic power output when the control source is maximally absorbing.

8.4 ACTIVE CONTROL OF ACOUSTIC RADIATION FROM MULTIPLE PRIMARY MONOPOLE SOURCES USING MULTIPLE CONTROL MONOPOLE SOURCES

In this section and subsequent sections (except for Section 8.15 where it is used to distinguish amplitudes from time varying quantities), the bar over a variable to indicate the complex amplitude will be omitted to simplify the notation, but it is implied in all sound pressure and volume velocity quantities and any quantity that does not explicitly indicate time dependence.

The examination of the use of a single harmonically oscillating monopole control source to attenuate the output from a similarly oscillating single primary monopole source undertaken in the previous two sections can be extended to include the use of multiple primary and control sources, as well as multiple error sensing locations. The general methodology used to examine these enlarged systems, to determine the optimum control source volume velocities for minimising sound pressure or acoustic power, is the same as that used to examine the simple two source arrangement. However, the single complex quantities of pressure and volume velocity will need to be replaced with vectors of the acoustic pressures at a set of error sensing locations, or volume velocities from a set of sources. This extension will provide an introduction to the problem of controlling sound radiation from a vibrating structure, considered later in this chapter, where sound radiation from a set of structural modes of vibration will be considered. Both the problems of minimising sound pressure at a set of discrete error sensing locations, and of minimising the total acoustic power output, will be considered in this section. The former of these problems will be discussed first.

One related point worth mentioning here is that it is (theoretically) possible to group a set of monopole sources in a compact region and arrange the amplitude and phasing such that an arbitrary, far-field acoustic pressure distribution is achieved. Therefore, given an unwanted sound field, a 'cancelling' field could be derived using a (possibly infinite) set of monopole sources. This concept, suggested in (Kempton, 1976), will not be discussed here, owing to its perceived lack of practicality. For more detail, the reader is referred to Kempton (1976) or Nelson and Elliott (1992).

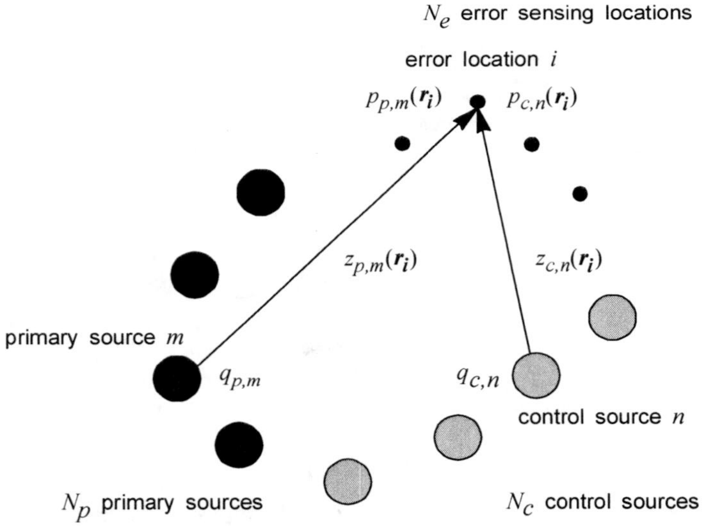

Figure 8.14 Multiple monopole primary sources/multiple monopole control sources system arrangement.

Consider the system arrangement shown in Figure 8.14, where there are N_p harmonically oscillating primary monopole sources, N_c harmonically oscillating control monopole sources, and N_e discrete error sensing locations where it is desired to minimise the acoustic pressure amplitude. The acoustic pressure at the ith error sensor location is, by superposition, the sum of contributions from the N_p primary sources and N_c control sources:

$$p(\boldsymbol{r}_i) = \sum_{m=1}^{N_p} p_{p,m}(\boldsymbol{r}_i) + \sum_{n=1}^{N_c} p_{c,n}(\boldsymbol{r}_i) \tag{8.4.1}$$

Using Equation (8.2.1), which defines the acoustic pressure at some location $\boldsymbol{r}$ relative to a monopole source, Equation (8.4.1) can be rewritten in terms of volume velocities:

$$p(\boldsymbol{r}_i) = \sum_{m=1}^{N_p} q_{p,m} z_{p,m}(\boldsymbol{r}_i) + \sum_{n=1}^{N_c} q_{c,n} z_{c,n}(\boldsymbol{r}_i) \tag{8.4.2}$$

where for the primary sources:

$$z_{p,m}(\boldsymbol{r}_i) = \frac{j\omega\rho_0 \mathrm{e}^{-jk|\boldsymbol{r}_i - \boldsymbol{r}_{p,m}|}}{4\pi|\boldsymbol{r}_i - \boldsymbol{r}_{p,m}|} \tag{8.4.3}$$

and for the control sources:

$$z_{c,n}(\boldsymbol{r}_i) = \frac{j\omega\rho_0 \mathrm{e}^{-jk|\boldsymbol{r}_i - \boldsymbol{r}_{c,n}|}}{4\pi|\boldsymbol{r}_i - \boldsymbol{r}_{c,n}|} \tag{8.4.4}$$

Equation (8.4.2) can be re-expressed in matrix form as:

$$p(\boldsymbol{r}_i) = \boldsymbol{z}_p^{\mathrm{T}}(\boldsymbol{r}_i)\boldsymbol{q}_p + \boldsymbol{z}_c^{\mathrm{T}}(\boldsymbol{r}_i)\boldsymbol{q}_c \tag{8.4.5}$$

where $\boldsymbol{q}_p$ and $\boldsymbol{q}_c$ are the $(N_p \times 1)$ vector of primary monopole source volume velocities and $(N_c \times 1)$ vector of control monopole source volume velocities respectively:

$$\boldsymbol{q}_p = \begin{bmatrix} q_{p,1} \\ q_{p,2} \\ \vdots \\ q_{p,N_p} \end{bmatrix}; \quad \boldsymbol{q}_c = \begin{bmatrix} q_{c,1} \\ q_{c,2} \\ \vdots \\ q_{c,N_c} \end{bmatrix} \tag{8.4.6a,b}$$

and $\boldsymbol{z}_p(\boldsymbol{r}_i)$ and $\boldsymbol{z}_c(\boldsymbol{r}_i)$ are the $(N_p \times 1)$ vector of primary source radiation transfer functions to the error location $\boldsymbol{r}_i$ and the $(N_c \times 1)$ vector of control source radiation transfer functions to the error location $\boldsymbol{r}_i$ respectively:

$$\boldsymbol{z}_p(\boldsymbol{r}_i) = \begin{bmatrix} z_{p,1}(\boldsymbol{r}_i) \\ z_{p,2}(\boldsymbol{r}_i) \\ \vdots \\ z_{p,N_p}(\boldsymbol{r}_i) \end{bmatrix}; \quad \boldsymbol{z}_c(\boldsymbol{r}_i) = \begin{bmatrix} z_{c,1}(\boldsymbol{r}_i) \\ z_{c,2}(\boldsymbol{r}_i) \\ \vdots \\ z_{c,N_c}(\boldsymbol{r}_i) \end{bmatrix} \tag{8.4.7a,b}$$

Equation (8.4.5) can be expanded to include consideration of the entire set of N_e error sensing locations:

$$\boldsymbol{p}_e = \boldsymbol{Z}_p \boldsymbol{q}_p + \boldsymbol{Z}_c \boldsymbol{q}_c \tag{8.4.8}$$

where $\boldsymbol{p}_e$ is the ($N_e \times 1$) vector of pressures at the error locations:

$$\boldsymbol{p}_e = \begin{bmatrix} p(\boldsymbol{r}_1) \\ p(\boldsymbol{r}_2) \\ \vdots \\ p(\boldsymbol{r}_{N_e}) \end{bmatrix} \tag{8.4.9}$$

and where $\boldsymbol{Z}_p$ and $\boldsymbol{Z}_c$ are the matrices of radiation transfer functions between the error locations and the primary and control source volume velocities respectively:

$$\boldsymbol{Z}_p = \begin{bmatrix} z_p^{\mathrm{T}}(\boldsymbol{r}_1) \\ z_p^{\mathrm{T}}(\boldsymbol{r}_2) \\ \vdots \\ z_p^{\mathrm{T}}(\boldsymbol{r}_{N_e}) \end{bmatrix} ; \quad \boldsymbol{Z}_c = \begin{bmatrix} z_c^{\mathrm{T}}(\boldsymbol{r}_1) \\ z_c^{\mathrm{T}}(\boldsymbol{r}_2) \\ \vdots \\ z_c^{\mathrm{T}}(\boldsymbol{r}_{N_e}) \end{bmatrix} \tag{8.4.10a,b}$$

The aim of the exercise in the first instance is to minimise the sound pressure at a set of error sensing locations, or to minimise the sum of the squared acoustic pressure amplitudes:

$$\sum_{i=1}^{N_e} |p(\boldsymbol{r}_i)|^2 = \sum_{i=1}^{N_e} p^*(\boldsymbol{r}_i) p(\boldsymbol{r}_i) \tag{8.4.11}$$

Equation (8.4.11) can be written in matrix form as:

$$\sum_{i=1}^{N_e} |p(\boldsymbol{r}_i)|^2 = \boldsymbol{p}_e^{\mathrm{H}} \boldsymbol{p}_e \tag{8.4.12}$$

where H is the Hermitian, or complex conjugate and transpose, of the matrix. Using Equation (8.4.8), this can be expanded as:

$$\begin{aligned} \boldsymbol{p}_e^{\mathrm{H}} \boldsymbol{p}_e &= \left[\boldsymbol{Z}_p \boldsymbol{q}_p + \boldsymbol{Z}_c \boldsymbol{q}_c\right]^{\mathrm{H}} \left[\boldsymbol{Z}_p \boldsymbol{q}_p + \boldsymbol{Z}_c \boldsymbol{q}_c\right] \\ &= \boldsymbol{q}_c^{\mathrm{H}} \boldsymbol{Z}_c^{\mathrm{H}} \boldsymbol{Z}_c \boldsymbol{q}_c + \boldsymbol{q}_c^{\mathrm{H}} \boldsymbol{Z}_c^{\mathrm{H}} \boldsymbol{Z}_p \boldsymbol{q}_p + \boldsymbol{q}_p^{\mathrm{H}} \boldsymbol{Z}_p^{\mathrm{H}} \boldsymbol{Z}_c \boldsymbol{q}_c + \boldsymbol{q}_p^{\mathrm{H}} \boldsymbol{Z}_p^{\mathrm{H}} \boldsymbol{Z}_p \boldsymbol{q}_p \end{aligned} \tag{8.4.13a,b}$$

Equation (8.4.13) can be written more conveniently as:

$$J_{prs} = \boldsymbol{p}_e^{\mathrm{H}} \boldsymbol{p}_e = \boldsymbol{q}_c^{\mathrm{H}} \boldsymbol{A}_{prs} \boldsymbol{q}_c + \boldsymbol{q}_c^{\mathrm{H}} \boldsymbol{b}_{prs} + \boldsymbol{b}_{prs}^{\mathrm{H}} \boldsymbol{q}_c + c \tag{8.4.14}$$

where

$$\boldsymbol{A}_{prs} = \boldsymbol{Z}_c^{\mathrm{H}} \boldsymbol{Z}_c \tag{8.4.15}$$

$$\boldsymbol{b}_{prs} = \boldsymbol{Z}_c^{\mathrm{H}} \boldsymbol{Z}_p \boldsymbol{q}_p \tag{8.4.16}$$

and

$$c = \boldsymbol{q}_p^{\mathrm{H}} \boldsymbol{Z}_p^{\mathrm{H}} \boldsymbol{Z}_p \boldsymbol{q}_p \tag{8.4.17}$$

Here, the subscript *prs* is being used to denote pressure minimisation. Note that c is the sum of the squared pressure amplitudes at the error sensing locations during primary excitation only, prior to the commencement of active control.

Equation (8.4.14) shows the error criterion J_{prs}, the minimisation of the squared acoustic pressure amplitudes at a set of discrete error locations, to be a real quadratic function of the *vector* of control source volume velocities. Hence the error surface, which is the error criterion plotted as a function of the real and imaginary components of the control source volume velocities, will be a $(2N_c+1)$-dimensional hyper-paraboloid whose axes are defined by the real and imaginary components of each complex volume velocity. This is a generalisation of the three-dimensional 'bowl' shown in Figure 8.2 for the case of $N_c = 1$. Being a quadratic function of $\boldsymbol{q}_c$, the error criterion of Equation (8.4.14) can be differentiated with respect to this quantity, and the resultant gradient expression set equal to zero to derive the vector of optimum control source volume velocities. Performing these operations on the real and imaginary parts separately gives:

$$\frac{\partial J_{prs}}{\partial \boldsymbol{q}_{c_{\mathrm{R}}}} = 2\boldsymbol{A}_{prs_{\mathrm{R}}} \boldsymbol{q}_{c_{\mathrm{R}}} + 2\boldsymbol{b}_{prs_{\mathrm{R}}} = 0 \tag{8.4.18a,b}$$

and

$$\frac{\partial J_{prs}}{\partial \boldsymbol{q}_{c_{\mathrm{I}}}} = 2\boldsymbol{A}_{prs_{\mathrm{I}}} \boldsymbol{q}_{c_{\mathrm{I}}} + 2\boldsymbol{b}_{prs_{\mathrm{I}}} = 0 \tag{8.4.19a,b}$$

Multiplying the imaginary part by j and adding it to the real part produces the following expression for the optimum vector of control source volume velocities:

$$\boldsymbol{q}_{c,p} = -\boldsymbol{A}_p^{-1} \boldsymbol{b}_p \tag{8.4.20}$$

The result of Equation (8.4.20) is directly comparable to Equation (8.2.12) for a system arrangement consisting of a single primary source, a single control source and a single error sensor. The difference is that Equation (8.4.20) uses matrix quantities and produces, as a result, the vector of optimum control source volume velocities for a number of control sources and error sensors.

Substitution of the result of Equation (8.4.20) into Equation (8.4.14) yields the expression for the minimum residual (controlled) sum of the squared sound pressures at the error locations:

$$\left\{ \boldsymbol{p}_e^{\mathrm{H}} \boldsymbol{p}_e \right\}_{\min} = c - \boldsymbol{b}_{prs}^{\mathrm{H}} \boldsymbol{A}_{prs}^{-1} \boldsymbol{b}_{prs} = c + \boldsymbol{b}_{prs}^{\mathrm{H}} \boldsymbol{q}_{c,p} \tag{8.4.21a,b}$$

It should be noted that if the operation of Equation (8.4.20) is to define a unique vector of control source volume velocities that will collectively minimise the sum of the squared acoustic pressure amplitudes at the error sensing locations, the matrix $\boldsymbol{A}_{prs}$ must be positive definite. For the case being considered here, that translates into the requirement that there be at least as many error locations as there are control sources. If there are fewer, then the matrix $\boldsymbol{A}_{prs}$ will be singular, resulting in an infinite number of 'optimum' control source volume velocity vectors. If there are an equal number of control sources and error locations then the vector of optimum control source volume velocities will completely null the acoustic pressure at the error locations. If there are more error locations than control sources, then the residual acoustic pressure at the error locations will probably not be equal to zero, although this may not have a negative influence upon the overall levels of acoustic power attenuation attained.

Before leaving this discussion of minimisation of the acoustic pressure at a set of discrete error locations, it is informative to have a brief qualitative look at the form of the matrix, $\boldsymbol{A}_{prs}$. Referring to the definition of this matrix given in Equation (8.4.15), it can be deduced that its formulation can be written heuristically as:

$$\boldsymbol{A}_{prs} = \begin{bmatrix} \text{control source 1 transfer functions} \\ \text{control source 2 transfer functions} \\ \cdots \\ \text{control source } N_c \text{ transfer functions} \end{bmatrix}^* \begin{bmatrix} \text{control source 1 transfer functions} \\ \text{control source 2 transfer functions} \\ \cdots \\ \text{control source } N_c \text{ transfer functions} \end{bmatrix}^{\mathrm{T}} \tag{8.4.22}$$

Viewed in this way, it can be deduced that the diagonal elements in $\boldsymbol{A}_{prs}$ are those terms that would be used in deriving the optimal control source volume velocities if the associated control source were the only one operating. The off-diagonal terms represent the required modification to, or influence upon, this solution, owing to the presence of other operating control sources. In other words, the off-diagonal terms are responsible for coupling the individually-optimal control source volume velocities. Consider, for example, a 2 primary source, 2 control source, 2 error sensor system. For this system, the matrix $\boldsymbol{A}$ will be:

$$\boldsymbol{A} = \frac{\mathrm{j}\omega\rho_0}{4\pi} \begin{bmatrix} \dfrac{r_{1,1} + r_{1,2}}{r_{1,1} r_{1,2}} & \dfrac{\mathrm{e}^{-\mathrm{j}k(r_{2,1} - r_{1,1})}}{r_{1,1} r_{2,1}} + \dfrac{\mathrm{e}^{-\mathrm{j}k(r_{2,2} - r_{1,2})}}{r_{2,2} r_{1,2}} \\ \dfrac{\mathrm{e}^{-\mathrm{j}k(r_{1,1} - r_{2,1})}}{r_{1,1} r_{2,1}} + \dfrac{\mathrm{e}^{-\mathrm{j}k(r_{1,2} - r_{2,2})}}{r_{2,2} r_{1,2}} & \dfrac{r_{2,1} + r_{2,2}}{r_{2,1} r_{2,2}} \end{bmatrix} \tag{8.4.23}$$

where the quantity $r_{a,b}$ is the distance between control source a and error sensor b. The diagonal terms are only concerned with a single source, while the off-diagonal terms couple the solution by considering the transfer functions between both sources and error sensors. If an attempt were made to obtain the set of optimal control source volume velocities by considering each control source individually, deriving the volume velocity of one source, which on its own would minimise the pressure at the error sensing locations, then deriving the next, and so on, the final result would be sub-optimal; it would be equivalent to forcing all of the off-diagonal terms in A_p to be equal to zero. For the optimal set of control source volume velocities to be derived, the problem must be considered holistically, taking into account the inherent coupling of the control sources, which is a result of mutual operation in the acoustic space, rather than individually. The quotation, 'no man is an island', is never more true than when applied to multiple control sources in an active control system.

Having now considered the minimisation of acoustic pressure squared at a number of discrete error locations, the second of the two error criteria, minimisation of total radiated acoustic power, will now be considered. It was stated in the previous section that acoustic power was equal to the integration of the real acoustic intensity radiating out of a sphere enclosing the noise sources as follows:

$$W = 2\pi r^2 \int_0^\pi \frac{|p(\boldsymbol{r})|^2}{2\rho_0 c_0} \sin\theta \, \mathrm{d}\theta \tag{8.4.24}$$

In a similar way to the single control source case, the acoustic pressure modulus squared term in Equation (8.4.23) can be expanded to re-express the total radiated power as a quadratic function of the vector of control source volume velocities. Equation (8.4.5) provides an expression for the acoustic pressure at some location in space during the operation of the N_p primary sources and N_c control sources. Using this, the pressure modulus squared at any location is:

$$\begin{aligned} |p(\boldsymbol{r})|^2 &= p(\boldsymbol{r})^* p(\boldsymbol{r}) \\ &= \boldsymbol{q}_c^{\mathrm{H}} \boldsymbol{z}_c^{*} \boldsymbol{z}_c^{\mathrm{T}} \boldsymbol{q}_c + \boldsymbol{q}_c^{\mathrm{H}} \boldsymbol{z}_c^{*} \boldsymbol{z}_p^{\mathrm{T}} \boldsymbol{q}_p + \boldsymbol{q}_p^{\mathrm{H}} \boldsymbol{z}_p^{*} \boldsymbol{z}_c^{\mathrm{T}} \boldsymbol{q}_c + \boldsymbol{q}_p^{\mathrm{H}} \boldsymbol{z}_p^{*} \boldsymbol{z}_p^{\mathrm{T}} \boldsymbol{q}_p \end{aligned} \tag{8.4.25a,b}$$

If Equation (8.4.25) is substituted into Equation (8.4.23), it is found that the total radiated acoustic power can be expressed as:

$$W = \boldsymbol{q}_c^{\mathrm{H}} \boldsymbol{A}_w \boldsymbol{q}_c + \boldsymbol{q}_c^{\mathrm{H}} \boldsymbol{b}_w + \boldsymbol{b}_w^{\mathrm{H}} \boldsymbol{q}_c + c \tag{8.4.26}$$

where

$$\boldsymbol{A}_w = \frac{\pi r^2}{\rho_0 c_0} \left\{ \int_0^\pi \boldsymbol{z}_c^{*} \boldsymbol{z}_c^{\mathrm{T}} \sin\theta \, \mathrm{d}\theta \right\} \tag{8.4.27}$$

$$\boldsymbol{b}_w = \frac{\pi r^2}{\rho_0 c_0} \left\{ \int_0^\pi \boldsymbol{z}_c^{*} \boldsymbol{z}_p^{\mathrm{T}} \sin\theta \, \mathrm{d}\theta \right\} \boldsymbol{q}_p \tag{8.4.28}$$

and

$$c = \frac{\pi r^2}{\rho_0 c_0} \boldsymbol{q}_p^{\mathrm{H}} \left\{ \int_0^\pi \boldsymbol{z}_p^{*} \boldsymbol{z}_p^{\mathrm{T}} \sin\theta \, \mathrm{d}\theta \right\} \boldsymbol{q}_p \tag{8.4.29}$$

From the form of Equation (8.4.26), where power is expressed as a quadratic function of the control source volume velocities $\boldsymbol{q}_c$, it is immediately apparent that the expression for the vector of control source volume velocities that will minimise the total radiated acoustic power will have the same form as Equation (8.4.21); that is,

$$\boldsymbol{q}_{c,w} = -\boldsymbol{A}_w^{-1}\boldsymbol{b}_w \tag{8.4.30}$$

which will result in a residual total acoustic power output, found by substituting Equation (8.4.29) into Equation (8.4.26), of:

$$W_{\min} = c - \boldsymbol{b}_w^{\mathrm{H}}\boldsymbol{A}_w^{-1}\boldsymbol{b}_w = c + \boldsymbol{b}_w^{\mathrm{H}}\boldsymbol{q}_{c,w} \tag{8.4.31a,b}$$

Note again that the term c is equal to the value of the error criterion during the operation of the primary sources only, in this case the primary source radiated acoustic power prior to the onset of active control. From Equation (8.4.31), the reduction in total acoustic power output is:

$$\Delta W = \frac{W_{\min}}{W_{p,\mathrm{unc}}} = 1 - c^{-1}\boldsymbol{b}_w^{\mathrm{H}}\boldsymbol{A}_w^{-1}\boldsymbol{b}_w = 1 + c^{-1}\boldsymbol{b}_w^{\mathrm{H}}\boldsymbol{q}_{c,w} \tag{8.4.32a–c}$$

For the problem being considered here, the form of the terms $\boldsymbol{A}_w$, $\boldsymbol{b}_w$ and c can be simplified by solving analytically the integral expressions in Equations (8.4.27) to (8.4.29). Consider first $\boldsymbol{A}_w$. The integration required for the m,nth term is, from Equations (8.4.4) and (8.4.27):

$$A_w(m,n) = \frac{\pi r^2}{\rho_0 c_0}\int_0^{\pi}\left(\frac{j\omega\rho_0 \mathrm{e}^{-jk|\boldsymbol{r}-\boldsymbol{r}_{c,m}|}}{4\pi|\boldsymbol{r}-\boldsymbol{r}_{c,m}|}\right)^*\left(\frac{j\omega\rho_0 \mathrm{e}^{-jk|\boldsymbol{r}-\boldsymbol{r}_{c,n}|}}{4\pi|\boldsymbol{r}-\boldsymbol{r}_{c,n}|}\right)\sin\theta\,\mathrm{d}\theta \tag{8.4.33}$$

As the integration is taking place in the far-field of the sources, the integration sphere can be approximated as having its origin midway along a line connecting the mth and nth control sources. By doing this, the geometry has been arranged so that it is the same as that shown in Figure 8.4, where $|\boldsymbol{r}-\boldsymbol{r}_{c,n}| \approx |\boldsymbol{r}-\boldsymbol{r}_{c,m}|$, and $|\boldsymbol{r}-\boldsymbol{r}_{c,m}| - |\boldsymbol{r}-\boldsymbol{r}_{c,n}| \approx d_{cm,cn}\cos\theta$, where $d_{cm,cn}$ is the separation distance between the mth and nth control sources. Making these approximations, Equation (8.4.33) can be written as:

$$A_w(m,n) = \frac{\omega^2\rho_0}{16\pi c_0}\int_0^{\pi} e^{-jk(d\cos\theta)}\sin\theta\,\mathrm{d}\theta \tag{8.4.34}$$

Using the standard results stated in Equation (8.3.8), $A_w(m,n)$ can now be evaluated as:

$$A_w(m,n) = \frac{\omega^2\rho_0}{8\pi c_0}\operatorname{sinc} kd_{cm,cn} \tag{8.4.35}$$

Note again that for the diagonal terms, where $d_{cm,cn} = 0$, sinc $d_{cm,cn} = 1.0$.

Using a similar procedure, $\boldsymbol{b}_w$ in Equation (8.4.27) can be simplified to:

$$\boldsymbol{b}_w = \boldsymbol{B}_w\boldsymbol{q}_p \tag{8.4.36}$$

where $\boldsymbol{B}_w$ is an $(N_c \times N_p)$ matrix, whose terms are defined by:

$$\boldsymbol{B}_w(m,n) = \frac{\omega^2 \rho_0}{8\pi c_0} \operatorname{sinc} k d_{cm,pn} \tag{8.4.37}$$

where $d_{cm,pn}$ is the separation distance between the mth control source and nth primary source. Finally, c can also be simplified to:

$$c = \boldsymbol{q}_p^{\mathrm{H}} \boldsymbol{C}_w \boldsymbol{q}_p \tag{8.4.38}$$

where $\boldsymbol{C}_w$ is an $(N_p \times N_p)$ matrix, the terms of which are defined by:

$$\boldsymbol{C}_w(m,n) = \frac{\omega^2 \rho_0}{8\pi c_0} \operatorname{sinc} k d_{pm,pn} \tag{8.4.39}$$

where $d_{pm,pn}$ is the separation distance between the mth and nth primary sources.

Before considering the results obtained thus far for minimising the total radiated acoustic power by the set of N_p primary sources and N_c control sources, it will again be worthwhile re-deriving the results using a near-field, rather than far-field, methodology (as was done in the previous section for the single primary source, single control source case; see also Nelson et al. (1987)). As was stated in the previous section, while it is not the most generally applicable methodology for analysing free-field radiation problems, particularly when a radiated structure is involved, the near-field methodology does provide a 'neat' means of solution for the monopole radiation problem.

It was stated in the previous section, in Equation (8.3.20), that the acoustic power output from a monopole source is equal to the real part of the product of the source volume velocity and the complex conjugate of the acoustic pressure at the source location:

$$W = \frac{1}{2} \mathrm{Re}\left\{ p(\boldsymbol{r}_q)^* q \right\} \tag{8.4.40}$$

Considering first the acoustic power output of the primary sources, the total power output of the group is:

$$W = \frac{1}{2} \sum_{i=1}^{N_p} \mathrm{Re}\left\{ p(\boldsymbol{r}_{pi})^* q_{pi} \right\} = \frac{1}{2} \mathrm{Re}\left\{ \boldsymbol{p}_p^{\mathrm{H}} \boldsymbol{q}_p \right\} \tag{8.4.41a,b}$$

where $\boldsymbol{p}_p$ is the $(N_p \times 1)$ vector of complex acoustic pressures at the primary source locations:

$$\boldsymbol{p}_p = \begin{bmatrix} p(\boldsymbol{r}_{p1}) \\ p(\boldsymbol{r}_{p2}) \\ \vdots \\ p(\boldsymbol{r}_{pN_p}) \end{bmatrix} \tag{8.4.42}$$

The acoustic pressures at the primary source locations, during the operation of the active control system, are the sum (superposition) of the pressures at that location attributable to the entire set of N_p primary sources and N_c control sources. Thus, $\boldsymbol{p}_p$ can be re-expressed in terms of the volume velocities of the entire set of sources, in a manner similar to that of Equation (8.4.8):

$$\boldsymbol{p}_p = \boldsymbol{Z}_{p/p}\boldsymbol{q}_p + \boldsymbol{Z}_{p/c}\boldsymbol{q}_c \tag{8.4.43}$$

where

$$\boldsymbol{Z}_{p/p} = \begin{bmatrix} z_p^{\mathrm{T}}(\boldsymbol{r}_{p1}) \\ z_p^{\mathrm{T}}(\boldsymbol{r}_{p2}) \\ \vdots \\ z_p^{\mathrm{T}}(\boldsymbol{r}_{pN_p}) \end{bmatrix} ; \quad \boldsymbol{Z}_{p/c} = \begin{bmatrix} z_c^{\mathrm{T}}(\boldsymbol{r}_{p1}) \\ z_c^{\mathrm{T}}(\boldsymbol{r}_{p2}) \\ \vdots \\ z_c^{\mathrm{T}}(\boldsymbol{r}_{pN_p}) \end{bmatrix} \tag{8.4.44a,b}$$

and $z_p(\boldsymbol{r}_i)$, $z_c(\boldsymbol{r}_i)$ are as defined in Equation (8.4.7). Substituting this expression into Equation (8.4.40), the total power output of the primary sources operating in the presence of the control sources is:

$$W_p = \frac{1}{2}\mathrm{Re}\left\{\boldsymbol{q}_p^{\mathrm{H}}\boldsymbol{Z}_{p/p}^{\mathrm{H}}\boldsymbol{q}_p + \boldsymbol{q}_p^{\mathrm{H}}\boldsymbol{Z}_{p/c}^{\mathrm{H}}\boldsymbol{q}_c\right\} \tag{8.4.45}$$

An identical set of steps can be undertaken to derive a matrix expression for the total control source acoustic power output when operating in the presence of the primary sources. The result of doing this is:

$$W_c = \frac{1}{2}\mathrm{Re}\left\{\boldsymbol{q}_c^{\mathrm{H}}\boldsymbol{Z}_{c/c}^{\mathrm{H}}\boldsymbol{q}_c + \boldsymbol{q}_c^{\mathrm{H}}\boldsymbol{Z}_{c/p}^{\mathrm{H}}\boldsymbol{q}_p\right\} \tag{8.4.46}$$

where

$$\boldsymbol{Z}_{c/c} = \begin{bmatrix} z_c^{\mathrm{T}}(\boldsymbol{r}_{c1}) \\ z_c^{\mathrm{T}}(\boldsymbol{r}_{c2}) \\ \vdots \\ z_c^{\mathrm{T}}(\boldsymbol{r}_{cN_c}) \end{bmatrix} ; \quad \boldsymbol{Z}_{c/p} = \begin{bmatrix} z_p^{\mathrm{T}}(\boldsymbol{r}_{c1}) \\ z_p^{\mathrm{T}}(\boldsymbol{r}_{c2}) \\ \vdots \\ z_p^{\mathrm{T}}(\boldsymbol{r}_{cN_c}) \end{bmatrix} \tag{8.4.47a,b}$$

and $z_p(\boldsymbol{r}_i)$, $z_c(\boldsymbol{r}_i)$ are again as defined in Equation (8.4.7). Combining Equations (8.4.44) and (8.4.45), the total acoustic power output of the system is:

$$\begin{aligned} W_{tot} &= W_p + W_c \\ &= \frac{1}{2}\mathrm{Re}\left\{\boldsymbol{q}_c^{\mathrm{H}}\boldsymbol{Z}_{c/c}^{\mathrm{H}}\boldsymbol{q}_c + \boldsymbol{q}_c^{\mathrm{H}}\boldsymbol{Z}_{c/p}^{\mathrm{H}}\boldsymbol{q}_p + \boldsymbol{q}_p^{\mathrm{H}}\boldsymbol{Z}_{p/c}^{\mathrm{H}}\boldsymbol{q}_c + \boldsymbol{q}_p^{\mathrm{H}}\boldsymbol{Z}_{p/p}^{\mathrm{H}}\boldsymbol{q}_p\right\} \end{aligned} \tag{8.4.48a,b}$$

Equation (8.4.48) can be re-expressed in a format more amenable to minimisation using quadratic optimisation techniques, which by now are well known to the reader, as:

$$W_{tot} = \boldsymbol{q}_c^{\mathrm{H}}\boldsymbol{A}_w\boldsymbol{q}_c + \boldsymbol{q}_c^{\mathrm{H}}\boldsymbol{b} + \boldsymbol{b}^{\mathrm{H}}\boldsymbol{q}_c + c \tag{8.4.49}$$

where

$$\boldsymbol{A}_w = \frac{1}{2}\mathrm{Re}\left\{\boldsymbol{Z}_{c/c}\right\} \tag{8.4.50}$$

$$\boldsymbol{b}_w = \frac{1}{2}\mathrm{Re}\left\{\boldsymbol{Z}_{p/c}^{\mathrm{H}}\,\boldsymbol{q}_p\right\} \tag{8.4.51}$$

and

$$c = \frac{1}{2}\boldsymbol{q}_p^{\mathrm{H}}\,\mathrm{Re}\left\{\boldsymbol{Z}_{p/p}^{\mathrm{H}}\right\}\boldsymbol{q}_p \tag{8.4.52}$$

However, the terms in these equations are exactly the same as those given in Equations (8.4.33), (8.4.35), and (8.4.37) respectively, derived using the far-field measure of acoustic power as a basis. Therefore, once again the quadratic expression is the same, whether a near-field or far-field measure of power is employed in its derivation.

The multiple source formulation developed in this section can now be used to examine the influence that control source and error sensor placement has upon the performance of the overall system.

8.5 EFFECT OF TRANSDUCER LOCATION

In the previous sections, expressions defining control source volume velocities that will minimise the total radiated acoustic power from a set of acoustic monopole sources, or that will minimise the squared acoustic pressure amplitude at a number of discrete error locations, have been derived. Being able to calculate this quantity is a first step in the design of an active control system. The next step is to use these quantities to optimally place the control sources and error sensors. The aim of this section is to use some of the results of the previous sections to demonstrate the effects that transducer placement has upon the performance of the monopole-based systems considered thus far. This will provide a 'primer' for the structural radiation discussion undertaken later in this chapter.

As mentioned in the introduction to this chapter and elsewhere in this book, the design of a feedforward active control system can be viewed as a hierarchical procedure aimed at optimising the performance of the four principal components of the system; introduction of the control signal, extraction of an error signal, acquisition of a primary disturbance-correlated reference signal for the controller, and design of the controller itself. If any one of these facets is improperly designed, then the system is doomed, as it is only as good as its weakest link. The first two of these facets are of principal interest here.

While it is true that all components must be optimised for the system to function, the placement of the control sources, to introduce the controlling disturbance into the system, is arguably the most important stage in designing the active control system. This is because the location of the control sources will place the absolute bound upon the level of acoustic power attenuation which can be achieved, regardless of the number of error sensors used to sense the residual sound field, and regardless of the quality and expense of the electronics used to drive the system. Unfortunately, optimising the control source locations is also one of the least straightforward aspects of designing an active control system. This is because acoustic power attenuation is not a linear function of control source location. Thus, while it is possible to determine directly the maximum levels of acoustic power attenuation that are possible with a given control source arrangement using the quadratic optimisation approach of the previous sections, it is not possible to use such an approach to determine the control source locations that will maximise these levels. Usually, to place the control sources, some form of (numerical) 'search' routine must be undertaken, using the calculation of acoustic power attenuation for a given control source arrangement (calculated using the quadratic optimisation procedure of the previous sections) to provide the error criterion. There are a few extremely simple monopole arrangements where an equation for power reduction in terms of control source location can be derived, and it is these that will be discussed here.

For the most simple system of all, where a single monopole control source is used to attenuate acoustic radiation from a single monopole primary source, the only location consideration is separation distance. It was shown in Section 8.3 that the effect which separation distance has upon the levels of acoustic power attenuation that can be achieved is quantified in the expression:

$$\frac{W_{\text{controlled}}}{W_{\text{uncontrolled}}} = 1 - \text{sinc}^2 kd \tag{8.5.1}$$

It is worthwhile emphasising again the extremely rapid reduction in acoustic power attenuation that accompanies an increase in the distance between the primary and control sources. Using the approximate measure of 10 dB reduction at a separation distance of 1/10 wavelength, consider that at 343 Hz, the control source must be with 100 mm of the primary source to provide this rather moderate amount of attenuation. If the separation distance is increased to 500 mm, no attenuation of the total radiated acoustic power is possible. Close source proximity is a must in active control of sound power.

The addition of a second control source can improve the acoustic power attenuation when the sources are in close proximity, but can do little to improve the result when the source separation distance is greater than one half of an acoustic wavelength. Consider the arrangement depicted in Figure 8.15, where two monopole control sources are used to attenuate the radiated acoustic power from a single monopole primary source. All sources are on a common axis, and the control sources are an equal distance from the primary source. The maximum levels of acoustic power attenuation for this arrangement are plotted as a function of separation distance in Figure 8.16. Note that while the result is better than that obtained using a single control source for 'small' separation distances, the improvement deteriorates to a negligible level when the separation distance exceeds one half of an acoustic wavelength.

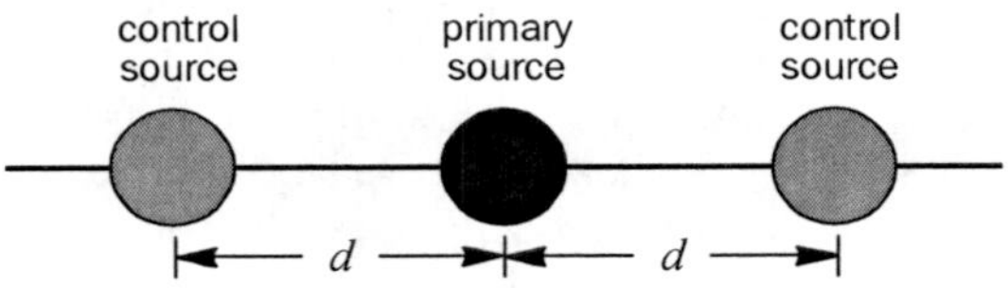

Figure 8.15 Single monopole primary source/dual monopole control source system arrangement.

A third case where the effect of control source location on acoustic power attenuation can be relatively easily calculated is that of a single monopole control source being used to attenuate acoustic radiation from a dipole-like pair of primary sources. The optimum control source volume velocity relationship can be calculated using the quadratic optimisation approach outlined in the previous section.

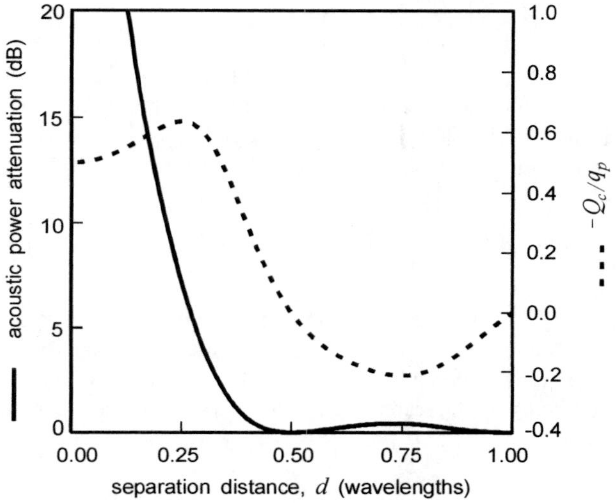

Figure 8.16 Optimum control source volume velocities and maximum acoustic power attenuation as a function of separation distance d for the single primary/dual control monopole source system.

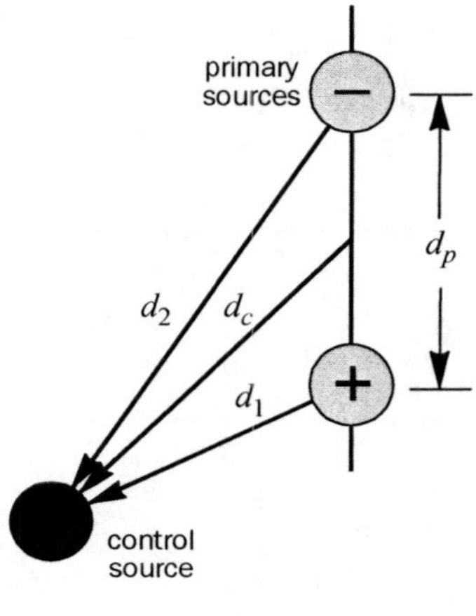

Figure 8.17 Dual primary source/monopole control source system arrangement.

For the arrangement shown in Figure 8.17, it can easily be deduced that:

$$a = \frac{\omega^2 \rho_0}{8\pi c_0}; \qquad b = \frac{\omega^2 \rho_0}{8\pi c_0} q_p(\text{sinc}\, kd_1 - \text{sinc}\, kd_2)$$

$$\text{and } c = \frac{\omega^2 \rho_0}{4\pi c_0} |q_p|^2 (1 - \text{sinc}\, kd_p) \tag{8.5.2a–c}$$

where q_p is the volume velocity of each primary source, taken to be equal in amplitude for the two sources but opposite in sign; d_1 and d_2 are the separation distances between the control source and the two primary sources; and d_p is the separation distance between the two primary sources. Substituting these definitions into Equation (8.4.31), for the case being considered here, the minimum achievable radiated acoustic power is:

$$W_{\text{min}} = \frac{\omega^2 \rho_0}{8\pi c_0} |q_p|^2 \left\{ 2(1 - \text{sinc}\, kd_p) - (\text{sinc}\, kd_1 - \text{sinc}\, kd_2)^2 \right\} \tag{8.5.3}$$

From Equation (8.5.3) it is clear that the optimum control source volume velocity is one that maximises the quantity $(\text{sinc}\, kd_1 - \text{sinc}\, kd_2)^2$, or maximises the difference between sinc kd_1 and sinc kd_2. The problem of determining the optimum control source location can therefore be solved heuristically, referring to the shape of the sinc function curve shown in Figure 8.18.

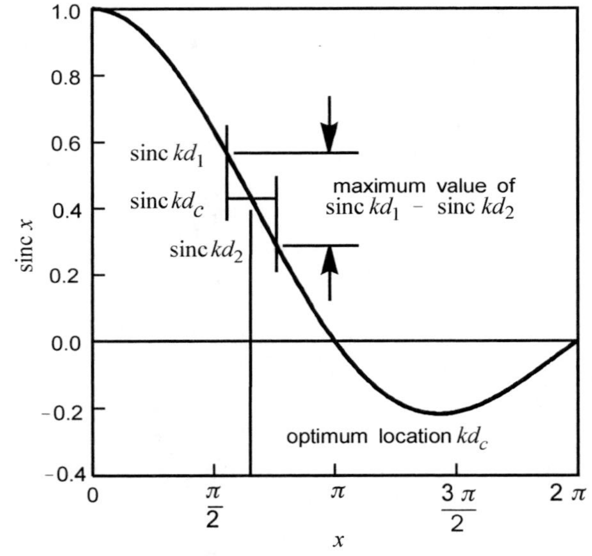

Figure 8.18 Plot of sinc function used to determine optimum control source location.

Optimally, the control source should be placed on the primary source axis (where $(d_1 - d_2)$ will be maximised), situated at a distance from the mid-point between the two sources such that the slope $(\text{sinc}\, kd_1 - \text{sinc}\, kd_2)$ is maximised, which for this separation distance and frequency is approximately $kd_c = 2\pi/3$. This trend is displayed clearly in Figure 8.19, which depicts the levels of acoustic power attenuation possible for a given control source location, where the primary source separation distance is 0.1 m and the excitation frequency is $\omega = 2\pi c_0$. Note that the optimum control source placements are at mirror image locations of $kd_c \approx \pm 2\pi/3$, and are not at the location of either primary source.

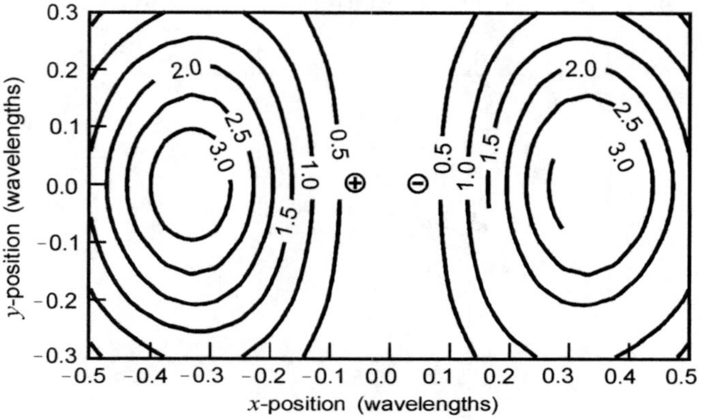

Figure 8.19 Maximum acoustic power attenuation as a function of control source location for a dipole primary source, primary source separation distance is 1/10 of a wavelength.

If the strengths of the two primary sources are unequal, the control source placement characteristics outlined above change. Consider again the arrangement depicted in Figure 8.17, but where the volume velocities of primary source 1 and 2 are αq_p and $-q_p$ respectively, where α can be viewed as a 'dipole modifying factor'. Working through the same analysis referenced previously, the terms *a*, *b* and *c* are now found to be equal to:

$$a = \frac{\omega^2 \rho_0}{8\pi c_0}$$

$$b = \frac{\omega^2 \rho_0}{8\pi c_0} q_p(\alpha \operatorname{sinc} kd_1 - \operatorname{sinc} kd_2) \tag{8.5.4a–c}$$

$$c = \frac{\omega^2 \rho_0}{4\pi c_0} |q_p|^2 (1 + \alpha^2 + (1 - \alpha)\operatorname{sinc} kd_p)$$

This leads to a minimum total acoustic power output of:

$$W_{\min} = \frac{\omega^2 \rho_0}{8\pi c_0} |q_p|^2 \left\{ (1 + \alpha^2 + (1 - \alpha)\operatorname{sinc} kd_p) - (\alpha \operatorname{sinc} kd_1 - \operatorname{sinc} kd_2)^2 \right\} \tag{8.5.5}$$

The influence of the different primary monopole source strengths can be qualified by considering Equation (8.5.5), where it can be seen that to achieve the lowest minimum acoustic power output, the term $(\alpha \operatorname{sinc} kd_1 - \operatorname{sinc} kd_2)^2$ must be maximised. In light of the characteristics of the sinc function, illustrated in Figure 8.20, it can be deduced that as α increases from a value of 1.0, the optimum control source location moves closer to the stronger source, so that the difference between (α sinc kd_1) and (sinc kd_2) is maximised. This difference also grows in size, meaning that larger values of attenuation are possible. This intuitively makes sense, as the monopole component of the sound field has increased, and the single control source will chiefly attenuate this component. Further, as α departs from a unity value, so too does the phenomenon of mirror image optimum control source locations.

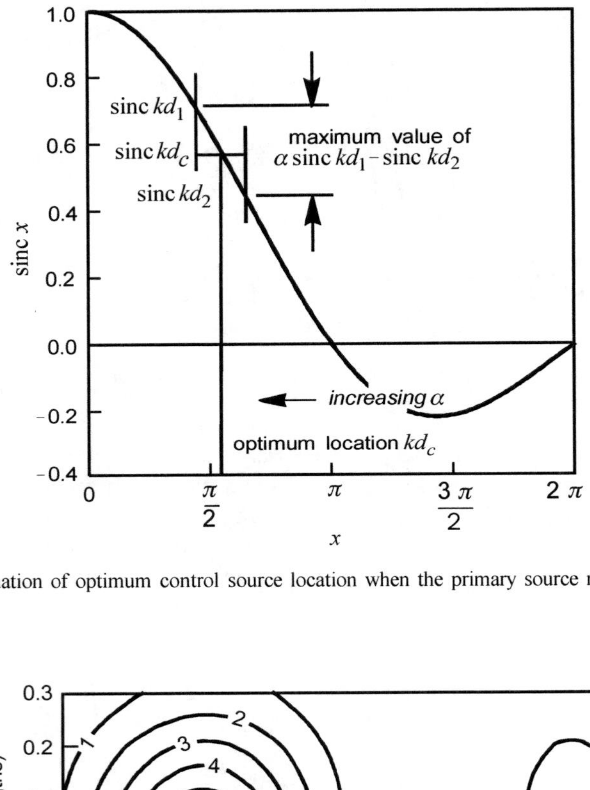

Figure 8.20 Evaluation of optimum control source location when the primary source monopoles are of unequal strengths.

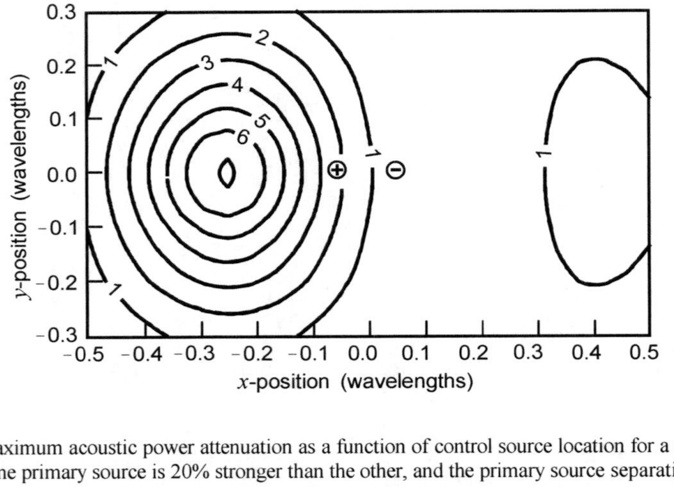

Figure 8.21 Maximum acoustic power attenuation as a function of control source location for a dipole-like primary source, where one primary source is 20% stronger than the other, and the primary source separation distance is 1/10 of a wavelength.

These qualitative views are mirrored in Figure 8.21, which illustrates the maximum levels of acoustic power attenuation as a function of control source location for a single sinusoidal primary source output with only a small deviation from equal source strengths, $\alpha = 1.2$. Here, the optimum control source location has moved from a value $kd_c = 2\pi/3$ to a position $kd_c = \pi/2$. Note also that the maximum levels of sound power attenuation have increased by a factor of two.

Once the control source location has been selected, the next step in the design of the active control system is to optimally locate the error sensor(s). For sinusoidal excitation, the control source output that will minimise the acoustic pressure amplitude at a specific location in space for a single control source and error sensor was given in Equation (8.2.5). For the optimum error microphone positions this volume-velocity relationship will be equal to the optimum for minimising the total radiated acoustic power, given in Equation (8.3.12). For primary source/control source arrangements that are closely spaced, therefore, the nearest optimum error sensor placements will run on a line perpendicular to the control source/primary source axis, centred at a position between the primary and control sources such that r_c/r_p = sinc kd, which will always be closer to the control source. Figure 8.22 illustrates the acoustic power attenuation resulting from minimising the acoustic pressure amplitude at a single location for harmonic excitation, with a primary source/control source separation distance of 0.1λ and an exciting frequency $\omega = 2\pi c_0$. The outlined characteristics of optimal error sensor locations are clearly evident.

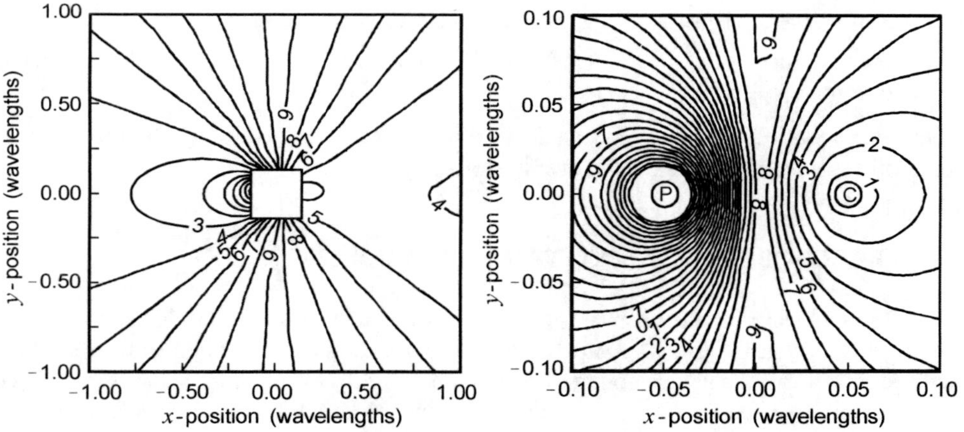

Figure 8.22 Acoustic power attenuation as a function of error sensor placement, single monopole primary and control sources separated by 1/10 wavelength.

There are two important points that are evident from this result. First, placing an error sensor in the near-field of the sources, especially the primary source, is a 'risky' pursuit; the acoustic power attenuation achieved when minimising the acoustic pressure at the sensor(s) varies rapidly with location. Second, the optimum error sensor location(s) are always at the locations of greatest acoustic pressure attenuation when the control source is generating the optimum volume-velocity relationship. Both of these results are general for the active control of free-field sound radiation.

8.5.1 Comparison of Near-Field Error Sensing Strategies

Notwithstanding the discussion in the previous section, in many cases it is not possible to locate error sensors in the far-field of the radiating structure. In these cases it is necessary to use some measure of the near-field as an error criterion for the minimisation of the radiated sound power. Qiu et al. (1998) compared results using eight different error criteria for sensing in the near-field of a monopole primary source and a dipole primary source, both with a monopole control source, which are listed below:

1. Potential energy density at a point (equivalent to the minimisation of the sound pressure at a point);
2. Kinetic energy density at a point (equivalent to the minimisation of the squared particle velocity at a point);
3. Total energy density at a point;
4. Mean active intensity in the radial direction at a point;
5. Sum of the potential energy density at a number of points through the near-field (equivalent to the minimisation of the sum of the squared pressure at a number of points throughout the near-field);
6. Sum of the kinetic energy density at a number of points through the near-field (equivalent to the minimisation of the sum of the squared particle velocity at a number of points throughout the near-field);
7. Sum of the total energy density over a number of points through the near-field; and
8. Sum of the mean active intensity in the radial direction over a number of points encompassing the sound sources.

They found that best results for near-field error sensing were obtained by minimising the total active sound intensity summed over a number of points.

8.6 REFERENCE SENSOR LOCATION CONSIDERATIONS

Thus far in this section, the problem of calculating the attenuation of a radiated sound field for a given ‘physical system’, control source(s) and error sensor(s), arrangement has been considered. In doing this, an optimum open-loop control law for minimising the radiated acoustic power or acoustic pressure at the error sensor location has been calculated and the resultant signal applied to the control source(s). The assumption behind this approach is that when a feedforward control system is attached to the physical system through the control sources and error sensors, it will do the same thing. However, feedforward control systems require a reference signal, which is correlated with the impending primary disturbance, from which to derive the control signal. When the unwanted (primary) disturbance is periodic, it is often possible to acquire this reference signal directly from the source of excitation, such as from a tachometer attached to an item of rotating machinery, although in general a sensor such as a microphone must be used to measure the primary and disturbance, providing the reference signal to the feedforward controller.

This, however, can introduce the new problems into the system design. Referring to the system arrangement shown in Figure 8.23, which depicts a single monopole control source being used to attenuate the acoustic radiation from a single monopole primary source with a single reference sensor and error sensor (microphones), it can be seen that feedback from the control source to the reference sensor can occur. From the equivalent block diagram in Figure 8.24, it is readily apparent that this feedback loop has the potential to drive the system unstable, purely from the geometry of the system. In this section, this problem, which was discussed at some length in Sections 6.15 to 6.17, will be re-investigated, and some criteria for maintaining system stability will be derived.

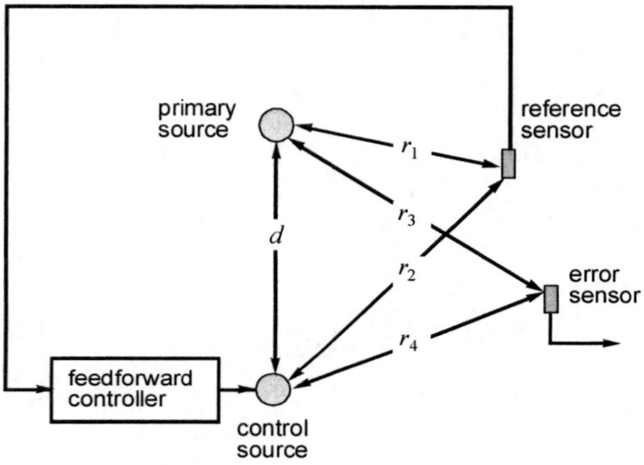

Figure 8.23 Single primary source, control source and error sensor system with feedforward control and reference sensor (single-channel feedforward active control system).

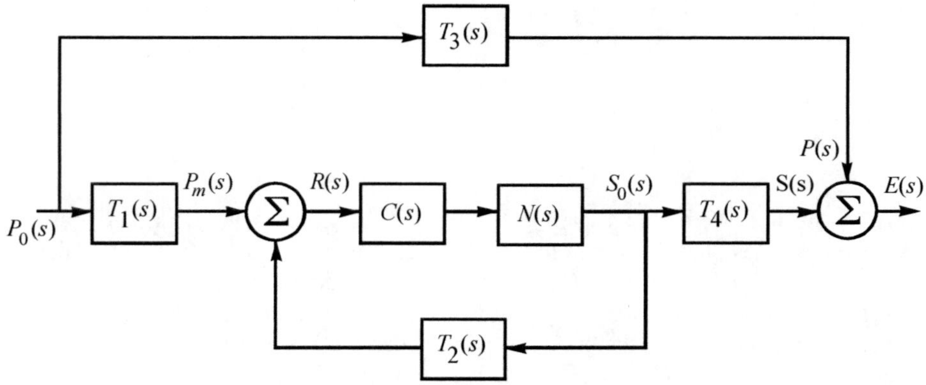

Figure 8.24 Block diagram of single-channel feedforward active control system shown in Figure 8.23.

8.6.1 Problem Formulation

For the control system to null the acoustic pressure at the error sensor, the control and primary signals at its location must be equal in amplitude and opposite in phase:

$$S(s) = -P(s) \tag{8.6.1}$$

From Figure 8.24, it can be seen that the primary source signal at the error sensor is equal to the primary source signal at the origin of the disturbance, $P_0(s)$, modified by the transfer function between this point and the error sensor, $T_3(s)$:

$$P(s) = T_3(s)P_0(s) \tag{8.6.2}$$

Similarly, the control source signal at the error sensor is equal to the control source signal $S_0(s)$, modified by the transfer function between this location and the error sensor, $T_4(s)$:

$$S(s) = T_4(s)S_0(s) \tag{8.6.3}$$

The control source signal can also be expressed in terms of the primary source signal as:

$$S_0(s) = \frac{C(s)N(s)T_1(s)}{1 - C(s)N(s)X_2(s)}P_0(s) \tag{8.6.4}$$

where $C(s)$ is the Laplace transform of the controller; $N(s)$ is the Laplace transform of the other 'peripheral' components of the control system, including the reference sensor and control source frequency-response characteristics, as well the frequency-response characteristics of any filters, amplifiers and other electronics; $T_1(s)$ is the transfer function between the primary disturbance origin and the reference sensor location; and $T_2(s)$ is the transfer function between the control source location and the reference sensor location. Substituting Equations (8.6.2), (8.6.3) and (8.6.4) into Equation (8.6.1), the frequency-response characteristics of the optimum controller, which will null the acoustic pressure at the error sensor, are defined by the relationship:

$$C(s) = \frac{T_3(s)}{N(s)[T_2(s)T_3(s) - T_1(s)T_4(s)]} \tag{8.6.5}$$

It can be observed from Equation (8.6.5) that feedback of the control signal to the reference sensor has placed poles in the optimum control source transfer function, the same result which was encountered in Chapter 6 when considering feedback to the reference sensor for a duct noise problem. Thus, feedback alters the result from the simple gain and phase changes that described the optimum control source characteristics without feedback to the reference sensor, which have been derived thus far in this chapter. There are several important conclusions that can be drawn from this result:

1. An adaptive feedforward control system based upon infinite impulse response (IIR) filters will be better suited to this problem than a system based upon finite impulse response (FIR) filters, owing to the IIR filter being a pole-zero architecture and the FIR filter being an all-zero architecture.
2. For some physical system arrangements, the required control system transfer function will be infinite.

3. It is possible that the feedback to the reference sensor, which forms a closed-loop system, will cause the control system to go unstable when an attempt to null the acoustic pressure at the error sensor is made. It is this last point that will be of interest here.

To examine the possibility of instability arising from feedback of the control signal to the reference sensor, only the closed-loop part of the system shown in Figure 8.25 need be considered. Using this arrangement, the control signal is now defined by the expression:

$$S(s) = C(s)N(s)[P_m(s) + T_2(s)S(s)] \tag{8.6.6}$$

where $P_m(s)$ is the primary source disturbance as measured by the reference sensor. This can be expressed in the more standard form of a closed-loop transfer function as:

$$\frac{S(s)}{P_m(s)} = \frac{C(s)N(s)}{1 + G(s)} \tag{8.6.7}$$

where $G(s)$ is defined by:

$$G(s) = -C(s)N(s)T_2(s) \tag{8.6.8}$$

From Equation (8.6.7), the roots of the characteristic equation $(1 + G(s))$ should all lie on the left-hand side of the imaginary axis for the system to be stable.

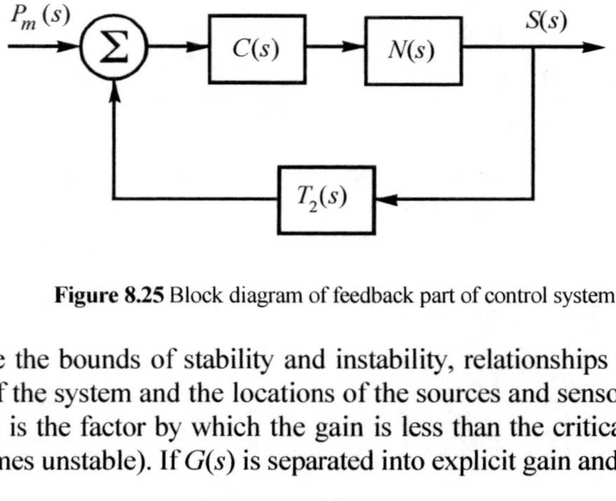

Figure 8.25 Block diagram of feedback part of control system.

To examine the bounds of stability and instability, relationships between the gain and phase margins of the system and the locations of the sources and sensors need to be derived. The gain margin is the factor by which the gain is less than the critical value (above which the system becomes unstable). If $G(s)$ is separated into explicit gain and phase components:

$$G(j\omega) = B(\omega)e^{j\theta(\omega)} \tag{8.6.9}$$

then the gain margin K_g is defined by the expression:

$$K_g = \frac{1}{B(\omega)} \quad \text{when } \theta(\omega) = -\pi \tag{8.6.10}$$

Phase margin is the additional phase, K_θ, required to make the system unstable, which is the amount by which the phase exceeds $-180°$ when the gain is equal to 1. For the problem being considered here, the phase margin is defined by:

$$K_\theta = \theta(\omega) + \pi \quad \text{when } B(\omega) = 1 \tag{8.6.11}$$

To calculate the gain and phase margins of the optimal system, the optimum controller transfer function of Equation (8.6.5) can be substituted into Equation (8.6.8). After some minor manipulation, the following is obtained (Tokhi and Leitch, 1991, 1992):

$$G(s) = \frac{1}{\dfrac{T_1(s)T_4(s)}{T_2(s)T_3(s)} - 1} \tag{8.6.12}$$

This can be simplified as:

$$G(j\omega) = \frac{1}{H(\omega)\mathrm{e}^{\,\mathrm{j}\varphi(\omega)} - 1} \tag{8.6.13}$$

where $H(\omega)$ and $\mathrm{e}^{j\varphi(\omega)}$ are gain and phase terms:

$$H(\omega)\mathrm{e}^{\,\mathrm{j}\varphi(\omega)} = \frac{T_1(\mathrm{j}\omega)T_4(\mathrm{j}\omega)}{T_2(\mathrm{j}\omega)T_3(\mathrm{j}\omega)} \tag{8.6.14}$$

Substituting Equation (8.6.14) into Equation (8.6.13) and the result into Equation (8.6.9), the gain and phase respectively of $G(s)$ are obtained as:

$$B(\omega) = \left[H^2(\omega) - 2H(\omega)\cos\varphi(\omega) + 1\right]^{-1/2} \tag{8.6.15}$$

and

$$\theta(\omega) = \tan^{-1}\frac{H(\omega)\sin\varphi(\omega)}{1 - H(\omega)\cos\varphi(\omega)} + 2m\pi, \quad m = 0, \pm 1, ... \tag{8.6.16}$$

The aim here is to examine system stability based upon the physical arrangement of the sources and sensors. To do this, the primary source is placed at a location (0,0,0), the control source at (x_s, y_s, z_s), the reference sensor at (x_r, y_r, z_r), and the error sensor at (x_e, y_e, z_e), as shown in Figure 8.26.

The various separation distances shown in the figure are now defined by the expressions:

$$r_1 = \left[x_r^2 + y_r^2 + z_r^2\right]^{1/2} \tag{8.6.17}$$

$$r_2 = \left[(x_r - x_s)^2 + (y_r - y_s)^2 + (z_r - z_s)^2\right]^{1/2} \tag{8.6.18}$$

$$r_3 = \left[x_e^2 + y_e^2 + z_e^2\right]^{1/2} \tag{8.6.19}$$

$$r_4 = \left[(x_e - x_s)^2 + (y_e - y_s)^2 + (z_e - z_s)^2\right]^{1/2} \tag{8.6.20}$$

and

$$d = \left[x_s^2 + y_s^2 + z_s^2\right]^{1/2} \tag{8.6.21}$$

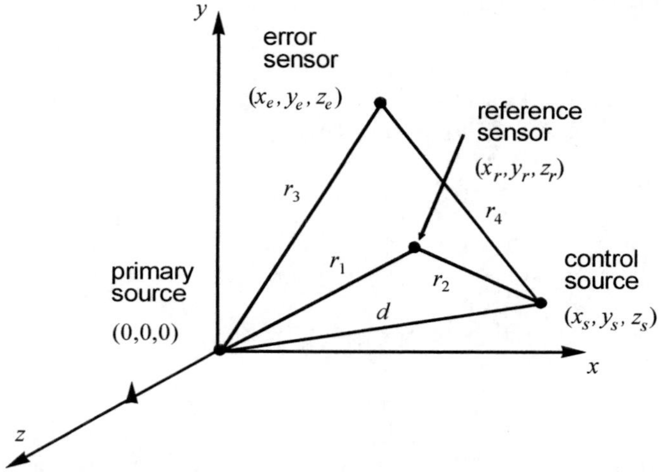

Figure 8.26 Source and sensor geometry for single-channel system.

Noting that in the acoustic medium, the transfer function between any two points is defined by:

$$T_i = \frac{e^{-jkr_i}}{r_i} \tag{8.6.22}$$

the gain and phase terms in Equation (8.6.14) respectively become (Tokhi and Leitch, 1992):

$$H(\omega) = \frac{r_2 r_3}{r_1 r_4} = \frac{\dfrac{r_3}{r_4}}{\dfrac{r_1}{r_2}} \tag{8.6.23a,b}$$

and

$$\varphi(\omega) = k\left[(r_3 - r_4) - (r_1 - r_2)\right] = k(r_{34} - r_{12}) \tag{8.6.24a,b}$$

where $r_{34} = r_3 - r_4$ and $r_{12} = r_1 - r_2$. Using these relationships, the effect that geometry has upon system stability can be examined.

8.6.2 Gain Margin

The gain margin will be examined first. To calculate gain margin, it is necessary to determine where the phase of the transfer function is 180°. Rewriting the phase relationship in Equation (8.6.16) as:

$$\frac{H(\omega)\sin\varphi(\omega)}{H(\omega)\cos\varphi(\omega) - 1} = \tan(2m + 1)\pi \tag{8.6.25}$$

it is apparent that for the phase of the transfer function to be 180°:

$$H(\omega)\sin\varphi(\omega) = 1 \quad \text{and} \quad H(\omega)\cos\varphi(\omega) - 1 < 0 \qquad (8.6.26a,b)$$

As $H(\omega)$ is a real positive number, this criterion becomes:

$$\begin{aligned} \sin\varphi(\omega) &= 0, \quad \cos\varphi(\omega) = 1 \text{ and } H(\omega) < 1 \\ \sin\varphi(\omega) &= 0, \quad \cos\varphi(\omega) = -1 \text{ and } H(\omega) > 0 \end{aligned} \qquad (8.6.27)$$

Therefore,

$$\varphi(\omega) = 2m\pi \quad \text{for } H(\omega) < 1, \quad (2m+1)\pi \quad \text{for } H(\omega) > 0 \qquad (8.6.28)$$

or

$$\varphi(\omega) = (2m+1)\pi \text{ for } H(\omega) \geq 1 \quad m\pi \text{ for } 0 < H(\omega) < 1 \qquad (8.6.29)$$

Substituting Equation (8.6.24) into Equation (8.6.29), the conditions under which the phase of $G(s)$ is −180°, where the gain margin can be evaluated, are found to be (Tokhi and Leitch, 1992):

$$r_{34} - r_{12} = \begin{cases} (2m+1)\dfrac{\lambda}{2} & \text{for } H(\omega) \geq 1 \\ m\dfrac{\lambda}{2} & \text{for } 0 \leq H(\omega) < 1 \end{cases} \qquad (8.6.30)$$

First the case where $\varphi(\omega)$ is an even multiple of π will be considered. Substituting $\varphi(\omega) = 2m\pi$ into Equation (8.6.15) produces the expression:

$$K_g = \left[H^2(\omega) - 2H(\omega) + 1\right]^{1/2} = |H(\omega) - 1| \qquad (8.6.31)$$

As $H(\omega)$ is less than 1 in this case (from Equation (8.6.29)):

$$K_g = 1 - H(\omega) \quad \text{for } 0 < H(\omega) < 1 \quad \text{and} \quad \varphi(\omega) = 2m\pi \qquad (8.6.32)$$

If $\varphi(\omega)$ is now an odd multiple of π, substituting $\varphi(\omega) = (2m+1)\pi$ into Equation (8.6.15) produces:

$$K_g = \left[H^2(\omega) + 2H(\omega) + 1\right]^{1/2} = |H(\omega) + 1| \qquad (8.6.33)$$

As $H(\omega)$ is greater than zero:

$$K_g = 1 + H(\omega) \quad \text{for } H(\omega) > 0 \quad \text{and} \quad \varphi(\omega) = (2m+1)\pi \qquad (8.6.34)$$

Combining Equations (8.6.32) and (8.6.34), the gain margin of the system is found to be:

$$K_g = \begin{cases} 1 - H(\omega) \text{ for } \varphi(\omega) = 2m\pi \text{ and } 0 < H(\omega) < 1 \\ 1 + H(\omega) \text{ for } \varphi(\omega) = (2m+1)\pi \text{ and } H(\omega) > 0 \end{cases} \qquad (8.6.35)$$

As mentioned previously, for the system to be stable, the gain margin must be greater than 1. In viewing the criteria in Equation (8.6.35), it is clear that this can only happen if $H(\omega) > 1$, in which case the gain margin is greater than 2 (Tohki and Leitch, 1992). Equation (8.6.23) shows $H(\omega)$ to be entirely defined by the location of the sources and sensors. Therefore, the requirement of $H(\omega) > 1$ can be used to examine what system arrangements will be stable, and what arrangements will be unstable.

Referring to Equation (8.6.23) and Figure 8.23, for $H(\omega)$ to be greater than 1:

$$\frac{r_3}{r_4} > \frac{r_1}{r_2} \tag{8.6.36}$$

In words, for the system to be stable, the ratio of the distance between the primary source and error sensor to the distance between the control source and error sensor must be greater than the ratio of the distance between the primary source and reference sensor and the distance between the control source and reference sensor. Further geometric consideration of this requirement can be found in Tohki and Leitch (1992).

8.6.3 Phase Margin

What remains now is to assess the phase margin. Referring to Equation (8.6.15), the transfer function gain $B(\omega)$ has a unity value when:

$$\left[H^2(\omega) - 2H(\omega)\cos\varphi(\omega) + 1\right]^{-1/2} = 1 \tag{8.6.37}$$

or

$$\cos\varphi(\omega) = \frac{H(\omega)}{2} \tag{8.6.38}$$

From the analysis of gain margin, it was found that the amplitude term $H(\omega)$ must be greater than 1 for the system to be stable. Based upon Equation (8.6.38), the limit on $H(\omega)$ is therefore,

$$1 \le H(\omega) \le 2 \tag{8.6.39}$$

Therefore, the range of $\varphi(\omega)$ is:

$$2m\pi - \frac{\pi}{3} \le \varphi(\omega) \le 2m\pi + \frac{\pi}{3} \quad \text{for } 1 \le H(\omega) \le 2 \tag{8.6.40}$$

Substituting Equation (8.6.24) into Equation (8.6.40) and simplifying yields the allowable distance difference $r_{34} - r_{12}$ as:

$$(6m - 1)\frac{\lambda}{6} \le r_{34} - r_{12} \le (6m + 1)\frac{\lambda}{6} \qquad \text{for } 1 \le H(\omega) \le 2 \tag{8.6.41}$$

Equation (8.6.46) provides the conditions under which $B(\omega) = 1$. To evaluate the resultant phase margin, note first that:

$$\sin\varphi(\omega) = \begin{cases} \dfrac{1}{2}\sqrt{4 - H^2(\omega)} & \text{for } 0 \le \varphi(\omega) \le \dfrac{(6m+1)\pi}{3} \\ -\dfrac{1}{2}\sqrt{4 - H^2(\omega)} & \text{for } \dfrac{(6m-1)\pi}{3} \le \varphi(\omega) < 0 \end{cases} \tag{8.6.42}$$

Substituting Equations (8.6.38) and (8.4.42) into Equation (8.6.16), the phase margin is found to be (Tohki and Leitch, 1992):

$$K_\theta = \begin{cases} \tan^{-1}\dfrac{H(\omega)\sqrt{4 - H^2(\omega)}}{2 - H^2(\omega)} + 2m\pi & \text{for } 0 \le \varphi(\omega) \le \dfrac{(6n+1)\pi}{3} \\ -\tan^{-1}\dfrac{H(\omega)\sqrt{4 - H^2(\omega)}}{2 - H^2(\omega)} + 2m\pi & \text{for } \dfrac{(6m-1)\pi}{3} \le \varphi(\omega) < 0 \end{cases} \tag{8.6.43}$$

Note that for values of $H(\omega)$ greater than 1, the phase margin is positive, indicating that the system is stable. Above a value of 2, however, the system may become unstable. Therefore, the conclusion that can be drawn is that for a feedforward system to be stable, the value of $H(\omega)$, defined in Equation (8.6.23) as purely a function of the location of the sources and sensors, must be between 1 and 2.

8.7 ACTIVE CONTROL OF HARMONIC SOUND RADIATION FROM PLANAR STRUCTURES: GENERAL PROBLEM FORMULATION

The work presented in the preceding sections of this chapter has been directed at the use of a set of idealised acoustic monopole control sources to attenuate the sound radiation from a second set of idealised monopole primary acoustic sources. Such a series of studies has two principal aims. First, there are many practical noise problems that can be viewed to a first approximation as originating from one or more point sources. However, possibly a more important reason for studying the relatively simple cases of the previous sections is to obtain insight, concerning both physical mechanisms and theoretical methodology, which can be used in the design of active control systems for other, possibly more complex and practical, free-field radiation problems. It is the philosophy behind this second reason which will be of use in the next sections, where the active control of harmonic sound radiation from planar, vibrating structures situated in an infinite baffle is examined. In the course of this study, the knowledge gained in the examination of the monopole problems will be used to develop a theoretical framework for obtaining optimum control source volume velocities for the minimisation of acoustic pressure at a set of discrete error locations, and for minimising the total radiated acoustic power. The exercise will then be repeated for the use of vibration sources, directly attached to the structure, to minimise the acoustic radiation. These theoretical formulations will be 'put to task' in the next section by applying them to the problem of controlling sound radiation from a vibrating plate. Finally, problems associated with controlling structural vibration with the aim of controlling the radiated sound field will be discussed.

Although a vibrating plane structure in an infinite baffle is an idealised case, it provides an opportunity to analytically study something resembling a real structural radiation problem, where the forced response of the structural modes is what is responsible for generating the sound field. In addition, there are many structures that can, to a first approximation, be represented by finite planar radiators in an infinite baffle. The results of the idealised study conducted in the next three sections will provide at least a qualitative feeling for what can be accomplished by active control, the means by which attenuation is attained, and the importance of various system parameters. Consideration of a vibrating structure also offers an opportunity to assess the potential of a different control arrangement than considered previously; the use of vibration control sources attached directly to structure, which will manipulate the velocity distribution of the structure in such a way as to attenuate the acoustic radiation.

To begin with, in this section, the theoretical framework required for examining the control of harmonic sound radiation from planar, vibrating structures using either acoustic or vibration control sources, will be developed. The development undertaken here is a generalisation of published work; see, for example, Deffayet and Nelson (1988) or Pan et al. (1992) for acoustic control source applications, or Fuller (1988, 1990) or Pan et al. (1992) for vibration source applications. As with the monopole cases discussed previously, there are two possible error criteria which can be used: minimisation of the acoustic pressure at a set of error locations, and minimisation of the total radiated acoustic power. The former of these will be examined first.

8.7.1 Minimisation of Acoustic Pressure at Discrete Locations Using Acoustic Monopole Sources

Consider an active control problem where a vibrating structure is responsible for the primary noise disturbance, and it is desired to use a set of N_c acoustic monopole sources to attenuate the acoustic pressure at a set of N_e error locations. The error criterion J_{prs} is the sum of the squared acoustic pressure amplitudes at the error locations:

$$J_{prs} = \sum_{i=1}^{N_e} |p(\boldsymbol{r}_i)|^2 = \sum_{i=1}^{N_e} p^*(\boldsymbol{r}_i)p(\boldsymbol{r}_i) \qquad (8.7.1a,b)$$

For the equivalent monopole problem of Section 8.3, the optimum control source volume velocity for minimising this error criterion was formulated by restating the criterion as a quadratic function of the vector of control source volume velocities, and applying quadratic optimisation techniques to obtain the optimal control source volume velocity vector. This would seem again to be the logical approach to take. To take this path, however, it is necessary to formulate a set of expressions for predicting the acoustic pressure at some location in space due to the vibration of the structure, which is the primary source.

The geometry that will be used for formulating this required expression is shown in Figure 8.27. A planar structure, situated in an infinite baffle, is subject to a harmonic exciting pressure field p_0 of the form $e^{j\omega t}$ at its interface with the fluid half-space. The fluid will be confined to being of relatively low density, which will allow the neglect of any fluid loading of the structure as it radiates. The velocity $v(\boldsymbol{x}, \omega)$ of the structure at frequency ω and location $\boldsymbol{x} = (x, y)$ due to this exciting force can be determined from the Green's function response equation (see Section 2.4, Equation (2.4.78)), where the integration is over the surface S of the structure:

$$v(\boldsymbol{x}, \omega) = \mathrm{j}\omega \int_S G_s(\boldsymbol{x}, \boldsymbol{x}_0, \omega)\, p_0(\boldsymbol{x}_0, \omega)\, \mathrm{d}\boldsymbol{x}_0 \tag{8.7.2}$$

where $G_s(\boldsymbol{x}, \boldsymbol{x}_0, \omega)$ is the structural Green's function, defined in Equation (2.4.76).

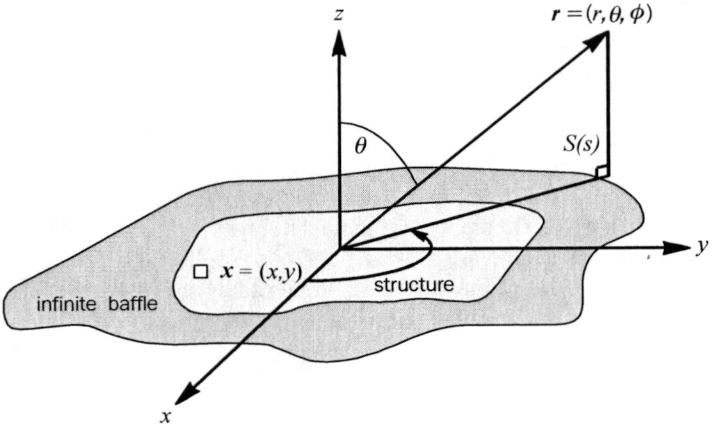

Figure 8.27 Geometry for the planar structure acoustic radiation problem formulation.

The exciting acoustic pressure field p_0 comprises two components: the blocked pressure and the radiation reaction pressure caused by the vibration of the structure. When the fluid medium surrounding the structure is of low density, so that the radiation reaction can pressure can be ignored, the exciting pressure field can be thought of as entirely due to the blocked pressure p_{bl} which is the pressure distribution from the primary disturbance:

$$p_0(\boldsymbol{x}_0, \omega) = p_{bl}(\boldsymbol{x}_0, \omega) \tag{8.7.3}$$

As harmonic excitation of frequency ω is assumed for the remainder of this section, the dependence of the acoustic pressure and structural response on ω is implicit and for convenience, the variable ω will not be included in brackets following structural velocity, acoustic pressure, Green's functions and impedance functions.

The velocity of the structure can be expanded in terms of an infinite set of *in vacuo* structural mode shape functions:

$$v(\boldsymbol{x}) = \sum_{i=1}^{\infty} v_i \psi_i(\boldsymbol{x}) \tag{8.7.4}$$

where v_i is the complex velocity amplitude of the ith structural mode. Substituting this expansion into Equation (8.7.2) enables the response of the structure to the primary exciting force to be expressed as:

$$\sum_{i=1}^{\infty} v_i \psi_i(\boldsymbol{x}) = \mathrm{j}\omega \int_S \left(\sum_{l=1}^{\infty} \frac{\psi_l(\boldsymbol{x})\, \psi_l(\boldsymbol{x}_0)}{m_l Z_l} \right) p_0(\boldsymbol{x}_0)\, \mathrm{d}\boldsymbol{x}_0 \tag{8.7.5}$$

where, the modal mass for mode l is defined by Equation (2.4.68) and the modal impedance Z_l is given by:

$$Z_l = \omega_l^2 - \omega^2 + j\omega_l^2\eta \tag{8.7.6}$$

From Equation (8.7.5), the response of a single lth mode can be written more compactly as:

$$Z_l m_l v_l = \Gamma_l \tag{8.7.7}$$

where Γ_l is the modal generalised force of the lth structural mode, which is based upon the blocked pressure only, and defined as:

$$\Gamma_l = \int_S p_{bl}(\boldsymbol{x}_0)\psi_l(\boldsymbol{x}_0)\,\mathrm{d}\boldsymbol{x}_0 = \int_S p_0(\boldsymbol{x}_0)\psi_l(\boldsymbol{x}_0)\,\mathrm{d}\boldsymbol{x}_0 \tag{8.7.8a,b}$$

From the form of Equation (8.7.5), it is apparent that when the forced response of the structure is expressed in terms of the normal modes of vibration, the problem is decoupled; that is, it can be expressed as the sum of an infinite set of (complex) scalar equations. This comes about as a direct result of neglecting fluid loading on the structure, assuming that it is radiating sound into air or some other low density fluid medium. If the fluid medium is of greater density, such as water, the assumption cannot be made, and the set of modal equations becomes coupled, thus complicating the analysis. This coupling is described, for example, by Davies (1971), Pope and Leibowitz (1974), Lomas and Hayek (1977), and Sandman (1977), as well as by Fahy (1985) and Junger and Feit (1986); active control of infinite fluid-loaded plates is described by Gu and Fuller (1991), and for finite plates by Meirovitch and Thangjitham (1990a) and Gu and Fuller (1993).

The response of the structure, expressed in Equation (8.7.7) as an infinite summation of *in vacuo* structural modes, can be *estimated* by summing over m structural modes only (do not confuse this modal index m with the modal mass m_l). Making this truncation, Equation (8.7.7) can be written in matrix form for the entire set of m structural modes as:

$$\boldsymbol{Z}_I \boldsymbol{v} = \boldsymbol{\Gamma} \tag{8.7.9}$$

where $\boldsymbol{v}$ is the $(m \times 1)$ vector of complex modal amplitudes, $\boldsymbol{Z}_I$ is the $(m \times m)$ diagonal matrix of modal input transfer functions, the (i,i)th element of which is:

$$z_I(i,i) = m_i Z_i \tag{8.7.10}$$

where Z_i is defined by Equation (8.7.6) and $\boldsymbol{\Gamma}$ is the $(m \times 1)$ vector of (blocked pressure) modal generalised forces. Thus, the amplitude of the structural modes under primary excitation can be expressed as:

$$\boldsymbol{v}_p = \boldsymbol{Z}_I^{-1}\boldsymbol{\Gamma}_p \tag{8.7.11}$$

where the subscript $\boldsymbol{p}$ denotes primary excitation.

Once the planar, infinitely baffled structure has been set into motion, the resultant acoustic pressure field can be calculated using the Rayleigh integral as follows:

$$p(\boldsymbol{r}) = \mathrm{j}\omega\rho_0 \int_S \frac{v(\boldsymbol{x})\mathrm{e}^{-\mathrm{j}kr}}{2\pi r} \mathrm{d}\boldsymbol{x} = \mathrm{j}\omega\rho_0 \int_S G(\boldsymbol{x},\boldsymbol{r}) v(\boldsymbol{x}) \,\mathrm{d}\boldsymbol{x}$$

$$= \mathrm{j}\omega\rho_0 \sum_{i=1}^{m} v_i \int_S G(\boldsymbol{x},\boldsymbol{r}) \psi_i(\boldsymbol{x}) \mathrm{d}\boldsymbol{x} \qquad (8.7.12\text{a–c})$$

where $G(\boldsymbol{x}, \boldsymbol{r})$ is the free-field Green's function of Equation (2.4.5), modified to account for the presence of the baffle and thus defined as:

$$G(\boldsymbol{x},\boldsymbol{r}) = \frac{\mathrm{e}^{-\mathrm{j}kR}}{2\pi R} \qquad (8.7.13)$$

where $R = |\boldsymbol{x} - \boldsymbol{r}|$ is the separation distance between the point on the structure at $\boldsymbol{x}$ and the point in the acoustic space at $\boldsymbol{r}$. Note here that Equation (8.7.12) is identical in form to Equation (8.2.1), which described the sound pressure at some location due to a monopole acoustic source, with the source volume velocity equal to:

$$q = 2 \int_S v(\boldsymbol{x}) \,\mathrm{d}\boldsymbol{x} \qquad (8.7.14)$$

The constant 2 is included because the sound field is generated using only half of the total fluid displacement of the structure, owing to the infinite baffle which blocks any contribution from the back of the plate.

Equation (8.7.12) shows that the total acoustic pressure at any location $\boldsymbol{r}$ in space is constructed from the sum of contributions from the m structural modes. The radiated sound pressure can therefore be expressed in matrix form as:

$$p(\boldsymbol{r}) = \boldsymbol{z}_{rad}^{\mathrm{T}} \boldsymbol{v} \qquad (8.7.15)$$

where $\boldsymbol{z}_{rad}$ is the $(m \times 1)$ vector of modal radiation transfer functions, the ith element of which is:

$$z_{rad}(i, \boldsymbol{r}) = \mathrm{j}\omega\rho_0 \int_S G(\boldsymbol{x}, \boldsymbol{r}) \psi_i(\boldsymbol{x}) \,\mathrm{d}\boldsymbol{x} = \frac{\mathrm{j}\rho_0\omega}{2\pi r} \int_S \mathrm{e}^{-\mathrm{j}kr} \psi_i(\boldsymbol{x}) \,\mathrm{d}\boldsymbol{x} \qquad (8.7.16\text{a,b})$$

Combining Equations (8.7.11) and (8.7.16) enables the radiated acoustic field under primary excitation to be evaluated as a function of the primary (pressure) forcing function:

$$p_p(\boldsymbol{r}) = \boldsymbol{z}_{rad}^{\mathrm{T}} \boldsymbol{v}_p = \boldsymbol{z}_{rad}^{\mathrm{T}} \boldsymbol{Z}_I^{-1} \boldsymbol{\Gamma}_p \qquad (8.7.17)$$

The problem of formulating a description of the sound field radiated by the control monopole acoustic sources will now be considered. The sound field radiated by the monopole sources in the arrangement being considered here will be somewhat different from that considered previously in this chapter, owing to the structure/infinite baffle arrangement, which will cause a reflection of the sound field. Making the assumption that the velocity distribution of the structure will be unaffected by the inclusion of the control sources in the system (another result of assuming a non-dense fluid medium), the total acoustic pressure at any location $\boldsymbol{r}$ in space during the operation of any particular control source can, in this instance, be modelled as coming from both the control source and its mirror image, as shown in Figure 8.28; the sound pressure radiated by each monopole control source can be determined by considering the source and its mirror image as two separate entities.

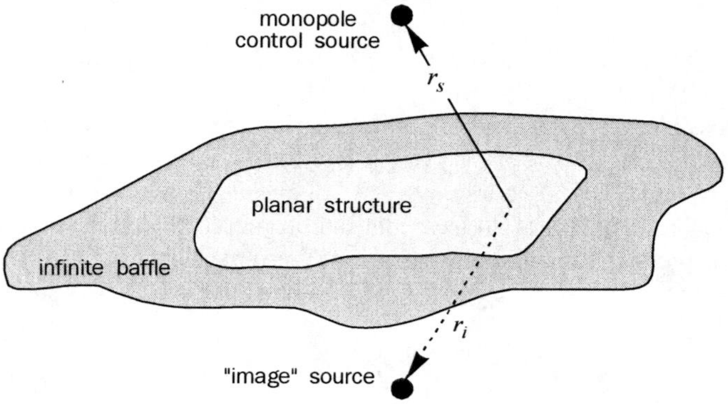

Figure 8.28 Monopole control source/image source arrangement.

Using the expression for the sound field at location produced by a monopole source given in Equation (8.2.1), the radiated sound field produced by any singular control source in the presence of the reflecting surface is:

$$p_c(\boldsymbol{r}) = \frac{\mathrm{j}\omega\rho_0 q_c \mathrm{e}^{-\mathrm{j}kr_s}}{4\pi r_s} + \frac{\mathrm{j}\omega\rho_0 q_c \mathrm{e}^{-\mathrm{j}kr_i}}{4\pi r_i} \tag{8.7.18}$$

where the subscripts s and i denote the distance from the location of the source and the mirror image respectively. This can be rewritten more compactly as:

$$p_c(\boldsymbol{r}) = \frac{\mathrm{j}\rho_0\omega q_c}{4\pi}\left(\frac{\mathrm{e}^{-\mathrm{j}kr_s}}{r_s} + \frac{\mathrm{e}^{-\mathrm{j}kr_i}}{r_i}\right) \tag{8.7.19}$$

If N_c monopole control sources are used, the total control source-generated acoustic field will be the superposition of the sound pressures radiated by each of the sources:

$$p_c(\boldsymbol{r}) = \sum_{n=1}^{N_c} \frac{\mathrm{j}\rho_0\omega q_{c,n}}{4\pi}\left(\frac{\mathrm{e}^{-\mathrm{j}kr_{n,s}}}{r_{n,s}} + \frac{\mathrm{e}^{-\mathrm{j}kr_{n,i}}}{r_{n,i}}\right) \tag{8.7.20}$$

This can be expressed in matrix form as:

$$p_c(\boldsymbol{r}) = \boldsymbol{z}_{mono}^{\mathrm{T}}(\boldsymbol{r})\boldsymbol{q}_c \tag{8.7.21}$$

where z_{mono} is the ($N_c \times 1$) vector of control source radiation transfer functions, which define the relationship between source volume velocity and radiated acoustic pressure, and $\boldsymbol{q}_c$ is the ($N_c \times 1$) vector of control source volume velocities. The nth element of z_{mono} is:

$$z_{mono}(\boldsymbol{r},n) = \frac{\mathrm{j}\rho_0\omega}{4\pi}\left(\frac{\mathrm{e}^{-\mathrm{j}kr_{n,s}}}{r_{n,s}} + \frac{\mathrm{e}^{-\mathrm{j}kr_{n,i}}}{r_{n,i}}\right) \tag{8.7.22}$$

where $r_{n,s}$ and $r_{n,i}$ are the distances from the nth actual and image source locations respectively.

Having derived matrix expressions for the sound pressure generated at any point in space due to the primary noise source (the radiating structure) and the monopole control sources, it is possible to express the error criterion, the sum of the squared acoustic pressure amplitudes at a discrete set of error sensing locations, as a quadratic function of control source volume velocities. The optimum set of volume velocities can then be calculated by means of the quadratic optimisation techniques already used in this chapter. The sum of the squared acoustic pressure amplitudes at the set of N_e error sensing locations can be expressed as:

$$J_p = \sum_{i=1}^{N_e} |p(\boldsymbol{r}_i)|^2 = \sum_{i=1}^{N_e} p^*(\boldsymbol{r}_i)p(\boldsymbol{r}_i) = \boldsymbol{p}_e^{\mathrm{H}}\boldsymbol{p}_e \tag{8.7.23a–c}$$

where $\boldsymbol{p}_e$ is the ($N_e \times 1$) vector of acoustic pressures at the error locations. As the system is linear, the total acoustic pressures are the superposition of the primary and control source generated pressures:

$$\boldsymbol{p}_e = \boldsymbol{p}_p + \boldsymbol{p}_c \tag{8.7.24}$$

where $\boldsymbol{p}_p$ and $\boldsymbol{p}_c$ are the ($N_e \times 1$) vectors of primary and control source generated pressures at the error locations respectively, so that:

$$J_p = \boldsymbol{p}_e^{\mathrm{H}}\boldsymbol{p}_e = \boldsymbol{p}_c^{\mathrm{H}}\boldsymbol{p}_c + \boldsymbol{p}_c^{\mathrm{H}}\boldsymbol{p}_p + \boldsymbol{p}_p^{\mathrm{H}}\boldsymbol{p}_c + \boldsymbol{p}_p^{\mathrm{H}}\boldsymbol{p}_p \tag{8.7.25a,b}$$

The vector of primary source pressures can be re-expressed using Equation (8.7.15) as:

$$\boldsymbol{p}_p = \boldsymbol{Z}_r\boldsymbol{v}_p \tag{8.7.26}$$

where $\boldsymbol{Z}_r$ is the ($N_e \times m$) matrix of modal radiation transfer functions between the m structural modes being used in the calculation and the N_e error locations, the rows of which are the transposed vectors of structural modal radiation transfer functions to each error location, defined in Equation (8.7.16):

$$\boldsymbol{Z}_r = \begin{bmatrix} z_{rad}^{\mathrm{T}}(\boldsymbol{r}_1) \\ z_{rad}^{\mathrm{T}}(\boldsymbol{r}_2) \\ \vdots \\ z_{rad}^{\mathrm{T}}(\boldsymbol{r}_{N_e}) \end{bmatrix} \tag{8.7.27}$$

The vector of control source pressures can be re-expressed in terms of Equation (8.7.21) as:

$$\boldsymbol{p}_c = \boldsymbol{Z}_m \boldsymbol{q}_c \tag{8.7.28}$$

where $\boldsymbol{Z}_m$ is the $(N_e \times N_c)$ matrix of radiation transfer functions between the N_c monopole control sources and the N_e error locations, the rows of which are the transposed vectors of control source radiation transfer functions to each error location, defined in Equation (8.7.22):

$$\boldsymbol{Z}_m = \begin{bmatrix} \boldsymbol{z}_{mono}^{\mathrm{T}}(\boldsymbol{r}_1) \\ \boldsymbol{z}_{mono}^{\mathrm{T}}(\boldsymbol{r}_2) \\ \vdots \\ \boldsymbol{z}_{mono}^{\mathrm{T}}(\boldsymbol{r}_{N_e}) \end{bmatrix} \tag{8.7.29}$$

Substituting Equations (8.7.26) and (8.7.28) into the error criterion as stated in Equation (8.7.25) enables the latter to be written as:

$$J_{prs} = \boldsymbol{q}_c^{\mathrm{H}} \boldsymbol{Z}_m^{\mathrm{H}} \boldsymbol{Z}_m \boldsymbol{q}_c + \boldsymbol{q}_c^{\mathrm{H}} \boldsymbol{Z}_m^{\mathrm{H}} \boldsymbol{Z}_r \boldsymbol{v}_p + \boldsymbol{v}_p^{\mathrm{H}} \boldsymbol{Z}_r^{\mathrm{H}} \boldsymbol{Z}_m \boldsymbol{q}_c + \boldsymbol{v}_p^{\mathrm{H}} \boldsymbol{Z}_r^{\mathrm{H}} \boldsymbol{Z}_r \boldsymbol{v}_p \tag{8.7.30}$$

or

$$J_{prs} = \boldsymbol{q}_c^{\mathrm{H}} \boldsymbol{A} \boldsymbol{q}_c + \boldsymbol{q}_c^{\mathrm{H}} \boldsymbol{b} + \boldsymbol{b}^{\mathrm{H}} \boldsymbol{q}_c + c \tag{8.7.31}$$

where

$$\boldsymbol{A} = \boldsymbol{Z}_m^{\mathrm{H}} \boldsymbol{Z}_m \tag{8.7.32}$$

$$\boldsymbol{b} = \boldsymbol{Z}_m^{\mathrm{H}} \boldsymbol{Z}_r \boldsymbol{v}_p \tag{8.7.33}$$

and

$$c = \boldsymbol{v}_p^{\mathrm{H}} \boldsymbol{Z}_r^{\mathrm{H}} \boldsymbol{Z}_r \boldsymbol{v}_p \tag{8.7.34}$$

The form of the error criterion given in Equation (8.7.31) is exactly the form that is needed, where it is expressed as a quadratic function of the vector of control source volume velocities, $\boldsymbol{q}_c$. The optimum set of volume velocities for minimising the sum of the squared sound pressure levels at N_e error locations can be found directly by differentiating Equation (8.7.31) with respect to $\boldsymbol{q}_c$ and setting the gradient equal to zero. The result of this exercise, as outlined in Section 8.4, is:

$$\boldsymbol{q}_{c,prs} = -\boldsymbol{A}^{-1} \boldsymbol{b} \tag{8.7.35}$$

where the subscripts ***c,prs*** on the control source volume velocity vector indicate that it is optimal with respect to minimising the acoustic pressure amplitude at a set of discrete error locations. (As outlined in Section 8.4, the matrix $\boldsymbol{A}$ must be positive definite for a unique solution to exist, and for the solution to be the optimum one.) Substituting this result back into Equation (8.7.31) produces the expression for the minimum sum of the squared acoustic pressures:

$$J_{prs,\min} = \left\{ \sum_{i=1}^{N_e} |p(\boldsymbol{r}_i)|^2 \right\}_{\min} = c - \boldsymbol{b}^{\mathrm{H}}\boldsymbol{A}^{-1}\boldsymbol{b} = c + \boldsymbol{b}^{\mathrm{H}}\boldsymbol{q}_{c,p} \tag{8.7.36a–c}$$

Note here again that the quantity c is equal to the value of the error criterion, the sum of the squared pressures, during operation of the primary noise source alone.

8.7.2 Minimisation of Total Radiated Acoustic Power Using Acoustic Monopole Sources

The procedure for calculating the control source volume velocity or force matrix that minimises the total radiated acoustic power is much the same as that outlined for minimising the sound pressure at a discrete point or points. The difference is that the matrices $\boldsymbol{A}$ and $\boldsymbol{b}$, and the scalar quantity c, must be modified to include the 'spatial integration' necessary in sound power calculations. There are two commonly employed approaches to calculating the control source volume velocities that will minimise the total radiated acoustic power of the system, both of which are aimed at expressing this error criterion as a quadratic function of control source output. The derived expression can then be differentiated with respect to control source volume velocities, and the resultant gradient equation set equal to zero to yield the optimum value. One approach to obtaining the desired quadratic expression is to construct the problem based upon a far-field measure of acoustic power, where the total radiated acoustic power is evaluated by integrating the far-field acoustic intensity over a hemisphere enclosing the radiating structure and control sources. Using the geometry shown in Figure 8.27, this can be written as:

$$W = \int_0^{2\pi}\int_0^{\pi/2} \frac{|p(\boldsymbol{r})|^2}{2\rho_0 c_0} |\boldsymbol{r}|^2 \sin\theta \, \mathrm{d}\theta \, \mathrm{d}\varphi \tag{8.7.37}$$

where the location $\boldsymbol{r}$ is defined by the spherical coordinates $\boldsymbol{r} = (r,\theta,\varphi)$. The squared acoustic pressure amplitude at any point is:

$$|p(\boldsymbol{r})|^2 = p^*(\boldsymbol{r})p(\boldsymbol{r}) \tag{8.7.38}$$

which, written explicitly in terms of the primary and control source contributions, is:

$$\begin{aligned} |p(\boldsymbol{r})|^2 &= \left(p_c(\boldsymbol{r}) + p_p(\boldsymbol{r})\right)^* \left(p_c(\boldsymbol{r}) + p_p(\boldsymbol{r})\right) \\ &= p_c^*(\boldsymbol{r})p_c(\boldsymbol{r}) + p_c^*(\boldsymbol{r})p_p(\boldsymbol{r}) + p_p^*(\boldsymbol{r})p_c(\boldsymbol{r}) + p_p^*(\boldsymbol{r})p_p(\boldsymbol{r}) \end{aligned} \tag{8.7.39a,b}$$

Using Equations (8.7.15) and (8.7.21), this can be re-expressed as:

$$\begin{aligned} |p(\boldsymbol{r})|^2 = {} & \boldsymbol{q}_c^{\mathrm{H}} \boldsymbol{z}_{mono}^* \boldsymbol{z}_{mono}^{\mathrm{T}} \boldsymbol{q}_c + \boldsymbol{q}_c^{\mathrm{H}} \boldsymbol{z}_{mono}^* \boldsymbol{z}_{rad}^{\mathrm{T}} \boldsymbol{v}_p \\ & + \boldsymbol{v}_p^{\mathrm{H}} \boldsymbol{z}_{rad}^* \boldsymbol{z}_{mono}^{\mathrm{T}} \boldsymbol{q}_c + \boldsymbol{v}_p^{\mathrm{H}} \boldsymbol{z}_{rad}^* \boldsymbol{z}_{rad}^{\mathrm{T}} \boldsymbol{v}_p \end{aligned} \tag{8.7.40}$$

If Equation (8.7.40) is now substituted into the defining Equation (8.7.37), the total radiated acoustic power can be written as a quadratic function of control source volume velocity:

$$W = \boldsymbol{q}_c^{\mathrm{H}} \boldsymbol{A} \boldsymbol{q}_c + \boldsymbol{q}_c^{\mathrm{H}} \boldsymbol{b} + \boldsymbol{b}^{\mathrm{H}} \boldsymbol{q}_c + c \tag{8.7.41}$$

where

$$\boldsymbol{A} = \int_0^{2\pi} \int_0^{\pi/2} \boldsymbol{z}_{mono}^* \frac{\boldsymbol{z}_{mono}^{\mathrm{T}}}{2\rho_0 c_0} |\boldsymbol{r}|^2 \sin\theta \, \mathrm{d}\theta \, \mathrm{d}\varphi \tag{8.7.42}$$

$$\boldsymbol{b} = \int_0^{2\pi} \int_0^{\pi/2} \frac{\boldsymbol{z}_{mono}^* \boldsymbol{z}_{rad}^{\mathrm{T}} \boldsymbol{v}_p}{2\rho_0 c_0} |\boldsymbol{r}|^2 \sin\theta \, \mathrm{d}\theta \, \mathrm{d}\varphi \tag{8.7.43}$$

and

$$c = \int_0^{2\pi} \int_0^{\pi/2} \frac{\boldsymbol{v}_p^{\mathrm{H}} \boldsymbol{z}_{rad}^* \boldsymbol{z}_{rad}^{\mathrm{T}} \boldsymbol{v}_p}{2\rho_0 c_0} |\boldsymbol{r}|^2 \sin\theta \, \mathrm{d}\theta \, \mathrm{d}\varphi \tag{8.7.44}$$

In this case, the quantity c is the acoustic power radiated by the primary source operating alone.

As before, differentiating Equation (8.7.41) with respect to $\boldsymbol{q}_c$ and setting the gradient equal to zero will produce an expression that defines the optimum control source volume velocities:

$$\boldsymbol{q}_{c,w} = -\boldsymbol{A}^{-1} \boldsymbol{b} \tag{8.7.45}$$

where the subscripts ***c,w*** on the control source volume velocity vector indicate that it is optimal with respect to minimising the total radiated acoustic power. Substituting this result back into Equation (8.7.41) produces the expression for the minimum radiated acoustic power:

$$W_{\min} = c - \boldsymbol{b}^{\mathrm{H}} \boldsymbol{A}^{-1} \boldsymbol{b} = c + \boldsymbol{b}^{\mathrm{H}} \boldsymbol{q}_{c,w} \tag{8.7.46}$$

In Sections 8.3 and 8.4, similar expressions were derived for the problem of controlling the total acoustic power output of a set monopole sources. In that exercise, it was found that the integrations, which are the equivalent of those stated in Equations (8.7.42) to (8.7.44), could be solved analytically, leading to a much more compact statement of the problem. Unfortunately, this will not normally be possible for the problem of controlling structural radiation. Consider the three terms in Equation (8.7.41) which require integration, and the specialised case of only one acoustic control source. Referring to Equation (8.7.22), the term $\boldsymbol{A}$ can be specialised in this instance as:

$$\boldsymbol{A} = \int_0^{2\pi} \frac{\left(\frac{\mathrm{j}\rho_0 \omega}{4\pi} \left\{ \frac{\mathrm{e}^{-\mathrm{j}kr_s}}{r_s} + \frac{\mathrm{e}^{-\mathrm{j}kr_i}}{r_i} \right\} \right)^* \left(\frac{\mathrm{j}\rho_0 \omega}{4\pi} \left\{ \frac{\mathrm{e}^{-\mathrm{j}kr_s}}{r_s} + \frac{\mathrm{e}^{-\mathrm{j}kr_i}}{r_i} \right\} \right)}{2\rho_0 c_0} |\boldsymbol{r}|^2 \sin\theta \, \mathrm{d}\theta \, \mathrm{d}\varphi \tag{8.7.47}$$

or

$$A = \frac{\omega^2 r^2}{16\pi} \int_0^{2\pi} \left\{ \frac{e^{-jkr_s}}{r_s} + \frac{e^{-jkr_i}}{r_i} \right\}^* \left\{ \frac{e^{-jkr_s}}{r_s} + \frac{e^{-jkr_i}}{r_i} \right\} \sin\theta \, d\theta \, d\varphi \tag{8.7.48}$$

Following on from the monopole work in Section 8.4, as the integration is being calculated in the far-field, the approximations $r_s \approx r_i \approx r$ and $r_i - r_s = z \cos\theta$ can be made. This allows Equation (8.7.48) to be restated, after some simple algebra, as:

$$A = \frac{\omega^2 \rho_0}{8\pi c_0} \int_0^{2\pi} \{1 + \cos(2kz\cos\theta)\} \sin\theta \, d\theta \tag{8.7.49}$$

Noting again that:

$$\int_0^{\pi/2} \cos(x \cos\theta) \sin\theta \, d\theta = \operatorname{sinc} x \tag{8.7.50}$$

enables the term in $\boldsymbol{A}$ to be evaluated as:

$$A = \frac{\omega^2 \rho_0}{8\pi c_0} (1 + \operatorname{sinc} 2kz) \tag{8.7.51}$$

From this, it can be seen that the problem is not in evaluating the $\boldsymbol{A}$ term. In fact, the problems that are encountered are in the evaluation of the $\boldsymbol{b}$ and c terms. In each of these terms the vector $\boldsymbol{z_{rad}}$ appears, the elements of which, from the definition of Equation (8.7.16), must themselves be evaluated through integration (recall that the terms in $\boldsymbol{z_{rad}}$ define the transfer function between the velocity of a given structural mode and the acoustic pressure at some location in space). It will rarely be possible to do this analytically, even with the consideration limited to far-field radiation and the most regular structural geometries. If such a result can be obtained, it will almost surely be a function of the angular coordinates, θ, φ (and of course radius r, which will be constant for the integration). To evaluate the terms in $\boldsymbol{b}$ and c, the result obtained for m $\boldsymbol{z_{rad}}$ must be multiplied by other spatially dependent terms and integrated. It can easily be deduced from this description of events that numerical techniques are usually required. At low frequencies, however, it is sometimes possible to formulate approximate solutions, which will be discussed shortly.

The second approach taken to obtain the desired quadratic equation for calculating the optimum set of control source volume velocities is to consider source power output directly, this quantity being equal to the integration of the near-field active intensity over the area of the surface:

$$W = \frac{1}{2} \operatorname{Re} \left\{ \int_S v(\boldsymbol{x}) \, p^*(\boldsymbol{x}) \, d\boldsymbol{x} \right\} \tag{8.7.52}$$

where Re denotes the real part of the expression, $v(\boldsymbol{x})$ is the velocity at some location $\boldsymbol{x}$ on the surface of the source, $p^*(\boldsymbol{x})$ is the complex conjugate of the acoustic pressure at this same location, and the integration is conducted over the surface of the source. Of course, both the near-field and far-field developments should lead to the same solution for the optimum control source output.

Consider again an active control system where a set of N_c acoustic monopole control sources is being used to attenuate the sound field radiated by a vibrating structure. The total acoustic power output W_t of this arrangement can be written in the form:

$$W_t = W_c + W_p \tag{8.7.53}$$

where the subscript t is used to denote a total (combined primary and control source) quantity. In Equation (8.7.53), W_c is the total acoustic power output of the control source distribution, equal to the sum of contributions from the N_c monopole control sources, the integral expression Equation (8.7.52) corresponding to the monopole control sources being replaced by multiplication of volume velocity with the complex conjugate of acoustic pressure at the monopole source location, to give:

$$W_c = \frac{1}{2}\mathrm{Re}\left\{ \sum_{i=1}^{N_c} q_{ci}\, p_t^*(\boldsymbol{r}_{ci}) \right\} \tag{8.7.54}$$

where q_{ci} is the volume velocity of the ith control source, located at a position $\boldsymbol{r}_{ci}$ in space. W_p in Equation (8.7.53) is the acoustic power output of the primary source (the radiating structure), given by:

$$W_p = \frac{1}{2}\,\mathrm{Re}\left\{ \int_S v_p(\boldsymbol{x})\; p_t^*(\boldsymbol{x})\; \mathrm{d}\boldsymbol{x} \right\} \tag{8.7.55}$$

As the total pressure at any location $\boldsymbol{r}$ can be considered to be the superposition of the primary and control source pressures at that location, Equation (8.7.53) can be expanded to:

$$W_t = \frac{1}{2}\mathrm{Re}\left\{ \sum_{i=1}^{N_c} q_{ci}\big[p_c(\boldsymbol{r}_{ci}) + p_p(\boldsymbol{r}_{ci})\big]^* + \int_S v(\boldsymbol{x})\big[p_c(\boldsymbol{x}) + p_p(\boldsymbol{x})\big]^* \mathrm{d}\boldsymbol{x} \right\} \tag{8.7.56}$$

Using Equations (8.7.17) and (8.7.21), the total acoustic power output of the system, defined in Equation (8.7.53), can be written in matrix form as:

$$W_t = \boldsymbol{q_c}^{\mathrm{H}} \boldsymbol{A} \boldsymbol{q_c} + \boldsymbol{b}^{\mathrm{H}} \boldsymbol{q_c} + \boldsymbol{q_c}^{\mathrm{H}} \boldsymbol{b} + c \tag{8.7.57}$$

where

$$\boldsymbol{A} = \frac{1}{2}\mathrm{Re}\left\{ \begin{bmatrix} z_{mono}^{\mathrm{T}}(\boldsymbol{r}_{c1}) \\ z_{mono}^{\mathrm{T}}(\boldsymbol{r}_{c2}) \\ \vdots \\ z_{mono}^{\mathrm{T}}(\boldsymbol{r}_{cN_c}) \end{bmatrix} \right\} \tag{8.7.58}$$

$$\boldsymbol{b} = \frac{1}{2}\mathrm{Re}\left\{\begin{bmatrix} z_{rad}^{\mathrm{T}}(\boldsymbol{r}_{c1}) \\ z_{rad}^{\mathrm{T}}(\boldsymbol{r}_{c2}) \\ \vdots \\ z_{rad}^{\mathrm{T}}(\boldsymbol{r}_{cN_c}) \end{bmatrix}\right\}\boldsymbol{v}_p = \frac{1}{2}\sum_{i=1}^{N_m}\left(\int_S \mathrm{Re}\left\{z_{mono}^*(\boldsymbol{x})\right\}\psi_i(\boldsymbol{x})\,\mathrm{d}\boldsymbol{x}\cdot v_{p,i}\right) \tag{8.7.59a,b}$$

$$c = \boldsymbol{v}_p^{\mathrm{H}}\boldsymbol{Z}_w\boldsymbol{v}_p \tag{8.7.60}$$

and the i,lth term of the matrix $\boldsymbol{Z}_w$, which will be referred to as the acoustic power transfer matrix in more in-depth studies later in this chapter, is defined by:

$$Z_w(i,l) = \frac{\omega\rho_0}{4\pi}\int_S\int_S \psi_i^*(\boldsymbol{x}_0)\frac{\sin kr}{r}\psi_l(\boldsymbol{x})\,\mathrm{d}\boldsymbol{x}\,\mathrm{d}\boldsymbol{x}_0 \tag{8.7.61}$$

Expressed in this 'standard' form, the vector of optimum control source volume velocities, calculated by differentiating Equation (8.7.57) with respect to $\boldsymbol{q}_c$ and setting the resultant gradient expression equal to zero, is found to be:

$$\boldsymbol{q}_{c_{\mathrm{opt}}} = -\boldsymbol{A}^{-1}\boldsymbol{b} \tag{8.7.62}$$

the use of which will result in a minimum acoustic power output from the controlled system of:

$$W_{\min} = c - \boldsymbol{b}^{\mathrm{H}}\boldsymbol{A}^{-1}\boldsymbol{b} \tag{8.7.63}$$

where c, defined in Equation (8.7.60), is the acoustic power output of the system in the absence of active control.

As mentioned in the previous far-field based, minimum-acoustic-power formulation, it would greatly simplify the exercise if the integrations in Equations (8.7.59) and (8.7.60) could be solved analytically. It was found in the far-field, acoustic-power formulation that this was not generally possible. This is also a problem in a near-field formulation of the problem. Consider first the matrix $\boldsymbol{A}$. The m,nth term in this real, symmetric matrix is the real part of the radiation transfer function between the mth and nth monopole control sources. From the definition given in Equation (8.7.58) this is:

$$A_{m,n} = \frac{\omega^2\rho_0}{8\pi c_o}\left(\mathrm{sinc}\,kr_{mn,s} + \mathrm{sinc}\,kr_{mn,i}\right) \tag{8.7.64}$$

where $r_{mn,s}$ and $r_{mn,i}$ are respectively the distance between the m and nth real sources (the same distance that exists between the two image sources) and the distance between an image and real source, defined as follows:

$$\begin{aligned} r_{mn,s} &= \sqrt{(x_m - x_n)^2 + (y_m - y_n)^2 + (z_m - z_n)^2} \\ r_{mn,i} &= \sqrt{(x_m - x_n)^2 + (y_m - y_n)^2 + (z_m + z_n)^2} \end{aligned} \tag{8.7.65}$$

Note that if a diagonal term of $\boldsymbol{A}$ is being evaluated, then the result is the radiation resistance of a monopole source near an infinite plane surface:

$$A_{m,m} = \frac{\omega^2 \rho_0}{8\pi c_0}\left(1 + \operatorname{sinc} 2kz_m\right) \tag{8.7.66}$$

This is the same expression as that derived in Equation (8.7.51) for the far-field formulation, as is to be expected.

The terms $\boldsymbol{b}$ and c are not as straightforward to evaluate, owing to the integrals contained within them. However, if the equations are reconsidered in the wavenumber domain, the problem lends itself to low-frequency approximations. Considering the former of these terms, in viewing Equation (8.7.59) it can be deduced that each term in the vector $\boldsymbol{b}$ describes the change in acoustic power output one source (a single control source or the vibrating structure) elicits in the other source (structure or control) per unit control source volume velocity (as reciprocity holds in free space the influence one source has upon the other will be mutual). Thus, each term in $\boldsymbol{b}$ is the result of the modal summation:

$$b_n = \sum_{i=1}^{m}\left\{\left[\frac{\omega\rho_0}{4\pi}\int_S \psi_i(\boldsymbol{x})\frac{\sin kr}{r}\,\mathrm{d}\boldsymbol{x}\right]\cdot v_i\right\} \tag{8.7.67}$$

where r is the distance between a point on the vibrating structure and the acoustic control source, $\psi_i(\boldsymbol{x})$ is the value of the ith mode shape function at location $\boldsymbol{x}$ on the structure, and v_i is the velocity amplitude of the ith mode. The integral expression in the square brackets in Equation (8.7.67) can be approximated in the low-frequency regime as follows. First, the (sin kr)/r term can be re-expressed using a MacLaurin series as:

$$\frac{\sin\, kr}{r} = \sum_{s=0}^{\infty}\frac{(-1)^s (kr)^{2s+1}}{r\,(2s+1)!} = \sum_{s=0}^{\infty}\frac{(-1)^s k^{2s+1} r^{2s}}{(2s+1)!} \tag{8.7.68a,b}$$

The separation distance r, defined in Equation (8.7.65), can be rewritten using a binomial expansion as:

$$r^{2s} = \left[(x - x_{cn})^2 + (y - y_{cn})^2 + z_{cn}^2\right]^s =$$

$$\sum_{m=0}^{s}\sum_{l=0}^{m}\sum_{p=0}^{2m-2l}\sum_{q=0}^{2l}\binom{s}{m}\binom{m}{l}\binom{2m-2l}{p}\binom{2l}{q} x_{cn}^{2m-2l-p} x^p y_{cn}^{2l-q} y^q z_{cn}^{2s-2m} \tag{8.7.69a,b}$$

where

$$\binom{a}{b} = \frac{a!}{b!\,(a-b)!} \tag{8.7.70}$$

and the monopole control source is at a location (x_{cn}, y_{cn}, z_{cn}). Therefore, the part of Equation (8.7.67) in curly brackets can be written as:

$$v_i \cdot \frac{\rho_0 c_0}{4\pi} \sum_{s=0}^{\infty} \frac{(-1)^s k^{2s+2}}{(2s+1)!} \sum_{m=0}^{s} \sum_{l=0}^{m} \sum_{p=0}^{2m-2l} \sum_{q=0}^{2l} \binom{s}{m} \binom{m}{l} \binom{2m-2l}{p} \binom{2l}{q} \times$$
$$x_{cn}^{2m-2l-p} y_{cn}^{2l-q} z_{cn}^{2s-2m} \int_S \varphi_i(\boldsymbol{x}) x^p y^q \, \mathrm{d}\boldsymbol{x} \tag{8.7.71}$$

The integral in Equation (8.7.71) can be re-expressed as a partial derivative by considering the problem in *k*-space (see Section 2.4). Doing this requires calculation of the Fourier transform of the mode shape function, defined by the expression:

$$\tilde{\psi}(k_x, k_y) = \int_S \psi(\boldsymbol{x}) e^{-jk_x x} e^{-jk_y y} \, \mathrm{d}\boldsymbol{x} \tag{8.7.72}$$

The moments of the mode shape function (the integral) in Equation (8.7.71) (in the plane $z = 0$) can be expressed as derivatives in *k*-space evaluated at $k_x = k_y = 0$, so that the integral in Equation (8.7.71) can be written in the form:

$$\int_S \psi(\boldsymbol{x}) x^a y^b \, \mathrm{d}\boldsymbol{x} = \mathrm{j}^{a+b} \left. \frac{\partial^{a+b} \tilde{\psi}(k_x, k_y)}{\partial^a k_x \, \partial^b k_y} \right|_{k_x = k_y = 0} \tag{8.7.73}$$

Thus, the part of Equation (8.7.67) in curly brackets can be approximated as:

$$v_i \cdot \frac{\rho_0 c_0}{4\pi} \sum_{s=0}^{\infty} \frac{(-1)^s k^{2s+2}}{(2s+1)!} \sum_{m=0}^{s} \sum_{l=0}^{m} \sum_{p=0}^{2m-2l} \sum_{q=0}^{2l} \binom{s}{m} \binom{m}{l} \binom{2m-2l}{p} \binom{2l}{q}$$
$$\times x_{cn}^{2m-2l-p} y_{cn}^{2l-q} z_{cn}^{2s-2m} \mathrm{j}^{p+q} \left. \left\{ \left(\frac{\partial}{\partial k_x} \right)^p \left(\frac{\partial}{\partial k_y} \right)^q \tilde{\psi}_i \right\} \right|_{k_x = k_y = 0} \tag{8.7.74}$$

This expression, while looking rather foreboding, in fact often provides a straightforward means of approximating the terms in the vector $\boldsymbol{b}$ as the transformed mode shape functions of many commonly studied structures can be approximated by an infinite series and the derivatives easily evaluated (see Williams (1983) for a related discussion with examples).

The terms in the matrix $\boldsymbol{Z}_w$ upon which the term c is based, can be evaluated using the same approach. Doing this, it is straightforward to show that Equation (8.7.61) can be expressed as:

$$Z_w(i, l) = \frac{\rho_0 c}{4\pi} \sum_{m=0}^{\infty} \frac{k^{2m+2}}{(2m+1)!} \sum_{l=0}^{m} \sum_{p=0}^{2m-2l} \sum_{q=0}^{2l} \binom{m}{l} \binom{2m-2l}{p} \binom{2l}{q}$$
$$\times \left. \left\{ \left(\frac{\partial}{\partial k_x} \right)^{2m-2l-p} \left(\frac{\partial}{\partial k_y} \right)^{2l-q} \tilde{\psi}_i^* \right\} \left\{ \left(\frac{\partial}{\partial k_x} \right)^p \left(\frac{\partial}{\partial k_y} \right)^q \tilde{\psi}_l \right\} \right|_{k_x = k_y = 0} \tag{8.7.75}$$

This approach has been used in calculating the radiation efficiencies for a variety of structural mode shape functions (Williams, 1983), as well as for calculating total acoustic power output of a structure (Snyder and Tanaka, 1992). Examples of the use of these equations will be given in the next section.

8.7.3 Minimisation of Acoustic Pressure at Discrete Locations Using Vibration Sources

Having derived a set of expressions that enable the determination of the optimal volume velocities for a set of monopole control sources used to attenuate radiation from an idealised vibrating structure, it is now of interest to examine an alternative means of achieving attenuation of the sound field radiated by a structure. When the sound field to be attenuated is produced by a vibrating structure, vibration control sources may be effective in providing the desired reduction in total radiated sound power, or acoustic pressure amplitude at a set of error sensing locations. Vibration control sources can provide this attenuation by altering the velocity distribution of the structure. To assess this capability, what is required is some means of calculating the optimum control source output with respect to the error criterion of interest: acoustic power or acoustic pressure.

Consider the system illustrated in Figure 8.29, where a force input is used to reduce the acoustic radiation from a vibrating structure. In practice, the force input will be provided by a vibration source, which is modelled as a point force for two reasons: first, it approximates the characteristics of a shaker attached to the structure at a point, such as through a stinger, and second, analytically it will prove to make the required calculations reasonably straightforward. (More complex piezoelectric ceramic vibration actuators are discussed later in Chapter 15; see also Fuller et al. (1991b), Clark et al. (1991), Clark and Fuller (1992c) and Koshigoe and Murdock (1993)). With a point source input to the structure, a quantity that can be manipulated in a manner analogous to the previous treatment of acoustic control source volume velocity is input force. A logical approach to this problem would therefore seem to be to attempt to express the error criterion as a quadratic function of the control source input forces, differentiate the expression with respect to this quantity, and set the resultant gradient expression equal to zero to obtain the optimum set of complex force amplitudes. This necessitates formulating expressions for both the primary and control source radiated acoustic pressure distribution, the latter in terms of input force. The required expressions for the primary source pressure distribution have already been derived and stated in Equation (8.7.17), so much of the required analysis has already been done.

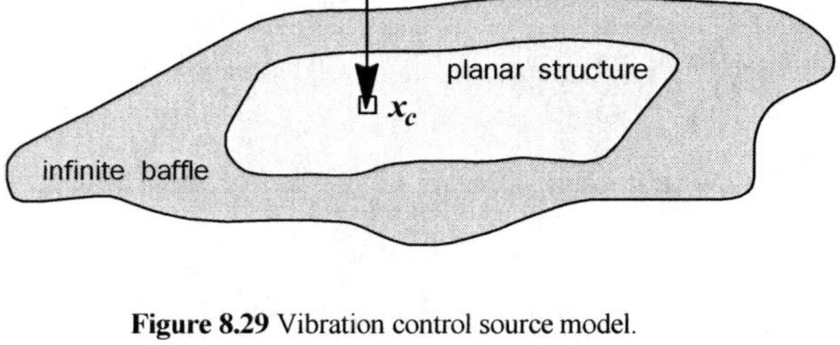

Figure 8.29 Vibration control source model.

To obtain an expression for the acoustic pressure radiated in response to the control point source input, note that the problem is virtually identical to that of determining the primary source radiated sound pressure. The only difference is that in the primary source case, the forcing function, the modal generalised force Γ defined in Equation (8.7.8), is assumed to be known either directly or implicitly as a result of knowledge of the existing vibration of the structure through Equation (8.7.11). If an expression for the modal generalised force in terms of the control source input force can be found, then the previous primary source directed derivation can be used to determine the control source pressure distribution.

Referring to Equation (8.7.8), to evaluate the modal generalised force, an expression for the force input per unit area, or blocked pressure, from the forcing function must be obtained. For a point force, this expression is:

$$p_{bl,c} = f_c \delta(\boldsymbol{x} - \boldsymbol{x}_c) \tag{8.7.76}$$

where f_c is the complex amplitude of the control force, $\delta(\boldsymbol{x} - \boldsymbol{x}_c)$ is a Dirac delta function, defined by:

$$\delta(\boldsymbol{x} - \boldsymbol{x}_c) = \begin{cases} 1 & \boldsymbol{x} = \boldsymbol{x}_c \\ 0 & \text{otherwise} \end{cases} \tag{8.7.77}$$

and $\boldsymbol{x}_c$ is the attachment location of the point control force. Substituting this into Equation (8.7.8), the lth modal generalised force due to the control force input is:

$$\Gamma_l = \int_S f_c \psi_l(\boldsymbol{x}) \delta(\boldsymbol{x} - \boldsymbol{x}_c) \,\mathrm{d}\boldsymbol{x} = f_c \psi_l(\boldsymbol{x}_c) \tag{8.7.78a,b}$$

In words, Equation (8.7.78) states that the modal generalised force resulting from the control point force input is equal to the amplitude of the control force multiplied by the value of the lth mode shape function at the application point (note again that harmonic time dependence is implicit in this relationship). If N_c control forces are to be used, then the total value of the lth modal generalised force is the superposition of the N_c individual contributions:

$$\Gamma_l = \sum_{i=1}^{N_c} f_{ci} \varphi_l(\boldsymbol{x}_i) \tag{8.7.79}$$

which can be written in matrix form as:

$$\Gamma_l = \boldsymbol{\psi}_{cl}^{\mathrm{T}} \boldsymbol{f}_c \tag{8.7.80}$$

where $\boldsymbol{\psi}_{cl}$ is the $(N_c \times 1)$ vector of values of the lth mode shape function at the control force input positions:

$$\boldsymbol{\psi}_{cl} = \begin{bmatrix} \psi_l(\boldsymbol{x}_1) \\ \psi_l(\boldsymbol{x}_2) \\ \vdots \\ \psi_l(\boldsymbol{x}_{N_c}) \end{bmatrix} \tag{8.7.81}$$

and $\boldsymbol{f}_c$ is the $(N_c \times 1)$ vector of complex control force amplitudes:

$$\boldsymbol{f}_c = \begin{bmatrix} f_{c1} \\ f_{c2} \\ \cdots \\ f_{cN_c} \end{bmatrix} \tag{8.7.82}$$

If m structural modes are being considered in the calculations, then the vector of modal generalised forces resulting from the control force inputs, $\boldsymbol{\Gamma}_c$, can be expressed as a product of two matrices:

$$\boldsymbol{\Gamma}_c = \boldsymbol{\Psi}_c \boldsymbol{f}_c \tag{8.7.83}$$

where $\boldsymbol{\Psi}_c$ is the $(m \times N_c)$ matrix of structural mode shape functions evaluated at the control source input locations, the rows of which are the transposed vectors of the mode shape functions evaluated at the control source locations defined in Equation (8.7.81):

$$\boldsymbol{\Psi}_c = \begin{bmatrix} \boldsymbol{\psi}_{c1}^{\mathrm{T}} \\ \boldsymbol{\psi}_{c2}^{\mathrm{T}} \\ \vdots \\ \boldsymbol{\psi}_{cm}^{\mathrm{T}} \end{bmatrix} \tag{8.7.84}$$

Equation (8.7.84) can now simply be substituted into the form of Equation (8.7.17) to produce an expression for the acoustic sound field at location $\boldsymbol{r}$ resulting from the control force input:

$$p_c(\boldsymbol{r}) = \boldsymbol{z}_{rad}^{\mathrm{T}} \boldsymbol{Z}_I^{-1} \boldsymbol{\Psi}_c \boldsymbol{f}_c \tag{8.7.85}$$

Now armed with an expression for the acoustic pressure that results from a given control force input, it is possible to consider expressing the error criterion as a quadratic function of the control force input, and using this expression to derive the optimum set of control forces. Consider first the minimisation of the sum of the squared acoustic pressure amplitudes at a discrete set of error sensing locations. This criterion was expressed in Equation (8.7.25) in terms of the vectors of sound pressures at the points of interest due to the control and primary sources operating alone. Using Equation (8.7.85), the $(N_e \times 1)$ vector of acoustic pressures, $\boldsymbol{p}_c$, due to the control force inputs will be:

$$\boldsymbol{p}_c = \boldsymbol{Z}_r \boldsymbol{Z}_I^{-1} \boldsymbol{\Phi}_c \boldsymbol{f}_c \tag{8.7.86}$$

where $\boldsymbol{Z}_r$ is the $(N_e \times m)$ matrix of modal radiation transfer functions between the m structural modes being used in the calculation and the N_e error locations, as defined in Equation (8.7.27). Using this expression, in conjunction with that for the vector of acoustic pressures generated during primary excitation, expressed in Equation (8.7.26), the error criterion when vibration control sources are being used can be expanded as:

$$J_p = \boldsymbol{q}_c^{\mathrm{H}} \boldsymbol{A} \boldsymbol{q}_c + \boldsymbol{q}_c^{\mathrm{H}} \boldsymbol{b} + \boldsymbol{b}^{\mathrm{H}} \boldsymbol{q}_c + c \tag{8.7.87}$$

where

$$\boldsymbol{A} = \boldsymbol{\Psi}_c^{\mathrm{H}} \left\{ \boldsymbol{Z}_I^{-1} \right\}^{\mathrm{H}} \boldsymbol{Z}_r^{\mathrm{H}} \boldsymbol{Z}_r \boldsymbol{Z}_I^{-1} \boldsymbol{\Psi}_c \tag{8.7.88}$$

$$\boldsymbol{b} = \boldsymbol{\Psi}_c^{\mathrm{H}} \left\{ \boldsymbol{Z}_I^{-1} \right\}^{\mathrm{H}} \boldsymbol{Z}_r^{\mathrm{H}} \boldsymbol{Z}_r \boldsymbol{v}_p \tag{8.7.89}$$

and

$$c = \boldsymbol{v}_p^{\mathrm{H}} \boldsymbol{Z}_r^{\mathrm{H}} \boldsymbol{Z}_r \boldsymbol{v}_p \tag{8.7.90}$$

The form of the error criterion given in Equation (8.7.87) is identical to that of Equation (8.7.31) developed for the use of acoustic control sources. Differentiating it and setting the resultant gradient expression equal to zero produces an expression for the vector of optimum control forces, $\boldsymbol{f}_c$:

$$\boldsymbol{f}_{c,prs} = -\boldsymbol{A}^{-1} \boldsymbol{b} \tag{8.7.91}$$

where the subscripts on the control force vector ***c, prs*** indicate that it is optimal with respect to minimising the acoustic pressure amplitude at a set of discrete error locations. Substituting this result into Equation (8.7.87) produces the expression for the minimum sum of the squared pressures:

$$J_{prs,\min} = \left\{ \sum_{i=1}^{N_e} |p(\boldsymbol{r}_i)|^2 \right\}_{\min} = c - \boldsymbol{b}^{\mathrm{H}} \boldsymbol{A}^{-1} \boldsymbol{b} = c + \boldsymbol{b}^{\mathrm{H}} \boldsymbol{f}_{c,prs} \tag{8.7.92a–c}$$

8.7.4 Minimisation of Total Radiated Acoustic Power Using Vibration Sources

The procedure for calculating the control source force inputs that minimise the total radiated acoustic power is much the same as that outlined previously for the use of acoustic control sources, where the total radiated acoustic power is expressed as a quadratic function of the control force inputs. It is possible to develop this expression based upon either a near-field or far-field measure of acoustic power. Both will be done here, beginning with the far-field case. Using Equation (8.7.85), it is straightforward to show, using the same steps taken in deriving Equation (8.7.41), that the total radiated acoustic power can be expressed as a quadratic function of the control force inputs as:

$$W = \boldsymbol{f}_c^{\mathrm{H}} \boldsymbol{A} \boldsymbol{f}_c + \boldsymbol{f}_c^{\mathrm{H}} \boldsymbol{b} + \boldsymbol{b}^{\mathrm{H}} \boldsymbol{f}_c + c \tag{8.7.93}$$

where

$$\boldsymbol{A} = \boldsymbol{\Psi}_c^{\mathrm{H}} \left\{ \boldsymbol{Z}_I^{-1} \right\}^{\mathrm{H}} \left\{ \int_0^{2\pi} \int_0^{\pi/2} \frac{\boldsymbol{z}_{rad}^{*} \boldsymbol{z}_{rad}^{\mathrm{T}}}{2\rho_0 c_0} |\boldsymbol{r}|^2 \sin\theta \, \mathrm{d}\theta \, \mathrm{d}\varphi \right\} \boldsymbol{Z}_I^{-1} \boldsymbol{\Psi}_c \tag{8.7.94}$$

$$b = \boldsymbol{\Psi}_c^{\mathrm{H}} \left\{ \boldsymbol{Z}_I^{-1} \right\}^{\mathrm{H}} \left\{ \int_0^{2\pi} \int_0^{\pi/2} \frac{\boldsymbol{z}_{rad}^* \boldsymbol{z}_{rad}^{\mathrm{T}}}{2\rho_0 c_0} |\boldsymbol{r}|^2 \sin\theta \, \mathrm{d}\theta \, \mathrm{d}\varphi \right\} \boldsymbol{v}_p \tag{8.7.95}$$

and

$$c = \boldsymbol{v}_p^{\mathrm{H}} \left\{ \int_0^{2\pi} \int_0^{\pi/2} \frac{\boldsymbol{z}_{rad}^* \boldsymbol{z}_{rad}^{\mathrm{T}}}{2\rho_0 c_0} |\boldsymbol{r}|^2 \sin\theta \, \mathrm{d}\theta \, \mathrm{d}\varphi \right\} \boldsymbol{v}_p \tag{8.7.96}$$

As before, differentiating Equation (8.7.93) with respect to $\boldsymbol{f}_c$ and setting the gradient expression equal to zero will produce an expression that defines the vector of optimum control force inputs:

$$\boldsymbol{f}_{c,w} = -\boldsymbol{A}^{-1}\boldsymbol{b} \tag{8.7.97}$$

where the subscripts ***c,w*** indicate that it is optimal with respect to minimising the total radiated acoustic power. Substituting this result back into Equation (8.8.95) produces the following expression for the minimum radiated acoustic power:

$$W_{\min} = c - \boldsymbol{b}^{\mathrm{H}}\boldsymbol{A}^{-1}\boldsymbol{b} = c + \boldsymbol{b}^{\mathrm{H}}\boldsymbol{f}_{c,w} \tag{8.7.98a,b}$$

As with the acoustic control source problem, obtaining an analytical solution for the terms in $\boldsymbol{A}$, $\boldsymbol{b}$ and c in Equations (8.7.94) to (8.7.96) is not, in general, a viable option. Rather, the spatial integrations must be performed numerically. However, as will be seen shortly, at low frequencies it may be possible to make some approximations.

The other (near-field) approach to deriving the necessary quadratic equation is to consider source power output directly (at the plate surface), this quantity being expressed previously in Equation (8.7.52). The total acoustic power output W_t of this linear structural radiation problem in response to the combination of primary and control excitation can be written in the form:

$$W_t = \frac{1}{2} \mathrm{Re} \left\{ \int_S v_t(\boldsymbol{x}) p_t^*(\boldsymbol{x}) \, \mathrm{d}\boldsymbol{x} \right\} \tag{8.7.99}$$

where t denotes a total (combined primary and control) quantity. Expanding the total acoustic pressure as the superposition of primary and control source generated components, Equation (8.7.99) becomes:

$$W_t = \frac{1}{2} \mathrm{Re} \left\{ \int_S v_c(\boldsymbol{x}) p_c^*(\boldsymbol{x}) \, \mathrm{d}\boldsymbol{x} + \int_S v_c(\boldsymbol{x}) p_p^*(\boldsymbol{x}) \, \mathrm{d}\boldsymbol{x} + \int_S v_p(\boldsymbol{x}) p_c^*(\boldsymbol{x}) \, \mathrm{d}\boldsymbol{x} + \int_S v_p(\boldsymbol{x}) p_p^*(\boldsymbol{x}) \mathrm{d}\boldsymbol{x} \right\} \tag{8.7.100}$$

Observe that the expansion produces a set of four expressions, from left to right describing the acoustic power output of the structure due to excitation by the control source distribution operating alone, the modification to this quantity due to the co-existing primary excited acoustic pressure field, the modification to the primary excited acoustic power output due to the co-existing control source generated acoustic pressure field, and the acoustic power output of the structure due to the sole operation of the primary forcing function.

For the planar structure under consideration here, the acoustic pressure $p(\boldsymbol{x})$ can be expressed in terms of the surface velocity using the Rayleigh integral as outlined in Equation (8.7.12). If Equation (8.7.14) is used to expand Equation (8.7.100), the result is a set of four expressions:

$$W_t = W_{cc} + W_{cp} + W_{pc} + W_{pp} \tag{8.7.101}$$

each with the form of the surface integral for acoustic power as described, for example, in Williams (1983). In the terms on the right-hand side of Equation (8.7.101), the first subscript refers to the surface velocity generating source distribution, and the second subscript the acoustic field generating source distribution. For example, W_{cp} is the component of total radiated acoustic power attributable to the combination of the surface velocity induced by the control source distribution (operating alone) and the acoustic field induced by the primary source distribution (also operating alone):

$$W_{cp} = \frac{\omega\rho_0}{4\pi}\,\mathrm{Re}\left\{\int_S\int_S v_p^*(\boldsymbol{x}_0)\,\frac{\sin kr}{r}\,v_c(\boldsymbol{x})\,\mathrm{d}\boldsymbol{x}_0\,\mathrm{d}\boldsymbol{x}\right\} \tag{8.7.102}$$

with similar relationships for all other terms. If the velocity at any point on the structure is expanded in terms of modal contributions, then Equation (8.7.102) can be expanded as:

$$W_{cp} = \frac{\omega\rho_0}{4\pi}\int_S\int_S\left(\sum_{i=1}^{m} v_{pi}^*(\boldsymbol{x}_0)\,\frac{\sin kr}{r}\,\sum_{l=1}^{m} v_{cl}(\boldsymbol{x})\,\mathrm{d}\boldsymbol{x}_0\,\mathrm{d}\boldsymbol{x}\right) \tag{8.7.103}$$

again with similar relationships for the other terms. Using Equations (8.7.11) and (8.7.83) it is straightforward to re-express Equation (8.7.101) in the standard form (as a quadratic function of the control force inputs) as:

$$W_t = \boldsymbol{f}_c^{\mathrm{H}}\boldsymbol{A}\boldsymbol{f}_c + \boldsymbol{b}^{\mathrm{H}}\boldsymbol{f}_c + \boldsymbol{f}_c^{\mathrm{H}}\boldsymbol{b} + c \tag{8.7.104}$$

where

$$\boldsymbol{A} = \boldsymbol{\Psi}_c^{\mathrm{H}}\left\{\boldsymbol{Z}_I^{-1}\right\}^{\mathrm{H}}\boldsymbol{Z}_w\boldsymbol{Z}_I^{-1}\boldsymbol{\Psi}_c \tag{8.7.105}$$

$$\boldsymbol{b} = \boldsymbol{\Psi}_c^{\mathrm{H}}\left\{\boldsymbol{Z}_I^{-1}\right\}^{\mathrm{H}}\boldsymbol{Z}_w\boldsymbol{v}_p \tag{8.7.106}$$

$$c = \boldsymbol{v}_p^{\mathrm{H}}\boldsymbol{Z}_w\boldsymbol{v}_p \tag{8.7.107}$$

and $\boldsymbol{Z}_w$ is the acoustic power transfer matrix as defined previously in Equation (8.7.61).

Expressed in this form, the vector of optimal control forces is by now well known to be equal to:

$$f_c = -A^{-1}b \tag{8.7.108}$$

In viewing Equations (8.7.105) to (8.7.107), it is apparent that to obtain an analytical solution for the optimum set of control force inputs, it will be necessary to obtain an analytical solution for the terms in the acoustic power transfer matrix Z_w. While this is not, in general, a viable option, it is often possible to obtain an approximate solution if the excitation frequency is 'low', as outlined in Equation (8.7.75). The next section will specialise this expression for a simply supported, baffled plate, and will show that the off-diagonal terms in the power transfer matrix can be calculated by some simple algebraic manipulation of the diagonal terms.

All of the tools have now been developed for examining the effectiveness, or otherwise, of applying active control to planar, vibrating structures situated in an infinite baffle. If acoustic control sources are of interest, then for a given arrangement of sources and a given frequency, use of Equations (8.7.32) to (8.7.35) will enable the derivation of a set of volume velocities that will minimise the acoustic pressure at a set of potential error sensing locations, or use of Equations (8.7.42) to (8.7.45) or (8.7.58) to (8.7.62) will produce a set of volume velocities that will minimise the total radiated acoustic power. If vibration control sources are of interest, then for a given arrangement of source and a given frequency, use of Equations (8.7.88) to (8.7.91) will enable the derivation of a set of input forces that will minimise the acoustic pressure at a set of potential error sensing locations. Alternatively, use of Equations (8.7.94) to (8.7.97) or Equations (8.7.105) to (8.7.108) will produce a set of input forces that will minimise the total radiated acoustic power. In the next section, the use of this methodology will be demonstrated by applying it to the specific problem of controlling sound radiation from a infinitely baffled, simply supported rectangular plate.

It is also possible to measure the vibration levels at a number of vibration sensors on the radiating structure that occur when the acoustic pressure at a set of potential error sensing locations is minimised. Then the microphones can be removed and the vibration of the structure can be controlled in such a way that the previous vibration levels are achieved at the vibration sensors. This process is called 'model reference control' and was demonstrated for radiation from a beam and a rectangular plate by Clark and Fuller (1992d) and Clark et al. (1993).

8.8 AN EXAMPLE: CONTROL OF SOUND RADIATION FROM A RECTANGULAR PLATE

In the previous section, generalised analytical models were developed that could be used to determine the optimum control source volume velocities and forces for minimising periodic sound radiation from baffled, planar vibrating structures. In this section a specific example will be considered to demonstrate how to specialise the general models for a specific problem. For this purpose, the case where the radiating 'structure' is a simply supported rectangular plate will be used. The reason for choosing this arrangement for a closer inspection of the active control problem is the relative simplicity of analytically modelling the system, which arises from its regular geometry. This simplicity correlates well with its use as a model structure in published literature. This specific example will be used several times in the sections that follow to demonstrate certain characteristics of the active control problem.

8.8.1 Specialisation for Minimisation of Acoustic Pressure at Discrete Locations

This example study begins by considering the problem of minimising the acoustic pressure amplitude at discrete locations in space. A quick critique of the main equations of interest for calculating the optimum control source volume velocities and forces that minimise acoustic pressure, Equations (8.7.32) to (8.7.35) and Equations (8.7.88) to (8.7.91), shows that the matrices that must be specialised are $\boldsymbol{\Psi}_c$, $\boldsymbol{Z}_I$ and $\boldsymbol{Z}_r$ (note that the terms in $\boldsymbol{Z}_m$, defined in Equation (8.7.29), are the same for all baffled, planar structure problems and can be traced back to Equation (8.7.19)). Starting with the first of these, the mode shape function of the lth mode of the rectangular plate, with the origin at the lower left corner, is:

$$\psi_l(\boldsymbol{x}) = \sin\frac{m_l \pi x}{L_x} \sin\frac{n_l \pi y}{L_y} \tag{8.8.1}$$

where m_l and n_l are the modal indices in the x- and y-directions respectively, and L_x, L_y are the plate dimensions. Therefore, the terms in the vector ψ_{cl}, defined in Equation (8.7.81), which is used in constructing the rows of $\boldsymbol{\Psi}_c$, as defined in Equation (8.7.84), are:

$$\psi_{cl} = \begin{bmatrix} \psi_l(\boldsymbol{x}_1) \\ \psi_l(\boldsymbol{x}_2) \\ \vdots \\ \psi_l(\boldsymbol{x}_{N_c}) \end{bmatrix} = \begin{bmatrix} \sin\dfrac{m_l \pi x_1}{L_x} \sin\dfrac{n_l \pi y_1}{L_y} \\ \sin\dfrac{m_l \pi x_2}{L_x} \sin\dfrac{n_l \pi y_2}{L_y} \\ \vdots \\ \sin\dfrac{m_l \pi x_{N_c}}{L_x} \sin\dfrac{n_l \pi y_{N_c}}{L_y} \end{bmatrix} \tag{8.8.2}$$

In words, the terms in the $(N_c \times 1)$ vector $\boldsymbol{\Psi}_{cl}$ are the values of the lth structural mode shape function evaluated at the application point of each of the N_c vibration control sources for a total of M modes. The $(M \times N_c)$ matrix $\boldsymbol{\Psi}_c$ is then,

$$\boldsymbol{\Psi}_c = \begin{bmatrix} \sin\dfrac{m_1 \pi x_1}{L_x} \sin\dfrac{n_1 \pi y_1}{L_y} & \cdots & \sin\dfrac{m_1 \pi x_{N_c}}{L_x} \sin\dfrac{n_1 \pi y_{N_c}}{L_y} \\ & \vdots & \\ \sin\dfrac{m_M \pi x_1}{L_x} \sin\dfrac{n_M \pi y_1}{L_y} & \cdots & \sin\dfrac{m_M \pi x_{N_c}}{L_x} \sin\dfrac{n_M \pi y_{N_c}}{L_y} \end{bmatrix} \tag{8.8.3}$$

To specialise the second of these matrices, $\boldsymbol{Z}_I$, what is required (from the definition in Equation (8.7.10)) are the modal input impedances and modal masses. The former of these requires calculation of the resonance frequencies associated with the plate modes. These are found from the solution to the wave equation derived in Chapter 4 and are (Junger and Feit, 1986):

$$\omega_l = \left(\frac{D}{m_s h}\right)^{1/2}\left\{\left(\frac{m_l\pi}{L_x}\right)^2 + \left(\frac{n_l\pi}{L_y}\right)^2\right\} \tag{8.8.4}$$

where *m*, *n* are the modal indices of the *l*th mode and *D* is the plate bending stiffness, given by:

$$D = EI = \frac{Eh^3}{12(1-\nu)^2} \tag{8.8.5}$$

I is the moment of inertia of the plate, m_s is the mass per unit area of the plate, *h* is the plate thickness, *E* is the modulus of elasticity, and ν is Poisson's ratio. The modal mass can be calculated by substituting the mode shape function of Equation (8.8.1) into the defining equation of modal mass, Equation (2.4.68). Assuming a plate which is homogeneous in both thickness and material density, this produces:

$$\begin{aligned} m_n &= \int_S m_s(\boldsymbol{x})\psi_l^2(\boldsymbol{x})\,\mathrm{d}\boldsymbol{x} \\ &= \int_0^\pi\int_0^\pi \rho_s h\left(\sin\frac{m_l\pi x}{L_x}\sin\frac{n_l\pi y}{L_y}\right)^2 \mathrm{d}x\,\mathrm{d}y = \frac{m_s A}{4} \end{aligned} \tag{8.8.6}$$

where *A* is the plate area, $A = xy$. Therefore, the (*l*,*l*)th term in the diagonal matrix $\boldsymbol{Z_I}$, as defined in Equation (8.7.10), is:

$$z_I(l,l) = m_{nl}Z_l = \frac{m_s A}{4}\left\{\omega_l^2(1 + \mathrm{j}\eta) - \omega^2\right\} \tag{8.8.7}$$

where Z_l is defined in Equation (8.7.6).

The final matrix to specialise is $\boldsymbol{Z_r}$, which requires calculation of the terms in the vector $\boldsymbol{z_{rad}}$. From the definition given in Equation (8.7.16), this requires evaluation of the integral expression:

$$z_{rad}(l,\boldsymbol{r}) = 2\mathrm{j}\omega\rho_0\int_S G(\boldsymbol{x_0},\boldsymbol{r})\psi_l(\boldsymbol{x_0})\,\mathrm{d}\boldsymbol{x_0} \tag{8.8.8}$$

using the modal shape function of Equation (8.8.1). It is not possible to evaluate this integral analytically for any arbitrary location $\boldsymbol{r}$ in space. However, if the problem is restricted to locations in the far-field of the structure, such that $r >> L_x$ and $r >> L_y$, an approximate solution can be found (Wallace, 1972):

$$z_{rad}(l,\boldsymbol{r}) \approx \frac{\mathrm{j}\omega\rho_0}{2\pi r}\mathrm{e}^{-\mathrm{j}kr}\frac{L_xL_y}{m_ln_l\pi^2}\left(\frac{-1^{m_l}\mathrm{e}^{-\mathrm{j}\alpha}-1}{\left(\dfrac{\alpha}{m_l\pi}\right)^2-1}\right)\left(\frac{-1^{n_l}\mathrm{e}^{-\mathrm{j}\beta}-1}{\left(\dfrac{\beta}{n_l\pi}\right)^2-1}\right) \tag{8.8.9}$$

where

$$\alpha = kL_x \sin\theta \cos\varphi ; \qquad \beta = kL_y \sin\theta \sin\varphi \tag{8.8.10}$$

The terms in the ($N_e \times M$) matrix $\boldsymbol{Z}_r$ are simply the evaluation of this expression for the combination of each of the M structural modes at the N_e error sensing locations.

With these specialisations, the terms in Equations (8.7.32) and (8.7.33), and Equations (8.7.88) and (8.7.89) can now be quantified. Using the relationships in Equations (8.7.35) and (8.7.88), the control inputs that will minimise the acoustic pressure amplitude at discrete locations can be calculated. It is then straightforward to calculate the residual acoustic field at any location by adding together (superposition) the primary and control source pressures. For the primary source, the acoustic pressure is defined by Equation (8.7.17), for acoustic control sources by Equation (8.7.21), and for vibration control sources by Equation (8.7.85). This enables a complete picture to be determined of what will happen during the application of an active control system that minimises the acoustic pressure at discrete error sensing locations in space.

Before using the above-mentioned model to examine some of the phenomena associated with the active control of acoustic radiation from a rectangular plate, it is of interest to compare the results predicted by the theoretical model with some obtained experimentally. There have been a number of theoretical and experimental investigations of this problem, as well as other related plate radiation problems published, including Fuller et al. (1989, 1991a, 1991b), Thomas et al. (1990), Hansen and Snyder (1991), and Metcalf et al. (1992). Here, results related to those of Pan et al. (1992) will be used. With this study, a simply supported rectangular steel plate of dimensions 380 mm × 300 mm, and either 2 mm or 9.5 mm thick, was situated in a baffle in an anechoic room, with a primary noise disturbance from a non-contacting electro-magnetic shaker mounted above the plate centre. The important resonance frequencies are listed in Table 8.1.

Table 8.1 Experimental and theoretical plate resonances.

Mode	1.9 mm Thick Plate		9.5 mm Thick Plate	
	Theoretical Resonance (Hz)	Experimental Resonance (Hz)	Theoretical Resonance (Hz)	Experimental Resonance (Hz)
(1,1)	86.3	88	418.5	444
(2,1)	185.8	187	900.6	920
(1,2)	245.9	244	1191.9	1196
(2,2)	345.4	343	1674.0	1688
(3,1)	351.6	349	1703.9	1692
(3,2)	511.1	—	2477.4	—
(1,3)	511.9	501	2481.0	2456
(4,1)	583.6	581	2828.8	2796
(2,3)	611.3	595	2963.0	2928
(4,2)	743.0	732	3602.2	—

The results presented here are the radiated sound field measured across a horizontal arc, 1.8 metres from the plate centre, a distance that can be considered as just in the far-field

(Beranek, 1986, p. 100), from $\boldsymbol{r}$ = (1.8 m, 90°, 180°) to (1.8 m, 90°, 0°). The primary disturbance will be modelled as a point force (a thorough discussion of the experimental conditions can be found in Pan et al. (1992)).

Consider first the use of an acoustic control source to attenuate radiation from the 2 mm thick plate at 338 Hz. The acoustic source used here is a horn driver located 20 mm in front of the centre of the plate, with a horn diameter of 50 mm at the exit (the acoustic control source will be modelled as a monopole source 20 mm from the front of the plate). Figure 8.30 illustrates the effect of minimising the sound pressure at $\boldsymbol{r}$ = (1.8 m, 50°, 180°). The comparison between theory and experiment shows that the general agreement is good, although diffraction around the horn driver has slightly altered the acoustic field, introducing an interference pattern into the result. This sort of deviation from an idealised model is to be expected, where practical transducers are not infinitely small. However, the ability of the model to predict the major characteristics of the residual sound field is clearly evident.

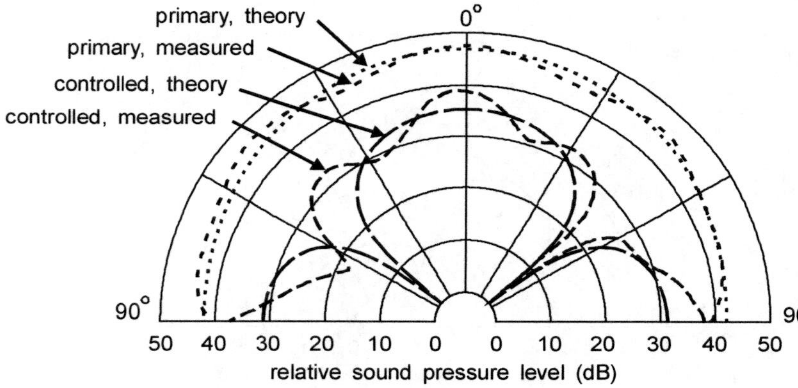

Figure 8.30 Theoretical and experimental acoustic pressure distributions for a single acoustic control source in the plate centre, error sensor at (1.8 m, 50°, 180°), 338 Hz.

It would be expected that this diffraction effect would become more pronounced as both the frequency and number of control sources, placed in front of the plate, are increased. To examine this, the plate thickness will be increased to 9.5 mm, the frequency of excitation to 1707 Hz, and the number of control sources to three. The two additional horn drivers are placed 100 mm in the x-direction on either side of the centre of the plate. To simplify the experiment, the magnitudes of each horn driver output are constrained to be equal, with the relative phases varying 0°/180°/0° across the plate. With this arrangement, a single error sensor will be adequate to implement the control system; had the three control sources been completely independent, a single error sensor would result in an under-determined system, and the overall experimental result somewhat unpredictable. To model this arrangement analytically, the term in the vector $\boldsymbol{z}_{mono}$ defined in Equation (8.7.22) can be modified to:

$$z_{mono} = \frac{\mathrm{j}\rho_0\omega}{4\pi}\left(\frac{\mathrm{e}^{-\mathrm{j}kr_{s1}}}{r_{s1}} + \frac{\mathrm{e}^{-\mathrm{j}kr_{i1}}}{r_{i1}} + \frac{\mathrm{e}^{-\mathrm{j}kr_{s2}}}{r_{s2}} + \frac{\mathrm{e}^{-\mathrm{j}kr_{i2}}}{r_{i2}} + \frac{\mathrm{e}^{-\mathrm{j}kr_{s3}}}{r_{s3}} + \frac{e^{-\mathrm{j}kr_{i3}}}{r_{i3}}\right) \tag{8.8.11}$$

The theoretical and experimental sound fields under primary excitation are illustrated in Figure 8.31. Observe that the diffraction has become worse, although the general characteristics still match quite well. The theoretical and experimental residual sound fields obtained when minimising the sound pressure at $\boldsymbol{r}$ = (1.8 m, 0°, 0°) are shown in Figure 8.32. Here, the interference pattern has become markedly worse, owing to multiple sound sources diffracting around multiple objects (horns). However, the average amplitudes of the predicted and measured residual sound fields, as well as the general characteristics of the directivity pattern, match quite well.

The use of vibration control sources to attenuate the radiated acoustic field will now be considered. For this purpose, vibration control will be applied using an electrodynamic shaker through a stinger attached to the plate 150 mm to the left of the plate centre. A plot of the theoretical and measured primary radiated and controlled residual sound pressure levels for excitation at 338 Hz and sound pressure minimisation at $\boldsymbol{r}$ = (1.8 m ,0°, 0°) are given in Figure 8.33. Here, the agreement between theory and experiment is very good, as the diffraction phenomenon associated with having the acoustic sources in front of the plate is removed.

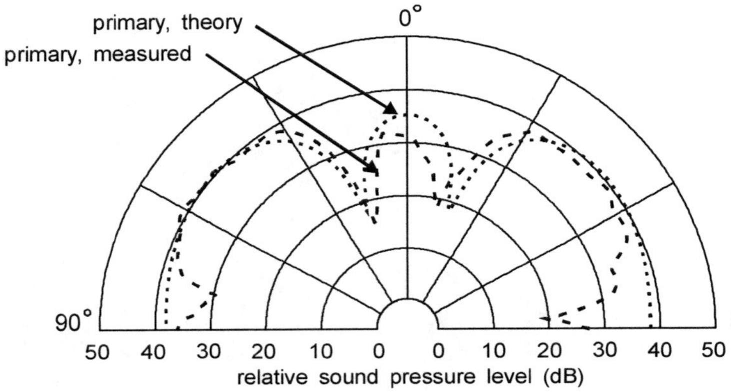

Figure 8.31 Theoretical and experimental primary source acoustic pressure distributions at 1707 Hz.

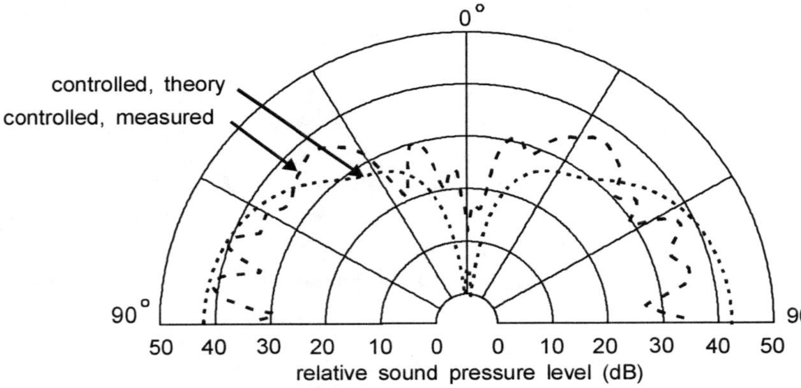

Figure 8.32 Theoretical and experimental residual pressure distributions for a three acoustic control sources, error sensor at (1.8 m, 0°, 0°), 1707 Hz.

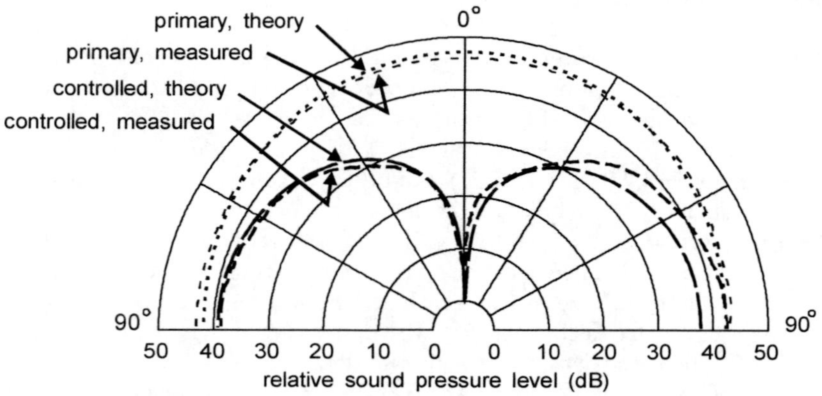

Figure 8.33 Theoretical and experimental pressure distributions for a vibration control source, error sensor at (1.8 m, 0°, 0°), 338 Hz.

One very useful type of vibration actuator is the piezoelectric ceramic patch, described in Chapter 15. These actuators are relatively small, and can be bonded directly to the plate surface to help create a 'smart structure', a structure with embedded actuators and sensors, and a control system to modify its response characteristics. As discussed in Chapter 15, these actuators can be used as vibration sources to attenuate structural acoustic radiation. The final experimental result to be discussed here involves the use of a piezoelectric crystal actuator, bonded to the centre of the 9.5 mm plate, to control the sound field at 1707 Hz. Theoretical and experimental results are as plotted in Figure 8.34 for the piezoelectric crystal actuator modelled as a simple point force. Although there is some evidence of diffraction (from the experimental baffle and microphone boom) which is a result of the relatively high test frequency, the correspondence between the theoretical and experimental results is very good. The conclusion that can be drawn from this and other tests is that the relatively simple theoretical model developed above is quite accurate in predicting the effect of applying active control using either acoustic or vibration control sources.

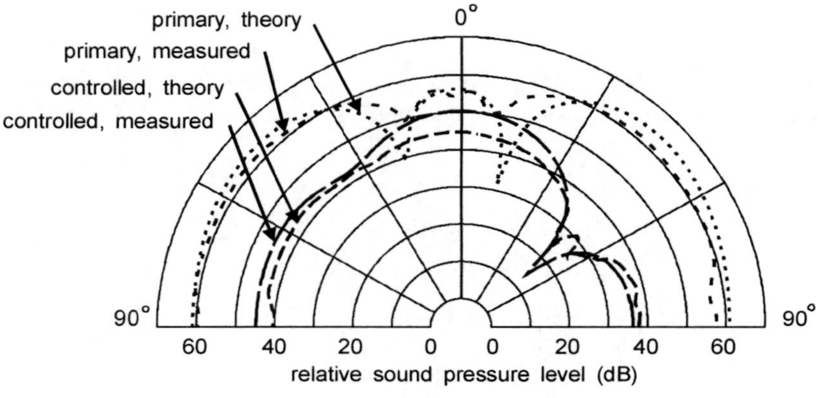

Figure 8.34 Theoretical and experimental pressure distributions for a piezoelectric ceramic path vibration control source, error sensor at (1.8 m , 50°, 180°), 1707 Hz.

8.8.2 Minimisation of Radiated Acoustic Power

The problem of minimising the acoustic power output of the system will now be considered. In general, to investigate this topic, the system specific quantities of the previous section (mode shape function, resonance frequencies and radiation transfer function) are used in Equations (8.7.42) to (8.7.45) and/or Equations (8.7.58) to (8.7.62) for acoustic control sources, or Equations (8.7.94) to (8.7.97) and/or Equations (8.7.105) to (8.7.108) for vibration control sources. The required integrations are done numerically to calculate the optimum control source outputs. From there it is straightforward to find the acoustic power attenuation using the minimum power expressions of Equations (8.7.46), (8.7.63), (8.7.98) or Equation (8.7.106), noting that the term c in each of these is the uncontrolled acoustic power output of the system. Later in this chapter, when aspects of designing an active control system are briefly discussed, this is the approach that will be taken. However, at low frequencies it is often possible to obtain approximate expressions for the acoustic power attenuation that do not require numerical integration. Such expressions can be quite useful when trying to paint a qualitative picture of the influence that certain parameters have upon system performance. The techniques required to obtain these approximate solutions were outlined in the previous section, where the problem was transformed into the wavenumber domain to change the integrals into partial derivatives. The current section will concentrate on obtaining these approximate solutions for the rectangular plate system, and demonstrate how to fill out the terms in Equations (8.7.58) to (8.7.62) and Equations (8.7.105) to (8.7.108). These solutions will be used in subsequent sections.

A general form for low-frequency approximate solutions for the elements in the vector $\boldsymbol{b}$, defined in Equation (8.7.59) for the problem of minimising radiated acoustic power using an acoustic control source, was derived in Equation (8.7.74), and for the acoustic power transfer matrix $\boldsymbol{Z}_w$ in Equation (8.7.75). To use these expressions, the Fourier transform of the mode shape function of the simply supported rectangular plate system must be obtained. The mode shape function of the simply supported plate was derived in Chapter 2 and given in Equations (2.3.151) and (2.3.152). It is separable in the x- and y-directions, as follows:

$$\psi_l(\boldsymbol{x}) = X_l(x)\,Y_l(y) = \sin\frac{m_l \pi x}{L_x}\,\sin\frac{n_l \pi y}{L_y} \tag{8.8.12a,b}$$

Taking the Fourier transform of the x-component of the mode shape function as outlined in Equation (8.7.72), if the modal index is an odd number, then,

$$\tilde{X}(x) = \frac{2\dfrac{m_l \pi}{L_x}\cos\dfrac{k_x L_x}{2}}{\left(\dfrac{m_l \pi}{L_x}\right)^2 - k_x^2} \tag{8.8.13}$$

and if the modal index is an even number:

$$\tilde{X}(x) = \mathrm{j}\frac{2\dfrac{m_l\pi}{L_x}\sin\dfrac{k_xL_x}{2}}{\left(\dfrac{m_l\pi}{L_x}\right)^2 - k_x^2} \tag{8.8.14}$$

with similar expressions for the y component of the mode shape function. In many cases, the solution for the required partial derivatives of the Fourier transformed mode shape functions is simplified if the transformed mode shape functions are first expressed as a series expansion (although this is of debatable value for the simple example being considered here). Working towards this goal, a general expression for the Fourier transformed, simply supported rectangular plate mode shape function is:

$$\tilde{\psi}(k_x,k_y) = \frac{4L_xL_y}{mn\pi^2} f(k_x)f(k_y) \tag{8.8.15}$$

where $k_x = k\sin\theta\cos\varphi$, $k_y = k\sin\theta\sin\varphi$, and for odd modes:

$$f(k_x) = \frac{\left(\dfrac{m_l\pi}{L_x}\right)^2\cos\dfrac{k_xL_x}{2}}{k_x^2 - \left(\dfrac{n_l\pi}{L_x}\right)^2} \tag{8.8.16}$$

while for even modes:

$$f(k_x) = \mathrm{j}\frac{\left(\dfrac{m_l\pi}{L_x}\right)^2\sin\dfrac{k_xL_x}{2}}{k_x^2 - \left(\dfrac{m_l\pi}{L_x}\right)^2} \tag{8.8.17}$$

with similar expressions for k_y, where the y-dimension is used in place of the x-dimension, and the y modal index in place of the x modal index. A MacLaurin expansion of Equations (8.8.16) and (8.8.17) can be easily taken. For odd modes, Equation (8.8.16) is expanded as:

$$f(k_x) = -\left(1 - \frac{(k_xL_x)^2}{8} + \cdots\right)\left(1 + \left(\frac{k_xL_x}{m_l\pi}\right)^2 + \cdots\right) \tag{8.8.18}$$

and for even modes Equation (8.8.17) is expanded as:

$$f(k_x) = -\mathrm{j}\left(\frac{k_xL_x}{2} - \frac{(k_xL_x)^3}{48} + \cdots\right)\left(1 + \left(\frac{k_xL_x}{m_l\pi}\right)^2 + \cdots\right) \tag{8.8.19}$$

The Fourier transformed mode shape functions of Equation (8.8.15), used in conjunction with the odd/even expansions of Equations (8.8.18) and (8.8.19), can be substituted into Equation (8.7.75) to calculate the terms in the acoustic power transfer matrix $\boldsymbol{Z}_w$. The mode shape functions can also be substituted into Equation (8.7.69) to calculate the term $\boldsymbol{b}$, used in the acoustic control source minimum acoustic power formulation, for the simply supported rectangular plate system. These can then be used in conjunction with Equations (8.7.58) to (8.7.62) and Equations (8.7.105) to (8.7.108) to provide an approximate calculation of the optimum control source outputs and Equations (8.7.63) and (8.7.109) for the residual radiated acoustic power at low frequencies.

The calculation of the terms in the acoustic power transfer matrix $\boldsymbol{Z}_w$ will be considered first. One point to note prior to undertaking this calculation is that only similar index-type modal combinations will produce a non-zero result ($\boldsymbol{Z}_w(i, l)$ will be non-zero if the modes i and l are both (odd-odd), (odd-even), (even-odd) or (even-even) modes). While this result is to be expected, it is not immediately obvious from the form of Equation (8.7.75). Upon substituting values into the equation, however, it is found that the inherent 'squareness' of the equation (the sum of the orders of the partial derivatives in each term must be an even number, equal to $2m$), combined with the differences in the form of Equations (8.7.16) and (8.7.17), Equation (8.7.16) is an even power expansion, Equation (8.7.17) is an odd power expansion) always combine to enforce this result. For this reason, only similar index-type modal combinations will be discussed here.

Consider first the calculation of the term $\boldsymbol{Z}_w(i,l)$ when i and l are both odd-odd index modes. To match the accuracy of the commonly stated results for radiation efficiency, also derived using low-frequency approximations, the infinite summation in Equation (8.7.75) can be truncated at $m = 1$ (this places a low-frequency constraint upon the solution, as for such a premature truncation of the infinite series to yield an accurate result, $(ka/m_l\pi) << 1$, $(kb/n_l\pi) << 1$). The term $\boldsymbol{Z}_w(i,l)$ for odd-odd modal combinations is:

$$\boldsymbol{Z}_{w_{o,o}}(i,l) = \frac{4\rho_0 c_0 k^2 L_x^2 L_x^2}{m_i n_i m_l n_l \pi^5}\left\{1 \quad \frac{k^2 L_x L_y}{24}\left[\left(1 - \frac{8}{(n_i\pi)^2}\right)\frac{L_x}{L_y}\right.\right.$$
$$\left.\left. + \left(1 - \frac{8}{(n_i\pi)^2}\right)\frac{L_y}{L_x} + \left(1 - \frac{8}{(m_l\pi)^2}\right)\frac{L_x}{L_y} + \left(1 - \frac{8}{(n_l\pi)^2}\right)\frac{L_y}{L_x}\right]\right\} \tag{8.8.20}$$

For even-odd mode types, in order to obtain the same accuracy it is necessary to consider terms in Equation (8.7.75) up to $m = 2$ (note here that the variable m is an index in a binomial expansion sum – when it has a subscript l, it is a modal index and when it has a subscript, n, it is a modal mass). Doing this, the terms in $\boldsymbol{Z}_w$ for even-odd mode combinations are:

$$\boldsymbol{Z}_{w_{e,o}}(i,l) = \frac{\rho_0 c_0 k^4 L_x^4 L_y^2}{3 m_i n_i m_l n_l \pi^5}\left\{1 - \frac{k^2 L_x L_y}{40}\left[\left(1 - \frac{24}{(m_i\pi)^2}\right)\frac{L_x}{L_y}\right.\right.$$
$$\left.\left. + \left(1 - \frac{8}{(n_i\pi)^2}\right)\frac{L_y}{L_x} + \left(1 - \frac{24}{(m_l\pi)^2}\right)\frac{L_x}{L_y} + \left(1 - \frac{8}{(n_l\pi)^2}\right)\frac{L_y}{L_x}\right]\right\} \tag{8.8.21}$$

Note also that this expression will hold for odd-even modes, as the power transfer matrix term $\boldsymbol{Z}_w(i,l)$ will be the same as in Equation (8.8.21), with the x and y modal indices and dimensions swapped. Finally, for even-even mode types, to obtain the desired accuracy it will be necessary to include terms in the infinite series of Equation (8.7.75) up to m=3. Doing this, the term $\boldsymbol{Z}_w(i,l)$ for even-even modal combinations is found to be:

$$\boldsymbol{Z}_{w_{e,e}}(i,l) = \frac{\rho_0 c_0 k^6 L_x^4 L_y^4}{60 m_i n_i m_l n_l \pi^5}\left\{1 \quad \frac{k^2 L_x L_y}{28}\left[\left(1 - \frac{24}{(m_i\pi)^2}\right)\frac{L_x}{L_y} + \left(1 - \frac{24}{(n_i\pi)^2}\right)\frac{L_y}{L_x} + \left(1 - \frac{24}{(m_l\pi)^2}\right)\frac{L_x}{L_y} + \left(1 - \frac{24}{(n_l\pi)^2}\right)\frac{L_y}{L_x}\right]\right\} \tag{8.8.22}$$

Approximate solutions will now be derived for the terms in the $\boldsymbol{b}$ vector used in the acoustic control source problem, defined in Equation (8.7.59), by filling out the terms in the expansion of Equation (8.7.74). Here, it is found that the number of terms in the expansion required to obtain the same accuracy as for the expressions above is the same for the same mode types. For odd-odd modes, the low-frequency approximate solution is:

$$b_i = v_i \frac{\rho_0 c_0 k^4 L_x L_y}{6 m_i n_i \pi^3}\left\{1 - \frac{k^2 L_x L_y}{40}\left[\frac{|2r|^2}{L_x L_y} + \left(1 - \frac{24}{(m_i\pi)^2}\right)\frac{L_x}{L_y} + \left(1 - \frac{8}{(n_i\pi)^2}\right)\frac{L_y}{L_x}\right]\right\} \tag{8.8.23}$$

where

$$|2\boldsymbol{r}|^2 = 4\left(\left(x - \frac{L_x}{2}\right)^2 + \left(y - \frac{L_y}{2}\right)^2 + z^2\right) \tag{8.8.24}$$

For even-odd and odd-even modes:

$$b_i = v_i \frac{\rho_0 c_0 k^4 L_x^2 L_y \left(x - \frac{L_x}{2}\right)}{6 m_i n_i \pi^3}\left\{1 - \frac{k^2 L_x L_y}{40}\left[\frac{|2r|^2}{L_x L_y} + \left(1 - \frac{24}{(m_i\pi)^2}\right)\frac{L_x}{L_y} + \left(1 - \frac{8}{(n_i\pi)^2}\right)\frac{L_y}{L_x}\right]\right. \tag{8.8.25}$$

Finally, for even-even modes:

$$b_i = v_i \frac{\rho_0 c_0 k^6 L_x^2 L_y^2 \left(x - \frac{L_x}{2}\right)\left(y - \frac{L_y}{2}\right)}{60 m_i n_i \pi^3} \times \left\{1 - \frac{k^2 L_x L_y}{28}\left[\frac{|2r|^2}{L_x L_y} + \left(1 - \frac{24}{(m_i\pi)^2}\right)\frac{L_x}{L_y} + \left(1 - \frac{24}{(n_i\pi)^2}\right)\frac{L_y}{L_x}\right]\right\} \tag{8.8.26}$$

As mentioned previously, one of the purposes of deriving low-frequency approximate solutions is to obtain a qualitative picture of the effects of system parameters on the ability of the system to provide acoustic power attenuation. For example, substituting the above solutions into Equations (8.7.58) to (8.7.62), it is found that with a single acoustic control source and a single structural mode, the leading terms in the acoustic power attenuation series expansion are, for an odd-odd mode (after Deffayet and Nelson, 1988):

$$\frac{W_{\min}}{W_{pri}} \approx \frac{k^2|\boldsymbol{r}|^2}{3} \tag{8.8.27}$$

for an even-odd or odd-even mode:

$$\frac{W_{\min}}{W_{pri}} \approx 1 - \frac{k^2\left(x - \frac{L_x}{2}\right)^2}{3}\left(1 - \frac{k^2|\boldsymbol{r}|^2}{5}\right) \tag{8.8.28}$$

and for an even-even mode:

$$\frac{W_{\min}}{W_{pri}} \approx 1 - \frac{k^4\left(x - \frac{L_x}{2}\right)^2\left(y - \frac{L_y}{2}\right)^2}{120}\left(1 - \frac{4k^2|\boldsymbol{r}|^2}{14}\right) \tag{8.8.29}$$

As a side note, one immediately obvious parameter having the greatest influence over the system performance is control source location. It can be observed from Equations (8.8.27) to (8.8.29) that with an odd-odd mode, the optimum control source location (for low-frequency excitation) is in the centre of the plate, while a control source at this location for the other mode types will be ineffective (Deffayet and Nelson, 1988).

8.9 ELECTRICAL TRANSFORMER NOISE CONTROL

Noise from large electrical transformers is characterised by single frequency components at twice, four times, six times and eight times the a.c. line frequency (50 Hz in Europe and Australia and 60 Hz in North America), although in some cases higher harmonics also contribute. When transformers are located close to residential communities, the characteristic low-frequency humming noise is often a cause of widespread complaints. This noise is caused by vibrations of the core (caused principally by magnetostriction) which are transmitted through the oil bath to the outer tank, which then vibrates and radiates noise. As it is extremely difficult to reduce the amplitude of the magnetostriction and thus control the problem at its source, traditional noise control involves the construction of a massive enclosure around the transformer. This enclosure must be cooled with forced air ventilation drawn into the enclosure and exhausted through silenced ductwork. This is extremely expensive and in many cases very inconvenient for maintenance and inspection purposes. For this reason, the prospect of active control has become increasingly attractive as practical automatic control systems have become more easily realised.

The first published attempt to control electrical transformer noise was reported by Conover in 1956. He used the voltage signal, from the low voltage side of the transformer, and a full wave rectifier to generate a reference signal. This reference signal was passed through narrow bandpass filters to isolate periodic noise at 120 Hz, 240 Hz and 360 Hz (see Figure 8.35). Each of these three signals was then passed through a variable gain amplifier and variable phase shifter before being recombined and fed to a single loudspeaker placed next to the transformer. The amplitudes and phases of the three individual signals were adjusted manually using the variable gain amplifier and phase shifter until the sound at an error microphone was minimised. Conover found that although he could achieve noise reductions of up to 25 dB at the error microphone, this reduction was restricted to a very small angle from the line joining the error microphone to the centre of the tank wall against which the loudspeaker was placed. In other directions, the overall noise level invariably increased with the control loudspeaker turned on. He also found that as the transformer noise varied considerably from day to day, the system had to be adjusted regularly. He suggested that this might be done automatically but did not follow this up.

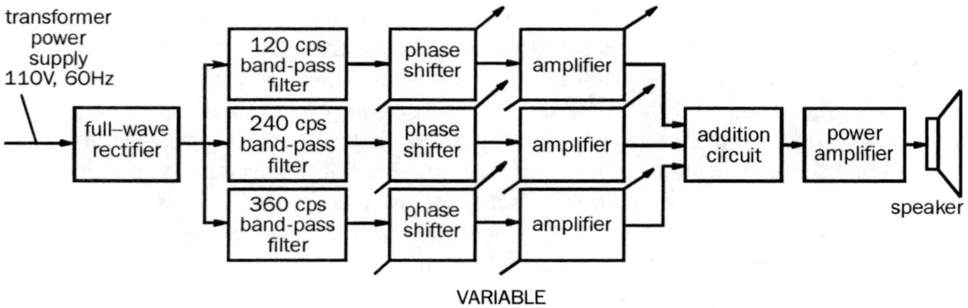

Figure 8.35 Conover's system for the control of electrical transformer noise.

More than 20 years later, Ross (1978) used an almost identical system to demonstrate control of transformer noise transmitted to an office adjacent to two transformers in a courtyard. It appears that Ross developed this independently, as no mention of Conover's work was made. At about the same time, Hesselmann (1978) and then Angevine (1981) demonstrated that global control could be achieved for a transformer in an anechoic room provided that the transformer was completely surrounded by loudspeakers. Angevine used independent single-channel controllers to minimise the sound level in the vicinity of each control source. He found that the attenuation was dependent on the number of control sources and that this dependence was stronger at lower frequencies. For example, increasing the number of control sources from 10 to 30 increased the attenuation at 125 Hz from 9 dB to 19 dB.

Berge et al. (1987, 1988) reported on an attempt to actively control electrical transformer noise using an approach similar to that of Conover (1956) except that the gains and phases of the individual frequency components (100 Hz and 200 Hz) were adjusted automatically to minimise an r.m.s. signal at a microphone, using an iterative algorithm that updated the gains and phases every 1 to 4 minutes. Rapid updating (every 12 seconds) seemed to degrade the performance for some unexplained reason. Using only one loudspeaker control source, they found that significant noise attenuation could only be achieved over a very narrow angle on either side of the error microphone. In addition, expected levels in other directions invariably increased when the active control system was turned on. Clearly, the single control source had a negligible effect on the radiation impedance 'seen' by the transformer tank and served only to provide an interfering sound field, which must increase noise levels in some locations if reductions are to be achieved at other locations.

Angevine (1992, 1993, 1994, 1995) reported some success using multiple loudspeaker sources in front of a transformer tank to achieve significant noise reductions over a wide area (15to 20 dB over an angle of 35° to 40°).

Another possible means of controlling electrical transformer noise is to use active force actuators on the tank to suppress the vibration of modes that contribute most to sound radiation (Mcloughlin et al., 1994; Brungardt et al., 1997).

Later work reported by Li (2000) involved the measurement of transfer functions between possible control source locations and error sensor locations for a large transformer tank of power, 160 MVA and dimensions, 4 m × 4 m × 4.6 m high, using both loudspeakers and control forces.

For the purpose of experimental transfer function measurements, the force was applied using a large impact hammer. However, when implementing an active noise control system, the force is applied using large piezoelectric actuators or inertial shakers (see Chapter 15 and Li, 2000). Transfer function measurements were made for 96 error sensor locations (at a 1 m distance from the transformer tank), 350 potential loudspeaker locations and 200 potential force actuator locations. Two quantities were measured at the error sensor location, mean square acoustic pressure and acoustic intensity, to determine whether there was any significant advantage in using sound intensity rather than squared acoustic pressure as the cost function for minimisation by the control system. The transfer function data were used to determine the minimum achievable mean square sound pressure or sound intensity over all 96 error sensors using sub-sets of 80 control sources derived from the 350 potential loudspeaker locations (adjacent to the transformer tank) and 200 actuator locations for which transfer function measurements were made. The procedure for doing this is described in Section 6.22. A genetic search algorithm, similar to that described in Section 6.20, was used to determine the optimal sets of control source locations for force actuators and loudspeakers that would produce the minimum of the cost function (mean squared sound pressure or mean sound intensity averaged over all error sensor locations). With 80 loudspeakers, the spacing was approximately 1/2 of a wavelength at 100 Hz and a wavelength at 200 Hz.

Results achieved by Li (2000) for the maximum achievable noise reductions, assuming a perfect adaptive controller and a reference signal highly correlated with the error sensor signals, are listed in Table 8.2. The data in the table correspond to results for the control source locations optimised separately for each tonal frequency. As this is unlikely to be practical, results were also calculated at 200 Hz and 300 Hz for the control source arrangement that was optimal for 100 Hz and these values are indicated in brackets. Although the near-field sound pressure minimisation results may not translate to a reduction in far-field sound pressure levels, minimisation of the active sound intensity will translate to a similar reduction in far-field sound levels. Achieving these global reductions in transformer sound radiation, for the size of transformer tested, required a 96 channel control system with 80 vibration control sources (or 80 loudspeakers) and 96 error sensors. Use of the same number of loudspeakers adjacent to the transformer tank and distributed around and over it was capable of achieving 5 dB to 7 dB less noise reduction (see Table 8.2).

In practice, it was found that the required vibration control forces could be achieved using piezoelectric actuators, 100 mm diameter and 2 mm thick, for 200 Hz and 300 Hz, but the forces required for 100 Hz could not be generated by these actuators. Li (2000) also investigated the use of inertial shakers as force actuators and concluded that even low-cost inertial shakers were capable of generating the required force input at 100 Hz. Due to the simple nature of the acoustic field radiated by electrical transformers, specialist multi-channel algorithms can be tailored to ensure fast adaptation of the cancellation path as well as the control filter. One such algorithm, the multi-channel waveform synthesis algorithm (Qiu et al., 2002), is described in Section 6.4.1.

Table 8.2 Estimated maximum achievable noise reductions for a large transformer tank and a control system with 80 control sources and 96 error sensors.

Frequency	SPL Reduction (dB)		Sound Intensity Reduction (dB)	
(Hz)	80 loudspeakers	80 force actuators	80 loudspeakers	80 force actuators
100	19.6	23.1	18.2	23.9
200	12.8 (11.5)	16.0 (13.6)	15.6	20.4
300	10.3 (8.5)	16.0 (14.9)	9.8	16.2

The idea of using vibration actuators to control enclosure wall vibration may be a generally efficient means of improving passive enclosure transmission loss at low tonal frequencies generated by rotating equipment inside.

8.10 A CLOSER LOOK AT CONTROL MECHANISMS AND A COMMON LINK AMONG ALL ACTIVE CONTROL SYSTEMS

The previous two sections were concerned with developing analytical models that can be used to assess the outcome of applying active control with a given system arrangement. What is yet to be done is to examine, in physical terms, how this outcome (acoustic power attenuation) is achieved; that is, what the actual control mechanisms are, and this is the focus of this current section.

Although the principal aim of the previous sections was to develop and give examples of general theoretical models that could be used in an analytical study or design exercise, the few results presented tend to leave one with the feeling that acoustic- and vibration-control-source based systems are orthogonal entities. Certainly the acoustic-control-source efforts could be viewed as an extension to the theoretical work undertaken earlier in the chapter on monopole source arrangements, with qualitative characteristics likely to be common. The efforts directed at the use of vibration control sources to suppress structural sound radiation are, however, unique to the structural radiation problem, with an apparently unique set of qualitative system characteristics (such as the influence of source location). It may seem unlikely that any important similarities between acoustic-source based control systems and vibration-source based control systems would exist. In fact, there is a common link between all active control systems, found at the fundamental level of control mechanisms and the 'control source' generated value of the global error criterion of interest. Here, active noise control systems will be considered specifically, although the conclusions drawn are not confined to the use of an acoustic power error criterion, but will exist with other global error criteria such as kinetic energy in vibrating structures.

Thus, this section will begin with the derivation of a 'common link', then progress on to a discussion of how this applies to acoustic- and vibration-source based systems. In doing so, a closer examination will be undertaken of how active control systems provide attenuation of the radiated acoustic field, especially the more complicated question of how vibration-source based systems provide attenuation of the radiated acoustic field by altering the velocity distribution of the structure to which they are attached.

8.10.1 Common Link

Consider the active noise control arrangement shown in Figure 8.36, where some generic control source distribution is being used to attenuate the acoustic field originating from some generic primary source distribution. The primary source is subject to harmonic excitation of the form $e^{j\omega t}$, which will again be left implicit in the equations, with no specifications yet made concerning the characteristics of the acoustic environment. In the development of a 'common link' it will prove insightful to examine the acoustic power output of the control source during application of active control. To do this, what is first required is calculation of the optimum control source output (volume velocity for acoustic control sources, or input force for vibration control sources) which will minimise the total radiated acoustic power of the system. This has been done previously in Sections 8.4 and 8.7 using two different approaches (near-field and far-field), both of which were aimed at expressing the error criterion (total radiated acoustic power) as a quadratic function of the control source output. This expression was then differentiated with respect to the control source output, and the resultant gradient equation set equal to zero to yield the optimum value. The process will be repeated here in a much more general way, using only the near-field formulation of the error criterion.

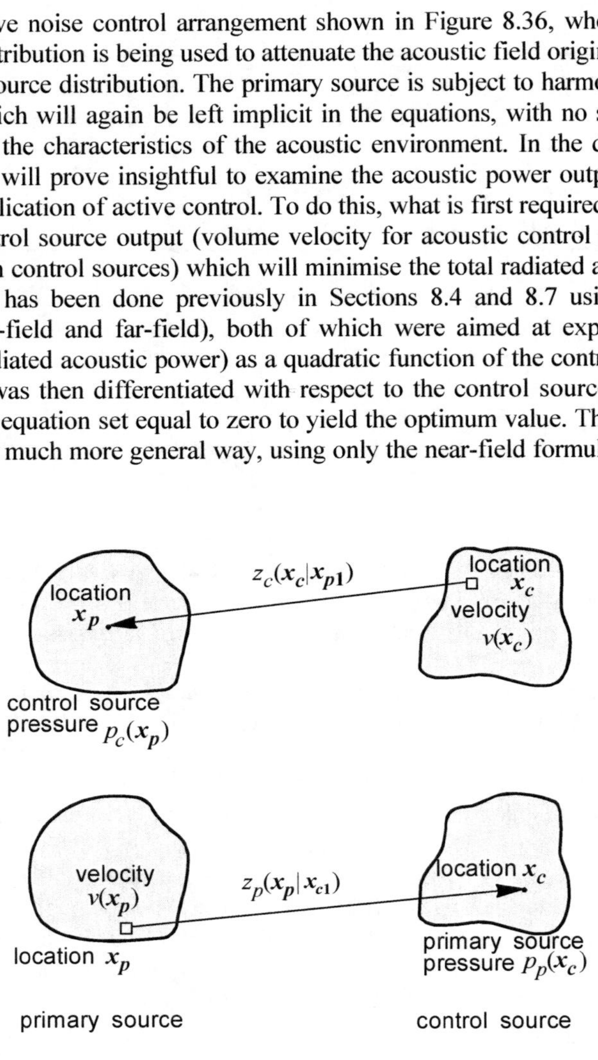

Figure 8.36 Geometry for a generic control source distribution being used to attenuate acoustic radiation from a generic primary source distribution.

Recalling that with a near-field formulation of the error criterion, the source acoustic power output is considered directly, and is equal to:

$$W = \frac{1}{2}\mathrm{Re}\left\{ \int_S v(\boldsymbol{x})\, p^*(\boldsymbol{x})\, \mathrm{d}\boldsymbol{x} \right\} \tag{8.10.1}$$

where Re denotes the real part of the expression, $v(\boldsymbol{x})$ is the velocity at some location $\boldsymbol{x}$ on the surface of the source, $p^*(\boldsymbol{x})$ is the complex conjugate of the acoustic pressure at this same location, and the integration is conducted over the surface S of the source. Returning to Figure 8.36, an expression for the total acoustic power output W_t of this arrangement can be expressed in the form:

$$W_t = W_c + W_p = \frac{1}{2}\mathrm{Re}\left\{ \int_{S_c} v(\boldsymbol{x}_c) p_t^*(\boldsymbol{x}_c)\, \mathrm{d}\boldsymbol{x}_c + \int_{S_p} v(\boldsymbol{x}_p) p_t^*(\boldsymbol{x}_p)\, \mathrm{d}\boldsymbol{x}_p \right\} \tag{8.10.2a,b}$$

where W_c is the acoustic power output of the control source, given by:

$$W_c = \frac{1}{2}\mathrm{Re}\left\{ \int_{S_c} v(\boldsymbol{x}_c)\, p_t^*(\boldsymbol{x}_c)\, \mathrm{d}\boldsymbol{x}_c \right\} \tag{8.10.3}$$

W_p is the acoustic power output of the primary source:

$$W_p = \frac{1}{2}\mathrm{Re}\left\{ \int_{S_p} v(\boldsymbol{x}_p)\, p_t^*(\boldsymbol{x}_p)\, \mathrm{d}\boldsymbol{x}_p \right\} \tag{8.10.4}$$

and the subscripts p, c and t are used to denote primary source, control source and total (combined primary and control source) quantities. As the total pressure at any location can be considered to be the superposition of the primary and control source pressures at that location, Equation (8.10.2b) can be expanded to:

$$W_t = \frac{1}{2}\mathrm{Re}\left\{ \int_{S_c} v(\boldsymbol{x}_c) \left(p_c(\boldsymbol{x}_c) + p_p(\boldsymbol{x}_c)\right)^* \mathrm{d}\boldsymbol{x}_c + \int_{S_p} v(\boldsymbol{x}_p) \left(p_c(\boldsymbol{x}_p) + p_p(\boldsymbol{x}_p)\right)^* \mathrm{d}\boldsymbol{x}_p \right\} \tag{8.10.5}$$

The control source generated acoustic pressure at a location $\boldsymbol{r}$ is related to the control source velocity distribution through an equation of the form:

$$p_c(\boldsymbol{r}) = \int_{S_c} v(\boldsymbol{x}_c)\, z_c(\boldsymbol{x}_c|\boldsymbol{r})\, \mathrm{d}\boldsymbol{x}_c = q_c \int_{S_c} z_v(\boldsymbol{x}_c)\, z_c(\boldsymbol{x}_c|\boldsymbol{r})\, \mathrm{d}\boldsymbol{x}_c \tag{8.10.6a,b}$$

where $z_c(\boldsymbol{x}_c|\boldsymbol{r})$ is the radiation transfer function between the control source velocity at location $\boldsymbol{x}_c$ and the acoustic pressure at location $\boldsymbol{r}$, q_c is the complex amplitude of the control source output, and $z_v(\boldsymbol{x}_c)$ is the transfer function between this overall amplitude and the specific velocity level of the control source at location $\boldsymbol{x}_c$. Note that q_c is completely general, and fills the role of either volume velocity or input force amplitude, considered separately in the derivations of Section 8.7. It could even represent a more 'unusual' quantity, such as the amplitude of a structural mode or the amplitude of some section of the vibrating structure, if it were of interest to find what the optimum value of these quantities, with respect to minimising the error criterion, would be.

Substituting Equation (8.10.6) into Equation (8.10.5), the total radiated acoustic power can be written as:

$$W_t = \frac{1}{2}\text{Re}\left\{ q_c \int_{S_c} z_v(\boldsymbol{x}_c^{\prime}) \left(q_c \int_{S_c} z_v(\boldsymbol{x}_c)\, z_c(\boldsymbol{x}_c|\boldsymbol{x}_c^{\prime})\, \mathrm{d}\boldsymbol{x}_c + p_p(\boldsymbol{x}_c^{\prime}) \right)^* \mathrm{d}\boldsymbol{x}_c^{\prime} \right. \\ \left. + \int_{S_p} v(\boldsymbol{x}_p) \left(q_c \int_{S_c} z_v(\boldsymbol{x}_c) z_c(\boldsymbol{x}_c|\boldsymbol{x}_p)\, \mathrm{d}\boldsymbol{x}_c + p_p(\boldsymbol{x}_p) \right)^* \mathrm{d}\boldsymbol{x}_p \right\} \tag{8.10.7}$$

or

$$W_t = q_c^* a q_c + \text{Re}\left\{ q_c b_1^* \right\} + \text{Re}\left\{ b_2 q_c^* \right\} + c \tag{8.10.8}$$

where

$$a = \frac{1}{2}\text{Re}\left\{ \int_{S_c} z_v(\boldsymbol{x}_c^{\prime}) \int_{S_c} z_v^*(\boldsymbol{x}_c)\, z_c^*(\boldsymbol{x}_c|\boldsymbol{x}_c^{\prime})\, \mathrm{d}\boldsymbol{x}_c\, \mathrm{d}\boldsymbol{x}_c^{\prime} \right\} \tag{8.10.9}$$

$$b_1^* = \frac{1}{2} \int_{S_c} z_v(\boldsymbol{x}_c^{\prime})\, p_p^*(\boldsymbol{x}_c^{\prime})\, \mathrm{d}\boldsymbol{x}_c^{\prime} \tag{8.10.10}$$

$$b_2 = \frac{1}{2} \int_{S_p} v(\boldsymbol{x}_p) \int_{S_c} z_v^*(\boldsymbol{x}_c)\, z_c^*(\boldsymbol{x}_c|\boldsymbol{x}_p)\, \mathrm{d}\boldsymbol{x}_c\, \mathrm{d}\boldsymbol{x}_p \tag{8.10.11}$$

and

$$c = \frac{1}{2}\text{Re}\left\{ \int_{S_p} v(\boldsymbol{x}_p)\, p_p^*(\boldsymbol{x}_p)\, \mathrm{d}\boldsymbol{x}_p \right\} \tag{8.10.12}$$

Equation (8.10.8) is the desired quadratic form, defining the total acoustic power output of the system, which can be used to derive the optimum value of q_c, the complex control source amplitude. It should again be noted that no specifications concerning the characteristics of the primary source distribution, control source distribution, or the acoustic environment have yet been made. In fact, it is possible to derive a quadratic expression of the form of Equation (8.10.8) for the majority of error criteria used in the development of active control systems for harmonic excitation problems.

To use Equation (8.10.8) to obtain the optimum control source output, it must first be rewritten in terms of its real and imaginary parts:

$$W_t = a q_{c\mathrm{R}}^2 + a q_{c\mathrm{I}}^2 + b_{1\mathrm{R}} q_{c\mathrm{R}} + b_{1\mathrm{I}} q_{c\mathrm{I}} + b_{2\mathrm{R}} q_{c\mathrm{R}} + b_{2\mathrm{I}} q_{c\mathrm{I}} + c \tag{8.10.13}$$

where the quantities a and c are defined in Equations (8.10.9) and (8.10.12) as being strictly real. This can then be differentiated with respect to the real and imaginary parts of the control source output, and the resultant gradient expressions set equal to zero to define the optimum values:

$$\frac{\partial W_t}{\partial q_{c\mathrm{R}}} = 2aq_{c\mathrm{R}} + b_{1\mathrm{R}} + b_{2\mathrm{R}} = 0 \; ; \qquad \frac{\partial W_t}{\partial q_{c\mathrm{I}}} = 2aq_{c\mathrm{I}} + b_{1\mathrm{I}} + b_{2\mathrm{I}} = 0 \tag{8.10.14a–d}$$

Recombining the real and imaginary parts, the optimum control source output is found to be defined by the relationship:

$$q_{c,\mathrm{opt}} = -\frac{1}{2}a^{-1}(b_1 + b_2) \tag{8.10.15}$$

Note that for the relationship of Equation (8.10.15) to actually define a unique value of the control source output that will minimise the total radiated acoustic power, the term a must be positive, which from consideration of the physical meaning of the terms in Equation (8.10.8) (which will be discussed next) will surely be the case.

It is worthwhile here to assign some physical meaning to the terms defined in Equations (8.10.9) to (8.10.12), specifically in terms of the acoustic power error criterion which is of most interest in this chapter. The simplest of these is the term c, which is the value of the error criterion (acoustic power output) of the primary source in the absence of active control. The term a can be viewed as the radiation resistance of the control source, the quantity that defines the acoustic power produced by the source, operating alone, for a given source output. From this it is apparent that a must always be positive, because if it were zero, the control source would be incapable of radiating any (real) acoustic power, regardless of the amplitude of its output; and if it were negative, it would simply absorb energy in response to operating (an acoustic black hole!). It is the quantities b_1 and b_2, however, which will prove to be the most important for the topic being considered here. The terms in Equation (8.10.8) in which b_1 and b_2 appear are the modifications to the control source and primary source acoustic power outputs respectively, due to the addition of the other sound source distributions into the acoustic environment. From a mathematical perspective, b_1 is simply equal to the integration of half the product of the primary source generated acoustic pressure and the control source velocity transfer function over the surface area of the control source, which when multiplied by the control source velocity amplitude, defines the change in control source acoustic power output. Heuristically, it can be viewed as representing the desire of the control system to absorb the primary source radiated power. Similarly, from a mathematical perspective, the term b_2 is equal to the integration over the primary source surface of half the value of the product of the velocity at some point on the surface of the primary source and the complex conjugate of the transfer function between the control source amplitude and the acoustic radiation to that point, which when multiplied by the control source velocity amplitude, defines the change in primary source acoustic power output. However, the heuristic interpretation of b_2 is that it represents the desire of the control system to suppress the primary source power output. Therefore, the bracketed part of Equation (8.10.15) can be viewed as the balance between wanting to absorb the primary source power output and wanting to suppress it. The final acoustic power output of the control source will be a result of this balance. In line with these interpretations, the a term can be viewed heuristically as simply a magnitude modulating term.

If the optimum control source output of Equation (8.10.15) is substituted into Equations (8.10.3) and (8.10.4), in terms of the quantities defined in Equations (8.10.9) to (8.10.12), the controlled power outputs of the primary and control sources are found to be:

$$W_c = \frac{1}{4} a^{-1} (b_1 + b_2)^* (b_1 + b_2) - \frac{1}{2} a^{-1} \mathrm{Re}\left\{(b_1 + b_2) b_1^*\right\} \tag{8.10.16}$$

and

$$W_p = -\frac{1}{2} a^{-1} \mathrm{Re}\left\{(b_1 + b_2)^* b_2\right\} + c \tag{8.10.17}$$

After some minor algebraic manipulation, the former of these, the acoustic power output of the control source, can be expressed as:

$$W_c = \frac{1}{4} a^{-1} \left(|b_2|^2 - |b_1|^2\right) \tag{8.10.18}$$

From this rearrangement, it can be deduced that the control source power output will be zero if:

$$|b_1| = |b_2| \tag{8.10.19}$$

or heuristically, if the desire to suppress the primary source radiation is equal to the desire to absorb it. On the surface this would appear an unlikely occurrence. It is, however, very common, as will be seen in the following analysis.

To give more credence to this line of heuristic thought, it is of interest to consider what happens if b_1 and b_2 are related by some multiplying factor, so that $b_1 = \gamma b_2$. Substituting this into Equation (8.10.18), the power output of the control source becomes:

$$W_c = \frac{1}{4} a^{-1} \left\{|b_2|^2 (1 - \gamma^2)\right\} \tag{8.10.20}$$

From this relationship, if γ is greater than 1, interpreted as the desire to absorb, b_1, being greater than the desire to suppress, b_2, the final power output of the control source will be negative and absorption will occur. Similarly, if γ is less than 1, interpreted as the desire to absorb, b_1, being less than the desire to suppress, b_2, the final power output of the control source will be positive.

To quantify the circumstances under which the desires to suppress and absorb are perfectly balanced, and hence the control source acoustic power output is equal to zero, further consideration of the terms b_1 and b_2 is necessary. The primary source acoustic pressure at the control source location, $p_p(\boldsymbol{x}_c)$, which appears in the term b_1 as defined in Equation (8.10.10), can be rewritten in terms of the velocity distribution of the primary source as follows:

$$p_p(\boldsymbol{x}_c) = \int_{S_p} v(\boldsymbol{x}_p)\, z_p(\boldsymbol{x}_p | \boldsymbol{x}_c)\, \mathrm{d}\boldsymbol{x}_p \tag{8.10.21}$$

where $z_p(\boldsymbol{x}_p | \boldsymbol{x}_c)$ is the radiation transfer function between the surface velocity at the location $\boldsymbol{x}_p$ on the primary source and control source location $\boldsymbol{x}_c$. Substituting this into Equation (8.10.10), the term b_1^* can be rewritten as:

$$b_1^* = \frac{1}{2} \int_{S_c} z_v(\boldsymbol{x}_c') \int_{S_p} v^*(\boldsymbol{x}_p)\, z_p^*(\boldsymbol{x}_p | \boldsymbol{x}_c')\, \mathrm{d}\boldsymbol{x}_p\, \mathrm{d}\boldsymbol{x}_c' \tag{8.10.22}$$

If the zero control source power output criterion of Equation (8.10.19) is expanded using this result in conjunction with Equation (8.10.11), then for zero control source acoustic power output, the following is obtained:

$$\left|\frac{1}{2}\int_{S_c} z_v(\boldsymbol{x}_c')^* \int_{S_p} v(\boldsymbol{x}_p) z_p(\boldsymbol{x}_p|\boldsymbol{x}_c')\,\mathrm{d}\boldsymbol{x}_p\,\mathrm{d}\boldsymbol{x}_c'\right| = \left|\frac{1}{2}\int_{S_p} v(\boldsymbol{x}_p)\int_{S_c} z_v(\boldsymbol{x}_c)^* z_c(\boldsymbol{x}_c|\boldsymbol{x}_p)^*\,\mathrm{d}\boldsymbol{x}_c\,\mathrm{d}\boldsymbol{x}_p\right| \tag{8.10.23}$$

The criterion of Equation (8.10.23) will only be satisfied if z_p and z_c are equal, real numbers. While it is unlikely that this will occur directly, there is a common situation that can occur which will enable the neglect of the imaginary components of these terms, and in so doing enable fulfilment of the criterion. To examine this, it is first necessary to combine the two middle terms of the acoustic power Equation (8.10.8), producing:

$$W_t = q_c^* a q_c + \mathrm{Re}\left\{q_c b_1^* + b_2 q_c^*\right\} + c \tag{8.10.24}$$

This newly combined term can be expressed, using Equations (8.10.11) and (8.10.22), as:

$$\mathrm{Re}\left\{q_c b_1^* + b_2 q_c^*\right\}$$
$$= \frac{1}{2}\int_{S_c}\int_{S_p}\mathrm{Re}\left\{\left(q_c z_v(\boldsymbol{x}_c)\right) v^*(\boldsymbol{x}_p)\, z_p^*(\boldsymbol{x}_p|\boldsymbol{x}_c) + \left(q_c z_v(\boldsymbol{x}_c)\right)^* v(\boldsymbol{x}_p)\, z_c^*(\boldsymbol{x}_c|\boldsymbol{x}_p)\right\}\mathrm{d}\boldsymbol{x}_p\,\mathrm{d}\boldsymbol{x}_c \tag{8.10.25}$$

Suppose for a moment that $z_p(\boldsymbol{x}_p|\boldsymbol{x}_c) = z_c(\boldsymbol{x}_c|\boldsymbol{x}_p)$. If this is the case, then the terms in the integral in Equation (8.10.25) have the form:

$$\mathrm{Re}\left\{de^* f^* + d^* e f^*\right\} = de^*\,\mathrm{Re}\{f\} + d^* e\,\mathrm{Re}\{f\} \tag{8.10.26}$$

Here, the term, f, is the transfer function $z_p(\boldsymbol{x}_p|\boldsymbol{x}_c) = z_c(\boldsymbol{x}_c|\boldsymbol{x}_p)$. This result means that if $z_p(\boldsymbol{x}_p|\boldsymbol{x}_c) = z_c(\boldsymbol{x}_c|\boldsymbol{x}_p)$, only the real part of these terms contributes to the acoustic power calculation, and so the imaginary components of these terms can be neglected without any alteration to the final result of the equation. If that is the case, then the criterion for zero control source acoustic power output is satisfied. However, stating that $z_p(\boldsymbol{x}_p|\boldsymbol{x}_c) = z_c(\boldsymbol{x}_c|\boldsymbol{x}_p)$ is simply stating that acoustic reciprocity exists. Therefore, the principal result of this section is that if acoustic reciprocity exists between each point on the primary source and each point on the control source, the acoustic power output of the control source during optimal operating conditions will be zero (Snyder and Tanaka, 1993a). (Note here again that only harmonic excitation is being discussed in this section, and hence in the preceding statement. Later in this chapter, random noise will be considered and a rather different result will be found.) The systems that are considered in this chapter all radiate into free space, where reciprocity exists, neglecting wind effects. Therefore, it can be stated that the acoustic power output of the control source under optimal conditions must be zero.

The question now arises, if there are multiple control sources, what is the acoustic power output of each source? Simply having a total acoustic power output of zero from the control source array does not in itself mean that contributions from the constituent elements will be zero. To answer this, it is necessary to start by generalising Equation (8.10.8) for N_c multiple, separate control sources as:

$$W_t = \mathrm{Re}\left\{\boldsymbol{q}_c^{\mathrm{H}}\boldsymbol{A}\boldsymbol{q}_c\right\} + \mathrm{Re}\left\{\boldsymbol{b}_1^{\mathrm{H}}\boldsymbol{q}_c\right\} + \mathrm{Re}\left\{\boldsymbol{q}_c^{\mathrm{H}}\boldsymbol{b}_2\right\} + c \tag{8.10.27}$$

where $\boldsymbol{q}_c$ is an N_c length vector of control source outputs, $\boldsymbol{A}$ is an $(N_c \times N_c)$ matrix of transfer functions, whose terms are defined by the expression:

$$A(i,l) = \frac{1}{2}\mathrm{Re}\left\{\int_{S_{ci}} z_v(\boldsymbol{x}_{ci}^{\prime}) \int_{S_{cl}} z_v^*(\boldsymbol{x}_{cl}^{\prime})\, z_c^*(\boldsymbol{x}_{cl}|\boldsymbol{x}_{ci}^{\prime})\,\mathrm{d}\boldsymbol{x}_{cl}\,\mathrm{d}\boldsymbol{x}_{ci}^{\prime}\right\} \tag{8.10.28}$$

$\boldsymbol{b}_1$ is an N_c length vector whose terms are defined by:

$$b_1(i) = \frac{1}{2}\int_{S_{ci}} z_v^*(\boldsymbol{x}_{ci}^{\prime})\, p_p(\boldsymbol{x}_{ci}^{\prime})\,\mathrm{d}\boldsymbol{x}_{ci}^{\prime} \tag{8.10.29}$$

and $\boldsymbol{b}_2$ is an N_c length vector whose terms are defined by:

$$b_2(i) = \frac{1}{2}\int_{S_p} v(\boldsymbol{x}_p) \int_{S_{ci}} z_v^*(\boldsymbol{x}_{ci})\, z_c^*(\boldsymbol{x}_{ci}|\boldsymbol{x}_p)\,\mathrm{d}\boldsymbol{x}_{ci}\,\mathrm{d}\boldsymbol{x}_p \tag{8.10.30}$$

Supposing that acoustic reciprocity exists in the system, then defining the variable:

$$\boldsymbol{b}_1 = \boldsymbol{b}_2^* = \boldsymbol{b} \tag{8.10.31a,b}$$

the total acoustic power output can be expressed as:

$$W_t = \boldsymbol{q}_c^{\mathrm{H}}\boldsymbol{A}\boldsymbol{q}_c + \boldsymbol{q}_c^{\mathrm{H}}\boldsymbol{b} + \boldsymbol{b}^{\mathrm{H}}\boldsymbol{q}_c + c \tag{8.10.32}$$

This is the standard expression of an error criterion as a quadratic function of control source output, which has been used several times already in this chapter. Based on the previous work, it is straightforward to write an expression defining the vector of optimum control source amplitudes:

$$\boldsymbol{q}_c = -\boldsymbol{A}^{-1}\boldsymbol{b} \tag{8.10.33}$$

The total acoustic power output of the control source array is defined by the first two terms in Equation (8.10.27) as follows:

$$W_c = \mathrm{Re}\left\{\boldsymbol{q}_c^{\mathrm{H}}\boldsymbol{A}\boldsymbol{q}_c\right\} + \mathrm{Re}\left\{\boldsymbol{b}_1^{\mathrm{H}}\boldsymbol{q}_c\right\} \tag{8.10.34}$$

From this total, the acoustic power output of a single ith control source is equal to:

$$W_{ci} = \mathrm{Re}\left\{\boldsymbol{q}_c^{\mathrm{H}}\boldsymbol{A}(\mathrm{col}\ i)\, q_c(i)\right\} + \mathrm{Re}\left\{b^*(i)\, q_c(i)\right\} \tag{8.10.35}$$

where $\boldsymbol{A}(\mathrm{col}\ i)$ is the ith column of the matrix $\boldsymbol{A}$, $b(i)$ is the ith element in the vector $\boldsymbol{b}$, and $q_c(i)$ is the complex amplitude of the ith control source, the ith element of the vector $\boldsymbol{q}_c$. If the vector of control source outputs is set equal to the optimum value defined in Equation (8.10.33), the acoustic power output of the ith control source is:

$$W_{ci} = \mathrm{Re}\left\{[-\boldsymbol{A}^{-1}\boldsymbol{b}]^{\mathrm{H}}\boldsymbol{A}(\mathrm{col}\ i)\, q_c(i)\right\} + \mathrm{Re}\left\{b(i)^* q_c(i)\right\} = 0 \tag{8.10.36a,b}$$

Therefore, the acoustic power output of each element in the control source array will also be zero.

It is important to note that this result has been derived on a completely general basis, without any specification about the type of primary or control source. In fact, this result could have been derived for any global error criterion that can be expressed as a quadratic function of the control source output, such as minimisation of structural kinetic energy or acoustic potential energy in an enclosed space. This result, therefore, provides a common link among all active control systems. The problem now is how to apply it to obtain a maximum understanding of the physical mechanisms responsible for providing attenuation in an active system. As will be seen, the concept is straightforward in an acoustic-source based system, but will require some 'lateral thinking' in a vibration-source based system.

8.10.2 Acoustic Control Source Mechanisms and the Common Link

Examination of this common link, and the relationship that it has to control mechanisms, will begin with consideration of acoustic-control-source based systems. With these systems, it is straightforward to apply the above result, as the control source is an obviously separate vibrating entity, which will have zero acoustic power output under ideal operating conditions. This result has already been derived in Section 8.4 in relation to systems with monopole primary and control sources. It will, however, be worth re-deriving the result in terms of the theory of this section before progressing on to consideration of the control of sound radiation from an infinitely baffled, planar structure using a monopole control source.

Consider first the two monopole arrangement shown in Figure 8.37. As stated in Section 8.2, the sound pressure distribution of a monopole source is defined as:

$$p(\boldsymbol{r}) = \frac{\mathrm{j}\omega\rho_0 q}{4\pi R}\,\mathrm{e}^{-\mathrm{j}kR} = \mathrm{j}\omega\rho_0 q\, G(\boldsymbol{r}_q, \boldsymbol{r}) \tag{8.10.37a,b}$$

where $\boldsymbol{r}_q$ is the location of the source, $R = |\,\boldsymbol{r}_q - \boldsymbol{r}\,|$ and G is the free space Green's function, defined in Section 2.4.1 and given by:

$$G(\boldsymbol{r}_q, \boldsymbol{r}) = \frac{\mathrm{e}^{-\mathrm{j}kR}}{kR} \tag{8.10.38}$$

The control source output that needs to be optimised in this instance, q_c in Equation (8.10.15), is its volume velocity.

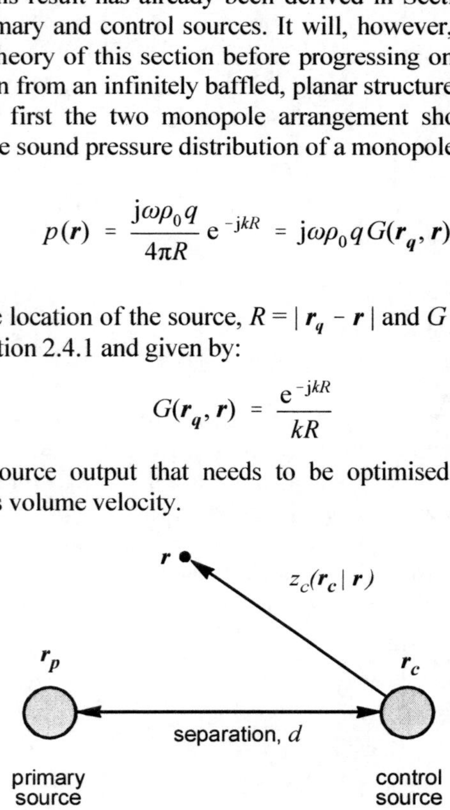

Figure 8.37 Single primary source/single control source arrangement.

The relationship of Equation (8.10.38) can be used to specialise the terms in a, b_1 and b_2 in Equations (8.10.9) to (8.10.11) to examine the power output of the control source. To do this, the term q_c in Equation (8.10.8) is set equal to the control source volume velocity amplitude, with a velocity transfer function z_v equal to 1.0. The radiation transfer function z_c is, from Equation (8.10.30):

$$z_c(\boldsymbol{x}_c|\boldsymbol{r}) = \mathrm{j}\omega\rho_0 \frac{\mathrm{e}^{-\mathrm{j}kR}}{4\pi R} = \mathrm{j}\omega\rho_0 G(\boldsymbol{x}_c, \boldsymbol{r}) \tag{8.10.39a,b}$$

with an identical expression for the primary source radiation transfer function z_p. For monopole sources, as the velocity is isolated to a single location $\boldsymbol{r}_q$ in space, the integrations can be replaced by multiplications of the values at the source locations. Taking these factors into account, the term a, which is the radiation resistance of a monopole source operating alone, and the terms b_1 and b_2 become:

$$a = \frac{1}{2}\mathrm{Re}\left\{\mathrm{j}\omega\rho_0 G(\boldsymbol{r}_c, \boldsymbol{r}_c)\right\} = \frac{\omega k\rho_0}{8\pi} \tag{8.10.40a,b}$$

$$b_1 = \frac{1}{2} z_p(\boldsymbol{r}_p|\boldsymbol{r}_c) q_p = \frac{1}{2}\mathrm{j}\omega\rho_0 G(\boldsymbol{r}_p, \boldsymbol{r}_c)\, q_p = \frac{\mathrm{j}\omega\rho_0 \mathrm{e}^{-\mathrm{j}k|\boldsymbol{r}_c - \boldsymbol{r}_p|}}{8\pi|\boldsymbol{r}_c - \boldsymbol{r}_p|} q_p \tag{8.10.41a–c}$$

and

$$b_2 = \frac{1}{2} z_c^*(\boldsymbol{r}_c|\boldsymbol{r}_p) q_p = -\frac{1}{2}\mathrm{j}\omega\rho_0 G^*(\boldsymbol{r}_c, \boldsymbol{r}_p) q_p = \frac{-\mathrm{j}\omega\rho_0 \mathrm{e}^{\mathrm{j}k|\boldsymbol{r}_p - \boldsymbol{r}_c|}}{8\pi|\boldsymbol{r}_p - \boldsymbol{r}_c|} q_p \tag{8.10.42a–c}$$

By comparing Equations (8.10.41) and (8.10.42) it is apparent that $z_p(\boldsymbol{r}_p|\boldsymbol{r}_c) = z_c(\boldsymbol{r}_c|\boldsymbol{r}_p)$, so that the imaginary parts of these transfer functions can be neglected in the overall acoustic power equation without altering the result; thus,

$$\begin{aligned} W_t &= q_c^* a q_c + \mathrm{Re}\left\{q_c b_1^*\right\} + \mathrm{Re}\left\{b_2 q_c^*\right\} + c \\ &= q_c^* a q_c + q_c \mathrm{Re}\left\{z_p(r_p|r_c)\right\} q_p^* + q_c^* \mathrm{Re}\left\{z_c(r_c|r_p)\right\} q_p + c \end{aligned} \tag{8.10.43a,b}$$

As was found in the theoretical development of this section, neglect of the imaginary components of these transfer functions, in conjunction with equality of their real parts, results in the control source acoustic power output being equal to zero. This can be shown by substituting a, as defined in Equation (8.10.40), and b_1 and b_2, as defined in Equations (8.10.41) and (8.10.42), with only the real parts of the radiation transfer functions, into Equation (8.10.18), producing the following result for the control source acoustic power output:

$$W_c = \frac{1}{4} a^{-1}\left(|b_2|^2 - |b_1|^2\right) = \frac{1}{4}\frac{8\pi}{\omega k\rho_0}\left(\left|\frac{\omega\rho_0 \sin kd}{8\pi d}\right|^2 - \left|\frac{\omega\rho_0 \sin kd}{8\pi d}\right|^2\right) = 0 \tag{8.10.44a–c}$$

This is the same result as was obtained in Section 8.3.

An important point to note is that it is the radiation transfer functions between the acoustic sources that are responsible for the final acoustic power output of the control source, and not the real part of the radiation impedance (radiation resistance) presented to the sources when they are operating alone. This quantity, which for the control source was shown to be equal to the term a in the total acoustic power Equation (8.10.8), can best be viewed as an amplitude modulating factor. This can be demonstrated by considering the arrangement depicted in Figure 8.38, where the control and primary monopole sources are operating in the presence of an infinite, rigid baffle (a problem considered by Snyder and Tanaka (1993a) and Cunefare and Shepard (1993)). The sources are arranged roll-offly in a line perpendicular to the baffle, with the control source, at a distance z_c, closer to the baffle than the primary source, at a distance z_p.

As has already been discussed, in the presence of an infinite rigid baffle, the sound field produced by a monopole source can be modelled as the superposition of sound fields radiated by the real source and a mirror image source in the absence of the baffle; thus the acoustic pressure at location $\boldsymbol{r}$ is given by:

$$p(\boldsymbol{r}) = \frac{j\omega\rho_0 q}{4\pi}\left(\frac{e^{-jkr_s}}{r_s} + \frac{e^{-jkr_i}}{r_i}\right) \tag{8.10.45}$$

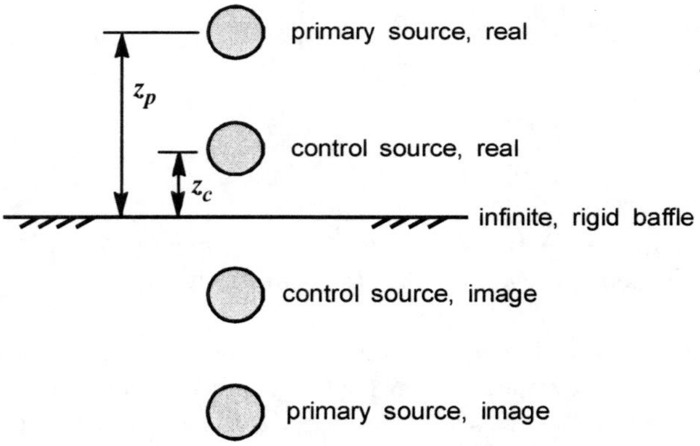

Figure 8.38 Single primary source/single control source in the presence of an infinite rigid baffle.

where the subscripts s and i are used to denote the real source and image source. The acoustic power output of a single source, operating in the presence of the baffle, can be calculated using the acoustic pressure distribution of Equation (8.10.45) in the defining Equation (8.10.1), specialised for the monopole case as follows:

$$\begin{aligned} W &= \frac{1}{2}\mathrm{Re}\left\{q\, p_t(\boldsymbol{r}_q)^*\right\} = |q|^2\frac{\omega\rho_0}{8\pi}\mathrm{Re}\left\{\frac{-je^{jkr_s}}{r_s} + \frac{-je^{jkr_i}}{r_i}\right\} \\ &= |q|^2\frac{\omega k\rho_0}{8\pi}(1 + \mathrm{sinc}\,2kz) \end{aligned} \tag{8.10.46a–c}$$

Comparing this result to Equation (8.10.40), which is the radiation resistance of a monopole source in free space, it can be deduced that the radiation resistance has been increased by the introduction of the baffle by a factor of (1 + sinc $2kz$). As such, the primary and control sources in the system illustrated in Figure 8.38 will have different radiation resistances, due to their different distances from the baffle. Theoretically, this should not make any difference to the final value of control source acoustic power output; it should still be equal to zero, as the radiation transfer functions between any two sources (real or image) will be the same. Only the amplitude of the control source output should change.

To test these theoretical predictions, it is necessary calculate the terms b_1 and b_2 as well as the term a. From Equation (8.10.46), the latter is:

$$a = \frac{\omega k \rho_0}{8\pi}(1 + \mathrm{sinc}\,2kz) \tag{8.10.47}$$

Using the description of acoustic pressure given in Equation (8.10.45), the terms b_1 and b_2 can be calculated as:

$$\begin{aligned} b_1 &= \frac{1}{2} z_p(\boldsymbol{r_p}|\boldsymbol{r_c})\, q_p = \frac{1}{2}\mathrm{j}\omega\rho_0 \left\{ G(\boldsymbol{r_{p,r}}, \boldsymbol{r_c}) + G(\boldsymbol{r_{p,i}}, \boldsymbol{r_c}) \right\} q_p \\ &= \frac{\mathrm{j}\omega\rho_0}{8\pi} \left\{ \frac{\mathrm{e}^{-\mathrm{j}k(z_p - z_c)}}{z_p - z_c} + \frac{\mathrm{e}^{-\mathrm{j}k(z_p + z_c)}}{z_p + z_c} \right\} q_p \end{aligned} \tag{8.10.48a–c}$$

and

$$\begin{aligned} b_2 &= \frac{1}{2} z_c(\boldsymbol{r_c}|\boldsymbol{r_p})^* q_p = -\frac{1}{2}\mathrm{j}\omega\rho_0 \left\{ G(\boldsymbol{r_{p,r}}, \boldsymbol{r_c}) + G(\boldsymbol{r_{p,i}}, \boldsymbol{r_c}) \right\}^* q_p \\ &= \frac{-\mathrm{j}\omega\rho_0}{8\pi} \left\{ \frac{\mathrm{e}^{\mathrm{j}k(z_p - z_c)}}{z_p - z_c} + \frac{\mathrm{e}^{\mathrm{j}k(z_p + z_c)}}{z_p + z_c} \right\} q_p \end{aligned} \tag{8.10.49a–c}$$

From Equations (8.10.48) and (8.10.49) it is apparent that $z_p(\boldsymbol{r_p}|\boldsymbol{r_c}) = z_c(\boldsymbol{r_c}|\boldsymbol{r_p})$, which again leads to a zero control source power output, the same result as for the previous system without the baffle.

Thus, the change in source (primary or control) radiation resistance has had no effect upon the final acoustic energy state of the control source. It has, however, altered the optimum value of the control source volume velocity from the unbaffled case, which was shown in Section 8.4 to be defined by the relationship $q_c = -q_p$ sinc kd, where d is the separation distance between the two sources. Substitution of Equations (8.10.47) to (8.10.49) into Equation (8.10.15) shows the optimal relationship in the presence of the baffle to be:

$$q_{c,\mathrm{opt}} = -q_p \left\{ \mathrm{sinc}(kz_p + kz_c) + \mathrm{sinc}(kz_p - kz_c) \right\} \tag{8.10.50}$$

Both the zero acoustic power output and the altered optimum control source volume velocity were predicted qualitatively by the general theory of the previous section.

Having seen how the 'common link' theory is applied to simple monopole source problems, it is now of interest to investigate the acoustic radiation from a vibrating structure, the system arrangement of which is shown in Figure 8.39.

The description of the sound field radiated by the control source is identical to that of the previous case, for the monopole sources operating in the presence of the baffle, stated in Equation (8.10.45). The sound field radiated by the structure is defined by the Rayleigh integral, in terms of surface velocity as:

$$p(r) = \frac{\mathrm{j}\omega\rho_0}{2\pi} \int_{S_p} v(\boldsymbol{x_p}) \frac{\mathrm{e}^{-\mathrm{j}kr}}{r}\, \mathrm{d}\boldsymbol{x_p} = \mathrm{j}2\omega\rho_0 \int_{S_p} v(\boldsymbol{x_p}) G(\boldsymbol{x_p}, \boldsymbol{r})\, \mathrm{d}\boldsymbol{x_p} \tag{8.10.51a,b}$$

where r is the distance between the location $\boldsymbol{x}_p$ on the structure and the location in space of interest.

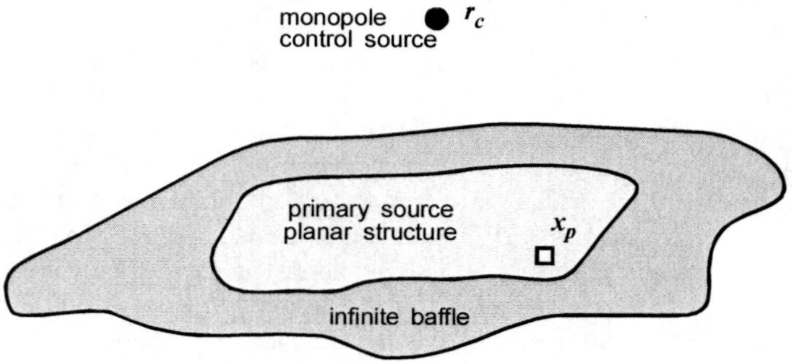

Figure 8.39 Monopole control source being used to attenuate acoustic radiation from a vibrating plate.

To calculate the control source volume velocity from the viewpoint of mapping the residual acoustic pressure field, or calculating the precise value of the acoustic power attenuation, Equation (8.10.51) is often specialised for the structure of interest and separated in terms of modal contributions to the sound field. This was the approach taken in Section 8.7, which was later specialised for the rectangular plate example in Section 8.8. This approach, however, is not necessary to demonstrate the point of interest here. Rather, Equation (8.10.47) can be used directly in the formulation of the terms b_1 and b_2, producing:

$$b_1 = \frac{1}{2}\int_{S_p} v(\boldsymbol{x}_p)\, z_p(\boldsymbol{x}_p|\boldsymbol{r}_c)\,\mathrm{d}\boldsymbol{x}_p = \frac{\mathrm{j}\omega\rho_0}{4\pi}\int_{S_p} v(\boldsymbol{x}_p)\frac{\mathrm{e}^{-\mathrm{j}k|\boldsymbol{r}_c-\boldsymbol{x}_p|}}{|\boldsymbol{r}_c-\boldsymbol{x}_p|}\,\mathrm{d}\boldsymbol{x}_p \tag{8.10.52a,b}$$

and

$$b_2 = \frac{1}{2}\int_{S_p} v(\boldsymbol{x}_p) z_c(\boldsymbol{r}_c|\boldsymbol{x}_p)^*\,\mathrm{d}\boldsymbol{x}_p$$

$$= -\frac{1}{2}\mathrm{j}\omega\rho_0\int_{S_p} v(\boldsymbol{x}_p)\left\{G(\boldsymbol{r}_{p,r},\boldsymbol{r}_c) + G(\boldsymbol{r}_{p,i},\boldsymbol{r}_c)\right\}^*\,\mathrm{d}\boldsymbol{x}_p = \frac{-\mathrm{j}\omega\rho_0}{4\pi}\int_{S_p} v(\boldsymbol{x}_p)\frac{\mathrm{e}^{\mathrm{j}k|\boldsymbol{x}_p-\boldsymbol{r}_c|}}{|\boldsymbol{x}_p-\boldsymbol{r}_c|}\,\mathrm{d}\boldsymbol{x}_p$$

(8.10.53a–c)

In viewing Equations (8.10.52) and (8.10.53), it is again apparent that, despite the fact that the primary source is a radiating structure and the control source is a monopole source, $z_p(\boldsymbol{r}_p|\boldsymbol{r}_c) = z_c(\boldsymbol{r}_c|\boldsymbol{r}_p)$, which is a direct result of acoustic reciprocity. As such, the control source acoustic power output will again be zero under optimal operating conditions.

The final item of business in this discussion is to equate the above results with the control mechanisms discussed previously. For acoustic control sources this is a very simple point, as it has been assumed that the primary and control sources have a constant volume velocity. Thus, the control mechanism must be one of pressure reduction at the face of both sources, or a reduction in radiation impedance seen by both the control source (to zero) and primary source. This control mechanism has been referred to previously in this book as one of acoustic power suppression, or source unloading.

8.10.3 Mechanism Prelude: A Vibration Source Example

While the determination of control mechanisms and application of the common link theory for systems using acoustic control sources are rather straightforward, the use of vibration control sources represents a much more complicated problem. Vibration control sources provide attenuation of the radiated acoustic field by altering the velocity distribution of the vibrating structure. Before examining this effect in terms of mechanisms and the common link theory, it will prove informative to briefly consider an example that will be considered as a simply supported rectangular plate of dimensions, 0.38 m × 0.30 m × 2 mm, excited in the four corners by point forces, at $x = 0.019$ m and 0.361 m, and $y = 0.015$ m and 0.285 m, with all forces equal in amplitude and phase. This is similar to the form of excitation that may be found on a rotating machinery cover plate. The excitation frequency is 350 Hz, which lies slightly below the fourth and fifth plate modes, the (2,2) and (3,1) modes, as shown in Table 8.3. This frequency was chosen for its ability to readily demonstrate all control mechanisms, as will be shown. The control source is a single point force that can be applied to any location on the plate.

Table 8.3 Theoretical plate resonance frequencies.

Mode	Theoretical Resonance Frequency (Hz)
(1,1)	88.1
(2,1)	189.6
(1,2)	250.9
(2,2)	352.4
(3,1)	358.7
(3,2)	521.6
(1,3)	522.3
(4,1)	595.5
(2,3)	623.8
(4,2)	758.4

Figure 8.40 illustrates the maximum achievable sound power attenuation as a function of control source location on the plate, showing four optimum control source locations, one on the horizontal centre-line, one at each of the two edges and one in the centre of the plate. It will prove interesting to examine the changes in the plate vibration characteristics that accompany the application of acoustic power-optimal control with a control source at the centre and then at one edge.

The effect of controlling the plate radiation, at the optimum location of (0.36 m, 0.15 m), upon the amplitudes and relative phases of the first 10 plate modes is evident in Figure 8.41. Here, it can be seen that the amplitudes of the nearly resonant (3,1) mode and the (1,1) mode are substantially reduced. These two modes, based on their velocity levels and radiation efficiencies at this frequency, have the greatest potential for sound power generation. Their reduction causes a significant reduction in the total radiated sound power, which is the primary control mechanism producing the 19.9 dB of attenuation achieved here. (One additional point to note concerning these results is the lack of excitation of any even numbered modes under the primary forcing function. This is due solely to the symmetric nature of the primary forcing function.)

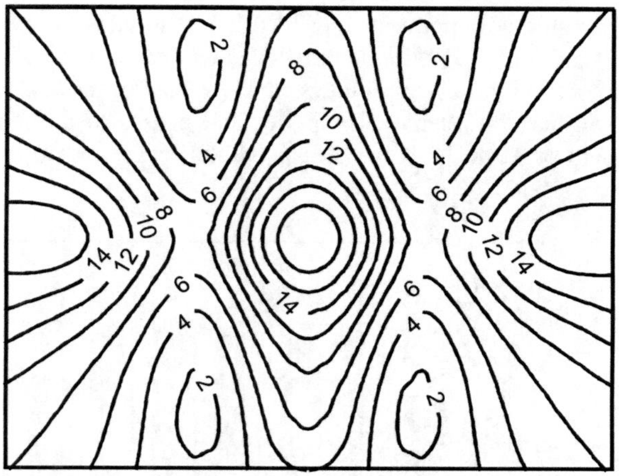

Figure 8.40 Maximum acoustic power attenuation (dB) as a function of vibration control source location on the plate, 350 Hz.

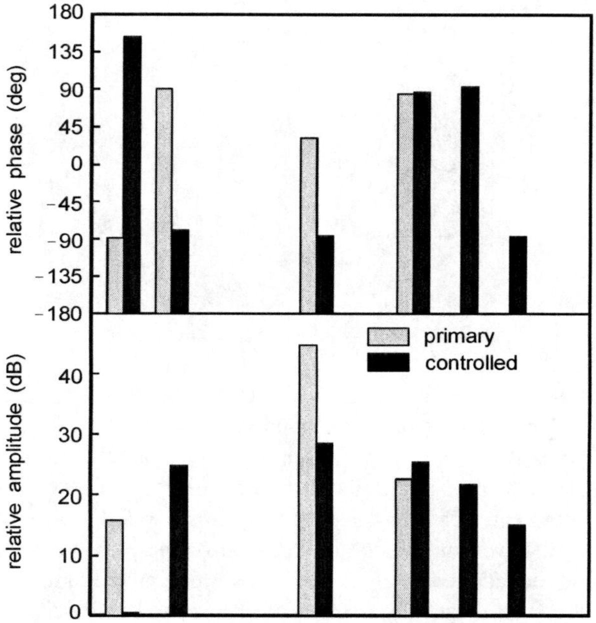

Figure 8.41 Plate modal amplitudes and phases before and after control, with the vibration control source near the plate edge.

The effect that using a vibration source at (0.19 m, 0.15 m) to control sound power radiation has upon the first 8 non-zero amplitude plate modes is shown in Figure 8.42. Here, the nearly resonant (3,1) mode, as well as the (1,3) mode, both have a reduction in amplitude of approximately 10 dB. This would appear to be offset, however, by an increase in the amplitude of the (1,1) mode of approximately 10 dB. Just in viewing these results, it would seem unlikely that the 22.7 dB reduction in radiated sound power would be achieved only by a reduction in the modal vibration amplitudes of the radiating modes. However, it can be observed in the figure that the relative amplitudes and phases of the modes have changed significantly. This will prove to be what is responsible for the extent of the sound power reduction achieved.

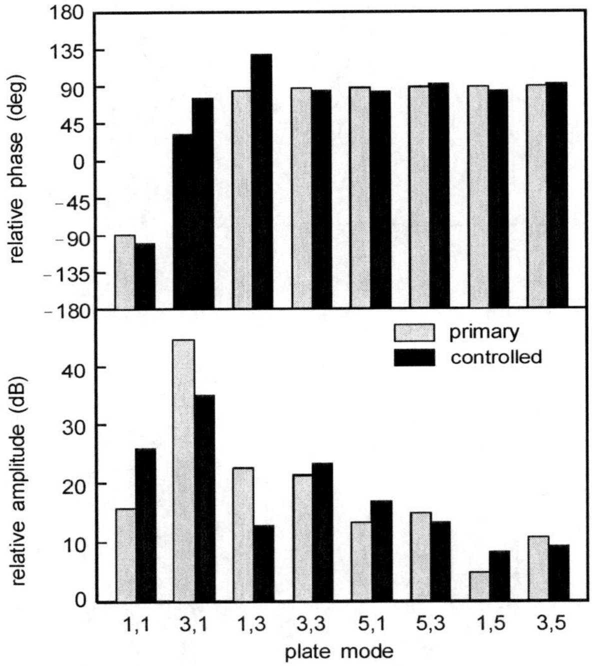

Figure 8.42 Plate modal amplitudes and phases before and after control, vibration control source in plate centre.

For the purposes of the discussion here, it can be viewed that there are two ways in which an alteration of the plate velocity distribution can provide a reduction in the radiated acoustic power. The first is simply a reduction in the velocity levels of the principal offending modes, as was the case in the first example. This type of control is implicitly modal control, and is an approach commonly taken in systems explicitly targeting a reduction in vibration level. The other way in which an alteration of the plate velocity distribution can provide a reduction in the radiated acoustic field is by reducing the radiation efficiency of the structure, without necessarily causing a reduction in the modal velocity levels (Snyder and Hansen, 1991a). This is possible in the above example problem because the plate modes are not orthogonal radiators, so that the total acoustic power output depends not only upon the velocity of individual modes, but also upon the relative phases of the modal velocities (cross-terms in the expression of acoustic power in terms of structural modes). Changing the amplitude and phase relationship between modes will therefore change the acoustic power output of the system. Minimising acoustic power by altering the relative complex velocities of the structural modes to minimise radiation efficiency has been termed 'modal rearrangement', or 'modal restructuring', as opposed to 'modal control'. (The details of the relationship between the modes of the rectangular plate system will be discussed in more detail in the next section.)

It should be noted at this point that these two control mechanisms, reduction of vibration level or reduction in overall radiation efficiency, have appeared under a variety of labels, the pros and cons of which can be debated. It can be viewed (Burdisso and Fuller, 1991, 1992) that during the application of active control, the 'controlled eigenfunctions', or controlled mode shape functions, of the structure are different from those of the uncontrolled structure, and that these different mode shape functions are less efficient radiators. This is one way of describing the control mechanism because when the active control system is engaged, all control-system referenced disturbances see a different set of boundary conditions, made up of the original system boundary conditions plus those forced by the active controller.

The application of active control can also be related to the wavenumber spectrum of the vibrating structure (Fuller and Burdisso, 1991; Clark and Fuller, 1992a; Wang and Fuller, 1992), where the radiating components of the spectrum are reduced (wavenumber domain considerations are discussed further in Chapter 2, Section 2.4).

Still another way to view what happens under the action of active control is that the amplitude of orthogonally radiating sets of structural modes decrease when the active control system is implemented, which may or may not reduce the overall vibration levels on the structure (this concept will be discussed further in the next section). However, from the standpoint of developing a physical understanding of what leads to a reduction in the radiated acoustic field upon application of vibration-source based active control, the simple concepts of a reduction in vibration levels, or a reduction in the overall radiation efficiency of the plate, seem to adequately describe the observed phenomena.

Figure 8.43 illustrates the mean square velocity levels of the plate before and after the application of active control for the second vibration source location at (0.19 m, 0.15 m).

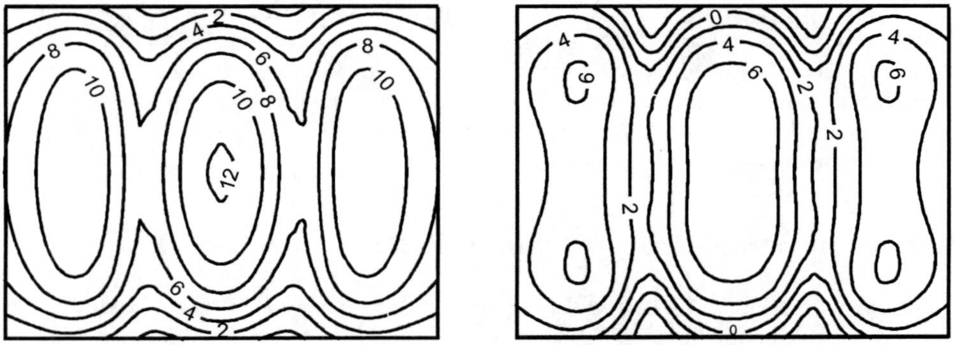

(a) (b)

Figure 8.43 Plate velocity distribution (dB): (a) during primary excitation; (b) during application of control input.

Note that an overall reduction in amplitude of approximately 6 dB has occurred. This is because the reduction in amplitude of the (3,1) and (1,3) modes is greater than the increase in amplitude of the (1,1) mode. This, however, is not enough to account for the 22.7 dB of radiated sound power attenuation. The remainder of the sound attenuation must therefore be due to modal rearrangement.

8.10.4 Control Sources and Sources of Control

Before relating the vibration control source results to the common link, it is worthwhile briefly considering the idea of a 'control source' in the light of physical 'sources of control'. Referring to the defining equation of acoustic power in Equation (8.10.1), there are two classes of physical mechanism, or sources of control, which will attenuate the radiated acoustic power of the system. The first of these are acoustic control mechanisms, where the in-phase pressure, and hence radiation impedance, on the surface of the control source and/or other radiators in the system is reduced (note that acoustic absorption falls within this definition). The second are velocity control mechanisms, where the velocity levels of the radiating sources are reduced, or the structural input impedance to the primary excitation is increased. It has already been demonstrated that an acoustic control source can be a source of acoustic control. It is also obvious that a vibration control source can be a source of structural vibration control. Less obvious is the fact that a vibration control source can also be a source of acoustic control through the modal rearrangement control mechanism, as a result of the vibration source being attached to the structure.

Structural vibration will often form inherent sources of acoustic control, areas of the structure whose acoustic radiation is responsible for reducing the acoustic power output of other areas on the same structure. One common example of this is seen in the areas of negative acoustic intensity, which frequently appear on the surface of vibrating structures, areas that obviously reduce the total acoustic power output of the structure. Vibration control sources are potential vehicles for optimising the attenuating effect of these inherent phenomena via modal rearrangement. Indeed, it is possible to actually *increase* the velocity levels of a structure while providing attenuation of the radiated sound field. Clearly, in this case the vibration control source has greatly reduced the overall radiation efficiency of the structure, an effect that can only come about by enhancing the effectiveness of the physical phenomena responsible for limiting the acoustic power radiated from the structure, the inherent sources of acoustic control. Thus, when the term 'control source' is used in the discussion here, it will be rather generic; vibration control sources attached to a structure operating as sources of velocity control, or vibration control sources acting via the structure as sources of acoustic control. With this idea in mind it is possible to apply the common link theory to vibration-source based systems.

8.10.5 Vibration Source Control Mechanisms and the Common Link

Vibration-source based systems will now be discussed, where a vibration control (point) force, attached directly to the vibrating structure, is used to modify the velocity distribution of the structure in such a way as to attenuate the radiated acoustic field. This situation represents a departure from the previous study in that the control source is no longer strictly acoustic nor distinct; both the primary and control radiation is from the same structure. It can, however, still be adequately represented by the previously outlined theory. To investigate the control source power output, the terms a, b_1 and b_2 must again be specialised for particular the active control arrangement, where the term q_c is now the control force amplitude, rather than volume velocity. While it is specifically the radiation transfer functions z_p and z_c that require calculation in order to demonstrate zero acoustic power output, it is also useful to specialise the control source velocity transfer function z_v to demonstrate how the control arrangement fits into the previously outlined theory.

The velocity transfer function $z_v(\boldsymbol{x}_c)$ is the transfer function between the control force output and the velocity at a location $\boldsymbol{x}_c$ on the structure. Expressing the velocity at this location in terms of the set of structural mode shape functions, the velocity transfer function can be defined using the following Green's function relationship for the velocity response:

$$v(\boldsymbol{x}_c) = \mathrm{j}\omega \int_{S_c} G_s(\boldsymbol{x}', \boldsymbol{x}_c) f_c \delta(\boldsymbol{x}' - \boldsymbol{x}_f)\, \mathrm{d}\boldsymbol{x}' = f_c \mathrm{j}\omega G_s(\boldsymbol{x}_c, \boldsymbol{x}_f) \tag{8.10.54a,b}$$

In Equation (8.10.54) $\boldsymbol{x}_f$ is the application point of the control force, $\delta(\boldsymbol{x}' - \boldsymbol{x}_f)$ is the Dirac delta function, and $G_s(\boldsymbol{x}_c, \boldsymbol{x}_f)$ is the structural Green's function. Therefore,

$$z_v(\boldsymbol{x}_c) = \mathrm{j}\omega G_s(\boldsymbol{x}_c, \boldsymbol{x}_f) \tag{8.10.55}$$

The radiated sound field of the planar structure is governed by the Rayleigh integral. In viewing this, it is clear that z_p and z_c both have the form:

$$z_c(\boldsymbol{x}_c|\boldsymbol{x}) = \mathrm{j}2\omega\rho_0 G_f(\boldsymbol{x}_c, \boldsymbol{x}) = \frac{\mathrm{j}\omega\rho_0 \mathrm{e}^{-\mathrm{j}k|\boldsymbol{x} - \boldsymbol{x}_c|}}{2\pi|\boldsymbol{x} - \boldsymbol{x}_c|} \tag{8.10.56a,b}$$

Therefore, terms b_1 and b_2 are:

$$b_1 = \frac{1}{2}\int_{S_c} z_v^*(\boldsymbol{x}_c') \int_{S_p} v(\boldsymbol{x}_p)\, z_p(\boldsymbol{x}_p|\boldsymbol{x}_c)\, \mathrm{d}\boldsymbol{x}_p\, \mathrm{d}\boldsymbol{x}_c'$$

$$= \frac{1}{2}\int_{S_c} z_v^*(\boldsymbol{x}_c') \int_{S_p} v(\boldsymbol{x}_p) \left\{ \frac{\mathrm{j}\omega\rho_0 \mathrm{e}^{-\mathrm{j}k|\boldsymbol{x}_p - \boldsymbol{x}_c'|}}{2\pi|\boldsymbol{x}_p - \boldsymbol{x}_c'|} \right\} \mathrm{d}\boldsymbol{x}_p\, \mathrm{d}\boldsymbol{x}_c' \tag{8.10.57a,b}$$

and

$$b_2 = \frac{1}{2}\int_{S_p} v(\boldsymbol{x}_p) \int_{S_c} z_v^*(\boldsymbol{x}_c)\, z_c^*(\boldsymbol{x}_c|\boldsymbol{x}_p)\, \mathrm{d}\boldsymbol{x}_c\, \mathrm{d}\boldsymbol{x}_p$$

$$= \frac{1}{2}\int_{S_p} v(\boldsymbol{x}_p) \int_{S_c} z_v(\boldsymbol{x}_c)^* \left\{ \frac{\mathrm{j}\omega\rho_0 \mathrm{e}^{-\mathrm{j}k|\boldsymbol{x}_c - \boldsymbol{x}_p|}}{2\pi|\boldsymbol{x}_c - \boldsymbol{x}_p|} \right\}^* \mathrm{d}\boldsymbol{x}_c\, \mathrm{d}\boldsymbol{x}_p \tag{8.10.58a,b}$$

In viewing these expressions for b_1 and b_2, it is clear that the radiation transfer functions, the bracketed part of the expressions, are equal, which leads to a control source power output equal to zero. This result, however, could easily have been anticipated from the fact that the control source (radiation) and primary source (radiation) are (from) the same object, the vibrating structure.

For vibration control sources, the control source acoustic power output that will be zero, the term W_c in Equation (8.10.3), is defined mathematically as the acoustic power output of the structure with a control source-induced velocity distribution operating in the total (combined primary and control) acoustic pressure field:

$$W_c = \frac{1}{2} \mathrm{Re}\left\{ q_c \int_{S_c} z_v(\boldsymbol{x}_c)\, p_t^*(\boldsymbol{x}_c)\, \mathrm{d}\boldsymbol{x}_c \right\} \tag{8.10.59}$$

However, it is informative to relate this result to 'real-world' phenomena, which would be witnessed upon implementation of a power-optimal feedforward active control system, using the potential control mechanisms outlined previously in this section. For vibration sources operating as sources of velocity control, the concept is straightforward: the overall velocity levels of the structure will be minimised. For vibration sources operating as sources of acoustic control, the meaning is more obscure, and is best elucidated by means of another example.

Consider a specific case where the vibrating 'structure' is again a simply supported, lightly damped, rectangular plate in an infinite baffle. The concept of intercellular cancellation can be used to explain the small value of radiation efficiency of this plate at low frequencies. Referring to the sketch of the cross-section of a three-index mode shown in Figure 8.44, at low frequencies the acoustic radiation from the centre nodal area can be viewed as 'cancelling' half of the acoustic radiation from each of the end nodal areas, a phenomenon that leads to an area of negative acoustic intensity in the centre of the plate (an inherent source of acoustic control). With this idea in mind, it is possible to determine analytically what centre nodal area amplitude, relative to the edge amplitudes, would minimise the total acoustic power output of the plate. This can be accomplished using the previously derived common link theory, where the control source velocity distribution is over the centre nodal area and the primary source velocity distribution is over the edge nodal areas.

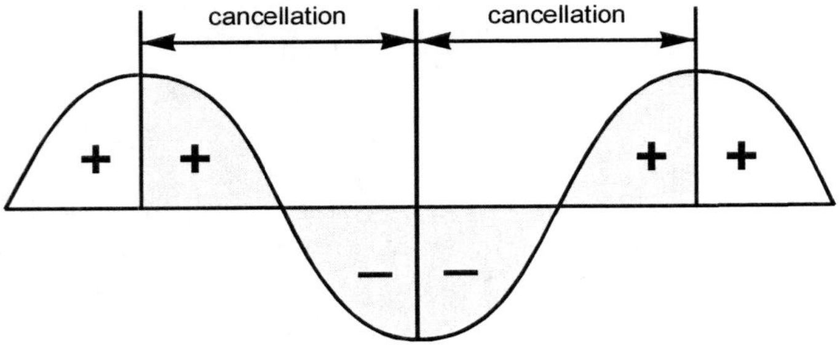

Figure 8.44 Intercellular cancellation in a simply supported rectangular plate three-index mode.

Working through the theory, the complex control source amplitude q_c introduced prior to Equation (8.9.3) simply becomes the complex amplitude of the centre nodal area, for which an optimum value can be derived once the remaining terms in Equation (8.9.11) are calculated. For example, for a 0.88 m × 1.80 m × 9 mm steel plate oscillating at 89 Hz, the optimum centre nodal area is found to be 1.66 times that of the edge nodal areas. More importantly for the discussion here, however, is that it follows that the inherent source of acoustic control, the area of negative acoustic intensity, will have an acoustic power output of zero when optimally tuned.

Figure 8.45 depicts the acoustic power output of the inherent source of acoustic control (which is represented by the areas of negative acoustic intensity or, in other words, negative acoustic power, or absorption), the 'primary source' (which is represented by the areas of positive acoustic intensity), and the total acoustic power output of the plate excited at 89 Hz, as a function of the ratio of the amplitudes of the centre and edge nodal areas is varied. Figure 8.46 depicts the acoustic intensity distribution on the plate corresponding to several specific cases. As illustrated in Figure 8.46, under normal conditions (where the amplitude ratio = 1), there is acoustic power flow out of the plate edges and into the centre of the plate; both positive and negative contributions to the total acoustic power output exist.

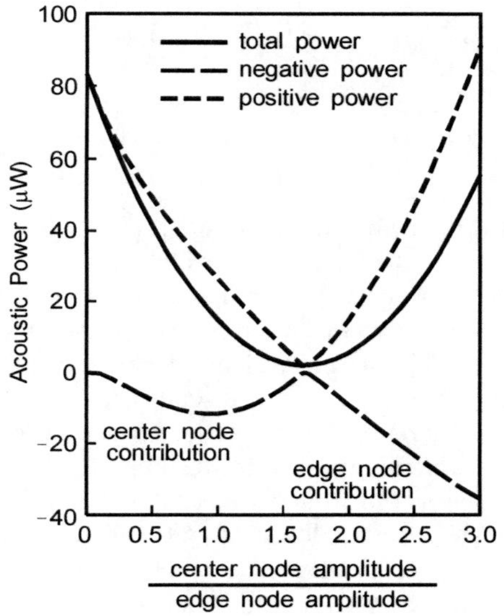

Figure 8.45 Acoustic power output of negative and positive intensity regions (labelled negative and positive power) on the plate, as a function of the ratio of the edge nodal area amplitudes to the centre nodal area amplitude.

As described previously, the centre nodal area can be viewed as attenuating the acoustic radiation from the outer nodal areas, performing the function of a source of acoustic control. Altering the relative amplitudes of the nodal areas alters this balance. If the inherent source of acoustic control, the area of negative intensity, is removed (amplitude ratio = 0), the acoustic power output of the plate increases, as expected. Conversely, if the magnitude of the amplitude ratio is substantially increased, the outer nodal areas become the acoustic power sinks. However, the most interesting result occurs at a value of amplitude ratio equal to 1.66, where the amplitude of the centre nodal area has increased relative to the edge nodal areas. At this point, the minimum acoustic power output from the plate occurs. More importantly for the topic of discussion here is the fact that at this point the centre nodal area, the inherent source of acoustic control, stops absorbing and actually has an acoustic power output of zero, with the radiation from the outer nodal areas greatly attenuated. This is the physical manifestation of the idea of zero control source acoustic power output when vibration sources are used as sources of acoustic control.

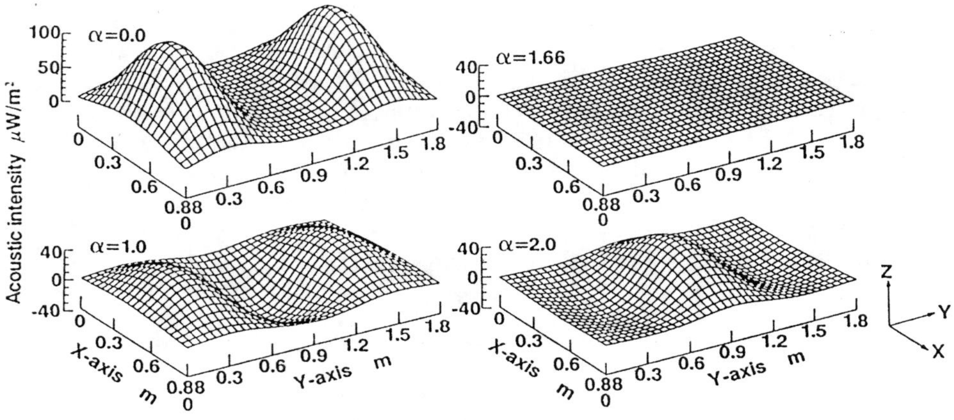

Figure 8.46 Acoustic intensity distributions on the plate surface for various ratios of the edge nodal area amplitudes to the centre nodal area amplitude.

While this discussion may seem somewhat fanciful, it is, in fact, possible to alter the relative velocities of the centre and edge nodal areas by superimposing a second odd-index mode, such as a 1-index mode, which will in turn alter the effectiveness of the intercellular cancellation phenomenon. Vibration sources can be used to alter the relative amplitude and phasing of the two modes (this is the modal rearrangement control mechanism), thereby becoming vehicles to 'tune' this inherent source of acoustic control. Because there is a single global minimum in the acoustic power error criterion, when the problem is formulated in terms of vibration control source force inputs, the resulting final state of the vibrating plate must be qualitatively the same as that found when the problem is formulated in terms of nodal areas in Equation (8.9.13): the amplitude of the centre nodal area will have increased relative to the amplitude of the edge nodal areas, and the acoustic power output of this source of acoustic control will be zero; the area of negative intensity will disappear.

To demonstrate this, consider again the 0.88 m × 1.80 m × 9 mm plate, subject to harmonic primary excitation at 89 Hz by a point force located at a position x = 0.44 m, y = 0.30 m, with the origin of the coordinate system at the lower left corner of the plate. A vibration control source, attached at the plate centre, x = 0.44m, y = 0.90 m, is used to attenuate the radiated acoustic field. For clarity of explanation, only two modes, the (1,1) and (1,3), will be used in the calculations presented here, although qualitatively the phenomena are the same as for the example of Section 8.10.3. The resonance frequency of the (1,1) mode is 35.2 Hz, and for the (1,3), it is 89.5 Hz. The vibration velocities of the two modes in response to both primary excitation and optimum controlled conditions are plotted in Figure 8.47.

The acoustic power attenuation obtained by the application of the optimum vibration control force is 10.8 dB. However, the change in modal velocities alone does not reflect this attenuation. In fact, the velocities of both modes have increased substantially. Rather than being due to a reduction in plate velocity, attenuation is the result of an acoustic mechanism-based phenomenon, arising from the change in the relative complex amplitudes of the modes. Figures 8.48(a) and 8.48(b) depict the plate velocity distribution before and after the application of active control. It can be seen that while the distribution pattern is the same, the relative amplitude of the centre nodal area has increased compared to that of the outside nodes, which is as expected.

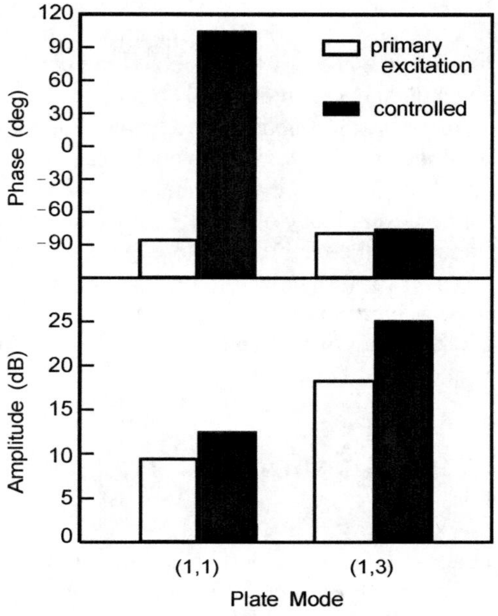

Figure 8.47 Amplitudes and phases of the (1,1) and (1,3) modes before and after application of active control.

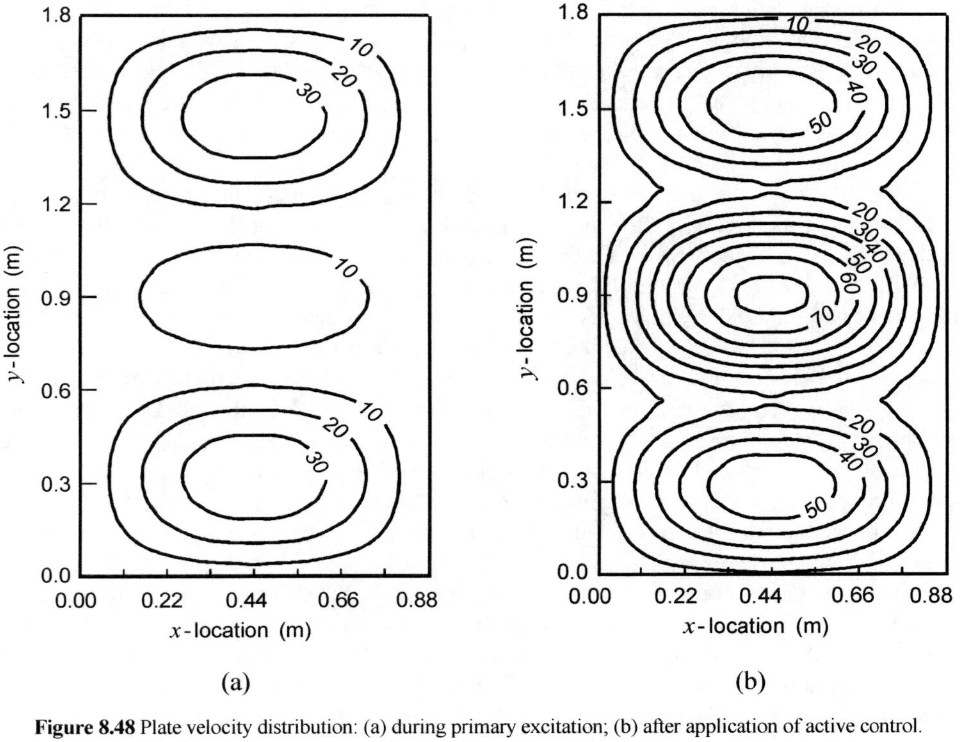

Figure 8.48 Plate velocity distribution: (a) during primary excitation; (b) after application of active control.

The final question to answer here is why the vibration control source has employed an acoustic-type rather than a velocity-type control mechanism. The reason is that with the primary source located towards the edge of the plate and the control source located in the centre, it is not possible to simultaneously reduce the amplitude of both modes, or to provide velocity (modal) control. However, the 'modal rearrangement' mechanism can be invoked, reducing the radiation efficiency of the plate. If the plate is very lightly damped and vibrating at resonance, the control effect would be a combination of modal control and modal rearrangement, a reduction in the amplitude of the (1,3) mode and subsequent increase in the amplitude of the (1,1) mode until the centre node to edge node amplitude ratio reached a value of 1.66. If the control source were moved to the plate edge, this would change, as the (1,3) and (1,1) modes can be reduced in amplitude simultaneously. Either way, one result is consistent: the acoustic power output of the 'control source', as defined in Section 8.9.4, is zero under optimally controlled conditions.

8.11 MINIMISING SOUND RADIATION BY MINIMISING ACOUSTIC RADIATION MODES

In structural/acoustic problems, the use of vibration control sources offers several potential advantages over the use of more conventional acoustic control sources, one of these being system compactness: somewhat bulky speakers and cabinets can be replaced by compact control sources such as surface mounted or embedded piezoelectric ceramic actuators (which are discussed further in Chapter 15), creating a 'smart structure' when driven by an appropriate control system. A potential hindrance to realising this advantage is the use of acoustic sensors, such as microphones, to provide an error signal to the control system. As observed in the monopole source studies in the beginning of this chapter, microphones should usually be placed in the far-field of the acoustic system to ensure that global sound attenuation accompanies the reduction in acoustic pressure at the sensor. It can be envisaged that an improvement would be had if vibration error sensors could be used instead of acoustic sensors.

The use of vibration error sensors, such as accelerometers, in systems targeting a reduction in structural vibration is commonplace. With these arrangements, the control system can be designed to explicitly reduce structural modal amplitudes, using what is termed a 'modal filter', which will resolve a set of measured vibration signals into a set of modal signals (modal velocity, displacement, etc.). A more novel approach which is discussed in Chapter 15 is the use of shaped piezoelectric polymer film sensors that discriminate between structural modes, thus eliminating the need for explicit error signal pre-filtering. However, as observed in the previous section, it does not always happen that minimising the radiated acoustic power reduces the modal velocity levels of the principal offending modes; in fact, a previous example in this chapter demonstrated an *increase* in structural velocity accompanied a reduction in radiated acoustic power. Clearly, a control system that is constrained to reducing structural velocity is inappropriate for the control of structural sound radiation.

This section is concerned with the problem of sensing some measure of structural vibration to provide a measure of the error criterion of interest, such as acoustic power or structural kinetic energy. Such a signal could be used as an error signal for either an adaptive feedforward active control system, or a feedback system, to reduce structural or acoustic disturbances. The development of the problem will be general, specifically targeting the questions of what should be sensed in terms of structural modes to maximise the levels of global attenuation of the error criterion of interest, and how should an adaptive feedforward active control system using vibration-based error signals be implemented. Application of the theory developed in this section to feedback control systems will be undertaken later in Section 8.15.

The idea of the analysis in this section is to express the vibration response of a structure in terms of modes that are orthogonal with respect to sound radiation. This means that reducing the amplitude of one such mode will be guaranteed to reduce the sound power radiated. Such a mode is called a 'radiation mode' and it is made up of weighted contributions of various structural vibration modes which are orthogonal in terms of structural vibration but not in terms of acoustic radiation. Thus, minimising an acoustic radiation mode will not necessarily lead to a reduction in the structural vibration level, just as minimising a structural vibration mode will not necessarily result in a reduction in the radiated sound power.

Methods to generate radiation modes generally involve the use of singular value decomposition of the Green's function and this will be discussed in detail later in this section. The eigenvalues associated with the singular value decomposition are interpreted as the modal radiation efficiencies and the eigenvectors represent the mode shapes. It was pointed out by Borgiotti (1990) that the radiation modes are real valued and therefore, in principle, allow delay-free sensing of acoustic radiation, which is particularly useful if feedback control is to be used to minimise the radiated sound power.

The real-time implementation of radiation mode minimisation to minimise radiated sound power is complicated by the fact that both the radiation mode shapes and the radiation efficiencies depend on frequency. However, Borgiotti and Jones (1994) demonstrated that minimising the radiation modal error signals obtained by weighting the sensor signals with the radiation mode shapes at the highest controlled frequency also gives good results at lower frequencies. The ability to obtain good results without using frequency weighting for free-field radiation was confirmed by Berkhoff (2000), although Cazzolato and Hansen (1998) showed that this may not be possible for sound radiation into an enclosed space. Berkhoff (2000) also showed that there was more than one possible choice of radiation mode shape function that could be used. In another paper, Berkhoff (2002) determined the radiation mode shapes and associated radiation efficiencies based on broadband sound radiation. The radiation mode shapes so obtained were optimum in an average sense over a pre-defined frequency band. The formulation presented by Berkhoff (2002) can be used for the identification of multiple radiation modes from experimental data, using microphones in the far-field.

The radiation modal analysis that is undertaken in the following sub-sections is based upon the idea of finding orthogonal groupings of structural modes that radiate as a set, and is based principally on the work published by Snyder and Tanaka (1993b) and Snyder et al. (1994a). For further discussion, the reader is referred to Baumann et al. (1991, 1992), Elliott and Johnson (1992, 1993) and Naghshineh and Koopmann (1992, 1993). Related work was also published by Keltie and Peng (1987), Borgiotti (1990), Cunefare (1991), Cunefare and Shepard (1993), Cunefare et al. (2001), Naghshineh et al. (1992), Berkhoff (2000, 2001 and 2002), Snyder and Burgan (2002), Burgan et al. (2002), Hill and Snyder (2007), Hill et al. (2002, 2005, 2007a, 2007b, 2008) and Snyder et al. (2004).

8.11.1 General Theory

Consider some generic structure, subject to harmonic excitation by an unspecified primary forcing function. The aim of active control is to globally attenuate some measure of the system response, which can normally be expressed as a quadratic function of the structural modal velocities. This measure may be vibrational, such as the kinetic energy of the structure, or acoustic, such as the radiated acoustic power or acoustic potential energy in a coupled enclosure problem. Each of these quantities, or global error criteria, can be expressed in the form:

$$J = \boldsymbol{v}^{\mathrm{H}} \boldsymbol{A} \boldsymbol{v} \tag{8.11.1}$$

where J is the global error criterion of interest, $\boldsymbol{v}$ is the vector of complex modal velocity amplitudes, H is the matrix Hermitian (conjugate transpose), and $\boldsymbol{A}$ is a symmetric transfer matrix (symmetric provided that reciprocity holds). Note that to write this relationship, it has been assumed that the infinite modal summation required to exactly decompose the structural velocity distribution has been truncated at some finite number of modes. The diagonal terms of the matrix $\boldsymbol{A}$ define the value of the error criterion that would result from the vibration of a single, isolated structural mode, and the off-diagonal terms define the modification to this value caused by the co-existence of the other structural modes.

As $\boldsymbol{A}$ is a symmetric matrix, it can be diagonalised by the orthonormal transformation:

$$\boldsymbol{A} = \boldsymbol{Q} \boldsymbol{\Lambda} \boldsymbol{Q}^{-1} \tag{8.11.2}$$

where $\boldsymbol{Q}$ is the (square) orthonormal transform matrix, the columns of which are the eigenvectors of the transfer matrix $\boldsymbol{A}$, and $\boldsymbol{\Lambda}$ is the diagonal matrix of associated eigenvalues. The symmetric form of $\boldsymbol{A}$ dictates that the eigenvectors in $\boldsymbol{Q}$ will be orthogonal. If the transformation of Equation (8.11.2) is substituted into Equation (8.11.1), the generic global error criterion expression becomes:

$$J = \boldsymbol{v}^{\mathrm{H}} \boldsymbol{A} \boldsymbol{v} = \boldsymbol{v}^{\mathrm{H}} \boldsymbol{Q} \boldsymbol{\Lambda} \boldsymbol{Q}^{-1} \boldsymbol{v} = \boldsymbol{v}'^{\mathrm{H}} \boldsymbol{\Lambda} \boldsymbol{v}' \tag{8.11.3a–c}$$

where the vector of transformed modal velocities, $\boldsymbol{v}'$, is defined by the expression:

$$\boldsymbol{v}' = \boldsymbol{Q}^{-1} \boldsymbol{v} \tag{8.11.4}$$

and use has been made of the following property of the orthonormal transform matrix:

$$\boldsymbol{Q}^{-1} = \boldsymbol{Q}^{\mathrm{T}} \tag{8.11.5}$$

The orthonormal transformation has decoupled the global error criterion expression, enabling it to be expressed as a weighted summation of the form:

$$J = \sum_{i=1}^{m} \lambda_i |v_i'|^2 \tag{8.11.6}$$

where λ_i is the ith eigenvalue of $\boldsymbol{A}$, v_i' is the ith transformed modal velocity of the vector $\boldsymbol{v}'$, and the summation is over the m structural modes that are being modelled.

The important point to note about the error criterion as expressed in Equation (8.11.6) is that v_i' is not necessarily equal to the velocity amplitude of the ith structural mode v_i, but is rather equal to some orthogonal grouping of modal contributions (with each group referred to as a 'radiation mode'), defining the ith principal axis of the error 'surface' (the plot of error criterion as a function of modal velocity), defined by the quadratic criterion of Equation (8.11.1). The concept of error surfaces is discussed in more depth in Chapter 6 in relation to adaptive signal processing, a problem that has the same qualitative characteristics. Consider, for example, some hypothetical problem where only two structural modes are being modelled and the error criterion is defined by the transfer matrix:

$$\boldsymbol{A} = \begin{bmatrix} 1.0 & 0.5 \\ 0.5 & 1.0 \end{bmatrix} \tag{8.11.7}$$

In this instance the orthonormal transform matrix $\boldsymbol{Q}$ and the eigenvalue matrix $\boldsymbol{\Lambda}$ are equal to:

$$\boldsymbol{Q} = \frac{1}{\sqrt{2}}\begin{bmatrix} 1.0 & 1.0 \\ 1.0 & -1.0 \end{bmatrix}; \quad \boldsymbol{\Lambda} = \begin{bmatrix} 1.5 & 0.0 \\ 0.0 & 0.5 \end{bmatrix} \tag{8.11.8a,b}$$

Therefore, the transformed velocities, as defined by Equation (8.11.4), are:

$$v_1' = \frac{1}{\sqrt{2}}\left(v_1 + v_2\right); \qquad v_2' = \frac{1}{\sqrt{2}}\left(v_1 - v_2\right) \tag{8.11.9a,b}$$

These transformed velocities defined the *m* principal axes of the ($m + 1 = 3$)-dimensional hyper-parabolic error surface associated with the quadratic (in modal velocity) error criterion of Equation (8.11.1).

The problem of implementing an adaptive feedforward active control system to attenuate the global error criterion of Equation (8.11.1) will now be considered, where a vibration control source is being used. The desired error signal, to be minimised by the control system, is some measure of structural velocity which can be decomposed into a set of orthogonal constituents by some electronic pre-processing operation such as modal filtering, or by the use of a custom sensor such as constructed from shaped piezoelectric polymer film. (How to shape a piezoelectric polymer film sensor to obtain a measurement of some desired quantity is discussed in more depth in Chapter 15. The aim of the exercise in this chapter is to derive the quantity to be measured.)

One approach that could be undertaken for decomposing the error signal is to base it upon the *in vacuo* mode shape functions of the structure. However, using this data in its 'raw' form may not produce the desired result, as strictly speaking the value of the global error criterion is based not solely upon (squared) individual *in vacuo* (untransformed) modal contributions but rather upon these terms in conjunction with modifications resulting from other co-existing modes (in other words, the weighting matrix $\boldsymbol{A}$ is not diagonal). Simplifying this holistic view by aiming to minimise only individual contributions can prove detrimental to performance, as when controlling acoustic radiation from a vibrating structure. A better approach is to resolve the orthogonal transformed modal velocities in $\boldsymbol{v}'$ and simply weight the contributions by the associated eigenvalues. Not only does this implicitly account for the off-diagonal, or cross-coupling terms in the transfer matrix $\boldsymbol{A}$, but it may also reduce the required number of error signal inputs to the controller to obtain a satisfactory result, as some eigenvalues may be so small as to make contributions to the error criterion by the associated transformed modal velocities negligible. This latter point is becoming evident even in the previous simple two-mode example, where one eigenvalue is three times the value of the other, indicating that measurement of the associated eigenvector v_1' is three times as important.

8.11.2 Minimising Vibration versus Minimising Acoustic Power

To obtain some insight into differences that minimising structural vibration has compared to minimising radiated acoustic power, it will prove useful to first consider the problem of minimising structural kinetic energy, which provides a measure of the vibration of a finite structure. Structural kinetic energy is defined here as:

$$E_k = \frac{1}{2} \int_S m_s(\boldsymbol{x}) v^2(\boldsymbol{x}) \, \mathrm{d}\boldsymbol{x} \tag{8.11.10}$$

where $m_s(\boldsymbol{x})$ and $v^2(\boldsymbol{x})$ are respectively the structural material mass per unit area and velocity at location $\boldsymbol{x}$ on the structure, and the integration is conducted over the surface area S of the structure. The velocity term in Equation (8.11.10) can be expanded in terms of modal contributions as follows:

$$v(\boldsymbol{x}) = \sum_{i=1}^{N_m} v_i \psi_i(\boldsymbol{x}) \tag{8.11.11}$$

where v_i is the complex velocity of the ith mode $\psi_i(\boldsymbol{x})$ is the value of the associated mode shape function at location $\boldsymbol{x}$, and there are N_m modes being modelled in the calculations. Substituting Equation (8.11.11) into Equation (8.11.10) and taking advantage of modal orthogonality enables structural kinetic energy to be expressed as:

$$E_k = \frac{1}{2} \int_S m_s(\boldsymbol{x}) \left(\sum_{i=1}^{N_m} v_i \psi_i(\boldsymbol{x}) \right)^* \left(\sum_{i=1}^{N_m} v_i \psi_i(\boldsymbol{x}) \right) \mathrm{d}\boldsymbol{x} = \frac{1}{2} \sum_{i=1}^{N_m} M_i |v_i|^2 \tag{8.11.12a,b}$$

where M_i is the ith modal mass, given by:

$$M_i = \int_S m_s(\boldsymbol{x}) \psi_i^2(\boldsymbol{x}) \, \mathrm{d}\boldsymbol{x} \tag{8.11.13}$$

and * denotes complex conjugate. Structural kinetic energy can therefore be written in the form of the global error criterion expression Equation (8.11.1):

$$J = E_k = \boldsymbol{v}^{\mathrm{H}} \boldsymbol{A} \boldsymbol{v} \tag{8.11.14a,b}$$

where $\boldsymbol{A}$ is a diagonal matrix of modal masses, with the elements defined as $A(i,i) = M_i/2$.

From the description of Equation (8.11.14), it is apparent that in this problem the orthonormal transform matrix $\boldsymbol{Q}$ of Equation (8.11.2) will simply be the identity matrix, and that the eigenvalues in $\boldsymbol{\Lambda}$ in Equation (8.11.2) are equal to the modal masses. The transformed modal velocities v_i' of Equation (8.11.4) are therefore simply equal to the *in vacuo* modal velocities v_i. In structures, where all of the modal masses are equal, such as simply supported rectangular plates, the eigenvalue weighting can be neglected in a practical implementation (because the weighting given to each input will be the same), so that the optimal error signals to provide to the adaptive control system for minimisation of the kinetic energy of the structure are simply the *in vacuo* structural modal velocities (this simple result will not, however, occur in general for structural acoustic problems, as will be seen in the next two sections).

A feedforward control system based on this idea for a general structure could look something like the arrangement shown in Figure 8.49, where a common adaptive controller, such as based upon an FIR filter and filtered-*x* LMS algorithm described in Chapter 6, is being employed to 'tune' the control force inputs to minimise the weighted set of structural modal velocity error signals, as measured by shaped piezoelectric polymer film sensors (used, for example, in Lee and Moon (1990), Collins et al. (1992), Clark and Fuller (1991, 1992b), and Snyder et al. (1993, 1994a, 1994b). For the kinetic energy case, the error signal weighting is frequency independent (defined by the eigenvalues) and simply equal to the modal masses (note again that with some structures, such as rectangular simply supported plate, all modal masses are the same and so the weighting can be neglected).

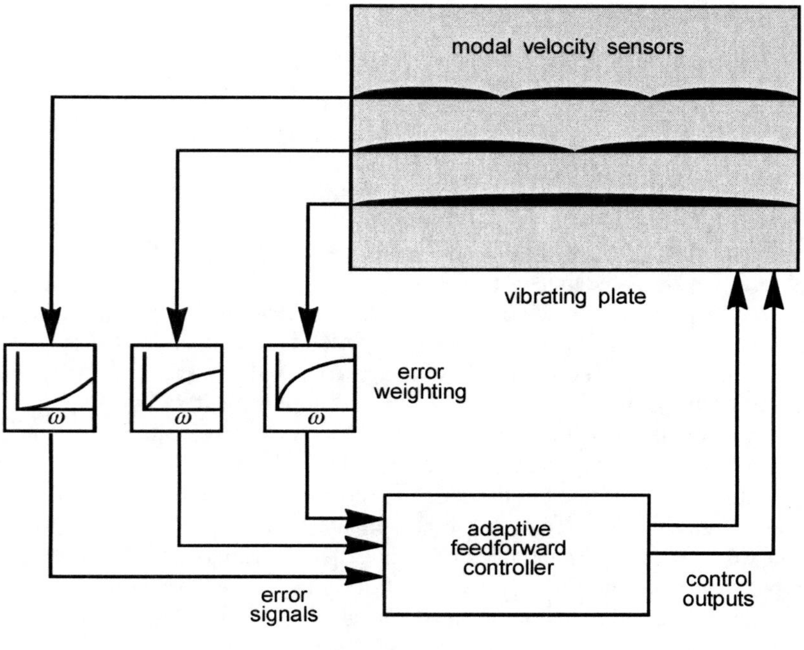

Figure 8.49 Feedforward control arrangement using modal velocity sensors.

This result can be contrasted to that obtained when considering minimisation of radiated acoustic power. With this latter problem, the transfer matrix $\boldsymbol{A}$ of Equation (8.11.2) is neither constrained to be diagonal nor to have equal magnitude terms, and as such, this both couples and weights the problem. Couples because the contribution of one *in vacuo* mode the error criterion is dependent upon the amplitudes of other *in vacuo* modes. To explicitly minimise radiated acoustic power, it is not sufficient to consider individual structural modal amplitudes; rather, cross-coupling between the modes (defined by the off-diagonal terms of $\boldsymbol{A}$) must also be taken into account. This difference between minimising structural velocity and acoustic power infers that any control approach aimed at minimising the sum of (possibly weighted) individual structural modal velocities with a view to minimising radiated acoustic power is fundamentally flawed. Rather, acoustic power attenuation will be achieved by minimising the weighted sum of the squared amplitudes of a set of coupled modal velocity groups, or by *altering* the velocity distribution. This alteration may result in reduced velocity levels for some or all structural modes, but this result is by no means guaranteed, even for resonant or nearly resonant modes.

8.11.3 Example: Minimising Radiated Acoustic Power from a Rectangular Plate

To consider this idea further it is best to pick a specific example. The one chosen here is again a baffled, simply supported rectangular plate. With this structure and an acoustic power error criterion, the transfer matrix $\boldsymbol{A}$ is equal to the acoustic power transfer matrix $\boldsymbol{Z}_w$, defined initially in Equation (8.7.61). As the acoustic power transfer matrix is real and symmetric, it can be diagonalised by an orthonormal transformation as defined in Equation (8.11.2), enabling the acoustic power to be expressed in the form of Equations (8.11.3) and (8.11.6), as follows:

$$W = \boldsymbol{v}^{\prime H}\boldsymbol{\Lambda}\boldsymbol{v}^{\prime} = \sum_{i=1}^{m} \lambda_i |v_i^{\prime}|^2 \tag{8.11.15}$$

where the m transformed structural modes used in the summation are as defined in Equation (8.11.4), with the orthonormal matrix $\boldsymbol{Q}$ containing the eigenvectors of the acoustic power transfer matrix $\boldsymbol{Z}_w$ (or $\boldsymbol{A}$). If the transformed modal velocities could be sensed directly and the signals weighted by the associated eigenvalue (which is effectively the radiation efficiency of the transformed mode), then the radiated acoustic power could be minimised by devising a control system that would minimise the set of weighted signals. In this case, a feedforward control system would again look something like the arrangement shown in Figure 8.49, where transformed modal velocities are measured with shaped piezoelectric film sensors, and the weighting factors are the eigenvalues associated with each transformed mode.

The orthonormal transformation described in the previous section enables the problem of controlling sound radiation from a vibrating plate to be described as a problem of minimising a weighted set of amplitudes of orthogonal contributors to the error criterion, thereby simplifying a potential feedforward control arrangement. However, there is still a (commonly encountered) problem that must be discussed: the power transfer matrix $\boldsymbol{Z}_w$, and hence its eigenvectors used in calculating the transformed modes and eigenvalues used in weighting the problem, is frequency dependent. This means that the constituents of a given transformed mode, which could be measured using a shaped piezoelectric polymer film sensor, will be frequency dependent. At first, this would appear to be a serious impediment to the implementation of optimally shaped sensors for control of sound radiation over anything but the narrowest of frequency bands. However, it is insightful to qualitatively consider the form of the frequency dependence of the constituent terms of the acoustic power transfer matrix $\boldsymbol{Z}_w$. At low frequencies (such that $(kL_x/m\pi) << 1$ and $(kL_y/n\pi) << 1$, where k is the acoustic wavenumber at the frequency of interest), the (α, β)th term in the matrix $\boldsymbol{Z}_w$ can be expressed in terms of modal radiation efficiencies as:

$$\boldsymbol{Z}_w(\alpha,\beta) = \frac{\rho_0 c_0 L_x L_y}{8}\,\frac{\left(1+(-1)^{M_\alpha+M_\beta}\right)\left(1+(-1)^{N_\alpha+N_\beta}\right)}{4}\left(\frac{\dfrac{M_\alpha N_\alpha}{M_\beta N_\beta}S_\alpha + \dfrac{M_\beta N_\beta}{M_\alpha N_\alpha}S_\beta}{2}\right) \tag{8.11.16}$$

where (M_α, N_α), (M_β, N_β) and S_α, S_β are the (x, y) modal indices and radiation efficiencies for modes α and β respectively, the latter found in work published by Wallace (1972). The important point to note about this result is that it shows that at low frequencies, the relative values of the terms in the acoustic power transfer matrix $\boldsymbol{Z}_w$ are related to each other by the relative values of the individual modal radiation efficiencies, while the absolute values are related to the radiation efficiencies. This is important because it means that while the eigenvectors of $\boldsymbol{Z}_w$, which define the constituents of the transformed modes, will vary with frequency, in the low-frequency regime, this variation can often be taken as insignificant when viewed in light of the attenuation of radiated acoustic power that can actually be obtained. This is because the ratio of the radiation efficiencies of any two modes may not vary greatly with frequency (see the graphs in Section 2.5). However, Equation (8.11.16) also shows that the eigenvalues of $\boldsymbol{Z}_w$ will vary markedly with frequency, as these will be related to the individual values of radiation efficiency (as stated, the eigenvalues are equivalent to the radiation efficiencies of the transformed modal velocities).

To investigate this further, a steel plate of dimensions 1.8 m × 0.88 m × 10.1 mm, subject to harmonic excitation within the frequency band 30 to 200 Hz will be studied. In this range there are seven modal resonances, as outlined in Table 8.4. If the acoustic power transfer matrix $\boldsymbol{Z}_w$ is calculated at any given frequency in this band using 30 structural modes and then decomposed as outlined in Equation (8.11.4), it is found that at low frequencies only three transformed acoustic power modes (sometimes called ‘radiation modes’) may be considered as ‘significant’, significance being defined by the associated eigenvalue which will weight the influence that the transformed mode has upon the calculation. These transformed modes are, in order of importance, based upon orthogonal groupings of (odd-odd), (even-odd), and (odd-even) structural modes, and will be referred to as transformed modes $w(o,o)$, $w(e,o)$ and $w(o,e)$ respectively. As the frequency increases, two more modes, based upon a second grouping of (odd-odd) modes and a grouping of (even-even) modes, become important. A plot of the variation in eigenvalues of these transformed modes as a function of frequency is given in Figure 8.50.

Table 8.4 Theoretically calculated plate resonances below 200 Hz.

Mode	Theoretical Resonance Frequency (Hz)
1,1	39.5
2,1	62.3
3,1	100.4
1,2	135.0
4,1	153.7
2,2	157.9
3,2	195.9

The interesting point to note here is that while the eigenvalues of these transformed modes vary greatly with frequency, the proportion of each *in vacuo* structural mode contributing to the transformed mode defined by the eigenvectors varies by a relatively small degree. Consider for example the $w(o,o)$ mode, which is shown in Figure 8.50 to have the greatest radiation potential. Table 8.5 lists the values of the non-zero contributions from the first 60 structural modes to this transformed mode at 30 Hz and 100 Hz. This mode is dominated by the (1,1), (3,1) and (1,3) structural modes, with the proportional contributions changing only slightly.

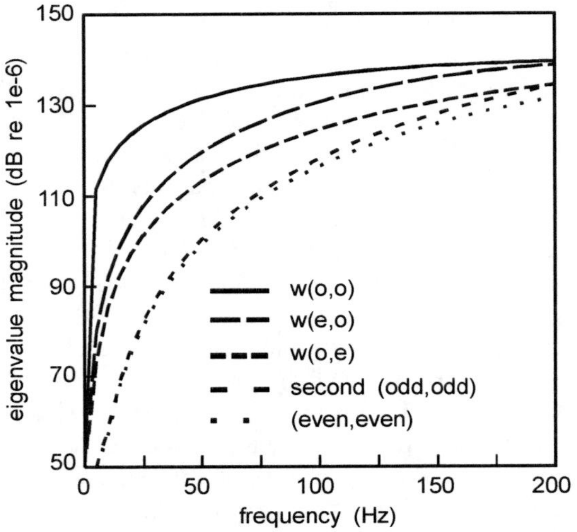

Figure 8.50 Amplitude of the five largest eigenvalues of the acoustic power transfer matrix as a function of frequency.

Table 8.5 Constituents of the first transformed mode *w(o,o)* as defined at 30 Hz and 100 Hz.

Mode	30 Hz Value	100 Hz Value
1,1	0.913	0.919
3,1	0.229	0.199
5,1	0.138	0.117
1,3	0.233	0.241
1,5	0.141	0.143

To give further insight into the physical significance of transformed modes, it is useful to consider the variation in acoustic power output as a function of the relative amplitudes of the (1,1) and (3,1) structural modes. Figure 8.51 depicts the acoustic power output of the described plate at 100 Hz in response to primary point excitation at (0.35 m, 0.64 m), where the amplitude of the (1,1) and (3,1) modes are varied relative to the original amplitude of the (1,1) mode, and the amplitudes of all other structural modes are kept constant. Observe that the principal axes of the quadratic surface are *not* parallel to the axes defined by the structural modal velocities, but rather, are skewed. The first transformed mode, $w(o,o)$, describes the steepest principal axis of the function (the axis along which the greatest variation in acoustic power is seen). Clearly, minimising the acoustic power amplitude along this axis (equivalent to minimising the $w(o,o)$ mode) will go a long way towards minimising the total acoustic power output of the plate. For interest, the other principal axis of the figure is described by the second orthogonal grouping of (odd-odd) structural modes mentioned previously.

Note also that by considering the structural radiation in terms of orthogonal transformed modes, it is perfectly clear that radiated acoustic power can be reduced while the vibration amplitude of the structure is increased. For example, consider two situations, where the amplitude of the (1,1) structural mode is 1.0 and the amplitude of the (3,1) structural mode is -1.0 and -5.0 respectively, while the amplitude of all other modes is 0.0. The kinetic energy of the structure is less in the first of these two cases. However, in viewing the constituents of the dominant transformed mode given in Table 8.5, it can be seen that the radiated acoustic power is lower for the second of these two cases (the sum of the products of the amplitudes and their proportions in the transformed mode are less), even though the kinetic energy is significantly increased. This phenomenon occurs because structural modes are *not* orthogonal radiators, whereas transformed modes, as defined in this section, are. Clearly, to minimise structural sound radiation, it is necessary to minimise the amplitudes of the weighted set of transformed vibration modes.

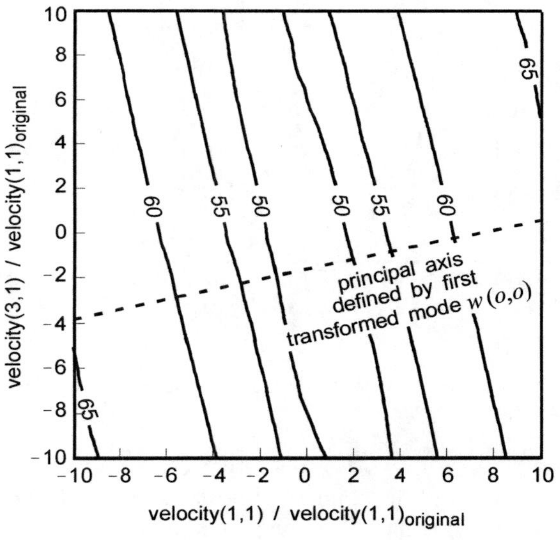

Figure 8.51 Acoustic power output as a function of the amplitudes of the (1,1) and (3,1) structure modes.

There are two questions that arise from the results of Table 8.5 which need to be addressed. First, from the standpoint of using measurements of the amplitudes of transformed modes to formulate a control strategy, how important are the relatively small variations in the proportional contributions of the constituents as a function of frequency? Consequently, is it possible to base a feedforward control arrangement on minimising transformed modes which are strictly only correct at a single frequency (calculated using the eigenvectors of the matrix $\boldsymbol{Z}_w$ at that frequency) and vary only the weighting factor (eigenvalue) when the frequency is changed? Second, what real advantage does the minimisation of transformed modal velocities have over the minimisation of ordinary structural modal velocities?

Illustrated in Figure 8.52 are the theoretical maximum levels of attenuation of acoustic power radiated from the previously described plate, as well as the attenuation achieved using the $w(o,o)$, and both the $w(o,o)$ and $w(o,e)$, transformed modal velocities, as defined at 30 Hz (strictly only optimal at this frequency) and weighted by the frequency-correct eigenvalue (the form of implementation shown in Figure 8.49), with a primary point force at (0.44 m,0.35 m), a control point force at (0.24 m,0.9m), and loss factor η = 0.02 for all modes. Note that at low frequencies, minimisation of a single transformed modal velocity achieves practically the optimal result, while the minimisation of the second transformed modal velocity extends the region of control over the entire bandwidth. These results suggest that the slight variations in the relative amplitudes of the constituents of the transformed structural modes are essentially negligible in light of the acoustic power attenuation that is achievable with an ideal feedforward control system.

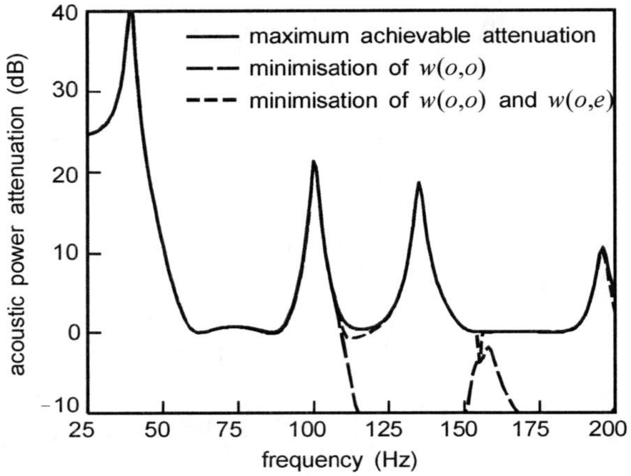

Figure 8.52 Maximum acoustic power attenuation and attenuation achieved by minimising the first, and first and third, transformed modes.

Figure 8.53 illustrates the levels of acoustic power attenuation that can be achieved by minimising the kinetic energy of the structure (calculated using all 30 modelled modes), and which can be obtained by minimising the amplitude of the (3,1) mode. Observe that the minimisation of the kinetic energy of the structure achieves almost the same result as the minimisation of the transformed modal velocities (a result which would not necessarily be found for a general structure). It falls short, however, in the vicinity of the (3,1) mode resonance frequency. This is a result of there being two ways to reduce radiated acoustic power, as has been discussed: reduction in structural vibration velocity, and an alteration of the relative amplitudes and temporal phasing of the vibration modes to enhance the inherent acoustic pressure-control mechanisms, thereby reducing the radiation efficiency of the structure. The additional control in this vicinity is achieved by minimising the amplitude of the transformed $w(o,o)$ mode, which effectively balances the decrease in the amplitude of the (3,1) structural mode against the resulting increase in the amplitude of the (1,1) mode, such that the centre nodal area of the (3,1) mode is essentially increased in amplitude (superposition of the (1,1) and (3,1) modes) while the edge nodal areas are decreased in amplitude. This enhances the effect of intercellular cancellation, the inherent acoustic control mechanism, as described in the previous section. It is not possible to achieve this result by simply minimising the velocity distribution of the structure, or by minimising the mode that contributes most to the radiated sound.

Note also that minimising the amplitude of the (3,1) mode only guarantees attenuation at the (3,1) resonance, while Figure 8.52 shows that minimisation of the single $w(o,o)$ transformed mode will provide attenuation at both the (1,1) and (3,1) resonances. Finally, it should again be mentioned that the relative quality of results obtained by minimising the kinetic energy of the structure (as compared to the maximum possible levels of acoustic power attenuation) are source location dependent, while those obtained by minimising the weighted set of transformed modal velocities are not.

Another point to consider is the importance of the eigenvalue weighting. Figure 8.54 illustrates the levels of acoustic power attenuation that are achieved by minimising the unweighted set and also the weighted set of transformed modal velocities. In comparing the data shown in the two curves, the results corresponding to the unweighted case are seen to be only slightly sub-optimal. This is because the resonances of the structure inherently weight the problem to some degree. At low frequencies, where the $w(o,o)$ transformed mode is dominant, the (1,1) and (3,1) modes are resonant. As the frequency increases, such that the $w(e,o)$ and $w(o,e)$ transformed modes become important, the (1,2) and (2,1) resonances occur. This characteristic is not globally applicable, however, and neglecting the eigenvalue weighting with other source placements and in other problems can seriously degrade the results. In the next chapter, it will be seen that this is especially true in coupled structural/acoustic problems.

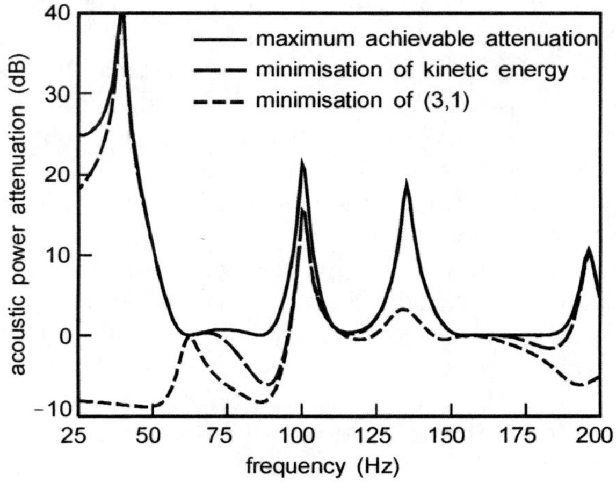

Figure 8.53 Maximum acoustic power attenuation and attenuation achieved by minimising the structural kinetic energy, and the amplitude of the (3,1) structural mode.

Active control systems targeting attenuation of acoustic radiation from vibrating plates have been implemented experimentally using shaped piezoelectric polymer film sensors to measure the transformed modes outlined in this section. For both broadband and tonal excitation, the attenuation in radiated acoustic power achieved was found to be near optimal. For further details, see Snyder et al. (1993, 1994a).

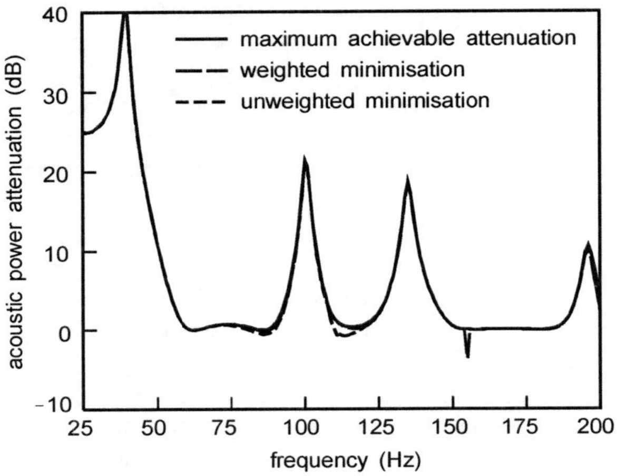

Figure 8.54 Maximum acoustic power attenuation and attenuation achieved by minimising the first three transformed modes with and without eigenvalue weighting.

8.11.4 Alternative Methods for Minimising Sound Radiation from Vibrating Structures

Since the early work of the early to mid-1990s, there have been a number of alternative schemes proposed for the provision of error signals to a controller, which when minimised will reduce the sound power radiated by a vibrating structure. One such scheme, called model reference control, has already been discussed at the end of Section 8.7.4.

In 2000, Berkhoff showed that spatial weighting functions other than radiation modes are possible and that it is possible to reduce the frequency dependence of such weighting functions. He also presented a scheme for determining the weighting functions for near-field pressure sensing rather than vibration sensing. In 2001, Berkhoff extended the radiation mode analysis from point sensors to the use of piezoelectric patch sensors. Methods for experimentally determining radiation mode shapes and radiation efficiencies for the first few important radiation modes were presented by Berkhoff et al. (2001). Cunefare et al. (2001) gave an excellent tutorial on the radiation mode formulation and examined the grouping of radiation modes with similar radiation efficiencies as a function of frequency.

In 1999, Cazzolato and Hansen presented an analysis showing how radiation mode control could be used to control sound transmission through a vibrating curved structure into a backing enclosure such that the potential energy in the enclosure is minimised (Cazzolato and Hansen, 1998, 1999).

8.12 SOME NOTES ON APPROACHING THE DESIGN OF AN ACTIVE CONTROL SYSTEM FOR SOUND RADIATION FROM A VIBRATING SURFACE

Having developed analytical models which enable us to examine free space active noise control systems, it is a good time to briefly discuss the problem of designing applicable feedforward active control systems. It will be assumed here that an 'incorruptible' reference signal is available for implementation of the control system (reference signal corruption was discussed in Section 8.6), and concentrate on the problem of optimising the control source/error sensor arrangement.

Although equations that facilitate calculation of the optimum control source outputs for a given control source/error sensor arrangement have been derived, a direct analytical means of determining the optimum control source and error sensor location has not been found. This is because the error criterion of interest, radiated acoustic power or the squared acoustic pressure amplitude at a set of discrete locations, is not a simple quadratic function of source and sensor location. Therefore, in this case, the system design tends to have the appearance of being largely a 'trial and error' process. It is possible, however, to give some structure to the task, and point out some potential pitfalls and shortcuts, which will be the aim of the discussion in this section.

8.12.1 Stepping through the Design of a System

Before beginning the design process, the system, and its response under primary excitation, must be characterised. The minimum requirement is that the primary sound field (amplitude and relative phase) on an imaginary test surface surrounding the structure in the far-field be measured at the frequency of interest using an appropriate measurement location distribution similar to that used when sound intensity is used to estimate sound power. If vibration control sources are to be used, the structural response must also be measured. If the structural response can be decomposed into its individual contributing modes, this often proves helpful in painting a qualitative picture of the influence of system variables on performance (see the example in Section 8.8.2). It is not absolutely necessary, however, as the optimum control source outputs and reduction in the global error criterion for a given control source arrangement can be calculated using the 'spatial', rather than 'modal', theoretical model developed in Section 8.10 for the common link study.

Once the system to be controlled has been characterised, the task is to choose the type of control source to be used. As demonstrated in the previous sections, the use of either acoustic or vibration control sources can produce significant levels of reduction in the overall sound power radiated into free space by a vibrating structure. The mechanisms by which they achieve this attenuation, and the influence that various geometric and structural/acoustic parameters have upon the magnitudes of these reductions, are different for the two control source types. In choosing the control source type for a particular application, it is sometimes helpful to think of the following points:

1. Acoustic control sources are the easier of the two types with which to design a system. No information about the structural response is required; only measurements of the primary radiated sound field at the frequency of interest are necessary, and this is a very distinct advantage for the free space radiation problem. One disadvantage is that the system is constrained to using an acoustic error sensor(s) (such as a microphone) to provide feedback to the electronic control system. Another disadvantage is that the levels of sound attenuation that can be attained per source may be less than those attained with vibration control source(s), especially if the plate dimensions are approximately equal to, or larger than, the acoustic wavelength at the frequency of interest, or if the plate is excited at a frequency close to one of its resonance

frequencies. Acoustic control sources can also be more obtrusive than some vibration sources, such as piezoelectric actuators.

2. Vibration control sources are more difficult to use than acoustic control sources for system design. A detailed description of the structural response, and also the resulting sound radiation field, must be known if an optimised design is to be undertaken. However, if this can be done, vibration control sources may exhibit an increased level of performance (on a per source basis) over acoustic control sources. The final system may also be more compact and unobtrusive for vibration control sources, especially if piezoelectric actuators and vibration error sensors can be used.

Once the system has been characterised, and the type and number of control sources chosen, the next step is to optimise the location of the control sources. This can be a particularly difficult exercise, as it is, in general, impossible to directly determine the optimum source locations due to them not being a linear function of sound power attenuation. A numerical search routine is therefore required to optimise the control source locations. What is required for this is a means to estimate of the maximum possible sound power attenuation for a given control source arrangement, which can be accomplished using the theoretical models already developed in this chapter, or estimated using the multiple regression approach outlined at the end of this section.

One problem with the use of a numerical search routine is that there may be local optima (minima) in the error surface. In this case, the starting point(s) for the search algorithm will influence the final result. The designer must be aware of this, restarting the procedure at several locations or by using a random search technique such as a genetic algorithm to determine the optimal starting location if it is not possible to choose a starting point based on 'common sense'. Such common sense guidelines would include initial placement of the control sources as close as possible to the antinodes of the modes contributing most to the sound radiation, and also placement of the vibration control sources in the least stiff parts of the structure so that control effort is minimised.

Once the control source location has been selected, the final step is to select the type and location of the error sensor(s). Regardless of the control source type, microphones sampling the far-field will always be effective (provided there are no background noise problems from other sources). With the control source placed, there is a straightforward rule of thumb for selecting the optimum error sensor location: error sensors should be placed at the location of greatest sound pressure difference between the primary radiated and optimally controlled (control source output optimised with respect to minimising acoustic power output) residual sound fields. A simple example of this is shown in Figure 8.55, which illustrates the primary and controlled acoustic fields in the plate mid-section, at a distance of 1.8 m, and the acoustic power attenuation achieved when minimising the acoustic pressure at a single point in this arc, for the 0.38 m × 0.30 m × 2 mm plate system described in Section 8.11.3 with a single point control force in the plate centre.

Note that the maximum levels of acoustic power attenuation are achieved with the error sensors at the locations of the greatest sound pressure difference between the primary and optimally controlled (minimum acoustic power) residual sound fields. This concept also applies to multiple control sources (Snyder and Hansen, 1991).

If vibration control sources are being used, then vibration error sensors can also be used. The problem of using a measure of structural vibration to minimise radiated acoustic power was considered in some depth in the previous section. Cazzolato and Hansen (1998, 1999) showed how the mode transformation method discussed above can be used to design distributed error sensors for controlling sound transmission through a panel into a coupled enclosure.

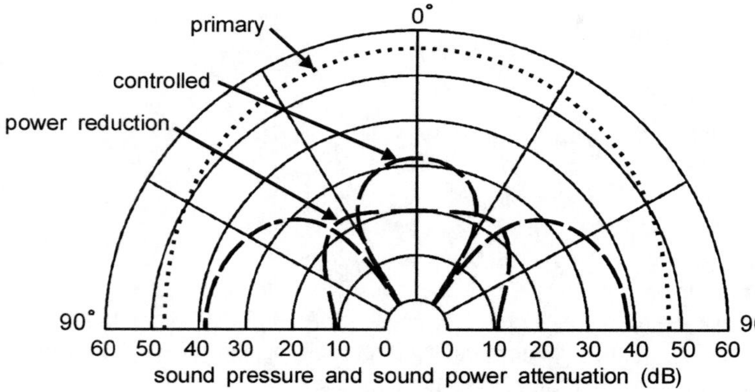

Figure 8.55 Primary and residual acoustic pressure distributions, plotted with acoustic power attenuation as a function of error microphone position.

8.12.2 Shortcut: Determination of the Optimum Control Source Amplitudes and Phases Using Multiple Regression

For some systems targeted for active control, it is not feasible to perform a detailed analysis in a reasonable length of time, owing to the complexity of the system. With others, perhaps all that is desired is some way of ranking the quality of potential control source/error sensor arrangements. One alternative to the detailed theoretical development of the previous section, which is readily amenable to the design of systems based upon measured data, is the use of multiple regression to determine the optimum control source amplitudes (Snyder and Hansen, 1991b). Multiple regression is a generalised linear least-squares technique, where several independent variables are used to predict the dependent variable of interest. Here, the dependent variable of interest is the 180° inverse of the primary sound field (which will provide the greatest level of acoustic power attenuation), while the independent variables are the control source transfer functions and volume velocities or forces.

From the previous far-field based developments it can be concluded that minimisation of the radiated sound power error criterion, used for determining the maximum possible attenuation with a given control source arrangement, is equivalent to minimising the average squared sound pressure over an imaginary hemisphere enclosing the source (for free-field radiation). This can be estimated as the minimisation of the sound pressure squared at a finite number of points on the hemisphere, given by:

$$\sum_{i=1}^{N} (p_{p,i} + p_{c,i})^2 \tag{8.12.1}$$

where the number of points N should be chosen so that Equation (8.12.1) is representative of the radiated acoustic power. For vibration control sources, the generated sound pressure at any point i can be expressed as:

$$p_{c,i} = z_{vt,i}^{\mathrm{T}} f \tag{8.12.2}$$

where, theoretically:

$$z_{vt,i}^{\mathrm{T}} = z_{rad}^{\mathrm{T}} Z_I^{-1} \boldsymbol{\Psi}_c \tag{8.12.3}$$

For acoustic control sources, $p_{c,i}$ is defined by:

$$p_{c,i} = z_{mono}^{\mathrm{T}}(\boldsymbol{r}_i)\boldsymbol{q}_c \tag{8.12.4}$$

Note, however, that measured data can be used for $p_{p,i}$ and the terms in $z_{vt,i}$ and $z_{mono,i}$. As both of these quantities are complex, they must be measured as transfer functions, so for pressure they must be measured relative to some reference signal. In doing this, note that the absolute value of the reference signal is unimportant in determining the 'quality' of a given control source placement, measured by the global attenuation levels.

Substituting these relations into Equation (8.12.1), the optimisation criterion for vibration control sources is that the following expression should be minimised:

$$\sum_{i=1}^{N} \left(p_{p,i} + \sum_{j=1}^{L} z_{vt,j,i}\, f_j \right)^2 \tag{8.12.5}$$

and for acoustic control sources, the following should be minimised:

$$\sum_{i=1}^{N} \left(p_{p,i} + \sum_{j=1}^{L} z_{mono,j,i}\, q_j \right)^2 \tag{8.12.6}$$

where the subscripts j, i denote the transfer function between the jth control source and the ith measurement point. Equations (8.12.5) and (8.12.6) can be written in matrix form. For vibration control sources, Equation (8.12.5) becomes:

$$\left| \boldsymbol{Z}_v \boldsymbol{f}_c - \{ -\boldsymbol{p}_p \} \right| \tag{8.12.7}$$

and for acoustic control sources, Equation (8.12.6) becomes:

$$\left| \boldsymbol{Z}_m \boldsymbol{q}_c - \{ -\boldsymbol{p}_p \} \right| \tag{8.12.8}$$

where

$$\boldsymbol{p}_p = \left[p_{p1} \;\; p_{p2} \;\; \cdots \;\; p_{pN} \right]^{\mathrm{T}} \tag{8.12.9}$$

$$\boldsymbol{Z}_v = \left[z_{vt,1} \;\; z_{vt,2} \;\; \cdots \;\; z_{vt,N} \right]^{\mathrm{T}} \tag{8.12.10}$$

and

$$\boldsymbol{Z}_m = \left[z_{mono,1} \;\; z_{mono,2} \;\; \cdots \;\; z_{mono,N} \right]^{\mathrm{T}} \tag{8.12.11}$$

The minimisation problems of Equations (8.12.7) and (8.12.8) can be solved by various methods, such as by the use of singular value decomposition, or by using one of many commercially available multiple regression software packages. The control forces and/or control volume velocities that result are those that are optimal (those that minimise the power radiation) for the given control source positions.

The level of power attenuation achieved using active control can be estimated as:

$$\Delta W = -10 \log_{10} \left(\sum_{i=1}^{N} \frac{(p_{p,i} + p_{c,i})^2}{(p_{p,i})^2} \right) \tag{8.12.12}$$

From Equations (8.12.7) and (8.12.8) it can be seen that the control source sound pressure $p_{c,i}$ desired is actually the estimated inverse of the primary source sound pressure, $-p_{p,i}$. Therefore, Equation (8.12.5) can be written as:

$$\Delta W = -10 \log_{10} \left(\sum_{i=1}^{N} \frac{(p_{p,i} - p'_{p,i})^2}{(p_{p,i})^2} \right) \tag{8.12.13}$$

where $'$ denotes an estimate. The denominator of Equation (8.12.13) is equivalent to the sum of the squares of the measured dependent variable SS_p, while the numerator is equivalent to the sum of squares of the residuals SS_{res}. Using these equivalences, the estimated acoustic power reduction for the given control source arrangement can be written as:

$$\Delta W = -10 \log_{10} \left(\frac{SS_{res}}{SS_p} \right) = -10 \log_{10} (1 - R^2) \tag{8.12.14a,b}$$

where R is the multiple correlation coefficient.

Therefore, the multiple correlation coefficient can be used to estimate the acoustic power reduction under optimum control for a given control source arrangement. As the number of measurement points increases, so does the accuracy of the estimate. For a very large number of points, this method becomes equivalent to the integration methods of the previous section for determining the optimum control source volume velocities or forces for a given control source and error sensor arrangement. The main advantages of this method are the speed of calculation and the ability to easily incorporate combined vibration and acoustic control sources. This makes it well suited to practical implementation in a multi-dimensional optimisation routine.

It should be noted that this technique is also suitable for determining the control source volume velocities or forces that minimise the sum of the squared sound pressure at a specific point or points. In this case, only the error sensing locations would be used as measurement points in the equations.

There are further advantages to the use of multiple regression in active control system design. One of these concerns the fact, mentioned earlier in this section, that the optimum error microphone locations are at the points of maximum difference between the primary and controlled sound fields. These points can be determined directly using a commercial multiple regression package. These packages usually produce, as part of their output data, a vector of residuals. These residuals are the difference between the measured quantity (pressure here) and the value predicted by the regression equation. The optimum error sensor locations are the locations with the smallest residual values.

8.13 ACTIVE CONTROL OF FREE-FIELD RANDOM NOISE

Analysis of feedforward active control of harmonic excitation is easily facilitated by considering it in the frequency domain, which produces an optimal control source output 'schedule', or control source output as a function of time, which is not constrained to be causal with respect to the primary disturbance. For a large number of problems targeted for active noise control, such an analytical approach and result is completely acceptable. There are instances, however, where a non-causal control source schedule is not practically realisable. One of these instances occurs when considering the active control of free-field random noise. To analyse such a problem, and constrain the solution for the optimum control source output to be causal, the problem must be developed using a time domain, rather than a frequency domain, approach. In this section an analysis for the problem of controlling the free-field radiation from a point acoustic (primary) monopole source by the introduction of a second point acoustic (control) monopole source will be developed. In so doing, some qualitative results that can be applied to the range of causally constrained problems will be developed.

8.13.1 Analytical Basis

Here, the active control of sound radiation from one (primary) monopole source by the introduction of a second (control) monopole source will be considered again, but in this instance the primary monopole oscillation will be random, rather than harmonic. As with all free-field radiation problems, there are two possible error criteria that immediately come to mind: control of acoustic pressure at a point or points, and control of total radiated acoustic power. It will prove to be more straightforward in this instance if the control of acoustic pressure at a single error sensing point is considered first. The starting point for the analysis is the inhomogeneous wave equation, which was originally outlined in Section 2.1:

$$\left(\nabla^2 - \frac{1}{c_0^2}\frac{\partial^2}{\partial t^2}\right) p(\boldsymbol{r},t) = -\rho_0 \frac{\partial Q(\boldsymbol{r},t)}{\partial t} \tag{8.13.1}$$

As monopole acoustic sources are being considered, the source strength Q can be taken as concentrated at a single location $\boldsymbol{r}_q$, such that $Q(\boldsymbol{r},t) = q(t)\,\delta(\boldsymbol{r} - \boldsymbol{r}_q)$. With this assumption, the solution for the acoustic pressure fluctuation at some location $\boldsymbol{r}_e$ due to the operation of the monopole source located at $\boldsymbol{r}_q$ is:

$$p(\boldsymbol{r}_e,t) = \frac{\rho_0}{4\pi r_{q/e}} \frac{\partial q\left(t - \dfrac{r_{q/e}}{c_0}\right)}{\partial t} = \frac{\rho_0}{4\pi r_{q/e}}\, \dot{q}\left(t - \frac{r_{q/e}}{c_0}\right) \tag{8.13.2}$$

where $r_{q/e} = |\,\boldsymbol{r}_e - \boldsymbol{r}_q\,|$. This expression is simply the time domain version of Equation (8.2.1), which describes the acoustic pressure radiated from a monopole source in the frequency domain, where the explicit time derivative in the time domain is equivalent to $j\omega$ in the frequency domain.

During operation of the active control system, the residual acoustic pressure at any location $\boldsymbol{r_e}$ ($\boldsymbol{e}$ denoting error sensor location) can be viewed as the superposition of the primary source and control source generated acoustic pressure at that location, as follows:

$$p(\boldsymbol{r_e},t) = \frac{\rho_0}{4\pi}\left(\frac{\dot{q}_p\left(t - \frac{r_{p/e}}{c_0}\right)}{r_{p/e}} + \frac{\dot{q}_c\left(t - \frac{r_{c/e}}{c_0}\right)}{r_{c/e}}\right) \tag{8.13.3}$$

where $r_{p/e} = |\boldsymbol{r_e} - \boldsymbol{r_p}|$ and $r_{c/e} = |\boldsymbol{r_e} - \boldsymbol{r_c}|$, with $\boldsymbol{r_p}$ and $\boldsymbol{r_c}$ being the locations of the primary and control monopole sources respectively. It is readily apparent from Equation (8.13.3) that the control source output that will minimise the acoustic pressure at the error sensing location is:

$$\dot{q}_{c,\mathrm{opt}}(t) = \frac{r_{c/e}}{r_{p/e}}\,\dot{q}_p\left(t - \frac{r_{p/e} - r_{c/e}}{c_0}\right) \tag{8.13.4}$$

This control output is, in theory, realisable in a causally constrained system if the error sensor location is closer to the control source than the primary source, such that $t-(r_{p/e} - r_{c/e})/c_0$ is always a point in time prior to the control output time t (in practice, the control source must be closer than the primary source to the error sensor by an amount greater than or equal to the time it takes for a signal to propagate through the electronic control, as discussed in Chapter 13). Conversely, if the error sensor is located at a point closer to the primary source than the control source, the constraints of causality make the realisation of the control source output described by Equation (8.13.4) a practical impossibility. However, it is possible to use knowledge of the statistical properties of the primary source output to *estimate* the primary source output at some time in the future based upon its past outputs. The form of the problem then becomes one of optimally estimating the primary source output at time $t - (r_{p/e} - r_{c/e})/c_0$, which can viewed as a Weiner filtering problem.

Weiner filter theory has been encountered previously in Chapter 6, for the purpose of deriving an optimal finite impulse response filter, the output of which most closely matches some desired output based solely on previous samples. This is identical to the problem being considered here. Consider the system arrangement shown in Figure 8.56. Here, the desired control source output, which will completely minimise the acoustic pressure amplitude at the error sensor, is as stated in Equation (8.13.4). The actual control source output is defined by the relationship:

$$\dot{q}_c(t) = \int_0^{\infty} h(\tau)\,\dot{q}_p(t - \tau)\,\mathrm{d}\tau \tag{8.13.5}$$

where $h(\tau)$ is the impulse response function of the control source, and the lower limit on the convolution integral has been set to zero for physical realisability.

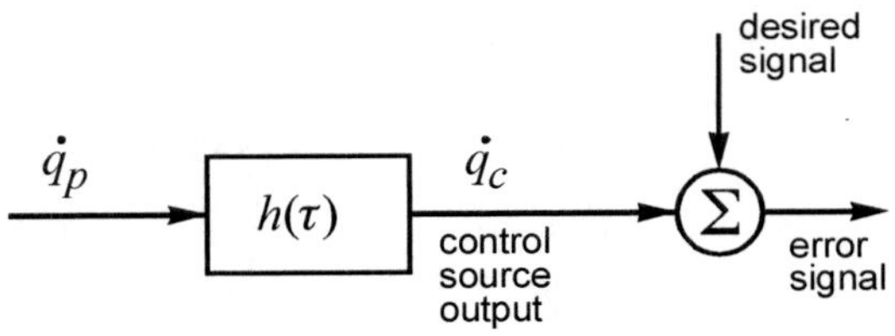

Figure 8.56 Derivation of optimal control source output, and hence filter impulse response function.

Therefore, what is desired is some optimal control source impulse function, h_{opt}, which satisfies the relationship:

$$\dot{q}_{c,\mathrm{opt}}(t) = \frac{r_{c/e}}{r_{p/e}} \dot{q}_p\left(t - \frac{r_{p/e} - r_{c/e}}{c_0}\right) = \int_0^\infty h_{\mathrm{opt}}(\tau)\, \dot{q}_p(t-\tau)\,\mathrm{d}\tau \tag{8.13.6}$$

The product of this desired output and the primary source volume velocity time derivative is:

$$\dot{q}_p(t)\left(\frac{r_{c/e}}{r_{p/e}}\right) \dot{q}_p\left(t - \frac{r_{p/e} - r_{c/e}}{c_0}\right) = \int_0^\infty h_{\mathrm{opt}}(\gamma)\, \dot{q}_p(t)\, \dot{q}_p(t-\tau-\gamma)\,\mathrm{d}\gamma \tag{8.13.7}$$

Taking expected values, this can be written as a cross-correlation relationship:

$$\left(\frac{r_{c/e}}{r_{p/e}}\right) R_{\dot{p}\dot{p}}\left(\tau - \frac{r_{p/e} - r_{c/e}}{c_0}\right) = \int_0^\infty h_{\mathrm{opt}}(\gamma)\, R_{\dot{p}\dot{p}}(\tau-\gamma)\,\mathrm{d}\gamma \tag{8.13.8}$$

or

$$\left(\frac{r_{c/e}}{r_{p/e}}\right) R_{\dot{p}\dot{p}}(\tau) = \int_0^\infty h_{\mathrm{opt}}(\gamma)\, R_{\dot{p}\dot{p}}\left(\tau + \frac{r_{p/e} - r_{c/e}}{c_0}\gamma\right)\mathrm{d}\gamma \tag{8.13.9}$$

where $R_{\dot{p}\dot{p}}$ is the auto-correlation of the primary source time derivative. Equation (8.13.6) is simply the Weiner–Hopf integral equation, the solution of which defines the optimum control source impulse response function, h_{opt}.

The solution to the Weiner–Hopf integral equation was formulated in Chapter 6 in terms of spectral density functions, which are based upon the Fourier transforms of the quantities in the equation. It will be advantageous here, however, to state the result in terms of Laplace transformed quantities, for ease of demonstrating an example. Referring to Figure 8.57, modelling the primary source output as a white noise sequence, v, passed through a filter with an impulse response function a, such that:

$$\dot{q}_p(t) = \int_0^\infty a(\tau)\, v(t-\tau)\,\mathrm{d}\tau \tag{8.13.10}$$

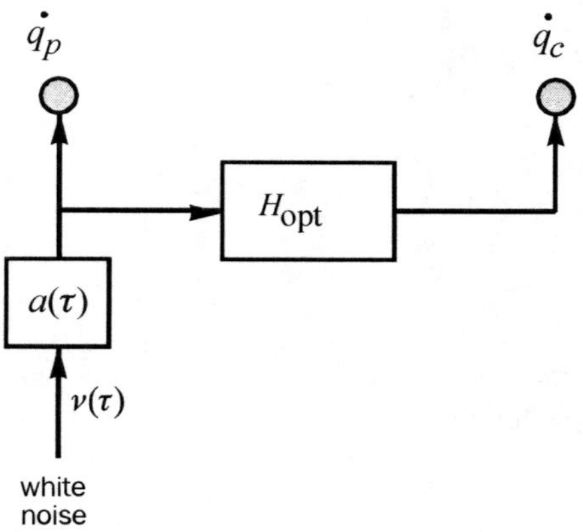

Figure 8.57 Single-channel active control system arrangement as a time domain filtering exercise.

The optimum control source transfer function $H_{opt}(s)$ which is found by solving the Weiner–Hopf integral equation using the primary source characteristics given in Equation (8.13.10) is:

$$H_{opt}(s) = \int_0^\infty h_{opt}(t)\,e^{-st}\,dt = -\frac{r_{c/e}}{r_{p/e}} A^{-1}(s)\,B^{-1}(s) \tag{8.13.11}$$

where

$$A(s) = \int_0^\infty a(t)\,e^{-st}\,dt \tag{8.13.12}$$

$$B(s) = \int_0^\infty a\left(t + \frac{r_{p/e} - r_{c/e}}{c_0}\right) e^{-st}\,dt \tag{8.13.13}$$

The use of Equation (8.13.11) in deriving a control source output schedule is best illustrated by example.

As an example, the case where the primary source volume velocity is modelled as passing white noise through a second-order filter will now be modelled. The characteristics of the second-order filter are defined by:

$$\alpha(s) = \frac{\omega_n^2}{s^2 + 2\zeta\omega_n s + \omega_n^2} \tag{8.13.14}$$

where ω_n and ζ are respectively the natural frequency and critical damping ratio of the system. (This response from this model is similar to the response of a structure excited by a broadband source where there are resonances in the excitation frequency band.) The time derivative of the primary source volume velocity is therefore modelled as white noise passed through a filter whose characteristics are the derivative of $\alpha(s)$:

$$A(s) = \frac{s\omega_n}{s^2 + 2\zeta\omega_n s + \omega_n^2} \tag{8.13.15}$$

The impulse response of the above transfer function is a standard table result:

$$a(t) = -\frac{\omega_n^3}{\omega_d} e^{-\zeta\omega_n t} \sin(\omega_d t - \varphi) \tag{8.13.16}$$

where ω_d is the damped natural frequency, $\omega_d = \omega\sqrt{1 - \zeta^2}$, and $\varphi = \tan^{-1}\left[\sqrt{(1 - \zeta^2)/\zeta}\right]$. The impulse response $b(t)$ is found to be:

$$b(t) = a(t + \Delta t) = -\frac{\omega_n^3}{\omega_d} e^{-\zeta\omega_n (t + \Delta t)} \sin(\omega_d t - \varphi + \Delta t) \tag{8.13.17}$$

where $\Delta t = (r_{p/e} - r_{c/e})/c_0$. Taking the Laplace transform of this produces (Nelson et al., 1988, 1990):

$$B(s) = -\frac{\omega_n^3}{\omega_d} e^{-\zeta\omega_n \Delta t} \left\{ \frac{s\left(\zeta\omega_n \sin\omega_d \Delta t - \omega_d \cos\omega_d \Delta t\right) + \omega_n^2 \sin\omega_d \Delta t}{s^2 + 2\zeta\omega_n s + \omega_n^2} \right\} \tag{8.13.18}$$

Therefore, using the result of Equation (8.13.11), the optimum control source transfer function is:

$$H_{\text{opt}}(s) = \frac{r_{c/e}}{r_{p/e}} \frac{e^{-\zeta\omega_n \Delta t}}{\omega_d} \left(\left(\zeta\omega_d \sin\omega_d \Delta t\right) + \frac{\omega_n^2}{s} \sin\omega_d \Delta t \right) \tag{8.13.19}$$

Taking the inverse Laplace transform of this, the optimum control source output is found to be:

$$\dot{q}_{c,\text{opt}}(t) = \frac{r_{c/e}}{r_{p/e}} \left\{ q_p(t) \left(\frac{e^{-\zeta\omega_n \Delta t}}{\omega_d} \omega_n^2 \sin\omega_d \Delta t \right) - \dot{q}_p(t) \left(\frac{e^{-\zeta\omega_n \Delta t}}{\omega_d} \left[\omega_d \cos\omega_d \Delta t - \zeta\omega_n \sin\omega_d \Delta t \right] \right) \right\} \tag{8.13.20}$$

8.13.2 Minimum Sound Pressure Amplitude at the Error Sensor

Having now derived an expression for determining the optimal control source output characteristics as a function of the primary noise statistics, the next step is to determine the sound pressure reduction that will be achieved at the error sensor when using this control signal. If the error sensor is closer to the control source than the primary source, such that the control schedule of Equation (8.13.4) can be implemented exactly, the residual sound pressure will, in theory, be zero. However, if the error sensor is closer to the primary source than the control source, such that the control source characteristics must be based on the optimal estimator derived in Equation (8.13.11), the residual sound pressure will probably not equal zero. The actual value can be easily quantified by considering the analysis of Weiner filtering given in Chapter 6. In this analysis it was shown that the minimum mean square error was equal to:

$$\xi_{\min} = \mathrm{E}\{y^2(t)\} - \mathrm{E}\{y(t)\hat{y}(t)\} \tag{8.13.21}$$

where y is the desired output, and $\hat{y}$ is the optimal estimate of the desired output. Therefore, by substituting the obtained (from Equation (8.13.11)) optimal estimate of the desired control source output into Equation (8.13.21), the residual sound pressure amplitude can be calculated. This again is best demonstrated by means of example. Consider the problem of calculating the minimum sound pressure amplitude at the error sensor for the control source output derived in the previous problem (from Nelson et al., 1988, 1990). To calculate the residual sound pressure amplitude at the error sensor location, note that the first term in Equation (8.13.21) is simply the squared sound pressure amplitude of the primary disturbance before the addition of active control, which can be denoted simply as J_p. Substituting the final result of the previous example into Equation (8.13.21), the minimum mean square pressure becomes:

$$J_{\min} = J_p - \mathrm{E}\left\{\frac{\rho_0 \dot{q}_p\left(t - r_{p/e}/c_0\right)}{4\pi r_{p/e}}\left(\frac{M\rho_0 \dot{q}_p\left(t - \frac{r_{c/e}}{c_0}\right)}{4\pi r_{c/e}} + \frac{N\rho_0 q_p\left(t - \frac{r_{c/e}}{c_0}\right)}{4\pi r_{c/e}}\right)\right\} \tag{8.13.22}$$

where

$$M = \frac{\mathrm{e}^{-\zeta\omega_n\Delta t}}{\omega_d}\left(\zeta\omega_n \sin\omega_d\Delta t - \omega_d \cos\omega_d\Delta t\right) \tag{8.13.23}$$

$$N = \frac{\mathrm{e}^{-\zeta\omega_n\Delta t}}{\omega_d}\,\omega_n^2 \sin\omega_d\Delta t \tag{8.13.24}$$

This can be simplified to:

$$J_{\min} = J_p - \left(\frac{\rho_0}{4\pi r_{p/e}}\right)^2 \left(M R_{\dot{p}\dot{p}}(\Delta t) + N R_{p\dot{p}}(\Delta t)\right) \tag{8.13.25}$$

where

$$R_{\dot{p}\dot{p}}(\Delta t) = \mathrm{E}\left\{\dot{q}_p(t)\,\dot{q}_p(t+\Delta t)\right\} \tag{8.13.26}$$

$$R_{p\dot{p}}(\Delta t) = \mathrm{E}\left\{q_p(t)\,\dot{q}_p(t+\Delta t)\right\} \tag{8.13.27}$$

The auto-correlation and cross-correlation functions given above can be solved for the description of the primary noise source characteristics given in the beginning of the previous example, giving (Bendat and Piersol, 1986, Chapter 6; Nelson et al., 1988, 1990):

$$R_{\dot{p}\dot{p}}(\Delta t) = \frac{e^{-\zeta\omega_n\Delta t}}{\omega_d}\frac{\omega_n^3}{8\zeta}\left(\omega_d\cos\omega_d\Delta t - \zeta\omega_n\sin\omega_d\Delta t\right) \tag{8.13.28}$$

and

$$R_{p\dot{p}}(\Delta t) = -\frac{e^{-\zeta\omega_n\Delta t}}{\omega_d}\frac{\omega_n^3}{8\zeta}\sin\omega_d\Delta t \tag{8.13.29}$$

The reduction in squared sound pressure amplitude at the error sensing location can be expressed as:

$$\Delta p = \frac{J_{min}}{J_p} \tag{8.13.30}$$

Using the previous equations, with some algebraic manipulation (Nelson, 1988), this can be expressed as:

$$\Delta p = 1 - \frac{e^{-2\zeta\omega_n\Delta t}}{\omega_d^2}\left(\omega_n^2\sin^2\omega_d\Delta t + (\omega_d\cos\omega_d\Delta t - \zeta\omega_n\sin\omega_d\Delta t)^2\right) \tag{8.13.31}$$

In viewing the description of the acoustic pressure reduction, it is apparent that the reduction that can be achieved is dependent on two parameters: the damping of the primary source characteristics (determining how flat the response is), and the parameter $\omega_n\Delta t$, which can be expressed as:

$$\omega_n\Delta t = \frac{\omega_n(r_{p/e} - r_{c/e})}{c_0} = \frac{2\pi(r_{p/e} - r_{c/e})}{\lambda_n} \tag{8.13.32}$$

which provides a direct representation of the effect of the difference in path length, from the primary and control sources to the error sensor, on the level of attenuation that can be achieved. As would be expected, as the difference in path length becomes small, or the damping of the primary source response decreases (such that there is a sharper peak, or the output becomes more 'tonal'), the levels of attenuation that can be achieved increase.

8.13.3 Minimisation of Total Radiated Acoustic Power

Having considered the problem of minimising acoustic pressure amplitude at a single point in space, the next case to consider is that of minimising the radiated acoustic power from the primary source/control source pair. For the time domain analysis, the appropriate error criterion is instantaneous acoustic power, defined by the far-field integration of instantaneous acoustic intensity over a surface enclosing the source pair:

$$W = \int_S I(\boldsymbol{r},t)\,dS \tag{8.13.33}$$

where the far-field instantaneous acoustic intensity is directly related to the pressure amplitude squared:

$$I(\boldsymbol{r},t) = \frac{|p(\boldsymbol{r},t)|^2}{2\rho_0 c_0} \tag{8.13.34}$$

As such, minimising the total radiated sound power is equivalent to minimising the squared sound pressure level spatially averaged over a surface enclosing the source. For the control of random noise, this means that the causally constrained control source output can be considered as a combination of exact 'signal delay' control, for the points on the intensity integration surface closer to the control source than the primary source, and optimal prediction control, for points on the integration surface closer to the primary source than the control source. However, rather than directly use the spatial integration to derive the optimum control source operating 'schedule', the near-field approach used in analysing the power radiated by a harmonic monopole source (in the beginning of this chapter) will be used again to derive the desired control source characteristics.

Expressed in the time domain, the time-averaged acoustic power output of a monopole source is:

$$W = \int_S \mathrm{E}\{p(\boldsymbol{r},t)v(\boldsymbol{r},t)\}\cdot\boldsymbol{n}\,\mathrm{d}S = \mathrm{E}\{p(\boldsymbol{r}_q,t)\,q(t)\} \tag{8.13.35}$$

However, as was mentioned in Section 2.4, the pressure at the source location $\boldsymbol{r}_q$ is a singular quantity. This problem can be overcome by rewriting Equation (8.13.35) using Equation (8.13.2) as follows:

$$W = \lim_{r/c_0 \to 0} \mathrm{E}\left\{\frac{\rho_0}{4\pi r}\dot{q}\left(t - \frac{r}{c_0}\right)q(t)\right\} \tag{8.13.36}$$

where r is the distance from the source. Substituting the expansion:

$$q(t) = q\left(t - \frac{r}{c_0}\right) + \left(\frac{r}{c_0}\right)\dot{q}\left(t - \frac{r}{c_0}\right) \tag{8.13.37}$$

into Equation (8.13.36), and noting that for a stationary random process:

$$\mathrm{E}\{\dot{x}(t)\,x(t)\} = 0 \tag{8.13.38}$$

Equation (8.13.36) can be evaluated to give an expression for the time-averaged power output from a monopole source, stated in the time domain as:

$$W = \frac{\rho_0}{4\pi c_0}\mathrm{E}\{\dot{q}^2(t)\} \tag{8.13.39}$$

Having now formulated an expression to overcome the problem of acoustic pressure singularity when evaluating the power output of a monopole source, it is now feasible to consider minimisation of the total radiated acoustic power. Using Equation (8.13.35), the total radiated power from the primary source/control source pair is:

$$W = \mathrm{E}\left\{\left(p_p(\boldsymbol{r_p},t) + p_c(\boldsymbol{r_p},t)\right) q_p(t) + \left(p_p(\boldsymbol{r_c},t) + p_c(\boldsymbol{r_c},t)\right) q_c(t)\right\} \tag{8.13.40}$$

where $p_p(\boldsymbol{r},t)$, $p_c(\boldsymbol{r},t)$ denote respectively the primary and control source generated sound pressures at location $\boldsymbol{r}$ and time t. Using Equation (8.13.39), Equation (8.13.40) can be re-expressed as:

$$W = \mathrm{E}\left\{\frac{\rho_0}{4\pi c_0}\dot{q}_p^2(t) + p_c(\boldsymbol{r_p},t)\,q_p(t) + p_p(\boldsymbol{r_c},t)\,q_c(t) + \frac{\rho_0}{4\pi c_0}\dot{q}_c^2(t)\right\} \tag{8.13.41}$$

The second and third terms in Equation (8.13.41) can be restated using Equation (8.13.2):

$$\mathrm{E}\left\{p_c(\boldsymbol{r_p},t)q_p(t)\right\} = \frac{\rho_0}{4\pi d}\mathrm{E}\left\{\dot{q}_c\left(t - \frac{d}{c_0}\right) q_p(t)\right\} = \frac{\rho_0}{4\pi d}\mathrm{E}\left\{\dot{q}_c(t)\, q_p\left(t + \frac{d}{c_0}\right)\right\} \tag{8.13.42a–c}$$

where d is the primary source/control source separation distance, and

$$\mathrm{E}\left\{p_p(\boldsymbol{r_c},t)\,q_c(t)\right\} = \frac{\rho_0}{4\pi d}\mathrm{E}\left\{\dot{q}_p\left(t - \frac{d}{c_0}\right) q_c(t)\right\} = -\frac{\rho_0}{4\pi d}\mathrm{E}\left\{q_p\left(t - \frac{d}{c_0}\right)\dot{q}_c(t)\right\} \tag{8.13.43a–c}$$

Substituting Equations (8.13.42) and (8.13.43) into Equation (8.13.41), and noting that the first term in Equation (8.13.41) is the acoustic power, W_p, radiated by the primary source in the absence of active control, produces the following expression for the total radiated acoustic power:

$$W = W_p + \frac{\rho_0}{4\pi c_0}\mathrm{E}\left\{\dot{q}_c^2(t) + \frac{c_0}{d}\dot{q}_c(t)\left[q_p\left(t + \frac{d}{c_0}\right) - q_p\left(t - \frac{d}{c_0}\right)\right]\right\} \tag{8.13.44}$$

This expression can be used to formulate a set of control source characteristics which will minimise the total radiated acoustic power.

To derive the optimum control source characteristics, it will be useful to assume that the time derivative of the control source volume velocity is related to the primary source volume velocity, rather than its time derivative as was assumed when minimising sound pressure at a point. With this assumption, the control source output can be expressed as:

$$\dot{q}_c(t) = \int_0^\infty h_w(\tau)\, q_p(t - \tau)\,\mathrm{d}\tau \tag{8.13.45}$$

Deriving the required Weiner–Hopf integral equation is not quite as simple as the 'intuitive' approach used when minimising sound pressure at a point, as the optimum control source output is not immediately apparent, and so variational methods will be employed for the task (Laning and Battin, 1956; Nelson et al., 1988). With these methods, it is assumed that the impulse response h_w can be written as:

$$h_w(t) = h_{w,\mathrm{opt}}(t) + \varepsilon h_w^{\prime}(t) \tag{8.13.46}$$

where $h_{w,\text{opt}}$ is the desired optimal impulse response function, ε is a real number, and $h_w^{/}$ is some arbitrary function of t. The approach taken to solving the problem is to re-express the power output in terms of ε, and then, noting that for the optimum impulse response function, $\varepsilon = 0$, differentiate the expression with respect to ε and set the result of the differentiation equal to zero. Substituting Equation (8.13.46) into Equation (8.13.45) produces:

$$W = W_p + \frac{\rho_0}{4\pi c_0}\left\{ \int_0^\infty\int_0^\infty \left(h_{w,\text{opt}}(\tau) + \varepsilon h_w^{/}(\tau)\right)\left(h_{w,\text{opt}}(\gamma) + \varepsilon h_w^{/}(\gamma)\right) R_{pp}(\tau - \gamma)\,\mathrm{d}\tau\,\mathrm{d}\gamma \right.$$

$$\left. + \frac{c_0}{d}\int_0^\infty \left(h_{w,\text{opt}}(\tau) + \varepsilon h_w^{/}(\tau)\right) R_{pp}\left(\tau + \frac{d}{c_0}\right)\mathrm{d}\tau - \frac{c_0}{d}\int_0^\infty \left(h_{w,\text{opt}}(\tau) + \varepsilon h_w^{/}(\tau)\right) R_{pp}\left(\tau - \frac{d}{c_0}\right)\mathrm{d}\tau \right\} \tag{8.13.47}$$

Expanding Equation (8.13.47) in terms of powers of ε, differentiating with respect to ε, and setting $(\partial W/\partial\varepsilon)_{\varepsilon=0} = 0$, produces the expression:

$$\int_0^\infty\int_0^\infty \left(h_{w,\text{opt}}(\tau) h_w^{/}(\tau) h_{w,\text{opt}}(\gamma)\right) R_{pp}(\tau-\gamma)\,\mathrm{d}\tau\mathrm{d}\gamma + \frac{c_0}{d}\int_0^\infty h_w^{/}(\tau)\left(R_{pp}\left(\tau + \frac{d}{c_0}\right) - R_{pp}\left(\tau - \frac{d}{c_0}\right)\right)\mathrm{d}\tau = 0 \tag{8.13.48}$$

Noting that $R_{pp}(\tau - \gamma) = R_{pp}(\gamma - \tau)$, so that the first term in Equation (8.13.48) can be expressed as:

$$\int_0^\infty\int_0^\infty 2h_w^{/}(\tau)\, h_{w,\text{opt}}(\gamma)\, R_{pp}(\tau-\gamma)\,\mathrm{d}\tau\,\mathrm{d}\gamma \tag{8.13.49}$$

Equation (8.13.48) can be rewritten as:

$$\int_0^\infty h_w^{/}(\tau)\left\{ \int_0^\infty 2h_{w,\text{opt}}(\tau)\, R_{pp}(\tau - \gamma)\,\mathrm{d}\gamma + \frac{c_0}{d}\left[R_{pp}\left(\tau + \frac{d}{c_0}\right) - R_{pp}\left(\tau - \frac{d}{c_0}\right)\right]\right\}\mathrm{d}\tau = 0 \tag{8.13.50}$$

As Equation (8.13.50) must hold for any arbitrary function $h_w{}'$, it can be deduced that the expression inside the curly brackets must be equal to zero. Therefore,

$$\int_0^\infty h_{w,\text{opt}}(\tau)\, R_{pp}(\tau-\gamma)\,\mathrm{d}\gamma + \frac{c_0}{2d}\left[R_{pp}\left(\tau + \frac{d}{c_0}\right) - R_{pp}\left(\tau - \frac{d}{c_0}\right)\right] = 0 \tag{8.13.51}$$

This is the Weiner–Hopf integral equation, which defines the optimum control source characteristics for minimising the total radiated acoustic power (Nelson et al., 1988, 1990).

Now that the desired form of the Weiner–Hopf integral equation has been derived, what remains is to solve it for the optimum control source impulse function. To do this, it will be assumed that the optimum impulse response function is a combination of a 'delay' control and optimal estimation, resulting in the following form for the solution:

$$h_{w,\text{opt}}(t) = \frac{c_0}{2d}\left[\delta\left(t - \frac{d}{c_0}\right) - h_{w,e}(t)\right] \tag{8.13.52}$$

where $\delta(t - d/c_0)$ is the pure time delay and $h_{w,e}$ is the impulse response for the estimation part of the impulse response function. Substituting this assumed form of solution into Equation (8.13.51) shows that $h_{w,e}$ must satisfy the expression:

$$\int_0^\infty h_{w,e}(\gamma)\, R_{pp}(\tau-\gamma)\, \mathrm{d}\gamma - R_{pp}\left(\tau + \frac{d}{c_0}\right) = 0 \qquad \tau \geq 0 \tag{8.13.53}$$

because

$$\frac{c_0}{2d}\left[\int_0^\infty \delta\left(\gamma - \frac{d}{c_0}\right) R_{pp}(\tau - \gamma)\, \mathrm{d}\gamma - R_{pp}\left(t - \frac{d}{c_0}\right)\right] = 0 \tag{8.13.54}$$

From Equation (8.13.53) it is clear that the impulse response function $h_{w,e}$ is the solution to the problem which requires optimal prediction of the primary source output at a time in the future corresponding to the propagation time delay due to the primary source/control source separation distance. Further, the pure delay component of the impulse response corresponds physically to one of the control mechanisms (absorption) seen previously. Substituting Equation (8.13.52) into Equation (8.13.45) shows that the optimum control source output characteristics are defined by (Nelson et al., 1988, 1990):

$$\dot{q}_{c,\text{opt}}(t) = \frac{c_0}{2d}\left(q_p\left(t - \frac{d}{c_0}\right) - \int_0^\infty h_{w,e}(\tau)\, q_p(t-\tau)\, \mathrm{d}\tau\right) \tag{8.13.55}$$

The first term in Equation (8.13.55) describes a control source output that maximises the absorption of acoustic power, while the second maximises suppression of acoustic power.

8.13.4 Calculation of the Minimum Power Output

Now that an expression has been derived for the optimum control source output with respect to minimising total radiated acoustic power, what remains is to find out what levels of attenuation it will provide. Substituting the optimum control source volume velocity time derivative of Equation (8.13.55) into Equation (8.13.41), which describes the total acoustic power output of the system, gives:

$$W_{\min} = \frac{\rho_0}{4\pi c_0} \mathrm{E}\left\{\dot{q}_p^2(t) + \dot{q}_{c,\text{opt}}^2(t) + \frac{c_0}{d}\dot{q}_{c,\text{opt}}(t)\left[q_p\left(t + \frac{d}{c_0}\right) - q_p\left(t - \frac{d}{c_0}\right)\right]\right\} \tag{8.13.56}$$

Equation (8.13.56) can be re-expressed as:

$$W_{\min} = \frac{\rho_0}{4\pi c_0} \mathrm{E}\left\{ \dot{q}_p^2(t) + \frac{c_0}{2d} \dot{q}_{c,\mathrm{opt}}(t) \left[q_p\left(t + \frac{d}{c_0}\right) - q_p\left(t - \frac{d}{c_0}\right) \right] \right.$$
$$\left. + \dot{q}_{c,\mathrm{opt}}^2(t) + \frac{c_0}{2d} \dot{q}_{c,\mathrm{opt}}(t) \left[q_p\left(t + \frac{d}{c_0}\right) - q_p\left(t - \frac{d}{c_0}\right) \right] \right\} \tag{8.13.57}$$

It can be shown that the term in the square brackets in Equation (8.13.57) is equal to zero by using the substitution:

$$\dot{q}_{c,\mathrm{opt}}(t) = \int_0^\infty h_{w,\mathrm{opt}}(\tau)\, q_p(t-\tau)\, \mathrm{d}\tau \tag{8.13.58}$$

and noting that:

$$\mathrm{E}\left\{ \dot{q}_{c,\mathrm{opt}}^2(t) + \frac{c_0}{2d} \dot{q}_{c,\mathrm{opt}}(t) \left[q_p\left(t + \frac{d}{c_0}\right) - q_p\left(t - \frac{d}{c_0}\right) \right] \right\} =$$
$$\int_0^\infty h_{w,\mathrm{opt}}(\tau) \left\{ \int_0^\infty h_{w,\mathrm{opt}}(\gamma)\, R_{pp}(\tau - \gamma)\, \mathrm{d}\gamma + \frac{c_0}{2d} \left[R_{pp}\left(\tau + \frac{d}{c_0}\right) - R_{pp}\left(\tau - \frac{d}{c_0}\right) \right] \right\} \tag{8.13.59}$$

an equality which is a result of the defining Weiner–Hopf integral equation. Therefore, the minimum radiated acoustic power is:

$$W_{\min} = \frac{\rho_0}{4\pi c_0} \mathrm{E}\left\{ \dot{q}_p^2(t) + \frac{c_0}{2d} \dot{q}_{c,\mathrm{opt}}(t) \left[q_p\left(t + \frac{d}{c_0}\right) - q_p\left(t - \frac{d}{c_0}\right) \right] \right\} \tag{8.13.60}$$

or

$$W_{\min} = W_p + \frac{\rho_0}{4\pi c_0} \mathrm{E}\left\{ \frac{c_0}{2d} \dot{q}_{c,\mathrm{opt}}(t) \left[q_p\left(t + \frac{d}{c_0}\right) - q_p\left(t - \frac{d}{c_0}\right) \right] \right\} \tag{8.13.61}$$

where W_p is the power radiated by the primary source in the absence of active control.

Therefore, the minimum acoustic power output of the system, as well as the control source operating schedule, is a function of both a desire to absorb the primary acoustic radiation that has already occurred, and the desire to suppress the acoustic radiation that will occur in the future.

8.14 ACTIVE CONTROL OF IMPACT ACCELERATION NOISE

Many of the prominent sources of industrial noise pollution in factories are machines concerned with the reshaping of a piece of material. Such machines normally perform their allotted task via some impulsive process, such as stamping, forging or riveting. The noise signature from such an operation is compiled from two separate generating mechanisms: the rapid acceleration or deceleration of some part of the workpiece and/or machine, and the pseudo-steady-state vibration of the machine and/or workpiece following the impact, as it dissipates its vibrational energy (Osman et al., 1974; Akay, 1978; Richards et al., 1979). The result of this is often a signature that is characterised by two separate parts; a short duration, high amplitude initial pulse due to the acceleration noise, and a much longer duration, lower amplitude 'ringing' following this. While it is the residual structural vibrations that are the primary contributors to the overall noise level (Akay, 1978), it is perhaps the initial acceleration noise pulse that holds the greatest potential for hearing damage. This is due to its short duration, which limits the ability of the middle ear mechanism to stiffen and thus partially protect the inner ear from the pressure pulse (Akay, 1978), resulting in the pressure pulses passing directly into the inner ear with their full amplitude and hearing damage potential.

In this section, the potential for applying active control to attenuate the sound generated by an impulsive process will be examined. The initial acceleration impulse noise portion of the noise signature will be specifically targeted, because feedback control of radiated acoustic power (probably the best approach for controlling the acoustic radiation from the residual vibration) will be discussed later in this chapter.

One of the characteristics of the acoustic pressure fluctuation caused by the initial acceleration is that it is essentially deterministic. Further, for elastic collisions in simple systems it can be shown that its amplitude is linearly correlated with the impact force (Koss and Alfredson, 1973; Koss, 1974), which can be measured easily with a load cell. These features suggest that the use of a feedforward active control system may be feasible, where the impact force, or machine acceleration, is provided to the control system as a reference signal. Alternatively, if the characteristics of successive impulses are constant, a 'preview' signal, such as the acceleration of the machine, can be used as a reference signal. The collisions that will be of relevance here will be limited to elastic collisions between two rigid bodies, such as the collision between the hammer and anvil in a forge. For these cases, the acoustic waveform generated by the initial acceleration is constructed from what is essentially a single sinusoidal pulse from each of the impacting bodies.

The active control of what is basically sinusoidal (harmonic) radiation from rigid bodies has already been considered in this chapter, in regard to control of acoustic radiation from monopole sources. However, the impact noise problems being considered here have a distinguishing characteristic: they require control inputs that are constrained to be causal with respect to the primary disturbance. This rules out the use of the frequency domain analysis techniques employed in the examination of the performance potential of feedforward control for the previously mentioned harmonic excitation problems, as these methods were shown to produce control output schedules that are not constrained to be causal with respect to the primary noise disturbance. Causally constrained systems were studied in the previous section using an analysis akin to Weiner filter theory, employing statistical expectation operators to develop a control source whose characteristics minimise the mean square value of the acoustic quantity of interest (power or pressure). However, as the problem being considered here is essentially deterministic, this approach is more complex than what is really needed.

Therefore, this section will begin with the development of a methodology for calculating the optimum control source output 'schedule' for controlling both the overall (spatially averaged) levels of the initial impact pulse, as well as minimising the acoustic pressure amplitude at prospective (discrete) error sensing locations. Following this, a very simple problem, the control of a single sinusoidal pulse from a monopole source will be considered.

8.14.1 Method for Obtaining Optimum Control Source Pressure Output Schedules

It is appropriate to begin the study of the problem of actively attenuating impact acceleration noise by developing a methodology to facilitate calculation of the optimum control source output schedule, or output as a function of time, which will minimise the particular error criterion of interest. This is either the total radiated acoustic power or the acoustic pressure at a given set of potential error sensing locations. The method that will be developed to calculate the desired control source output schedules is similar to both the previously developed frequency and time domain techniques in that it is least-squares based, but is best viewed as a linear regression approach to solving the problem.

Consider the active control problem illustrated in Figure 8.58, where some generic control source distribution is being used to actively attenuate the acoustic radiation from some generic primary source distribution. The error criterion that will be examined first in the time domain analysis of the problem is instantaneous radiated acoustic power, calculated as the integration of instantaneous acoustic intensity I over the surface of a volume enclosing both the primary and control source distribution. Instantaneous acoustic intensity can be expressed as:

$$I(\boldsymbol{r},t) = \frac{|p(\boldsymbol{r},t)|^2}{\rho_0 c_0} \tag{8.14.1}$$

where $p(\boldsymbol{r}, t)$ is the acoustic pressure at location $\boldsymbol{r}$ at time t. This can be approximated at any given time by minimising the squared acoustic pressure amplitude at a discrete number of points, N_n, distributed evenly over the surface of the enclosing space as follows:

$$\text{minimizing } W \approx \text{minimizing } \sum_{n=1}^{N_n} |p(\boldsymbol{r}_n,t)|^2 \tag{8.14.2}$$

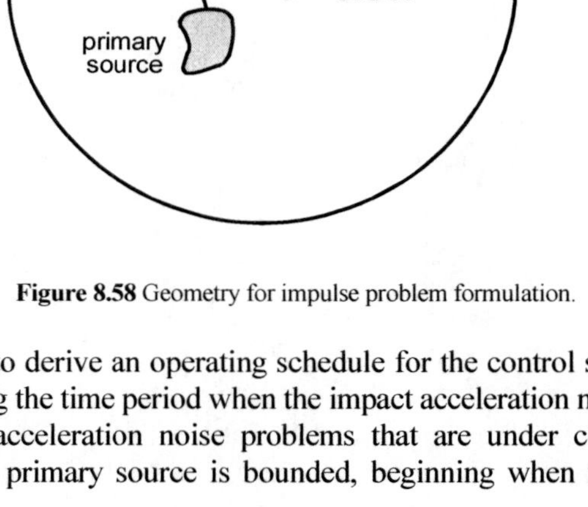

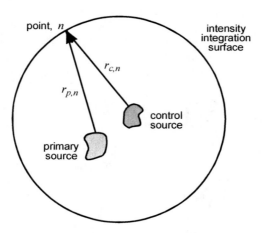

Figure 8.58 Geometry for impulse problem formulation.

The objective now is to derive an operating schedule for the control source distribution that will achieve this during the time period when the impact acceleration noise is present.

For the impact acceleration noise problems that are under consideration here, the operating time of the primary source is bounded, beginning when the impact occurs and finishing when the impacted body returns to the point of equilibrium. The duration of the non-zero contribution to the acoustic pressure at all points on the integration surface from the impact is therefore also bounded, beginning at time $t_{p,i}$, when the pressure pulse initially contacts the closest point on the enclosing surface and finishing at time $t_{p,f}$, when the trailing edge of the pulse is exiting from the most distant point. The object of applying active control is therefore to minimise the sum of the squared acoustic pressure amplitudes at all points on the enclosing surface throughout the period $(t_{p,i}, t_{p,f})$. As with the discretisation of the continuous spatial integration, the minimisation of pressure squared during this time can be approximated as the problem of minimising the pressure squared at discrete time intervals during the period $(t_{p,i}, t_{p,f})$, such that the error criterion used to analytically determine the optimum control source operating schedule is:

$$\text{minimise} \sum_{\gamma=0}^{N_t-1} \sum_{n=1}^{N_n} |p(r_n, t = t_{p,i} + \gamma \Delta t_p)|^2 \tag{8.14.3}$$

where N_t is the total number of points in time between $t_{p,i}$ and $t_{p,f}$ which are considered, and each point is incremented by Δt_p from the one before, where

$$\Delta t_p = \frac{t_{p,i} - t_{p,f}}{N_t - 1} \tag{8.14.4}$$

During operation of the active noise control system, the acoustic pressure $p(\boldsymbol{r}_n,t)$ at any location $\boldsymbol{r}_n$ and time t can be considered as the superposition of the primary and control source contributions as follows:

$$p(\boldsymbol{r}_n, t) = p_p(\boldsymbol{r}_n, t) + p_c(\boldsymbol{r}_n, t) \tag{8.14.5}$$

where the subscripts p, c denote primary and control respectively. The control sources of interest here will in practice be loudspeakers, but will be modelled to a good approximation at the frequencies of interest as acoustic monopole sources. The time domain description of the pressure fluctuation at some distance r from an acoustic monopole source due to its operation can be found from the inhomogeneous wave equation:

$$\left(\nabla^2 - \frac{1}{c_0^2} \frac{\partial^2}{\partial t^2} \right) p(\boldsymbol{r},t) = -\rho_0 \frac{\partial Q(\boldsymbol{r},t)}{\partial t} \tag{8.14.6}$$

Concentrating the source strength Q at a single location $\boldsymbol{r}_q$, such that $Q(\boldsymbol{r}, t) = q(t)\delta(\boldsymbol{r}-\boldsymbol{r}_q)$, enables a solution to be written for the acoustic pressure at some point $\boldsymbol{r}$ resulting from the monopole source as (see Equation (8.13.2)):

$$p(\boldsymbol{r},t) = \frac{\rho_0}{4\pi r} \frac{\partial q\left(t - \dfrac{r}{c_0} \right)}{\partial t} = \frac{\rho_0}{4\pi r} \dot{q}\left(t - \frac{r}{c_0} \right) \tag{8.14.7a,b}$$

From Equation (8.14.7), it can be concluded that the control source operating schedule, which will define the contribution of the control source to the acoustic pressure on the intensity integration surface, will in fact be a record of the time derivative of the control source volume velocity $\dot{q}$ as a function of time.

If there are N_c multiple, separate control sources, then Equation (8.14.5) can be further expanded as:

$$p(\boldsymbol{r_n}, t) = p_p(\boldsymbol{r_p}, t) + \sum_{j=1}^{N_c} p_{cj}(\boldsymbol{r_n}, t) \tag{8.14.8}$$

The acoustic pressure at the set of N_n points on the intensity integration surface at time t can be written as the sum of vectors:

$$\boldsymbol{p}(t) = \boldsymbol{p}_p(t) + \sum_{j=1}^{N_c} \boldsymbol{p}_{cj}(t) \tag{8.14.9}$$

where $\boldsymbol{p}$, $\boldsymbol{p}_p$ and $\boldsymbol{p}_{cj}$ are respectively the ($N_n \times 1$) vectors of the total acoustic pressure, the primary source contributed acoustic pressure and the jth control source contributed acoustic pressure, defined as follows:

$$\boldsymbol{p}(t) = \begin{bmatrix} p(\boldsymbol{r}_1, t) \\ p(\boldsymbol{r}_2, t) \\ \vdots \\ p(\boldsymbol{r_n}, t) \end{bmatrix}; \quad \boldsymbol{p}_p(t) = \begin{bmatrix} p_p(\boldsymbol{r}_1, t) \\ p_p(\boldsymbol{r}_2, t) \\ \vdots \\ p_p(\boldsymbol{r_n}, t) \end{bmatrix}; \quad \boldsymbol{p}_{cj} = \begin{bmatrix} p_{cj}(\boldsymbol{r}_1, t) \\ p_{cj}(\boldsymbol{r}_2, t) \\ \vdots \\ p_{cj}(\boldsymbol{r_n}, t) \end{bmatrix} \tag{8.14.10a–c}$$

Considered over the entire time period ($t_{p,i}$, $t_{p,f}$), the set of total, primary and control sound pressures can therefore be expressed in matrix form as:

$$\boldsymbol{P} = \boldsymbol{P}_p + \sum_{j=1}^{N_c} \boldsymbol{P}_{cj} = \boldsymbol{P}_p + \boldsymbol{P}_c \tag{8.14.11}$$

where

$$\boldsymbol{P} = \begin{bmatrix} \boldsymbol{p}(t_0) \\ \boldsymbol{p}(t_1) \\ \vdots \\ \boldsymbol{p}(t_{N_t - 1}) \end{bmatrix}; \quad \boldsymbol{P}_p = \begin{bmatrix} \boldsymbol{p}_p(t_0) \\ \boldsymbol{p}_p(t_1) \\ \vdots \\ \boldsymbol{p}_p(t_{N_t - 1}) \end{bmatrix}; \quad \boldsymbol{P}_{cj} = \begin{bmatrix} \boldsymbol{p}_{cj}(t_0) \\ \boldsymbol{p}_{cj}(t_1) \\ \vdots \\ \boldsymbol{p}_{cj}(t_{N_t - 1}) \end{bmatrix}; \quad \boldsymbol{P}_c = \sum_{j=1}^{N_c} \boldsymbol{P}_{cj} \tag{8.14.12a–d}$$

Using Equation (8.14.11), the error criterion of Equation (8.14.3) can be written simply in matrix form as:

$$\text{minimise } \boldsymbol{P}^{\mathrm{T}}\boldsymbol{P} = \boldsymbol{P}_c^{\mathrm{T}}\boldsymbol{P}_c + \boldsymbol{P}_c^{\mathrm{T}}\boldsymbol{P}_p + \boldsymbol{P}_p^{\mathrm{T}}\boldsymbol{P}_c + \boldsymbol{P}_p^{\mathrm{T}}\boldsymbol{P}_p \tag{8.14.13}$$

To use this form of the error criterion to derive the optimum control source operating schedules, note that the (non-zero) control source operating time, like the primary source operating time, will be finite, confined within the period $(t_{c,i}, t_{c,f})$. Also, note that the control source operating period $(t_{c,i}, t_{c,f})$ is not constrained to be equivalent to the primary source operating period $(t_{p,i}, t_{p,f})$ in either starting or finishing times, or in total duration. By discretising this duration into N_{tc} separate periods, the matrix $\boldsymbol{P}_c$ can be written as:

$$\boldsymbol{P}_c = \boldsymbol{ZS} \tag{8.14.14}$$

where $\boldsymbol{S}$ is the $(N_c N_{tc} \times 1)$ vector of control source output schedules, given by:

$$\boldsymbol{S} = \left[\text{control source 1 schedule} \ \vdots \ \cdots \ \vdots \ \text{control source } N_c \text{ schedule}\right]^{\mathrm{T}}$$

$$= \left[\dot{q}_{c1}(\tau_0) \ \ \dot{q}_{c1}(\tau_1) \ \cdots \ \dot{q}_{c1}(\tau_{N_{tc}-1}) \ \vdots \ \cdots \ \vdots \ \dot{q}_{cN_c}(\tau_0) \ \ \dot{q}_{cN_c}(\tau_1) \ \cdots \ \dot{q}_{cN_c}(\tau_{N_{tc}-1})\right]^{\mathrm{T}}$$

(8.14.15a,b)

The quantity $\dot{q}_{cj}(\tau)$ is the time derivative of the volume velocity generated by control source j at control source operating time τ, and $\boldsymbol{Z}$ is the $(N_n N_t \times N_c N_{tc})$ matrix of space and time dependent transfer functions between these control source volume velocity derivatives, and the point locations and times on the intensity integration surface, defined as follows:

$$\boldsymbol{Z} = \left[z_c1(\tau_0) \ \ z_c1(\tau_1) \ \cdots \ z_c1(\tau_{N_{tc}-1}) \ \vdots \ \cdots \ \vdots \ z_{cN_c}(\tau_0) \ \ z_{cN_c}(\tau_1) \ \cdots \ z_{cN_c}(\tau_{N_{tc}-1})\right] \tag{8.14.16}$$

where

$$z_{cj}(\tau) = \left[z_{cj,\tau}(\boldsymbol{r}_1,t_0) \ \cdots \ z_{cj,\tau}(\boldsymbol{r}_{N_n},t_0) \ \vdots \ \cdots \ \vdots \ z_{cj,\tau}(\boldsymbol{r}_1,t_{N_1-1}) \ \cdots \ z_{cj,\tau}(\boldsymbol{r}_{N_n},t_{N_t-1})\right]^{\mathrm{T}} \tag{8.14.17}$$

and, from Equation (8.14.7), for monopole sources:

$$z_{cj,\tau}(\boldsymbol{r}_1,t) = \begin{cases} \dfrac{\rho_0}{4\pi r} & \text{if } \ \tau = t - \dfrac{r_1}{c_0} \\ 0 & \text{otherwise} \end{cases} \tag{8.14.18}$$

Note that for each point location $\boldsymbol{r}_n$ on the intensity integration surface, there will be only one non-zero transfer function between that point at a given time t and the jth control source output schedule in the matrix $\boldsymbol{S}$, $\dot{q}_j(\tau)$, corresponding to time $\tau = t - r/c_0$, and hence only one non-zero entry in the columns of the transfer function matrix $\boldsymbol{Z}$.

Using Equation (8.14.14), Equation (8.14.13) can be written in the standard quadratic form as:

$$\boldsymbol{P}^{\mathrm{T}}\boldsymbol{P} = \boldsymbol{S}^{\mathrm{T}}\boldsymbol{AS} + \boldsymbol{S}^{\mathrm{T}}\boldsymbol{b} + \boldsymbol{b}^{\mathrm{T}}\boldsymbol{S} + c \tag{8.14.19}$$

where

$$\boldsymbol{A} = \boldsymbol{Z}^{\mathrm{T}}\boldsymbol{Z}, \ \boldsymbol{b} = \boldsymbol{Z}^{\mathrm{T}}\boldsymbol{P}_p, \ c = \boldsymbol{P}_p^{\mathrm{T}}\boldsymbol{P}_p \tag{8.14.20a–c}$$

Expressed in this form, the vector of optimum control source operating schedules, $\boldsymbol{S}_{\mathrm{opt}}$, is defined by the expression:

$$S_{\mathrm{opt}} = -A^{-1}b \tag{8.14.21}$$

The minimum value of the error criterion, the sum of the squared acoustic pressure amplitudes over all points and times is therefore,

$$P^{\mathrm{T}}P = c - b^{\mathrm{T}}A^{-1}b \tag{8.14.22}$$

Note that from Equation (8.14.20), the term c is the value of the error criterion during primary excitation only.

This procedure for deriving the optimum control source output pressure schedules is analogous to the frequency domain methods used previously in this chapter, and is identical to the procedures used in a multiple regression problem. The dependent variable in this case is $-\boldsymbol{P}_p$, which is what is desired to achieve the maximum levels of attenuation of the radiated acoustic power, and the independent variables are the terms in the matrix of transfer functions $\boldsymbol{Z}$. Hence the optimum control source operating schedules can be easily calculated using commercially available multiple regression software as described previously in this chapter.

Note finally that the methodology developed here for deriving the optimum control source operating schedules with respect to minimising the total radiated acoustic power can also be used to derive similar schedules for minimising the acoustic pressure amplitude at discrete error sensing locations. To do this, the set of N_n distributed measuring points on the intensity integration sphere should simply be replaced with the error sensing locations of interest.

8.14.2 Example: Control of a Sinusoidal Pulse from a Single Source

The model problem that will be considered here is the control of a single sinusoidal pulse from one acoustic (primary) monopole source by the introduction of a second acoustic (control) monopole source. This model is a good starting point for a study of feedforward active control of impact acceleration noise for three reasons. First, it is a very simple case, making its implementation, in the theoretical framework presented in the previous section, a straightforward exercise. Second, it can be viewed as half of the acceleration noise problem, which to a first approximation is the emission of single sinusoidal acoustic pulses from two impacting bodies. Third, the active control of harmonic sound radiation from one monopole acoustic source by a second monopole acoustic source has already been studied in this chapter, so that the previously obtained results provide a basis against which to compare results derived here.

The system to being studied in this section is illustrated in Figure 8.59. Here, a single monopole control source is being used to control the acoustic radiation from a single monopole primary source. The primary source generated acoustic pressure that will be arriving at location $\boldsymbol{r}_n$ at time t' is defined using Equation (8.14.7) as follows:

$$p_p(\boldsymbol{r}_n, t') = \frac{\rho_0}{4\pi r_{n,p}} \dot{q}_p\left(t - \frac{r_{n,p}}{c_0}\right) \tag{8.14.23}$$

where for the single sinusoidal pulse problem being considered in this section $\dot{q}_p$ is defined by:

$$\dot{q}_p(t) = \begin{cases} \dot{q}_p \sin \omega t & \text{if} \quad 0 < t < \dfrac{2\pi}{\omega} \\ 0 & \text{otherwise} \end{cases} \tag{8.14.24}$$

Without loss of generality, $\dot{q}_p$ will be set equal to 1.0, so that the results can be viewed as normalised.

For comparison, recall from the work in Section 8.3 that when the arrangement shown in Figure 8.59 is subjected to harmonic excitation, the minimum acoustic power output is produced when the control source volume velocity relationship is:

$$q_{c,\text{opt}} = -q_p \operatorname{sinc} kd \tag{8.14.25}$$

where k is the acoustic wavenumber and d is the source separation distance. With this optimal value of control source volume velocity, the reduction in total radiated acoustic power from the initial state of primary source operation only was found to be:

$$\Delta W = 1 - \operatorname{sinc}^2 kd \tag{8.14.26}$$

The first, and possibly most important, question that must be answered is how much attenuation of the total radiated acoustic power of this single pulse can actually be achieved. Figure 8.60 illustrates the maximum levels of acoustic power reduction that can be achieved as a function of primary source/control source separation distance, plotted with the equivalent curve for the harmonic excitation case. The first point to note is that the levels of attenuation of the single pulse acoustic power fall off rapidly as the separation distance d between the primary and control sources increases, conservatively following the ($1 - \operatorname{sinc}^2 kd$) relationship of the harmonic excitation case and crossing the 10 dB level at approximately 1/10 wavelength separation. However, a perhaps more surprising result is that it is always possible to achieve slightly greater levels of acoustic power attenuation when controlling a single sinusoidal pulse, using a causally constrained control signal, than would be similarly attained when controlling harmonic excitation with an unconstrained control signal.

To explain why the performance potential has been improved, note that as with the previously studied frequency domain cases, the total acoustic power output of the system arrangement shown in Figure 8.59 is the sum of four terms (relating to the four terms in Equation (8.14.13)), which can be expressed as:

$$W_{\text{total}} = W_c + \Delta W_c + \Delta W_p + W_p \tag{8.14.27}$$

where W_c, W_p are the acoustic power outputs of the control and primary sources respectively, which would be measured if they were operating on their own, and ΔW_c, ΔW_p are the respective modifications to these power outputs arising from the existence of the other source in the acoustic domain. For significant (or even any) acoustic power attenuation to be achieved, the sum of these modification terms must be a negative value, as the acoustic power outputs of the sources operating on their own will always be positive values. For the arrangement being considered here, applying the analysis of the previous section (see Equation (8.13.44)), the sum of ΔW_c and ΔW_p is equal to:

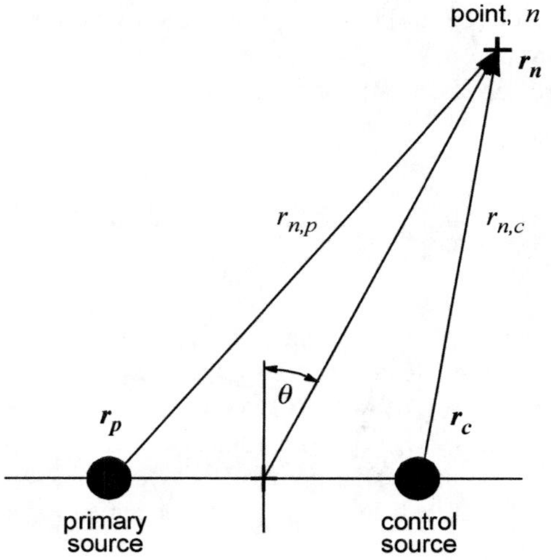

Figure 8.59 Geometry for single primary source/single control source system.

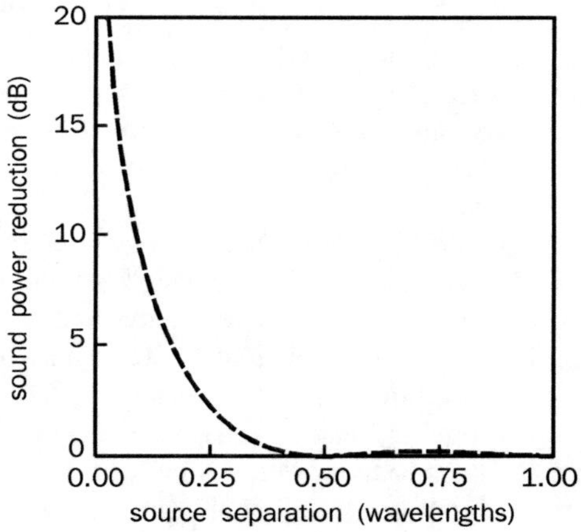

Figure 8.60 Acoustic power attenuation as a function of source separation distance.

$$\Delta W_c + \Delta W_p = \mathrm{E}\left\{\dot{q}_s(t)\,\frac{\rho_0}{4\pi d}\,q_p\left(t + \frac{d}{c_0}\right)\right\} + \mathrm{E}\left\{q_s(t)\,\dot{q}_p\left(t - \frac{d}{c_0}\right)\frac{\rho_0}{4\pi d}\right\} \tag{8.14.28}$$

where E{ } is the statistical expectation operator. In viewing this equation, it can be deduced that the optimum control source output at time t will be based upon two criteria: producing a pressure that will arrive at the primary source at time $(t + d/c_0)$ and be out of phase with its volume velocity, hence reducing its power output (primary source suppression); and producing a volume velocity that will be out of phase with the impinging primary source pressure, generated at an earlier time ($t - d/c_0$) (self-suppression or absorption). Similar to the conclusion that was derived in the common link theory derivation of Section 8.9, these two criteria represent a compromise, rather than a coalition. This can be seen more clearly by restating the second term on the right-hand side of Equation (8.14.28) in terms of $\dot{q}_s$, to produce:

$$\Delta W_c + \Delta W_p = \mathrm{E}\left\{\dot{q}_s(t)\,\frac{\rho_0}{4\pi d}\,q_p\left(t + \frac{d}{c_0}\right)\right\} - \mathrm{E}\left\{\dot{q}_s(t)\,q_p\left(t - \frac{d}{c_0}\right)\frac{\rho_0}{4\pi d}\right\} \tag{8.14.29}$$

With harmonic excitation, it was found in Sections 8.4 and 8.9 that this compromise leads, under optimal conditions, to a control source whose characteristic is zero acoustic power output, and a primary source with a reduced acoustic power output level. As the separation distance approaches $\lambda/2$, this compromise results in an ever decreasing control source volume velocity output, until at a separation distance of $\lambda/2$ the difference between the primary source output at ($\tau + d/c_o$) and ($\tau - d/c_0$) becomes zero; hence the best control output is no control output.

With the impact acceleration noise problem, however, the control source is *not* constrained to provide a control source output that results in a compromise between primary source acoustic power suppression and self-acoustic power suppression during the time periods ($0 < t < d/c_0$) and ($2\pi/\omega - d/c_0 < t$), owing to the finite operating time of the primary source; the control source is able to concentrate solely on initially suppressing the future primary source power output, and is similarly free to absorb some of the acoustic energy of trailing part of the pulse. Therefore, the improved performance is a direct result of the finite operating period of the primary noise disturbance. This trend manifests itself in different control source schedules for the single sinusoidal pulse and harmonic excitation problems, as illustrated in Figure 8.61.

As expected, the two control source schedules are identical for the time period ($d/c_0 < t < 2\pi/\omega - d/c_0$). However, before and after this period, the volume velocity outputs are different; they are based solely on the primary source output at times ($\tau + d/c_0$) and ($\tau - d/c_0$) respectively. It is in these periods that an increased control effect is facilitated. This is particularly apparent at a separation distance of $\lambda/2$, when the optimum control source output for the harmonic excitation problem is zero, but non-zero for the single sinusoidal pulse case. If, for example, the problem were extended to a pulse consisting of two sinusoidal cycles, the control output would follow the zero output schedule of the harmonic excitation case for the period ($d/c_0 < t < 4\pi/\omega - d/c_0$) and be non-zero outside this period.

One other point to notice in regard to the control schedules of Figure 8.61 is the non-zero starting point of the single pulse schedules. This is clearly a practical impossibility, the result of which will be a slightly reduced level of acoustic power attenuation actually achieved. It should also be noted here that when practically implementing a system, the time delays inherent in the analogue anti-aliasing and reconstruction filters in a digital control system, as well as the digital (sampling) delay, further reduces the ability of the system to output a pulse ‘quickly’ (it takes a finite time for a reference signal to pass through the controller). Therefore, if a preview control system is not used, the practical results will be further deteriorated from the theoretical maximum levels of attenuation. However, it can be concluded that the levels of acoustic power attenuation that can be achieved when controlling a single sinusoidal pulse from an acoustic point source are at least as high, probably slightly higher, than can be obtained when controlling harmonic excitation.

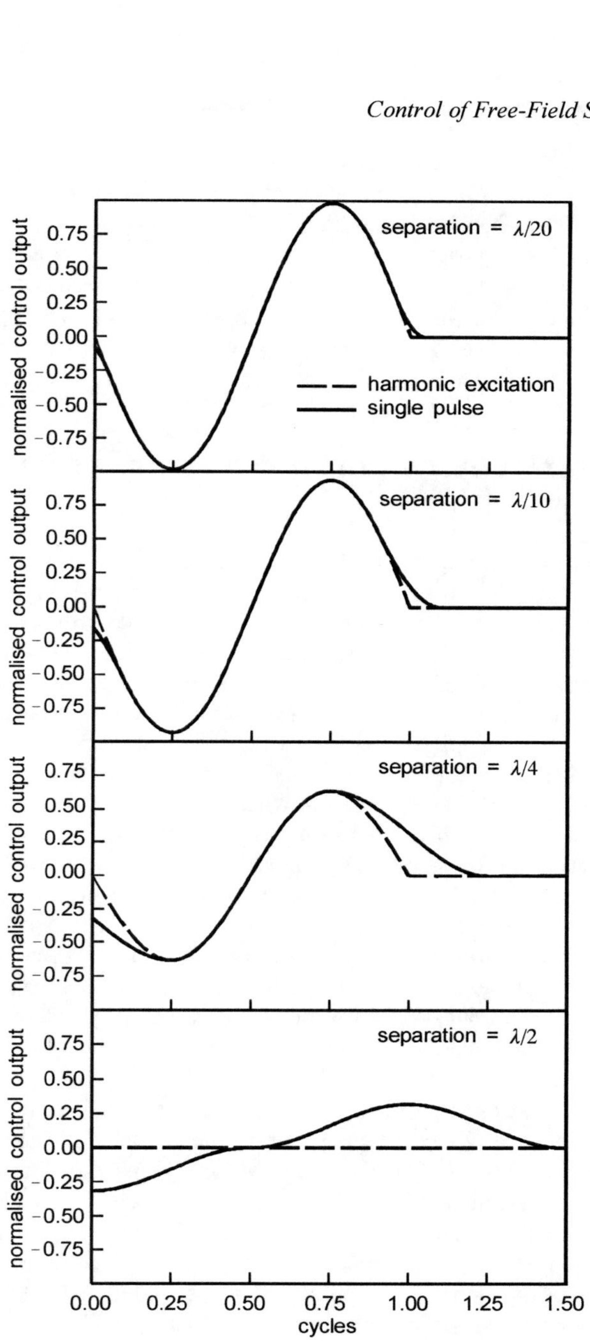

Figure 8.61 Control source operating schedules as a function of source separation distance; solid line is for a single pulse, dashed line is for harmonic excitation.

One additional point to mention is that the above result may not hold when the primary source is essentially a multipole source, and a single control source is used. In these instances, the attenuation of the single pulse may be significantly less than the attenuation achieved during harmonic excitation or the attenuation achieved when the primary source is an impulsive monopole.

Having now seen how much acoustic power attenuation is actually possible for a given control source/primary source arrangement, the next step is to consider the inclusion of an error microphone in the system to make it practically realisable, and to optimise its placement. For comparative purposes, recall that in Section 8.5, it was found that for primary source/control source arrangements that are closely spaced and subjected to harmonic excitation, the nearest optimum error sensor placements will be on a line perpendicular to the control source/primary source axis, and centred at a position between the primary and control sources such that r_c/r_p = sinc *kd*, which will always be closer to the control source.

8.15 FEEDBACK CONTROL OF SOUND RADIATION FROM VIBRATING STRUCTURES

Thus far, this chapter has considered the design of the 'physical' part of active control systems, the idea being to develop models that enable the quantification of how much attenuation of the radiated acoustic field is possible for a given control source/error sensor placement. These models can then be used to optimise sensor and control source placements for a particular problem, and to study the mechanisms underlying the active attenuation. To implement an active control system using the desired control source/error sensor arrangement, it has been assumed that a feedforward control system, such as that outlined in Chapter 6, will be used. There are a range of active control problems where this approach to implementation is not desirable, such as cases where a reference signal is not (readily) available. In these instances, a feedback control strategy can be used.

Design of feedback control systems was reviewed in Chapter 5, and will receive further attention in Chapter 11 where feedback control of flexible structures will be discussed. Many of the design requirements outlined in these chapters are relevant to feedback control of structural/acoustic radiation into free space. In fact, this topic can be viewed as a specific case of feedback control of flexible structures, where the optimal control state weighting matrix is chosen to reflect radiated acoustic power, or acoustic pressure in a particular direction. It may also be advantageous to transform the system states to better reflect the characteristics of structural/acoustic radiation.

In this section, feedback control of structural/acoustic radiation will be briefly considered. Once developed, it will be possible to apply most of the discussion in Chapter 11 to this active control problem with specific state weighting and transformation matrices. Here, the focus will be on systems that explicitly target a reduction in radiated acoustic power, such as considered by Baumann et al. (1991, 1992) and Snyder et al. (1994b). Formulating the problem to explicitly reduce acoustic pressure at certain locations in space requires only minor modifications to the procedure to be outlined; for this problem, the reader is referred to Meirovitch and Thangjitham (1990a, 1990b) and Meirovitch (1993).

8.15.1 Derivation of Structural State Equations

Derivation of the state equations required for constructing a control law aimed at attenuating radiated acoustic power from a vibrating structure will be undertaken in two phases: derivation of the state equations for a damped, distributed parameter system, and modification of these equations to reflect the characteristics of acoustic power output. Derivation of the first of these, the state equations describing the motion of the distributed parameter system, will be discussed more fully in Chapter 11, but will be repeated here in brief for completeness.

The motion of a damped, flexible (distributed parameter) system can be expressed as the partial differential equation:

$$m_s(\boldsymbol{x})\ddot{w}(\boldsymbol{x},t) + c_d\dot{w}(\boldsymbol{x},t) + \kappa w(\boldsymbol{x},t) = u(\boldsymbol{x},t) \tag{8.15.1}$$

where $w(\boldsymbol{x},t)$ is the displacement at location $\boldsymbol{x}$ on the structure at time t in response to the applied force distribution u (having units of force per unit area, or pressure). Here, m_s is the surface density (mass per unit area), κ is a time-invariant, non-negative differential operator, and c_d is defined as:

$$c_d = 2\zeta\kappa^{1/2} \tag{8.15.2}$$

where ζ is the (non-negative) critical damping ratio of the system. The eigenvalue problem associated with this differential equation is:

$$\kappa\psi_n = \lambda_n m \psi_n \tag{8.15.3}$$

where λ_n is the nth eigenvalue and ψ_n is the associated eigenfunction (mode shape function). The eigenvalues are related to the undamped natural frequencies by:

$$\lambda_n = \omega_n^2 \tag{8.15.4}$$

The eigenfunctions are orthogonal, satisfying the relation:

$$\int_S m_s(\boldsymbol{x})\psi(\boldsymbol{x})\psi_n(\boldsymbol{x})\,\mathrm{d}\boldsymbol{x} = \delta(m-n) \tag{8.15.5}$$

where $\delta(m-n)$ is a Kronecker delta function.

The displacement at any location $\boldsymbol{x}$ on the structure can be expressed as a sum of modal contributions as follows:

$$w(\boldsymbol{x},t) = \sum_{i=1}^{\infty} x_i(t)\psi_i(\boldsymbol{x}) \tag{8.15.6}$$

where $x_i(t)$ is the amplitude of the ith mode at time t, and $\Psi_i(\boldsymbol{x})$ is the value of the associated mode shape function at location $\boldsymbol{x}$ on the structure. If this modal expansion is substituted into the equation of motion, Equation (8.15.1), which is then multiplied by Ψ_n and integrated over the surface of the structure, the following equation is produced that describes the behaviour of the nth mode:

$$\ddot{x}_n(t) + 2\zeta\omega_n\dot{x}(t) + \omega_n^2 x_n(t) = f_n(t) \tag{8.15.7}$$

where $f_n(t)$ is the nth modal generalised force given by:

$$f_n(t) = \int_S \psi_n(\boldsymbol{x})u(\boldsymbol{x},t) \tag{8.15.8}$$

Equation (8.15.7) can be expressed in state variable form as:

$$\dot{\boldsymbol{x}}_n(t) = \boldsymbol{A}_n \boldsymbol{x}_n(t) + \boldsymbol{f}_n(t) \qquad (8.15.9)$$

where

$$\boldsymbol{x}_n(t) = \begin{bmatrix} x_n(t) \\ \dot{x}_n(t) \end{bmatrix}; \quad \boldsymbol{A}_n = \begin{bmatrix} 0 & 1 \\ -\omega_n^2 & -2\zeta\omega_n \end{bmatrix}; \quad \boldsymbol{\gamma}_n(t) = \begin{bmatrix} 0 \\ f_n(t) \end{bmatrix} \qquad (8.15.10a–c)$$

To calculate control forces which will minimise the radiated acoustic power, it is necessary to divide the modal generalised force into two components, one due to the primary disturbance f_p and one due to the control force input f_c:

$$f_n = f_{n,c} + f_{n,p} \qquad (8.15.11)$$

where n is the mode of interest. For the control input modal generalised force, it is useful to express the force distribution $u(\boldsymbol{x}, t)$ as the product of some force amplitude $f(t)$, and a transfer function $z_f(\boldsymbol{x})$ to the location $\boldsymbol{x}$ (having units of force per unit area per unit force):

$$u(\boldsymbol{x},t) = f(t) z_f(\boldsymbol{x}) \qquad (8.15.12)$$

such that the control modal generalised force is equal to:

$$f_{n,c} = f(t) \int_S \psi_n(\boldsymbol{x}) \, z_f(\boldsymbol{x}) \, \mathrm{d}\boldsymbol{x} = f(t) \psi_{g,n} \qquad (8.15.13a,b)$$

where $\Psi_{g,n}$ is the nth modal generalised force transfer function, given by:

$$\psi_{g,n} = \int_s \psi_n(\boldsymbol{x}) \, z_f(\boldsymbol{x}) \, \mathrm{d}\boldsymbol{x} \qquad (8.15.14)$$

For the simplest control force (a point force), the modal generalised force is described by:

$$f_{c,n} = f(t) \int_S \psi_n(\boldsymbol{x}) \, \delta(\boldsymbol{x} - \boldsymbol{x}_f) \, \mathrm{d}\boldsymbol{x} = f(t) \psi_n(\boldsymbol{x}_f) \qquad (8.15.15)$$

where $\boldsymbol{x}_f$ is the location of the point force input to the structure and $\delta(\boldsymbol{x} - \boldsymbol{x}_f)$ is a Dirac delta function. Therefore, the modal generalised force transfer function is simply equal to the value of the mode shape function at the control force application point. If there are N_c control sources, then the control source modal generalised force can be written as the matrix expression:

$$f_{c,n} = \boldsymbol{B}_n \boldsymbol{u}(t) \qquad (8.15.16)$$

where

$$\boldsymbol{B}_n = \begin{bmatrix} 0 & 0 & \cdots & 0 \\ \psi_{gn,1} & \psi_{gn,2} & \cdots & \psi_{gn,N_c} \end{bmatrix} ; \quad \boldsymbol{u}(t) = \begin{bmatrix} f_1(t) \\ f_2(t) \\ \vdots \\ f_{N_c}(t) \end{bmatrix} \tag{8.15.17a,b}$$

Therefore, the state-space equation of motion for mode n in Equation (8.15.9) can be expressed as:

$$\dot{\boldsymbol{x}}_n(t) = \boldsymbol{A}_n \boldsymbol{x}_n(t) + \boldsymbol{B}_n \boldsymbol{u}(t) + \boldsymbol{f}_{p,n} \tag{8.15.18}$$

Returning to consideration of the entire system, if there are N_m structural modes being modelled, the system equations of motion of the modelled modes can be written in state variable form as:

$$\dot{\boldsymbol{x}}(t) = \boldsymbol{A}\boldsymbol{x}(t) + \boldsymbol{B}\boldsymbol{u}(t) + \boldsymbol{f}_p(t) \tag{8.15.19}$$

where

$$\boldsymbol{x}(t) = \begin{bmatrix} x_1(t) \\ x_2(t) \\ \vdots \\ x_{N_m}(t) \end{bmatrix} ; \quad \boldsymbol{A} = \begin{bmatrix} \boldsymbol{A}_1 & & & \\ & \boldsymbol{A}_2 & & \\ & & \ddots & \\ & & & \boldsymbol{A}_{N_m} \end{bmatrix} ; \quad \boldsymbol{B} = \begin{bmatrix} \boldsymbol{B}_1 \\ \boldsymbol{B}_2 \\ \vdots \\ \boldsymbol{B}_{N_m} \end{bmatrix} ; \quad \boldsymbol{f}_p(t) = \begin{bmatrix} f_{p,1}(t) \\ f_{p,2}(t) \\ \vdots \\ f_{p,N_m}(t) \end{bmatrix}$$

(8.15.20a–d)

Note that $\boldsymbol{A}$ is not truly a diagonal matrix, but rather a diagonal arrangement of modal matrices $\boldsymbol{A}_i$, defined in Equation (8.15.10).

The state equation describing the system response, Equation (8.15.18), is the same equation that will be encountered in Chapter 11 when considering the control of vibrating structures.

8.15.2 Modification for Acoustic Radiation

The modal equation of motion of Equation (8.15.19) can easily be used to derive control laws aimed at attenuating the velocity levels of the structure. However, as has been seen already in this chapter, this is not necessarily the best approach to take when considering the minimisation of acoustic radiation. To restate the control problem in a form more amenable to calculation of a control law for the problem of acoustic radiation, it is necessary to first consider the characteristics of the radiation of acoustic power from the vibrating structure. Acoustic power output for harmonic excitation can be calculated by integrating the real part of the acoustic intensity I over the surface of the vibrating structure as follows:

$$W = \int_S \mathrm{Re}\{I(\boldsymbol{x})\}\,\mathrm{d}\boldsymbol{x} = \int_S \frac{1}{2}\mathrm{Re}\{\bar{p}(\boldsymbol{x})\bar{u}^*(\boldsymbol{x})\}\,\mathrm{d}\boldsymbol{x} = \int_S \frac{1}{2}A_p(\boldsymbol{x})\,\boldsymbol{A}_u(\boldsymbol{x})\cos\theta_r\,\mathrm{d}\boldsymbol{x} \qquad (8.15.21\text{a–c})$$

where A_p and $\boldsymbol{A}_u$ are real amplitudes of the acoustic pressure and particle velocity respectively, and where $\bar{u}(\boldsymbol{x}) = \bar{\dot{w}}(\boldsymbol{x})$ is the complex surface velocity amplitude at a location $\boldsymbol{x}$ on the structure, $p(\boldsymbol{x})$ is the complex acoustic pressure at this same location, and S denotes that integration is over the surface of the vibrating structure. The bar above a quantity denotes its complex amplitude. For the time domain formulation being undertaken here, however, it is more appropriate to consider the integration of instantaneous acoustic intensity I_i defined as:

$$I_i(\boldsymbol{x},t) = \mathrm{Re}\{p(\boldsymbol{x},t)\,\dot{w}(\boldsymbol{x},t)\} \qquad (8.15.22)$$

Note that acoustic intensity I is equal to the temporal average of the instantaneous acoustic intensity:

$$I(\boldsymbol{x}) = \frac{1}{T}\int_0^T I_i(\boldsymbol{x},t)\,\mathrm{d}t \qquad (8.15.23)$$

From this relationship it is clear that minimisation of the spatial integral (over the surface of the vibrating structure) of instantaneous acoustic intensity at all points in time t will have the effect of minimising radiated acoustic power. Therefore, the acoustic power-related error criterion J_w to be used here is the spatial integral:

$$J_w(t) = \int_S \mathrm{Re}\{p(\boldsymbol{x},t)\,\dot{w}(\boldsymbol{x},t)\}\,\mathrm{d}\boldsymbol{x} = \int_S [\mathrm{Re}\{p(\boldsymbol{x},t)\}\,\mathrm{Re}\{\dot{w}(\boldsymbol{x},t)\}]\,\mathrm{d}\boldsymbol{x} \qquad (8.15.24\text{a,b})$$

Equation (8.15.24) is a direct result of multiplying two vector quantities together and is explained in Section 2.5.3 above Equation (2.5.87).

To simplify the following discussion, consideration will be limited here to infinitely baffled, planar structures. The methodology used in deriving the expressions, however, can be adapted to other geometries.

If the analysis is restricted to planar structures for simplicity, the acoustic pressure $p(\boldsymbol{x}, t)$ can be expressed in terms of the structural surface velocity using the Rayleigh integral as follows:

$$p(\boldsymbol{x},t) = \frac{\mathrm{j}\omega\rho_0}{2\pi}\int_S \dot{w}(\boldsymbol{x}',t)\,\frac{\mathrm{e}^{-\mathrm{j}k|\boldsymbol{x}-\boldsymbol{x}'|}}{|\boldsymbol{x}-\boldsymbol{x}'|}\,\mathrm{d}\boldsymbol{x}' = \frac{\mathrm{j}\omega\rho_0}{2\pi}\int_S \dot{w}(\boldsymbol{x}',t)\,\frac{\mathrm{e}^{-\mathrm{j}kr}}{r}\,\mathrm{d}\boldsymbol{x}' \qquad (8.15.25\text{a,b})$$

where k is the acoustic wavenumber and r is the distance between the source location $\boldsymbol{x}'$ and the receiver location $\boldsymbol{x}$:

$$r = |\boldsymbol{x}-\boldsymbol{x}'| = \sqrt{(x-x')^2 + (y-y')^2} \qquad (8.15.26\text{a,b})$$

Substituting Equation (8.15.25) into Equation (8.15.24) produces the following integral expression for the acoustic power error criterion:

$$J_w(t) = \frac{\rho_0\omega}{2\pi} \int_S \int_S \dot{w}(\boldsymbol{x}',t) \frac{\sin kr}{r} \dot{w}(\boldsymbol{x},t) \, \mathrm{d}\boldsymbol{x}' \mathrm{d}\boldsymbol{x} \tag{8.15.27}$$

Expanding the velocity in terms of modal contributions enables Equation (8.15.27) to be written as:

$$J_w(t) = \frac{\rho_0\omega}{2\pi} \int_S \int_S \left(\sum_{i=1}^{N_m} \dot{x}_i(t) \psi_i(\boldsymbol{x}') \right) \frac{\sin kr}{r} \left(\sum_{l=1}^{N_m} \dot{x}_l(t) \psi_l(\boldsymbol{x}) \right) \mathrm{d}\boldsymbol{x}' \mathrm{d}\boldsymbol{x} \tag{8.15.28}$$

This notation can be simplified to the matrix expression:

$$J_w(t) = \boldsymbol{v}(t)^{\mathrm{T}} \boldsymbol{Z}_w \boldsymbol{v}(t) \tag{8.15.29}$$

where $\boldsymbol{v}(t)$ is the $(N_m \times 1)$ vector of modal velocity amplitudes:

$$\boldsymbol{v}(t) = \begin{bmatrix} \dot{x}_1(t) \\ \dot{x}_2(t) \\ \vdots \\ \dot{x}_{N_m}(t) \end{bmatrix} \tag{8.15.30}$$

and $\boldsymbol{Z}_w$ is a 'power transfer matrix', which has been discussed previously in this chapter, the terms of which are defined by:

$$Z_w(i,l) = \frac{\rho_0\omega}{2\pi} \int_S \int_S \psi_i(\boldsymbol{x}') \frac{\sin kr}{r} \psi_l(\boldsymbol{x})^{\mathrm{T}} \, \mathrm{d}\boldsymbol{x}' \mathrm{d}\boldsymbol{x} \tag{8.15.31}$$

One point to note here is that the same analysis can be undertaken using modal displacements in place of modal velocities, the end result having the same form as Equation (8.15.29) with the displacement power transfer matrix related to the velocity power transfer matrix by:

$$\boldsymbol{Z}_{w,\,\text{displacement}} = \omega^2 \boldsymbol{Z}_{w,\,\text{velocity}} \tag{8.15.32}$$

As observed in Section 8.11 in the discussion of the differences between the control of structural vibration, and the resultant acoustic radiation, the power transfer matrix $\boldsymbol{Z}_w$ will not, in general, be diagonal, as structural modes do not radiate independently of each other. It may therefore not be appropriate to design a control law that aims to explicitly minimise modal amplitudes. This conclusion covers both unweighted and weighted summations, such as where the structural modes with the greatest radiation efficiency are preferentially treated.

The relationship in Equation (8.15.29), as well as the similar relationship relating displacement to acoustic radiation through Equation (8.15.32), will enable an optimal control problem to be formulated where the error criterion is explicitly a measure of radiated acoustic power. It should be noted that it is possible to develop the same form of equations to describe acoustic radiation to particular points in space, and to develop control laws to minimise this measure of acoustic radiation (Meirovitch and Thangjitham, 1990a, 1990b; Meirovitch, 1993). This problem is a straightforward modification of the one above and will not be explicitly studied here.

8.15.3 Problem Statement in Terms of Transformed Modes

To aid in the formulation of the control law, it will be useful to decouple Equation (8.15.29) so that the acoustic power error criterion can be expressed as the (weighted) sum of squared contributions. To do this, the approach outlined in Section 8.11 is adopted, noting that the power transfer matrix $\boldsymbol{Z}_w$ is real symmetric and so can be diagonalised by the orthonormal transformation:

$$\boldsymbol{Z}_w = \boldsymbol{T}\boldsymbol{\Lambda}_w\boldsymbol{T}^{-1} \tag{8.15.33}$$

where $\boldsymbol{T}$ is the orthonormal transform matrix, the columns of which are the eigenvectors of the power transfer matrix $\boldsymbol{Z}_w$, and $\boldsymbol{\Lambda}_w$ is the diagonal matrix of associated eigenvalues, where the subscript w is used to denote eigenvalues associated with the acoustic power problem as opposed to eigenvalues associated with the vibration problem. Note that the transformation in Equation (8.15.33) will produce the same eigenvectors for both the displacement and velocity formulation of the power transfer matrix, but different eigenvalues related by:

$$\boldsymbol{\Lambda}_{w,\,\text{displacement}} = \omega^2\boldsymbol{\Lambda}_{w,\,\text{velocity}} \tag{8.15.34}$$

The form of $\boldsymbol{Z}_w$ dictates that the eigenvectors in $\boldsymbol{T}$ will be orthogonal. If the transformation of Equation (8.15.33) is substituted into Equation (8.15.29), the acoustic power error criterion expression becomes:

$$J_w = \boldsymbol{v}(t)^{\mathrm{T}}\boldsymbol{Z}_w\boldsymbol{v}(t) = \boldsymbol{v}(t)^{\mathrm{T}}\boldsymbol{T}\boldsymbol{\Lambda}_w\boldsymbol{T}^{-1}\boldsymbol{v}(t) = \boldsymbol{v}'(t)^{\mathrm{T}}\boldsymbol{\Lambda}_w\boldsymbol{v}'(t) \tag{8.15.35a–c}$$

where the vector of transformed modal velocities, $\boldsymbol{v}'$, is defined by the expression:

$$\boldsymbol{v}'(t) = \boldsymbol{T}^{-1}\boldsymbol{v}(t) \tag{8.15.36}$$

and use has been made of the property of the orthonormal transform matrix:

$$\boldsymbol{T}^{-1} = \boldsymbol{T}^{\mathrm{T}} \tag{8.15.37}$$

Observe that the orthonormal transformation has decoupled the global error criterion expression, so that it is now equal to a weighted summation of the form:

$$J_w(t) = \sum_{i=1}^{N_m}\lambda_{wi}\left(\dot{x}_i'(t)\right)^2 \tag{8.15.38}$$

where λ_{wi} is the ith eigenvalue of $\boldsymbol{Z}_w$, $v_i'(t)$ is the ith transformed modal velocity of the vector $\boldsymbol{v}'(t)$, and the summation is over the N_m structural modes being modelled. It is also straightforward to show that a similar relationship can be developed using transformed modal displacements, with the related error criterion defined by the expression:

$$J_{w,\text{displacement}}(t) = \sum_{i=1}^{N_m} \omega^2 \lambda_{wi} \left(x_i'(t)\right)^2 \tag{8.15.39}$$

with the transformed modal displacement related to modal displacement in the same way that the transformed modal velocity is related to modal velocity, as defined in Equation (8.15.36).

The orthonormal transformation matrix of Equation (8.15.33) can be used to cast the state-space modal equations of motion, Equation (8.15.19), into a form that will facilitate calculation of the control law for minimising the radiated acoustic power. The only problem is that the dimension of $\boldsymbol{x}$ is $(2N_m \times 1)$, as opposed the $(N_m \times 1)$ dimension of the orthonormal transformation matrix $\boldsymbol{T}$. Observe, however, that an expanded orthonormal transformation matrix $\boldsymbol{T}_e$ can be constructed as:

$$\boldsymbol{T}_e = \begin{bmatrix} T(1,1) & 0 & T(1,2) & 0 & T(1,3) & 0 & \cdots \\ 0 & T(1,1) & 0 & T(1,2) & 0 & T(1,3) & \cdots \\ T(2,1) & 0 & T(2,2) & 0 & T(2,3) & 0 & \cdots \\ 0 & T(2,1) & 0 & T(2,2) & 0 & T(2,3) & \cdots \\ T(3,1) & 0 & T(3,2) & 0 & T(3,3) & 0 & \cdots \\ 0 & T(3,1) & 0 & T(3,2) & 0 & T(3,3) & \cdots \\ & & & \vdots & & & \end{bmatrix} \tag{8.15.40}$$

which will still have the orthonormal property, $\boldsymbol{T}_e^{-1} = \boldsymbol{T}_e^{\mathrm{T}}$. This expanded orthonormal matrix can be applied to Equation (8.15.19) as:

$$\boldsymbol{T}_e^{-1}\dot{\boldsymbol{x}}(t) = \boldsymbol{T}_e^{-1}\boldsymbol{A}\,\boldsymbol{T}_e\boldsymbol{T}_e^{-1}\boldsymbol{x}(t) + \boldsymbol{T}_e^{-1}\boldsymbol{B}\boldsymbol{u}(t) + \boldsymbol{T}_e^{-1}\boldsymbol{f}_p(t) \tag{8.15.41}$$

or

$$\dot{\boldsymbol{x}}'(t) = \boldsymbol{A}'\boldsymbol{x}'(t) + \boldsymbol{B}'\boldsymbol{u}(t) + \boldsymbol{f}_p'(t) \tag{8.15.42}$$

where the transformed quantities, denoted by $'$, are defined by the relationships:

$$\boldsymbol{x}'(t) = \boldsymbol{T}_e^{-1}\boldsymbol{x}(t);\;\; \boldsymbol{A}' = \boldsymbol{T}_e^{-1}\boldsymbol{A}\boldsymbol{T}_e;\;\; \boldsymbol{B}' = \boldsymbol{T}_e^{-1}\boldsymbol{B};\;\; \boldsymbol{f}_p' = \boldsymbol{T}_e^{-1}\boldsymbol{f}_p \tag{8.15.43a–d}$$

Observe now that with transformed modal velocities and displacements as the system states, the acoustic power output of the system will be minimised by minimising the error criterion:

$$J_w = \boldsymbol{x}'^{\mathrm{T}}(t)\boldsymbol{Q}_w(\omega)\boldsymbol{x}'(t) \tag{8.15.44}$$

where $\boldsymbol{Q}_w(\omega)$ is a $(2N_m \times 2N_m)$ diagonal matrix, the terms of which are defined by:

$$Q_w(\omega) = \begin{bmatrix} \omega^2\lambda_1 & 0 & 0 & 0 & \cdots \\ 0 & \lambda_1 & 0 & 0 & \cdots \\ 0 & 0 & \omega^2\lambda_2 & 0 & \cdots \\ 0 & 0 & 0 & \lambda_2 & \cdots \\ & & \vdots & & \end{bmatrix} \tag{8.15.45}$$

The problem of suppressing radiated acoustic power from the vibrating structure can now be expressed as the optimal control problem:

$$J = \int_0^\infty \left(x'^{\mathrm{T}}(t)\, Q_w(\omega)\, x'(t) + u^{\mathrm{T}}(t) R u(t) \right) \mathrm{d}t \tag{8.15.46}$$

Note that the weighting matrix $Q_w(\omega)$ in the optimal control error criterion is frequency dependent.

As was shown in the previous section, the state transformations of this section are not strictly required in order to formulate the optimal control problem in terms of minimising radiated acoustic power. From Equations (8.15.29) and (8.15.32) it is easily shown that the optimal control error criterion could be re-expressed in terms of untransformed states, structural modal displacements and velocities, as:

$$J = \int_0^\infty \left(x^{\mathrm{T}}(t) Q(\omega) x(t) + u^{\mathrm{T}}(t) R u(t) \right) \mathrm{d}t \tag{8.15.47}$$

where $Q(\omega)$ is defined as:

$$Q = \begin{bmatrix} \omega^2 Z_w(1,1) & 0 & \omega^2 Z_w(1,2) & 0 & \omega^2 Z_w(1,3) & 0 & \cdots \\ 0 & Z_w(1,1) & 0 & Z_w(1,2) & 0 & Z_w(1,3) & \cdots \\ \omega^2 Z_w(2,1) & 0 & \omega^2 Z_w(2,2) & 0 & \omega^2 Z_w(2,3) & 0 & \cdots \\ 0 & Z_w(2,1) & 0 & Z_w(2,2) & 0 & Z_w(2,3) & \cdots \\ \omega^2 Z_w(3,1) & 0 & \omega^2 Z_w(3,2) & 0 & \omega^2 Z_w(3,3) & 0 & \cdots \\ 0 & Z_w(3,1) & 0 & Z_w(3,2) & 0 & Z_w(3,3) & \cdots \\ & & & \vdots & & & \end{bmatrix} \tag{8.15.48}$$

The question then arises as to what advantage there is in expressing the problem in terms of transformed states. In addition to the computational benefits that arise from decoupling the problem, the chief advantage is a simplified means of model reduction. As discussed in Section 8.11, at low frequencies, only a few transformed modes need to be viewed as 'important', enabling a significantly reduced order model of the problem to be used. The transformed modes are described by orthogonal combinations of structural modes, and could therefore be measured using a set of modal filters (Meirovitch and Baruh, 1983; Morgan,

1992; see also Chapter 11) to resolve the structural modes, with these values combined in accordance with the eigenvectors of the power transfer matrix $\boldsymbol{Z}_w$ to produce the transformed modes. This approach only partially takes advantage of the benefit of formulating the problem in terms of transformed modes, as while the number of transformed modes required to model the problem is low, the number of structural modes required to accurately construct each transformed mode is somewhat higher. If custom sensors, such as piezoelectric polymer films, were used to measure the transformed modes, however, only a few measurements would be required. Experimental implementation of this approach can be found in Snyder et al. (1994b).

Finally, as has been noted, the state weighting matrix that is required for the optimal control problem to explicitly consider acoustic power minimisation, is frequency dependent. Perhaps the most straightforward way to account for this frequency dependence is to augment the state vector with additional variables whose response is related to the frequency dependence of the error criterion. This is described in more detail by Baumann et al. (1991, 1992). A general discussion of the problem of frequency-weighted error criteria in optimal control problems is discussed by Gupta (1980), Anderson and Mingori (1985), Moore and Mingori (1987), and Anderson and Moore (1990).

REFERENCES

Akay, A. (1978). A review of impact noise. *Journal of the Acoustical Society of America*, **64**, 977–987.

Anderson, B.D.O. and Mingori, D.L. (1985). Use of frequency dependence in linear quadratic control problems to frequency shape robustness. *Journal of Guidance, Control, and Dynamics*, **8**, 397–401.

Anderson, B.D.O. and Moore, J.B. (1990). *Optimal Control, Linear Quadratic Methods*. Prentice Hall, Englewood Cliffs, NJ.

Angevine, O.L. (1981). Active acoustic attenuation of electric transformer noise. *Proceedings of Inter-Noise '81*, 303–306.

Angevine, O.L. (1992). Active cancellation of the hum of large electric transformers. In *Proceedings of Inter-Noise '92*, 313–316.

Angevine, O.L. (1993). Active control of hum from large power transformers: the real world. In *Proceedings of the Second conference on Recent Advances in Active Control of Sound and Vibration*, Virginia Polytechnic Institute and State University, Blacksburg, VA, 279–290.

Angevine, O.L. (1995). The prediction of transformer noise. *Sound and Vibration*, October, 16–18.

Angevine, O.L. (1995). Active systems for attenuation of noise. *International Journal of Active Control*, **1**, 65–78.

Baumann, W.T., Saunders, W.R. and Robertshaw, H.H. (1991). Active suppression of acoustic radiation from impulsively excited structures. *Journal of the Acoustical Society of America*, **90**, 3202–3208.

Baumann, W.T., Ho, F.S. and Robertshaw, H.H. (1992). Active structural acoustic control of broadband disturbances. *Journal of the Acoustical Society of America*, **92**, 1998–2005.

Bendat, J.S. and Piersol, A.G. (1986). *Random Data*. John Wiley & Sons, New York.

Beranek, L.L. (1986). *Acoustics*. Acoustical Society of America, New York.

Berge, T., Pettersen, O.K.O. and Sorsdal, S. (1987). Active noise cancellation of transformer noise. In *Proceedings of Inter-Noise '87*, 537–540.

Berge, T., Pettersen, O.K.O. and Sorsdal, S. (1988). Active cancellation of transformer noise: field measurements. *Applied Acoustics*, **23**, 309–320.

Berkhoff, A. P. (2000). Sensor scheme design for active structural acoustic control. *Journal of the Acoustical Society of America*, **108**, 1037–1045.

Berkhoff, A. P. (2001). Piezoelectric sensor configuration for active structural acoustic control, *Journal of Sound and Vibration*, **246**, 175–183.

Berkhoff, A. P., Sarajlic, E, Cazzolato, B.S. and Hansen, C.H. (2001). Inverse and reciprocity methods for experimental determination of radiation modes. In *Proceedings of ICSV8,* Hong Kong, July, 1629–1637.

Berkhoff, A. P. (2002). Broadband radiation modes: estimation and active control. *Journal of the Acoustical Society of America*, **111**, 1295–1305.

Berry, A. (2001). Advanced sensing strategies for the active control of vibration and structural radiation. *Noise Control Engineering Journal*, **49**, 54–65.

Borgiotti, G.V. (1990). The power radiated by a vibrating body in an acoustic fluid and its determination from boundary measurements. *Journal of the Acoustical Society of America*, **88**, 1884–1893.

Borgiotti, G.V. and Jones, K.E. (1994). Frequency independence property of radiation spatial filters. *Journal of the Acoustical Society of America*, **96**, 3516–3524.

Brungardt, K., Vierengel, J. and Weissman, K. (1997). Active structural acoustic control of the noise from power transformers. In *Proceedings of Noise-Con 97*, pp. 173–182.

Burdisso, R.A. and Fuller, C.R. (1991). Eigenproperties of feedforward controlled flexible structures. *Journal of Intelligent Material Systems and Structures*, **2**, 494–507.

Burdisso, R.A. and Fuller, C.R. (1992). Theory of feedforward controlled system eigenproperties. *Journal of Sound and Vibration*, **153**, 437–451 (also, comments and reply, *Journal of Sound and Vibration*, **163**, 363–371).

Burgan, N.C., Snyder, S.D., Tanaka, N. and Zander, A.C. (2002a). A generalised approach to modal filtering for active noise control. Part I. Vibration sensing. *IEEE Sensors*, **2**, 577–589.

Burgemeister, K.A. and Hansen, C.H. (1993). Use of a secondary perforated panel to actively control the sound radiated by a heavy structure. In *Proceedings of ASME, Winter Annual Meeting*, 93-WA/NCA-2, New Orleans, LA.

Cazzolato, B.S. and Hansen, C.H. (1998). Active control of sound transmission using structural error sensing. *Journal of the Acoustical Society of America*, **104**, 2878–2889.

Cazzolato, B.S. and Hansen, C.H. (1999). Structural radiation mode sensing for active control of sound radiation into enclosed spaces. *Journal of the Acoustical Society of America*, **106**, 3732–3735.

Clark, R.L. and Fuller, C.R. (1991). Control of sound radiation with adaptive structures. *Journal of Intelligent Material Systems and Structures*, **2**, 431–452.

Clark, R.L., Fuller, C.R. and Wicks, A.L. (1991). Characterization of multiple piezoceramic actuators for structural excitation. *Journal of the Acoustical Society of America*, **90**, 346–357.

Clark, R.L. and Fuller, C.R. (1992a). Active structural acoustic control with adaptive structures including wavenumber considerations. *Journal of Intelligent Material Systems and Structures*, **3**, 296–315.

Clark, R.L. and Fuller, C.R. (1992b). Modal sensing of efficient acoustic radiators with PVDF distributed sensors in active structural acoustic approaches. *Journal of the Acoustical Society of America*, **91**, 3321–3329.

Clark. R.L. and Fuller, C.R. (1992c). Experiments on active control of structurally radiated sound using multiple piezoceramic actuators. *Journal of the Acoustical Society of America*, **91**, 3313–3320.

Clark. R.L. and Fuller, C.R. (1992d). A model reference approach for implementing active structural acoustic control. *Journal of the Acoustical Society of America*, **92**, 1534–1544.

Clark. R.L., Gibbs, G.P. and Fuller, C.R. (1993). An experimental study for implementing model reference active structural acoustic control. *Journal of the Acoustical Society of America*, **93**, 3258–3264.

Collins, S.A., Padilla, C.E., Notestine, R.J., von Flotow, A.H., Schmitz, E. and Ramey, M. (1992). Design, manufacture, and application to space robots of distributed piezoelectric film sensors. *Journal of Guidance, Control, and Dynamics*, **15**, 396–403.

Conover, W.B. (1956). Fighting noise with noise. *Noise Control*, **2**, 78–82.

Cunefare, K.A. (1991). The minimum multi-modal radiation efficiency of baffled finite beams. *Journal of the Acoustical Society of America*, **90**, 2521–2529.

Cunefare, K.A. and Shepard, S. (1993). The active control of point acoustic sources in a half-space. *Journal of the Acoustical Society of America*, **93**, 2732–2739.

Cunefare, K.A., Currey, M. N., Johnson, M. E., Elliott, S. J. (2001). The radiation efficiency grouping of free-space acoustic radiation modes. *Journal of the Acoustical Society of America*, **109**, 203–215.

Davies, H.G. (1971). Low frequency random excitation of water-loaded rectangular plates. *Journal of Sound and Vibration*, **15**, 107–126.

Deffayet, C. and Nelson, P.A. (1988). Active control of low-frequency harmonic sound radiated by a finite panel. *Journal of the Acoustical Society of America*, **84**, 2192–2199.

Elliott, S.J. and Johnson, M.E. (1992). *A Note on the Minimisation of Radiated Sound Power*. ISVR Tecnical Note 1992.

Elliott, S.J. and Johnson, M.E. (1993). Radiation modes and the active control of sound power. *Journal of the Acoustical Society of America*, **94**, 2194–2204.

Elliott, S.J. and Nelson, P.A. (1986). The implications of causality in active control. In *Proceedings of Inter-Noise '86*, 583–588.

Elliott, S.J., Joseph, P., Nelson, P.A. and Johnson, M.E. (1991). Power output minimisation and power absorption in the active control of sound. *Journal of the Acoustical Society of America*, **90**, 2501–2512.

Fahy, F. (1985). *Sound and Structural Vibration: Radiation, Transmission, and Response*. Academic Press, London.

Ffowcs Williams, J.E. (1984). Anti-sound. *Proceedings of the Royal Society of London, Series A*, **395**, 63–88.

Fuller, C.R. (1988). Analysis of active control of sound radiation from elastic plates by force inputs. In *Proceedings of Inter-Noise '88*, 1061–1064.

Fuller, C.R. (1990). Active control of sound transmission/radiation from elastic plates by vibration inputs. I. Analysis. *Journal of Sound and Vibration*, **136**, 1–15.

Fuller, C.R. and Burdisso, R.A. (1991). A wavenumber domain approach to the active control of sound and vibration. *Journal of Sound and Vibration*, **148**, 355–360.

Fuller, C.R., Hansen, C.H. and Snyder, S.D. (1989). Active control of structurally radiated noise using piezoceramic actuators. In *Proceedings of Inter-Noise '89*, 509–511.

Fuller, C.R., Hansen, C.H. and Snyder, S.D. (1991a). Experiments on active control of sound radiation from a panel using a piezo ceramic actuator. *Journal of Sound and Vibration*, **150**, 179–190.

Fuller, C.R., Hansen, C.H. and Snyder, S.D. (1991b). Active control of sound radiation from a vibrating panel by sound sources and vibration inputs: An experimental comparison. *Journal of Sound and Vibration*, **145**, 195–216.

Gu, Y. and Fuller, C.R. (1991). Active control of sound radiation due to subsonic wave scattering from discontinuities on fluid loaded plates. I. Far-field pressure. *Journal of the Acoustical Society of America*, **90**, 2020–2026.

Gu, Y. and Fuller, C.R. (1993). Active control of sound radiation from a fluid-loaded rectangular uniform plate. *Journal of the Acoustical Society of America*, **93**, 337–345.

Gupta, N.K. (1980). Frequency-shaped loop functionals: Extensions of linear-quadratic-Gaussian design methods. *Journal of Guidance and Control*, **3**, 529–535.

Hansen, C.H. and Snyder, S.D. (1991). Effect of geometric and structural/acoustic variables on the active control of sound radiation from a vibrating surface. In *Proceedings of Recent Advances in Active Control of Sound and Vibration*, 487–506.

Hansen, C.H., Snyder, S.D. and Fuller, C.R. (1989). Reduction of noise radiated by a vibrating rectangular panel by active sound sources and active vibration sources: a comparison. In *Proceedings of Noise and Vibration '89*, Singapore, 16–18 August, E50–E51.

Hesselmann, N. (1978). Investigation of noise reduction on a 100 kVA transformer tank by means of active methods. *Applied Acoustics*, **11**, 27–34.

Hill, S.G. and Snyder, S.D. (2007). Acoustic based modal filtering of orthogonal radiating functions for active noise control. Part I. Theory and simulation. *Mechanical Systems and Signal Processing*, **21**, 1815–1838.

Hill, S.G. Snyder, S.D. and Cazzolato, B.S. (2005). Deriving time-domain models of structural-acoustic radiation into free space, *Mechanical Systems and Signal Processing*, **19**, 1015–1033.

Hill, S.G. Snyder, S.D. and Tanaka, N. (2007a). Practical implementation of an acoustic-based modal filtering sensing technique for active noise control. *Applied Acoustics*, **68**, 1400–1426.

Hill, S.G., Tanaka, N and Snyder, S.D. (2007b). Acoustic based modal filtering of orthogonal radiating functions for active noise control. Part II. Implementation. *Mechanical Systems and Signal Processing*, **21**, 1937–1952.

Hill, S.G., Tanaka, N and Snyder, S.D. (2008). Acoustic based sensing of orthogonal radiating functions for three-dimensional noise sources: background and experiments. *Journal of Sound and Vibration*, **318**, 1050–1076.

Hill, S.G., Snyder, S.D., Cazzolato, B.S., Tanaka, N. and Fukuda, R. (2002). A generalised approach to modal filtering for active noise control. Part II. Acoustic sensing. *IEEE Sensors*, **2**, 590–596.

Junger, M.C. and Feit, D. (1986). *Sound, Structures, and Their Interaction*. MIT Press, Cambridge, MA.

Keltie, R.F. and Peng, H. (1987). Acoustic power radiated from point-forced thin elastic plates. *Journal of Sound and Vibration*, **112**, 45–52.

Kempton, A.J. (1976). The ambiguity of acoustic sources: a possibility for active control? *Journal of Sound and Vibration*, **48**, 475–483.

Knyazev, A.S. and Tartakovskii, B.D. (1967). Abatement of radiation from flexurally vibrating plates by means of active local vibration dampers. *Soviet Physics Acoustics*, **13**, 115–116.

Koshigoe, S. and Murdock, J.W. (1993). A unified analysis of both active and passive damping for a plate with piezoelectric transducers. *Journal of the Acoustical Society of America*, **93**, 346–355.

Koss, L.L. and Alfredson, R.J. (1973). Transient sound radiated by spheres undergoing an elastic collision. *Journal of Sound and Vibration*, **27**, 59–75.

Koss, L.L. (1974). Transient sound from colliding spheres: normalised results. *Journal of Sound and Vibration*, **36**, 541–553.

Laning, J.H. and Battin, R.H. (1956). *Random Processes in Automatic Control*. McGraw-Hill, New York.

Lee, C.K. and Moon, F.C. (1990). Modal sensors/actuators. *Journal of Applied Mechanics*, **57**, 396–403.

Levine, H. (1980a). On source radiation. *Journal of the Acoustical Society of America*, **68**, 1199–1205.

Levine, H. (1980b). A note on sound radiation from distributed sources. *Journal of Sound and Vibration*, **68**, 203–207.

Levine, H. (1980c). Output of acoustical sources. *Journal of the Acoustical Society of America*, **67**, 1935–1946.

Li, X. (2000). *Physical System Design for the Active Control of Electrical Transformer Noise*. Ph.D thesis, University of Adelaide.

Lomas, N.S. and Hayek, S.I. (1977). Vibration and acoustic radiation of elastically supported rectangular plates. *Journal of Sound and Vibration*, **52**, 1–25.

McLoughlin, M., Hildebrand, S. and Hu, Z. (1994). A novel active transformer quietening system. In *Proceedings of Inter-Noise '94*.

Meirovitch, L. and Baruh, H. (1983). On the problem of observation spillover in self-adjoint distributed-parameter systems. *Journal of Optimization Theory and Applications*, **39**, 269–291.

Meirovitch, L. and Thangjitham, S. (1990a). Control of sound radiation from submerged plates. *Journal of the Acoustical Society of America*, **88**, 402–407.

Meirovitch, L. and Thangjitham, S. (1990b). Active control of sound radiation pressure. *Journal of Vibration and Acoustics*, **112**, 237–244.

Metcalf, V.L., Fuller, C.R., Silcox, R.J. and Brown, D.E. (1992). Active control of sound transmission/radiation from elastic plates by vibrational inputs. II. Experiments. *Journal of Sound and Vibration*, **153**, 387–402.

Moore, J.B. and Mingori, D.L. (1987). Robust Frequency-Shaped LQ Control. *Automatica*, **23**, 641–646.

Morgan, D.R. (1991). An adaptive modal-based active control system. *Journal of the Acoustical Society of America*, **89**, 248–256.

Naghshineh, K. and Koopmann, G.H. (1992). A design method for achieving weak radiator structures using active vibration control. *Journal of the Acoustical Society of America*, **92**, 856–870.

Naghshineh, K. and Koopmann, G.H. (1993). Active control of sound power using acoustic basis functions as surface velocity filters. *Journal of the Acoustical Society of America*, **93**, 2740–2752.

Naghshineh, K., Koopmann, G.H. and Belegundu, A.D. (1992). Material tailoring of structures to achieve a minimum radiation condition. *Journal of the Acoustical Society of America*, **92**, 841–855.

Nelson, P.A. and Elliott, S.J. (1986). The minimum power output of a pair of free field monopole sources. *Journal of Sound and Vibration*, **105**, 173–178.

Nelson, P.A. and Elliott, S.J. (1992). *Active Control of Sound*. Academic Press, London.

Nelson, P.A., Curtis, A.R.D. and Elliott, S.J. (1985). Quadratic optimization problems in the active control of free and enclosed sound fields. *Proceedings of the Institute of Acoustics*, **7**, 45–53.

Nelson, P.A., Curtis, A.R.D. and Elliott, S.J. (1986). On the active absorption of sound. In *Proceedings of Inter-Noise '86*, 601–606.

Nelson, P.A., Curtis, A.R.D., Elliott, S.J. and Bullmore, A.J. (1987). The minimum power output of free field point sources and the active control of sound. *Journal of Sound and Vibration*, **116**, 397–414.

Nelson, P.A., Hammond, J.K., Joseph, P. and Elliott, S.J. (1988). *The Calculation of Causally Constrained Optima in the Active Control of Sound*. ISVR Technical Report 147.

Nelson, P.A., Hammond, J.K., Joseph, P. and Elliott, S.J. (1990). Active control of stationary random sound fields. *Journal of the Acoustical Society of America*, **87**, 963–975.

Osman, A., Sadek, M. and Knight, W. (1974). Noise and vibration analysis of an impact forming machine. *Journal of Engineering in Industry*, **February**, 233–240.

Pan, J., Snyder, S.D., Hansen, C.H. and Fuller, C.R. (1992). Active control of far field sound radiated by a rectangular panel: a general analysis. *Journal of the Acoustical Society of America*, **91**, 2056–2066.

Pope, L.D. and Leibowitz, R.C. (1974). Intermodal coupling coefficients for a fluid-loaded rectangular plate. *Journal of the Acoustical Society of America*, **56**, 408–415.

Qiu, X., Li, X., Ai, Y. and Hansen, C.H. (2002). A waveform synthesis algorithm for active control of transformer noise: implementation. *Applied Acoustics*, **63**, 467–479.

Richards, E.J., Westcott, M.E. and Jeyapalan, R.K. (1979). Prediction of impact noise, I. Acceleration noise. *Journal of Sound and Vibration*, **62**, 547–575.

Ross, C.F. (1978). Experiments on the active control of transformer noise. *Journal of Sound and Vibration*, **61**, 473–480.

Sandman, B.E. (1977). Fluid-loaded vibration of an elastic plate carrying a concentrated mass. *Journal of the Acoustical Society of America*, **61**, 1502–1510.

Snyder, S.D. (1991). *A Fundamental Study of Active Noise Control System Design*. PhD thesis, University of Adelaide.

Snyder, S.D and Burgan, N.C. (2002). An acoustic-based modal filtering approach to sensing system design for active control of structural acoustic radiation: theoretical development. *Mechanical Systems and Signal Processing*, **16**, 123–139.

Snyder, S.D and Hansen, C.H. (1991a). Mechanisms of active noise control using vibration sources. *Journal of Sound and Vibration*, **147**, 519–525.

Snyder, S.D. and Hansen, C.H. (1991b). Using multiple regression to optimise active noise control system design. *Journal of Sound and Vibration*, **148**, 537–532.

Snyder, S.D. and Tanaka, N. (1993a). To absorb or not to absorb: control source power output in active noise control systems. *Journal of the Acoustical Society of America*, **94**, 185–195.

Snyder, S.D. and Tanaka, N. (1993b). On feedforward active control of sound and vibration using vibration error signals. *Journal of the Acoustical Society of America*, **94**, 2181–2193.

Snyder, S.D. and Tanaka, N. (1995). Calculating total acoustic power output using modal radiation efficiencies. *Journal of the Acoustical Society of America*, **97**, 1702–1709.

Snyder, S.D., Tanaka, N. and Hansen, C.H. (1993). Shaped vibration sensors for feedforward control of structural radiation. In *Proceedings of Recent Advances in Active Control of Sound and Vibration*, 177–188.

Snyder, S.D., Tanaka, N. and Kikushima, Y. (1996). The use of optimally shaped piezoelectric film sensors in the active control of free field structural radiation. Part 2. Feedback control. *ASME Journal of Vibration and Acoustics*, **118**, 112–121.

Snyder, S.D., Tanaka, N. and Kikushima, Y. (1995). The use of optimally shaped piezoelectric film sensors in the active control of free field structural radiation. Part 1. Feedforward control. *ASME Journal of Vibration and Acoustics*, **117**, 311–322.

Snyder, S.D., Hill, S.G., Burgan, N.C., Tanaka, N. and Cazzolato, B.S. (2004). Acoustic-centric modal filter design for active noise control. *Control Engineering Practice*, **12**, 1055–1064.

Thomas, D.R., Nelson, P.A. and Elliott, S.J. (1990). Experiments on the active control of the transmission of sound through a clamped rectangular plate. *Journal of Sound and Vibration*, **139**, 351–355.

Thi, J., Unver, E. and Zuniga, M. (1991). Comparison of design approaches in sound radiation suppression. In *Proceedings of Recent Advances in Active Control of Sound and Vibration*, Virginia Polytechnic Institute and State University, Blacksburg, VA, 534–551.

Tokhi, M.O. and Leitch, R.R. (1991a). Design of active noise control systems operating in three-dimensional non-dispersive propagation medium. *Noise Control Engineering Journal*, **36**, 41–53.

Tokhi, M.O. and Leitch, R.R. (1991b). The robust design of active noise control systems based on relative stability measures. *Journal of the Acoustical Society of America*, **90**, 334–345.

Walker, L.A. (1976). Characteristics of an active feedback system for the control of plate vibrations. *Journal of Sound and Vibration*, **46**, 157–176.

Wallace, C.F. (1972). Radiation resistance of a rectangular plate. *Journal of the Acoustical Society of America*, **51**, 946–952.

Wang, B.T. and Fuller, C.R. (1992). Near-field pressure, intensity, and wave-number distributions for active structural acoustic control of plate radiation: theoretical analysis. *Journal of the Acoustical Society of America*, **92**, 1489–1498.

Wang, B.T. Dimitradis, E.K., and Fuller, C.R. (1990). Active control of structurally radiated noise using multiple piezo electric actuators. In *Proceedings of AIAA SDM Conference AIAA paper* **90-1172-CP**, 2409–2416.

Williams, E.G. (1983). A series expansion of the acoustic power radiated from planar sources. *Journal of the Acoustical Society of America*, **73**, 1520–1524.

CHAPTER 9

Active Control of Enclosed Sound Fields

9.1 INTRODUCTION

In this chapter a third 'general group' of active noise control problems will be examined: that of controlling sound fields in enclosed spaces. This group of problems includes many commonly encountered systems, such as automobile interiors, aircraft cabins, ship cabins and rooms in houses. Initial consideration of active control of enclosed sound fields was undertaken in the 1950s, with the aim being to achieve both global and local sound attenuation using an 'electronic sound absorber' (Olson and May, 1953; Olson, 1956). This device was basically a single speaker and microphone placed in close proximity of each other, with a feedback control system that could drive the speaker to either reduce the pressure at the microphone or absorb part of the incident acoustic field. Little became of this device at its time of inception, possibly due to limitations in electronics technology.

Over the next 30 years there appears to have been little work done on the active attenuation of enclosed sound fields. There was, however, research conducted on the related area of 'assisted resonance' in concert halls (Parkin and Morgan, 1965, 1971). This is basically active noise control where the aim is not to null the sound field, but rather to modify the reverberation characteristics of the enclosed space. Once again, this technology does not seem to have found widespread use.

Since the early 1980s, interest in the active control of enclosed sound fields has increased dramatically, owing in part to parallel advances in micro-processor technology that makes possible the implementation of such systems. Passenger vehicles in particular have become targets of active noise control research. Advances in engineering materials have led to an increase in the strength to weight ratio, particularly in aircraft and automobiles. It has also led to the development of more fuel efficient, yet sometimes louder (in the case of modern turbo-prop engines (Magliozzi, 1984)), propulsion systems. Passengers do not, however, expect modern advances to result in any decrease in creature comfort; on the contrary, they expect to be pampered with ever-increasing comfort. This has led to a problem for structural acousticians, as their old ally, mass, is being eliminated before their very eyes. Active noise control is viewed by many as a possible means to control low-frequency noise without substantially increasing the mass of these sleek, modern carriers. There have been numerous experimental studies on implementations in aircraft (Zalas and Tichy, 1984; Dorling et al., 1989; Elliott et al., 1989a, 1990; Salikuddin and Ahuja, 1989; Simpson et al., 1989, 1991, Ross, 1999, Hinchliffe et al., 2002, Gorman et al., 2004), automobiles (Oswald, 1984; Elliott et al., 1988a, Park et al., 2002, 2004), and in tractor cabins (Nadim and Smith, 1983, Faber and Sommerfeldt, 2004).

Simply introducing a few transducers connected to an electronic controller into an enclosed sound field will not, however, guarantee satisfactory results, even if it is physically possible to substantially attenuate a given enclosed sound field. The work presented in this section is aimed at developing methods of analysis that will assist in optimally implementing active control systems for this class of problem. The following sections concentrate on developing methods for calculating the maximum levels of attenuation that are possible for a given control source arrangement, predicting the levels of attenuation that will be achieved by minimising the disturbance at discrete error sensing locations, examining the mechanisms by which active control provides attenuation, and developing a qualitative feeling for how effective active noise control will be for a given problem.

The first part of the chapter will concentrate on feedforward control of low-frequency harmonic excitation, first for sound sources in rigid enclosures and then for sound transmission into coupled enclosures. As in the previous section, the analysis will be directed at obtaining a set of governing equations in the form of matrix expressions, which can be manipulated simply, to provide optimum control source outputs for a variety of error criteria using quadratic optimisation techniques (see, for example, Piraux and Nayroles, 1980; Nelson et al., 1985, 1987). These models will then be used to examine control mechanisms and the effect of source/sensor arrangements for feedforward control implementations. Control at higher frequencies will then be considered, where the modal density in the enclosed space is high. Finally, there will be a limited discussion relating to the application of feedback control to the problem.

9.2 CONTROL OF HARMONIC SOUND FIELDS IN RIGID ENCLOSURES AT DISCRETE LOCATIONS

A good starting point for study of the control of enclosed sound fields is the simplest problem, the active control of sound fields in rigid walled enclosures at discrete locations. By specifying 'rigid' walls, all of the enclosure boundaries are assumed to be locally reacting, meaning that wave motion in the enclosing structure is not possible and therefore the enclosed sound field is entirely the result of sound sources contained within it or on its boundaries. (It should be noted that even in very thick walled enclosures such as reverberation rooms, this assumption may not be strictly valid, especially at low frequencies (Pan and Bies, 1990a, 1990b), but it is still commonly employed, yielding reasonably accurate results.)

The aim of this section is to develop relationships for the volume velocities of acoustic control sources, which will result in the minimisation of acoustic pressure at discrete locations in the enclosed space, locations that may be potential error sensing points. As in the previous two chapters, the approach that will be taken is to formulate the governing equations that describe the enclosed sound field as matrix expressions, then describe the error criterion in terms of these matrices, which will greatly simplify the task of calculating the control source output. In the section following this, the global minimisation of the rigidly enclosed sound field, and means for obtaining the best possible result for a given control source arrangement, will be discussed.

As was outlined earlier in this book, the sound field within an enclosed space is composed of an infinite set of acoustic modes φ. Referring to Figure 9.1, the sound pressure $p_i(\boldsymbol{r}_e)$ at some location $\boldsymbol{r}_e$ in the enclosure due to the ith mode, excited by some acoustic source distribution contained within the enclosed space, is defined by the relationship:

$$p_i(\boldsymbol{r}_e) = \mathrm{j}\rho_0\omega\int_V \frac{\varphi_i(\boldsymbol{r})\varphi_i(\boldsymbol{r}_e)}{\Lambda_i(\kappa_i^2 - k^2)} s(\boldsymbol{r})\,\mathrm{d}\boldsymbol{r} \tag{9.2.1}$$

where ρ_0 is the density of the acoustic medium, ω is the excitation frequency, $s(\boldsymbol{r})$ is the acoustic source strength in units of volume velocity per unit volume, $\varphi_i(\boldsymbol{r})$ is the value of the acoustic mode shape $\imath$ at location $\boldsymbol{r}$ in the enclosed space, $\varphi_\imath(\boldsymbol{r}_e)$ is the value of acoustic mode shape $\imath$ at location $\boldsymbol{r}_e$ in the enclosed space, and $\Lambda_\imath$ is the volume normalisation of acoustic mode $\imath$ given by:

$$\Lambda_\imath = \int_V \varphi_\imath^2(\boldsymbol{r})\ \mathrm{d}\boldsymbol{r} \tag{9.2.2}$$

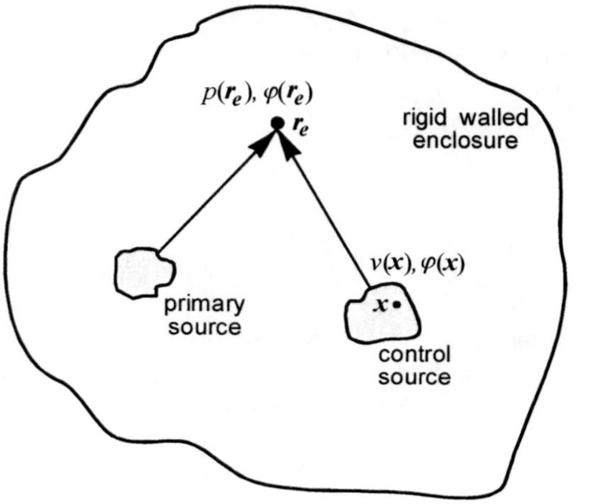

Figure 9.1 Active control arrangement in a rigid walled enclosure.

The quantity $\kappa_\imath$ is the eigenvalue associated with acoustic mode $\imath$, k is the acoustic wavenumber at the excitation frequency, and the integrations in Equations (9.2.1) and (9.2.2) are all over the volume of the enclosed space, V. Acoustic damping will be included in the system models developed here, so that $\kappa_\imath$ is a complex eigenvalue of the cavity, defined as $\kappa_\imath = \omega_\imath / [c_0(1 - j2\zeta_\imath)]$, where the imaginary component defines the acoustic modal damping, $\zeta_\imath$ being the viscous damping ratio of the mode.

Consider, for example, a system where there is a single primary monopole source and a single control monopole source operating in the enclosed space, and the aim of the active control system is to minimise the $\imath$th modal sound pressure at the error sensing location $\boldsymbol{r}_e$. From Equation (9.2.1), the modal sound pressure at location $\boldsymbol{r}_e$ resulting from the operation of a monopole sound source at location $\boldsymbol{r}_q$ is:

$$p_\imath(\boldsymbol{r}_e) = \mathrm{j}\rho_0\omega \int_V \frac{\varphi_\imath(\boldsymbol{r}_q)\varphi_\imath(\boldsymbol{r}_e)}{\Lambda_\imath(\kappa_\imath^2 - k^2)}\, q\delta(\boldsymbol{r} - \boldsymbol{r}_q)\,\mathrm{d}\boldsymbol{r} \tag{9.2.3}$$

where the source strength $s(\boldsymbol{r})$ is equal to the product of the monopole source volume velocity amplitude q and a Dirac delta function $\delta(\boldsymbol{r} - \boldsymbol{r}_q)$, where the delta function has units of L^{-3}, or reciprocal volume. Evaluating the integral in Equation (9.2.3), the modal sound pressure is found to be:

$$p_\imath(\boldsymbol{r}_e) = \mathrm{j}\rho_0\omega q \frac{\varphi_\imath(\boldsymbol{r}_q)\varphi_\imath(\boldsymbol{r}_e)}{\Lambda_\imath(\kappa_\imath^2 - k^2)} \tag{9.2.4}$$

As the system considered here is linear, the sound pressure at location $\boldsymbol{r}_e$ in space during operation of both the primary and control sources will be equal to the superposition of the two components:

$$p_\iota(\boldsymbol{r}_e) = p_{p,\iota}(\boldsymbol{r}_e) + p_{c,\iota}(\boldsymbol{r}_e) \tag{9.2.5}$$

Expanding Equation (9.2.5) using Equation (9.2.4), it is easily shown that the modal sound pressure at $\boldsymbol{r}_e$ can be completely nullified by the choice of control source volume velocity,

$$q_c = -q_p \frac{\varphi_\iota(\boldsymbol{r}_p)}{\varphi_\iota(\boldsymbol{r}_c)} \tag{9.2.6}$$

where $\boldsymbol{r}_p$ and $\boldsymbol{r}_c$ are the locations of the primary and control source monopoles respectively. Note that in the relationship given in Equation (9.2.6), there is no dependency upon error location $\boldsymbol{r}_e$, because for a single mode, nullifying the acoustic pressure at one location is equivalent to nullifying the acoustic pressure at all locations.

The utility of the result of Equation (9.2.6) is somewhat dubious, as it is unlikely that consideration of a single mode in an enclosed space will be adequate to represent any physical system. Theoretically, exact modelling of an enclosed sound field requires consideration of an infinite number of acoustic modes:

$$p(\boldsymbol{r}_e) = \mathrm{j}\rho_0\omega \int_V \sum_{\iota=1}^{\infty} \frac{\varphi_\iota(\boldsymbol{r})\varphi_\iota(\boldsymbol{r}_e)}{\Lambda_\iota(\kappa_\iota^2 - k^2)} s(\boldsymbol{r})\,\mathrm{d}\boldsymbol{r} = \mathrm{j}\rho_0\omega \int_V G_a(\boldsymbol{r}|\boldsymbol{r}_e)s(\boldsymbol{r})\,\mathrm{d}\boldsymbol{r} \tag{9.2.7a,b}$$

where $G_a(\boldsymbol{r}|\boldsymbol{r}_e)$ is the Green's function of the interior space, defined in Chapter 2 as:

$$G_a(\boldsymbol{r}|\boldsymbol{r}_e) = \sum_{\iota=1}^{\infty} \frac{\varphi_\iota(\boldsymbol{r})\varphi_\iota(\boldsymbol{r}_e)}{\Lambda_\iota(\kappa_\iota^2 - k^2)} \tag{9.2.8}$$

In enclosures of low modal density (low-frequency excitation), sufficiently accurate results can be obtained by truncating the infinite summation at a relatively low number of modes, N_n. The discussion in this section will be restricted to systems of low modal density, explicitly truncating the acoustic modal summation at N_n modes. A discussion of the effects of higher modal density on the qualitative results will be presented in Section 9.9.

Observe now that the integral expression in Equation (9.2.7) can be rewritten, using the truncated summation, as:

$$p(\boldsymbol{r}_e) = \mathrm{j}\rho_0\omega \sum_{\iota=1}^{N_n} \frac{\varphi_\iota(\boldsymbol{r}_e)}{\Lambda_\iota(\kappa_\iota^2 - k^2)} \int_V \varphi_\iota(\boldsymbol{r})s(\boldsymbol{r})\,\mathrm{d}\boldsymbol{r} \tag{9.2.9}$$

Expressed in this way, the integral in Equation (9.2.9) can be viewed as a modal generalised volume velocity, analogous to the structural modal generalised forces encountered in the previous chapter. It will be useful to further re-express this integral by writing the acoustic source strength distribution $s(\boldsymbol{r})$ as the product of a complex volume velocity amplitude q and a transfer function $z_v(\boldsymbol{r})$ to that point:

$$\int_V \varphi_\imath(\boldsymbol{r}) s(\boldsymbol{r})\,\mathrm{d}\boldsymbol{r} = q\int_V \varphi_\imath(\boldsymbol{r}) z_v(\boldsymbol{r})\,\mathrm{d}\boldsymbol{r} = q\varphi_{g,\imath} \tag{9.2.10a,b}$$

where $\varphi_{g,\imath}$ is the modal generalised volume velocity transfer function for the $\imath$th acoustic mode, which describes the volume velocity excitation of the mode for a unit amplitude source volume velocity:

$$\varphi_{g,\imath} = \int_V \varphi_\imath(\boldsymbol{r}) z_v(\boldsymbol{r})\,\mathrm{d}\boldsymbol{r} \tag{9.2.11}$$

If, for example, the source of interest was sufficiently small to be modelled as an acoustic monopole, the modal generalised volume velocity would be:

$$q\varphi_{g,\imath} = \int_V \varphi_\imath(\boldsymbol{r}) q\delta(\boldsymbol{r} - \boldsymbol{r_q})\,\mathrm{d}\boldsymbol{r} = q\varphi_\imath(\boldsymbol{r_q}) \tag{9.2.12a,b}$$

so that the modal generalised volume velocity transfer function $\varphi_{g,\imath}$ is simply the value of the mode shape at the monopole source location:

$$\varphi_{g,\imath} = \int_V \varphi_\imath(\boldsymbol{r}) \delta(\boldsymbol{r} - \boldsymbol{r_q})\,\mathrm{d}\boldsymbol{r} = \varphi_\imath(\boldsymbol{r_q}) \tag{9.2.13a,b}$$

It is also straightforward to simplify Equation (9.2.11) for other sources of regular geometry, such as rectangular pistons (as were considered in Chapter 7 for plane wave sound propagation in air handling ducts). This enables explicit modelling of the effects of control source size.

An alternative representation of the sound field in an enclosure has been derived by Nelson and Elliott (1992). Their theoretical formulation is also based upon the assumption that the sound field can be expressed as the sum of acoustic modal contributions. For a truncated set of acoustic modes, the sound pressure at a position $\boldsymbol{r_e}$ due to a source distribution within the volume of an enclosure is given by (Nelson and Elliott, 1992):

$$p(\boldsymbol{r_e}) = \sum_{\imath=1}^{N_n} a_\imath \varphi_\imath(\boldsymbol{r_e}) \tag{9.2.14}$$

where the coefficients $a_\imath$ are the complex modal amplitudes given by:

$$a_\imath = \frac{\omega\rho_0 c_0^2}{V[2\zeta_\imath\omega_\imath\omega + \mathrm{j}(\omega^2 - \omega_\imath^2)]}\int_V \varphi_\imath(\boldsymbol{r}) s(\boldsymbol{r})\,\mathrm{d}V \tag{9.2.15}$$

If the enclosure is excited by an acoustic monopole at location $\boldsymbol{r}_p$ and a second monopole is introduced at $\boldsymbol{r}_c$ to control the sound field, the coefficients $a_\imath$ become:

$$a_\imath = \frac{\omega\rho_0 c_0^2}{V[2\zeta_\imath\omega_\imath\omega + \mathrm{j}(\omega^2 - \omega_\imath^2)]}\left[q_p\varphi_\imath(\boldsymbol{r}_p) + q_c\varphi_\imath(\boldsymbol{r}_c)\right] \tag{9.2.16}$$

Equation (9.2.16) clearly shows that the modal sound pressure at position $\boldsymbol{r}_e$ can be completely nullified if the control source volume velocity q_c is set according to Equation (9.2.6).

Returning to Equation (9.2.9), by truncating the modal summation it is possible to express the total acoustic pressure at some location $\boldsymbol{r}_e$ in the enclosed space as the product of a set of matrices. Defining an $(N_n \times 1)$ vector of modal generalised volume velocity transfer functions $\boldsymbol{\varphi}_g$:

$$\boldsymbol{\varphi}_g = \begin{bmatrix} \varphi_{g,1} \\ \varphi_{g,2} \\ \vdots \\ \varphi_{g,N_n} \end{bmatrix} \tag{9.2.17}$$

an $(N_n \times N_n)$ diagonal matrix of acoustic modal transfer functions $\boldsymbol{Z}_a$, the terms of which describe the complex amplitude of the acoustic modes for a unit modal generalised volume velocity:

$$Z_a(i,i) = \frac{\mathrm{j}\rho_0\omega}{\Lambda_i(\kappa_i^2 - k^2)} \tag{9.2.18}$$

and an $(N_n \times 1)$ vector of acoustic mode shape values at the location $\boldsymbol{r}_e$ in the enclosed space $\boldsymbol{\varphi}(\boldsymbol{r}_e)$:

$$\boldsymbol{\varphi}(\boldsymbol{r}_e) = \begin{bmatrix} \varphi_1(\boldsymbol{r}_e) \\ \varphi_2(\boldsymbol{r}_e) \\ \vdots \\ \varphi_{N_n}(\boldsymbol{r}_e) \end{bmatrix} \tag{9.2.19}$$

the acoustic pressure at $\boldsymbol{r}_e$ can be written as:

$$p(\boldsymbol{r}_e) = \boldsymbol{\varphi}^{\mathrm{T}}(\boldsymbol{r}_e)\, \boldsymbol{Z}_a\, \boldsymbol{\varphi}_g\, q \tag{9.2.20}$$

As has been seen in the previous two chapters, expressing the acoustic pressure in matrix form enables straightforward calculation of the control source volume velocity that will minimise the acoustic pressure amplitude, consisting of the sum of contributions from all N_n acoustic modes being modelled, at the error location of interest in the enclosed space.

As the systems being examined here are linear, the total acoustic pressure at any location in the enclosed space during operation of the active control system will be equal to the sum of the primary and control components:

$$p(\boldsymbol{r}_e) = p_c(\boldsymbol{r}_e) + p_p(\boldsymbol{r}_e) \tag{9.2.21}$$

Therefore, the squared amplitude of the acoustic pressure at $\boldsymbol{r}_e$ during operation of the control and primary source distributions is equal to:

$$|p(\boldsymbol{r}_e)|^2 = p(\boldsymbol{r}_e)^* p(\boldsymbol{r}_e) = [p_c(\boldsymbol{r}_e) + p_p(\boldsymbol{r}_e)]^* [p_c(\boldsymbol{r}_e) + p_p(\boldsymbol{r}_e)] \tag{9.2.22a,b}$$

Using Equation (9.2.20), this can be rewritten as:

$$|p(\boldsymbol{r}_e)|^2 = q_c^* a q_c + q_c^* b + b^* q_c + c \tag{9.2.23}$$

where

$$a = \boldsymbol{\varphi}_{gc}^{\mathrm{H}} \boldsymbol{Z}_a^{\mathrm{H}} \boldsymbol{Z}_p \boldsymbol{Z}_a \boldsymbol{\varphi}_{gc} \tag{9.2.24}$$

$$b = \boldsymbol{\varphi}_{gc}^{\mathrm{H}} \boldsymbol{Z}_a^{\mathrm{H}} \boldsymbol{Z}_p \boldsymbol{Z}_a \boldsymbol{\varphi}_{gp}\, q_p \tag{9.2.25}$$

$$c = q_p^* \boldsymbol{\varphi}_{gp}^{\mathrm{H}} \boldsymbol{Z}_a^{\mathrm{H}} \boldsymbol{Z}_p \boldsymbol{Z}_a \boldsymbol{\varphi}_{gp}\, q_p \tag{9.2.26}$$

and where

$$\boldsymbol{Z}_p = \boldsymbol{\varphi}(\boldsymbol{r}_e)\boldsymbol{\varphi}^{\mathrm{T}}(\boldsymbol{r}_e) \tag{9.2.27}$$

In Equations (9.2.24) to (9.2.26), the subscripts p and c denote primary and control source related quantities. One point to note briefly here is that $\boldsymbol{Z}_p$ is a real, symmetric matrix, effectively a weighting matrix for minimising the acoustic modal amplitudes. This concept will be discussed further in Section 9.7 in relation to determining the optimum error sensor location.

Equation (9.2.23) shows the squared acoustic pressure amplitude at any location $\boldsymbol{r}_e$ in the enclosed space to be a real quadratic function of the complex control source volume velocity q_c. This is seen more easily by rewriting the equation in terms of its real and imaginary parts:

$$|p(\boldsymbol{r}_e)|^2 = a q_{c\mathrm{R}}^2 + 2 b_{\mathrm{R}} q_{c\mathrm{R}} + a q_{c\mathrm{I}}^2 + 2 b_{\mathrm{I}} q_{c\mathrm{I}} + c \tag{9.2.28}$$

where the subscripts I and R denote real and imaginary components respectively. As has been discussed previously, this means that a plot of the squared acoustic pressure amplitude as a function of the real and imaginary components of the control source volume velocity forms a 'bowl', as was shown in Figure 8.2. The bottom of the bowl defines the optimum control source volume velocity that will minimise the error criterion, which is the squared acoustic pressure amplitude. This value of volume velocity can be found by differentiating Equation (9.2.23) with respect to its real and imaginary components, and setting the resultant gradient expression equal to zero. Doing this:

$$\frac{\partial |p(\boldsymbol{r}_e)|^2}{\partial q_{c\mathrm{R}}} = 2 a q_{c\mathrm{R}} + 2 b_{\mathrm{R}} = 0 \tag{9.2.29a,b}$$

$$\frac{\partial |p(\boldsymbol{r}_e)|^2}{\partial q_{c\mathrm{I}}} = 2 a q_{c\mathrm{I}} + 2 b_{\mathrm{I}} = 0 \tag{9.2.30a,b}$$

Putting together the real, Equation (9.2.29), and imaginary, Equation (9.2.30), components of the criteria, through multiplying the latter by j and adding it to the former, produces the following expression for the optimum control source volume velocity:

$$q_{c,p} = -a^{-1}b \tag{9.2.31}$$

where the subscript (c, p) denotes control source volume velocity derived with respect to minimising a pressure error criterion. If this control source volume velocity is substituted back into Equation (9.2.23), the minimum squared pressure amplitude is found to be:

$$|p(\boldsymbol{r}_e)|^2_{\min} = c - b^* a^{-1} b = c + b^* q_{c,p} \tag{9.2.32a,b}$$

As has been seen before, the term c is the value of the error criterion, the squared acoustic pressure amplitude at r_e, during primary excitation only.

This formulation for the optimum single control source output q_c that will minimise the acoustic pressure at a single error sensing location $\boldsymbol{r}_e$ in the enclosed space, can easily be extended to include multiple control sources and multiple error sensors. The ($N_e \times 1$) vector of sound pressures at the set of N_e discrete error sensing locations $\boldsymbol{p}_e$, resulting from the operation of the set of N_s acoustic sources in the enclosed space, can be expressed as:

$$\boldsymbol{p}_e = \boldsymbol{\Phi}_e \boldsymbol{Z}_a \boldsymbol{\Phi}_g \boldsymbol{q} \tag{9.2.33}$$

where $\boldsymbol{\Phi}_e$ is the ($N_e \times N_n$) matrix of acoustic mode shape functions evaluated at the set of error sensing locations:

$$\boldsymbol{\Phi}_e = \begin{bmatrix} \boldsymbol{\varphi}(\boldsymbol{r}_1)^{\mathrm{T}} \\ \boldsymbol{\varphi}(\boldsymbol{r}_2)^{\mathrm{T}} \\ \vdots \\ \boldsymbol{\varphi}(\boldsymbol{r}_{N_e})^{\mathrm{T}} \end{bmatrix} \tag{9.2.34}$$

$\boldsymbol{\Phi}_g$ is the ($N_n \times N_s$) matrix of modal generalised volume velocity transfer functions for the N_s sources:

$$\boldsymbol{\Phi}_g = \begin{bmatrix} \boldsymbol{\varphi}_{g1} & \boldsymbol{\varphi}_{g2} & \cdots & \boldsymbol{\varphi}_{gN_s} \end{bmatrix} \tag{9.2.35}$$

and $\boldsymbol{q}$ is the ($N_s \times 1$) vector of complex source velocity amplitudes:

$$\boldsymbol{q} = \begin{bmatrix} q_1 \\ q_2 \\ \vdots \\ q_{N_s} \end{bmatrix} \tag{9.2.36}$$

Using these quantities, the vector of acoustic pressure amplitudes at the N_e error sensing locations during combined operation of N_p primary and N_c control sources is:

$$\boldsymbol{p}_e = \boldsymbol{p}_{ce} + \boldsymbol{p}_{pe} = \boldsymbol{\Phi}_e \boldsymbol{Z}_a \boldsymbol{\Phi}_{gc} \boldsymbol{q}_c + \boldsymbol{\Phi}_e \boldsymbol{Z}_a \boldsymbol{\Phi}_{gp} \boldsymbol{q}_p \tag{9.2.37a,b}$$

where $\boldsymbol{p}_{pe}$ and $\boldsymbol{p}_{ce}$ are the ($N_e \times 1$) vectors of acoustic pressures at the error sensing locations during primary and control excitation only respectively. The sum of the squared acoustic pressure amplitudes at the error sensing locations is therefore,

$$\sum_{i=1}^{N_e} |p(\boldsymbol{r}_{ei})|^2 = \boldsymbol{p}_e^{\mathrm{H}} \boldsymbol{p}_e = [\boldsymbol{p}_{ce} + \boldsymbol{p}_{pe}]^{\mathrm{H}} [\boldsymbol{p}_{ce} + \boldsymbol{p}_{pe}] \tag{9.2.38a,b}$$

which can be written more compactly as:

$$\sum_{i=1}^{N_e} |p(\boldsymbol{r}_{ei})|^2 = \boldsymbol{q}_c^{\mathrm{H}} \boldsymbol{A} \boldsymbol{q}_c + \boldsymbol{q}_c^{\mathrm{H}} \boldsymbol{b} + \boldsymbol{b}^{\mathrm{H}} \boldsymbol{q}_c + c \tag{9.2.39}$$

where

$$\boldsymbol{A} = \boldsymbol{\Phi}_{gc}^{\mathrm{H}} \boldsymbol{Z}_a^{\mathrm{H}} \boldsymbol{Z}_p \boldsymbol{Z}_a \boldsymbol{\Phi}_{gc} \tag{9.2.40}$$

$$\boldsymbol{b} = \boldsymbol{\Phi}_{gc}^{\mathrm{H}} \boldsymbol{Z}_a^{\mathrm{H}} \boldsymbol{Z}_p \boldsymbol{Z}_a \boldsymbol{\Phi}_{gp} \boldsymbol{q}_p \tag{9.2.42}$$

$$c = \boldsymbol{q}_p^{\mathrm{H}} \boldsymbol{\Phi}_{gp}^{\mathrm{H}} \boldsymbol{Z}_a^{\mathrm{H}} \boldsymbol{Z}_p \boldsymbol{Z}_a \boldsymbol{\Phi}_{gp} \boldsymbol{q}_p \tag{9.2.41}$$

and where $\boldsymbol{Z}_p$ is now based upon multiple error sensor locations:

$$\boldsymbol{Z}_p = \boldsymbol{\Phi}_e^{\mathrm{T}} \boldsymbol{\Phi}_e \tag{9.2.43}$$

Equation (9.2.39) shows this expanded error criterion, the minimisation of the squared acoustic pressure amplitudes at a set of discrete error locations, to be a real quadratic function of the *vector* of control source volume velocities. As discussed in Section 8.4, this means that the error surface, or error criterion plotted as a function of the real and imaginary components of the control source volume velocities, will be a ($2N_c$+1)-dimensional hyper-paraboloid, which is a generalisation of the three-dimensional 'bowl' shown in Figure 8.2 for the case of $N_c = 1$. Being a quadratic function of $\boldsymbol{q}_c$, the error criterion of Equation (9.2.39) can again be differentiated with respect to this quantity, and the resultant gradient expression set equal to zero to derive the vector of optimum control source volume velocities. Performing these operations on the real and imaginary parts separately:

$$\frac{\partial \boldsymbol{p}_e^{\mathrm{H}} \boldsymbol{p}_e}{\partial \boldsymbol{q}_{c_{\mathrm{R}}}} = 2\boldsymbol{A}_{\mathrm{R}} \boldsymbol{q}_{c_{\mathrm{R}}} + 2\boldsymbol{b}_{\mathrm{R}} = 0 \tag{9.2.44a,b}$$

$$\frac{\partial \boldsymbol{p}_e^{\mathrm{H}} \boldsymbol{p}_e}{\partial \boldsymbol{q}_{c_{\mathrm{I}}}} = 2\boldsymbol{A}_{\mathrm{I}} \boldsymbol{q}_{c_{\mathrm{I}}} + 2\boldsymbol{b}_{\mathrm{I}} = 0 \tag{9.2.45a,b}$$

multiplying the imaginary part by j and adding it to the real part produces the expression for the optimum vector of control source volume velocities:

$$\boldsymbol{q}_{c,p} = -\boldsymbol{A}^{-1} \boldsymbol{b} \tag{9.2.46}$$

The result of Equation (9.2.46) is simply an extension of Equation (9.2.31), derived for a system arrangement consisting of a single primary and control source and a single error sensor. Substituting the result of Equation (9.2.46) back into Equation (9.2.39) produces the expression for the minimum residual (controlled) sum of the squared sound pressures at the error locations:

$$[\boldsymbol{p}_e^{\mathrm{H}}\boldsymbol{p}_e]_{\min} = c - \boldsymbol{b}^{\mathrm{H}}\boldsymbol{A}^{-1}\boldsymbol{b} = c + \boldsymbol{b}^{\mathrm{H}}\boldsymbol{q}_{c,p} \tag{9.2.47a,b}$$

It should again be pointed out that if the relationship of Equation (9.2.47) is to define a unique vector of control source volume velocities that will collectively minimise the sum of the squared acoustic pressure amplitudes at the error sensing locations, the matrix $\boldsymbol{A}$ must be positive definite. For the case being considered here, this often translates into the requirement that there be at least as many error locations as there are control sources. If there are less, then the matrix $\boldsymbol{A}$ may be singular, and hence there will be an infinite number of optimum control source volume velocity vectors (or optimal control source configurations). If there are an equal number of control sources and error locations, then the vector of optimum control source volume velocities will often completely null the acoustic pressure at the error locations. If there are more error locations than control sources, then the residual acoustic pressure at the error locations will probably not be equal to zero, although this may not have a negative influence upon the overall levels of acoustic power attenuation attained.

A specific example of minimising acoustic pressure in a rigid walled enclosure will be presented at the end of the next section, after expressions have been derived for determining the best possible result for the minimisation of acoustic potential energy in the enclosed space for a given control source arrangement.

9.3 GLOBAL CONTROL OF SOUND FIELDS IN RIGID ENCLOSURES

As pointed out throughout this book, simply minimising the sound field at discrete locations does not guarantee the best result in terms of global attenuation of the sound field throughout the enclosure. To determine the maximum levels of global sound attenuation that are possible for a given source arrangement, the problem must be reformulated in terms of some global error criterion.

In the previous two chapters, the global error criterion that has been used is real acoustic power, as it is this quantity that will propagate away from the sound source and become a nuisance to the surrounding population. When considering enclosed sound fields, however, acoustic power is not a particularly useful basis for examination, as for a given volume velocity source, the only real acoustic power output will be a by-product of acoustic damping. This is evident in the defining equation for acoustic pressure, Equation (9.2.1), where the acoustic pressure in the absence of damping is seen to be 90° out of phase with the velocity (recall that acoustic power is defined by the product of the in-phase components of pressure and velocity). A more suitable global measure of acoustic pressure when considering enclosed sound fields is acoustic potential energy E_p:

$$E_p = \frac{1}{4\rho_0 c_0^2}\int_V |p(\boldsymbol{r})|^2 \,\mathrm{d}\boldsymbol{r} \tag{9.3.1}$$

which is explicitly based upon the volume integration of the squared acoustic pressure amplitudes in the enclosed space. To give some further physical meaning to acoustic potential energy, note that the acoustic pressure term in Equation (9.3.1) can be expanded as the sum of modal contributions as follows:

$$p(\boldsymbol{r}) = \sum_{\imath=1}^{\infty} p_\imath \varphi_\imath(\boldsymbol{r}) \tag{9.3.2}$$

where $p_\imath$ is the complex pressure amplitude of the $\imath$th acoustic mode. Substituting this expansion into Equation (9.3.1):

$$E_p = \frac{1}{4\rho_0 c_0^2} \int_V \left(\sum_{\imath=1}^{\infty} p_\imath \varphi_\imath(\boldsymbol{r})\right)^* \left(\sum_{\imath=1}^{\infty} p_\imath \varphi_\imath(\boldsymbol{r})\right) \mathrm{d}V \tag{9.3.3}$$

Acoustic modes in a rigid enclosure are orthogonal, so that:

$$\int_V \varphi_i(\boldsymbol{r}) \varphi_\imath(\boldsymbol{r})\, d\boldsymbol{r} = \begin{cases} \Lambda_\imath & i = \imath \\ 0 & i \neq \imath \end{cases} \tag{9.3.4}$$

where $\Lambda_\imath$ is the volume normalisation of the $\imath$th acoustic mode. Using this property, the expression for acoustic potential energy given in Equation (9.3.3) can be simplified to:

$$E_p = \frac{1}{4\rho_0 c_0^2} \sum_{\imath=1}^{\infty} \Lambda_\imath |p_\imath|^2 \tag{9.3.5}$$

Equation (9.3.5) shows that minimising acoustic potential energy in the enclosed space is equivalent to minimising the weighted (by mode normalisation) sum of the squares of the acoustic modal amplitudes. Therefore, to formulate expressions for the control source volume velocities that will minimise the enclosed sound field, the acoustic modal amplitudes themselves must be considered, rather than their values at specific error sensing locations.

From Equation (9.2.1) in the previous section, it is apparent that the amplitude of the $\imath$th acoustic mode resulting from the operation of some acoustic source distribution located within the enclosed space is defined by the equation:

$$p_\imath = \mathrm{j}\rho_0 \omega \int_V \frac{\varphi_\imath(\boldsymbol{r})}{\Lambda_\imath(\kappa_\imath^2 - k^2)} s(\boldsymbol{r})\, \mathrm{d}\boldsymbol{r} \tag{9.3.6}$$

Considering a truncated set of acoustic modes, the ($N_n \times 1$) vector of acoustic modal amplitudes, $\boldsymbol{p}$, resulting from the operation of a single velocity source can be expressed in matrix form as follows using the quantities defined in the previous section:

$$\boldsymbol{p} = \boldsymbol{Z_a}\, \boldsymbol{\varphi_g}\, q \tag{9.3.7}$$

The sum of the squared acoustic modal amplitudes during operation of single control and primary source is therefore equal to:

$$\sum_{\imath=1}^{N_n} |p_\imath|^2 = \sum_{\imath=1}^{N_n} p_\imath^* p_\imath = [\boldsymbol{p_c} + \boldsymbol{p_p}]^{\mathrm{H}} [\boldsymbol{p_c} + \boldsymbol{p_p}] \tag{9.3.8a,b}$$

Using Equation (9.3.8), the acoustic potential energy for this single control and primary source arrangement can therefore be expressed as:

$$E_p = \frac{1}{4\rho_0 c_0^2} \sum_{i=1}^{N_n} |p_i|^2 = q_c^* a q_c + q_c^* b + b^* q_c + c \tag{9.3.9a,b}$$

where

$$a = \boldsymbol{\varphi}_{gc}^{\mathrm{H}} \boldsymbol{Z}_a^{\mathrm{H}} \boldsymbol{Z}_E \boldsymbol{Z}_a \boldsymbol{\varphi}_{gc} \tag{9.3.10}$$

$$b = \boldsymbol{\varphi}_{gc}^{\mathrm{H}} \boldsymbol{Z}_a^{\mathrm{H}} \boldsymbol{Z}_E \boldsymbol{Z}_a \boldsymbol{\varphi}_{gp} q_p \tag{9.3.11}$$

and

$$c = q_p^* \boldsymbol{\varphi}_{gp}^{\mathrm{H}} \boldsymbol{Z}_a^{\mathrm{H}} \boldsymbol{Z}_E \boldsymbol{Z}_a \boldsymbol{\varphi}_{gp} q_p \tag{9.3.12}$$

where the terms of the acoustic potential energy-based diagonal weighting matrix $\boldsymbol{Z}_E$ are defined by:

$$Z_E(i,i) = \frac{\Lambda_i}{4\rho_0 c_0} \tag{9.3.13}$$

Observing that Equation (9.3.9) is identical in form to Equation (9.2.23) of the previous section, the optimum control source volume velocity, with respect to providing global reduction of the enclosed sound field, is defined by the same relationship given in Equation (9.2.31):

$$q_{c,E} = -a^{-1} b \tag{9.3.14}$$

where the subscript (c, E) denotes a control source output, which is optimal with respect to minimising acoustic potential energy. Substituting this relationship back into Equation (9.3.9), the minimum value of acoustic potential energy is found to be:

$$E_{p,\min} = c - b^* a^{-1} b = c + b^* q_{c,E} \tag{9.3.15a,b}$$

where the term c is again the value of the error criterion (acoustic potential energy in the enclosed space) under primary excitation only.

There is an interesting observation that can be made here in regard to the phasing of the control source volume velocity relative to the primary source when minimising acoustic potential energy. If the modal generalised volume velocity transfer functions for the primary and control sources are both real, which will usually be the case (for example when both sources are modelled as acoustic monopoles), then as the quantity a in Equation (9.3.14) is real (as defined in Equation (9.3.10)) and the quantity b will be equal to some real number multiplied by the primary source amplitude, the control source will always be either in-phase or 180° out of phase with the primary source. This is a generic trait of active noise control systems, where when using one source to control the output of a second equivalent source, the optimum result will be obtained when the sources are precisely in or out of phase.

Using Nelson and Elliott's (1992) theoretical formulation of the harmonic sound field within an enclosure given in Equation (9.2.14), the acoustic potential energy can be written as:

$$E_p = \frac{V}{4\rho_0 c_0^2} \sum_{\imath=1}^{N_n} |a_\imath|^2 \tag{9.3.16}$$

where $a_\imath$ are the complex modal amplitude coefficients. For a single primary and control source at $\boldsymbol{r}_p$ and $\boldsymbol{r}_c$ respectively, the coefficients $a_\imath$ can be written in the form:

$$a_\imath = a_{p\imath} + B_\imath q_c \tag{9.3.17}$$

where

$$a_{p\imath} = \frac{\omega \rho_0 c_0^2 \varphi_\imath(\boldsymbol{r}_p)}{V\left[2\zeta_\imath \omega_\imath \omega + \mathrm{j}(\omega^2 - \omega_\imath^2)\right]} q_p \tag{9.3.18}$$

and

$$B_\imath = \frac{\omega \rho_0 c_0^2 \varphi_\imath(\boldsymbol{r}_c)}{V\left[2\zeta_\imath \omega_\imath \omega + \mathrm{j}(\omega^2 - \omega_\imath^2)\right]} \tag{9.3.19}$$

It follows from Equations (9.3.16) and (9.3.17) that the total time-averaged acoustic potential energy is given by:

$$E_p = \frac{V}{4\rho_0 c_0^2} \sum_{\imath=1}^{N_n} |a_{p\imath} + B_\imath q_c|^2 \tag{9.3.20}$$

Expanding Equation (9.3.20) gives the acoustic potential energy E_p as a quadratic function of q_c according to:

$$E_p = q_c^* a q_c + q_c^* b + b^* q_c + c \tag{9.3.21}$$

where in this case:

$$a = \frac{V}{4\rho_0 c_0^2} \sum_{\imath=1}^{N_n} |B_\imath|^2 \tag{9.3.22}$$

$$b = \frac{V}{4\rho_0 c_0^2} \sum_{\imath=1}^{N_n} B_\imath^* a_{p\imath} \tag{9.3.23}$$

and

$$c = \frac{V}{4\rho_0 c_0^2} \sum_{\imath=1}^{N_n} |a_{p\imath}|^2 \tag{9.3.24}$$

The minimum value of the acoustic potential energy function in Equation (9.3.21) is given by Equation (9.3.15) and occurs when the optimum control source volume velocity is according to Equation (9.3.14):

$$q_{c,E} = \frac{-\sum_{i=1}^{N_n} B_i^* a_{pi}}{\sum_{i=1}^{N_n} |B_i|^2} \tag{9.3.25}$$

The previous single primary and control source formulation given in Equations (9.3.1) to (9.3.15) can easily be expanded to incorporate multiple control and primary sources. Again using quantities defined in the previous section, the acoustic potential energy in the enclosure with this expanded system arrangement is:

$$E_p = \frac{1}{4\rho_0 c_0^2} \sum_{i=1}^{N_n} |p_i|^2 = \boldsymbol{q}_c^{\mathrm{H}} \boldsymbol{A} \boldsymbol{q}_c + \boldsymbol{q}_c^{\mathrm{H}} \boldsymbol{b} + \boldsymbol{b}^{\mathrm{H}} \boldsymbol{q}_c + c \tag{9.3.26a,b}$$

where

$$\boldsymbol{A} = \boldsymbol{\Phi}_{gc}^{\mathrm{H}} \boldsymbol{Z}_a^{\mathrm{H}} \boldsymbol{Z}_E \boldsymbol{Z}_a \boldsymbol{\Phi}_{gc} \tag{9.3.27}$$

$$\boldsymbol{b} = \boldsymbol{\Phi}_{gc}^{\mathrm{H}} \boldsymbol{Z}_a^{\mathrm{H}} \boldsymbol{Z}_E \boldsymbol{Z}_a \boldsymbol{\Phi}_{gp} \boldsymbol{q}_p \tag{9.3.28}$$

and

$$c = \boldsymbol{q}_p^{\mathrm{H}} \boldsymbol{\Phi}_{gp}^{\mathrm{H}} \boldsymbol{Z}_a^{\mathrm{H}} \boldsymbol{Z}_E \boldsymbol{Z}_a \boldsymbol{\Phi}_{gp} \boldsymbol{q}_p \tag{9.3.29}$$

As Equation (9.3.26) is identical in form to Equation (9.2.39), the previous result of Equation (9.2.46) can be used to simply write the expression for the vector of optimum control source outputs as:

$$\boldsymbol{q}_{c,E} = -\boldsymbol{A}^{-1}\boldsymbol{b} \tag{9.3.30}$$

Substituting this result into Equation (9.3.26) produces the following expression for the minimum acoustic potential energy:

$$E_{p,\min} = c - \boldsymbol{b}^{\mathrm{H}}\boldsymbol{A}^{-1}\boldsymbol{b} = c + \boldsymbol{b}^{\mathrm{H}}\boldsymbol{q}_{c,E} \tag{9.3.31a,b}$$

Before considering a simple example of control of a rigidly enclosed sound field, there is one final point that should be discussed, that of physical mechanisms. It has been noted in previous chapters that when (constant volume velocity) acoustic sources are used in an active control system, two control mechanisms are possible: acoustic power suppression, where the control source(s) mutually unload the primary source such that the total acoustic power output is reduced, and acoustic power absorption, where the control source absorbs some of the primary source acoustic power output. However, it has already been pointed out that when considering the active control of sound fields in an enclosed space, acoustic power is not a particularly useful basis for examination, as for a given velocity source, the only real acoustic power output will be a by-product of acoustic damping. This is the reason that acoustic potential energy, based upon volume-averaged acoustic pressure levels, as opposed to acoustic power, was used as a global error criterion. Therefore, when examining control mechanisms in a rigid enclosure, it is more appropriate to examine the individual source contributions to acoustic potential energy.

It was shown in Equation (9.3.5) that acoustic potential energy is equal to the weighted sum of squared acoustic modal pressure amplitudes. This sum can be expressed in a manner analogous to total acoustic power as the sum of primary and control source contributions:

$$E_p = \frac{1}{4\rho_0 c_0^2} \sum_{\imath=1}^{n} \Lambda_\imath |p_\imath|^2 = E_{pp} + E_{pc} \tag{9.3.32}$$

where E_{pp} and E_{pc} are the contributions to acoustic potential energy by the primary source and control source distributions respectively, which are equal to:

$$E_{p(p/c)} = \frac{1}{4\rho_0 c_0^2} \sum_{\imath=1}^{n} \Lambda_\imath \operatorname{Re}\left\{p_{\imath,p/c}\, p_{\imath,tot}^*\right\} \tag{9.3.33}$$

Here, *p*/*c* denotes primary or control source, *tot* denotes total (sum of primary and control source contributions), and Re denotes the real part of the expression. The acoustic modal amplitude resulting from some source distribution is defined in Equation (9.3.7) for single sources, and for multiple sources it can easily be extended to:

$$\boldsymbol{p} = \boldsymbol{Z}_a \boldsymbol{\Phi}_g \boldsymbol{q} \tag{9.3.34}$$

(where the terms must be specialised for either control sources or primary sources). Thus, for a single control source, single primary source system, it is straightforward to show that the control source contribution to the acoustic potential energy can be expressed as:

$$E_{pc} = q_c^* a q_c + \operatorname{Re}\left\{q_c b^*\right\} \tag{9.3.35}$$

Similarly, the primary source contribution to the potential energy of the enclosed space is:

$$E_{pp} = \operatorname{Re}\left\{q_c^* b\right\} + c \tag{9.3.36}$$

If the optimum value of control source output, as defined in Equation (9.3.14), is substituted into Equation (9.3.34), which defines the control source acoustic potential energy, it is found that under optimum conditions:

$$E_{pc} = q_{c,E}^* a q_{c,E} + \left\{q_{c,E} b^*\right\} = (-a^{-1}b)^* a(-a^{-1}b) + \operatorname{Re}\left\{(-a^{-1}b)b^*\right\} = 0 \tag{9.3.37a,b}$$

That is, the contribution made by the control source to the acoustic potential energy in the enclosed space is zero; it makes neither a negative nor positive contribution. The control mechanism, which is optimally employed, must the be one of pure suppression, where a reduction in acoustic potential energy is achieved by a mutual unloading of the primary and control sources, resulting in a zero contribution by the control source and a reduced contribution by the primary source to the acoustic potential energy. Note that if the optimum control source output of Equation (9.3.14) is substituted into Equation (9.3.35), which defines the primary source contribution to acoustic potential energy, the result is:

$$E_{pp} = \mathrm{Re}\left\{ q_{c,E}^{*} b + c \right\} = c - b^{*} a^{-1} b \tag{9.3.38a,b}$$

which is the same as the minimum acoustic potential energy of the system, as outlined in Equation (9.3.15). The residual acoustic potential energy can be entirely traced to the primary source distribution.

This result can be easily expanded to included consideration of multiple sources. In terms of the quantities defined in Equations (9.3.27) and (9.3.28), Equation (9.3.32) can be specialised for a set of control sources as:

$$E_{pc} = \boldsymbol{q_c}^{\mathrm{H}} \boldsymbol{A} \boldsymbol{q_c} + \mathrm{Re}\left\{ \boldsymbol{b}^{\mathrm{H}} \boldsymbol{q}_c \right\} \tag{9.3.39}$$

If the vector of optimal control source outputs as defined in Equation (9.3.30) is now inserted into this relation, noting that $\boldsymbol{A}$ is a Hermitian matrix, the acoustic potential energy attributable to the control source array is:

$$E_{p,c} = [-\boldsymbol{A}^{-1}\boldsymbol{b}]^{\mathrm{H}} \boldsymbol{A} [-\boldsymbol{A}^{-1}\boldsymbol{b}] + \mathrm{Re}\left\{ \boldsymbol{b}^{\mathrm{H}} [-\boldsymbol{A}^{-1}\boldsymbol{b}] \right\} = 0 \tag{9.3.40a,b}$$

That is, the acoustic potential energy attributable to the control source array is zero, as may have been expected. Once again, the control mechanism is one of pure suppression.

Simply having a total contribution to the acoustic potential energy of zero from the control source array does not in itself mean that contributions from the constituent elements will necessarily be zero. However, note that the acoustic potential energy attributable to the operation of a single, *i*th, control source can be expressed as:

$$E_{pc}(i) = \mathrm{Re}\left\{ \boldsymbol{q_{c,E}}^{\mathrm{H}} \boldsymbol{A}(\mathrm{col}\ i) q_{c,E}(i) \right\} + \mathrm{Re}\left\{ b^{*}(i) q_{c,E}(i) \right\} \tag{9.3.41}$$

where $\boldsymbol{A}(\mathrm{col}\ i)$ is the *i*th column of the matrix $\boldsymbol{A}$, $b(i)$ is the *i*th element in the vector $\boldsymbol{b}$, and $q_{c,E}(i)$ is the optimum volume velocity of the *i*th control source, the *i*th element of the vector $\boldsymbol{q_{c,E}}$. If the vector of optimum control source volume velocities is again expanded, then the result is:

$$E_{pc}(i) = \mathrm{Re}\left\{ [-\boldsymbol{A}^{-1}\boldsymbol{b}]^{\mathrm{H}} \boldsymbol{A}(\mathrm{col}\ i) q_{c,E}(i) \right\} + \mathrm{Re}\left\{ b(i)^{*} q_{c,E}(i) \right\} = 0 \tag{9.3.42a,b}$$

It has therefore just been shown that the acoustic potential energy contribution of each element in the control source array will also be zero. Note that this is analogous to the free space result of the previous section, where it was shown that if reciprocity exists between each point of the primary source distribution (in this instance the vibrating structure) and each point on the control source array, then the acoustic power output of the control sources, under optimal conditions, will be zero.

Example 9.1

Application of the above methodology to analyse the active control of sound fields in rigid walled enclosures has been undertaken for a number of problems (Bullmore, 1988; Bullmore et al., 1987, 1990; Elliott et al., 1987, Stanef et al., 2004, Wareing et al., 2011). As a simple example of the active control of a rigidly enclosed sound field, the problem of low-frequency sound attenuation in a rigid walled rectangular enclosure of dimensions 3.43 m × 0.1 m × 0.1 m will be chosen. This is essentially the problem considered at length by Curtis et al. (1985, 1987, 1990). To fit this problem into the theoretical models that have been developed thus far, only a knowledge of the acoustic mode shape functions and resonance frequencies is required. For a rectangular enclosure, the acoustic mode shape function is:

$$\varphi_{l,m,n}(x,y,z) = \cos\frac{l\pi x}{L_x}\cos\frac{m\pi y}{L_y}\cos\frac{n\pi z}{L_z} \tag{9.3.43}$$

where l, m, n and L_x, L_y, L_z are the modal indices and enclosure dimensions in the x- , y- and z-directions respectively, with the origin of the coordinate system in the lower left corner of the enclosure. The resonance frequencies associated with these acoustic modes are defined by the expression (Bies and Hansen, 2009):

$$\omega_{l,m,n} = c_0\pi\sqrt{\left(\frac{l}{L_x}\right)^2 + \left(\frac{m}{L_y}\right)^2 + \left(\frac{n}{L_z}\right)^2} \tag{9.3.44}$$

The acoustic mode normalisation associated with this mode shape function is:

$$\Lambda_{l,m,n} = \int_V \varphi^2_{l,m,n}(\boldsymbol{r})\,\mathrm{d}\boldsymbol{r} = \frac{V}{8}\varepsilon_l\varepsilon_m\varepsilon_n,$$

$$\varepsilon_{l,m,n} = \begin{cases} 1 & l,m,n \neq 0 \\ 2 & l,m,n = 0 \end{cases} \tag{9.3.45a–c}$$

In considering the dimensions of the example enclosure in light of the defining expression for resonance frequency, it is apparent that at low frequencies the problem is essentially one-dimensional, with the modal indices varying only in the x-direction ((0,0,0), (1,0,0), (2,0,0), etc.). A sketch of the enclosure geometry and the x-direction variation for several low-frequency resonant modes is shown in Figure 9.2. Also, for an enclosure of these dimensions, the low order resonance frequencies will be spaced at 50 Hz intervals, the (0,0,0) at 0 Hz, (1,0,0) at 50 Hz, (2,0,0) at 100 Hz, etc. This example will include the use of a single monopole primary source, located at $x = 0$ (note that for the low order modes, the y- and z-locations are irrelevant), a single control source and a single error sensor.

The first point of interest is how much attenuation of acoustic potential energy is possible, and why. Figure 9.3 illustrates the levels of acoustic potential energy reduction possible for a single control source located at the end of the enclosure $x = L_x$ as a function of frequency for two levels of acoustic damping: 0.005 (light damping) and 0.05 (moderate damping). The data display a number of attenuation peaks corresponding to the resonance frequencies of the modes. To interpret these results, there are two points that must be noted. First, the mode shapes are ordered such that with this source arrangement it is impossible to

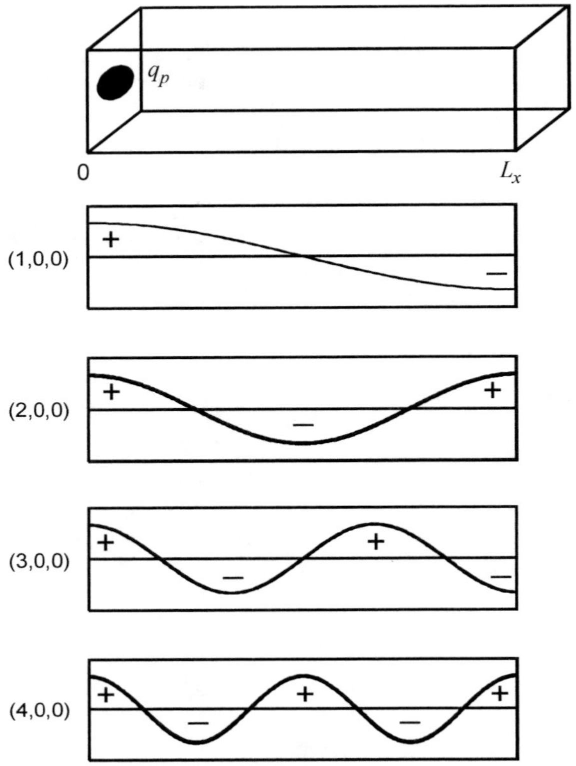

Figure 9.2 One-dimensional rectangular enclosure with several low-frequency mode shape functions.

simultaneously reduce the amplitude of two sequential (in resonance frequency) modes with this primary source/control source arrangement; a reduction in the amplitude of one mode will always cause an increase in the amplitude of the neighbouring (in resonance frequency) modes.

Second, it was shown previously in this section that acoustic potential energy is equal to the weighted sum of squared acoustic modal amplitudes, values of which are illustrated in Figures 9.4 and 9.5 for the low-frequency modes as a function of frequency for the two damping rates. At and near the resonance frequencies, a single mode is clearly dominant in this summation, a mode that will have a very small input impedance, defined by the quantity $(\kappa_i^2 - k^2)$ in the denominator of the governing Equation (9.2.1). This mode can be significantly decreased in amplitude with only a relatively small spillover effect (the increase in the amplitude of the two neighbouring modes), so that a large attenuation of acoustic potential energy is possible.

As the damping of the enclosure increases, the amplitude of the resonant mode decreases (due to the increase in the modal input impedance), and so too does the reduction in acoustic potential energy. This trend is apparent when comparing the data in Figures 9.3, 9.4 and 9.5. At frequencies away from resonance frequencies, however, the summation is not dominated by a single mode, but rather is very 'multi-modal'. Decreasing the amplitude of the most dominant mode will increase the amplitude of the next most important in the summation by a relatively significant amount. The end result is a very reduced potential for attenuation.

Clearly, a single control source is only effective at controlling a resonant response where a single acoustic mode is largely responsible for the sound field. As the number of modes increases, so too does the number of control sources required to control them.

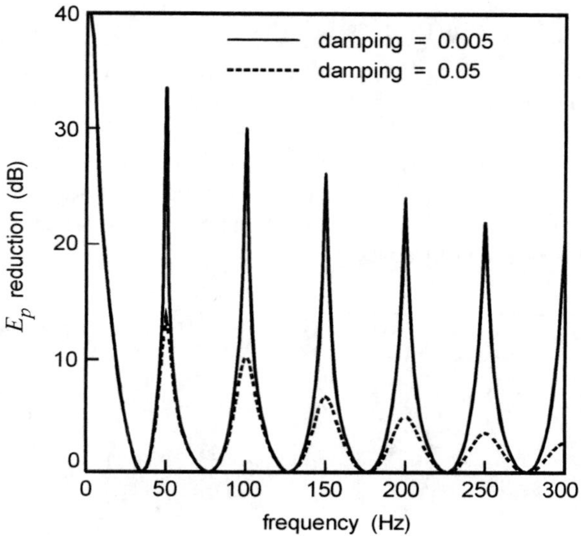

Figure 9.3 Maximum possible acoustic potential energy reduction as a function of frequency with a single control source at $x = L_x$ with acoustic damping of 0.005 and 0.05.

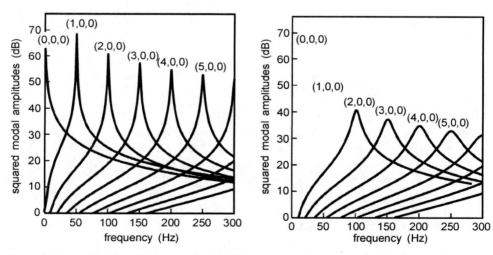

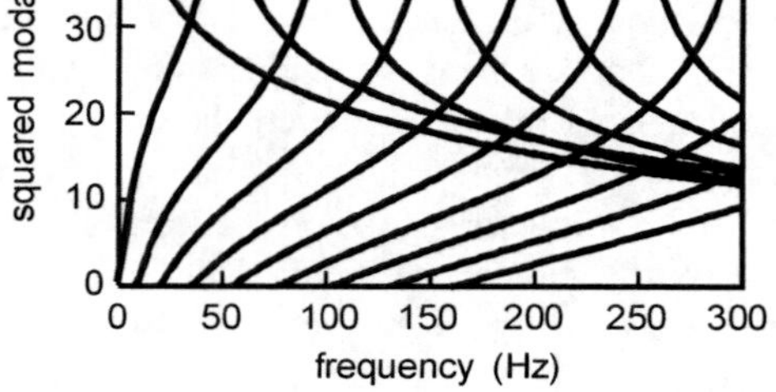

Figure 9.4 Squared modal amplitudes as a function of frequency, damping = 0.005.

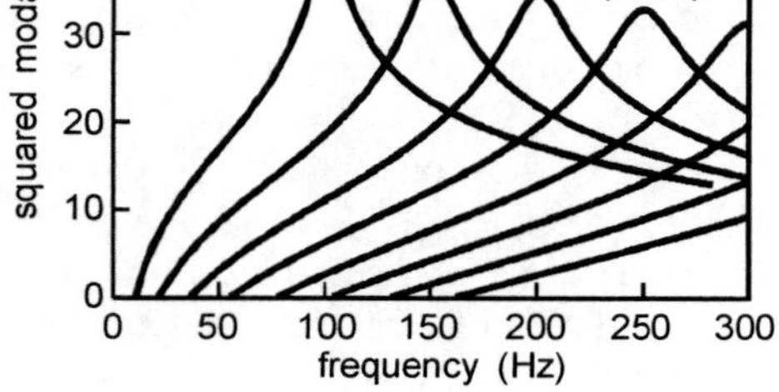

Figure 9.5 Squared modal amplitudes as a function of frequency, damping = 0.05.

The location of the control source plays an important role in the level of attenuation that can be obtained. Figure 9.6 shows the maximum levels of acoustic potential energy attenuation as a function of control source position for two frequencies: 100 Hz where the resonant (2,0,0) mode dominates the acoustic potential energy summation, and 125 Hz, where two modes, the (2,0,0) and (3,0,0), have equal importance. For both of these cases, the maximum levels of attenuation are achieved with the control sources near the primary source position, as at this position it will be possible to attenuate the amplitude of all modes simultaneously. At 100 Hz, significant levels of attenuation are possible with the control source well away from the primary source, as a single mode dominates the response. The troughs in the attenuation curve are at the nodal points (zero amplitude) of the resonant mode. The 125 Hz case does not fare so well for control source locations away from the primary source, however, as two modes must be reduced in amplitude simultaneously to obtain a satisfactory result.

It was pointed out previously that when a single control source is used to control an identical primary source, it will either be exactly in or out of phase with it. Figure 9.7 shows the ratio of control source to primary source volume velocity when the control source is at a position $x = L_x$, as a function of frequency.

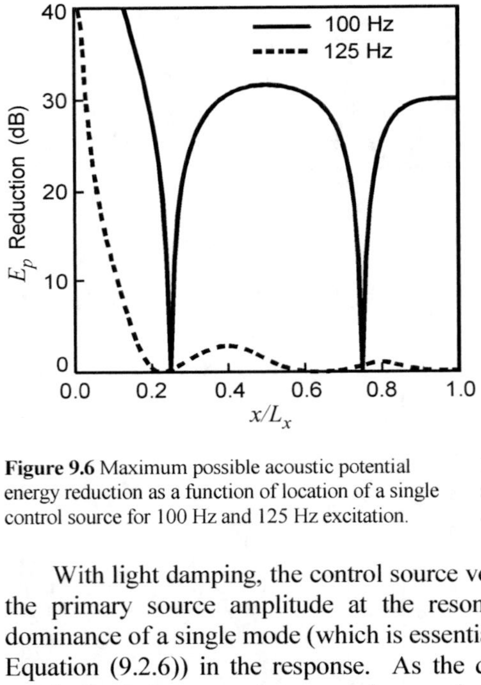

Figure 9.6 Maximum possible acoustic potential energy reduction as a function of location of a single control source for 100 Hz and 125 Hz excitation.

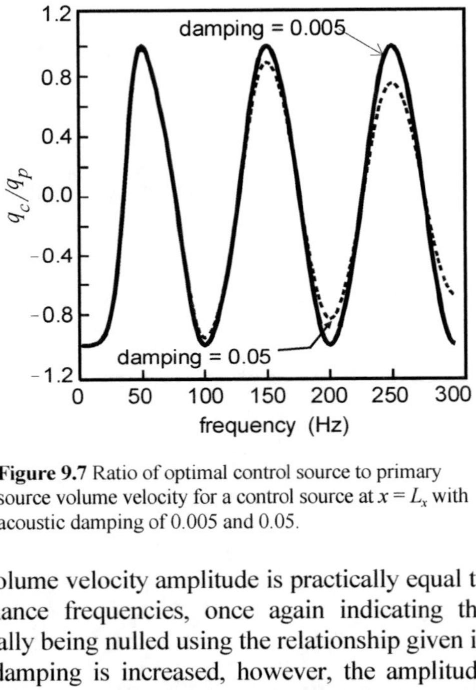

Figure 9.7 Ratio of optimal control source to primary source volume velocity for a control source at $x = L_x$ with acoustic damping of 0.005 and 0.05.

With light damping, the control source volume velocity amplitude is practically equal to the primary source amplitude at the resonance frequencies, once again indicating the dominance of a single mode (which is essentially being nulled using the relationship given in Equation (9.2.6)) in the response. As the damping is increased, however, the amplitude required of the control source for maximum potential energy reduction decays with increasing frequency. This is due to a reduction in the amplitude of the resonance peaks in the response, and hence a reduction in the dominance of a single acoustic mode in the result. For both levels of damping, the optimum control source output is zero at some frequency between the resonance peaks. At this point, a decrease in one modal amplitude will be exactly offset by an increase in its neighbour; hence it is best to do nothing at all.

It was shown theoretically that under optimum conditions, the contribution of the acoustic control source(s) to the acoustic potential energy in the enclosure will be zero. Away from the optimum this will not be the case. Figures 9.8 and 9.9 illustrate the acoustic potential energy, and the primary and control source contributions, as a function of phase when the volume velocity amplitudes are fixed to be equal.

In Figure 9.8, data for a frequency of 100 Hz are plotted. In this figure, the primary and control source contributions are identical, and are shown plotted as one line. This is at the resonance frequency of a lightly damped mode, so that ideally the volume velocity amplitudes should practically be equal. Observe that in this instance there are only positive contributions to acoustic potential energy, and no negative. This is also the case at 125 Hz, the data for which are shown in Figure 9.9. Here, there is a net increase in acoustic potential energy for all phase differences, as the decrease in amplitude of one mode is offset by an increase in another. This result is different from that seen in the previous chapter for free-field radiation, where it was seen that altering the phase difference between two closely spaced monopole sources of equal amplitude will cause one source to begin absorbing acoustic power.

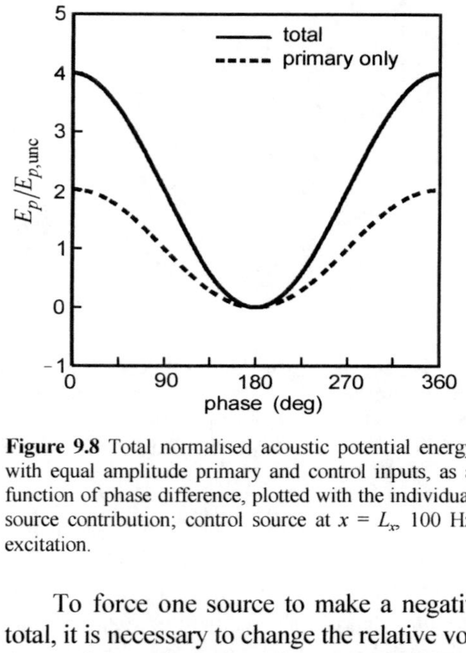

Figure 9.8 Total normalised acoustic potential energy with equal amplitude primary and control inputs, as a function of phase difference, plotted with the individual source contribution; control source at $x = L_x$, 100 Hz excitation.

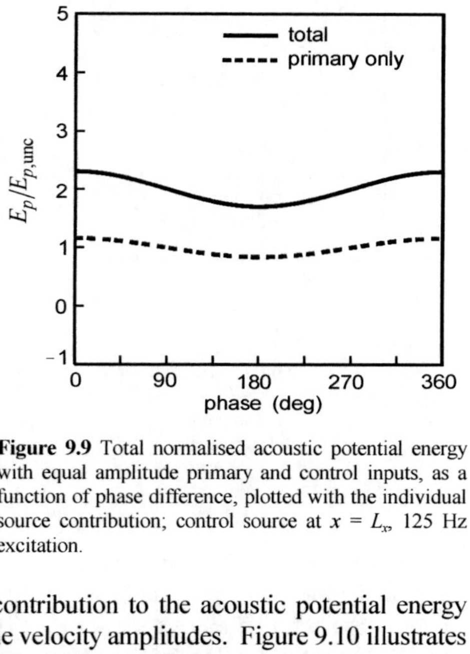

Figure 9.9 Total normalised acoustic potential energy with equal amplitude primary and control inputs, as a function of phase difference, plotted with the individual source contribution; control source at $x = L_x$, 125 Hz excitation.

To force one source to make a negative contribution to the acoustic potential energy total, it is necessary to change the relative volume velocity amplitudes. Figure 9.10 illustrates the total acoustic potential energy, and the primary and control source contributions, for a control source mounted at $x = L_x$, operating out of phase with the primary source at 100 Hz, as a function of control source amplitude. Note that at low amplitudes, the control source is making a negative contribution to the total, or 'absorbing' acoustic potential energy (in fact, in a real source it is simply using the pressure field to help 'push' the source, reducing its mechanical impedance). At high control source amplitudes, these roles have reversed. Only at one point, the optimum amplitude ratio of approximately 1.0 for this frequency, is the control source neither producing nor absorbing, and it is at this point the maximum attenuation of acoustic potential energy occurs.

Having now seen how much attenuation is maximally possible for a given control source arrangement, it is time to consider including a microphone into the system to make it practically realisable. Figure 9.11 illustrates the levels of acoustic potential energy reduction with a control source at $x = L_x$ as function of error microphone location for two frequencies: 100 Hz, which is dominated by the resonant response of the (2,0,0) mode, and 125 Hz, the response at which is not dominated by a single mode. For the 100 Hz case, significant levels of attenuation can be achieved at any error microphone location away from the nodes of the resonant mode, as the resonant mode dominates the pressure field as well as the acoustic potential energy. For the 125 Hz case, however, at most locations there is a significant increase in the sound field when minimising acoustic pressure at a single point. These results are reflected in the sound field plots shown in Figures 9.12 and 9.13. Observe that while there is a reduction in acoustic pressure at the error sensor for the 125 Hz case, there is an overall increase in the sound pressure level or acoustic potential energy.

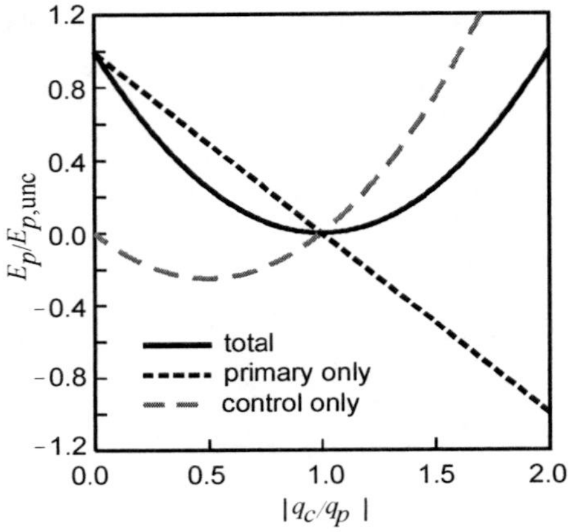

Figure 9.10 Total and individual source normalised acoustic potential energy with out of phase primary and control inputs as a function of volume velocity amplitude ratio, control source at $x = L_x$, 100 Hz.

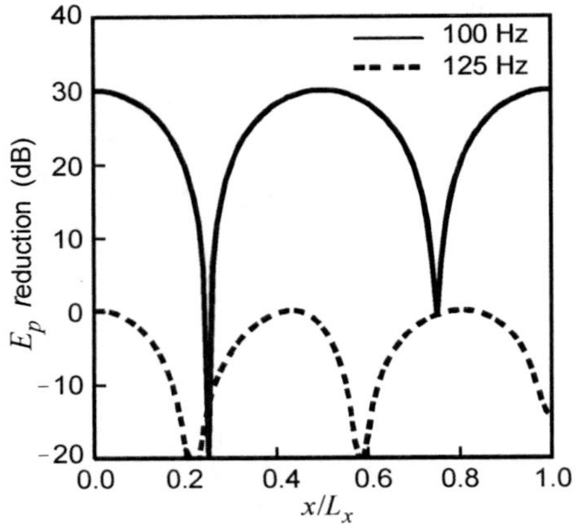

Figure 9.11 Acoustic potential energy reduction resulting from minimising the output of a single error sensor as a function of location of error sensor location for a single point control source at $x = L_x$ for 100 Hz and 125 Hz excitation.

While considering residual sound fields, it is insightful to compare the residual sound fields under the action of an optimally adjusted control source, shown in Figure 9.14 for the 100 Hz case and Figure 9.13 for the 125 Hz case, with the plot of acoustic potential energy reduction as a function of error sensor position given in Figure 9.11. Note that the locations of maximum pressure difference between the primary and optimal residual fields (determined that the acoustic potential energy can be accurately measured) are also the locations at which a single error sensor should be placed to obtain the maximum possible reduction in acoustic potential energy as a result of minimising the acoustic pressure at one point. This is a general result, to be discussed further in Section 9.7, which can be used to optimally place acoustic error sensors once a control source arrangement has been decided upon.

A conclusion that can be drawn from this short example is that active control of sound fields in lightly damped enclosed spaces is most effective at resonances of the acoustic space. In these instances, the problem is essentially the control of a single mode, which increases the potential for attenuation, as well as making the placement of control sources and error sensors a more forgiving task. For multiple modes to be controlled, the number of control sources and error sensors can be increased; however, the potential for attenuation is never as great as at a resonance frequency.

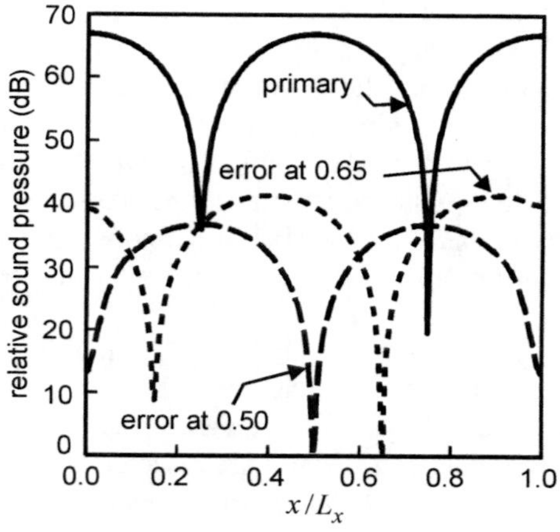

Figure 9.12 Primary and controlled sound fields resulting from minimising acoustic pressure at a single location, control source at $x = L_x$, 100 Hz. Illustrated are the primary sound field, the controlled sound field with an error sensor at $x/L_x = 0.50$, and the controlled sound field with an error at $x/L_x = 0.65$.

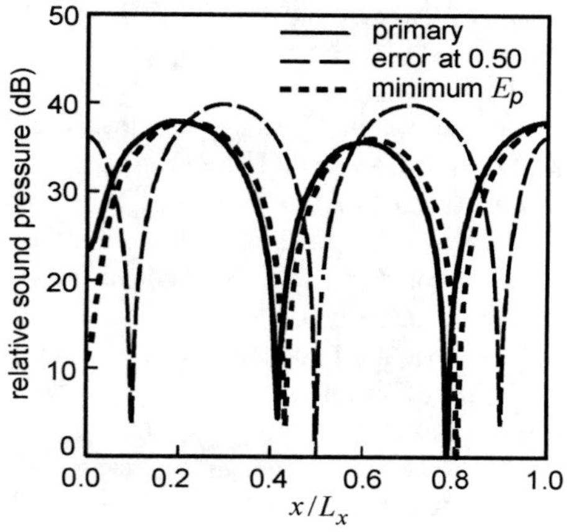

Figure 9.13 Primary and controlled sound fields resulting from minimising acoustic pressure at a single location, control source at $x = L_x$, 125 Hz. Illustrated are the primary sound field, the controlled sound field with an error sensor at $x/L_x = 0.50$, and the controlled sound field with an error at $x/L_x = 0.65$.

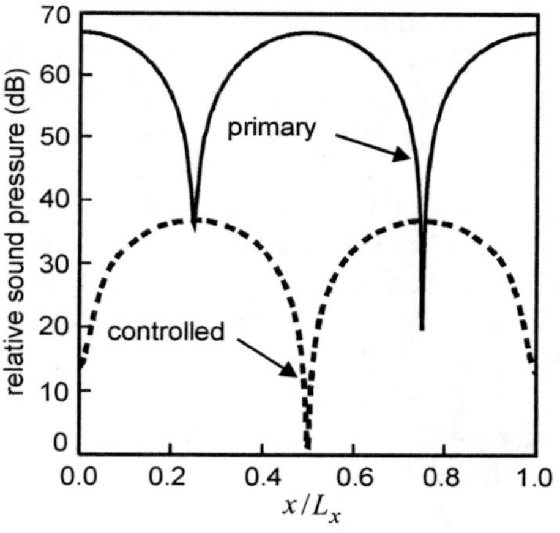

Figure 9.14 Primary and controlled sound fields resultant from minimising acoustic potential energy, control source at $x = L_x$, 100 Hz.

9.4 CONTROL OF SOUND FIELDS IN COUPLED ENCLOSURES AT DISCRETE LOCATIONS

The discussion in the previous two sections has concentrated on the active control of harmonic sound fields in rigid walled enclosures, where discrete sound sources either contained within the enclosed space or on its boundary were solely responsible for the sound field. Many practical systems, however, have enclosed sound fields that actually result from the vibration of part or all of the enclosing structure. A common example of this is an aircraft, where the interior (cabin) noise is generated by the vibration of the fuselage. While the analysis of the previous two sections provides useful qualitative results for coupled structural/acoustic systems such as this, in many ways it falls short. An obvious example of this is the fact that it is not possible to examine the use of vibration sources for controlling sound transmission and vibration sensors for providing an error signal unless the vibration of the enclosing structure is included in the model. The aim of the next two sections is to extend the work of the previous two sections by considering the active control of harmonic sound fields in coupled enclosures, where the vibration of the surrounding structure is responsible for the noise disturbance.

Several methods for modelling coupled enclosures for active control have been utilised in the past (Bullmore et al., 1986; Fuller, 1986; Pan et al., 1990). The one to be developed here is based upon modal coupling theory (Pope, 1971; Fahy, 1985; Pan and Bies, 1990a; Pan et al., 1990; Snyder and Hansen, 1994a). Modal coupling theory uses the premise that there is 'weak' modal coupling between the vibrating structural and acoustic modes, such that the *in vacuo* mode shapes of the structure and the mode shapes of a rigidly enclosed space can be used to determine the coupled system response. If the fluid medium is non-dense, such as air, and the structure is not 'small', weak coupling can normally be assumed, and this approach will provide an accurate model of the system response. The advantage of using modal coupling theory is that, theoretically, any weakly coupled system can be considered, the only requirement being a knowledge of the structural and acoustic resonance frequencies, the associated mode shape functions and a calculation of the coupling between them.

The analysis in this section will develop the basic equations required for modelling the coupled system response, then go on to apply them to the problem of minimising the acoustic field at discrete error sensing locations. The section that follows will extend this analysis to consider minimisation of acoustic potential energy in the enclosed space, as well as several 'hybrid' error criteria. Specific examples using the theoretical models developed here and in the next section will be given in Section 9.6, during a discussion of the control mechanisms, and in Section 9.7, during a discussion of the influence of control source and error sensor location. As in the previous studies of rigid walled enclosures, the aim of the analysis outlined here is to derive an equation defining acoustic pressure at a point as a matrix expression. From this equation it will be relatively easy to derive a relationship for the control source outputs that will minimise the pressure.

The first step in the analysis is to derive an expression for the structural velocity levels that result from a given external forcing function. Once these are known, the resulting interior sound field can be examined. Referring to Figure 9.15, the velocity at any location $\vec{x}$ on the vibrating structure is defined by the relationship:

$$v(\boldsymbol{x}) = \mathrm{j}\omega \int_S G_s(\boldsymbol{x}'|\boldsymbol{x})[p_{ext}(\boldsymbol{x}') - p(\boldsymbol{x}')]\,\mathrm{d}\boldsymbol{x}' \tag{9.4.1}$$

where ω is the frequency of interest in rad/s, p_{ext} and p are the external and internal acoustic pressures respectively, and $\boldsymbol{x}$ is a location on the vibrating structure. $G_s(\boldsymbol{x}'|\boldsymbol{x})$ is the structural Green's function, defined as:

$$G_s(\boldsymbol{x}'|\boldsymbol{x}) = \sum_{i=1}^{\infty} \frac{\psi_i(\boldsymbol{x}')\psi_i(\boldsymbol{x})}{M_i Z_i} \tag{9.4.2}$$

In the Green's function of Equation (9.4.2), $\psi_i(\boldsymbol{x})$ is the ith mode shape of the structure evaluated at location $\boldsymbol{x}$ on the structure, and M_i and Z_i are the modal mass and *in vacuo* structural input impedance of the ith mode, defined as:

$$M_i = \int_S m(\boldsymbol{x})\psi_i^2(\boldsymbol{x})\,\mathrm{d}\boldsymbol{x}; \qquad Z_i = (\omega_i^2 + \mathrm{j}\eta_i\omega_i^2 - \omega^2) \tag{9.4.3a,b}$$

where $m(\boldsymbol{x}) = \rho_s(\boldsymbol{x})h(\boldsymbol{x})$ is the surface density at location $\boldsymbol{x}$, $\rho_s(\boldsymbol{x})$ and $h(\boldsymbol{x})$ are respectively the density and thickness of the structure at $\boldsymbol{x}$, ω_i is the resonance frequency of the ith structural mode, and η_i is its associated hysteretic loss factor. For the analysis outlined here, a more useful set of quantities than surface velocity at certain locations on the structure are modal velocity amplitudes v_i. Equation (9.4.1) can be re-expressed in terms of these by expanding the surface velocity term in terms of modes as:

$$\sum_{i=1}^{\infty} v_i\psi_i(\boldsymbol{x}) = \mathrm{j}\omega\int_S G_s(\boldsymbol{x}'|\boldsymbol{x})[p_{ext}(\boldsymbol{x}') - p(\boldsymbol{x}')]\,\mathrm{d}\boldsymbol{x}' \tag{9.4.4}$$

To simplify the notation, a single structural mode, mode r will be considered, having a velocity at location $\boldsymbol{x}$ of:

$$v_r\psi_r(\boldsymbol{x}) = \mathrm{j}\omega\left\{\int_S \frac{\psi_r(\boldsymbol{x}')\psi_r(\boldsymbol{x})}{M_r Z_r}[p_{ext}(\boldsymbol{x}') - p(\boldsymbol{x}')]\,\mathrm{d}\boldsymbol{x}'\right\} \tag{9.4.5}$$

The analysis will then be extended at the end to included consideration of the entire set of modes to be modelled.

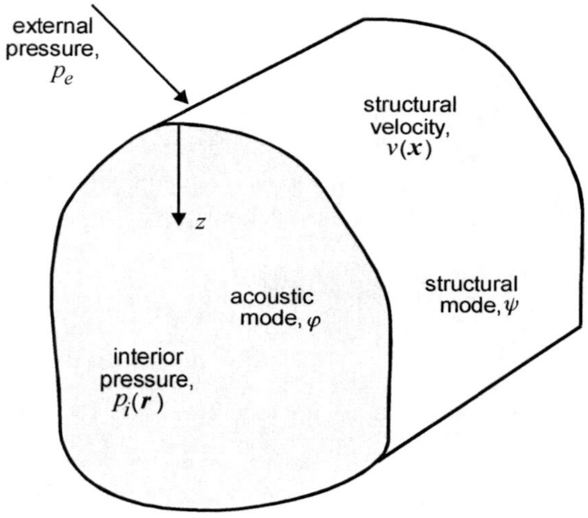

Figure 9.15 Sound transmission into a coupled enclosure.

To derive a matrix-based expression for structural velocity, the bracketed integral expression in Equation (9.4.5) must be simplified. For the first term, this is done simply by rewriting it as:

$$\int_S \frac{\psi_r(\mathbf{x}')\psi_r(\mathbf{x})}{M_r Z_r} p_{ext}(\mathbf{x}')\,\mathrm{d}\mathbf{x}' = \frac{\gamma_r \psi_r(\mathbf{x})}{M_r Z_r} \tag{9.4.6}$$

where γ_r is the rth modal generalised force:

$$\gamma_r = \int_S p_{ext}(\mathbf{x}')\psi_r(\mathbf{x}')\,\mathrm{d}\mathbf{x}' \tag{9.4.7}$$

The simplification of the second term in the brackets, however, will require a bit more work.

Equation (9.2.1), defining acoustic pressure in the enclosed space resulting from some volume velocity source distribution, can be specialised for the case where the source distribution is the vibrating enclosing structure by replacing acoustic mode shape functions with structural mode shape functions, as the pressure of interest is adjacent to the structure surface:

$$p(\mathbf{x}') = \mathrm{j}\rho_0\omega \int_S \sum_{\imath=1}^{\infty} \frac{\psi_\imath(\mathbf{x}'')\psi_\imath(\mathbf{x}')}{\Lambda_\imath(\kappa_\imath^2 - k^2)} v(\mathbf{x}'')\,\mathrm{d}\mathbf{x}'' \tag{9.4.8}$$

If the velocity term $v(\mathbf{x}'')$ is expanded in terms of structural modes, this can be written as:

$$p(\mathbf{x}') = \mathrm{j}\rho_0\omega \int_S \sum_{\imath=1}^{\infty} \frac{\psi_\imath(\mathbf{x}'')\psi_\imath(\mathbf{x}')}{\Lambda_\imath(\kappa_\imath^2 - k^2)} \sum_{i=1}^{\infty} v_i\psi_i(\mathbf{x}'')\,\mathrm{d}\mathbf{x}'' \tag{9.4.9}$$

or

$$p(\mathbf{x}') = \mathrm{j}\rho_0\omega S \sum_{\imath=1}^{\infty} \frac{\psi_\imath(\mathbf{x}')}{\Lambda_\imath(\kappa_\imath^2 - k^2)} \sum_{i=1}^{\infty} v_i B_{\imath,i} \tag{9.4.10}$$

where S is the surface area of the structure and $B_{\imath,i}$ is the non-dimensional modal coupling coefficient between the $\imath$th acoustic mode and ith structural mode:

$$B_{\imath,i} = \frac{1}{S}\int_S \psi_i(\mathbf{x}'')\varphi_\imath(\mathbf{x}'')\,\mathrm{d}\mathbf{x}'' \tag{9.4.11}$$

Equation (9.4.10) can be used to expand the second term in the bracketed integral expression in Equation (9.4.5) as follows:

$$\int_S \frac{\psi_r(\mathbf{x}')\psi_r(\mathbf{x})}{M_r Z_r} p(\mathbf{x}')\,\mathrm{d}\mathbf{x} = \mathrm{j}\rho_0\omega S \int_S \frac{\psi_r(\mathbf{x}')\psi_r(\mathbf{x})}{M_r Z_r} \sum_{\imath=1}^{\infty} \frac{\psi_\imath(\mathbf{x}')}{\Lambda_\imath(\kappa_\imath^2 - k^2)} \sum_{i=1}^{\infty} v_i B_{\imath,i}\,\mathrm{d}\mathbf{x} \tag{9.4.12}$$

Equation (9.4.12) can be further simplified by again using the modal coupling coefficient defined in Equation (9.4.11):

$$\int_S \frac{\psi_r(\boldsymbol{x}')\psi_r(\boldsymbol{x})}{M_r Z_r} p(\boldsymbol{x}')\,\mathrm{d}\boldsymbol{x} = \mathrm{j}\rho_0 \omega S^2 \frac{\psi_r(\boldsymbol{x})}{M_r Z_r} \sum_{\imath=1}^{\infty} \sum_{i=1}^{\infty} \frac{v_i B_{\imath,r} B_{\imath,i}}{\Lambda_\imath(\kappa_\imath^2 - k^2)} \tag{9.4.13}$$

Substituting the simplified expressions of Equations (9.4.6) and (9.4.13) into Equation (9.4.5), and dividing through by the mode shape term, $\psi(\boldsymbol{x})$, produces the defining expression for the rth modal velocity amplitude in response to an external exciting force:

$$v_r = \mathrm{j}\frac{\omega \gamma_r}{M_r Z_r} + \frac{\rho_0 \omega^2 S^2}{M_r Z_r} \sum_{\imath=1}^{\infty} \sum_{i=1}^{\infty} \frac{v_i B_{\imath,r} B_{\imath,i}}{\Lambda_\imath(\kappa_\imath^2 - k^2)} \tag{9.4.14}$$

This can be restated as:

$$\left(\mathrm{j}\rho_0 S^2 \omega \left(\sum_{\imath=1}^{\infty} \frac{B_{\imath,r} B_{\imath,r}}{\Lambda_\imath(\kappa_\imath^2 - k^2)} \right) - \frac{\mathrm{j} M_r Z_r}{\omega} \right) v_r + \sum_{i=1,\neq r}^{\infty} \mathrm{j}\rho_0 S^2 \omega \left(\sum_{\imath=1}^{\infty} \frac{B_{\imath,r} B_{\imath,i}}{\Lambda_\imath(\kappa_\imath^2 - k^2)} \right) v_i = \gamma_{r,p} \tag{9.4.15}$$

To write Equation (9.4.15) as a matrix expression, it is necessary to truncate the infinite summation of structural modes to N_m modes, and the infinite summation of acoustic modes to N_n modes. As was discussed in Section 9.2, such a truncation effectively limits the utility of the analysis to low frequencies, where a 'manageable' number of modes can be used in the calculations and a sufficiently accurate answer still obtained. With these truncations, the (velocity) response of the structure to the external forcing function can be written as:

$$\boldsymbol{v} = \boldsymbol{Z}_I^{-1} \boldsymbol{\gamma} \tag{9.4.16}$$

where $\boldsymbol{v}$ is the ($N_m \times 1$) vector of structural modal velocities resulting from the external forcing function, $\boldsymbol{Z}_I$ is the ($N_m \times N_m$) structural modal input impedance matrix, the terms of which are defined by the expressions:

$$Z_I(u,u) = \mathrm{j}\rho_0 S^2 \omega \sum_{\imath=1}^{n} \frac{B_{\imath,u} B_{\imath,u}}{\Lambda_\imath(\kappa_\imath^2 - k^2)} - \frac{\mathrm{j} M_u Z_u}{\omega} = \text{diagonal terms} \tag{9.4.17}$$

$$Z_I(u,v) = \mathrm{j}\rho_0 S^2 \omega \sum_{\imath=1}^{n} \frac{B_{\imath,u} B_{\imath,v}}{\Lambda_\imath(\kappa_\imath^2 - k^2)} = \text{off–diagonal terms} \tag{9.4.18}$$

where u,v refer to the uth and vth structural modes respectively, $Z_I(u,u)$ represents the velocity response of mode u to a unit excitation force, $Z_I(u,v)$ represents the response of mode v to a unit response of mode u, and $\boldsymbol{\gamma}$ is the ($N_m \times 1$) vector of modal generalised forces, the ith element of which is:

$$\gamma(i) = \gamma_i = \int_S \psi_i(\boldsymbol{x})\, p_{ext}(\boldsymbol{x})\, \mathrm{d}\boldsymbol{x} \tag{9.4.19}$$

Having derived expressions describing the structural velocity levels for a given external forcing function, attention must now be directed back to the acoustic space to derive expressions describing the resulting sound field. If the structural velocity term in Equation (9.4.8) is expanded using the modelled set of N_m modes, the internal pressure distribution can be expressed as:

$$p(\boldsymbol{r}) = \mathrm{j}\rho_0\omega \int_S \frac{\varphi_\imath(\boldsymbol{x})\varphi_\imath(\boldsymbol{r})}{\Lambda_\imath(\kappa_\imath^2 - k^2)} v(\boldsymbol{x})\, \mathrm{d}\boldsymbol{x} = \mathrm{j}\rho_0\omega \sum_{\imath=1}^{N_n} \frac{\varphi_\imath(\boldsymbol{r})}{\Lambda_\imath(\kappa_\imath^2 - k^2)} \sum_{i=1}^{N_m} v_i \int_S \psi_i(\boldsymbol{x})\varphi_\imath(\boldsymbol{x})\, \mathrm{d}\boldsymbol{x} \tag{9.4.20a,b}$$

This can be written more compactly using the modal coupling coefficient defined in Equation (9.4.11) as:

$$p(\boldsymbol{r}) = \mathrm{j}\rho_0\omega S \sum_{\imath=1}^{N_n} \frac{\varphi_\imath(\boldsymbol{r})}{\Lambda_\imath(\kappa_\imath^2 - k^2)} \sum_{i=1}^{N_m} v_i B_{\imath,i} \tag{9.4.21}$$

Expressed in this way, Equation (9.4.21) can also be written as a matrix expression:

$$p(\boldsymbol{r}) = \boldsymbol{\varphi}^{\mathrm{T}}(\boldsymbol{r})\boldsymbol{p} = \boldsymbol{\varphi}^{\mathrm{T}}(\boldsymbol{r})\, \boldsymbol{Z}_a\, \boldsymbol{B}\, \boldsymbol{v} = \boldsymbol{\varphi}^{\mathrm{T}}(\boldsymbol{r})\, \boldsymbol{Z}_a\, \boldsymbol{B}\, \boldsymbol{Z}_I^{-1}\, \boldsymbol{\gamma} \tag{9.4.22a–c}$$

Equation (9.4.22) is the target of this analysis, expressing acoustic pressure $p(\boldsymbol{r})$ at any location in the enclosed space resulting from vibration of the structure. The quantity $\boldsymbol{p}$ is the $(N_n \times 1)$ vector of complex acoustic modal amplitudes, and $\boldsymbol{Z}_a$ is the $(N_n \times N_n)$ diagonal matrix of acoustic modal radiation transfer functions, the terms of which were originally defined in Equation (9.2.18) by the relation:

$$Z_a(\imath,\imath) = \frac{\mathrm{j}\rho_0\omega}{\Lambda_\imath(\kappa_\imath^2 - k^2)} \tag{9.4.23}$$

and $\boldsymbol{B}$ is the $(N_m \times N_n)$ matrix of modal coupling coefficients, the elements of which are defined by:

$$B(\imath,i) = SB_{\imath,i} \tag{9.4.24}$$

Before using Equation (9.4.22) to derive control source relationships for minimising the acoustic pressure at discrete error sensing locations, it is interesting to compare this expression with Equation (9.2.33), the equivalent relationship for the use of multiple discrete velocity sources in rigid walled enclosures.

For a single error sensing location, the first two matrix terms, $\boldsymbol{\varphi}^{\mathrm{T}}(\boldsymbol{r}_e)$ and $\boldsymbol{Z}_a$, are the same. If the terms in the matrix $\boldsymbol{B}$ are considered in light of the definition of the modal coupling coefficient given in Equation (9.4.11), it can be seen that they are equivalent to the terms in the matrix of modal generalised volume velocity transfer functions $\boldsymbol{\Phi}_g$ used in the discrete source formulation, the terms of which are defined in Equation (9.2.11). That is, the terms in $\boldsymbol{B}$ define the volume velocity excitation of the acoustic modes in the enclosed space, in response to a unit velocity amplitude for each structural mode. In Equation (9.4.22), each structural mode is treated in a manner analogous to a single discrete source in a rigid walled enclosure, treatment that can be viewed as a 'modal' approach to the problem. Later in this chapter a different approach will be discussed, one that can be viewed as a 'spatial' approach to the problem using boundary element methods, where each node on a grid of locations placed over the structure is viewed as a discrete sound source. Each of these methods has pros and cons. If mode shapes can be found, the modal approach often lends itself to an enhanced physical understanding of the physical phenomena employed by active control systems. However, for complicated systems modelled using finite element methods, a spatial approach may make more sense computationally.

One point that should be mentioned briefly here is that structural/acoustic modal coupling is far more selective in terms of what acoustic modes will be excited by a given structural mode than the modal generalised volume velocity transfer functions were in determining what acoustic modes will be excited by, for example, a monopole source. In the latter case, a source must be located on a nodal line of an acoustic mode for it not to be excited. Modal coupling can be selective to the point of 'allowing' a structural mode to effectively excite only a single acoustic mode in a given frequency band, as is the case for a finite cylinder (considered in more depth in Section 9.7).

Figure 9.16 illustrates a coupled and uncoupled structural/acoustic modal pair. In fact, for systems of regular geometry, often a necessary but not sufficient requirement for coupling is that both structural and acoustic modes be either symmetric or anti-symmetric about some line of symmetry in the structure. In Figure 9.16, only the anti-symmetric acoustic mode couples with the anti-symmetric structural mode. This characteristic of structural/acoustic modal coupling becomes important when assessing control mechanisms, as will be discussed in Section 9.7.

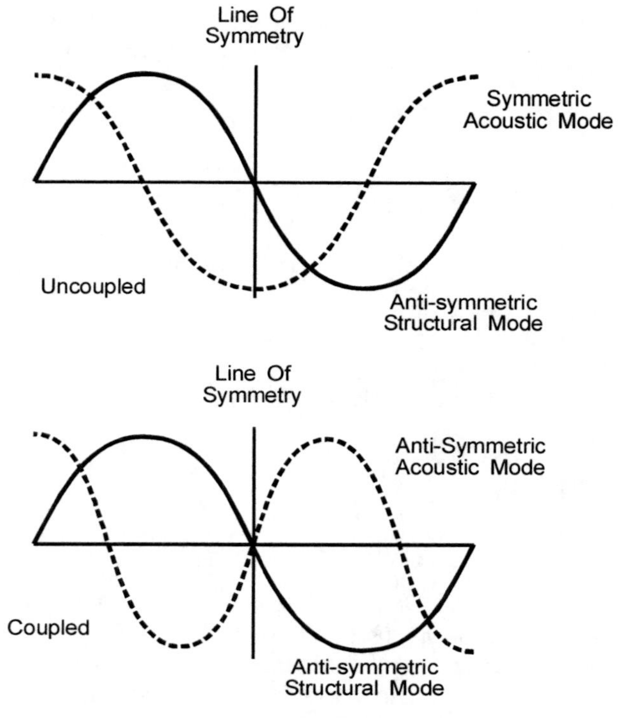

Figure 9.16 Example of structural/acoustic modal coupling.

The problem of minimising the sound pressure at discrete error sensing locations in the enclosed space will now be considered. There are two types of control source available to achieve this objective: discrete acoustic control sources, as were considered in Section 9.2 with rigid walled enclosures, and vibration control sources, attached directly to the vibrating structure. The former acoustic control sources will be considered first.

The squared amplitude of the acoustic pressure at some error sensing location $\boldsymbol{r}_e$ during operation of the control and primary source distributions, is equal to:

$$|p(\boldsymbol{r}_e)|^2 = p(\boldsymbol{r}_e)^* p(\boldsymbol{r}_e) = [p_c(\boldsymbol{r}_e) + p_p(\boldsymbol{r}_e)]^* [p_c(\boldsymbol{r}_e) + p_p(\boldsymbol{r}_e)] \tag{9.4.25a,b}$$

where the subscripts p and c denote primary and control source related quantities respectively.

Here, the primary noise source is the vibrating structure, for which the governing acoustic pressure field relationship is given in Equation (9.4.22). For acoustic control sources, in many weakly coupled structural/acoustic systems, sufficient accuracy is attained by considering the enclosure to be rigid walled, neglecting the structural/acoustic interaction when formulating the sound field present in the enclosure during operation of the control sources alone. The exception to this is at a structural resonance where neglecting this interaction may result in some discrepancy between the theoretical and experimental results. The interaction will be neglected here for brevity; those interested in including it can refer to the work of Pan (1992).

Neglecting any structural/acoustic interaction enables use of the previously derived relationship for the sound field in a rigid walled enclosure, Equation (9.2.20), to describe the pressure field generated by a single control source. Using Equations (9.4.22) and (9.2.20), the squared acoustic pressure amplitude at $\vec{r}_e$ can be re-expressed as:

$$|p(\boldsymbol{r}_e)|^2 = q_c^* a q_c + q_c^* b + b^* q_c + c \tag{9.4.26}$$

where

$$a = \boldsymbol{\varphi}_{gc}^{\mathrm{H}} \boldsymbol{Z}_a^{\mathrm{H}} \boldsymbol{Z}_p \boldsymbol{Z}_a \boldsymbol{\varphi}_{gc} \tag{9.4.27}$$

$$b = \boldsymbol{\varphi}_{gc}^{\mathrm{H}} \boldsymbol{Z}_a^{\mathrm{H}} \boldsymbol{Z}_p \boldsymbol{Z}_a \boldsymbol{B} \boldsymbol{v}_p \tag{9.4.28}$$

$$c = \boldsymbol{v}_p^{\mathrm{H}} \boldsymbol{B}^{\mathrm{T}} \boldsymbol{Z}_a^{\mathrm{H}} \boldsymbol{Z}_p \boldsymbol{Z}_a \boldsymbol{B} \boldsymbol{v}_p \tag{9.4.29}$$

and where

$$\boldsymbol{Z}_p = \boldsymbol{\varphi}(\boldsymbol{r}) \boldsymbol{\varphi}^{\mathrm{T}}(\boldsymbol{r}) \tag{9.4.30}$$

In Equations (9.4.28) and (9.4.29), the structural modal velocities resulting from the primary excitation $\boldsymbol{v}_p$ have been explicitly used in the equations. Alternatively, the primary acoustic modal amplitudes or structural input impedance and modal generalised force could be used, as shown in Equation (9.4.22). Note also that when confronted with a practical problem, measured data could easily be used in these primary source related matrix quantities.

Equation (9.4.26) is in what is now a fairly standard form, so that the following relationship can be written for the single optimum control source output directly:

$$q_{c,p} = -a^{-1}b \tag{9.4.31}$$

which will result in the minimised squared pressure amplitude at the error sensing location of:

$$|p(\boldsymbol{r}_e)|^2_{\min} = c - b^* a^{-1} b \tag{9.4.32}$$

For a single error sensor and single control source, this minimum value will be theoretically equal to zero.

As with the rigid walled enclosure, this result can easily be extended to include consideration of multiple control sources and error sensing positions. Using quantities defined in Section 9.2, the sum of squared acoustic pressure amplitudes at a set of error sensing locations can be expressed as:

$$\sum_{i=1}^{N_e} |p(\boldsymbol{r}_e)|^2 = \boldsymbol{q}_c^{\mathrm{H}} \boldsymbol{A} \boldsymbol{q}_c + \boldsymbol{q}_c^{\mathrm{H}} \boldsymbol{b} + \boldsymbol{b}^{\mathrm{H}} \boldsymbol{q}_c + c \tag{9.4.33}$$

where

$$\boldsymbol{A} = \boldsymbol{\Phi}_{gc}^{\mathrm{H}} \boldsymbol{Z}_a^{\mathrm{H}} \boldsymbol{Z}_p \boldsymbol{Z}_a \boldsymbol{\Phi}_{gc} \tag{9.4.34}$$

$$\boldsymbol{b} = \boldsymbol{\Phi}_{gc}^{\mathrm{H}} \boldsymbol{Z}_a^{\mathrm{H}} \boldsymbol{Z}_p \boldsymbol{Z}_a \boldsymbol{B} \boldsymbol{v}_p \tag{9.4.35}$$

$$c = \boldsymbol{v}_p^{\mathrm{H}} \boldsymbol{B}^{\mathrm{T}} \boldsymbol{Z}_a^{\mathrm{H}} \boldsymbol{Z}_p \boldsymbol{\varphi}_e \boldsymbol{Z}_a \boldsymbol{B} \boldsymbol{v}_p \tag{9.4.36}$$

and where $\boldsymbol{Z}_p$ is now based upon multiple error sensing locations and is defined as:

$$\boldsymbol{Z}_p = \boldsymbol{\Phi}_e^{\mathrm{T}} \boldsymbol{\Phi}_e \tag{9.4.37}$$

Equation (9.4.33) is again in standard form for multiple source problems, with the vector of optimum control source outputs defined by the relationship

$$\boldsymbol{q}_{c,p} = c - \boldsymbol{b}^{\mathrm{H}} \boldsymbol{A}^{-1} \boldsymbol{b} \tag{9.4.38}$$

The sum of the residual squared acoustic pressure amplitudes under optimal conditions, found by substituting Equation (9.4.38) into Equation (9.4.33), is:

$$\left(\sum_{i=1}^{N_e} |p(\boldsymbol{r}_e)|^2 \right)_{\min} = c - \boldsymbol{b}^{\mathrm{H}} \boldsymbol{A}^{-1} \boldsymbol{b} \tag{9.4.39}$$

The second control source option, vibration sources, will now be considered. As the acoustic pressure at the error sensor(s) due to operation of these sources is a result of structural radiation, the previous analysis, culminating in Equation (9.4.22), will be used in formulating both the control and primary source contributions. As the systems that are of interest here are linear, it is possible to simply arithmetically add the two contributions in the analysis. What is required first is to re-express the modal generalised force term γ to directly reflect the control source input. To do this, note that the rth modal generalised force, as defined in Equation (9.4.7), can be rewritten for a single vibration source as:

$$\gamma_r = \int_S p_{ext}(\boldsymbol{x})\psi_r(\boldsymbol{x})\,\mathrm{d}\boldsymbol{x} = f_c\int_S z_f(\boldsymbol{x})\psi_r(\boldsymbol{x})\,\mathrm{d}\boldsymbol{x} \tag{9.4.40a,b}$$

where f_c is a complex force amplitude, and $z_f(\boldsymbol{x})$ is a transfer function between this force and the pressure at location $\boldsymbol{x}$. This can be simplified to:

$$\gamma_r = f_c\psi_{g,r} \tag{9.4.41}$$

where $\psi_{g,r}$ is a modal generalised force transfer function, defined as:

$$\psi_{g,r} = \int_S z_f(\boldsymbol{x})\psi_r(\boldsymbol{x})\,\mathrm{d}\boldsymbol{x} \tag{9.4.42}$$

If, for example, the control source were a point force f_c, the modal generalised force would be:

$$f_c\psi_{g,r} = \int_S \psi_r(\boldsymbol{x})f_c\,\delta(\boldsymbol{x}-\boldsymbol{x}_f)\,\mathrm{d}\boldsymbol{x} = f_c\psi(\boldsymbol{r}_f) \tag{9.4.43a,b}$$

where f_c is the complex amplitude of the force, input at a location $\boldsymbol{x}_f$, and $\delta(\boldsymbol{x} - \boldsymbol{x}_f)$ is a Dirac delta function. In this case, the modal generalised force transfer function $\psi_{g,r}$ is simply the value of the mode shape at the point force input location, as follows:

$$\psi_{g,r} = \int_S \psi_r(\boldsymbol{x})\,\delta(\boldsymbol{x}-\boldsymbol{x}_f)\,\mathrm{d}\boldsymbol{x} = \psi_r(\boldsymbol{x}_f) \tag{9.4.44a,b}$$

Substituting this expansion in Equation (9.4.22) gives the acoustic field generated by the vibration control source, which is defined by the relationship:

$$p(\boldsymbol{r}) = \boldsymbol{\varphi}^{\mathrm{T}}(\boldsymbol{r})\,\boldsymbol{Z_a}\,\boldsymbol{B}\,\boldsymbol{Z_I}^{-1}\,\boldsymbol{\psi_{gc}}\,f_c \tag{9.4.45}$$

where $\boldsymbol{\psi_{gc}}$ is the $(N_m \times 1)$ vector of control source modal generalised force transfer functions. Using Equation (9.4.22) for the primary source and Equation (9.4.45) for the control source, the squared acoustic pressure amplitude at $\boldsymbol{r_e}$ as defined in Equation (9.4.25) can be expressed as:

$$|p(\boldsymbol{r_e})|^2 = f_c^* a f_c + f_c^* b + b^* f_c + c \tag{9.4.46}$$

where

$$a = \boldsymbol{\psi_{gc}}^{\mathrm{H}}\left\{\boldsymbol{Z_I}^{-1}\right\}^{\mathrm{H}}\boldsymbol{B}^{\mathrm{T}}\,\boldsymbol{Z_a}^{\mathrm{H}}\,\boldsymbol{Z_p}\,\boldsymbol{Z_a}\,\boldsymbol{B}\,\boldsymbol{Z_I}^{-1}\boldsymbol{\psi_{gc}} \tag{9.4.47}$$

$$b = \psi_{gc}^{\mathrm{T}} \left\{ \boldsymbol{Z}_I^{-1} \right\}^{\mathrm{H}} \boldsymbol{B}^{\mathrm{T}} \boldsymbol{Z}_a^{\mathrm{H}} \boldsymbol{Z}_p \boldsymbol{Z}_a \boldsymbol{B} \boldsymbol{Z}_I^{-1} \boldsymbol{v}_p \tag{9.4.48}$$

and

$$c = \boldsymbol{v}_p^{\mathrm{H}} \boldsymbol{B}^{\mathrm{T}} \boldsymbol{Z}_a^{\mathrm{H}} \boldsymbol{Z}_p \boldsymbol{Z}_a \boldsymbol{B} \boldsymbol{v}_p \tag{9.4.49}$$

where $\boldsymbol{Z}_p$ is as defined in Equation (9.4.30). Equation (9.4.46) is written in standard form, this time as a quadratic function of vibration control source amplitude. This enables the following relationship to be written for the optimum control force:

$$f_{c,p} = -a^{-1} b \tag{9.4.50}$$

which results in the minimised squared pressure amplitude at the error sensing location of:

$$|p(r_e)|^2_{\min} = c - b^* a^{-1} b \tag{9.4.51}$$

Again, the subscripts p and c denote primary and control source respectively, and for a single error sensor and single control source, this minimum value will be theoretically equal to zero.

This analysis can also be extended easily to include consideration of multiple sources and error sensing positions. Defining the ($N_c \times 1$) vector of control forces, $\boldsymbol{f}_c$:

$$\boldsymbol{f}_c = \begin{bmatrix} f_1 \\ f_2 \\ \vdots \\ f_{N_c} \end{bmatrix} \tag{9.4.52}$$

and the ($N_m \times N_c$) matrix whose columns are the control source modal generalised force transfer function vectors, $\boldsymbol{\Psi}_{gc}$:

$$\boldsymbol{\Psi}_{gc} = \left[\psi_{g1} \;\; \psi_{g2} \;\; \cdots \;\; \psi_{gN_c} \right] \tag{9.4.53}$$

The sum of squared acoustic pressure amplitudes at a set of error sensing locations can be expressed as:

$$\sum_{i=1}^{N_e} |p(\boldsymbol{r}_e)|^2 = \boldsymbol{f}_c^{\mathrm{H}} \boldsymbol{A} \boldsymbol{f}_c + \boldsymbol{f}_c^{\mathrm{H}} \boldsymbol{b} + \boldsymbol{b}^{\mathrm{H}} \boldsymbol{f}_c + c \tag{9.4.54}$$

where

$$\boldsymbol{A} = \boldsymbol{\Psi}_{gc}^{\mathrm{H}} \left\{ \boldsymbol{Z}_I^{-1} \right\}^{\mathrm{H}} \boldsymbol{B}^{\mathrm{T}} \boldsymbol{Z}_a^{\mathrm{H}} \boldsymbol{Z}_p \boldsymbol{Z}_a \boldsymbol{B} \boldsymbol{Z}_I^{-1} \boldsymbol{\Psi}_{gc} \tag{9.4.55}$$

$$\boldsymbol{b} = \boldsymbol{\Psi}_{gc}^{\mathrm{H}} \left\{ \boldsymbol{Z}_I^{-1} \right\}^{\mathrm{H}} \boldsymbol{B}^{\mathrm{T}} \boldsymbol{Z}_a^{\mathrm{H}} \boldsymbol{Z}_p \boldsymbol{Z}_a \boldsymbol{B} \boldsymbol{v}_p \tag{9.4.56}$$

and

$$c = \boldsymbol{v}_p^{\mathrm{H}} \boldsymbol{B}^{\mathrm{T}} \boldsymbol{Z}_a^{\mathrm{H}} \boldsymbol{Z}_p \boldsymbol{Z}_a \boldsymbol{B} \boldsymbol{v}_p \tag{9.4.57}$$

where $\boldsymbol{Z}_p$ is now based upon multiple error sensors, as defined in Equation (9.4.37). Equation (9.4.54) is in standard form for multiple source problems, expressed as a quadratic function of vibration control source output. The vector of optimum control source outputs therefore is defined by the relationship:

$$\boldsymbol{f}_{c,p} = -\boldsymbol{A}^{-1}\boldsymbol{b} \tag{9.4.58}$$

The sum of the residual squared acoustic pressure amplitudes under optimal conditions, found by substituting Equation (9.4.58) into Equation (9.4.54), is:

$$\left(\sum_{i=1}^{N_e} |p(\boldsymbol{r}_e)|^2 \right)_{\min} = c - \boldsymbol{b}^{\mathrm{H}}\boldsymbol{A}^{-1}\boldsymbol{b} \tag{9.4.59}$$

9.5 MINIMISATION OF ACOUSTIC POTENTIAL ENERGY IN COUPLED ENCLOSURES

The relationships developed in the previous section facilitate calculation of the sound field in a coupled enclosure, as well as the acoustic or vibration control source outputs that will minimise the acoustic pressure at discrete locations in the field. What remains to be done is to extend these analyses to include consideration of the best possible result; minimisation of acoustic potential energy in the enclosed space. In addition, this section will consider a few 'hybrid' control source and error sensor arrangements, such as the use of both acoustic and vibration control sources and the minimisation of the velocity levels of the enclosing structure. The first task, however, is the minimisation of acoustic potential energy.

In Section 9.3, Equation (9.3.5), it was shown that acoustic potential energy is equal to the weighted sum of squared acoustic modal amplitudes. Then a vector relationship describing the acoustic modal amplitudes resulting from the operation of a discrete acoustic source (see Equation (9.3.7)) was derived. What is required here then, is a matrix expression for the vector of acoustic modal amplitudes, $\boldsymbol{p}$, resulting from excitation by the surrounding structure. Using Equations (9.4.21) and (9.4.22), this can be written directly as:

$$\boldsymbol{p} = \boldsymbol{Z}_a \boldsymbol{B} \boldsymbol{v} = \boldsymbol{Z}_a \boldsymbol{B} \boldsymbol{Z}_I^{-1} \boldsymbol{\gamma} \tag{9.5.1a,b}$$

Considering first the use of acoustic control sources, the acoustic potential energy in the enclosure can be written in matrix form as:

$$E_p = \frac{1}{4\rho_0 c_0^2} \sum_{i=1}^{N_n} \Lambda_i |p_i|^2 = \boldsymbol{q}_c^{\mathrm{H}}\boldsymbol{A}\boldsymbol{q}_c + \boldsymbol{q}_c^{\mathrm{H}}\boldsymbol{b} + \boldsymbol{b}^{\mathrm{H}}\boldsymbol{q}_c + c \tag{9.5.2a,b}$$

where

$$\boldsymbol{A} = \boldsymbol{\Phi}_{gc}^{\mathrm{H}} \boldsymbol{Z}_a^{\mathrm{H}} \boldsymbol{Z}_E \boldsymbol{Z}_a \boldsymbol{\Phi}_{gc} \tag{9.5.3}$$

$$b = \boldsymbol{\Phi}_{gc}^{\mathrm{H}} \, \boldsymbol{Z}_a^{\mathrm{H}} \, \boldsymbol{Z}_E \, \boldsymbol{Z}_a \, \boldsymbol{B} \, \boldsymbol{v}_p \tag{9.5.4}$$

and

$$c = \boldsymbol{v}_p^{\mathrm{H}} \, \boldsymbol{B}^{\mathrm{T}} \, \boldsymbol{Z}_a^{\mathrm{H}} \, \boldsymbol{Z}_E \, \boldsymbol{Z}_a \, \boldsymbol{B} \, \boldsymbol{v}_p \tag{9.5.5}$$

where the elements of the diagonal weighting matrix $\boldsymbol{Z}_E$ are defined by:

$$Z_E(i,i) = \frac{\Lambda_i}{4\rho_0 c_0^2} \tag{9.5.6}$$

Equation (9.5.2) is in standard form for multiple source problems, with the vector of optimum control source outputs defined by the relationship:

$$\boldsymbol{q}_{c,E} = -\boldsymbol{A}^{-1}\boldsymbol{b} \tag{9.5.7}$$

The minimum value of acoustic potential energy under optimal conditions, found by substituting Equation (9.5.7) into Equation (9.5.2), is:

$$E_{p,\min} = c - \boldsymbol{b}^{\mathrm{H}}\boldsymbol{A}^{-1}\boldsymbol{b} \tag{9.5.8}$$

A virtually identical set of equations can be written for the use of vibration control sources to minimise acoustic potential energy in the coupled enclosure. Using Equation (9.5.1) in conjunction with the expansion of the modal generalised force described in Equation (9.4.41), the acoustic potential energy in the enclosure can be written as a quadratic function of the vector of vibration control source outputs as:

$$E_p = \frac{1}{4\rho_0 c_0^2} \sum_{l=1}^{N_n} \Lambda_l |p_l|^2 = \boldsymbol{f}_c^{\mathrm{H}} \boldsymbol{A} \boldsymbol{f}_c + \boldsymbol{f}_c^{\mathrm{H}} \boldsymbol{b} + \boldsymbol{b}^{\mathrm{H}} \boldsymbol{f}_c + c \tag{9.5.9a,b}$$

where

$$\boldsymbol{A} = \boldsymbol{\Psi}_{gc}^{\mathrm{T}} \left\{\boldsymbol{Z}_I^{-1}\right\}^{\mathrm{H}} \boldsymbol{B}^{\mathrm{T}} \, \boldsymbol{Z}_a^{\mathrm{H}} \, \boldsymbol{Z}_E \, \boldsymbol{Z}_a \, \boldsymbol{B} \, \boldsymbol{Z}_I^{-1} \, \boldsymbol{\Psi}_{gc} \tag{9.5.10}$$

$$\boldsymbol{b} = \boldsymbol{\Psi}_{gc}^{\mathrm{T}} \left\{\boldsymbol{Z}_I^{-1}\right\}^{\mathrm{H}} \boldsymbol{B}^{\mathrm{T}} \, \boldsymbol{Z}_a^{\mathrm{H}} \, \boldsymbol{Z}_E \, \boldsymbol{Z}_a \, \boldsymbol{B} \, \boldsymbol{v}_p \tag{9.5.11}$$

and c and $\boldsymbol{Z}_E$ are as defined in Equations (9.5.5) and (9.5.6) respectively. Written in this form, the relationship defining the vector of optimum control source outputs can be written directly as:

$$\boldsymbol{f}_{c,E} = -\boldsymbol{A}^{-1}\boldsymbol{b} \tag{9.5.12}$$

leading to a minimum value of acoustic potential energy equal to:

$$E_{p,\,\min} = c - \boldsymbol{b}^{\mathrm{H}}\boldsymbol{A}^{-1}\boldsymbol{b} \tag{9.5.13}$$

In addition to the basic system arrangements involving the use of either acoustic or vibration control sources to minimise the sound pressure at discrete points or throughout the enclosed space, there are a number of 'hybrid' control source and error sensor arrangements that may be worthwhile investigating in some situations. The first of these is the use of both acoustic and vibration control sources to minimise acoustic pressure or acoustic potential energy. The analytical model required to describe this is a simple extension of those presented in the previous two sections, where the control source generated sound pressure is now the summation of that generated by the vibration control source(s), $p_{cv}(\boldsymbol{r}_e)$, and the acoustic control source(s), $p_{ca}(\boldsymbol{r}_e)$:

$$p_c(\boldsymbol{r}_e) = p_{cv}(\boldsymbol{r}_e) + p_{ca}(\boldsymbol{r}_e) \tag{9.5.14}$$

Using the matrices outlined previously in Section 9.4, the vector of acoustic pressures at a set of error sensor locations in the enclosure can be expressed as:

$$\boldsymbol{p}_{ce} = \left[\boldsymbol{\Phi}_e \boldsymbol{Z}_a \boldsymbol{B} \boldsymbol{Z}_I^{-1} \boldsymbol{\Psi}_{gc} \;\middle|\; \boldsymbol{\Phi}_e \boldsymbol{Z}_a \boldsymbol{\Phi}_{gc}\right] \begin{bmatrix} \boldsymbol{f}_c \\ \boldsymbol{q}_c \end{bmatrix} \tag{9.5.15}$$

Using this relationship, the sum of the squared sound pressures at a set of error sensing locations is:

$$\sum_{i=1}^{I} |p(\boldsymbol{r}_i)|^2 = \begin{bmatrix} \boldsymbol{f}_c \\ \boldsymbol{q}_c \end{bmatrix}^{\mathrm{H}} \boldsymbol{A} \begin{bmatrix} \boldsymbol{f}_c \\ \boldsymbol{q}_c \end{bmatrix} + \begin{bmatrix} \boldsymbol{f}_c \\ \boldsymbol{q}_c \end{bmatrix}^{\mathrm{H}} \boldsymbol{b} + \boldsymbol{b}^{\mathrm{H}} \begin{bmatrix} \boldsymbol{f}_c \\ \boldsymbol{q}_c \end{bmatrix} + c \tag{9.5.16}$$

where

$$\boldsymbol{A} = \left[\boldsymbol{\Phi}_e \boldsymbol{Z}_a \boldsymbol{B} \boldsymbol{Z}_I^{-1} \boldsymbol{\Psi}_{gc} \;\middle|\; \boldsymbol{\Phi}_e \boldsymbol{Z}_a \boldsymbol{\Phi}_{gc}\right]^{\mathrm{H}} \left[\boldsymbol{\Phi}_e \boldsymbol{Z}_a \boldsymbol{B} \boldsymbol{Z}_I^{-1} \boldsymbol{\Psi}_{gc} \;\middle|\; \boldsymbol{\Phi}_e \boldsymbol{Z}_a \boldsymbol{\Phi}_{gc}\right] \tag{9.5.17}$$

$$\boldsymbol{b} = \left[\boldsymbol{\Phi}_e \boldsymbol{Z}_a \boldsymbol{B} \boldsymbol{Z}_I^{-1} \boldsymbol{\Psi}_{gc} \;\middle|\; \boldsymbol{\Phi}_e \boldsymbol{Z}_a \boldsymbol{\Phi}_{gc}\right]^{\mathrm{H}} \boldsymbol{\Phi}_e^{\mathrm{T}} \boldsymbol{Z}_a \boldsymbol{B} \boldsymbol{v}_p \tag{9.5.18}$$

and c is as defined in Equation (9.4.36). The optimum set of control forces and volume velocities has the same form as in the previous sections, and is:

$$\begin{bmatrix} \boldsymbol{f}_c \\ \boldsymbol{q}_c \end{bmatrix}_{opt} = -\boldsymbol{A}^{-1} \boldsymbol{b} \tag{9.5.19}$$

If minimisation of acoustic potential energy is the error criterion of interest, the same form of equation as stated in Equation (9.5.16) is used, with

$$\boldsymbol{A} = \begin{bmatrix} \boldsymbol{A}_v & \boldsymbol{A}_{vq}^{\mathrm{H}} \\ \boldsymbol{A}_{vq} & \boldsymbol{A}_q \end{bmatrix} \tag{9.5.20}$$

$$b = \left[Z_a B Z_I^{-1} \Psi_{gc} \;\; | \;\; Z_a \Phi_{gc} \right]^{\mathrm{H}} Z_a B v_p \tag{9.5.21}$$

and with c as defined in Equation (9.5.5). Here, A_v is the matrix A for vibration control sources operating alone, as defined in Equation (9.5.10), A_q is the matrix A for acoustic control sources operating alone, as defined in Equation (9.5.3), and A_{vq} is:

$$A_{vq} = \Phi_{gc}^{\mathrm{T}} Z_a^{\mathrm{H}} Z_E Z_a B Z_I^{-1} \Psi_{gc} \tag{9.5.22}$$

The next error criterion that may be of interest is the minimisation of structural velocity levels at discrete points, or the global error criterion of kinetic energy of the structure, when vibration control sources are used. As with sound pressure, the surface velocity at any location can be considered as the sum of the primary generated and control source generated velocities:

$$v(x) = v_p(x) + v_c(x) \tag{9.5.23}$$

In terms of the matrices defined previously, the velocity levels at any point on the structure resulting from the primary and control sources can be expressed as:

$$v_p(x) = \psi^{\mathrm{T}}(x) v_p \tag{9.5.24}$$

and

$$v_c(x) = \psi^{\mathrm{T}}(x) Z_I^{-1} \Psi_{gc} f_c \tag{9.5.25}$$

where $\psi(x)$ is the $(N_m \times 1)$ vector of structural mode shape functions evaluated at location, x. Using these expressions, the sum of squared velocity levels at a set of error sensing points can be expressed as:

$$\sum_{i=1}^{I} |v(x_i)|^2 = f_c^{\mathrm{H}} A f_c + f_c^{\mathrm{H}} b + b^{\mathrm{H}} f_c + c \tag{9.5.26}$$

where

$$A = \Psi_{gc}^{\mathrm{H}} \left\{ Z_I^{-1} \right\}^{\mathrm{H}} Z_v Z_I^{-1} \Psi_{gc} \tag{9.5.27}$$

$$b = \Psi_{gc}^{\mathrm{H}} \left\{ Z_I^{-1} \right\}^{\mathrm{H}} Z_v v_p \tag{9.5.28}$$

$$c = v_p^{\mathrm{H}} Z_v v_p \tag{9.5.29}$$

and where Z_v is the weighting matrix based upon minimising the velocity levels at discrete locations, defined as:

$$Z_v = \Psi_e^{\mathrm{T}} \Psi_e \tag{9.5.30}$$

and $\boldsymbol{\Psi}_e$ is the $(N_m \times \iota)$ matrix whose columns are the $(N_m \times 1)$ structural mode shape function vectors $\boldsymbol{\psi}^{\mathrm{T}}(\boldsymbol{x})$ evaluated at the N_e error sensing locations. By inspection, it can be deduced that the optimum set of control forces, to minimise structural velocity levels, are found by using the expression:

$$\boldsymbol{f}_{c,v} = -\boldsymbol{A}^{-1}\boldsymbol{b} \tag{9.5.31}$$

If minimisation of the kinetic energy E_k of the structure:

$$E_k = \int_S \rho_s(\boldsymbol{x}) h(\boldsymbol{x}) v^2(\boldsymbol{x}) \,\mathrm{d}\boldsymbol{x} \tag{9.5.32}$$

is desired, the problem can be examined in the same way as the acoustic potential energy problem. Taking advantage of modal orthogonality, it can be shown that to minimise this quantity, the matrix $\boldsymbol{Z}_v$ in Equation (9.5.30) must be replaced with a diagonal weighting matrix $\boldsymbol{Z}_k$, the terms of which are defined by:

$$Z_k(i,i) = M_i \tag{9.5.33}$$

where M_i is the modal mass of the ith structural mode.

As was done with the acoustic potential energy in Section 9.3, it will be useful here to give some further physical meaning to the concept of minimising the kinetic energy of the structure. The velocity term in the definition of structural kinetic energy, given in Equation (9.5.32), can be expanded as the sum of modal contributions as follows:

$$v(\boldsymbol{x}) = \sum_{i=1}^{N_m} v_i \psi_i(\boldsymbol{x}) \tag{9.5.34}$$

where v_i is the complex amplitude of the ith structural mode. Therefore,

$$E_k = \int_S \rho_s(\boldsymbol{x}) h(\boldsymbol{x}) \left(\sum_{i=1}^{N_m} v_i \psi_i(\boldsymbol{x}) \right)^* \left(\sum_{i=1}^{N_m} v_i \psi_i(\boldsymbol{x}) \right) \mathrm{d}\boldsymbol{x} \tag{9.5.35}$$

Observe, however, that by using modal orthogonality, this can be simplified to:

$$E_k = \sum_{i=1}^{N_m} M_i |v_i|^2 \tag{9.5.36}$$

Thus, minimising the structural kinetic energy is equivalent to minimising the weighted (by modal mass) sum of the squares of the structural modal amplitudes.

An extension to the use of vibration sources to control surface velocity levels is the use of vibration sources to control both surface velocity levels and interior acoustic levels, as may be encountered when both accelerometers and microphones are used as error sensors. The set of acoustic pressures and surface velocities under both primary and control excitation can be expressed in matrix form as:

$$\begin{bmatrix} \boldsymbol{p} \\ \boldsymbol{v} \end{bmatrix}_p = \begin{bmatrix} \boldsymbol{\Phi}_e^{\mathrm{T}} \boldsymbol{Z}_a \boldsymbol{B} \boldsymbol{v}_p \\ \boldsymbol{\Psi}_e^{\mathrm{T}} \boldsymbol{v}_p \end{bmatrix} = \begin{bmatrix} \boldsymbol{\Phi}_e^{\mathrm{T}} \boldsymbol{Z}_a \boldsymbol{B} \\ \boldsymbol{\Psi}_e^{\mathrm{T}} \end{bmatrix} \boldsymbol{v}_p \tag{9.5.37a,b}$$

and

$$\begin{bmatrix} \boldsymbol{p} \\ \boldsymbol{v} \end{bmatrix}_c = \begin{bmatrix} \boldsymbol{\Phi}_e^{\mathrm{T}} \boldsymbol{Z}_a \boldsymbol{B} \boldsymbol{Z}_I^{-1} \boldsymbol{\Psi}_{gc} \\ \boldsymbol{\Psi}_e^{\mathrm{T}} \boldsymbol{Z}_I^{-1} \boldsymbol{\Psi}_{gc} \end{bmatrix} \boldsymbol{f}_c = \begin{bmatrix} \boldsymbol{\Phi}_e^{\mathrm{T}} \boldsymbol{Z}_a \boldsymbol{B} \\ \boldsymbol{\Psi}_e^{\mathrm{T}} \end{bmatrix} \boldsymbol{Z}_I^{-1} \boldsymbol{\Psi}_{gc} \boldsymbol{f}_c \tag{9.5.38a,b}$$

Therefore, the sum of the squared sound pressures and surface velocities at the set of error sensing points is:

$$\begin{bmatrix} \boldsymbol{p} \\ \boldsymbol{v} \end{bmatrix}^{\mathrm{H}} \begin{bmatrix} \boldsymbol{p} \\ \boldsymbol{v} \end{bmatrix} = \boldsymbol{f}_c^{\mathrm{H}} \boldsymbol{A} \boldsymbol{f}_c + \boldsymbol{f}_c^{\mathrm{H}} \boldsymbol{b} + \boldsymbol{b}^{\mathrm{H}} \boldsymbol{f}_c + c \tag{9.5.39}$$

where $\boldsymbol{A}$, $\boldsymbol{b}$ and c are as given in Equations (9.5.27) to (9.5.29), with $\boldsymbol{Z}_v$ now equal to:

$$\boldsymbol{Z}_v = \begin{bmatrix} \boldsymbol{\Phi}_e^{\mathrm{T}} \boldsymbol{Z}_a \boldsymbol{B} \\ \boldsymbol{\Psi}_e^{\mathrm{T}} \end{bmatrix}^{\mathrm{H}} \begin{bmatrix} \boldsymbol{\Phi}_e^{\mathrm{T}} \boldsymbol{Z}_a \boldsymbol{B} \\ \boldsymbol{\Psi}_e^{\mathrm{T}} \end{bmatrix} \tag{9.5.40}$$

One point to note when modelling systems utilising 'mixed medium' error sensors is that in a practical implementation these transducers will have different amplifications, which will inherently introduce some weighting into the problem. This may have an influence upon the overall control obtained by the system, and must be considered in the analytical modelling stage. This can be done by incorporating scalar weighting factors into the $\boldsymbol{Z}_v$ matrix of Equation (9.5.36), producing:

$$\boldsymbol{Z}_w = \begin{bmatrix} w_a \boldsymbol{\Phi}_e^{\mathrm{T}} \boldsymbol{Z}_a \boldsymbol{B} \\ w_v \boldsymbol{\Psi}_e^{\mathrm{T}} \end{bmatrix}^{\mathrm{H}} \begin{bmatrix} w_a \boldsymbol{\Phi}_e^{\mathrm{T}} \boldsymbol{Z}_a \boldsymbol{B} \\ w_v \boldsymbol{\Psi}_e^{\mathrm{T}} \end{bmatrix} \tag{9.5.41}$$

where w_a and w_v are scalar weighting factors for the minimisation of the acoustic and vibration error sensors respectively. The scalar weighting factors may be replaced with vector quantities to allow more importance to be placed on some error sensors than others, enabling the sound levels to be reduced further at critical locations.

9.5.1 Multiple Regression as a Shortcut

To implement the previous analytical models in a numerical search routine to optimise the placement of control sources/error sensors can be a lengthy process, and may not be practical in some instances due to the complexity of the structural/acoustic system, nor desired if the aim of the analysis is simply to discriminate between possible design options. The process can, however, be simplified using multiple regression (Snyder and Hansen, 1991, 1994a).

This is a generalised linear least-squares technique, where several independent variables are used to predict the dependent variable of interest. Here, the dependent variable of interest is the 180° inverse of the primary sound field (which will provide the greatest level of acoustic attenuation), while the independent variables are the control source transfer functions (dependent only upon the position of the sources and sensors) and volume velocities or forces. As will be seen shortly, one advantage to this approach is the ease with which measured data can be included in the design process.

Referring to Equation (9.3.1), it can be concluded that minimisation of the acoustic potential energy is equivalent to minimising the average squared sound pressure in the enclosure or, using a finite number of points, is equivalent to minimising:

$$\sum_{i=1}^{N} \left| p_p(\boldsymbol{r}_i) + p_c(\boldsymbol{r}_i) \right|^2 \tag{9.5.42}$$

where the number of points N at which the pressure is to be minimised should ideally be chosen so that Equation (9.5.42) is representative of the acoustic potential energy; if the aim is simply to discriminate between design options, the number of points considered can be substantially reduced. For vibration control sources, the control generated sound pressure at any point i can be expressed as:

$$p_c(\boldsymbol{r}_i) = \boldsymbol{z}_{vt}^{\mathrm{T}}(\boldsymbol{r}_i)\boldsymbol{f}_c \tag{9.5.43}$$

where the terms in $\boldsymbol{z}_{vt}(\boldsymbol{r}_i)$ could be calculated analytically as:

$$\boldsymbol{z}_{vt}^{\mathrm{T}}(\boldsymbol{r}_i) = \boldsymbol{\varphi}^{\mathrm{T}}(\boldsymbol{r}_i)\boldsymbol{Z}_a \boldsymbol{B} \boldsymbol{Z}_I^{-1} \boldsymbol{\Psi}_{gc} \tag{9.5.44}$$

or measured data could be used. For acoustic control sources, $p_c(\boldsymbol{r}_i)$ is defined by the expression:

$$p_c(\boldsymbol{r}_i) = \boldsymbol{z}_{at}^{\mathrm{T}} \boldsymbol{q}_c \tag{9.5.45}$$

where analytically:

$$\boldsymbol{z}_{at}^{\mathrm{T}} = \boldsymbol{\varphi}^{\mathrm{T}}(\boldsymbol{r})\boldsymbol{Z}_a \boldsymbol{\Phi}_{gc} \tag{9.5.46}$$

Alternatively, measured data could again be used. Note also that the primary generated sound pressure, as well as the transfer functions between the response at the control source locations and any point on the structure or in the acoustic space, can either be calculated using the previously outlined analytical methods or measured *in situ*.

Substituting these relations into Equation (9.5.42) gives the optimisation criterion for L vibration control sources; that is, the following expression should be minimised:

$$\sum_{i=1}^{N} \left| p_p(\boldsymbol{r}_i) + \sum_{l=1}^{L} z_{vt}(\boldsymbol{r}_{i,l}) f_l \right|^2 \tag{9.5.47}$$

and for L acoustic control sources, the following should be minimised:

$$\sum_{i=1}^{N} \left| p_p(\boldsymbol{r}_i) + \sum_{l=1}^{L} z_{at}(\boldsymbol{r}_{i,l}) q_l \right|^2 \tag{9.5.48}$$

Equations (9.5.47) and (9.5.48) can be written in matrix form. For vibration control sources, Equation (9.5.47) becomes:

$$|\boldsymbol{Z}_{vt}\boldsymbol{f}_c - (-\boldsymbol{p}_p)|^2 \tag{9.5.49}$$

where $\boldsymbol{Z}_{vt}$ is the $(N \times N_c)$ vector of transposed transfer function vectors $\boldsymbol{z}_{vt}^{\mathrm{T}}(\boldsymbol{r}_i)$ between the control force inputs and the error sensing locations. For acoustic control sources, Equation (9.5.48) becomes:

$$|\boldsymbol{Z}_{at}\boldsymbol{q}_c - (-\boldsymbol{p}_p)|^2 \tag{9.5.50}$$

where $\boldsymbol{Z}_{at}$ is the $(N \times N_c)$ vector of transposed transfer function vectors $\boldsymbol{z}_{at}^{\mathrm{T}}(\boldsymbol{r}_i)$ between the control volume velocities and the error sensing locations.

Equations (9.5.49) and (9.5.50) can be solved by various methods, such as by the use of singular value decomposition or by using one of many commercially available multiple regression software packages. The control forces and/or control volume velocities that result are those that are optimal for the given control source/error sensor positions. (It should be noted that a combination of acoustic and vibration control sources can be optimised in the same manner.)

The level of acoustic potential energy attenuation achieved using active noise control can be estimated as:

$$\Delta W = -10\log_{10}\left(\frac{\sum_{i=1}^{N}|p_p(\boldsymbol{r}_i) + p_c(\boldsymbol{r}_i)|^2}{\sum_{i=1}^{N}|p_p(\boldsymbol{r}_i)|^2}\right) \tag{9.5.51}$$

From Equations (9.5.47) and (9.5.48), it can be seen that the desired control source sound pressure $p_c(\boldsymbol{r}_i)$ is actually the estimated inverse of the primary source sound pressure $-p_p(\boldsymbol{r}_i)$. Therefore, Equation (9.5.51) can be written as:

$$\Delta W = -10\log_{10}\left(\frac{\sum_{i=1}^{N}|p_p(\boldsymbol{r}_i) - \hat{p}_p(\boldsymbol{r}_i)|^2}{\sum_{i=1}^{N}|p_p(\boldsymbol{r}_i)|^2}\right) \tag{9.5.52}$$

where ^ denotes estimate. For periodic sound, the mean (complex) sound pressure is zero (for both the real and imaginary parts). Thus, Equation (9.5.52) can be expressed as:

$$\Delta W = -10\log_{10}\left(\frac{\sum_{i=1}^{N}|p_p(\boldsymbol{r}_i) - \hat{p}_p(\boldsymbol{r}_i)|^2}{\sum_{i=1}^{N}|p_p(\boldsymbol{r}_i) - \tilde{p}_p(\boldsymbol{r}_i)|^2}\right) \tag{9.5.53}$$

where ~ denotes the mean value. Expressed in this form, the denominator of Equation (9.5.53) is equivalent to the sum of the squares of the measured dependent variable, SS_p, while the numerator is equivalent to the sum of squares of the residuals, SS_{res}. Using this notation, the estimated acoustic potential energy reduction for the given control source arrangement can be written as:

$$\Delta W = -10\log_{10}\left(\frac{SS_{res}}{SS_p}\right) \tag{9.5.54}$$

or

$$\Delta W = -10\log_{10}\left(1 - R^2\right) \tag{9.5.55}$$

where R is the multiple correlation coefficient (Tabachnick and Fidell, 1989). Thus, the multiple correlation coefficient can be used to estimate the acoustic power reduction under optimum control for a given control source arrangement. As the number of data points increases, so does the accuracy of the estimate. For a very large number of points, this method becomes equivalent to the quadratic optimisation methods of the previous section for determining the optimum control source volume velocities or forces for a given control source and error sensor arrangement.

The main advantages to be had in using a multiple regression approach are the speed of calculation and the ease with which practical transfer function measurements can be incorporated into the design process. An active control system could be optimally designed using all measured data and a multiple regression routine as follows:

1. A grid of prospective control source and error sensor locations can be laid out, and the transfer functions between each of these measured at the frequency(s) of interest. These measurements will comprise the terms in Equations (9.5.44) and (9.5.46). If the aim of the analysis is to discriminate between potential design choices, the grid can be coarse.
2. Next, the transfer function between some arbitrary reference point and each potential error sensor location can be measured under primary excitation to make up the terms in $\boldsymbol{p}_p$.
3. Different control source locations, or combinations of locations, can be inserted into the multiple regression routine. This could be done either manually or by using one of the 'stepwise' regression procedures commonly available. The multiple convergence coefficient R^2 can be used to estimate the acoustic potential energy reduction, as outlined in Equation (9.5.55).

One point to note concerning the implementation of the outlined multiple regression routine is that the majority of terms (acoustic pressures, forces, volume velocities and impedances) are complex, and therefore cannot be directly incorporated into a commercial package. Rather, the real and imaginary components of the acoustic pressure at each location must be considered separately, doubling the size of the problem. Consider, for example, Equation (9.5.49). For one pressure point and one control source, this can be written as:

$$(z_{vt\mathrm{R}} + \mathrm{j}z_{vt\mathrm{I}})(f_{c\mathrm{R}} + \mathrm{j}f_{c\mathrm{I}}) = (-p_{p\mathrm{R}} - \mathrm{j}p_{p\mathrm{I}}) \tag{9.5.56}$$

where the subscripts R and I refer to real and imaginary components respectively. Equation (9.5.56) can be written in a matrix form suitable for implementation as:

$$\begin{bmatrix} z_{vtR} & -z_{vtI} \\ z_{vtI} & z_{vtR} \end{bmatrix} \begin{bmatrix} f_{cR} \\ f_{cI} \end{bmatrix} = \begin{bmatrix} -p_{pR} \\ -p_{pI} \end{bmatrix} \tag{9.5.57}$$

Besides those already mentioned, there is one further advantage to using multiple regression. As has been already seen in Section 9.3 and will be discussed further in Section 9.7, the optimum error sensor locations are at the points of maximum sound pressure difference between the primary and optimally controlled residual sound fields. These points can be determined directly using a commercial multiple regression package, and the error sensors located accordingly. These packages, as part of their output data, generally produce a vector of residuals, which are the differences between the measured quantities (pressure here), and the values predicted by the regression equation. The point of minimum acoustic pressure in the residual sound field will be the point associated with the smallest residual value.

9.6 CALCULATION OF OPTIMAL CONTROL SOURCE VOLUME VELOCITIES USING BOUNDARY ELEMENT METHODS

In Sections 9.4 and 9.5, equations were developed that enabled calculation of the sound field in a coupled enclosure, as well as the control source outputs that would minimise this field. At that time it was noted that this was essentially a modal approach to the problem, where each structural mode was being treated as a discrete velocity source. In this section, another approach to the problem will be considered briefly: a spatial approach, where individual points on the enclosure boundary are explicitly considered in the calculation. This approach is based upon boundary element methods, and may be the only viable analytical methodology when the enclosure of interest has a complex geometry.

There are two commonly utilised boundary element formulations for predicting acoustic response: direct boundary element methods, based directly upon the Helmholtz integral equations, and indirect boundary element methods, based upon Huygen's principle. The one that will be considered here is the latter, the indirect boundary element, such as developed in Mollo and Bernhard (1987, 1989, 1990). Those interested in formulations based upon the direct boundary element method are directed to Cunefare and Koopmann (1991).

The indirect boundary element method is essentially a numerical implementation of Huygen's principle, where a fictitious distribution of sources on the boundary is used in the calculations. Details of the development of this methodology can be found in various publications (Brebbia and Walker, 1980; Banerjee and Butterfield, 1981; Bernhard et al., 1986). The indirect boundary element method produces a matrix equation for the acoustic pressure at a number of points as follows:

$$\boldsymbol{p}_f = \boldsymbol{S\alpha} + \boldsymbol{Tq} \tag{9.6.1}$$

where $\boldsymbol{p}_f$ is the $(N_p \times 1)$ vector of acoustic pressures at the N_p field points, $\boldsymbol{\alpha}$ is the vector of boundary conditions associated with each element, $\boldsymbol{q}$ is the vector of volume velocities of any acoustic point sources present in the domain, and $\boldsymbol{S}$ and $\boldsymbol{T}$ are boundary element matrices.

It has been shown several times in this chapter that the acoustic potential energy in an enclosed space is equal to the weighted sum of squared acoustic modal amplitudes. However, in the boundary element formulation of the problem of minimising the enclosed sound field, the analysis is not concerned explicitly with modes, but rather acoustic pressure at discrete points. Therefore, acoustic potential energy must be approximated by considering the squared acoustic pressure amplitude at discrete points, similar to the approach taken in the multiple regression formulation outlined in Section 9.5. In this instance, a weighting term will be included for generalisation, such that the error criterion to be minimised, J_p, is the weighted sum of squared acoustic pressure amplitudes at the field points:

$$J_p = \sum_{i=1}^{N_p} w_i |p_i|^2 \tag{9.6.2}$$

where w_i is the weighting factor associated with point p_i. Note that inclusion of the weighting term enables either local or global control to be investigated, as some of the weighting factors can be set equal to zero. Equation (9.6.2) can be written more compactly in matrix form as:

$$J_p = \boldsymbol{p}_f^{\mathrm{H}} \boldsymbol{W} \boldsymbol{p}_f \tag{9.6.3}$$

where $\boldsymbol{W}$ is a diagonal matrix of weighting terms.

To investigate active noise control, the last term in Equation (9.6.1) can be partitioned into primary and control source components, producing:

$$\boldsymbol{p}_f = \boldsymbol{S\alpha} + \boldsymbol{T}_p \boldsymbol{q}_p + \boldsymbol{T}_c \boldsymbol{q}_c \tag{9.6.4}$$

where the subscripts p and c denote primary and control source related quantities respectively. Using Equation (9.6.4), the error criterion can be expressed as:

$$J_p = \boldsymbol{q}_c^{\mathrm{H}} \boldsymbol{A} \boldsymbol{q}_c + \boldsymbol{b}^{\mathrm{H}} \boldsymbol{q}_c + \boldsymbol{q}_c^{\mathrm{H}} \boldsymbol{b} + c \tag{9.6.5}$$

where

$$\boldsymbol{A} = \boldsymbol{T}_c^{\mathrm{H}} \boldsymbol{W} \boldsymbol{T}_c \tag{9.6.6}$$

$$\boldsymbol{b} = \boldsymbol{T}_c^{\mathrm{H}} \boldsymbol{W} \{ \boldsymbol{S\alpha} + \boldsymbol{T}_p \boldsymbol{q}_p \} \tag{9.6.7}$$

and

$$c = \{ \boldsymbol{S\alpha} + \boldsymbol{T}_p \boldsymbol{q}_p \}^{\mathrm{H}} \boldsymbol{W} \{ \boldsymbol{S\alpha} + \boldsymbol{T}_p \boldsymbol{q}_p \} \tag{9.6.8}$$

Equation (9.6.5) is in exactly the same form as the error criterion expressions that were developed previously for multiple control source problems. As the matrix $\boldsymbol{A}$ is positive definite, the direct solution for the optimum control source volume velocities can be written as:

$$\boldsymbol{q}_{c,B} = -\boldsymbol{A}^{-1} \boldsymbol{b} \tag{9.6.9}$$

where the subscript $\boldsymbol{c,B}$ is used to denote the optimum control source volume velocities with respect to a boundary element problem. If Equation (9.6.9) is expanded using Equations (9.6.6) and (9.6.7), the vector of optimum control source volume velocities is found to be:

$$\boldsymbol{q}_{c,B} = -\{ \boldsymbol{T}_c^{\mathrm{H}} \boldsymbol{W} \boldsymbol{T}_c \}^{-1} \boldsymbol{T}_c^{\mathrm{H}} \boldsymbol{W} \{ \boldsymbol{S\alpha} + \boldsymbol{T}_p \boldsymbol{q}_p \} \tag{9.6.10}$$

Substituting Equation (9.6.9) into Equation (9.6.5), the minimum value of the error criterion is found to be:

$$J_{p,\,\min} = c - \boldsymbol{b}^{\mathrm{H}}\boldsymbol{A}^{-1}\boldsymbol{b} \tag{9.6.11}$$

Note again that the approach taken in calculating the vector of optimum control source volume velocities is the same whether a spatial or modal approach to the problem formulation is taken.

Examples of the implementation of this boundary element approach for examining active control in enclosed spaces can be found in various published work (Mollo and Bernhard, 1987, 1989, 1990, Bai and Chang, 1996, Brancati and Aliabadi, 2012).

9.7 CONTROL MECHANISMS

The analytical models developed in Sections 9.4 and 9.5 enable calculation of a number of important parameters in the design of active noise control systems, such as how much attenuation of acoustic potential energy is possible for a given control source arrangement, and how much attenuation will result by minimising the disturbance (vibrational or acoustic) at the error sensors. These models will be used in this section to consider a more fundamental point, that of how attenuation of the enclosed sound field is achieved physically. What are the physical mechanisms? From what has been discussed so far, attenuation can be the result of two possible classes of physical mechanism, or sources of control: acoustic control mechanisms, where the acoustic pressure is reduced without necessarily a reduction in the velocity levels of the coupled structural modes, and velocity control mechanisms, where the velocity amplitudes of the coupled structural modes are explicitly reduced.

It is intuitively obvious that acoustic control sources can be sources of acoustic control, and vibration control sources can be sources of velocity control. Perhaps not so obvious is the fact that vibration sources can also be sources of acoustic control, altering, but not necessarily reducing, the velocity levels of the coupled structural modes while decreasing the acoustic pressure in the enclosed space. The focus of this section will be on an expansion of these general concepts. As the physical mechanisms by which acoustic control sources achieve active sound attenuation are less complicated than those of vibration control sources, they will be considered first.

9.7.1 Acoustic Control Source Mechanisms

As was discussed in Section 9.3 in relation to active control in rigid enclosures, while acoustic power is a suitable basis quantity for examination of global sound attenuation in free-field structural radiation problems, it is not particularly useful for the examination of enclosed sound fields. This is because any real acoustic power flow will only be as a result of acoustic damping, as evident in Equation (9.2.1), and will not in general reflect the global sound pressure levels in the enclosed space. This is why acoustic potential energy was chosen as a global error criterion in Sections 9.2 and 9.4, and it is this latter quantity that will be used as the basis of examination of acoustic control mechanisms.

It has been noted in previous chapters that when acoustic control mechanisms are employed, two outcomes are possible: suppression, where the control source(s) mutually unload the primary source such that the acoustic radiation is reduced, and absorption, where the control source absorbs some of the primary source acoustic radiation. Therefore, to elucidate acoustic control mechanisms employed in transmission problems, what must be examined is the contribution made by the noise sources to the total acoustic potential energy under optimal operating conditions. It was outlined in Section 9.3 that acoustic potential energy is equal to the weighted sum of squared acoustic modal pressure amplitudes, quantities which themselves can be expressed as the sum of the primary source, E_{pp}, and control source, E_{pc}, contributions. These contributions are defined by the relationship:

$$E_{p(p/c)} = \frac{1}{4\rho_0 c_0^2} \sum_{l=1}^{N_n} \Lambda_l \operatorname{Re}\left\{p_{l,p/c}\, p_{l,tot}^*\right\} \tag{9.7.1}$$

where *p*/*c* denotes primary or control source, *tot* denotes total (sum of primary and control source contributions), and Re denotes the real part of the expression. The vector of acoustic modal amplitudes resulting from primary excitation generated by vibration of the surrounding structure were defined previously in Equation (9.5.1) as:

$$\boldsymbol{p}_p = \boldsymbol{Z}_a \boldsymbol{B} \boldsymbol{v}_p \tag{9.7.2}$$

Once again, neglecting any structural/acoustic interaction from operation of the acoustic control sources alone, the resulting acoustic modal amplitudes are as given in Equation (9.3.7):

$$\boldsymbol{p}_c = \boldsymbol{Z}_a \boldsymbol{\Phi}_{gc} \boldsymbol{q}_c \tag{9.7.3}$$

If Equations (9.7.2) and (9.7.3) are substituted into Equation (9.7.1), it is straightforward to show that the control source contribution to the acoustic potential energy can be expressed as:

$$E_{pc} = \boldsymbol{q}_c^{\mathrm{H}} \boldsymbol{A} \boldsymbol{q}_c + \operatorname{Re}\left\{\boldsymbol{b}^{\mathrm{H}} \boldsymbol{q}_c\right\} \tag{9.7.4}$$

where the terms $\boldsymbol{A}$ and $\boldsymbol{b}$ are those used in the formulation of the vector of optimum acoustic control source outputs, defined in Equations (9.5.3) and (9.5.4) respectively. Noting that $\boldsymbol{A}$ is a symmetric matrix, if the vector of optimal acoustic control source outputs, as defined in Equation (9.5.7), is now inserted into Equation (9.7.4), the acoustic potential energy attributable to the control source array is:

$$E_{pc} = [-\boldsymbol{A}^{-1}\boldsymbol{b}]^{\mathrm{H}} \boldsymbol{A} [-\boldsymbol{A}^{-1}\boldsymbol{b}] + \operatorname{Re}\left\{\boldsymbol{b}^{\mathrm{H}}[-\boldsymbol{A}^{-1}\boldsymbol{b}]\right\} = 0 \tag{9.7.5a,b}$$

That is, the acoustic potential energy attributable to the control source array is zero. This is exactly the same result as derived for the use of acoustic control sources in rigid walled enclosures, in Section 9.3. This means that the acoustic control mechanism under ideal conditions is one of pure suppression, where the control source itself is neither providing a positive contribution to the acoustic potential energy total nor absorbing any of the primary source acoustic radiation. This fact is confirmed by considering the primary source acoustic potential energy contribution under optimally controlled conditions. If Equations (9.7.2) and

(9.7.3) are again substituted into Equation (9.7.1) to calculated the primary source acoustic potential energy, it is straightforward to show that the result can be expressed as:

$$E_{pp} = \mathrm{Re}\left\{\boldsymbol{q}_c^{\mathrm{H}}\boldsymbol{b}\right\} + c = \mathrm{Re}\left\{b^{\mathrm{H}}\boldsymbol{q}_c\right\} + c \qquad (9.7.6\mathrm{a,b})$$

where c is as defined in Equation (9.5.5). Substituting the vector of optimum control source outputs into Equation (9.7.6) produces:

$$E_{pp} = \mathrm{Re}\left\{-\boldsymbol{b}^{\mathrm{H}}\boldsymbol{A}^{-1}\boldsymbol{b}\right\} + c = c - \boldsymbol{b}^{\mathrm{H}}\boldsymbol{A}^{-1}\boldsymbol{b} \qquad (9.7.7\mathrm{a,b})$$

which is identical to the expression for the minimum acoustic potential energy of the system, as defined in Equation (9.5.8). The primary source is the only non-zero contributor to this total.

As was noted in Section 9.3, simply having a total contribution of zero to the acoustic potential energy from the control source array does not necessarily mean that contributions from the constituent elements will be zero. The acoustic potential energy attributable to a single acoustic control source in the array can be expressed as:

$$E_{pc}(i) = \mathrm{Re}\left\{\boldsymbol{q}_c^{\mathrm{H}}\boldsymbol{A}(\mathrm{col}\ i)q_c(i)\right\} + \mathrm{Re}\left\{b^*(i)q_c(i)\right\} \qquad (9.7.8)$$

where $\boldsymbol{A}(\mathrm{col}\ i)$ is the ith column of the matrix $\boldsymbol{A}$ defined in Equation (9.5.3), and $b(i)$ is the ith element in the vector $\boldsymbol{b}$ as defined in Equation (9.5.4). If the vector of optimum control source volume velocities is again expanded, then the result is:

$$E_{pc}(i) = \mathrm{Re}\left\{[-\boldsymbol{A}^{-1}\boldsymbol{b}]^{\mathrm{H}}\boldsymbol{A}(\mathrm{col}\ i)q_c(i)\right\} + \mathrm{Re}\left\{b(i)^*q_c(i)\right\} = 0 \qquad (9.7.9\mathrm{a,b})$$

Therefore the acoustic potential energy contribution of each element in the control source array must also be zero, as was the case for the rigid walled enclosure problem.

The fact that the control mechanism employed when using optimally tuned acoustic control sources is one of suppression infers under what general conditions the maximum levels of acoustic potential energy attenuation per control source will be obtained. In most coupled structural/acoustic systems, the coupling characteristics for low-frequency excitation will be 'multi-modal'; that is, any given acoustic mode will couple to several structural modes that are significantly excited, and vice versa. At an acoustic resonance, a single acoustic mode will dominate the acoustic pressure distribution in the enclosure, and hence the levels of acoustic potential energy. A single acoustic control source can easily excite this same single mode, and hence quite easily provide a significant reduction in pressure over the entire surface of the structure (unloading over the surface of the structure). At a structural resonance, however, several acoustic modes may be significantly excited. For the control source distribution to similarly (in relative phase and amplitude) excite the acoustic modes, control source placement and number can become critical. Therefore, generally speaking, acoustic control sources are recommended for controlling acoustic resonances.

The effect of an acoustic control source is shown in Figure 9.17, which illustrates the maximum levels of acoustic potential energy attenuation as a function of frequency for the case where a normally incident plane wave primary disturbance impinges upon a rectangular panel/cavity system (described in more detail shortly). The single acoustic control source is located in the bottom and centre of the enclosure, at a position (0.434, 0.575, 0.0) m, where the box interior dimensions are (0.868 × 1.15 × 1.0) m and the top panel is steel, 6 mm thick. Not surprisingly, the greatest levels of attenuation are achieved at acoustic resonances.

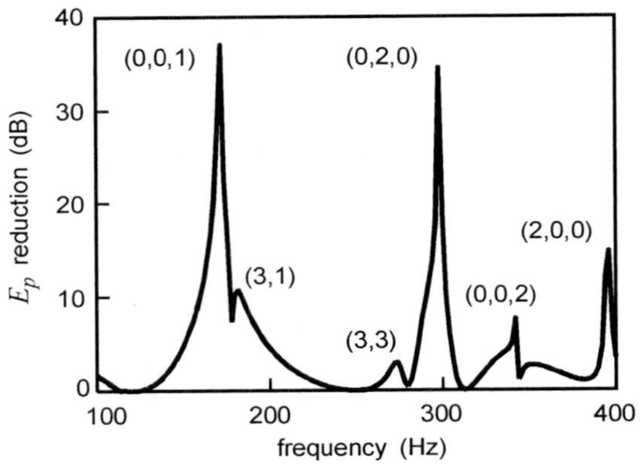

Figure 9.17 Maximum levels of acoustic potential energy reduction for rectangular enclosure system as a function of frequency. Locations of acoustic modal resonances are represented by three integers in brackets and structural modes are represented by two integers.

9.7.2 Vibration Control Source Mechanisms

Attention will now be directed to the mechanisms employed by vibration control sources in providing attenuation of the acoustic potential energy in the coupled enclosure. Vibration control sources provide a reduction in the levels of acoustic potential energy in the enclosed space by altering the velocity distribution of the enclosing structure. From the viewpoint of active control, the alteration manifests itself in two different ways: a reduction in the amplitudes of the dominant coupled structural modes, and an alteration in the relative amplitudes and phases of the coupled structural modes. Each of these effects has the potential to provide attenuation, and may do so individually or in partnership with the other. For this reason, the active control of sound transmission into enclosed spaces is more complicated (in terms of control mechanisms) when vibration control sources are used than when acoustic control sources are used, and a closer examination of the response of the coupled structural/acoustic system is required to obtain a qualitative overview of what is physically happening. A reduction in acoustic potential energy is equivalent to a reduction in the weighted (by mode normalisation) sum of the squared acoustic modal amplitudes. Therefore, to obtain this overview, the excitation of a single acoustic mode by the enclosing structure will be examined.

It was shown in Section 9.4, Equation (9.4.21), that the amplitude of the lth acoustic mode, excited by the vibration of the enclosing structure, is defined by the relationship:

$$p_l = \frac{j\rho_0 \omega S}{\Lambda_l(\kappa_l^2 - k^2)} \sum_{i=1}^{\infty} \beta_{l,i} v_i \tag{9.7.10}$$

If the infinite summation in this relationship is truncated at N_m structural modes, the expression defining the complex amplitude of the $\imath$th acoustic mode can be written as the product of two vectors as follows:

$$p_\imath = \boldsymbol{z}_{a,\imath}^{\mathrm{T}} \boldsymbol{v} \tag{9.7.11}$$

where $\boldsymbol{v}$ is the $(N_m \times 1)$ vector of structural modal velocity amplitudes, and $\boldsymbol{z}_{a,\imath}$ is the $(N_m \times 1)$ vector of radiation transfer functions between the structural modes and the $\imath$th acoustic mode, the terms of which are defined by:

$$z_{a,\imath}(i) = \frac{j\rho_0 \omega S}{\Lambda_\imath(\kappa_\imath^2 - k^2)} \beta_{\imath,i} \tag{9.7.12}$$

From this, the squared amplitude of the $\imath$th acoustic mode is defined by the matrix expression:

$$|p_\imath|^2 = \boldsymbol{v}^{\mathrm{H}} \boldsymbol{z}_{a,\imath}^{*} \boldsymbol{z}_{a,\imath}^{\mathrm{T}} \boldsymbol{v} \tag{9.7.13}$$

Minimising the acoustic potential energy in the enclosed space is equivalent to minimising a weighted set of equations of the form of Equation (9.7.13).

Consider the problem of minimising the amplitude of the $\imath$th acoustic mode in light of the characteristics of its excitation by the structure, as defined in Equation (9.7.12). If the modal coupling characteristics and excitation frequency of the system are such that only one structural mode is essentially responsible for excitation of the acoustic mode, then attenuation will be provided by reducing the amplitude of that single coupled structural mode, hence by velocity, or modal, control. This control characteristic applies to systems with extremely selective modal coupling characteristics and to systems excited near the resonance frequency of one of the (lightly damped) coupled structural modes. If, however, there are two or more structural modes that have the potential to significantly excite the acoustic mode of interest, there are two ways in which attenuation can be provided. The first is simply to reduce the amplitude of this set of structural modes, the mechanism employed again being one of velocity, or modal, control. The second way in which attenuation can be provided, is to alter the relative amplitudes and phases of these structural modes, such that the overall excitation of the acoustic mode, which is the sum of contributions from the structural modes, is reduced. The mechanism employed in this instance is one of acoustic control, where the acoustic pressure is reduced without necessarily a reduction in the velocity of the structure. In contrast to modal control, this mechanism was labelled as 'modal rearrangement' in the previous chapter, where the relative complex amplitudes of the structural modes are altered, not necessarily reduced.

When there are two or more structural modes capable of significant excitation of the acoustic mode, there is then some element of choice in the control mechanism employed. The choice taken is dependent upon the characteristics of the primary forcing function and the control source arrangement. If there are as many or more control sources as there are 'significant' structural modes, then it is usually possible to provide modal control. If there are less, then the control mechanism employed is highly location dependent. If it is possible to reduce the amplitude of all structural modes simultaneously, the control mechanism employed will be velocity, or modal, control. If it is possible to simply reduce the total energy transfer from the structural modes into the acoustic mode by altering the relative complex amplitudes of the structural modes, then acoustic control, or modal rearrangement, will be employed. Often the final result will be a combination of modal control and modal rearrangement. It should be emphasised that the mechanism employed is a function of the physical arrangement of the control source and error sensors; the attached electronic control control system simply aims to attenuate the unwanted disturbance at the error sensor(s) and has no effect upon the mechanism employed to do it.

To study these qualitative characteristics further, it is useful to conduct a brief analytical investigation of several sound transmission problems. For this study, three separate coupled structural/acoustic systems are considered: a rectangular panel/cavity system, a finite circular cylinder and a finite circular cylinder with an integral longitudinal partition (floor). This combination of systems is selected for two reasons. First, solutions for the eigenvalues (natural frequencies) and eigenvectors (mode shapes) of these systems are obtained relatively easily, with the rectangular and plain cylindrical systems presenting an analytically tractable problem, and the cylinder with floor requiring only a relatively simple numerical approach. Second, these three systems represent a range of structural/acoustic modal coupling characteristics, the nature and importance of which will become apparent. The first task in this study, however, is to specialise the theory of the previous two sections to the three outlined systems.

9.7.3 Specialisation of Theory for the Rectangular Enclosure Case

The first system of interest is the rectangular enclosure (often referred to as a panel/cavity system) depicted in Figure 9.18, having four rigid walls and a rigid bottom, with a flexible top comprising a simply supported rectangular panel. This arrangement has been considered

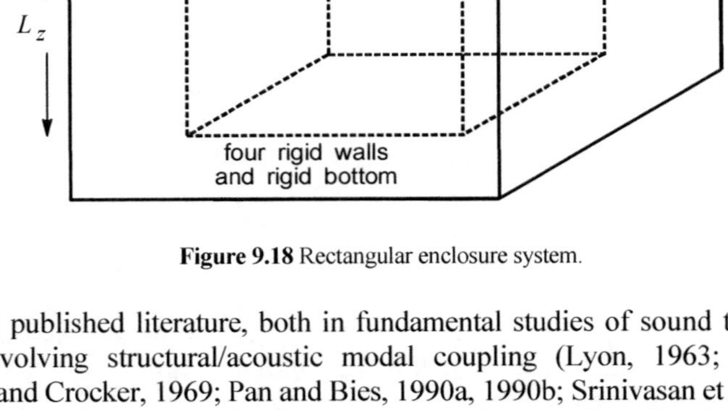

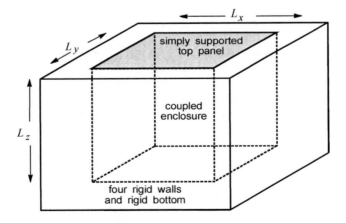

Figure 9.18 Rectangular enclosure system.

extensively in published literature, both in fundamental studies of sound transmission into enclosures involving structural/acoustic modal coupling (Lyon, 1963; Pretlove, 1966; Bhattacharya and Crocker, 1969; Pan and Bies, 1990a, 1990b; Srinivasan et al., 2007), and in fundamental studies of active control of sound transmission using vibration sources (Pan et al., 1990; Pan and Hansen, 1991a, 1991b; Snyder and Hansen, 1994b; Kim and Brennan,

2000; Gardonio et al., 2004; Al-Bassyiouni and Balachandran, 2005). The reason for this is its simply geometry, which facilitates a clear study of physical phenomena. For the purposes of the following analysis, the simple geometry will be doubly beneficial, as it leads to what can be viewed as 'multi-modal' structural acoustic coupling at low frequencies.

As was outlined in Section 9.4 in the beginning of the development of the coupled structural/acoustic equations, the assumption of weak modal coupling implies that the response of the coupled structural/acoustic system can be described by the *in vacuo* response of the structural component, the rigid walled response of the acoustic component, and the coupling between the two. For the system shown in Figure 9.18, the *in vacuo* structural mode shape functions of the simply supported rectangular top panel, and the acoustic mode shape functions of the rigid walled rectangular enclosure are respectively:

$$\psi_{u,v}(x,y) = \sin\frac{u\pi x}{L_x}\sin\frac{v\pi y}{L_y} \tag{9.7.14}$$

and

$$\varphi_{l,m,n}(x,y,z) = \cos\frac{l\pi x}{L_x}\cos\frac{m\pi y}{L_y}\cos\frac{n\pi z}{L_z} \tag{9.7.15}$$

where u and v are the structural modal indices; l, m and n are the acoustic modal indices; and L_x, L_y and L_z are the (inside) dimensions of the enclosure. The origin of the coordinate system is the lower left corner of the panel shown in the figure. The structural and acoustic resonance frequencies associated with these modes are respectively:

$$\omega_{u,v} = \left(\frac{D}{\rho_s h}\right)^{½}\left(\left(\frac{m\pi}{L_x}\right)^2 + \left(\frac{n\pi}{L_y}\right)^2\right) \tag{9.7.16}$$

and

$$\omega_{l,m,n} = c_0\pi\sqrt{\left(\frac{l}{L_x}\right)^2 + \left(\frac{m}{L_y}\right)^2 + \left(\frac{n}{L_z}\right)^2} \tag{9.7.17}$$

where D is the panel bending stiffness, given by:

$$D = EI = \frac{Eh^3}{12(1 - \nu^2)} \tag{9.7.18}$$

E is the modulus of elasticity, I is the cross-sectional second moment of area, h is the panel thickness and ν is Poisson's ratio. The modal mass of the top panel and the acoustic mode normalisation are respectively:

$$M_i = \int_S \rho_s(\boldsymbol{x})h(\boldsymbol{x})\psi_i^2(\boldsymbol{x})\,\mathrm{d}\boldsymbol{x} = \frac{\rho_s hS}{4} \tag{9.7.19}$$

and

$$\Lambda_{l,m,n} = \int_V \varphi_{l,m,n}^2(\boldsymbol{r})\,\mathrm{d}\boldsymbol{r} = \frac{V}{8}\varepsilon_l\varepsilon_m\varepsilon_n 1; \quad \varepsilon_{l,m,n} = \begin{cases} 1 & l,m,n \neq 0 \\ 2 & l,m,n = 0 \end{cases} \tag{9.7.20a,b}$$

where S and V are the surface area of the top panel and enclosed volume of the cavity respectively.

Once the characteristic eigenvalues and eigenvectors of the structural and acoustic subsystems are known, all that remains to be done to enable calculation of the system response is to define the coupling between them. Substituting the structural and acoustic mode shape functions outlined in Equations (9.7.14) and (9.7.15) into the modal coupling relation, Equation (9.4.11), shows the coupling between acoustic mode (l,m,n) and structural mode (u,v) to be defined by (Pan and Bies, 1990a):

$$B_{(u,v),(l,m,n)} = \frac{1}{S}\int_S \psi_{u,v}(\boldsymbol{x})\varphi_{l,m,n}(\boldsymbol{x})\,\mathrm{d}\boldsymbol{x} = \begin{cases} (-1)^n \dfrac{uv[(-1)^{l+u}-1][(-1)^{m+v}-1]}{\pi^2[l^2-u^2][m^2-v^2]} & l \neq u,\; m \neq v \\ 0 & \text{otherwise} \end{cases} \tag{9.7.21a,b}$$

Note that, from Equation (9.7.21), all odd index panel modes couple with even index acoustic modes, and vice versa. Even at low frequencies, this means that several significantly excited or excitable panel modes (such as the (1,1), (1,3) and (3,1)) will be coupled to the same set of acoustic modes, each transferring energy from the structure into the acoustic space. This is what was meant by 'multi-modal' coupling at low frequencies, a characteristic that will be important in determining what physical mechanism is responsible for attenuation of the acoustic potential energy in the enclosed space.

Equations (9.7.14) to (9.7.21) can be used in the previously outlined theory to determine the acoustic pressure at a point, the acoustic potential energy in the enclosure, or the optimum control forces/volume velocities for a given error criterion. For example, if it is desired to know the acoustic pressure at some point $\boldsymbol{r}$ in the enclosure under primary excitation, where the primary forcing function and hence the modal generalised force matrix γ_p (the terms of which are defined in Equation (9.4.7)), are known, Equation (9.4.22) can be used, where $\varphi(\vec{r})$ is now the vector of acoustic mode shape functions defined in Equation (9.7.15) and evaluated at the point $\vec{r}$ of interest. The terms in $\boldsymbol{Z}_a$, as outlined in Equation (9.4.23), are evaluated using the acoustic mode resonance frequencies determined from Equation (9.7.17), and $\boldsymbol{Z}_I$, whose terms are defined in Equations (9.4.17) and (9.4.18), is evaluated using Equations (9.7.14) to (9.7.21).

9.7.4 Specialisation of Theory for the Finite Length Circular Cylinder Case

The second system of interest, shown in Figure 9.19, is a finite length circular cylinder, the ends of which are assumed to be simply supported and sufficiently capped such that the sound transmission through them can be ignored. This arrangement is also popular in published literature, owing to its resemblance to an aircraft fuselage. It has been examined previously in some depth using a modal coupling theoretical approach in the context of describing general sound transmission (Pope et al., 1980, 1982), in which the methodology was found to be quite accurate in its ability to predict the sound transmission. It has also been considered by a number of researchers in active control, using both acoustic and vibration control sources (for vibration control source studies, see for example, Fuller and Jones, 1987; Jones and Fuller, 1987; Snyder and Hansen, 1994b; for acoustic control source studies, see for example, Bullmore et al., 1986; Silcox et al., 1987).

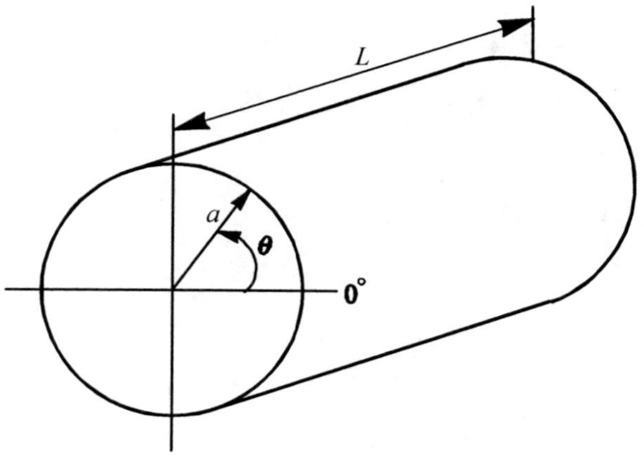

Figure 9.19 Plain cylindrical enclosure system.

As was done in the previous section for the rectangular enclosure system, the quantities that are required to specialise the general theory to the circular cylinder case are the *in vacuo* response of the structure, the rigid walled response of the enclosed acoustic space, and the coupling between the two. The structural and acoustic mode shape functions for these are respectively:

$$\psi_{u,v}(z,\theta) = \sin\frac{u\pi z}{L}\left(\cos(v\theta) + \sin(v\theta)\right) \tag{9.7.22}$$

and

$$\varphi_{q,n,s}(r,\theta,z) = \cos\frac{q\pi z}{L}\, J_n\left(\frac{\gamma_{ns}r}{a}\right)\left(\cos(n\theta) + \sin(n\theta)\right) \tag{9.7.23}$$

where u,v are the axial and circumferential structural modal indices; L is the cylinder length; q,n,s, are the axial, circumferential, and radial acoustic modal indices respectively; a is the cylinder radius; J_n is a Bessel function of the first kind of order n; and γ_{ns} is the value of the sth zero of the derivative of the Bessel function of order n:

$$J_n^{\prime}(\gamma_{ns}) = 0 \tag{9.7.24}$$

Note that both sine and cosine functions are required to describe the angular distribution of the structural and acoustic modes on the cylinder.

The resonance frequencies associated with the structural mode shape functions can be found from the characteristic equation (Leissa, 1973):

$$(\Omega^2)^3 - (K_2+\tilde{h}\,\Delta K_2)(\Omega^2)^2 + (K_1+\tilde{h}\,\Delta K_1)(\Omega^2) - (K_0+\tilde{h}\,\Delta K_0) = 0 \tag{9.7.25}$$

where

$$\Omega^2 = \frac{\rho_s(1-\nu^2)}{E}a^2\omega_{u,v}^2 = \frac{a^2\omega_{u,v}^2}{c_s} \tag{9.7.26a,b}$$

$$\tilde{h} = h^2/(12a^2) \tag{9.7.27}$$

and c_s is the speed of sound in the material. The constants K_0, K_1 and K_2 in Equation (9.7.25) are from the Donnell–Mustari shell theory, and are defined as:

$$K_0 = \frac{1-\nu}{2}\left((1-\nu^2)\lambda^4 + \tilde{h}(\nu^2+\lambda^2)^4\right)$$

$$K_1 = \frac{1-\nu}{2}\left((3+2\nu)\lambda^2 + \nu^2 + (\nu^2+\lambda^2)^2 + \frac{3-\nu}{1-\nu}\tilde{h}(\nu^2+\lambda^2)^3\right) \tag{9.7.28a–c}$$

$$K_2 = 1 + \frac{3-\nu}{2}(\nu^2+\lambda^2) + \tilde{h}(\nu^2+\lambda^2)^2$$

where

$$\lambda = \frac{u\pi a}{L} \tag{9.7.29}$$

ΔK_0, ΔK_1, and ΔK_2 are the modifying constants of the Goldenveizer–Novozhilov/Arnold–Warburton shell theory, defined as:

$$\Delta K_0 = \frac{1-\nu}{2}\left(4(1-\nu^2)\lambda^4 + 4\lambda^2\nu^2 + \nu^4 - 2(2-\nu)(2+\nu)\lambda^4 - 8\lambda^2\nu^4 - 2\nu^6\right)$$

$$\Delta K_1 = 2(1-\nu)\lambda^2 + \nu^2 + 2(1-\nu)\lambda^4 - (2-\nu)\lambda^2\nu^2 - \frac{3+\nu}{2}\nu^4 \tag{9.7.30a–c}$$

$$\Delta K_2 = 2(1-\nu)\lambda^2 + \nu^2$$

Goldenveizer–Novozhilov/Arnold–Warburton shell theory will be used in the current study because of its previously demonstrated ability to accurately predict the modal response of a thin, circular finite length shell of the type considered here (Pope et al., 1980, 1982).

The cubic characteristic Equation (9.7.25) has three roots for each set of structural modal indices, corresponding to three resonance frequencies, associated respectively with motion predominantly in each of the radial, tangential and axial directions. The first root is the one associated with the radial response of the dominant resonance (see the discussion in Section 2.3), and will be the only one used in the calculations presented here, as it is the flexure of the cylinder that is responsible for the interior sound generation.

The modal mass associated with each structural mode (assuming that the structure has uniform material properties) is found from:

$$M_i = \int_S \rho_s(\boldsymbol{x})h(\boldsymbol{x})\psi_{u,v}^2(\boldsymbol{x})\,\mathrm{d}S = \frac{\rho_s hS}{4} \tag{9.7.31}$$

where S is the surface area of the cylinder (excluding the ends) and h is the wall thickness. The resonance frequencies of the acoustic modes are found from:

$$\omega_{q,n,s} = c_0\sqrt{\gamma_{n,s}^2 + \left(\frac{q\pi}{L}\right)^2} \tag{9.7.32}$$

The associated acoustic mode normalisation term is:

$$\Lambda_{q,n,s} = \int_V \varphi^2_{q,n,s}(\boldsymbol{r})\,\mathrm{d}\boldsymbol{r} = \frac{\gamma^2_{ns} - n^2}{2\gamma^2_{n,s}}\frac{J^2_n(\gamma_{n,s})}{2}\pi a^2 L \varepsilon_q \varepsilon_n \tag{9.7.33}$$

where

$$\varepsilon_q,\ \varepsilon_n = \begin{cases} 2 & q,n = 0 \\ 1 & q,n > 0 \end{cases} \tag{9.7.34}$$

The final term to evaluate is the coupling factor between the structural and acoustic modes. Substituting the acoustic and structural mode shape functions of Equations (9.7.22) and (9.7.23) into the coupling relation yields:

$$B_{(u,v)(q,n,s)} = \begin{cases} \dfrac{J_n(\gamma_{n,s})}{2L}\,\dfrac{\varepsilon_n \kappa_u [1 - (-1)^{u+q}]}{\kappa_u^2 - \kappa_q^2}\ ; & n = v\ ; \quad \dfrac{u+q}{2} \neq \text{integer} \\ 0 & \text{otherwise} \end{cases} \tag{9.7.35}$$

where

$$\kappa_u = \frac{u\pi}{L}\ ; \qquad \kappa_q = \frac{q\pi}{L} \tag{9.7.36a,b}$$

Note that the modal coupling characteristics of the circular cylinder, defined by Equation (9.7.35), are much more selective than those of the rectangular enclosure given in Equation (9.7.21). For the circular enclosure, the structural and acoustic modes will only couple if the circumferential modal indices match, and the axial modal indices are an odd/even combination. For example, if the acoustic mode of interest is the (1,2,1) mode, only the (2,2), (4,2), ... structural modes will couple to it, modes whose resonance frequencies are far apart. From a practical standpoint this means that effectively only one structural mode is coupled to one acoustic mode when the excitation frequency is 'low', which is what will be considered here. This has implications for the physical control mechanism, to be considered shortly.

9.7.5 Specialisation of General Model for the Cylinder with Floor System

Use of the previously developed analytical models is not confined to systems with analytically tractable characteristic equations, with readily describable mode shape functions and resonance frequencies; rather, the analytical models are valid for any weakly coupled structural/acoustic system. One example of a system which is of interest theoretically and experimentally, but which does not have analytically tractable solutions for the mode shape functions and resonance frequencies, is the finite length circular cylinder with an integral longitudinal partition, or floor, shown in Figure 9.20. This arrangement is a more accurate representation of an aircraft fuselage than the plain circular cylinder considered in the previous section and has been used as the structural model for aircraft interior acoustics in a number of studies (Bullmore et al., 1990; Cazzolato and Hansen, 1998; Missaoui and Cheng, 1999; Li et al., 2002, 2004; Simpson and Hansen, 2006). In this section, the circular cylinder with floor system will be shown to be slightly different to the plain circular cylinder with respect to the physical control mechanisms available to reduce the acoustic potential energy in the enclosed space.

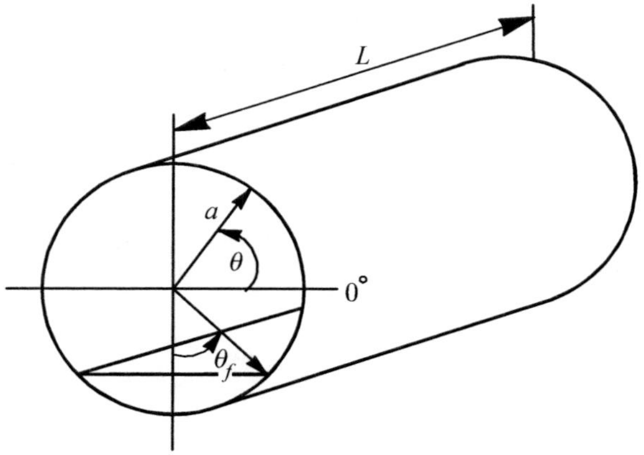

Figure 9.20 Cylindrical enclosure with floor.

As with the two previous cases, what is required to study the active control of sound transmission into this system is a description of the response of the structural and acoustic subsystems, and the coupling between them. The response of the structural subsystem can be determined using component mode synthesis (Peterson and Boyd, 1978). Using this method, the basis function for the total structural mode shape is a combination of the mode shape basis functions for the plain cylinder and the floor (plate) structural components. Simply supported boundary conditions are assumed for the longitudinal end of both the cylinder and floor structure, so that the axial mode shape function of the structure is:

$$\varphi_u = \sin\frac{u\pi z}{L} \tag{9.7.37}$$

The acoustic mode shape functions can be found by using a finite difference implementation of the Helmholtz equation (Pope and Wilby, 1982; Pope et al., 1983). The acoustic mode shape functions are then represented as eigenvectors of a set of values at a discrete set (grid) of points. Linear interpolation can be used to assess the value of the mode shape at any off-grid point. The axial mode shape function is equal to that of a one-dimensional enclosure:

$$\varphi_q = \cos\frac{q\pi z}{L} \tag{9.7.38}$$

Coupling between the structural and acoustic modes can be evaluated numerically. When doing this evaluation, it is found that the coupling characteristics become much more 'multi-modal' at low frequencies, similar to a rectangular enclosure, as the floor angle θ_f increases. By this, it is meant that most horizontally symmetric and anti-symmetric structural modes will couple, to some extent, to most horizontally symmetric and anti-symmetric acoustic modes respectively, subject to the axial constraint of an odd/even combination of structural and acoustic axial modal indices. As a result, vibration control sources may have an expanded scope for employing both control mechanisms, as will now be demonstrated.

9.7.6 Examination of Mechanisms

As was outlined in Section 9.4, a weakly coupled enclosure has the majority of the overall system energy associated with the individual responses of the structural and acoustic systems, with relatively little associated with the interaction between the two. Within this definition, therefore, it its possible to define two regimes of coupled system response; cavity controlled response, where the majority of the total system energy is associated with the response of the enclosed acoustic space (such as near an acoustic resonance) and structure-controlled response, where the majority of the total system energy is associated with the response of the structure (such as near a structural resonance). The distinction between these two regimes becomes important in qualifying the control mechanisms associated with vibration control sources.

Consider first the case of a structure-controlled response. If it is predominantly the response of the structure that is creating the noise problems, it would appear intuitively obvious that the best way to overcome the problem is to eliminate the principal offending structural mode or modes. This is, in fact, what can occur. Consider again the rectangular panel system mentioned at the end of the discussion of acoustic control mechanisms. Figure 9.21 illustrates the (theoretical) response of the dominant modes of the top panel of the rectangular enclosure system for an excitation frequency of 120 Hz, close to the (1,3) resonance of the panel, before and after the application of optimum vibration control at a panel location (x = 0.434, y = 0.9) m to attenuate the sound transmission of a normally incident plane wave. Clearly, the amplitudes of the dominant modes have been reduced, which is reflected in a 23.1 dB reduction in acoustic potential energy. This reduction in the amplitudes of the dominant structural modes will not, however, always be the only control mechanism with a structure-controlled response. Figure 9.22 illustrates the change in the amplitudes and phases of these same modes when the control source has been moved to the centre of the top panel. As before, the nearly resonant (1,3) panel mode has been reduced in amplitude. However, this reduction is much less than that achieved when the control source was located in its previous position, and is offset to some degree by the increase in amplitude of the (1,1) mode. Despite this, there is still a reduction of 22.6 dB in the enclosure acoustic potential energy.

From the standpoint of assessing the physical control mechanism, an important aspect of the structural modal amplitudes and phases of Figure 9.22 is the alteration in the relative amplitudes and phases of the (1,1), (1,3) and, to a lesser extent, the (3,1) modes. As was outlined earlier in this section, the modal coupling characteristics of the rectangular enclosure are such that odd index structural modes couple with even index acoustic modes, and vice versa, so that all three of these structural modes couple to the same set of acoustic modes. By altering the relative amplitudes and phases of these modes, it is possible to reduce the total excitation of the (coupled) acoustic modes, which is the summation of the contributions from each (coupled) structural mode. This modal rearrangement control mechanism occurs when the vibration source is placed at the panel centre, because the transfer function between the control force input and the two dominant structural modes, the (1,1) and (1,3) modes, differs in phase by a significant amount (greater than 90°), and so the amplitudes of both of these structural modes could not be reduced concurrently.

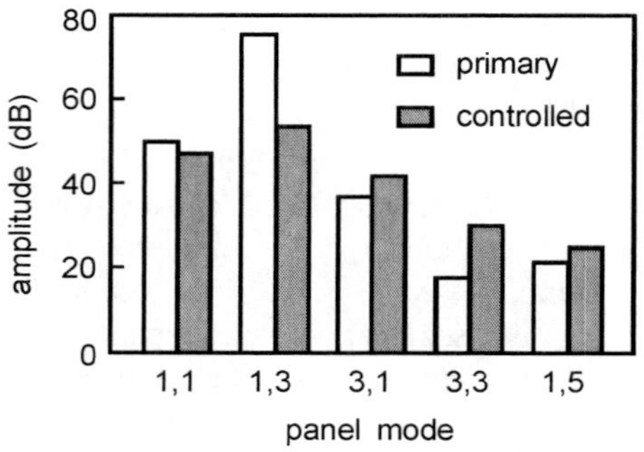

Figure 9.21 Change in structural modal amplitudes of the top panel of the rectangular enclosure system with a point control force applied off-centre at 120 Hz.

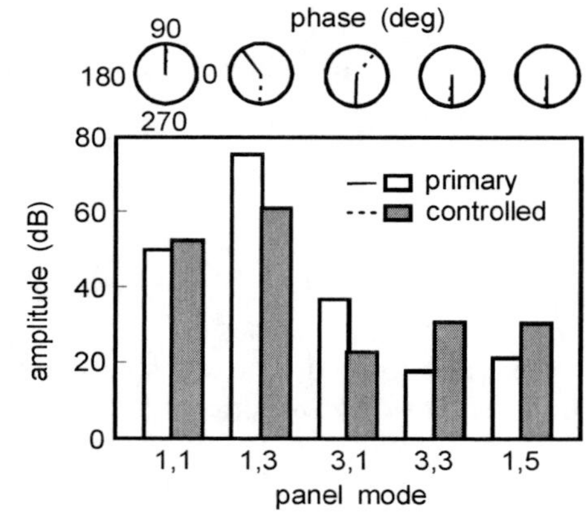

Figure 9.22 Change in structural modal amplitudes of the top panel of the rectangular enclosure system with a point control force applied on-centre at 120 Hz.

With this rectangular enclosure system, the modal rearrangement control mechanism can be even more evident with an acoustic cavity controlled response. Figures 9.23 and 9.24 illustrate the phase and amplitudes of the dominant structural and acoustic modes for the active control of sound transmission from a normally incident plane wave at 171 Hz, near the (0,0,1) acoustic cavity resonance frequency, using a control force at the panel centre. Here, there is no significant decrease in the amplitude of any of the structural modes, while the nearly resonant (0,0,1) acoustic mode has decreased in amplitude, leading to a reduction in acoustic potential energy of 14.9 dB.

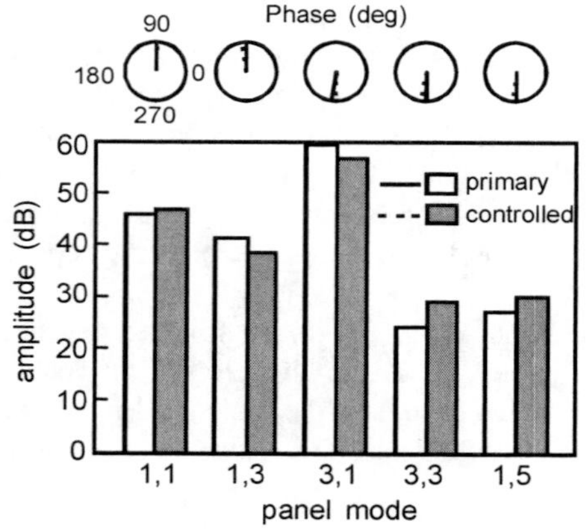

Figure 9.23 Change in structural modal amplitudes of the top panel of the rectangular enclosure system with a point control force applied on centre at 171 Hz.

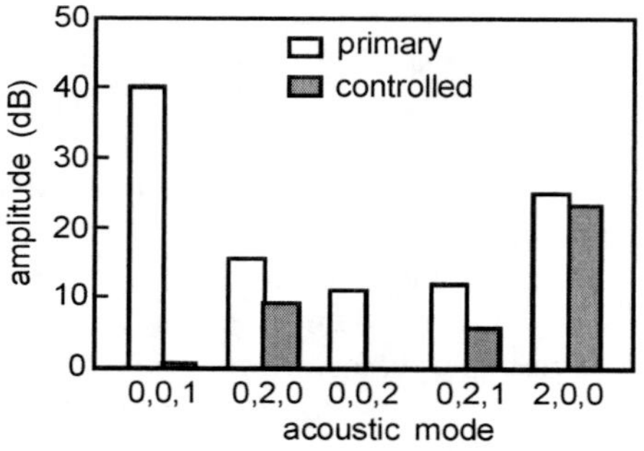

Figure 9.24 Change in acoustic modal amplitudes of the rectangular enclosure system with a point control force applied on centre at 171 Hz.

It is important to note that the reason modal rearrangement is able to provide global sound attenuation in this system is that there are several structural modes with 'similar' magnitude input impedances coupled to a single acoustic mode; in other words, modal rearrangement works because there are several modes available to 'rearrange'. This seemingly obvious point has implications for the active control of coupled structural/acoustic systems other than the rectangular enclosure; systems such as a cylindrical enclosure.

Consider a simply supported aluminium cylindrical enclosure of 1.2 m length, 0.254 m diameter and 1.6 mm wall thickness, with a normally incident plane wave primary forcing function, and a control force located at (z = 0.6 m, θ = 0°). Figures 9.25 and 9.26 show the amplitudes of the dominant structural and acoustic modes for an excitation frequency of 272 Hz, near the (1,2) structural resonance. Here, the amplitude of this nearly resonant structural mode has been reduced, with a corresponding reduction in the (0,2,1) acoustic mode that is coupled to it, and an overall reduction of 17.7 dB in the acoustic potential energy.

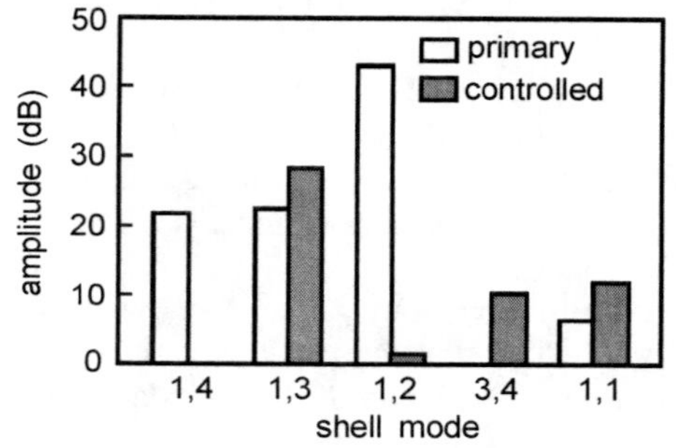

Figure 9.25 Change in structural modal amplitudes of the plain cylindrical enclosure system with a point control force applied at the mid-section, 272 Hz.

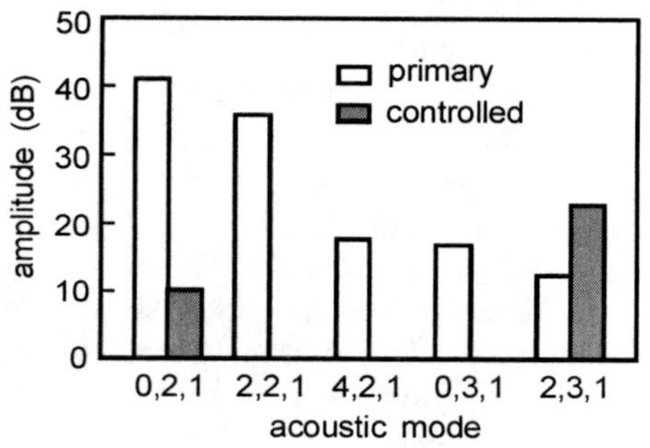

Figure 9.26 Change in acoustic modal amplitudes of the plain cylindrical enclosure system with a point control force applied at the mid-section, 272 Hz.

It is shown in Figures 9.27 and 9.28 that the modal amplitudes are the same for control at a frequency of 656 Hz, near the (0,2,1) acoustic resonance. Again, modal amplitude control is the physical mechanism that provides the 11.1 dB reduction in acoustic potential energy. In fact, modal amplitude control is the *only* physical mechanism (for vibration control sources) that will provide global sound attenuation for a circular cylinder. This is because of its structural/acoustic modal coupling characteristics, which at low frequencies essentially limits coupling to single structural mode/acoustic mode pairs. With only a single structural mode exciting a given acoustic mode, it is not possible to use modal rearrangement.

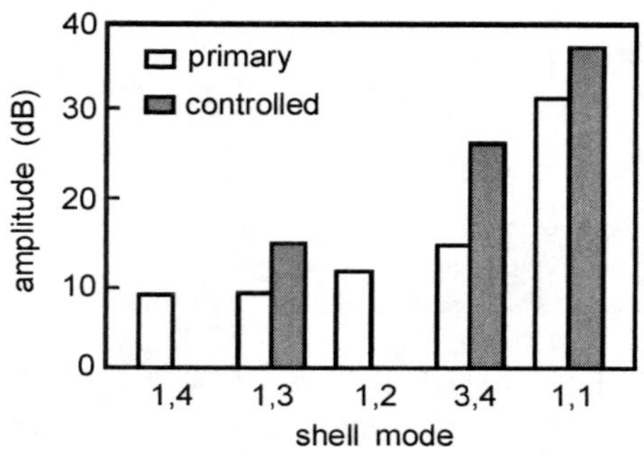

Figure 9.27 Change in structural modal amplitudes of the plain cylindrical enclosure system with a point control force applied at the mid-section, 656 Hz.

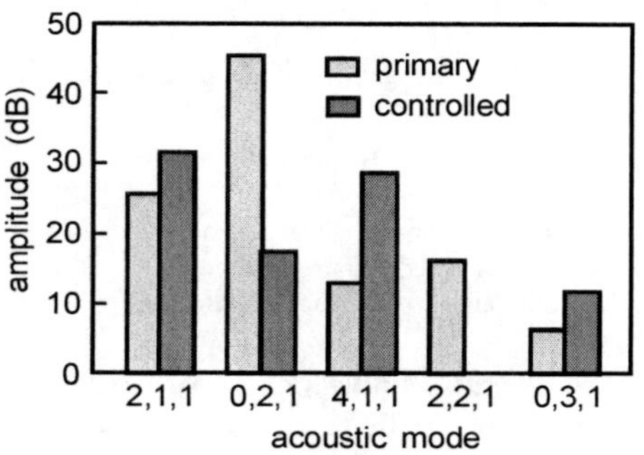

Figure 9.28 Change in acoustic modal amplitudes of the plain cylindrical enclosure system with a point control force applied at the mid-section, 656 Hz.

The final point to note here is that while a reduction in structural modal amplitudes is the physical mechanism that provides global sound attenuation in a circular cylinder when vibration sources are used, it does not necessarily follow that the overall levels of shell vibration will decrease when point forces are used. This fact is illustrated in Figures 9.29 and 9.30, which plot the maximum possible levels of acoustic potential energy reduction that can be achieved for a normally incident plane wave primary forcing function, controlled by a single vibration (point) control force at ($z = 0.6$, $\theta = 0°$), as a function of frequency, and the corresponding change in mean square velocity levels. Clearly, the overall velocity levels increase in many cases. This spillover effect results from the (point) control force inadvertently exciting higher-order structural modes, which do not drive the interior sound field efficiently, while controlling the lower order structural modes that are largely responsible for the interior sound field. Hence, this effect has little impact upon the controlled sound field.

In a practical system such as an aircraft or automobile, then, will modal rearrangement be a mechanism that can provide global sound attenuation? The answer is that it probably will, as most practical systems do not exhibit such ideal single mode coupling as found in the cylindrical enclosure system. In fact, the characteristics of the circular cylinder can be altered by simply including a longitudinal 'floor' in the enclosure.

To illustrate this, consider the introduction, into the previously considered cylindrical enclosure, of a 1.6 mm longitudinal partition, mounted so that the two longitudinal edges subtend an angle ($2\theta_f$ in Figure 9.20) of 90° with the central axis of the cylinder. Let the excitation frequency now be 465 Hz, which is near the cut-on (axial modal index $q = 0$) of the circumferential acoustic mode shape shown in Figure 9.31. Figures 9.32 and 9.33 illustrate the sound fields in a cross-section midway along the cylinder length before and after control with a point force located in the same position ($z = 0.6$ m, $\theta = 0°$) as the previous (plain cylinder) example. The reduction in sound pressure levels is readily apparent, and is matched by the 23.1 dB reduction in acoustic potential energy.

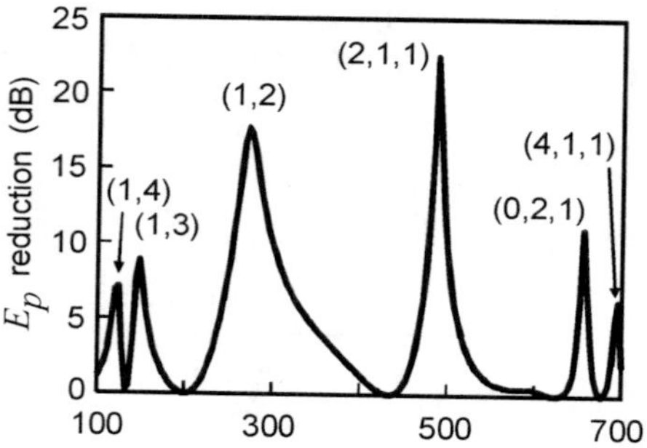

Figure 9.29 Maximum achievable levels of acoustic potential energy reduction for the plain cylindrical enclosure system as a function of frequency.

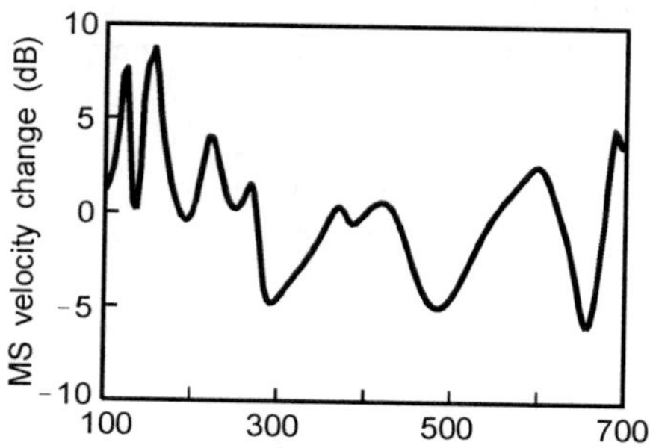

Figure 9.30 Change in overall structural velocity levels corresponding to the maximum acoustic potential energy reduction as a function of frequency.

This reduction, however, is not matched by a similar reduction in the principal coupled structural modes that are illustrated in Figure 9.34. Many of the amplitudes of the dominant coupled structural modes do, in fact, increase, which is accompanied by an increase in the overall structural velocity levels, demonstrating that modal rearrangement is the control mechanism at work here.

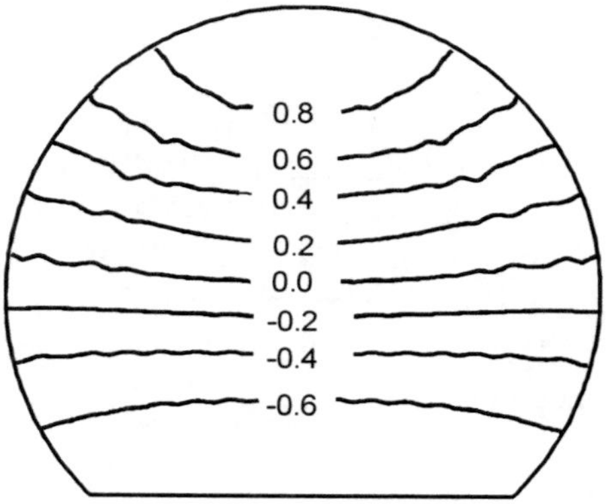

Figure 9.31 Cross-section of the acoustic mode shape resonant at 465 Hz for the cylinder with floor system.

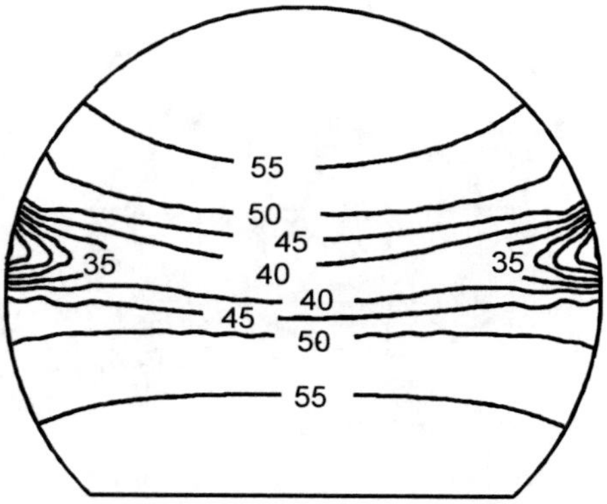

Figure 9.32 Primary sound field at the mid-section of the cylinder with floor under primary excitation, 465 Hz.

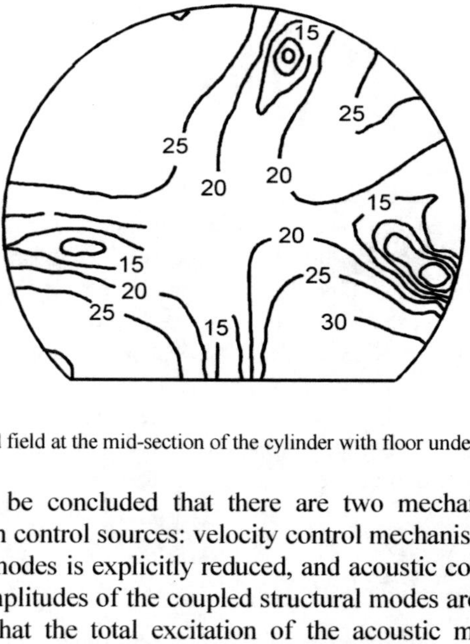

Figure 9.33 Controlled sound field at the mid-section of the cylinder with floor under primary excitation, 465 Hz.

Therefore, it can be concluded that there are two mechanisms that are generally employable by vibration control sources: velocity control mechanisms, where the velocity of the coupled structural modes is explicitly reduced, and acoustic control mechanisms, where the relative complex amplitudes of the coupled structural modes are altered in relative phase and magnitude, such that the total excitation of the acoustic mode is reduced, without necessarily providing a reduction in any of the individual structural modes. The choice of mechanism is dependent upon the modal coupling characteristics of the system and the placement of the primary and control sources.

9.8 INFLUENCE OF CONTROL SOURCE AND ERROR SENSOR ARRANGEMENT

As discussed in previous chapters, it has proved thus far not possible to directly (analytically) optimise the design of active control systems, including those targeting sound transmission

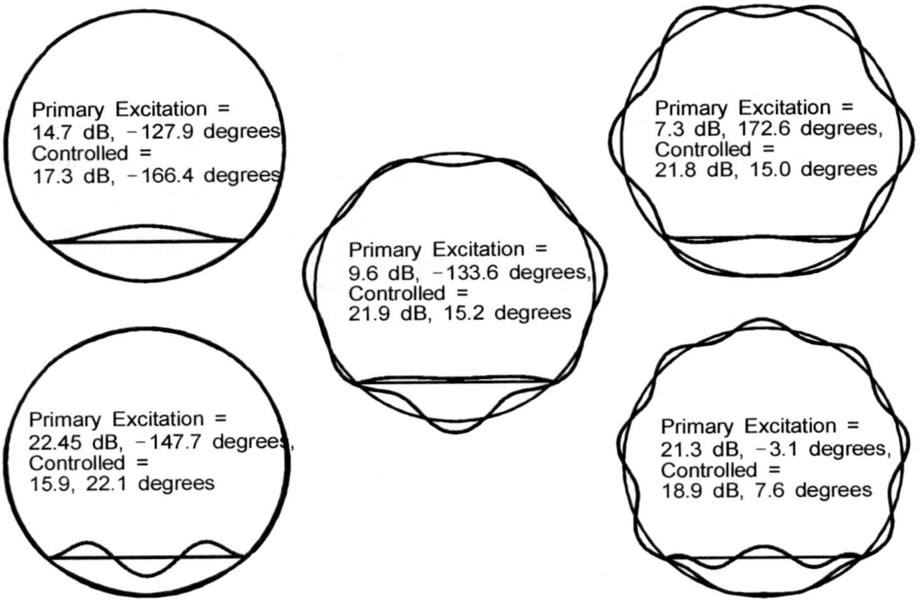

Figure 9.34 Change in amplitude and phase of the dominant structural modes under control, 465 Hz. For all the modes shown in cross-section, the axial modal index $m = 1$.

problems. This is because the maximum level of reduction in acoustic potential energy is not a linear function of control source location, and because the optimum error sensor locations are dependent upon the location of the control sources. Therefore, analytical models that have been developed thus far in this chapter, models that enable the prediction of the maximum levels of reduction in acoustic potential energy possible for a given control source arrangement, and the control source volume velocities/forces required to minimise the sound pressure/structural velocity at potential error sensing locations, must be implemented in some form of a numerical search routine to optimise the physical arrangement of the system. This section will examine some of the implications of control source and error sensor placement, the aim being to obtain some qualitative results that will aid in the optimisation process.

9.8.1 Control Source/Error Sensor Type

When approaching the design of an active system to control sound transmission into a coupled structural/acoustic system, the first variables to consider are the types of control source and error sensor. In terms of control 'potential', a good general rule is that acoustic control sources are best at controlling acoustic resonances, while vibration control sources are best at controlling structural resonances. The reason is simply that in each of these cases the control source will be concerned principally with a single mode, which makes for a more forgiving design exercise. With most coupled structural/acoustic systems, a structural resonance will drive several acoustic modes, while at an acoustic resonance the acoustic mode will be driven by several structural modes. Therefore, when a control source of one medium is used to attenuate a resonance of another medium, it must contend with several modes, complicating the design problem. Thus, the best choice of control source is dependent upon which type of coupled modes (structurally dominated or acoustically dominated) contribute most to the interior sound field in the frequency range of interest.

Kim and Brennan (2000) compared the use of a single point-force actuator and a single acoustic piston source in controlling sound transmission into a coupled structural/acoustic system. In their study, a rectangular enclosure (panel/cavity system) was considered. As expected, the structural actuator was found to be more effective in controlling sound transmission at the resonance frequency of a structurally dominant mode while the acoustic source was more effective at an acoustic resonance. As the enclosed sound field is governed by both acoustic and structural modes, a hybrid approach employing both acoustic and structural sources was found to be the most effective in minimising transmitted sound and also reduced the required actuator control effort.

There are other factors that need to be considered when selecting a control source type for controlling sound transmission into a coupled structural/acoustic system. From the standpoint of peace of mind regarding structural integrity (a point often considered important by aviation authorities!), and ease of installation and modification, acoustic control sources are often a better choice. On the other hand, piezoelectric vibration sources are much lighter than acoustic sources (speakers), and require less power to generate the required control signal.

A study by Hirsch and Sun (1998) has shown that if an array of acoustic sources is placed on the structural-acoustic boundary of an enclosure in such a way as to emulate the structural radiation patterns, then the acoustic sources can in fact outperform structural actuators in suppressing the interior noise levels. This was found to occur even when the enclosed sound field was dominated by structural modes.

In general with structural/acoustic problems, the choice of error sensor type is influenced by the choice of control source type. Obviously, acoustic control sources require acoustic error sensors. With vibration control sources, however, there is a choice, but caution must be exercised when using discrete vibration error sensors, as it has been shown that the maximum attenuation of acoustic potential energy can result in an increase in the vibration levels of the structure (during the employment of control mechanisms involving modal rearrangement). As will be seen in the next section, the use of distributed vibration sensors to provide an error signal is also subject to some difficulty, leading to the conclusion that acoustic error sensors may, in most cases, be the better option when attempting to minimise sound radiation.

9.8.2 Effect of Control Source Arrangement/Numbers

The next questions to consider are how critical is the placement of control sources in achieving the maximum levels of acoustic potential energy reduction, and whether there is only one (global) optimum arrangement or several local optima, or 'minima'. The answer to the first of these questions is 'very', and the answer to the second is 'there usually are several' (Pan and Hansen, 1991b; Snyder and Hansen, 1994b).

Figure 9.35 shows the maximum levels of acoustic potential energy reduction that can be obtained as a function of the location of a single control force on the top panel of the rectangular enclosure studied in the previous section, for a plane wave primary disturbance at 190 Hz. This figure shows several 'optimal' control source placements, which are actually dominated by different control mechanisms (the edge optima achieve control by employment of velocity (modal control) mechanisms, while the centre optimum relies on acoustic (modal rearrangement) mechanisms). The general locations of these optima are not surprising; they are at anti-nodes of the (1,3) structural mode, which dominates the acoustic transmission at this frequency.

It is important to note here is that if control sources are simply placed in 'convenient' locations in the system (aircraft, automobile, etc.), the level of global sound attenuation achieved may not be as much as is possible, and may in fact be discouraging. Ideally, the control sources should be placed using a numerical search routine, with the error criterion being the minimisation of the acoustic potential energy or the minimisation of the sound pressure level at a number of critical locations, or a combination of both, the latter effect achieved by using the potential energy minimisation procedure and applying a higher weighting to error sensors at critical locations. However, if this is not possible, potential design arrangements can be discriminated between by using a multiple regression approach with limited measured data, as was outlined in Section 9.4.

The next point to consider is how many control sources to use. As with location, this is not an easy question to answer. Heuristically, the best control will be achieved if the control sources are capable of 'duplicating' the response of the system to the primary disturbance. Determining the number of control sources required to do this is analogous to determining the length of a digital filter required to model a dynamic system, which was discussed briefly in Chapter 6. The basic way of determining the number of variables (sources/filter stages) in both of these is simply to include an additional one, re-optimise the arrangement, and see what the improvement in the result there is. If the improvement is not 'significant', the number of control sources can be considered adequate. It may also be tempting to simply include a very large number of control sources, with the aim of compensating for non-ideal placement. This temptation should be avoided, as increasing the number of control sources (and error sensors) will simply make it hard for the electronic control system to keep up with changes in the operating environment, as the increased computational load and the required decrease in the adaptive algorithm convergence coefficient (discussed in Chapter 13) will decrease its reaction time.

9.8.3 Effect of Error Sensor Location

The data illustrated in Figure 9.35 clearly demonstrate that the locations of the active control source(s) have a very significant influence upon the levels of reduction in acoustic potential energy that can be achieved. This is also true for the placement of the error sensor(s). Figure 9.36 illustrates the levels in acoustic potential energy reduction that can be achieved by minimising the sound pressure at a single point in the previously studied rectangular enclosure, in the z = 0 plane, with a vibration control source at (0.4, 0.5) and a normally incident plane wave primary disturbance at 190 Hz. As can be seen, there is a single optimum error sensor location in this plane.

A quantitative assessment of the characteristics of an optimum error sensor arrangement can be obtained by comparing the governing equations of acoustic potential energy and the sum of squared acoustic pressure amplitudes at a number of discrete locations. For a rigid walled enclosure, comparing Equations (9.2.39) and (9.3.26) shows that for the problem of minimising acoustic pressure to be equivalent to the problem of minimising acoustic potential energy:

$$\boldsymbol{Z}_p = \boldsymbol{Z}_E \tag{9.8.1}$$

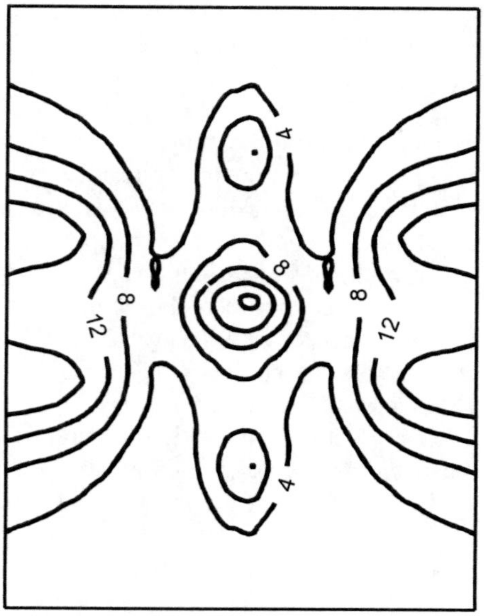

Figure 9.35 Maximum levels of acoustic potential energy reduction as a function of vibration control source location on the top of the rectangular enclosure system, 190 Hz.

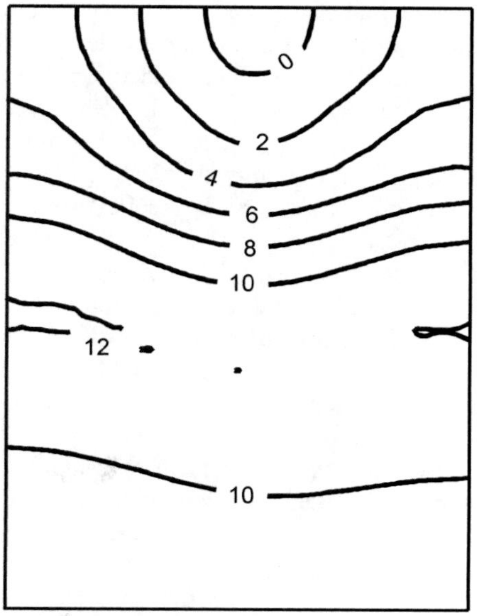

Figure 9.36 Acoustic potential energy reduction achieved by minimising the sound pressure level at one point on the bottom of the rectangular enclosure, vibration control source at (0.4, 0.5), 190 Hz.

where these can be viewed as weighting matrices for the two problems respectively. The same conclusion can be arrived at for the problem of controlling sound transmission into a coupled enclosure, by comparing Equation (9.4.33) and Equation (9.5.2) for an acoustic control source, and Equations (9.4.54) and (9.5.9) for a vibration control source. If the terms in these matrices are considered, it can be concluded that for an optimal set of N_e error sensors:

$$\sum_{i=1}^{N_e} \begin{bmatrix} \varphi_1^2(\boldsymbol{r}_i) & \varphi_i(\boldsymbol{r}_i)\varphi_2(\boldsymbol{r}_i) & \cdots & \varphi_1(\boldsymbol{r}_i)\varphi_{N_n}(\boldsymbol{r}_i) \\ \varphi_2(\boldsymbol{r}_i)\varphi_1(\boldsymbol{r}_i) & \varphi_2^2(\boldsymbol{r}_i) & \cdots & \varphi_2(\boldsymbol{r}_i)\varphi_{N_n}(\boldsymbol{r}_i) \\ & & \vdots & \\ \varphi_{N_n}(\boldsymbol{r}_i)\varphi_1(\boldsymbol{r}_i) & \varphi_{N_n}(\boldsymbol{r}_i)\varphi_2(\boldsymbol{r}_i) & \cdots & \varphi_{N_n}^2(\boldsymbol{r}_i) \end{bmatrix} \rightarrow \begin{bmatrix} \Lambda_1 & 0 & \cdots & 0 \\ 0 & \Lambda_2 & \cdots & 0 \\ & & \vdots & \\ 0 & 0 & \cdots & \Lambda_{N_n} \end{bmatrix} \tag{9.8.2}$$

It is possible to use this as a criterion for assessing the effect of error sensor placement for a number of simple systems (Bullmore, 1988).

There is, however, a simpler way to optimally place error sensors. Considering the data shown in Figure 9.36, it is possible to make a prediction of the optimal error sensor location by looking at the difference between the primary and optimally controlled sound fields for the given control source arrangement as shown in Figure 9.37 for the $z = 0$ plane. Comparing this with the optimal error sensor location plot, it is apparent that the optimal error sensor location is at the point of greatest acoustic pressure reduction under optimum (in terms of acoustic potential energy reduction) control. This intuitively sensible result was seen previously in Section 9.3 when considering a simple one-dimensional example. This characteristic is especially advantageous when using multiple regression to optimise the control source location. The points of maximum difference between the primary and optimally controlled residual sound fields can be determined directly using the residuals vector produced as an output by commercial multiple regression packages. These residuals are the difference between the measured quantity (pressure here) and the value predicted by the regression equation. The point of minimum acoustic pressure in the residual sound field is the point with the smallest residual value. Illustrated in Figure 9.38 is a plot of the 'normalised' residuals for the arrangement under discussion, where the control force has been optimised using multiple regression:

$$\text{normalized residual} = \frac{p_{\text{residual}}}{|p_{pri}|} \qquad (9.8.3)$$

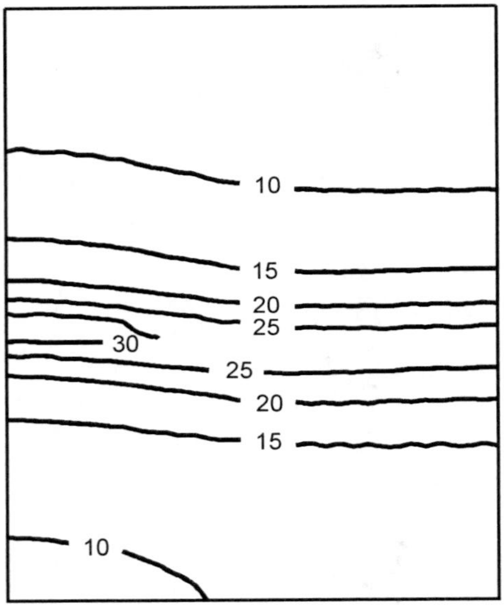

Figure 9.37 Attenuation of the sound field under optimal vibration control applied at (0.4, 0.5) in the bottom of the rectangular enclosure, 190 Hz.

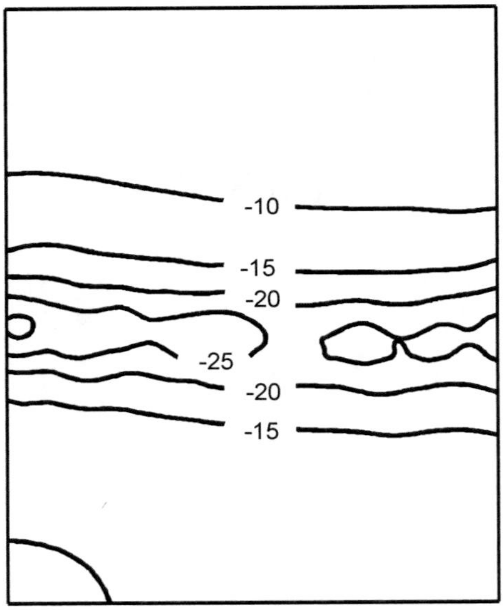

Figure 9.38 Multiple regression residuals under optimal vibration control applied at (0.4, 0.5) in the bottom of the rectangular enclosure, 190 Hz.

9.9 CONTROLLING VIBRATION TO CONTROL SOUND TRANSMISSION

In the previous section, it was pointed out that the levels of acoustic potential energy that are achieved by a practical system are highly dependent upon the location of discrete error sensors used in the system. In this section, the implementation of a system with another form of error signal will be considered, where sets of structural modes are used as error signals in vibration control-source-based systems. Sets of modes could be resolved using 'modal filters' (discussed in Chapter 11), where a modal decomposition is performed on a set of discrete error signals prior to their use by the electronic control system. Alternatively, shaped piezoelectric polymer film sensors could be used to spatially decompose the structural vibration into the desired sets of structural modes (see Chapter 15).

The problem of what function of structural vibration to measure and minimise that will reduce the associated acoustic field was considered in Chapter 8 for controlling sound radiation into free space. It was outlined there that most global error criteria can be expressed in the form:

$$J = \boldsymbol{v}^{\mathrm{H}}\boldsymbol{A}\boldsymbol{v} \tag{9.9.1}$$

where J is the global error criterion of interest and $\boldsymbol{A}$ is some real, symmetric, positive definite transfer matrix. As such, $\boldsymbol{A}$ can be diagonalised by the orthonormal transformation:

$$\boldsymbol{A} = \boldsymbol{Q}\boldsymbol{\Lambda}\boldsymbol{Q}^{-1} \tag{9.9.2}$$

where $\boldsymbol{Q}$ is the orthonormal transform matrix, the (square) column vector of eigenvectors of the transfer matrix $\boldsymbol{A}$; and $\boldsymbol{\Lambda}$ is the diagonal matrix of associated eigenvalues. The form of $\boldsymbol{A}$ dictates that the eigenvectors in $\boldsymbol{Q}$ will be orthogonal. If the transformation of Equation (9.9.2) is substituted into Equation (9.9.1), the global error criterion expression becomes:

$$J = \boldsymbol{v}^{\mathrm{H}}\boldsymbol{A}\boldsymbol{v} = \boldsymbol{v}^{\mathrm{H}}\boldsymbol{Q}\boldsymbol{\Lambda}\boldsymbol{Q}^{-1}\boldsymbol{v} = \boldsymbol{v}'^{\mathrm{H}}\boldsymbol{\Lambda}\boldsymbol{v}' \tag{9.9.3a–c}$$

where the vector of transformed modal velocities $\boldsymbol{v}'$ is defined by the expression:

$$\boldsymbol{v}' = \boldsymbol{Q}^{-1}\boldsymbol{v} \tag{9.9.4}$$

and use has been made of the property of the orthonormal transform matrix:

$$\boldsymbol{Q}^{-1} = \boldsymbol{Q}^{\mathrm{T}} \tag{9.9.5}$$

The orthonormal transformation has decoupled the global error criterion expression, so that it is now equal to a weighted summation of the form:

$$J = \sum_{i=1}^{N_m} \lambda_i |v_i'|^2 \tag{9.9.6}$$

where λ_i is the ith eigenvalue of $\boldsymbol{A}$ and v_i' is the ith transformed modal velocity. As discussed, it is the transformed modal velocities that ideally should be measured and used as error signals, where the measured signals are each weighted by the associated eigenvalues prior to use by the electronic control system (Snyder and Tanaka, 1992). The utility of this approach for the problem of controlling sound transmission into coupled enclosures is what will be investigated here.

With the control of sound transmission into coupled enclosures, the global error criterion used so far is acoustic potential energy in the enclosure, E_p, shown in Equation (9.3.5) to be equal to the weighted sum of squared acoustic modal amplitudes. From the result of Equations (9.4.20) and (9.4.21), it is straightforward to show that the amplitude of the ith acoustic mode can be expressed as:

$$p_i = \frac{\mathrm{j}\rho_0\omega}{\Lambda_i(\kappa_i^2 - k^2)} \int_S \sum_{\imath=1}^{N_m} \psi_\imath(\boldsymbol{x})\varphi_i(\boldsymbol{x}) v_\imath \, \mathrm{d}\boldsymbol{x} = \frac{\mathrm{j}\rho_0\omega S}{\Lambda_i(\kappa_i^2 - k^2)} \sum_{\imath=1}^{N_m} \beta_{i,\imath} v_\imath \tag{9.9.7a,b}$$

where $\beta_{i,\imath}$ is the non-dimensional modal coupling coefficient for the ith acoustic mode and the $\imath$th structural mode, and is defined in Equation (9.4.11). Equation (9.9.7) can be written as a matrix expression, where the complex amplitude of the ith acoustic mode is equal to:

$$p_i = \boldsymbol{z}_{a,i}^{\mathrm{T}} \boldsymbol{v} \tag{9.9.8}$$

where $\boldsymbol{v}$ is the $(N_m \times 1)$ vector of structural modal velocity amplitudes and $\boldsymbol{z}_{a,i}$ is the $(N_m \times 1)$ vector of radiation transfer functions between the structural modes and the ith acoustic mode, the terms of which are defined by:

$$z_{a,i}(\imath) = \frac{\mathrm{j}\rho_0\omega S}{\Lambda_i(\kappa_i^2 - k^2)} \beta_{i,\imath} \tag{9.9.9}$$

Therefore, the vector of N_n acoustic modal amplitudes, $\boldsymbol{p}$, is defined by the expression:

$$\boldsymbol{p} = \boldsymbol{Z_a v} \tag{9.9.10}$$

where $\boldsymbol{Z_a}$ is the $(N_n \times N_m)$ matrix of radiation transfer vectors:

$$\boldsymbol{Z_a} = \begin{bmatrix} \boldsymbol{z}_{\boldsymbol{a1}}^{\mathrm{T}} \\ \boldsymbol{z}_{\boldsymbol{a2}}^{\mathrm{T}} \\ \vdots \\ \boldsymbol{z}_{\boldsymbol{a}N_m}^{\mathrm{T}} \end{bmatrix} \tag{9.9.11}$$

Equation (9.9.10) can be substituted into Equation (9.3.5), defining acoustic potential energy as the weighted sum of squared acoustic modal amplitudes, to express the acoustic potential energy in the form of Equation (9.9.1) as:

$$J = E_p = \boldsymbol{v}^{\mathrm{H}} \boldsymbol{A} \boldsymbol{v} \tag{9.9.12a,b}$$

where the transfer matrix $\boldsymbol{A}$ is defined by the relationship:

$$\boldsymbol{A} = \boldsymbol{Z_a^* W Z_a^{\mathrm{T}}} \tag{9.9.13}$$

such that the (m,n)th term is equal to:

$$A(m,n) = \sum_{i=1}^{N_n} \frac{\rho_0 k^2 S^2}{4\Lambda_i |(\kappa_i^2 - k^2)|^2} \beta_{i,m} \beta_{i,n} \tag{9.9.14}$$

As with radiated acoustic power, the defining relationship for the transfer matrix $\boldsymbol{A}$ given in Equation (9.9.14) shows that it is not constrained to be diagonal. This suggests that the normal structural modes are not orthogonal contributors to the radiated sound field, and minimising the amplitude of the structural modes will not necessarily reduce the interior acoustic potential energy. The amount of modal cross-coupling, represented by the importance of the off-diagonal terms of $\boldsymbol{A}$, will be dictated by the modal coupling characteristics of the system. If the coupling is very specific, to the point where a given acoustic mode is coupled to a single structural mode, then the transfer matrix $\boldsymbol{A}$ will be diagonal. The orthonormal transform matrix $\boldsymbol{Q}$ will then be the identity matrix, and the eigenvalues will be equal to:

$$\lambda_m = \sum_{i=1}^{N_n} \frac{\rho_0 k^2 S^2}{4\Lambda_i |(\kappa_i^2 - k^2)|^2} \beta_{i,m}^2 \tag{9.9.15}$$

It has already been shown that this is essentially the case for low-frequency excitation of a simply supported circular cylinder, where for a structural mode to couple to an acoustic mode, the circumferential modal indices must match and the axial modal indices must form and odd/even pair. This means that while a given acoustic mode, for example with circumferential index n and an odd axial index, will couple to the infinite set of structural modes $(2,n)$, $(4,n)$, ..., the separation (in frequency) of the resonance frequencies of the structural modes dictate that at low frequencies acoustic modes will often be driven by essentially a single structural mode. Most practical systems, however, will have a significant degree of cross-coupling, where several structural modes are coupled to a single acoustic mode, so that even at low frequencies (with a very limited set of modes being considered), the transfer matrix $\boldsymbol{A}$ will not be diagonal.

The values of the terms in the transfer matrix are frequency dependent, governed by both the actual frequency of excitation and the admittance $(\kappa_i^2 - k^2)^{-1}$ of the acoustic modes at the excitation frequency. The eigenvalues of the transfer matrix will therefore peak at resonances of the acoustic modes to which the transformed modal velocities are coupled. This will have significant implications for the practicality of implementing the type of control system developed in Chapter 8 for minimising acoustic radiation into free space using measurements of structural vibration, as will be discussed shortly.

To further investigate the problem of sensing vibration to control acoustic potential energy in a coupled enclosure, it will again be useful to specialise the problem. The system chosen is a rectangular cavity/panel system shown in Figure 9.18, with a description of the governing equations given in Section 9.7. Here, a simply supported rectangular panel of dimensions 1.8 m × 0.88 m × 9 mm is placed on top of a rigid box of 1.0 m depth, with vibration of the panel responsible for excitation of the enclosed sound field. The first ten structural and acoustic modal resonance frequencies are listed in Table 9.1. Recall that for this arrangement, there is a significant degree of cross-coupling in this system, as the only requirement for a structural and acoustic mode to couple is that the modal indices in both *x*- and *y*-directions form an odd/even (structural/acoustic) pair (the modal coupling relationship is defined in Equation (9.7.21)). This means that the transfer matrix, $\boldsymbol{A}$, will not be diagonal, even with a low number modal truncation at low frequencies.

For the calculations outlined here, 30 acoustic modes and 30 structural modes will be used, and acoustic and structural loss factors of 0.01 will be used in formulating the transfer matrix $\boldsymbol{A}$. As for the case in Chapter 8, when decomposing the transfer matrix in the frequency range up to 250 Hz, it is found that only a limited number of eigenvalues (four) are of significant amplitude. These eigenvalues are plotted as a function of frequency in Figure 9.39. The transformed modal velocities associated with these are based predominantly upon the (1,1), (2,1), (3,1) and (1,2) structural modes respectively. The first two of these, normalised to an amplitude of 1.0, are plotted in Figures 9.40 and 9.41.

Table 9.1 Resonance frequencies of first 10 structural and acoustic modes.

Structural Mode	Resonance Frequency (Hz)	Acoustic Mode	Resonance Frequency (Hz)
(1,1)	35.17	(0,0,0)	0.00
(2,1)	55.52	(1,0,0)	95.28
(3,1)	89.44	(0,0,1)	171.50
(1,2)	120.33	(2,0,0)	190.56
(4,1)	136.94	(0,1,0)	194.89
(2,2)	140.68	(1,0,1)	196.19
(3,2)	174.60	(1,1,0)	216.93
(5,1)	198.00	(2,0,1)	256.37
(4,2)	222.09	(0,1,1)	259.60
(1,3)	262.25	(2,1,0)	272.57

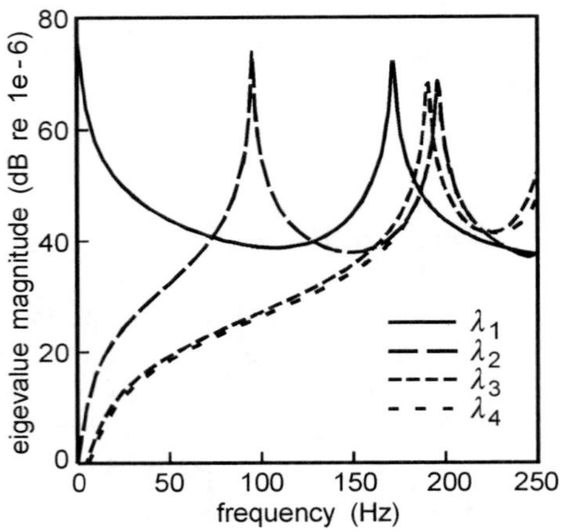

Figure 9.39 Eigenvalues of the first four (most important) transformed modal velocities.

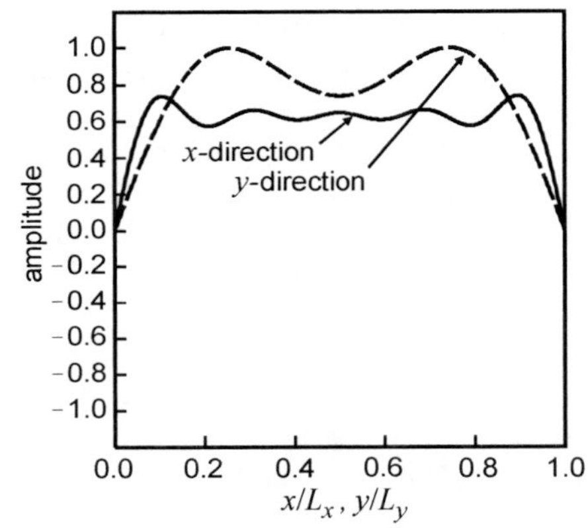

Figure 9.40 First transformed mode shape.

Note the similarity between these transformed modal velocities and those obtained for the problem of radiated acoustic power in Chapter 8. Further, it is found that while both the eigenvalues and eigenvectors are frequency dependent, the variation in the eigenvectors with frequency is relatively small, once again opening the possibility of implementing a system using, as error signals, transformed modal velocities that are based upon eigenvectors which are strictly correct at only one frequency, such as could be measured using a shaped piezoelectric polymer film sensor, weighting these measurements by the frequency correct eigenvalue to provide an error signal and achieving near-optimum results (as was done for free-field radiation in the previous chapter).

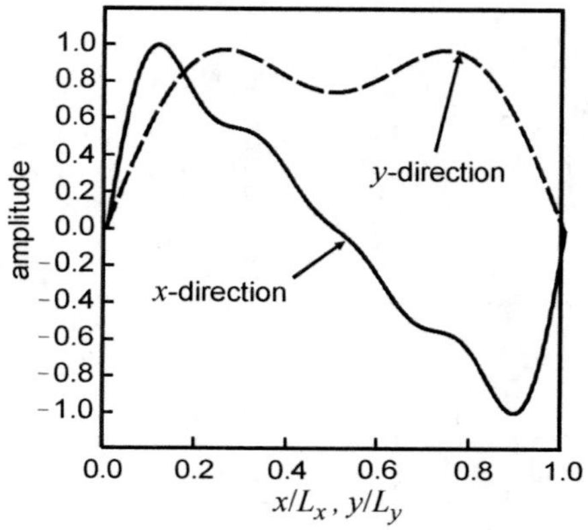

Figure 9.41 Second transformed mode shape.

It is interesting to note that while the eigenvalues shown in Figure 9.39 do peak at the resonance frequencies of the acoustic modes to which the transformed modal velocities are coupled, the coupling is extremely selective. For example, the first eigenvector, based predominantly upon the (1,1) structural mode, peaks at the (0,0,*n*) acoustic resonances *only*, where $n = 0, 1, ...,$ it does *not* peak at (2,0,*n*) acoustic resonances, even though all of the structural modes in the eigenvector are (odd-odd) index modes, which will individually couple to this acoustic mode. Similarly, the eigenvalues associated with the third eigenvector, based predominantly upon the (3,1) structural mode, peak at resonances of (2,0,*n*) acoustic modes *only*, and not at other (even-even,*n*) acoustic resonances. Decomposition of the transfer matrix $\boldsymbol{A}$ produces eigenvectors defining transformed modal velocities that drive a single *x,y* index set of acoustic modes. Therefore, when the modal density in the enclosed space is low, such as at low frequencies in relatively small enclosures, only a few transformed modal velocities need to be considered. As the modal density increases, however, more transformed modal velocities become important and the size of the control problem begins to increase.

Illustrated in Figure 9.42 are the maximum achievable levels of acoustic potential energy reduction, as a function of frequency, with a primary point force at a location (x = 0.35 m, y = 0.64 m) relative to the lower left corner, and a control point force placed symmetrically at (x = 1.45 m, y = 0.24 m). Figure 9.43 shows the acoustic potential energy reduction that results from minimising the first and first to fourth transformed modal velocities. In this calculation the transformed modal velocities were based upon eigenvectors calculated at 30 Hz, and weighted in the calculation using the frequency correct eigenvalues shown in Figure 9.39. As may have been expected, minimising the amplitude of the first transformed modal velocity produces good results at resonances of its constituent structural modes, such as the (1,1) and (3,1), and at acoustic resonances of the form (0,0,*n*).

Minimising the weighted set of four transformed modal velocities achieves practically maximal attenuation over the frequency range of interest.

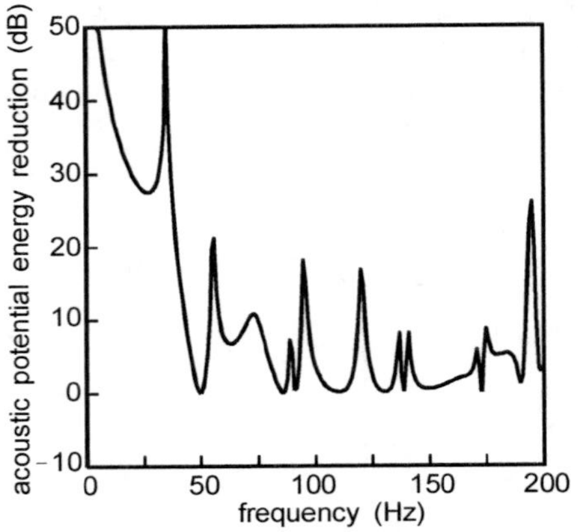

Figure 9.42 Maximum levels of acoustic potential energy reduction, primary source at (0.35 m, 0.64 m) and control source at (1.45 m, 0.24 m).

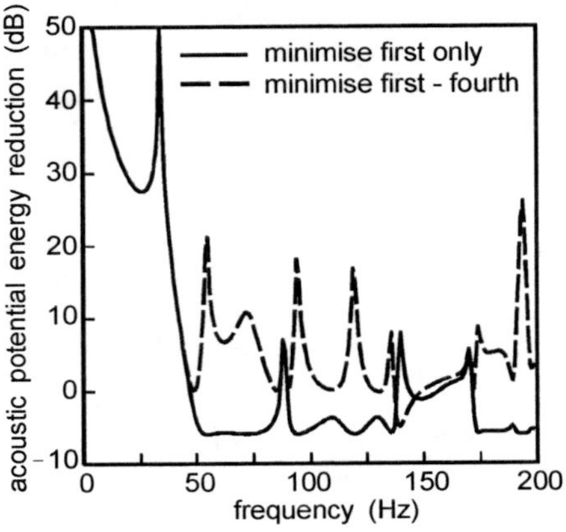

Figure 9.43 Acoustic potential energy reduction achieved by minimising the weighted sum of transformed modal velocities, primary source at (0.35 m, 0.64 m) and control source at (1.45 m, 0.24 m). Shown are the results from minimisation of first transformed modal velocity, and minimisation of first to fourth transformed modal velocities.

Figure 9.44 illustrates the levels of attenuation that result from minimising the kinetic energy of the vibrating panel, or the structural velocity amplitudes. While this approach produces good results near structural resonances, it fails to provide adequate levels of attenuation at acoustic resonances. At acoustic resonances, the sound field will not reflect (in magnitude) the structural velocity distribution, owing to the disproportionate values of acoustic modal admittance $(\kappa_i^2 - k^2)^{-1}$. Clearly, minimising structural velocity with the aim of attenuating the enclosed acoustic field is not, in general, a sound approach.

Having now seen that the approach of minimising the sum of transformed modal velocity amplitudes, calculated using the eigenvectors of the transfer matrix $\boldsymbol{A}$, which are strictly only correct for a single frequency (as could be measured by a shaped sensor), and weighted by the frequency correct eigenvalue, is theoretically capable of producing the desired control effect, the question of practicality again arises. Unfortunately, for control of sound transmission into coupled enclosures, this may present a problem.

Figure 9.45 illustrates the levels of acoustic potential energy attenuation that can be achieved using the set of four transformed modal velocities, but with the eigenvalue weighting neglected. While the attenuation is still satisfactory at many frequencies, the overall result is far less impressive than that obtained with the weighting included. One issue of practicality is therefore the ability to construct eigenvalue 'filters' with the frequency characteristics shown in Figure 9.39, equivalent to the problem of constructing a set of filters that will mirror the response spectrum of a specific set of acoustic modes. While this is possible, it is not as straightforward as the construction of what are essentially high-pass (eigenvalue) filters for the control of acoustic power radiation into free space, as outlined in the previous chapter. There is also the problem of increasing modal density requiring an increasing number of transformed modal velocities to be measured. Therefore, while it is feasible to use structural velocity measurements as error signals in adaptive feedforward control systems targeting the attenuation of sound transmission into coupled enclosures using vibration control sources, practicality must be assessed very much on a per-problem (system/frequency range) basis.

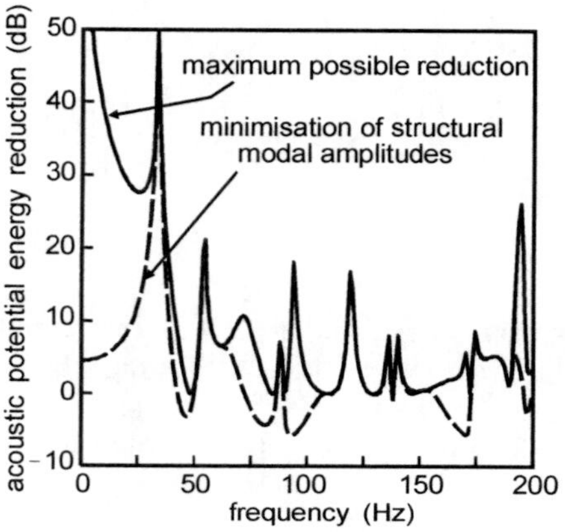

Figure 9.44 Acoustic potential energy reduction achieved by minimising structural modal amplitudes, primary source at (0.35 m, 0.64 m) and control source at (1.45 m, 0.24 m). Shown are the maximum possible reduction and the result obtained from minimisation of structural modal amplitudes.

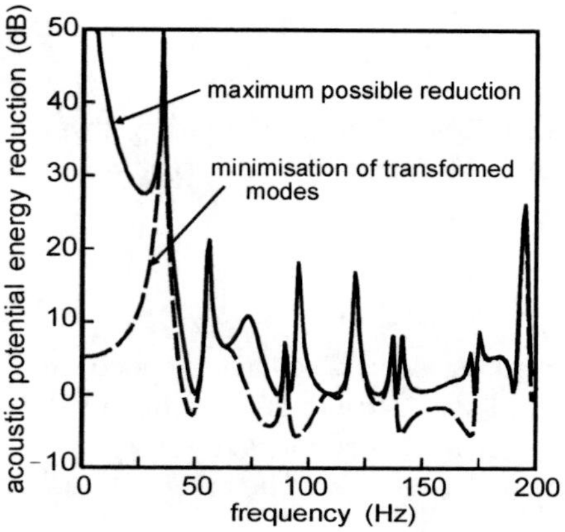

Figure 9.45 Acoustic potential energy reduction achieved by minimising the unweighted set of four transformed modal velocities, primary source at (0.35 m, 0.64 m) and control source at (1.45 m, 0.24 m).

The previous analysis has shown that it is possible to calculate a set of orthonormal structural modes that can be used as the error signals in an active control system to minimise sound transmission into an enclosure. In a number of publications, including those by Cazzolato and Hansen (1998, 1999), these orthonormal sets are referred to as structural radiation modes. Cazzolato and Hansen (1998) investigated structural error sensing to minimise sound transmission through the walls of a stiffened cylinder. Their study discussed issues regarding practical implementation of an active control system minimising sound transmission using structural error sensing, and some of the results from this work are presented below.

As stated earlier, it is transformed modal velocities $\boldsymbol{v}'$, referred to as the amplitudes of radiation modes by Cazzolato and Hansen (1998), that should be measured and used as the error signals by the active control system. Cazzolato and Hansen (1998) showed that a least-squares approach can be used to extract the transformed modal velocities from measurements of structural vibration. When using a number of discrete error sensors to measure the structural velocity distribution, the structural velocity levels at the N_e error sensor locations are given by:

$$\boldsymbol{v}_e = \boldsymbol{\Psi}_e \boldsymbol{v} \tag{9.9.16}$$

where $\boldsymbol{v}_e$ is the $(N_e \times 1)$ vector of the velocity levels measured at the sensors and $\boldsymbol{\Psi}_e$ is the $(N_e \times N_m)$ structural mode shape matrix at the sensor locations. It follows from Equation (9.9.16) that the structural modal velocity vector is:

$$\boldsymbol{v} = \boldsymbol{\Psi}_e^{-1} \boldsymbol{v}_e \tag{9.9.17}$$

Substituting Equation (9.9.17) into (9.9.4) gives an expression for the transformed modal velocities as a function of the velocity levels at the error sensor locations:

$$\boldsymbol{v}' = \boldsymbol{Z}_t \boldsymbol{v}_e \tag{9.9.18}$$

where $\boldsymbol{Z}_t$ is the modal filter matrix that relates the error sensor vibration velocity levels to the transformed modal velocities and is given by:

$$\boldsymbol{Z}_t = \boldsymbol{Q}^{\mathrm{T}} \boldsymbol{\Psi}_e^{-1} \tag{9.9.19}$$

The elements of each row of the modal filter matrix in Equation (9.9.19) represent a weighting value which, when applied to the signal from the error sensors measuring vibration and summed for all sensors, will provide a measure of the transformed modal velocities.

Using Equation (9.9.4), the expression for the structural velocity levels at the error sensors given by Equation (9.9.16) can be written as:

$$\boldsymbol{v}_e = \boldsymbol{\Psi}_e \boldsymbol{Q} \boldsymbol{Q}^{-1} \boldsymbol{v} = [\boldsymbol{\Psi}_e \boldsymbol{Q}] \boldsymbol{v}' \tag{9.9.20}$$

where $\boldsymbol{\Psi}_e \boldsymbol{Q}$ is the mode shape matrix of the transformed modal velocities at the discrete error sensor locations.

In a practical system, an array of accelerometers could be used to measure the velocity levels on the structural surface. The modal filters $\boldsymbol{Z}_t$ given by Equation (9.9.19) decompose the velocity measurements into the transformed modal velocities. The matrix of eigenvalues, $\boldsymbol{\Lambda}$, can then be used to weight the transformed modal velocities according to Equation (9.9.6) and these would then be input to the controller.

As stated earlier, one issue with practical implementation of an active noise control system employing structural error sensing is that the eigenvalue and eigenvector matrices of transfer matrix $\boldsymbol{A}$ are frequency dependent. To overcome this, the discrete error sensors can be replaced with shaped piezoelectric polymer film sensors optimised for operation at a single frequency. Shaped sensors effectively provide the modal filtering, meaning that modal filters, $\boldsymbol{Z}_t$, are not required. As the eigenvectors vary slowly with frequency, this approach is effective over the frequency range where the transformed mode shape variation is small.

Fixing operation at a single frequency is performed by normalising the eigenvector matrix $\boldsymbol{Q}$ at the fixed frequency to produce a new eigenvector matrix $\boldsymbol{Q}_f = \boldsymbol{KQ}$, where $\boldsymbol{K}$ is a square but highly diagonal correction matrix. Using this approach, only the frequency dependent eigenvalues in matrix $\boldsymbol{\Lambda}$ need to be implemented digitally, and a shaped sensor whose shape is optimised for a selected frequency can be used as the modal filter.

9.10 INFLUENCE OF MODAL DENSITY

Thus far, this chapter has been explicitly concerned with systems subject to low-frequency excitation, where the response can be accurately modelled using a relatively small number of acoustic modes. It has been shown that in such systems, active control is most effective at suppressing system resonances, where the response of the dominant mode can be reduced without having a significant spillover effect (non-resonant modes are inadvertently increased in amplitude in the control process). In this section, the applicability of these results to higher frequency excitation will be considered, where the response of the system is unlikely to be dominated by a single mode, and where the reduction in amplitude of one set of modes is likely to be offset by a similar increase in the amplitude of another set of modes. Specifically, a criterion for predicting the level of global attenuation that is maximally possible based upon the nature of the system response, will be developed.

A parameter that is useful in developing such a criterion is modal overlap, $M(\omega)$, which quantifies the likely number of resonance frequencies of other modes lying within the 3 dB bandwidth of a given modal resonance. This parameter is equal to the product of the average modal 3 dB bandwidth $\Delta\omega$ and modal density $n(\omega)$, which is the average number of resonance frequencies per unit frequency:

$$M(\omega) = \Delta\omega\; n(\omega) \tag{9.10.1}$$

Assuming viscous damping, the 3 dB bandwidth of the $\imath$th mode is:

$$\Delta\omega = 2\zeta_\imath\omega_\imath \tag{9.10.2}$$

where $\zeta_\imath$ and $\omega_\imath$ are the critical damping ratio and resonance frequency of the $\imath$th acoustic mode respectively. Therefore, modal density can be re-expressed as:

$$M(\omega_\imath) = 2\zeta_\imath\omega_\imath n(\omega_\imath) \tag{9.10.3}$$

It is possible (Tohyama and Suzuki, 1987; Elliott, 1989) to derive an expression for the average reduction in acoustic potential energy for a rigid walled system containing a single primary source and a single control source remote from it, as an explicit function of modal density. This will provide a qualitative picture of how well active noise control will work (globally) at higher frequencies, away from somewhat isolated modal resonances. The related problem of local control is discussed in the next section.

As has been discussed several times, acoustic potential energy is equal to the weighted sum of squared acoustic modal amplitudes:

$$E_p = \frac{1}{4\rho_0 c_0^2}\sum_{\imath=1}^{\infty}\Lambda_\imath|p_\imath|^2 \tag{9.10.4}$$

A defining relationship for acoustic modal amplitudes resulting from excitation by a volume velocity source can be written, using Equations (9.3.6) and (9.2.12), as:

$$p_\imath = \frac{\mathrm{j}\rho_0\omega}{\Lambda_\imath(\kappa_\imath^2 - k^2)}\varphi_{g,\imath}q \tag{9.10.5}$$

where q is the complex volume velocity amplitude of the source and $\varphi_{g,\imath}$ is the modal generalised volume velocity transfer function for the $\imath$th mode. For the examination here, it will be useful to state explicitly the complex eigenvalue $\kappa_\imath$ in terms of both resonance frequency and damping:

$$p_\imath = \frac{\mathrm{j}\rho_0 c_0{}^2\omega}{\Lambda_\imath\left(\omega_\imath^2(1 - \mathrm{j}2\zeta) - \omega^2\right)}\varphi_{g,\imath}q \tag{9.10.6}$$

Expanded in this way, it is possible to write the relation for acoustic potential energy as:

$$E_p \geq \sum_{\imath=1}^{\infty} \frac{\rho_0 c_0^2 \omega^2}{4\Lambda_\imath\left(4\zeta_\imath\omega_\imath^2\omega^2 + (\omega^2 - \omega_\imath^2)\right)^2}|\varphi_{g,\imath}q_\imath|^2 \tag{9.10.7}$$

Consider now a system where there is a single primary source and single control source contained within a rigid walled enclosure, such that the total volume velocity excitation of any mode is equal to the sum of the primary and control source components. If it is now assumed that at the excitation frequency ω, there is a single dominant acoustic mode, mode d, then the acoustic potential energy can be re-expressed as the sum of dominant, E_{pd}, and residual E_{pr}, modal components as:

$$E_p = E_{pd} + E_{pr} \approx D|\varphi_{gp,d}q_p + \varphi_{gc,d}q_c|^2 + \varepsilon_p|q_p|^2 + \varepsilon_c|q_c|^2 \tag{9.10.8a,b}$$

where

$$D = \frac{\rho_0 c_0^2 \omega^2}{4\Lambda_d\left(4\zeta_d^2\omega_d^2\omega^2 + (\omega^2 - \omega_d^2)^2\right)} \tag{9.10.9}$$

$$\varepsilon_p = \sum_{\imath=1,\neq d}^{\infty} \frac{\rho_0 c_0^2 \omega^2}{4\Lambda_\imath\left(4\zeta_\imath^2\omega_\imath^2\omega^2 + (\omega^2 - \omega_\imath^2)^2\right)}|\varphi_{gp,\imath}|^2 \tag{9.10.10}$$

and

$$\varepsilon_c = \sum_{\imath=1,\neq d}^{\infty} \frac{\rho_0 c_{0^2} \omega^2}{4\Lambda_\imath\left(4\zeta_\imath{}^2\omega_\imath^2\omega^2 + (\omega^2 - \omega_\imath^2)^2\right)}|\varphi_{gc,\imath}|^2 \tag{9.10.11}$$

Here, it has been assumed that, as the primary and control sources are remote from each other, modal cross-terms of the form $\varphi_{gp,\imath}\,\varphi_{gc,\imath}$ will have random signs and so their sum over all modes will be small and can be ignored. Using Equations (9.10.8) to (9.10.11), the acoustic potential energy can be expressed in terms of the dominant and residual modal components as:

$$E_p = q_c^* a q_c + q_c^* b + b^* q_c + c \tag{9.10.12}$$

where

$$a = D\varphi_{gc,d}^2 + \varepsilon_c \tag{9.10.13}$$

$$b = D\varphi_{gc,d}\varphi_{gp,d}q_p \tag{9.10.14}$$

and

$$c = (D\varphi_{gp,d}^2 + \varepsilon_p)|q_p|^2 \tag{9.10.15}$$

Based on previous work, the optimum control source volume velocity derived from a quadratic expression of the standard form used in Equation (9.10.12) is defined by the relationship:

$$q_{c,E} = -a^{-1}b = -\frac{D\varphi_{gc,d}\varphi_{gp,d}}{D\varphi_{gc,d}^2 + \varepsilon_c} q_p \qquad (9.10.16a,b)$$

Observe that if only the dominant mode existed, Equation (9.10.16) would be simply:

$$q_{c,E} = -a^{-1}b = -\frac{\varphi_{gp,d}}{\varphi_{gc,d}} q_p \qquad (9.10.17a,b)$$

which is the same as Equation (9.2.7), when minimising the amplitude of a single mode.

Similarly based on previous work, the reduction in acoustic potential energy is defined by the expression:

$$\frac{E_{p,\min}}{E_{orig}} = \frac{c - b^* a^{-1} b}{c} \qquad (9.10.18)$$

This can expanded using Equations (9.10.13) to (9.10.15) as:

$$\frac{E_{p,\min}}{E_{p,orig}} = 1 - \frac{D^2\varphi_{gc,d}^2\varphi_{gp,d}^2}{(D\varphi_{gc,d}^2 + \varepsilon_c)(D\varphi_{gp,d}^2 + \varepsilon_p)} \qquad (9.10.19)$$

Making the simplifying assumptions that $\varphi_{gc,d} \approx \varphi_{gp,d} \approx 1$ and $\varepsilon_p \approx \varepsilon_c = \varepsilon$, Equation (9.10.19) can be written as:

$$\frac{E_{p,\min}}{E_{p,orig}} \approx 1 - \frac{D^2}{(D + \varepsilon)^2} = 1 - \left(1 + \frac{\varepsilon}{D}\right)^{-2} \qquad (9.10.20a,b)$$

If $\varepsilon/D << 1$, such as at an isolated acoustic resonance, then,

$$\frac{E_{p,\min}}{E_{p,orig}} = \frac{2\varepsilon}{D} \qquad (9.10.21)$$

The result of Equation (9.10.21) defines an explicit, albeit approximate, relationship between acoustic potential energy reduction and the excitation characteristics of the enclosure in terms of dominant and residual modes. What remains now is to re-express this relationship in terms of modal density, which will require consideration of a specific type of enclosure. The one that will be chosen for the analysis undertaken here is a rigid walled rectangular enclosure.

In a rectangular enclosure, the infinite set of acoustic modes can be divided into three categories: axial modes, which are essentially one-dimensional (such as the (1,0,0) mode); tangential, which are essentially two-dimensional (such as the (1,1,0) mode); and oblique, which are essentially three-dimensional (such as the (1,1,1) mode). The modal density of axial modes in a rectangular enclosure is (Morse, 1948):

$$n(\omega) = \frac{L}{\pi c_0} \tag{9.10.22}$$

where L is the length of the enclosure over which the modal indices are varying. The modal density of tangential modes is (Morse, 1948):

$$n(\omega) = \frac{\omega S}{2\pi c_0^2} \tag{9.10.23}$$

where S is the area of the two-dimensional plane over which the modal indices are varying. The modal density of oblique modes is (Morse, 1948):

$$n(\omega) = \frac{\omega^2 V}{2\pi^2 c_0^3} \tag{9.10.24}$$

where V is the volume of the enclosure. Using Equation (9.10.3), the modal overlap of these modal categories is then,

$$M(\omega) = \frac{2\zeta\omega L}{\pi c_0} \tag{9.10.25}$$

for axial modes;

$$M(\omega) = \frac{\zeta\omega^2 S}{\pi c_0^2} \tag{9.10.26}$$

for tangential modes; and

$$M(\omega) = \frac{\zeta\omega^3 V}{\pi c_0^3} \tag{9.10.27}$$

for oblique modes. Responses that are dominated by oblique (from Tohyama and Suzuki, 1987) and axial (from Elliott, 1989) modes will be considered here.

Consider first the oblique mode dominated response. If the excitation frequency is set equal to the resonance frequency of the dominant mode d, then from Equation (9.10.9):

$$D \geq \frac{\rho_0 c_0^2}{2\zeta_d^2 \omega^2 V} \tag{9.10.28}$$

where use is made of the fact that for oblique modes, the normalisation term Λ is equal to $V/8$, as defined in Equation (9.3.45a). It will be assumed that the level of excitation of the other, residual, acoustic modes is equal to the diffuse field value. Assuming that $\zeta = \zeta_i$ for each mode, this is equal to (after Nelson et al., 1987):

$$\varepsilon = \frac{\omega\rho_0}{4\pi c_0 \zeta} \tag{9.10.29}$$

Assuming that ζ_d is also equal to ζ, the residual to dominant mode ratio in Equation (9.10.20) is equal to:

$$\frac{\varepsilon}{D} \approx \frac{\zeta\omega^3 V}{2\pi c_0^3} = \frac{\pi}{2}M(\omega) \tag{9.10.30a,b}$$

where $M(\omega)$ is the oblique field modal overlap, as given in Equation (9.10.27). Substituting this relationship into Equation (9.10.20), the reduction in acoustic potential energy resulting from suppression of the dominant acoustic mode is:

$$\frac{E_{p,\min}}{E_{p,orig}} = 1 - \left(1 + \frac{\pi}{2}M(\omega)\right)^{-2} \tag{9.10.31}$$

If the modal density $M(\omega) << 1$, such as at low frequencies, this can be further simplified to:

$$\frac{E_{p,\min}}{E_{p.orig}} \approx \pi M(\omega) \tag{9.10.32}$$

In an enclosure dominated by axial modes, the result is slightly different. As was noted in the simple example in Section 9.3, axial modes are uniformly spaced in natural frequency, the frequencies following the relationship:

$$\omega_\imath = \frac{\imath c_0 \pi}{L} \tag{9.10.33}$$

where L is the length of the enclosure over which the modal indices are varying. The term ε governing residual mode acoustic potential energy can therefore be approximated by examining the acoustic potential energy in the enclosure contributed by the two modes that are immediately preceding and following a resonant mode, such that $\omega_\imath = \omega \pm c_0\pi/2L$. Assuming that $c_0\pi/2L << \omega$ and $2\zeta\omega << c_0\pi/L$, the residual mode acoustic potential energy is governed by:

$$\varepsilon = \frac{\rho_0 L^2}{2\pi^2 V} \tag{9.10.34}$$

With the excitation frequency equal to the resonance frequency of the dominant mode, the term D governing the acoustic potential energy contributed by the dominant mode is as stated in Equation (9.10.28). The ratio of residual mode to dominant mode acoustic potential energy for the an axial mode dominated system is therefore,

$$\frac{\varepsilon}{D} = \frac{8\zeta^2\omega^2 L^2}{\pi^2 c_0^2} = 2M^2(\omega) \tag{9.10.35a,b}$$

where the modal overlap $M(\omega)$ is now that of an axial mode system, as outlined in Equation (9.10.25). Substituting this into Equation (9.10.20), the reduction in acoustic potential energy is:

$$\frac{E_{p,\min}}{E_{p,orig}} \approx 1 - \left(1 + 2M^2(\omega)\right)^{-2} \tag{9.10.36}$$

If the modal overlap $M(\omega) << 1$ this can be simplified to:

$$\frac{E_{p,\min}}{E_{p,orig}} \approx 4M^2(\omega) \tag{9.10.37}$$

In comparing Equations (9.10.31) and (9.10.32) for oblique modes with Equations (9.10.36) and (9.10.37) for axial modes, it can be seen that if the modal overlap is small, there is a significant difference in the effectiveness of active control in the two systems. However, as the modal overlap increases, the systems tend toward the same asymptotic result:

$$\frac{E_{p,\min}}{E_{p,orig}} \to 1 \text{ for } M(\omega) \geq 1 \tag{9.10.38}$$

These results are evident in Figure 9.46, which depicts acoustic potential energy attenuation as a function of modal overlap for both mode types.

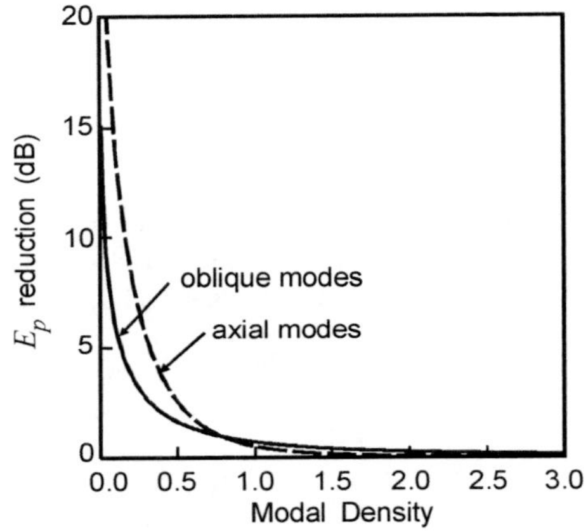

Figure 9.46 Maximum attenuation of acoustic potential energy as a function of modal density for a sound field composed of oblique modes and a sound field composed of axial modes.

To demonstrate the effectiveness of these predictive formulae, it is interesting to apply them to the simple one-dimensional example discussed in Section 9.3. Figures 9.47 and 9.48 illustrate the predicted attenuation using Equation (9.10.36) with the previously calculated (exact) results for damping ratios of 0.005 and 0.05. The prediction is seen to envelop the maximum results, occurring at the resonance frequencies of the acoustic modes (the assumption made in deriving the criteria) very well.

These results demonstrate that the usefulness of (single source) active noise control is limited to essentially isolated resonance problems. The exception to this is when there is a single compact primary noise source, and the control source is placed in close proximity to it. To arrive at this latter conclusion, note that the expression for the control source volume velocity, which is optimal with respect to minimising acoustic potential energy and stated in Equation (9.3.14), can be rewritten as:

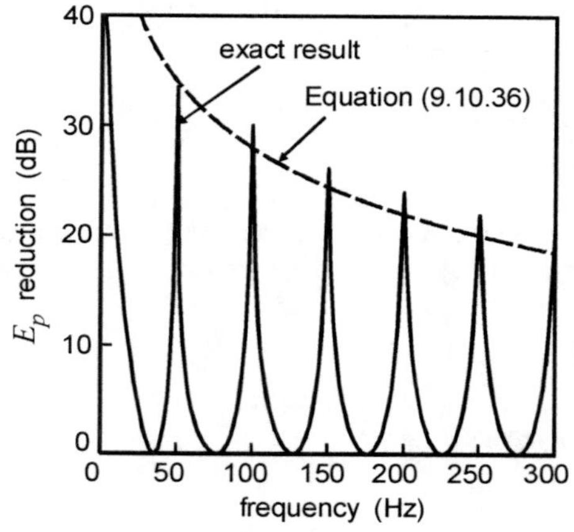

Figure 9.47 Maximum acoustic potential energy reduction as a function of frequency with a single control source at $x = L_x$, damping of 0.005. Shown are the exact result and the estimated result using Equation (9.10.36).

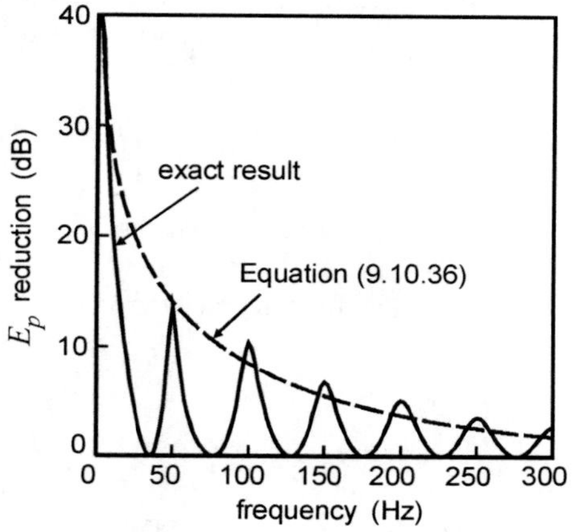

Figure 9.48 Maximum acoustic potential energy reduction as a function of frequency with a single control source at $x = L_x$, damping of 0.05. Shown are the exact result and the estimated result using Equation (9.10.36).

$$q_{c,E} = -q_p \frac{\displaystyle\sum_{\imath=1}^{\infty} \frac{\omega^2 \varphi_\imath(\boldsymbol{r_c})\varphi_\imath(\boldsymbol{r_p})}{(\omega_\imath^2 - \omega^2)^2 + (2\zeta_\imath\omega_\imath\omega)^2}}{\displaystyle\sum_{\imath=1}^{\infty} \frac{\omega^2 \varphi_\imath^2(\boldsymbol{r_p})}{(\omega_\imath^2 - \omega^2)^2 + (2\zeta\omega_\imath\omega)^2}} = -q_p \frac{\displaystyle\sum_{\imath=1}^{N_m} |Y_\imath(\omega)|^2 \varphi_\imath(\boldsymbol{r_c})\varphi_\imath(\boldsymbol{r_p})}{\displaystyle\sum_{\imath=1}^{\infty} |Y_\imath(\omega)|^2 \varphi_\imath^2(\boldsymbol{r_p})} \tag{9.10.39a,b}$$

where

$$|Y_\imath(\omega)|^2 = \frac{\omega^2}{(\omega_\imath^2 - \omega^2)^2 + (2\zeta\omega_\imath\omega)^2} \tag{9.10.40}$$

In writing this expression, consideration is again restricted to oblique modes. At high values of modal overlap, where the excitation is above the Schroeder frequency, the infinite summations in Equation (9.10.39) can be approximated by an integral expression. With this approximation, the numerator of Equation (9.10.39) becomes:

$$\sum_{\imath=1}^{\infty} |Y_\imath(\omega)|^2 \varphi_\imath(\boldsymbol{r_p})\varphi_\imath(\boldsymbol{r_c}) \approx \int_0^{\infty} \langle |Y_\imath|^2 \rangle_u \langle \varphi_\imath(\boldsymbol{r_p})\varphi_\imath(\boldsymbol{r_c}) \rangle_u \, n(u) \, \mathrm{d}u \tag{9.10.41}$$

where the continuous variable u has replaced the discrete variable $\omega_\imath$; the symbol $< >_u$ means that the value in the angular brackets is averaged over a small bandwidth centred on u; and $n(u)$ is the modal density at u. The first of the terms in the integral expression of Equation (9.10.41) is evaluated as:

$$\langle |Y_\imath(\omega)|^2 \rangle_u = \frac{\omega^2}{(u^2 - \omega^2)^2 + (2\zeta u \omega)^2} \tag{9.10.42}$$

where the critical damping ratio is now assumed to be constant for all modes. The second term in the integral expression is evaluated as (Nelson et al., 1987):

$$\langle \varphi_i(\boldsymbol{r}_p)\varphi_i(\boldsymbol{r}_c)\rangle_u = \operatorname{sinc}\frac{ud}{c_0} \tag{9.10.43}$$

where d is the separation distance between the sources. As only oblique modes are being considered, the modal density term in Equation (9.10.39) can be expanded using Equation (9.10.24), so that the integral expression can be written as:

$$\sum_{i=1}^{N_m} |Y_i(\omega)|^2\varphi_i(\boldsymbol{r}_p)\varphi_i(\boldsymbol{r}_c) = \frac{\omega^2 V}{2\pi^2 c_0^3}\int_0^\infty \frac{\omega^2}{(u^2-\omega^2)^2 + (2\zeta u\omega)^2}\operatorname{sinc}\frac{ud}{c_0}\,du \tag{9.10.44}$$

With the further assumption of light damping, this integral can be evaluated as (Nelson et al., 1987):

$$\sum_{i=1}^{N_m} |Y_i(\omega)|^2\varphi_i(\boldsymbol{r}_p)\varphi_i(\boldsymbol{r}_c) = \frac{\omega^2 V}{8\pi c_0^3\zeta}\operatorname{sinc} kd \tag{9.10.45}$$

Using the same series of steps on the denominator as were just used for the numerator of Equation (9.10.39), it is found that:

$$\sum_{i=1}^{N_m} |Y_i(\omega)|^2\,\varphi_i^2(\boldsymbol{r}_p) = \frac{\omega^2 V}{8\pi c_0^3\zeta} \tag{9.10.46}$$

Therefore, the relationship defining the control source volume velocity, which is optimal with respect to minimising the acoustic potential energy in the enclosure, is:

$$q_{c,E} = -q_p \operatorname{sinc} kd \tag{9.10.47}$$

which is exactly the same as the expression obtained for minimising the radiated acoustic power for the same system arrangement operating in free space. At high modal density, only control of the direct field of the source is possible, as is the case in free space, which is why the two solutions are the same. It follows that the maximum attenuation of acoustic potential energy is defined by the relationship:

$$\frac{E_{p,\min}}{E_{p,orig}} = 1 - \operatorname{sinc}^2 kd \tag{9.10.48}$$

Thus, as the control source becomes remote from the primary source, such that $kd \geq \pi$, global attenuation of the sound field becomes impossible. This provides an explicit analytical demonstration that global control of enclosed sound fields of high modal density is *only* possible with closely spaced compact noise sources, and is virtually impossible in the general sense.

9.11 CONTROL OF SOUND AT A POINT IN ENCLOSURES WITH HIGH MODAL DENSITIES

While it may not always be possible to obtain significant levels of global sound attenuation in an enclosed space using active control, especially when the modal density of the sound field is high, it is usually theoretically possible to reduce the acoustic pressure at one or more single points. This problem has been considered previously in Sections 9.2 and 9.4 for systems operating in low modal density environments. In this section, the problem for a system operating in a high modal density environment will be considered, examining both local and global effects of the control strategy. Rather than consider a discrete set of modal contributions, as was done for the low modal density studies of Sections 9.2 and 9.4, statistical operators will be employed to examine the sound field. For simplicity, the discussion will be restricted to a system with a single monopole primary source, single monopole control source, and single (acoustic) error sensing location.

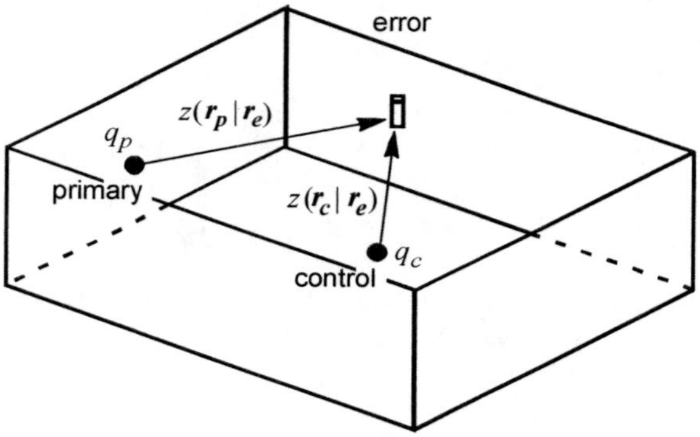

Figure 9.49 System geometry.

Consider the problem as shown in Figure 9.49, where the output characteristics of the control source are to be adjusted to minimise the acoustic pressure amplitude at the error sensing location. The acoustic pressure at the error sensing location $p(\boldsymbol{r}_e)$ is equal to the sum of primary and control contributions:

$$p(\boldsymbol{r}_e) = p_c(\boldsymbol{r}_e) + p_p(\boldsymbol{r}_e) = q_c z(\boldsymbol{r}_c|\boldsymbol{r}_e) + q_p z(\boldsymbol{r}_p|\boldsymbol{r}_e) \qquad (9.11.1a,b)$$

where q_c and q_p are the volume velocities of the control and primary sources respectively, and $z(\boldsymbol{r}_c|\boldsymbol{r}_e)$ and $z(\boldsymbol{r}_p|\boldsymbol{r}_e)$ are the transfer functions between the control and primary source locations and the error sensing location. It is obvious that the acoustic pressure at the error sensing location can be driven to zero by a control source satisfying the simple relationship:

$$q_c = -q_p \frac{z(\boldsymbol{r}_p|\boldsymbol{r}_e)}{z(\boldsymbol{r}_c|\boldsymbol{r}_e)} \qquad (9.11.2)$$

What is of interest here, however, is the development of expressions that will facilitate examination of the overall effect of applying active control.

Consider the acoustic pressure at some location $\boldsymbol{r}_q$, and its relationship to the acoustic pressure at a point some distance $\Delta \boldsymbol{r}$ away, as shown in Figure 9.50.

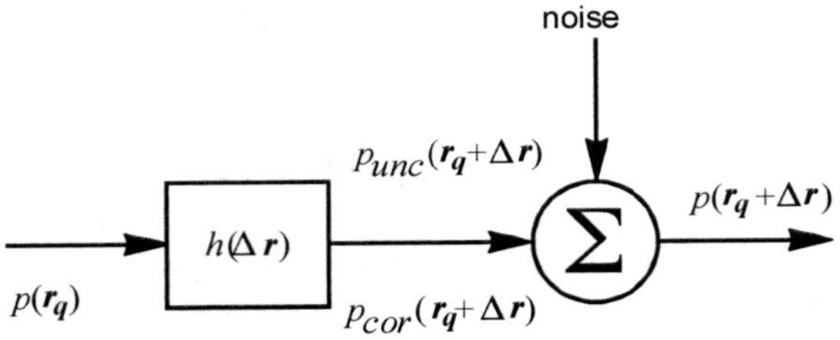

Figure 9.50 Relationship between acoustic pressures $p(\boldsymbol{r}_q)$ and $p(\boldsymbol{r}_q+\Delta\boldsymbol{r})$.

It is possible to view the sound field at $(\boldsymbol{r}_q+\Delta\boldsymbol{r})$ as composed of two parts, one perfectly correlated with the acoustic pressure at $\boldsymbol{r}_q$, the other perfectly uncorrelated (Elliott et al., 1989):

$$p(\boldsymbol{r}_q+\Delta\boldsymbol{r}) = p_{\text{cor}}(\boldsymbol{r}_q+\Delta\boldsymbol{r}) + p_{\text{unc}}(\boldsymbol{r}_q+\Delta\boldsymbol{r}) \tag{9.11.3}$$

where the subscripts cor and unc denote the part of the acoustic pressure that is perfectly correlated and perfectly uncorrelated with the sound field at $\boldsymbol{r}_q$ respectively. Note that by this definition, the cross-correlation between $p(\boldsymbol{r}_q)$ and $p_{\text{unc}}(\boldsymbol{r}_q+\Delta\boldsymbol{r})$ is equal to:

$$\text{E}\left\{p(\boldsymbol{r}_q)\, p_{\text{unc}}(\boldsymbol{r}_q+\Delta\boldsymbol{r})\right\} = 0 \tag{9.11.4}$$

where E{ } denotes expected value of the variable in the curly brackets. In the linear system, the correlated part of $p(\boldsymbol{r}_q+\Delta\boldsymbol{r})$ is related to the pressure at $p(\boldsymbol{r}_q)$ by a simple linear relationship:

$$p_{\text{cor}}(\boldsymbol{r}_q+\Delta\boldsymbol{r}) = h(\Delta\boldsymbol{r})p(\boldsymbol{r}_q) \tag{9.11.5}$$

where $h\,(\Delta\boldsymbol{r})$ is a linear transformation operator, in this case a scalar quantity. Therefore, the cross-correlation between the acoustic pressure $p(\boldsymbol{r}_q)$ and $p_{\text{cor}}(\boldsymbol{r}_q+\Delta\boldsymbol{r})$ is equal to:

$$\text{E}\left\{p(\boldsymbol{r}_q)p_{\text{cor}}(\boldsymbol{r}_q+\Delta\boldsymbol{r})\right\} = h(\Delta\boldsymbol{r})\ \text{E}\left\{p^2(\boldsymbol{r}_q)\right\} \tag{9.11.6}$$

It can be shown (Cook et al., 1955; Morrow, 1971) that in a diffuse field:

$$h(\Delta\boldsymbol{r}) = \text{sinc}\ k\Delta r \tag{9.11.7}$$

where Δr is the length of $\Delta\boldsymbol{r}$. Noting that:

$$\text{E}\left\{p_{\text{cor}}(\boldsymbol{r}_q+\Delta\boldsymbol{r})\ p_{\text{unc}}(\boldsymbol{r}_q+\Delta\boldsymbol{r})\right\} = 0 \tag{9.11.8}$$

it is possible to write the relationship:

$$\text{E}\left\{p^2(\boldsymbol{r}_q+\Delta\boldsymbol{r})\right\} = \text{E}\left\{p^2_{\text{cor}}(\boldsymbol{r}_q+\Delta\boldsymbol{r})\right\} + \text{E}\left\{p^2_{\text{unc}}(\boldsymbol{r}_q+\Delta\boldsymbol{r})\right\} \tag{9.11.9}$$

as

$$\left\{p_{\text{cor}}(\boldsymbol{r}_q+\Delta\boldsymbol{r})\right\} = \text{E}\left\{p(\boldsymbol{r}_q)\right\}\text{sinc}\ k\Delta r \tag{9.11.10}$$

it is possible to express the expected pressure at any location as:

$$\mathrm{E}\{p^2\} = \mathrm{E}\{p^2\}\operatorname{sinc}^2 k\Delta r + \mathrm{E}\{p_{\mathrm{unc}}^2(\boldsymbol{r}_q + \Delta\boldsymbol{r})\} \tag{9.11.11}$$

so that:

$$\{p_{\mathrm{unc}}^2(\boldsymbol{r}_q + \Delta\boldsymbol{r})\} = \mathrm{E}\{p^2(\boldsymbol{r}_q)\}\,(1 - \operatorname{sinc}^2 k\Delta r) \tag{9.11.12}$$

It is of interest to consider what happens to the sound field when the acoustic pressure at the error sensing location $\boldsymbol{r}_e$ is minimised. The acoustic pressure at some point $\Delta\boldsymbol{r}$ away from the error sensing location is equal to the superposition of the primary and control source contributions:

$$p(\boldsymbol{r}_e + \Delta\boldsymbol{r}) = p_p(\boldsymbol{r}_e + \Delta\boldsymbol{r}) + p_c(\boldsymbol{r}_e + \Delta\boldsymbol{r}) \tag{9.11.13}$$

The primary and control source acoustic pressures can each be divided into two components, one perfectly correlated with the acoustic pressure at the error sensing location $\boldsymbol{r}_e$ and one perfectly uncorrelated, so that:

$$p_p(\boldsymbol{r}_e + \Delta\boldsymbol{r}) = p_{p,\mathrm{cor}}(\boldsymbol{r}_e + \Delta\boldsymbol{r}) + p_{p,\mathrm{unc}}(\boldsymbol{r}_e + \Delta\boldsymbol{r}) \tag{9.11.14}$$

where

$$p_{p,\mathrm{cor}}(\boldsymbol{r}_e + \Delta\boldsymbol{r}) = p_p(\boldsymbol{r}_e)\operatorname{sinc} k\Delta r$$

$$\mathrm{E}\{p_{p,\mathrm{unc}}^2(\boldsymbol{r}_e + \Delta\boldsymbol{r})\} = \mathrm{E}\{p_p^2(\boldsymbol{r}_e)\}(1 - \operatorname{sinc}^2 k\Delta r) \tag{9.11.15a,b}$$

and

$$p_c(\boldsymbol{r}_e + \Delta\boldsymbol{r}) = p_{c,\mathrm{cor}}(\boldsymbol{r}_e + \Delta\boldsymbol{r}) + p_{c,\mathrm{unc}}(\boldsymbol{r}_e + \Delta\boldsymbol{r}) \tag{9.11.16}$$

where

$$p_{c,\mathrm{cor}}(\boldsymbol{r}_e + \Delta\boldsymbol{r}) = p_c(\boldsymbol{r}_e)\operatorname{sinc} k\Delta r$$

$$\mathrm{E}\{p_{c,\mathrm{unc}}^2(\boldsymbol{r}_e + \Delta\boldsymbol{r})\} = \mathrm{E}\{p_c^2(\boldsymbol{r}_e)\}\,(1 - \operatorname{sinc}^2 k\Delta r) \tag{9.11.17a,b}$$

If it is now assumed that the primary and control sources are remote from each other, compared to the wavelength of sound, then,

$$\mathrm{E}\{p_{p,\mathrm{unc}}(\boldsymbol{r}_e + \Delta\boldsymbol{r})\, p_{c,\mathrm{unc}}(\boldsymbol{r}_e + \Delta\boldsymbol{r})\} = 0 \tag{9.11.18}$$

If the acoustic pressure at the error sensing location is driven to zero, then,

$$p_c(\boldsymbol{r}_e) = -p_p(\boldsymbol{r}_e) \tag{9.11.19}$$

and so

$$p_{c,\mathrm{cor}}(\boldsymbol{r}_e + \Delta\boldsymbol{r}) = -p_{p,\mathrm{cor}}(\boldsymbol{r}_e + \Delta\boldsymbol{r}) \tag{9.11.20}$$

The squared acoustic pressure level at $(\boldsymbol{r}_e + \Delta\boldsymbol{r})$ is equal to:

$$\mathrm{E}\left\{p^2(\boldsymbol{r_e}+\Delta\boldsymbol{r})\right\} = \mathrm{E}\left\{\left(p_{c,\mathrm{cor}}(\boldsymbol{r_e}+\Delta\boldsymbol{r}) + p_{c,\mathrm{unc}}\boldsymbol{r_e}+\Delta\boldsymbol{r}) + p_{p,\mathrm{cor}}(\boldsymbol{r_e}+\Delta\boldsymbol{r}) + p_{p,\mathrm{unc}}(\boldsymbol{r_e}+\Delta\boldsymbol{r})\right)^2\right\} \tag{9.11.21}$$

Using the previous outlined relationships, this can be simplified to (Elliott et al., 1989):

$$\mathrm{E}\left\{p^2(\boldsymbol{r_e}+\Delta\boldsymbol{r})\right\} = \left(\left\{p_p^2\right\}+\left\{p_c^2\right\}\right)\left(1 - \mathrm{sinc}^2 k\Delta r\right) \tag{9.11.22}$$

where $\mathrm{E}\{p^2\}$ is the expected squared pressure amplitude when the source is operating alone.

Equation (9.11.22) shows that the squared acoustic pressure amplitude rapidly increases away from the error sensing location, tending towards the sum of the individual primary and control source quantities. For example, if a reduction of 10 dB is required at a certain point, it must be within a distance of approximately 1/10 wavelength from the error sensor (cancellation point). Generally speaking, the sound field away from the error sensor increases in amplitude when the acoustic pressure is minimised at a single point in a diffuse sound field (with all sources and sensors remote from each other). This qualitative idea is evident in Figure 9.51, which is the result of a computer simulation examining the pressure field around a single error sensor for a single primary and control source placed randomly in an enclosure containing a diffuse sound field (from Elliott et al., 1988). The data shown represents the average for 200 simulations, and clearly shows the extreme localisation of the 'zone of quiet'. Observe that away from the error sensor, the sound field has *increased* in amplitude.

It is interesting to note that this result is not exactly repeatable, although qualitatively the same characteristics could always be expected. To explain this, note that Equation (9.11.22) shows the mean square acoustic pressure away from the error sensing location to be equal to the sum of the mean square values of primary and control source generated pressure. Thus, the increase in level of the mean square value of the sound field is dependent upon the ratio of the control to primary source volume velocities under optimum (cancelling) conditions. Equation (9.11.2) shows this ratio to be

$$\frac{|q_c|^2}{|q_p|^2} = \frac{|z(\boldsymbol{r_p}|\boldsymbol{r_e})|^2}{|z(\boldsymbol{r_c}|\boldsymbol{r_p})|^2} \tag{9.11.23}$$

If the amplitude of the transfer function $z(\boldsymbol{r_c}|\boldsymbol{r_e})$ between the control source and error sensor is very small, such that the control source is not well coupled to the error sensor, then the amplitude of the control source volume velocity will be very large compared to that of the primary source, and the increase in the amplitude of the pressure field well away from the point of minimisation will be correspondingly large.

The probability density function of the real and imaginary components of acoustic pressure at any point in a diffuse sound field will be Gaussian. The probability density function of the ratio of Equation (9.11.23) will therefore be chi-squared with two degrees of freedom, described by the relationship

$$f(x) = \frac{1}{(1+x)^2} \tag{9.11.24}$$

where x is the variable of interest, which in this case is the ratio of the mean square volume velocities in Equation (9.11.23). Figure 9.52 illustrates the probability density function associated with the simulation result of Figure 9.51, along with the chi-squared probability density function with two degrees of freedom (from Elliott et al., 1989). The correspondence is obvious. The point of this discussion, however, is that large increases in the mean square acoustic pressure away from the cancellation point can occur, with single frequency excitation, due to poor control source placement. As the quality of the control source placement will not be so readily apparent in sound fields of high modal density as in sound fields of low modal density when the sources and sensors are widely dispersed, and for broadband control the source placement is likely to be poor at some frequencies, there is no utility in implementing such an active control arrangement in systems of high modal density with the aim of sound attenuation over anything other than a very small spatial region.

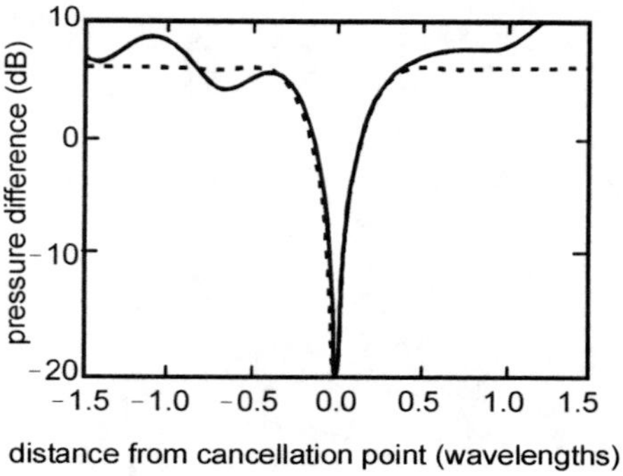

Figure 9.51 Average value of mean square pressure difference as a function of distance from the cancellation point (after Elliott et al., 1988). Shown are theoretical and computer simulated results.

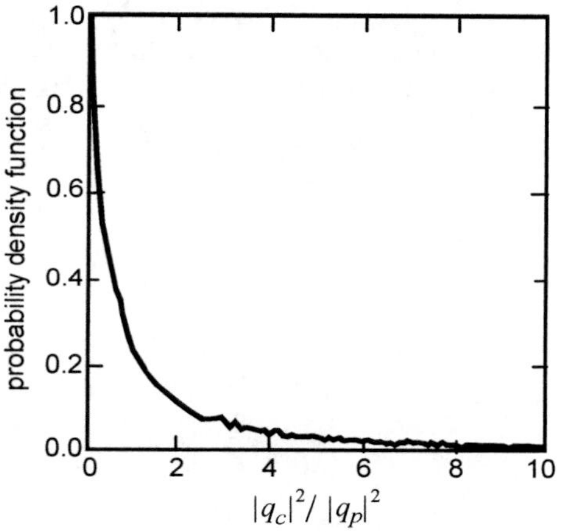

Figure 9.52 Probability density function for the ratio of control to primary mean square source volume velocities (after Elliott et al., 1988). Shown are the results of computer simulation, and a chi-squared distribution with two degrees of freedom.

One approach that can be taken to improve these results, at least from the standpoint of guarding against substantial increases in acoustic potential energy in the enclosed space when minimising the acoustic pressure at a single location, is to place the error microphone directly in front of the control source as illustrated in Figure 9.53. With this arrangement, the control source volume velocity required to minimise the acoustic pressure at the error sensor is greatly reduced, and is largely independent of the statistical properties of the sound field. This, in fact, is one of the modes of operation of an 'electronic sound reducer' proposed in 1953 (Olson and May, 1953), although the control approach taken was feedback rather than feedforward. Part of the utility of this device was envisaged as being able to provide local acoustic control for a variety of arrangements, such as near an occupant's head in an automobile or aircraft, shown in the sketch in Figure 9.54. The previous analysis can be extended to examine how effective such an arrangement could be.

To do this, the transfer function between the control source and error sensor is first expressed as the sum of two components, that due to direct radiation and that due to the reverberant field:

$$z(\boldsymbol{r}_c|\boldsymbol{r}_e) = z_d(\boldsymbol{r}_c|\boldsymbol{r}_e) + z_r(\boldsymbol{r}_c|\boldsymbol{r}_e) \tag{9.11.25}$$

where the subscripts d and r are used to denote the direct and reverberant field components respectively. Substituting this into Equation (9.11.2), the optimum control source volume velocity becomes:

$$q_c = -q_p \frac{z(\boldsymbol{r}_p|\boldsymbol{r}_e)}{z_d(\boldsymbol{r}_c|\boldsymbol{r}_e) + z_r(\boldsymbol{r}_c|\boldsymbol{r}_e)} \tag{9.11.26}$$

With the error microphone placed in the vicinity of the control source, the direct field transfer function will dominate Equation (9.11.26), being much greater than either the primary source transfer function or the reverberant field component of the control source transfer function. This is why the control source volume velocity required to minimise the acoustic pressure at the error microphone will be greatly reduced, which will reduce the problem of greatly increased sound pressure levels away from the error sensor.

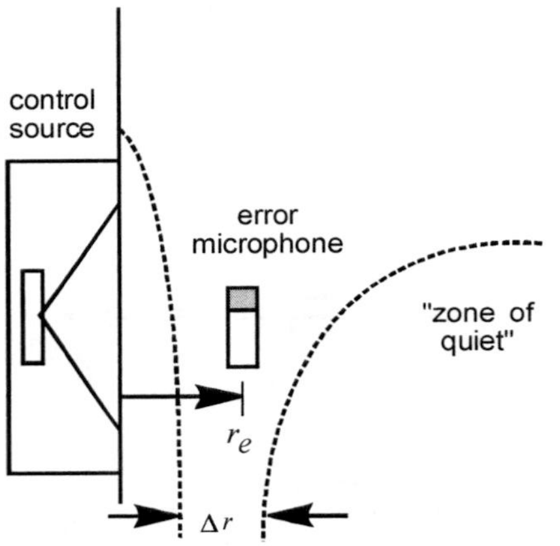

Figure 9.53 Near-field local control arrangement.

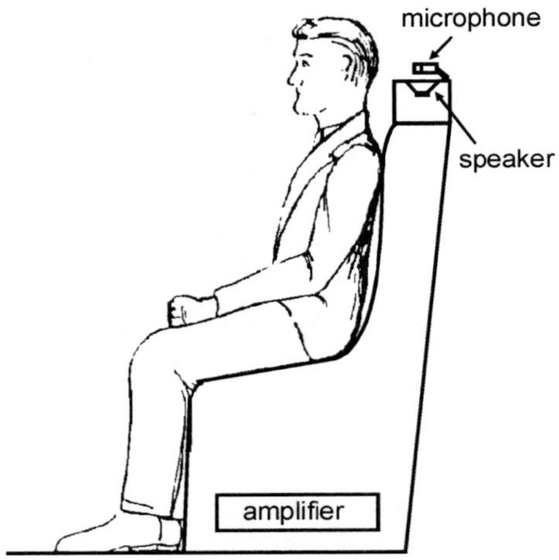

Figure 9.54 One application of an 'electronic sound reducer' (after Olson and May, 1953).

As the direct component of the control source transfer function will be dominant in its immediate vicinity, to a first approximation, the control source generated reverberant field component can be neglected when examining the sound field in this area. The acoustic pressure a small distance $\Delta\boldsymbol{r}$ away from the error sensor can therefore be approximated as:

$$p(\boldsymbol{r}_e + \Delta\boldsymbol{r}) \approx p_p(\boldsymbol{r}_e + \Delta\boldsymbol{r}) + z_d(\boldsymbol{r}_c|\boldsymbol{r}_e + \Delta\boldsymbol{r})q_c \tag{9.11.27}$$

This can be rewritten as:

$$p(\boldsymbol{r}_e + \Delta\boldsymbol{r}) \approx p_p(\boldsymbol{r}_e + \Delta\boldsymbol{r}) - \frac{z_d(\boldsymbol{r}_c|\boldsymbol{r}_e + \Delta\boldsymbol{r})}{z_d(\boldsymbol{r}_c|\boldsymbol{r}_e)} p_p(\boldsymbol{r}_e) \tag{9.11.28}$$

The change in mean square pressure amplitude at the point $\Delta\boldsymbol{r}$ away from the error sensor is therefore,

$$\frac{\mathrm{E}\left\{|p(\boldsymbol{r}_e+\Delta\boldsymbol{r})|^2\right\}}{\mathrm{E}\left\{|p_p(\boldsymbol{r}_e)|^2\right\}} = 1 - 2\ \mathrm{Re}\left\{\frac{z_d(\boldsymbol{r}_c|\boldsymbol{r}_e+\Delta\boldsymbol{r})}{z_d(\boldsymbol{r}_c|\boldsymbol{r}_e)}\right\}\mathrm{sinc}\ \Delta\boldsymbol{r} + \left|\frac{z_d(\boldsymbol{r}_c|\boldsymbol{r}_e+\Delta\boldsymbol{r})}{z_d(\boldsymbol{r}_c|\boldsymbol{r}_e)}\right|^2 \tag{9.11.29}$$

As consideration is restricted to small distances $\Delta\boldsymbol{r}$, the term sinc $\Delta\boldsymbol{r}$ in Equation (9.11.29) can be approximated as equal to 1.0. The transfer function term $z(\boldsymbol{r}_c|\boldsymbol{r}_e+\Delta\boldsymbol{r})$ can be expanded as a Taylor series about $\boldsymbol{r}_e$, retaining only the first term because the distance $\Delta\boldsymbol{r}$ is small, as:

$$z_d(\boldsymbol{r}_c|\boldsymbol{r}_e+\Delta\boldsymbol{r}) \approx z_d(\boldsymbol{r}_c|\boldsymbol{r}_e) + \nabla z_d(\boldsymbol{r}_c|\boldsymbol{r}_e)\cdot\Delta\boldsymbol{r} \tag{9.11.30}$$

Substituting this expansion into Equation (9.11.29) and using the approximation for small distances:

$$\mathrm{E}\left\{|p_p(\boldsymbol{r}_e)|^2\right\} \approx \mathrm{E}\left\{|p_p(\boldsymbol{r}_e+\Delta\boldsymbol{r})|^2\right\} \tag{9.11.31}$$

produces the expression (after Joseph et al., 1989):

$$\frac{\mathrm{E}\left\{|p(\boldsymbol{r}_e+\Delta\boldsymbol{r})|^2\right\}}{\mathrm{E}\left\{|p_p(\boldsymbol{r}_e+\Delta\boldsymbol{r})|^2\right\}} \approx \left|\frac{\nabla z_d(\boldsymbol{r}_c|\boldsymbol{r}_e)\cdot\Delta\boldsymbol{r}}{z_d(\boldsymbol{r}_c|\boldsymbol{r}_e)}\right|^2 \tag{9.11.32}$$

This result shows that, to a first approximation, the change (hopefully reduction) in the mean square acoustic pressure amplitude at a location in the vicinity of the error sensor is dependent upon the ratio of the pressure gradient to the absolute value of the pressure in the near-field of the control source as a result of the following relationship:

$$\left|\frac{\nabla z_d(\boldsymbol{r}_c|\boldsymbol{r}_e)\cdot\Delta\boldsymbol{r}}{z_d(\boldsymbol{r}_c|\boldsymbol{r}_e)}\right|^2 = \left|\frac{\nabla p_d(\boldsymbol{r}_c)\cdot\Delta\boldsymbol{r}}{p_d(\boldsymbol{r}_c)}\right|^2 \tag{9.11.33}$$

This result can be further simplified by re-expressing the pressure gradient in terms of particle velocity (for harmonic pressure) as:

$$\nabla p_d(\boldsymbol{r}_c) = -\mathrm{j}\omega\rho_0 u_d(\boldsymbol{r}_c) \tag{9.11.34}$$

where $u_d(\boldsymbol{r}_c)$ is the particle velocity associated with the direct acoustic pressure. Assuming that the particle velocity is evaluated in the same direction as $\Delta\boldsymbol{r}$, substituting Equation (9.11.34) into Equation (9.11.33), and expressing the result in terms of specific acoustic impedance:

$$z_d(\boldsymbol{r}_c) = \frac{u_d(\boldsymbol{r}_c)}{p_d(\boldsymbol{r}_c)} \tag{9.11.35}$$

produces the following expression:

$$\frac{\mathrm{E}\left\{|p(\boldsymbol{r}_e+\Delta\boldsymbol{r})|^2\right\}}{\mathrm{E}\left\{|p_p(\boldsymbol{r}_e+\Delta\boldsymbol{r})|^2\right\}} \approx \left|\frac{\rho_0 c_0}{z_d(\boldsymbol{r}_c)}\right|^2 k^2\Delta\boldsymbol{r}^2 \tag{9.11.36}$$

This equation can be used to predict the change in the pressure field away from the error sensing location.

As an example, consider the simple case of a monopole control source, for which the specific acoustic impedance some distance $\Delta \boldsymbol{r}$ away is:

$$z(\Delta \boldsymbol{r}) = \rho_0 c_0 \frac{\mathrm{j}k\Delta r}{1 + \mathrm{j}k\Delta r} \tag{9.11.37}$$

Substituting this into Equation (9.11.36) produces the following expression for the change in acoustic pressure in the vicinity of the error microphone:

$$\frac{\mathrm{E}\left\{|\, p(\boldsymbol{r}_e + \Delta \boldsymbol{r})\,|^2\right\}}{\mathrm{E}\left\{|\, p_p(\boldsymbol{r}_e + \Delta \boldsymbol{r})\,|^2\right\}} \approx (1 + k^2 r_e^2)\left(\frac{\Delta r}{r_e}\right)^2 \tag{9.11.38}$$

where r_e is the distance between the monopole control source and the error sensor. For small separations between the control source and error sensor, this can be approximated as:

$$\frac{\mathrm{E}\left\{|\, p(\boldsymbol{r}_e + \Delta \boldsymbol{r})\,|^2\right\}}{\mathrm{E}\left\{|\, p_p(\boldsymbol{r}_e + \Delta \boldsymbol{r})\,|^2\right\}} \approx \left(\frac{\Delta r}{r_e}\right)^2 < \varepsilon \tag{9.11.39}$$

where ε is the mean square pressure reduction. Therefore, a 10 dB zone of quiet will encircle the control source for a radial distance equal to twice the value of Equation (9.11.39), or

$$\Delta r = 2r_e\sqrt{\varepsilon} \approx 0.6r_e \tag{9.11.40a,b}$$

As with the case of a separated control source and error sensor, the area of pressure reduction is extremely localised; it is, in fact, even smaller than before. Recall from Figure 9.51 that with the separated arrangement, a 10 dB zone of quiet existed at a radial distance of approximately 1/10 wavelength from the error microphone. Figure 9.55 depicts the size of the 10 dB zone of quiet around the error sensor as a function of frequency for various separations of the error sensor from the 0.05 m piston control source (see Elliott and David, 1992). These are, in fact, bounded by the 1/10 wavelength criterion, being much less than this amount for small separation distances. Clearly, the advantage of locating the error microphone close the control source is the reduction in the problem of significant increases in acoustic pressure away from the error sensor, and not an enhancement in the size of the 'zone of quiet'.

It is also worth mentioning that the size of the zone of quiet can be enlarged by employing an acoustic energy density cost function. Acoustic energy density is formed by the sum of acoustic potential energy density and acoustic kinetic energy density, and in practice is calculated using the weighted sum of squared pressure and squared particle velocity. Acoustic energy density represents the total energy at a point and has been found to be an effective cost function for both local and global active noise control applications (Sommerfeldt and Nashif, 1994; Sommerfeldt and Parkins, 1994). In addition, minimising the acoustic energy density overcomes the observability problems associated with reducing the acoustic pressure alone. Elliott and Garcia-Bonito (1995) investigated the local control of both pressure and pressure gradient in a pure tone diffuse sound field with two secondary sources. Minimising both the pressure and pressure gradient along a single axis produced a far-field zone of quiet over a distance of 1/2 a wavelength, in the direction of pressure gradient measurement. This is considerably larger than the zone of quiet obtained by minimising the pressure alone (in which the 10 dB zone of quiet was limited to 1/10 of a wavelength). This size increase can be explained by the fact that the average size of the zone of quiet is a function of the spatial correlation properties of the sound field, and the squared sum of the pressure and pressure gradient cross-correlation functions extend over a larger region compared to the squared pressure correlation function alone.

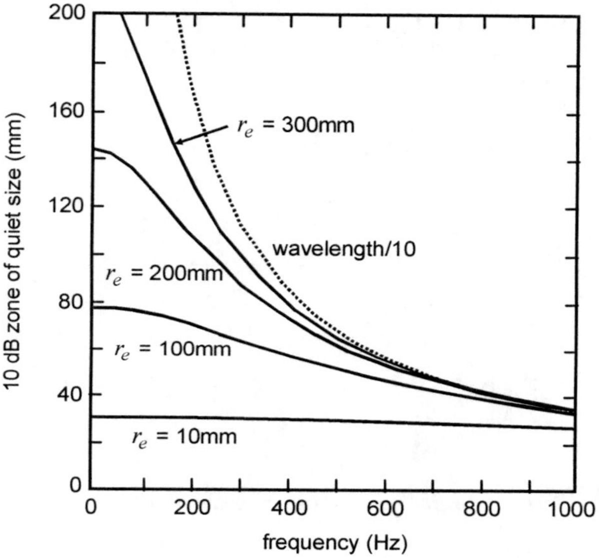

Figure 9.55 The axial extent of a 10 dB minimum zone of quiet generated by a 0.05 m radius piston control source as a function of frequency for various control source/error sensor separation distances (after Elliott and David, 1992).

9.12 STATE-SPACE MODELS OF ACOUSTIC SYSTEMS

While there has been a substantial amount of work conducted on the application of feedforward active control systems to attenuate enclosed sound fields, the application of feedback control systems to the problem has received less attention. In this section, this latter problem will be considered to a limited degree by simply formulating state-space models of the enclosed sound field, which can be used to formulate control laws. The formulation is very similar to that used to examine distributed parameter (structural) systems, which will be considered later in Chapter 11.

The starting point for the development of a state-space model is the inhomogeneous wave equation, derived in Chapter 2. This can be stated in terms of acoustic pressure as:

$$\left(\frac{1}{c_0^2}\frac{\partial^2}{\partial t^2} + \nabla^2\right) p(r,t) = \rho_0 \frac{\partial}{\partial t} s(\boldsymbol{r}_s, t) \tag{9.12.1}$$

where s is the acoustic volume velocity distribution in the enclosure. For an enclosed acoustic space, the acoustic pressure at any location $\boldsymbol{r}$ can be expressed as contributions from an infinite set of acoustic modes, φ:

$$p(\boldsymbol{r},t) = \sum_{n=1}^{\infty} p_n(t)\,\varphi_n(\boldsymbol{r}) \tag{9.12.2}$$

where p_n is the complex amplitude of the nth acoustic mode. The eigenvalue problem associated with the (self-adjoint) acoustic modes is:

$$\nabla^2\varphi_n(\boldsymbol{r}) = \lambda_n\varphi_n(\boldsymbol{r}) \tag{9.12.3}$$

where λ_n is the nth eigenvalue, and the acoustic mode shape functions will be ortho-normalised in such a way that:

$$\int_V \varphi_m(\boldsymbol{r})\varphi_n(\boldsymbol{r})\,\mathrm{d}\boldsymbol{r} = \delta(m-n) \tag{9.12.4}$$

where δ is a Kronecker delta function:

$$\delta(m-n) = \begin{cases} 1 & m=n \\ 0 & m\neq n \end{cases} \tag{9.12.5}$$

Note that the relationship between the eigenvalue λ_n and the natural frequency ω_n of the nth mode is:

$$\lambda_n = \omega_n^2 \tag{9.12.6}$$

Substituting the modal expansion of acoustic pressure into the wave equation produces:

$$\frac{1}{c_0^2}\sum_{n=1}^{\infty}\frac{\partial^2 p_n(t)}{\partial t^2}\varphi_n(\boldsymbol{r}) + \sum_{n=1}^{\infty}\lambda_n p_n(t)\varphi_n(\boldsymbol{r}) = \rho_0\frac{\partial}{\partial t}s(\boldsymbol{r}_s,t) \tag{9.12.7}$$

Multiplying Equation (9.12.7) through by the mth acoustic mode shape, φ_m, and integrating over the enclosed space produces the equation governing the response of the mth mode as:

$$\frac{1}{c_0^2}\frac{\partial^2 p_m(t)}{\partial t} + \lambda_m p_m(t) = \rho_0\alpha_m(t) \tag{9.12.8}$$

or

$$\frac{\partial^2 p_m(t)}{\partial t} + c_0^2\omega_m^2 p_m(t) = \rho_0 c_0^2\alpha_m(t) \tag{9.12.9}$$

where $\alpha_m(t)$ in the modal generalised volume velocity of the mth acoustic mode:

$$\alpha_m(t) = \int_V \varphi_m(\boldsymbol{r})s(\boldsymbol{r},t)\,\mathrm{d}\boldsymbol{r} \tag{9.12.10}$$

Using Equation (9.12.10), the response of the mth acoustic mode can be written in state-space form as:

$$\dot{\boldsymbol{x}}_m(t) = \boldsymbol{A}_m\boldsymbol{x}_m(t) + \boldsymbol{\alpha}_m(t) \tag{9.12.11}$$

where

$$x_m(t) = \begin{bmatrix} p_m(t) \\ \dot{p}_m(t) \end{bmatrix}; \quad A_m = \begin{bmatrix} 0 & 1 \\ -c_0^2\omega_m^2 & 0 \end{bmatrix}; \quad \alpha_m(t) = \begin{bmatrix} 0 \\ \rho_0 c_0^2 \alpha_m(t) \end{bmatrix} \tag{9.12.12a–c}$$

To formulate a control law for attenuating the enclosed sound field, it is necessary to explicitly state the matrix $\boldsymbol{\alpha}_m$ in terms of a control input and a transfer function. To do this, the volume velocity distribution is expressed as the product of some amplitude q and a transfer function to the location of interest, z_q, as follows:

$$s(\boldsymbol{r},t) = q(t) z_q(\boldsymbol{r}) \tag{9.12.13}$$

Substituting this product into Equation (9.12.10), the modal generalised volume velocity can be written as:

$$\alpha_m(t) = q(t) \int_V \varphi_m(\boldsymbol{r}) z_q(\boldsymbol{r}) \,\mathrm{d}\boldsymbol{r} = q(t)\varphi_{g,m} \tag{9.12.14}$$

where $\varphi_{g,m}$ is the mth modal generalised volume velocity transfer function given by:

$$\varphi_{g,m} = \int_V \varphi_m(\boldsymbol{r}) z_q(\boldsymbol{r}) \,\mathrm{d}\boldsymbol{r} \tag{9.12.15}$$

This term was encountered previously in Section 9.2, where it was shown in Equation (9.2.13) that for the simplest source arrangement, a monopole acoustic source, the modal generalised volume velocity transfer function was simply equal to the value of the acoustic mode shape function at the monopole location. Using the expansion of Equation (9.12.14), the state-space Equation (9.12.11) can be rewritten as:

$$\dot{\boldsymbol{x}}_m(t) = \boldsymbol{A}_m \boldsymbol{x}_m(t) + \boldsymbol{B}_m \boldsymbol{u}(t) \tag{9.12.16}$$

where

$$\boldsymbol{B} = \begin{bmatrix} 0 & 0 & \cdots & 0 \\ \varphi_{g,1} & \varphi_{g,2} & \cdots & \varphi_{g,N_c} \end{bmatrix}; \quad \boldsymbol{u}(t) = \begin{bmatrix} q_1(t) \\ q_2(t) \\ \vdots \\ q_{N_c}(t) \end{bmatrix} \tag{9.12.17a,b}$$

where N_c sources are being used in the system. If the pressure field is being modelled using N_n acoustic modes, then the state-space description of the system can be written as:

$$\dot{\boldsymbol{x}}(t) = \boldsymbol{A}\boldsymbol{x}(t) + \boldsymbol{B}\boldsymbol{u}(t) \tag{9.12.18}$$

where

$$\boldsymbol{x}(t) = \begin{bmatrix} \boldsymbol{x}_1(t) \\ \boldsymbol{x}_2(t) \\ \vdots \\ \boldsymbol{x}_{N_n}(t) \end{bmatrix}; \quad \boldsymbol{A} = \begin{bmatrix} \boldsymbol{A}_1 & & & \\ & \boldsymbol{A}_2 & & \\ & & \ddots & \\ & & & \boldsymbol{A}_{N_n} \end{bmatrix}; \quad \boldsymbol{B} = \begin{bmatrix} \boldsymbol{B}_1 \\ \boldsymbol{B}_2 \\ \vdots \\ \boldsymbol{B}_{N_n} \end{bmatrix} \tag{9.12.19a–c}$$

Note that $\boldsymbol{A}$ is not a diagonal matrix, but rather a diagonal arrangement of modal state vectors, defined in Equation (9.12.12).

Acoustic damping can be added to this model by rewriting the wave equation as:

$$\left(\frac{1}{c^2}\frac{\partial^2}{\partial t^2} + \zeta\nabla^2\frac{\partial}{\partial t} + \nabla^2\right) p(\boldsymbol{r},t) = \rho_0\frac{\partial}{\partial t}s(\boldsymbol{r}_s,t) \tag{9.12.20}$$

where ζ is the critical damping ratio. Going through the same steps as before, the damped modal equation of motion is found to be:

$$\frac{\partial^2 p_m(t)}{\partial t} + \zeta c_0^2\omega_m^2\frac{\partial p_m(t)}{\partial t} + c_0^2\omega_m^2 p_m(t) = \rho_0 c_0^2\alpha_m(t) \tag{9.12.21}$$

which can be written in the state-space form of Equation (9.12.11) with the modified state matrix as:

$$\boldsymbol{A}_m = \begin{bmatrix} 0 & 1 \\ -c_0^2\omega_m^2 & -\zeta c_0^2\omega_m^2 \end{bmatrix} \tag{9.12.22}$$

This damped state matrix can then be used in place of the undamped state matrices in $\boldsymbol{A}$, as defined in Equation (9.12.19).

If microphones are used to measure the sound field at discrete locations, the acoustic pressure at each location can be expressed as the sum of modal contributions:

$$p(\boldsymbol{r},t) = \sum_{i=1}^{N_n} p_i(t)\varphi_i(\boldsymbol{r}) \tag{9.12.23}$$

If N_e error sensors are being used, the vector of acoustic pressures at their locations, the output equation, can be expressed as:

$$\boldsymbol{y}(t) = \boldsymbol{C}\boldsymbol{x}(t) \tag{9.12.24}$$

where $\boldsymbol{y}(t)$ is the vector of pressures:

$$\boldsymbol{y}(t) = \begin{bmatrix} p(\boldsymbol{r}_1,t) \\ p(\boldsymbol{r}_2,t) \\ \vdots \\ p(\boldsymbol{r}_{N_e},t) \end{bmatrix} \tag{9.12.25}$$

and $\boldsymbol{C}$ is the output matrix given by:

$$\boldsymbol{C} = \begin{bmatrix} \varphi_1(\boldsymbol{r}_1) & 0 & \varphi_2(\boldsymbol{r}_1) & 0 & \cdots & \varphi_{N_n}(\boldsymbol{r}_1) & 0 \\ \varphi_1(\boldsymbol{r}_2) & 0 & \varphi_2(\boldsymbol{r}_2) & 0 & \cdots & \varphi_{N_n}(\boldsymbol{r}_2) & 0 \\ & & & \vdots & & & \\ \varphi_1(\boldsymbol{r}_{N_e}) & 0 & \varphi_2(\boldsymbol{r}_{N_e}) & 0 & \cdots & \varphi_{N_n}(\boldsymbol{r}_{N_e}) & 0 \end{bmatrix} \tag{9.12.26}$$

An example of the application of the state Equations (9.12.18) and (9.12.24) in formulating a control law for attenuating the sound field in a rigid walled rectangular enclosure was given by Dohner and Shoureshi (1989), where an LQG approach was taken.

9.12.1 Feedback Control of Sound Transmission into a Launch Vehicle

Recently, Griffin et al. (1999, 2001) investigated feedback control of sound transmission into a spacecraft launch vehicle fairing by controlling structural vibration. A launch vehicle fairing is a flexible, thin-walled structure that houses the payload. When subjected to an acoustic or structural disturbance, the fairing vibrates, producing an acoustic response in its interior. During launch, the payload housed within the fairing is exposed to damagingly high acoustic loads from a number of sources including aerodynamic buffeting, shocks during stage separations and structural vibrations induced by the rocket motors. The sound levels within the fairing can exceed 140 dB, with the levels at low frequencies (10 to 200 Hz) reaching 130 dB (Lane et al., 2001b).

Griffin et al. (1999, 2001) developed a MIMO radiation state-space model of a rocket fairing that relates the out-of-plane structural vibration modes to the time varying acoustic pressure in the fairing interior. The modal-interaction method was used in the analysis whereby the fairing structural response was described using its *in-vacuo* mode shapes and modal frequencies while the acoustic response of the fairing interior was modelled using the rigid-walled mode shapes and resonance frequencies.

Feedback control to minimise sound transmission into the fairing was investigated in numerical simulations using linear quadratic Gaussian (LQG) control laws and twelve collocated displacement sensors and proof-mass actuators. The sensors and actuators were optimally located on the fairing to best control the structural modes responsible for the dominant interior acoustic modes. The passive effect of the proof-mass actuators was found to reduce the spatially averaged sound pressure level in the fairing interior by 3.5 dB over the 0 to 300 Hz frequency range. Active control with performance limits on the control effort achieved only a marginal improvement on the open-loop response, with an overall reduction of 4.2 dB over the 0 to 300 Hz frequency range. This slight increase in noise attenuation is not likely to be worth the added mass and computational complexity introduced by the active control system.

Lane et al. (2001a) simulated active structural acoustic control of low-frequency noise transmitted into a launch vehicle fairing with monolithic piezoceramic actuators. Using the fully coupled state-space fairing model developed by Griffin et al. (1999, 2001) and linear quadratic regulator (LQR) control laws, the piezoceramic patch actuators were found to reduce the spatially averaged interior fairing noise levels by up to 2.5 dB over the 0 to 300 Hz frequency range. Later, Lane et al. (2003) presented simulated results of active structural acoustic control in a launch vehicle fairing using single-crystal piezoelectric actuators. An overall attenuation of more than 10 dB was achieved over the 0 to 300 Hz bandwidth with an LQR controller. Single-crystal actuators were found to be much more effective in the control of fairing interior noise than the monolithic actuators used in the authors' earlier study, and this was attributed to the higher strain coefficient of the single-crystal actuators.

Lane et al. (2001b) also conducted an experimental investigation on feedback control of low-frequency sound in a rocket fairing interior. Experiments were conducted on a full-scale composite fairing model with sixteen collocated moving coil actuators and microphones and an additional sixteen microphones for monitoring the control system performance. System identification was first conducted using an eigen system realisation algorithm (Juang, 1994) to determine a state-space model of the open-loop system. This state-space model was then used to design the control system employing H_2 feedback control laws. In experiments, low-frequency acoustic modes were reduced by 6 to 12 dB with minor spillover and a spatially averaged noise reduction of 3 dB was achieved throughout the fairing interior over the 20 to 200 Hz frequency range. It was hypothesised that higher levels of noise attenuation could be achieved by optimising the number and placement of actuators.

9.13 AIRCRAFT INTERIOR NOISE

9.13.1 Introduction

It is well known and generally accepted that interior noise levels in general and commercial aviation propeller driven aircraft are unacceptable. Unfortunately, the weight penalty resulting from the use of passive control measures to reduce the noise to acceptable levels is even more unacceptable. Passive control is usually associated with the attachment to the fuselage interior surface of multi-layer trim panels, an example of which might be made up as follows: a layer of lightweight damping material to the fuselage, a central weighted layer separated from the fuselage by a fibreglass blanket and a stiff decorative inner trim panel separated from the central weighted layer by another fibreglass blanket (Silence, 1991).

Another form of passive control, which is also characterised by a significant weight penalty, is the use of tuned vibration dampers acting on the fuselage to reduce vibration at the fundamental and first harmonic of the propeller blade passing frequency (Emborg and Halvorsen, 1988). Experimental tests on a SAAB340 aircraft demonstrated in-flight noise reductions of up to 6 dB(A) at seat locations close to the plane of the propellers. Clearly, this type of control must be used together with some form of trim panel control as well.

Existing interior noise problems in propeller driven aircraft together with the resurgence of interest in the development of more fuel efficient, high speed turbo propeller aircraft has generated considerable interest in the application of active control to reduce interior noise levels with a small weight penalty. Available data (Wilby et al., 1985; Patrick, 1986; Wilby and Wilby, 1987) indicate that 20 to 30 dB of noise reduction is required for the first three or four propeller blade passage harmonics and 10 to 20 dB for the next five or six to reduce them to levels resulting from turbulent boundary layer excitation of the fuselage.

As might be concluded from the preceding discussion, active control is particularly suited to controlling tonal noise such as generated by a propeller. Researchers are also working on reducing jet engine noise by placing acoustic control sources in the periphery of an extension to the shroud (or cowling) at the inlet and exit of the engine. However, control of noise generated by action of the turbulent boundary layer on the fuselage is considerably more difficult, although progress is likely to be made by applying adaptive feedback control to vibration actuators on the fuselage to minimise a cost function proportional to interior noise levels. Such a cost function could be derived from appropriately shaped PVDF film, which could be configured to detect only those structural modes that contribute significantly to interior noise levels.

In recent years, there has been a major thrust from a European consortium named the Advanced Study of Active Noise Control in Aircraft, ASANCA (Borchers et al., 1992; Van der Auweraer, 1993; Emborg and Ross, 1993; Borchers et al. 1997; Johansson et al., 1999a; Tsahalis et al., 2000) to implement active noise control systems in several medium size passenger aircraft. In these studies, testing and theoretical work has been undertaken for a number of different aircraft including the Dornier 228 and 328, Saab 340, ATR 42, and Fokker 100. They were selected to identify general noise control information and demonstrate that active noise control was feasible in aircraft that have large differences in acoustic character.

The ASANCA consortium implemented active noise control using a fixed array of acoustic control sources, located in the interior trim and ceiling of the aircraft (Borchers et al., 1992; Van der Auweraer, 1993; Emborg and Ross, 1993). A fuselage test cell of a Dornier 228 was subsequently used to experimentally verify the analytical results (Hackstein et al., 1992). Reductions of 15 and 16 dB were achieved for the blade passage tone and first harmonic respectively.

Additional flight tests (Emborg and Ross, 1993) were performed using a similar system in a Saab 340 aircraft. A total of 48 error sensors and 24 loudspeakers were installed. Reductions in this aircraft were 10 dB and 3 dB for the blade passage tone and first harmonic respectively. These results were not as good as predicted for the Dornier 228 aircraft, but this is probably as a result of differences in the fuselage structure and the presence of cabin furniture and passengers.

Following this work, the first commercial active noise control system for propeller induced noise was developed by Ultra Electronics (England) and Saab Aircraft (Sweden). The first commercial aircraft to be fitted with this system were the Saab 340 and the Saab 2000 in 1994 (Ross and Purver, 1997). The active noise control system in the Saab 340 uses 48 error microphones and 24 loudspeakers, while the active noise control system in the Saab 2000 uses 72 error microphones and 48 loudspeakers (Emborg, 1998). Ultra Electronics now has a large number of active noise and vibration control systems in operation and these are implemented in a wide variety of turboprop aircraft including the KingAir 90, 200, 300 and 350, Twin Commander, Q400 and USAF C130–H3 aircraft (Ross, 1999; Hinchliffe et al., 2002; Gorman et al., 2004). The more recent systems rely on actuators attached to ribs and stringers on the fuselage rather than loudspakers as control sources.

As part of ASANCA, Johansson et al. (1999b) investigated the reduction of propeller induced noise in a Saab 340 fuselage test cell using narrowband dual-reference feedforward active noise and vibration control. Using this approach, the controller was synchronised to both propellers using a separate reference signal from each engine. Experiments were conducted with twelve error microphones, five loudspeakers, three shakers and twenty microphones that monitor system performance. Under stationary flight conditions with fully synchronised propellers, the mean attenuation in the fundamental blade passing frequency was 18 dB. This attenuation level reduced by 3 to 6 dB when the propellers were unsynchronised so that beating occurred.

In an earlier series of flight tests, Elliott et al. (1990a) used 26 configurations involving up to 16 loudspeakers and 32 microphones to actively reduce in-flight interior noise levels in a BAe748 propeller driven aircraft. They reported significant reductions (of the order of 10 dB) in noise levels at the fundamental, second and third harmonics (88 Hz, 176 Hz and 264 Hz respectively) of the propeller blade passing frequency, which resulted in a 7 dB(A) reduction at forward seat locations near the plane of the propellers. However, reductions further from the propeller plane were much less.

Simpson et al. (1991) reported similar reductions using active vibration control sources acting on the fuselage structure of the aft section of a McDonnell Douglas DC-9. In this case, the aircraft section was on the ground and engine noise was simulated using shakers attached to engine pylons.

More recently, Palumbo et al. (2001) investigated active structural acoustic control of propeller induced noise in a Raython Aircraft Company 1900D. The control system employed 21 inertial force actuators mounted on the aircraft frame and 32 microphones. In flight tests, an attenuation of 10, 2.5 and 3 dB was achieved in the mean microphone response at the blade passing frequency and its second and third harmonics respectively.

Other experimental tests on actual aircraft have involved in-flight tests on the ATR-42 (Paonessa et al., 1993), and laboratory tests on a Handley Page 137 Jetstream III fuselage (Warner and Bernhard, 1987) and an Airbus A400M test cell (Breitbach and Sachau, 2009). Noise reductions were up to 12 dB for the ATR-42, between 12 and 22 dB for the Handley Page fuselage, and up to 20 dB for the Airbus A400M test cell.

At very low frequencies characterised by the first or second harmonic of a typical propeller blade passing frequency (below about 160 Hz), it is feasible to attempt noise reductions using local cancellation which does not result in global noise reductions but only reductions in specific areas, usually at the expense of increases in other areas (Elliott et al., 1990a). A single-channel feedback system may be used for example to produce a zone of quiet around a headrest in a seat. Elliott (1991) showed by computer simulation that it is possible to produce a spherical zone of quiet of diameter equal to one-tenth of a wavelength around a single microphone with a control source mounted in close proximity (for example, in a headrest). In this zone, noise levels can be expected to be 10 dB quieter on average than they were with just the primary field. However, this technique is only suitable for very low frequencies (at 200 Hz, $\lambda/10 \approx 170$ mm) and is less preferable than the option of using feedforward control to achieve global noise reductions. Later work has shown that this zone of quiet can be extended to half a wavelength in one direction if two control sources and error sensors are used.

It is of interest to note in passing that of the many patents for active noise control devices, a number are directed particularly at aircraft interior noise. One (Fuller, 1987) addresses the use of vibration actuators on the fuselage to reduce interior noise levels and another (Warnaka and Zalas, 1985) addresses the use of acoustic sources within the cabin.

9.13.2 Analytical Modelling

Although a significant proportion of the energy exciting the fuselage and thus generating interior noise in an aircraft is structure-borne from the engine mounts and the wings to the fuselage, most of the analytical modelling work has been directed at predicting the effects of excitation of the fuselage by an external pressure field. However, as will be shown, it is relatively straightforward to include the structure-borne energy transmission in the acoustic model, provided the forcing function at the attachment points of the wing to the fuselage can be estimated. This forcing function is made up of a combination of mechanical energy at the engine mounts and aerodynamic pressures imposed by the propeller wake on the wing, the latter being much more important for high speed turboprop aircraft (Wilby, 1990).

As pointed out by Wilby (1989), large errors in predictions are obtained if the analytical model is too greatly simplified; for example, if the fuselage is modelled as a simple cylinder as was done by Bullmore et al. (1990). This is because the presence of the floor destroys the axi-symmetric nature of the interior acoustic modes, thus allowing more than one structural mode to couple with each acoustic mode, resulting in a second mechanism available for acoustic control, as described in Section 9.7. If the fuselage is modelled as a simple cylinder, then this second control mechanism is not analytically possible, and large prediction errors can occur at some frequencies.

As was made clear earlier in this chapter, the first step in developing an analytical model that can be used for the design of an active noise control system, is to develop an accurate model capable of predicting local as well as average fuselage vibration levels and interior noise levels. The cylinder with floor model discussed earlier in this chapter is adequate for investigating mechanisms of active control and likely achievable noise reductions, but it is inadequate for designing an active control system for an actual aircraft or for accurately predicting expected noise reductions due to active control. The same may be said of the detailed analytical model described by Wilby (1989) and Pope (1990), called PAIN (propeller aircraft interior noise), which is a combination of closed form and numerical approaches. In PAIN, the structural model is a cylinder with a floor, and its resonance frequencies and mode shapes are characterised in closed form by smearing the effects of the stiffeners over the entire structural skin. This is done by calculating an equivalent average EI (where E is Young's modulus of elasticity and I is the second moment of area (or area moment of inertia)) for the skin and stiffeners and also an effective structural mass per unit area of the structural cross-section averaged over the skin and stiffeners. Finite difference methods are used to calculate the modal characteristics of the interior space (cylinder with a floor). To simplify the model with little loss of accuracy, the modal parameters of the structure are calculated for the fuselage in a vacuum, those for the interior space are calculated assuming a rigid structural boundary, and the results are coupled together using a similar approach to that outlined earlier in this chapter.

The propeller noise field in the PAIN model is obtained either analytically or from test data, and involves the determination of amplitude and phase data describing the acoustic pressure associated with the exterior acoustic field at all exterior locations on the fuselage. Interior acoustic treatments (trim panels) have been included as an integral part of the pain model.

Perhaps the most promising modelling approach for practical fuselage active noise control system design is to use separate finite element analyses to determine the modal characteristics of both the *in-vacuo* structure and the rigidly enclosed acoustic space (Unruh and Dobosz, 1988; Martin, 1993). Once found, these can be coupled, and the effect of active control predicted using the same procedure as used for the cylinder with a floor earlier in this chapter. Use of finite element analysis enables discrete modelling of important structural and acoustic characteristics that influence the overall fuselage response. In some previous studies (Borchers et al., 1992 – Dornier 228 aircraft; Van der Auweraer, 1993 – Dornier 228 aircraft; Emborg and Ross, 1993 – SAAB 340 aircraft; Paxton et al., 1993 – MD-80 jet aircraft; Green, 1992 – SAAB 2000 aircraft; Monaco et al., 2008 – ATR42/72 aircraft), the structure was modelled as a small length of fuselage adjacent to the propeller, where the sound transmission is at its highest. No consideration was given to the effect of wing attachments and the potential for structure-borne excitation via that path, and the excitation input into the structure was primarily represented as a propeller blade passage pressure distribution on the fuselage surface. Modelling a small length of fuselage is computationally expedient, but the results can be of limited use in evaluating the effectiveness of global active noise control systems.

The structural modal parameters required for the analysis approach outlined above can be found easily by using most commercially available FEA software packages. Some commercial software packages do not have acoustic elements, so solutions for acoustical modal properties may be obtained with structural FEA packages by using an appropriate displacement–pressure or stress–pressure analogy as described by Lamancusa (1988). In summary, a structural finite element analysis package can be used to solve an acoustic problem using the displacement–pressure analogy by taking the following steps.

1. Set Poisson's ratio $\nu = 0$.
2. Set density $\rho = 1.205$ kg/m^3.
3. Set elastic modulus $E = \rho c^2$ (1.42×10^5 Pa).
4. Set modulus of rigidity $G = E$.
5. Set structural displacements equal to zero in all but one direction. Displacement in the remaining non-zero direction is acoustic pressure. Any direction is acceptable (as pressure is a scalar), except in an axi-symmetric problem it must be along the axis of symmetry.
6. At a free surface, the pressure is set equal to zero.
7. Along rigid walls, no boundary conditions are needed.
8. Where the applied acoustic pressure is known, give the selected degree-of-freedom a forced displacement equal to the applied pressure.
9. At a surface where the normal component of displacement is known, apply an external force $F = -S\ddot{a}_n(t)$, where S is the area of the moving surface and $\ddot{a}_n(t)$ is the normal component of surface acceleration.

Herdic et al. (2005) have developed a full finite element model of a Cessna Citation II fuselage section including the fuselage skin, frames, stringers and window sections. The structural response of the fuselage and its interior sound field were also examined in experiments. Predictions with the numerical model showed a high degree of accuracy but this was only achieved by developing the numerical model with the assistance of the experimental results.

One of the important steps in the development of an analytical model is to accurately define the forcing function acting on the fuselage, which is a combination of the exterior acoustic pressure field and the structure-borne forces at the wing attachment points. The exterior acoustic field can be determined to a reasonable accuracy by on-ground measurements, or to a lesser accuracy by approximating the propeller as a pair of dipoles whose axes intersect at right angles (Mahan and Fuller 1986) and which are 90° out of phase with one another. If a counter-clockwise rotating source is to be simulated, then each of the four monopoles making up the two dipoles must lag its clockwise neighbour by 90°. While the monopoles themselves remain motionless, they will produce a combined dipole type directivity pattern that rotates in the counter-clockwise direction with an angular frequency equal to the angular frequency of oscillation of the dipoles. Note that the rotation of the directivity pattern through one cycle does not represent one rotation of the propeller; rather, it represents one cycle of the fundamental, or a harmonic, of the complex acoustic pressure field generated by the motion of an individual propeller blade past the fuselage. The centre of the dipoles is between the propeller hub and the fuselage, at about 60% of the distance from the propeller hub to the propeller tip, in the plane of the propeller. In one example, Mahan and Fuller (1986) found that good results were obtained for a spacing of 10% of the fuselage radius between individual monopoles making up each of the two dipoles.

In summary, an analytical model to predict interior noise levels would be constructed using the following steps.

1. Development of a finite element model to determine *in vacuo* structural modal parameters.

2. Development of a finite element model to determine rigid boundary acoustic mode shapes.
3. Generation of the acoustic field forcing function and the structural forcing function at the wing attachment points.
4. Determination of the structural response to the forcing functions using experimentally determined or estimated structural loss factors.
5. Determination of the interior acoustic field using the coupled structural/acoustic modal analysis outlined in Section 9.4.
6. Optimisation of control source types, locations and strengths, as well as error sensor types, locations and weighting factors using the procedure outlined in Section 9.7.

9.13.3 Control Sources and Error Sensors

There are a number of practical alternatives that can be used for control sources and error sensors. Acoustic control sources can be either loudspeakers or horn drivers placed at appropriate locations in the aircraft cabin. It is also possible to use ductwork to direct the control sound to locations where it is needed and where there may not necessarily be room for a loudspeaker. A special purpose lightweight speaker for aircraft has been developed (Warnaka et al., 1992) using a rare earth permanent magnet encapsulated by steel to reduce stray magnetic fields. Loudspeakers designed specifically for use in aircraft are also commercially available (for example Misco Loudspeakers). Generally, loudspeakers should be capable of generating approximately 105 to 110 dB at 1 m at frequencies to be controlled. In practice, the effectiveness of the loudspeakers is limited by the available cone excursion, rather than the required input power, which is usually between 1 and 4 W.

Vibration control sources, directed at controlling the fuselage vibration in a way that will minimise interior noise, can be thin piezoceramic crystals bonded to the fuselage, or piezoceramic stack actuators or magnetostrictive rod actuators, the latter two being mounted between the fuselage wall and the interior trim panel or between flanges on fuselage ribs. Each of these actuator types is discussed in detail in Chapter 15. One possible disadvantage associated with using vibration actuators on the fuselage is that some energy will spill over into structural modes, which are not excited by the primary source. Also, minimisation of the interior acoustic field does not necessarily result in lower structural vibration levels and in some cases these levels may increase, thus increasing the potential for acoustic fatigue failure. Nevertheless, vibration control sources are more effective than acoustic sources in reducing interior noise levels dominated by structural controlled modes, although acoustic sources are more effective than vibration sources in reducing sound levels dominated by acoustic controlled modes. Thus, a practical system would probably include both types of source with the fuselage vibration level as one of the error inputs, which would be assigned a weighting to allow a balance to be obtained between interior noise level reductions and fuselage vibration level increases. Another way of minimising fuselage vibration levels would be to use a 'control effort' error signal made up of the sum of the squares of all of the signals driving the control actuators, thus ensuring that the vibrational energy into the fuselage is minimised. This would also have the effect of ensuring that one control actuator was not driven significantly harder than the others; it would assist in the control effort being divided more equally between control actuators. If necessary, a similar error signal could be derived for acoustic sources.

Practical error sensors include interior mounted microphones and PVDF film (see Chapter 16) bonded to the fuselage. There is even a possibility that appropriate shaping of the PVDF sensor may allow a single vibration error signal to be obtained, which is proportional to the average radiated interior noise levels. Signals from both vibration and acoustic error sensors could be appropriately weighted to place more or less importance on interior noise levels in specific locations.

9.14 AUTOMOBILE INTERIOR NOISE

With the ever-increasing drive for greater fuel efficiency and lower manufacturing costs, there is considerable interest in reducing the weight of passenger cars. Thus, there is a need for the development of appropriate technology to control engine noise, exhaust noise and road noise, so that noise levels in the passenger compartment will not be increased as the vehicle weight is reduced. One approach with great potential for low-frequency noise reduction is active control. The use of active mufflers to control exhaust noise was discussed in Chapter 7, and the use of active vibration isolation to control low-frequency engine noise will be discussed in Chapter 12. Here, the discussion will be restricted to the active control of noise generated by the tyre-road interaction. This noise is random in nature, and physical constraints limit the extent of control that is achievable to typically 5 to 10 dB, depending on the particular vehicle being tested. The active systems discussed in this section have a very different purpose from the low bandwidth active suspension systems discussed in Chapter 12. Here, the aim is to minimise noise levels in the passenger compartment in the audio frequency range, whereas the purpose of an active suspension is to improve the ride and reduce vibration levels at sub-audio frequencies.

Considerable effort in a number of countries has been focussed on the application of active control of road noise in automobile passenger compartments. Guicking et al. (1991) used measured road noise data and simulation techniques to estimate the likely benefit of active control. They found that it was important to limit control effort to avoid non-linear distortion in the loudspeakers. Since 1988, researchers at the Institute of Sound and Vibration Research (ISVR) have been publishing the results of their joint efforts with Lotus Engineering on active control of road noise (Elliott et al., 1988a; Sutton et al., 1990; Elliott et al., 1990b; Sutton et al., 1991, 1994). Their early simulation work using measured data showed that a relatively short processing delay in the controller could be tolerated, but after that, further delays resulted in a significant degradation in performance. In later work, the research team demonstrated a reduction of 7 dB(A) in the frequency range 100 to 200Hz on a Citroen AX sedan travelling at 60 km h^{-1}, and published guidelines on optimal locations for reference sensors and control loudspeakers (Sutton et al., 1994).

The active noise control technologies developed by the ISVR and Lotus Engineering for road noise cancellation have been detailed in a review paper by Mackay and Kenchington (2004). In the same time period, Bernhard and his co-workers, working on a similar and parallel program to that of the ISVR team, have published a number of papers (summarised by Bernhard, 1995) outlining their work and have demonstrated results that are similar to those achieved by the ISVR team. Perhaps the most important contribution of Bernhard's team has been the development of design techniques for the optimal location of reference sensors.

An example of a commercially available, mass produced active road noise cancellation system is one implemented in the Honda Accord station wagon released in June 2000 (Sano et al., 2001). In this system, a fixed feedback controller reduces noise in the front seats using the front door speakers and a feedforward controller minimises noise in the rear seats using the rear door speakers. When the vehicle is driven at 50 km h^{-1} over a rough surface, narrowband road noise centred at 40 Hz is reduced by 10 dB in the front seats. An important aspect of this active noise control system is the use of the front and rear door audio loudspeakers which significantly reduces the system cost, allowing mass production.

Road noise is transmitted into the passenger compartment through the wheel spindle and then through various suspension components to the body panels, which vibrate and radiate sound. Although the spectrum is generally broadband, it usually also includes predictable components due to the repeating nature of the tread vibration pattern, which is even apparent on rough roads. As all of the energy is transmitted through the wheel spindle, this seems the obvious location for a reference sensor to be used with a feedforward active control system. However, as discussed by Bernhard (1995), often more than one reference sensor per wheel is needed to obtain good results. In agreement with this, Couche and Fuller (1999) found that in practice, significant road noise cancellation requires two reference sensors per wheel. They investigated feedforward control of simulated road noise in the cabin of a sport utility vehicle. To simulate road noise, one wheel was excited by a roller covered with aggregate. Using two reference sensors on the wheel spindle, two conventional loudspeakers and two error sensors in the vehicle interior, attenuation of greater than 9 dB was measured at the error sensors between 100 and 500 Hz.

In addition to the number and locations of reference sensors, other system parameters that must be considered in the active control system design are the number and optimal locations of control loudspeakers in the passenger compartment, the number and optimal locations of microphone error sensors, the length and type of adaptive control filters, and algorithm convergence speed to maintain stability.

Bernhard (1995) summarised the optimal characteristics of many of the parameters just mentioned and these results will be summarised here. First he indicated the importance of the coherence between the reference and error sensors, as the maximum theoretically achievable noise reduction assuming an ideal controller is:

$$NR = 10\log_{10}(1 - \gamma^2) \tag{9.14.1}$$

where γ is the coherence for a single input, single output system and the multiple coherence for a multiple input, single output system. The optimum number of reference sensors was found to be either two or three, depending on the vehicle; and the optimum locations were very vehicle specific and were chosen to maximise the coherence between the reference and error sensors (Kompella, 1992), while at the same time minimising the coherence between one reference sensor and another. Good coherence between two or more reference sensors results in very slow convergence of the adaptive filters and poor tracking ability of the active noise control system. The reference sensors must also be located such that it is possible for the adaptive system to realise the optimal filter with the number of filter taps available.

As shown earlier in this chapter, one control source is required for each significant mode in the passenger compartment that must be controlled, and the best locations are at the antinodes of the enclosure response. The passenger compartment response of most automobiles is dominated at low frequencies (up to 150 Hz for small cars and up to 200 Hz for large cars) by four modes (Bernhard, 1995) with antinodes at the corners of the compartment. Thus, it is a practical alternative to use the four loudspeakers already present for the vehicle sound system as control sources, even if this requires some small change to the location of the speakers. With four control sources, four error microphones are needed and the best location for these is likely to be in the headrests of the seats, one for each seat. Previous research has indicated that between 5 and 7 dB reduction in road noise may be achieved in automobile passenger compartments in the frequency range 50 to 150 Hz using a four-channel feedforward active control system.

Bernhard (1995) found that FIR filters perform as well as IIR filters with the same number of coefficients, and that the optimal filter length varied between 300 and 600 taps for each control channel.

Park et al. (2002, 2004) showed that two independent feedforward control systems can be used simultaneously to overcome the computational limitations on broadband control and attenuate road noise over a large frequency range in a passenger vehicle. In their study, the first control system was designed to attenuate low-frequency noise (between 30 and 300 Hz) while the second targeted noise in the mid frequency region (between 250 and 500 Hz). Using loudspeakers and conformal actuators, dual frequency control with the two control systems achieved 5.6 dB of attenuation at the error sensors distributed throughout the cabin interior over the 30 to 500 Hz frequency range. This demonstrates the viability of using multiple control systems simultaneously to extend the frequency range of road noise cancellation compared to a single control system alone.

REFERENCES

Al–Bassyiouni, M. and Balachandran, B. (2005). Sound transmission through a flexible panel into an enclosure: structural-acoustics model. *Journal of Sound and Vibration* **284**, 467–486.

Bai, M.R. and Chang, S. (1996). Active noise control of enclosed harmonic fields by using BEM-based optimization techniques. *Applied Acoustics*, **48**, 15–32.

Banerjee, P.K. and Butterfield, R. (1981). *Boundary Element Methods in Engineering Science*. McGraw-Hill, London.

Bernhard, R.J. (1995). Active control of road noise inside automobiles. In *Proceedings of Active 95*, Institute of Noise Control Engineering, pp. 21–32.

Bernhard, R.J., Gardner, B.K., Mollo, C.G. and Kipp, C.R. (1986). Prediction of sound fields in cavities using boundary element methods. *AIAA Paper*, AIAA-86-1864.

Bhattacharya, M.C. and Crocker, M.J. (1969). Forced vibration of a panel and radiation of sound into a room. *Acustica*, **22**, 275–294.

Borchers, I.U., Emborg, U., Sollo, A., Waterman, E.H., Palliard, J., Larsen, P.N., Venet, G., Göransson, P. and Martin, V. (1992). Advanced study for active noise control in aircraft. In *Proceedings of the 4th NASA/SAE/DLR Aircraft Interior Noise Workshop*. Friedrichshafen, Germany, 19–20 May.

Brancati, A. and Aliabadi, M.H. (2012). Boundary element simulations for local active noise control using an extended volume. *Engineering Analysis with Boundary Elements*, **36**, 190–202.

Brebbia, C.A. and Walker, S. (1980). *Boundary Element Techniques in Engineering*. Newnes–Butterworths, London.

Breitbach, H. and Sachau, D. (2009). Active acoustic solutions for a high speed turboprop transport aircraft. In *Proceedings of Active 2009*. Ottawa, Canada, 20–22 August.

Bullmore, A.J. (1988). *The Active Minimisation of Harmonic Enclosed Sound Fields with Particular Application to Propeller Induced Cabin Noise*. PhD thesis, University of Southampton.

Bullmore, A.J., Nelson, P.A., Curtis, A.R.D. and Elliott, S.J. (1987). The active minimisation of harmonic enclosed sound fields. Part II. A computer simulation. *Journal of Sound and Vibration*, **117**, 15–33.

Bullmore, A.J., Nelson, P.A. and Elliott, S.J. (1990). Theoretical studies of the active control of propeller-induced cabin noise. *Journal of Sound and Vibration*, **140**, 191–217.

Cazzolato, B.S. and Hansen, C.H. (1999). Structural radiation mode sensing for active control of sound radiation into enclosed spaces. *Journal of the Acoustical Society of America*, **106**, 3732–3735.

Cazzolato, B.S. and Hansen, C.H. (1998). Active control of sound transmission using structural error sensing. *Journal of the Acoustical Society of America* **104**, 2878–2889.

Cook, R.K., Waterhouse, R.V., Berendt, R.D., Edelman, S. and Thompson, M.C. (1955). Measurement of correlation coefficients in reverberant sound fields. *Journal of the Acoustical Society of America*, **27**, 1072–1077.

Couche, J. and Fuller, C.R. (1999). Active control of power train and road noise in the cabin of a sport utility vehicle with advanced speakers. In *Proceedings of Active 99.* Fort Lauderale, FL, 2–4 December, pp. 609–620.

Cunefare, K.A. and Koopmann, G.H. (1991). Global optimum active noise control: surface and far-field effects. *Journal of the Acoustical Society of America*, **90**, 365–373.

Curtis, A.R.D., Elliott, S.J. and Nelson, P.A. (1985). Active control of one-dimensional enclosed sound fields. In *Proceedings of Inter-Noise 85*, 579–582.

Curtis, A.R.D., Nelson, P.A., Elliott, S.J. and Bullmore, A.J. (1987). The active suppression of acoustic resonances. *Journal of the Acoustical Society of America*, **81**, 624–631.

Curtis, A.R.D., Nelson, P.A. and Elliott, S.J. (1990). Active reduction of a one-dimensional enclosed sound field: an experimental investigation of three control strategies. *Journal of the Acoustical Society of America*, **88**, 2265–2268.

Dohner, J.L. and Shoureshi, R. (1989). Modal control of acoustic plants. *Journal of Vibration, Acoustics, Stress, and Reliability in Design*, **111**, 326–330.

Dorling, C.M., Eatwell, G.P., Hutchins, S.M., Ross, C.F. and Sutcliffe, S.G. (1989). A demonstration of active noise reduction in an aircraft cabin. *Journal of Sound and Vibration*, **128**, 358–360.

Elliott, S.J. (1989). *The Influence of Modal Overlap in the Active Control of Sound and Vibration.* ISVR Memorandum, 695.

Elliott, S.J. (1991). Active control of enclosed sound fields. In *Notes for a Short Course on Active Control of Noise and Vibration*, Ed. C.H. Hansen. Department of Mechanical Engineering, University of Adelaide, GPO Box 498 Adelaide, South Australia, 9.1–9.38.

Elliott, S.J. and David, A. (1992). A virtual microphone arrangement for local active sound control. In *Proceedings of the First International Conference on Motion and Vibration Control (MOVIC)*, 1027–1031.

Elliott, S.J., Curtis, A.R.D., Bullmore, A.J. and Nelson, P.A. (1987). The active minimisation of harmonic enclosed sound fields. Part III. Experimental verification. *Journal of Sound and Vibration*, **117**, 35–58.

Elliott, S.J., Stothers, I.M., Nelson, P.A., McDonald, A.M., Quinn, D.C. and Saunders, T. (1988a). The active control of engine noise inside cars. In *Proceedings of Inter-Noise '88*, 987–990.

Elliott, S.J., Joseph, P., Bullmore, A.J. and Nelson, P.A. (1988b). Active cancellation at a point in a pure tone diffuse sound field. *Journal of Sound and Vibration*, **120**, 183–189.

Elliott, S.J., Nelson, P.A., Stothers, I.M. and Boucher, C.C. (1989). Preliminary results of in-flight experiments on the active control of propeller-induced cabin noise. *Journal of Sound and Vibration*, **128**, 355–357.

Elliott, S.J., Nelson, P.A., Stothers, I.M. and Boucher, C.C. (1990a). In-flight experiments on the active control of propeller-induced cabin noise. *Journal of Sound and Vibration*, **140**, 219–238.

Elliott, S.J., Nelson, P.A. and Sutton, P.J. (1990b). The active control of low frequency engine and road noise inside automotive interiors. *Active Noise and Vibration Control,* ASME publication NCA-8, 125–129.

Elliott, S.J. and Garcia-Bonito, J. (1995). Active cancellation of pressure and pressure gradient in a diffuse sound field. *Journal of Sound and Vibration*, **186**, 696–704.

Emborg, U. and Halvorsen, W.C. (1988). Development of tuned dampers for control of bladepass tones in cabin of SAAB 340. In *Proceedings of the 3rd SAE/NASA Aircraft Interior Noise Workshop.* Hampton, Virginia, April, 11–12.

Emborg, U. (1998). Cabin noise control in the Saab 2000 high-speed turboprop aircraft. In *Proceedings of ISMA 23*, pp. 13–25.

Emborg, U., and Ross, C.F., (1993) Active control in the Saab 340. In *Proceedings of the Recent Advances in Active Control of Sound and Vibration*. Blacksburg, VA, April, 100–109.

Faber, B.M and Sommerfeldt, S.D. (2004). Global control in a mock tractor cabin using energy density. In *Proceedings of Active 04*. Williamsburg, Virginia, 20–22 September.

Fahy, F. (1985). *Sound and Structural Vibration: Radiation, Transmission, and Response*. Academic Press, London.

Fuller, C.R. (1987). Apparatus and method for global noise reduction. US Patent 4 715 559.

Fuller, C.R. (1986). Analytical model for investigation of interior noise characteristics in aircraft with multiple propellers including synchrophasing. *Journal of Sound and Vibration*, **109**, 141–156.

Fuller, C.R. and Jones, J.D. (1987). Experiments on the reduction of propeller induced interior noise by active control of cylinder vibration. *Journal of Sound and Vibration*, **112**, 389–395.

Gardonio, P., Bianchi, E. and Elliott, S.J. (2004). Smart panel with multiple decentralized units for the control of sound transmission. Part 1. Theoretical predictions. *Journal of Sound and Vibration* **274**, 163–192.

Gorman, J., Hinchliffe, R. and Sothers, I. (2004). Active control on the flight deck of a C130 Hercules. In *Proceedings of Active 04*. Williamsburg Virginia, 20–22 September.

Green, I.S. (1992). Vibro-acoustic FE analyses of the Saab 2000 aircraft. In *Proceedings of the 4th NASA/SAE/DLR Aircraft Interior Noise Workshop*. Friedrichshafen, Germany, 44–69.

Griffin, S., Lane, S.A., Hansen, C. and Cazzolato, B. (2001). Active structural-acoustic control of a rocket fairing using proof-mass actuators. *Journal of Spacecraft and Rockets*, **38**, 219–225.

Griffin, S., Hansen, C. and Cazzolato, B. (1999). Feasibility of feedback control of transmitted sound into a launch vehicle fairing using structural sensing and proof mass actuators. AIAA paper 99-1529. *AIAA/ASME/ASCE/AHS/ASC Structures, Sturctural Dynamics and Materials Conference and Exhibit*. St. Louis, MO, 12–15 April.

Guicking, D. Bronzel, M. and Bohm, W. (1991). Active adaptive noise control in cars. In *Proceedings of Recent Advances in Active Control of Sound and Vibration*. Technomic Publishing Co., Inc., 657–670.

Guy, R.W. (1979). The response of a cavity backed panel to external airborne excitation: a general analysis. *Journal of the Acoustical Society of America*, **65**, 719–731.

Hackstein H.J., Borchers, I.U., Renger, K. and Vogt, K., (1992). The Dornier 328 acoustic test cell (ATC) for interior noise tests and selected test results. In *Proceedings of the Recent Advances in Active Control of Sound and Vibration*. Blacksburg, VA, April, pp. 35-43.

Herdic, P.C., Houston, B.H., Marcus, M.H., Williams, E.G. and Baz, A.M. (2005). The vibro-acoustic response and analysis of a full-scale aircraft fuselage section for interior noise reduction. *Journal of the Acoustical Society of America* **117**, 3667–3678.

Hinchliffe, R., Scott, I., Purver, M. and Stothers, I. (2002). Tonal active control in production on a large turbo-prop aircraft. In *Proceedings of Active 02*. ISVR Southampton, UK, 15–17 July.

Hirsch, S.M. and Sun, J.Q. (1998). Numerical studies of acoustic boundary control for interior sound suppression. *Journal of the Acoustical Society of America*, **104**, 2227–2235.

Johansson, S., Claesson, I., Nodebo, S. and Sjosten, P. (1999a). Evaluation of multiple reference active noise control algorithms on Dornier 328 aircraft data. *IEEE Transactions on Speech and Audio Processing*, **7**, 473–477.

Johansson, S., Persson, P. and Claesson, I. (1999b). Active control of propeller-induced noise in an aircraft mock-up. In *Proceedings of Active 99*. Ft Lauderdale, FL, 2–4 December.

Jones, J.D. and Fuller, C.R. (1987). Active control of sound fields in elastic cylinders by multi-control forces. *AIAA Paper*, AIAA-87-2707.

Juang, J. (1994). *Applied System Identification*, 1st ed., Prentice Hall, Englewood Cliffs, NJ.

Junger, M.C. and Feit, D. (1986). *Sound, Structures, and Their Interaction*. MIT Press, Cambridge, MA.

Kim, S.M. and Brennan, M.J. (2000) Active control of harmonic sound transmission into an acoustic enclosure using both structural and acoustic sources. *Journal of the Acoustical Society of America*, **107**, 2523–2534.

Kompella, M.S. (1992). *Improved Multiple Input/Multiple Output Modeling Procedures with Consideration of Statistical Information*. PhD thesis, Purdue University.

Lamancusa, J.S. (1988). Acoustic finite element modelling using commercial structural analysis programs. *Noise Control Engineering Journal*, **30**, 65–71.

Lane, S.A., Griffin, S. and Leo, D. (2001a). Active structural acoustic control of a launch vehicle fairing using monolithic piezoceramic actuators. *Journal of Intelligent Material Systems and Structures*, **12**, 795–806.

Lane, S.A., Griffin, S. and Leo, D. (2003). Active-structural-acoustic control of composite fairings using single-crystal piezoelectic actuators. *Smart Materials and Structures*, **12**, 96–104.

Lane, S.A., Kemp, J.D., Griffin, S. and Clark, R.L. (2001b). Active acoustic control of a rocket fairing using spatially weighted transducer arrays. *Journal of Spacecraft and Rockets*, **38**, 112–119.

Leissa, A.W. (1973). *Vibration of Shells*. NASA SP-288.

Li, D.S., Cheng, L. and Gosselin, G.M. (2004). Optimal design of PZT actuators in active structural acoustic control of a cylindrical shell with floor partition. *Journal of Sound and Vibration*, **269**, 569–588.

Li, D.S., Cheng, L. and Gosselin, G.M. (2002). Analysis of structural acoustic coupling of a cylindrical shell with an internal floor partition. *Journal of Sound and Vibration*, **250**, 903–921.

Lyon, R.H. (1963). Noise reduction of rectangular enclosures with one flexible wall. *Journal of the Acoustical Society of America*, **35**, 1791–1797.

Mackay, A. and Kenchington, S. (2004). Active control of noise and vibration – a review of automotive applications. In *Proceedings of Active 04*. Williamsburg, VA, 20–22 September.

Magliozzi, B. (1984). Advanced turboprop noise: a historical review. *AIAA paper*, AIAA 84-226.

Mahan, J.R. and Fuller, C.R. (1986). A propeller model for studying trace velocity effects on interior noise. *Journal of Aircraft*, **23**, 142–147.

Martin, V. (1993). Small-scale vibro-acoustic modelling for active noise control in aircraft. In *Proceedings of Noise–93*, St. Petersburg, Russia, Vol. 2, 195–200.

Missaoui, J. and Cheng, L. (1999). Vibroacoustic analysis of a finite cylindrical shell with internal floor partition. *Journal of Sound and Vibration*, **226**, 101–123.

Mollo, C.G. and Bernhard, R.J. (1987). *A Generalised Method for Optimization of Active Noise Controllers in Three-Dimensional Spaces*. AIAA paper, AIAA-87-2705.

Mollo, C.G. and Bernhard, R.J. (1989). Generalized method of predicting optimal performance of active noise controllers. *AIAA Journal*, **27**, 1473–1478.

Mollo, C.G. and Bernhard, R.J. (1990). Numerical evaluation of the performance of active noise control systems. *Journal of Vibration and Acoustics*, **112**, 230–236.

Monaco, E., Lecce, L., Natale, C., Pirozzi, S. and May, C. (2008). Active noise control in turbofan aircraft: theory and experiments. In *Proceedings of Active 08*. Paris, France, 29 June–4 July.

Morrow, C.T. (1971). Point to point correlation of sound pressures in reverberant chambers. *Journal of Sound and Vibration*, **16**, 29–42.

Morse, P.M. (1948). *Vibration and Sound*. McGraw-Hill, New York (reprinted by the Acoustical Society of America, 1981).

Nadim, M. and Smith, R.A. (1983). Synchronous adaptive cancellation in vehicle cabins. In *Proceedings of Inter-Noise 83*, 461–464.

Nelson, P.A., Curtis, A.R.D. and Elliott, S.J. (1985). Quadratic optimisation problems in the active control of sound. *Proceedings of the Institute of Acoustics*, **7**, 45–53.

Nelson, P.A., Curtis, A.R.D., Elliott, S.J. and Bullmore, A.J. (1987). The active minimisation of harmonic enclosed sound fields. Part I. Theory. *Journal of Sound and Vibration*, **117**, 1–13.

Nelson, P.A. and Elliott, S.J. (1992). *Active Control of Sound*. Academic Press, London.

Olson, H.F. (1956). Electronic control of noise, vibration, and reverberation. *Journal of the Acoustical Society of America*, **28**, 966–972.

Olson, H.F. and May, E.G. (1953). Electronic sound absorber. *Journal of the Acoustical Society of America*, **25**, 1130–1136.

Oswald, L.J. (1984). Reduction of diesel noise inside passenger compartments using active, adaptive noise control. In *Proceedings of Inter-Noise 84*, 483–488.

Palumbo, D., Cabell, R., Sullivan, B. and Cline, J. (2001). Flight test of active structural acoustic noise control system. *Journal of Aircraft*, **38**, 277–284.

Pan, J. (1992). The forced response of an acoustic-structural coupled system. *Journal of the Acoustical Society of America*, **91**, 949–956.

Pan, J. and Bies, D.A. (1990a). The effect of fluid-structural coupling on sound waves in an enclosure-theoretical part. *Journal of the Acoustical Society of America*, **87**, 691–707.

Pan, J. and Bies, D.A. (1990b). The effect of fluid-structural coupling on sound waves in an enclosure – experimental part, *Journal of the Acoustical Society of America*, **87**, 708–717.

Pan, J., Hansen, C.H. and Bies, D.A. (1990). Active control of noise transmission through a panel into a cavity. Part I. Analytical study. *Journal of the Acoustical Society of America*, **87**, 2098–2108.

Pan, J. and Hansen, C.H. (1991a). Active control of noise transmission through a panel into a cavity. Part II. Experimental study. *Journal of the Acoustical Society of America*, **90**, 1488–1492.

Pan, J. and Hansen, C.H. (1991b). Active control of noise transmission through a panel into a cavity. Part III. Effect of actuator location, *Journal of the Acoustical Society of America*, **90**, 1493–1501.

Paonessa, A., Sollo, A., Paxton, M. Purver, M. and Ross, C.F. (1993). Experimental active control of sound in the ATR 42. In *Proceedings of Noise-Con '93*, 225–230.

Park, C.G., Fuller, C.R. and Long, J.T. (2004). On-road demonstration of noise control in a passenger automobile. Part 2. In *Proceedings of Active 04*. Williamsburg, Virginia, 20–22 September.

Park, C.G., Fuller, C.R. and Kidner, M. (2002). Evaluation and demonstration of advanced active noise control in a passenger automobile. In *Proceedings of Active 02*. ISVR Southampton, UK, 15–17 July, pp. 275–284.

Parkin, P.H. and Morgan, K. (1965). Assisted resonance. *Journal of Sound and Vibration*, **2**, 464.

Parkin, P.H. and Morgan, K. (1971). Assisted resonance in the Royal Festival Hall, London: 1965–1969. *Journal of Sound and Vibration*, **15**, 127–141.

Patrick, H.V.L. (1986). Cabin noise characteristics of a small propeller powered aircraft. AIAA paper. In *Proceedings of the 10th Aeroacoustics Conference*, Seattle, 9–11 July, p. 1096.

Paxton, M., Purver, M., Ross, C.F., Baptist, M., Lang, M.A., May, D.N. and Simpson, M.A. (1993). Active control of sound in a MD-80. In *Proceedings of the Recent Advances in Active Control of Sound and Vibration*. Blacksburg, VA, April, 67–73.

Peterson, M.R. and Boyd, D.E. (1978). Free vibrations of circular cylinders with longitudinal, interior partitions. *Journal of Sound and Vibration*, **60**, 45–62.

Piraux, J. and Nayroles, B. (1980). A theoretical model for active noise attenuation in three-dimensional space. In *Proceedings of Inter-Noise 80*, 703–706.

Pope, L.D. (1971). On the transmission of sound through finite closed shells: statistical energy analysis, modal coupling, and non-resonant transmission. *Journal of the Acoustical Society of America*, **50**, 1004–1018.

Pope, L.D. (1990). On the prediction of propeller tone sound levels and pressure gradients in an aeroplane cabin. *Journal of the Acoustical Society of America*, 88, 2755–2765.

Pope, L.D. and Wilby, E.G. (1982). *Analytical Prediction of the Interior Noise for Cylindrical Models of Aircraft Fuselages for Prescribed Exterior Noise Fields, Phase II*. NASA CR-165869.

Pope, L.D., Rennison, D.C. and Wilby, E.G. (1980). *Analytical Prediction of the Interior Noise for Cylindrical Models of Aircraft Fuselages for Prescribed Exterior Noise Fields*. NASA CR-159363.

Pope, L.D., Rennnison, D.C., Willis, C.M. and Mayes, M.H. (1982). Development and validation of preliminary analytical models for aircraft noise prediction. *Journal of Sound and Vibration*, **82**, 541–575.

Pope, L.D., Wilby, E.G., Willis, C.M. and Mayes, W.H. (1983). Aircraft interior noise models: sidewall trim, stiffened structures, and cabin acoustics with floor partition. *Journal of Sound and Vibration*, **89**, 371–417.

Pretlove, A.J. (1966). Forced vibrations of a rectangular panel backed by a closed rectangular cavity. *Journal of Sound and Vibration*, **3**, 252–261.

Ross, C.F. (1999). Active Noise Control in Aircraft. In *Proceedings of the Sixth International Congress on Sound and Vibration*. Technical University of Denmark, Denmark, 5–8 July, pp. 1611–1618.

Ross, C.F. and Purver, M.R.J. (1997). Active cabin noise control. In *Proceedings of Active 97*. Budapest, Hungary, 21–23 August.

Salikuddin, M. and Ahuja, K.K. (1989). Application of localised active control to reduce propeller noise transmitted through fuselage surface. *Journal of Sound and Vibration*, **133**, 467–481.

Sano, H., Inoue, t., Takahashi, A., Terai, K. and Nakamura, Y. (2001). Active control system for low-frequency road noise combined with an audio system. *IEEE Transactions on Speech and Audio Processing*, **9**, 755–763.

Silcox, R.J., Fuller, C.R. and Lester, H.C. (1987). *Mechanisms of Active Control in Cylindrical Fuselage Structures*. AIAA Paper, AIAA-87-2703.

Silence, J.L. (1991). Meeting the noise control needs of today's aircraft. *Noise and Vibration Worldwide*, July, 14–16.

Simpson, M.A., Luong, T.M., Swinbanks, M.A., Russell, M.A. and Leventhall, H.G. (1989). Full scale demonstration tests of cabin noise reduction using active noise control. In *Proceedings of Inter-Noise 89*, 459–462.

Simpson, M.A., Luong, T.M., Fuller, C.R. and Jones, J.D. (1991). Full-scale demonstration tests of cabin noise reduction using active vibration control. *Journal of Aircraft*, **28**, 208–215.

Simpson, M.T. and Hansen, C.H. (2006) Simultaneous noise and vibration control using active structural acoustic control inside an enclosed stiffened cylinder with floor structure. In *Proceedings of Active 06.* Adelaide Australia, 18–20 September.

Snyder, S.D. and Hansen, C.H. (1991). Using multiple regression to optimize active noise control system design. *Journal of Sound and Vibration*, **148**, 537–542.

Snyder, S.D. and Hansen, C.H. (1994a). The design of systems to actively control periodic sound transmission into enclosed spaces. Part 1. Analytical models. *Journal of Sound and Vibration*, **170**, 433–449.

Snyder, S.D. and Hansen, C.H. (1994b). The design of systems to actively control periodic sound transmission into enclosed spaces. Part 2. Mechanisms and trends, *Journal of Sound and Vibration*, **170**, 451–472.

Snyder, S.D. and Tanaka, N. (1992). On feedforward active control of sound and vibration using vibration error signals. *Journal of the Acoustical Society of America*, **94**, 2181–2193.

Sommerfeldt, S. D. and Nashif, P. J. (1994). An adaptive filtered-*x* algorithm for energy based active control. *Journal of the Acoustical Society of America*, **96**, 300–305.

Sommerfeldt, S. D. and Parkins, J. W. (1994). An evaluation of active noise attenuation in rectangular enclosures. In *Proceedings of Inter-Noise 94*, Yokohama, 1351–1356.

Srinivasan, P., Morrey, D., Durodola, J.F., Morgans, R.C. and Howard, C.Q. (2007). A comparison of structural-acoustic coupled reduced order models (ROMS): Modal coupling and implicit moment matching via Arnoldi. In *Proceedings of the 14th International Congress on Sound and Vibration.* Cairns Australia, 9–12 July.

Stanef, D.A., Hansen, C.H. and Morgans, R.C. (2004). Active control analysis of mining vehicle cabin noise using finite element modelling. *Journal of Sound and Vibration*, **227**, 277–297.

Sutton, T.J., Elliott, S.J. and Nelson, P.A. (1990). The active control of road noise inside automobiles. In *Proceedings of Inter-Noise '90*, 689–695.

Sutton, T.J., Elliott, S.J. and Moore, I. (1991). Use of non-linear controllers in the active attenuation of road noise inside cars. In *Proceedings of Recent Advances in Active Control of Sound and Vibration.* Technomic Publishing Co. Inc., 682–690.

Sutton, T.J., Elliott, S.J. McDonald, A.M. and Saunders, T.J. (1994). Active control of road noise inside vehicles. *Noise Control Engineering Journal*, **42**, 137–147.

Tabachnick, B.G. and Fidell, L.S. (1989). *Using Multivariante Statistics*. Harper and Row, New York.

Tohyama, M. and Suzuki, A. (1987). Active power minimisation of a sound source in a closed space. *Journal of Sound and Vibration*, **119**, 562–564.

Tsahalis, D.T., Katsikas, S.K. and Manolas, D.A. (2000). A genetic algorithm for optimal positioning of actuators in active noise control: results from the ASANCA project. *Aircraft Engineering and Aerospace Technology*, **72**, 252–258.

Unruh, J.F. and Dobosz, S.A. (1988). Fuselage structural-acoustic modelling for structure-borne interior noise transmission. *Journal of Vibration, Acoustics, Stress and Reliability in Design*, **110**, 226–233.

Van der Auweraer, H., Otte, D., Venet, G. and Catalifaud, J. (1993). Aircraft interior sound field analysis in view of active control: results from the ASANCA project. In *Proceedings of Noise-Con '93*, 219–224.

Wareing, R.R., Wilson, C.L., Hansen, C.H. and Pearce, J.R. (2011). Active control of low frequency noise in a simulated mining vehicle cabin. Part 2. In *Proceedings of Noise-Con 2011*. Portland, OR, 25–27 July.

Warnaka, G.E. and Zalas, J.M. (1985). Active attenuation of noise in a closed structure. US Patent 4 562 589.

Warnaka, G.E., Kleinle, M., Tsangaris, P., Oslac, M.J. and Moskow, H.J. (1992). A lightweight loudspeaker for aircraft communications and active noise control. In *Proceedings of the 4th MASA/SAE/DLR Aircraft Interior Noise Workshop*. Friedrichshafen, Germany, 19–20 May.

Warner, J.V. and Bernhard, R.J. (1987). Digital control of sound fields in three-dimensional enclosures. AIAA Paper 87-2706. In *Proceedings of the 11th Aeroacoustics Conference*. Palo Alto, CA, 19–21 October.

Wilby, J.F. (1989). Noise transmission into propeller-driven aeroplanes. *The Shock and Vibration Digest*, **21**, 3–10.

Wilby, J.F. and Wilby, E.G. (1987). Measurements of propeller noise in a light turboprop aeroplane. AIAA paper 87-2737, In *Proceedings of the 11th Aeroacoustics conference*. Palo Alto, CA, 19–21 October.

Wilby, J.F., McDaniel, C.D. and Wilby, E.G. (1985). In-Flight Acoustic Measurements on a Light Twin-engined Turboprop Aeroplane. NASA-CR 178004.

Zalas, J.M. and Tichy, J. (1984). *Active Attenuation of Propeller Blade Passage Noise.* NASA CR-172386.

CHAPTER 10

Feedforward Control of Vibration in Beams and Plates

The vibration control of structures to minimise their noise radiation was discussed in Chapters 8 and 9, and the control of global structural vibration levels will be discussed in Chapter 11. In this chapter, the control of vibration transmission along structural elements is discussed. Such control may be for the purpose of reducing the sound radiated from another structure attached to the structural element, or it may be for the purpose of reducing vibration levels in an attached structure. Examples of the former case include the control of vibration transmission along a strut connecting the gearbox to the cabin in a helicopter (Sutton, 1997) or the control of vibratory energy transmitted along a submarine hull excited by the propeller pressure field. An example of the latter case is the control of vibratory power transmitted along a support arm to a space telescope.

Structural vibration may be described in terms of waves of different types travelling in different directions, or in terms of vibration modes. If the transmission of vibratory energy is to be controlled, then a wave description is more appropriate; whereas if overall structural vibration levels are to be controlled, a modal description gives better results. The control of modal vibration using feedforward control has not been discussed extensively in the literature, although Clark (1994) described a methodology for approaching such problems. However, as this chapter is concerned with the control of transmission of vibratory energy rather than global vibration control, a wave description is used in the analysis. All possible wave types (flexural, longitudinal and torsional (or shear)) are considered, because although longitudinal and torsional waves do not generally in themselves contribute significantly to sound radiation or feelable structural vibration, they are easily scattered into flexural waves at structural junctions and discontinuities, and these waves are generally associated with significant sound radiation and often feelable vibration.

As the topic being considered here is vibration transmission, it is assumed that it is possible to obtain a measure of the incoming disturbance. For periodic signals, it is not important when the measurement is obtained relative to the controller output; it is only important that it reflect the frequency content of the signal to be controlled. However, for random disturbances, the measurement must be made sufficiently far ahead of the control actuator for the wave propagation time from the measurement point to the control actuator to be greater than the delay through the control system. Otherwise the controller will be non-causal. Achieving causality can be very difficult for longitudinal waves which, in common structural materials, propagate at a speed that is commonly an order of magnitude faster than flexural waves. For flexural waves, the propagation speed is much less, but even for these waves, the non-propagating, evanescent field associated with vibration sources can result in error signals not being proportional to the propagating power transmission. This problem will be discussed in more detail later on in this chapter.

If it is not possible to obtain a measure of the incoming disturbance in sufficient time, then it is necessary to use feedback control and this is discussed in Chapter 11. Feedback control effectively adds damping, stiffness or mass to a structure (at the force input location), thus effectively changing system resonance frequencies. It can be very effective in reducing overall vibration levels in a reverberant structure, and also effective in controlling random propagating disturbances. However, if the disturbance is periodic, the performance achieved even for a reverberant structure is not as good as that obtained using feedforward control, because in the feedback system the error signal (which is diminished by the action of the controller) is used directly to obtain the control signal; thus, large gains in the feedback loop are needed to obtain a substantial reduction in the error signal and this comes at some cost to system stability margins. On the other hand, in the feedforward system, the error signal is used as the quantity to be minimised in a control algorithm and is not directly related to the control signal. When feedback control is used with a single force actuator to control the propagation of travelling waves in a beam, large feedback gains effectively result a pinned or simply supported condition at the force location. This results in control of the component of the bending wave associated with transverse displacement, but provides no control for the component associated with rotation. As the two components are associated with equal energies, the maximum reduction in total propagating power that can be obtained is 3 dB. However, when displacement feedback (as opposed to acceleration, velocity or force feedback) is used, there is one frequency dependent value of the feedback gain that will result in total suppression of the travelling wave. Interestingly, this value is the same as the optimal feedforward gain with the error sensor co-located with the control source (or force). This is discussed in more detail in Chapter 11.

Both feedforward and feedback systems are governed by laws of controllability and observability, and this is especially important in reverberant structures, where the vibration may be characterised by modes. A particular mode or wave type is controllable only if the control actuator is located so that it can drive that mode or configured so that it can drive that particular wave type. The same comments can be made about the law of observability and its applicability to the sensor used by the controller to derive the control signal.

It is interesting to note the differences in control mechanisms associated with the different types of control and input disturbance. For feedback control, the mechanism, as mentioned previously, effectively involves artificially increasing the structural stiffness, mass and/or damping, thus resulting in a smaller response to or a reflection of an incoming disturbance. On the other hand, with feedforward control, the mechanism can be any one or all of three: reflection, absorption or suppression. For periodic excitation, the likely mechanisms are absorption of the propagating wave power by the control actuator or suppression of the primary source power generation by effectively changing the impedance it sees. For random noise, the power is either absorbed by the control actuator or reflected back from whence it came. In this case, causality effects prevent the control source from affecting the impedance of the primary source and thus suppressing the power generation unless, of course, the control source is located in the near-field of the primary source.

The active control of beam vibration is the simplest form of active control of structural vibration, being a similar type of problem to active control of plane waves in ducts, with the added complication that the flexural wavespeed is frequency dependent, thus making it more difficult to implement control for random noise disturbances. Active control of beam vibration was first discussed by Scheuren (1985) who considered the feedforward control of flexural waves in an infinite beam using a single control force. To control both the propagating flexural wave component and the evanescent near-field, two control forces have been used (Mace, 1987; Redman-White et al., 1987; Scheuren 1988; McKinnell, 1988, 1989).

In addition to the control of flexural waves in infinite beams, effort has also been directed at the control of longitudinal and torsional waves (Pan and Hansen, 1991), as these wave types can be converted to the more efficiently radiating and more easily feelable flexural waves at structural junctions and discontinuities. The effect of error sensor location and type of cost function (power or vibration level) on the maximum achievable control of flexural waves has also been examined (Pan and Hansen, 1993a), as has the control of flexural wave propagation in finite beams with various end conditions (Pan and Hansen, 1993b).

Although the preceding analyses were mostly undertaken using harmonic primary excitation, McKinnell (1989) showed that the results are equally applicable to a random noise disturbance provided that the control source is in the far-field of the primary source and provided that the time required to process the reference signal to obtain the desired control source signal is less than the time taken for the wave to propagate from the reference signal transducer to the control source. This effectively imposes an upper limit on the frequency of the waves which can be controlled, at least for dispersive type waves such as flexural waves. If an accelerometer or other form of vibration sensor is used as the error sensor, a low-frequency limit will also exist, which is governed by the frequency at which the wavelength of the travelling waves is sufficiently long that the error sensor encroaches on the near-field of the control source. When this happens, the maximum achievable control is substantially reduced (Pan and Hansen, 1993a). In practice, for random noise excitation, a higher low-frequency limit results from lack of coherence between the reference (or detection) and error signals (Elliott et al., 1990).

If the overall structural vibration level rather than vibratory power transmission is to be suppressed, then a modal description of the structure is the more appropriate one, whether feedback or feedforward control is used. Feedback control for this case is discussed in Chapter 11, and feedforward control follows a similar approach to that used for minimising sound radiation as discussed in Chapter 8 or for minimising the potential energy in an enclosure as discussed in Chapter 9. The main difference is that the structural case involves the minimisation of structural kinetic energy approximated using a number of vibration transducers, whilst the acoustic case involves minimisation of the acoustic potential energy approximated using a number of microphones. Thus, this aspect will not be considered further here.

The remainder of this chapter begins with a discussion of the feedforward control of vibratory power transmission in an infinitely long beam, with flexural, longitudinal and torsional waves all being taken into account. This discussion is followed by an analysis of the feedforward control of flexural waves in beams with various end conditions, which is followed by an analysis of the feedforward control of vibratory power transmission in a semi-infinite plate. The minimisation of the vibratory power transmission at the error sensor is compared with the minimisation of flexural wave vibration amplitude in terms of effectiveness in controlling vibration downstream of the error sensor. The effect of control source/primary source separation and error sensor/control source separation on vibration levels both upstream and downstream of the primary source is also considered. Note that the downstream direction is defined as the direction from the primary source to the control source. Means for measuring power transmission of all three wave types in a beam simultaneously are also discussed.

An advantage of feedforward systems is that the physical system can be analysed separately from the electronic control system. Thus, for any given control source and error sensor arrangement, it is possible to determine the maximum achievable reduction in vibration level or vibratory power transmission, assuming an ideal controller. Therefore, this chapter will only be concerned with the analysis of the physical system. Design of the feedforward controller is discussed in Chapters 6 and 13, and feedback control of structural vibration is discussed in Chapter 11.

10.1 INFINITE BEAM

Using feedforward control to actively control any one particular wave type in a beam is analogous to controlling plane waves propagating in an air duct as discussed in Chapter 7. For longitudinal and torsional waves on the beam, the only added problem is the greatly increased near-field size for a particular excitation frequency due to the increased wavelengths, with the result that the error sensor must be much further from the control source to escape its near-field influence. Another difference arises for the case of flexural waves which are characterised as dispersive; that is, their propagation speed increases with increasing frequency.

A typical feedforward control system arrangement for controlling a single wave type on an infinite beam using accelerometer reference and error sensors is illustrated in Figure 10.1(a), and the equivalent block diagram is shown in Figure 10.1(b) (Elliott et al., 1990). The quantity $\ddot{w}(x)$ is the acceleration measured by an accelerometer at location x. Note that the feedback path $F(\mathrm{j}\omega)$ shown in the block diagram is part of the physical system, not the electronic control system, and represents the effect of the control source on the reference signal input into the electronic controller. The transfer function or frequency response $P(\mathrm{j}\omega)$ represents the transfer function of the path from the reference (or detection) sensor along the beam to the error sensor and includes the electromechanical transfer functions of the reference and error transducers. Similarly, $C(\mathrm{j}\omega)$ represents the transfer function from the controller electrical output through the control source, along the beam to the error sensor and through the error sensor to the error sensor output, also taking into account the electro-mechanical transfer functions of the control source and error transducer (referred to as the cancellation path transfer function). The electronic control system is represented by $T(\mathrm{j}\omega)$ and the adaptive algorithm which adjusts $T(\mathrm{j}\omega)$ to minimise the error signal. Note that $R(\omega)$, $Y(\omega)$ and $E(\omega)$ are Fourier transforms of the reference signal $r(t)$, the controller output signal $y(t)$, and the error signal $e(t)$. The contribution of the primary source to the reference (or detection) sensor is $I(\mathrm{j}\omega)$, while the contribution to the error sensor is $D(\omega)$. Using the frequency domain for system analysis is purely a convenience, as the time domain equivalents (impulse responses) of frequency domain transfer functions cannot simply be multiplied together; they must be convolved as explained in Chapter 6, thus making algebraic manipulations more complex.

For an infinite beam, where the dynamic effects of the actuators and sensors are ignored, the transfer functions discussed above are defined formally as:

$$C(\mathrm{j}\omega) = \left.\frac{E(\omega)}{Y(\omega)}\right|_{I(\omega)=0} = \left.\frac{\ddot{w}(x_e)}{F_c}\right|_{\ddot{w}(x_p)=0} = \frac{\mathrm{j}\omega^2}{4EIk_b^3}\left[\mathrm{e}^{-\mathrm{j}k_b(x_e - x_c)} - \mathrm{j}\mathrm{e}^{-k_b(x_e - x_c)}\right] \quad (10.1.1\text{a–c})$$

$$F(\mathrm{j}\omega) = \left.\frac{R(\omega)}{Y(\omega)}\right|_{I(\omega)=0} = \left.\frac{\ddot{w}(x_r)}{F_c}\right|_{\ddot{w}(x_p)=0} = \frac{\mathrm{j}\omega^2}{4EIk_b^3}\left[\mathrm{e}^{-\mathrm{j}k_b(x_r - x_c)} - \mathrm{j}\mathrm{e}^{-k(x_r - x_c)}\right] \quad (10.1.2\text{a–c})$$

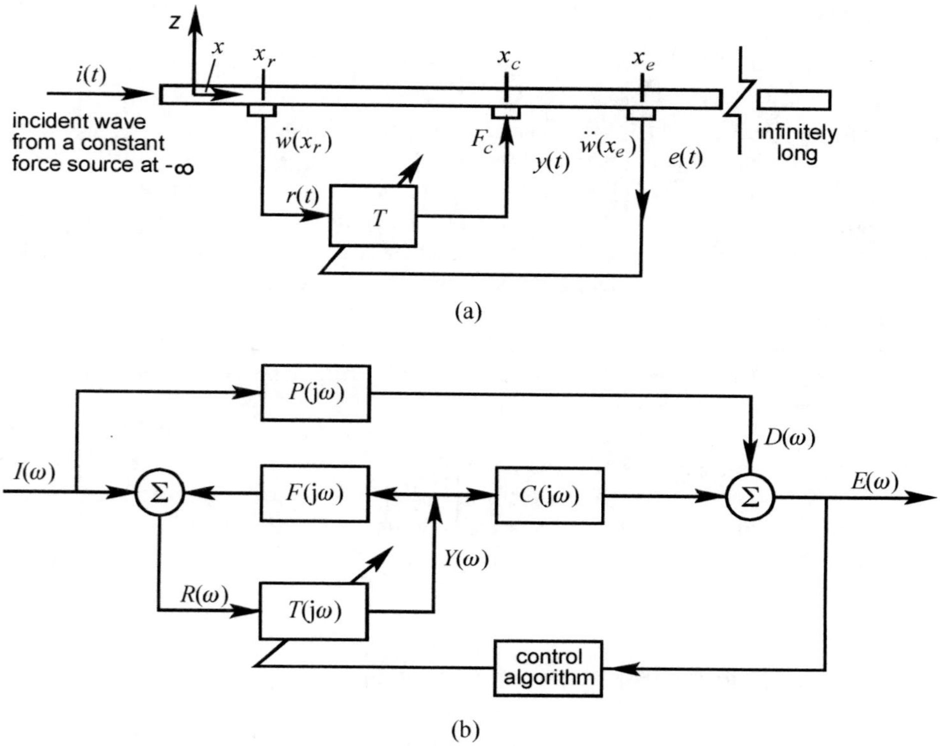

Figure 10.1 Feedforward control of wave propagation in an infinite beam: (a) physical system arrangement; (b) equivalent electrical block diagram.

$$P(\mathrm{j}\omega) = \left.\frac{E(\omega)}{R(\omega)}\right|_{T(\mathrm{j}\omega)=0} = \left.\frac{\ddot{w}(x_e)}{\ddot{w}(x_r)}\right|_{F_c=0} = \mathrm{e}^{-\mathrm{j}k_b(x_e - x_r)} \tag{10.1.3a–c}$$

From Figure 10.1(b), the total error signal is:

$$E(\omega) = I(\omega)\left[P(\mathrm{j}\omega) + \frac{C(\mathrm{j}\omega)\,T(\mathrm{j}\omega)}{1 - T(\mathrm{j}\omega)\,F(\mathrm{j}\omega)}\right] \tag{10.1.4}$$

If there is no measurement noise, the error signal $E(\omega)$ can be made zero if:

$$T(\mathrm{j}\omega) = \frac{-P(\mathrm{j}\omega)}{C(\mathrm{j}\omega) - P(\mathrm{j}\omega)\,F(\mathrm{j}\omega)} \tag{10.1.5}$$

For control of single frequency noise, $T(\mathrm{j}\omega)$ can be modelled in the time domain using a transversal filter with two coefficients or taps. These two taps, however, may have very large weight values, thus presenting numerical difficulties during implementation, which can be overcome by using a few more taps (up to about 10). For control of broadband noise, two orders of magnitude more taps were found to be necessary (Elliott et al., 1990). In practice, the filter must be made adaptive to account for changes in the physical system transfer functions that usually occur relatively slowly with time. When controlling broadband noise, the maximum attenuation in vibration level at the error sensor at frequency ω has been shown (Elliott et al., 1990) to be $10\log_{10}\left(\frac{1}{1-\gamma^2_{xe}(\omega)}\right)$, where γ^2_{xe} is the coherence between the controller reference input signal and error input signal at frequency ω.

In practice, the transfer functions $P(\omega)$, $F(\omega)$ and $C(\omega)$ of Equation (10.1.5) may be measured by driving the control actuators with random noise, with all other input sources deactivated. Changes in $C(\omega)$ over the lifetime of the controller can cause it to go unstable, and changes in the other transfer functions result in decreases in controller performance. Ways of overcoming this problem include using an adaptive filter and algorithm which includes an on-line estimate of $C(\omega)$ as discussed in Chapter 6. Note that the inclusion of the feedback path $F(\omega)$ only affects the desired controller transfer function for optimal control. It does not affect the maximum achievable controllability with an assumed ideal controller as discussed in the remainder of this chapter.

The transfer functions also clearly depend on the wave type being controlled; thus, it is clear that this single-channel controller cannot be used to control more than one wave type simultaneously. Use of a two-channel feedforward controller to control a single frequency flexural and longitudinal wave was reported by Fuller et al. (1990), and the simultaneous control of two single frequency flexural waves and one longitudinal wave using a three channel controller was reported by Clark et al. (1992).

10.1.1 Flexural Wave Control: Minimising Vibration

In this section an analysis of flexural waves propagating along an infinite beam is undertaken with the idea of examining the effect of error sensor location on the maximum achievable reduction in vibration levels downstream from the error sensor. The coordinate system used for the analysis is shown in Figure 10.2. It will be assumed that the excitation forces act normal to the *x*-*y* plane, in the *z*-direction. A harmonic primary point force of magnitude F_p and zero relative phase acts at $x = 0$. It is assumed that the system to be controlled is linear; that is, the principle of superposition holds, whereby the response at a point on the beam under the simultaneous action of a number of forces is equal to the sum of the responses due to each individual force. Here, and in subsequent sections, Euler beam theory will be used; that is, the effects of shear and rotary inertia will be neglected. The consequences of this were discussed in Section 2.3.

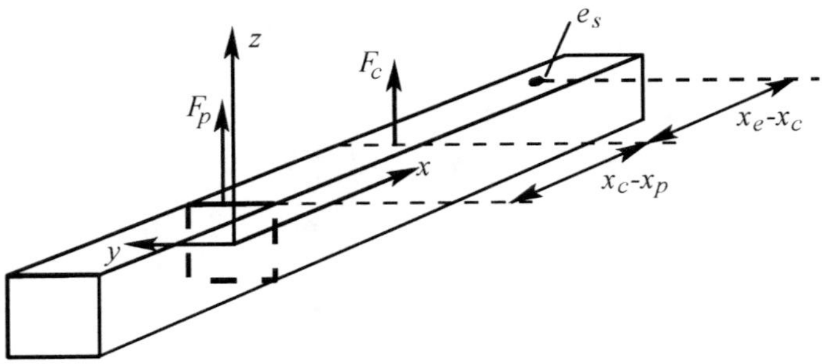

Figure 10.2 Coordinate system for flexural wave propagation in a beam.

As discussed in Chapter 2, the general solution for flexural wave propagation in a beam for flexural vibration in the *z*-direction is (omitting time dependence for convenience):

$$\bar{w}_z(x) = A_1 e^{-jk_b x} + A_2 e^{jk_b x} + A_3 e^{-k_b x} + A_4 e^{k_b x} \tag{10.1.6}$$

Considering only waves propagating in the positive *x*-direction, $A_2 = A_4 = 0$.

If a point force F_p is applied at $x = 0$, the symmetry constrains the slope to be zero at that point. That is,

$$\frac{\partial \bar{w}_z(x)}{\partial x} = 0 \quad \text{at } x = 0 \tag{10.1.7}$$

Thus,

$$-jk_b A_1 e^{-jk_b x} - k_b A_3 e^{-k_b x}\Big|_{x=0} = 0 \tag{10.1.8}$$

or

$$A_1 = jA_3 \tag{10.1.9}$$

Thus, Equation (10.1.6) becomes:

$$\bar{w}_z(x) = A_1 \left(e^{-jk_b x} - je^{-k_b x}\right) \tag{10.1.10}$$

Following the argument used in Section 2.5 to derive the point impedance for a beam, A_1 is given by

$$A_1 = \frac{-jF_p}{4EIk_b^3} \tag{10.1.11}$$

Thus, Equation (10.1.10) can be written as:

$$\bar{w}_z(x) = \frac{-F_p}{4EIk_b^3}\left(je^{-jk_b x} + e^{-k_b x}\right) \tag{10.1.12}$$

If instead of being applied at the origin, the primary force is applied at x_p, and if an additional (control) force is applied at x_c, the equation for the displacement at any location *x* can be derived from Equation (10.1.12) and written as:

$$\bar{w}_z(x) = F_p \beta_p + F_c \beta_c \tag{10.1.13}$$

where the coefficients β_p and β_c are defined as:

$$\beta_{p,c} = -\frac{1}{4EIk_b^3}\left(je^{-jk_b|x - x_{p,c}|} + e^{-k_b|x - x_{p,c}|}\right) \tag{10.1.14}$$

Remember that the first term in Equation (10.1.14) is the propagating component and the second term is the non-propagating or evanescent component. The control force required to make $w(x)$ equal to zero is then,

$$F_c = -F_p \frac{\beta_p(x_e)}{\beta_c(x_e)} \tag{10.1.15}$$

The ratio of controlled to uncontrolled displacement amplitude downstream of the control source is:

$$R = \left| \frac{(F_p\beta_p + F_c\beta_c)}{F_p\beta_p} \right| = \left| 1 + \frac{F_c\beta_c(x)}{F_p\beta_p(x)} \right| \tag{10.1.16a,b}$$

By substituting Equation (10.1.15) into this, the following is obtained:

$$R = \left| 1 - \frac{\beta_p(x_e)\beta_c(x)}{\beta_c(x_e)\beta_p(x)} \right| \tag{10.1.17}$$

The β_p and β_c factors can then be replaced by their definitions and the limit as x approaches infinity is taken so that the expression for the reduction in far-field residual vibration is:

$$\lim_{x \to \infty} R = \frac{\left| 1 - \mathrm{e}^{-k_b(x_c - x_p)(1-\mathrm{j})} \right|}{\left| \mathrm{j}\mathrm{e}^{k_b(x_e - x_c)(1+\mathrm{j})} + 1 \right|} \tag{10.1.18}$$

From the equation, it is clear that the reduction in far-field acceleration level depends on the separation between the control source and the primary source as well as the separation between the control source and the error sensor (at which the acceleration is minimised). If the distance between the control and primary source locations is large compared with a wavelength; that is, if $k_b(x_c - x_p) >> 1$, the numerator of Equation (10.1.18) approaches unity. If in addition, the distance between the error sensor and the control source locations is large compared with a wavelength; that is, if $k_b(x_e - x_c) >> 1$, the final expression for the downstream residual vibration becomes:

$$\lim_{x \to \infty} R \approx \mathrm{e}^{-k_b(x_e - x_c)} \tag{10.1.19}$$

For harmonic excitation, the net vibrational power transmission in the beam is proportional to the square of the far-field acceleration level and so the power transmission reduction is proportional to $\frac{20}{\log_e 10} k_b(x_e - x_c)$ dB.

The acceleration distribution along the beam may be calculated from Equation (10.1.13) for both the controlled condition where the control source is given by Equation (10.1.15) and the uncontrolled condition. In the example shown in Figure 10.3(a), a unit primary force is applied at $x_p = 0$, the error sensor is placed in the far-field of the control source, and the control source is also in the far-field of the primary source. The dotted line represents the effect of the primary force acting alone, while the solid line shows the effect of adding at x_c the optimal control force defined by Equation (10.1.15). Clearly, the (constant) residual acceleration level on the right of the error sensor at x_e is very much smaller than the uncontrolled level. For this particular case, with the primary force at $x_p = 0$, and the control force at $x_c/\lambda_b = 2.12$ ($\lambda_b = 0.4823$ m is the flexural wavelength), the reduction in vibration level is about 110 dB. Of course, it would not be feasible to achieve this with a practical control system. Nevertheless, the calculation does indicate the maximum theoretically possible reduction. Between the control source location and the error sensor, the vibration level decreases exponentially from the uncontrolled level to the residual level. As required in the analysis, the vibration amplitude at the error sensor ($x_e/\lambda_b = 4.0$) is zero ($-\infty$ dB). When the control source is placed in the near-field of the primary source, the resulting reduction in vibration levels downstream of the error sensor is slightly greater, as expected by inspection of Equation (10.1.18).

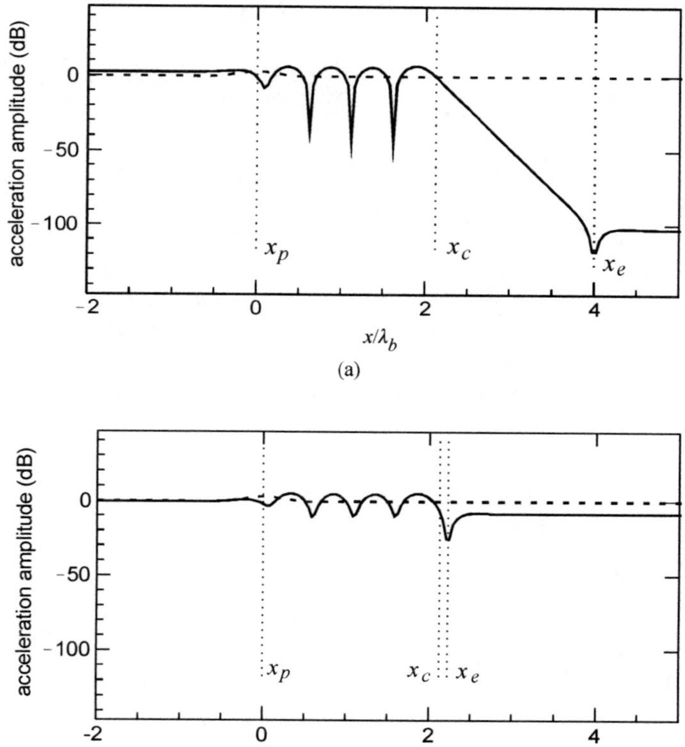

(b)

Figure 10.3 Distribution of acceleration amplitude in an anechoically terminated beam of 25 mm × 50 mm section with the primary source at $x_p/\lambda_p = 0$ and the control source at $x_s/\lambda_b = 2.12$.
----- primary force only, -------- controlled.
(a) Error sensor at $x_e/\lambda_b = 4.0$; (b) error sensor at $x_e/\lambda_b = 2.22$.

For the results shown in Figure 10.3(b), the error sensor is in the near-field of the control source. The residual acceleration level on the right of the error sensor is only 9 dB lower than the uncontrolled level. This is much less than that shown in Figure 10.3(a) and indicates the importance of placing a vibration error sensor in the far-field of the control source if the power transmission is to be controlled adequately using an acceleration sensor.

Between the primary and control sources there is a standing wave pattern in both figures. Because the applied forces represent structural discontinuities with finite rather than infinite or zero impedance, the primary and control sources appear at neither the nodes nor the anti-nodes of these standing wave patterns. It was shown by Brennan et al. (1992) that these standing waves can be eliminated by using two control sources rather than one, as they can be configured to suppress the wave propagating towards the primary source as well as the downstream propagating wave. Two control sources can also be used to suppress the near (or evanescent) field downstream of the control source as well as the downstream propagating wave. If three control sources are used, it is possible to suppress all of the waves mentioned above simultaneously (Brennan et al., 1992); and if four closely-spaced control sources are used, then it is possible to suppress the evanescent field on the upstream side of the control sources as well. Note that as the number of closely spaced (0.1λ) control sources increases, the total control force effort increases exponentially (Brennan et al., 1992). Also, the required control force becomes prohibitively large as the control source separation approaches zero or an integer multiple of half wavelengths.

10.1.2 Flexural Wave Control: Minimising Power Transmission

It is of interest to discover what control improvements are possible if the error sensor is capable of measuring structural power transmission in the beam and providing the control algorithm with a measure of this quantity rather than just vibration amplitude.

The flexural wave power being transmitted along the beam through any one cross-sectional location is made up of an active component and a reactive component as discussed in Section 2.5. As shown by Equation (2.5.202), the active or time-averaged power transmission is:

$$P_{Bza} = \mathrm{Re}\left\{ \frac{\mathrm{j}\omega EI_{yy}}{2} \left(\bar{w}_z^*(x) \frac{\partial^3 \bar{w}_z(x)}{\partial x^3} - \frac{\partial \bar{w}_z^*(x)}{\partial x} \frac{\partial^2 \bar{w}_z(x)}{\partial x^2} \right) \right\} \tag{10.1.20}$$

and the amplitude of the reactive component is:

$$P_{Bzr} = \mathrm{Im}\left\{ \frac{\mathrm{j}\omega EI_{yy}}{2} \left(\bar{w}_z^*(x) \frac{\partial^3 \bar{w}_z(x)}{\partial x^3} - \frac{\partial \bar{w}_z^*(x)}{\partial x} \frac{\partial^2 \bar{w}_z(x)}{\partial x^2} \right) \right\} \tag{10.1.21}$$

Using Equation (10.1.13), Equation (10.1.20) can be written in quadratic form in terms of the control force F_c as:

$$P_{Bza} = R_e\left\{ F_c^* a_B F_c + b_{1B} F_c + F_c^* b_{2B} + c_B \right\} \tag{10.1.22}$$

where c_B is the power due to the primary force only and is found by setting $F_c = 0$ in Equation (10.1.13) and substituting the result into Equation (10.1.20) to give:

$$c_B = \frac{\mathrm{j}\omega EI_{yy}}{2} F_p^2 \left(\beta_p^* \frac{\partial^3 \beta_p}{\partial x^3} - \frac{\partial \beta_p^*}{\partial x} \frac{\partial^2 \beta_p}{\partial x^2} \right) \tag{10.1.23}$$

where β_p is defined by Equation (10.1.14).

The quantities a_B, b_{1B} and b_{2B} in Equation (10.1.22) are defined as follows:

$$a_B = \frac{j\omega EI_{yy}}{2}\left[\beta_c^* \frac{\partial^3 \beta_c}{\partial x^3} - \frac{\partial \beta_c^*}{\partial x}\frac{\partial^2 \beta_c}{\partial x^2}\right] \tag{10.1.24}$$

$$b_{1B} = \frac{j\omega EI_{yy}F_p}{2}\left[\beta_p^* \frac{\partial^3 \beta_c}{\partial x^3} - \frac{\partial \beta_p^*}{\partial x}\frac{\partial^2 \beta_c}{\partial x^2}\right] \tag{10.1.25}$$

$$b_{2B} = \frac{j\omega EI_{yy}F_p}{2}\left[\beta_c^* \frac{\partial^3 \beta_p}{\partial x^3} - \frac{\partial \beta_c^*}{\partial x}\frac{\partial^2 \beta_p}{\partial x^2}\right] \tag{10.1.26}$$

In the preceding four equations, care must be exercised when differentiating β, as the results are different for $x > x_p$ or $x > x_c$ than they are for $x < x_p$ or $x < x_c$.

Equation (10.1.22) is the familiar quadratic minimisation problem, which in this case is solved by differentiating the real part of the power in Equation (10.1.22) with respect to both the real and imaginary parts of the control force and setting each result to zero. The result is the optimal control force F_c, given by (Pan and Hansen, 1991) as:

$$F_c = -\frac{1}{2\mathrm{Re}\{a_B\}}(b_{1B}^* + b_{2B}) \tag{10.1.27}$$

If the power is minimised in the far-field of the control and primary sources, then $b_{1B}^* = b_{2B}$. Substituting Equations (10.1.23) to (10.1.26) into Equation (10.1.27) gives (for minimising far-field power transmission):

$$F_c = -F_p \mathrm{e}^{-jk_b(x_c - x_p)} \tag{10.1.28}$$

Substituting Equation (10.1.14) into Equation (10.1.22) (with $F_c = 0$) and taking the limit as $x \to \infty$, gives the following result for the far-field flexural wave power transmission with no control:

$$P_{Bzau} = \frac{F_p^2 \omega}{16EI_{yy}k_b^3} \tag{10.1.29}$$

The controlled power transmission to the right of the error sensor location x_e is found by substituting Equations (10.1.23) to (10.1.27) into Equation (10.1.22) and is equal to zero. This is in contrast to the result obtained in Equation (10.1.18) for acceleration control where it was shown that the power transmission downstream of the error sensor was dependent on both the separation between the control source and primary source (weakly) and the separation between the error sensor and the control source (strongly).

Between the primary and control sources, the reduction in power transmission due to the control force is found to be:

$$\frac{P_{Bza}}{P_{Bzau}} = 2\sin\left[k_b(x_c - x_p)\right]\mathrm{e}^{-k_b(x_c - x_p)} \tag{10.1.30}$$

To the left of the primary source the reduction in power transmission is:

$$\frac{P_{Bza}}{P_{Bzau}} = 2\cos\left[2k_b(x_c - x_p)\right] - 2 \tag{10.1.31}$$

Equations (10.1.30) and (10.1.31) are zero if the following condition is satisfied:

$$x_c - x_p = n\frac{\pi}{k_b} = n\frac{\lambda_b}{2} \tag{10.1.32a,b}$$

Thus, best results are obtained for power being transmitted on both sides of the primary source and between the control source and primary source if the separation between the latter two sources is an integer multiple of half the structural wavelength. When this occurs, the primary source power is completely suppressed at the primary source with no absorption at the control source. Similar conclusions were drawn in Chapter 7 in the discussion of waves propagating in air ducts.

The relative effectiveness of a vibration error sensor compared with a power transmission (or intensity) sensor for the active control of power transmission in a beam is indicated in Figure 10.4. In part (a), the maximum achievable power transmission control is plotted as a function of frequency for an aluminium beam of cross-sectional dimensions 50 mm × 25 mm, terminated anechoically, with the error sensor located 1 m from the control source and the control source located 1 m from the primary source.

It was shown by Pan and Hansen (1993a) that the reactive (or near) field is 20 dB below the active or propagating field at distances greater than 0.73 λ_b from a point force source. This corresponds to $k_b(x_e - x_c) = 4.6$. From Figure 10.4(b) it can be seen that the influence of the near-field at this point is negligible. As mentioned earlier, if the error sensor is an accelerometer or velocity or displacement sensor, and is placed in the near-field of a control source, it will be much less effective in controlling the downstream power transmission than if it were in the far-field of the control source. This places a lower frequency limit on the vibration that can be controlled (as the near-field grows in size as the frequency of flexural waves decreases). Elliott and Billet (1993) showed that for practically sized beams, this low-frequency limit is below the limit imposed by other constraints such as obtaining a reference signal that is coherent with the vibration to be controlled. Elliott and Billet (1993) also pointed out a high-frequency limitation, which is associated with the increasing flexural wavespeed as the disturbance frequency is increased and the fixed, frequency independent, finite time taken by the controller to produce a control signal, once it receives a reference signal. This is clearly not a real limit for periodic vibration, but it is very important for random vibration, which requires the controller to be causal. In the same article, it was also shown that the bandwidth of controllability increases exponentially as the controller processing time is reduced. The group delay τ_g associated with a flexural wave travelling a distance L along a beam is obtained using Equation (2.3.58) and is given by:

$$\tau_g = L/c_g = \frac{L}{2\omega^{1/2}}\left[\frac{m}{EI}\right]^{1/4} \tag{10.1.33}$$

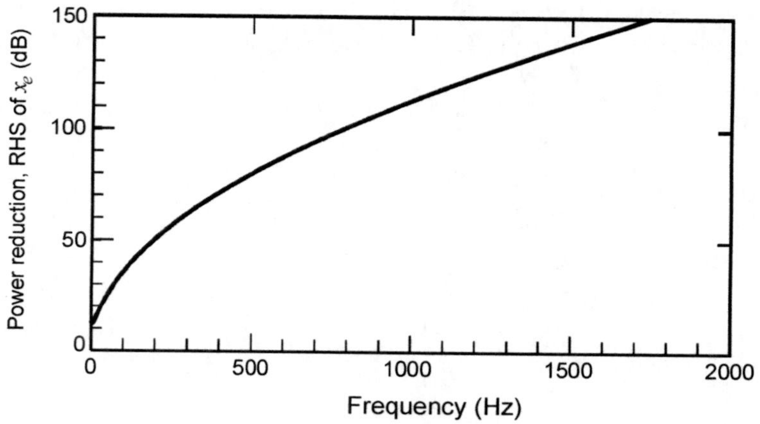

(a)

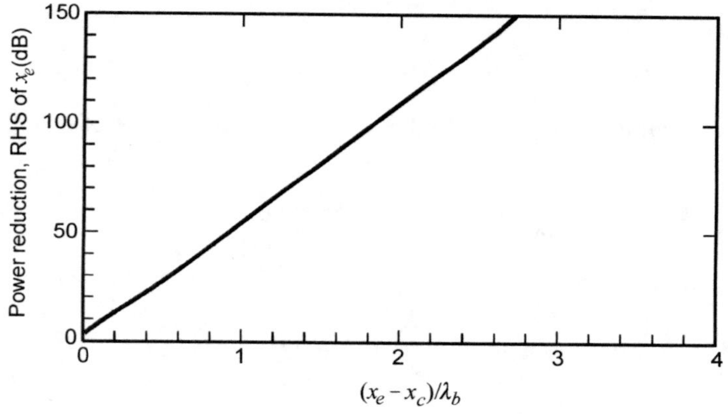

(b)

Figure 10.4 Maximum achievable reduction in real power flow (and for field vibration level) for an infinitely long beam excited by a point force with an error sensor at which the acceleration is minimised located at $x = x_e$: (a) as a function of frequency for $(x_e - x_c) = 1$ m, and $(x_c - x_p) = 1$ m; (b) as a function of the product of the ratio of the separation between the error sensor and control source, and the wavelength, for $(x_c - x_p)/\lambda_b = 1$ m.

An alternative approach to the active control of structural power transmission is to use the frequency domain. This approach is described in detail by Pereira and Arruda (2000).

10.1.3 Simultaneous Control of All Wave Types: Power Transmission

In this section, expressions for the power being transmitted along an infinite beam in the form of flexural, longitudinal and torsional waves are investigated for the purpose of determining the optimal control forces to control these wave types and the maximum control achievable assuming an ideal controller. Another reason for pursuing this analysis is to investigate the spillover of energy into other wave types when attempts are made to actively control flexural wave power with a control force that does not act exactly normal to the beam surface. Pan and Hansen (1991) have analysed the active control of all four wave types (two flexural, one torsional and one longitudinal) in a thick beam, and some experimental results have been presented by Clark et al. (1992). Brennan et al. (1992) reported a means of controlling the three wave types by using three magnetostrictive actuators mounted within a hollow tube used to model a helicopter strut. The analysis to follow is based principally on the work reported by Pan and Hansen (1991).

Consider the rectangular section beam shown in Figure 10.5 excited by a primary force at some angle α_p from the x-axis in the x-z plane and angle φ_p from the x-axis in the x-y

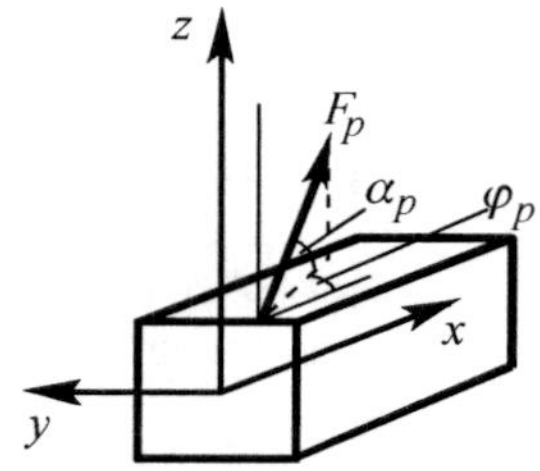

Figure 10.5 Beam excited by a force at an arbitrary location and orientation.

plane. The force is located at (x_p, y_p, z_p) and can be represented in terms of unit vectors along the x-, y- and z-axes. Thus,

$$F_{pL} = F_p \delta(x - x_p) = \left(F_{px}\boldsymbol{i} + F_{py}\boldsymbol{j} + F_{pz}\boldsymbol{k}\right)\delta(x - x_p) \qquad (10.1.34a,b)$$

where the Dirac delta function is used as a mathematical convenience so that the force can be represented as a force per unit length along the x-direction. Equation (10.1.34) indicates that the force is zero everywhere except at $x = x_p$. The line moment per unit length about the origin of the $x = x_p$ plane resulting from the action of F_{pL} is given by:

$$M_{pL} = \boldsymbol{\sigma_p} \times \boldsymbol{F_{pL}} = M_{px}\boldsymbol{i} + M_{py}\boldsymbol{j} + M_{pz}\boldsymbol{k} \qquad (10.1.35a,b)$$

where

$$\boldsymbol{\sigma_p} = x_p\boldsymbol{i} + y_p\boldsymbol{j} + z_p\boldsymbol{k} \qquad (10.1.36)$$

Each wave type will now be considered separately.

10.1.3.1 Longitudinal Waves

Using the wave Equation (2.3.8) derived in Section 2.3 for free vibration and including the external force, the following displacement equation is obtained for the a beam of cross-sectional area S and material Young's modulus E:

$$\frac{\partial^2 \xi_x}{\partial x^2} - \frac{1}{c_L^2}\frac{\partial^2 \xi_x}{\partial t^2} = -\frac{1}{ES}F_{pL}\boldsymbol{i} \qquad (10.1.37)$$

10.1.3.2 Torsional Waves

Using the analysis of Section 2.3, Equations (2.3.20) and (2.3.21), the torsional wave equation can be written as:

$$\frac{\partial^2 \theta_x}{\partial x^2} - \frac{1}{c_s^2}\frac{\partial^2 \theta_x}{\partial t^2} = -\frac{1}{GJ} M_{pL} \boldsymbol{i} \tag{10.1.38}$$

10.1.3.3 Flexural Waves

The forced wave equation for flexural waves corresponding to displacements in the y-direction can be derived from:

$$\frac{\partial^4 w_y}{\partial x^4} + \frac{1}{c_{by}^2}\frac{\partial^2 w_y}{\partial t^2} = \frac{1}{EI_{zz}}\left[F_{pL}\boldsymbol{j} - \frac{\partial M_{pL}}{\partial x}\boldsymbol{k}\right] \tag{10.1.39}$$

where

$$c_{by}^4 = \frac{\omega^2 EI_{zz}}{m_b} \tag{10.1.40}$$

m_b is the mass per unit length of the beam, E is Young's modulus of the beam material and I_{zz} is the second moment of area of the beam cross-section about the z-axis.

The forced wave equation corresponding to displacements in the z-direction is:

$$\frac{\partial^4 w_z}{\partial x^4} + \frac{1}{c_{bz}^2}\frac{\partial^2 w_z}{\partial t^2} = \frac{1}{EI_{yy}}\left[F_{pL}\boldsymbol{k} + \frac{\partial M_{pL}}{\partial x}\boldsymbol{j}\right] \tag{10.1.41}$$

where

$$c_{bz}^4 = \frac{\omega^2 EI_{yy}}{m_b} \tag{10.1.42}$$

If a control force F_c is now introduced at location (x_c, y_c, z_c), the combined effect of both the control force and primary force can be taken into account in the preceding equations by replacing F_{pl} with $F_{cL} + F_{pL}$, M_{pL} with $M_{pL} + M_{cL}$, and $\frac{\partial M_{pL}}{\partial x}$ with $\frac{\partial M_{pL}}{\partial x} + \frac{\partial M_{cL}}{\partial x}$.

With both the primary force and the control force included, the solutions to the preceding equations of motion are (Pan and Hansen, 1991):

$$w_{x0}(x,\omega,t) = \left(F_{px}\beta_{pL} + F_{cx}\beta_{cL}\right)\mathrm{e}^{\mathrm{j}\omega t} \tag{10.1.43}$$

$$w_{y0}(x,\omega,t) = \left(F_{py}\beta_{pby} + F_{cy}\beta_{cby} + M_{pz}\beta_{pbyM} + M_{cz}\beta_{cbyM}\right)\mathrm{e}^{\mathrm{j}\omega t} \tag{10.1.44}$$

$$w_{z0}(x,\omega,t) = \left(F_{pz}\beta_{pbz} + F_c\beta_{cbz} + M_{py}\beta_{pbzM} + M_{cy}\beta_{cbzM}\right)\mathrm{e}^{\mathrm{j}\omega t} \tag{10.1.45}$$

$$\theta_x(x,\omega,t) = \left(M_{px}\beta_{pT} + M_{cx}\beta_{cT}\right)e^{j\omega t} \tag{10.1.46}$$

The β coefficients corresponding to the primary excitation force are defined as follows. The subscript 0 in Equations (10.1.43) to (10.1.45) indicates that the displacements are those of the centre of the beam cross-section at any particular location x.

$$\beta_{pL} = \frac{-j e^{-jk_L|x-x_p|}}{2ESk_L} \tag{10.1.47}$$

$$\beta_{pby} = \frac{-j e^{-jk_{by}|x-x_p|} - e^{-k_{by}|x-x_p|}}{4EI_{zz}k_{by}^3} \tag{10.1.48}$$

$$\beta_{pbz} = \frac{-j e^{-jk_{bz}|x-x_p|} - e^{-k_{bz}|x-x_p|}}{4EI_{yy}k_{bz}^3} \tag{10.1.49}$$

$$\beta_{pT} = \frac{-j e^{-jk_T|x-x_p|}}{2GJk_T} \tag{10.1.50}$$

$$\beta_{pbyM} = \frac{(x-x_p)\left(e^{-jk_{by}|x-x_p|} - e^{-k_{by}|x-x_p|}\right)}{4EI_{zz}k_{by}^3\,|x-x_p|} \tag{10.1.51}$$

$$\beta_{pbzM} = \frac{(x-x_p)\left(e^{-jk_{bz}|x-x_p|} - e^{-k_{bz}|x-x_p|}\right)}{4EI_{yy}k_{bz}^3\,|x-x_p|} \tag{10.1.52}$$

Similar expressions hold for the β coefficients with the c subscript, which refers to the control forces and moments. In this latter case, x_p is replaced by x_c.

The wavenumbers k_{bz}, k_{by}, k_L and k_T correspond respectively to bending waves with displacements along the y-axis, longitudinal waves and torsional waves. They are defined as follows:

$$k_{bz}^4 = \frac{\omega^2 m_b}{EI_{yy}} = \left(\frac{\omega}{c_{bz}}\right)^4 \tag{10.1.53a,b}$$

$$k_{by}^4 = \frac{\omega^2 m_b}{EI_{zz}} = \left(\frac{\omega}{c_{by}}\right)^4 \tag{10.1.54a,b}$$

$$k_L = \omega\sqrt{\frac{\rho}{E}} = \frac{\omega}{c_L} \tag{10.1.55a,b}$$

$$k_T = \omega\sqrt{\frac{\rho}{G}} = \frac{\omega}{c_T} \tag{10.1.56a,b}$$

where J in Equation (10.1.50) is the polar second moment of area of the beam cross-section, and in the above equations, m_b is the mass per unit area of the beam, G is the shear modulus or modulus of rigidity of the beam material and E is its modulus of elasticity.

The general displacement vector for the centre of the beam can be written as:

$$\boldsymbol{W}_0 = \left(\xi_{x0},\ w_{y0},\ w_{z0},\ \theta_x,\ \theta_y,\ \theta_z\right)^{\mathrm{T}} \tag{10.1.57}$$

Note that θ_y and θ_z are defined as:

$$\theta_y = -\frac{\partial w_{z0}}{\partial x} \tag{10.1.58}$$

$$\theta_z = \frac{\partial w_{y0}}{\partial x} \tag{10.1.59}$$

The time independent part of the general displacement vector can be written, by combining Equations (10.1.43) to (10.1.55), as:

$$\bar{\boldsymbol{W}}_0 = \boldsymbol{a}_p \boldsymbol{Q}_p + \boldsymbol{a}_c \boldsymbol{Q}_c \tag{10.1.60}$$

where the generalised forces are defined as:

$$\boldsymbol{Q}_{p,c} = \left[F_x,\ F_y,\ F_z,\ M_x,\ M_y,\ M_z\right]^{\mathrm{T}}_{p,c} = \boldsymbol{R}_{p,c} F_{p,c} \tag{10.1.61a,b}$$

$$\begin{aligned}\boldsymbol{R}_{c,p} = [&\cos\alpha_{p,c}\cos\varphi_{p,c},\ \cos\alpha_{p,c}\sin\varphi_{p,c},\ \sin\alpha_{p,c},\ y_{p,c}\sin\alpha_{p,c} - z_{p,c}\cos\alpha_{p,c}\sin\varphi_{p,c},\\ &z_{p,c}\cos\alpha_{p,c}\cos\varphi_{p,c},\ -y_{p,c}\cos\alpha_{p,c}\sin\varphi_{p,c}]^{\mathrm{T}}\end{aligned} \tag{10.1.62}$$

(the subscript p corresponds to the primary force and $\boldsymbol{R}_p$, while the subscript c corresponds to the control force and $\boldsymbol{R}_c$).

The influence coefficient matrix $\boldsymbol{a}_p$ is defined as:

$$\boldsymbol{a}_p = \begin{bmatrix} \beta_{pL} & 0 & 0 & 0 & 0 & 0 \\ 0 & \beta_{pby} & 0 & 0 & 0 & \beta_{pbyM} \\ 0 & 0 & \beta_{pbz} & 0 & \beta_{pbzM} & 0 \\ 0 & 0 & 0 & \beta_T & 0 & 0 \\ 0 & 0 & \beta'_{pbz} & 0 & \beta'_{pbzM} & 0 \\ 0 & \beta'_{pby} & 0 & 0 & 0 & \beta'_{pbyM} \end{bmatrix} \tag{10.1.63}$$

where the prime denotes differentiation with respect to x and $\boldsymbol{a}_s$ is defined by replacing the subscript p with the subscript s in Equation (10.1.58).

Although the preceding equations have been derived for a single point primary force and a single point control force, they can be extended easily to the case of multiple discrete point forces or distributed forces. In the former case, the right-hand side of Equation (10.1.60) becomes a summation over all the point forces and in the latter case it becomes an integration over the region of the distributed force.

Equation (2.5.220) gives the following for the time-averaged power transmission in the beam:

$$P_a = \tfrac{1}{2}\,\mathrm{Re}\left\{ \mathrm{j}\omega\, \bar{\boldsymbol{W}}_0^{\mathrm{H}} \boldsymbol{\Lambda} \bar{\boldsymbol{W}}_0 \right\} \tag{10.1.64}$$

where $\bar{\boldsymbol{W}}_0$ is defined in Equation (10.1.60) and where Λ is a diagonal matrix defined as:

$$\Lambda = \begin{bmatrix} ES\dfrac{\partial}{\partial x} & 0 & 0 & 0 & 0 & 0 \\ 0 & -EI_{zz}\dfrac{\partial^3}{\partial x^3} & 0 & 0 & 0 & 0 \\ 0 & 0 & -EI_{yy}\dfrac{\partial^3}{\partial x^3} & 0 & 0 & 0 \\ 0 & 0 & 0 & -GJ\dfrac{\partial}{\partial x} & 0 & 0 \\ 0 & 0 & 0 & 0 & EI_{yy}\dfrac{\partial}{\partial x} & 0 \\ 0 & 0 & 0 & 0 & 0 & EI_{zz}\dfrac{\partial}{\partial x} \end{bmatrix} \tag{10.1.65}$$

Using Equations (10.1.60), (10.1.61), (10.1.62) and (10.1.64), it is possible to write the equation for the time-averaged power transmission in terms of the control force amplitude F_c. That is,

$$P_a = \bar{F}_c^{*} a_B \bar{F}_c + b_{1B}\bar{F}_c + \bar{F}_c^{*} b_{2B} + c_B \tag{10.1.66}$$

where

$$a_B = \boldsymbol{R}_c^{\mathrm{T}} \boldsymbol{a}_c^{\mathrm{H}} \boldsymbol{\Lambda} \boldsymbol{a}_c \boldsymbol{R}_c \tag{10.1.67}$$

$$b_{1B} = \boldsymbol{Q}_p \boldsymbol{a}_p^{\mathrm{H}} \boldsymbol{\Lambda} \boldsymbol{a}_c \boldsymbol{R}_c \tag{10.1.68}$$

$$b_{2B} = \boldsymbol{R}_c^{\mathrm{T}} \boldsymbol{a}_c^{\mathrm{H}} \boldsymbol{\Lambda} \boldsymbol{a}_p \boldsymbol{Q}_p \tag{10.1.69}$$

$$c_B = \boldsymbol{Q}_p \boldsymbol{a}_p^{\mathrm{T}} \boldsymbol{\Lambda} \boldsymbol{a}_p \boldsymbol{Q}_p \tag{10.1.70}$$

Differentiating Equation (10.1.66) with respect to the real and imaginary parts of the control force and setting both results to zero gives the following for the optimum control force:

$$\bar{F}_c^{\text{opt}} = -\frac{1}{2\text{Re}\{a_B\}}\left(b_{1B}^* + b_{2B}\right) \tag{10.1.71}$$

Substituting Equation (10.1.71) into Equation (10.1.66) provides an expression for the minimum achievable real power transmission, assuming an ideal electronic controller and some form of ideal power transmission (or intensity) sensor. In practice, it is difficult to construct a power transmission sensor of sufficient accuracy, and vibration sensors are often used instead. These give good results if placed in the far-field of the control source where power transmission is proportional to vibration velocity squared.

Pan and Hansen (1991) showed that where a single control force is used, its angular orientation (α_c, φ_c) is extremely important. For example, if it is desired to control flexural waves, effective control of total power transmission without a significant amount of spillover into other wave types can only be achieved if the control force orientation is within 1° or 2° of the normal to the beam surface. The net result is that one control force is needed for each wave type that is to be controlled. It is not possible to control successfully two or more wave types with only one control source as the optimum orientation and relative phase is very much dependent on the particular wave type to be controlled.

Control of three waves (two flexural and one longitudinal) simultaneously using piezoceramic crystal actuators bonded to a beam was demonstrated by Clark et al. (1992). The error sensor configuration consisted of six accelerometers which were configured to allow the vibration level corresponding to each wave type to be determined.

Flexural wave vibration amplitudes were measured as shown in Figure 2.37, using two accelerometers located on the beam surfaces at $y = \pm L_y/2$. The outputs from these accelerometers were pre-amplified and subtracted using an analogue computer to obtain a voltage proportional to the flexural wave component characterised by displacement in the y-direction. Two additional accelerometers located on the beam surface at $z = \pm L_z/2$ provided a signal proportional to the flexural wave component characterised by displacement in the z-direction. To measure the longitudinal wave component, two accelerometers were located on the beam at $z = 0$, $y = \pm L_y/2$. The accelerometers were mounted on their side as shown in Figure 2.41, and their outputs were summed in an analogue computer to give a voltage proportional to the amplitude of the longitudinal waves.

The accelerometer cross-axis sensitivity can be a problem in cases where the flexural wave amplitude is large compared to the longitudinal wave amplitude, as this results in the longitudinal wave amplitude measurement being contaminated significantly by the flexural wave.

The controller configuration used for the simultaneous minimisation of the three wave types is discussed in detail by Clark et al. (1992). Means of determining the amplitudes of each wave type at various locations along the beam and the power transmission corresponding to each wave type before and after control was discussed in Section 2.5.5.

10.1.4 Effect of Damping

As the beam is terminated anechoically and no reverberant field exists, the results obtained for the maximum achievable reductions in vibration level with active control are affected by the structural damping characteristics of the beam.

10.2 FINITE BEAMS

As mentioned previously, it is often important to control the transmission of vibratory power in structural elements such as beams and plates to minimise vibration in, or sound radiation from, attached structures. In the previous section, an analysis was presented that enabled the effect of active control on the reduction of power transmission in an infinite (or anechoically terminated) beam to be calculated.

Here, the more practical problem of active control of vibratory power and vibration level in finite length beams with arbitrary terminations, as shown in Figure 10.6, is examined, following closely the work of Hansen et al. (1993), Pan and Hansen (1993b), and Young and Hansen (1994). Only control of flexural waves is discussed, as the analysis including all wave types is not easily managed. Support conditions at the end of the beam will be defined in terms of force and moment impedances, and referred to as boundary impedances, which may be defined in terms of force and moment impedances and are analogous to the concept of point impedance discussed in Section 2.5.

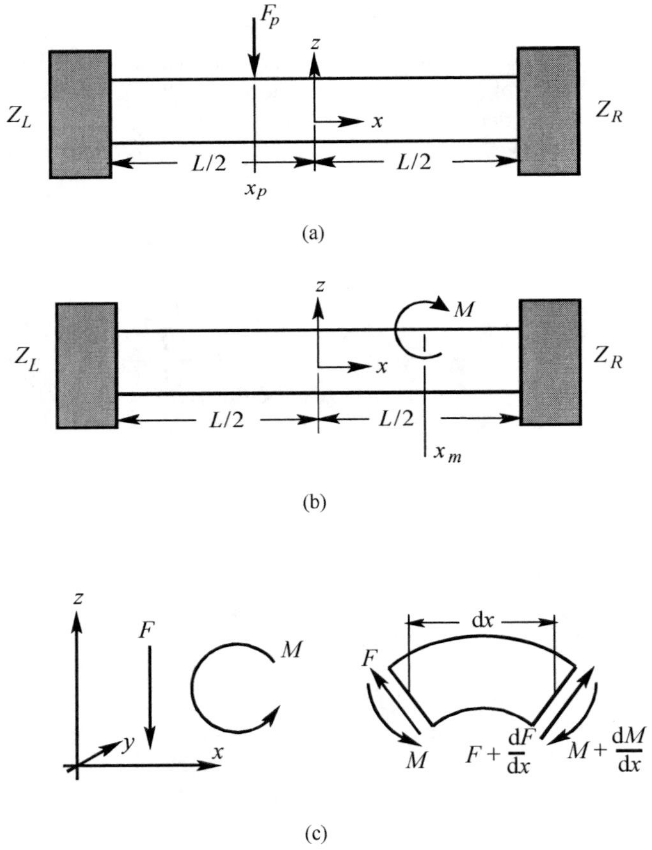

Figure 10.6 Finite beam model: (a) point force excitation; (b) line moment excitation; (c) sign conventions.

In Section 2.5, the concept of impedance to a point force was discussed assuming that no applied moment was acting on the beam. However, in the presence of an applied moment, there will exist coupling between the moment and the force impedance, as the moment will generate a lateral displacement as well as a rotation. Similarly, a pure applied force will generate rotation as well as lateral displacement. Thus, it is not possible to write the boundary impedance for a beam in terms of simple force and moment impedances. Rather, a fully coupled impedance matrix formulation must be used. The meaning of this will be made clearer by the following analysis.

The local lateral beam velocity and angular rotational velocity generated as a result of an applied moment M_e and an applied force F_e may be written as:

$$\dot{w} = \frac{M_e}{Z_{mf}} + \frac{F_e}{Z_f} \tag{10.2.1}$$

$$\dot{\theta} = \frac{M_e}{Z_m} + \frac{F_e}{Z_{fm}} \tag{10.2.2}$$

where Z_{mf} and Z_{fm} ($Z_{mf} \neq Z_{fm}$) are the coupling impedances, and Z_f and Z_m are the force and moment impedances respectively.

At the beam boundary, the moment and force applied to the beam by the supports is equal to the internal moment M and shear force Q respectively. Making these substitutions and rearranging allows Equations (10.2.1) and (10.2.2) to be written as:

$$\begin{bmatrix} Q \\ M \end{bmatrix} = \begin{bmatrix} Z_{f\dot{w}} & Z_{f\dot{\theta}} \\ Z_{m\dot{w}} & Z_{m\dot{\theta}} \end{bmatrix} \begin{bmatrix} \dot{w} \\ \dot{\theta} \end{bmatrix} \tag{10.2.3}$$

where

$$\begin{bmatrix} Z_{f\dot{w}} & Z_{f\dot{\theta}} \\ Z_{m\dot{w}} & Z_{m\dot{\theta}} \end{bmatrix} = \begin{bmatrix} 1/Z_m & 1/Z_{fm} \\ 1/Z_{mf} & 1/Z_f \end{bmatrix}^{-1} \tag{10.2.4}$$

Coefficients of the impedance matrix of Equation (10.2.3) corresponding to various beam end conditions are listed in Table 10.1. Note that for all of the conditions listed, the cross-coupling terms from the matrix of Equation (10.2.3) are zero. In the analysis to follow, dissipation of energy within the beam may be taken into account by replacing Young's modulus of elasticity E by $E(1+j\eta)$, where η is the beam loss factor at the frequency of interest.

10.2.1 Equivalent Boundary Impedance for an Infinite Beam

A finite beam can be made to behave as an infinite beam by applying appropriate boundary impedances to its ends. The appropriate impedance matrix is derived as follows. In an infinite beam with a flexural wave characterised by displacement in the *z*-direction, travelling toward the right and generated by a sole point force excitation, the displacement in the *z*-direction can be written as:

Table 10.1 Impedances corresponding to standard terminations.

End Condition	Representation	Boundary Condition	Impedance
Simply supported		$w = 0$ $\frac{\partial^2 w}{\partial x^2} = 0$	$Z_{f\dot{w}} = \infty$ $Z_{m\dot{\theta}} = 0$
Fixed		$w = 0$ $\frac{\partial w}{\partial x} = 0$	$Z_{f\dot{w}} = \infty$ $Z_{m\dot{\theta}} = \infty$
Free	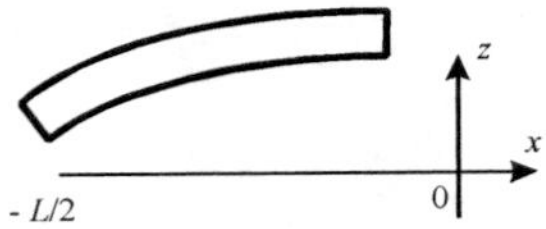	$\frac{\partial^2 w}{\partial x^2} = 0$ $\frac{\partial w^3}{\partial x^3} = 0$	$Z_{f\dot{w}} = 0$ $Z_{m\dot{\theta}} = 0$
Deflected spring		$\frac{\partial^2 w}{\partial x^2} = 0$ $EI_{yy}\frac{\partial^3 w}{\partial x^3} = -K_D w$	$Z_{f\dot{w}} = \mathrm{j}\frac{K_D}{\omega}$ $Z_{m\dot{\theta}} = 0$
Torsion spring		$w = 0$ $EI_{yy}\frac{\partial^2 w}{\partial x^2} = K_T\frac{\partial w}{\partial x}$	$Z_{f\dot{w}} = \infty$ $Z_{m\dot{\theta}} = -\mathrm{j}\frac{K_T}{\omega}$
Mass		$\frac{\partial^2 w}{\partial x^2} = 0$ $EI_{yy}\frac{\partial^3 w}{\partial x^3} = -m\frac{\partial^2 w}{\partial t^2}$	$Z_{f\dot{w}} = -\mathrm{j}\omega m$ $Z_{m\dot{\theta}} = 0$
Dashpot	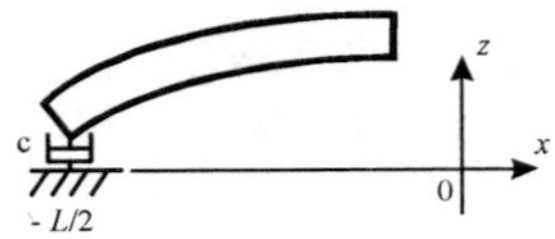	$\frac{\partial^2 w}{\partial x^2} = 0$ $EI_{yy}\frac{\partial^3 w}{\partial x^3} = -c\frac{\partial w}{\partial t}$	$Z_{f\dot{w}} = -c$ $Z_{m\dot{\theta}} = 0$

$$w(x,t) = \left(B_1 e^{-jk_b x} + B_3 e^{-k_b x}\right) e^{j\omega t} \tag{10.2.5}$$

so that:

$$\dot{w}(x,t) = \left(j\omega B_1 e^{-jk_b x} + j\omega B_3 e^{-k_b x}\right) e^{j\omega t} \tag{10.2.6}$$

and

$$\dot{\theta}(x,t) = -\frac{\partial \dot{w}(x,t)}{\partial x} = \left(-\omega k_b B_1 e^{-jk_b x} + j\omega k_b B_3 e^{-k_b x}\right) e^{j\omega t} \tag{10.2.7a,b}$$

where B_1 and B_3 are arbitrary constants.

Equations (10.2.6) and (10.2.7) can be written in a matrix form as:

$$\begin{bmatrix} \dot{w}(x,t) \\ \dot{\theta}(x,t) \end{bmatrix} = \begin{bmatrix} j\omega & j\omega \\ -\omega k_b & j\omega k_b \end{bmatrix} \begin{bmatrix} B_1 e^{-jk_b x} e^{j\omega t} \\ B_3 e^{-k_b x} e^{j\omega t} \end{bmatrix} \tag{10.2.8}$$

which can be inverted to give:

$$\begin{bmatrix} B_1 e^{-jk_b x} e^{j\omega t} \\ B_3 e^{-k_b x} e^{j\omega t} \end{bmatrix} = \begin{bmatrix} \dfrac{(1-j)}{2\omega} & \dfrac{-(1-j)}{2\omega k_b} \\ -\dfrac{(1+j)}{2\omega} & \dfrac{(1-j)}{2\omega k_b} \end{bmatrix} \begin{bmatrix} \dot{w}(x,t) \\ \dot{\theta}(x,t) \end{bmatrix} \tag{10.2.9}$$

where the dot denotes differentiation with respect to time.

Equation (10.2.5) can be differentiated (with respect to x) two more times to give the bending moment M and the shear force Q, and the result can be written in matrix form as:

$$\begin{bmatrix} Q(x,t) \\ M(x,t) \end{bmatrix} = -EI_{yy} \begin{bmatrix} -jk_b^3 & k_b^3 \\ -k_b^2 & k_b^2 \end{bmatrix} \begin{bmatrix} B_1 e^{-jk_b x} e^{j\omega t} \\ B_3 e^{-k_b x} e^{j\omega t} \end{bmatrix} \tag{10.2.10}$$

where E is Young's modulus of elasticity and I_{yy} is the second moment of area of the beam cross-section about the y-axis.

The column vector on the right-hand side of Equation (10.2.10) can be replaced with the right-hand side of Equation (10.2.9). Thus,

$$\begin{bmatrix} Q(x,t) \\ M(x,t) \end{bmatrix} = \boldsymbol{Z_R} \begin{bmatrix} \dot{w}(x,t) \\ \dot{\theta}(x,t) \end{bmatrix} \tag{10.2.11}$$

where

$$\boldsymbol{Z_R} = \begin{bmatrix} (1+j)EI_{yy}k_b^3/\omega & -EI_{yy}k_b^2/\omega \\ EI_{yy}k_b^2/\omega & -(1-j)EI_{yy}k_b/\omega \end{bmatrix} \tag{10.2.12}$$

so that the boundary bending moment and shear force are expressed as the product of the impedance matrix and a column vector containing the velocity and angular velocity of the beam. The quantity $\boldsymbol{Z_R}$ is known as the wave impedance matrix. By a similar process of considering a flexural wave travelling from a source toward the left in an infinite beam, the left wave impedance matrix is:

$$\boldsymbol{Z_L} = \begin{bmatrix} -(1+\mathrm{j})EI_{yy}k_b^3/\omega & -EI_{yy}k_b^2/\omega \\ EI_{yy}k_b^2/\omega & (1-\mathrm{j})EI_{yy}k_b/\omega \end{bmatrix} \tag{10.2.13}$$

Note that $\boldsymbol{Z_R} \neq \boldsymbol{Z_L}$ and $\boldsymbol{Z_R} \neq -\boldsymbol{Z_L}$. By using the wave impedance matrix $\boldsymbol{Z_R}$ (or $\boldsymbol{Z_L}$) as the right (or left) boundary impedance matrix in the finite beam model, the boundary is effectively removed and the beam becomes semi-infinite. An infinite beam can be modelled by using both the left and right wave impedance matrices as boundary impedances at opposite ends of a finite length beam. This use of wave impedance matrices to model the termination of a finite length beam produces numerical results identical to those obtained by evaluating the expression derived from the analysis of an infinite beam.

10.2.2 Response to a Point Force

The purpose of this section is to determine the response to a simple harmonic point force excitation applied at $x = x_0$ of a finite beam with left and right boundary conditions specified as impedance matrices $\boldsymbol{Z_L}$ and $\boldsymbol{Z_R}$ as shown in Figure 10.6(a). The applied point force produces a discontinuity in the shear force function $Q(x)$ and so it is necessary to calculate two sets of constant coefficients A_1, A_2, A_3 and A_4 on the left, and B_1, B_2, B_3 and B_4 on the right of the applied force, so that:

$$w_1(x) = A_1 \mathrm{e}^{-\mathrm{j}k_b x} + A_2 \mathrm{e}^{\mathrm{j}k_b x} + A_3 \mathrm{e}^{-k_b x} + A_4 \mathrm{e}^{k_b x} \tag{10.2.14}$$

and

$$w_2(x) = B_1 \mathrm{e}^{-\mathrm{j}k_b x} + B_2 \mathrm{e}^{\mathrm{j}k_b x} + B_3 \mathrm{e}^{-k_b x} + B_4 \mathrm{e}^{k_b x} \tag{10.2.15}$$

At the common boundary $x = x_0$:

$$w_1(x_0) = w_2(x_0) \tag{10.2.16}$$

$$w_1'(x_0) = w_2'(x_0) \tag{10.2.17}$$

$$w_1''(x_0) = w_2''(x_0) \tag{10.2.18}$$

$$w_1'''(x_0) = w_2'''(x_0) - \frac{F_e}{EI_{yy}} \tag{10.2.19}$$

where the prime denotes differentiation with respect to x.

From Equation (10.2.3) the left-hand boundary condition of the beam at $x = x_L$ can be written as:

$$\begin{bmatrix} Q(x_L,t) \\ M(x_L,t) \end{bmatrix} = \begin{bmatrix} Z_{Lf\dot{w}} & Z_{Lf\dot{\theta}} \\ Z_{Lm\dot{w}} & Z_{Lm\dot{\theta}} \end{bmatrix} \begin{bmatrix} \dot{w}_1(x_L,t) \\ \dot{\theta}_l(x_L,t) \end{bmatrix} \tag{10.2.20}$$

By using Equations (2.3.35) and (2.3.36) to replace the bending moment and shear force with a derivative of the displacement function, the following is obtained:

$$\begin{bmatrix} Z_{Lf\dot{w}} & Z_{Lf\dot{\theta}} \\ Z_{Lm\dot{w}} & Z_{Lm\dot{\theta}} \end{bmatrix} \begin{bmatrix} \dot{w}_1(x_L,t) \\ \dot{\theta}_l(x_L,t) \end{bmatrix} + EI_{yy} \begin{bmatrix} -w_1'''(x_L,t) \\ w_1''(x_L,t) \end{bmatrix} = 0 \tag{10.2.21}$$

Similarly, for the right-hand boundary of the beam at $x = x_r$:

$$\begin{bmatrix} Z_{Rf\dot{w}} & Z_{Rf\dot{\theta}} \\ Z_{Rm\dot{w}} & Z_{Rm\dot{\theta}} \end{bmatrix} \begin{bmatrix} \dot{w}_2(x_R,t) \\ \dot{\theta}_2(x_R,t) \end{bmatrix} + EI_{yy} \begin{bmatrix} -w_2'''(x_R,t) \\ w_2''(x_R,t) \end{bmatrix} = 0 \tag{10.2.22}$$

where

$$\dot{\theta}_1(x,t) = -\frac{\partial \dot{w}_1(x,t)}{\partial x} \; ; \qquad \dot{\theta}_2(x,t) = -\frac{\partial \dot{w}_2(x,t)}{\partial x} \tag{10.2.23a,b}$$

Equations (10.2.14) and (10.2.15) are then differentiated to produce expressions for w_1, $w_2, \dot{\theta}_1$, $\dot{\theta}_2$, w_1', w_1'', w_1''', w_2', w_2'' and w_2''', which contain the unknown coefficients A_1, A_2, A_3, A_4, B_1, B_2, B_3 and B_4. These expressions can be substituted into Equations (10.2.16), (10.2.17), (10.2.18), (10.2.19), (10.2.21) and (10.2.22), and the equations can be combined into a single system of linear equations to obtain:

$$\boldsymbol{a X} = \boldsymbol{F} \tag{10.2.24}$$

where

$$\boldsymbol{X} = [A_4, A_3, A_2, A_1, B_4, B_3, B_2, B_1]^{\mathrm{T}} \tag{10.2.25}$$

$$\boldsymbol{F} = \left[0, 0, 0, 0, 0, 0, 0, \frac{F_e}{EI_{yy} k_b^3}\right]^{\mathrm{T}} \tag{10.2.26}$$

and

$$
\boldsymbol{\alpha} = \left[\begin{array}{cccc}
\mathrm{j}\omega\beta_L(Z_{Lm\dot{w}} - k_b Z_{Lm\dot{\theta}} + H_2) & \frac{\mathrm{j}\omega(Z_{Lm\dot{w}} + k_b Z_{Lm\dot{\theta}} + H_2)}{\beta_L} & \mathrm{j}\omega\beta_L^{\mathrm{j}}(Z_{Lm\dot{w}} - \mathrm{j}k_b Z_{Lm\dot{\theta}} - H_2) & \frac{\mathrm{j}\omega(Z_{Lm\dot{w}} + \mathrm{j}k_b Z_{Lm\dot{\theta}} - H_2)}{\beta_L^{\mathrm{j}}} \\
\mathrm{j}\omega\beta_L(Z_{Lf\dot{w}} - k_b Z_{Lf\dot{\theta}} - H_1) & \frac{\mathrm{j}\omega(Z_{Lf\dot{w}} + k_b Z_{Lf\dot{\theta}} + H_1)}{\beta_L} & \mathrm{j}\omega\beta_L^{\mathrm{j}}(Z_{Lf\dot{w}} - \mathrm{j}k_b Z_{Lf\dot{\theta}} + \mathrm{j}H_1) & \frac{\mathrm{j}\omega(Z_{Lf\dot{w}} + \mathrm{j}k_b Z_{Lf\dot{\theta}} - \mathrm{j}H_1)}{\beta_L^{\mathrm{j}}} \\
0 & 0 & 0 & 0 \\
0 & 0 & 0 & 0 \\
\beta_0 & 1/\beta_0 & \beta_0^{\mathrm{j}} & 1/\beta_0^{\mathrm{j}} \\
\beta_0 & -1/\beta_0 & \mathrm{j}\beta_0^{\mathrm{j}} & -j/\beta_0^{\mathrm{j}} \\
\beta_0 & 1/\beta_0 & -\beta_0^{\mathrm{j}} & -1/\beta_0^{\mathrm{j}} \\
-\beta_0 & 1/\beta_0 & \mathrm{j}\beta_0^{\mathrm{j}} & -\mathrm{j}/\beta_0^{\mathrm{j}}
\end{array}\right.
$$

$$
\begin{array}{cccc}
0 & 0 & 0 & \\
0 & 0 & 0 & \\
\mathrm{j}\omega\beta_R(Z_{Rm\dot{w}} - k_b Z_{Rm\dot{\theta}} + H_2) & \frac{\mathrm{j}\omega(Z_{Rm\dot{w}} + k_b Z_{Rm\dot{\theta}} + H_2)}{\beta_R} & \mathrm{j}\omega\beta_R^{\mathrm{j}}(Z_{Rm\dot{w}} - \mathrm{j}k_b Z_{Rm\dot{\theta}} - H_2) & \mathrm{j}\omega(Z \\
\mathrm{j}\omega\beta_R(Z_{Rf\dot{w}} - k_b Z_{Rf\dot{\theta}} - H_1) & \frac{\mathrm{j}\omega(Z_{Rf\dot{w}} + k_b Z_{Rf\dot{\theta}} + H_1)}{\beta_R} & \mathrm{j}\omega\beta_R^{\mathrm{j}}(Z_{Rf\dot{w}} - \mathrm{j}k_b Z_{Rf\dot{\theta}} + \mathrm{j}H_1) & \mathrm{j}\omega(Z \\
-\beta_0 & -1/\beta_0 & -\beta_0^{\mathrm{j}} & \\
-\beta_0 & 1/\beta_0 & -\mathrm{j}\beta_0^{\mathrm{j}} & \\
-\beta_0 & -1/\beta_0 & \beta_0^{\mathrm{j}} & \\
\beta_0 & -1/\beta_0 & -\mathrm{j}\beta_0^{\mathrm{j}} &
\end{array}
\tag{10.2.27}
$$

where $\beta_L = \mathrm{e}^{k_b x_L}$, $\beta_0 = \mathrm{e}^{k_b x_0}$, $\beta_R = \mathrm{e}^{k_b x_R}$, $\beta_L^{\mathrm{j}} = \mathrm{e}^{\mathrm{j}k_b x_L}$, $\beta_0^{\mathrm{j}} = \mathrm{e}^{\mathrm{j}k_b x_0}$, $\beta_R^{\mathrm{j}} = \mathrm{e}^{\mathrm{j}k_b x_R}$, $H_1 = EI_{yy}k_b^3/(\mathrm{j}\omega)$ and $H_1 = EI_{yy}k_b^2/(\mathrm{j}\omega)$.

The solution vector $\boldsymbol{X}$ characterises the response of a finite length beam to a point force simple harmonic excitation.

10.2.3 Response to a Concentrated Line Moment

Here, the response to a simple harmonic concentrated line moment across the beam applied at $x = x_0$, is considered, as shown in Figure 10.6(b), with left and right boundary conditions specified as impedance matrices $\boldsymbol{Z_L}$ and $\boldsymbol{Z_R}$. Following the sign conventions shown in Figure

10.6(c) (which are consistent with those used in Section 2.3), the equation of motion for the flexural vibration of the beam shown in Figure 10.6(b) is:

$$EI_{yy}\frac{\partial^4 w}{\partial x^4} + \rho S\frac{\partial^2 w}{\partial t^2} = \frac{dM_e}{dx}\,\delta(x - x_0)\,\mathrm{e}^{j\omega t} \tag{10.2.28}$$

where M_e is the amplitude of a harmonic point moment applied at $x = x_0$, and $\delta(x - x_0)$ is the Dirac delta function. The equations describing the displacement on the left and right side of the applied moment are identical to Equations (10.2.14) and (10.2.15) for a point force excitation.

The first two boundary conditions at the common boundary between the left and right sides of the beam are the same as Equations (10.2.16) and (10.2.17) for a point force. The other two equations are slightly different and are:

$$w_1''(x_0,t) = w_2''(x_0,t) - \frac{M_e}{EI_{yy}} \tag{10.2.29}$$

$$w_1'''(x_0,t) = w_2'''(x_0,t) \tag{10.2.30}$$

The solution procedure is the same as for a point force excitation; in fact, Equations (10.2.20) to (10.2.27) for a point force also describe the solution for an applied moment except that the force vector of Equation (10.2.26) is replaced by a moment vector defined as:

$$\boldsymbol{M} = \left[0,\,0,\,0,\,0,\,0,\,0,\,-\frac{M_e}{EI_{yy}k_b^2},\,0\right]^{\mathrm{T}} \tag{10.2.31}$$

10.2.4 Active Vibration Control with a Point Force

In the following analysis, a primary point force acts at $x = x_p$, a control point force acts at $x = x_c$ and an error sensor is located at $x = x_e$ as shown in Figure 10.7. The cost function to be minimised is the beam flexural displacement amplitude squared at the location of the error sensor. The total beam response may be considered as the sum of the responses due to the primary and control forces, each of which may be calculated separately.

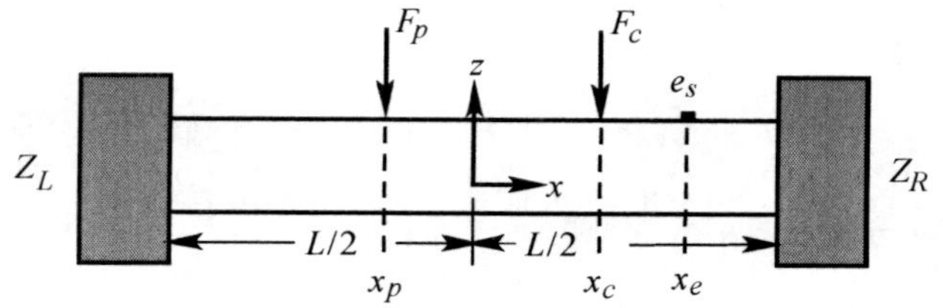

Figure 10.7 Control force and error sensor configuration for point force control of a beam with arbitrary impedance terminations.

The boundary condition equation for the primary force is:

$$\boldsymbol{\alpha}_p \boldsymbol{X}_p = \boldsymbol{F}_p \tag{10.2.32}$$

or

$$\boldsymbol{X}_p = \boldsymbol{\alpha}_p^{-1} \boldsymbol{F}_p \tag{10.2.33}$$

where the primary force vector $\boldsymbol{F}_p = [0, 0, 0, 0, 0, 0, 0, F_p/(EI_{yy}k_b^3)]^{\mathrm{T}}$, $\boldsymbol{X}_p$ is the boundary eigenvector for the primary force and $\boldsymbol{\alpha}_p$ is the matrix of boundary condition coefficients for the primary force. The boundary condition equation for the control force $\boldsymbol{F}_c$ is:

$$\boldsymbol{\alpha}_c \boldsymbol{X}_c = \boldsymbol{F}_c \tag{10.2.34}$$

or

$$\boldsymbol{X}_c = \boldsymbol{\alpha}_c^{-1} \boldsymbol{F}_c \tag{10.2.35}$$

where the control force vector $\boldsymbol{F}_c = [0, 0, 0, 0, 0, 0, 0, F_c/(EI_{yy}k_b^3)]^{\mathrm{T}}$, $\boldsymbol{X}_c$ is the boundary eigenvector for the control force and $\boldsymbol{\alpha}_c$ is the matrix of boundary condition coefficients for the control force. In the following analysis the dependence of all variables on x and t will be implicit and the dependence will not be shown by following the variable with (x, t).

At the error sensor location x_e, the displacement due to the primary force is:

$$w_p = \boldsymbol{X}_p^{\mathrm{T}} \boldsymbol{E} \tag{10.2.36}$$

and the displacement due to the control force is:

$$w_c = \boldsymbol{X}_c^{\mathrm{T}} \boldsymbol{E} \tag{10.2.37}$$

where

$$\boldsymbol{E} = \begin{bmatrix} 0 & 0 & 0 & 0 & \mathrm{e}^{k_b x_e} & \mathrm{e}^{-k_b x_e} & \mathrm{e}^{\mathrm{j}k_b x_e} & \mathrm{e}^{-\mathrm{j}k_b x_e} \end{bmatrix}^{\mathrm{T}} \tag{10.2.38}$$

By adding Equations (10.2.36) and (10.2.37), the total displacement can be written as:

$$w = w_p + w_c = \boldsymbol{X}_p^{\mathrm{T}} \boldsymbol{E} + \boldsymbol{X}_c^{\mathrm{T}} \boldsymbol{E} \tag{10.2.39a,b}$$

and by substituting Equations (10.2.33) and (10.2.35) into Equation (10.2.39):

$$\begin{aligned} w &= [\boldsymbol{\alpha}_p^{-1} \boldsymbol{F}_p]^{\mathrm{T}} \boldsymbol{E} + [\boldsymbol{\alpha}_c^{-1} \boldsymbol{F}_c]^{\mathrm{T}} \boldsymbol{E} \\ &= \frac{F_p}{EI_{yy}k_b^3} [\boldsymbol{\alpha}_p^{-1}]_{i,8}^{\mathrm{T}} \boldsymbol{E} + \frac{F_c}{EI_{yy}k_b^3} [\boldsymbol{\alpha}_c^{-1}]_{i,8}^{\mathrm{T}} \boldsymbol{E} \end{aligned} \tag{10.2.40a,b}$$

The optimal control force $\boldsymbol{F}_c$ for the primary force $\boldsymbol{F}_p$ may be found by letting $w = 0$ in Equation (10.2.40). By writing the transpose of the eighth column of the inverse of $\boldsymbol{\alpha}_p$ as $\boldsymbol{P} = (\boldsymbol{\alpha}_p^{-1})^{\mathrm{T}}_{i,8}$ and the transpose of the eighth column of the inverse of $\boldsymbol{\alpha}_c$ as $\boldsymbol{C} = (\boldsymbol{\alpha}_c^{-1})^{\mathrm{T}}_{i,8}$, the optimal control force can be written as:

$$F_c = -\frac{\boldsymbol{PE}}{\boldsymbol{CE}} F_p \tag{10.2.41}$$

for an error sensor in either the near-field or the far-field of the control force. If the wave impedances from Equations (10.2.12) and (10.2.13) are substituted into Equation (10.2.22), the numerical result is the same as that obtained by using the expression derived in Section 10.1 for an infinite beam, which is:

$$F_c = -\frac{\beta_p}{\beta_c} F_p \tag{10.2.42}$$

Equation (10.2.41) is clearly more complicated than Equation (10.2.42) because it takes into account not only the boundary conditions (boundary impedances), but also the relative locations of the primary and control sources, and error sensor in relation to the ends of the beam.

10.2.4.1 Effect of Boundary Impedance

Acceleration distributions along the beam for both optimally controlled and uncontrolled cases corresponding to a harmonic primary forcing frequency of 1000 Hz (which does not correspond to a beam resonance) applied at $x_p/\lambda_b = 0$, a point control force at $x_c/\lambda_b = 2.24$ and an error sensor at $x_e/\lambda_b = 4.48$, (where $\lambda_b = 0.48$ m is the wavelength of the flexural wave in the beam), for a range of boundary impedances are shown in Figure 10.8.

For the dB ordinate scale in Figure 10.8 and all the following figures, the reference level is the uncontrolled infinite beam acceleration produced by the primary source in the far-field. By optimally controlled, it is meant that the single control source has been driven in such a way as to obtain the maximum achievable vibration reduction at the error sensor. Because the error sensor is not near the point of application of a force or a discontinuity, near-field effects on the beam response at the error sensor are expected to be numerically negligible.

For convenience, the analysis begins with a number of simple boundary conditions. Figure 10.8(a) shows the forced response of a free-free beam, Figure 10.8(b) shows the forced response of a pinned–pinned beam and Figure 10.8(c) shows the forced response of a fixed-fixed beam. From these figures, it can be seen that the free-free beam has the same acceleration distribution and potential for vibration reduction as the fixed-fixed beam except within about half a wavelength of the ends of the beam which are at $(x - x_p)/\lambda_b = \pm 10.32$. The pinned-pinned beam has a different acceleration distribution from that of the free-free and fixed-fixed beams; in particular, between each end and the adjacent source, the vibration level is about 15 dB lower. Also, the vibration level of the controlled beam between the primary and control sources is about 5 dB lower than on the left of the primary source. In practice, the extent of the reductions shown in the figures will not be realised. However, it is useful to present them to gain insight into the effect of the termination impedances on the beam controllability.

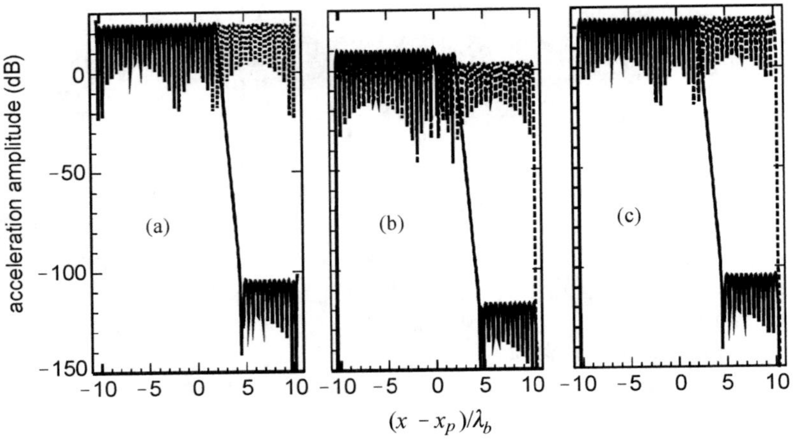

Figure 10.8 Controlled and uncontrolled distribution of beam acceleration amplitude as a function of boundary impedances. Acceleration level minimised at the error sensor $(x_e - x_p)/\lambda_b = 4.48$, with $x_p/\lambda_b = 0$ and $(x_c - x_p)/\lambda_b = 2.24$. The beam extends from $(x - x_p)/\lambda_b = -10.0$ to $(x - x_p)/\lambda_b = 10.0$. The forcing frequency is 1000 Hz.
-------- controlled,
------ uncontrolled.
Three standard boundary conditions are: (a) $Z_{f\omega} = 0$ and $Z_{m\dot{\theta}} = 0$ (free-free); (b) $Z_{f\omega} = \infty$ and $Z_{m\dot{\theta}} = 0$ (pinned-pinned); (c) $Z_{f\omega} = \infty$ and $Z_{m\dot{\theta}} = \infty$ (fixed-fixed).

10.2.4.2 Effect of Control Force Location

Figures 10.9(a), 10.9(c) and 10.9(e) show the maximum achievable mean attenuation of acceleration level in the far-field, upstream of the primary force, for free-free, pinned-pinned and infinite-infinite beams respectively. Control of the fixed-fixed beam produces similar attenuation as the free-free beam, so the results for the fixed-fixed beam are not shown in this figure. In these examples, the mean attenuation of acceleration is shown as a function of the distance $(x_c - x_p)$ between the primary and control forces. The forcing frequency is 1000 Hz (which is non-resonant) and the error sensor is two wavelengths downstream of the control force such that $(x_e - x_c)/\lambda_b = 2.0$. From these figures, it can be seen that the acceleration upstream of the primary force is maximally reduced if the separation between the primary and control forces is an integer multiple of half of a wavelength. The attenuation minima for the free beam occur at odd multiples of a quarter wavelength separation between the control and primary forces. The pinned beam has a different acceleration distribution to that of the free-free beam; the vibration level is about 20 dB lower and its minima are not located at control and primary separations of multiples of a quarter wavelength. For the infinite beam, the peaks in attenuation also occur at half wavelength intervals and are at about the same level as those for the pinned-pinned beam.

Figures 10.9(b), 10.9(d) and 10.9(f) show the corresponding maximum achievable mean attenuation downstream of the error sensor, assuming an ideal feedforward controller. For the free beam (Figure 10.9(b)), the maxima are at integer multiples of a half wavelength and the minima are at odd integer multiples of a quarter wavelength. The attenuation for the pinned beam (Figure 10.9(d)) is greatest at integer multiples of a quarter of a wavelength and is minimal at odd integer multiples of one-eighth of a wavelength. The infinite beam (Figure 10.9(f)) produces a constant maximum achievable mean attenuation of about 110 dB for the

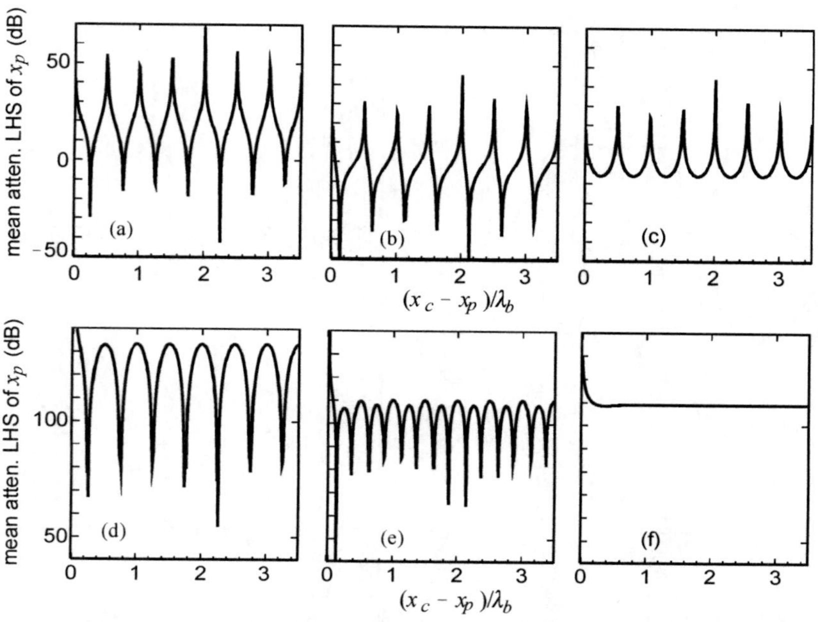

Figure 10.9 Effect of control force location x_c/λ_b on the mean attenuation of acceleration. The location of the primary force and the excitation frequency are the same as in Figure 10.9, with fixed $(x_e - x_c)/\lambda_b = 2.0$: (a) Mean attenuation of acceleration upstream of the primary force (free-free); (b) mean attenuation of acceleration upstream of the primary force (pinned-pinned); (c) mean attenuation of acceleration upstream of the primary force (infinite-infinite); (d) mean attenuation of acceleration downstream of the error sensor (free-free); (e) mean attenuation of acceleration downstream of the error sensor (pinned-pinned); (f) mean attenuation of acceleration downstream of the error sensor (infinite-infinite).

particular location which was used for the error sensor. The maximum attenuation would approach $-\infty$ as the separation between the control force and error sensor approached ∞. For the case of the infinite beam, the independence of the separation between the primary and control forces ensures that good control is possible. In contrast, the extreme sensitivity of the finite beams to control force location indicates that it would be far more difficult to achieve satisfactory control if the primary force location or excitation frequency are not fixed. This implies that good control of a broadband signal, using an ideal feedforward controller, would only occur over narrowbands separated by very narrow bands where control is poor, unless multiple control forces and error sensors are used.

10.2.4.3 Effect of Error Sensor Location

Figure 10.10(a) shows the maximum achievable mean attenuation of acceleration level in the far-field, upstream of the primary force, as a function of both control force and error sensor locations. Each curve in the figure indicates the attenuation for a fixed but different value of control force location. The forcing frequency is 1000 Hz and the ends of the beam are free. This figure shows that the attenuation depends on the separation $(x_c - x_p)$, but not on the error sensor location, provided that the error sensor is in the far-field of the control and primary sources. However, the nature of the dependence on $(x_c - x_p)$ is not clear.

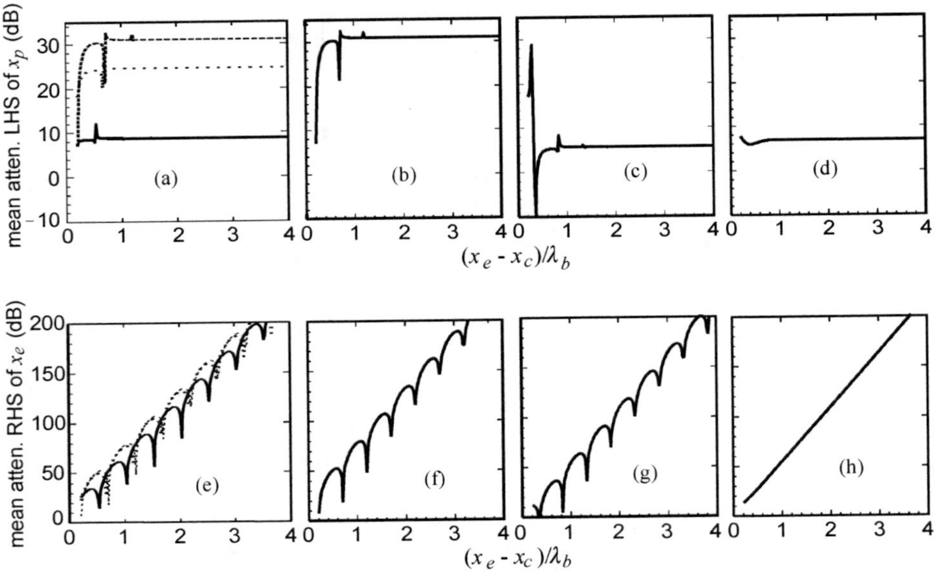

Figure 10.10 Effect of error sensor location $(x_e - x_c)/\lambda_b$ on the mean attenuation of acceleration for various separations between the primary and control forces. The location of primary force and the excitation frequency are the same as in Figure 10.9: (a) Mean attenuation of acceleration upstream of the primary force with varying $(x_c - x_p)/\lambda_b$ (free-free). (b) mean attenuation of acceleration upstream of the primary force with $(x_c - x_p)/\lambda_b = 1.0$ (free-free); (c) mean attenuation of acceleration upstream of the primary force with $(x_c - x_p)/\lambda_b = 1.0$ (pinned-pinned); (d) mean attenuation of acceleration upstream of the primary force with $(x_c - x_p)/\lambda_b = 1.0$ (infinite-infinite); (e) mean attenuation of acceleration downstream of the error sensor with varying $(x_c - x_p)/\lambda_b$ (free-free); (f) mean attenuation of acceleration downstream of the error sensor with $(x_c - x_p)/\lambda_b = 1.0$ (free-free); (g) mean attenuation of acceleration downstream of the error sensor with $(x_c - x_p)/\lambda_b = 1.0$ (pinned-pinned); (h) mean attenuation of acceleration downstream of the error sensor with $(x_c - x_p)/\lambda_b = 1.0$ (infinite-infinite).

\- - - - - - - - - - $(x_c - x_p)/\lambda_b = 1.0$ (part (a) only);

.... $(x_c - x_p)/\lambda_b = 2.0$ (parts (a) and (e) only);

———————— $(x_c - x_p)/\lambda_b = 0.2$.

Figure 10.10(b) shows the corresponding maximum achievable mean attenuation downstream of the error sensor. The average reduction in dB is clearly a linearly increasing function of $(x_e - x_c)/\lambda_b$. For each curve in this figure it is possible to draw an 'upper bound' straight line, which is a tangent to every lobe of the curve. The vertical distance between these straight lines is the same as the distance between the horizontal parts of the corresponding curves in Figure 10.10(a).

Figures 10.10(c), 10.10(e) and 10.10(g) show the maximum achievable mean attenuation of acceleration level on the left-hand side of the primary force for free-free, pinned-pinned and infinite-infinite beams. The control force is located at $(x_c - x_p)/\lambda_b = 1.0$. These figures show that the mean attenuation is different for each of the different boundary conditions. As in Figure 10.10(a), the attenuation upstream of primary force in all three beams does not depend on the error sensor location if the error sensor is in the far-field of the control and primary forces. Figures 10.10(d), 10.10(f) and 10.10(h) show the corresponding mean attenuation downstream of the error sensor. For the free beam (Figure 10.10(d)), minima in the curve correspond to separations between the control force and error sensor which are odd integer multiples of one-quarter of a wavelength. As the error sensor is moved further downstream, local maxima in attenuation are encountered at intervals of half a wavelength separation between the control force and error sensor. In the case of the pinned beam (Figure 10.10(f)), although the distance between successive extrema is the same, their actual locations are different from those of the free beam. In the control force far-field of the infinite beam (Figure 10.10(h)), the attenuation is simply proportional to the distance between the error sensor and the control force.

10.2.4.4 Effect of Forcing Frequency

Figure 10.11(a) shows the maximum achievable mean attenuation of acceleration level in the far-field, upstream of the primary force, as a function of the forcing frequency. Figure 10.11(b) shows the corresponding maximum achievable mean attenuation downstream of the error sensor. In these examples, the boundary force and moment impedances are zero. In both cases, the acceleration is reduced maximally at frequencies for which the separation between the primary and control forces is an integer multiple of half of a wavelength. In Figure 10.11(b), the average attenuation clearly increases with increasing frequency. The performance is poor at low frequencies because the error sensor is in the near-field of the control force.

Figures 10.11(c) and 10.11(d) show the maximum achievable mean attenuation of vibration in an infinite beam as a function of frequency. There are six sharp peaks in Figure 10.11(c).

10.2.4.5 Summary of Control Results Using a Single Control Force

From the preceding results and those of Pan and Hansen (1993b), it can be concluded that although reflective boundaries reduce the maximum control achievable, it is still possible to achieve high levels of vibration reduction over a range of termination impedances and harmonic excitation frequencies with a single error sensor and a single control force. The extent of achievable control with a feedforward controller is strongly dependent upon excitation frequency, control force location and error sensor location.

10.2.5 Minimising Vibration Using a Piezoceramic Actuator and an Angle Stiffener

Figure 10.12(b) shows the resultant forces and moments applied to the beam by the angle stiffener and piezoceramic stack shown in Figure 10.12(a), with a primary force F_p at $x = x_p$. Control forces F_1 and F_2 act at $x = x_1$ and $x = x_2$ respectively, with the concentrated moment M_1 also acting at $x = x_1$. An error sensor is located at axial location $x = x_e$. The boundary condition equation for the primary (excitation) point force was shown in Section 10.2.2 to be:

$$\boldsymbol{\alpha}_p \boldsymbol{X}_p = \boldsymbol{F}_p \tag{10.2.43}$$

or

$$\boldsymbol{X}_p = \boldsymbol{\alpha}_p^{-1} \boldsymbol{F}_p \tag{10.2.44}$$

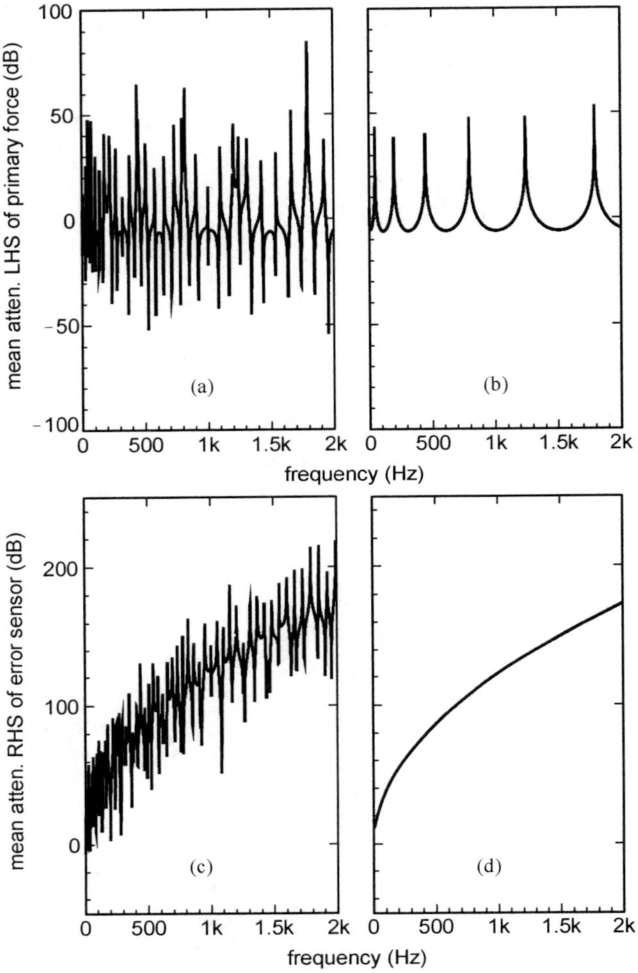

Figure 10.11 Mean attenuation of acceleration as a function of forcing frequency. The location of primary and control forces and location of the error sensor are the same as in Figure 10.9: (a) mean attenuation of acceleration upstream of the primary force (free-free); (b) mean attenuation of acceleration upstream of the primary force (infinite-infinite); (c) mean attenuation of acceleration downstream of the error sensor (free-free); (d) mean attenuation of acceleration downstream of the error sensor (infinite-infinite).

At each of these peaks, the distance between control and primary forces is an integer multiple of half of a wavelength. The number of peaks is limited to six because there are no reflections from boundaries (that is, no resonances). In Figure 10.11(d), the attenuation is proportional to the square root of frequency because the attenuation is proportional to $(x_e - x_c)/\lambda_b$ and $\lambda_b \propto f^{-1/2}$. More detailed results are discussed by Pan and Hansen (1993b).

where $\boldsymbol{F_p} = \left[0,0,0,0,0,0,0\ \frac{F_p}{k_b^3 EI_{yy}}\right]^{\mathrm{T}}$, α_p is the matrix of boundary condition coefficients for the primary force and $\boldsymbol{X}_p$ is the boundary eigenvector for the primary force. Similarly:

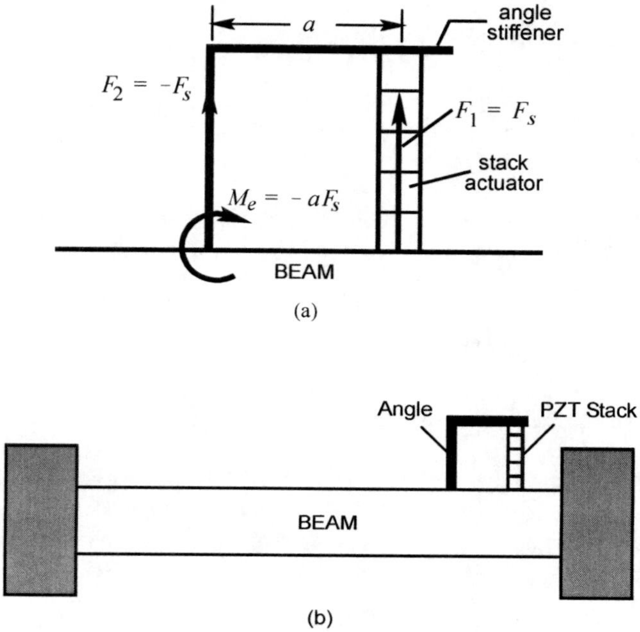

Figure 10.12 Model for analysing the response of a finite length beam to excitation with a piezoceramic stack: (a) stack and angle model; (b) beam model with stack actuator.

$$\boldsymbol{X}_1 = \boldsymbol{\alpha}_1^{-1} \boldsymbol{F}_1 \tag{10.2.45}$$

$$\boldsymbol{X}_2 = \boldsymbol{\alpha}_2^{-1} \boldsymbol{F}_2 \tag{10.2.46}$$

where $\boldsymbol{X}_1$ and $\boldsymbol{X}_2$ are the boundary eigenvectors for the two control forces. In addition:

$$\boldsymbol{X}_m = \boldsymbol{\alpha}_m^{-1} \boldsymbol{M} \tag{10.2.47}$$

where $\boldsymbol{X}_m$ is the boundary eigenvector for the control moment and $\boldsymbol{M}$ is defined by Equation (10.2.31). At the error sensor ($x = x_e$), the displacement due to each force and the moment is given by, for $z = p$, 1, 2 and m:

$$w_z = \boldsymbol{X}_z^{\mathrm{T}} \boldsymbol{E} \tag{10.2.48}$$

where $\boldsymbol{E}$ is defined by Equation (10.2.38).

Using superposition to sum the displacements defined in Equation (10.2.47), the total displacement is:

$$w = w_p + w_1 + w_2 + w_m = \boldsymbol{X}_p^{\mathrm{T}} \boldsymbol{E} + \boldsymbol{X}_1^{T} \boldsymbol{E} + \boldsymbol{X}_2^{\mathrm{T}} \boldsymbol{E} + \boldsymbol{X}_m^{\mathrm{T}} \boldsymbol{E} \tag{10.2.49a,b}$$

By substituting Equations (10.2.44) to (10.2.47) into Equation (10.2.49):

$$w = [\boldsymbol{\alpha}_p^{-1}\boldsymbol{F}_p]^{\mathrm{T}}\boldsymbol{E} + [\boldsymbol{\alpha}_1^{-1}\boldsymbol{F}_1]^{\mathrm{T}}\boldsymbol{E} + [\boldsymbol{\alpha}_2^{-1}\boldsymbol{F}_2]^{\mathrm{T}}\boldsymbol{E} + [\boldsymbol{\alpha}_m^{-1}\boldsymbol{M}]^{\mathrm{T}}\boldsymbol{E}$$

$$= \frac{F_p}{k_b^3 EI_{yy}}[\boldsymbol{\alpha}_p^{-1}]_{i,8}^{\mathrm{T}}\boldsymbol{E} + \frac{F_1}{k_b^3 EI_{yy}}[\boldsymbol{\alpha}_1^{-1}]_{i,8}^{\mathrm{T}}\boldsymbol{E} + \frac{F_2}{k_b^3 EI_{yy}}[\boldsymbol{\alpha}_2^{-1}]_{i,8}^{\mathrm{T}}\boldsymbol{E} + \frac{M_e}{k_b^2 EI_{yy}}[\boldsymbol{\alpha}_m^{-1}]_{i,7}^{\mathrm{T}}\boldsymbol{E} \tag{10.2.50a,b}$$

Setting $w = 0$ to find the optimal control force, defining the transpose of the eighth column of the inverse of $\boldsymbol{\alpha}_p$ as $P = [\boldsymbol{\alpha}_p^{-1}]_{i,8}^{\mathrm{T}}$ (and similarly for $\boldsymbol{A} = [\boldsymbol{\alpha}_1^{-1}]_{i,8}^{\mathrm{T}}$, and $\boldsymbol{B} = [\boldsymbol{\alpha}_2^{-1}]_{i,8}^{\mathrm{T}}$), and defining the transpose of the seventh column of the inverse of $\boldsymbol{\alpha}_m$ as $\boldsymbol{C} = [\boldsymbol{\alpha}_m^{-1}]_{i,7}^{\mathrm{T}}$, Equation (10.2.50) can be rewritten as:

$$\boldsymbol{AE}F_1 + \boldsymbol{BE}F_2 + k_b\boldsymbol{CE}M_e = -\boldsymbol{PE}F_p \tag{10.2.51}$$

Analysis of the forces applied by the stack and angle stiffener gives $F_1 = -F_2 = F_c$ and $M_e = -aF_c$, where F_c is the compressive force applied by the piezoceramic stack (see Figure 10.12(b)) and is positive as shown. The optimal control force F_c can be written as:

$$F_c = -\frac{\boldsymbol{PE}}{\boldsymbol{AE} - \boldsymbol{BE} - k_b a\boldsymbol{CE}}F_p \tag{10.2.52}$$

The discussion in the following sections examines the effect of varying the forcing frequency, control force location, error sensor location and stiffener flange length, a (see Figure 10.12) on the active control of flexural harmonic vibration in beams with two sets of end conditions, infinite and pinned. Although these two end conditions yield substantially different results, the trends observed for the pinned beam are similar for free or clamped ends or any end conditions resulting in a substantial reflected wave amplitude. End conditions which result in absorption of a substantial part of the incoming flexural wave would exhibit trends between those shown by the infinite and pinned beams.

In the following discussion, the control force amplitude is expressed as a fraction or multiple of the primary force amplitude, and the control force phase is relative to the primary force phase. For convenience, the primary excitation force is always located at $x = 0$. It is also assumed that in all cases the control force has been optimised to minimise the acceleration level at the error sensor. The acceleration reference level is the far-field acceleration level with just the primary excitation force acting.

10.2.5.1 Effect of Variations in Forcing Frequency, Stiffener Flange Length and Control Location on the Control Force

The discussion that follows examines the effect of varying forcing frequency, control force location (which is defined here as the location of the angle/beam joint), error sensor location, and stiffener flange length on the active control of vibration in beams with simply supported ends. For interest, results are compared to those obtained for beams of infinite length. The beam parameters (including location of the control force, primary force and error sensor) are listed in Table 10.2.

Table 10.2 Beam parameters used in the analysis.

Parameter	Value
Beam length, L_x	5.0 m
Beam width, L_y	0.05 m
Beam height, L_z	0.025 m
Young's modulus, E	71.1 GPa
Excitation force location, x_0	0.0 m
Frequency, f	1000 Hz
Wavelength, λ	0.4824 m

Control force amplitudes are expressed as multiples of the primary force amplitude, and the acceleration amplitude dB scale reference level is the far-field uncontrolled infinite-beam acceleration produced by the primary force acting alone. In all cases, the control force is assumed to be optimally adjusted to minimise the acceleration at the error sensor location.

Figure 10.13 shows the effect of varying the forcing frequency on the magnitude of the control force required to minimise the beam vibration at the error sensor location for beams with four different end conditions: simply supported, free, fixed and infinite. In each case, the two ends of the beam have the same end conditions. The control force is located 1 m from the primary force and the error sensor 2 m from the primary force, as indicated by Table 10.2. The minima on the curves for the simply supported beam occur at resonance frequencies, when control is easier. At these frequencies, the control force amplitude is small but non-zero. The difference in control effort required to control resonant and non-resonant response is associated with the variation from mode to mode in the phase of the optimal control force input required with the result that decreasing the response of one mode can increase the response of another. At resonance, only one mode dominates the response with the result that only a small force is required to achieve control. However, off-resonance the response is a result of the combined effect of more than one mode, generally having widely differing optimal control force phases.

The maxima in Figure 10.13 occur when the relative spacing between primary and control forces is given by $x = (c + n/2)\lambda$ for integer n and constant c. This is verified in Figure 10.14, which shows the control force magnitude as a function of separation between the primary and control forces, with a constant error sensor location−control force location separation of 1 m (2.07λ) and an excitation frequency of 1000 Hz.

The maxima occur because of the difficulty in controlling the flexural vibration when the controller is placed at a node of the standing wave caused by reflection from the terminations. The constant, c, represents the distance (in wavelengths) between the primary force and the first node in the standing wave in the direction of the control force. This constant changes with frequency, and for $f = 1000$ Hz it is approximately zero.

Interestingly, the phase of the optimal control force relative to the primary force is zero (or 180°) for the simply supported beam and varies between 0° and 180° as a function of frequency for the infinite beam. This is because the response of a lightly damped beam can be modelled as the sum of many modal responses for which the control force location is either 0 or 180° out of phase with the primary force location. This also makes control of a finite beam difficult with a single control force off-resonance, because decreasing the response of one mode requires the oppositely phased signal to that required by the two modes nearest in resonance frequency to the first. Thus, the resulting control only incrementally increases as the control force magnitude increases drastically. Conversely at resonance frequencies, where the beam response is dominated by a single resonant mode, control is much easier.

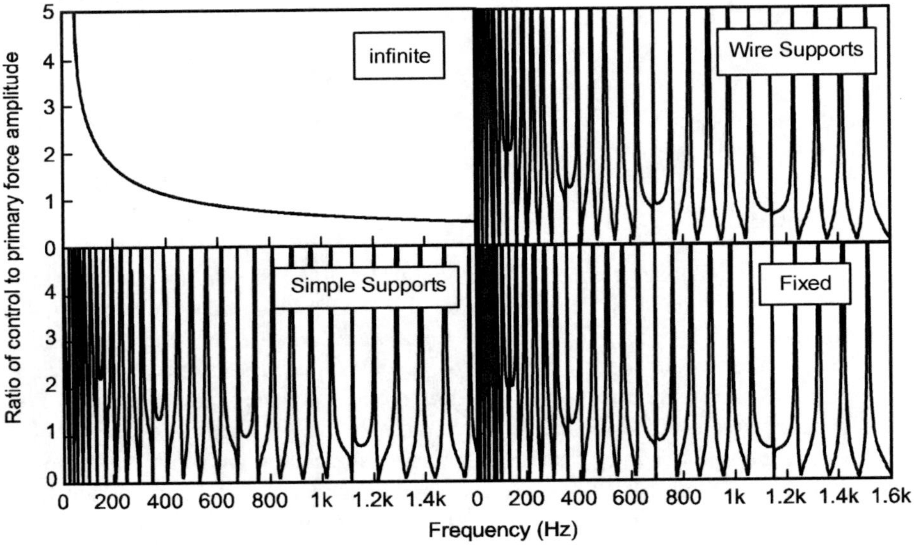

Figure 10.13 Control force amplitude for optimal feedforward control as a function of frequency for various beam end conditions: minima occur at beam resonances; maxima occur at $x_c - x_p = (c + n/2)\lambda$, where c is the frequency dependent distance between x_p and the first node in the standing wave.

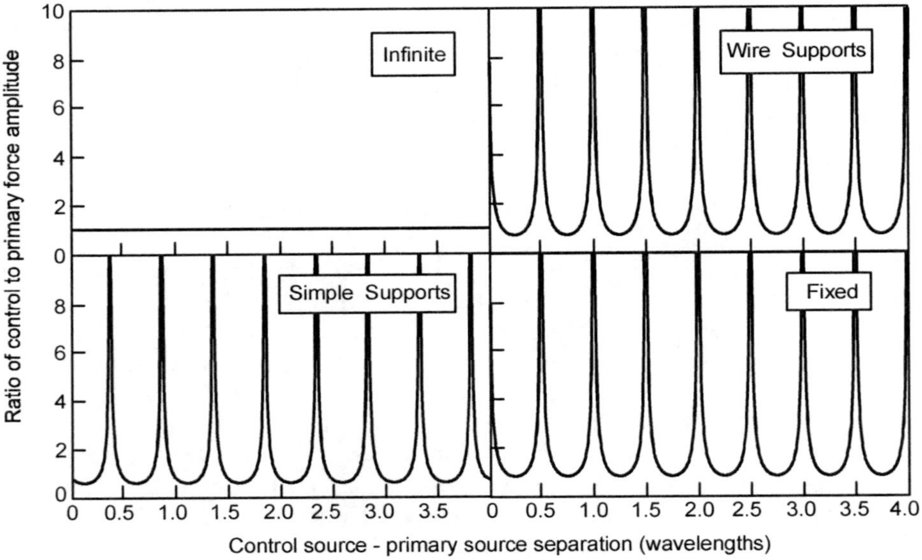

Figure 10.14 Control force amplitudes for optimal feedforward control as a function of the separation between the control and primary forces $(x_c - x_p)/\lambda_b$.

Figure 10.15 shows the control force magnitude plotted as a function of increasing stiffener flange length in wavelengths a/λ (see Figure 10.12(b) for a definition of the stiffener flange length). The exponentially shaped decrease in control force magnitude with increasing stiffener flange length can be attributed to the increasing size of the flange relative to the flexural wavelength. When the wavelength is large compared to the stiffener flange length, the two control forces operating in opposite directions tend to cancel. This effect can also be seen in Figure 10.13(a) (the infinite beam case) where the relative control force amplitude is plotted as a function of frequency, and to a lesser extent in Figure 10.13(b).

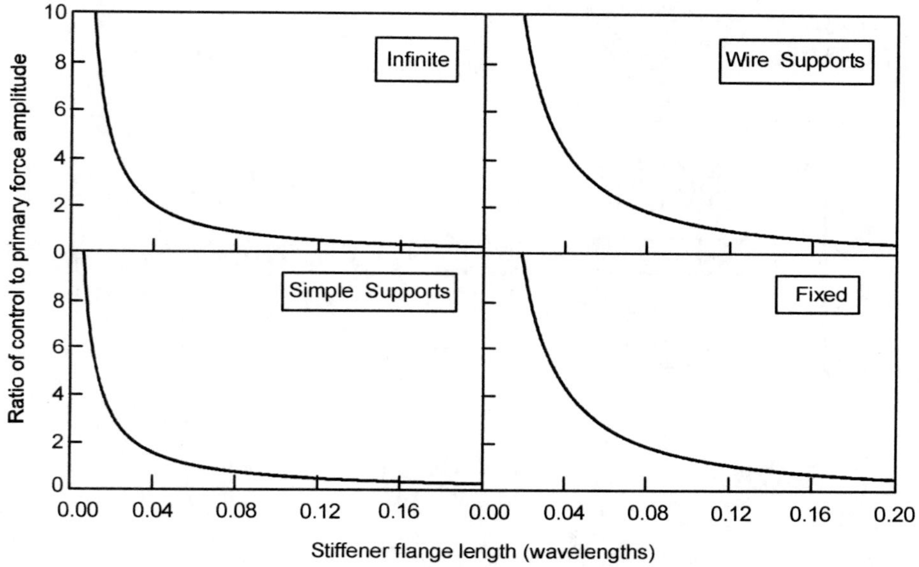

Figure 10.15 Control force amplitudes for optimal feedforward control as a function of the stiffener flange length.

10.2.5.2 Acceleration Distribution for Controlled and Uncontrolled Cases

Figure 10.16 shows the acceleration amplitude distribution for a controlled beam, excited at 1000 Hz using the control, primary and error sensor locations given in Table 10.2. The curves dip to a minimum at the error sensor location ($x = 4.14\lambda$) where acceleration has been minimised. In both cases, the theoretically achievable reduction in acceleration amplitude downstream of the error sensor, using an ideal feedforward controller, is over 100 dB. For the simply supported beam, the acceleration amplitude is also reduced upstream of the primary force. However, the amount of this reduction or increase in acceleration amplitude upstream of the primary force depends on the control force location, with the maximum attenuation upstream of the primary force being achieved when the control force-primary force separation is $(0.23 + n/2)\lambda$. This value is independent of the beam length and the excitation frequency, and noting also that the maxima occur for the infinite beam as well as the finite beams, it can be concluded that the maxima are a result of a modal beam response between the primary and control forces only. Note that the constant of 0.23 arises as the control force location has been defined at the point of attachment of the angle stiffener to the beam. However, the effective point of action of the control force is at a location between the point of attachment of the angle stiffener to the beam and the piezoceramic actuator stack, such that the constant would be 0.25.

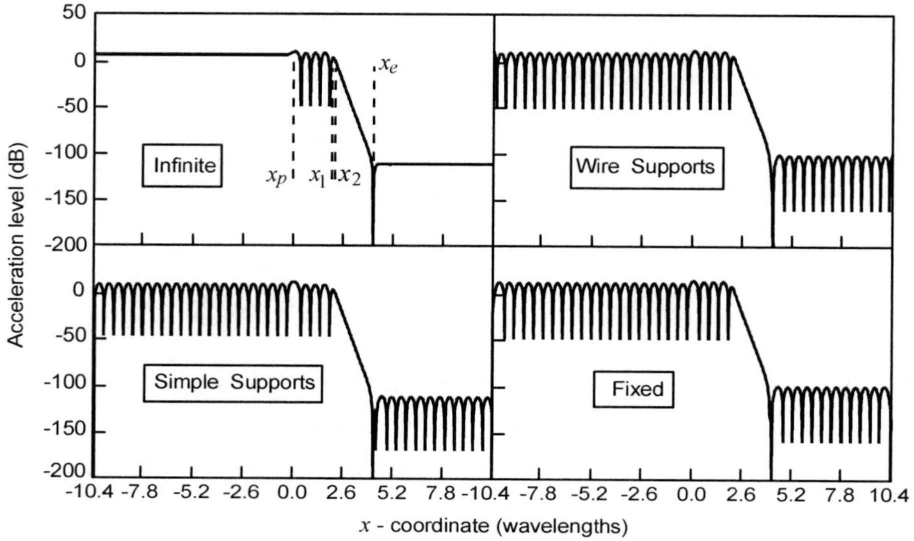

Figure 10.16 Acceleration distribution along the optimally controlled beam.

10.2.5.3 Effect of Control Location on Attenuation of Acceleration Level

Figure 10.17 shows the mean attenuation of acceleration level downstream of the error sensor as a function of separation between primary and control forces. The control force locations giving the best results upstream of the primary force also give high attenuation downstream of the error sensor. Every second minimum occurs at a location corresponding to a maximum in the control force magnitude (see Figure 10.14), and again these are separated by half a wavelength. The odd-numbered minima occur at control force–primary force separations of $(d + n/2)\lambda$, where d is a constant dependent on frequency.

10.2.5.4 Effect of Error Sensor Location on Attenuation of Acceleration Level

Figure 10.18 shows the mean attenuation downstream of the error sensor as a function of the separation between the control force and error sensor. It can be seen that downstream of the error sensor, mean attenuation increases with increasing separation between the error sensor and control location at the rate of around 50 dB per wavelength separation. The minima in the curves for the simply supported beam correspond to separations in the error sensor and control location of $(d + n/2)\lambda$; $n = 1, 2, ...,$ where d is the constant dependent on the frequency previously defined.

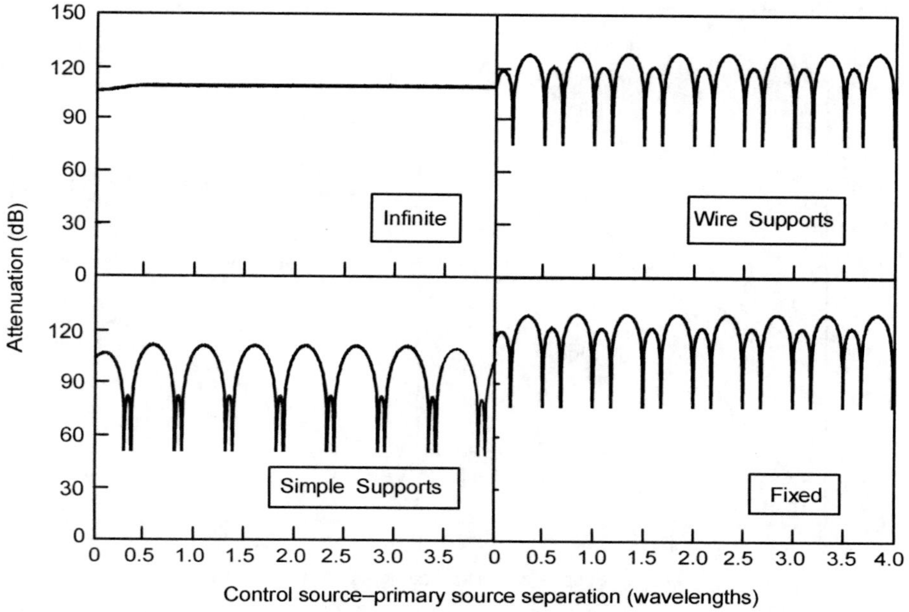

Figure 10.17 Mean, maximum achievable attenuation of acceleration level for an optimally controlled beam, downstream of the error sensor as a function of separation between the primary and control forces. Each even minimum occurs at a minimum in the control force amplitude (see Figure 10.14). Each odd minimum occurs at $(d + n/2)\lambda,\ n = 1, 2,\ldots$.

10.2.6 Determination of Beam End Impedances

In practice, rarely will beams be characterised by the ideal end impedances listed in Table 10.1. Thus, to be able to predict the effect of active control on a beam terminated with an arbitrary impedance, it is necessary to determine the end impedances by some sort of measurement. Fortunately, it is not necessary to determine the fully coupled beam end impedance matrix directly; rather, an equivalent impedance matrix containing only two uncoupled terms instead of four terms can be shown to give the same beam displacements and derivatives as the actual four-element impedance matrix, at all locations on the beam. In the following paragraphs it will be shown how this two-element impedance matrix characterising each end of the beam can be determined using data from four accelerometers mounted on the beam. It is not necessary for the two ends of the beam to be terminated in the same way.

Using Equation (10.2.4), beam end conditions may be characterised by impedance matrices Z_L and Z_R, corresponding respectively to the left and right ends of the beam as follows:

$$\boldsymbol{Z_L} = \begin{bmatrix} Z_{Lf\dot{w}} & Z_{Lf\dot{\theta}} \\ Z_{Lm\dot{w}} & Z_{Lm\dot{\theta}} \end{bmatrix} = \begin{bmatrix} \dfrac{1}{Z_{Lm}} & \dfrac{1}{Z_{Lfm}} \\ \dfrac{1}{Z_{Lmf}} & \dfrac{1}{Z_{Lf}} \end{bmatrix}^{-1} \tag{10.2.53a,b}$$

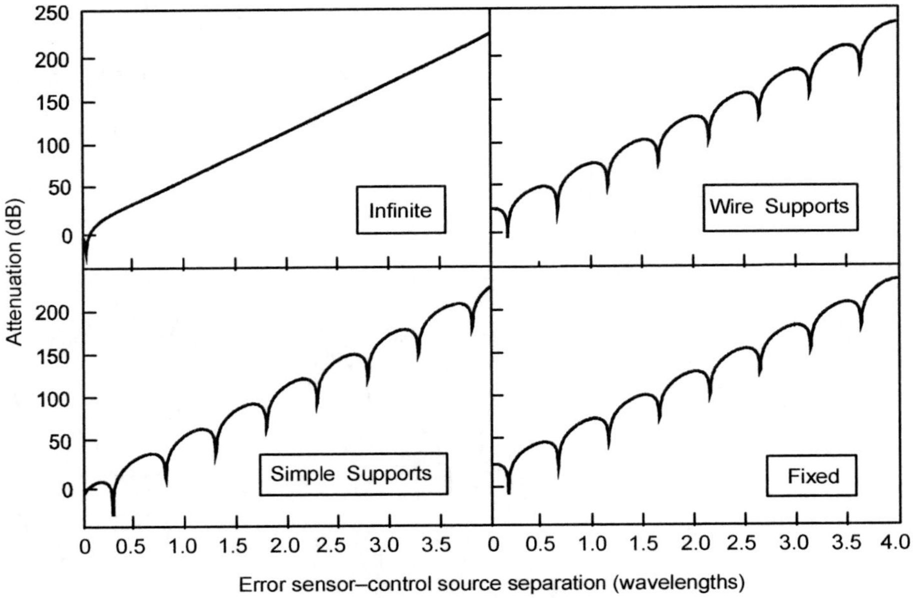

Figure 10.18 Mean, maximum achievable attenuation of acceleration level for an optimally controlled beam, downstream of the error sensor as a function of separation between the control force and error sensor. Minima occur at $(d + n/2)\lambda, n = 1, 2, ..., \infty$.

$$\boldsymbol{Z_R} = \begin{bmatrix} Z_{Rf\dot{w}} & Z_{Rf\dot{\theta}} \\ Z_{Rm\dot{w}} & Z_{Rm\dot{\theta}} \end{bmatrix} = \begin{bmatrix} \dfrac{1}{Z_{Rm}} & \dfrac{1}{Z_{Rfm}} \\ \dfrac{1}{Z_{Rmf}} & \dfrac{1}{Z_{Rf}} \end{bmatrix}^{-1} \tag{10.2.54a,b}$$

Equation (10.2.24) may be generalised for force or moment excitation by replacing the vector $\boldsymbol{F}$ by $\boldsymbol{B}$, where for moment excitation $\boldsymbol{B} = \boldsymbol{M}$ and for force excitation $\boldsymbol{B} = \boldsymbol{F}$, where $\boldsymbol{M}$ is defined by Equation (10.2.31) and $\boldsymbol{F}$ is defined by Equation (10.2.26). Substituting $\boldsymbol{B}$ for $\boldsymbol{F}$ in Equation (10.2.24) and inverting the result gives the following:

$$\boldsymbol{X} = \boldsymbol{\alpha}^{-1}\boldsymbol{B} \tag{10.2.55}$$

Because the vector $\boldsymbol{B}$ consists of a non-zero element in either the seventh row or eighth row only, the only columns of importance in the inverse matrix $\boldsymbol{\alpha}^{--1}$ are the seventh and eight columns, as all other columns will be multiplied by a zero element on the vector $\boldsymbol{B}$. The practical implication of this result is that only the larger elements of the first four rows of the $\boldsymbol{\alpha}$ matrix will affect the solution vector $\boldsymbol{X}$. It is proposed that the accuracy of the solution vector $\boldsymbol{X}$ can be maintained with the diagonal elements $Z_{m\dot{w}}$ and $Z_{f\dot{\theta}}$ of the impedance matrix set to zero. This simplification is justified by examples rather than by formal proof, as inverting the complex matrix $\boldsymbol{\alpha}$ symbolically is not practicable.

Once the impedance matrix has been approximated by the equivalent matrix with just two unknowns, determining the unknown equivalent impedance of a given beam termination from experimental data is possible. Beginning with the beam shown in Figure 10.6(a), the unknown termination at the left-hand end may be described by the equivalent impedance matrix:

$$\boldsymbol{Z_L} = \begin{bmatrix} Z_{L1} & 0 \\ 0 & Z_{L2} \end{bmatrix} \sim \begin{bmatrix} Z_{Lf\dot{w}} & Z_{Lf\dot{\theta}} \\ Z_{Lm\dot{w}} & Z_{Lm\dot{\theta}} \end{bmatrix} \tag{10.2.56a,b}$$

The right-hand termination may be such that the impedance values are known, or it may be the same unknown termination used on the left-hand end, in which case (following the sign conventions given in Figure 10.6(c)) the equivalent impedance matrix $\boldsymbol{Z_R}$ is given by:

$$\boldsymbol{Z_R} = \begin{bmatrix} Z_{R1} & 0 \\ 0 & Z_{R2} \end{bmatrix} = \begin{bmatrix} -Z_{L1} & 0 \\ 0 & -Z_{L2} \end{bmatrix} \tag{10.2.57a,b}$$

The method that follows will be the same regardless of whether a known or unknown termination is used at the right-hand end.

Setting the coupling terms in Equation (10.2.56(b)) equal to zero for the equivalent case, and using Equations (10.2.3) and (10.2.54):

$$Z_{L1} = \frac{Q}{\dot{w}} \tag{10.2.58}$$

and

$$Z_{L2} = \frac{M}{\dot{\theta}} \tag{10.2.59}$$

For harmonic signals, $\dot{w} = j\omega w$ and $\dot{\theta} = -j\omega w'$. Replacing the bending moment and shear force with derivatives of the displacement function (see Equations (2.3.35) and (2.3.36)) gives:

$$Z_{L1} = \frac{EI_{yy}\bar{w}_1'''(x_L)}{j\omega\bar{w}_1(x_L)} \tag{10.2.60}$$

$$Z_{L2} = \frac{EI_{yy}\bar{w}_1''(x_L)}{j\omega\bar{w}_1'(x_L)} \tag{10.2.61}$$

where the bar over a variable denotes amplitude. Note that the time dependence has been omitted as it is identical in the numerator and denominator and thus cancels out.

All that remains is to find the displacement and derivatives required in Equations (10.2.60) and (10.2.61). The accelerations, $\bar{a}_1, \bar{a}_2, \bar{a}_3, ..., \bar{a}_n$, are measured (both amplitude and phase relative to an arbitrary, fixed reference signal) at n locations, $x_1, x_2, x_3, ..., x_n$, such that $x_L < x_i < x_0$. For $i = 1$ to n:

$$\bar{a}_{ie}(x_i) = -\omega^2 \bar{w}_{ie}(x_i) = -\omega^2\left(A_{1e}\,\mathrm{e}^{-\mathrm{j}k_b x_i} + A_{2e}\mathrm{e}^{\mathrm{j}k_b x_i} + A_{3e}\mathrm{e}^{-k_b x_i} + A_{4e}\mathrm{e}^{k_b x_i}\right) \qquad (10.2.62\text{a,b})$$

where the subscript e denotes experimentally obtained values. Note that the acceleration $\bar{a}_{ie}(x_i)$ at location (x_i) is complex, being defined in terms of a magnitude and a phase. In matrix form:

$$\begin{bmatrix} \dfrac{-\bar{a}_{1e}}{\omega^2} \\ \dfrac{-\bar{a}_{2e}}{\omega^2} \\ \dfrac{-\bar{a}_{3e}}{\omega^2} \\ \vdots \\ \dfrac{-\bar{a}_{ne}}{\omega^2} \end{bmatrix} = \begin{bmatrix} \beta_1 & \beta_1^{-1} & \beta_1^{\mathrm{j}} & \beta_1^{-\mathrm{j}} \\ \beta_2 & \beta_2^{-1} & \beta_2^{\mathrm{j}} & \beta_2^{-\mathrm{j}} \\ \beta_3 & \beta_3^{-1} & \beta_3^{\mathrm{j}} & \beta_3^{-\mathrm{j}} \\ \vdots & \vdots & \vdots & \vdots \\ \beta_n & \beta_n^{-1} & \beta_n^{\mathrm{j}} & \beta_n^{-\mathrm{j}} \end{bmatrix} \begin{bmatrix} A_{4e} \\ A_{3e} \\ A_{2e} \\ A_{1e} \end{bmatrix} \qquad (10.2.63)$$

where $\beta_i = \mathrm{e}^{k_b x_i}$, $\beta_i^{-1} = \mathrm{e}^{-k_b x_i}$, $\beta_i^{\mathrm{j}} = \mathrm{e}^{\mathrm{j}k_b x_i}$ and $\beta_i^{-\mathrm{j}} = \mathrm{e}^{-\mathrm{j}k_b x_i}$. Rearranging gives:

$$\begin{bmatrix} A_{4e} \\ A_{3e} \\ A_{2e} \\ A_{1e} \end{bmatrix} = \begin{bmatrix} \beta_1 & \beta_1^{-1} & \beta_1^{\mathrm{j}} & \beta_1^{-\mathrm{j}} \\ \beta_2 & \beta_2^{-1} & \beta_2^{\mathrm{j}} & \beta_2^{-\mathrm{j}} \\ \beta_3 & \beta_3^{-1} & \beta_3^{\mathrm{j}} & \beta_3^{-\mathrm{j}} \\ \vdots & \vdots & \vdots & \vdots \\ \beta_n & \beta_n^{-1} & \beta_n^{\mathrm{j}} & \beta_n^{-\mathrm{j}} \end{bmatrix}^{-1} \begin{bmatrix} -\dfrac{\bar{a}_{1e}}{\omega^2} \\ -\dfrac{\bar{a}_{2e}}{\omega^2} \\ -\dfrac{\bar{a}_{3e}}{\omega^2} \\ \vdots \\ -\dfrac{\bar{a}_{ne}}{\omega^2} \end{bmatrix} \qquad (10.2.64)$$

where the superscript '-1' represents' the generalised inverse or pseudo-inverse of the matrix. The matrix is better conditioned if the accelerometer spacings are random rather than uniform. Equation (10.2.63) represents a system of n equations in four unknowns. If $n = 4$, the system is determined, but the solution $[A_{4e}, A_{3e}, A_{2e}, A_{1e}]$ is extremely sensitive to errors in the measured accelerations. However, if an overdetermined system ($n > 4$) is used, the error is dramatically reduced, as will now be shown.

Let $\bar{w}_e$ be the displacement amplitude calculated from the constants A_{1e}, A_{2e}, A_{3e} and A_{4e}. For the beam described by the parameters of Table 10.2, with end conditions modelled as infinite, the error induced in the displacement $\bar{w}_e$, given an initial error in the accelerations (a_i, $i = 1 - n$) of 10%, is plotted as a function of the number of accelerometers n in Figure 10.19. For this case, $n \geq 10$ provides a satisfactory accuracy.

Once calculated, A_{1e}, A_{2e}, A_{3e} and A_{4e} are substituted into Equation (10.2.14) and differentiation carried out to find $\bar{w}_{1e}(x_L)$, $\bar{w}_{1e}^{\prime}(x_L)$, $\bar{w}_{1e}^{\prime\prime}(x_L)$ and $\bar{w}_{1e}^{\prime\prime\prime}(x_L)$. Equations (10.2.60) and (10.2.61) can then be used to find the equivalent impedances Z_{L1} and Z_{L2}, which may be substituted into the $\boldsymbol{\alpha}$ matrix to find the solution vector at any other location along the beam with and without control.

10.2.6.1 Accuracy of the Approximation

It should be noted that use of the 'equivalent' impedance matrix obtained by eliminating the diagonal elements from the impedance matrix is only valid for analysis similar to that followed in this chapter. It is not claimed that the resulting impedance matrix closely approximates the real impedance values of the termination in general circumstances. However, the numerical answers for all derivatives of displacement (and hence acceleration, etc.), calculated at any point along the beam, are correct to eight significant figures or are insignificant (less than 10^{-16}) when compared to the corresponding derivatives obtained by using the 'exact' impedance matrix. This is clearly well within the accuracy available from the experimental measurements!

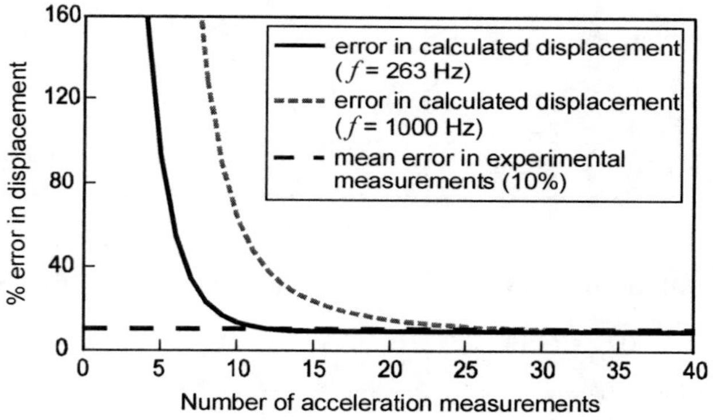

Figure 10.19 Error induced in the calculated controlled beam response for an accelerometer measurement error under primary excitation of 10% (in both the real and imaginary parts).

The accuracy of this method has been tested with a variety of cases. The 'exact' impedance matrices and the corresponding approximations calculated using the method described are given in Table 10.3. All of these examples utilise a right-hand impedance corresponding to an ideally infinite beam, and approximations are made for the various test cases at the left-hand end of the beam. The parameters characterising the beam are those given in Table 10.2. For all the examples, all derivatives of displacement calculated using the 'exact' and approximate impedance matrices are either identical to at least eight significant figures, or are insignificant (smaller than the reference by a factor of 10^{15} or more). It would be expected that the simplification might fail when the original matrix had large elements on the diagonal, but this is not the case, as shown by the first two examples. The third example shows an approximation for an impedance matrix with four complex elements. Examples 4, 5 and 6 show the exact and approximate matrices for the ideal free, fixed and infinite beam impedances respectively.

Note that if the measurements were taken at a large distance from the beam termination such that the terms A_{3e} and A_{4e} were very small, the matrix in Equation (10.2.64) would be ill-conditioned and the steps outlined in Section 10.2.7 below would have to be undertaken.

Table 10.3 Impedance matrices and corresponding approximations.

Example No.	'Exact' Matrix $[Z_L]$	Corresponding Approximation (with zero diagonal elements)
1	$\begin{bmatrix} 10^{100} + 10^{100j} & 0 \\ 0 & -10^{100} - 10^{100j} \end{bmatrix}$	$\begin{bmatrix} 0 & 2.44\times10^{15} - 5.00\times10^{16}j \\ -1.61\times10^{19} - 9.30\times10^{18}j & 0 \end{bmatrix}$
2	$\begin{bmatrix} 10^{100} & 0 \\ 0 & -10^{100} \end{bmatrix}$	$\begin{bmatrix} 0 & 1.01\times10^{17} - 4.28\times10^{16}j \\ -2.07\times10^{19} - 1.27\times10^{18}j & 0 \end{bmatrix}$
3	$\begin{bmatrix} 0 & 0 \\ 0 & 0 \end{bmatrix}$	$\begin{bmatrix} 0 & -1.5\times10^{-15} - 2.8\times10^{-16}j \\ 5.6\times10^{-14} - 1.3\times10^{-13}j & 0 \end{bmatrix}$
4	$\begin{bmatrix} 0 & 10^{100} \\ 10^{100} & 0 \end{bmatrix}$	$\begin{bmatrix} 0 & 5.00\times10^{15} - 8.13\times10^{16}j \\ -1.38\times10^{19} - 8.48\times10^{17}j & 0 \end{bmatrix}$
5	$\begin{bmatrix} 124.99 & 9.596 - 9.596j \\ -1628.0 - 1628.0j & -124.99 \end{bmatrix}$	$\begin{bmatrix} 0 & 9.596 \\ -1628.0 + 1.35\times10^{-13}j & 0 \end{bmatrix}$
6	$\begin{bmatrix} 1 + 2j & 3 + 4j \\ 5 + 6j & 7 + 8j \end{bmatrix}$	$\begin{bmatrix} 0 & 4.963 + 6.101j \\ 222.396 + 10.889j & 0 \end{bmatrix}$

10.2.7 Measuring Amplitudes of Waves Travelling Simultaneously in Opposite Directions

Examination of Equation (10.2.64) reveals that the analysis of the previous section can also be applied to determining the amplitudes of the two flexural waves (A_{1e} and A_{2e}) propagating in opposite directions, and the amplitudes of the non-propagating near-field components (A_{3e} and A_{4e}). If the measurements are made in the far-field of any vibration sources, beam discontinuities or boundaries such that A_{3e} and A_{4e} are very small, then the matrix requiring inversion in Equation (10.2.64) will be ill-conditioned and the results for A_{1e} and A_{2e} will be inaccurate. In this situation, better results are obtained by setting A_{3e} and A_{4e} equal to zero, removing A_{3e} and A_{4e} from the left side vector and removing the terms β_i and β_i^{-1} from the matrix.

Longitudinal waves may be measured in a similar way to flexural waves, provided the guidelines outlined in Chapter 2, Section 2.5.4.1.5 for obtaining accurate measurements are followed closely, especially if more than one wave type is present. In fact, determining the amplitudes of two longitudinal waves propagating in opposite directions is less prone to error than doing the same for flexural waves, as there are no near-field effects to contend with and the waves may be described using:

$$\bar{\xi}_x(x) = A_1 e^{-jkx} + A_2 e^{jkx} \tag{10.2.65}$$

where the complex wave amplitudes A_1 and A_2 are determined as for flexural waves. Similar procedures can be followed to determine the amplitudes of two torsional waves travelling in opposite directions. As mentioned elsewhere in this book, the interest in longitudinal and torsional waves is not so much due to their direct contribution to feelable (or potentially damaging) vibration or sound radiation, but rather because their energy can be converted to the more problematic flexural waves at structural junctions and discontinuities, and especially at locations where there is effectively an eccentric mass.

10.3 ACTIVE CONTROL OF VIBRATION IN A SEMI-INFINITE PLATE

The active control of vibratory power transmission along a semi-infinite plate is of interest from the point of view of reducing vibration levels in, or sound radiation from, the plate or reducing the same quantities in an attached structure. To make the analytical problem tractable, only the more important flexural waves will be considered here. This does not mean that in a practical situation longitudinal waves could not be important; indeed, conversion of longitudinal wave energy into the more 'feelable' and more efficiently radiating flexural waves is a common phenomenon at structural junctions and discontinuities.

It is interesting to note that the problem of flexural waves in a plate is similar in some respects to higher-order mode propagation in a two-dimensional air duct. The model considered here is a semi-infinite plate, free at one end, anechoically terminated at the other and simply supported along the other two sides. Thus, the analysis describes the vibration in terms of bending waves travelling from the free end to the anechoically terminated end. These waves consist of the fundamental wave and various higher-order waves involving reflections at the simply supported edges in much the same way as plane wave and higher-order mode propagation occur in an air duct.

10.3.1 Response of a Semi-Infinite Plate to a Line of Point Forces Driven in-Phase

The plate considered here lies in the x-y plane with a free edge at $x = 0$ and simply supported edges at $y = 0$ and $y = L_y$ as shown in Figure 10.20.

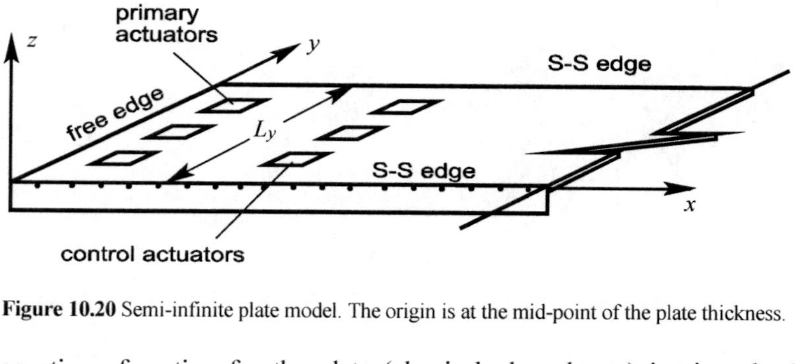

Figure 10.20 Semi-infinite plate model. The origin is at the mid-point of the plate thickness.

The equation of motion for the plate (classical plate theory) is given by Equqtion (2.3.101). The external force per unit area excitation represented by q in that equation is an array of n equally spaced point forces spanning the plate along a line parallel to the y-axis. At this stage, all of these forces at locations (x_0, y_i, i = 1, n) will be assumed to be driven in-phase and with the same complex amplitude, F_0, so that from Equation (2.3.124):

$$q = F_0 \sum_{i=1}^{n} \delta(x - x_0)\delta(y - y_i) \qquad (10.3.1)$$

The two sides (y = 0 and $y = L_y$) are simply supported, and so the following harmonic series solution in y can be assumed for the vibrational displacement:

$$w(x, y, t) = \sum_{m=1}^{\infty} w_m(x) \sin\frac{m\pi x}{L_y} \mathrm{e}^{\,\mathrm{j}\omega t} \qquad (10.3.2)$$

where m is the mode number. Each eigen function $w_m(x)$ can be expressed in terms of unknown constants A_1, A_2, A_3 and A_4 and modal wavenumbers k_{im} as follows:

$$\bar{w}_m(x) = A_1 \mathrm{e}^{k_{1m}x} + A_2 \mathrm{e}^{k_{2m}x} + A_3 \mathrm{e}^{k_{3m}x} + A_4 \mathrm{e}^{k_{4m}x} \qquad (10.3.3)$$

To find the modal wavenumbers (that is, eigenvalues), the homogeneous form of Equation (2.3.101) is multiplied by $\sin\frac{m\pi y}{L_y}$ and integrated with respect to y over the width the plate (that is, from 0 to L_y), to give:

$$\frac{\mathrm{d}^4 \bar{w}_m(x)}{\mathrm{d}x^4} - 2\left(\frac{m\pi}{L_y}\right)^2 \frac{\mathrm{d}^2 \bar{w}_m(x)}{\mathrm{d}x^2} + \left[\left(\frac{m\pi}{L_y}\right)^4 - \frac{\rho h\omega^2}{D_h}\right]\bar{w}_m(x) = 0 \qquad (10.3.4)$$

where $D_h = Eh\,/[12(1 - \nu^2)$ is the flexural rigidity of the plate. Solutions to this ordinary differential equation are assumed to be of the form $\mathrm{e}^{k_m x}$, and its characteristic equation is:

$$k_m^4 - 2\left(\frac{m\pi}{L_y}\right)^2 k_m^2 + \left(\frac{m\pi}{L_y}\right)^4 - \frac{\rho h\omega^2}{D_h} = 0 \qquad (10.3.5)$$

which has the roots:

$$k_{1m,2m} = \pm\sqrt{\left(\frac{m\pi}{L_y}\right)^2 + \sqrt{\frac{\rho h\omega^2}{D_h}}} \qquad (10.3.6)$$

$$k_{3m,4m} = \pm\sqrt{\left(\frac{m\pi}{L_y}\right)^2 - \sqrt{\frac{\rho h\omega^2}{D_h}}} \qquad (10.3.7)$$

On each side of the applied force, the eigen function $\bar{w}_m(x)$ is then a different linear combination of the terms $\mathrm{e}^{k_{1m}x}$ (with i = 1, 2, 3, 4). For $x < x_0$:

$$\bar{w}_{1m}(x) = A_1 e^{k_{1m}x} + A_2 e^{k_{2m}x} + A_3 e^{k_{3m}x} + A_4 e^{k_{4m}x} \tag{10.3.8}$$

and for $x > x_0$:

$$\bar{w}_{2m}(x) = B_2 e^{k_{2m}x} + B_4 e^{k_{4m}x} \tag{10.3.9}$$

The phenomenon of wave propagation in this plate is very similar to the propagation of longitudinal waves in an air duct. Below a certain cut-on frequency, the flexural wave propagation will be one-dimensional as in a beam or as longitudinal plane waves in an air duct. Above the cut-on frequency, higher-order modes, characterised by wave reflections from the two simply supported plate edges, begin to propagate, resulting in a very non-uniform vibration field on the plate. As the frequency becomes higher, more modes cut-on and the vibration field on the plate becomes even more complex. The cut-on frequencies of the higher-order modes can be obtained from Equation (10.3.8) by determining the frequency when the wavenumber first becomes imaginary. Thus, the mth higher-order mode will cut-on when the excitation frequency is given by:

$$f = \frac{\pi}{2}\left(\frac{m}{L_y}\right)^2 \sqrt{\frac{D_h}{\rho h}} \tag{10.3.10}$$

Note that in Equation (10.3.9), the coefficients B_1 and B_3 have been omitted because, for a semi-infinite plate, there is no boundary to produce reflected waves with a negative horizontal velocity component. At the junction $x = x_0$, the following boundary conditions must be satisfied:

$$\bar{w}_{1m} = \bar{w}_{2m} \tag{10.3.11}$$

$$\frac{\partial \bar{w}_{1m}}{\partial x} = \frac{\partial \bar{w}_{2m}}{\partial x} \tag{10.3.12}$$

$$\frac{\partial^2 \bar{w}_{1m}}{\partial x^2} = \frac{\partial^2 \bar{w}_{2m}}{\partial x^2} \tag{10.3.13}$$

and

$$\frac{\partial^3 \bar{w}_{1m}}{\partial x^3} - \frac{\partial^3 \bar{w}_{2m}}{\partial x^3} = -\frac{2F_0}{L_y D_h} \sum_{i=1}^{n} \sin\frac{m\pi y_i}{L_y} \tag{10.3.14}$$

For the free edge at $x = 0$, the expression:

$$-D_h \left(\frac{\partial^2 \bar{w}}{\partial x^2} + \nu \frac{\partial^2 \bar{w}}{\partial y^2} \right) \Bigg|_{x=0} = 0 \tag{10.3.15}$$

given by Leissa (1960) can be used to express the bending moment boundary condition:

$$M_x(0, y) = 0 \tag{10.3.16}$$

in terms of displacement, with the following result:

$$\sum_{m=1}^{\infty}\left[k_{1m}^2 A_1 + k_{2m}^2 A_2 + k_{3m}^2 A_3 + k_{4m}^2 A_4 - \nu\left(\frac{m\pi}{L_y}\right)^2 (A_1 + A_2 + A_3 + A_4)\right]\sin\frac{m\pi y}{L_y} = 0 \qquad (10.3.17)$$

By again multiplying both sides by $\sin\frac{m\pi y}{L_y}$ and integrating from $y = 0$ to L_y, the following is obtained:

$$\left[k_{1m}^2 - \nu\left(\frac{m\pi}{L_y}\right)^2\right]A_1 + \left[k_{2m}^2 - \nu\left(\frac{m\pi}{L_y}\right)^2\right]A_2 + \left[k_{3m}^2 - \nu\left(\frac{m\pi}{L_y}\right)^2\right]A_3 + \left[k_{4m}^2 - \nu\left(\frac{m\pi}{L_y}\right)^2\right]A_4 = 0 \qquad (10.3.18)$$

The free edge condition also requires that the net vertical force at $x = 0$ be zero (Leissa, 1960). Thus,

$$V_x = Q_x + \frac{\partial M_{xy}}{\partial y} = 0 \qquad (10.3.19a,b)$$

which can be expressed in terms of displacement as follows:

$$V_x(x,y)\big|_{x=0} = -D_h\left[\frac{\partial^3 w}{\partial x^3} + (2-\nu)\frac{\partial^3 w}{\partial x \partial y^2}\right]\Bigg|_{x=0} = 0 \qquad (10.3.20a,b)$$

Thus,

$$\left[k_{1m}^3 - (2-\nu)\left(\frac{m\pi}{L_y}\right)^2 k_{1m}\right]A_1 + \left[k_{2m}^3 - (2-\nu)\left(\frac{m\pi}{L_y}\right)^2 k_{2m}\right]A_2 + \left[k_{3m}^3 - (2-\nu)\left(\frac{m\pi}{L_y}\right)^2 k_{3m}\right]A_3 + \left[k_{4m}^3 - (2-\nu)\left(\frac{m\pi}{L_y}\right)^2 k_{4m}\right]A_4 = 0 \qquad (10.3.21)$$

Equations (10.3.11) to (10.3.14), (10.3.18) and (10.3.21) can be written as a 6 × 6 matrix equation as follows:

$$\begin{bmatrix} k_{1m}^2 - \nu H & k_{2m}^2 - \nu H & k_{3m}^2 - \nu H & k_{4m}^2 - \nu H & 0 & 0 \\ k_{1m}^3 - (2-\nu)Hk_{1m} & k_{2m}^3 - (2-\nu)Hk_{2m} & k_{3m}^3 - (2-\nu)Hk_{3m} & k_{4m}^3 - (2-\nu)Hk_{4m} & 0 & 0 \\ \mathrm{e}^{k_{1m}x_0} & \mathrm{e}^{k_{2m}x_0} & \mathrm{e}^{k_{3m}x_0} & \mathrm{e}^{k_{4m}x_0} & -\mathrm{e}^{k_{2m}x_0} & -\mathrm{e}^{k_{4m}x_0} \\ k_{1m}\mathrm{e}^{k_{1m}x_0} & k_{2m}\mathrm{e}^{k_{2m}x_0} & k_{3m}\mathrm{e}^{k_{3m}x_0} & k_{4m}\mathrm{e}^{k_{4m}x_0} & -k_{2m}\mathrm{e}^{k_{2m}x_0} & -k_{4m}\mathrm{e}^{k_{4m}x_0} \\ k_{1m}^2\mathrm{e}^{k_{1m}x_0} & k_{2m}^2\mathrm{e}^{k_{2m}x_0} & k_{3m}^2\mathrm{e}^{k_{3m}x_0} & k_{4m}^2\mathrm{e}^{k_{4m}x_0} & -k_{2m}^2\mathrm{e}^{k_{2m}x_0} & -k_{4m}^2\mathrm{e}^{k_{4m}x_0} \\ k_{1m}^3\mathrm{e}^{k_{1m}x_0} & k_{2m}^3\mathrm{e}^{k_{2m}x_0} & k_{3m}^3\mathrm{e}^{k_{3m}x_0} & k_{4m}^3\mathrm{e}^{k_{4m}x_0} & -k_{2m}^3\mathrm{e}^{k_{2m}x_0} & -k_{4m}^3\mathrm{e}^{k_{4m}x_0} \end{bmatrix} \begin{bmatrix} A_1 \\ A_2 \\ A_3 \\ A_4 \\ B_2 \\ B_4 \end{bmatrix} = \begin{bmatrix} 0 \\ 0 \\ 0 \\ 0 \\ 0 \\ -\dfrac{2F_0}{L_y D_h} \displaystyle\sum_{i=1}^{n} \sin\frac{m\pi y_i}{L_y} \end{bmatrix} \tag{10.3.22}$$

which can be written as $\boldsymbol{\alpha X} = \boldsymbol{F}$. For each value of m, the solution $\boldsymbol{X}$ of the 6 × 6 system of equations is an eigenvector which describes a travelling wave mode shape and amplitude. The modal wavenumbers k_{1m}, k_{2m}, k_{3m} and k_{4m} required by Equation (10.3 22) are calculated from Equations (10.3.6) and (10.3.7), and the quantity, $H = \left(\dfrac{m\pi}{L_y}\right)^2$. The plate response amplitude at any location (x, y) due to the row of in-phase harmonic point forces is:

$$\bar{w} = \bar{F}_0 \bar{w}_{0\text{-}f} \tag{10.3.23}$$

where $w_{0\text{-}f}$ is the response amplitude to unit force amplitude excitation, which is obtained by solving Equation (10.3.22) and substituting the results for A_1, A_2, A_3, A_4, B_2 and B_4 into Equations (10.3.8) and (10.3.2) or Equations (10.3.9) and (10.3.2), depending upon the location at which the plate response is to be evaluated. Finally, the response of the plate to unit primary and unit control force excitation is required. By expressing the solution for the eigenvectors as:

$$\boldsymbol{X} = \boldsymbol{\alpha}^{-1}\boldsymbol{F} \tag{10.3.24}$$

the response amplitude at (x, y) is:

$$\bar{w}_{0\text{-}f}(x, y) = \sum_{m=1}^{\infty} \frac{2}{L_y D_h} \left[\sum_{i=1}^{n} \sin\frac{m\pi y_i}{L_y}\right] \left[(\boldsymbol{\alpha})^{-1}\right]_{i,6}^{\mathrm{T}} \boldsymbol{E_m} \sin\frac{m\pi y}{L_y} \tag{10.3.25}$$

where n is the number of forces in the array, m is the mode number and $\left[(\boldsymbol{\alpha})^{-1}\right]^{\mathrm{T}}_{i,6}$ is the sixth column of the inverse of the coefficient matrix $\boldsymbol{\alpha}$ from Equation (10.3.22).

For $x < x_0$:

$$\boldsymbol{E}_m = \left[\mathrm{e}^{k_{1m}x} \;\; \mathrm{e}^{k_{2m}x} \;\; \mathrm{e}^{k_{3m}x} \;\; \mathrm{e}^{k_{4m}x} \;\; 0 \;\; 0\right]^{\mathrm{T}} \tag{10.3.26}$$

and for $x > x_0$:

$$\boldsymbol{E}_m = \left[0 \;\; 0 \;\; 0 \;\; 0 \;\; \mathrm{e}^{k_{2m}x} \;\; \mathrm{e}^{k_{4m}x}\right]^{\mathrm{T}} \tag{10.3.27}$$

10.3.2 Minimisation of Acceleration with a Line of in-Phase Control Forces

If the plate is excited by a line of in-phase primary point forces of complex amplitude $\bar{F}_p$, located at $x = x_p$, the plate response at any location (x, y) is found by using Equations (10.3.23) and (10.3.25) and is:

$$\bar{w}_p(x,y) = \bar{F}_p \bar{w}_{p-f}(x,y) = \bar{F}_p \sum_{m=1}^{\infty} \frac{2}{L_y D_h} \left[\sum_{i=1}^{n_p} \sin\frac{m\pi y_i}{L_y}\right] \left[(\boldsymbol{\alpha}_p)^{-1}\right]^{\mathrm{T}}_{i,6} \boldsymbol{E}_m \sin\frac{m\pi y}{L_y} \tag{10.3.28a,b}$$

where $\bar{w}_{p-f}(x,y)$ is the plate response amplitude at location (x, y) to unit primary force excitation, n_p is the number of primary forces, and $\boldsymbol{\alpha}_p$ is similar to $\boldsymbol{\alpha}$ with $\bar{F}_0$ replaced by $\bar{F}_p$ and x_0 replaced by x_p. Similarly, if a line of control forces of complex amplitude $\bar{F}_c$ is placed at $x = x_c$, the plate response due to these acting alone is:

$$\bar{w}_c(x,y) = \bar{F}_c \bar{w}_{c-f}(x,y) = \bar{F}_c \sum_{m=1}^{\infty} \frac{2}{L_y D_h} \left[\sum_{i=1}^{n_c} \sin\frac{m\pi y_i}{L_y}\right] \left[(\boldsymbol{\alpha}_c)^{-1}\right]^{\mathrm{T}}_{i,6} \boldsymbol{E}_m \sin\frac{m\pi y}{L_y} \tag{10.3.29a,b}$$

where n_c is the number of control actuators. The total plate response amplitude due to primary and control forces acting together is then,

$$\bar{w} = \bar{w}_p + \bar{w}_c = \bar{F}_p \bar{w}_{p-f} + \bar{F}_c \bar{w}_{c-f} \tag{10.3.30a,b}$$

The optimal control force amplitude $\bar{F}_c$ for minimising the acceleration along the width of the plate at a constant x may be found by integrating the mean square of the displacement amplitude defined in Equation (10.3.30), and setting the partial derivatives of the integration with respect to the real and imaginary components of the control force equal to zero. Thus,

$$\bar{F}_c = -\bar{F}_p \frac{\int_0^{L_y} \bar{w}_{p-f} \bar{w}^*_{c-f}\, \mathrm{d}y}{\int_0^{L_y} |\bar{w}_{c-f}|^2\, \mathrm{d}y} \tag{10.3.31}$$

10.3.3 Minimisation of Acceleration with a Line of *n* Independently Driven Control Forces

If the plate is driven by an array of in-phase primary point forces in a line at $x = x_p$ and n independent control point forces in a line at $x = x_c$, the total plate response may be written as:

$$\bar{w} = \bar{w}_p + \bar{w}_c = \bar{F}_p \bar{w}_{p-f} + \bar{F}_{c1}\bar{w}_{c-f1} + \bar{F}_{c2}\bar{w}_{c-f2} + \cdots + \bar{F}_{cn}\bar{w}_{c-fn} \tag{10.3.32a,b}$$

The quantities $\bar{w}_{c-f1}, \bar{w}_{c-f2}, \ldots, \bar{w}_{c-fn}$ are each calculated in a similar way to w_{c-f} in Equation (10.3.29), except that n_c is set equal to one in each case.

The optimal control force amplitudes for minimising acceleration may be found by integrating the mean square of the displacement defined in Equation (10.3.32) and setting the partial derivatives of the integration with respect to each of the real and imaginary components of the control forces equal to zero. The result is an optimal set of control forces as follows:

$$\begin{bmatrix} \bar{F}_{c1} \\ \bar{F}_{c2} \\ \vdots \\ \bar{F}_{cn} \end{bmatrix} =$$

$$- \begin{bmatrix} \int_0^{L_y} |\bar{w}_{c-f1}|^2 \, dy & \int_0^{L_y} \bar{w}^*_{c-f1}\bar{w}_{c-f2} \, dy & \cdots & \int_0^{L_y} \bar{w}^*_{c-f1}\bar{w}_{c-fn} \, dy \\ \int_0^{L_y} \bar{w}_{c-f1}\bar{w}^*_{c-f2} \, dy & \int_0^{L_y} |\bar{w}_{c-f2}|^2 \, dy & \cdots & \int_0^{L_y} \bar{w}^*_{c-f2}\bar{w}_{c-fn} \, dy \\ \vdots & \vdots & \vdots & \vdots \\ \int_0^{L_y} \bar{w}_{c-f1}\bar{w}^*_{c-fn} \, dy & \int_0^{L_y} \bar{w}_{c-f2}\bar{w}^*_{c-fn} \, dy & \cdots & \int_0^{L_y} |\bar{w}_{c-fn}|^2 \, dy \end{bmatrix}^{-1} \begin{bmatrix} \int_0^{L_y} \bar{w}_{p-f}\bar{w}^*_{c-f1} \, dy \\ \int_0^{L_y} \bar{w}_{p-f}\bar{w}^*_{c-f2} \, dy \\ \vdots \\ \int_0^{L_y} \bar{w}_{p-f}\bar{w}^*_{c-fn} \, dy \end{bmatrix} \bar{F}_p \tag{10.3.33}$$

10.3.4 Power Transmission

From Section 2.5, the expression for the x component of flexural wave vibration intensity (or power transmission per unit width) in a plate is:

$$P_x(t) = -\left[\dot{w}Q_x - \frac{\partial \dot{w}}{\partial x}M_x - \frac{\partial \dot{w}}{\partial y}M_{xy}\right] \tag{10.3.34}$$

Note that here the contribution from longitudinal and shear waves is assumed to be zero. The total instantaneous flexural wave power transmission through a section at constant x is then given by:

$$P_x(t) = -\int_0^{L_y} \left[\dot{w} Q_x - \frac{\partial \dot{w}}{\partial x} M_x - \frac{\partial \dot{w}}{\partial y} M_{xy} \right] \mathrm{d}y \tag{10.3.35}$$

or, for a single frequency:

$$P_x(t) = -\int_0^{L_y} \left[\mathrm{j}\omega w Q_x - \mathrm{j}\omega \frac{\partial w}{\partial x} M_x - \mathrm{j}\omega \frac{\partial w}{\partial y} M_{xy} \right] \mathrm{d}y \tag{10.3.36}$$

In Equations (10.3.35) and (10.3.36), all displacements, forces and moments are time dependent quantities and also dependent on the coordinate location (x, y) on the plate.

As discussed in Section 2.5, for single frequency excitation, the real (or active) part of the power transmission along the plate is the product of the real part of the force term with the real part of the velocity term for each pair of terms in Equation (10.3.36) and the result is time averaged. Thus, the active power transmission for harmonic excitation is given by:

$$P_{xa} = -\frac{1}{2} \int_0^{L_y} \mathrm{Re} \left\{ [\, \mathrm{j}\omega \bar{w}]^* \bar{Q}_x - \left[\mathrm{j}\omega \frac{\partial \bar{w}}{\partial x} \right]^* \bar{M}_x - \left[\mathrm{j}\omega \frac{\partial \bar{w}}{\partial y} \right]^* \bar{M}_{xy} \right\} \mathrm{d}y \tag{10.3.37}$$

where the quantities $\bar{Q}_x$, $\bar{M}_x$ and $\bar{M}_{xy}$ have been defined in Section 2.3.

For one line of primary actuators and a second line of control actuators parallel to the y-axis, the resulting total power transmission can be expressed in terms of the primary and control forces, using superposition.

10.3.5 Minimisation of Power Transmission with a Line of in-Phase Point Control Forces

The total power transmission resulting from a line of in-phase point primary forces and a line of in-phase point control forces acting together can be found by substituting Equations (10.3.28) and (10.3.29) into Equation (10.3.30) and the result into Equation (10.3.37). The shear forces and moments may also be expressed in terms of the displacements associated with each excitation force using from Section 2.3 and Equations (10.3.28), (10.3.29) and (10.3.30) to give:

$$\bar{M}_x = \frac{Eh^3}{12(1 - \nu^2)} \left[\bar{F}_p \frac{\partial^2 \bar{w}_{p-f}}{\partial x^2} + \bar{F}_c \frac{\partial^2 \bar{w}_{c-f}}{\partial x^2} + \nu \left(\bar{F}_p \frac{\partial^2 \bar{w}_{p-f}}{\partial y^2} + \bar{F}_c \frac{\partial^2 \bar{w}_{c-f}}{\partial y^2} \right) \right] \tag{10.3.38}$$

$$\bar{M}_{xy} = -\frac{Eh^3}{12(1 + \nu)} \left[\bar{F}_p \frac{\partial^2 \bar{w}_{p-f}}{\partial x \partial y} + \bar{F}_c \frac{\partial^2 \bar{w}_{c-f}}{\partial x \partial y} \right] \tag{10.3.39}$$

$$\begin{aligned} \bar{Q}_x &= \frac{\partial \bar{M}_x}{\partial x} - \frac{\partial \bar{M}_{xy}}{\partial y} \\ &= -\frac{Eh^3}{12(1 - \nu^2)} \left[\bar{F}_p \frac{\partial^3 \bar{w}_{p-f}}{\partial x^3} + \bar{F}_c \frac{\partial^3 \bar{w}_{c-f}}{\partial x^3} + \bar{F}_p \frac{\partial^3 \bar{w}_{p-f}}{\partial x \partial y^2} + \bar{F}_c \frac{\partial^3 \bar{w}_{c-f}}{\partial x \partial y^2} \right] \end{aligned} \tag{10.3.40a,b}$$

where ν is Poisson's ratio and E is Young's modulus of elasticity. The preceding three equations and Equation (10.3.30) can be substituted into the expression for the power transmission through any plate cross-section at axial location x and Equation (10.3.37) to give:

$$P_{xa} = \frac{1}{2}\int_0^{L_y} \mathrm{Re}\left\{ \bar{F}_c \bar{F}_c^* A + \bar{F}_c B \bar{F}_p^* + \bar{F}_c^* C \bar{F}_p + \bar{F}_p \bar{F}_p^* D \right\} \mathrm{d}y \tag{10.3.41}$$

where

$$A = \frac{\mathrm{j}\omega E h^3}{12(1-\nu^2)}\left[-\left(\frac{\partial^3 \bar{w}_{c-f}}{\partial x^3} + \frac{\partial^3 \bar{w}_{c-f}}{\partial x \partial y^2} \right) \bar{w}_{c-f}^* + \left(\frac{\partial^2 \bar{w}_{c-f}}{\partial x^2} + \nu \frac{\partial^2 \bar{w}_{c-f}}{\partial y^2} \right) \frac{\partial \bar{w}_{c-f}^*}{\partial x} + (1-\nu) \frac{\partial^2 \bar{w}_{c-f}}{\partial x \partial y} \frac{\partial \bar{w}_{c-f}^*}{\partial y} \right] \tag{10.3.42}$$

$$B = \frac{\mathrm{j}\omega E h^3}{12(1-\nu^2)}\left[-\left(\frac{\partial^3 \bar{w}_{c-f}}{\partial x^3} + \frac{\partial^3 \bar{w}_{c-f}}{\partial x \partial y^2} \right) \bar{w}_{p-f}^* + \left(\frac{\partial^2 \bar{w}_{c-f}}{\partial x^2} + \nu \frac{\partial^2 \bar{w}_{c-f}}{\partial y^2} \right) \frac{\partial \bar{w}_{p-f}^*}{\partial x} + (1-\nu) \frac{\partial^2 \bar{w}_{c-f}}{\partial x \partial y} \frac{\partial \bar{w}_{p-f}^*}{\partial y} \right] \tag{10.3.43}$$

$$C = \frac{\mathrm{j}\omega E h^3}{12(1-\nu^2)}\left[-\left(\frac{\partial^3 \bar{w}_{p-f}}{\partial x^3} + \frac{\partial^3 \bar{w}_{p-f}}{\partial x \partial y^2} \right) \bar{w}_{c-f}^* + \left(\frac{\partial^2 \bar{w}_{p-f}}{\partial x^2} + \nu \frac{\partial^2 \bar{w}_{p-f}}{\partial y^2} \right) \frac{\partial \bar{w}_{c-f}^*}{\partial x} + (1-\nu) \frac{\partial^2 \bar{w}_{p-f}}{\partial x \partial y} \frac{\partial \bar{w}_{c-f}^*}{\partial y} \right] \tag{10.3.44}$$

The partial derivatives of the active (real) power transmission with respect to the real and imaginary components of the control force are:

$$\frac{\partial P_{xa}}{\partial \bar{F}_r} = \frac{1}{2}\int_0^{L_y} \mathrm{Re}\left\{ \bar{F}_c^* A + \bar{F}_c A + \bar{F}_p^* B + \bar{F}_p C \right\} \mathrm{d}y \tag{10.3.45}$$

$$\frac{\partial P_{xa}}{\partial \bar{F}_j} = \frac{1}{2}\int_0^{L_y} \mathrm{Re}\left\{ \mathrm{j}\bar{F}_c^* A - \mathrm{j}\bar{F}_c A + \mathrm{j}\bar{F}_p^* B - \mathrm{j}\bar{F}_p C \right\} \mathrm{d}y \tag{10.3.46}$$

respectively, where $\bar{F}_c = \bar{F}_r + \mathrm{j}\bar{F}_i$. The optimal force is found by requiring that both of these derivatives are zero. That is,

$$\int_0^{L_y} \mathrm{Re}\left\{ \bar{F}_c^* A + \bar{F}_c A + \bar{F}_p^* B + \bar{F}_p C \right\} \mathrm{d}y = 0 \tag{10.3.47}$$

$$\int_0^{L_y} \mathrm{Re}\left\{ \mathrm{j}\bar{F}_c^* A - \mathrm{j}\bar{F}_c A + \mathrm{j}\bar{F}_p^* B - \mathrm{j}\bar{F}_p C \right\} \mathrm{d}y = 0 \tag{10.3.48}$$

The result is:

$$\bar{F}_r^{\mathrm{opt}} = -\frac{\int_0^{L_y} \mathrm{Re}\left\{ \bar{F}_p^* B + \bar{F}_p C \right\} \mathrm{d}y}{2\int_0^{L_y} \mathrm{Re}\{A\}\, \mathrm{d}y} = -\frac{\int_0^{L_y} \mathrm{Re}\left\{ \bar{F}_p^* B \right\} \mathrm{d}y + \int_0^{L_y} \mathrm{Re}\left\{ \bar{F}_p C \right\} \mathrm{d}y}{2\int_0^{L_y} \mathrm{Re}\{A\}\, \mathrm{d}y} \tag{10.3.49a,b}$$

$$\bar{F}_i^{\mathrm{opt}} = -\frac{\int_0^{L_y} \mathrm{Re}\left\{ \mathrm{j}\bar{F}_p^* B - \mathrm{j}\bar{F}_p C \right\} \mathrm{d}y}{2\int_0^{L_y} \mathrm{Re}\{A\}\, \mathrm{d}y} = -\frac{-\int_0^{L_y} \mathrm{Im}\left\{ \bar{F}_p^* B \right\} \mathrm{d}y + \int_0^{L_y} \mathrm{Im}\left\{ \bar{F}_p C \right\} \mathrm{d}y}{2\int_0^{L_y} \mathrm{Re}\{A\}\, \mathrm{d}y} \tag{10.3.50a,b}$$

or

$$\bar{F}_c^{\mathrm{opt}} = \bar{F}_r^{\mathrm{opt}} + \mathrm{j}\bar{F}_i^{\mathrm{opt}} = -\frac{\int_0^{L_y} B^*\, \mathrm{d}y + \int_0^{L_y} C\, \mathrm{d}y}{2\int_0^{L_y} \mathrm{Re}\{A\}\, \mathrm{d}y} \bar{F}_p \tag{10.3.51a,b}$$

Equation (10.3.51) is an expression for the control force which minimises the power transmission along the plate, and is of the same form (apart from the integration over y) as the expression for the optimal force required to control the power transmission in a beam (see Section 10.1). The optimum plate power transmission is given by:

$$P_{xa}^{\mathrm{opt}} = -\frac{1}{2}\left[\frac{\left| \int_0^{L_y} B\, \mathrm{d}y \right|^2 + 2\mathrm{Re}\left\{ \int_0^{L_y} B\, \mathrm{d}y \int_0^{L_y} C\, \mathrm{d}y \right\} + \left| \int_0^{L_y} C\, \mathrm{d}y \right|^2}{4\int_0^{L_y} \mathrm{Re}\{A\}\, \mathrm{d}y} - \int_0^{L_y} \mathrm{Re}\{D\}\, dy \right] |\bar{F}_p|^2 \tag{10.3.52}$$

where the power transmission has been obtained by integrating the intensity over a plate cross-section. The differential of Equation (10.3.52) with respect to x and y, which is the intensity, is a strongly varying function of location, but by conservation of energy, the total power transmission through a particular plate cross-section must be independent of the location and shape of the cross-section surface. It is assumed here that it is possible in practice to actually measure the total power transmission through a particular plate cross-section. In practice, this may be approximated by making a number of point measurements across the width of the plate and averaging the results. Pan and Hansen (1995) show that the required number of measurements to achieve an acceptable accuracy can be very small (as low as three in some cases).

10.3.6 Minimisation of Power Transmission with a Line of n Independently Driven Point Control Forces

For this case, the plate response is given by Equation (10.3.32) which can be substituted into Equation (10.3.37) to give an expression for the total power being transmitted past a line across the plate at an axial location x, which can be written in matrix form as:

$$P_{xa} = \frac{1}{2}\int_0^{L_y} \mathrm{Re}\left\{\bar{\boldsymbol{F}}^{\mathrm{H}} \boldsymbol{A} \bar{\boldsymbol{F}}\right\} \mathrm{d}y \tag{10.3.53}$$

where

$$\bar{\boldsymbol{F}} = [\bar{F}_p \;\; \bar{F}_{c1} \;\; \bar{F}_{c2} \;\; \cdots \;\; \bar{F}_{cn}]^{\mathrm{T}} \tag{10.3.54}$$

and

$$\boldsymbol{A} = \begin{bmatrix} A(1,1) & A(1,2) & A(1,3) & \cdots & A(1,n+1) \\ A(2,1) & A(2,2) & A(2,3) & \cdots & A(2,n+1) \\ A(3,1) & A(3,2) & A(3,3) & \cdots & A(3,n+1) \\ \vdots & \vdots & \vdots & \vdots & \vdots \\ A(n+1,1) & A(n+1,2) & A(n+1,3) & \cdots & A(n+1,n+1) \end{bmatrix} \tag{10.3.55}$$

The coefficients $A(i, j)$ ($i = 1, n+1; j = 1, n+1$) of matrix $\boldsymbol{A}$ result from the product of terms in Equation (10.3.37), each of which contains contributions from the $n+1$ different force elements of Equation (10.3.54). Thus, $A(i,j)$ is the product of the contribution to the first part of each term in Equation (10.3.37) due to the ith element of $\bar{\boldsymbol{F}}$ with the contribution to the second part of each term due to the jth element of $\bar{\boldsymbol{F}}$. For example:

$$\begin{aligned} A(1,1) = \frac{\mathrm{j}\omega E h^3}{12(1-\nu^2)} \Bigg[& -\left(\frac{\partial^3 \bar{w}_{p\text{-}f}}{\partial x^3} + \frac{\partial^3 \bar{w}_{p\text{-}f}}{\partial x \partial y^2}\right) \bar{w}^*_{p\text{-}f} \\ & + \left(\frac{\partial^2 \bar{w}_{p\text{-}f}}{\partial x^2} + \nu \frac{\partial^2 \bar{w}_{p\text{-}f}}{\partial y^2}\right) \frac{\partial \bar{w}^*_{p\text{-}f}}{\partial x} + (1-\nu) \frac{\partial^2 \bar{w}_{p\text{-}f}}{\partial x \partial y} \frac{\partial \bar{w}^*_{p\text{-}f}}{\partial y} \Bigg] \end{aligned} \tag{10.3.56}$$

$$\begin{aligned} A(2,1) = \frac{\mathrm{j}\omega E h^3}{12(1-\nu^2)} \Bigg[& -\left(\frac{\partial^3 \bar{w}_{p\text{-}f}}{\partial x^3} + \frac{\partial^3 \bar{w}_{p\text{-}f}}{\partial x \partial y^2}\right) \bar{w}^*_{c\text{-}f1} \\ & + \left(\frac{\partial^2 \bar{w}_{p\text{-}f}}{\partial x^2} + \nu \frac{\partial^2 \bar{w}_{p\text{-}f}}{\partial y^2}\right) \frac{\partial \bar{w}^*_{c\text{-}f1}}{\partial x} + (1-\nu) \frac{\partial^2 \bar{w}_{p\text{-}f}}{\partial x \partial y} \frac{\partial \bar{w}^*_{c\text{-}f1}}{\partial y} \Bigg] \end{aligned} \tag{10.3.57}$$

$$A(n+1,1) = \frac{j\omega E h^3}{12(1-\nu^2)}\left[-\left(\frac{\partial^3 \bar{w}_{p-f}}{\partial x^3} + \frac{\partial^3 \bar{w}_{p-f}}{\partial x \partial y^2} \right) \bar{w}^*_{c-fn} + \left(\frac{\partial^2 \bar{w}_{p-f}}{\partial x^2} + \nu \frac{\partial^2 \bar{w}_{p-f}}{\partial y^2} \right) \frac{\partial \bar{w}^*_{c-fn}}{\partial x} + (1-\nu) \frac{\partial^2 \bar{w}_{p-f}}{\partial x \partial y} \frac{\partial \bar{w}^*_{c-fn}}{\partial y} \right] \tag{10.3.58}$$

By defining $\bar{F}_{c1} = \bar{F}_{r1} + j\bar{F}_{i1}$, $\bar{F}_{c2} = \bar{F}_{r2} + j\bar{F}_{i2}$, $\bar{F}_{cn} = \bar{F}_{rn} + j\bar{F}_{in}$, the partial derivatives of the active (real) power transmission with respect to each real and imaginary control force may be written as:

$$\frac{\partial P_{xa}}{\partial \bar{F}_{r1}} = \frac{1}{2}\int_0^{L_y} \mathrm{Re}\Big\{ \bar{F}^*_p A(1,2) + \bar{F}_p A(2,1) + \bar{F}_{c1} A(2,2) + \bar{F}^*_{c1} A(2,2) + \bar{F}^*_{c2} A(3,2) + \bar{F}_{c2} A(2,3) + \cdots + \bar{F}^*_{cn} A(n+1,2) + \bar{F}_{cn} A(2,n+1) \Big\} \, dy \tag{10.3.59}$$

$$\frac{\partial P_{xa}}{\partial \bar{F}_{i1}} = \frac{1}{2}\int_0^{L_y} \mathrm{Re}\Big\{ j\bar{F}^*_p A(1,2) - j\bar{F}_p A(2,1) + j\bar{F}^*_{c1} A(2,2) - j\bar{F}_{c1} A(2,2) + j\bar{F}^*_{c2} A(3,2) - j\bar{F}_{c2} A(2,3) + \cdots + j\bar{F}^*_{cn} A(n+1,2) - j\bar{F}_{cn} A(2,n+1) \Big\} \, dy \tag{10.3.60}$$

$$\frac{\partial P_{xa}}{\partial \bar{F}_{r2}} = \frac{1}{2}\int_0^{L_y} \mathrm{Re}\Big\{ \bar{F}^*_p A(1,3) + \bar{F}_p A(3,1) + \bar{F}^*_{c1} A(2,3) + \bar{F}_{c1} A(3,2) + \bar{F}^*_{c2} A(3,3) + \bar{F}^*_{c2} A(3,3) + \cdots + \bar{F}^*_{cn} A(n+1,3) + \bar{F}_{cn} A(3,n+1) \Big\} \, dy \tag{10.3.61}$$

$$\frac{\partial P_{xa}}{\partial \bar{F}_{i2}} = \frac{1}{2}\int_0^{L_y} \mathrm{Re}\Big\{ j\bar{F}^*_p A(1,3) - j\bar{F}_p A(3,1) + j\bar{F}^*_{c1} A(2,3) - j\bar{F}_{c1} A(3,2) + j\bar{F}^*_{c2} A(3,3) - j\bar{F}_{c2} A(3,3) + \cdots + j\bar{F}^*_{cn} A(n+1,3) - j\bar{F}_{cn} A(3,n+1) \Big\} \, dy \tag{10.3.62}$$

$$\frac{\partial P_{xa}}{\partial \bar{F}_{rn}} = \frac{1}{2}\int_0^{L_y} \mathrm{Re}\Big\{ \bar{F}^*_p A(1,n+1) + \bar{F}_p A(n+1,1) + \bar{F}^*_{c1} A(2,n+1) + \bar{F}_{c1} A(n+1,2) + \bar{F}^*_{c2} A(3,n+1) + \bar{F}_{c2} A(n+1,3) + \bar{F}^*_{cn} A(n+1,n+1) + \bar{F}_{cn} A(n+1,n+1) \Big\} \, dy \tag{10.3.63}$$

$$\frac{\partial P_{xa}}{\partial \bar{F}_{in}} = \frac{1}{2}\int_0^{L_y} \text{Re}\Big\{ \text{j}\bar{F}_p^* A(1,n+1) - \text{j}\bar{F}_p A(n+1,1) + \text{j}\bar{F}_{c1}^* A(2,n+1) - \text{j}\bar{F}_{c1} A(n+1,2)$$
$$+ \text{j}\bar{F}_{c2}^* A(3,n+1) - \text{j}\bar{F}_{c2} A(n+1,3) + \cdots + \text{j}\bar{F}_{cn}^* A(n+1,n+1) - \text{j}\bar{F}_{cn}^* A(n+1,n+1)\Big\}\,\text{d}y \tag{10.3.64}$$

An optimum set of control forces corresponding to a minimum power transmission is achieved when each of the derivatives is zero. The matrix form of the system Equations (10.3.59) to (10.3.64) is:

$$\begin{bmatrix} \bar{F}_{c1} \\ \bar{F}_{c2} \\ \vdots \\ \bar{F}_{cn} \end{bmatrix} = - \begin{bmatrix} \int_0^{L_y}[A^*(2,2) + A(2,2)]\,\text{d}y & \int_0^{L_y}[A^*(3,2) + A(2,3)]\,\text{d}y & \cdots & \int_0^{L_y}[A^*(n+1,2) + A(2,n+1)]\,\text{d}y \\ \int_0^{L_y}[A^*(2,3) + A(3,2)]\,\text{d}y & \int_0^{L_y}[A^*(3,3) + A(3,3)]\,\text{d}y & \cdots & \int_0^{L_y}[A^*(n+1,3) + A(3,n+1)]\,\text{d}y \\ \vdots & \vdots & \vdots & \vdots \\ \int_0^{L_y}[A^*(2,n+1) + A(n+1,2)]\,\text{d}y & \int_0^{L_y}[A^*(3,n+1) + A(n+1,3)]\,\text{d}y & \cdots & \int_0^{L_y}[A^*(n+1,n+1) + A(n+1,n+1)]\,\text{d}y \end{bmatrix}^{-1} \left(\begin{bmatrix} \int_0^{L_y} A^*(1,2)\,\text{d}y \\ \int_0^{L_y} A^*(1,3)\,dy \\ \vdots \\ \int_0^{L_y} A^*(1,n+1)\,\text{d}y \end{bmatrix} + \begin{bmatrix} \int_0^{L_y} A(2,1)\,\text{d}y \\ \int_0^{L_y} A(3,1)\,\text{d}y \\ \vdots \\ \int_0^{L_y} A(n+1,1)\,\text{d}y \end{bmatrix} \right) \bar{F}_p \tag{10.3.65}$$

By comparing Equation (10.3.41) with Equation (10.3.53) and Equation (10.3.51) with Equation (10.3.65), it can be seen that the single force changes to a force vector when in-phase force control changes to independent force control. The expression for power transmission given by Equation (10.3.53) not only includes each force term, but also includes coupling force terms, which makes independent force control much more complex than in-phase force control to analyse. As has been shown in the analysis, the procedure can be used for any number n of independently controlled forces.

The matrix that must be inverted in Equation (10.3.65) is ill-conditioned for the case where the location of power transmission minimisation is in the near-field of the control force array; thus, more reliable results are obtained if the point of minimisation is always in the far-field of the control force.

10.3.7 Numerical Results

Numerical results obtained using the preceding analysis for a semi-infinite plate indicate that it is extremely difficult to achieve a significant degree of vibratory power transmission control using either a single control force or a row of in-phase control forces (Pan and Hansen, 1995). Although control force locations which result in significant (greater than 15 to 20 dB) power transmission reductions do exist for any particular frequency, the locations are extremely frequency dependent and for higher frequencies (above the cut-on frequency of the third cross-mode) the locations are very small in area. As the frequency dependency of the location can also be viewed as speed of sound (or temperature) sensitivity, it may be concluded that the use of a single control force or line of in-phase control forces to control vibratory power transmission in a semi-infinite plate is impractical. This is demonstrated by the plots in Figures 10.21(a) and (b), for which the control force location is optimum at 210 Hz. The figures show how the maximum achievable reduction in power transmission varies with frequency for a single primary and control force (Figure 10.2(a)) and for a row of equal number of primary and control forces (Figure 10.2(b)). As expected, good control is achieved only over very narrow frequency ranges in both cases, although the row of control forces performs somewhat better. It can be seen from Figure 10.21(b), that the difference in using six control forces compared to three is very small and in fact the results for six and twelve primary and control forces are almost identical to the three force case.

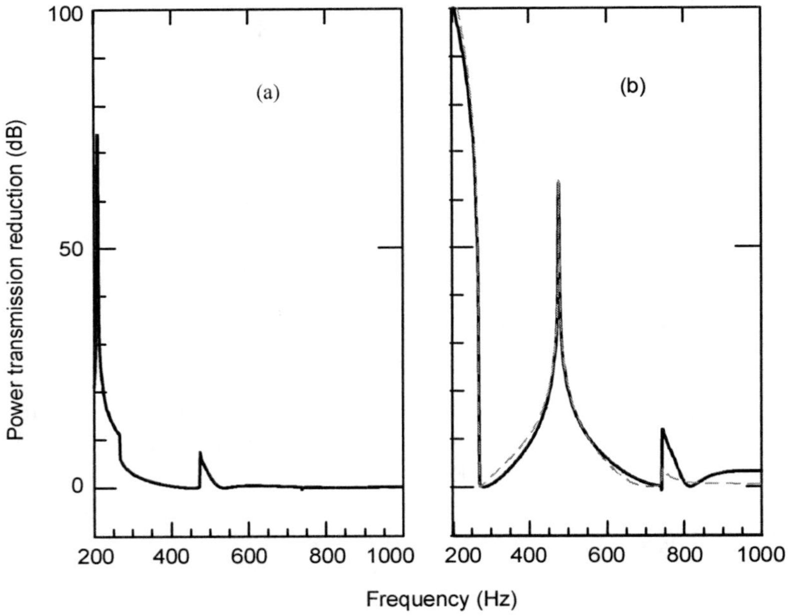

Figure 10.21 Power flow reduction as a function of frequency with the control forces located at the optimum location for 210 Hz (from Pan and Hansen, 1995): (a) a single off-centre primary and control force; (b) equal numbers of primary and control forces driven in-phase.

——— a single primary force with a single control force in part (a) and 3 primary forces with 3 control forces in part (b);

— — — — — 6 primary forces and 6 control forces.

In contrast to the results for the row of in-phase control forces, a row of three independently driven control forces produces large reductions in power transmission over a wide frequency range as indicated in Figure 10.22.

As an aside, it is of interest to note that the far-field vibration generated by a pair of piezoceramic crystals placed on opposite sides of a thin plate or beam can be simulated by a point force of a specified amplitude (Pan and Hansen, 1993b).

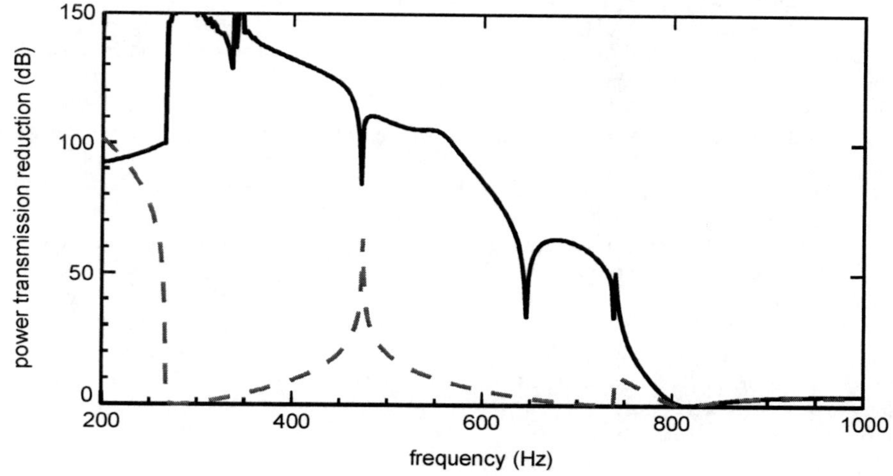

Figure 10.22 Power flow reduction for three in-phase, uniform-amplitude primary forces and three independent control forces, as a function of frequency with the control forces located at the optimum location for 210 Hz.
— — — — — — 3 primary forces and 3 control forces driven in-phase;
———————— 3 primary forces driven in-phase and 3 control forces driven independently.

REFERENCES

Brennan, M.J., Elliott, S.J. and Pinnington, R.J. (1992). Active control of vibrations transmitted through struts. In *Proceedings of First International MOVIC Conference*. Japan, 605–609.

Clark, R.L. (1994). Adaptive feedforward modal space control. *Journal of the Acoustical Society of America*, **95**, 2989.

Clark, R.L., Pan, J and Hansen, C.H. (1992). An experimental study of the active control of multiple wave types in a beam. *Journal of the Acoustical Society of America*, **92**, 871–876.

Elliott, S.J., Stothers, I.M. and Billet, L. (1990). The use of adaptive filters in the active control of flexural waves in structures. In *Active Noise and Vibration Control.* ASME publication NCA. **8**, 161–166.

Elliott, S.J. and Billet, L. (1993). Adaptive control of flexural waves propagating in a beam. *Journal of Sound and Vibration*, **163**, 295–310.

Fuller, C.R., Gibbs, C.P. and Silcox, R.J. (1990). Simultaneous control of flexural and extensional waves in beams. *Journal of Intelligent Material Systems and Structures*, **1**, 235–247.

Hansen, C.H., Young, A.J. and Pan, X. (1993). Active control of harmonic vibration in beams with arbitrary end conditions. In *Proceedings of the Second Conference on Recent Advances in Active Control of Sound and Vibration.* Technomic Publishing Co., Blacksburg, VA, 487–506.

Mace, R.B. (1987). Active control of flexural vibrations. *Journal of Sound and Vibration*, **114**, 253–270.

McKinnell, R.J. (1988). Active vibration isolation by cancellation of bending waves. *Proceedings of the Institute of Acoustics*, **10**, 581–588.

McKinnell, R.J. (1989). Active vibration isolation by cancelling bending waves. *Proceedings of the Royal Society of London*, **A421**, 357–393.

Pan, J. and Hansen, C.H. (1991). Active control of total vibratory power flow in a beam. 1. Physical system analysis. *Journal of the Acoustical Society of America*, **89**, 200–209.

Pan, X. and Hansen, C.H. (1993a). Effect of error sensor type and location on the active control of beam vibration. *Journal of Sound and Vibration*, **165**, 497–510.

Pan, X. and Hansen, C.H. (1993b). Effect of end conditions on the active control of beam vibration. *Journal of Sound and Vibration*, **168**, 429–448.

Pan, X. and Hansen, C.H. (1995). Active control of vibratory power transmission along a semi-infinite plate. *Journal of Sound and Vibration*, **184**, 585-610.

Pereira, A.K.A. and Arruda, R.F. (2000). Active control of the structural intensity in beams using a frequency domain adaptive method. *Journal of Intelligent Material systems and Structures*, **11**, 3–13.

Redman–White, W., Nelson, P.A. and Curtis, A.R.D. (1987). Experiments on the active control of flexural wave power flow. *Journal of Sound and Vibration*, **112**, 187–191.

Scheuren, J. (1985). Active control of bending waves in beams. In *Proceedings of Inter-Noise '85.* Institute of Noise Control Engineering, 591–595.

Scheuren, J. (1988). Non-reflecting termination for bending waves in beams by active means. In *Proceedings of Inter-Noise '88.* Institute of Noise Control Engineering, 1065–1068.

Scheuren, J. (1989). Iterative design of bandlimited FIR filters with gain constraints for active control of wave propagation. In *Proceedings of the International Conference on Acoustics Speech and Signal Processing.* IEEE, paper A2.3.

Sutton T.J., Elliott S.J., Brennan M.J., Heron K.H. and Jessop D.A.C (1997). Active isolation of multiple structural waves on a helicopter gearbox support strut. *Journal of Sound and Vibration*, **205**, 81–101.

Young, A.J. and Hansen, C.H. (1994). Control of flexural vibration in a beam using a piezoceramic actuator and an angle stiffener. *Journal of Intelligent Material Systems and Structures*, **5**, 536–549.

CHAPTER 11

Feedback Control of Flexible Structures Described in Terms of Modes

11.1 INTRODUCTION

In an attempt to put bounds on the amount of material included here, this chapter, perhaps more than others in this book, has been limited in its scope, specifically to consideration of structures that are not too large or flexible.

Active vibration control of flexible structures (using feedback techniques) is a topic that has been vastly researched and published in recent years. Significant effort has been directed at the problem of vibration attenuation in large flexible structures, with a focus on large space structures. The dynamic response of such structures is typically characterised by fundamental modal resonance frequencies in the order of 1 Hz, and a (very) large number of lightly damped, poorly modelled modes in the bandwidth of most ('practically' formulated) control systems. Therefore, there is a limit to which the response of such a structure, and hence the controller design for the problem, can be based upon system modes (von Flotow, 1988).

Good starting points for the study of large space structure control include texts edited by Atluri and Amos (1988), and Junkins (1990). The series of texts edited by Leondes are also valuable sources of information. Review papers by Balas (1982), Nurre *et al.* (1984) and Hallauer (1990) are also useful. More recent useful texts on structural vibration control include Fuller et al. (1996), Preumont (2002) and Moheimani et al. (2003).

In contrast to the work cited above for large space structures, the remainder of this chapter will be concerned with the vibration control of structures described in terms of system modes. This subject description limits the applicability of the analysis to low-frequency controllers for 'small' structures and/or 'stiff' structures, the responses of which are largely characterised by well spaced resonances in the controller bandwidth. These response characteristics are similar to the response characteristics of the dynamic systems that have been considered thus far in this book, such as the low-frequency response of coupled structural/acoustic enclosures.

In the following sections, the development of feedback control systems will be considered based upon a knowledge of the modes of the target structure. The treatment will begin with the development of feedback laws for this 'modal control', and then go on to a consideration of issues such as stability, model reduction, and sensor and actuator placement. Much of the work outlined in this chapter follows on directly from the discussion of Chapter 5, the review of modern control theory, which should be referred to for the foundations of the following analyses.

11.2 MODAL CONTROL

The mathematical modelling approach, based on a modal description of a vibrating structure to examine feedback control of its vibration, has been used for many years, with early references found in, for example, Balas (1978a, 1978b, 1982), Meirovitch and Baruh (1982) and Meirovitch and Oz (1980a). Early related works on modal control include Simon and Mitter (1968) and Porter and Crossley (1972).

As mentioned in the introduction, there is a practical limit to any control approach formulated on the basis of modal analysis. This limit is approached as the maximum frequency in the controller bandwidth increases. Increasing frequency results in increasing modal density, including higher-order modes in the controller bandwidth. These higher-order modes are most susceptible to model errors, especially when finite element models are used, and the increasing number of poorly known modes is a potential disaster for the control system designer. Analysis and feedback control techniques based upon wave propagation have been put forward (for example, MacMartin and Hall, 1991; Miller and von Flotow, 1989; Miller et al., 1990; von Flotow, 1986; von Flotow and Schafer, 1986), but are outside the scope of this text (however, wave propagation analysis was used in Chapter 10 for the analysis of feedforward control systems for structures, and the interested reader is referred to that chapter).

In formulating the control laws in this section, there are several practical constraints that will be imposed (Athans, 1970):

1. A finite (small) number of sensors and actuators is available;
2. The controller must be finite (small) in dimension and
3. There is some degree (usually small) of damping in the system.

These constraints will influence the controller design, and implicitly the first two constraints will be responsible for a few problems (spillover), while the last has the potential to enhance the stability of the controller.

11.2.1 Development of the Governing Equations

The systems that are of interest in this chapter, mechanically flexible structures, are distributed parameter systems, where the system parameters are spatially dependent. Such systems are, in theory, infinite-dimensional, which means that in practice for a model to have good fidelity, it must be large. As will be seen later in this chapter, some of the fundamental problems associated with the active control of flexible structures arise from the use of a (relatively) small-dimension controller to control a large-dimension system.

The response of each mode of a multi-modal structure can be described by the partial differential equation:

$$m(\boldsymbol{r})\frac{\partial^2 w(\boldsymbol{r},t)}{\partial t^2} + c_d\frac{\partial w(\boldsymbol{r},t)}{\partial t} + kw(\boldsymbol{r},t) = f(\boldsymbol{r},t) \tag{11.2.1}$$

where $w(\boldsymbol{r}, t)$ is the displacement of the structure at some location $\boldsymbol{r}$ at time t, $m(\boldsymbol{r})$ is the effective modal mass influencing the motion at location $\boldsymbol{r}$, c_d and k describe the modal damping and stiffness of the structure respectively, and $f(\boldsymbol{r}, t)$ is the applied force at location $\boldsymbol{r}$ at time t.

The damping of a flexible structure, is often small and not well known, so it is common to assume that $c_d = 0$. When the effects of damping are to be included in the model, c_d will be represented here by proportional damping, a class of viscous damping where the damping operator is a linear combination of the stiffness operator and mass distribution (as discussed in Chapter 4):

$$c_d = \alpha_1 m + \alpha_2 k \tag{11.2.2}$$

For the discussion here, it will be assumed that:

$$c_d = 2\zeta\sqrt{k/m} \tag{11.2.3}$$

where ζ is a small (positive) damping coefficient.

For the analytical development being undertaken in this section, the control sources will be modelled as point forces, so that the applied force distribution can be written as the sum of a disturbance input and N_s control inputs $f_{c,i}(t)$:

$$f(\boldsymbol{r},t) = \sum_{i=1}^{N_s} b_i(\boldsymbol{r}) f_{c,i}(t) + \text{disturbances} \tag{11.2.4}$$

where $b_i(\boldsymbol{r})$ is an influence function value at location $\boldsymbol{r}$. For point force excitation, the influence function is commonly taken to be a Dirac delta function.

The response of the structure to the applied force distribution is measured by a set of linearly independent sensors, the output of each of which is defined by the expression:

$$y(t) = \int_S \left[g(\boldsymbol{r}) w(\boldsymbol{r},t) + e(\boldsymbol{r}) \frac{\partial w(\boldsymbol{r},t)}{\partial t} \right] \mathrm{d}x \tag{11.2.5}$$

where g is an influence function defining the measurement of the structure displacement, e is an influence function defining the velocity measurement, and integration is over the surface of the structure, S. If the measurement is from a point displacement sensor, where the influence function $g(\boldsymbol{r})$ is defined by the product of a gain and a Dirac delta function, the output from the sensor is defined by the expression:

$$y(t) = w(\boldsymbol{r},t) g \tag{11.2.6}$$

where g is the gain of the displacement sensing system. Similarly, if the measurement is from a point velocity sensor, where the influence function $e(\boldsymbol{r})$ is also defined by the product of a gain and a Dirac delta function, the output from the sensor is defined by the expression:

$$y(t) = \dot{w}(\boldsymbol{r},t) e \tag{11.2.7}$$

where e is the gain of the velocity sensing system.

The domain of the stiffness operator k is assumed to contain all smooth functions satisfying the boundary conditions of the structure, and which have the usual inner product (*,*) and norm $\| \; \|$. For the structures of interest here, k is assumed to have a discrete spectrum, with isolated resonance frequencies defined by the eigenvalue problem associated with the governing partial differential Equation (11.2.1):

$$k\psi_i = \lambda_i m \psi_i \tag{11.2.8}$$

where λ_i is the ith eigenvalue and ψ_i is the associated eigenvector. The resonance frequencies ω are related to the eigenvalues of Equation (11.2.8) by:

$$\omega_i = \lambda_i^{1/2} \tag{11.2.9}$$

while the mode shape functions are defined by the eigenvectors. It is usual to assume that the stiffness operator is self-adjoint, such that all eigenvectors (mode shape functions) are orthogonal. In this chapter it will also be assumed that the mode shape functions are mass normalised, to be consistent with literature on this topic. This means that the mass and stiffness terms can be normalised respectively as:

$$\int_S m(\boldsymbol{r})\psi_i(\boldsymbol{r})\psi_l(\boldsymbol{r})\ \mathrm{d}S = \delta_{i,l} \tag{11.2.10}$$

and

$$\int_S \psi_i(\boldsymbol{r})k\psi_l(\boldsymbol{r})\ \mathrm{d}S = \lambda_i\delta_{i,l} = \omega_i^2\delta_{i,l} \tag{11.2.11}$$

where $\delta_{i,l}$ is the Kronecker delta function, defined by:

$$\delta_{i,l} = \begin{cases} 0 & i \neq l \\ 1 & i = l \end{cases} \tag{11.2.12}$$

It is also usual to assume that the stiffness operator is positive definite, such that all resonance frequencies are greater than zero. For the discussion in this chapter, it will be assumed that the resonance frequencies (and hence eigenvalues) are ordered as $0 \le \omega_1 \le \omega_2 \le \ldots$.

Having defined several properties of the systems of interest, the responses of which are governed by the partial differential Equation (11.2.1), it is straightforward to re-express Equation (11.2.1) as a set of modal equations, where each equation defines the response of a particular structural mode. The structural displacement amplitude at time t and location $\boldsymbol{r}$, $w(\boldsymbol{r}, t)$, can be expressed as an infinite sum of modal contributions:

$$w(\boldsymbol{r},t) = \sum_{i=1}^{\infty} z_i(t)\psi_i(\boldsymbol{r}) \tag{11.2.13}$$

where $\psi_i(\boldsymbol{r})$ is the value of the ith mode shape function at location $\boldsymbol{r}$ and $z_i(t)$ is the modal amplitude at time t. By substituting this expansion into Equation (11.2.1), multiplying both sides by ψ_i, and integrating over the surface of the structure, S, the governing equation of motion can be re-expressed as an infinite sum of modal contributions:

$$\sum_{i=1}^{\infty} \left[\ddot{z}_i(t) + 2\zeta\omega_i\dot{z}_i(t) + \omega_i^2 z_i(t) = f_i(t)\right] \tag{11.2.14}$$

where $f_i\,(t)$ is a modal generalised force:

$$f_i(t) = \int_S f(\boldsymbol{r},t)\psi_i(r)\ \mathrm{d}r \tag{11.2.15}$$

It can be seen that each term in the infinite summation has the same form as the equation of motion for a single-degree-of-freedom system discussed in Chapter 4. Thus, individual modes may be considered as behaving as single degree of freedom systems characterised by a resonance frequency and damping coefficient. Also, as the control inputs are assumed to be coming from point sources, the control source contribution to the total modal generalised force can be written as:

$$f_{c,i}(t) = \sum_{\iota=1}^{N_s} f_c(\boldsymbol{r}_\iota, t)\psi_i(\boldsymbol{r}_\iota) \tag{11.2.16}$$

where $\boldsymbol{r}_\iota$ is the location at which control force, $f_c(\boldsymbol{r}_\iota)$ is applied.

The system outputs are now also defined in terms of structural modes. With the assumption that a measurement is done by point displacement sensor, the output of the ith mode is defined by:

$$y_i(t) = z_i(t)\psi_i(\boldsymbol{r})g \tag{11.2.17}$$

Similarly, with the assumption that a measurement is done by point velocity sensor, the output of the ith mode is defined by:

$$y_i(t) = \dot{z}_i(t)\psi_i(\boldsymbol{r})e \tag{11.2.18}$$

11.2.2 Discrete Element Model Development

It is often common to use finite element methods (FEMs), or some other form of discretisation method, to model the response of a structure. A typical approximation of the governing partial differential Equation (11.2.1) can be obtained from the relationship:

$$w(\boldsymbol{r}, t) \approx \sum_{i=1}^{N_\theta} q_i(t)\theta_i(\boldsymbol{r}) \tag{11.2.19}$$

where $\theta_i(\boldsymbol{r})$ is a (known) local approximation of the structure, and $q_i(t)$ is the amplitude of the ith discrete element (note that if finite elements were being used, then the value of the local approximation of the structure at $\boldsymbol{r}$ would be zero for all elements except the element actually located at $\boldsymbol{r}$). Using Equation (11.2.19), a well-known approximation of Equation (11.2.1) is (Balas, 1990):

$$\boldsymbol{M}\ddot{\boldsymbol{q}} + \boldsymbol{C}\dot{\boldsymbol{q}} + \boldsymbol{K}\boldsymbol{q} = \boldsymbol{f} \tag{11.2.20}$$

where $\boldsymbol{q}$ is the vector of element displacements:

$$\boldsymbol{q} = \begin{bmatrix} q_1 & q_2 & \cdots & q_{N_\theta} \end{bmatrix}^{\mathrm{T}} \tag{11.2.21}$$

$\boldsymbol{M}$ is a mass matrix, the i,ι element of which is defined by:

$$\boldsymbol{M}(i,\iota) = \int_S m(\boldsymbol{r})\theta_i(\boldsymbol{r})\theta_\iota(\boldsymbol{r})\, \mathrm{d}\boldsymbol{r} \tag{11.2.22}$$

where $m(\boldsymbol{r})$ is the mass per unit area at $\boldsymbol{r}$, $\boldsymbol{C}$ is a damping matrix, the i,ι element of which is defined by the relationship:

$$\boldsymbol{C}(i,\iota) = \int_S \theta_i(\boldsymbol{r})c_d\theta_\iota(\boldsymbol{r})\, \mathrm{d}\boldsymbol{r} \tag{11.2.23}$$

and $\boldsymbol{K}$ is a stiffness matrix, the i, $\imath$ element of which is defined by the relationship:

$$\boldsymbol{K}(i,\imath) = \int_S \theta_i(\boldsymbol{r}) k \theta_\imath(\boldsymbol{r})\, \mathrm{d}\boldsymbol{r} \tag{11.2.24}$$

The control force contribution to the force on the structure is defined by:

$$\boldsymbol{f} = \boldsymbol{B}\boldsymbol{f}_c \tag{11.2.25}$$

where the $i,\imath$ element of the control input matrix $\boldsymbol{B}$ is defined by the expression:

$$\boldsymbol{B}(i,\imath) = \int_S \theta_i(\boldsymbol{r}) b_\imath(\boldsymbol{r})\, \mathrm{d}\boldsymbol{r} \tag{11.2.26}$$

where $b_\imath$ is the influence function for the $\imath$th control source.

The system output associated with Equation (11.2.20), which is the finite element approximation of Equation (11.2.5), is given by the expression:

$$\boldsymbol{y} = \boldsymbol{G}\boldsymbol{q} + \boldsymbol{E}\dot{\boldsymbol{q}} \tag{11.2.27}$$

where $\boldsymbol{G}$ and $\boldsymbol{E}$ are displacement and velocity output matrices, defined by the relationships:

$$\boldsymbol{G}(i,\imath) = \int_S \theta_i(\boldsymbol{r}) g_\imath(\boldsymbol{r})\, \mathrm{d}\boldsymbol{r} \tag{11.2.28}$$

$$\boldsymbol{E}_0(i,\imath) = \int_S \theta_i(\boldsymbol{r}) e_\imath(\boldsymbol{r})\, \mathrm{d}\boldsymbol{r} \tag{11.2.29}$$

This system model is sometimes referred to as a configuration space model, with the displacement vector referred to as the configuration vector. For complicated structures, it is more likely that the system model will be formulated in this lumped parameter form than in the (exact) distributed parameter form of Equation (11.2.1).

The finite element model of system response can also be restated in terms of system modes, using the generalised eigenvalue problem associated with Equation (11.2.20):

$$\boldsymbol{K}\boldsymbol{\psi}_i = \lambda_i \boldsymbol{M}\boldsymbol{\psi}_i \tag{11.2.30}$$

where λ_i is the ith eigenvalue, and ψ_i is the associated eigenvector. Transformation from element coordinates to modal coordinates, z, is accomplished using the transformation matrix $\boldsymbol{\Psi}$:

$$\boldsymbol{q} = \boldsymbol{\Psi}\boldsymbol{z} \tag{11.2.31}$$

where the columns of $\boldsymbol{\Psi}$ are the eigenvectors defined by Equation (11.2.30):

$$\boldsymbol{\Psi} = \begin{bmatrix} \boldsymbol{\psi}_1 & \boldsymbol{\psi}_2 & \cdots & \boldsymbol{\psi}_{N_\theta} \end{bmatrix} \tag{11.2.32}$$

and N_θ is the number of modes considered in the model. As both the mass matrix $\boldsymbol{M}$ and stiffness matrix $\boldsymbol{K}$ are non-negative and symmetric, the transforms:

$$\boldsymbol{\Psi}^{\mathrm{T}} \boldsymbol{M} \boldsymbol{\Psi} = \boldsymbol{I} \tag{11.2.33}$$

and

$$\boldsymbol{\Psi}^{\mathrm{T}} \boldsymbol{K} \boldsymbol{\Psi} = \boldsymbol{\Lambda} = \mathrm{diag}[\lambda_1 \ \lambda_2 \ \cdots \ \lambda_{N_\theta}] \tag{11.2.34}$$

exist, and so Equation (11.2.20) can be transformed into 'approximate' (discrete models of continuous) modal coordinates as:

$$\ddot{z} + 2\xi\omega\dot{z} + \omega^2 z = \tilde{\boldsymbol{B}}\boldsymbol{f}_c \tag{11.2.35}$$

where

$$\tilde{\boldsymbol{B}} = \boldsymbol{\Psi}^{\mathrm{T}}\boldsymbol{B} \tag{11.2.36}$$

The output equation is similarly transformed to:

$$y = \tilde{\boldsymbol{G}}z + \tilde{\boldsymbol{E}}\dot{z} \tag{11.2.37}$$

where

$$\tilde{\boldsymbol{G}} = \boldsymbol{G\Psi}; \qquad \tilde{\boldsymbol{E}} = \boldsymbol{E\Psi} \tag{11.2.38a,b}$$

It should be noted that the accuracy of the mode shape functions derived from a finite element model decreases as the mode number increases. In general, the number of useable modes from the calculation is an order of magnitude less than the number of finite elements used in the calculation.

11.2.3 Transformation into State-Space Form

It is straightforward to re-express the modal equations of motion in a state-space format. To do so, the state vector can be defined as:

$$\boldsymbol{x}_i = \begin{bmatrix} z_i \\ \dot{z}_i \end{bmatrix} \tag{11.2.39}$$

Assuming point forces are used for control, and ignoring the (primary source) disturbance inputs in this development (which, as discussed in Chapter 5, is commonly done in the development of state-space models), the response of a single mode in Equation (11.2.14) can be written as:

$$\dot{\boldsymbol{x}}_i(t) = \boldsymbol{A}_i x_i(t) + \boldsymbol{B}_i \boldsymbol{f}_c(t) \tag{11.2.40}$$

where

$$\boldsymbol{A}_i = \begin{bmatrix} 0 & 1 \\ -\omega_i^2 & -2\zeta\omega_i \end{bmatrix}; \quad \boldsymbol{B}_i = \begin{bmatrix} 0 & \cdots & 0 \\ \psi_i(\boldsymbol{r}_1) & \cdots & \psi_i(\boldsymbol{r}_{N_c}) \end{bmatrix} \tag{11.2.41a,b}$$

and N_c is the number of control inputs, situated at locations $\boldsymbol{r}_1,\ldots,\boldsymbol{r}_N$. The output of this mode, as measured by point displacement and velocity sensors, is:

$$y = Gx \tag{11.2.42}$$

where y is the vector of measured outputs, and

$$G = \begin{bmatrix} \psi_1(r_1) & 0 \\ \vdots & \\ \psi_1(r_{N_d}) & 0 \\ 0 & \psi_1(r_1) \\ \vdots & \\ 0 & \psi_1(r_{N_v}) \end{bmatrix} \tag{11.2.43}$$

where there are N_d displacement sensors and N_v velocity sensors being used.

Equation (11.2.40) describes the response of a single mode of vibration. This can be expanded to the set of N_m modelled modes as:

$$\dot{x} = Ax + Bf_c(t) \tag{11.2.44}$$

where the state vector is now defined as:

$$x = \begin{bmatrix} z_1 & z_2 & \cdots & z_{N_m} & \dot{z}_1 & \dot{z}_2 & \cdots & \dot{z}_{N_m} \end{bmatrix}^{\mathrm{T}} \tag{11.2.45}$$

and

$$A = \begin{bmatrix} \mathbf{0} & I \\ -\Lambda & -2\zeta\Lambda^{1/2} \end{bmatrix}; \quad B = \begin{bmatrix} \mathbf{0} \\ B^* \end{bmatrix} = \begin{bmatrix} 0 & \cdots & 0 \\ & \vdots & \\ 0 & \cdots & 0 \\ \psi_1(r_1) & \cdots & \psi_I(r_{N_c}) \\ & \vdots & \\ \psi_{N_m}(r_1) & \cdots & \psi_{N_m}(r_{N_c}) \end{bmatrix} \tag{11.2.46a–c}$$

In Equation (11.2.46), I is an $(N_m \times N_m)$ identity matrix, Λ is the $(N_m \times N_m)$ diagonal matrix of system eigenvalues, and B^* is the lower, non-zero part of B. Also, the damping coefficient is shown as the same for all modes, but this need not be the case. The output equation is now:

$$y = Gx \tag{11.2.47}$$

where, if it is again assumed that displacement and velocity sensors are being used:

$$G = \begin{bmatrix} G_d & \mathbf{0} \\ \mathbf{0} & G_v \end{bmatrix} = \begin{bmatrix} \psi_1(r_1) & \cdots & \psi_{N_m}(r_1) & 0 & \cdots & 0 \\ \vdots & & & & & \\ \psi_1(r_{N_d}) & \cdots & \psi_{N_m}(r_{N_d}) & 0 & \cdots & 0 \\ 0 & \cdots & 0 & \psi_1(r_1) & \cdots & \psi_{N_m}(r_1) \\ \vdots & & & & & \\ 0 & \cdots & 0 & \psi_1(r_{N_v}) & \cdots & \psi_{N_m}(r_{N_v}) \end{bmatrix} \tag{11.2.48a,b}$$

One final point to note is that in the absence of damping, the eigenvalues of the system are purely real; any complex component to the eigenvalues arises from the damping. From the discussion in Chapter 5, where it was pointed out that the response of a system mode is described by an exponential with a time constant proportional to the associated eigenvalue, it can be concluded that the reverberation time of many such systems is extremely long. Also, the stability margins of the systems are small. The former provides an impetus for applying active control to flexible structures; the latter dictates that care in system design is required.

11.2.4 Model Reduction

The number of modes required to model the response of a structure with good fidelity may be quite large, and so it is generally impractical to consider all modelled modes in the development of a control strategy. Therefore, a reduced order model (ROM) of the system must be formulated in order to derive a control strategy. A discussion of some of the ways to derive the 'best' reduced order model for a given system will be undertaken later in this chapter. In this section it will be assumed that some criteria have been applied to the set of modelled modes to facilitate partitioning into controlled (modelled in controller design) and residual or uncontrolled (unmodelled in controller design) mode subsets, and then the control strategy will be formulated solely from consideration of the controlled modes.

If the set of modelled modes is divided into N controlled modes and R uncontrolled modes, the state vector containing the displacements and velocities of the controlled modes is defined as $\boldsymbol{x}_N$, and the state vector containing the displacements and velocities of the uncontrolled modes is defined as $\boldsymbol{x}_R$, then,

$$\dot{\boldsymbol{x}}_N = \boldsymbol{A}_N \boldsymbol{x}_N + \boldsymbol{B}_N \boldsymbol{f}_c \tag{11.2.49}$$

and

$$\dot{\boldsymbol{x}}_R = \boldsymbol{A}_R \boldsymbol{x}_R + \boldsymbol{B}_R \boldsymbol{f}_c \tag{11.2.50}$$

where

$$\boldsymbol{A}_N = \begin{bmatrix} \mathbf{0} & \boldsymbol{I}_N \\ -\boldsymbol{\Lambda}_N & -2\zeta\boldsymbol{\Lambda}_N^{1/2} \end{bmatrix} \tag{11.2.51a}$$

$$\boldsymbol{B}_N = \begin{bmatrix} \mathbf{0} \\ \boldsymbol{B}_N^* \end{bmatrix} = \begin{bmatrix} 0 & \cdots & 0 \\ & \vdots & \\ 0 & \cdots & 0 \\ \psi_1(\boldsymbol{r}_1) & \cdots & \psi_1(\boldsymbol{r}_{N_c}) \\ & \vdots & \\ \psi_N(\boldsymbol{r}_1) & \cdots & \psi_N(\boldsymbol{r}_{N_c}) \end{bmatrix} \tag{11.2.51b,c}$$

and

$$A_R = \begin{bmatrix} \mathbf{0} & I_R \\ -\Lambda_R & -2\zeta\Lambda_R^{1/2} \end{bmatrix} \tag{11.2.52a}$$

$$B_R = \begin{bmatrix} \mathbf{0} \\ B_R^* \end{bmatrix} = \begin{bmatrix} 0 & \cdots & 0 \\ & \vdots & \\ 0 & \cdots & 0 \\ \psi_1(r_1) & \cdots & \psi_1(r_{N_c}) \\ & \vdots & \\ \psi_R(r_1) & \cdots & \psi_R(r_{N_c}) \end{bmatrix} \tag{11.2.52b,c}$$

Similarly, the system output is partitioned as:

$$y = y_N + y_R = G_N x_N + G_R x_R \tag{11.2.53}$$

where

$$G_N = \begin{bmatrix} G_{Nd} & \mathbf{0} \\ \mathbf{0} & G_{Nv} \end{bmatrix} = \begin{bmatrix} \varphi_1(r_1) & \cdots & \varphi_N(r_1) & 0 & \cdots & 0 \\ \vdots & & & & & \\ \psi_1(r_{N_d}) & \cdots & \psi_N(r_{N_d}) & 0 & \cdots & 0 \\ 0 & \cdots & 0 & \psi_1(r_1) & \cdots & \psi_N(r_1) \\ \vdots & & & & & \\ 0 & \cdots & 0 & \psi_1(r_{N_v}) & \cdots & \psi_N(r_{N_v}) \end{bmatrix} \tag{11.2.54a,b}$$

and

$$G_R = \begin{bmatrix} G_{Rd} & \mathbf{0} \\ \mathbf{0} & G_{Rv} \end{bmatrix} = \begin{bmatrix} \psi_1(r_1) & \cdots & \psi_R(r_1) & 0 & \cdots & 0\sigma \\ \vdots & & & & & \\ \psi_1(r_{N_d}) & \cdots & \psi_R(r_{N_d}) & 0 & \cdots & 0 \\ 0 & \cdots & 0 & \psi_1(r_1) & \cdots & \psi_R(r_1) \\ \vdots & & & & & \\ 0 & \cdots & 0 & \psi_1(r_{N_v}) & \cdots & \psi_R(r_{N_v}) \end{bmatrix} \tag{11.2.55a,b}$$

11.2.5 Modal Control

Following development of a state-space model of a flexible structure, the next step is to develop feedback control laws to change the dynamic response of the structure in some desired way. In this section, control laws will be developed based upon the reduced order model, and in the process, the requirements for controllability and observability will be investigated. In the section which follows, the effect of considering only some of the structural modes in development of the control law will be evaluated.

Design of a feedback control system for flexible structures using state-space methods follows the methodology reviewed in Chapter 5. Here, the active control system consists of two parts: a state observer or estimator, which estimates the values of the system states at any time t, and a control law, or regulator, which filters the state estimates to produce control input signals. For the discussion here, the control law will consist of static gains, so that the control inputs are derived from linear combinations of the system states.

If the output signal-to-noise ratio is high, the state observer can be deterministic, as discussed in Section 5.9. However, if the signal-to-noise ratio is not high, the state observer will need to be a Kalman filter, discussed in Section 5.11. Either way, the state observer has the form:

$$\frac{\mathrm{d}\hat{\boldsymbol{x}}_N}{\mathrm{d}t} = \boldsymbol{A}_N\hat{\boldsymbol{x}}_N + \boldsymbol{B}_N\boldsymbol{u} + \boldsymbol{L}_N(\boldsymbol{y} - \boldsymbol{G}\hat{\boldsymbol{x}}_N) \tag{11.2.56}$$

where ^ denotes an estimated (or observed) quantity, N denotes quantities associated with the controlled modes in the reduced order model, and the time dependence of the state, control and output vectors has been dropped for clarity of notation.

The estimation error $\boldsymbol{e}_N$, defined as the difference between the actual state value and estimated state value:

$$\boldsymbol{e}_N = \hat{\boldsymbol{x}}_N - \boldsymbol{x}_N \tag{11.2.57}$$

is defined by the relationship:

$$\dot{\boldsymbol{e}}_N = \left(\boldsymbol{A}_N - \boldsymbol{L}_N\boldsymbol{G}_N\right)\boldsymbol{e}_N + \boldsymbol{L}_N\boldsymbol{G}_R\boldsymbol{x}_R \tag{11.2.58}$$

If the residual system contribution is ignored, this becomes:

$$\dot{\boldsymbol{e}}_N = \left(\boldsymbol{A}_N - \boldsymbol{L}_N\boldsymbol{G}_N\right)\boldsymbol{e}_N \tag{11.2.59}$$

The result of Equation (11.2.59) is applicable to systems employing a deterministic observer; for those employing a Kalman filter or optimal observer, there is an additional white noise term, as outlined in Section 5.11. Thus, the equations governing the observer for the reduced-order system are the same as those governing a fully modelled system.

The control law of interest is (negative) linear state variable feedback, defined by the relationship:

$$\boldsymbol{f}_c = -\boldsymbol{K}_N\boldsymbol{x}_N \tag{11.2.60}$$

As discussed in Chapter 5, if the system is controllable and deterministic, then the control gains can be calculated using a pole placement approach. If the system is controllable, the eigenvalues can (ignoring practicalities) be arbitrarily placed, giving the system any desired damping and stiffness characteristics.

In practice, the control law actually implemented uses estimated state values rather than the actual ones:

$$\boldsymbol{f}_c = -\boldsymbol{K}_N\hat{\boldsymbol{x}}_n \tag{11.2.61}$$

To quantify the effect which this has upon the system, the state equation can be augmented with the estimator error of Equation (11.2.59):

$$z_N = \begin{bmatrix} x_N \\ e_N \end{bmatrix} \tag{11.2.62}$$

It is straightforward to show, using the same series of steps taken in Section 5.12, that the overall response of the controller (which uses only the reduced-model modes) is governed by the expression:

$$\dot{z}_N = \begin{bmatrix} A_N - B_N K_N & B_N K_N \\ 0 & A_N - L_N G_N \end{bmatrix} z_N \tag{11.2.63}$$

As discussed in Chapter 5, the form of Equation (11.2.63) shows that the eigenvalues of the system are a combination of those of the controlled system with full (of the reduced order model) state feedback and those of the observer; the use of an observer does not change the location of the poles of the controlled system, but rather adds its own poles. This is the deterministic separation principle. A similar result exists for systems implementing an optimal (Kalman filter) observer, known as the stochastic separation principle.

The control gains could also be calculated using a steady-state optimal control approach, where the control law is chosen to minimise the performance index:

$$J_N = \int_0^\infty \left[x_N^{\mathrm{T}}(t) Q_n x_N(t) + f_c^{\mathrm{T}}(t) R_N f_c(t) \right] \mathrm{d}t \tag{11.2.64}$$

As the total energy in the controlled modes is given by the sum, $E_N(t)$, of the kinetic and potential energies of the (normalised) system:

$$E_N(t) = \frac{1}{2} \sum_{i=1}^{N} \left[\dot{z}_i^2 + \omega_i^2 z_i^2(t) \right] \tag{11.2.65}$$

where N is the number of controlled modes, a common choice for the state weighting matrix is:

$$Q_N = \frac{1}{2} \begin{bmatrix} \Lambda_N & 0 \\ 0 & I_N \end{bmatrix} \tag{11.2.66}$$

If the optimal control law is implemented with the estimated state, the controller is, in general, suboptimal. The increase in the performance index which results is (Bongiorno and Youla, 1968):

$$\Delta J = \int_0^\infty e_N^{\mathrm{T}}(t) K_N^{\mathrm{T}} R_N K_N e_N(t) \; \mathrm{d}t \tag{11.2.67}$$

As discussed in Section 5.8, deriving the optimal control gains normally involves solving a matrix Riccati equation which is the same order as the state equation, which for the systems of interest here is $(2N \times 2N)$. This order can be reduced by taking advantage of the form of the state equations, as will be seen shortly.

When the system response is described in terms of 'modal' state equations, it is straightforward to assess controllability and observability. From the discussion in Chapter 5, for an LTI system to be controllable, the controllability matrix for the pair $(\boldsymbol{A}_N, \boldsymbol{B}_N)$, defined as:

$$\begin{bmatrix} \boldsymbol{B}_N & \boldsymbol{A}_N\boldsymbol{B}_N & \cdots & \boldsymbol{A}_N^{2N-1}\boldsymbol{B}_N \end{bmatrix} \tag{11.2.68}$$

must be full rank, which in this case is rank $2N$. Referring to the definition of $\boldsymbol{A}_N$ given in Equation (11.2.51), and ignoring the damping term, this is equivalent to the controllability matrix for the pair $(\boldsymbol{\Lambda}_N, \boldsymbol{B}_N)$, defined as:

$$\begin{bmatrix} \boldsymbol{B}_N^* & \boldsymbol{\Lambda}_N\boldsymbol{B}_N^* & \cdots & \boldsymbol{\Lambda}_N^{N-1}\boldsymbol{B}_N^* \end{bmatrix} \tag{11.2.69}$$

being rank N (Balas, 1978a,b). As $\boldsymbol{\Lambda}$ is diagonal, this means that each row of $\boldsymbol{B}_N^*$ must have a non-zero entry. In other words, for the system to be controllable, at least one control actuator must be at a non-zero location of each mode shape function. This criterion is a useful one for guiding control source placement, which will be discussed further, later in this chapter.

Similar to the criterion for controllability, for the system to be observable, the observability matrix for the pair $(\boldsymbol{A}_N, \boldsymbol{G}_N)$, defined as:

$$\begin{bmatrix} \boldsymbol{G}_N \\ \boldsymbol{G}_N\boldsymbol{A}_N \\ \vdots \\ \boldsymbol{G}_N\boldsymbol{A}_N^{2N-1} \end{bmatrix} \tag{11.2.70}$$

must be full rank, which is $2N$. When only displacement sensors are used, from the definition of $\boldsymbol{A}_N$ given in Equation (11.2.51), this means that the observability matrix for the pair $(\boldsymbol{\Lambda}_N, \boldsymbol{G}_{Nd})$, defined as:

$$\begin{bmatrix} \boldsymbol{G}_{Nd} \\ \boldsymbol{G}_{Nd}\boldsymbol{\Lambda}_N \\ \vdots \\ \boldsymbol{G}_{Nd}\boldsymbol{\Lambda}_N^{N-1} \end{bmatrix} \tag{11.2.71}$$

must be of rank N (Balas, 1978a,b). As $\boldsymbol{\Lambda}$ is diagonal, this means that each column of $\boldsymbol{G}_{Nd}$ must have a non-zero entry. In other words, for the system to be observable, at least one displacement sensor must be at a non-zero location of each mode shape function. This criterion is a useful one for guiding sensor placement, which will be discussed further, later in this chapter.

When velocity sensors are used, the observability result differs slightly from that obtained with displacement sensors. Here, the requirement is that the observability matrix for the pair $(\boldsymbol{\Lambda}_N, \boldsymbol{G}_{Nv})$, defined as:

$$\begin{bmatrix} \boldsymbol{G}_{Nv} \\ \boldsymbol{G}_{Nv}\boldsymbol{\Lambda}_N \\ \vdots \\ \boldsymbol{G}_{Nv}\boldsymbol{\Lambda}_N^{N-1} \end{bmatrix} \tag{11.2.72}$$

be of rank N, and also that the matrix $\boldsymbol{\Lambda}_N$ be of rank N (Balas, 1978a,b). This follows from the fact that, with velocity sensors, for $(\boldsymbol{A}_N, \boldsymbol{G}_N)$ to be observable, then both $(\boldsymbol{\Lambda}_N, \boldsymbol{G}_{Nv})$ and $(\boldsymbol{\Lambda}_N, \boldsymbol{G}_{Nv}\boldsymbol{\Lambda}_N)$ must be rank N. The second criterion is equivalent to $\boldsymbol{\Lambda}_N$ being of rank N when $(\boldsymbol{\Lambda}_N, \boldsymbol{G}_{Nv})$ is observable. The requirement that $\boldsymbol{\Lambda}_N$ be of rank N means that if there are rigid body modes (zero eigenvalues), a system that has only velocity sensors will not be observable; displacement sensors are (additionally or solely) required.

The above controllability and observability results hold for systems that have non-repeated eigenvalues. Therefore, if each eigenvalue of the modelled system has unit multiplicity, the system can be made controllable and observable with a single control actuator and sensor, provided that they are located away from a zero-value location of any of the mode shape functions included in the modelled system. If, however, there are repeated eigenvalues (as may be the case when rigid body modes are present), then the diagonal block of $\boldsymbol{\Lambda}_N$ associated with the repeated eigenvalue, and the associated blocks of $\boldsymbol{B}_N$ and $\boldsymbol{G}_N$, must form an observable and controllable subsystem. This will only be true if the rank of the associated blocks in $\boldsymbol{B}_N$ and $\boldsymbol{G}_N$ is equal to the eigenvalue multiplicity (Chen, 1970, p. 191). Therefore, a single (point) actuator and sensor cannot produce an observable and controllable system when there are repeated eigenvalues. In fact, the number of sensors required is equal to, or greater than, the eigenvalue multiplicity (Balas, 1978a,b).

11.2.6 Spillover

Thus far, only the controlled part of the system has been considered, and the influence which the residual (uncontrolled) part of the system has upon controller performance has been ignored. This would be acceptable if both the residual input matrix $\boldsymbol{B}_R$ and residual output matrix $\boldsymbol{G}_R$ were zero. However, in general they are not. Referring to Figure 11.1, the residual modes are usually excited to some degree by the active control system, and the sensor signal usually includes some measure of the response of the residual (uncontrolled) modes. These two effects, referred to as control spillover and observation spillover respectively can have a detrimental effect upon system performance.

To examine this, the augmented state vector χ is defined as:

$$\chi = \begin{bmatrix} z_N \\ x_R \end{bmatrix} \tag{11.2.73}$$

where z_N is as defined in Equation (11.2.62). Using Equations (11.2.49), (11.2.50), (11.2.58) and (11.2.61), it is straightforward to show that the closed-loop system equations satisfy the relationship:

$$\dot{\chi} = \begin{bmatrix} \boldsymbol{H}_{11} & \boldsymbol{H}_{12} \\ \boldsymbol{H}_{21} & \boldsymbol{H}_{22} \end{bmatrix} \chi \tag{11.2.74}$$

where

$$H_{11} = \begin{bmatrix} A_N - B_N K_N & -B_N K_N \\ 0 & A_N - L_N G_N \end{bmatrix} \tag{11.2.75}$$

$$H_{12} = \begin{bmatrix} 0 \\ L_N G_R \end{bmatrix} \tag{11.2.76}$$

$$H_{21} = \begin{bmatrix} -B_R K_N & -B_R K_N \end{bmatrix} \tag{11.2.77}$$

and $H_{22} = A_R$.

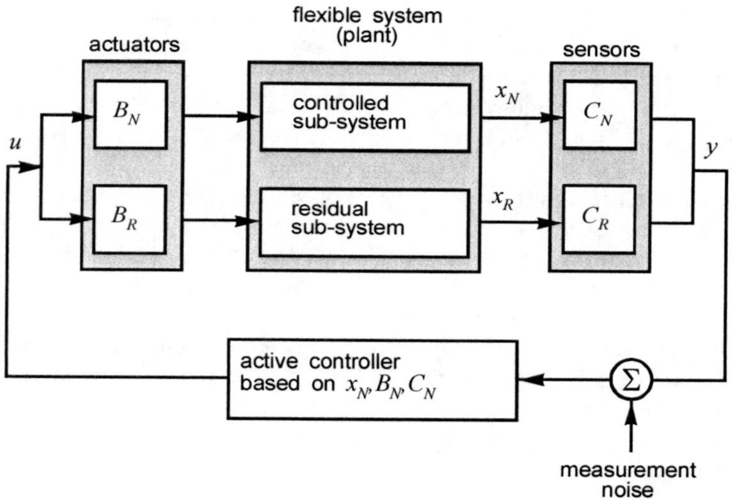

Figure 11.1 Block diagram of active control system, illustrating where control and observation spillover arise (after Balas, 1978a,b).

If there is no observation spillover (in other words, the sensors measure only the controlled system), then $G_R = 0$ and the system response is identical to that considered in the previous discussion where the effects of the residual (uncontrolled) system are neglected. In this case, the poles of the closed-loop system are those of the modelled system under full state feedback control, $A_N - B_N K_N$, the modelled system observer $A_N - L_N G_N$, and those of the residual system A_R.

The response of the residual system when there is control spillover but not observation spillover is defined by the expression:

$$\dot{x}_R = A_R x_R + B_R K_N [x_N + e_N] \tag{11.2.78}$$

It is because the control input is purely ‘external’, not a function of the residual system state, that the poles of the residual system are unchanged. The conclusion that can be drawn is that when observation spillover is not present, while control spillover may degrade the overall response characteristics of the controlled system, it cannot destabilise the system (Balas, 1978a,b).

The next step is to consider what happens when observation spillover is present. To do this, the state matrix $\boldsymbol{H}$ is first divided as:

$$\boldsymbol{H} = \boldsymbol{H}_c + \boldsymbol{H}_0 \tag{11.2.79}$$

where

$$\boldsymbol{H}_c = \begin{bmatrix} \boldsymbol{H}_{11} & \mathbf{0} \\ \boldsymbol{H}_{21} & \boldsymbol{H}_{22} \end{bmatrix}; \quad \boldsymbol{H}_0 = \begin{bmatrix} \mathbf{0} & \boldsymbol{H}_{12} \\ \mathbf{0} & \mathbf{0} \end{bmatrix} \tag{11.2.80a,b}$$

It can be observed that $\boldsymbol{H} = \boldsymbol{H}_c$ in the absence of observation spillover (control spillover only), and that $\boldsymbol{H}_0$ describes the observation spillover. From this it can be concluded that the poles of $\boldsymbol{H}$ are (continuous) functions of the (real) parameters of $\boldsymbol{H}_0$, and hence $\boldsymbol{G}_R$, and hence the degree of observation spillover. The conclusion that can be drawn then is that observation spillover can destabilise the controlled system (Balas, 1978a,b). Usually the poles of the residual system can be expected to be most susceptible to this effect, especially when the damping in the system is low (or negligible), because the controller and observer will have some degree of stability margin (recall that it has been shown that the controlled system poles are the combination of the poles of the controller, observer and residual system).

The question is then, how to avoid observation spillover. One way would be to locate the sensors on the zero locations of the residual mode shape functions. This, however, is normally impractical, especially when one considers that the residual modes usually include high-frequency resonances with rapid spatial variations in mode shape. A common way to avoid observation spillover is to filter the sensor signal, to remove the resonances of the uncontrolled modes (or only to measure the resonances of the controlled modes). This can be done by using bandpass filters, or even phase locked loops (Balas, 1978a,b). Another way to avoid the destabilising effects of observation spillover is by using collocated sensors and actuators, which will be discussed further, later in this chapter.

Example 11.1

As a numerical example of modal control, the active control of vibration in a simply supported beam (from Balas, 1978a,b) will be considered. In this example, it will be assumed that the beam has no damping, so that the response of the beam is governed by the Euler-Bernoulli partial differential equation:

$$m\frac{\partial^2 w(x,t)}{\partial t^2} + EI\frac{\partial^4 w(x,t)}{\partial x^4} = f(x,t) \tag{11.2.81}$$

where $w(x,t)$ is the displacement of the (one-dimensional) beam at x at time t, $f(x,t)$ is the applied force distribution, E and I are the modulus of elasticity and moment of inertia of the beam respectively, m is the mass per unit length of the beam, and $0 \le x \le L$, where L is the beam length. The simply supported boundary conditions are described by:

$$w(0,t) = w(L,t) = 0 \tag{11.2.82a,b}$$

and

$$\frac{\partial^2 w(0,t)}{\partial x^2} = \frac{\partial^2 w(L,t)}{\partial x^2} = 0 \tag{11.2.83a,b}$$

The mode shape functions are the beam are described by:

$$\psi_i(x) = \sin\frac{i\pi x}{L} \tag{11.2.84}$$

For simplicity, the quantities E, I, m and L will be set equal to one. The associated resonance frequencies are described by:

$$\omega_i = \left(\frac{i\pi}{L}\right)^2 \tag{11.2.85}$$

The beam/active controller arrangement is shown in Figure 11.2. The beam is controlled by a single point control force at $x_c = L/6$, so that the applied force distribution is modelled by $f(x, t) = f(t)\,\delta(x_c - x)$. The beam output is measured by a single (unity gain) point displacement sensor at $x_e = 5/6L$, so that the beam output is $y = w(x_e,t)$. For this simple example, the transducer dynamics and system noise will be ignored.

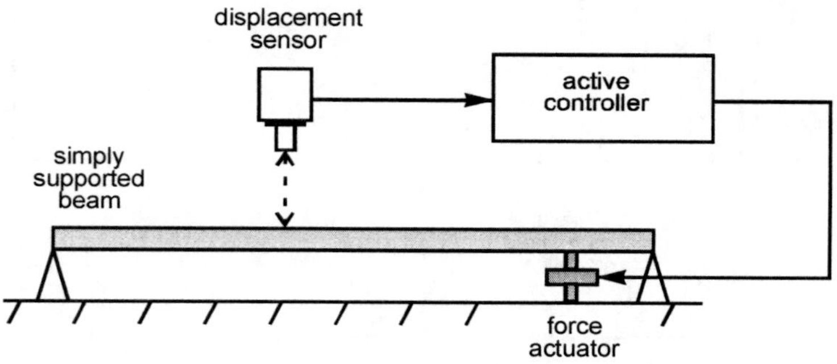

Figure 11.2 Arrangement for beam simulation (after Balas, 1978a,b).

The active controller in this example is designed to control the first three beam modes. With three controlled modes, the state observer must be six-dimensional, and, since noise in the system is ignored, it is deterministic. The control gains are calculated from an optimal control law that minimises the (unweighted) energy in the first three modes. Therefore, with the state matrix:

$$\boldsymbol{x} = \begin{bmatrix} z_1 & z_2 & z_3 & \dot{z}_1 & \dot{z}_2 & \dot{z}_3 \end{bmatrix}^{\mathrm{T}} \tag{11.2.86}$$

the weighting matrices in the optimal control law are:

$$\boldsymbol{Q} = \begin{bmatrix} \boldsymbol{\Lambda}_N & \boldsymbol{0} \\ \boldsymbol{0} & \boldsymbol{I} \end{bmatrix}; \quad \boldsymbol{R} = 0.1\boldsymbol{I} \tag{11.2.87a,b}$$

where $\boldsymbol{I}$ is the identity matrix. The resulting control gains are (Balas, 1978a,b):

$$\boldsymbol{k} = -[0.5448 \quad 5.018 \quad 18.303 \quad 3.162 \quad 3.1597 \quad 3.156] \tag{11.2.88}$$

The observer gains are chosen to place the observer poles to the left of the controlled system poles, using a pole placement routine. The resulting observer poles are (Balas, 1978a,b):

$$l = [31.7821 \quad -95.626 \quad 160.9996 \quad -118.304 \quad 837.444 \quad 11388.99]^T \tag{11.2.89}$$

In the calculations, the beam is given an initial displacement such that the first three modes have unity displacement, while the other modes are unperturbed. The sensor output is shown in Figure 11.3, the control input in Figure 11.4, and the energy in the controlled modes in

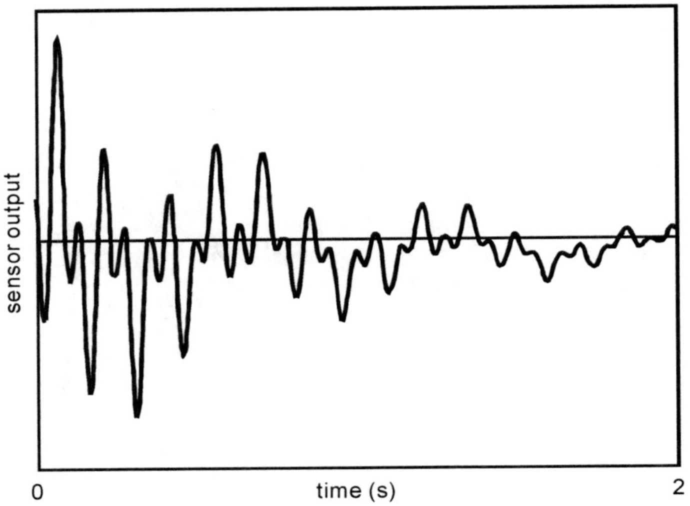

Figure 11.3 Sensor output (after Balas, 1978a,b).

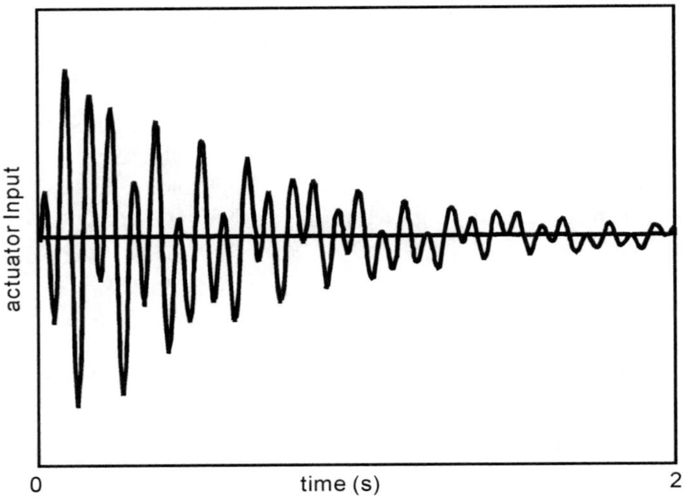

Figure 11.4 Actuator input (after Balas, 1978a,b).

Figure 11.5. It can be observed that the response of the controlled beam modes decays at a rate which is slightly faster than if each mode had one per cent critical damping. The energy in the fourth (residual) mode is shown in Figure 11.6, where it can be seen that the energy in this mode is increasing over time. In fact, this mode is now unstable. This is a result of observation spillover. The observation spillover-induced instability could be eliminated by either filtering the sensor signal, or in this case by adding a small amount of passive damping, which would shift the residual pole slightly to the left, increasing its stability margin and hence providing some tolerance to spillover effects.

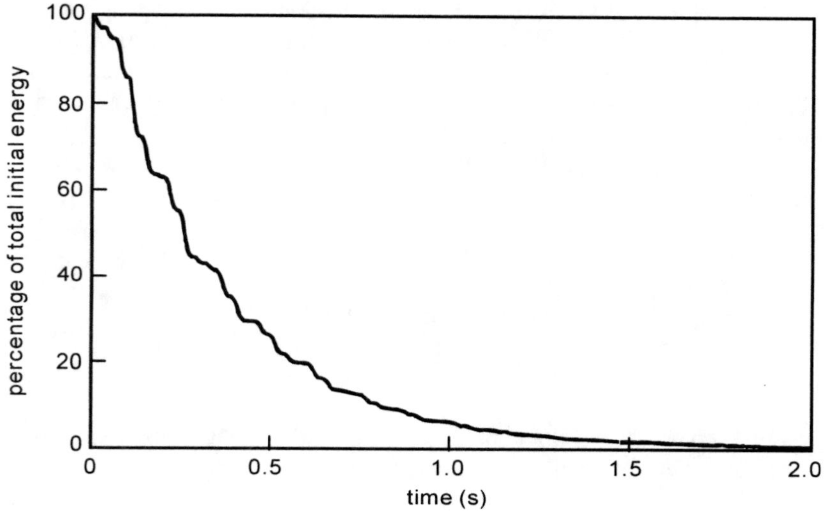

Figure 11.5 Energy in controlled modes during application of active control (after Balas, 1978a,b).

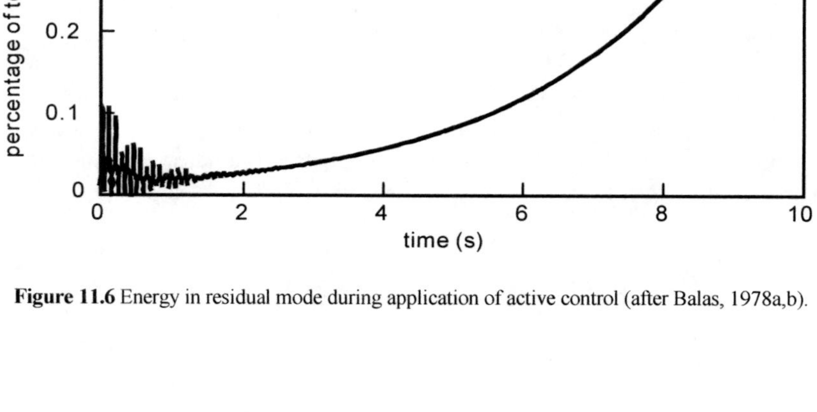

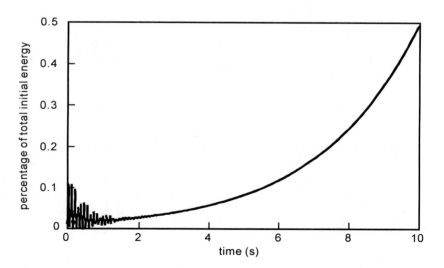

Figure 11.6 Energy in residual mode during application of active control (after Balas, 1978a,b).

11.2.7 Optimal Control Gains for Second-Order Matrix Equations

Thus far in this section, the 'basics' of modal control have been considered, with the analysis based on state equations that describe the response of the system in terms of system modes. These equations can be used to calculate optimal control gains for attenuating the system response to the unwanted disturbance. Calculation of these gains can be carried out using the methods outlined in Chapter 5, which involves solving a matrix Riccati equation of dimension ($2N \times 2N$), where there are N modes in the reduced order system model.

In this section, the calculation of optimal control gains for systems described by second-order matrix equations will be considered. These equations are a generalisation of the form of the modal state equations which were considered in the preceding section. Specifically, it will be found that there are characteristics of the coefficient matrices which can be used to reduce the problem to one of solving three ($N \times N$) matrix equations, which may significantly reduce the computational burden.

From Chapter 5, the steady-state optimal control input for a system described in state-space notation has the form:

$$\boldsymbol{f}_c = -\boldsymbol{K}\boldsymbol{x} \tag{11.2.90}$$

where the gain matrix $\boldsymbol{K}$ is defined by the relationship:

$$\boldsymbol{K} = \boldsymbol{R}^{-1}\boldsymbol{B}^{\mathrm{T}}\boldsymbol{P} \tag{11.2.91}$$

where $\boldsymbol{R}$ is the control effort weighting matrix, $\boldsymbol{B}$ is the input matrix, and the matrix $\boldsymbol{P}$ satisfies the steady-state matrix Riccati equation:

$$\boldsymbol{A}^{\mathrm{T}}\boldsymbol{P} + \boldsymbol{P}\boldsymbol{A} - \boldsymbol{P}\boldsymbol{B}\boldsymbol{R}^{-1}\boldsymbol{B}^{\mathrm{T}}\boldsymbol{P} + \boldsymbol{Q} = 0 \tag{11.2.92}$$

where $\boldsymbol{A}$ is the state matrix.

The (full) state and input matrices for the systems of interest here are defined in Equation (11.2.46), with the reduced order matrices defined in Equation (11.2.51). Note that modal state equations provide a 'lumped' approximation of the equations governing the response of the distributed parameter system, and are a (truncated) set of second-order matrix equations. The method of solution for the optimal control gains to be outlined here is valid for any such lumped parameter system, for which the equations of motion can be written as a set of simultaneous second-order differential equations:

$$\boldsymbol{M}\ddot{\boldsymbol{q}} + \boldsymbol{C}\dot{\boldsymbol{q}} + \boldsymbol{K}\boldsymbol{q} = \boldsymbol{B}\boldsymbol{f}_c \tag{11.2.93}$$

where $\boldsymbol{M}$, $\boldsymbol{C}$ and $\boldsymbol{K}$ are ($N \times N$) mass, damping, and stiffness matrices respectively, $\boldsymbol{q}$ is an N-dimensional displacement vector and $\boldsymbol{f}_c$ is an N-dimensional control force vector (there are N modelled modes). When expressed in state variable format, the state vector, state matrix and input matrix have the form:

$$\boldsymbol{x} = \begin{bmatrix} \boldsymbol{q} \\ \dot{\boldsymbol{q}} \end{bmatrix}; \quad \boldsymbol{A} = \begin{bmatrix} \boldsymbol{0} & \boldsymbol{I} \\ -\boldsymbol{M}^{-1}\boldsymbol{K} & -\boldsymbol{M}^{-1}\boldsymbol{C} \end{bmatrix}; \quad \boldsymbol{B} = \begin{bmatrix} \boldsymbol{0} \\ \boldsymbol{M}^{-1}\boldsymbol{B}' \end{bmatrix} \tag{11.2.94a–c}$$

If the terms in Equation (11.2.94) are compared with those of the (normalised) modal state equations given in Equation (11.2.46), it is found that $\boldsymbol{M}^{-1}\boldsymbol{K} = \boldsymbol{\Lambda}$, $\boldsymbol{M}^{-1}\boldsymbol{C} = 2\zeta\boldsymbol{\Lambda}^{1/2}$, and $\boldsymbol{M}^{-1}\boldsymbol{B}' = \boldsymbol{B}^*$.

If the state weighting matrix used in the optimal control problem is diagonal, then,

$$\boldsymbol{Q} = \begin{bmatrix} \boldsymbol{Q}_1 & 0 \\ 0 & \boldsymbol{Q}_2 \end{bmatrix} \tag{11.2.95}$$

The form of the state and input matrices can be used to put the matrix Riccati equation in a convenient form for solution. As discussed previously in this section, this is commonly the case in active noise and vibration control problems, where the weighting matrix $\boldsymbol{Q}$ is chosen such that $\boldsymbol{x}^{\mathrm{T}}\boldsymbol{Q}\boldsymbol{x}$ represents the total system energy (in this generalised case, by setting $\boldsymbol{Q}_1 = \boldsymbol{K}$, $\boldsymbol{Q}_2 = \boldsymbol{M}$).

To simplify the optimal control problem, the matrix $\boldsymbol{P}$ in the Riccati equation is first divided into four ($N \times N$) sub-matrices as follows:

$$\boldsymbol{P} = \begin{bmatrix} \boldsymbol{P}_{11} & \boldsymbol{P}_{12} \\ \boldsymbol{P}_{12}^{\mathrm{T}} & \boldsymbol{P}_{22} \end{bmatrix} \tag{11.2.96}$$

Using the notation of Equations (11.2.60) and (11.2.62), the ($2N \times 2N$) Riccati Equation (11.2.58) can be expressed as the following set of ($N \times N$) matrix equations (Kwak and Meirovitch, 1993):

$$-\boldsymbol{Q}_1 + \boldsymbol{K}\boldsymbol{M}^{-1}\boldsymbol{P}_{12}^{\mathrm{T}} + \boldsymbol{P}_{12}\boldsymbol{M}^{-1}\boldsymbol{K} + \boldsymbol{P}_{12}\boldsymbol{M}^{-1}\boldsymbol{B}^*\boldsymbol{R}^{-1}\boldsymbol{B}^{*\mathrm{T}}\boldsymbol{M}^{-1}\boldsymbol{P}_{12}^{\mathrm{T}} = 0 \tag{11.2.97}$$

$$-\boldsymbol{P}_{11} + \boldsymbol{K}\boldsymbol{M}^{-1}\boldsymbol{P}_{22} + \boldsymbol{P}_{12}\boldsymbol{M}^{-1}\boldsymbol{C} + \boldsymbol{P}_{12}\boldsymbol{M}^{-1}\boldsymbol{B}^*\boldsymbol{R}^{-1}\boldsymbol{B}^{*\mathrm{T}}\boldsymbol{M}^{-1}\boldsymbol{P}_{22} = 0 \tag{11.2.98}$$

$$-\boldsymbol{P}_{11} + \boldsymbol{P}_{22}\boldsymbol{M}^{-1}\boldsymbol{K} + \boldsymbol{C}\boldsymbol{M}^{-1}\boldsymbol{P}_{12}^{\mathrm{T}} + \boldsymbol{P}_{22}\boldsymbol{M}^{-1}\boldsymbol{B}^*\boldsymbol{R}^{-1}\boldsymbol{B}^{*\mathrm{T}}\boldsymbol{M}^{-1}\boldsymbol{P}_{12}^{\mathrm{T}} = 0 \tag{11.2.99}$$

$$\begin{aligned} &-\boldsymbol{Q}_2 - \boldsymbol{P}_{12} - \boldsymbol{P}_{12}^{\mathrm{T}} + \boldsymbol{C}\boldsymbol{M}^{-1}\boldsymbol{P}_{22} + \boldsymbol{P}_{22}\boldsymbol{M}^{-1}\boldsymbol{C} + \\ &\boldsymbol{P}_{22}\boldsymbol{M}^{-1}\boldsymbol{B}^*\boldsymbol{R}^{-1}\boldsymbol{B}^{*\mathrm{T}}\boldsymbol{M}^{-1}\boldsymbol{P}_{22} = 0 \end{aligned} \tag{11.2.100}$$

Note again that these equations can be restated using the modal state-space equation notation of Equation (11.2.46) by setting $\boldsymbol{M}^{-1}\boldsymbol{K} = \boldsymbol{\Lambda}$, $\boldsymbol{M}^{-1}\boldsymbol{C} = 2\xi\boldsymbol{\Lambda}^{1/2}$ and $\boldsymbol{M}^{-1}\boldsymbol{B}' = \boldsymbol{B}^*$.

Owing to the form of the input matrix $\boldsymbol{B}$, not all of the submatrices of $\boldsymbol{P}$ are needed to derive the optimum control gains. If Equation (11.2.91) is expanded, then,

$$\boldsymbol{K} = \boldsymbol{R}^{-1}\begin{bmatrix} \boldsymbol{0} & \boldsymbol{B}'^{\mathrm{T}}\boldsymbol{M}^{-1} \end{bmatrix} \begin{bmatrix} \boldsymbol{P}_{11} & \boldsymbol{P}_{12} \\ \boldsymbol{P}_{12}^{T} & \boldsymbol{P}_{22} \end{bmatrix} = \boldsymbol{R}^{-1}\boldsymbol{B}'^{\mathrm{T}}\boldsymbol{M}^{-1}\begin{bmatrix} \boldsymbol{P}_{12}^{\mathrm{T}} & \boldsymbol{P}_{22} \end{bmatrix} \tag{11.2.101a,b}$$

From this it is apparent that the sub-matrix $\boldsymbol{P}_{11}$ is not used in the calculation of the control gains, although it would be calculated if the Riccati Equation (11.2.92) were simply solved as written. If this term is eliminated from Equations (11.2.98) and (11.2.99), the following matrix equation is obtained (Kwak and Meirovitch, 1993):

$$P_{22}M^{-1}K - KM^{-1}P_{22} + CM^{-1}P_{12}^{\mathrm{T}} - P_{12}M^{-1}D$$
$$+ P_{22}M^{-1}B^{*}R^{-1}B^{*\mathrm{T}}M^{-1}P_{12}^{\mathrm{T}} - P_{12}M^{-1}B^{*}R^{-1}B^{*T}M^{-1}P_{22} = 0 \quad (11.2.102)$$

The $(N \times N)$ matrix Equations (11.2.97), (11.2.100) and (11.2.102) can be solved for $\boldsymbol{P}_{12}$ and $\boldsymbol{P}_{22}$ to facilitate calculation of the optimum control gains as an alternative to solving the full $(2N \times 2N)$ matrix Riccati equation. In designs that involve a large number of modelled modes, this may present a (significant) computational savings.

On the surface, it may appear from Equation (11.2.97), which contains only $\boldsymbol{P}_{12}$ and has the appearance of an $(N \times N)$ Riccati equation, that it is possible to solve for $\boldsymbol{P}_{12}$ independently of $\boldsymbol{P}_{22}$. This is not, however, the case, as $\boldsymbol{P}_{12}$ is not in general a symmetric matrix. However, an iterative approach can be used to solve for the two variables (Kwak and Meirovitch, 1993).

To derive an iterative procedure, note first that the matrix $\boldsymbol{P}_{12}$ can be expressed as the sum of a symmetric and skew-symmetric matrix (Noble and Daniel, 1977):

$$P_{12} = \bar{P}_{12} + \tilde{P}_{12} \quad (11.2.103)$$

where – denotes the symmetric matrix, and ˜ denotes the skew-symmetric matrix. Substituting the expansion of Equation (11.2.103) into Equation (11.2.97) yields the relationship:

$$-Q_1 + KM^{-1}\bar{P}_{12} + \bar{P}_{12}M^{-1}K + \bar{P}_{12}M^{-1}B^{*}R^{-1}B^{*\mathrm{T}}M^{-1}\bar{P}_{12}$$
$$- KM^{-1}\tilde{P}_{12} + \tilde{P}_{12}M^{-1}K + \tilde{P}_{12}M^{-1}B^{*}R^{-1}B^{*\mathrm{T}}M^{-1}\bar{P}_{12} \quad (11.2.104)$$
$$- \bar{P}_{12}M^{-1}B^{*}R^{-1}B^{*\mathrm{T}}M^{-1}\tilde{P}_{12} - \tilde{P}_{12}M^{-1}B^{*}R^{-1}B^{*\mathrm{T}}M^{-1}\tilde{P}_{12} = 0$$

With a view to solving a Riccati equation for the symmetric part of $\boldsymbol{P}_{12}$, Equation (11.2.104) can be rewritten as:

$$-(Q_1 + S) + KM^{-1}\bar{P}_{12} + \bar{P}_{12}M^{-1}K + \bar{P}_{12}M^{-1}B^{*}R^{-1}B^{*\mathrm{T}}M^{-1}\bar{P}_{12} = 0 \quad (11.2.105)$$

where $\boldsymbol{S}$ is a symmetric matrix defined by the expression:

$$S = -\tilde{P}_{12}\left(M^{-1}K + M^{-1}B^{*}R^{-1}B^{*\mathrm{T}}M^{-1}\bar{P}_{12}\right)$$
$$+ \left(KM^{-1} + \bar{P}_{12}M^{-1}B^{*}R^{-1}B^{*\mathrm{T}}M^{-1}\right)\tilde{P}_{12} \quad (11.2.106)$$
$$+ \tilde{P}_{12}M^{-1}B^{*}R^{-1}B^{*\mathrm{T}}M^{-1}\tilde{P}_{12}$$

For a given symmetric matrix $\boldsymbol{S}$, the solution of Equation (11.2.105) is a solution of an $(N \times N)$ matrix Riccati equation.

If the expansion of Equation (11.2.103) is substituted into Equation (11.2.100), a second Riccati equation is obtained as follows:

$$-(Q_2 + 2\bar{P}_{12}) + DM^{-1}P_{22} + P_{22}M^{-1}D + P_{22}M^{-1}B^{*}R^{-1}B^{*\mathrm{T}}M^{-1}P_{22} = 0 \quad (11.2.107)$$

Finally, if Equation (11.2.103) is substituted into Equation (11.2.102), the Lyapunov equation is obtained:

$$\begin{aligned}
&-\tilde{P}_{12}\left(M^{-1}D + M^{-1}B^{*}R^{-1}B^{*T}M^{-1}P_{22}\right) \\
&-\left(DM^{-1} + P_{22}M^{-1}B^{*}R^{-1}B^{*T}M^{-1}\right)\tilde{P}_{12} \\
&+P_{22}\left(M^{-1}K + M^{-1}B^{*}R^{-1}B^{*T}M^{-1}\bar{P}_{12}\right) \\
&-\left(KM^{-1} + \bar{P}_{12}M^{-1}B^{*}R^{-1}B^{*T}M^{-1}\right)P_{22} \\
&\quad + DM^{-1}\bar{P}_{12} - \bar{P}_{12}M^{-1}D = 0
\end{aligned} \tag{11.2.108}$$

An iterative procedure for solving for P_{12} and P_{22} can now be devised as follows (Kwak and Meirovitch, 1993), starting with some initial guess of S:

1. Insert matrix S into Equation (11.2.105) and solve for the symmetric part of P_{12}.
2. Insert the just-computed value of P_{12} into Equation (11.2.107) and solve this Riccati equation for P_{22}.
3. Insert both the symmetric part of P_{12} and P_{22} into Equation (11.2.108) and solve for the skew-symmetric matrix P_{12}.
4. Insert both the symmetric and skew-symmetric parts of P_{12} into Equation (11.2.106) and calculate a new value of S.
5. Repeat steps (1) through (4) until the matrices have converged.

A numerical example of this procedure can be found in Kwak and Meirovitch (1993).

11.2.8 Brief Note on Passive Damping

One of the constraints cited in the beginning of this section was that at least a small amount of passive damping is required to be present in structures that are to be controlled. Passive damping improves controller the stability (Hughes and Abdel-Rahman, 1979; Spanos, 1989; Grandhi, 1990; Rao et al., 1990; Gueler et al., 1993). In some instances, it can also reduce the number of sensors and actuators required to establish controllability and observability (Hughes and Skelton, 1980a).

Consider the desired controller response outlined in Figure 11.7. Here, there are two requirements for the controller, the first of which is gain stabilisation of the response of the unmodelled, or poorly modelled, modes outside the controller bandwidth (Gueler et al., 1993). The loop gain of a flexible structure peaks at structural resonances, where the structural response is controlled only by its damping. As there is an inverse relationship between damping and response at resonance, it is obvious that any structure without damping cannot be gain stabilised. As the resonance frequency increases, the effectiveness of (viscous) passive damping also increases, which leads to the small high-frequency damping requirement shown in Figure 11.8.

The second requirement of a controller is phase stabilisation of the system response within the controller bandwidth. This requirement follows from the fact that to achieve high bandwidth control of a flexible structure, the structural dynamics must be compensated, particularly at resonances. Often, the controllers implemented to do this effectively notch filter the response at structural resonances, cancelling the structure's poles and zeroes with the controller's zeroes and poles respectively. For this to be achieved, an accurate model of the response of the target structure is needed. In an undamped system, uncertainty in this model can lead to instability.

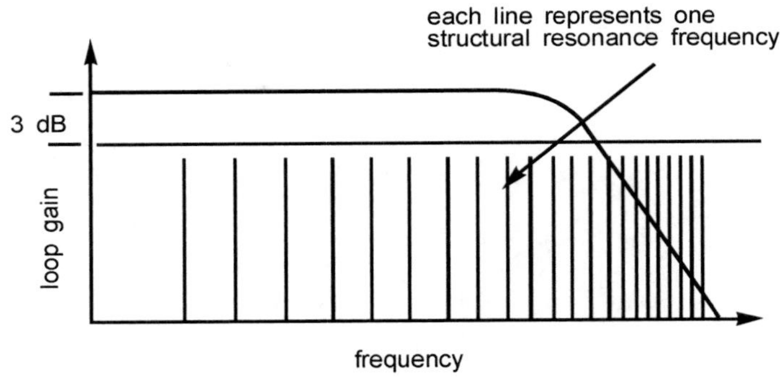

Figure 11.7 Figurative depiction of problem statement for bandwidth to include many poorly modelled, lightly damped, closely spaced modes (after Gueler et al., 1993).

To demonstrate this, consider Figure 11.9, which illustrates the departure of the root locus of a system with a pole above or below a zero, a situation that may exist in a compensated, undamped system where there was uncertainty in the system model. If it is assumed that the remainder of the plant dynamics is responsible for a phase lag of $-90°$, the departure angle will be at $180°$ when the pole is above the zero, but $0°$ when the zero is above the pole. This means that if the dynamic system has no (or very little) damping, in the second instance the system will become unstable as the compensator gain is increased.

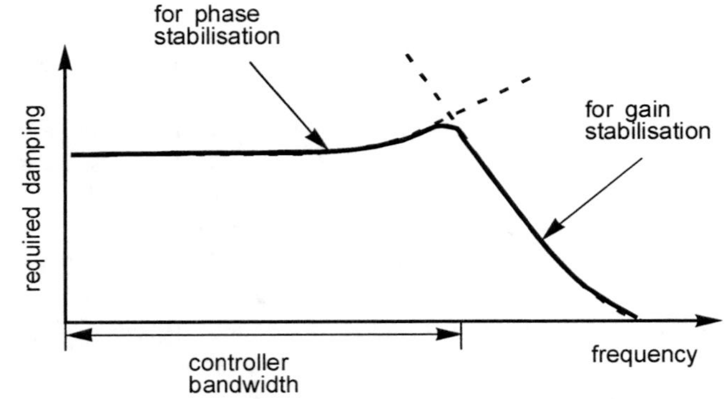

Figure 11.8 Required level of passive damping to meet problem specification (after Gueler et al., 1993).

If it is assumed that the poles migrate to the zeroes in a pattern that approximates a semi-circle, the amount of damping that is required to avoid instability can be estimated. As the radius of the semi-circle is dependent upon the pole-zero separation, the level of damping required to assure stability robustness is (Gueler et al., 1993):

$$\zeta = \frac{|\omega_z - \omega_n|}{\omega_z + \omega_n} \qquad (11.2.109)$$

where ω_z is the zero frequency, and ω_n is the resonance (pole) frequency. This level of passive damping will ensure that the pole does not migrate across the imaginary axis for any gain value.

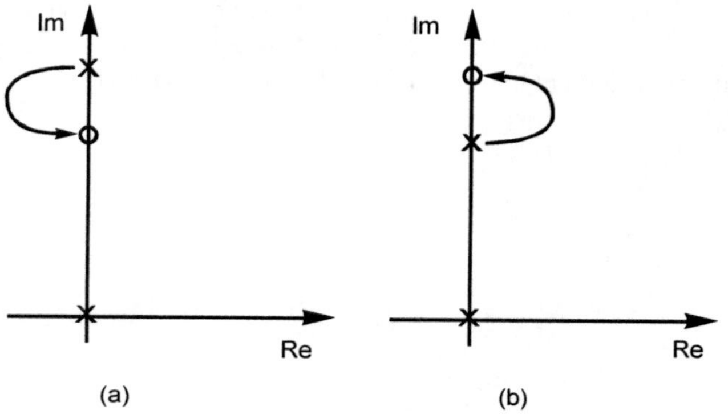

Figure 11.9 Departure angles of root locus of a single oscillatory system as a result of uncertainty in pole location (after Gueler et al., 1993): (a) pole above zero; (b) zero above pole.

To derive a relationship for the required degree of passive damping, which is perhaps more readily applicable, it is necessary to consider the system response in the vicinity of the resonance of an isolated structural mode, which is governed by the relationship:

$$G(s) = \frac{1}{s^2 + 2\zeta\omega_n s + \omega_n^2} \qquad (11.2.110)$$

where $s = j\omega$.

The phase angle of this transfer function follows the relationship:

$$\theta(\omega) = -\tan^{-1}\frac{2\zeta\omega_n\omega}{\omega_n^2 - \omega^2} \qquad (11.2.111)$$

Therefore, at resonance, the change in phase with respect to frequency is given by:

$$\frac{d\theta}{d\omega} = -\frac{1}{\zeta\omega_n} \qquad (11.2.112)$$

Therefore, the phase change as the system passes through resonance will be sharp if the damping is low. If the uncertainty in resonance frequency is defined as $\delta\omega = \omega_n - \omega_{actual}$, then a first-order approximation of the uncertainty in phase angle near resonance is given by (Gueler et al., 1993):

$$\delta\theta = -\frac{\delta\omega}{\zeta\omega_n} \qquad (11.2.113)$$

This relationship states that, given an uncertainty in resonance frequency, the uncertainty in phase is inversely proportional to the damping.

It is possible to use the relationship of Equation (11.2.113) to quantify the degradation in closed-loop stability resulting from imperfect pole-zero cancellation for a specific problem (see Gueler et al., 1993). The important point to note, however, which is evident from the sketches in Figure 11.9, is that if there is no passive damping in the system, the phase excursion can be 180°; passive damping reduces this value.

For a (local) phase margin of θ_{pm}, which from the definitions in Chapter 5 is the amount of additional phase lag required to make the system unstable at a given frequency, the permissible amount of uncertainty in plant natural frequency is given by:

$$\delta\omega \leq \theta_{pm}\zeta\omega_n \tag{11.2.114}$$

Therefore, the amount of passive damping needed, given a pole-zero mismatch of $\delta\omega$ and a phase margin θ_{pm}, is (Gueler et al., 1993):

$$\zeta \geq \frac{\delta\omega}{\theta_{pm}\omega_n} \tag{11.2.115}$$

This expression leads to the damping requirement for phase stabilisation in the controller bandwidth sketched in Figure 11.8.

Further examples of the need for passive damping to avoid controller instability can be found in Gueler et al. (1993).

11.3 INDEPENDENT MODAL SPACE CONTROL

In the previous section, the design of active control systems for flexible structures was considered using state-space design methods, where the system states are modal displacements and velocities. The general problem framework fits into that reviewed in Chapter 5, where the control and observer gains can be calculated using either pole placement or optimal (linear quadratic) methodologies. There can, however, be computational problems with this approach when a large number of modes are included in the control calculations; even using the iterative approach outlined in the previous section, where solution of the matrix Riccati equation was reduced from the solution of a $(2N \times 2N)$ matrix equation to the solution of three $(N \times N)$ matrix equations, may not reduce the computational burden sufficiently.

Computational problems can arise with a large system design because the control system is 'coupled'; that is, because the actuators and sensors excite and measure multiple structural modes, a control input derived from consideration of one mode will excite additional modes (controlled and residual) upon application. Thus, feedback control destroys the independence of the open-loop modal equations. Because of this coupling, the various controller design equations (such as the Riccati equation) will have cross-terms describing the influence that a control input derived from one state has upon other states, which limits the extent to which 'shortcuts' can be taken in the design process.

One approach to simplifying control design, then, would be to somehow eliminate the extensive cross-coupling between the states (or modes). If the independence of the modal equations could be maintained during implementation of a feedback control system (closed-loop), then the control law could be derived as a set of independent control laws, one for each mode. This basic idea is behind a form of control referred to as independent modal-space control (IMSC) (Bennighof and Meirovitch, 1988; Meirovitch and Baruh, 1982; Meirovitch and Bennighof, 1986; Meirovitch and Oz, 1980a, 1980b, 1980c; Meirovitch et al., 1977, 1979, 1983).

11.3.1 Control Law Development

To develop the independent modal space control (IMSC) methodology, the first consideration is the control input to the structure, which is the root cause of control system coupling. If a set of point actuators is used in the control system, then the control input from each can be modelled using a Dirac delta function as follows:

$$f(\boldsymbol{r},t) = f_c(t)\delta(\boldsymbol{r}-\boldsymbol{r}_c) \tag{11.3.1}$$

where $f_c(t)$ is the control input at time t, and $\boldsymbol{r}_c$ is the location of the control actuator on the structure. If there are N_s control actuators in the system, then the ith modal generalised force, defined in the previous section in Equation (11.2.15), is simply:

$$f_i(t) = \sum_{\imath=1}^{N_s} f_{c,\imath}(t)\psi_i(\boldsymbol{r}_\imath) \tag{11.3.2}$$

The relationship in Equation (11.3.2) implies that all control inputs may excite all modelled structural modes (provided they are not located on a zero of the mode shape function). This is where the system coupling arises upon implementation of closed-loop control.

To simplify the development, the controlled system equations will be written below in a slightly different manner than was done in the previous section. The response of the controlled system can still be written as the state-space equation:

$$\dot{\boldsymbol{x}}_N = \boldsymbol{A}_N \boldsymbol{x}_N + \boldsymbol{B}_N u \tag{11.3.3}$$

However, this time the system states, which are still modal displacements and velocities, are arranged as:

$$\boldsymbol{x}_N = \begin{bmatrix} z_1 & \dot{z}_1 & \cdots & z_N & \dot{z}_N \end{bmatrix}^{\mathrm{T}} \tag{11.3.4}$$

which puts the state matrix in a block diagonal form as:

$$\boldsymbol{A}_N = \begin{bmatrix} \boldsymbol{A}_{N1} & 0 & \cdots & 0 \\ 0 & \boldsymbol{A}_{N2} & \cdots & 0 \\ & & \ddots & \\ 0 & 0 & \cdots & \boldsymbol{A}_{NN} \end{bmatrix} \tag{11.3.5}$$

where, for a lightly damped system:

$$\boldsymbol{A}_{Ni} = \begin{bmatrix} 0 & 1 \\ -\omega_{Ni}^2 & -2\zeta\omega_{Ni} \end{bmatrix} \tag{11.3.6}$$

and where N is the number of modelled modes, which in this instance is the same as the number of modes to be controlled. For N_s discrete control (point) inputs,

$$\boldsymbol{Bf}_c = \boldsymbol{f} = \begin{bmatrix} 0 & 0 & \cdots & 0 \\ \psi_1(\boldsymbol{r}_1) & \psi_1(\boldsymbol{r}_2) & \cdots & \psi_1(\boldsymbol{r}_{N_s}) \\ & & \vdots & \\ 0 & 0 & \cdots & 0 \\ \psi_N(\boldsymbol{r}_1) & \psi_N(\boldsymbol{r}_2) & \cdots & \psi_N(\boldsymbol{r}_{N_s}) \end{bmatrix} \begin{bmatrix} u_1 \\ u_2 \\ \vdots \\ u_{N_s} \end{bmatrix} = \begin{bmatrix} 0 \\ f_1 \\ \vdots \\ 0 \\ f_N \end{bmatrix} \tag{11.3.7}$$

where $\boldsymbol{f}$ is the vector of modal generalised forces.

The basic idea behind IMSC is to restate the problem as a set of independent modal equations, thereby eliminating coupling, which will simplify the controller design exercise. To do this, the controller can be designed in terms of the modal generalised forces in the vector $\boldsymbol{f}$, then the result transformed to obtain the actual control inputs $\boldsymbol{f}_c$:

$$\boldsymbol{f}_c = \boldsymbol{B}_N^{-1}\boldsymbol{f} \tag{11.3.8}$$

where $\boldsymbol{B}_N^{-1}$ is a pseudo inverse in general, and an actual inverse if the control matrix $\boldsymbol{B}_N$ is square ($\boldsymbol{B}_N$ will be square if the number of control inputs is equal to the number of controlled modes).

To see how designing for a single mode will simplify the controller design, a pole placement exercise is considered first. When considering a single mode, the control input is defined by the expression:

$$\boldsymbol{bf}_c = -\boldsymbol{bk}_i\boldsymbol{x}_i = -\begin{bmatrix} 0 \\ 1 \end{bmatrix} \begin{bmatrix} k_{i1} & k_{i2} \end{bmatrix} \begin{bmatrix} z_i \\ \dot{z}_i \end{bmatrix} \tag{11.3.9}$$

Using the state matrix outlined in Equation (11.3.4) specialised for a single mode, the characteristic equation for this mode is:

$$\left| s\boldsymbol{I} - \boldsymbol{A}_{Ni} + \boldsymbol{bk}_i\boldsymbol{x}_i \right| = \begin{vmatrix} s & -1 \\ \omega_i^2 + k_{i1} & s + 2\zeta\omega_i + k_{i2} \end{vmatrix} \tag{11.3.10}$$

or

$$s^2 + (k_{i2} + 2\zeta\omega_i)s + (\omega_i^2 + k_{i1}) = 0 \tag{11.3.11}$$

If the desired location of the poles for this mode defines the desired characteristic equation:

$$\gamma = s^2 + \alpha_1 s + \alpha_2 \tag{11.3.12}$$

then the control gains are simply defined by the relationships:

$$k_{i1} = \alpha_2 - \omega_i^2$$
$$k_{i2} = \alpha_1 - 2\zeta\omega_i \tag{11.3.13}$$

Optimal control problems are also more straightforward when designing for a single modal state equation, and deriving a modal generalised force. When the modal equations are independent, the cost function associated with the optimal control problem is simply the sum of the cost functions of the individual modes:

$$J = \sum_{i=1}^{N} J_i \tag{11.3.14}$$

With steady-state optimal control, these modal cost functions are defined by the relationship:

$$J_i = \int_0^{\infty} \left[\boldsymbol{x}_i^{\mathrm{T}}(\tau) \boldsymbol{Q} \boldsymbol{x}_i(\tau) + \boldsymbol{f}_{c,i}^{\mathrm{T}}(\tau) \boldsymbol{R} \boldsymbol{f}_{c,i}(\tau) \right] \mathrm{d}\tau \tag{11.3.15}$$

where $\boldsymbol{Q}$ is the state weighting matrix, and $\boldsymbol{R}$ is the control effort weighting matrix.

For steady-state optimal control, the control signal for the ith modal 'system' is defined by the relationship:

$$\boldsymbol{f}_{c,i} = -\boldsymbol{K}\boldsymbol{x}_i \tag{11.3.16}$$

where the gain matrix, $\boldsymbol{K}_i$ is defined by the relationship:

$$\boldsymbol{K}_i = \boldsymbol{R}^{-1} \boldsymbol{B}_i^{\mathrm{T}} \boldsymbol{P}_i \tag{11.3.17}$$

where $\boldsymbol{R}$ is the control effort weighting matrix, $\boldsymbol{B}_i$ is the input matrix for the ith mode and $\boldsymbol{P}_i$ satisfies the matrix Riccati equation:

$$\boldsymbol{A}_i^{\mathrm{T}} \boldsymbol{P}_i + \boldsymbol{P}_i \boldsymbol{A}_i - \boldsymbol{P}_i \boldsymbol{B}_i \boldsymbol{R}^{-1} \boldsymbol{B}_i^{\mathrm{T}} \boldsymbol{P}_i + \boldsymbol{Q} = 0 \tag{11.3.18}$$

where $\boldsymbol{Q}$ is the state weighting matrix.

When considering a modal generalised force, the following relationship is useful:

$$\boldsymbol{f}_i = \begin{bmatrix} 0 \\ f_i \end{bmatrix} = -\boldsymbol{K}_i \boldsymbol{x}_i = -\begin{bmatrix} k_{11} & k_{12} \\ k_{21} & k_{22} \end{bmatrix} \begin{bmatrix} z_i \\ \dot{z}_i \end{bmatrix} \tag{11.3.19}$$

From this relationship it is apparent the $k_{11} = k_{12} = 0$. For this relationship to hold, the control effort weighting matrix must have the form (Oz and Meirovitch, 1980a):

$$\boldsymbol{R} = \begin{bmatrix} \infty & 0 \\ 0 & R_{22} \end{bmatrix} \tag{11.3.20}$$

If minimising system energy is of interest, such that the state weighting matrix is:

$$\boldsymbol{Q} = \begin{bmatrix} \omega^2 & 0 \\ 0 & 1 \end{bmatrix} \tag{11.3.21}$$

(so that $\boldsymbol{x}^{\mathrm{T}}\boldsymbol{Q}\boldsymbol{x}$ reflects the system kinetic energy), substituting values into Equations (11.3.17) and (11.3.18), the gain matrix is then (Meirovitch and Baruh, 1982):

$$\boldsymbol{K}_i = \begin{bmatrix} 0 & 0 \\ \omega_i^2\left(\sqrt{1+R_{22}^{-1}} - 1\right) & \left[2\omega_i\left(\sqrt{1+R_{22}^{-1}} - 1\right) + \omega_i^2 R_{22}^{-1}\right]^{1/2} \end{bmatrix} \tag{11.3.22}$$

Solving a set of these equations can be considerably faster than solving a single large Riccati equation (see Meirovitch et al., 1983). When implementing this control law, the closed-loop poles are (Meirovitch and Baruh, 1982):

$$\lambda_i = -\frac{\omega_i}{2}\left(2\sqrt{1+R_{22}^{-1}} - 1 + R_{22}^{-1}\right)^{1/2} \pm \frac{\omega_i}{2}\left(2\sqrt{1+R_{22}^{-1}} - 1 - R_{22}^{-1}\right)^{1/2} \tag{11.3.23}$$

Once the control gains have been derived for each modelled mode, the result must be transformed back into the form of the original state equation. For a control input defined by:

$$\boldsymbol{f}_c = -\boldsymbol{K}\boldsymbol{x} \tag{11.3.24}$$

where $\boldsymbol{K}$ is, in general, a $(2N \times m)$ matrix of gains for $2N$ states (N modes) and m control inputs, the transformation requires solving the relationship:

$$-\boldsymbol{B}\boldsymbol{K}\boldsymbol{x} = \begin{bmatrix} 0 & 0 & \cdots & 0 \\ \psi_1(\boldsymbol{r}_1) & \psi_1(\boldsymbol{r}_2) & \cdots & \psi_1(\boldsymbol{r}_m) \\ & & \vdots & \\ 0 & 0 & \cdots & 0 \\ \psi_n(\boldsymbol{r}_1) & \psi_n(\boldsymbol{r}_2) & \cdots & \psi_n(\boldsymbol{r}_m) \end{bmatrix} \begin{bmatrix} k_{1,1} & k_{1,2} & \cdots & k_{1,2n} \\ & & \vdots & \\ k_{m,1} & k_{m,2} & \cdots & k_{m,2n} \end{bmatrix} \begin{bmatrix} x_1 \\ \dot{x}_1 \\ \vdots \\ x_n \\ \dot{x}_n \end{bmatrix} =$$

$$= \begin{bmatrix} 0 \\ f_1 \\ \vdots \\ 0 \\ f_n \end{bmatrix} = \begin{bmatrix} 0 & 0 & \cdots & 0 & 0 \\ k_{11} & k_{12} & \cdots & 0 & 0 \\ & & \vdots & & \\ 0 & 0 & \cdots & k_{n1} & k_{n2} \end{bmatrix} \begin{bmatrix} x_1 \\ \dot{x}_1 \\ \vdots \\ x_n \\ \dot{x}_n \end{bmatrix} \tag{11.3.25a–c}$$

Defining the terms:

$$\boldsymbol{B}' = \begin{bmatrix} \psi_1(\boldsymbol{r}_1) & \psi_1(\boldsymbol{r}_2) & \cdots & \psi_1(\boldsymbol{r}_m) \\ & & \vdots & \\ \psi_n(\boldsymbol{r}_1) & \psi_n(\boldsymbol{r}_2) & \cdots & \psi_n(\boldsymbol{r}_m) \end{bmatrix} ; \quad \boldsymbol{K}' = \begin{bmatrix} k_{11} & k_{12} & \cdots & 0 & 0 \\ & & \ddots & & \\ 0 & 0 & \cdots & k_{n1} & k_{n2} \end{bmatrix} \tag{11.3.26a,b}$$

solving Equation (11.3.25) is equivalent to solving the matrix expression:

$$\boldsymbol{B}'\boldsymbol{K} = \boldsymbol{K}' \tag{11.3.27}$$

The control gain matrix $\boldsymbol{K}$ is therefore defined by the expression:

$$\boldsymbol{K} = \boldsymbol{B}^{-1}\boldsymbol{K}' \tag{11.3.28}$$

where $\boldsymbol{B}^{-1}$ is the left inverse of the matrix $\boldsymbol{B}'$ (simply an inverse if the number of control inputs is equal to the number of modelled modes).

Numerical examples of the implementation of independent modal space control can be found in the references listed at the beginning of this chapter.

It should be pointed out here that while independent modal space control has the advantage of simplifying the control law calculation, it is not without some drawbacks. Possibly chief among these arises from the operation in Equation (11.3.28); the method is dependent upon calculation of inverses of matrices that can be large and ill-conditioned. The ill-conditioning problem is a particular problem when the structure and transducer arrangement have some form of geometric symmetry.

Finally, as derivation of feedback control gains using the independent modal-space control methodology requires a matrix inversion, implementation is made easier if the number of control actuators is set equal to the number of modes to be controlled (otherwise a pseudo-inverse is required). Alternatively, if a small number of actuators is used to control a larger number of modes, some form of time-sharing (controlling a smaller number of modes at any given time) can be used (Baz et al., 1989; Baz and Poh, 1990). This is referred to as the modified independent modal space control (MIMSC) method.

11.3.2 Modal Filters

To implement the independent modal-space control methodology, a measure of the displacement and velocity of the system modes is required. It is possible to use the observers outlined in the previous section for this purpose (which provided estimates of modal displacements and velocities). In doing so, it would be important again to take precautions to avoid the destabilising effects of observation spillover. A second method for isolating the (resonant) response of a particular structural mode would be to use a bandpass filter, with the passband centre frequency set equal to the resonance frequency of the mode of interest (Hallauer et al., 1982). However, this technique can be difficult to implement when the resonant mode of interest is not sufficiently isolated from the resonance frequencies of other modes, particularly a problem with higher-order modes. A further alternative is to use 'modal filters'. These could be implemented using shaped piezoelectric polymer film sensors (Lee and Moon, 1990), as discussed in Chapter 8 for controlling structural/acoustic radiation into free space. They could also be implemented as spatial filters, which resolve modal displacements and velocities from a set of discrete sensor measurements (Meirovitch and Baruh, 1982, 1985; Morgan, 1991). It is this last technique that is of interest here.

The concepts underlying the construction of modal filters are the same as those used to formulate the modal control problem. As discussed previously, if a set of modes is orthogonal (self-adjoint), then the displacement of the *i*th mode can be extracted from a known structural displacement distribution through the relationship:

$$z_i(t) = \int_S m(\boldsymbol{r})\psi_i(\boldsymbol{r})w(\boldsymbol{r},t)\,\mathrm{d}\boldsymbol{r} \tag{11.3.29}$$

where integration is over the surface of the structure. Similarly, the velocity of the *i*th mode can be extracted from a known structural velocity distribution through the relationship:

$$\dot{z}_i(t) = \int_S m(\boldsymbol{r})\psi_i(\boldsymbol{r})\dot{w}(\boldsymbol{r},t)\,\mathrm{d}\boldsymbol{r} \tag{11.3.30}$$

Usually, structural displacement and velocity sensors are discrete transducers. Therefore, Equations (11.3.29) and (11.3.30) could be implemented by interpolating the discrete measurements to estimate the continuous displacement and velocity distributions, and the integrations performed numerically (Meirovitch and Oz, 1982). However, there is a second approach to modal filtering, which is better suited to real-time implementation. Considering only displacement (development of the problem for velocity sensing would be identical), the estimated displacement distribution of the structure, based upon a measurement of displacement at N_e discrete locations, is defined by:

$$\hat{w}(\boldsymbol{r},t) = \sum_{j=1}^{N_e} G(\boldsymbol{r},\boldsymbol{r}_j)w(\boldsymbol{r}_j,t) \tag{11.3.31}$$

where $G(\boldsymbol{r},\boldsymbol{r}_j)$ is the interpolation function between the measurement position $\boldsymbol{r}_j$ and the location $\boldsymbol{r}$. If Equation (11.3.31) is multiplied through by $m(\boldsymbol{r})\psi_i(\boldsymbol{r})$ and integrated over the surface of the structure, an estimate of the displacement of mode *i* based upon a set of discrete displacement measurements is obtained as follows:

$$z_i = \sum_{j=1}^{N_e} f_{ij}\,w(\boldsymbol{r},t) \tag{11.3.32}$$

The important point to note about this equation is that the estimate of modal displacements from a set of discrete displacement measurements has become a simple input-output problem; in other words, a simple (modal) filtering problem. The coefficients, f_{ij} are fixed for a given set of structural mode shape functions and sensor locations, and can be calculated prior to implementing the control system.

It is possible to derive an expression for the filter coefficients f_{ij} by first solving for the interpolation functions $G(\boldsymbol{r},\boldsymbol{r}_j)$, a task that can be accomplished using either a Rayleigh–Ritz or finite element approach. For the present purposes, however, the problem will be tackled using a least-squares approach. What is desired is to form linear combinations of discrete sensor signals to extract the displacement response of a given structural mode, which is a spatial filtering operation. The coefficients in the ith spatial filter $\boldsymbol{f}_i$ can be expressed as a vector as follows:

$$\boldsymbol{f}_i = \left[f_{i,1} \; f_{i,2} \; \cdots \; f_{i,N_e} \right]^{\mathrm{T}} \tag{11.3.33}$$

It will be assumed for the moment that the number of modes to be resolved, N, is less than or equal to the number of sensors, N_e. If the spatial filter $\boldsymbol{f}_i$ is to resolve the ith structural mode, then the desired response for the filter is defined by the relationship:

$$\boldsymbol{f}_i^{\mathrm{T}} \boldsymbol{\psi}_j = \begin{cases} 1 & i=j \\ 0 & i \neq j \end{cases} \tag{11.3.34}$$

where $\boldsymbol{\psi}_j$ is the vector of mass-normalised mode shape values (for mode j) at the sensing locations:

$$\boldsymbol{\psi}_j = \left[\psi_j(\boldsymbol{r}_1) \; \psi_j(\boldsymbol{r}_2) \; \cdots \; \psi(\boldsymbol{r}_{N_e}) \right]^{\mathrm{T}} \tag{11.3.35}$$

If the filter and mode shape vectors are assembled into two matrices:

$$\boldsymbol{\Psi} = \left[\boldsymbol{\psi}_1 \; \boldsymbol{\psi}_2 \; \cdots \; \boldsymbol{\psi}_N \right] ; \quad \boldsymbol{F} = \left[\boldsymbol{f}_1 \; \boldsymbol{f}_2 \; \cdots \; \boldsymbol{f}_N \right] \tag{11.3.36a,b}$$

the desired set of modal filters satisfies the relationship (Morgan, 1991):

$$\boldsymbol{F} = \boldsymbol{\Psi} \left[\boldsymbol{\Psi}^{\mathrm{T}} \boldsymbol{\Psi} \right]^{-1} \tag{11.3.37}$$

Note that $\boldsymbol{F}^{\mathrm{T}}$ is the generalised, or Moore-Penrose, inverse of the mode shape value matrix $\boldsymbol{\Psi}$.

In most real-world situations, the total number of modes excited in a structure is greater than the number of sensors, as well as the number of modes in the reduced order model. As discussed previously in this chapter, this can lead to problems of observation spillover. One approach to overcoming this problem would be to consider more modes than simply those in the reduced order model when designing the modal filters, and design the filters to minimise the cross-response between the modes. Therefore, for the ith modal filter, the following should be minimised:

$$\sum_{j=1}^{N} \left(\boldsymbol{f}_i^{\mathrm{T}} \boldsymbol{\psi}_j \right)^2 \tag{11.3.38}$$

subject to the constraint:

$$\boldsymbol{f}_i^{\mathrm{T}} \boldsymbol{\psi}_j = 1 \tag{11.3.39}$$

Using vector differentiation with the method of Lagrange multipliers, the solution to the problem posed in Equations (11.3.38) and (11.3.39) is (Morgan, 1991):

$$F = \left[\Psi\Psi^{\mathrm{T}}\right]^{-1}\Psi\Lambda \tag{11.3.40}$$

where Λ is a diagonal scaling matrix calculated to satisfy the constraint Equation (11.3.39). Note that apart from the scaling matrix, the solution in Equation (11.3.40) is the generalised inverse of the transposed mode shape value matrix Ψ^{T}.

There are several points worth mentioning here. First, it is possible to weight the above least-squares problem to take into account differences in the excitation of individual modes, if they are known (Morgan, 1991). Second, it is possible to estimate modal velocities using only modal displacement measurements, and vice versa, using a 'modal' observer. To understand this, consider the *i*th modal state equation (as it would appear in an independent modal space control problem):

$$\dot{x}_i(t) = A_i x_i + b f_i(t) \tag{11.3.41}$$

where

$$x_i = \begin{bmatrix} z_i \\ \dot{z}_i \end{bmatrix}; \quad A_i = \begin{bmatrix} 0 & 1 \\ -\omega_i^2 & -2\zeta\omega_i \end{bmatrix}; \quad b = \begin{bmatrix} 0 \\ 1 \end{bmatrix} \tag{11.3.42a–c}$$

Consider the case where only modal displacement measurements are available (from modal filters using discrete displacement sensors). The modal output equation is then,

$$y_i(t) = c x_i(t) \tag{11.3.43}$$

where

$$c = \begin{bmatrix} 1 & 0 \end{bmatrix} \tag{11.3.44}$$

The question is now, how can the modal velocities be recovered? To this end, consider a modal observer, described by (Meirovitch and Baruh, 1985) as:

$$\frac{\mathrm{d}\hat{x}_i(t)}{\mathrm{d}t} = A_i\hat{x}_i(t) + b f_i(t) + l_i[y_i(t) - \hat{y}_i(t)] \tag{11.3.45}$$

where $\hat{x}$ is the estimated state, and l_i is the observer gain vector

$$l_i = \begin{bmatrix} l_{i,1} & l_{i,2} \end{bmatrix} \tag{11.3.46}$$

To assist in assigning the modal observer gains, a modal error vector e_i can be defined as:

$$e_i(t) = \hat{x}_i(t) - x_i(t) \tag{11.3.47}$$

The evolution of the error vector is governed by the expression:

$$\dot{e}_i(t) = \frac{\mathrm{d}\hat{x}_i(t)}{\mathrm{d}t} - \dot{x}_i = (A_i - l_i c)e_i(t) \tag{11.3.48a,b}$$

Therefore, the observer gains can be assigned such that the poles of $(\boldsymbol{A}_i - \boldsymbol{l}_i\boldsymbol{c})$ have negative real parts that allow the observer to converge as quickly as possible. While the observer outlined above is deterministic, in a noisy environment a Kalman filter could also be used.

The final point to make here concerns the locations of discrete sensors for modal filtering applications. When attempting to numerically solve the integral Equation (11.3.29) directly, it seems logical to locate the discrete sensors in a regular manner. However, when attempting to implement modal filters in the manner described above, regular spacing of discrete sensors, especially on a structure with regular geometry, can lead to matrix conditioning problems. Usually, a more randomised placement, perhaps within certain bounds, is desirable.

Example 11.2

As a simple example of modal filter design (taken from Morgan (1991)), consider the problem where there are three modes and two sensors. This situation could arise where there are two modes in the reduced order model (that is, two controlled modes), but three modelled modes. Suppose that the (2 sensor × 3 mode) mode shape value matrix is equal to:

$$\boldsymbol{\Phi} = \begin{bmatrix} 0 & \sqrt{3}/2 & 1 \\ 1 & 1/2 & 0 \end{bmatrix} \tag{11.3.49}$$

Therefore, the modal filter matrix is:

$$\boldsymbol{F} = \left[\boldsymbol{\Phi}\boldsymbol{\Phi}^{\mathrm{T}}\right]^{-1}\boldsymbol{\Phi}\boldsymbol{\Lambda} = \frac{1}{8}\begin{bmatrix} -\sqrt{3} & 2\sqrt{3} & 5 \\ 7 & 2 & -\sqrt{3} \end{bmatrix}\boldsymbol{\Lambda} \tag{11.3.50a,b}$$

To satisfy the constraint, Equation (11.3.39), the diagonal scaling matrix $\boldsymbol{\Lambda}$ is:

$$\boldsymbol{\Lambda} = \begin{bmatrix} 8/7 & 0 & 0 \\ 0 & 2 & 0 \\ 0 & 0 & 8/5 \end{bmatrix} \tag{11.3.51}$$

so that the modal filter matrix, the columns of which define the three modal filters, is:

$$\boldsymbol{F} = \begin{bmatrix} -\sqrt{3}/7 & \sqrt{3}/2 & 1 \\ 1 & 1/2 & -\sqrt{3}/5 \end{bmatrix} \tag{11.3.52}$$

The measurements of modal displacement and/or velocity required to implement a modal control system could now be obtained by taking the measurements (displacement and/or velocity) from the two sensors and multiplying them by the coefficients in the columns of $\boldsymbol{F}$.

11.4 MODEL REDUCTION

The controllers that have been considered thus far in this chapter are of the same order as the systems themselves (feedback from each state). Given this fact, it is then worthwhile considering whether a smaller (lower-order) controller could be designed that would achieve approximately the same performance. Lower-order controllers are usually preferable to higher-order controllers for reasons of reduced complexity, which leads to simplified implementation and enhanced physical insight.

The are three basic ways in which the reduction of the order of a controller could be approached: approximation of the dynamic system (plant) by a reduced order model prior to the controller design, such that the latter process is based entirely upon the reduced order model; reduction of the (full order) controller size after it has been designed, by consideration of the whole of the dynamic system; and formulation of the controller design problem with order constraints imposed from the start. The first and third of these methods are referred to in various texts as direct methods, while the second is referred to as an indirect method. In this section, only the reduction of the model prior to control system design method will be considered. For a thorough discussion of techniques for reducing the size of the controller after its formulation, refer to Anderson and Moore (1990); a comparison of several techniques can also be found in Liu and Anderson (1986). For a thorough discussion of controllers with order constraints imposed upon them, refer to Bernstein and Hyland (1990). Also, for a thorough discussion of the model reduction problem that will be briefly discussed here, the interested reader is referred to Gawronski and Juang (1990).

The approach of reducing the size of a system model prior to the design of the feedback control system is one which has been used in the previous sections in this chapter. It is also one which general practitioners of control try to avoid, for a number of reasons. One objection is that it contradicts the (heuristic, but generally correct) notion that if there is any approximation in the design process, it should be postponed as long as possible. In fact, in early work on control of distributed parameter systems, a practical constraint placed upon controller design was that 'the distributed nature of the system should be retained until numerical results are required' (Athans, 1970). A second objection arises from the idea (Anderson and Moore, 1990) that:

> *what constitutes satisfactory approximation of the plant necessarily involves the controller: it is the closed-loop behaviour that one is ultimately interested in, and it is clear that a controller design could yield a situation in which very big variations in the open-loop plant in a limited frequency range had little effect on the closed-loop performance, while rather small variations in another frequency range could dramatically affect the closed-loop behaviour. Now since the definition of a good plant approximation involves the controller, and since the controller is not known at the time of approximation, one is caught in a logical loop.*

Despite these objections, it is often the case that the system model must be reduced prior to controller design simply because its size is simply unmanageable in the calculation procedure. For systems described by large order finite element models, common techniques for model reduction are the Guyan-Irons reduction method (Craig, 1981), or application of component mode synthesis techniques (Craig, 1987). When the dynamic system is described in terms of system modes, a common technique for model reduction is modal cost analysis (Skelton and Hughes, 1980; Skelton and Yousuff, 1982,1983; Skelton et al., 1990; Yousuff et al., 1985; Yousuff and Breida, 1992).

11.4.1 Modal Cost Analysis

The concept behind modal cost analysis can be summarised as follows. Given a dynamic system, described in terms of its normal modes such that the system output is:

$$y = \sum_{i=1}^{n} \left(c_i z_i + e_i \dot{z}_i \right) \tag{11.4.1}$$

where

$$\ddot{z}_i + 2\zeta_i \omega_i \dot{z}_i + \omega_i^2 z_i = f_i \tag{11.4.2}$$

find a subset of r system modes, the output of which is:

$$y_r = \sum_{i=1}^{r} \left(c_i z_i + e_i \dot{z}_i \right) \tag{11.4.3}$$

such that the error between the full order model and reduced order model, defined as the difference between the output of the two:

$$e(t) = y(t) - y_r(t) \tag{11.4.4}$$

is minimised. For a given number of modes in the subset r, the optimal set of constituents that would minimise the model error δJ is defined as:

$$\delta J = \lim_{t \to \infty} \mathrm{E} \left\{ \frac{1}{t} \int_0^t e^{\mathrm{T}}(\tau)\, e(\tau)\, \mathrm{d}\tau \right\} \tag{11.4.5}$$

For a given system, the model error is usually computed after the reduced order model is specified. The model reduction process then takes on the form of an iterative procedure, where different combinations of systems modes are trialled, and the 'best' set retained as the reduced order model. To simplify this process, component cost analysis (referred to as modal cost analysis when the system is described in terms of its normal modes) can be employed. Modal cost analysis uses a modified version of Equation (11.4.5) for the criterion for model reduction and aims to minimise the cost error ΔJ, defined as:

$$\Delta J = J - J_r \tag{11.4.6}$$

where

$$J = \lim_{t \to \infty} \mathrm{E} \left\{ \frac{1}{t} \int_0^t y^{\mathrm{T}}(\tau) y(\tau)\, \mathrm{d}\tau \right\} \tag{11.4.7}$$

and

$$J_r = \lim_{t \to \infty} \mathrm{E} \left\{ \frac{1}{t} \int_0^t y_r^{\mathrm{T}}(\tau) y_r(\tau)\, \mathrm{d}\tau \right\} \tag{11.4.8}$$

If the reduced-order model is optimal, it will satisfy the orthogonality condition (Wilson, 1974):

$$\langle e, y_r \rangle = \lim_{t\to\infty} \mathrm{E}\left\{ \frac{1}{t} \int_0^t e^{\mathrm{T}}(\tau) y_r(\tau)\, \mathrm{d}\tau \right\} = 0 \tag{11.4.9a,b}$$

where $\langle\ \rangle$ represents an inner product. The model error in Equation (11.4.5) can be expressed in terms of the cost error of Equation (11.4.6) and the inner product in Equation (11.4.9) as:

$$\delta J = \Delta J - 2\langle e, y_r \rangle \tag{11.4.10}$$

It can be surmised that minimising the cost error of Equation (11.4.10) is a satisfactory approach provided that the reduced-order model is known to be at least near-optimal. However, there is still the problem that the reduced-order model needs to be known before the error criterion can be calculated. To overcome this problem, modal cost analysis uses a predicted cost error, defined as:

$$\Delta J' = J - J_r' \tag{11.4.11}$$

where $'$ denotes a predicted value. J_r' is calculated from:

$$J_r' = \sum_{i \in r} J_i' \tag{11.4.12}$$

where J_i' is the ith modal cost, defined as:

$$J_i' = \frac{1}{2} \lim_{t\to\infty} \mathrm{E}\left\{ \frac{1}{t} \int_0^t \frac{\partial[y^{\mathrm{T}}(\tau) y(\tau)]}{\partial z_i(\tau)} z_i(\tau)\, \mathrm{d}\tau \right\} \tag{11.4.13}$$

In the more general component cost analysis (Skelton and Yousuff, 1983), the modal state z_i is replaced by the state of some component on the structure.

As the modal costs can be calculated prior to calculation of the error criterion, the predicted cost error can be calculated before the formulation of the reduced-order model. Therefore, modal cost analysis aims to derive an 'optimal' reduced-order model by minimising the predicted cost error defined in Equation (11.4.11).

One problem with modal cost analysis is that, except in special circumstances, there is no guarantee that the predicted cost error is equal to the model error, and therefore the result may not be the optimal reduced-order model. It is, however, possible to calculate the model error directly in much the same manner as the predicted cost error (Yousuff and Breida, 1992). First, a modal output is defined as:

$$y_i = c_i z_i + e_i \dot{z}_i \tag{11.4.14}$$

such that the system output is simply defined by the expression:

$$y = \sum_{i \in N_n} y_i \tag{11.4.15}$$

Second, the residual modes are defined as the difference between the total modal set and the reduced-order subset as follows:

$$N_{res} = N_n - N_r \tag{11.4.16}$$

Then,

$$e = \sum_{i \in N_{res}} y_i \tag{11.4.17}$$

From this, the model error is:

$$\delta J = \sum_{i \in N_{res}} \sum_{j \in N_{res}} y_{ij} \tag{11.4.18}$$

where

$$y_{ij} = \lim_{t \to \infty} \mathrm{E}\left\{ \frac{1}{t} \int_0^t y_i^{\mathrm{T}}(\tau) y_j(\tau)\, \mathrm{d}\tau \right\} \tag{11.4.19}$$

This relationship can be used in the modal cost analysis procedure. Improved results obtained using this approach are found in Yousuff and Breida (1992).

The relationship between the 'maximum possible' disturbance attenuation and that provided by a control system designed using a reduced-order model is directly related to the cost function, J. If the cost is 'small', then the performance of the control system designed using a reduced order model can be expected to be good.

One final point to mention is that even if a reduced order model is used in the control law formulation, it may be necessary to reduce its order further for implementation. In that case, the various procedures outlined in Anderson and Moore (1990) for controller reduction are necessary.

11.4.2 Optimal Truncated Model

In this section, ways are outlined in which the analytical approximation terms can be developed when it is desired to model modes only within a certain frequency band. In Section 11.4.2.1, classical optimal truncation was considered, which is where modes below a certain frequency are all modelled, and an approximation term is developed to account for all frequencies above this. In Section 11.4.2.2, an extension to the classical approach is discussed, in which the modes are modelled only in a specific frequency band of interest, and the approximation terms account for frequencies both above and below this frequency band. In Section 11.4.2.3, the method by which the corrected truncation model for a specified frequency band can be adjusted is outlined with particular application to optimal $\mathcal{H}_\infty$ and $\mathcal{H}_2$ controllers. More details of the analysis presented here are given by Barrault et al. (2007).

11.4.2.1 Classical Optimal Truncation for Low Frequencies

To capture the structural response of a complex system, $\boldsymbol{G}(s)$, in a time-efficient manner, the modal model must be truncated. Traditionally, only the $\boldsymbol{M}$ lowest frequency modes are modelled, resulting in a reduced order model $\boldsymbol{G}_r(s)$ with residual dynamics $\boldsymbol{G}_d(s)$:

$$\begin{aligned} \boldsymbol{G}(s) &= \sum_{i=1}^{\infty} \frac{\boldsymbol{F}_i}{s^2 + 2\zeta_i\omega_i s + \omega_i^2} \\ &= \sum_{i=1}^{M} \frac{\boldsymbol{F}_i}{s^2 + 2\zeta_i\omega_i s + \omega_i^2} + \sum_{i=M+1}^{\infty} \frac{\boldsymbol{F}_i}{s^2 + 2\zeta_i\omega_i s + \omega_i^2} \\ &= \boldsymbol{G}_r(s) + \boldsymbol{G}_d(s) \end{aligned} \tag{11.4.20a–c}$$

where $\boldsymbol{F}_i$ is the external force matrix, i is the mode number, ζ_i is the ith damping ratio and ω_i is the ith natural frequency.

Ignoring the effect of the residual dynamics from the higher-order modes may result in erroneous estimates of the structural response, degrading the performance of the control system. Thus, higher-order modes should be approximated through some other means and a zero-order correction term $\boldsymbol{K}_d$ applied to account for these. The overall structural response is then estimated as:

$$\hat{\boldsymbol{G}}(s) = \boldsymbol{G}_r(s) + \boldsymbol{K}_d \tag{11.4.21}$$

The optimal correction term, $\boldsymbol{K}_d$ is evaluated by minimising the $\mathcal{H}_2$ norm (r.m.s. of the impulse response) of the following cost function:

$$J = \left\| W(s)\left(G(s) - \hat{G}(s)\right) \right\|_2^2 \tag{11.4.22}$$

where $W(s)$ is a perfect low-pass filter with a cut-off between $\pm(\omega_M + \omega_{M+1})/2$. Differentiating the cost function J the optimal correction term is found to be (Moheimani and Clark, 2000; Moheimani et al., 2003; Barrault, 2006):

$$\boldsymbol{K}_d = \frac{1}{2\omega_c} \sum_{i=M+1}^{\infty} \frac{\boldsymbol{F}_i}{\omega_i} \log_e \left(\frac{\omega_i + \omega_c}{\omega_i - \omega_c} \right) \tag{11.4.23}$$

11.4.2.2 Optimal Truncation for Specified Frequency Band

Classical truncation is based on the assumption that the lower order modes dominate the system response. There are, however, cases where this is not the case, and there are a number of modes within a higher frequency band that contribute significantly to the response. In such cases, it is preferable to truncate around a specific bandwidth, modelling only modes within that bandwidth, and incorporating optimal model approximation terms to account for modes at both lower and higher frequencies. Barrault et al. (2007) developed a procedure for calculating these terms:

$$\begin{aligned} G(s) &= \sum_{i=1}^{\infty} \frac{F_i}{s^2 + 2\zeta_i\omega_i s + \omega_i^2} \\ &= \sum_{i=1}^{m_1-1} \frac{F_i}{s^2 + 2\zeta_i\omega_i s + \omega_i^2} + \sum_{i=m_1}^{m_2} \frac{F_i}{s^2 + 2\zeta_i\omega_i s + \omega_i^2} \\ &\quad + \sum_{i=m_2+1}^{\infty} \frac{F_i}{s^2 + 2\zeta_i\omega_i s + \omega_i^2} \\ &= G_l(s) + G_r(s) + G_d(s) \end{aligned} \tag{11.4.24}$$

where m_1 and m_2 are the lowest and highest modes within the bandwidth to be fully modelled and $G_l(s)$ and $G_d(s)$ are the lower and higher-order mode residual dynamics respectively. As in classical optimal truncation, the higher-order modes are approximated through a zero-order correction term K_d. The lower-order modes are approximated via a second-order correction term K_l/ω^2. Thus, in the frequency domain ($s = j\omega$):

$$\hat{G}(\omega) = \frac{K_l}{\omega^2} + G_r(\omega) + K_d \tag{11.4.25}$$

The optimal correction terms, K_d and K_l are evaluated by minimising the $\mathcal{H}_2$ norm of the following cost function:

$$J = \left\| W(\omega)\left(G(\omega) - \hat{G}(\omega)\right) \right\|_2^2 \tag{11.4.26}$$

where $W(\omega)$ is a perfect bandpass filter with unit value in $[-\omega_c, -\omega_a]$ and $[\omega_c, \omega_a]$, where $\omega_c = (\omega_{m1} + \omega_{m1+1})/2$ and $\omega_a = (\omega_{m2} + \omega_{m2+1})/2$. Differentiating the cost function J gives the low-frequency optimal correction term as (Barrault et al., 2007):

$$K_l = \frac{1}{\beta}\left(\Gamma_i - \left(\frac{\omega_c - \omega_a}{\omega_c\omega_a}\right) K_d\right) \tag{11.4.27}$$

where

$$\Gamma_i = \sum_{\substack{i=1 \\ i \notin [m_1, m_2]}}^{\infty} \frac{X_i}{\omega_i^2} + \left(\frac{\omega_c - \omega_a}{\omega_c\omega_a}\right) \sum_{\substack{i=1 \\ i \notin [m_1, m_2]}}^{\infty} \frac{F_i}{\omega_i^2} \tag{11.4.28}$$

$$X_i = \frac{F_i}{\omega_i^2} \log_e\left(\frac{(\omega_c + \omega_i)|\omega_a - \omega_i|}{|\omega_c - \omega_i|(\omega_a + \omega_i)}\right) \tag{11.4.29}$$

$$\beta = \frac{\omega_c^3 + \omega_a^3}{3\omega_c^3\omega_a^3} \tag{11.4.30}$$

and the high-frequency optimal correction term is:

$$\boldsymbol{K}_d = \frac{1}{\gamma}\left(\sum_{\substack{i=1 \\ i \notin [m_1, m_2]}}^{\infty} \mathbf{X}_i - \left(\frac{\omega_c - \omega_a}{\omega_c \omega_a} \right) \frac{\boldsymbol{\Gamma}_i}{\beta} \right) \tag{11.4.31}$$

where

$$\gamma = \omega_c - \omega_a - \frac{1}{\beta}\left(\frac{\omega_c - \omega_a}{\omega_c \omega_a} \right)^2 \tag{11.4.32}$$

11.4.2.3 Optimisation for Robust Control Design

The state-space model state and output equations for a discrete time-invariant system are:

$$\boldsymbol{x}(k+1) = \boldsymbol{A}\boldsymbol{x}(k) + \boldsymbol{B}\boldsymbol{u}(k) \tag{11.4.33}$$

and

$$\boldsymbol{y}(k) = \boldsymbol{C}\boldsymbol{x}(k) + \boldsymbol{D}\boldsymbol{u}(k) \tag{11.4.34}$$

where $\boldsymbol{x}$ is the $(n \times 1)$ state vector, $\boldsymbol{u}$ is the $(r \times 1)$ vector of r system inputs, $\boldsymbol{y}$ is the $(m \times 1)$ output vector, $\boldsymbol{A}$ is the $(n \times n)$ state matrix, $\boldsymbol{B}$ is the $(n \times r)$ input matrix, $\boldsymbol{C}$ is the $(m \times n)$ output matrix, and $\boldsymbol{D}$ is the $(m \times r)$ direct transmission matrix. The transfer function estimate matrix can be expressed in compact state-space notation form as:

$$\hat{\boldsymbol{G}}(s) = \left[\begin{array}{cc|c} \boldsymbol{A} & \boldsymbol{0} & \boldsymbol{B} \\ \boldsymbol{0} & \boldsymbol{A}_l & \boldsymbol{B}_l \\ \hline \boldsymbol{C} & \boldsymbol{C}_l & \boldsymbol{K}_d \end{array}\right] = \begin{bmatrix} \hat{\boldsymbol{A}} & \hat{\boldsymbol{B}} \\ \hat{\boldsymbol{C}} & \boldsymbol{K}_d \end{bmatrix} \tag{11.4.35}$$

where subscript $\boldsymbol{l}$ denotes states that are implied by the $\boldsymbol{K}_l$ correction term, and the matrices $\boldsymbol{A}_l$, $\boldsymbol{B}_l$ and $\boldsymbol{C}_l$ are given as:

$$\boldsymbol{A}_l = \begin{bmatrix} \boldsymbol{0} & \boldsymbol{I} \\ \boldsymbol{0} & \boldsymbol{0} \end{bmatrix}; \quad \boldsymbol{B}_l = \begin{bmatrix} \boldsymbol{0} \\ \boldsymbol{I} \end{bmatrix}; \quad \boldsymbol{C}_l = \begin{bmatrix} \boldsymbol{K}_l & \boldsymbol{0} \end{bmatrix} \tag{11.4.36a–c}$$

To design a $\mathcal{H}_2$ or $\mathcal{H}_\infty$ controller, the associated matrix:

$$\boldsymbol{H} = \begin{bmatrix} \hat{\boldsymbol{A}} - j\omega\boldsymbol{I} & \hat{\boldsymbol{B}} \\ \hat{\boldsymbol{C}} & \boldsymbol{K}_d \end{bmatrix} \tag{11.4.37}$$

must have full column rank (Zhou et al., 1996). However, this condition is not met due to the component of $\boldsymbol{A}_l$ in the $\boldsymbol{A}$ matrix (see Equation (11.4.36). The corrected truncation model for a specified frequency band must therefore be adjusted for use in optimal $\mathcal{H}_\infty$ and $\mathcal{H}_2$

controllers. Barrault et al. (2007) overcame this by replacing the $\boldsymbol{A}_l$ matrix with an adjusted version as follows:

$$\boldsymbol{A}_l \approx \boldsymbol{A}_{adj} = \begin{bmatrix} \boldsymbol{0} & \boldsymbol{I} \\ -\omega_1^2 \boldsymbol{I} & -2\zeta_1\omega_1 \boldsymbol{I} \end{bmatrix} \tag{11.4.38a,b}$$

where ζ_1 and ω_1 are the fundamental mode damping ratio and natural frequency respectively.

11.5 EFFECT OF MODEL UNCERTAINTY

Real control systems have inherent uncertainties that lead to differences between the plant model for which the controller is designed and the real system to which the controller is applied. These uncertainties must be addressed to ensure that performance and stability of the control system remain within reasonable bounds.

One way of dealing with uncertainty is through the use of stochastic control theory (a comprehensive introduction of which is given by Lewis et al. (2007)), whereby probability distributions are used to represent model uncertainties. Stochastic control theory can, however, sometimes lead to results that are erroneous (Rollins, 1999), and thus may not be appropriate in certain cases such as those where safety is of importance.

Here, means by which how uncertainty can be accounted for are discussed, using robust stability and performance analysis. The interested reader is directed to Zhou et al. (1996) and Skogestad and Postlethwaite (2005) for more in-depth descriptions of this topic.

11.5.1 Modelling of Unstructured Uncertainties

Uncertainties can be either unstructured or structured. Unstructured uncertainties are generic uncertainties associated with the model as a whole. High-frequency errors due to model truncation and other effects can only be adequately represented as unstructured uncertainties. Structured uncertainties are those uncertainties that are functions of specific parameters (parametric uncertainty) of the nominal model; for example, damping. Low-frequency errors due to uncertainties in the nominal model dynamics are generally represented as structured uncertainties, which can be treated individually, or combined into some form of unstructured uncertainty – in the latter case, so that methods of analysis used for unstructured uncertainty problems can be applied.

Unstructured uncertainties are usually accounted for in the form of an additive or multiplicative uncertainty. Consider first introducing a feed-through term to the nominal model ($\boldsymbol{G}_n$) via an additive uncertainty ($\boldsymbol{E}_a$):

$$\boldsymbol{G}_a(s) = \boldsymbol{G}_n(s) + \boldsymbol{E}_a(s)\Delta(s) \tag{11.5.1}$$

where $\boldsymbol{E}_a$ is the upper bound of the additive uncertainties being considered, and $\Delta(j\omega) \leq 1$ for all frequencies.

As an example, recall that the use of model truncation to make a control problem more tractable was discussed in Section 11.4. Neglecting out-of-bandwidth modes may, however, introduce uncertainty as these neglected modes can contribute to the dynamics within the control frequency bandwidth of interest. Inclusion of a few additional modes can be an effective means of reducing the error, but some error will still remain as long as there is truncation. Any other dynamics that are not captured in the nominal model will also lead to unmodelled dynamics uncertainty. In this case, the frequency dependent error bounds on the additive uncertainty can be determined by considering the differences between the full-order and truncated models. Moheimani et al. (2003) discuss techniques that use an additive uncertainty to account for the loss of information due to unmodelled dynamics.

Consider now introducing a multiplicative uncertainty ($\boldsymbol{E}_m$) to the nominal model ($\boldsymbol{G}_n$):

$$\boldsymbol{G}_m(s) = \boldsymbol{G}_n(s)[1 + \boldsymbol{E}_m(s)\Delta(s)] \tag{11.5.2}$$

where $\boldsymbol{E}_m$ is the upper bound of the multiplicative uncertainties being considered, and $\Delta(\mathrm{j}\omega) \leq 1$ for all frequencies. The upper bound is a function of frequency and tends to be much greater at higher frequencies, where the nominal plant model is less accurate.

Unstructured uncertainties can also be treated in ways other than simply additive or multiplicative (for example, feedback uncertainties); but as the approaches are similar, they are not discussed here. It should be noted that modelling uncertainties as unstructured is a conservative approach. This is because it assumes that every plant in the uncertainty set is realistically possible when in reality this is generally not true. Determining the upper bound on the additive and multiplicative uncertainties for a given system is not a trivial task. Zhou et al. (1996) and Skogestad and Postlethwaite (2005) provide more details on this process.

11.5.2 Robust Stability and Performance

Robust performance control aims to determine the bounds on uncertainties, and achieve robust performance, and robust stability. Here, robust stability will be discussed first, and this will be followed by a discussion of robust performance.

11.5.2.1 Robust Stability

A system is said to achieve robust stability if the controller provides internal stability (that is, all internal signals as well as the output are bounded when the input is bounded in magnitude) for all plants in the uncertainty set. For a system with robust stability, robust performance is also achieved if all the performance objectives are satisfied for all plants in the uncertainty set.

When determining the stability of the closed-loop system in the presence of uncertainty, it is assumed that the open-loop system and nominal system with feedback are stable (that is, the nominal system has been previously designed to meet stability requirements). The conditions for robust stability for the system in the presence of uncertainty can then be determined from the Nyquist stability criterion (the small gain theorem could also be used instead; see, for example, Zhou et al. (1996) for further discussion).

The closed-loop for all plants in the uncertainty set will be stable only if the open-loop transfer function does not encircle −1 on a Nyquist plot (Skogestad and Postlethwaite, 2005). Considering the case of multiplicative uncertainty, the open-loop transfer function of the plant with uncertainty is:

$$\boldsymbol{L}_m(s) = \boldsymbol{K}(s)\boldsymbol{G}_n(s)[1 + \boldsymbol{E}_m(s)\Delta(s)] \tag{11.5.3}$$

and hence:

$$1 + \boldsymbol{L}_m(s) = 1 + \boldsymbol{L}_n(s) + \boldsymbol{K}(s)\boldsymbol{G}_n(s)\boldsymbol{E}_m(s)\Delta(s) \tag{11.5.4}$$

where $\boldsymbol{K}(s)$ represents the controller gain and $\boldsymbol{L}_n(s) = \boldsymbol{K}(s)\boldsymbol{G}_n(s)$ is the nominal open-loop transfer function. The open-loop transfer function with multiplicative uncertainty $\boldsymbol{L}_m(s)$ will thus not encircle -1 if:

$$|1 + \boldsymbol{L}_n(s)| > |\boldsymbol{K}(s)\boldsymbol{G}_n(s)\boldsymbol{E}_m(s)| \tag{11.5.5}$$

which can be rearranged to give:

$$|\boldsymbol{T}(s)| = \left|\frac{\boldsymbol{G}_n(s)\boldsymbol{K}(s)}{1 + \boldsymbol{G}_n(s)\boldsymbol{K}(s)}\right| < \frac{1}{\boldsymbol{E}_m(s)} \tag{11.5.6}$$

where $\boldsymbol{T}(s)$ is the complementary sensitivity function (discussed in Section 5.5.3).

A similar argument for the case of additive uncertainty yields a robust stability criterion of:

$$|\boldsymbol{S}(s)| = \left|\frac{1}{1 + \boldsymbol{G}(s)\boldsymbol{K}(s)}\right| < \frac{1}{\boldsymbol{K}(s)\boldsymbol{E}_a(s)} \tag{11.5.7}$$

where $\boldsymbol{S}(s)$ is the sensitivity function (discussed in Section 5.5.3).

11.5.2.2 Robust Performance

It was previously discussed in Section 5.5.3 that there must be some trade-off in control loop gains because, for good tracking and good disturbance rejection, a large feedback gain is desirable but for good sensor noise rejection, a small feedback gain is desirable. Additionally, gains must be limited in certain frequency ranges to ensure that saturation of control actuators does not occur. Generally, it is desirable to have good tracking and disturbance rejection at low frequencies (small sensitivity function $\boldsymbol{S}$), and good sensor noise rejection at higher frequencies (small complementary sensitivity function $\boldsymbol{T}$).

A system with uncertainty will achieve robust performance if it is able to achieve its performance objectives for all plants in the uncertainty set. If, for example, the nominal plant sensitivity function is considered to be the performance indicator and a frequency-dependent upper bound is enforced on its magnitude, the following criterion results:

$$|\boldsymbol{S}(\mathrm{j}\omega)| < \left|\frac{1}{\boldsymbol{M}(\mathrm{j}\omega)}\right| \Rightarrow |\boldsymbol{M}(\mathrm{j}\omega)\boldsymbol{S}(\mathrm{j}\omega)| < 1 \quad \text{for all } \omega \tag{11.5.8}$$

For robust performance, this condition must be satisfied for all plants in the uncertainty set; that is, $|\boldsymbol{M}(\mathrm{j}\omega)\boldsymbol{S}_m(\mathrm{j}\omega)| < 1$ for all frequencies and all possible uncertainties, where $\boldsymbol{S}_m(\mathrm{j}\omega)$ is the sensitivity function of the plant with uncertainty. This can again be done by considering a graphical representation on a Nyquist plot, or via algebraic derivation. Both will lead to the following criterion for robust performance for multiplicative uncertainty:

$$|\boldsymbol{M}(\mathrm{j}\omega)\boldsymbol{S}(\mathrm{j}\omega)| + |\boldsymbol{E}_m(\mathrm{j}\omega)\,\boldsymbol{T}(\mathrm{j}\omega)| < 1 \tag{11.5.9}$$

Note that this condition must be satisfied for all plants within the uncertainty set.

11.6 EXPERIMENTAL DETERMINATION OF THE SYSTEM MODEL THROUGH SUBSPACE MODEL IDENTIFICATION

Conventional techniques for vibration control usually assume either a sufficiently accurate analytical full system model or full system simulation. Such assumptions can be valid for simple structures (for example, a flexible beam); however, practical systems will often be far more complex. These complex systems cannot always be analytically modelled accurately and numerical simulations that are practicable can often over-simplify the system. Experimental determination of a mathematical full system model (using data obtained from the real system) is one means of overcoming this problem.

Consider the generic state-space representation of a system with actuators and sensors:

$$\dot{\boldsymbol{x}}(t) = \boldsymbol{A}\boldsymbol{x}(t) + \boldsymbol{B}_a v_a(t) \tag{11.6.1}$$

$$\boldsymbol{v}_s(t) = \boldsymbol{C}_s\boldsymbol{x}(t) + \boldsymbol{D}_{as}\boldsymbol{v}_a(t) \tag{11.6.2}$$

where, for a real system $\boldsymbol{x}$ is the system state, $\boldsymbol{v}_a$ represents the actuator inputs and $\boldsymbol{v}_s$ represents the sensor outputs, $\boldsymbol{A}$ is the system matrix, $\boldsymbol{B}_a$ is the actuator input matrix, $\boldsymbol{C}_s$ is the sensor output matrix, and $\boldsymbol{D}_{as}$ is the direct transmission matrix relating the actuator outputs to the sensor inputs. Subspace model identification (SMI) algorithms can use measured actuator inputs and sensor outputs to determine the order n of the system, and thus can be used to estimate $\boldsymbol{A}$, $\boldsymbol{B}_a$, $\boldsymbol{C}_s$ and $\boldsymbol{D}_{as}$ using linear algebra. Several of these algorithms, which incorporate an LQ factorisation, a singular value decomposition and a least-squares solution, are outlined by Overschee and Moor (1996), Haverkamp (2001) and Verhaegen and Verdult (2007). The most suitable algorithm depends upon the physical system under consideration; for example, a system assuming white measurement noise should be analysed using a different algorithm to one assuming coloured measurement noise.

Barrault et al. (2008) successfully applied SMI for experimental determination of the system dynamics of a shell structure using shaker actuators and accelerometer sensors (experimental results are detailed in this paper). Barrault's system can be represented in discrete state-space as:

$$\hat{\boldsymbol{x}}(k+1) = \boldsymbol{A}\hat{\boldsymbol{x}}(k) + \boldsymbol{B}_a\boldsymbol{v}_a(k) + \boldsymbol{K}\boldsymbol{e}(k) \tag{11.6.3}$$

$$\hat{\boldsymbol{v}}_s(k) = \boldsymbol{C}_s\hat{\boldsymbol{x}}(k) + \boldsymbol{D}_{as}\boldsymbol{v}_a(k) \tag{11.6.4}$$

$$\boldsymbol{v}_s(k) = \hat{\boldsymbol{v}}_s(k) + \boldsymbol{e}(k) \tag{11.6.5}$$

where $\boldsymbol{K}$ is a Kalman matrix filter gain, $\boldsymbol{e}$ is zero-mean white noise, and ^ represents the minimum variance estimate. As the system equations are a combination of deterministic and stochastic terms, and because it is assumed that white noise and process noise combine to produce the output error, the Past Output Multivariable Output-Error State-space (PO-MOESP) SMI algorithm was applied to measured actuator inputs and sensor outputs in order to determine the $\boldsymbol{K}$ matrix and the system matrices up to a similarity transformation (that is, the matrices estimated from the experiment do not necessarily have the correct individual elements, but when used in analysis they produce the same results as matrices with the correct elements). Stability of the controllable modes, observability of the state and sensor output matrices, and spectrally rich continuous actuator inputs are assumed.

Barrault (2006) describes the application of the PO-MOESP SMI algorithm for estimating the deterministic part of the model (that is, estimating the system matrices $\boldsymbol{A}$, $\boldsymbol{B}_a$, $\boldsymbol{C}_s$ and $\boldsymbol{D}_{as}$ up to a similarity transformation). Note that their description follows the rigorous discussion of the algorithm given by Haverkamp (2001).

Once the system matrices have been estimated experimentally, further processes need to be undertaken in order to apply these in a robust controller. Overschee and Moor (1996) and Haverkamp (2001) outline algebraic procedures for estimating the stochastic part of the model (specifically the $\boldsymbol{K}$ matrix). Barrault (2006) and Barrault et al. (2008) discuss the methodologies that they applied subsequent to using the sub-space model identification outlined here for controlling vibration of their shell structure and also show experimental results. Chapter 5 and specialist control books also provide further detail on the steps from taking the state estimates to constructing a working controller.

11.7 SENSOR AND ACTUATOR PLACEMENT CONSIDERATIONS

In the control law development outlined in this chapter, the approach taken has basically been one of assuming that the location of sensors and actuators on a structure is given, and the problem has been one of formulating an optimal control law (with reference to some performance objectives) given these location constraints. It is intuitively obvious, however, that sensor and actuator placement will have a large influence upon the ultimate performance of the control system. For example, in a poor system design, an actuator could be placed on or near the nodal line of a mode which is to be controlled, the result being that either an excessively large force is required for control at best, or that the mode is uncontrollable at worst. Similarly, a sensor could be placed on or near a nodal line of a mode to be controlled, meaning that the signal-to-noise ratio is poor at best, or the mode is unobservable (and hence uncontrollable) at worst.

A large number of strategies for optimising the placement of sensors and actuators have been developed in recent years, largely based upon the idea of minimising some performance index associated with transducer placement. For (probably) the majority of these, the performance index is based upon some measure of controllability and observability, derived from minimum energy considerations (see, for example, Hughes and Skelton, 1980b; Arbel, 1981; Vander Velde and Carignan, 1984; Hac and Liu, 1992; Maghami and Joshi, 1993). In this section, some of the concepts associated with sensor and actuator placement, based upon energy considerations, will be outlined, and will be seen to be intimately connected to ideas of controllability and observability. A more complete tutorial-like discussion of actuator and sensor placement issues can be found in an article by Baruh (1992).

11.7.1 Actuator Placement

In placing actuators on a structure, the aim here is to excite the structure with the minimum control effort for the various operating conditions. The operating conditions of particular interest are transient response recovery and attenuation of persistent excitation.

Heuristically, the criterion of minimum control effort for actuator placement can be seen to take into account the metric of controllability, because if a mode is uncontrollable, the control effort required is infinite. However, simply assessing controllability in itself is not necessarily a good criterion for actuator placement; an actuator can be extremely close to a nodal line of a mode of interest and the system will still be (theoretically) controllable, although the control effort required for the control may be in excess of what is practically achievable.

The response of the systems of interest here are again described in terms of system modes, where the governing equation for each mode is:

$$\ddot{z}_i(t) + 2\zeta\omega_i\dot{z}_i(t) + \omega_i^2 z_i(t) = f_i(t) \tag{11.7.1}$$

This can be expressed in state-space form as:

$$\dot{\boldsymbol{x}} = \boldsymbol{A}\boldsymbol{x} + \boldsymbol{B}\boldsymbol{u} \tag{11.7.2}$$

In this instance, however, the state vector and matrix will be written in a slightly different form. Defining the state vector (for N modelled modes) as:

$$\boldsymbol{x} = \begin{bmatrix} x_1 & \omega_1\dot{x}_1 & \cdots & x_N & \omega_N\dot{x}_N \end{bmatrix}^{\mathrm{T}} \tag{11.7.3}$$

the state matrix is in block diagonal form, given by:

$$\boldsymbol{A} = \begin{bmatrix} \boldsymbol{A}_1 & & \\ & \ddots & \\ & & \boldsymbol{A}_N \end{bmatrix}, \quad \boldsymbol{A}_i = \begin{bmatrix} 0 & \omega_i \\ -\omega_i & -2\xi\omega_i \end{bmatrix} \tag{11.7.4}$$

Accordingly, the control input matrix for N_c control forces is now defined as:

$$\boldsymbol{B} = \begin{bmatrix} 0 & \cdots & 0 \\ \psi_1(\boldsymbol{r}_1) & \cdots & \psi_1(\boldsymbol{r}_{N_c}) \\ & \vdots & \\ 0 & \cdots & 0 \\ \psi_N(\boldsymbol{r}_1) & \cdots & \psi_N(\boldsymbol{r}_{N_c}) \end{bmatrix} \tag{11.7.5}$$

11.7.1.1 Transient Excitation

The transient excitation case will be considered first, where the system is subject to some perturbation and the control system is to return it to its original state. To guide the actuator placement, the problem can be defined as one of returning the perturbed system to equilibrium at some time t_f, subject to minimising the energy criterion:

$$J = \int_0^{t_f} \boldsymbol{u}^{\mathrm{T}}(\tau)\boldsymbol{u}(\tau)\,\mathrm{d}\tau \tag{11.7.6}$$

where $\boldsymbol{u}(\tau)$ is the vector of actuator forces at time, τ.

As discussed in Chapter 5, the optimal solution to this problem is given by Kalman et al. (1962) as:

$$\boldsymbol{u}_{opt}(t) = -\boldsymbol{B}^{\mathrm{T}} e^{A(t_f - t)} \boldsymbol{P}^{-1}(t_f)(e^{A\mathrm{T}}\boldsymbol{x}(0) - \boldsymbol{x}(t_f)) \tag{11.7.7}$$

where $\boldsymbol{P}$ is the controllability grammian, defined originally in Chapter 5 as:

$$\boldsymbol{P}(t) = \int_0^{t} \mathrm{e}^{A(\tau)}\boldsymbol{B}\boldsymbol{B}^{\mathrm{T}}\mathrm{e}^{A^{\mathrm{T}}(\tau)}\,\mathrm{d}\tau \tag{11.7.8}$$

The minimum energy associated with this optimal result is (Kalman et al., 1962):

$$J_{\min} = \left[\mathrm{e}^{A(t_f)}\boldsymbol{x}(0) - \boldsymbol{x}(t_f)\right]^{\mathrm{T}} \boldsymbol{P}^{-1}(t_f) \left[\mathrm{e}^{A(t_f)}\boldsymbol{x}(0) - \boldsymbol{x}(t_f)\right] \tag{11.7.9}$$

It can be observed that, from Equation (11.7.7), if the controllability grammian is small (such that the inverse is large), the optimum control forces will be large. Associated with this, the minimum value of the energy criterion, defined in Equation (11.7.9), will be large. Therefore, it is desirable to make the controllability grammian as large as possible (this defines a relationship between controllability and energy considerations). Referring to the definition of the controllability grammian given in Equation (11.7.8), it can be seen that it is dependent upon the control input matrix $\boldsymbol{B}$, which, from Equation (11.7.5), is dependent upon actuator location. This dependency will enable a definition of some degree of quality for a given actuator arrangement: a 'good' actuator arrangement is one that maximises the norm (square root of the sum of the squares of all elements) of the controllability grammian.

As an example of how controllability can vary with actuator position is shown in Figure 11.10, which illustrates the degree of controllability of a free-free beam with three modelled modes. It can be observed that at the nodal location for each mode, the controllability goes to zero, which is to be expected. The maximum controllability is at the ends of the beam, where all modes have an antinode.

The final problem associated with using controllability to guide actuator placement arises because the controllability grammian is dependent upon the time designated for return of the system to its equilibrium state following perturbation, t_f. To overcome this problem, the steady-state solution to the controllability grammian relationship, where t_f is infinite, will be considered. In this case, for asymptotically stable systems, the controllability grammian satisfies the Lyapunov equation:

$$\boldsymbol{AP} + \boldsymbol{PA}^{\mathrm{T}} + \boldsymbol{BB}^{\mathrm{T}} = 0 \tag{11.7.10}$$

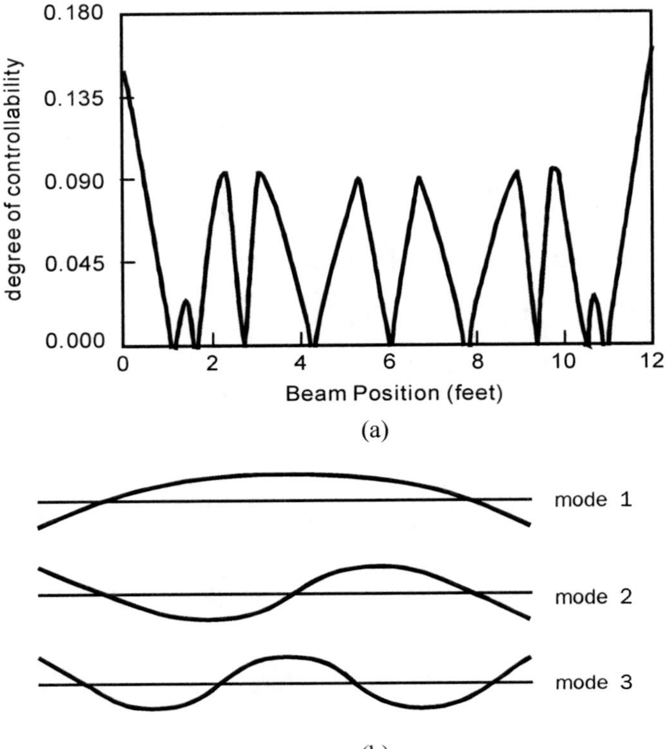

Figure 11.10 (a) Degree of controllability for a free-free beam (after Vander Velde and Carignan, 1984); (b) sketch of first three mode shapes for a free-free beam.

Because of the special form of the state matrix $\boldsymbol{A}$, a closed form solution for the Lyapunov Equation (11.7.10) can be found for system models of any order N (Gawronski and Juang, 1990). Further, if the damping is small, the controllability grammian is essentially block diagonal (Gawronski and Juang, 1990):

$$\boldsymbol{P} \approx \begin{bmatrix} P_1 & & \\ & \ddots & \\ & & P_N \end{bmatrix} \tag{11.7.11}$$

where

$$P_i = \begin{bmatrix} \dfrac{\sum_{l=1}^{N_c} \psi_i^2(\boldsymbol{r}_l)}{4\zeta\omega_i} & 0 \\ 0 & \dfrac{\sum_{l=1}^{N_c} \psi_i^2(\boldsymbol{r}_l)}{4\zeta\omega_i} \end{bmatrix} \tag{11.7.12}$$

11.7.1.2 Persistent Excitation

For persistent excitation, the objective of actuator placement is to arrive at an arrangement that maximises the influence which the control system has upon the steady-state behaviour. This idea can be quantified by the statement that the energy transmitted by the actuators to the structure as a whole, as well as to the individual modes, should be maximised for a given control effort limitation. To quantify this, it is supposed that the actuator signals are mutually uncorrelated white noise processes of unit intensity and have a covariance matrix defined by:

$$\mathrm{E}\left\{\boldsymbol{u}(t)\boldsymbol{u}^{\mathrm{T}}(t)\right\} = \boldsymbol{I}\delta(t-\tau) \tag{11.7.13}$$

Under these conditions, the covariance matrix of the system state:

$$\mathrm{E}\left\{\boldsymbol{x}(t)\boldsymbol{x}^{\mathrm{T}}(t)\right\} = \boldsymbol{X}(t) \tag{11.7.14}$$

is a steady-state solution of the Lyapunov Equation (11.7.10) (Hac and Liu, 1992). Further, the mean value of the total energy of the modelled modes is (Hac and Liu, 1992):

$$\mathrm{E}\{E(t)\} = \sum_{i=1}^{N} \frac{\sum_{l=1}^{N_c} \psi_i^2(\boldsymbol{r}_l)}{4\zeta\omega_i} \tag{11.7.15}$$

where E is the expectation operator and $E(t)$ is the system energy at time t. Therefore, the total energy is the sum of the diagonal elements of the controllability grammian divided by two.

From consideration of the transient and persistent excitation cases, two different performance criteria arise. The transient response case performance criteria are dependent upon the product of the diagonal elements of the controllability grammian, while the persistent excitation case is based upon the sum of the diagonal elements. Therefore, in placing control actuators, it is possible to use these criteria separately, or combine them. One proposed form of combination is (Hac and Liu, 1992):

$$J = \left(\sum_{i=1}^{2N} \lambda_j\right) \times \sqrt[2N]{\prod_{i=1}^{2N} \lambda_i} \tag{11.7.16}$$

where λ_i is the ith eigenvalue of the controllability grammian, which is a diagonal element when the controllability grammian is diagonal.

11.7.2 Sensor Placement

Design of the sensing system can be viewed as the complement of the design of the actuating system. Rather than being interested in maximising the response of the structure as a whole, as well as its individual modes, to a given control input, for the sensing problem the aim is to maximise the (sensor) signal power for a given excitation amplitude of the structure. As before, both transient excitation and persistent excitation conditions will be considered.

11.7.2.1 Transient Excitation

For the transient excitation problem, for simplicity it will be assumed that only displacement sensors are being used in the sensing system (a similar development is possible with velocity sensors). In this case the system output is:

$$\boldsymbol{y}(t) = \boldsymbol{C}\boldsymbol{x}(t) \tag{11.7.17}$$

where the output matrix $\boldsymbol{C}$ is defined by:

$$\boldsymbol{C} = \begin{bmatrix} \psi_1(\boldsymbol{r}_1) & \cdots & \psi_1(\boldsymbol{r}_{N_e}) \\ 0 & \cdots & 0 \\ & \vdots & \\ \psi_N(\boldsymbol{r}_1) & \cdots & \psi_N(\boldsymbol{r}_{N_e}) \\ 0 & \cdots & 0 \end{bmatrix} \tag{11.7.18}$$

where there are N_e sensors in the system. If the system is perturbed at time $t = 0$, starting with state $\boldsymbol{x}(0)$, the output is defined by the relationship:

$$\boldsymbol{y}(\tau) = \boldsymbol{C}\mathrm{e}^{\boldsymbol{A}(\tau)}\boldsymbol{x}(0) \tag{11.7.19}$$

The system output energy, which is to be maximised with the sensor placement, is defined by:

$$J = \int_0^{t_f} \boldsymbol{y}^{\mathrm{T}}(\tau)\boldsymbol{y}(\tau)\, \mathrm{d}\tau \tag{11.7.20}$$

Substituting Equation (11.7.19) into Equation (11.7.20), the system output energy can be written as:

$$J = \boldsymbol{x}^{\mathrm{T}}(0)\boldsymbol{M}(t_f)\boldsymbol{x}(0) \tag{11.7.21}$$

where $\boldsymbol{M}$ is the observability grammian which, from the definition in Chapter 5, Equation (5.6.23), can be written as:

$$\boldsymbol{M}(t_f) = \int_0^{t_f} \mathrm{e}^{\boldsymbol{A}^{\mathrm{T}}(\tau)}\boldsymbol{C}^{\mathrm{T}}\boldsymbol{C}\mathrm{e}^{\boldsymbol{A}(\tau)}\, \mathrm{d}\tau \tag{11.7.22}$$

As an example of how the observability grammian can vary with sensor position, Figure 11.11 illustrates the degree of observability for the same free-free beam used in Figure 11.10. It can be observed that again at the nodal location for each mode, the observability goes to zero. The maximum observability is again at the edge of the beam, where all modes have an anti-node.

As with the controllability grammian in the previous section, to overcome the time dependency in Equation (11.7.21) (as t_f is usually not defined *a priori*), the steady-state case will be used. Here, for asymptotically stable systems, the controllability grammian satisfies the Lyapunov equation:

$$\boldsymbol{A}^{\mathrm{T}}\boldsymbol{M} + \boldsymbol{M}\boldsymbol{A} + \boldsymbol{C}^{\mathrm{T}}\boldsymbol{C} = 0 \tag{11.7.23}$$

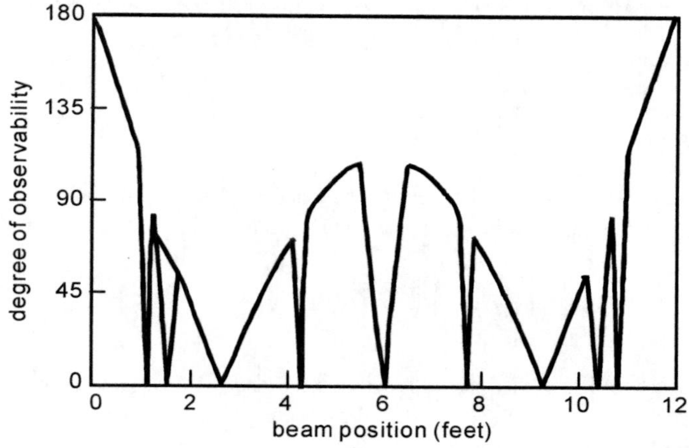

Figure 11.11 Degree of observability for a free-free beam (after Vander Velde and Carignan, 1984).

Because of the special form of the state matrix $\boldsymbol{A}$, a closed form solution for the Lyapunov Equation (11.7.23) can again be found. Further, if the damping is small, the observability grammian is essentially a block diagonal as follows:

$$\boldsymbol{M} \approx \begin{bmatrix} M_1 & & \\ & \ddots & \\ & & M_N \end{bmatrix} \tag{11.7.24}$$

where (Hac and Liu, 1992):

$$M_i = \begin{bmatrix} \dfrac{\sum_{\iota=1}^{N_e} \psi_i^2(\boldsymbol{r}_\iota)}{4\zeta\omega_i^3} & 0 \\ 0 & \dfrac{\sum_{\iota=1}^{N_e} \psi_i^2(\boldsymbol{r}_\iota)}{4\zeta\omega_i^3} \end{bmatrix} \tag{11.7.25}$$

In this case, the system output energy is (Hac and Liu, 1992):

$$\boldsymbol{x}^{\mathrm{T}}(0)\boldsymbol{M}\boldsymbol{x}(0) = \sum_{i=1}^{N} \frac{\sum_{\iota=1}^{N_e} \psi_i^2(\boldsymbol{r}_\iota)}{4\zeta\omega_i^3}[x_{2i-1}^2(0) + x_{2i}^2(0)] \tag{11.7.26}$$

Clearly, to maximise the system output energy, both the individual diagonal terms, as well as their sum, must be maximised. This can be accomplished by maximising the norm of the observability grammian. This result is the complement of the actuator problem, where it was found that the norm of the controllability grammian must be maximised.

11.7.2.2 Persistent Excitation

For the persistent excitation problem, it will be assumed that a white noise process excites all modes with equal strength. In this case, the mean square value of the system output is:

$$J = \mathrm{E}\left\{ \boldsymbol{y}^{\mathrm{T}}(t)\boldsymbol{y}(t) \right\} = tr\left[(\boldsymbol{C}^{\mathrm{T}}\boldsymbol{C})\boldsymbol{X}(t) \right] \tag{11.7.27a,b}$$

where $\boldsymbol{X}(t)$ is the previously defined covariance matrix of the state vector. If the eigenvalues of the state matrix are well spaced, so that the spectrum contains isolated resonances and the damping is small, then the covariance matrix is approximately diagonal. As a result, the mean square value of the system output is (Hac and Liu, 1992):

$$J = \mathrm{E}\left\{ \boldsymbol{y}^{\mathrm{T}}(t)\boldsymbol{y}(t) \right\} = \sum_{i=1}^{N} \frac{\sum_{\iota=1}^{N_e} \psi_i^2(\boldsymbol{r}_\iota)}{4\zeta\omega_i^3} \tag{11.7.28}$$

It can be seen that this result is the complement of the actuator placement result for persistent excitation given in Equation (11.7.15).

From the previous analyses of the transient and persistent excitation cases, it can be seen that the transient response case performance criteria is dependent upon the product of the diagonal elements of the controllability grammian, while the persistent excitation case is based upon the sum of the diagonal elements. Therefore, as in placing control actuators, when placing sensors it is possible to use these criteria separately, or combine them. This combination can be the same as outlined in Equation (11.7.16), but where λ_i is now the ith eigenvalue of the observability grammian. Because the placement criteria are complementary, when velocity sensors and force actuators are used, their optimal placements coincide.

11.7.3 Additional Comments

There are a few additional points that need to be made in respect to the above developments. First, as the sensor and actuator placement exercise is dependent upon both the sum and product of the eigenvalues of the controllability and observability grammians, it is important that the eigenvalues be properly scaled (of similar value) to avoid any bias in the result. By using $[z_i, \omega_i z_i]$ as system states, rather than $[z_i, \dot{z}_i]$, both eigenvalues associated with a given mode are equal. If different states are used, different eigenvalues result.

The development above was only concerned with sensing and actuating modes in the reduced order system model. The development could be expanded to account for the problem of control and observation spillover. To do this, the criterion of Equation (11.7.16) could be formulated for both controlled modes and residual modes. The criterion used in actuator and sensor placement would then be equal to the controlled mode value minus some fraction of the residual mode value, the factor weighting the importance of overcoming spillover problems. Further consideration of actuator placement and spillover can be found in Lindberg and Longman (1982).

Finally, one other consideration in sensor and actuator placement that has been studied is the ability to recognise component failures, and minimise the resulting reduction in controllability and observability which will arise (Vander Velde and Carignan, 1984; Baruh and Choe, 1990a, 1990b; Baruh, 1992). One approach to doing this is to assign a probability density function to the state of failures of components, and use this to derive a measure of the expected value of the degree of controllability and/or observability over a given operating period. A further description of this can be found in Vander Velde and Carignan (1984).

11.7.4 Collocated Sensors and Actuators

Thus far, this chapter has been concerned with control strategies that are essentially suitable for controlling a low number of critical structural modes, where the risk of performance-reducing control spillover and destabilising observation spillover, resulting from the presence of untargeted modes in the structure, exists. There is, however, another control approach where it can be guaranteed that all vibration modes will remain stable when the active control system is in operation, at the expense of control performance. That technique is referred to as low authority control, and is commonly implemented using velocity feedback (Balas, 1979; Aubrun, 1980; Joshi, 1981, 1986, 1989; Schulz and Heimbold, 1983; Martinovic et al., 1990; Creamer and Junkins, 1991; Hanagud et al., 1992; Clark *et al.*, 1998; Preumont, 2002). Low authority control specifically aims to augment the damping of a structure, and by using co-located (usually written 'collocated' in the literature) sensors and actuators, this can be done in a stable fashion. Here, this result will be briefly described; a thorough study of collocated (dissipative) controllers is provided by Joshi (1989).

The following discussion will begin with the open-loop modal equation:

$$\ddot{\boldsymbol{z}}(t) + 2\zeta\boldsymbol{\Lambda}^{1/2}\dot{\boldsymbol{z}}(t) + \boldsymbol{\Lambda}\boldsymbol{z}(t) + \boldsymbol{B}\boldsymbol{f}(t) \tag{11.7.29}$$

where $\boldsymbol{z}$ is an $(n \times 1)$ vector of modal displacement amplitudes, n is the number of modelled modes, $\boldsymbol{\Lambda}$ is a diagonal matrix of open-loop eigenvalues, such that $\boldsymbol{\Lambda}^{1/2}$ is a diagonal matrix of resonance frequencies, $\boldsymbol{f}$ is an $(N_c \times 1)$ vector of control (point) force inputs, and $\boldsymbol{B}$ is an $(n \times N_c)$ matrix of influence functions, where the i,jth term is the value of the ith mode shape function at the location of the jth control force. The output equation associated with this is:

$$\boldsymbol{y}(t) = \boldsymbol{C}\boldsymbol{\Psi}\boldsymbol{z}(t) + \boldsymbol{E}\boldsymbol{\Psi}\dot{\boldsymbol{z}}(t) \tag{11.7.30}$$

where $\boldsymbol{y}$ is an $(N_e \times 1)$ vector of N_e sensor outputs, $\boldsymbol{C}$ and $\boldsymbol{E}$ are $(N_e \times N_e)$ diagonal gain matrices, and $\boldsymbol{\Psi}$ is an $(N_e \times n)$ matrix whose i,jth term is the value of the jth mode shape function at the ith sensor location.

Several assumptions will now be made. First, the number of sensors N_e will be set equal to the number of control sources, N_c. Second, only point velocity sensors will be used in the system. Third, the control sources and error sensors will be collocated. With these assumptions, $\boldsymbol{C} = \boldsymbol{0}$, $\boldsymbol{E} = \boldsymbol{I}$ and $\boldsymbol{\Psi} = \boldsymbol{B}^{\mathrm{T}}$. Therefore, the output Equation (11.7.30) becomes:

$$y(t) = \boldsymbol{B}^{\mathrm{T}}\dot{\boldsymbol{z}}(t) \tag{11.7.31}$$

With direct velocity feedback, the control law has the form:

$$\boldsymbol{f}(t) = -\boldsymbol{K}\boldsymbol{y}(t) \tag{11.7.32}$$

where $\boldsymbol{K}$ is an $(N_e \times N_e)$ symmetric, non-negative gain matrix.

One of the interesting, and attractive, features of the control approach outlined in Equation (11.7.32) is that it is energy dissipative (Balas, 1979), and therefore always stable (similar to the Lyapunov stability criterion discussed in Chapter 5). Further, if there are no zero resonance frequencies in $\boldsymbol{\Lambda}$ (no rigid body modes, so that $\boldsymbol{\Lambda}$ is positive definite), then the closed-loop system is stable for any damping coefficient $\zeta \geq 0$, and asymptotically stable if either $\zeta > 0$ or $\zeta \geq 0$ and $\boldsymbol{BK}^{1/2}$ is non-singular (Balas, 1979). To prove this, the total system energy $E(t)$ is defined as:

$$E(t) = 1/2\left(\dot{\boldsymbol{z}}^{\mathrm{T}}\dot{\boldsymbol{z}} + \boldsymbol{z}^{\mathrm{T}}\boldsymbol{\Lambda}\boldsymbol{z}\right) \tag{11.7.33}$$

Using Equation (11.7.29), the following can be obtained:

$$\dot{E}(t) = \dot{\boldsymbol{z}}^{\mathrm{T}}\ddot{\boldsymbol{z}} + \dot{\boldsymbol{z}}^{\mathrm{T}}\boldsymbol{\Lambda}\boldsymbol{z} = -2\zeta\dot{\boldsymbol{z}}^{\mathrm{T}}\boldsymbol{\Lambda}^{1/2}\dot{\boldsymbol{z}} + \dot{\boldsymbol{z}}^{\mathrm{T}}\boldsymbol{B}\boldsymbol{f} \tag{11.7.34a,b}$$

From Equations (11.7.31) and (11.7.32):

$$\boldsymbol{f} = -\boldsymbol{K}\boldsymbol{B}^{\mathrm{T}}\boldsymbol{z} \tag{11.7.35}$$

Therefore,

$$\dot{E}(t) = -\dot{\boldsymbol{z}}^{\mathrm{T}}\left(2\zeta\boldsymbol{\Lambda}^{1/2} + \boldsymbol{B}\boldsymbol{K}\boldsymbol{B}^{\mathrm{T}}\right)\dot{\boldsymbol{z}} \tag{11.7.36}$$

Note, however, that $\boldsymbol{\Lambda}^{1/2} \geq \boldsymbol{0}$ and $\boldsymbol{BKB}^{\mathrm{T}} \geq \boldsymbol{0}$, and so the derivative of the total system energy, defined in Equation (11.7.36), must be negative. Therefore, the value of the total system energy must be decreasing, so the system is dissipative. If $\boldsymbol{\Lambda} > 0$, then the total system energy is positive while its first derivative is negative for all non-zero states $(z, \dot{z})$, then the total system energy $E(t)$ is a Lyapunov function. Therefore, from Equation (11.7.34) the closed-loop system is stable for any $\zeta \geq 0$, and asymptotically stable if the matrix $\boldsymbol{W}=2\,\zeta\boldsymbol{\Lambda}^{1/2} + \boldsymbol{BKB}^{\mathrm{T}}$ is positive definite. This is true if either $\zeta > 0$ or $\zeta \geq 0$, and $\boldsymbol{BK}^{1/2}$ is non-singular, as stated above (Balas, 1979).

The important significance of this result is that for point-collocated sensors and actuators, neglecting transducer dynamics, system stability is maintained regardless of the number of modes in the model (to be controlled), and regardless of the inaccuracy of the knowledge of parameters. The spillover problem is completely avoided.

If the system response does include contributions from rigid body modes, such that $\boldsymbol{\Lambda}$ contains zero eigenvalues, then a sufficient condition for stability is that the energy in the zero resonance frequency modes remain constant (that is, the control actuators do not excite the rigid body modes). If the zero resonance frequency modes are damped, then the system is asymptotically stable (Balas, 1979).

When velocity feedback is used with point-collocated sensors and actuators, the resulting closed-loop transfer function has alternating poles and zeroes on the imaginary axis (Martin and Bryson, 1980; Lee and Speyer, 1993; Preumont, 2002). With this arrangement, the system can always be stabilised (with lead compensation) (Martin and Bryson, 1980). Unfortunately, sensor and actuator dynamics, if unmodelled, can alter this pole/zero arrangement, and lead to system instability (Goh and Caughey, 1985; Inman, 1990). As discussed in Section 11.2, a small amount of passive damping can add stability robustness, and help to alleviate this problem.

Although they provide excellent performance and stability characteristics when used with velocity feedback control, point-collocated sensor pairs such as accelerometers and shakers can often be relatively heavy and bulky, and thus significant research has been undertaken into using 'matched' (or 'dual') distributed smart material (for example, piezoelectric) transducers. A simple configuration would be to have distributed actuator and sensor elements of identical shape positioned in-line on either side of the structure. However, distributed actuators will yield both in-plane and out-of-plane forces (Gibbs and Fuller, 1992), and thus the collocated distributed transducers described here will experience in-plane coupling, which can cause instability of the velocity feedback control system (Lee, 2011; Lee et al., 2002). Double sensor-actuator pairs (a distributed actuator bonded directly to a sensor of the same shape, attached to one side of the structure, with an identical configuration on the other side of the structure) have been suggested by Yang and Huang (1998) as a means of eliminating in-plane coupling; however, is has been shown experimentally by Lee et al. (2003) that in-plane coupling may not be eliminated if there is direct inter-layer coupling between each sensor and actuator pair. The in-plane coupling problem should be overcome by using distributed sensoriactuators (for which the same piezoelectric element is used for both sensing and actuation) on either side of the structure; however, true sensoriactuators are difficult to design and implement (see Section 15.12.3).

Wave-absorbing controllers (von Flotow and Schafer, 1986), developed for large space structure control for which a description of the system response in terms of modes is undesirable, bear a distinct resemblance to the collocated controllers described here. In fact, the wave-absorbing compensators designed by von Flotow and Schafer (1986) can be viewed as velocity feedback controllers, modified to have a slightly different phase, with the gain increasing with frequency.

11.8 CENTRALISED VERSUS DECENTRALISED AND DISTRIBUTED CONTROL

Traditional multi-channel vibration control systems use a centralised control system, whereby a single controller is used to process all sensor inputs and generate all controller output signals. Whilst this can work well for smaller systems, centralised controllers become computationally inefficient and can exhibit excessive wiring and mass once the number of actuators and sensors becomes large; that is, they suffer from scalability problems. As such, *decentralised* and *distributed* control systems are seen as potential alternatives. Centralised and decentralised controllers (and in some cases distributed controllers) have been compared and contrasted directly for both active vibration control and also control of sound radiation from vibrating structures by several researchers (for example, Elliott, 2005; Engels et al., 2006; Frampton et al., 2010). The ideas here could thus also be applied to sound radiation applications such as those discussed in Chapter 8.

For decentralised control, the inputs and outputs are separated into paired groups, which are incorporated into individual controllers that independently work to achieve the performance objective. These controllers can potentially exhibit limited performance as well as stability issues if there is significant coupling between neighbouring controllers, and thus use of inherently robust control methodologies is desirable to avoid this. Low-authority control using collocated and dual sensors and actuators has been shown to be particularly effective for decentralised control (Frampton et al., 2010).

Consider a decentralised vibration controller consisting of numerous independent collocated force actuator and velocity sensor pairs. It has previously been discussed in Section 11.7.4 how, individually, these controllers are dissipative and exhibit excellent performance and stability characteristics. This concept holds for the case here, and thus each individual controller can be designed and integrated without concern for disrupting global stability. Such a controller does, however, have the problem that if the actuators and sensors are not completely dual and collocated, high-frequency stability problems can result (Schiller et al., 2010). One should also be reminded that this controller, and any other low-authority controller will also intrinsically sacrifice performance for robustness, which may be problematic if high performance is desired.

Each individual collocated actuator/sensor pair must be carefully positioned such that each pair can affect all of the modes that are to be controlled. Global control can be achieved by adjusting individual gains to meet specific performance indices (for example, minimisation of the kinetic energy of the structure); however, unlike fully centralised control, decentralised velocity controllers cannot directly control the amplitudes of each of the modes that one wishes to control.

The term 'distributed controller' refers to configurations that lie in between the centralised and fully decentralised configurations; that is, they consist of a set of localised controllers that can communicate with one another to a certain extent. Examples include hierarchical control and clustering. They are designed to achieve a global control performance comparable to centralised controllers, whilst still retaining the scalability advantages of decentralised control systems.

The relative effectiveness of each of these types of control is strongly dependent upon the number of controllers, the number of modes to be controlled, and coupling between the system states; that is, the relative amplitudes of the centralised gain matrix diagonal terms versus the amplitudes of the off-diagonal terms. If the gain matrix is dominated by the diagonal terms, then cross-coupling between different sensor-actuator pairs is minimal and the optimal centralised controller is practically decentralised. As such, similar performance in these cases can be obtained using either centralised or decentralised methods and a decentralised controller would likely be preferred due to its relative simplicity. If, however, the gain matrix has significant off-diagonal terms (that is, it has 'colouration'), centralised or distributed controllers should out-perform fully decentralised controllers. Frampton et al. (2010) discuss possible scenarios that may lead to colouration of the feedback gain matrix.

11.8.1 Biologically Inspired Control

One means of controlling the vibrational response of a distributed elastic system via numerous distributed control actuators is a method termed 'biologically inspired control' (in reference to biological muscle control), which was introduced by Fuller and Carneal (1993). This approach controls only one actuator (the 'master' actuator) directly from the centralised main controller, whilst the control inputs to the remaining actuators are determined by localised learning rules. Linear quadratic control theory was used to determine the input voltage into the 'master' actuator through minimisation of a vibration energy density cost function which is dependent on the superposition of the system responses due to an input disturbance force and all of the control actuators. Although Fuller and Carneal's control system used a feedforward approach, the ideas could also be applied to a feedback system.

The local learning rule employed by Fuller and Carneal (1993) was to compare the effect on the cost function of applying voltages to the control actuator immediately neighbouring the 'master' actuator with signals dependent on the 'master' actuator voltage: in-phase with the 'master' actuator voltage, out-of-phase with the 'master' actuator voltage, and zero input. The voltage yielding a minimum cost function was selected as the control input to the second actuator. This method was then repeated for all other actuators in order of proximity, resulting in a distributed actuator control system. The method was successfully applied analytically to a simply supported beam with a harmonic point force excitation, and was also experimentally validated for narrow-band excitation in a later publication (Carneal and Fuller, 1995).

REFERENCES

Anderson, B.D.O. and Moore, J.B. (1990). *Optimal Control, Linear Quadratic Methods*. Prentice Hall, Englewood Cliffs, NJ.

Arbel, A. (1981). Controllability measures and actuator placement in oscillatory systems. *International Journal of Control*, **33**, 565–574.

Athans, M. (1970). Toward a practical theory of distributed parameter systems. *IEEE Transactions on Automatic Control*, **AC-15**, 245–247.

Atluri, S.N. and Amos, A.K. (1988). *Large Space Structures: Dynamics and Control*. Springer-Verlag, Berlin.

Aubrun, J.N. (1980). Theory of control of structures by low-authority controllers. *Journal of Guidance and Control*, **3**, 444–451.

Balas, M.J. (1978a). Feedback control of flexible systems. *IEEE Transactions on Automatic Control*, **AC-23**, 673–679.

Balas, M.J. (1978b). Active control of flexible systems. *Journal of Optimization Theory and Applications*, **25**, 415–436.

Balas, M.J. (1979). Direct velocity feedback control of large space structures. *Journal of Guidance and Control*, **2**, 252–253.

Balas, M.J. (1982). Trends in large space structure control theory: fondest hopes, wildest dreams. *IEEE Transactions on Automatic Control*, **AC-27**, 522–535.

Balas, M.J. (1990). Low order control of linear finite-element models of large flexible structures using second-order parallel architectures, in *Mechanics and Control of Large Flexible Structures*, J.L. Junkins, Ed. AIAA Press, Washington, DC.

Barrault, G.F.G. (2006). *Controle ativo de vibrações de baixas e altas freqüências e ruído radiado de estruturas complexas*. PhD thesis, Universidade Federal de Santa Catarina.

Barrault, G., Halim, D., Hansen, C. and Arcanjo, L. (2007). Optimal truncated model for vibration control design within a specified bandwidth. *International Journal of Solids and Structures*, **44**, 4673–4689.

Barrault, G., Halim, D., Hansen, C. and Arcanjo, L. (2008). High frequency spatial vibration control for complex structures. *Applied Acoustics*, **69**, 933–944.

Baruh, H. (1992). Placement of sensors and actuators in structural control. In *Control and Dynamic Systems*, Vol. 52, C.T. Leondes, Ed. Academic Press, San Diego, CA.

Baruh, H. and Choe, K. (1990a). Sensor placement in structural control. *Journal of Guidance, Control, and Dynamics*, **13**, 524–533.

Baruh, H. and Choe, K. (1990b). Reliability issues in structural control. In *Control and Dynamic Systems*, Vol. 32, C.T. Leondes, Ed. Academic Press, San Diego, CA.

Baz, A., Poh, S. and Studer, P. (1989). Modified independent modal space control method for active control of flexible systems. *Proceedings of the Institution of Mechanical Engineers*, **203**,103–112.

Baz, A. and Poh, S. (1990). Experimental implementation of the modified independent modal space control method. *Journal of Sound and Vibration*, **139**, 133–149.

Benhabibm R.J., Iwens, R.P. and Jackson, R.L. (1981). Stability of large space structure control systems using positivity concepts. *Journal of Guidance and Control*, **4**, 487–494.

Bennighof, J.K. and Meirovitch, L. (1988). Active vibration control of a distributed system with moving support. *Journal of Vibration, Acoustics, Stress, and Reliability in Design*, **110**, 246–253.

Bernstein, D.S. and Hyland, D.C. (1990). Optimal projection approach to robust fixed-structure control design. In *Mechanics and Control of Large Flexible Structures*, J.L. Junkins, Ed. AIAA Press; Washington, DC.

Bongiorno, J. and Youla, D. (1968). On observers in multi-variable control systems. *International Journal of Control*, **8**, 221–243.

Carneal, J.P. and Fuller, C.R. (1995). A biologically inspired controller. *Journal of the Acoustical Society of America*, **98**, 386–396.

Chen, C. (1970). *Introduction to Linear System Theory*. Holt, Rinehart and Winston, New York.

Choe, K. and Baruh, H. (1992). Actuator placement in structural control. *Journal of Guidance, Control, and Dynamics*, **15**, 40–48.

Clark, R.L., Saunders, W.R. and Gibbs, G.P. (1998). *Adaptive Structures*. John Wiley & Sons, New York.

Craig, R.R., Jr. (1981). *Structural Dynamics: An Introduction to Computer Methods*. John Wiley & Sons, New York.

Craig, R.R., Jr. (1987). A review of time-domain and frequency-domain component-mode synthesis techniques. *International Journal of Analytical and Experimental Modal Analysis*, **2**, 59–72.

Creamer, N.G. and Junkins, J.L. (1991). Low-authority eigenvalue placement for second-order structural systems. *Journal of Guidance, Control, and Dynamics*, **14**, 698–701.

Elliott, S.J. (2005). Distributed control of sound and vibration. *Noise Control Engineering Journal*, **53**, 165–180.

Engels, W.P., Baumann, O.N. and Elliott, S.J. (2006). Centralized and decentralized control of structural vibration and sound radiation. *Journal of the Acoustical Society of America*, **119**, 1487–1495.

Frampton, K.D., Baumann, O.N. and Gardonio, P. (2010). A comparison of decentralized, distributed and centralized vibro-acoustic control. *Journal of the Acoustical Society of America*, **128**, 2798–2806.

Fuller, C.R. and Carneal, J.P. (1993). A biologically inspired control approach for distributed elastic systems. *Journal of the Acoustical Society of America*, **93**, 3511–3513.

Fuller, C.R., Elliott, S.J. and Nelson, P.A. (1993). *Active Control of Vibration*. Academic Press Limited, London.

Gawronski, W. and Juang, J. (1990). Model reduction for flexible structures. In *Control and Dynamics*, Vol. 36, C.T. Leondes, Ed. Academic Press, San Diego, CA, 143–222.

Gibbs, G.P. and Fuller, C.R. (1992). Excitation of thin beams using asymmetric piezoelectric actuators. *Journal of the Acoustical Society of America*, **92**, 3221–3227.

Goh, C.J. and Caughey, T.K. (1985). On the problem caused by finite actuator dynamics in the collocated control of large space structures. *International Journal of Control*, **41**, 787–802.

Grandhi, R.V. (1990). Optimum design of space structures with active and passive damping. *Engineering With Computers*, **6**, 117–183.

Gueler, R., von Flotow, A.H. and Vos, D.W. (1993). Passive damping for robust feedback control of flexible structures. *Journal of Guidance, Control and Dynamics*, **16**, 662–667.

Hac, A. and Liu, L. (1992). Sensor and actuator location in motion control of flexible structures. In *Proceedings of First International Conference on Motion and Vibration Control (MOVIC)*, Yokohama, 86–91.

Hallauer, W.L., Jr. (1990). Recent literature on experimental structural dynamics and control research. In *Mechanics and Control of Large Flexible Structures*, J.L. Junkins, Ed. AIAA Press, Washington, DC.

Hallauer, W.L., Jr., Skidmore, G.R. and Mesquita, L.C. (1982). Experimental-theoretical study of active vibration control. In *Proceedings of the International Modal Analysis Conference*, 39–45.

Hanagud, S., Obal, M.W. and Calise, A.J. (1992). Optimal vibration control by the use of piezoceramic sensors and actuators. *Journal of Guidance, Control, and Dynamics*, **15**, 1199–1206.

Haverkamp, B. (2001). *State space identification: theory and practice*. PhD thesis, System and Control Engineering Group, Delft University of Technology.

Hughes, P.C. and Skelton, R.E. (1980a). Controllability and observability of linear matrix-second-order systems. *Journal of Applied Mechanics*, **47**, 415–420.

Hughes, P.C. and Abdel–Rahman, T.M. (1979). Stability of proportional-plus-derivative-plus-integral control of flexible spacecraft. *Journal of Guidance and Control*, **2**, 499–503.

Hughes, P.C. and Skelton, R.E. (1980b). Controllability and observability for flexible spacecraft. *Journal of Guidance and Control*, **3**, 452–459.

Inman, D.J. (1990). Control/structure interaction: effects of actuator dynamics. In *Mechanics and Control of Large Flexible Structures*, J.L. Junkins, Ed. *AIAA Progress in Astronautics and Aeronautics*, 507–533.

Joshi, S.M. (1981). *A Controller Design Approach for Large Flexible Space Structures*. NASA Contractor Report, CR-165717.

Joshi, S.M. (1986). Robustness properties of collocated controllers for flexible spacecraft. *Journal of Guidance, Control, and Dynamics*, **9**, 85–91.

Joshi, S.M. (1989). *Control of Large Flexible Space Structures*. Springer-Verlag, Berlin.

Junkins, J.L. Ed. (1990). *Mechanics and Control of Large Flexible Structures*. AIAA Press, Washington, DC.

Kwak, M.K. and Meirovitch, L. (1993). An algorithm for the computation of optimal control gains for second order matrix equations. *Journal of Sound and Vibration*, **166**, 45–54.

Lee, C.-K. and Moon, F.C. (1990). Modal sensors/actuators. *Journal of Applied Mechanics*, **57**, 434–441.

Lee, Y.J. and Speyer, J.L. (1993). Zero locus of a beam with varying sensor and actuator locations. *Journal of Guidance, Control, and Dynamics*, **16**, 21–25.

Lee, Y.-S., (2011). Comparison of collocation strategies of sensor and actuator for vibration control. *Journal of Mechanical Science and Technology*, **25**, 61–68.

Lee, Y.-S., Elliott, S.J. and Gardonio, P. (2003). Matched piezoelectric double sensor/actuator pairs for beam motion control. *Smart Materials and Structures*, **12**, 541–548.

Lee, Y.-S., Gardonio, P. and Elliott, S.J. (2002). Coupling analysis of a matched piezoelectric sensor and actuator pair for vibration control of a smart beam. *Journal of the Acoustical Society of America*, **111**, 2715–2726.

Leondes, C.T. (1990). *Control and Dynamic Systems*. Academic Press, San Diego, CA.

Lewis, F.L., Xie, L. and Popa, D. (2007). *Optimal and Robust Estimation: With an Introduction to Stochastic Control Theory*. CRC Press, Boca Raton, FL.

Lindberg, R.E. and Longman, R.W. (1982). *Optimization of Actuator Placement via Degree of Controllability Criteria Including Spillover Considerations*. AIAA Paper, 82-1435.

Liu, Y. and Anderson, B.D.O. (1986). Controller reduction via stable factorization and balancing. *International Journal of Control*, **44**, 507–531.

MacMartin, D.G. and Hall, S.R. (1991). Structural control experiments using an H_∞ power flow approach. *Journal of Sound and Vibration*, **148**, 223–241.

Maghami, P.G. and Joshi, S.M. (1993). Sensor-actuator placement for flexible structures with actuator dynamics. *Journal of Guidance, Control, and Dynamics*, **16**, 301–307.

Martin, G.D. and Bryson, A.E., Jr. (1980). Attitude control of a flexible spacecraft. *Journal of Guidance and Control*, **3**, 37–41.

Martinovic, Z.N., Schamel, G.C., Hafta, R.T. and Hallauer, W.L. (1990). Analytical and experimental investigation of output feedback vs linear quadratic regulator. *Journal of Guidance, Control, and Dynamics*, **13**, 160–167.

Meirovitch, L. and Baruh, H. (1982). Control of self-adjoint distributed parameter systems. *Journal of Guidance, Control, and Dynamics*, **5**, 60–66.

Meirovitch, L. and Baruh, H. (1985). The implementation of modal filters for control of structures. *Journal of Guidance, Control, and Dynamics*, **8**, 707–716.
Meirovitch, L., Baruh, H. and Oz, H. (1983). A comparison of control techniques for large flexible systems. *Journal of Guidance, Control, and Dynamics*, **6**, 302–310.
Meirovitch, L. and Bennighof, J.K. (1986). Modal control of travelling waves in flexible structures. *Journal of Sound and Vibration*, **111**, 131–144.
Meirovitch, L. and Oz, H. (1980a). Modal-space control of large flexible spacecraft possessing ignorable coordinates. *Journal of Guidance and Control*, **3**, 569–577.
Meirovitch, L. and Oz, H. (1980b). Modal-space control of distributed gyroscopic systems. *Journal of Guidance and Control*, **3**, 140–150.
Meirovitch, L. and Oz, H. (1980c). Optimal modal-space control of flexible gyroscopic systems. *Journal of Guidance and Control*, **3**, 218–226.
Meirovitch, L., van Landingham, H.F. and Oz, H. (1977). Control of spinning flexible spacecraft by modal synthesis. *Acta Astronautica*, **4**, 985–1010.
Meirovitch, L., van Landingham, H.F. and Oz, H. (1979). Distributed control of spinning flexible spacecraft. *Journal of Guidance and Control*, **2**, 407–415.
Miller, D.W. and von Flotow, A.H. (1989). Power flow in structural networks. *Journal of Sound and Vibration*, **128**, 145–162.
Miller, D.W., Hall, S.R. and von Flotow, A.H. (1990). Optimal control of power flow at structural junctions. *Journal of Sound and Vibration*, **140**, 475–497.
Moheimani, S.O.R. and Clark, R.L. (2000). Minimizing the truncation error in assumed modes models of structures. *ASME Journal of Vibration and Acoustics*, **122**, 331–335.
Moheimani, S.O.R., Halim, D. and Fleming, A.J. (2003). Spatial control of vibration: theory and experiments. World Scientific, Singapore.
Morgan, D.R. (1991). An adaptive modal-based active control system. *Journal of the Acoustical Society of America*, **89**, 248–256.
Nurre, G.S., Ryan, R.S., Scofield, H.N. and Sims, J.L. (1984). Dynamics and control of large space structures. *Journal of Guidance, Control, and Dynamics*, **7**, 514–526.
Noble, B. and Daniel, J.W. (1977). *Applied Linear Algebra*, 2nd ed. Prentice Hall, Englewood Cliffs, NJ.
Overschee, P.V. and Moor, B.D. (1996). *Subspace Identification for Linear Systems*. Kluwer Academic Publishers, Massachusetts.
Porter, B. and Crossley, T.R. (1972). *Modal Control: Theory and Applications*. Taylor and Francis, London.
Preumont, A. (2002). *Vibration Control of Active Structures: An Introduction*. Kluwer Academic Publisher, Dordrecht, London.
Rao, S.S., Pan, T. and Venkayya, V.B. (1990). Robustness improvements of actively controlled structures through structural modification. *AIAA Journal*, **28**, 353–361.
Rollins, L. (1999). *Robust Control Theory*. Carnegie Mellon University, 18-849b Dependable Embedded Systems, Spring 1999.
Schiller, N.H., Cabell, R.H. and Fuller, C.R. (2010). Decentralized control of sound radiation using iterative loop recovery. *Journal of the Acoustical Society of America*, **128**, 1729–1737.
Schulz, G. and Heimbold, G. (1983). Dislocated actuator/sensor positioning and feedback design for flexible structures. *Journal of Guidance, Control, and Dynamics*, **6**, 361–367.
Simon, J.D. and Mitter, K. (1968). A theory of modal control. *Information and Control*, **13**, 316–353.
Skelton, R.E. and Hughes, P.C. (1980). Modal cost analysis of linear matrix second order systems. *Journal of Dynamic Systems, Measurement, and Control*, **102**, 151–158.
Skelton, R.E. and Yousuff, A. (1982). Component cost analysis of large-scale systems. In *Control and Dynamic Systems*, Vol. 18, C.T. Leondes, Ed. Academic Press, San Diego, CA.
Skelton, R.E. and Yousuff, A. (1983). Component cost analysis of large-scale systems. *International Journal of Control*, **37**, 285–304.
Skelton, R.E., Singh, R. and Ramakrishnan, J. (1990). Component model reduction by component cost analysis, in *Mechanics and Control of Large Flexible Structures*, J.L. Junkins, Ed. AIAA Press, Washington, DC.

Spanos, J.T. (1989). Control-structure interaction in precision pointing servo loops. *Journal of Guidance, Control, and Dynamics*, **12**, 256–263.

Vander Velde, W.E. and Carignan, C.R. (1984). Number and placement of control system components considering possible failures. *Journal of Guidance, Control, and Dynamics*, **6**, 703–709.

Verhaegen, M. and Verdult, V. (2007). Filtering and system identification: A least squares approach. *Cambridge University Press*, New York.

von Flotow, A.H. (1986). Disturbance propagation in structural networks. *Journal of Sound and Vibration*, **106**, 433–450.

von Flotow, A.H. (1988). The acoustic limit of control of structural dynamics, in *Large Space Structures: Dynamics and Control*, S.N. Atluri and A.K. Amos, Eds. Springer-Verlag, Berlin.

von Flotow, A.H. and Schafer, B. (1986). Wave-absorbing controllers for a flexible beam. *Journal of Guidance, Control, and Dynamics*, **9**, 673–680.

Wilson, D.A. (1974). Model reduction for multivariable systems. *International Journal of Control*, **20**, 57–64.

Yousuff, A. and Breida, M. (1992). Model reduction of mechanical systems. *Journal of Guidance, Control, and Dynamics*, **16**, 408–410.

Yousuff, A., Wagie, D.A. and Skelton, R.E. (1985). Linear systems approximations via covariance equivalent realizations. *Journal of Mathematical Analysis and Applications*, **106**, 91–114.

Yang, S.Y. and Huang, W.H. (1998). Is a collocated piezoelectric sensor/actuator pair feasible for an intelligent beam. *Journal of Sound and Vibration*, **216**, 529–538.

Zhou, K., Doyle, J.C. and Glover, K. (1996). *Robust and Optimal Control*. Prentice Hall, Englewood Cliffs, NJ.

CHAPTER 12

Vibration Isolation

12.1 INTRODUCTION

Active vibration isolation involves the use of an active system to reduce the transmission of vibration from one body or structure to another. A broader definition would also include the reduction of vibration of a machine or structure by an active vibration absorber. Passive vibration isolation is covered adequately in many textbooks (see, for example, Bies and Hansen, 2009) and will not be discussed here. In the analyses discussed in this chapter, a constant force (or infinite impedance) source is assumed. That is, it is assumed that the driving force is independent of the structure and does not change significantly if the dynamics of the structure change. Although this idealised case is not often found in practice, the constant force assumption simplifies complex analyses, and the results obtained are indicative of what can be achieved in many practical cases.

Active vibration isolation systems are usually significantly more complex and expensive than their passive counterparts, which consist of steel or rubber springs and dashpots (illustrated diagrammatically in Figure 12.1(a)) and which have been in use for many years. One may well ask what advantages are offered by active systems which justify their increased cost and complexity. Of course, the main advantages are better static stability of the supported equipment and better performance, especially at low frequencies, which in many cases makes an active system the only feasible choice. Active systems can also be used to minimise vibration at critical locations on a flexible support structure, at some distance from the isolator attachment point and for some applications this is a distinct advantage. Active systems also have the capability of adjusting to changes in machine operating conditions (and thus vibration excitation frequencies) without any outside intervention. Another important advantage of active systems is that they can dissipate energy as well as supply it, although a major disadvantage in addition to cost and complexity is the requirement of an external power source and in many cases the need for numerous sensors as well as actuators, which could have durability problems.

Active systems have been used in the past to isolate optical systems from support structure vibrations, vehicle cabins from tyre vibrations generated by an uneven road surface, space telescopes from vibrations generated by driving equipment, vehicles from engine induced vibrations, helicopter cabins from rotor gearbox vibration and the ground from vibrations generated by heavy machinery. Thus, in some cases it is desirable to isolate vibrations of an item of equipment from a support structure and in other cases it is desirable to isolate equipment from a vibrating support, the latter often being referred to as base excitation.

Sometimes, active systems are used either in parallel with or in series with passive isolators. Such systems are often referred to as semi-active and consist of two main types. The first type, often used for suspension systems of luxury vehicles, involves control of the system damping, generally by varying the orifice size in a hydraulic damper. It is shown diagrammatically in Figure 12.1(b) for a single-degree-of-freedom system. The second type of semi-active system involves the use of a force actuator driven by a control system. There are four ways of implementing this type of control, as illustrated in Figure 12.1. As shown in the figure, the control force may be either in series with or in parallel with the passive elements. Alternatively, the control force may act only on the vibrating body or only on the support structure. Each of the semi-active systems shown in Figure 12.1 requires a control system to drive it. However, this has been omitted from the figures for clarity.

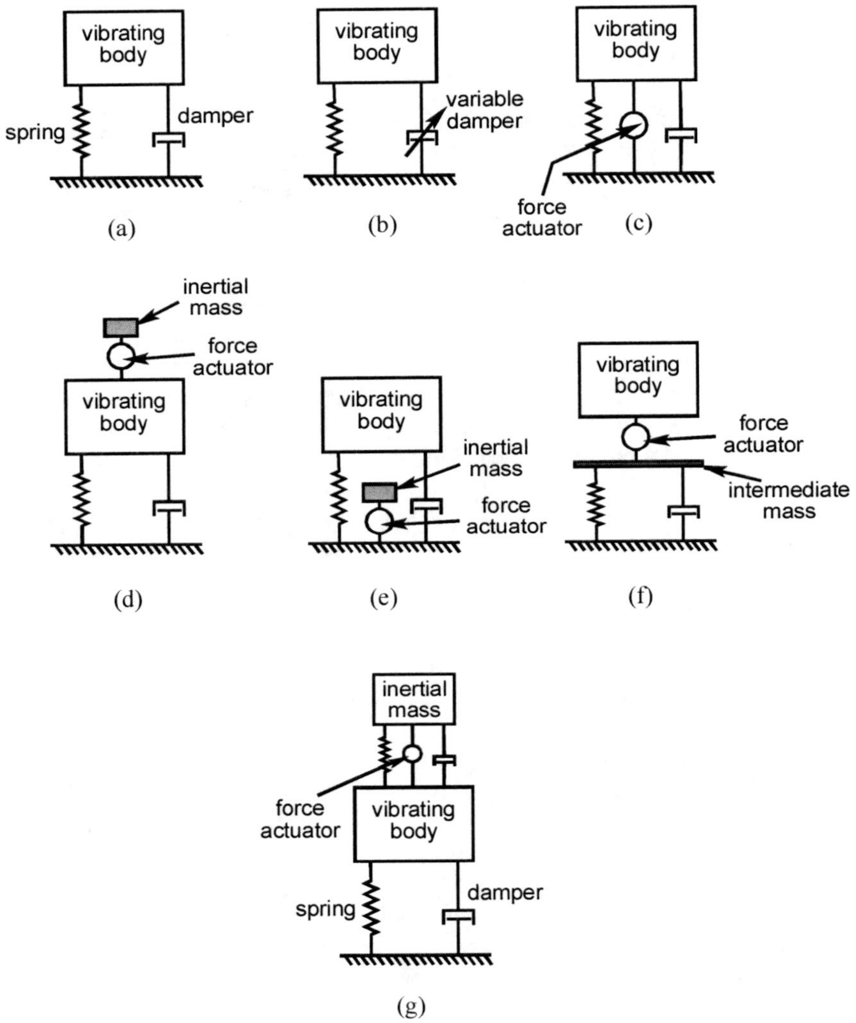

Figure 12.1 Passive and semi-active vibration isolation: (a) passive system showing a conventional spring and damper; (b) semi-active system with a variable damper; (c) semi-active system with a force on both the vibrating body and the support structure; (d) semi-active system with a control force applied only to the vibrating body; (e) semi-active system with a control force applied only to the structure to be isolated from the vibration source; (f) semi-active system with the force actuator in series with the passive elements; (g) semi-active vibration absorber.

Each of the systems illustrated in Figure 12.1 is characterised by advantages and disadvantages. The advantage of the variable damper shown in (b) is its relative simplicity and low cost. In some cases its performance is comparable to systems containing a force actuator, but in many other cases the performance of a system such as this is not as good. The improved performance of the system shown in (c) at low frequencies is sometimes offset by the reduced performance at high frequencies, due to high-frequency force transmission through the actuator as a result of the limited actuator bandwidth, especially if hydraulic actuators are used. In practice, low-frequency performance is also often limited by the high displacement requirement of the actuators, which often precludes the use of magnetostrictive or piezoceramic materials. Instead, hydraulic or electromagnetic actuators with their associated weight problems and in the case of hydraulic actuators, their associated hydraulic supply inconvenience, must be used.

With the force actuator acting either on the vibrating body or support structure, the loss in high-frequency performance is no longer a problem, as the system will be equivalent to a passive system at worst. However, for large machines, a large actuator with a large inertial mass will be needed to provide the required control force and in some cases this may be impractical. However, if the frequency range over which control is desired is limited to very low frequencies or to a very narrow band of frequencies, the actuator force requirements can be significantly reduced (Tanaka and Kikushima, 1988); and as many practical actuators are characterised by a relatively high force capability and low displacement capability, this may be the preferred configuration in many cases.

The configuration shown in Figure 12.1(f) also does not suffer from the loss in high-frequency performance as a result of limited actuator bandwidth. However, the actuator must be large enough to support the weight of the vibrating machine and passive suspension system. The main advantage of this configuration is that the active system is isolated from the dynamics of the supporting structure, which is an important advantage in terms of control system stability if the supporting structure is non-rigid.

In cases where it is desired to limit the vibration of a specific machine or structure at a single frequency, an active vibration absorber can be attached to it as shown in Figure 12.1(g). The advantage of the active vibration absorber over the traditional passive one is that it can be adjusted to track variations in the exciting frequency resulting from speed variations of the machine causing the vibration problem.

Fully active systems contain no passive elements and are usually implemented in one of two ways as shown in Figure 12.2. Again, the control system used to drive the actuator and the required vibration sensors have been omitted for clarity. These will be discussed later in this chapter.

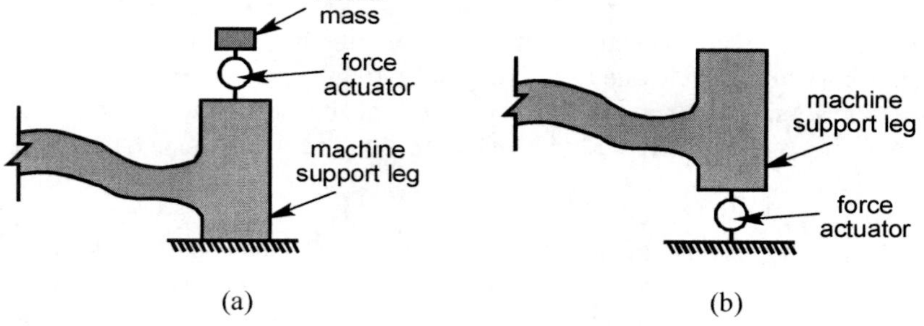

Figure 12.2 Active vibration isolation: (a) force actuator acting only on the vibrating machine support leg; (b) actuator between the vibrating machine and the support structure and supporting the full weight of the machine.

The two configurations shown in Figure 12.2 may be used to minimise vibration transmission from the machine to its supporting structure or to minimise vibration transmission from the structure to the equipment mounted on it, which may be an optical device, such as a telescope or microscope.

The advantage of the system shown in part (a) of the figure is that the machine can be supported very rigidly so that there is no risk of machine sway. However, for large machines, a large actuator and seismic mass will be needed, as the force actuator must exert sufficiently large dynamic forces on the support point to counteract the forces produced by the machine. For the configuration shown in part (b) of the figure, the force actuator must be capable of supporting the weight of the machine in a stable manner. A further disadvantage of this arrangement is that sometimes the force actuator will cause excessive vibrations in the machine as it attempts to minimise the vibration in the supporting structure.

In implementing active vibration isolation systems, great care must be taken to ensure that all vibration transmission paths are accounted for, because in some cases, control of one transmission path and not others could result in an increase in vibration level at locations where it should be reduced (Ross and Yorke, 1987). This is because in some instances, vibration energy arriving at a location by way of more than one transmission path can destructively interfere; removal of one of the transmission paths can reduce the destructive interference and result in an increase in vibration level. This is especially true for periodic vibration signals such as those generated by rotating or reciprocating machinery. Other likely transmission paths (or flanking paths) that may also need to be controlled are horizontal and rotational vibration transmission through the mounts, airborne acoustic waves exciting the support structure, and vibration transmission through pipes and other fittings directly connecting the engine to the support structure. This is a very good reason for including all vibration types (vertical, horizontal and rotational) in the analyses of Sections 12.4 to 12.7.

Identification of all flanking paths is especially important when vibration isolation or structural vibration control is used to reduce structurally radiated noise into free space or into an enclosure, such as a car interior, for example. For this example, the main vibration transmission path would be through the engine and gearbox isolation mounts, but flanking paths also exist via other mechanical attachments to the engine and via acoustic paths from the air intake and exhaust. An example of interior noise control in a car was presented by Quinn (1992) who demonstrated the effectiveness of controlling the main vibration transmission path with an active engine isolator, and using two interior loudspeakers to further reduce the interior noise. His system also used an accelerometer error sensor, four interior mounted error microphones, and a feedforward multi-channel control system, and was directed mainly at controlling periodic noise at the engine firing frequency.

Although Quinlan (1992) showed that it was feasible to substantially isolate low-frequency periodic engine vibration from the vehicle body using a single isolator with an active element acting in only one direction, generally three translational and three rotational vibration transmission components must be included. This is especially true for isolation of higher frequency noise which is a problem when isolating submarine equipment platforms from the hull.

The design of an isolation mount is often a compromise between good vibration isolation and acceptable static rigidity. Use of an active element in the system helps to overcome this trade-off. To reduce the complexity of the mount, passive elements can be designed to provide good isolation in all directions but the one along which the transmission of vibration is the greatest, and this is the direction addressed by the active element. Jenkins et al. (1990, 1993) demonstrated such a mount which involved the use of an intermediate air bag to remove the shear and rotational components of the vibration. This is discussed in more detail in Section 12.4.

12.1.1 Feedforward versus Feedback Control

As discussed earlier in this book, there are two fundamentally different approaches that have been used in the past for implementing active noise and vibration control systems: feedforward and feedback control. Feedforward control involves feeding a signal related to the disturbance input into the controller, which then generates a signal to drive a control actuator in such a way as to cancel the disturbance. On the other hand, feedback control uses a signal derived from the system response to a disturbance, which is amplified, passed through a compensator circuit and used to drive the control actuator to cancel the residual effects occurring after the initial disturbance has passed.

Feedforward control has been shown to provide better results than feedback control, provided that a reference signal well correlated with the disturbance input is available to the controller. Obtaining such a signal is especially simple for periodic excitation such as that generated by a rotating machine, or in cases where the active isolator is sufficiently far from the vibration source that the control system has time to respond to a measure of the incoming disturbance before it arrives at the actuator. For this latter case, it is possible to isolate random as well as periodic noise. However, in many cases for which active vibration isolation is applicable, it is not possible to obtain the required measure of the disturbance input and a feedback control system must be used, such as for active vehicle suspensions and for active isolation of sensitive equipment from a vibrating support structure.

As described in Chapter 6, practical implementation of a feedforward system for noise or vibration control usually requires the system to be adaptive so that it can adapt to changes in system parameters such as speed of sound in the structure, which changes with temperature. An adaptive feedforward isolation system is best implemented using a digital filter, the weights of which are updated by an algorithm that uses as its inputs, the primary disturbance signal and the signal from one or more error sensors that measure residual vibratory power transmission into or vibration on the structure to be controlled. The measured disturbance signal is passed through the filter into the control actuator to produce the force required to minimise the residual vibration in the structure.

An inherent disadvantage of feedback control systems is their tendency to go unstable if the feedback gain is set too high, but a high feedback gain is needed for good performance of active noise and vibration control systems. Another problem is that as the controller takes effect, the feedback signal is reduced in magnitude until it is no longer capable of providing an adequate signal, thus limiting the potential performance of a feedback system. Thus, feedforward systems are preferred whenever it is possible to obtain a signal related to the disturbance input as, unlike feedback systems, they do not modify the dynamic response of the system being controlled, and are inherently much more stable. However, in many cases, a feedback system is the only feasible type, and care must be taken to limit the feedback gain so the system remains stable over the whole range of possible system inputs and variations in the dynamics of the system being controlled (due to ambient temperature changes, for example).

12.1.2 Flexible versus Stiff Support Structures

When the support structure is flexible, it is often characterised by vibration modes in the frequency range for which vibration isolation is required. In this case, a passive isolation system can be quite ineffective, especially at frequencies corresponding to resonance frequencies of the supporting structure. Even with an active isolation system (which will typically involve more than one force actuator), it may not be possible to drive the structural response to zero at each mount point. Power can be transmitted to the structure by moments as well as by forces along axes at 90° to the isolator axis. Thus, for a machine supported on a number of isolators, the control forces used to drive the actuators are not independent and must be derived using a multi-channel control system. Horizontal control forces may also sometimes be required. Thus, in many cases, the control system to drive each active isolator cannot be designed independently; usually, a multi-channel controller must be designed. Of course, this is not necessarily so for the case of a vehicular active suspension system where it is quite possible for each of the four wheel suspensions to act independently.

In the remainder of this chapter, simple feedback active isolators are discussed first of all, beginning with a single-degree-of-freedom system and covering base excited systems as well as vibration absorbers. Isolation of a machine from a flexible sub-structure is also discussed. Following a discussion of feedback control, feedforward control of single-degree-of-freedom systems and multi-degree-of-freedom systems involving flexible subsystems is considered.

It should be noted that there have been significant advances in controller algorithm development over the years, with a plethora of both simple and complex techniques published in the literature. Discussion and comparison of these techniques is beyond the scope of this book, although some of the techniques are very briefly mentioned by name when discussing vibration isolation applications.

12.2 FEEDBACK CONTROL

As discussed in Section 12.1, feedback control is the best control approach in situations where it is not possible to sample the incoming disturbance soon enough for a feedforward control system to be effective. This section will begin with a discussion of the vibration response of a single-degree-of-freedom system consisting of a mass supported on a spring and damper connected to a rigid foundation. The system will be excited by a single force acting on the mass and may be modelled as a second-order system (by using a second-order differential equation). The effect on the response of the mass, of applying various types of feedback to drive a control force, will then be examined. Next, the case where the foundation is free to move will be discussed and the transmission of vibration from the foundation to the mass will be examined. This will be followed by an examination of the control of vibration of the mass by using a vibration absorber tuned by feedback control. Finally, a special case involving the active isolation of vibration from both a rigid and flexible substructure through a single mount will be discussed.

Cases involving active vibration isolation of rigid bodies from flexible substructures using feedforward control will be discussed in Section 12.4.

12.2.1 Single-Degree-of-Freedom Passive System

A single-degree-of-freedom passive isolation system is shown in Figure 12.3. For now, attention will be restricted to the motion of the mass. The transmission of force into the foundation will be considered in Section 12.2.5.

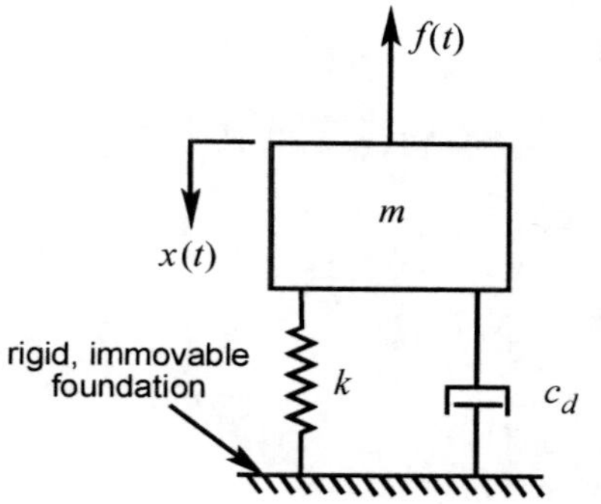

Figure 12.3 Single-degree-of-freedom passive isolation system.

The second-order differential equation that describes the motion of mass m is:

$$m\ddot{x}(t) + c_d\dot{x}(t) + kx(t) = f(t) \tag{12.2.1}$$

where the $\cdot$ denotes derivative with respect to time and $\cdot\cdot$ denotes the double derivative with respect to time.

Taking the Laplace transform ($s = j\omega$) gives:

$$m\left[s^2X(s) - sx(0) - \dot{x}(0)\right] + c_d\left[sX(s) - x(0)\right] + kX(s) = F(s) \tag{12.2.2}$$

Assuming zero initial conditions ($x(0) = \dot{x}(0) = 0$), Equation (12.2.2) becomes:

$$\left[ms^2 + c_d s + k\right]X(s) = F(s) \tag{12.2.3}$$

and the transfer function relating displacement to force is:

$$H(s) = \frac{X(s)}{F(s)} = \frac{1}{ms^2 + c_d s + k} \tag{12.2.4a,b}$$

The system natural frequency may be defined as:

$$\omega_0 = \sqrt{k/m} \tag{12.2.5}$$

and the viscous damping ratio as:

$$\zeta = c_d/2m\omega_0 \tag{12.2.6}$$

Using these definitions, Equation (12.2.4) can be written as:

$$H(s) = \frac{1}{m}\frac{1}{s^2 + 2\zeta\omega_0 s + \omega_0^2} \tag{12.2.7}$$

To find the response of the system to a unit impulse, the transfer function is expanded into partial fractions as follows:

$$H(s) = \frac{A}{(s+p_1)} - \frac{A}{(s+p_2)} \tag{12.2.8}$$

where

$$A = \frac{1}{2m\omega_0\sqrt{\zeta^2 - 1}} \tag{12.2.9}$$

$$p_1 = \omega_0\left(\zeta - \sqrt{\zeta^2 - 1}\right) \tag{12.2.10}$$

$$p_2 = \omega_0\left(\zeta + \sqrt{\zeta^2 - 1}\right) \tag{12.2.11}$$

Transforming back into the time domain, the following is obtained:

$$h(t) = A\mathrm{e}^{-\left(\zeta - \sqrt{\zeta^2 - 1}\right)\omega_0 t} - A\mathrm{e}^{-\left(\zeta + \sqrt{\zeta^2 - 1}\right)\omega_0 t} \tag{12.2.12}$$

The magnitude of the quantity $h(t)/A$ is shown in Figure 12.4 as a function of time for various values of the viscous damping ratio ζ.

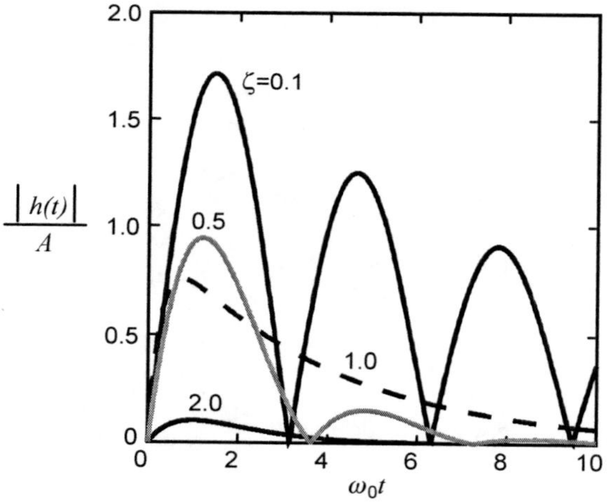

Figure 12.4 Single-degree-of-freedom system impulse response for various values of damping ratio.

Inspection of Equation (12.2.8) shows that if the poles, p_1 and p_2, have positive real parts, they will lie in the left half of the *s*-plane plot (where Re$\{s\}$ is the horizontal axis and Im$\{s\}$ is the vertical axis) and thus the system will be stable. This will be true provided that the damping ratio $\zeta > 0$, which is a property of all actual passive isolation systems.

If the substitution $\mathrm{j}\omega = s$ is made in Equation (12.2.7), the frequency-response function $H(\mathrm{j}\omega)$ is obtained:

$$H(\mathrm{j}\omega) = \frac{1}{m}\frac{1}{\omega_0^2 - \omega^2 + 2\mathrm{j}\omega_0\omega\zeta} \tag{12.2.13}$$

or in dimensionless form:

$$m\omega_0^2 H(\mathrm{j}\omega) = kH(\mathrm{j}\omega) = \frac{1}{1 - (\omega/\omega_0)^2 + 2\mathrm{j}\zeta\omega/\omega_0} \tag{12.2.14a,b}$$

The transfer function can also be written as:

$$H(\mathrm{j}\omega) = |H(\mathrm{j}\omega)|\,\mathrm{e}^{\mathrm{j}\theta} \tag{12.2.15}$$

where

$$|H(\mathrm{j}\omega)| = \frac{1/k}{\left[\left[1 - (\omega/\omega_0)^2\right]^2 + \left[(2\zeta\omega/\omega_0)^2\right]\right]^{½}} \tag{12.2.16}$$

and the relative phase angle between the force and displacement is given by:

$$\theta = \tan^{-1}\left[\frac{-2\zeta\omega/\omega_0}{1 - (\omega/\omega_0)^2}\right] \tag{12.2.17}$$

The modulus and phase angle of $H(\mathrm{j}\omega)$ for various values of the critical damping ratio are plotted in Figure 12.5.

12.2.2 Feedback Control of Single-Degree-of-Freedom System

The dynamics of the system discussed in the previous section can be modified by adding a control force proportional to the displacement, velocity or acceleration (or a combination of these) to the vibrating mass. This is called feedback control.

With a control force $f_c(t)$ acting on the system, the equation of motion can be written as

$$m\ddot{x}(t) + c_d\dot{x}(t) + kx(t) = f(t) + f_c(t) \tag{12.2.18}$$

If the acceleration, velocity and displacement of the mass are detected and fed back through gains, K_a, K_v and K_d to obtain $f_c(t)$, then,

$$f_c(t) = -\left[K_a\ddot{x}(t) + K_v\dot{x}(t) + K_d x(t)\right] \tag{12.2.19}$$

A block diagram illustrating this feedback control arrangement is shown in Figure 12.6(b) and the physical system is shown in Figure 12.6(a).

In practice, typical feedback control systems use some combination of acceleration, velocity, displacement or force feedback. In Figure 12.6, the dynamics of the suspended mass on its own is represented by:

$$G(s) = \frac{1}{ms^2 + c_d s + k} \tag{12.2.20}$$

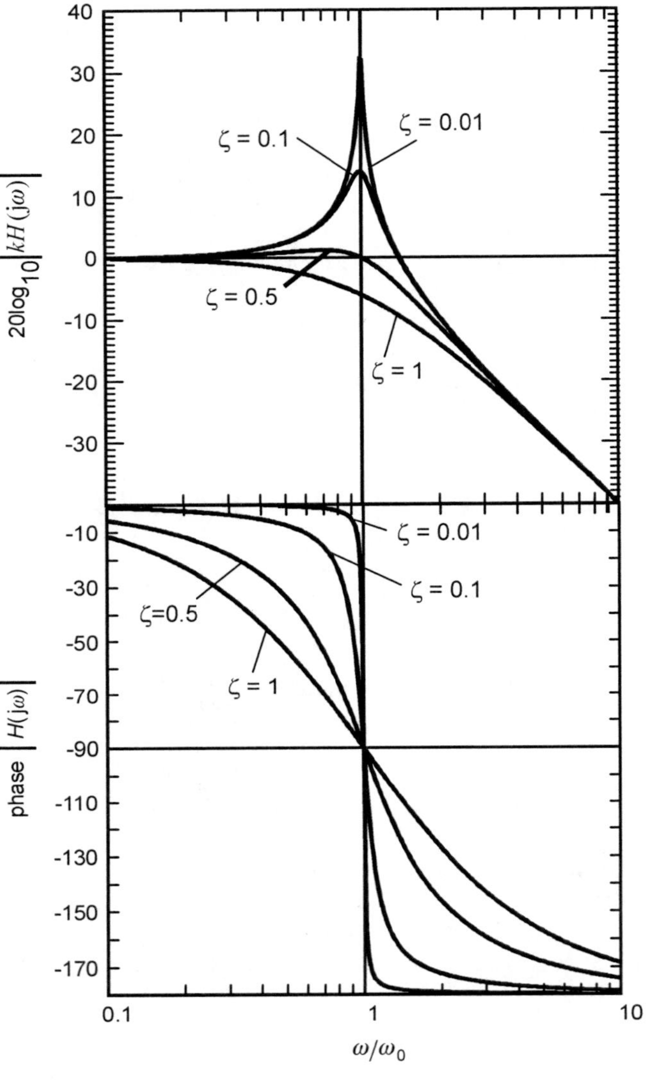

Figure 12.5 Frequency response of a single-degree-of-freedom passive isolation system.

and the feedback control force in the *s* domain is found by taking the Laplace transform of Equation (12.2.19) with zero initial conditions. Thus,

$$F_c(s) = (K_a s^2 + K_v s + K_d)X(s) = B(s)X(s) \qquad (12.2.21a,b)$$

The frequency response is given by:

$$H(s) = \frac{X(s)}{F(s)} = \frac{G(s)}{1 + G(s)B(s)} = \frac{1}{(m + K_a)s^2 + (c_d + K_v)s + k + K_d} \qquad (12.2.22a–c)$$

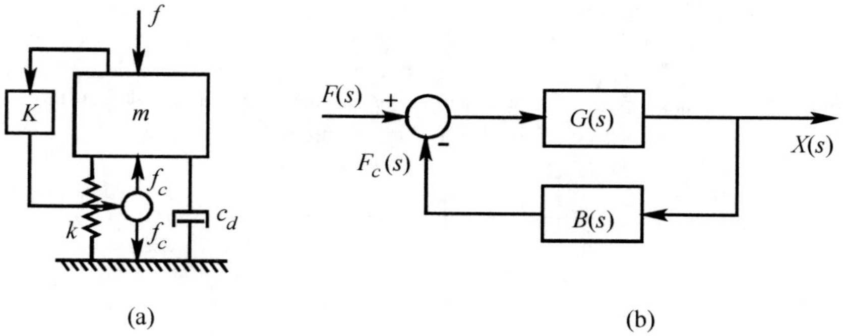

Figure 12.6 Feedback control of a single-degree-of-freedom isolation system: (a) physical system; (b) block diagram.

The time domain equivalent of this equation which could also have been derived by substituting Equation (12.2.19) into Equation (12.2.18), is:

$$(m + K_a)\ddot{x}(t) + (c_d + K_v)\dot{x}(t) + (k + K_d)x(t) = f(t) \tag{12.2.23}$$

In real physical systems, there is a finite time delay between receiving the signal from the vibration sensor, processing it, feeding it to the vibration actuator and propagating again to the vibration sensor. This affects the system stability and it can be shown that because of this phenomenon, velocity feedback systems are generally the most inherently stable (Fuller et al., 1994).

12.2.2.1 Displacement Feedback

The new resonance frequency and damping ratio for displacement feedback only (K_v, $K_a = 0$) are:

$$\omega_0' = \sqrt{\frac{k + K_d}{m}} = \omega_0\sqrt{1 + K_d/k} \tag{12.2.24a,b}$$

$$\zeta' = \frac{c_d}{2\sqrt{(k + K_d)m}} \tag{12.2.25}$$

If $K_a = K_v = 0$, use is made of Equations (12.2.5) and (12.2.6) and the substitution $s = \mathrm{j}\omega$ is made in Equation (12.2.22c), the following can be written for the frequency response with displacement feedback:

$$kH(\mathrm{j}\omega) = \frac{1}{1 + K_d/k - (\omega/\omega_0)^2 + 2\mathrm{j}\zeta(\omega/\omega_0)} \tag{12.2.26}$$

where ω_0 and ζ are the same quantities as used in Equation (12.2.14). Alternatively, ω_0' and ζ' may be substituted for ω_0 and ζ respectively in Equation (12.2.14).

The normalised modulus of the closed-loop frequency-response function $|H(j\omega)|$ is plotted in Figure 12.7 showing the effect of varying the displacement feedback gain K_d for various values of K_d/k, with $\zeta = 0.05$. It can be seen from the figure that increasing the displacement feedback increases low-frequency isolation but decreases high-frequency isolation, and that the crossover from low to high-frequency behaviour is dependent on the stiffness of the passive isolation system.

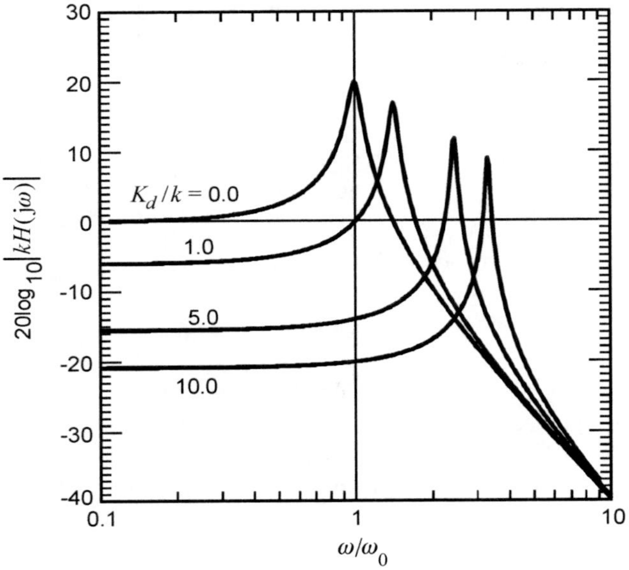

Figure 12.7 Effect of displacement feedback on the response of a single-degree-of-freedom system for $\zeta = 0.05$.

12.2.2.2 Velocity Feedback

In this case, the new resonance frequency and damping are given by:

$$\omega_0' = \sqrt{\frac{k}{m}} = \omega_0 \tag{12.2.27a,b}$$

$$\zeta' = \frac{c_d + K_v}{2\sqrt{km}} = \zeta(1 + K_v/c_d) \tag{12.2.28a,b}$$

Letting $K_v = K_d = 0$, using Equations (12.2.5) and (12.2.6) and substituting $s = j\omega$ in Equation (12.2.22c) allows the following to be written for the frequency response with velocity feedback:

$$kH(j\omega) = \frac{1}{1 - (\omega/\omega_0)^2 + 2j\zeta(1 + K_v/c_d)(\omega/\omega_0)} \tag{12.2.29}$$

Alternatively, ω_0' and ζ' may be substituted for ω_0 and ζ respectively in Equation (12.2.14).

The normalised modulus of the closed-loop frequency-response function of Equation (12.2.29) is plotted in Figure 12.8, and shows the effect of varying the velocity feedback gain for various values of K_v/c_d with $\zeta = 0.05$. It can be seen from the figure that increasing velocity feedback effectively increases the system damping, thus increasing the isolation in the region of the system resonance with little effect at low and high frequencies.

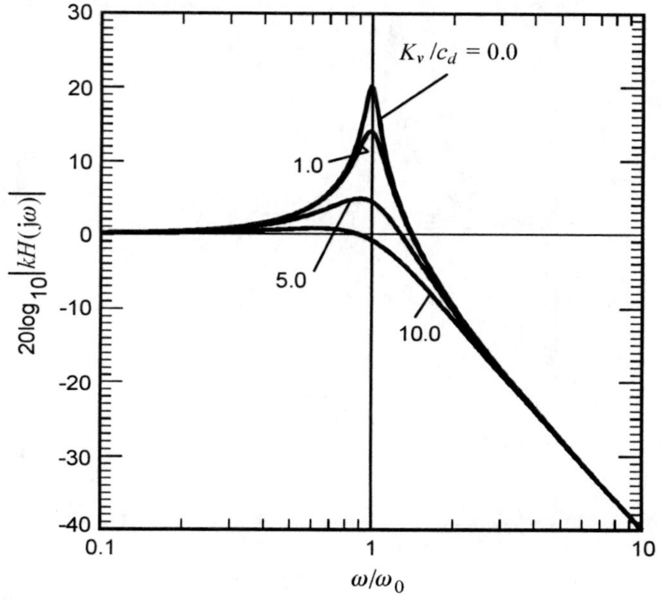

Figure 12.8 Effect of velocity feedback on the response of a single-degree-of-freedom system for $\zeta = 0.05$.

12.2.2.3 Acceleration Feedback

In this case, the new system resonance frequency and damping are given by:

$$\omega_0' = \sqrt{\frac{k}{m + K_a}} \tag{12.2.30}$$

$$\zeta' = \frac{c_d}{2\sqrt{k(m + K_a)}} \tag{12.2.31}$$

Letting $K_a = K_d = 0$, using Equations (12.2.5) and (12.2.6) and substituting $s = j\omega$ in Equation (12.2.22c) allows the following to be written for the frequency response with acceleration feedback:

$$kH(j\omega) = \frac{1}{1 - (\omega/\omega_0)^2(1 + K_a/m) + 2j\zeta(\omega/\omega_0)} \tag{12.2.32}$$

Alternatively, ω_0' and ζ' may be substituted for ω_0 and ζ respectively in Equation (12.2.14).

The normalised modulus of the closed-loop frequency-response function of Equation (12.2.32) is plotted in Figure 12.9, and shows the effect of varying the acceleration feedback gain for various values of K_a/m with $\zeta = 0.05$. It can be seen from the figure that increasing acceleration feedback decreases low-frequency isolation but increases high-frequency isolation.

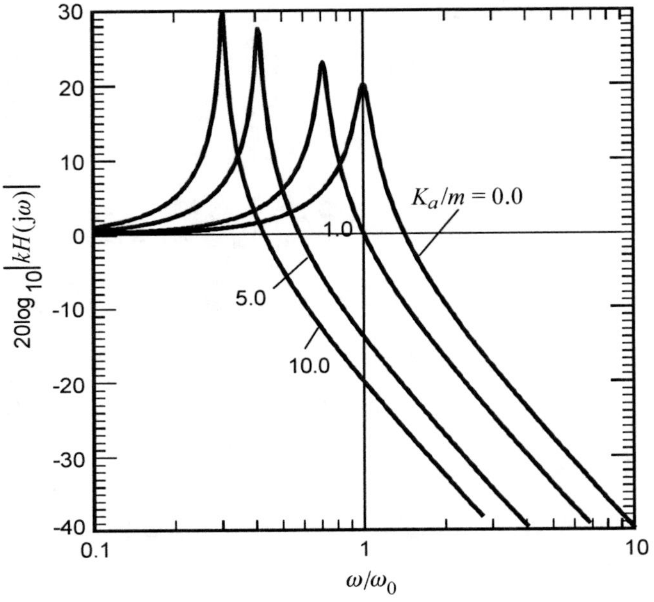

Figure 12.9 Effect of acceleration feedback on the response of a single-degree-of-freedom system for $\zeta = 0.05$.

Thus, it can be seen that adding feedback to the single-degree-of-freedom system actually changes the system parameters (resonance frequency and damping). It will be shown in Section 12.4 that feedforward control does not do this. This characteristic of the two different types of control was discussed in detail in Chapter 6.

It can also be seen that all types of feedback control can either reduce the frequency response over the entire frequency range of interest or reduce it at some desired frequency by moving the system resonance frequency and increasing the effective damping ratio.

12.2.2.4 Theoretical Closed-Loop Stability

It is of interest to determine the bounds on the gains K_d, K_v and K_a for stability considerations. The new pole locations with the addition of all three types of feedback can be expressed in terms of the new resonance frequency and damping ratio. That is, poles occur when:

$$s = -\omega_0\left(\zeta' \pm \sqrt{\zeta^2 - 1}\right) \tag{12.2.33}$$

where

$$\omega_0 = \frac{\sqrt{k + K_d}}{m} \tag{12.3.34}$$

and

$$\zeta = \frac{c_d + K_v}{2\sqrt{(k + K_d)(m + K_a)}} \tag{12.2.35}$$

Thus, the poles will lie in the negative half-plane (that is, the system will be stable) provided all of the following conditions are satisfied:

$$K_a + m > 0, \qquad K_v + c_d > 0, \qquad K_d + k > 0 \tag{12.2.36a,b,c}$$

In all practical cases mass, damping coefficient, stiffness and controller gains will be positive, and thus all three control methodologies are unconditionally stable.

12.2.2.5 Closed-Loop Instabilities in Practical Systems

It is important to point out that, although unconditional stability is achieved in theory, instabilities caused by instrumentation can, and likely will, arise when these methodologies are applied to physical systems. The reasons for this are that high-pass filters, which are used to remove the DC gain in the feedback signal, will result in a phase advance at low frequencies; and at high frequencies a phase lag can occur due to the limited bandwidth of the actuators. Both of these effects can influence the system stability and thus will limit the allowable control gains. The low-frequency instabilities are of particular importance as active control applications often tend to be in the lower frequency range. Here, the findings of Brennan et al. (2007) are summarised. They investigated both of these issues analytically, deriving simple formulae that govern the frequencies and gains at which systems will become unstable.

Signals for acceleration, velocity and displacement control are usually measured using an accelerometer. For velocity and displacement control, the measured absolute acceleration is then electronically integrated (once for velocity and twice for displacement). At frequencies well above the high-pass filter corner frequency, this process is indeed a pure integrator. However, at lower frequencies, closer to the high-pass filter corner frequency, the process is influenced by the characteristics of the high-pass filter. The corner frequency of a high-pass filter is the frequency at which the signal power is attenuated by 3 dB. The region above the corner frequency is called the passband and the region below is called the stopband.

Additional high-pass filters can exist in both sensor conditioning amplifiers and in power amplifiers. Thus, acceleration, displacement and velocity feedback may have up to two, three or four high-pass filters respectively. The critical frequency (frequency at which the system becomes unstable) and maximum gain that can be applied before the system becomes unstable are specified in Table 12.1 for zero through four high-pass filters (each assumed to have the same corner frequency) for each of acceleration, velocity and displacement control.

Practical systems experience phase-lags due to transducer dynamics, low-pass filters (if present) and digitalisation (if applicable). The resulting critical frequency

(frequency at which the system becomes unstable) and maximum gain that can be applied before the system becomes unstable, where the phase lags are modelled as a pure time delay in the control system, are specified in Table 12.2.

Table 12.1 Approximate maximum feedback gains and critical frequencies for a system with high-pass filters assuming low damping and a high-pass filter corner frequency, ω_c, much less than the system natural frequency (after Brennan et al., 2007).

	Acceleration Feedback		Velocity Feedback		Displacement Feedback	
High-Pass Filters	$K_{a,\max}$	Critical Frequency	$K_{v,\max}$	Critical Frequency	$K_{d,\max}$	Critical Frequency
0	Unconditionally stable		Unconditionally stable		Unconditionally stable	
1	$2m\zeta\left(\dfrac{\omega_0}{\omega_c}\right)$	$\dfrac{\omega_0}{\sqrt{1+2\zeta(\omega_0/\omega_c)}}$	Unconditionally stable		Unconditionally stable	
2	$m\zeta\left(\dfrac{\omega_0}{\omega_c}\right)$	$\dfrac{\omega_0}{\sqrt{1+\zeta(\omega_0/\omega_c)}}$	$2\sqrt{km}\left(\dfrac{\omega_0}{\omega_c}\right)$	ω_c	Unconditionally stable	
3	—	—	$\dfrac{8}{9}\sqrt{km}\left(\dfrac{\omega_0}{\omega_c}\right)$	$\sqrt{3}\,\omega_c$	$8k$	$\dfrac{\omega_c}{\sqrt{3}}$
4	—	—	—	—	$4k$	ω_c

Table 12.2 Approximate maximum feedback gains and critical frequencies for a system with phase lag (time delay) assuming low damping and a time delay, T_d, much less than the system's natural period, T_0 (after Brennan et al., 2007).

Acceleration Feedback		Velocity Feedback		Displacement Feedback	
$K_{a,\max}$	Critical Frequency	$K_{v,\max}$	Critical Frequency	$K_{d,\max}$	Critical Frequency
m	$\dfrac{\omega_0}{2(T_d/T_0)}$	$\dfrac{\sqrt{km}}{4(T_d/T_0)}$	$\dfrac{\omega_0}{4(T_d/T_0)}$	$\dfrac{\zeta k}{\pi(T_d/T_0)}$	$\omega_0\left(1+\dfrac{\zeta}{2\pi(T_d/T_0)}\right)$

The analysis of the closed-loop response for systems with high-pass filters and phase lags, outlined by Brennan et al. (2007), showed that of the three control strategies discussed, velocity feedback is the most robust to instrumentation related instabilities. From Table 12.1 it can be observed that higher gains for velocity feedback can be achieved if the system natural frequency is well above the high-pass filter corner frequency, and thus in practical systems, the corner frequency should be chosen to be as low as practicable (this argument also holds for practical systems with acceleration feedback). From Tables 12.1 and 12.2 it is observed that higher system damping will result in higher allowable gains for acceleration and displacement feedback.

12.2.3 Base Excited Second-Order System

For this system, the support is no longer considered fixed; it is free to move, although it is constrained to move vertically. It is of interest to investigate how the various types of feedback can be used to control vibration transmission from the support to the mass, for the system shown in Figure 12.10.

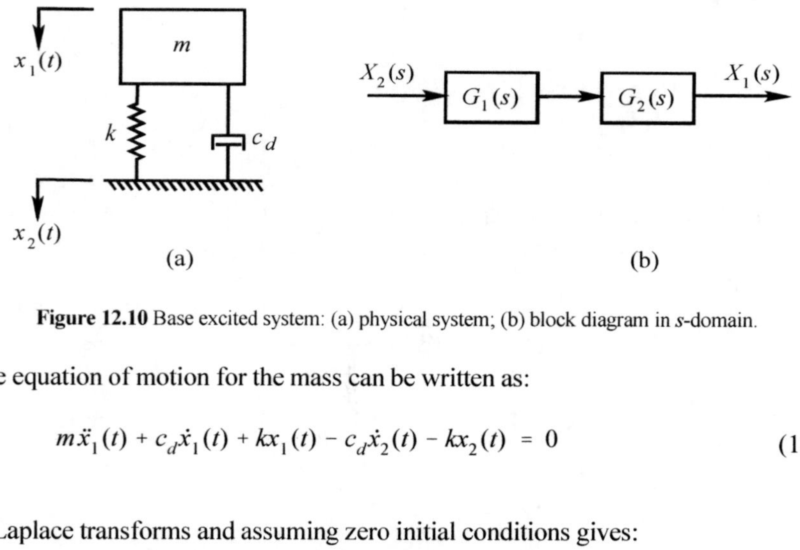

Figure 12.10 Base excited system: (a) physical system; (b) block diagram in *s*-domain.

The equation of motion for the mass can be written as:

$$m\ddot{x}_1(t) + c_d\dot{x}_1(t) + kx_1(t) - c_d\dot{x}_2(t) - kx_2(t) = 0 \tag{12.2.37}$$

Taking Laplace transforms and assuming zero initial conditions gives:

$$\left[ms^2 + c_d s + k\right]X_1(s) - [c_d s + k]X_2(s) = 0 \tag{12.2.38}$$

or in terms of the block diagram shown in Figure 12.10(b):

$$X_1(s) - G_1(s)G_2(s)X_2(s) = 0 \tag{12.2.39}$$

and thus the transfer function $H(s)$ relating the displacement of the mass to the displacement of the base can be written as:

$$H(s) = \frac{X_1(s)}{X_2(s)} = G_1(s)G_2(s) = \frac{c_d s + k}{ms^2 + c_d s + k} \tag{12.2.40a,b}$$

or in terms of the system resonance frequency and damping:

$$H(s) = \frac{2\zeta\omega_0 s + \omega_0^2}{s^2 + 2\zeta\omega_0 s + \omega_0^2} \tag{12.2.41}$$

where ω_0 and ζ are defined by Equations (12.2.5) and (12.2.6) respectively.

Using $s = \mathrm{j}\omega$, Equation (12.2.41) can be written as:

$$H(\mathrm{j}\omega) = \frac{2\mathrm{j}\zeta(\omega/\omega_0) + 1}{1 + 2\mathrm{j}\zeta(\omega/\omega_0) - (\omega/\omega_0)^2} = |H(\mathrm{j}\omega)|\,\mathrm{e}^{\,\mathrm{j}\theta} \tag{12.2.42a,b}$$

where

$$|H(\mathrm{j}\omega)| = \sqrt{\frac{1 + (2\zeta\omega/\omega_0)^2}{\left[1 - (\omega/\omega_0)^2\right]^2 + [2\zeta\omega/\omega_0]^2}} \tag{12.2.43}$$

and

$$\theta = \tan^{-1}\left[\frac{2\zeta\omega/\omega_0}{(\omega_0/\omega)^2 - 1 + 4\zeta^2}\right] \tag{12.2.44}$$

The normalised magnitude of $H(\mathrm{j}\omega)$ is shown in Figure 12.11 as a function of ω/ω_0 for various values of the damping ratio ζ. Note the reduced high-frequency performance as a result of damping.

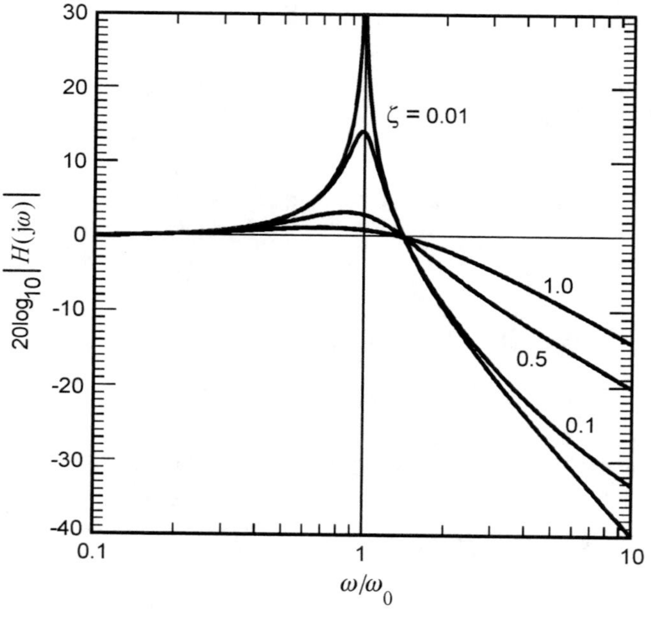

Figure 12.11 Effect of damping on the isolation of a base excited system.

12.2.3.1 Relative Displacement Feedback

The block diagram shown in Figure 12.12(b) represents the physical system shown in Figure 12.12(a).

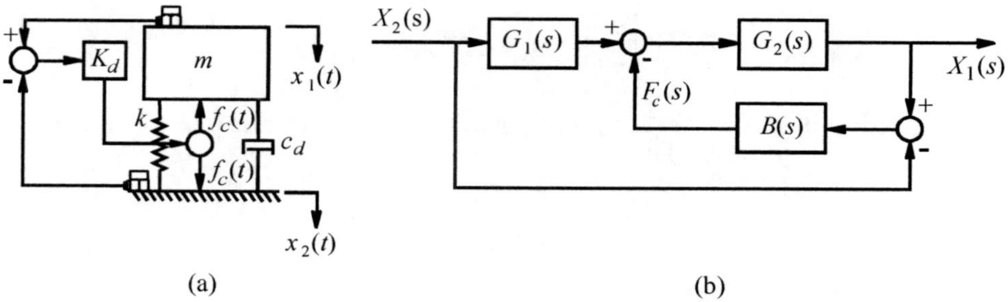

Figure 12.12 Relative displacement feedback for base excited system: (a) physical system; (b) block diagram.

Here:

$$G_1(s) = c_d s + k \tag{12.2.45}$$

$$G_2(s) = \frac{1}{ms^2 + c_d s + k} \tag{12.2.46}$$

$$B(s) = K_d \tag{12.2.47}$$

$$F_c(s) = B(s)X_1(s) \tag{12.2.48}$$

and

$$H(s) = \frac{X_1(s)}{X_2(s)} = \frac{G_1(s)G_2(s) + B(s)G_2(s)}{1 + G_2(s)B(s)} = \frac{G_1(s) + B(s)}{\dfrac{1}{G_2(s)} + B(s)} \qquad (12.2.49\text{a,b,c})$$

Substituting Equations (12.2.45) to (12.2.47) into Equation (12.2.49) gives:

$$H(s) = \frac{c_d s + k + K_d}{ms^2 + c_d s + k + K_d} \tag{12.2.50}$$

In the time domain, the control force is:

$$f_c(t) = -K_d(x_1(t) - x_2(t)) \tag{12.2.51}$$

and the system equation of motion is:

$$m\ddot{x}_1(t) + c_d\dot{x}_1(t) + (k + K_d)x_1(t) - c_d\dot{x}_2(t) - (k + K_d)x_2(t) = 0 \tag{12.2.52}$$

Note that the same results are obtained for the motion of the mass (but not the support) if the control reacts against the support or if an inertial mass is used for it to react against. The new system resonance frequency and damping ratio are:

$$\omega_0' = \sqrt{\frac{k + K_d}{m}} = \omega_0\sqrt{1 + K_d/k} \tag{12.2.53a,b}$$

$$\zeta' = \frac{c_d}{2\sqrt{(k + K_d)m}} = \frac{\zeta}{\sqrt{1 + K_d/k}} \tag{12.2.54a,b}$$

Using Equations (12.2.5) and (12.2.6) and substituting $s = \mathrm{j}\omega$ in Equation (12.2.50), the following can be written for the frequency response with relative displacement feedback:

$$H(\mathrm{j}\omega) = \frac{1 + K_d/k + 2\mathrm{j}\zeta(\omega/\omega_0)}{1 + K_d/k - (\omega/\omega_0)^2 + 2\mathrm{j}\zeta(\omega/\omega_0)} \tag{12.2.55}$$

Alternatively, ω_0' and ζ' may be substituted for ω_0 and ζ respectively in Equation (12.2.42).

The modulus of the closed-loop frequency-response function of Equation (12.2.55) is plotted in Figure 12.13, and shows the effect of varying the relative displacement feedback gain for various values of K_d/k, with $\zeta = 0.05$. It can be seen from the figure that increasing relative displacement feedback increases the effective system resonance frequency, but has little effect on the system damping, resulting in increased low-frequency isolation but decreased high-frequency isolation.

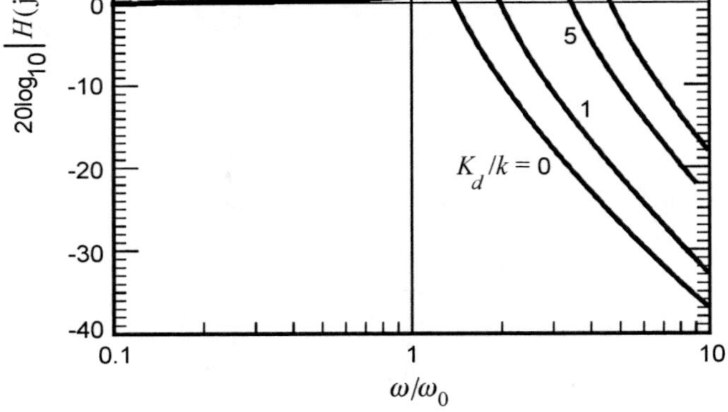

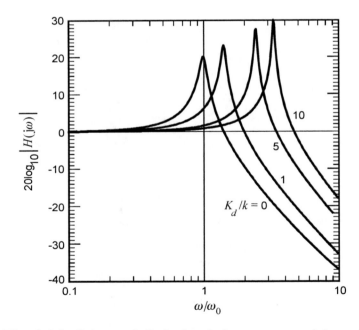

Figure 12.13 Effect of relative displacement feedback gain on the frequency response of a base excited system for $\zeta = 0.05$.

12.2.3.2 Absolute Displacement Feedback

Again, the block diagram shown in Figure 12.14(b) is an *s*-plane representation of the physical system shown in Figure 12.14(a), with all of the quantities defined as in Equations (12.2.45) to (12.2.48).

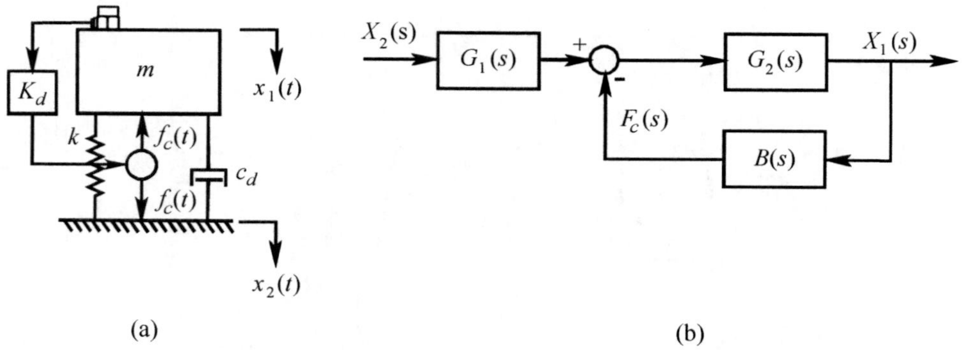

Figure 12.14 Absolute displacement feedback.

In this case:

$$H(s) = \frac{G_1(s)G_2(s)}{1 + G_2(s)B(s)} = \frac{G_1(s)}{\dfrac{1}{G_2(s)} + B(s)} \tag{12.2.56a,b}$$

or

$$H(s) = \frac{c_d s + k}{ms^2 + c_d s + k + K_d} \tag{12.2.57}$$

In the time domain, the control force $f_c(t)$ and equation of motion of the mass are:

$$f_c(t) = -K_d x_1(t) \tag{12.2.58}$$

and

$$m\ddot{x}_1(t) + c_d\dot{x}_1(t) + (k + K_d)x(t) - c_d\dot{x}_2(t) - kx_2(t) = 0 \tag{12.2.59}$$

It is not possible to define an equivalent damping ratio and resonance frequency as was done for relative displacement feedback. However, using Equations (12.2.5) and (12.2.6) and substituting $s = \mathrm{j}\omega$ in Equation (12.2.57), the following can be written for the frequency response with absolute displacement feedback:

$$H(\mathrm{j}\omega) = \frac{1 + 2\mathrm{j}\zeta(\omega/\omega_0)}{1 + K_d/k - (\omega/\omega_0)^2 + 2\mathrm{j}\zeta(\omega/\omega_0)} \tag{12.2.60}$$

The modulus of the closed-loop frequency-response function of Equation (12.2.60) is plotted in Figure 12.15 and shows the effect of varying the absolute displacement feedback gain for various values of K_d/k, with $\zeta = 0.05$. It can be seen from the figure that increasing the feedback gain increases the system natural frequency, as it effectively increases the stiffness of the system. This results in increased low-frequency isolation as K_d is increased and is much more effective than relative displacement feedback.

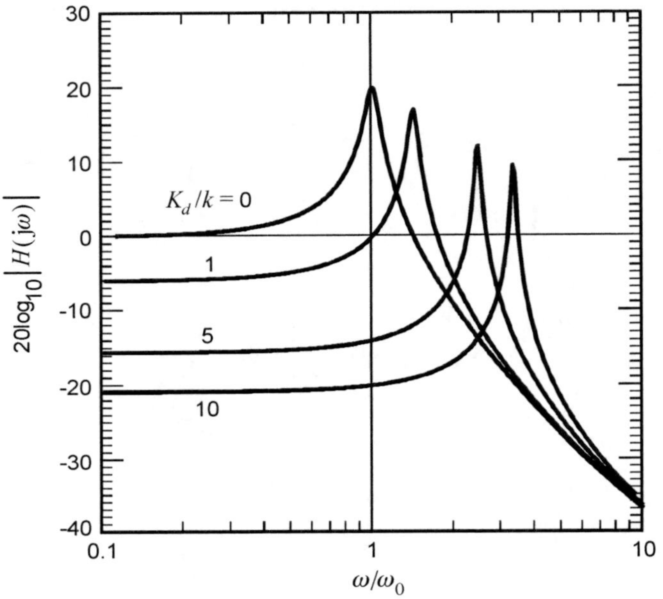

Figure 12.15 Effect of absolute displacement feedback gain on the frequency response of a base excited system.

12.2.3.3 Relative Velocity Feedback

The arrangement for this is similar to that shown in Figure 12.12 for relative displacement feedback, except that K_d is replaced by $K_v s$. Using Equation (12.2.49), the following is obtained for the transfer function:

$$H(s) = \frac{X_1(s)}{X_2(s)} = \frac{(c_d + K_v)s + k}{ms^2 + (c_d + K_v)s + k} \tag{12.2.61a,b}$$

In this case, the control force in the time domain is given by:

$$f_c(t) = -K_v\left[\dot{x}_1(t) - \dot{x}_2(t)\right] \tag{12.2.62}$$

and the system equation of motion is:

$$m\ddot{x}_1(t) + (c_d + K_v)\dot{x}_1(t) + kx_1(t) - (c_d + K_v)\dot{x}_2(t) - kx_2(t) = 0 \tag{12.2.63}$$

Of course, these equations could have been derived in the reverse order by writing down the equation of motion from inspection of the physical system and then Laplace transforms could have been taken to obtain Equation (12.2.61).

The new resonance frequency and damping ratio are:

$$\omega_0' = \omega_0 \tag{12.2.64}$$

$$\zeta' = \frac{c_d + K_v}{2\sqrt{km}} \tag{12.2.65}$$

Thus, Equation (12.2.61) can be written in terms of the system resonance frequency and critical damping ratio by replacing ω_0 and ζ in Equation (12.2.42) with ω_0' and ζ'.

Alternatively, using Equations (12.2.5) and (12.2.6) and substituting $s = j\omega$ in Equation (12.2.61), the following can be written for the frequency response with relative velocity feedback:

$$H(j\omega) = \frac{1 + 2j\zeta(\omega/\omega_0)(1 + K_v/c_d)}{1 - (\omega/\omega_0)^2 + 2j\zeta(\omega/\omega_0)(1 + K_v/c_d)} \tag{12.2.66}$$

The modulus of the closed-loop frequency-response function of Equation (12.2.66) is plotted in Figure 12.16 and shows the effect of varying the relative velocity feedback gain for various values of K_v/c_d, with $\zeta = 0.05$. It can be seen from the figure that increasing the feedback gain increases isolation in the region of the system natural frequency, but reduces high-frequency isolation.

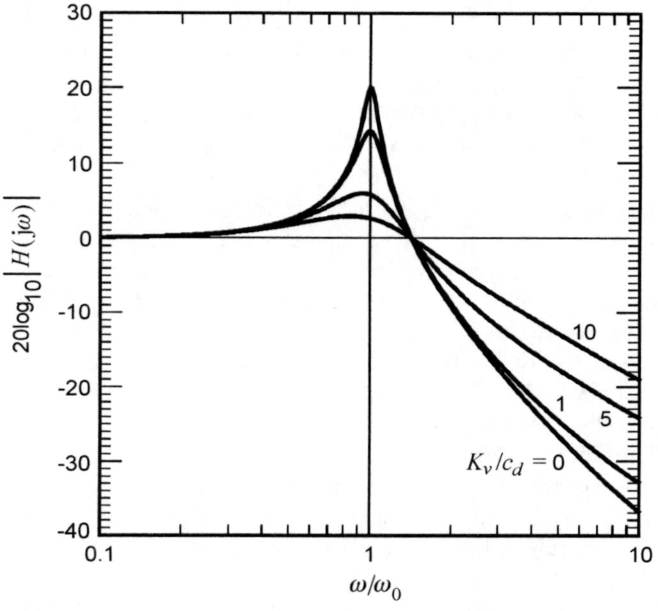

Figure 12.16 Effect of relative velocity feedback gain on the frequency response of a base excited system for $\zeta = 0.05$.

12.2.3.4 Absolute Velocity Feedback

The arrangement for this is similar to that shown in Figure 12.14 for absolute displacement feedback, except that K_d is replaced by K_v in Figure 12.14(a) (and the displacement transducer is replaced by a velocity transducer), and $B(s)$ of Figure 12.14(b) is now equal to $K_v s$. Using these substitutions in Equation (12.2.56), the following may be written for the transfer function:

$$H(s) = \frac{X_1(s)}{X_2(s)} = \frac{c_d s + k}{ms^2 + (c_d + K_v)s + k} \qquad (12.2.67a,b)$$

In the time domain, the control force is:

$$f_c = -K_v \dot{x}_1(t) \qquad (12.2.68)$$

and the equation of motion of the mass is:

$$m\ddot{x}_1(t) + (c_d + K_v)\dot{x}_1(t) + kx_1(t) - c_d\dot{x}_2(t) - kx_2(t) = 0 \qquad (12.2.69)$$

Using Equations (12.2.5) and (12.2.6) and substituting $s = \mathrm{j}\omega$ in Equation (12.2.67), the the following can be written for the frequency response with absolute velocity feedback:

$$H(\mathrm{j}\omega) = \frac{1 + 2\mathrm{j}\zeta(\omega/\omega_0)}{1 - (\omega/\omega_0)^2 + 2\mathrm{j}\zeta(\omega/\omega_0)(1 + K_v/c_d)} \qquad (12.2.70)$$

The modulus of the closed-loop frequency-response function of Equation (12.2.70) is plotted in Figure 12.17 and shows the effect of varying the absolute velocity feedback gain for various values of K_v/c_d, with $\zeta = 0.05$. It can be seen from the figure that increasing the feedback gain increases the isolation in the vicinity of the system resonance frequency without significantly affecting the isolation at other frequencies.

For the case where the base is perfectly rigid, the additional damping provided by feedback of absolute velocity can be considered equivalent to applying damping to the mass only by way of a sky-hook, whereby a passive damper is attached to an inertial base (see Figure 12.18). In real systems, the base off which the control reacts will have some degree of flexibility, and thus the reactive force can also influence the velocity of the system mass. However, as long as the system remains rigid and there are no internal resonances in the passive mounts, both of which tend to be true within the low-frequency range for which a controller is normally designed, the controller will retain stability when a flexible base structure is used (Elliott et al. 2001).

An alternative means of achieving absolute velocity feedback control is to react the control force not against the support base, but against an inertial mass actuator. This has the advantage of being entirely self-contained. However, as discussed in Section 15.7, in such a set-up there will be a phase shift associated with the actuator response that can lead to instability if the actuator resonance frequency is not significantly lower than the resonance frequency of the structure being controlled.

It should be noted that although the method discussed here is termed *absolute velocity feedback*, in practice absolute acceleration is usually measured and integral control then used to feedback the velocity; however, mathematically these are equivalent.

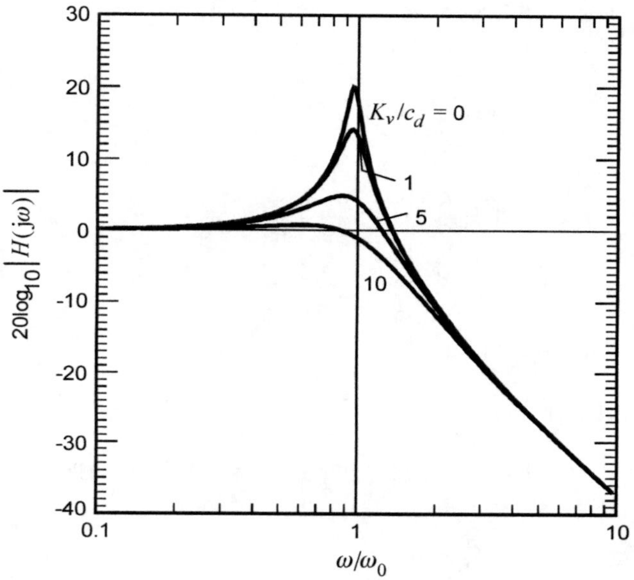

Figure 12.17 Effect of absolute velocity feedback gain on the frequency response of a base excited system for $\zeta = 0.05$.

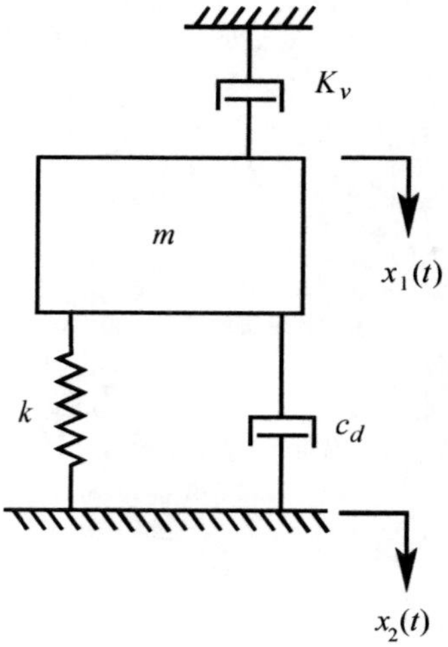

Figure 12.18 The 'sky-hook' damper.

12.2.3.5 Relative Acceleration Feedback

This arrangement is similar to that shown in Figure 12.12 for relative displacement feedback except that in part (a), K_d is replaced by K_a (and the displacement transducers are replaced with acceleration transducers) and in part (b), $B(s)$ is equal to $K_a s^2$.

Then, using these substitutions and Equation (12.2.49):

$$H(s) = \frac{X_1(s)}{X_2(s)} = \frac{K_a s^2 + c_d s + k}{(K_a + m)s^2 + c_d s + k} \tag{12.2.71a,b}$$

In this case, the control force in the time domain is given by:

$$f_c(t) = -K_a\left[\ddot{x}_1(t) - \ddot{x}_2(t)\right] \tag{12.2.72}$$

and the equation of motion of the mass is:

$$(m + K_a)\ddot{x}_1(t) + c_d\dot{x}_1(t) + kx_1(t) - K_a\ddot{x}_2(t) - c_d\dot{x}_2(t) - kx_2(t) = 0 \tag{12.2.73}$$

The new system resonance frequency and critical damping ratio are:

$$\omega_0' = \sqrt{\frac{k}{K_a + m}} \tag{12.2.74}$$

$$\zeta' = \frac{c_d}{2\sqrt{k(K_a + m)}} \tag{12.2.75}$$

Using Equations (12.2.5) and (12.2.6) and substituting $s = \mathrm{j}\omega$ in Equation (12.2.71), the following can be written for the frequency response with relative acceleration feedback:

$$H(\mathrm{j}\omega) = \frac{1 + 2\mathrm{j}\zeta(\omega/\omega_0) - (K_a/m)(\omega/\omega_0)^2}{1 - (\omega/\omega_0)^2(1 + K_a/m) + 2\mathrm{j}\zeta(\omega/\omega_0)} \tag{12.2.76}$$

Alternatively, ω_0' and ζ' may be substituted for ω_0 and ζ respectively in Equation (12.2.42).

The modulus of the closed-loop frequency-response function of Equation (12.2.76) is plotted in Figure 12.19 and shows the effect of varying the relative acceleration feedback gain for various values of K_a/m, with $\zeta = 0.05$. It can be seen from the figure that increasing the feedback gain increases the isolation in the vicinity of the system resonance frequency but decreases the isolation achieved at other frequencies.

12.2.3.6 Absolute Acceleration Feedback

The arrangement for this is similar to that shown in Figure 12.14 for absolute displacement feedback except that K_d is replaced by K_a (and the velocity transducer is replaced by an accelerometer) in Figure 12.14(a) and $B(s)$ of Figure 12.14(b) is now equal to $K_a s^2$. Using these substitutions in Equation (12.2.49), the following may be written for the transfer function:

$$H(s) = \frac{X_1(s)}{X_2(s)} = \frac{c_d s + k}{(K_a + m)s^2 + c_d s + k} \quad (12.2.77a,b)$$

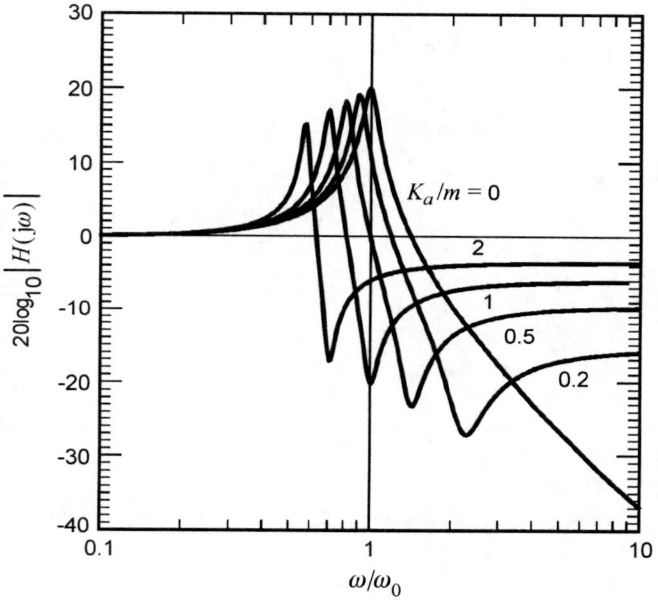

Figure 12.19 Effect of relative acceleration feedback gain on the frequency response of a base excited system for $\zeta = 0.05$.

The system equation of motion for this case is:

$$(m + K_a)\ddot{x}_1(t) + c_d\dot{x}_1(t) + kx_1(t) - c_d\dot{x}_2(t) - kx_2(t) = 0 \quad (12.2.78)$$

Using Equations (12.2.5) and (12.2.6) and substituting $s = j\omega$ in Equation (12.2.77), the following can be written for the frequency response with absolute acceleration feedback:

$$H(j\omega) = \frac{1 + 2j\zeta(\omega/\omega_0)}{1 - (\omega/\omega_0)^2(1 + K_a/m) + 2j\zeta(\omega/\omega_0)} \quad (12.2.79)$$

The modulus of the closed-loop frequency-response function of Equation (12.2.79) is plotted in Figure 12.20 and shows the effect of varying the absolute acceleration feedback gain for various values of K_a/m, with $\zeta = 0.05$. It can be seen from the figure that increasing the feedback gain is equivalent to adding mass to the system, resulting in a movement of the resonance frequency to lower frequencies, thus producing higher isolation at higher frequencies at the expense of lower isolation at low frequencies.

12.2.3.7 Force Feedback

Here, the case will be considered where the total force transmitted by the base into the suspension system will be used as the feedback variable (see Figure 12.21).

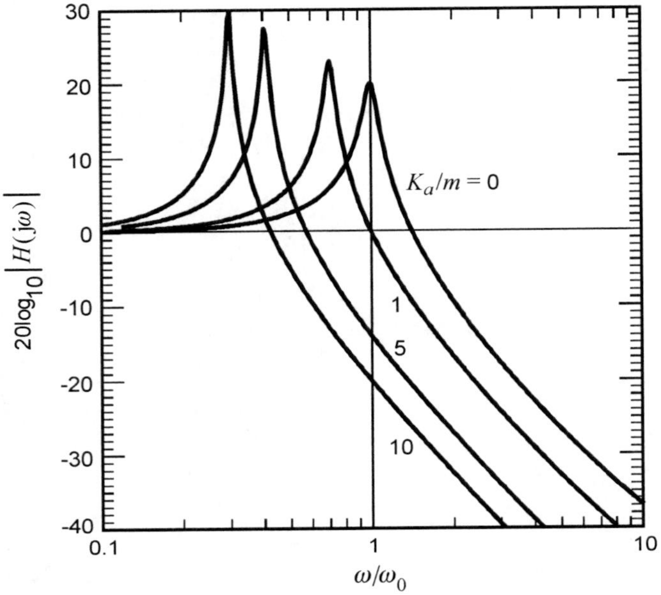

Figure 12.20 Effect of absolute acceleration feedback gain on the frequency response of a base excited system for $\zeta = 0.05$.

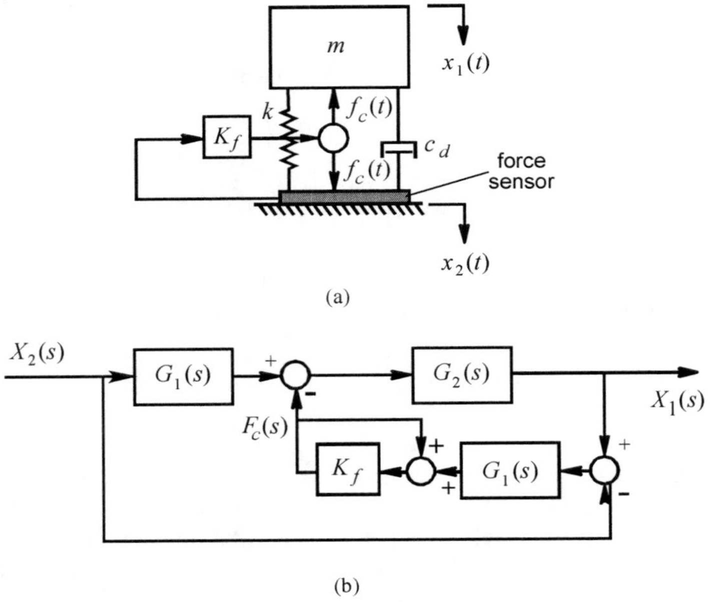

Figure 12.21 Force feedback for a base excited system.

From Figure 12.21, the transfer function in the frequency domain is given by:

$$H(s) = \frac{X_1(s)}{X_2(s)} = \frac{G_1(s) + \dfrac{G_1(s)K_f}{1-K_f}}{\dfrac{1}{G_2(s)} + \dfrac{G_1(s)K_f}{1-K_f}} = \frac{c_d s + k}{(1-K_f)ms^2 + c_d s + k} \tag{12.2.80a,b,c}$$

where $G_1(s)$ and $G_2(s)$ are defined by Equations (12.2.45) and (12.2.46).

The dynamic equations describing the system in Figure 12.21 are:

$$m\ddot{x}_1 + c_d(\dot{x}_1 - \dot{x}_2) + k(x_1 - x_2) + K_f f_T(t) = 0 \tag{12.2.81}$$

$$c_d(\dot{x}_1 - \dot{x}_2) + k(x_1 - x_2) + K_f f_T(t) = f_T \tag{12.2.82}$$

where the transmitted force f_T and the control force f_c are related by $f_c(t) = K_f f_T(t)$.

The new system resonance frequency and critical damping ratio are given by:

$$\omega_0' = \sqrt{\frac{k}{m(1-K_f)}} \tag{12.2.83}$$

$$\zeta' = \frac{c_d}{2\sqrt{km(1-K_f)}} \tag{12.2.84}$$

It can be seen from the above equations that force feedback effectively increases the suspension critical damping ratio.

Using Equations (12.2.5) and (12.2.6) and substituting $s = \mathrm{j}\omega$ in Equation (12.2.80c), the following can be written for the frequency response with force feedback (force acting on the base and suspended mass simultaneously):

$$H(\mathrm{j}\omega) = \frac{1 + 2\mathrm{j}\zeta(\omega/\omega_0)}{1 - (\omega/\omega_0)^2(1-K_f) + 2\mathrm{j}\zeta(\omega/\omega_0)} \tag{12.2.85}$$

Alternatively, ω_0' and ζ' may be substituted for ω_0 and ζ respectively in Equation (12.2.42).

The modulus of the closed-loop frequency-response function of Equation (12.2.85) is plotted in Figure 12.22 and shows the effect of varying the force feedback gain for various values of K_f/m, with $\zeta = 0.05$. It can be seen from the figure that increasing the feedback gain effectively increases the damping and stiffness of the system. This results in an improved low-frequency performance which extends also to high frequencies for high values of K_f/m. As will be discussed later, this extension to high frequencies will not occur in practice because the system becomes unstable if K_f/m is greater than one.

If the system were configured so that the control force acted only on the suspended mass and not the base, then the innermost feedback loop of Figure 12.21(b) would be eliminated and Equation (12.2.80) would become:

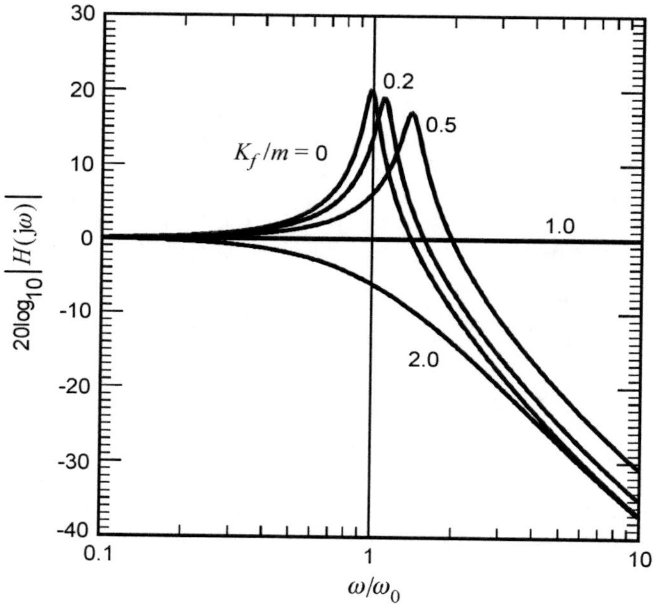

Figure 12.22 Effect of force feedback gain on the frequency response of a base excited system for $\zeta = 0.05$.

$$H(s) = \frac{c_d s + k}{\dfrac{m}{1 + K_f} s^2 + c_d s + k} \tag{12.2.86}$$

The modulus of the frequency-response function would be the same as before except that ω_0' and ζ' would be defined as:

$$\omega_0' = \sqrt{\frac{k(1 + K_f)}{m}} \tag{12.2.87}$$

$$\zeta' = \frac{c_d}{2\sqrt{\dfrac{km}{1 + K_f}}} \tag{12.2.88}$$

Using Equations (12.2.5) and (12.2.6) and substituting $s = j\omega$ in Equation (12.2.86), the following can be written for the frequency response with force feedback (force acting only on the suspended mass):

$$H(j\omega) = \frac{1 + 2j\zeta(\omega/\omega_0)}{1 - (\omega/\omega_0)^2 \dfrac{1}{(1 + K_f)} + 2j\zeta(\omega/\omega_0)} \tag{12.2.89}$$

Alternatively, ω_0' and ζ' may be substituted for ω_0 and ζ respectively in Equation (12.2.42).

The modulus of the closed-loop frequency-response function of Equation (12.2.89) is plotted in Figure 12.23 and shows the effect of varying the force (on only the suspended mass) feedback gain for various values of K_f/m, with $\zeta = 0.05$. It can be seen from the figure that increasing the feedback gain effectively increases the damping and stiffness of the system. This results in an improved low-frequency performance which, in contrast to the previous case, does not extend to high frequencies for high values of K_f/m.

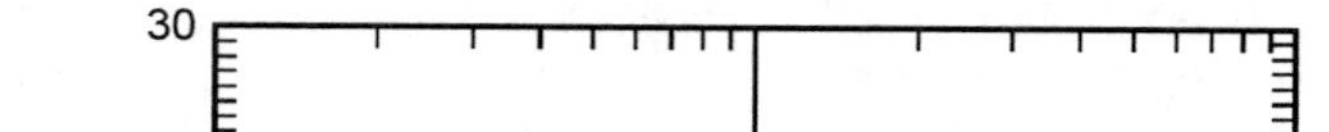

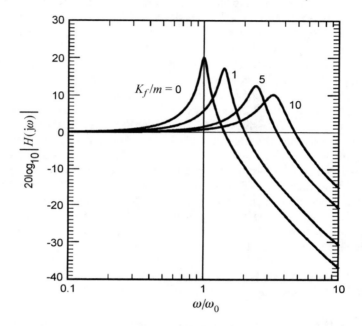

Figure 12.23 Effect of force feedback gain on the frequency response of a base excited system with the control force only acting on the suspended mass for $\zeta = 0.05$.

12.2.3.8 Integral Force Feedback

In Section 12.2.3.4 it was discussed how absolute velocity feedback control using an integrated absolute acceleration signal could be used to implement a sky-hook damper system. Similar results can be achieved by measuring force with a force sensor and integrating the resulting signal. In this case the arrangement is similar to that shown in Figure 12.21 for pure force feedback, except that the integral controller is incorporated by replacing the K_f term with $-K_f/s$. The frequency domain transfer function thus becomes:

$$H(s) = \frac{X_1(s)}{X_2(s)} = \frac{c_d s + k}{ms^2 + (c_d + K_f m)s + k} \qquad (12.2.90a,b)$$

which is the same as the absolute velocity feedback transfer function of Equation (12.2.67b) except that the absolute velocity gain K_v is here replaced with $K_f m$ (that is, the control gain is multiplied by the mass m that relates the total force to acceleration).

Preumont et al. (2002) have shown that when the physical system being controlled is a rigid structure, integral absolute acceleration (absolute velocity) and integral force feedback have equivalent open-loop transfer functions, which exhibit alternating poles and zeroes, guaranteeing controller stability. However, if the system is flexible (for example, it has one or more flexible appendages), integral force and integral absolute acceleration feedback control strategies exhibit differing characteristics. Preumont et al. (2002) show that if a force sensor is used, even when the structure is flexible, alternating poles and zeroes are still produced in the open-loop transfer function, guaranteeing controller stability; however, in cases where the structure has flexible modes which interfere with the isolation system, absolute velocity (integral absolute acceleration) feedback will not have alternating poles and zeroes in the open-loop transfer function, and thus stability is no longer guaranteed.

Although integral force feedback systems theoretically exhibit superior stability characteristics, absolute velocity feedback control can provide better attenuation at the internal equipment's resonance frequency. Thus, absolute velocity feedback can potentially offer practical advantages, as long as conditions under which it suffers from instabilities can be avoided. Elliott et al. (2004) have determined the conditions under which such systems will become unstable, and thus the conditions that must be met to avoid instability. For the case when the base system is considered to have a mass-controlled dynamic behaviour within the frequency range of interest (that is, input mobility of $Y_b = v_b/f_b = 1/(m_b s)$, where v_b is the base velocity at the point of attachment and f_b is the force transmitted by the actuator and mount), unconditional stability using absolute velocity feedback will be achieved as long as the natural frequency of the dynamic equipment structure is higher than that of the isolation system. For the case when both the equipment and base are flexible, absolute velocity feedback appears to remain stable for gain amplitudes that are within the bounds of what would be required to achieve sufficient vibration isolation.

12.2.3.9 Effects of Isolator Mass

Up to now the analysis has ignored the mass of the vibration isolator. This assumption should not be problematic at lower frequencies, but may not be adequate for higher frequency isolation. In general, isolators have some form of distributed mass, damping and stiffness, which can introduce high-frequency isolator resonances.

Recently, Yan et al. (2010) modelled a distributed parameter isolator in order to investigate the effect of isolator resonances on control performance and stability. They considered absolute velocity feedback control due to its high performance and stability characteristics. They found that high-frequency isolator resonances were not attenuated by the controller and also that instabilities could occur if the system's base were flexible. Two means for improving (but not completely rectifying) the system's stability characteristics were proposed and implemented: increasing the base mass and including a lead compensator in the control loop.

12.2.3.10 Closed-Loop Stability of the Base Excited System

As discussed in Section 12.2.1, the system will be stable provided that all of its poles are on the left of the $j\omega$ axis in the s-plane. For the cases discussed, this condition is met provided that all of the conditions expressed in Equation (12.2.36) are satisfied. Additionally, for the force feedback system, K_f/m must be less than one for the case of the force acting on the suspended mass and the base.

It was discussed in Section 12.2.5 how for a single-degree-of-freedom system, closed-loop instabilities due to instrumentation will affect system stability. The key points arising from this earlier discussion are equally applicable to base excited systems.

All of the simple feedback compensation circuits $B(s)$ (consisting only of a simple gain) discussed so far are the simplest form of feedback control possible. Better performance over a wider frequency band can be achieved in practice using more complicated compensator circuits $B(s)$, depending upon the application, and this is where a large research effort is currently concentrated. Some examples of more complex controllers have been discussed in Chapter 11.

Another consideration that has not been detailed here is that physical control actuators have limits on their magnitudes and rates. Any real control system must therefore take into account the saturation limits of actuators, which will in turn limit the allowable control gains. Some vibration isolation systems may experience mainly small disturbances, with only occasional large disturbances. In such cases it would be desirable to design a controller that exhibits high performance when disturbances are small, but is still able to handle large disturbances without saturating the actuators. One method of achieving this is through the use of 'anti-windup' strategies, whereby the basic control system is designed only considering the smaller nominal disturbances, and an 'anti-windup' compensator is then implemented to ensure that the control system can handle larger disturbances without the actuators saturating. Teel et al. (2006) give an overview of published 'anti-windup' strategies and experimentally validate one such strategy on an industrial active vibration isolation system.

12.2.4 Multiple-Mount Vibration Isolation

So far this chapter has been concerned with vibration isolation using feedback control through a single active mount. The single-mount concepts of feedback vibration isolation discussed in Sections 12.2.2 and 12.2.3 can be extended to systems with multiple mounts using multi-channel controllers for isolation of equipment from base vibration. Such systems have been successfully developed using decentralised controllers with absolute velocity feedback for a two-mount system (Serrand and Elliott, 2000) with improvements implemented over the following few years for a four-mount system (Kim et al., 2001; Huang et al., 2003; and Brennan et al., 2006), with both rigid and flexible base structures.

Using absolute velocity feedback control with decentralised controllers means that each controller operates as a single-input, single-output independent channel with the sensor output fed directly into the corresponding actuator, resulting in a relatively simple controller. Note that these controllers could theoretically be considered collocated (see Section 11.7.4) if the base structure were rigid; but for a flexible base, additional dynamics are transmitted to the equipment via the mounting system and thus the collocated assumption no longer holds. Further descriptions of velocity feedback controllers, collocated actuators and sensors, and decentralised controllers, including potential limitations, are given in Chapter 11 (Sections 11.7.4 and 11.8).

If the base structure is rigid, unconditional stability can be achieved for both rigid and flexible equipment; however, in a practical system, actuator-sensor pairs tend not to be perfectly collocated, and the effects of this on controller stability should be considered (Huang et al., 2003). Additionally, the potential instabilities caused by high-pass filters in physical instrumentation (discussed in Section 12.2.2.5) apply equally to multi-channel systems. Thus, ensuring that high-pass filters have a very low corner frequency ensures that high gains can be implemented without destabilising the system (Brennan et al., 2006).

When the base structure is flexible, unconditional theoretical stability no longer holds. However, Elliott et al. (2004) have shown that in practice, even when the base structure is flexible, *single-mount* systems will most likely retain good stability characteristics. The multiple-mount experimental results of Huang et al. (2003) and Brennan et al. (2006) have also displayed good stability characteristics for the systems they considered.

12.2.5 Tuned Vibration Absorbers

The term *tuned vibration absorber* refers to a mass that is suspended from a structure using some form of spring/damper system and tuned to a specific frequency for the purpose of vibration control (see Figure 12.24). Vibration absorbers can be used in two distinct ways. Here, the terminology of Bonello et al. (2005) is adopted. The term 'tuned mass damper' is used to refer to a vibration absorber that is tuned to reduce the vibration amplitude of a structure or machine vibrating around the resonance frequency and the term 'tuned vibration neutraliser' to refer to a device that is tuned to target non-resonant vibration due to forced excitation at a non-resonance frequency.

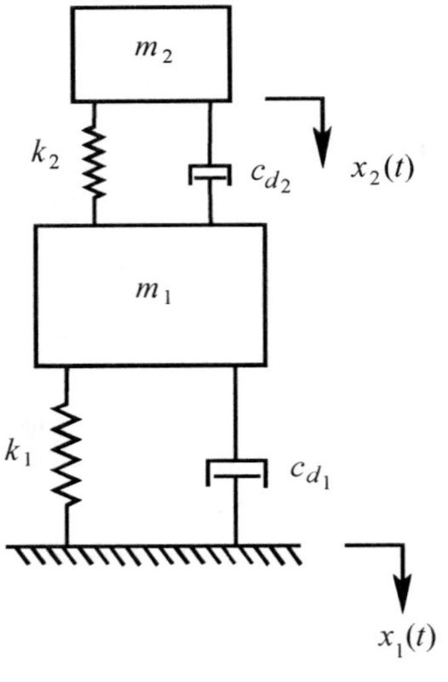

Figure 12.24 Vibration absorber, $\zeta_2 = C_2/[2(k_2 m_2)^{½}]$.

12.2.5.1 Tuned Mass Damper

A tuned mass damper is tuned to a frequency that is usually slightly less than the original system's natural frequency. If no damping is used, the effect of the absorber is to change the system from one characterised by a single resonance to one characterised by two resonances, the latter two appearing on either side (in the frequency domain) of the original. Although the system response at the original resonance frequency is substantially reduced, the response at the two new resonance frequencies is generally much larger than it was before. As the amount of damping is increased, the response at the two new resonances frequencies decreases and the total system response at the old original resonance frequency increases until a point is reached at which they are essentially equal, thus defining what is known as the optimum critical damping ratio ζ_2 for the suspended mass.

Note that the response at the two new resonance frequencies is not necessarily the same; in fact, it will only be the same if the ratio of the stiffness k_2 to stiffness k_1 is optimised.

Values of the stiffness ratio and critical damping ratio for 'optimum' performance of the absorber are (Den Hartog, 1956; Davies, 1965):

$$\frac{k_2}{k_1} = \frac{m_1 m_2}{(m_1 + m_2)^2} \tag{12.2.91a}$$

$$\zeta_2^2 = \frac{3(m_2/m_1)^3}{8(1 + m_2/m_1)^3} \tag{12.2.91b}$$

and the predicted amplitude of response of the mass m_1 in the frequency range including and between the two system resonances is:

$$x_1/d = \sqrt{1 + 2m_1/m_2} \tag{12.2.92}$$

where d is the static deflection of the original mass m_1.

Den Hartog (1956) derived the above equations on the basis that there was no existing damping in the vibrating machine support system (that is, ζ_1 was assumed to be zero). Soom and Lee (1983) showed that as ζ_1 increases from 0 to 0.1, the true optimum for ζ_2 increases from that given by Equation (12.2.91b) by between 2% (for a mass ratio of 0.5) and 7% (for a mass ratio of 0.1), and the true optimum stiffness ratio (k_2/k_1) decreases from that given by Equation (12.2.91a) by between 6% (for a mass ratio of 0.1) and 10% (for a mass ratio of 0.5). Soom and Lee (1983) also showed that there would be little benefit in adding a vibration absorber to a system that already had a critical damping ratio ζ_1 greater than 0.2.

Note that in some cases the so-called 'optimum' solution may not be the best. For example, if a machine is being excited by a single frequency which is very close to or coincides with the resonance frequency of the machine on its mounts, it is desirable to reduce the system response by as much as possible at the exciting frequency; the response at other frequencies is not so important. In this case, it is better to use a value of ζ_2 smaller than the optimum, which will have the effect of reducing the system response at the original resonance frequency at the expense of increasing it at the two new resonance frequencies.

Regardless of the stiffness ratio or critical damping ratio ζ_2, the separation between the two new resonance frequencies and the original resonance frequency is dependent solely on the mass ratio m_2/m_1. The larger this ratio, the greater will be the frequency separation. Determination of the actual frequencies can be accomplished using published charts (Bies and Hansen, 2009).

The equation of motion for the mass m_2 in the system shown in Figure 12.24 can be written as:

$$m_2\left[\ddot{x}_2(t) - \ddot{x}_1(t)\right] + c_{d_2}\left[\dot{x}_2(t) - \dot{x}_1(t)\right] + k\left[x_2(t) - x_1(t)\right] = -m_2\ddot{x}_1(t) \tag{12.2.93}$$

Putting $z(t) = x_2(t) - x_1(t)$, and dividing through by m_2 gives:

$$\ddot{z}(t) + 2\zeta_2\omega_{02}\dot{z}(t) + \omega_{02}^2 z(t) = -\ddot{x}_1(t) \tag{12.2.94}$$

where

$$\omega_{02} = \sqrt{k_2/m_2} \tag{12.2.95}$$

and

$$\zeta_2 = \frac{c_{d_2}}{2\sqrt{k_2 m_2}} \tag{12.2.96}$$

Equation (12.2.94) describes a second-order system, having the resonance frequency and critical damping ratio of the supplementary absorber system, but whose output is the relative displacement of the two masses and whose input is the acceleration of the main mass.

Equation (12.2.94) can be solved to give the required output, $z(t) = x_2(t) - x_1(t)$, for a given input acceleration, damping ratio and resonance frequency of the secondary system. If the optimum values for damping ratio and stiffness ratio are required, then these can be used in the equation to find the corresponding optimum value of $z(t)$.

Multiple structural resonances excited by a broadband force can be controlled by installing vibration absorbers tuned to each mode to be controlled (that is, one absorber for each mode).

12.2.5.2 Adaptive, Semi-Active and Active Tuned Mass Dampers

The system in Figure 12.24 is a passive tuned mass damper and, as such, cannot be re-tuned once installed. There are, however, many situations in which it is desirable to be able to alter the mass damper's dynamic characteristics post-installation.

Consider a situation where either it were difficult to obtain accurate estimates of m_1 and k_1 of the original system, or a situation where parameters of the original system were to change over time. The former would make it difficult to optimise values of k_2 and ζ_2 of the absorber at the design stage, and the latter could result in a system that is initially optimal but will not work effectively once the primary system changes. These problems could be overcome by using values of k_2 and ζ_2 that were close to the initial desired values and tuning the absorber with a feedback control system.

In an early implementation reported by Lund (1980), Equation (12.2.94) was implemented on an analogue computer such that the desired output $z(t)$ was calculated for a given input acceleration of the main mass (see Figure 12.25).

Values of ω_{02} and ζ_2 can be made adjustable in the computer model so that they can be manually adjusted to optimise the response of the main mass. The control force $f_c(t)$ can then be determined so that it modifies the stiffness and damping of the attached absorber to the values used in the computer model. This is accomplished by driving the control force (as shown in Figure 12.25) with a control signal proportional to the difference between the actual measured value of $z(t)$ and that desired according to the computer model solution of Equation (12.2.94). A system such as the one just described has been used to reduce wind induced sway in tall buildings; an example of a specific installation is described by Petersen (1980).

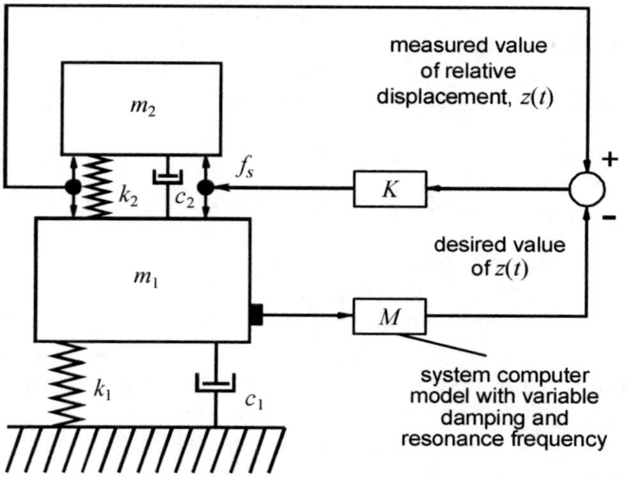

(a)

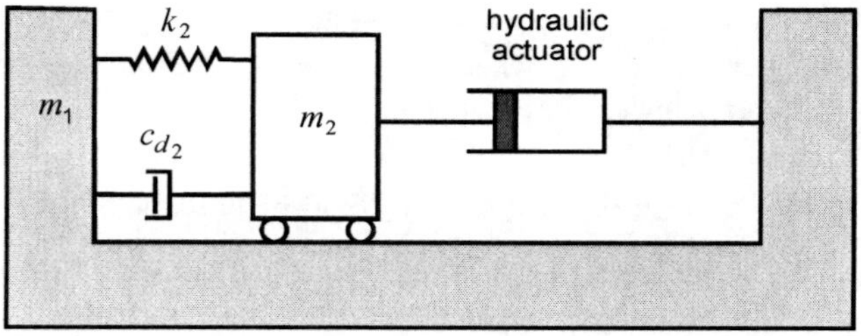

(b)

Figure 12.25 Active tuning of a vibration absorber: (a) for a vertically mounted machine; (b) to control building vibrations.

Adding a control force in parallel with the original passive mass damper as described above is an example of a fully *active* tuned mass damper. Although fully active systems can theoretically produce excellent results, one potential disadvantage of such systems is that the controller can require significant amounts of energy, and poor implementation can result in controller instabilities. As such, they are generally not preferred over *semi-active* and *adaptive* systems.

Adaptive and *semi-active* tuned mass dampers have parameters (for example, stiffness, damping and/or mass) that can be changed with time. They do not apply energy directly into the system to be controlled and thus can avoid instability problems whilst still exhibiting many of the performance characteristics of fully active systems. The term 'adaptive' is used to refer to systems that have parameters which can be adjusted over time, but are effectively constant over one vibration cycle. The term 'semi-active' is used to refer to systems where the parameter adjustment is rapid relative to the vibration cycle. Adaptive tuned mass dampers most commonly have variable stiffness elements whilst semi-active systems most commonly have variable dampers.

Sun et al. (1995) and Kela and Vähäoja (2009) have reviewed different types of active, semi-active and adaptive tuned mass dampers, as well as hybrid tuned mass dampers (combination of passive and active absorbers). There are also several papers in which semi-active vibration isolation developments have been comprehensively reviewed (for example, Jalili 2002; and Soong and Spencer, 2002).

12.2.5.3 Adaptive Tuned Vibration Neutraliser

A tuned vibration neutraliser is of a similar construction to a tuned mass damper, but as already stated, differs from it in that a vibration neutraliser targets non-resonant vibration due to forced excitation at a non-resonance frequency. The vibration neutraliser frequency is made equal to the forcing frequency that is causing the undesirable structural vibration.

A simple tuned vibration neutraliser will effectively act as a notch filter, and thus will isolate against forced excitation over only a very narrow bandwidth. In many practical applications, it is expected that the forcing frequency will drift; and if this occurs, the tuned vibration neutraliser will no longer effectively isolate against the forced excitation; in fact, vibration levels may actually increase above what they would have been for the host structure with no vibration neutraliser attached. For this reason, there has been concerted effort in the development of adaptive tuned vibration neutralisers; that is, tuned vibration neutralisers with a real-time variable stiffness element. Variable-length beam elements and changes of shape are commonly used to achieve variable stiffness, although this is only one of many means; Brennan (2006) and Kela and Vähäoja (2009) reviewed several other mechanisms to achieve a variable stiffness element.

12.2.6 Vibration Isolation of Equipment from a Rigid or Flexible Support Structure

The discussion until now in this chapter has been concerned with the reduction in the response of a mass suspended on a spring and damper system attached to a foundation that may or may not move. In practice, such systems include optical equipment mounted on vibration isolators and vibration absorbers attached to structures, flexibly mounted equipment or buildings. However, there still remains a large class of vibration isolation problems that have not yet been considered, and which are characterised by the common requirement that vibration of a support structure caused by vibrating equipment attached to it is to be reduced by interposing an isolation system between the vibration source and the structure.

An ideal vibration isolation mount would provide high stiffness below some limiting frequency so that low-frequency loads (including the weight) can be transmitted. Above the limiting frequency, the isolator should have zero stiffness so that vibrations at these higher frequencies are not transmitted. It is very difficult to approximate the ideal mount with a passive system; thus the attractiveness of an active system. The discussion in this section will only be concerned with a feedback system; isolation of a machine from a support structure using feedforward control is considered in Sections 12.4 onwards.

12.2.6.1 Rigid Support Structure

Returning to the single-degree-of-freedom system discussed in Section 12.2.1, the force transmitted through the isolation system to the support will be determined and then used as a measure of the vibratory energy transmitted.

The equation of motion for the mass is given by Equation (12.2.1) and from Figure 12.3 it can be seen clearly that the force transmitted into the structure is given by:

$$f_T(t) = c_d\dot{x}(t) + kx(t) \tag{12.2.97}$$

Using Equations (12.2.1) and (12.2.97), the following is obtained for the ratio of transmitted force to the force acting on the mass:

$$\frac{f_T(t)}{f(t)} = \frac{c_d\dot{x}(t) + kx(t)}{m\ddot{x}(t) + c_d\dot{x}(t) + kx(t)} \tag{12.2.98}$$

Taking Laplace transforms and zero initial conditions gives:

$$H(s) = \frac{F_T(s)}{F(s)} = \frac{c_d s + k}{ms^2 + c_d s + k} \tag{12.2.99a,b}$$

which is identical to Equation (12.2.40) for the displacement transmission for a base excited system.

In the following sections, the effect of various types of feedback on the force transmission will be examined: displacement, velocity, acceleration and force feedback.

12.2.6.1.1 Displacement feedback

The equations of motion for the system in Figure 12.26(a) are:

$$m\ddot{x}(t) + c_d\dot{x}(t) + (k + K_d)x = f(t) \tag{12.2.100}$$

$$c_d\dot{x}(t) + (k + K_d)x = f_T(t) \tag{12.2.101}$$

For the equivalent s-plane system shown in Figure 12.26(b):

$$X(s) = \frac{F(s)G_1(s)}{1 + B(s)G_s(s)} \tag{12.2.102}$$

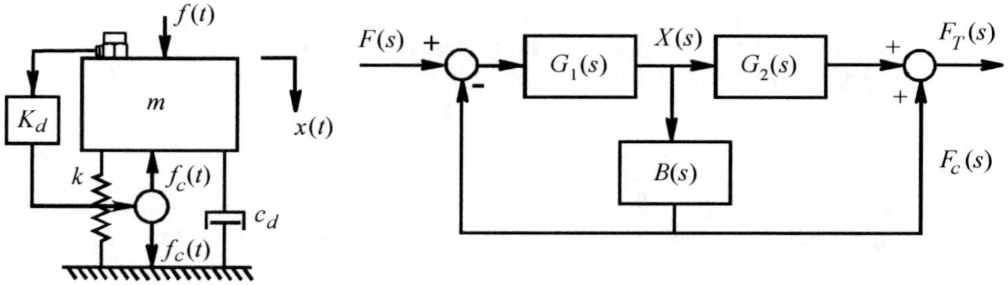

(a) (b)

Figure 12.26 Absolute displacement feedback for control of transmitted force $f_T(t)$: (a) physical system; (b) frequency domain block diagram.

Also,

$$F_T(s) = X(s)\left[G_2(s) + B(s)\right] = \frac{F(s)\left[G_2(s) + B(s)\right]}{\dfrac{1}{G_1(s)} + B(s)} \tag{12.2.103a,b}$$

Thus,

$$H(s) = \frac{F_T(s)}{F(s)} = \frac{G_2(s) + B(s)}{\dfrac{1}{G_1(s)} + B(s)} = \frac{c_d s + k + K_d}{ms^2 + c_d s + k + K_d} \tag{12.2.104a–c}$$

where $G_1(s)$ and $G_2(s)$ are defined by Equations (12.2.45) and (12.2.46) respectively, and $B(s) = K_d$ is defined as Equation (12.2.50) for the relative displacement feedback of a base excited system.

If the control force only acted on the base then,

$$H(s) = \frac{c_d s + k + K_d}{ms^2 + c_d s + k} \tag{12.2.105}$$

Using Equations (12.2.5) and (12.2.6) and substituting $s = \mathrm{j}\omega$ in Equation (12.2.105), the following can be written for the frequency response with absolute displacement feedback (force acting only on the base):

$$H(\mathrm{j}\omega) = \frac{1 + K_d/k + 2\mathrm{j}\zeta(\omega/\omega_0)}{1 - (\omega/\omega_0)^2 + 2\mathrm{j}\zeta(\omega/\omega_0)} \tag{12.2.106}$$

The modulus of the closed-loop frequency-response function of Equation (12.2.106) is plotted in Figure 12.27 and shows the effect of varying the displacement feedback gain for various values of K_d/k, with $\zeta = 0.05$. It can be seen from the figure that increasing the negative feedback gain effectively increases the isolation at all frequencies without affecting the system resonance frequency. However, it is unlikely that this would result in a stable system.

The case of the control force acting only on the mass is the same as for absolute displacement feedback of the base excited system plotted in Figure 12.15.

The case of the control force acting on the mass and reacting on the base is the same as for relative displacement feedback of the base excited system shown in Figure 12.12.

12.2.6.1.2 Velocity feedback

Again, similar results are obtained as were obtained for displacement transmission of the base excited system discussed earlier, except for the case where the control force acted only on the base. In this case, the force transmission frequency-response function is:

$$H(s) = \frac{(c_d + K_v)s + k}{ms^2 + c_d s + k} \tag{12.2.107}$$

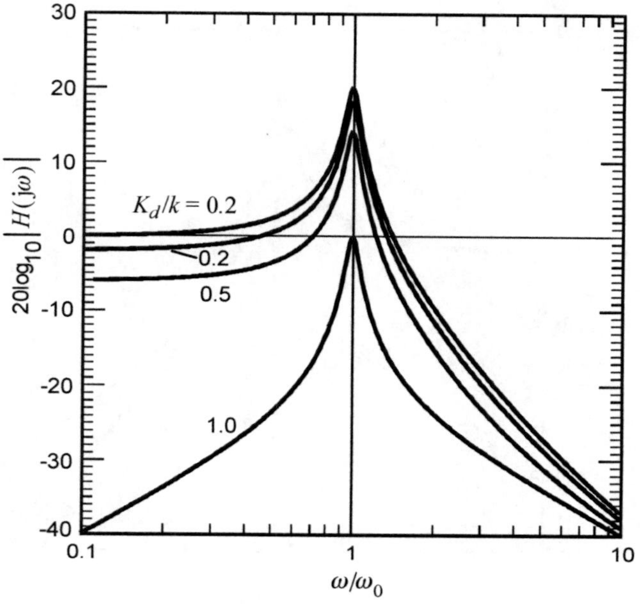

Figure 12.27 Modulus of the frequency response for force transmission into a rigid base with displacement feedback, and with the control force acting only on the base for $\zeta = 0.05$.

and $B(s)$ of Figure 12.22(b) is:

$$B(s) = K_v s \tag{12.2.108}$$

Using Equations (12.2.5) and (12.2.6) and substituting $s = j\omega$ in Equation (12.2.107), the following can be written for the force transmission frequency response with velocity feedback (force acting only on the base):

$$H(j\omega) = \frac{1 + 2j\zeta(\omega/\omega_0)(1 + K_v/C)}{1 - (\omega/\omega_0)^2 + 2j\zeta(\omega/\omega_0)} \tag{12.2.109}$$

The modulus of the closed-loop frequency-response function of Equation (12.2.109) is plotted in Figure 12.28 and shows the effect of varying the velocity feedback gain for various values of K_v/C, with $\zeta = 0.05$. It can be seen from the figure that increasing the negative feedback gain increases the isolation only slightly at high frequencies (provided the gain K_v/C is less than one).

12.2.6.1.3 Acceleration feedback

Again similar results are obtained as were obtained for the base excited system, except where the control force acts only on the base, the force transmission is given by:

$$H(s) = \frac{K_a s^2 + c_d s + k}{ms^2 + c_d s + k} \tag{12.2.110}$$

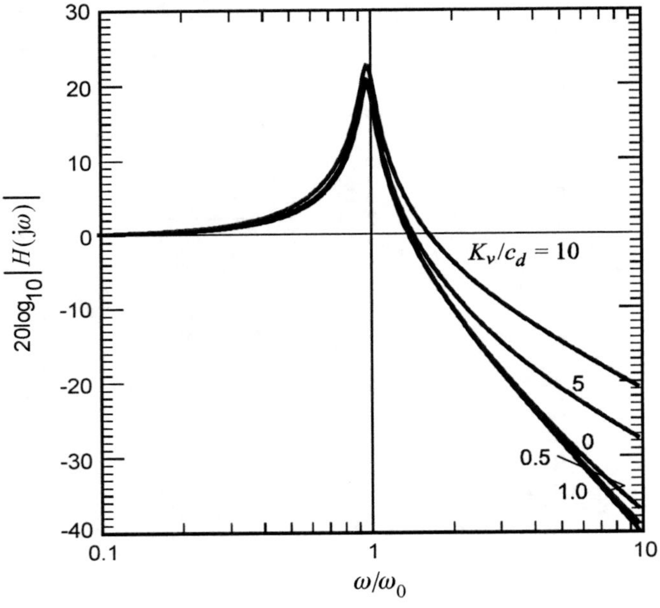

Figure 12.28 Modulus of the frequency response for the force transmission into a rigid base with velocity feedback and with the control force acting only on the base for $\zeta = 0.05$.

$B(s)$ of Figure 12.25(b) is given by:

$$B(s) = K_a s^2 \tag{12.2.110}$$

Using Equations (12.2.5) and (12.2.6) and substituting $s = j\omega$ in Equation (12.2.110), the following can be written for the force transmission frequency response with acceleration feedback (force acting only on the base):

$$H(j\omega) = \frac{1 + 2j\zeta(\omega/\omega_0) - (K_a/m)(\omega/\omega_0)^2}{1 - (\omega/\omega_0)^2 + 2j\zeta(\omega/\omega_0)} \tag{12.2.112}$$

The modulus of the closed-loop frequency-response function of Equation (12.2.112) is plotted in Figure 12.29 and shows the effect of varying the acceleration feedback gain for various values of K_a/m, with $\zeta = 0.05$. It can be seen from the figure that increasing the feedback gain increases the isolation at frequencies immediately above and immediately below the system resonance frequency with little effect at this frequency. Isolation at high frequencies is also markedly reduced.

12.2.6.1.4 Force feedback

For this case, $G_1(s)$ and $G_2(s)$ are as defined in Equations (12.2.45) and (12.2.46), and $B(s) = K_f$.

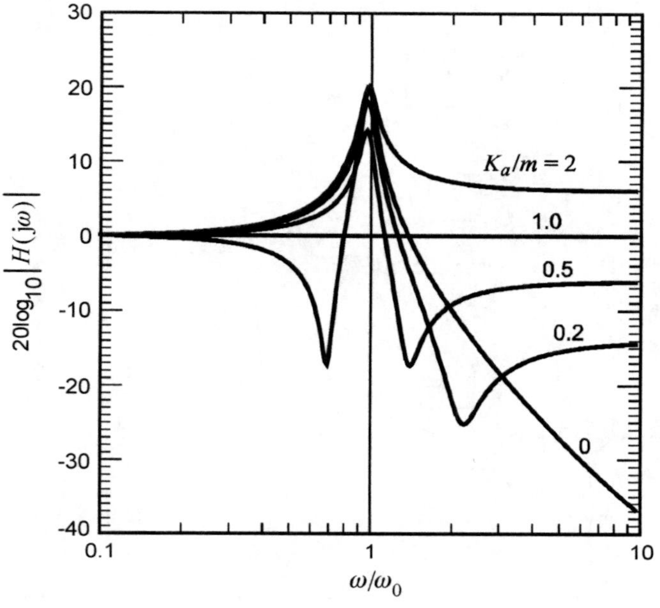

Figure 12.29 Modulus of the frequency response for force transmission into a rigid base with acceleration feedback and with the force acting only on the base for $\zeta = 0.05$.

The equations of motion for the system shown in Figure 12.30 are:

$$m\ddot{x}(t) + c_d\dot{x}(t) + kx(t) + K_f f_T(t) = f(t) \tag{12.2.113}$$

$$c_d\dot{x}(t) + kx(t) + K_f f_T(t) = f_T(t) \tag{12.2.114}$$

For the equivalent s-plane model shown in Figure 12.30:

$$F_T(s) = F(s)G_1(s)G_2(s) - F_T(s)B(s)G_1(s)G_2(s) + F_T(s)B(s) \tag{12.2.115}$$

or

$$H(s) = \frac{F_T(s)}{F(s)} = \frac{G_s(s)}{\dfrac{1}{G_1(s)} + B(s)G_2(s) - \dfrac{B(s)}{G_1(s)}} = \frac{c_d s + k}{ms^2(1 - K_f) + c_d s + k} \tag{12.2.116a–c}$$

Using Equations (12.2.5) and (12.2.6) and substituting $s = j\omega$ in Equation (12.2.116c), the following can be written for the force transmission frequency response with force feedback (force acting on the base and the mass):

$$H(j\omega) = \frac{1 + 2j\zeta(\omega/\omega_0)}{1 - (\omega/\omega_0)^2(1 - K_f) + 2j\zeta(\omega/\omega_0)} \tag{12.2.117}$$

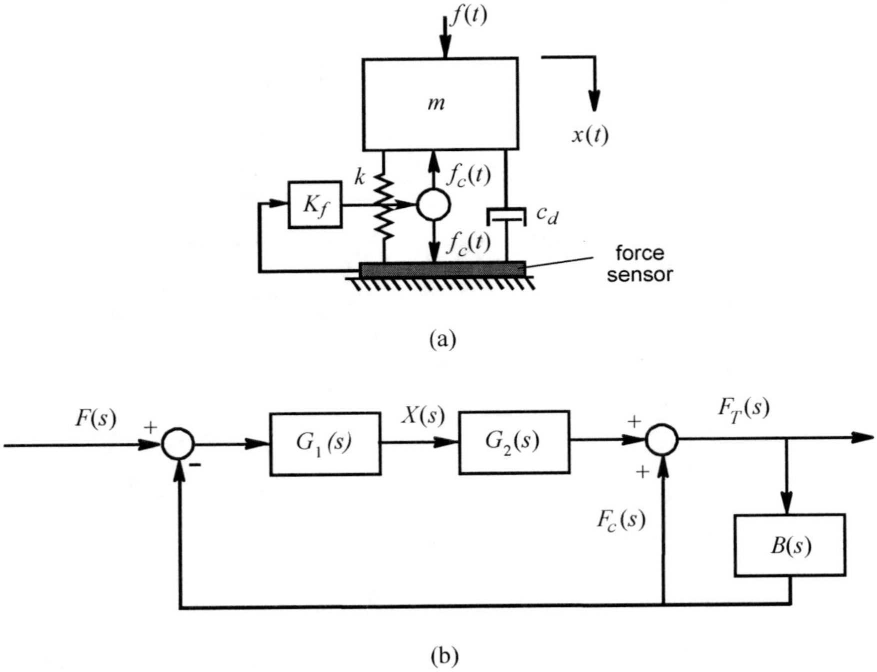

Figure 12.30 Force feedback for control of transmitted force: (a) physical system; (b) frequency domain block diagram.

The modulus of the closed-loop frequency-response function of Equation (12.2.117) is plotted in Figure 12.31 and shows the effect of varying the force feedback gain for various values of K_f/m, with $\zeta = 0.05$. It can be seen from the figure that increasing the feedback gain effectively increases the system resonance frequency and critical damping ratio which results in increases in the isolation achieved at low frequencies at the expense of reduced isolation at high frequencies. It is likely that the system would be unstable for values of the gain K_f/m greater than one. Note that K_f represents the amplitude of the control force as a fraction of the transmitted force in the absence of control.

12.2.6.1.5 Force feedback: Control force applied only to mass

Figure 12.32 shows the situation of the control force only applied to the mass. For this case, $G_1(s)$ and $G_2(s)$ are as defined in Equations (12.2.45) and (12.2.46), and $B(s) = K_f$.

Figure 12.32(a) shows the force actuator located between the vibrating mass and the passive isolation system. This is usually referred to as a 'series' system, as the active and passive elements are in series between the vibrating body and the support. It is assumed that the deformation of the force actuator is very small compared to the movement of the mass. For this case, it is clear that the force actuator must be capable of supporting the weight of the vibrating mass. This is in contrast to the situation shown in Figure 12.30(a), where it can be seen that the force actuator does not have to support the mass of the machine but it must be strong enough to overcome the stiffness of the stiffness element k so that it can develop an actuation force. This latter system is referred to as a parallel system, as the active and passive elements are mounted in parallel between the vibrating body and the support structure.

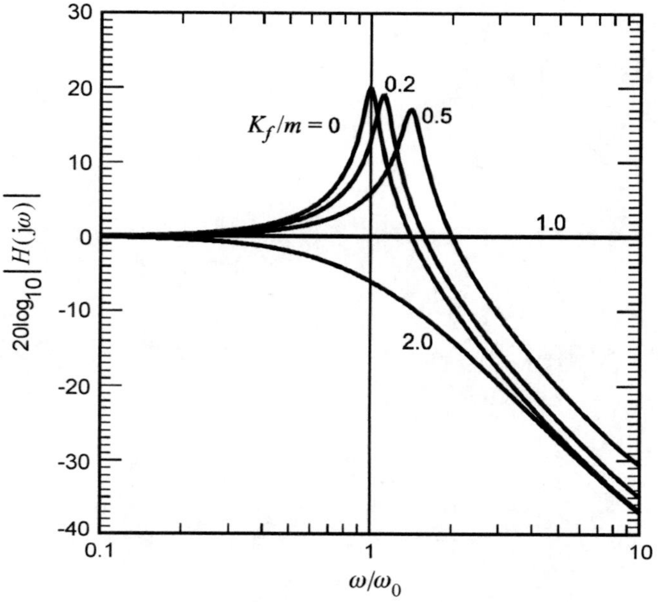

Figure 12.31 Modulus of the frequency response for force transmission into a rigid base with force feedback and the control force acting on the vibrating body and the rigid base for $\zeta = 0.05$.

In the configuration shown in Figure 12.32(b) the force is applied only to the vibrating rigid body. In dynamic terms this is identical to the configuration shown in Figure 12.32(a), provided that the force actuator is very rigid compared to the passive suspension. Such a transducer might be a piezoceramic stack or a magnetostrictive rod (see Chapter 15).

In practice, the control force shown in Figure 12.32(b) could be applied by a shaker mounted on the vibrating body of mass m and driving a second inertial mass attached only to the shaker.

The equations of motion for the system shown in Figure 12.32 are:

$$m\ddot{x}(t) + c_d\dot{x}(t) + kx(t) + K_f f_T(t) = f(t) \tag{12.2.118}$$

$$c_d\dot{x}(t) + kx(t) = f_T(t) \tag{12.2.119}$$

For the equivalent s-plane block diagram shown in Figure 12.32(c):

$$F_T(s) = F(s)G_1(s)G_2(s) - F_T(s)B(s)G_1(s)G_2(s) \tag{12.2.120}$$

$$H(s) = \frac{F_T(s)}{F_s} = \frac{G_1(s)G_2(s)}{1 + G_1(s)G_2(s)B(s)} = \frac{c_d s + k}{ms^2 + (1 + K_f)(c_d s + k)} \tag{12.2.121a–c}$$

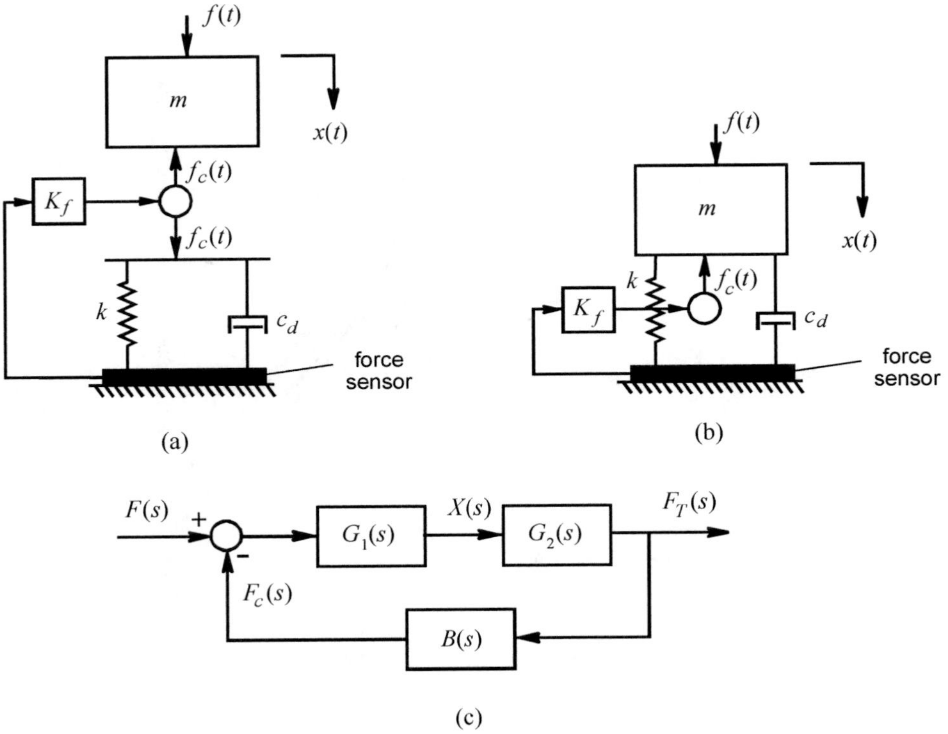

Figure 12.32 Force feedback on vibrating body for control of force transmission to a rigid foundation (sometimes a rigid mass is placed between the actuator and the passive system): (a) physical system 1; (b) physical system 2; (c) frequency domain block diagram.

Using Equations (12.2.5) and (12.2.6) and substituting $s = \mathrm{j}\omega$ in Equation (12.2.121c), the following can be written for the force transmission frequency response with force feedback (force acting only on the mass):

$$H(\mathrm{j}\omega) = \frac{1 + 2\mathrm{j}\zeta(\omega/\omega_0)}{1 - (\omega/\omega_0)^2(1 + K_f) + 2\mathrm{j}\zeta(\omega/\omega_0)(1 + K_f)} \tag{12.2.122}$$

The modulus of the closed-loop frequency-response function of Equation (12.2.122) is plotted in Figure 12.33 and shows the effect of varying the force feedback gain for various values of K_f/m, with $\zeta = 0.05$. It can be seen from the figure that increasing the feedback gain effectively decreases the system resonance frequency and increases the critical damping ratio, which results in increases in the isolation achieved at high frequencies at the expense of reduced isolation at low frequencies. It is likely that the system would be unstable for values of the gain K_f/m greater than one.

12.2.6.1.6 Force feedback: Control force applied only to support

The series system shown in Figure 12.34(a) is similar to that shown in Figure 12.32(a), except that the control force is now acting on the structure rather than the vibrating body. For this case, $G_1(s)$ and $G_2(s)$ are as defined in Equations (12.2.45) and (12.2.46), and $B(s) = K_f$.

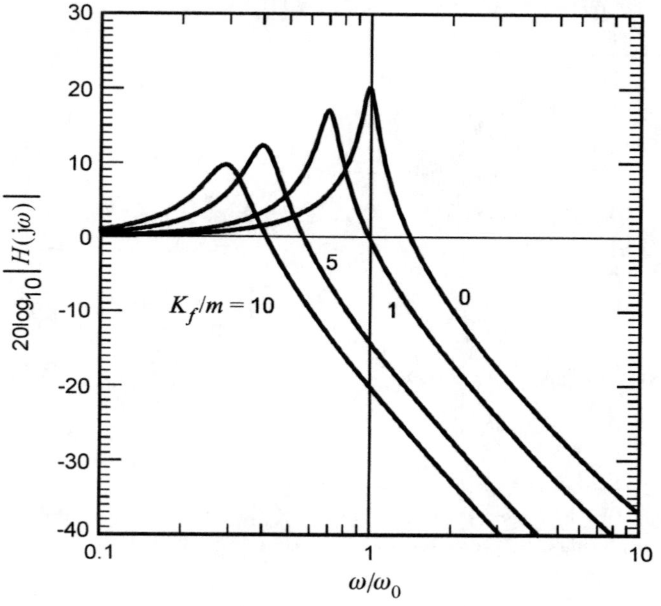

Figure 12.33 Modulus of the frequency response for force transmission into a rigid base with force feedback and the control force acting on the rigid body only.

The equations of motion for the system shown in Figure 12.34 are:

$$m\ddot{x}(t) + c_d\dot{x}(t) + kx(t) = f(t) \tag{12.2.123}$$

$$c_d\dot{x}(t) + kx(t) + K_f f_T(t) = f_T(t) \tag{12.2.124}$$

In the s-domain :

$$\frac{F_T(s)}{F(s)} = \frac{G_1(s)G_2(s)}{1 + K_f} = \frac{c_d s + k}{(ms^2 + c_d s + k)(1 + K_f)} \tag{12.2.125a,b}$$

Using Equations (12.2.5) and (12.2.6) and substituting $s = \mathrm{j}\omega$ in Equation (12.2.125b), the following can be written for the force transmission frequency response with force feedback (force acting only on the base):

$$H(\mathrm{j}\omega) = \frac{1 + 2\mathrm{j}\zeta(\omega/\omega_0)}{1 - (\omega/\omega_0)^2(1 - K_f) + 2\mathrm{j}\zeta(\omega/\omega_0)(1 - K_f)} \tag{12.2.126}$$

The modulus of the closed-loop frequency-response function of Equation (12.2.126) is plotted in Figure 12.35 and shows the effect of varying the force feedback gain for various values of K_f/m, with $\zeta = 0.05$. It can be seen from the figure that increasing the feedback gain effectively increases the system resonance frequency and decreases the critical damping ratio, which results in an increase in the isolation achieved at low frequencies at the expense of reduced isolation at high frequencies. It is likely that the system would be unstable for values of the gain K_f/m greater than one.

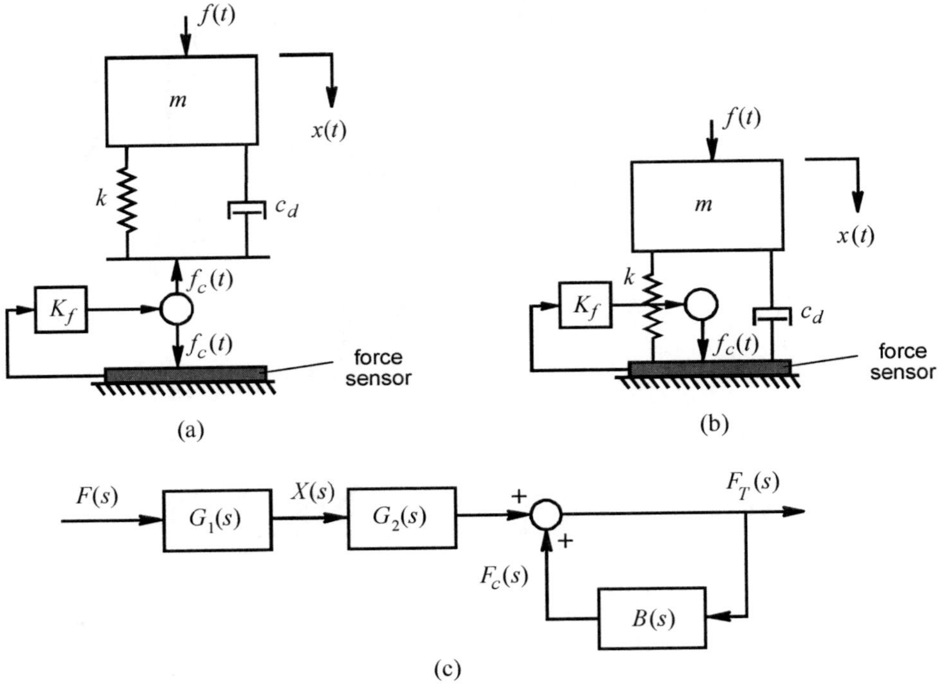

Figure 12.34 Force feedback on the support structure for control of force transmission from a vibrating rigid body into the structure: (a) physical system 1; (b) physical system 2; (c) frequency domain block diagram.

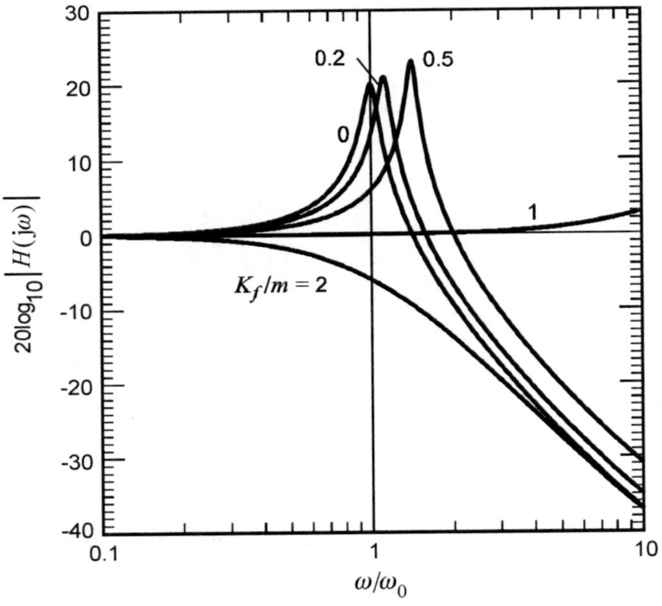

Figure 12.35 Modulus of the frequency response for force transmission into a rigid base with force feedback and the control force acting only on the rigid base for $\zeta = 0.05$.

12.2.6.1.7 Flexible support structure

Consider the simple system shown in Figure 12.36. The equation of motion for the mass m is:

$$m\ddot{x}_1(t) + c_d\dot{x}_1(t) + kx_1(t) - c_d\dot{x}_2(t) - kx_2(t) = f(t) \tag{12.2.127}$$

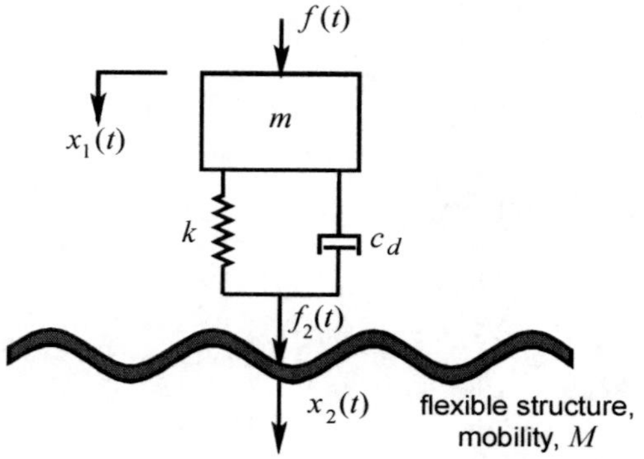

Figure 12.36 Vibration isolation of a vibrating mass from a flexible structure.

Taking Laplace transforms and zero initial conditions gives:

$$(ms^2 + c_d s + k)X_1(s) - (c_d s + k)X_2(s) = F(s) \tag{12.2.128}$$

What is desired is the transfer function between the input force $F(s)$ acting on the isolated mass, and the structural response $X_2(s)$. Thus, the quantity $X_1(s)$ needs to be eliminated from the equation. This is done by defining the structural point mobility $M(s)$ as:

$$M(s) = \frac{sX_2(s)}{F_2(s)} \tag{12.2.129}$$

where $F_2(s)$ is the force input to the structure, defined as:

$$F_2(s) = (c_d s + k)X_1(s) - (c_d s + k)X_2(s) \tag{12.2.130}$$

Substituting Equation (12.2.130) into Equation (12.2.129) and rearranging gives:

$$X_1(s) = X_2(s)\left[1 + \frac{s}{M(s)(c_d s + k)}\right] \tag{12.2.131}$$

Substituting Equation (12.2.131) into Equation (12.2.128) gives:

$$(ms^2 + c_d s + k)\left(1 + \frac{s}{M(s)(c_d s + k)}\right)X_2(s) - (c_d s + k)X_2(s) = F(s) \tag{12.2.132}$$

Rearranging gives:

$$\frac{X_2(s)}{F(s)} = \left[ms^2 + \frac{ms^3 + c_d s^2 + ks}{M(s)(c_d s + k)}\right]^{-1} \tag{12.2.133}$$

or

$$\frac{X_2(s)}{F(s)} = \frac{M(s)(c_d s + k)}{(s)(m c_d s^3 + mks^2) + ms^3 + c_d s^2 + ks} \tag{12.2.134}$$

The point mobility of a flexible structure may be modelled as (Scribner et al., 1990):

$$M(s) = \frac{sA \prod_{i=1}^{\infty} \left(1 - \frac{s}{z_i}\right)}{\prod_{i=1}^{\infty} \left(1 - \frac{s}{p_i}\right)} \tag{12.2.135}$$

where z_i and p_i are complex zeroes and poles and A is a real constant. The function in Equation (12.2.135) is guaranteed to have alternating poles and zeroes on the imaginary axis in the s-plane for the case of an undamped structure (Gevarter, 1970). On the other hand, for hysteretic structural damping, the poles and zeroes will be to the left of the imaginary axis. However, Scribner et al. (1990) show that, provided the structure is lightly damped, it is not very important from the point of view of controller design what the damping mechanism is, whether hysteretic, viscous or proportional viscous.

As Equation (12.2.134) is very complex, the control of the structural vibration by using a force driven by simple velocity, acceleration or displacement feedback is inadequate from either the performance or stability point of view. In this case, it is necessary to feed back either structural velocity, acceleration, displacement or driving force through a compensation circuit, which is more complicated than a simple gain, to the control force transducer. The number of terms used in Equation (12.2.125) to derive the optimal controller is dependent upon the upper frequency limit desired for control. In practice, it is extremely difficult to accurately determine the poles and zeroes of Equation (12.2.135), thus making it extremely difficult to design a robust controller with good performance over a wide frequency range, which is what would be desired in an ideal vibration isolation system. Such a system, which requires a detailed knowledge of the structural dynamics for its design, is referred to as a 'broadband' controller. However, as shown by Scribner et al. (1990), a narrowband controller can be designed without the need to know the details of the structural dynamics. Such a controller results in a mount with low stiffness over narrow frequency ranges, which are harmonically related. Thus, using this type of feedback controller, it is feasible to achieve good performance and stability for control of vibration transmission at a machine operating frequency and its harmonics, for example. With a suitably designed compensator circuit, it is possible for the feedback controller to track changes in the machine operating frequency (Scribner et al., 1990). The same authors also state that it is better to use force transmitted into the structure as the control variable, as better performance is possible and also it has the effect of controlling vibration transmission from the mounted machine to the structure and also from the structure to the mounted machine. On the other hand, velocity, acceleration or displacement feedback will only control vibration transmission to the part of the system on which the transducer is mounted.

It is of interest to examine the isolation system and compensator circuit used by Scribner et al. (1990). The physical system is shown in Figure 12.37(a) and the control block diagram in Figure 12.37(b).

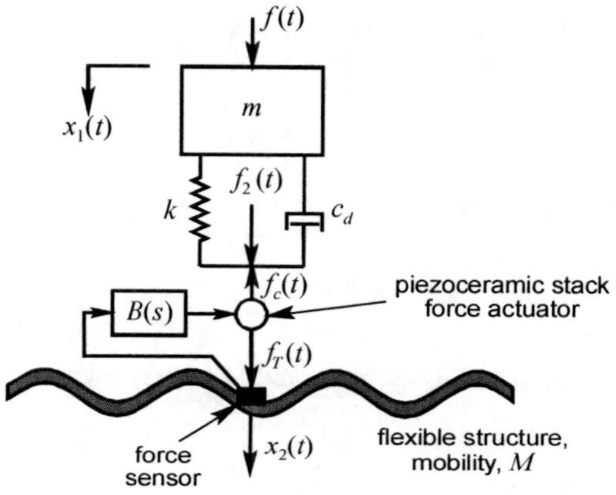

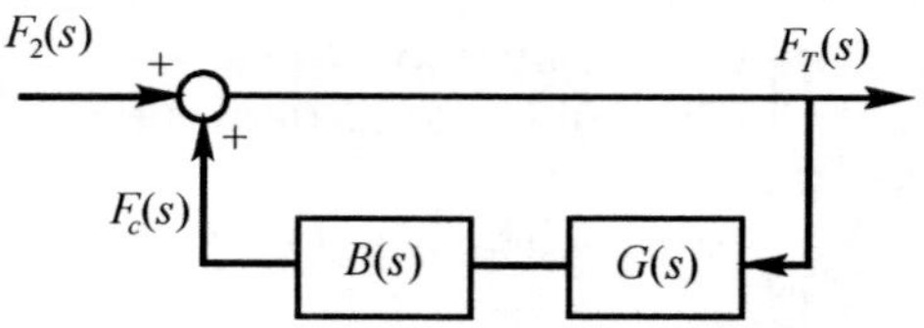

Figure 12.37 Vibration isolation of a rigid body from a flexible support structure: (a) physical arrangement; (b) control block diagram.

The quantity $G(s)$ represents the transfer function between the voltage into the piezo force generator and the voltage out of the force sensor, $F_2(s)$ is the disturbance input from the machine to the structure and $B(s)$ is the compensator transfer function.

From Figure 12.37(a) it can be seen that a series system (see Section 12.1) is being considered and from Figure 12.37(b) the relationship between the disturbance input force $F_2(s)$ (with the actuator turned off) and the structural vibration output force $F_T(s)$ is:

$$F_T(s) = F_2(s) \frac{1}{1 + B(s)G(s)} \tag{12.2.136}$$

which shows that the product, $B(s)\, G(s)$, must be large for $F_2(s)$ to have an insignificant effect on $F_T(s)$. The loop must also be stable. A compensator that satisfies these requirements is given by:

$$B(s) = \frac{1}{s^2 + 2\zeta_c \omega_c s + \omega_c^2} \tag{12.2.137}$$

where the compensator damping ζ_c must be low for high performance and the compensator resonance frequency ω_c should be set to the fundamental machine operating frequency.

However, Scribner et al. (1990) report a more clever compensation circuit (see Figure 12.38), which also allows the compensator to track the machine operating frequency ω.

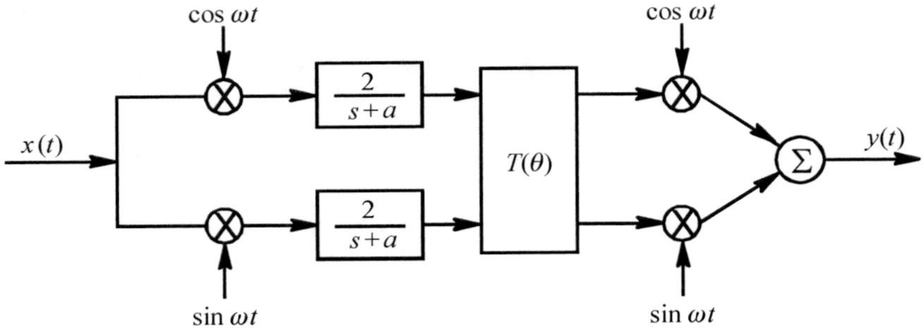

Figure 12.38 Frequency tracking analogue compensator circuit (after Scribner et al., 1990).

The signals $\sin\omega t$ and $\cos\omega t$ are derived from the fundamental machine excitation frequency ω (usually the machine rotational speed). These signals are multiplied with the input $x(t)$ and the result is low-pass filtered with a pole at $s = -a$. (Note that a should be about two orders of magnitude smaller than the machine operating frequency.) If $a = 0$, the multiplication process is equivalent to a Fourier transform of the time signal $x(t)$. As the pole a is so small in magnitude, the outputs from the low-pass filters are very close to d.c. and these two outputs are then multiplied by an orthonormal rotation matrix $T(\theta)$ given by:

$$T(\theta) = \begin{bmatrix} \cos\theta & -\sin\theta \\ \sin\theta & \cos\theta \end{bmatrix} \tag{12.2.138}$$

After the two signals have been rotated in-phase, they are converted back to the time domain by again multiplying by $\cos\omega t$ and $\sin\omega t$ respectively. The two results are then summed together to give the compensator output. The frequency response of the compensator in the s-domain is:

$$C(s) = \frac{Y(s)}{X(s)} = \frac{(s+a)\cos\theta + \omega\sin\theta}{(s+a)^2 + \omega^2} \tag{12.2.139a,b}$$

which has a zero at $s = -a - \omega\tan\theta$ and poles at $s = -a \pm j\omega$.

Note that as the system performance is best at the poles and worst at the zeroes, θ should be chosen so that the zero occurs between the fundamental and first harmonic of the machine operational frequency, and a should be made very small (100 times less than the machine operating frequency).

Although feedback control has been shown to be robust and effective over very narrow frequency ranges for controlling vibration transmission through an isolator to a flexible structure, it is very difficult to implement it over a wide frequency band. To do this, a very accurate identification of the structural dynamics is needed and as this usually changes over time, the identification should be carried out 'on-line'. This complicates the control problem to such an extent that a considerable amount of research is necessary before broadband feedback control will be practically implemented in vibration isolation systems. As feedforward systems are more effective than feedback systems for controlling periodic disturbances which can be converted to an electrical signal, the use of feedback controllers, even for narrowband control of vibration isolation systems is limited to those cases where it is not possible to measure the disturbance signal adequately. Feedback controllers are also the only option when the disturbance signal is a transient or random noise which cannot be measured sufficiently far enough ahead in time to allow a feedforward controller to produce a compensation signal.

Although feedback control over a narrow frequency range has been demonstrated for the control of vibration transmission along a single axis and at a single point into a structure, a considerable amount of research effort is needed to develop these ideas to control vibration translations along and rotations around multiple axes simultaneously and the control of vibration into several points on the structure simultaneously. Provided that all translations and rotations can be controlled at each structural excitation point, it may be possible to control each point with an independent control system, although control of all points with a single multi-channel controller may be more effective, as cross-coupling between mounts is likely to be important.

12.2.6.2 Use of an Intermediate Mass

To improve the isolation efficiency of passive systems, an intermediate rigid, massive plate can be placed between the vibrating body and the structure from which it is to be isolated. The mass of the plate needs to be sufficiently large for the additional resonance frequency introduced into the system by its presence to be well below the frequency range in which isolation is required. However, the mass should be no larger than necessary to minimise the actuation force required. Use of an active system to control the vibration of the intermediate mass can reduce the amount of mass required and can improve the low-frequency performance of the existing system (Ross et al., 1988). As pointed out by Ross, et al., it is important to control the rotations and horizontal translations, as well as the vertical translations of the mass to ensure that no vibration is transmitted to the support structure. Implementation of this idea in practice requires a minimum of six inertial shakers and six transducers, which can measure the six possible degrees of freedom (rotational and translational) of the rigid mass. In many cases the degree of freedom involving rotation about the vertical axis can be ignored.

As the intermediate rigid mass is poorly coupled to the vibration source and the rigid body, Ross et al. (1988) claim that each mount can be controlled independently of the others in a multi-mount system that is isolating a mass from a flexible structure. Whether or not this is true in practice remains to be seen. Also, the use of an intermediate mass decoupling the force actuator from the dynamics of the structure allows a more stable feedback control system to be realised. In practice, better performance has been achieved using feedback of the motion of the intermediate mass rather than the difference in motion between the mass and the structure as would be sensed by a gap sensor (Blackwood and von Flotow, 1993).

Although Ross et al. (1988) claim that the arrangement shown in Figure 12.39 can control rotational as well as translational motion, it is difficult to imagine how this may be implemented in practice, as it is clear that the shakers cannot move the mass to the left horizontally at the same time as they rotate it clockwise about its centre. In fact, horizontal motion to the left is also associated with anti-clockwise rotation. However, the addition of a third shaker shown as a dashed outline in Figure 12.39 will allow the simultaneous control of horizontal, vertical and rotational motion. A schematic layout of how a feedback controller may be implemented to control one horizontal translational, one rotational and the vertical translational degree-of-freedom is shown in Figure 12.40.

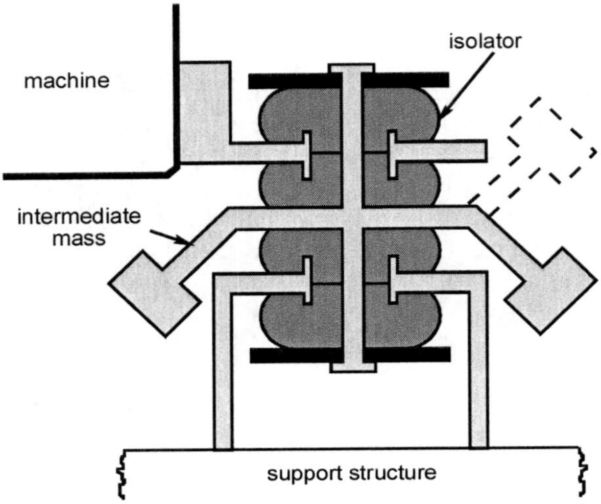

Figure 12.39 Active vibration isolation mount with an intermediate mass (after Ross et al., 1988) (The shaker shown as a dashed outline is not shown by Ross et al., and the additional damper at the bottom of the mount shown by Ross et al. has been omitted here for clarity.)

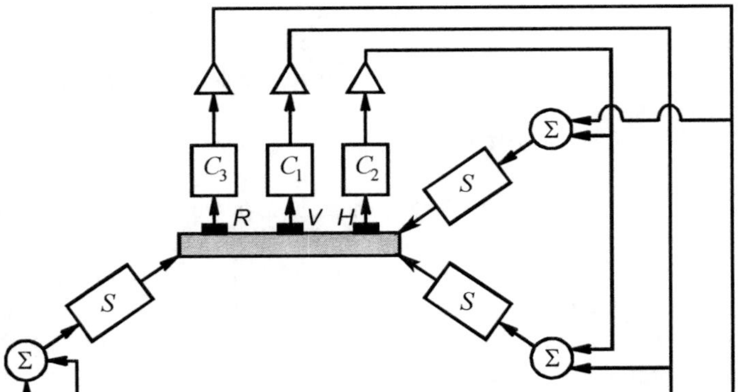

Figure 12.40 Possible implementation of a feedback system to control three degrees of freedom of the rigid mass of Figure 12.39. $\sum$ = sum, C_1, C_2, C_3 = compensation filters, ▷ = amplifier, S = inertial shaker, and V, R and H are the vertical, rotational and horizontal motion transducers respectively.

Control of five degrees of freedom with five inertial shakers would also be possible, although the arrangement would be even more complex than that shown in Figure 12.40, as the output from each sensor would be used to drive four of the five shakers, with a different set of four being driven by each transducer signal.

Stability problems associated with the arrangement shown in Figure 12.40 would be very difficult to overcome unless the interaction of the three transducer signals could be made almost negligible. This would require horizontal motion transducers that were insensitive to vertical and rotational motion. Similar requirements would be necessary for the vertical and rotational motion sensors. In addition, the frequency response of the control shakers would have to be well matched so that the vertical motion feedback does not generate horizontal or rotational motion, and vice versa for horizontal and rotational motion feedback. These problems have so far prevented practical implementation of the system to active isolation of motion other than vertical translation.

12.3 APPLICATIONS OF FEEDBACK CONTROL

In this section, various applications of feedback vibration isolation will be discussed, including the following topics:

1. Vehicle suspension systems (active and semi-active);
2. Transmission of vibration from rigidly mounted machinery to a stiff support structure;
3. Transmission of vibration from a support structure to an item of sensitive equipment mounted on it;
4. Vibration of tall buildings;
5. Isolation of machinery from flexible structures (including engine mounts);
6. Isolation of drive-train and rotor vibration from a helicopter cabin.

Many of these applications are also amenable to feedforward control, provided that a reference signal correlated with the disturbance can be obtained sufficiently far in advance for the controller to calculate the control signals so that they arrive at the error sensors at the same time as the primary disturbance used to obtain the reference signal. For periodic disturbances, feedforward control may indeed be preferable, as the reference signal need not be obtained in advance of the controller output signal to satisfy causality constraints. Feedforward control will be discussed in Sections 12.4 to 12.9.

12.3.1 Vehicle Suspension Systems

As mentioned in Section 12.1, vehicle suspension systems can be split into four categories: passive, semi-active, slow-active and fully active. Only the latter three types will be considered here. The main difference between fully active and semi-active systems is that a fully active system can put power into the suspension as well as dissipate it, whereas a semi-active system can only dissipate power, usually by varying the system damping.

Semi-active systems may be further subdivided into the continuous type and the on/off type. In the former system, the damping force is continuously variable during the time when it is being applied and in the latter case, it is constant during this time. Both types spend some of the time in the 'off' state, when no damping force is applied. The criterion determining when the damping force will be applied and when it will not be applied is known as the control law.

Similar control laws describing the control damping force are used for fully active and continuously variable semi-active systems, except that when the fully active system is supplying power, the semi-active system switches off.

Fully active systems generally use hydraulic actuators to generate the required forces in the suspension system. In practical systems, the actuator is usually placed in parallel with a passive spring so that it does not have to support the weight of the vehicle, thus greatly reducing the actuator static force requirements. The actuator needs to have a large bandwidth to minimise the transmission of high-frequency as well as low-frequency loads to the vehicle. This translates into an expensive actuator. One way around this is to use a low bandwidth actuator in series with a passive spring (slow-active system). In this system, the passive spring provides the required isolation at high frequencies while the actuator provides vibration control at low frequencies (usually below 3 Hz).

A disadvantage of high bandwidth actuators in active suspensions is the high power consumption (typically 10 kW for motor vehicles), although the power requirements for low bandwidth actuators are less (typically 2 to 3 kW) (Wendel and Stecklein, 1991). It is possible to reduce the power requirements of an active suspension by using one hydraulic pump for each suspension unit and driving all pumps with a common shaft. Thus, if an individual suspension unit must release fluid at a particular time, its pump will be displaced in the opposite direction, thus acting as a hydraulic motor which can help drive another pump which is required to supply fluid (as they are on the same drive shaft) (Wendel and Stecklein, 1991). However, the power requirements for semi-active systems are considerably smaller than either fully active system, as power is only needed to drive the sensors, damper valves and the controller, and is not needed to provide the control force. Because they only dissipate energy, semi-active systems are more failsafe and considerably less expensive than fully active systems. This is because semi-active systems do not require hydraulic pumps, accumulators, high performance filters, pipework actuators and servo-valves, just a rapidly adjustable damper (Hine and Pearce, 1988). Also to be considered are the running costs of a fully active system, which typically consumes a few kilowatts of power.

Although many journal papers concentrate on specifying the performance of an active suspension system in terms of ride quality (or the transmission of vibration from the wheel to the body), there are a number of other important parameters including those outlined below, which must also be considered.

1. Handling: this is pitch and roll of the vehicle body as a result of cornering and braking manoeuvres.
2. Road holding: this is the contact force between the tyres and the road. Clearly, this force should be as large as possible at all times.
3. Suspension travel: the allowable limit of suspension travel in any vehicle design will affect the performance achievable from the suspension systems.
4. Static deflection resulting from a variable payload.

Although passive suspensions can theoretically perform almost as well as active systems in terms of ride quality, they would do so at the expense of unacceptable performance for the four parameters discussed above. Active suspensions, on the other hand, do a much better job of optimising all performance parameters, although it is not possible to achieve the actual optimum for each parameter simultaneously. The optimisation ability of semi-active systems lies between that of passive and fully active systems.

Note that with the exception of very small damping ratios ζ, when more damping is added by a passive or on/off active suspension system, the ride quality is degraded but handling, road holding and suspension travel are improved. This is in contrast to the continuous variable semi-active or fully active systems where increasing the damping improves the ride quality (Miller, 1988a). This is caused because the damping force used in the latter systems is proportional to absolute velocity of the car body rather than the relative velocity between the car body and the wheel. Thus, the trade-off between ride comfort and handling that exists for the passive and on/off semi-active systems does not exist for the continuously variable semi-active and fully active systems. Note also that fully active systems can be designed to be self-levelling; that is, they can keep the vehicle on an even keel during cornering, etc. (Sharp and Hassan, 1986; Gao et al. 2006). In a practical system, however, some provision must be made for the wheel-to-body relative displacement feedback to be dominant at low frequencies, to ensure that the limit of the actuator stroke is not reached.

The great virtue of a fully active system is its ability to adapt to variable operating conditions, and to employ the full suspension working space (allowable suspension travel) to satisfy the ride comfort requirements and the four parameters listed previously. On the other hand, the major weakness of a suspension system that contains non-adjustable control forces is that it can be ideal for only one operating condition (or road type), and will probably be far from ideal in conditions differing from the one for which it was designed. As such, non-linear controllers that can adapt to different road conditions can offer potential advantages.

Note that only feedback (and not feedforward) control systems are generally used for vehicle systems because of the difficulty in measuring the incoming disturbance. However, some feedback systems have been designed using a preview of the incoming disturbance to improve performance (Foag, 1988; Marzbanrad et al., 2004).

Many papers have been published in the vehicle literature on 'intelligent' suspension systems. The idea of automatic (or active) suspensions for vehicles was first suggested by Federspeil-Labrosse in 1954. Since that time, servo-mechanisms have been introduced to provide the control forces (Rockwell and Lawther, 1964; Kimica, 1965), and the simple one-degree-of-freedom analytical model has been extended to a more complicated two-degrees-of-freedom model which has been analysed using state-space techniques and optimal linear control theory by Thompson (1979). Semi-active suspensions were introduced in the early 1970s (Crosby and Karnopp, 1973), and since that time there have been numerous papers published comparing the relative merits of active and semi-active suspension systems and describing a number of variations aimed to improve performance.

It is not the intention here to devote the considerable space needed for a comprehensive literature review. The work done prior to 1987 has been covered in an excellent review by Sharp and Crolla (1987). For a long time, controller design and performance predictions were based on a quarter car model which treated each wheel suspension as independent from all the others. Indeed, Sharp and Crolla (1987) showed that this simple model is capable of adequately representing the vehicle for calculations of ride comfort. To include vehicle handling as well as ride comfort, it has been assumed that it is sufficient to use suspension travel in the performance function to be minimised. However, to properly model the handling aspects, it is necessary to use more complex half or full car models (depending on whether pitch, roll or both is to be modelled) and degrees of freedom in addition to simple vertical movement at each suspension point (Todd, 1990; Shannan and Vanderploeg, 1989; Malek and Hedrick, 1985; El-Madany et al., 1987; Crolla and Abdel-Hady, 1991; Hrovat, 1997; Cao et al. 2008). Half or full car models have also been used for those cases involving a feedforward (or preview) part in the control system (Hrovat, 1991; Marzbanrad et al., 2004; Gopala Rao and Narayanan, 2008; Prabakar et al., 2009).

In almost all cases, whether they be full, half or quarter car models, each wheel suspension is modelled as a two-degrees-of-freedom system, as shown in Figure 12.41(b), where the upper mass, spring and damper represent the wheel mass and tyre stiffness. This allows the wheel dynamics to be included in the model and introduces an additional constraint of maximum tyre deflection, which is not included in the single degree-of-freedom model shown in Figure 12.41(a), in which the wheel dynamics are excluded.

Reducing tyre deflection reduces the potential for wheel hop and results in improved vehicle handling. Since single-degree-of-freedom models do not include the wheel hop constraint, the results obtained using these models give an upper limit to the possible suspension ride performance that can be achieved with the two-degrees-of-freedom model (Tseng and Hrovat, 1990; Karnopp, 1989; Thompson, 1976). However, the single-degree-of-freedom system is sometimes used to demonstrate trends for cases where only the maximum achievable ride performance is of interest; for example, when different control strategies are being compared (Shen et al., 2006; Alanoly and Sankar, 1987), or where the two-degrees-of-freedom problem becomes too complex to be manageable (Karnopp, 1989; Narayanan and Raju, 1990).

To obtain a truly accurate representation of passenger comfort, it is also necessary to include the seat dynamics to produce a three-degrees-of-freedom model as shown in Figure 12.41(c). However, this is rarely done in the literature (Hennecke and Zieglmeier, 1988; Craighead, 1988). In the latter case, the seat was replaced by an ambulance stretcher.

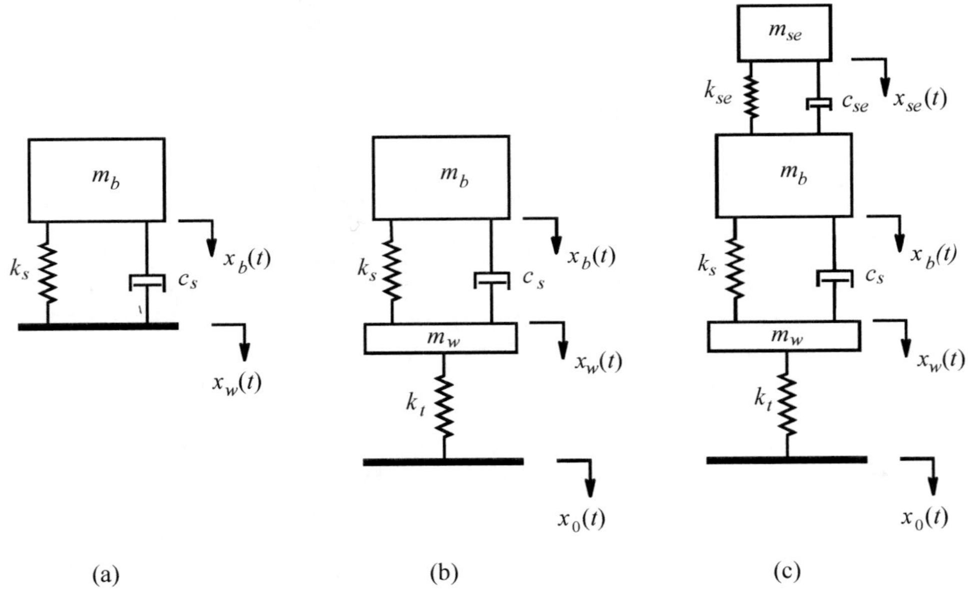

Figure 12.41 Quarter car suspension model: (a) single-degree-of-freedom model (subscript *b* refers to the vehicle body, *s* refers to the suspension, *w* to the wheel and *t* to the tyre); (b) two-degrees-of-freedom model (includes wheel dynamics); (c) three-degrees-of-freedom model (includes seat dynamics, where the subscript *se* refers to the seat).

Similar models to those shown in Figure 12.41(c) have also been used to design active suspensions for truck cabs (El-Madany and El-Razaz, 1988).

Using other approaches, systems have been developed for maximising passenger comfort using active or semi-active control of the seat suspension only (Cheok et al., 1989;

Stein, 1997). Junker and Seewald (1984) also investigated an active seat suspension system, but for a tractor.

In virtually all suspension models discussed in the literature, the tyre damping is assumed to be zero, as indicated by the absence of a damper in parallel with k_t in Figure 12.41(b). Levitt and Zorka (1991) showed that the effect of including a small non-zero ($\approx$ 0.02) value for the tyre damping ratio ζ_w is to reduce the calculated vehicle body acceleration by up to 30%. Thus, it is clear that if tyre damping is excluded from the model, the calculated car body accelerations (vertical direction) will be over-estimates of actual values, resulting in a conservative suspension design.

Automotive manufacturers have expended considerable effort in the development of both semi-active and active suspension systems. They have been incorporated into many production vehicles; however, there is still room for improvement in the development of future systems.

Applications of active suspensions to vehicles other than automobiles have also been reported in the literature. Significant research has been undertaken on active and semi-active suspensions for trains (Williams, 1985; Pollard and Simons, 1984, Okamoto et al., 1987; Gimenez et al., 1988; Goodall and Kortum, 1990; Goodall and Mei, 2006; Wang and Liao, 2009a; Wang and Liao 2009b).

Wheeled robot active suspension systems have been considered by Tani et al. (1990), various aspects of tractor active suspension systems have been considered by Stayner (1988), and Bluethmann et al. (2010) have considered active suspension systems for lunar crew rovers.

According to Crolla (1988), the procedure used to model vehicle suspensions generally involves the following stages:

1. Development of the equations of motion for the quarter car model, together with equations describing the (active or passive) force producing elements;
2. Derivation of appropriate (sometimes optimal) control laws;
3. Use of a random disturbance input at the tyre to calculate performance;
4. Repetition of the calculations for a range of road surfaces and vehicle speeds. As the suspension for most vehicles must be optimised to operate within a certain maximum vertical movement of the suspension, the performance of the various types is generally compared for a fixed maximum allowed vertical movement, usually ±(8 to 10) cm.

In the sections to follow, the first two steps of the above procedure will be outlined for active and semi-active systems using the principles of feedback control system analysis discussed in Chapter 5. However, the detailed results for various disturbance inputs will not be discussed; these are covered adequately in the various references mentioned above, so only general results will be discussed here. The discussion of vehicle suspensions will conclude with brief overviews of maglev (magnetic levitation) vehicle suspension vibration control and then sensors and actuators for suspension control.

12.3.1.1 Fully Active Suspensions

Fully active suspensions may be divided into two types: fast-active (often referred to just as fully active), where the control actuator may be mounted in parallel with a passive spring (Figure 12.42(a)) (or in some cases there may be no passive spring), and slow-active, where the control actuator is mounted in series with a passive spring (see Figure 12.42(b)). Slow-active systems will be discussed in Section 12.3.1.2. The system shown in Figure 12.42(a) can be represented schematically as shown in Figure 12.43.

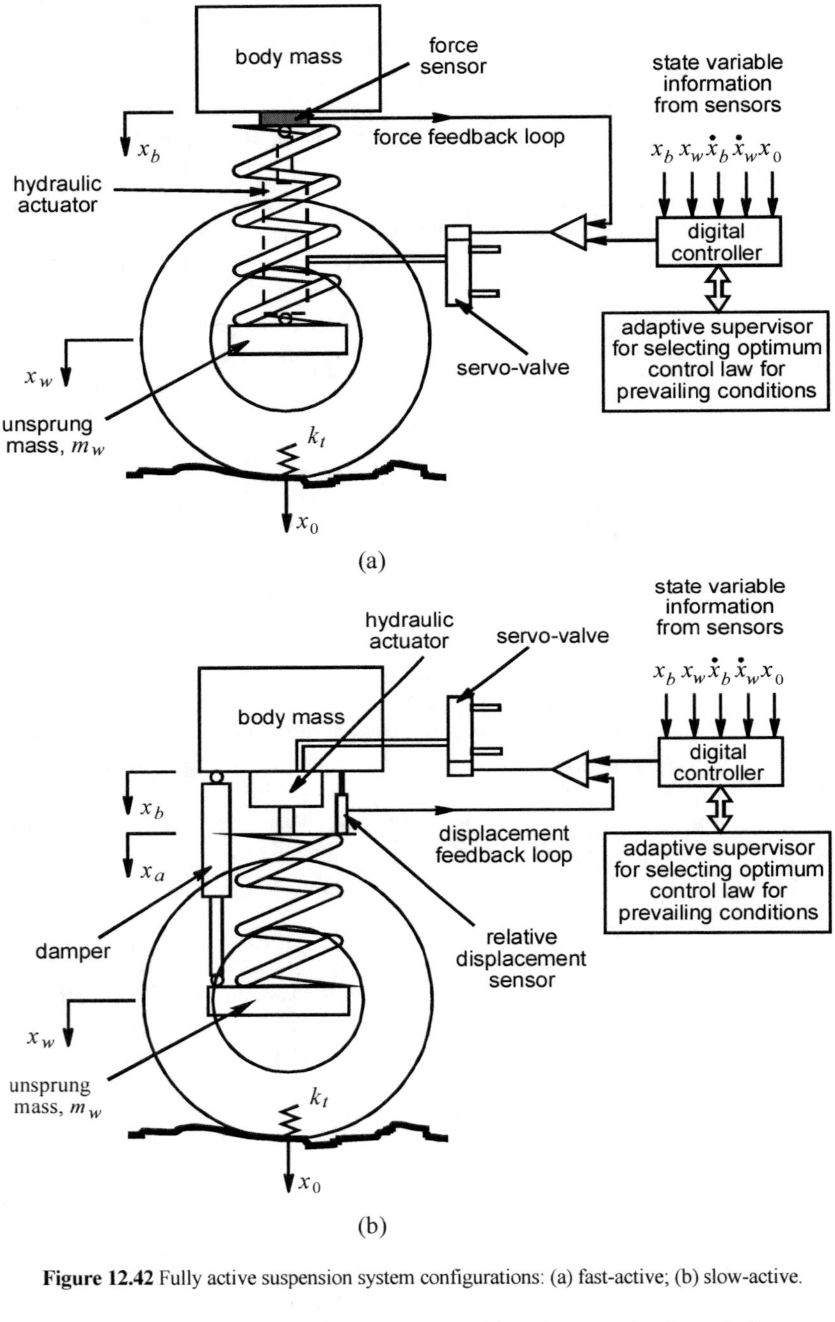

Figure 12.42 Fully active suspension system configurations: (a) fast-active; (b) slow-active.

In many technical papers, positive x is considered upwards, but similar results are obtained regardless of the convention used. The equation of motion for the vehicle body is:

$$m_b\ddot{x}_b + k_s(x_b - x_w) + f_d - f = 0 \tag{12.3.1}$$

If there is no spring in parallel with the actuator f_d, then $k_s = 0$.

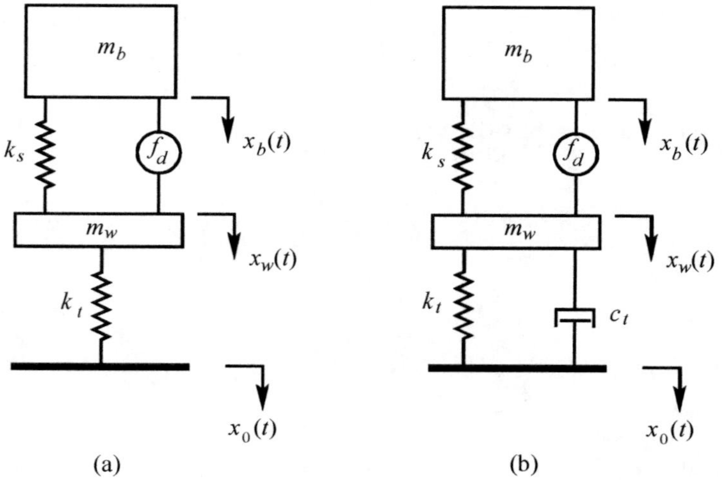

Figure 12.43 Schematic representation of a two-degrees-of-freedom quarter car model active suspension: (a) excluding tyre damping; (b) including tyre damping.

In Equation (12.3.1) and the equations to follow, the time varying nature of the quantities has been omitted to simplify the notation. The quantity x_b is the displacement of the car body, x_w is the displacement of the wheel, x_0 is the absolute displacement of the tyre surface, m_b is the mass of the quarter car body, m_w is the mass of the wheel, f is the disturbance force as a result of cornering or braking and f_d is the active control force.

The equation of motion for the wheel mass shown in Figure 12.43(a) is:

$$m_w\ddot{x}_w - c_s(\dot{x}_b - \dot{x}_w) - k_s(x_b - x_w) - f_d + c_t(\dot{x}_w - \dot{x}_0) + k_t(x_w - x_0) = 0 \tag{12.3.2}$$

In its full feedback form, the control force demand is given by Crolla (1988) as:

$$f_d = a_1(x_b - x_0) + a_2(x_w - x_0) + a_3\dot{x}_b + a_4\dot{x}_w \tag{12.3.3}$$

Due to the difficulty in measuring the relative displacement between the car body and the road, and the wheel and the road, the damping force is often calculated using only partial state feedback of the form:

$$f_d = a_3\dot{x}_b + a_4\dot{x}_w \tag{12.3.4}$$

with results obtained being almost as good as those obtained using full state feedback.

An alternative control law which is very similar to Equation (12.3.4) is given by Miller (1988a) as:

$$f_d = c_{on}\dot{x}_b + c_{off}(\dot{x}_b - \dot{x}_w) \tag{12.3.5}$$

where c_{on}, c_{off}, a_3 and a_4 are determined using optimal control theory as discussed later. The reason for expressing the control law as shown in Equation (12.3.5) is to demonstrate the similarity between fully active and semi-active systems, which will be discussed in Section 12.3.1.3. The effective damping ratios corresponding to c_{on} and c_{off} are given by Miller (1988a) as:

$$\zeta_{on} = \frac{c_{on}}{2\sqrt{k_s m_b}} \quad ; \quad \zeta_{off} = \frac{c_{off}}{2\sqrt{k_s m_b}} \tag{12.3.6a,b}$$

Thus, Equation (12.3.5) can be written as:

$$\frac{f_d}{m_b} = 2\zeta_{on}\omega_b\dot{x}_b + 2\zeta_{off}\omega_b(\dot{x}_b - \dot{x}_w) \tag{12.3.7}$$

where

$$\omega_b = \sqrt{\frac{k_s}{m_b}} \tag{12.3.8}$$

Equation (12.3.7) is identical to Equation (12.3.4), where

$$a_3 = 2m_b\omega_b(\zeta_{on} + \zeta_{off}) \tag{12.3.9}$$

and

$$a_4 = -2m_b\omega_b\zeta_{off} \tag{12.3.10}$$

The second term in Equation (12.3.7) corresponds to a passive damping element in the system having a damping ratio ζ_{off}. The first term in Equation (12.3.7) corresponds to feedback of absolute velocity (the 'sky-hook' damper discussed in Section 12.2) with a feedback gain of $2m_b\zeta_{on}\,\omega_b$, which results in a continuously variable damping force f_d.

Another control law, which is similar to that given in Equation (12.3.3), is given by Tseng and Hrovat (1990) and El-Madany (1990) as:

$$f_d = a_1(x_b - x_w) + a_2(x_w - x_0) + a_3\dot{x}_b + a_4\dot{x}_w \tag{12.3.11}$$

If the state variables $x_1 = (x_b - x_w)$, $x_2 = (x_w - x_0)$, $x_3 = \dot{x}_b$ and $x_4 = \dot{x}_w$, are then substituted, Equation (12.3.11) can be written as:

$$f_d = a_1x_1 + a_2x_2 + a_3x_3 + a_4x_4 \tag{12.3.12}$$

In terms of these state variables, the equations of motion, for the system shown in Figure 12.43, can be written in matrix form as:

$$\dot{\boldsymbol{x}} = \boldsymbol{A}\boldsymbol{x} + \boldsymbol{B}f_d + \boldsymbol{D}\dot{x}_0 \tag{12.3.13}$$

where the state vector is:

$$\boldsymbol{x} = [x_1\ x_2\ x_3\ x_4]^{\mathrm{T}} \tag{12.3.14}$$

and

$$\boldsymbol{A} = \begin{bmatrix} 0 & 0 & 1 & -1 \\ 0 & 0 & 0 & 1 \\ -k_s/m_b & 0 & 0 & 0 \\ k_s/m_w & -k_t/m_w & 0 & 0 \end{bmatrix} \tag{12.3.15}$$

$$\boldsymbol{B} = \begin{bmatrix} 0 & 0 & -1/m_b & 1/m_w \end{bmatrix}^{\mathrm{T}} \tag{12.3.16}$$

$$\boldsymbol{D} = \begin{bmatrix} 0 & -1 & 0 & 0 \end{bmatrix}^{\mathrm{T}} \tag{12.3.17}$$

If tyre damping were included, as shown in Figure 12.43(b), the matrices could be written as:

$$\boldsymbol{A} = \begin{bmatrix} 0 & 0 & 1 & -1 \\ 0 & 0 & 0 & 1 \\ -\omega_b^2 & 0 & 0 & 0 \\ \omega_b^2/\rho & -\omega_w^2 & 0 & -2\zeta_t\omega_w \end{bmatrix} \tag{12.3.18}$$

$$\boldsymbol{B} = \begin{bmatrix} 0 & 0 & -1/m_b & 1/m_w \end{bmatrix}^{\mathrm{T}} \tag{12.3.19}$$

$$\boldsymbol{D} = \begin{bmatrix} 0 & -1 & 0 & 2\zeta_t\omega_t \end{bmatrix}^{\mathrm{T}} \tag{12.3.20}$$

The variables in Equations (12.3.18) to (12.3.20) are defined as:

$$\omega_w^2 = k_t/m_w \tag{12.3.21}$$

$$\zeta_t = \frac{c_t}{2\sqrt{k_t m_w}} \tag{12.3.22}$$

$$\omega_b^2 = k_s/m_b \tag{12.3.23}$$

and $\rho = m_w/m_b$ is the ratio of wheel mass to body mass. Note that the inertia force f of Equation (12.3.1), caused by cornering or braking has been excluded from the matrix equations.

Values of m_b, m_w, k_s, ζ_t and k_t typical of automobiles are 250 kg, 25 kg, 20,000 $\mathrm{N^{-1}m}$, 0.02 and 120,000 $\mathrm{N^{-1}m}$ respectively, and the controller bandwidth typically needed is 15 Hz.

The control actuator is usually required to be sufficiently powerful to produce damping ratios given by Miller (1988a) as:

$$\frac{c_s}{2\sqrt{k_s m_b}} \geq 1 \tag{12.3.24}$$

The vehicle excitation x_0 is the apparent vertical roadway movement as a result of the vehicle's forward motion along an uneven road surface. It is generally accepted that the input excitation due to surface irregularities may be considered as a Gaussian random process with zero mean and single-sided power spectral density given by El-Madany and Abduljabbar 1989) as:

$$G_{x0}(\omega) = \frac{2\alpha V\sigma^2}{\pi} \frac{1}{\alpha^2 V^2 + \omega^2} \tag{12.3.25}$$

where σ^2 is the variance of road irregularities, α is a coefficient depending on the shape of these irregularities and V is the vehicle forward speed. For a concrete road, $\alpha = 0.15$ m^{-1} and $\sigma = 0.87 \times 10^{-2}$ m (El-Madany and Abduljabbar, 1989).

The spectrum of Equation (12.3.25) can be generated from a white noise process by using a filter of the form:

$$\dot{x}_0 = -\alpha V x_0 + \xi \tag{12.3.26}$$

where ξ is a zero mean Gaussian white noise process with the expected value given by:

$$\mathrm{E}\{\xi(t)\xi(t-\tau)\} = 2\alpha V\sigma^2 \delta(\tau) = q_f \delta(\tau) \tag{12.3.27a,b}$$

Equations (12.3.13) and (12.3.26) can be combined into a single matrix equation describing the vehicle motion in state vector form as follows:

$$\dot{\boldsymbol{x}} = \boldsymbol{A}\boldsymbol{x} + \boldsymbol{B}f_d + \boldsymbol{D}\xi \tag{12.3.28}$$

where the new state vector is:

$$\boldsymbol{x} = [x_1 \; x_2 \; x_3 \; x_4 \; x_0]^{\mathrm{T}} \tag{12.3.29}$$

and

$$\boldsymbol{A} = \begin{bmatrix} 0 & 0 & 1 & -1 & 0 \\ 0 & 0 & 0 & 1 & -1 \\ -\omega_b^2 & 0 & 0 & 0 & 0 \\ 0 & -\omega_w^2 & \omega_b^2/\rho & -2\zeta_t\omega_w & 2\zeta_t\omega_w \\ 0 & 0 & 0 & 0 & -\alpha V \end{bmatrix} \tag{12.3.30}$$

$$\boldsymbol{D} = [0 \;\; 0 \;\; 0 \;\; 0 \;\; 1]^{\mathrm{T}} \tag{12.3.31}$$

$$\boldsymbol{B} = [0 \;\; 0 \;\; 1/m_b \;\; 1/m_w \;\; 0]^{\mathrm{T}} \tag{12.3.32}$$

The optimal control force is that force which will result in the minimisation of a quadratic performance index. One performance index that considers the magnitude of the control force and all state variables, except the road surface profile x_0, is (Yue et al., 1989):

$$J = \mathrm{E}\left\{q_1 x_1^2 + q_2 x_2^2 + q_3 x_3^2 + q_4 x_4^2 + q_5 x_5^2 + r f_d^2\right\} \tag{12.3.33}$$

where $x_5 = \dot{x}_3$ and where E is the expectation operator; q_1 is the performance index that represents the importance of suspension deflection; q_2 represents tyre deflection; q_3 represents car body velocity; q_4 represents wheel velocity; q_5, which is an additional state variable, represents car body acceleration, which is related to passenger comfort; and r represents the importance of minimising the control force.

The performance index J may also be written as (Yoshimura et al., 1986; Yoshimura and Ananthanarayana, 1987):

$$J = \lim_{T \to \infty} \frac{1}{T} \int_0^T E\left[q_1 x_1^2 + q_2 x_2^2 + q_3 x_3^2 + q_4 x_4^2 + q_5 x_5^2 + r f_d^2\right] \mathrm{d}t \tag{12.3.34}$$

$$= \lim_{T \to \infty} \frac{1}{T} \int_0^T E\left[\boldsymbol{x}^{\mathrm{T}} \boldsymbol{Q} \boldsymbol{x} + \boldsymbol{F}^{\mathrm{T}} \boldsymbol{R} \boldsymbol{F}\right] \mathrm{d}t \tag{12.3.35}$$

where

$$\boldsymbol{Q} = \begin{bmatrix} q_1 & 0 & 0 & 0 & 0 \\ 0 & q_2 & 0 & 0 & 0 \\ 0 & 0 & q_3 & 0 & 0 \\ 0 & 0 & 0 & q_4 & 0 \\ 0 & 0 & 0 & 0 & q_5 \\ 0 & 0 & 0 & 0 & 0 \end{bmatrix} \tag{12.3.36}$$

$$\boldsymbol{x} = \left[x_1,\ x_2,\ x_3,\ x_4,\ x_5\ x_0\right]^{\mathrm{T}} \tag{12.3.37}$$

$$\boldsymbol{F} = f_d = a_1 x_1 + a_2 x_2 + a_3 x_3 + a_4 x_4 + a_5 x_5 + a_0 x_0 \tag{12.3.38a,b}$$

$$\boldsymbol{R} = r \tag{12.3.39}$$

The single element matrices $\boldsymbol{F}$ and $\boldsymbol{R}$ would have more elements if more control actuators were used (Yoshimura and Ananthananyana, 1987). Note that the control force f_d must be a function of all state variables included in the state vector, $\boldsymbol{x}$.

A simpler and more commonly used performance index can be derived by combining the last two terms of Equation (12.3.33) into one. This is justifiable, as the control force input can be used as a measure of ride comfort. Also, the velocities of the wheel and car body are not of much interest in terms of suspension performance. Thus, Equation (12.2.33) can be simplified to (El-Madany and Abduljabbar, 1989; Tseng and Hrovat, 1990):

$$J = \mathrm{E}\left\{q_1 x_1^2 + q_2 x_2^2 + r f_d^2\right\} \tag{12.3.40}$$

The problem of minimising J is often referred to as the linear quadratic regulator problem (or LQ control), and the required values of a_1, a_2, a_3, a_4 and a_0 to minimise J can be obtained by using optimal control theory, which involves solving the associated Riccati equation (see Chapter 5). Following the standard procedure (Tseng and Hrovat, 1990), the following can be written for the solution for a_1, a_2, a_3, a_4 and a_0:

$$[a_1 \; a_2 \; a_3 \; a_4 \; a_0]^{\mathrm{T}} = \boldsymbol{R}^{-1}\boldsymbol{B}^{\mathrm{T}}\boldsymbol{P} \tag{12.3.41}$$

giving the optimal control force, $f_d^* = a_1^* x_1 + a_2^* x_2 + a_3^* x_3 + a_4^* x_4 + a_0^* x_0$, as:

$$f_d^*(t) = \boldsymbol{R}^{-1}\boldsymbol{B}^{\mathrm{T}}\boldsymbol{P}\boldsymbol{x}(t) \tag{12.3.42}$$

where the asterisk denotes an optimal quantity and where the vector $\boldsymbol{x}$ is defined by Equation (12.3.29), the vector $\boldsymbol{B}$ is defined by Equation (12.3.32), and where $\boldsymbol{P} = \boldsymbol{P}^{\mathrm{T}}$ is the symmetric positive definite solution of the following algebraic Riccati equation:

$$\boldsymbol{PA} + \boldsymbol{A}^{\mathrm{T}}\boldsymbol{P} - \boldsymbol{PBR}^{-1}\boldsymbol{B}^{\mathrm{T}}\boldsymbol{P} + \boldsymbol{Q} = 0 \tag{12.3.43}$$

where the matrix $\boldsymbol{A}$ is defined by Equation (12.3.30) and

$$\boldsymbol{Q} = \begin{bmatrix} q_1 & 0 & 0 & 0 & 0 \\ 0 & q_2 & 0 & 0 & 0 \\ 0 & 0 & 0 & 0 & 0 \\ 0 & 0 & 0 & 0 & 0 \\ 0 & 0 & 0 & 0 & 0 \end{bmatrix} \tag{12.3.44}$$

Sometimes it is not feasible to directly measure all of the state feedback variables needed to evaluate the performance function J. Assuming that it is possible to measure at least one variable, it is possible to estimate the others using a Kalman filter (see Chapter 5), as follows. Equation (12.3.28) may be rewritten as:

$$\frac{\partial \hat{\boldsymbol{x}}}{\partial t} = \boldsymbol{A}\hat{\boldsymbol{x}} + \boldsymbol{B}f_d + \boldsymbol{L}\left(\boldsymbol{x}_m - \boldsymbol{C}_m\hat{\boldsymbol{x}}\right) \tag{12.3.45}$$

where the ^ indicates an estimated quantity and where $\boldsymbol{x}_m$ is the vector of measured variables. Usually, this vector contains only one variable, which is the suspension deflection. Thus,

$$\boldsymbol{x}_m = \left[(x_1 + v_1) \; 0 \; 0 \; 0 \; 0\right]^{\mathrm{T}} = \boldsymbol{C}_m\boldsymbol{x} + v_1 \tag{12.3.46a,b}$$

where

$$\boldsymbol{C}_m = [1,\; 0,\; 0,\; 0 \;\; 0] \tag{12.3.47}$$

and where v_1 represents measurement noise, which is assumed white and which has an expected value of:

$$\mathrm{E}\left\{v_1(t)\, v_1(t-\tau)\right\} = \mu\delta(\tau) \tag{12.3.48}$$

where δ is the Dirac delta function and μ is a constant.

The optimal control problem is to choose the control force gains, a_1, a_2, a_3, a_4 and a_0 (see Equation (12.3.38)) and the Kalman filter gains $\boldsymbol{L} = [h_1, h_2, h_3, h_4, h_0]^{\mathrm{T}}$ to minimise the performance index J. This is referred to as linear quadratic Gaussian control (or LQG control), and the overall compensator transfer function (in the frequency domain) between the active force $f_d(\mathrm{j}\omega)$ and the sensor output $x_m(\mathrm{j}\omega)$ is:

$$\frac{f_d(\mathrm{j}\omega)}{x_m(\mathrm{j}\omega)} = -\boldsymbol{G}\left(\boldsymbol{I} - \boldsymbol{A} + \boldsymbol{B}\,\boldsymbol{G} + \boldsymbol{L}\,\boldsymbol{C}_m\right)^{-1}\boldsymbol{L} \tag{12.3.49}$$

where $\boldsymbol{G} = (a_1, a_2, a_3, a_4, a_0)^{-1}$ and I is the identity matrix.

This LQG design procedure is guaranteed to produce a stable closed-loop controller design if the system response is observable by x_m and controllable by f_d. The LQG system applied to a typical vehicle suspension has a performance characteristic approaching that for full feedback control.

The optimal control force is found by replacing $\boldsymbol{x}(t)$ in Equation (12.3.42) with $\hat{\boldsymbol{x}}(t)$. Thus,

$$f_d^*(t) = \boldsymbol{R}^{-1}\boldsymbol{B}^{\mathrm{T}}\boldsymbol{P}\hat{\boldsymbol{x}}(t) \tag{12.3.50}$$

where $\boldsymbol{P}$ is the solution of Equation (12.3.43).

The Kalman filter gains $\boldsymbol{L}$ are found from:

$$\boldsymbol{L} = \boldsymbol{P}_f\boldsymbol{C}_m^{\mathrm{T}}\mu^{-1} \tag{12.3.51}$$

where $\boldsymbol{P}_f$ is the solution of the following Riccati equation:

$$\boldsymbol{P}_f\boldsymbol{A} + \boldsymbol{A}^{\mathrm{T}}\boldsymbol{P}_f - \boldsymbol{P}_f\boldsymbol{C}_m{}^{\mathrm{T}}\mu^{-1}\boldsymbol{C}_m\boldsymbol{P}_f + \boldsymbol{D}\,\boldsymbol{Q}_f\boldsymbol{D}^{\mathrm{T}} = 0 \tag{12.3.52}$$

where $\boldsymbol{D}$ is defined by Equation (12.3.31), $\boldsymbol{A}$ by Equation (12.3.30), $\boldsymbol{C}_m$ by Equation (12.3.47), and μ by Equation (12.3.48).

The quantity $\boldsymbol{Q}_f$ is a diagonal matrix containing a single non-zero element in the fifth row. That is,

$$\boldsymbol{Q}_f = \begin{bmatrix} 0 & 0 & 0 & 0 & 0 \\ 0 & 0 & 0 & 0 & 0 \\ 0 & 0 & 0 & 0 & 0 \\ 0 & 0 & 0 & 0 & 0 \\ 0 & 0 & 0 & 0 & q_f \end{bmatrix} \tag{12.3.53}$$

where q_f is defined by Equations (12.3.26) and (12.3.27).

The layout of the optimal controller and Kalman filter is shown in block diagram form in Figure 12.44. As discussed in Chapter 5, this type of approach is more prone to stability problems than if a state estimator is not used and less state variables are included in the system model.

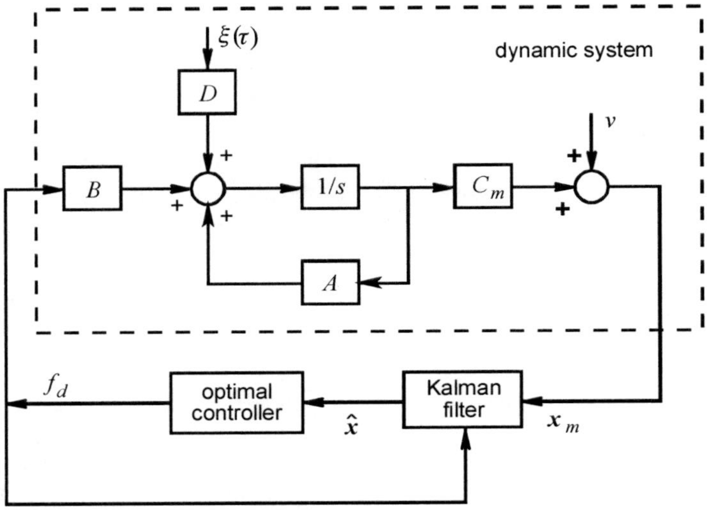

Figure 12.44 Block diagram of an active suspension with partial state feedback.

Another technique used to determine the optimum control force demand is referred to as model reference adaptive control (MRAC), in which the disturbance and vibration of the vehicle is reduced to a level that is determined by an ideal conceptual suspension system-a reference model. This is done by comparing the actual suspension response with the model response to the same impact, and adjusting the force signal to the control actuator to minimise the difference between the two. Sunwoo et al. (1991) used this technique with a 'sky-hook' damper (see Section 12.2) as the reference model. The two advantages of using MRAC over the standard active control method are that it adapts well to variations in suspension characteristics and the sprung load, and the characteristics of the reference model can be changed quickly to suit the desired ride condition by simply changing the computer algorithm used to determine the required actuator input.

Yet another innovative approach to a fully active system is the use of MRAC concepts with a neural network controller that can establish its own control laws without requiring a complete knowledge of the system dynamics (Cheok and Huang, 1989). Although this approach is more difficult to implement, it promises superior performance over other active systems.

The stability of control systems with various linear quadratic (LQ) control laws determined using optimal control theory is discussed by Ulsoy and Hrovat (1990) and Yue et al. (1989), who compare the relative performance of various control laws including full state feedback, feedback of absolute velocity of the vehicle body ('sky-hook' damper as discussed in Section 12.2) and a linear quadratic Gaussian (LQG) regulator using suspension deflection $(x_b - x_w)$ as the control variable to be minimised. They found that of the latter three types, the LQG regulator gave the best trade-off between ride quality, suspension travel and road holding constraints.

Suspension systems are inherently non-linear. Hrovat (1997), who presented a review of the application of optimal control techniques (focussing on linear quadratic controllers) to active suspensions, pointed out that although low-order linear models can be adequate for suspension control synthesis and analysis, more complex models would be required in order to fully understand the effects of non-linearities. Thus, there has been a focus towards developing controllers that can adequately account for non-linearities due to force saturation, force non-linearity and model uncertainty (Cao et al., 2011).

Due to its ability to handle parametric and modelling uncertainties and external disturbances in non-linear systems, sliding mode control is one technique that has shown promise here. With this technique, the controller applies a discontinuous feedback control signal, whereby the feedback gains switch between values depending upon the current position in the state space.

Yoshimura et al. (2001) applied sliding mode control to a quarter car model, finding that it compared favourably to a linear active suspension based on LQ control. Sam et al. (2004), also modelling a quarter car suspension, and compared a proportional integral sliding-mode controller to an LQG controller, finding the sliding mode controller to perform better and be more robust. Sliding-mode controllers traditionally suffer from chattering induced by high-speed switching of the controller signal. Chattering can potentially damage the suspension components and is therefore undesirable. Yagiz et al. (2000) developed a non-chattering sliding mode controller and later incorporated a fuzzy logic controller (Yagiz et al., 2008) in order to improve performance. Application of their control methodologies to a half car suspension showed that ride comfort could be improved with the inclusion of a fuzzy logic component. It was noted that the required power consumption and suspension working limits were realistically feasible.

Sliding-mode controllers have also been integrated with fuzzy logic controllers to overcome the disadvantages of traditional sliding-mode controllers. Cao et al. (2008) discuss this integration in their review of 'intelligent' methodologies for adaptive active suspension control. As well as fuzzy logic controllers, Cao et al. also comprehensively review neural networks and genetic algorithms and combinations of these three methods for dealing with non-linearities and conflicting performance metrics (which will be discussed in the next paragraph) in active suspension control. Due to space limitations, these techniques will not be discussed further here; however, they are and will continue to be important in active suspension control research, and thus it is recommended that the reader consult Cao et al.'s paper if they have interests in computational intelligence methodologies for active suspension control.

As discussed earlier, not only ride quality, but other performance metrics such as handling, road holding and suspension travel need to be considered in the design of an active suspension. These metrics will often be contradictory, for example, in order to improve ride quality (minimise body acceleration), greater suspension limits are required. Traditional active suspension designs were conservative, ensuring that suspension travel limits were not reached, even in the roughest conditions. It would, however, be better to focus on ensuring that suspension limits are not reached only when near the suspension limits (stiff suspension), and ensuring that body acceleration is minimised at other times (soft suspension). Non-linear techniques have been developed specifically for this purpose. For example, Lin and Kanellakopoulos (1997) designed a non-linear back-stepping controller in which the magnitude of the suspension travel governed the effective bandwidth of a non-linear filter with a wheel displacement input. This enabled the suspension stiffness to be adapted to road conditions such that ride quality and suspension travel performance metrics could be emphasised as appropriate. Fialho and Balas (2002) considered linear parameter-varying (LPV) gain-scheduling combined with non-linear back-stepping techniques, their motivation being that LPV has advantages over back-stepping techniques, including the requirement for less states to be fed back, more rapid switching between suspension settings, and the possibility of incorporation of robustness and sensor noise into problem formulation.

$\mathcal{H}^\infty$ control techniques can be designed to be robust and to attenuate disturbances effectively and thus have generated much interest in active suspension control. Some of these techniques use fixed performance metric weightings when there are no road conditions available. Chen and Guo (2005) thus suggest a constrained $\mathcal{H}^\infty$ control scheme in which improved ride quality could be achieved by capturing hard constraints, which are handled in a natural and explicit fashion, and constrained variables are allowed to remain free, as long as they remain within given bounds.

12.3.1.1.1 Preview control

The improved performance as a result of including a feedforward part (or preview control) with the feedback controller has been investigated by a number of authors. It was first proposed by Bender in 1968, who used a sensor to detect the road profile ahead of the vehicle to provide a signal to a controller designed to minimise the effect of the profile (or disturbance) input at the tyre/road contact point. Bender used Weiner filter theory to find the optimum controller, employing similar techniques to those outlined in Section 7.2. Similar techniques were used by Sasaki et al. (1976) and Iwata and Nakano (1976) on a two-dimensional model. The main difficulty with this approach is the implementation of the optimal controller, which is not guaranteed to be causal. This type of problem was discussed in relation to controlling random noise in ducts in Section 7.2.

Tomizuka (1975, 1976) derived an optimal feedback controller that had a feedforward component consisting of a weighted sum of future road elevations taken over a preview distance; however, the application of this theory in practice was hindered by the complicated mathematics, requiring solution of a large set of regressive matrix equations to determine the optimal controller. Thompson et al. (1980) approached the problem as a simple, linear quadratic regulator with a vector disturbance input. The preview sensor only senses the distance to the road at a point ahead of the wheel; thus, to determine the disturbance input, an estimate of the body displacement is necessary. This can be obtained by double integration of the signal from an accelerometer mounted on the body using integrators with feedback compensation to minimise low-frequency noise problems (Thompson et al., 1980). The filtering of the preview sensor to provide the delayed signal for use in the controller could be achieved using an FIR filter with an adaptive algorithm for updating the weights, although this technique has not yet been reported in the literature. Thompson et al. (1980) used a multi-stage fixed delay line, and relied on changing the point at which the input is fed to the filter from the first stage to a later stage as the vehicle speed increases (to account for the shorter required delay).

Foag (1988) showed that the use of preview significantly reduces the required control forces and actuator bandwidth compared to active systems without preview. As an example of a quarter car model analysis, the work of Yoshimura and Ananthanarayana (1991) will now be considered. They approached the preview problem slightly differently to their predecessors by including it in an extended Kalman filter, thus allowing a controller to be developed that would work for an irregular road surface. The dynamic single wheel model with preview control is illustrated in Figure 12.45.

The analysis is similar to that outlined previously for the Kalman filter, except for the presence of an additional measured input in the measured state vector $\boldsymbol{x}_m$, which is the quantity $x_b - x_{02}$. Equations (12.3.45) and (12.3.46) still describe this system except that the variables are more complex, and are defined as follows:

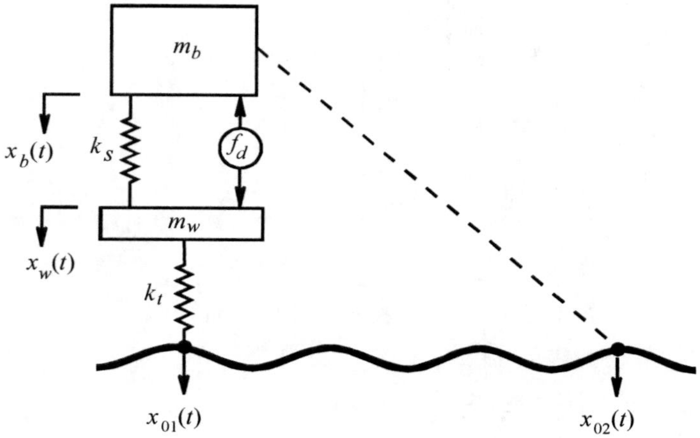

Figure 12.45 Single wheel model with preview.

$$\hat{\boldsymbol{x}} = [\hat{x}_1 \;\; \hat{x}_2 \;\; \hat{x}_3 \;\; \hat{x}_4 \;\; \hat{x}_{01} \;\; \hat{x}_{02}]^{\mathrm{T}} \tag{12.3.54}$$

$$\boldsymbol{x_m} = [x_{m_1} \;\; x_{m_2}]^{\mathrm{T}} \tag{12.3.55}$$

$$\boldsymbol{C_m} = \begin{bmatrix} 1 & 0 & 0 & 0 & 0 & 0 \\ 1 & 1 & 0 & 0 & 1 & -1 \end{bmatrix} \tag{12.3.56}$$

$$\boldsymbol{A} = \begin{bmatrix} 0 & 0 & 1 & -1 & 0 \\ 0 & 0 & 0 & 1 & -1 \\ -\omega_b^2 & 0 & 0 & 0 & 0 \\ 0 & -\omega_w^2 & -\omega_b^2/\rho & -2\zeta_t\omega_w & 2\zeta_t\omega_w \\ 0 & 0 & 0 & 0 & \alpha V \\ 0 & 0 & 0 & 0 & 0 \end{bmatrix} \tag{12.3.57}$$

$$\boldsymbol{B} = [0 \;\; 0 \;\; 1/m_b \;\; 1/m_w \;\; 0 \;\; 0]^{\mathrm{T}} \tag{12.3.58}$$

$$\boldsymbol{v} = \begin{bmatrix} v_1 \\ v_2 \end{bmatrix} \tag{12.3.59}$$

$$\boldsymbol{L} = \begin{bmatrix} l_{11} & l_{12} \\ l_{21} & l_{22} \\ l_{31} & l_{32} \\ l_{41} & l_{42} \\ l_{011} & l_{012} \\ l_{021} & l_{022} \end{bmatrix} \tag{12.3.60}$$

The elements of the measurement noise vector $\boldsymbol{v}$ (which replaces v_1 in Equation (12.3.46b) have expected values of:

$$\left.\begin{aligned} \mathrm{E}\left\{v_1(t)\, v_1(t-\tau)\right\} &= \mu_1 \delta(\tau) \\ \mathrm{E}\left\{v_2(t)\, v_2(t-\tau)\right\} &= \mu_2 \delta(\tau) \\ \mathrm{E}\left\{v_1(t)\, v_2(t-\tau)\right\} &= 0 \end{aligned}\right\} \tag{12.3.61a–c}$$

The surface profile at locations 1 and 2 can be expressed as follows:

$$\dot{x}_{01} = -\alpha V x_{01} + xi_1 \tag{12.3.62}$$

$$\dot{x}_{02} = -\alpha V x_{02} + \xi_2 \tag{12.3.63}$$

where the expected values of ξ are:

$$\left.\begin{aligned} \mathrm{E}\left\{\xi_1(t)\xi_1(t-\tau)\right\} &= 2\alpha V\sigma^2\delta(\tau) = q_f\delta(\tau) \\ \mathrm{E}\left\{\xi_2(t)\xi_2(t-\tau)\right\} &= 2\alpha V\sigma^2\delta(\tau) = q_f\delta(\tau) \\ \mathrm{E}\left\{\xi_1(t)\xi_2(t-\tau)\right\} &= 2\alpha V\sigma^2\delta(\tau-p) = q_f\delta(\tau-p) \end{aligned}\right\} \tag{12.3.64a–c}$$

where p is the preview time or the time taken for the vehicle to travel from x_{01} to x_{02}.

The optimal control force coefficients, a_1, a_2, a_3, a_4, a_{01} and a_{02}, are found as before, using Equation (12.3.42) with $\boldsymbol{R}$ defined by Equation (12.3.39) and $\boldsymbol{B}$ defined by Equation (12.3.58). The symmetric matrix $\boldsymbol{P}$ is the solution of Equation (12.3.43) with the matrix $\boldsymbol{A}$ given by Equation (12.3.57) and the matrix $\boldsymbol{Q}$ given by:

$$\boldsymbol{Q} = \begin{bmatrix} q_1 & 0 & 0 & 0 & 0 & 0 \\ 0 & q_2 & 0 & 0 & 0 & 0 \\ 0 & 0 & 0 & 0 & 0 & 0 \\ 0 & 0 & 0 & 0 & 0 & 0 \\ 0 & 0 & 0 & 0 & 0 & 0 \\ 0 & 0 & 0 & 0 & 0 & 0 \end{bmatrix} \tag{12.3.65}$$

The Kalman filter gains are found by using Equation (12.3.51), where Equation (12.3.56) is used for $\boldsymbol{C}_m$ and the following is used for $\boldsymbol{\mu}$:

$$\boldsymbol{\mu} = [\mu_1 \ \mu_2]^{\mathrm{T}} \tag{12.3.66}$$

$\boldsymbol{P}_f$ is a solution of Equation (12.3.52) with $\boldsymbol{A}$, $\boldsymbol{C}_m$ and $\boldsymbol{\mu}$ defined by Equations (12.3.57), (12.3.56) and (12.3.62) respectively. The vector $\boldsymbol{D}$ in Equation (12.3.52) is given by:

$$\boldsymbol{D} = [0 \ 0 \ 0 \ 0 \ 1 \ 1]^{\mathrm{T}} \tag{12.3.67}$$

The matrix $\boldsymbol{Q}_f$ in Equation (12.3.52) is defined for this case as:

$$\boldsymbol{Q}_f = \begin{bmatrix} 0 & 0 & 0 & 0 & 0 & 0 \\ 0 & 0 & 0 & 0 & 0 & 0 \\ 0 & 0 & 0 & 0 & 0 & 0 \\ 0 & 0 & 0 & 0 & 0 & 0 \\ 0 & 0 & 0 & 0 & q_f & 0 \\ 0 & 0 & 0 & 0 & 0 & q_f \end{bmatrix} \tag{12.3.68}$$

where q_f is defined by Equations (12.3.62), (12.3.63) and (12.3.64).

In practice, the preview signal can be derived from an ultrasonic, infrared or radar device. However, one of the main difficulties associated with the practical application of preview control is locating a sensor sufficiently far in front of the vehicle to provide sufficient delay at high vehicle speeds to enable the controller to develop the low-frequency actuator signals it needs. One way around this problem is to use preview only for the back wheels, with the disturbance input to the front wheels acting as the preview input to the controller for the back wheels (Hrovat, 1991). This is not as effective as providing it to both front and back wheels but it offers significant performance advantages over active suspensions with no preview. Analysis of such a system required a more complex two-dimensional, four-degree-of-freedom half vehicle model compared to the one-dimensional two-degrees-of-freedom quarter vehicle model used by Thompson et al. (1980).

Amongst many others, Thompson and Pearce (1998) modelled a half car model and solved a preview optimisation problem for a road input modelled as a unit step and then modelled as random irregularities. Marzbanrad et al. (2004) used a half car model and modelled road roughness as white noise. They incorporated a preview sensor on the front bumper and assumed that only relative velocities in the front and rear suspensions can be measured (that is, that not all states variable are measured), and considered the effect of noise in measurement of the state variables. Using stochastic optimal control theory and estimating their full state model using an observer similar to a Kalman filter, they found that preview control successfully reduced their performance indices. Marzbanrad et al. (2002) also applied preview control to a full car model.

Gopala Rao and Narayanan (2008) considered preview control of a half car model, but instead of assuming a linear model, they modelled the spring component of their suspension as a hysteretic type non-linear element (using a Bouc-Wen model). They note (in agreement with previous authors) that, although increases in performance are initially observed as preview distance increases, saturation beyond a certain preview distance is reached. Since control effort also increases with preview distance, the preview distance should therefore be carefully chosen.

The common conclusion of work reported so far has been that suspensions with preview control outperform similar systems without preview control.

12.3.1.2 Slow-Active Systems

As mentioned in the previous section, the actuator bandwidth required for a fully active suspension is typically 15 Hz, and achieving this is expensive. Considerable interest has developed in ways of reducing this bandwidth requirement without reducing performance. One idea is to use the actuator in series with passive springs so that the actuator controls low-frequency (less than 3 Hz) motion while the passive spring controls the higher frequency motion in the range where the actuator is essentially very stiff. Such a system is illustrated in Figure 12.42(b) and shown schematically in Figure 12.46 (see also Sharp and Hassan, 1987b). Here, the description concentrates on a quarter car model, although a more complex half or full car model could be used if more rigorous detail of the handling aspects were desired (see, for example, El-Madany and Qarmoush, 2011).

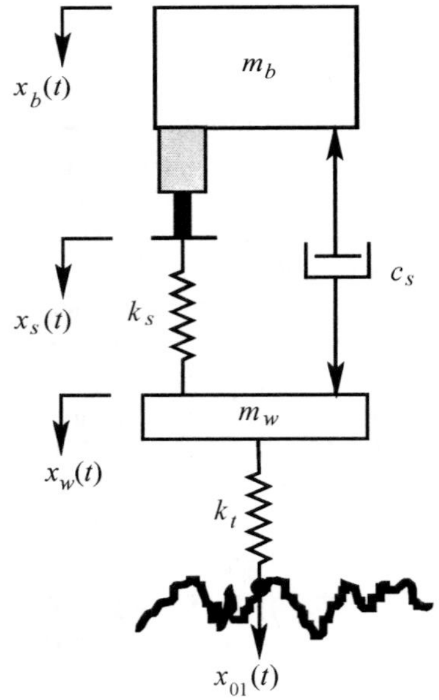

Figure 12.46 Schematic of a slow-active suspension system.

For the system shown in Figure 12.46, the control law is now based on displacement demand for the control actuator rather than force demand, as was used for the fully active system. In this case the equations of motion are:

$$m_b \ddot{x}_b + k_s(x_s - x_w) + c_s(\dot{x}_b - \dot{x}_w) = 0 \tag{12.3.69}$$

$$m_w\ddot{x}_w - k_s(x_s - x_w) - c_s(\dot{x}_b - \dot{x}_w) + k_t(x_w - x_0) = 0 \tag{12.3.70}$$

and the control law is (Crolla, 1988):

$$(x_b - x_s) = a_1 x_b + a_2 x_w + a_3 \dot{x}_b + a_4 \dot{x}_w \tag{12.3.71}$$

If fewer state measurements are available, then simpler control laws have been shown to work well. For example (Wilson et al., 1986):

$$(x_b - x_s) = a_1(\dot{x}_b - \dot{x}_s) \tag{12.3.72}$$

or

$$(x_b - x_s) = a_1\dot{x}_b - a_2\dot{x}_w \tag{12.3.73}$$

Optimal control theory can then be used to evaluate the unknown constants, a_1 a_2, etc., using the same approach as was taken in the last section for fully active systems. The performance obtained with these control laws and the arrangement shown in Figure 12.46 is very close to that for a fully active system, provided that the slow actuator has a bandwidth of at least 3 Hz (Sharp and Hassan, 1987).

The main disadvantage of the slow-active system just described is the high level of force demanded from the actuator unless a means can be found to support the vehicle weight by a spring in parallel with the actuator. This is simple to model but impractical to implement with metallic spring elements (due to excessive space demands) but it is possible to implement at some cost premium with air or hydro-pneumatic suspension systems.

12.3.1.3 Semi-Active Damping Suspension Systems

Semi-active damping systems follow similar control laws to a fully active system, shown schematically in Figure 12.43, except during times when the latter are required to supply power, the semi-active system damping is set to zero or a very small value. Thus, semi-active systems supply no power but simply adjust the damping coefficient of an otherwise passive damper.

During the period when an equivalent fully active damper would be absorbing power, the semi-active damper is set to its 'on' state and during the time when the equivalent fully active damper would be supplying power to the suspension, the semi-active damper is set to its 'off' state.

The semi-active suspension control literature is vast. Recent works that delve into some of the finer details include those by Guglielmino et al. (2008) and Savaresi et al. (2010). No attempt is made to cover all material here: instead, the focus is on some of the major developments in classical control approaches.

Karnopp (1974) developed the semi-active sky-hook damping technique, which can be realised by the following control law:

$$\text{Type 0} \quad f_d = \begin{cases} c\dot{x}_b; & \dot{x}_b(\dot{x}_b - \dot{x}_w) > 0 \\ 0; & \dot{x}_b(\dot{x}_b - \dot{x}_w) \le 0 \end{cases} \tag{12.3.74a,b}$$

Although the ideal damping in the 'off' state is zero (as for the fundamental type 0 damper), if only ride quality is to be maximised, it should be a little greater than zero to optimise the trade-off between ride quality and road holding (Miller, 1988a). For the type 0 system, two finite states are achieved as follows:

$$\text{Type 0 (2-states)} \quad f_d = \begin{cases} c_{\max}\dot{x}_b; & \dot{x}_b(\dot{x}_b - \dot{x}_w) > 0 \\ c_{\min}\dot{x}_b; & \dot{x}_b(\dot{x}_b - \dot{x}_w) \le 0 \end{cases} \tag{12.3.75a,b}$$

In the following discussion, a finite value of passive damping will usually be assumed.

Semi-active dampers may be divided essentially into two basic types: the simple on/off type where the damping in the 'on' state is constant but considerably greater than the damping in the 'off' state; and the continuously variable type where the damping in the 'on' state is continuously variable, and in the 'off' state it is constant and much smaller than in the 'on' state. Note that a constant damping coefficient results in a damping force proportional to the relative velocity between the vehicle wheel and body – a property that is characteristic of a passive damper, where the damping force f_d is:

$$f_d = c_{\text{off}}(\dot{x}_b - \dot{x}_w) = 2\zeta_{\text{off}} m_b \omega_b (\dot{x}_b - \dot{x}_w) \tag{12.3.76a,b}$$

An example of a continuously variable damper is characterised by the following control law (here called type 1), which comes directly from Equation (12.3.7) for a fully active damper with portal state feedback (Miller, 1988a):

$$\text{Type 1} \quad \frac{f_d}{m_b} = \begin{cases} 2\zeta_{\text{on}}\omega_b\dot{x}_b + 2\zeta_{\text{off}}\omega_b(\dot{x}_b - \dot{x}_w); & \dot{x}_b(\dot{x}_b - \dot{x}_w) > 0 \\ 2\zeta_{\text{off}}\omega_b(\dot{x}_b - \dot{x}_w); & \dot{x}_b(\dot{x}_b - \dot{x}_w) < 0 \end{cases} \tag{12.3.77a,b}$$

An example of a simple on/off damper will be referred to as type 2, and has the following control law (Miller, 1988a):

$$\text{Type 2} \quad \frac{f_d}{m_b} = \begin{cases} 2\zeta_{\text{on}}\omega_b(\dot{x}_b - \dot{x}_w); & \dot{x}_b(\dot{x}_b - \dot{x}_w) > 0 \\ 2\zeta_{\text{off}}\omega_b(\dot{x}_b - \dot{x}_w); & \dot{x}_b(\dot{x}_b - \dot{x}_w) < 0 \end{cases} \tag{12.3.78a,b}$$

It can be seen that, for both types, the semi-active damping ratio takes on two discrete and fixed values, depending on the value of the condition function $\dot{x}_b(\dot{x}_b - \dot{x}_w)$. The disadvantage of the first type is that a servo-valve of high bandwidth is required to provide the continuously variable damping coefficient (usually done by varying the size of the orifice in a passive damper on a continuous basis). For the second type, the damping coefficient remains constant in each state but equal to a different value for each state, thus avoiding the need for a high bandwidth servo-valve.

Types 0, 1 and 2 discussed above suffer from the disadvantage that they require a measurement of the absolute velocity of the vehicle, which can only be obtained by integrating the signal from an accelerometer mounted on the vehicle (and applying a high-pass filter to remove the DC offset). For the low-frequency signals of interest, this integration is difficult and hampered by hardware limitations. As such, numerous adaptations that do not rely on an absolute velocity measurement have been proposed. Several of these will be discussed here.

Rakheja and Sankar (1985) developed an alternative to the simple on/off damper (type 2) described above. This will be referred to here as a type 3 control law and is based on the following logic.

When the spring force and damping force act in the same direction, the damping force tends to increase the acceleration of the suspended mass; thus, to optimise ride comfort, a very small or zero damping force is preferred for this part of the cycle. When the spring force and damping force act in opposite directions, the force on the mass can be made zero if the magnitude of the damping force is made the same as the spring force. This logic is reflected in the condition function for the type 3 damper, the control law for which is:

$$\text{Type 3} \qquad \frac{f_d}{m_b} = \begin{cases} 2\zeta_{\text{on}}\omega_b(\dot{x}_b - \dot{x}_w)\,|\dot{x}_b - \dot{x}_w|; & (x_b - x_w)(\dot{x}_b - \dot{x}_w) > 0 \\ 2\zeta_{\text{off}}\omega_b(\dot{x}_b - \dot{x}_w)\,|\dot{x}_b - \dot{x}_w|; & (x_b - x_w)(\dot{x}_b - \dot{x}_w) < 0 \end{cases} \qquad (12.3.79\text{a,b})$$

It can be seen from Equation (12.3.79) that the only measurements that are required are the relative velocity and relative displacement between the vehicle body and the wheel (which could be measured by only one sensor).

Using the same reasoning as used to derive the type 3 on/off system, Alanoly and Sankar (1987) developed a continuously variable semi-active system which shall be referred to here as a type 4 system. The control law for this system is:

$$\text{Type 4} \qquad \frac{f_d}{m_b} = \begin{cases} \alpha_{\text{on}}\omega_b^2(x_b - x_w); & (x_b - x_w)(\dot{x}_b - \dot{x}_w) < 0 \\ 0; & (x_b - x_w)(\dot{x}_b - \dot{x}_w) > 0 \end{cases} \qquad (12.3.80\text{a,b})$$

where the coefficient $\alpha_{\text{on}} > 0$ is the gain of the semi-active controller. Although not mentioned by the authors, it is possible that a better trade-off between ride quality and road holding could be achieved if the mass normalised damping force in the 'off' state were made equal to $\alpha_{\text{off}}\,\omega_b^2\,(x_b - x_w)$, where α_{off} is much smaller than α_{on}.

Although the type 4 system requires a similar performing servo-valve to the type 1 system, it does not require a measurement of the absolute vertical velocity of the vehicle body, and has a similar overall performance.

Elimination of the need for a high bandwidth (>15 Hz) servo-valve and its associated cost penalty by using either the type 2 or type 3 on/off semi-active system comes at a considerable performance disadvantage, resulting in a suspension with a performance about midway between a passive and fully active system. On the other hand, use of either type 1 or type 4 continuously variable semi-active system results in a performance very close to that achieved by a fully active system (Alanoly and Sankar, 1987; El-Madanay and Abduljabbar, 1989).

Another fundamental difference between the passive and on/off semi-active systems on the one hand and the continuously variable semi-active and fully active on the other, is that in the former case, increasing the system damping beyond some very small value reduces ride quality, whereas in the latter case, increasing the damping increases ride quality. Note that high damping is desirable to maximise handling and vehicle stability.

Thus, the trade-off between ride quality and vehicle handling, which characterises passive and on/off type semi-active suspensions, does not occur for continuously variable semi-active or fully active systems. Similarly, it has been shown (Miller, 1988a) that increasing the damping for the passive and on/off semi-active systems reduces road holding performance (as the body and wheel bounce together on the tyre stiffness resulting in large variations in tyre contact force). On the other hand, increasing the damping in fully active or continuously variable semi-active systems improves road holding. However, it is difficult to match the road holding performance of a passive system with an active or semi-active system, although the performance of a semi-active system comes close with the addition of damping in the 'off' state (which results in a small reduction in ride quality).

Aside from type 1 variable damping, another method of achieving variable damping is by using the sky-hook linear approximation damper control law:

$$\text{Type 5} \quad c = \begin{cases} \underset{[c_{\text{off}},\, c_{\text{on}}]}{\textbf{sat}} \left(\dfrac{\alpha c_{\text{on}}(\dot{x}_b - \dot{x}_w) + (1-\alpha)c_{\text{off}}\dot{x}_b}{(\dot{x}_b - \dot{x}_w)} \right); & \dot{x}_b(\dot{x}_b - \dot{x}_w) \geq 0 \\ c_{\text{off}} \quad ; & \dot{x}_b(\dot{x}_b - \dot{x}_w) < 0 \end{cases} \qquad (12.3.81\text{a,b})$$

where **sat** is a saturation operator and α is a tuning parameter within the range [0,1]. Savaresi et al. (2003) suggest $\alpha = 0.5$ as a typical value, as it divides the damping coefficient equally between wheel and body). It should be noted that the limiting case of $\alpha = 1$ is equivalent to two-state sky-hook damping. This control law is suitable with dampers that can be continuously varied (for example, magnetorheological dampers).

Shen et al. (2006) developed two alternatives for the on/off dampers. To overcome the measurement of absolute acceleration required for the sky-hook principle, they instead suggested that the jerk of the vehicle (obtained by differentiating the filtered body acceleration) could instead be used:

$$\text{Type 0 modification} \quad \zeta_b = \begin{cases} \zeta_{on}; & \dddot{x}_b(\dot{x}_b - \dot{x}_w) < 0 \\ 0; & \dddot{x}_b(\dot{x}_b - \dot{x}_w) > 0 \end{cases} \qquad (12.3.82\text{a,b})$$

Shen et al. (2006) also suggested a simplified control law based on Rakheja and Sankar's (1985) and Alanoly and Sankar's (1987) methods. Alanoly and Sankar's method requires that the damping force follow the spring force. This can be difficult to implement in practice and thus Shen et al. (2006) suggested a simplified control law (which they termed a *modified Rakheja–Sankar method*) that sets damping as simply either on or off:

$$\text{Type 3 modification} \quad \zeta_b = \begin{cases} \zeta_{on}; & (x_b - x_w)(\dot{x}_b - \dot{x}_w) < 0 \\ 0; & (x_b - x_w)(\dot{x}_b - \dot{x}_w) > 0 \end{cases} \qquad (12.3.83\text{a,b})$$

Shen et al. experimentally verified both of these methodologies.

Motivated by the desire to decrease excessive tyre force (that is, improve roadholding characteristics), Valasek et al. (1997) introduced groundhook control, which is similar to sky-hook control, but differs in that the damper is between the unsprung mass (wheel mass) and the ground. The control law is:

$$\text{Type 6} \quad f_d = \begin{cases} c_{on}(\dot{x}_b - \dot{x}_w); & \dot{x}_w(\dot{x}_b - \dot{x}_w) < 0 \\ 0; & \dot{x}_w(\dot{x}_b - \dot{x}_w) \geq 0 \end{cases} \tag{12.3.84a,b}$$

Several variations to groundhook control have been suggested. As an example, although the specific application was different (tuned vibration absorbers), Koo et al. (2004) compared on/off and continuous versions of velocity-based and displacement-based groundhook controllers, finding that of the four controllers, on/off displacement based control performed the best.

Savaresi et al. (2005) proposed an acceleration-driven damping (ADD) control law to minimise the vertical body acceleration (that is, improve comfort) in cases where no road information (preview) is available:

$$\text{Type 7} \quad c = \begin{cases} c_{\text{on}}; & \ddot{x}_b(\dot{x}_b - \dot{x}_w) \geq 0 \\ c_{\text{off}}; & \ddot{x}_b(\dot{x}_b - \dot{x}_w) < 0 \end{cases} \tag{12.3.85a,b}$$

Although better comfort is achieved through ADD, road-holding characteristics actually deteriorate.

Savaresi and Spelta (2007) compared ADD with traditional sky-hook control, revealing that they perform best in different frequency regions: sky-hook provides quasi-optimal performance around the body resonance, whilst ADD provides quasi-optimal performance above the body resonance. To take advantage of both of these characteristics, Savaresi and Spelta (2007) developed a mixed sky-hook–ADD control algorithm and showed that its performance was close to that of a globally-optimal control algorithm.

Morselli and Zanasi (2008) pointed out that the ADD control algorithm can result in oscillations in the desired damping value (chatter). To provide results comparable to ADD, without the oscillation problem, they therefore proposed what is termed a power driven damper control algorithm:

$$\text{Type 8} \quad c = \begin{cases} c_{\text{on}} \; ; & k(x_b - x_w)(\dot{x}_b - \dot{x}_w) + c_{\text{on}}(\dot{x}_b - \dot{x}_w)^2 < 0 \\ c_{\text{off}} \; ; & k(x_b - x_w)(\dot{x}_b - \dot{x}_w) + c_{\text{off}}(\dot{x}_b - \dot{x}_w)^2 \geq 0 \\ \dfrac{c_{\text{on}} + c_{\text{off}}}{2} \; ; & x_b - x_w \neq 0 \;\wedge\; (\dot{x}_b - \dot{x}_w) = 0 \\ -\dfrac{k(x_b - x_w)}{(\dot{x}_b - \dot{x}_w)} \; ; & \text{for all other cases} \end{cases} \tag{12.3.86a,b}$$

At this stage it should be mentioned that many modern control theory approaches (for example, $\mathcal{H}^{\infty}$ controllers) to semi-active suspension damping have been developed. However, these are beyond the scope of this book and the interested reader is directed to Savaresi et al. (2010), who discuss some of these methodologies.

Suspension travel in a passive system is reduced by increasing the damping, but in all types of active system, the travel is increased as the damping increases beyond some very small value. Thus, the performance of active and semi-active suspensions is limited by the allowable space in which they can work (usually ±10 cm).

In summary, the best design trade-offs for a vehicle result in ζ for a passive system equal to 0.3, ζ_{on} and ζ_{off} for an on/off semi-active system equal to 0.4 to 0.6 and 0.1 to 0.2 respectively and for a continuously variable semi-active system, ζ_{on} and ζ_{off} should be equal to 1.0 or greater and 0.1 to 0.2 respectively. The desired values for a fully active suspension are similar to those for a continuously variable semi-active system.

Note that the theoretical performance of active and semi-active suspensions is somewhat reduced in practice due to inaccuracies in the hardware that converts velocities and displacements to electrical signals and electrical signals to damping coefficients. These problems are discussed in detail by Miller (1988b).

In summary, it can be said that all suspensions provide trade-offs between ride comfort, handling, road holding and suspension travel. However, active and continuously variable semi-active computer controlled suspensions can provide better compromises between the various requirements than are possible with purely passive systems. As the performance of a continuously variable semi-active suspension can be made almost as good as a fully active system (and much better than a passive system), there is some doubt whether the slightly improved performance of the fully active or slow-active systems can be justified in terms of the increased cost, number of components and complexity involved.

12.3.1.4 Semi-Active Systems Incorporating Variable Stiffness

Semi-active suspension systems have traditionally been focussed on only the damping element; however, one disadvantage of only varying the damping is that it can be difficult to reduce vibration levels at frequencies less than the natural frequency of the car and suspension. A semi-active system based on optimal control theory, incorporating a variable damper and a spring stiffness that could be varied over three distinct values, was developed by Youn and Hać (1995). The system compared favourably to a semi-active system with variable damping and fixed stiffness.

Conventional methods for implementing variable stiffness are far more complex than for implementing variable damping. To overcome this, Liu et al. (2008) incorporated a variable stiffness element using two controllable dampers along with two fixed stiffness springs. One damper was used to provide variable damping for the overall system. The second damper and one spring were aligned in parallel (a Voigt element), and this element was in series with the other spring, yielding an element with an overall effective stiffness that could be controlled via varying the damping coefficient of the second damper.

12.3.1.5 Switchable Damper

A switchable damper is different from an on/off type semi-active damper in that the switching of the damper force from one value to another occurs over a relatively long time period, whereas the semi-active damper is switched every hundred milliseconds or so.

Even though the switchable damper may stay in one setting for a relatively long time, it is capable of changing from one setting to another very quickly (15 to 20 ms). This changeover speed is necessary for safety considerations, as an excessively soft suspension can be disastrous during severe cornering or sudden vehicle direction changes.

Switchable dampers currently being considered have three settings: soft, medium and hard. The setting that is automatically selected depends upon the roughness of the road and the severity of the driver's steering actions (which are related to the vehicle speed as well as the steering wheel movement). In some cases, it is possible to switch between settings manually, but in this case, the soft setting cannot be made as soft as in the automation-only switching case, for obvious safety reasons. One particular switchable damper design is discussed by Meller and Frühauf (1988).

12.3.1.6 Maglev Vehicle Suspensions

Interest in magnetic levitation vibration problems has increased significantly over the years due to developments in Maglev (magnetic levitation) trains, which use magnetic forces to suspend and guide the vehicle. Suspensions are usually either electromagnetic (using electromagnets) or electrodynamic (using permanent or superconductor magnets), with hybrid versions also being developed. The former are more common, but are inherently unstable, and thus require control to stabilise the suspension system. Interactions between the control forces and a flexible or irregular guideway can result in coupled vibration problems. These vibration problems are generally described in three categories: (1) moving vehicle-bridge coupled vibrations; (2) stationary (or low speed) vehicle-guideway self-excited vibration; and (3) vehicle-guideway interaction as a result of track irregularities. Feedback vibration control algorithms have been employed for all of these. Zhou et al. (2010) provide an extensive review of system modelling and proposed/tested control algorithms for each of these categories.

Zhou et al. (2010) highlighted the stationary vehicle-guideway self-excited instability problem as something that 'should be solved urgently'. They thus focussed their subsequent efforts in this area, developing a system using time delay control with a virtual tuned mass damper, which uses an electromagnetic force to emulate the force of a real TMD (Zhou et al., 2011a), and developing an LMS cancellation algorithm with phase correction (Zhou et al., 2011b; Zhou et al., 2011c).

12.3.1.7 Sensors and Actuators

A large part of the effort in making active vehicle suspension systems practically realisable has been expended in the development of suitable low-cost sensors and actuators (Moore, 1988; Parker and Lau, 1988; Decker et al., 1988; Shiozaki et al., 1991; Hattori et al., 1990; Asano et al., 1991; Satoh and Osada, 1991; Akatsu et al., 1990) and a considerable effort has also been expended in developing ways of monitoring the health of these sensors and actuators and maintaining system performance in the event of component failures (de Benito and Eckert, 1990; Litkouhi and Boustany, 1988).

Sensors that have been developed include piezoceramic force and acceleration transducers, potentiometric position transducers and inductance position transducers, while actuator development has mainly focussed on various hydraulic or air actuator valve designs, (for semi-active suspensions), piezoceramic actuators and low-cost hydraulic actuators and associated servo-valves or solenoid valves (for fully active suspensions), and electro-rheological and magnetorheological fluids. An electro-rheological actuator (described in detail in Section 15.13) is a fluid formed by doping a dielectric base fluid with semi-conductor particles. On application of an electric field of sufficient strength, the particles form chains that link across the electrodes. The resistance to shear of these particles can be varied by controlling the strength of the electric field, thus forming a damping element suitable for a semi-active suspension (Stanway et al., 1989; Sproston and Stanway, 1992; Sproston et al., 1994; Choi et al., 1999; Choi et al., 2002).

Magnetorheological fluids are, however, preferred over electro-rheological fluids for semi-active suspension technologies due to superior performance characteristics. Magnetorheological fluids (described in Section 15.14) include suspended iron particles, and are similar to electro-rheological fluids, with the main difference being that a magnetic field is used to control the fluid viscosity. A good description is provided by Yao et al. (2002), who designed and built a magnetorheological damper, performed a parameter estimation of this, developed a quarter car suspension model incorporating the magnetorheological damper, and performed a semi-active sky-hook control simulation. Suspension systems with magnetorheological dampers are currently available commercially, although research into such systems continues to evolve (for example, Shen et al., 2006; Yu et al., 2009; Dong et al., 2010; Metered et al., 2010).

12.3.2 Rigid Mount Active Isolation

In cases where low-frequency machine movement cannot be tolerated, it is sometimes necessary to be able to connect a machine rigidly to a support foundation and at the same time reduce the forces acting at the support point due to the machine vibration. In these cases, a reduction in the vibratory force acting on the mount can be achieved by using a shaker attached to the support point and driving an inertial mass.

A system similar to that shown in Figure 12.47 was developed using a hydraulic actuator (and the slightly more complex control system needed) by Tanaka and Kikushima in 1988, who demonstrated a working system on an actual forge hammer with spectacular results.

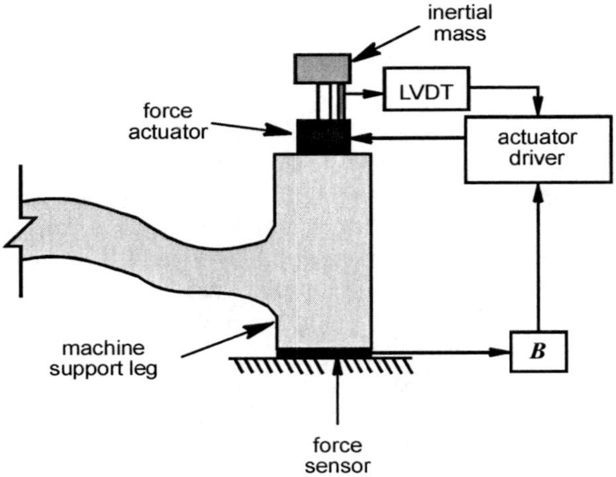

Figure 12.47 An active isolator for a rigid machine mount.

The purpose of the LVDT (linear variable differential transformer – see Chapter 15) was to sense the relative displacement of the actuator rod in the cylinder, so that it did not gradually drift to one extreme or another. Various forms of the optimal compensator ***B*** are discussed by Tanaka and Kikushima (1988).

12.3.3 Vibration Isolation of High-Precision Equipment

The problem of isolating high-precision equipment from vibrations in its support structure is similar to the vehicle suspension system problem, except that the frequency range of interest usually extends higher. Additionally, the more complex the sensitive equipment, the more difficult it can be to control.

Successful vibration isolation of a space telescope optical system from its structural support was reported by Kaplow and Velman in 1980, who used a simple form of velocity feedback control to achieve the desired result. Active control techniques have also been used to isolate not just telescopes, but also other high-precision equipment, such as interferometers and particle accelerators.

Commercial active vibration isolation assemblies are commonly available for small to medium size sensitive equipment. As an example, a space application will be considered here. The reason for this choice is that space applications tend to have sensitive systems, and thus control of vibrations in these systems requires extreme precision over a very large frequency range. Gravity gradient forces and atmospheric drag (in low earth orbits) can result in vibrations with frequencies of less than 0.001 Hz, whilst steady-state machine sources and some transient sources can create excitations of over 1 Hz. Specifically, vibration isolation systems for microgravity experiments will be discussed here.

12.3.3.1 Vibration Isolation Systems for Microgravity Experiments

There are many proposed active isolation systems for microgravity experiments. Early systems include those by Fenn et al. (1990), who developed an integrated six-degrees-of-freedom magnetic isolator, controlled with a non-linear feedback system to isolate microgravity experiments from space station movements and vibration. Stampleman and von Flotow (1990) developed a mount intended for a similar purpose using PVDF film controlled with a simple feedback system incorporating acceleration feedback and Jones et al., 1990, developed a six-degrees-of-freedom microgravity experiment isolator using simple displacement feedback.

By the mid-1990s, microgravity vibration isolation prototypes were starting to be flight-tested, with three flight-tested on space shuttle flights prior to the turn of the century. These systems are described in detail by Grodsinsky and Whorton (2001). An overview of some directly pertinent aspects is provided here. The first microgravity isolation system aboard a shuttle test-flight was STABLE (suppression of transient accelerations by levitation) in 1995. MIM (microgravity isolation mount) and ARIS (active rack isolation system) were tested over the following couple of years. The ARIS isolation system uses one single control system to isolate a set of racks that house a series of experiments. The control system had to be modifiable to cope with different items being changed in and out of the racks. The uncertain rack dynamics necessitated a controller with a limited bandwidth. The STABLE and MIM isolation systems provided vibration isolation at a component level directly to the payload, and thus there was no need to deal with uncertain rack dynamics. As such, the controllers could be more specific and higher bandwidth with superior disturbance rejection. Such a system does have mass, complexity and economic disadvantages, however, in that a dedicated isolation controller is required for each individual payload.

STABLE was designed to be robust, integrating a six-degrees-of-freedom, low-bandwidth relative position PID feedback controller with a high-bandwidth acceleration feedback control loop. Flight data showed that it was able to reduce accelerations by at least one order of magnitude.

MIM used six-degrees-of-freedom levitation provided by Lorentz force actuators. It was tested with several different control algorithms (both classical and optimal) using relative position, payload orientation and acceleration as input states. Initial flight data showed some low-frequency issues, which were found to result from attraction between the actuator magnets and ferromagnetic casing enveloping nearby position sensors, and the controllers were thus adjusted to overcome this problem.

ARIS incorporated six-degrees-of-freedom, low-authority position feedback blended with acceleration feedback. Flight data showed some promising results. Development of the ARIS system has continued into the twenty-first century and ARIS systems have been installed on two racks of the international space station.

Due to the stringent requirements of microgravity active isolation systems, modern control approaches to this problem continue to be proposed and developed (for example, $\mathcal{H}^\infty$ (Fialho, 2000), mixed $\mathcal{H}_2/\mu$ (Whorton, 2005), and linear-parameter varying (Mehendale et al., 2009)).

12.3.4 Vibration Reduction in Tall Buildings

Excessive vibration (or low-frequency sway) can be induced in tall buildings by either seismic activity or wind. Active structural control technologies can be employed to control vibration, and hence improve safety, of tall buildings as well as other civil structures (for example, suspension and cable-stay bridges). Korkmaz (2011) presents a recent cross-disciplinary comprehensive review of historical advances and future challenges in this area.

Active tendon control (using tensioned cables) and tuned mass dampers (or tuned vibration absorber, as discussed in Section 12.2.4) are both feasible control options (individually or in combination) for tall building vibration control (Abdel-Rohman, 1987). Tuned mass dampers are especially common for controlling earthquake-induced vibrations.

Fully active, hybrid and semi-active mass damper systems have been successfully installed in numerous tall buildings in Japan. Spencer and Nagarajaiah (2003) summarise many of these applications. The control laws used with active tuned mass dampers have continued to be adapted and developed over the years. For example, Wang and Lin (2007) have considered variable structure control and fuzzy sliding mode control for earthquake control, finding the latter to be more practical due to its lower force and hence energy requirements.

Active tendon control uses tensioned cables in the wall of each level of the building as shown in Figure 12.48. The tension in the cables (or active tendons) is controlled by hydraulic actuators, thus reducing building sway. Although the control forces required to counteract wind loading may exceed those that would be required using tuned mass dampers, the tensioned cable controller can be more efficient (Korkmaz, 2011). Traditional means of controlling these tensions using a feedback control system are discussed by Abdel-Rohman (1987) and Samali et al. (1985). Other controllers have since been developed. A recently proposed methodology (Lin et al., 2010) uses $\mathcal{H}^\infty$ direct output feedback control for the design of an active tendon system for irregular buildings (taking into account torsional coupling that results from the building's irregular shape) subject to earthquake excitation.

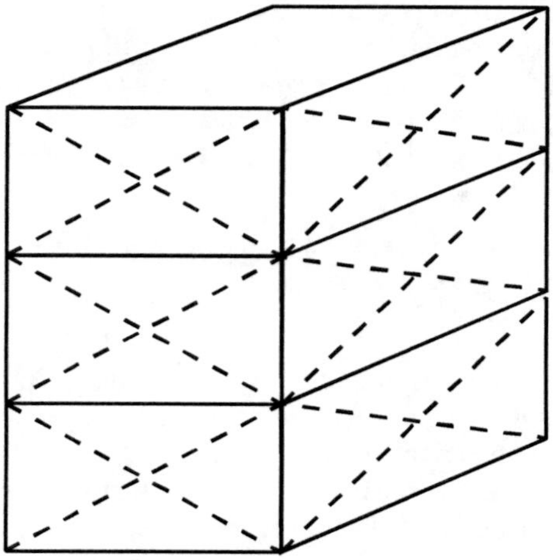

Figure 12.48 Use of tensioned cables to control building sway.

12.3.5 Active Isolation of Machinery from Flexible Structures: Engine Mounts

This topic was discussed in detail in Section 12.2.5. However, one special case that has not been discussed previously is that of active or semi-active isolation of an engine from a flexible support structure (car) as discussed by Graf and Shoureshi (1987). Engine mounts need to be able to provide engine support whilst simultaneously effecting vibration isolation. Ideally, damping and stiffness should be both frequency and amplitude dependent to provide optimal vibration isolation. Engine mounts are similar in principle to a semi-active suspension in that the mount damping is made adjustable by using an adjustable orifice through which hydraulic fluid passes, as shown in Figure 12.49.

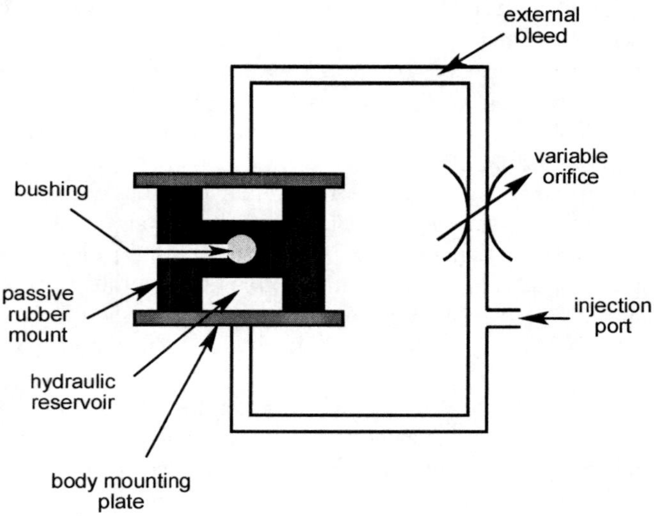

Figure 12.49 Semi-active engine mount.

The bushing set in the mount is used to connect the mount to the engine, and the vehicle body is attached to the mounting plates. The two fluid reservoirs are filled with fluid of density ρ and viscosity μ. The compliance of the reservoir volume is c_2, the length of the external bleed tube is l and the bleed tube diameter is d. The effective damping ratio of the mount is then given by Graf and Shoureshi (1987), as:

$$\zeta = \frac{32\mu l}{d^3\sqrt{2\pi\rho / c_2}} \tag{12.3.87}$$

Semi-active engine mounts improve low-frequency vibration isolation performance. Higher frequency control can be achieved through active techniques. Numerous active engine mounts have been described; most commonly, some form of actuator is used in parallel with a conventional passive mount. The actuation force has been provided by several means, including, for example, electromagnetic (Muller et al., 1994) and piezoelectric (Ushijima and Kumakawa, 1993). Yu et al. (2001) provide a review of developments in passive, semi-active and active engine mounts up until the year 2000.

Semi-active engine mounts have also been developed that use electro-rheological fluids (see Section 15.13) to provide the damping forces (for example, Sproston and Stanway, 1992; Sproston et al., 1994; Choi et al., 1999; Choi et al., 2002).

Arzanpour et al. (2008) have proposed semi-active engine bushings and mounts that incorporate magnetorheological fluids (discussed in Section 15.14) for use with variable displacement engines. Their motivation stemmed from variable displacement engines exhibiting changing load conditions (due to deactivation of some cylinders in light load conditions) which could potentially exceed desired vibration levels using existing isolators. They chose to investigate magnetorheological fluids over electro-rheological fluids for energy consumption and design reasons. Following on from this, Mansour et al. (2011a) developed a semi-active engine mount prototype that has an auxiliary magnetorheological chamber fitted within a hydraulic mount for use with variable displacement engines. The same group (Mansour et al., 2011b) has also numerically and experimentally investigated engine mounts for variable displacement engines that use an electromagnetic actuator retrofitted in a passive engine mount.

Numerous control algorithms for mounts have been proposed beyond the traditional standard controllers, for example (but not limited to), controllers that use linear quadratic Gaussian control (Choi et al., 2008) neural network algorithms (Darsivan et al., 2009) and $\mathcal{H}_2$ loop shaping (Olsson, 2006).

Although feedforward control has yet to be discussed in this chapter, it should be noted that feedforward techniques for engine mounts have been considered. For example, in an effort to design a low-cost engine mount, Lee and Lee (2009) developed a model-based feedforward algorithm that utilised a vibration estimation algorithm and a current shaping controller. Application of their methodologies to a prototype showed some promise, with reductions in transmission of an engine idle shake fundamental frequency when compared to a passive system.

Although active engine mount technology still has potential for improvements, they have, in the past few years, been incorporated by several manufacturers into production vehicles. It is anticipated that demand for this will increase and, as such, active engine mount technology will continue to develop.

12.3.6 Helicopter Vibration Control

Helicopters in flight experience a periodic variation in air loading on their rotor blades, inducing periodic forces. Transmission of these forces and moments from the helicopter rotor hub to the airframe results in a prominent source of vibration.

Active control of rotor and gearbox induced vibrations in a helicopter fuselage has been a subject of much interest since first investigated by Bies (1968).

A common approach for vibration control in helicopters is rotor-based control, which includes higher harmonic control (HHC) and individual blade control (IBC). Here, methods such as trailing edge flap control and active twist control are included in the definition of IBC. HHC is the oldest of these methodologies, and as such, early developments in active helicopter vibration control were seen in this area. An overview of the HHC technique will be given here with some specific details provided for some of the early developments in this area. A more general overview of some of the different IBC methodologies. For more detailed descriptions of the full scope of research that has been undertaken in these areas, the reader is directed to dedicated review papers (for example, Friedmann and Millot (1995), Kessler (2011a), and Kessler (2011b)). It should be noted at this point that although significant successful developments have occurred in the rotor-based HHC and IBC methods, they have yet to be applied to production helicopters.

HHC involves using a feedback control system to oscillate the rotor blades in pitch at harmonic frequencies greater than the fundamental rotational frequency of the rotor. The method was first suggested by Shaw (1967) and subsequent wind tunnel and flight tests were reported (McHugh and Shaw, 1978; Shaw and Albion, 1980). Some subsequent work in this area was reported by Nguyen and Chopra (1990), Prasad (1991), and Hall and Wereley (1992). These authors based their work on the use of a feedback control system driving an actuator to minimise vibration levels at one or more locations on the airframe.

Shaw's control algorithm may be classified essentially as a classical narrowband disturbance rejection compensator. Its integral action, rather than a detailed knowledge of helicopter dynamics, is responsible for the harmonic disturbance rejection achieved. The algorithm requires a knowledge of the transfer function between the control force input on the rotor swashplate and the locations on the fuselage where vibration is to be minimised. At each step in the adaptation process, the vibration signal measured at the minimisation points is harmonically analysed. The resulting vector of vibration amplitudes for each harmonic to be controlled is then multiplied by the inverse of the transfer function matrix (which is divided into sine and cosine terms) to give a vector of sine and cosine components of the harmonic control force input required for the next step of the adaptation process. Thus, the sine and cosine components of the *n*th harmonic at the measurement locations on the fuselage are:

$$\boldsymbol{w} = \boldsymbol{T}\boldsymbol{F} + \boldsymbol{w}_0 \tag{12.3.88}$$

where $\boldsymbol{w}$ is the vector of vibration amplitudes, $\boldsymbol{T}$ is the (assumed constant with time) transfer function matrix, $\boldsymbol{F}$ is the control input vector and $\boldsymbol{w}_0$ is the vector of vibration amplitudes with no control input. The resulting control law is:

$$\boldsymbol{w}_{n+1} = \boldsymbol{w}_n - \boldsymbol{T}^{-1}\boldsymbol{w}_n \tag{12.3.89}$$

For a single input, single output system, the transfer function matrix may be written as:

$$\boldsymbol{T} = \begin{bmatrix} T_{cc} & T_{cs} \\ T_{sc} & T_{ss} \end{bmatrix} \tag{12.3.90}$$

Details describing how this algorithm may be implemented in discrete- or continuous-time are given by Hall and Wereley (1992). These authors indicate that implementation of the algorithm in the continuous-time domain offers significant performance and stability advantages over discrete-time implementation.

A different method of implementing HHC was suggested by Gupta and Du Val (1982) and Du Val et al. (1984) and was essentially a linear quadratic regulator with frequency shaped cost functions (see Chapter 5). This approach provides similar performance results to those discussed above and is eloquently summarised by Hall and Wereley (1992).

Patt et al. (2005) and Kessler (2011a) reviewed significant recent developments in the HHC literature (Patt et al. also provide a detailed discussion and comparison of three versions of the HHC algorithm: classical, adaptive and 'relaxed'). One inherent disadvantage of HHC (Jonas, 2009; Kessler, 2011a) is that vibration will be controlled only at frequencies corresponding to harmonics (mN_b-1)T, mN_bT, and $(mN_b+1)T$, where m is a positive integer ($m = 1, 2, 3, ...$), N_b is the number of blades and T is the period of revolution. If, for example, a rotor with four blades were considered, the controllable frequencies would be 3T, 4T, 5T, 7T, 8T, 9T, etc. (that is, the 2T, 6T, etc. harmonics would not be controllable). Additionally, Kessler (2011a) point out that HHC has problems achieving simultaneous noise and vibration reduction.

A different method of reducing helicopter fuselage vibration is to use an active isolation system to isolate the gearbox and rotor from the fuselage. King (1988) reported on such a system used on a Westland Helicopter in the UK achieving fuselage vibration reductions of 15 to 25 dB. The active isolation system consisted of four hydraulic actuators (mounted in parallel with passive rubber elements so that the passive system supported the entire static load), located between the airframe and a raft structure, which carried the gearbox and rotor. Ten vibration sensors were used on the fuselage to provide the feedback signal to the electronic controller that was used to drive the four actuators. Two types of controller were investigated. The first was a time domain controller designed to minimise fuselage modal vibration using the IMSC method discussed in Chapter 11. The second and preferred system was a frequency domain controller that acted to suppress vibration transmission at the dominant rotor blade passing frequency only. Note that although the frequency domain controller was more robust, it had a very slow response time (≈ one second), limiting the vibration reduction capability during fast manoeuvres. The frequency domain approach involved optimising a performance function J (see Chapter 5) which includes control effort as well as fuselage vibration response, as follows:

$$J = \boldsymbol{Y}^{\mathrm{T}}\boldsymbol{C}^{\mathrm{T}}\boldsymbol{Y} + \boldsymbol{X}^{\mathrm{T}}\boldsymbol{D}\boldsymbol{X} \tag{12.3.91}$$

where $\boldsymbol{Y}$ is the complex vector of fuselage vibration measurement, $\boldsymbol{X}$ is the complex vector of actuator forces, and $\boldsymbol{C}$ is a weighting matrix which allows for the possibility of certain fuselage locations being more important than others in terms of the need for vibration reduction. The second term in Equation (12.3.91) involving the weighting matrix $\boldsymbol{D}$ is included to allow limiting the actuator commands within their practical restraints. This type of term in a performance function is often referred to as the control effort term and is commonly used to share the control effort between actuators as evenly as possible, as well as keeping the required effort within practical bounds.

The equation of motion relating the fuselage vibration to the control actuator inputs is:

$$\boldsymbol{Y} = \boldsymbol{TX} + \boldsymbol{B} \tag{12.3.92}$$

where $\boldsymbol{T}$ is a transfer matrix relating the actuator loads to the fuselage vibration and $\boldsymbol{B}$ is a vector representing the uncontrolled vibration at the measurement locations represented in $\boldsymbol{Y}$.

Minimisation of the performance function of Equation (12.3.91) gives for the optimal complex actuator forces:

$$\boldsymbol{X} = -\left[\boldsymbol{T}^{\mathrm{T}}\boldsymbol{CT} + \boldsymbol{D}\right]^{-1}\boldsymbol{T}^{\mathrm{T}}\ \boldsymbol{C}\ \boldsymbol{B} \tag{12.3.93}$$

Note that all of the terms in the matrices discussed above are complex; that is, they are described by amplitude and phase.

A schematic representation of the control system used by King (1988), which incorporated the performance function just discussed, is shown in Figure 12.51. In this figure the incoming signals from the vibration sensors on the fuselage are first converted to the frequency domain using a fast Fourier transform (FFT) and then the blade passing frequency component is extracted in the signal processing section. The transfer matrix $\boldsymbol{T}$ is estimated by the parameter estimator, and then the optimal actuator forces needed to minimise the performance function are calculated by the optimal controller. Force feedback from the actuators is used to compensate for the actuator dynamics, as indicated by the actuator loop closure block in the figure. The required actuator signals, calculated in the frequency domain, are then converted to time domain signals using an inverse DFT and then applied to the control actuators.

An important group of methods for controlling rotor-induced vibration that were mentioned at the start of this section but have not yet been described are collectively termed individual blade control (IBC). Although not as simple to implement as HHC, IBC does not suffer from the problems that are inherent to swashplate HHC methods (Corl, 2009; Kessler, 2011b) and thus the area has garnered significant research effort. The definition of IBC can differ between authors. Many deem IBC to include all the methods where the actuators are located in the rotating frame, whilst some use it to refer to only blade-root actuation systems (and thus classify methods such as active trailing edge flaps as distinct). The former of these two definitions will be assumed here.

Three well-studied methods of individual blade control are root pitch actuation (traditional IBC), active blade twist, and active trailing edge flaps. Blade-root actuation, or root pitch control, replaces the pushrods between the swashplate and the blade pitch horn with hydraulic actuators. The method is similar to HHC in that the entire rotor blade is oscillated with relatively high frequencies at its root; however, in IBC, each rotor blade is controlled by an individual feedback loop. Rotor blade excitation can be at any desired combination of harmonics; however, IBC is mechanically more complex than HHC. Additionally, since both mechanisms actuate the entire blade at its root, required actuation forces can be high and load distributions may not be ideal.

Active blade twist is a relatively recent method that uses distributed piezoceramic actuators integrated directly into the rotor blades which generate an active twist along the span of the rotor. They have the advantage of not requiring moving parts; however, they have the disadvantage of requiring a large number of distributed actuators.

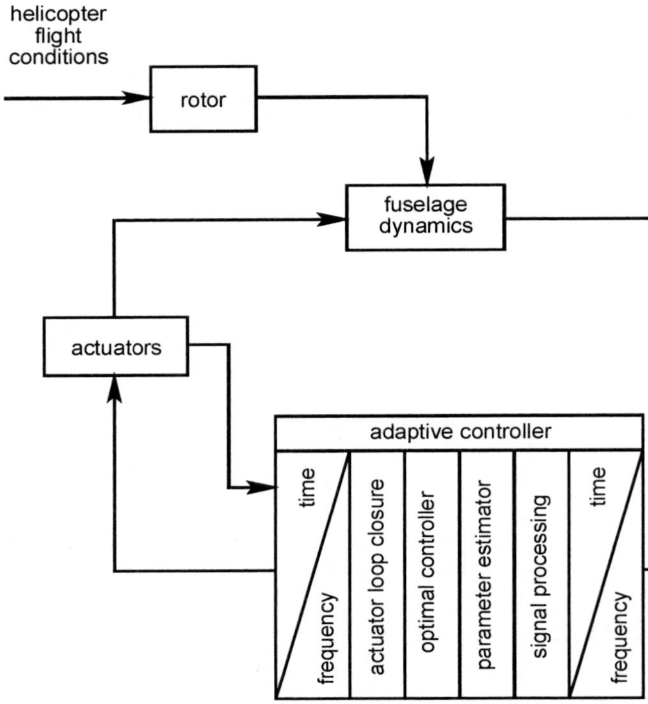

Figure 12.51 Schematic of King's 1988 system for active control of helicopter vibrations.

Active trailing edge flaps use actuators located in the blades, which drive trailing edge flaps that oscillate with small deflections at harmonics of the blade passage frequency in order to alter the aerodynamics. The flap actuators must be small, light and efficient and thus significant research effort has been expended on the flap actuation mechanism, with smart materials being a specific focus (Chopra, 2000). Piezoelectric actuators meet most of the requirements of a flap actuator, but provide limited stroke. As such, amplification methods have been developed to overcome this problem (Kim, 2007). Flap actuator positioning has also been considered, with multi-flap configurations (multiple flaps on each blade instead of an individual flap) showing promise (Kim, 2007). Corl (2009) identified active trailing edge flaps as showing the most progress of *on-blade* actuation techniques toward potential future implementation on production helicopters.

12.4 FEEDFORWARD CONTROL: BASIC SDOF SYSTEM

Feedforward control is the best approach for vibration isolation when the disturbance signal can be measured before the control actuator needs to act or when it is periodic (such as that generated by a rotating machine). Here, the discussion will begin by examining feedforward control of a single-degree-of-freedom system, which involves control of vibration transmission from a rigid mass to a stiff support. This will be followed in subsequent sections by a discussion of feedforward control of periodic vibration from a rigid body into a flexible beam, from a rigid body into a flexible panel, and from a rigid body into a flexible cylinder.

For the single-degree-of-freedom system illustrated in Figure 12.52, three different control force cases will be considered: the control force acting on the rigid body, the control force acting on the support structure and the control force acting on the rigid body but reacting against the support structure (Nelson et al., 1987; Nelson, 1991).

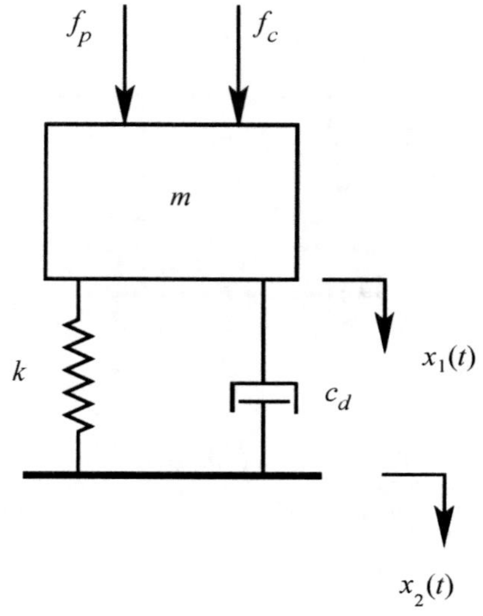

Figure 12.52 Single-degree-of-freedom system with the control force acting on the rigid body.

12.4.1 Control Force Acting on the Rigid Body

For the system shown in Figure 12.52, the equation of motion is:

$$m\ddot{x}_1 + c_d(\dot{x}_1 - \dot{x}_2) + k(x_1 - x_2) = f_p + f_c \tag{12.4.1}$$

For a periodic disturbing force f_p of frequency ω:

$$(-\omega^2 m + \mathrm{j}\omega c_d + k)x_1 - (\mathrm{j}\omega c_d + k)x_2 = f_p + f_c \tag{12.4.2}$$

It can be seen from Equation (12.4.2) (and intuitively) that the control force f_c which ensures that both x_1 and x_2 are zero is given by $f_c = -f_p$.

12.4.2 Control Force Acting on the Support Structure

For the system shown in Figure 12.53, let the mobility of the support structure at the excitation frequency be M, such that:

$$\mathrm{j}\omega x_2 = Mf \tag{12.4.3}$$

where f is the force acting on the support structure.

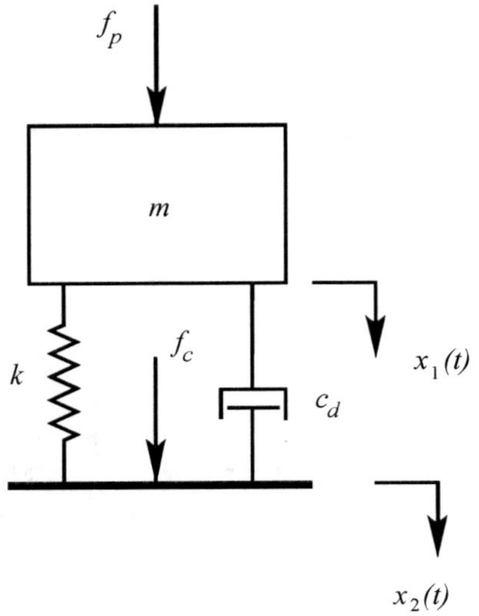

Figure 12.53 Single-degree-of-freedom system with the control force acting on the support structure.

The total force acting on the support structure is:

$$f = f_c + k(x_1 - x_2) + c_d(\dot{x}_1 - \dot{x}_2) \tag{12.4.4}$$

Thus, Equation (12.4.3) can be written as:

$$\mathrm{j}\omega x_2 = M\big(f_c + k(x_1 - x_2) + c_d(\dot{x}_1 - \dot{x}_2)\big) \tag{12.4.5}$$

The equation of motion for the rigid mass m is:

$$(-\omega^2 m + \mathrm{j}\omega c_d + k)x_1 - (\mathrm{j}\omega c_d + k)x_2 = f_p \tag{12.4.6}$$

Setting $x_p = 0$ in Equations (12.4.5) and (12.4.6) gives:

$$f_c = f_p \frac{-(\mathrm{j}\omega c_d + k)}{-\omega^2 m + \mathrm{j}\omega c_d + k} \tag{12.4.7}$$

Note that the control force to produce zero displacement of the support structure is independent of the support structure mobility.

Expressed in terms of the system resonance frequency ω_0 and damping ratio ζ, Equation (12.4.7) can be written as:

$$f_c = f_p \left[\frac{2\mathrm{j}\zeta\omega_0\omega + \omega_0^2}{\omega^2 - 2\mathrm{j}\zeta\omega_0\omega - \omega_0^2} \right] \tag{12.4.8}$$

and the magnitude is given by:

$$\left|\frac{f_c}{f_p}\right| = \sqrt{\frac{1 + (2\zeta\omega/\omega_0)^2}{\left[(\omega/\omega_0)^2 - 1\right]^2 + (2\zeta\omega/\omega_0)^2}} \tag{12.4.9}$$

which is identical to Equation (12.2.43), which describes the magnitude of (x_1/x_2) for a base excited system.

12.4.3 Control Force Acting on the Rigid Body and Reacting on the Support Structure

For the system shown in Figure 12.54, the equation for the support structure velocity is identical to Equation (12.4.5), but the equation of motion for the rigid body is:

$$(-\omega^2 m + \mathrm{j}\omega c_d + k)x_1 - (\mathrm{j}\omega c_d + k)x_2 = f_p - f_c \tag{12.4.10}$$

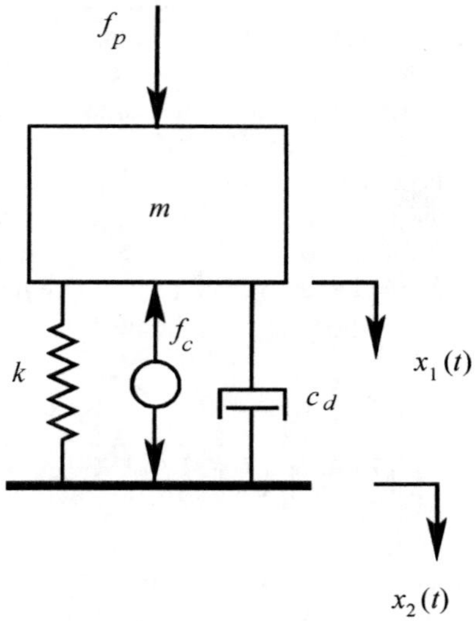

Figure 12.54 Single-degree-of-freedom system with the control force acting on the rigid body and reacting on the support structure.

Setting $x_2 = 0$ in Equations (12.4.5) and (12.4.8) gives:

$$f_c = f_p \frac{\mathrm{j}\omega c_d + k}{\omega^2 m} \tag{12.4.11}$$

Expressed in terms of the system resonance frequency ω_0 and damping ratio ζ, Equation (12.4.11) is:

$$f_c = f_p \frac{2\mathrm{j}\zeta\omega\omega_0 + \omega_0^2}{\omega^2} \tag{12.4.12}$$

and the amplitude ratio is:

$$\left|\frac{f_c}{f_p}\right| = \sqrt{\frac{1+(2\zeta\omega/\omega_0)^2}{(\omega/\omega_0)^2}} \tag{12.4.13}$$

12.4.4 Summary

It is important to note that in all three of the cases discussed above, the displacement of the support structure is reduced to zero by the control force. However, for a system consisting of multiple connections between the rigid body and support structure, with a control force at each mount, it is generally not possible to obtain a unique solution for the control forces required to reduce the amplitude to zero at each support point on the structure. In fact, because of the interaction or cross-coupling between forces acting at the base of each mount, it is possible for the control forces to 'fight' one another so that zero net displacement or force at the base of a mount may not be possible. Thus, where multiple mounts are used, it is preferable to use a multi-channel control system and total vibratory power transmission into the support structure through all support points as the cost function to be minimised, as it results in smaller required control forces and sometimes better overall reductions in structural vibration levels. Note also that feedforward control does not change the system resonance frequency and damping as did feedback control, which was discussed in Section 12.2.

It is interesting to plot Equations (12.4.9) and (12.4.13) as a function of ω/ω_o for various values of ζ to examine what sort of control force is necessary in each case. This is done in Figure 12.55; and in Figure 12.56, the control forces required using the latter two control strategies discussed above are compared as a function of ω/ω_0 for $\zeta = 0.05$. For the first strategy, the control force is simply equal to the inverse of the primary excitation force in all cases.

12.5 FEEDFORWARD CONTROL: SINGLE ISOLATOR BETWEEN A RIGID BODY AND A FLEXIBLE BEAM

In this section, a theoretical model will be developed to calculate the optimal feedforward control force required to minimise the vibratory power transmission from a rigid body through an isolator into a flexible beam as a function of excitation frequency. As there is only one attachment point, the minimum power transmission will be zero, as will be the force and displacement at the base of the isolator when the control force is optimal. However, in practice, the controller is never able to achieve the exact optimal control force; thus there will always be some residual power transmitted into the supporting beam. Thus, the analysis outlined here for deriving the optimal control force is useful for investigating the effects of the control forces varying slightly from the theoretical optimum.

The analysis begins with the development of the equations of motion for each of the three parts of the total system: the rigid body, the isolator and the beam. These are then combined into a single matrix equation for the total system which allows the vibratory power transmission into the beam to be calculated as a function of control force phase and magnitude. Quadratic optimisation is then used, as described in Chapter 8, to find the optimal control force. Once the optimal control force for a particular system configuration is found, the maximum achievable reduction in power transmission into the beam assuming an ideal feedforward controller is calculated.

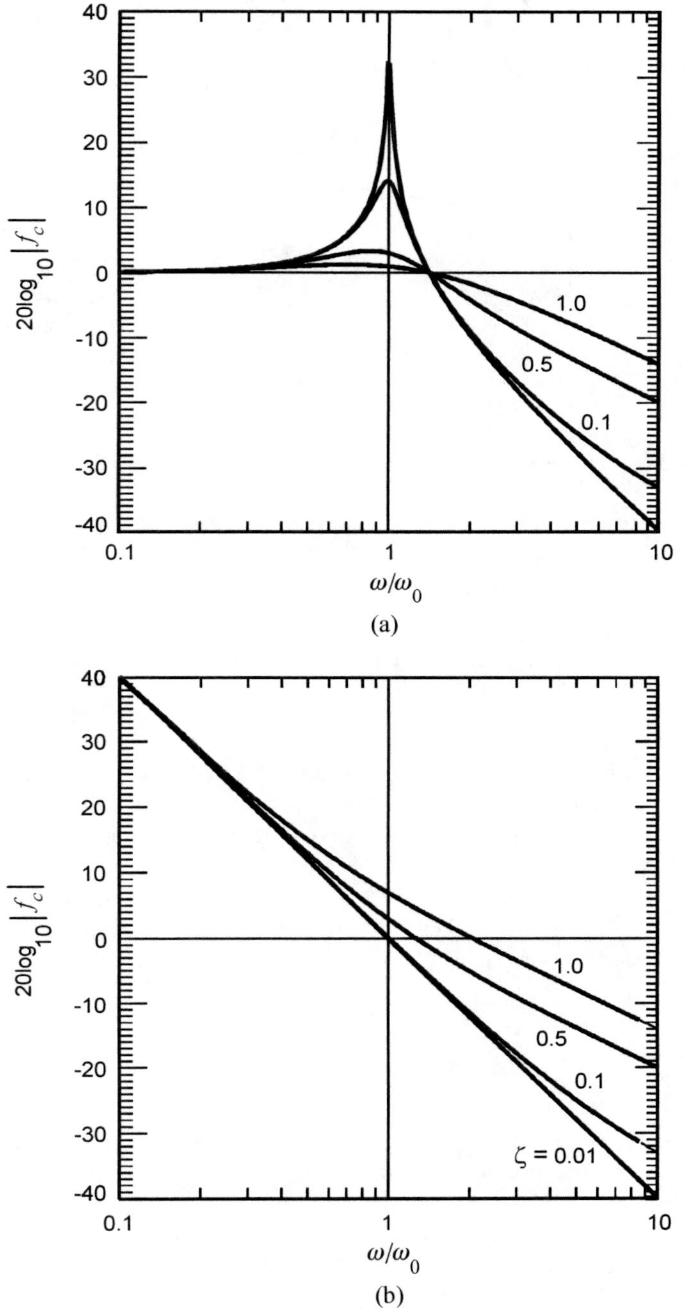

Figure 12.55 Effect of damping ratio and excitation frequency on the magnitude of the required control force for a single-degree-of-freedom system: (a) control force acting only on the support structure (Figure 12.52); (b) control force acting on the rigid body and reacting on the support structure (Figure 12.54).

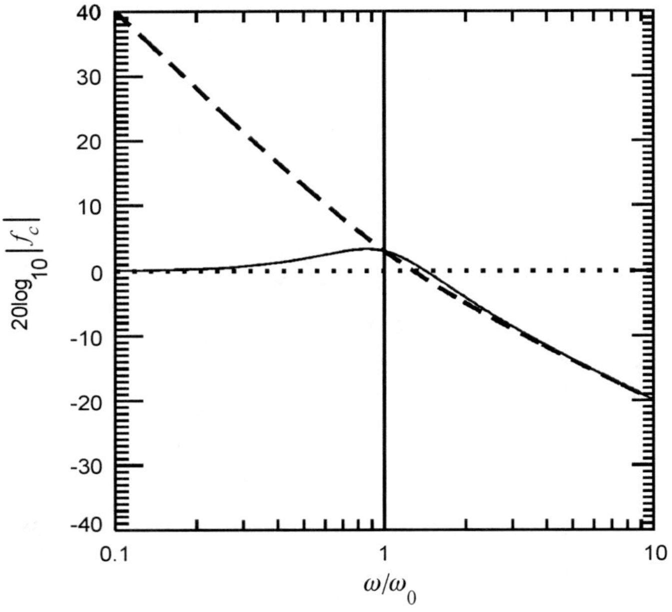

Figure 12.56 Effect of control strategy on the control force required to achieve optimal control for a damping ratio ζ = 0.05 as a function of ω/ω_0.
——— control force acting only on the support structure;
– – – control force acting on the rigid body and reacting on the support structure;
• • • • control force acting on the rigid body.

In the following analysis the superscripts and subscripts 0 refer to the centre of gravity of the rigid body, *t* refers to the top of the isolators, *b* refers to the bottom of the isolators, *J* refers to the *J*th isolator, *I* refers to the *I*th beam vibration mode and *c* refers to the control force.

In this case, the primary excitation force is not simply a vertical force, but a combination of all possible translations and rotations (that is, it is arbitrary). Thus, the primary excitation force may be written as:

$$\boldsymbol{Q_0} = \left[F_x F_y F_z M_x M_y M_z\right]^{\mathrm{T}} \tag{12.5.1}$$

To be consistent with published literature, the vertical direction is now the *z*-direction, the coordinate *x* is in the direction of the beam axis and *y* is the coordinate across the beam as shown in Figure 12.57.

The quantities labelled *F* and *M* in Equation (12.5.1) refer to force and moment inputs respectively on the rigid body. Note that another rigid body is shown between the beam and the lower end of the suspension in Figure 12.57. This more closely simulates the practical implementation of a suspension system such as this, where the lower part of the actuator and spring support are well represented by a rigid mass. In this case, the rigid body is modelled as a point mass loading on the beam and included in the beam equation of motion.

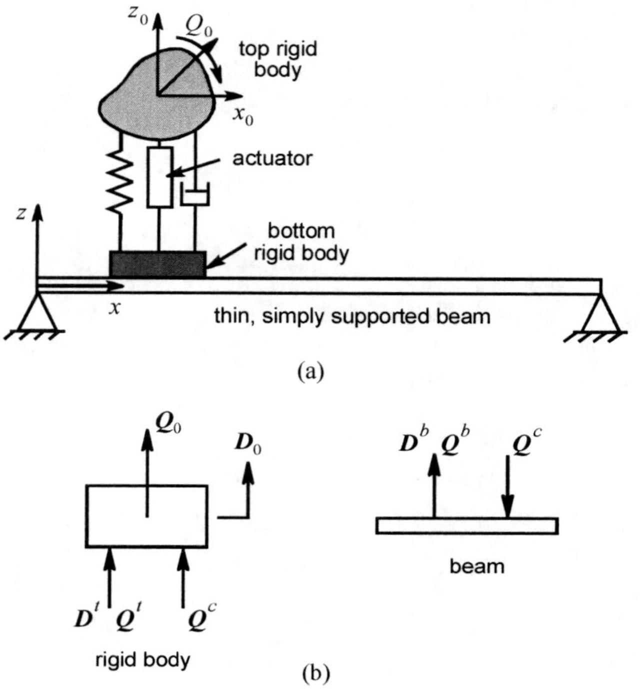

Figure 12.57 Theoretical model for a single isolator between a rigid body and a flexible beam: (a) beam-isolator model; (b) sign conventions.

12.5.1 Rigid Body Equation of Motion

Referring to Figure 12.57, the equation of motion of the rigid body can be written in matrix form as (Pan et al., 1993):

$$\boldsymbol{Z}\boldsymbol{D}_0 = \boldsymbol{Q}_0 + \boldsymbol{R}^t\boldsymbol{Q}^t \tag{12.5.2}$$

where $\boldsymbol{Z} = -\omega^2\boldsymbol{m}_0$, ω is the primary force excitation frequency and $\boldsymbol{m}_0$ is the diagonal inertia matrix given by:

$$\boldsymbol{m}_0 = \begin{bmatrix} m & & & & & \\ & m & & & & \\ & & m & 0 & & \\ & & & I_x & & \\ & 0 & & & I_y & \\ & & & & & I_z \end{bmatrix} \tag{12.5.3}$$

where m is the mass of the body and I_x, I_y and I_z are the moments of inertia of the body around its x-, y- and z-axes respectively. $\boldsymbol{D_0}$ in Equation (12.5.2) is the complex displacement matrix of the centre of gravity of the rigid body and is given in terms of the three linear displacements x_0, y_0, z_0, and the three angular rotational displacements θ_{0x}, θ_{0y}, and θ_{0z} as follows:

$$\boldsymbol{D_0} = [x_0 \; y_0 \; z_0 \; \theta_{x0} \; \theta_{y0} \; \theta_{z0}]^{\mathrm{T}} \tag{12.5.4}$$

The quantity $\boldsymbol{Q}^t = [F_x^t, F_y^t \; F_z^t \; M_x^t \; M_y^t \; M_z^t]^{\mathrm{T}}$ is the elastic force matrix for the passive mount, which acts at the point (x_1^t, y_1^t, z_1^t), where the top of the mount is connected to the rigid body. The quantity $\boldsymbol{R}^t$ is the matrix that accounts for coupling between moments and translational displacements, and is defined as follows:

$$\boldsymbol{R}^t = \begin{bmatrix} 1 & 0 & 0 & 0 & 0 & 0 \\ 0 & 1 & 0 & 0 & 0 & 0 \\ 0 & 0 & 1 & 0 & 0 & 0 \\ 0 & -z^t & y^t & 1 & 0 & 0 \\ z^t & 0 & -x^t & 0 & 1 & 0 \\ -y^t & x^t & 0 & 0 & 0 & 1 \end{bmatrix} \tag{12.5.5}$$

The complex stiffness matrix of the mount (which will be used later) is defined as:

$$\boldsymbol{K} = \begin{bmatrix} k_x & 0 & 0 & 0 & 0 & 0 \\ 0 & k_y & 0 & 0 & 0 & 0 \\ 0 & 0 & k_z & 0 & 0 & 0 \\ 0 & 0 & 0 & k_{\theta x} & 0 & 0 \\ 0 & 0 & 0 & 0 & k_{\theta y} & 0 \\ 0 & 0 & 0 & 0 & 0 & k_{\theta_z} \end{bmatrix} \tag{12.5.6}$$

where k_x, ··· $k_{\theta z}$ are complex stiffness coefficients of the mount corresponding to each degree of freedom:

$$k_x = k_x^0(1 + \mathrm{j}\eta_x)\,; \quad k_y = k_y^0(1 + \mathrm{j}\eta_y)\,; \; \cdots ; \; k_{\theta z} = k_{\theta z}^0(1 + \mathrm{j}\eta_{\theta z}) \tag{12.5.7a–c}$$

where k_x^0, ..., $k_{\theta z}^0$ are the stiffness coefficients and η_x, ..., $\eta_{\theta x}$ are the structural loss factors. For a cylindrical mount, the stiffness coefficients can be calculated according to Table 12.3, where E, v, L and a are respectively Young's modulus of elasticity and Poisson's ratio for the material, the length of the mount and the section radius of the mount.

Table 12.3 Stiffness coefficients of a cylindrical isolator.

k_x^0	k_y^0	k_z^0	$k_{\theta x}^0$	$k_{\theta y}^0$	$k_{\theta z}^0$
$\dfrac{3\pi E a^4}{4L^3}$	$\dfrac{3\pi E a^4}{4L^3}$	$\dfrac{\pi E a^2}{L}$	$\dfrac{\pi E a^4}{4L}$	$\dfrac{\pi E a^4}{4L}$	$\dfrac{\pi E a^4}{4(1+\nu)L}$

12.5.2 Supporting Beam Equation of Motion

For the thin beam considered here, with its thickness far smaller than its width and its width far smaller than its length, the transverse vibration in the z-direction dominates its motion. Errors arising from ignoring other motions, such as lateral vibration in the y-direction, rotational vibration around the x-direction and twisting vibration around the z-direction, are expected to be insignificant, as the beam is sufficiently thin and the base of the mount is attached at $y_b = 0$.

Force and moment components in the force vector $\boldsymbol{Q}^b = [F_{bx}, F_{by}, F_{bz}, M_{bx}, M_{by}, M_{bz}]^T$ for the mount acting on the beam are assumed to be point actions, so that Dirac delta functions and their differentials may be used for describing the forces and moments in terms of quantities per unit area. The point actions that generate the lateral vibration of the beam only include three such items:

$$F_{bz}\delta(x - x_b)\ ;\quad \frac{h}{2}F_{bx}\frac{\partial\delta(x - x_b)}{\partial x}\quad \text{and}\ M_{by}\frac{\partial\delta(x - x_b)}{\partial x}$$

Thus, the transverse displacement, $w(x, t: x_b)$, of the beam can be determined by using the following equation (see Section 2.3):

$$\rho S\frac{\partial^2 w}{\partial t^2} + EI_b\frac{\partial^4 w}{\partial x^4} = F_{bz}\delta(x - x_b) + (\frac{h}{2}F_{bx} + M_{by})\frac{\partial\delta(x - x_b)}{\partial x} + m^b\omega^2 w\,\delta(x - x_b) \qquad (12.5.8)$$

where m^b is the mass of the isolator base.

Boundary conditions at both simply supported ends of the beam are:

$$w = \frac{\partial^2 w}{\partial x^2} = 0 \qquad (12.5.9a,b)$$

where $F_{bx}, \cdots, M_{bz}$ are the elements of the force matrix $\boldsymbol{Q}^b$; and ρ, S and I_b are respectively the density of the material, the beam cross-sectional area and the second moment of area of the beam section about the neutral plane parallel with the z-axis. The location of the beam excitation point, or elastic mount support point, is $(x_b, 0, h/2)$. Damping is included on a modal basis.

For harmonic force excitation, the lateral displacement w in the z-direction for the simply supported beam can be written as (see Chapter 2):

$$w = \left[\sin\frac{\pi x}{L_b},\ \sin\frac{2\pi x}{L_b},\ \cdots,\ \sin\frac{p\pi x}{L_b}\right]\boldsymbol{w_p} \qquad (12.5.10)$$

where N is the number of modes included in the analysis, $\sin(i\pi x/L_b)$, $(i = 1, \cdots, N)$ are the mode shape functions, and $\boldsymbol{w}_p = [w_1, w_2, \cdots, w_P]^{\mathrm{T}}$ is the modal complex amplitude coefficient matrix, which depends upon x_b and ω. Substituting Equation (12.5.10) into Equation (12.5.8), and using the orthogonal property of the mode shape functions, gives:

$$m_i(\omega_i^2 - \omega^2 + j\eta_i\omega_i^2)w_i$$

$$= \left[-\frac{i\pi h}{2L_b}\cos\frac{i\pi x_b}{L_b}, 0, \sin\frac{i\pi x_b}{L_b}, 0, -\frac{i\pi}{L_b}\cos\frac{i\pi x_b}{L_b}, 0 \right] \boldsymbol{Q}^b + \sum_{i'=1}^{N} C_{i'i} w_{i'} \qquad (i = 1, \cdots, N)$$

(12.5.11)

where L_b and h are the length and thickness of the beam respectively, and ω_i, $(i = 1, \cdots, N)$ are the resonance frequencies of the beam, given by:

$$\omega_i = \left(\frac{i^2\pi^2}{L_b^2} \right) \sqrt{\frac{EI_b}{\rho S}} \qquad (12.5.12)$$

The quantity E is Young's modulus of elasticity, I_b is the second moment of area of the beam cross-section, and m_i is the modal mass for mode i, which is equal to half of the beam mass for each mode:

$$m_i = \frac{1}{2}\rho A L_b \qquad (i = 1, \cdots, N) \qquad (12.5.13)$$

The quantity η_i is the modal loss factor for mode i and is related to the 60 dB modal decay time T_i by:

$$\eta_i = \frac{4.4\pi}{T_i \omega_i} \qquad (12.5.14)$$

$C_{i',i}$ $(i', i = 1, N)$ is the concentrated mass contribution given by:

$$C_{i',i} = m^b \omega^2 \sin\frac{i\pi x_b}{L_b} \sin\frac{i'\pi x_b}{L_b} \qquad (12.5.15)$$

where m^b is the mass of the isolator base.

Equation (12.5.11) can be expressed compactly by making use of matrix notation. Thus,

$$\boldsymbol{Z}_p \boldsymbol{w}_p = \boldsymbol{R}^b \boldsymbol{Q}^b \qquad (12.5.16)$$

where

$$\boldsymbol{Z}_p = \begin{bmatrix} \chi_1 - C_{11} & -C_{12} & \cdots & -C_{1N} \\ -C_{21} & \chi_2 - C_{22} & & \vdots \\ \vdots & & \ddots & \\ -C_{N1} & \cdots & & \chi_N - C_{NN} \end{bmatrix} \qquad (12.5.17)$$

where

$$\chi_i = m_i(\omega_i^2 + j\eta_i\omega_i^2 - \omega^2) \tag{12.5.18}$$

and

$$\boldsymbol{R}^b = \begin{bmatrix} \frac{-h\pi}{2L_b}\cos\frac{\pi x_b}{L_b} & 0 & \sin\frac{\pi x_b}{L_b} & 0 & -\frac{\pi}{L_b}\cos\frac{\pi x_b}{L_b} & 0 \\ \frac{-2h\pi}{2L_b}\cos\frac{2\pi x_b}{L_b} & 0 & \sin\frac{2\pi x_b}{L_b} & 0 & -\frac{2\pi}{L_b}\cos\frac{2\pi x_b}{L_b} & 0 \\ \vdots & \vdots & \vdots & \vdots & \vdots & \vdots \\ \frac{-Nh\pi}{2L_b}\cos\frac{N\pi x_b}{L_b} & 0 & \sin\frac{N\pi x_b}{L_b} & 0 & -\frac{N\pi}{L_b}\cos\frac{N\pi x_b}{L_b} & 0 \end{bmatrix} \tag{12.5.19}$$

Using elastic thin plate theory (as the beam is much wider than it is thick), the elements of the displacement matrix $\boldsymbol{D}^b = [\zeta_x, \zeta_y, w, \theta_x, \theta_y, \theta_z]^{\mathrm{T}}$ at the elastic mount support point can be described in terms of the lateral displacement w as follows:

$$\xi_x = -\frac{h}{2}\frac{\partial w}{\partial x};\ \xi_y = -\frac{h}{2}\frac{\partial w}{\partial y} = 0;\ \theta_x = \frac{\partial w}{\partial y} = 0;\ \theta_y = -\frac{\partial w}{\partial x};\ \theta_z = 0 \tag{12.5.20 a–g}$$

Substituting Equation (12.5.10) into Equation (12.5.20) gives:

$$\boldsymbol{D}^b = (\boldsymbol{R}^b)^{\mathrm{T}}\boldsymbol{w}_p \tag{12.5.21}$$

12.5.3 System Equation and Power Transmission

The matrices describing the forces and moments acting on the rigid body $\boldsymbol{Q}^t = [F_{tx}, F_{ty}, F_{tz}, M_{tx}, M_{ty}, M_{tz}]^{\mathrm{T}}$ and on the support beam $\boldsymbol{Q}^b = [F_{bx}, F_{by}, F_{bz}, M_{bx}, M_{by}, M_{bz}]^{\mathrm{T}}$ are both proportional to the mount's linear and angular deformations which are described by the relative displacement of its top and bottom ends. Thus,

$$\boldsymbol{Q}^t = -\boldsymbol{Q}^b = \boldsymbol{K}(\boldsymbol{D}^b - \boldsymbol{D}^t) \tag{12.5.22a,b}$$

where $\boldsymbol{D}^t$ and $\boldsymbol{D}^b$ are respectively the displacement matrices for the top and bottom ends of the mount. It is easily shown that:

$$\boldsymbol{D}^t = (\boldsymbol{R}^t)^{\mathrm{T}}\boldsymbol{D}_0 \tag{12.5.23}$$

Selecting $\boldsymbol{D}_0$ (defined in Equation (12.5.4)) and $\boldsymbol{w}_p$ (defined following 12.5.10)) as the unknowns and substituting Equation (12.5.22) into Equations (12.5.2) and (12.5.6), the system can finally be expressed by the following complex linear equations with a symmetrical coefficient matrix:

$$\begin{bmatrix} A_{11} & A_{12} \\ A_{21} & A_{22} \end{bmatrix} \begin{bmatrix} D_0 \\ w_p \end{bmatrix} = \begin{bmatrix} Q_0 \\ \mathbf{0} \end{bmatrix} \tag{12.5.24}$$

where $\mathbf{0}$ is a $(P \times 1)$ zero vector and

$$A_{11} = Z + R^t K (R^t)^{\mathrm{T}} \tag{12.5.25}$$

$$A_{12} = (A_{21})^{\mathrm{T}} = -R^t K (R^b)^{\mathrm{T}} \tag{12.5.26a,b}$$

$$A_{22} = Z_p + R^b K (R^b)^{\mathrm{T}} \tag{12.5.27}$$

When $Q_0 = 0$, and all loss factors are assumed to be zero, Equation (12.5.24) represents the free vibration of the coupled system. Solving the eigenvalue problem of the following coefficient matrix which, in this case, is a real symmetrical matrix:

$$A = \begin{bmatrix} A_{11} & A_{12} \\ A_{21} & A_{22} \end{bmatrix} = A_1 - \omega^2 A_2 \tag{12.5.28a,b}$$

gives $N + 6$ resonance frequencies and corresponding mode shape vectors for the coupled system. When a force actuator is connected in parallel with the passive mount, the resulting active control force vector Q_c acts on both the upper rigid body and the support beam simultaneously, with the same amplitudes but opposite phases, and Equation (12.5.24) becomes:

$$\begin{bmatrix} A_{11} & A_{12} \\ A_{21} & A_{22} \end{bmatrix} \begin{bmatrix} D_0 \\ w_p \end{bmatrix} = \begin{bmatrix} Q_0 + R^t Q^c \\ -R^b Q^c \end{bmatrix} \tag{12.5.29}$$

A positive control force is defined here as one which acts at the bottom of the isolator in the direction of positive displacement (and in the opposite direction at the top of the isolator).

The time-averaged power transmission, P_0, into the support beam, which is the quantity that must be controlled, can be calculated by using the following equation:

$$P_0 = \mathrm{Re}\left\{-\frac{1}{2}\mathrm{j}\omega(Q^b - Q^c)^{\mathrm{H}} D^b\right\} = \frac{\omega}{2}\mathrm{Im}\left\{(Q^b - Q^c)^{\mathrm{H}} D^b\right\}$$
$$= -\frac{\omega}{2}\mathrm{Im}\left\{(D^b)^{\mathrm{H}}(Q^b - Q^c)\right\} \tag{12.5.30a–c}$$

where the superscript H represents the transpose and conjugate of a matrix. Note that removing j from the right-hand side of Equation (12.5.30a) and taking the imaginary part causes a sign change, as does changing the transpose conjugate from the force matrix to the displacement matrix. Positive power transmission is defined as in the direction of positive force and displacement, and is actually out of the beam. This is the reason for the minus sign in Equation (12.5.30a). Because only one mount is being considered, the power transmission P_0 corresponding to the optimal control force vector should be zero.

12.5.4 Optimum Control Force and Minimum Power Transmission

If the driving force matrix remains unchanged, each element of the matrices, $\boldsymbol{D}^t$, $\boldsymbol{D}^b$ and $\boldsymbol{Q}^b$ will generally be a linear function of the control force matrix elements. Therefore, the output power P_0 into the supporting beam is generally a quadratic function of these control variables and depends upon the driving frequency ω. For a specified driving frequency, the magnitude of the power transmission reduction will depend upon system parameters. Analytically, it is possible to select these parameters so that the power transmission or power reduction reaches an optimal value in a given frequency range. Substituting Equation (12.5.21) for $\boldsymbol{D}^b$ and Equation (12.5.25) for $\boldsymbol{D}^t$, Equation (12.5.22) can be used to write $(\boldsymbol{D}^b)^{\mathrm{H}}(\boldsymbol{Q}^b - \boldsymbol{Q}^c)$ as:

$$(\boldsymbol{D}^b)^{\mathrm{H}}(\boldsymbol{Q}^b - \boldsymbol{Q}^c) = -(\boldsymbol{w}_p)^{\mathrm{H}}\boldsymbol{R}^b\left[\boldsymbol{K}(\boldsymbol{R}^b)^{\mathrm{T}}\boldsymbol{w}_p - \boldsymbol{K}(\boldsymbol{R}^t)^{\mathrm{T}}\boldsymbol{D}_0 + \boldsymbol{Q}^c\right] \tag{12.5.31}$$

Using Equation (12.5.29):

$$\boldsymbol{D}_0 = \boldsymbol{B}_{11}\left[\boldsymbol{Q}_0 + \boldsymbol{R}^t\boldsymbol{Q}^c\right] - \boldsymbol{B}_{12}\boldsymbol{R}^b\boldsymbol{Q}^c \tag{12.5.32}$$

$$\boldsymbol{w}_p = \boldsymbol{B}_{21}\left[\boldsymbol{Q}_0 + \boldsymbol{R}^t\boldsymbol{Q}^c\right] - \boldsymbol{B}_{22}\boldsymbol{R}^b\boldsymbol{Q}^c \tag{12.5.33}$$

where the matrices $\boldsymbol{B}_{11}$, $\boldsymbol{B}_{12}$, $\boldsymbol{B}_{21}$ and $\boldsymbol{B}_{22}$ are the sub-matrix elements of the inverse of system matrix $\boldsymbol{A}$, as follows:

$$\begin{bmatrix} \boldsymbol{B}_{11} & \boldsymbol{B}_{12} \\ \boldsymbol{B}_{21} & \boldsymbol{B}_{22} \end{bmatrix} = \begin{bmatrix} \boldsymbol{A}_{11} & \boldsymbol{A}_{12} \\ \boldsymbol{A}_{21} & \boldsymbol{A}_{22} \end{bmatrix}^{-1} \tag{12.5.34}$$

$$\boldsymbol{B}_{11} = \left[\boldsymbol{A}_{11} - \boldsymbol{A}_{12}\boldsymbol{A}_{22}^{-1}\boldsymbol{A}_{21}\right]^{-1} \tag{12.5.35}$$

$$\boldsymbol{B}_{22} = \left[\boldsymbol{A}_{22} - \boldsymbol{A}_{21}\boldsymbol{A}_{11}^{-1}\boldsymbol{A}_{12}\right]^{-1} \tag{12.5.36}$$

$$\boldsymbol{B}_{12} = -\boldsymbol{A}_{11}^{-1}\boldsymbol{A}_{12}\left[\boldsymbol{A}_{22} - \boldsymbol{A}_{21}\boldsymbol{A}_{11}^{-1}\boldsymbol{A}_{12}\right]^{-1} \tag{12.5.37}$$

$$\boldsymbol{B}_{21} = -\boldsymbol{A}_{22}^{-1}\boldsymbol{A}_{21}\left[\boldsymbol{A}_{11} - \boldsymbol{A}_{12}\boldsymbol{A}_{22}^{-1}\boldsymbol{A}_{21}\right]^{-1} \tag{12.5.38}$$

Substituting Equation (12.5.32) and Equation (12.5.33) into Equation (12.5.31) gives the following quadratic form of $(\boldsymbol{D}^b)^{\mathrm{H}}(\boldsymbol{Q}^b - \boldsymbol{Q}_c)$:

$$(\boldsymbol{D}^b)^{\mathrm{H}}(\boldsymbol{Q}^b - \boldsymbol{Q}^c) = (\boldsymbol{Q}^c)^{\mathrm{H}}\boldsymbol{a}\boldsymbol{Q}^c + (\boldsymbol{Q}^c)^{\mathrm{H}}\boldsymbol{b}_1 + \boldsymbol{b}_2\boldsymbol{Q}^c + c \tag{12.5.39}$$

where

$$\boldsymbol{a} = \boldsymbol{G}_1\boldsymbol{K}\boldsymbol{G}_2 - \boldsymbol{G}_1 \tag{12.5.40}$$

$$b_1 = G_1 K G_3 \tag{12.5.41}$$

$$b_2 = G_4 K G_2 - G_4 \tag{12.5.42}$$

$$c = G_4 K G_3 \tag{12.5.43}$$

where the matrices G_1, G_2, G_3 and G_4 are defined as:

$$G_1 = (R^t)^T B_{21}{}^H R^b - (R^b)^T B_{22}{}^H R^b \tag{12.5.44}$$

$$G_2 = (R^t)^T B_{11} R^t - (R^t)^T B_{12} R^b - (R^b)^T B_{21} R^t + (R^b)^T B_{22} R^b \tag{12.5.45}$$

$$G_3 = \left[(R^t)^T B_{11} - (R^b)^T B_{21}\right] Q_0 \tag{12.5.46}$$

$$G_4 = Q_0{}^T B_{21}{}^H R^b \tag{12.5.47}$$

Thus, the output power P_0 in Equation (12.5.28) can be expressed as the following real quadratic function:

$$\begin{aligned} P_0 &= -\frac{\omega}{2}\mathrm{Im}\left\{(D^b)^H(Q^b - Q^c)\right\} \\ &= -\frac{\omega}{2}\left\{(q^c)^T\begin{bmatrix} a^i & a^r \\ -a^r & a^i\end{bmatrix} q^c + (q^c)^T\begin{bmatrix} b_1^i \\ -b_1^r\end{bmatrix} + \begin{bmatrix} b_2^i & b_2^r\end{bmatrix} q^c + c^i\right\} \end{aligned} \tag{12.5.48a,b}$$

or in terms of the following equivalent expression, with a symmetrical coefficient matrix for the quadratic term:

$$P_0 = -\frac{\omega}{2}\left\{(q^c)^T \alpha q^c + (q^c)^T \beta + \beta^T q^c + c^i\right\} \tag{12.5.49}$$

where

$$\alpha = \alpha^T = \frac{1}{2}\begin{bmatrix} a^i + (a^i)^T & a^r - (a^r)^T \\ -a^r + (a^r)^T & a^i + (a^i)^T \end{bmatrix} \tag{12.5.50a,b}$$

and

$$\beta = \frac{1}{2}\begin{bmatrix} (b_2^i)^T + b_1^i \\ (b_2^r)^T - b_1^r \end{bmatrix} \tag{12.5.51}$$

and the real matrices a^r, a^i, b_1^r, ··· represent respectively the real and imaginary parts of the complex matrices a, b_1, b_2 and the constant c. Clearly, the real output power for the uncontrolled case where $q^c = 0$ is given by $P_0^{\,l} = -\frac{1}{2}\omega c^i$. The real control 'force' vector, q^c, of twelve elements consists of real Q^{cr} and imaginary Q^{ci} parts of the control force vector Q^c; thus,

$$\boldsymbol{q}^c = \begin{bmatrix} \boldsymbol{Q}^{cr} \\ \boldsymbol{Q}^{ci} \end{bmatrix} \tag{12.5.52}$$

A reduction in output power as a result of the action of the control forces will occur when the relationship $\Delta P = P_0^{\prime} - P_0 < 0$ is satisfied (as positive power flow is defined upwards, out of the beam). In other words, control force vectors satisfying the following inequality can result in a reduction of the output power:

$$(\boldsymbol{q}^c)^{\mathrm{T}}\boldsymbol{\alpha}\boldsymbol{q}^c + (\boldsymbol{q}^c)^{\mathrm{T}}\boldsymbol{\beta} + \boldsymbol{\beta}^{\mathrm{T}}\boldsymbol{q}^c < 0 \tag{12.5.53}$$

From Chapter 8, the quadratic function of Equation (12.5.49) has a minimum (which should be equal to zero) given by:

$$\boldsymbol{P}_{\min} = \frac{\omega}{2}\left\{\boldsymbol{\beta}^{\mathrm{T}}\boldsymbol{\alpha}^{-1}\boldsymbol{\beta} - c^{i}\right\} \tag{12.5.54}$$

corresponding to an optimum control force vector of:

$$\boldsymbol{q}_{\mathrm{opt}}^{c} = -\boldsymbol{\alpha}^{-1}\boldsymbol{\beta} \tag{12.5.55}$$

where the coefficient matrix $\boldsymbol{\alpha}$ is a positive definite matrix.

The quantity $P_{\min}$ is theoretically equal to zero for an ideal controller, but the preceding analysis allows the characteristics of a non-ideal controller to be taken into account.

12.6 FEEDFORWARD CONTROL: MULTIPLE ISOLATORS BETWEEN A RIGID BODY AND A FLEXIBLE PANEL

In many cases, for example in submarines, equipment is isolated from its final support structure using a two-stage isolation system, which consists of an intermediate structure inserted between the machine and the support structure, which is shown in its simplest form in Figure 12.58.

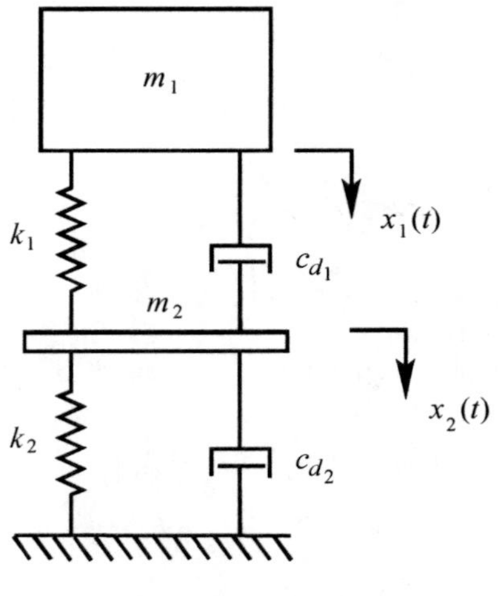

Figure 12.58 Two-stage vibration isolation system.

In a submarine, for example, the intermediate mass m_2 is the machinery support platform, and the base support shown in Figure 12.58 is the hull. However, in this case, the intermediate structure cannot be modelled simply as a mass as it is itself a multi-modal structure, with many resonances in the frequency range of interest which serve to reduce the effectiveness of the double mounting arrangement (which otherwise would be much more effective than the single-degree-of-freedom system with no intermediate mass). Thus, there is considerable interest in devising means of reducing the transmission of vibratory energy into the intermediate structure. One possibility is to use active control, and the purpose of this section is to investigate the performance improvement that can be achieved in terms of reduction in the response of the intermediate structure by the use of active control. To begin, it will be assumed that the final support structure is rigid; but in the final section of this chapter, the use of a flexible cylinder as the final support structure and minimisation of the vibratory power transmission into it will be examined.

In the work discussed in this section, a periodic excitation force will be assumed, as the analysis is directed towards the use of a feedforward active control system. To begin, only the transmission of vertical excitation forces into a clamped edge flexible panel will be considered. This will be extended later to include horizontal forces and moments.

12.6.1 Vertical Excitation Forces Only

This problem was first investigated by Jenkins et al. (1988, 1993), who examined the problem of four active isolators between a rigid body and an edge clamped flexible rectangular panel using finite element analysis. The active isolators consisted of an electromagnetic active element in parallel with a cylindrical flexible passive element with some internal damping, as shown in Figure 12.59.

Only vertical input forces from the rigid body m_1 were assumed, and the purpose of the active isolators was to minimise the resulting flexural vibration levels in the flexible support panel, so the cost function J to be minimised was:

$$J = \sum_{n=1}^{N} |w_n|^2 = \boldsymbol{W}^{\mathrm{H}}\boldsymbol{W} \tag{12.6.1a,b}$$

The total displacement at any point on the panel is a linear superposition of the displacements $\boldsymbol{W}_p$ due to the primary excitation force and those due to the control forces $\boldsymbol{f}_c$ so that:

$$\boldsymbol{W} = \boldsymbol{W}_p + \boldsymbol{R}\boldsymbol{f}_c \tag{12.6.2}$$

where $\boldsymbol{R}$ is the complex receptance matrix relating the displacement at any point on the panel to an input force at any other point.

Thus, the cost function to be minimised can be written as:

$$J = (\boldsymbol{W}_p + \boldsymbol{R}\boldsymbol{f}_c)^{\mathrm{H}}(\boldsymbol{W}_p + \boldsymbol{R}\boldsymbol{f}_c) \tag{12.6.3}$$

Rearranging gives:

$$J = \boldsymbol{f}_c^{\mathrm{H}}\boldsymbol{R}^{\mathrm{H}}\boldsymbol{R}\boldsymbol{f}_c + \boldsymbol{f}_c^{\mathrm{H}}\boldsymbol{R}^{\mathrm{H}}\boldsymbol{W}_p + \boldsymbol{W}_p^{\mathrm{H}}\boldsymbol{R}\boldsymbol{f}_c + \boldsymbol{W}_p^{\mathrm{H}}\boldsymbol{W}_p \tag{12.6.4}$$

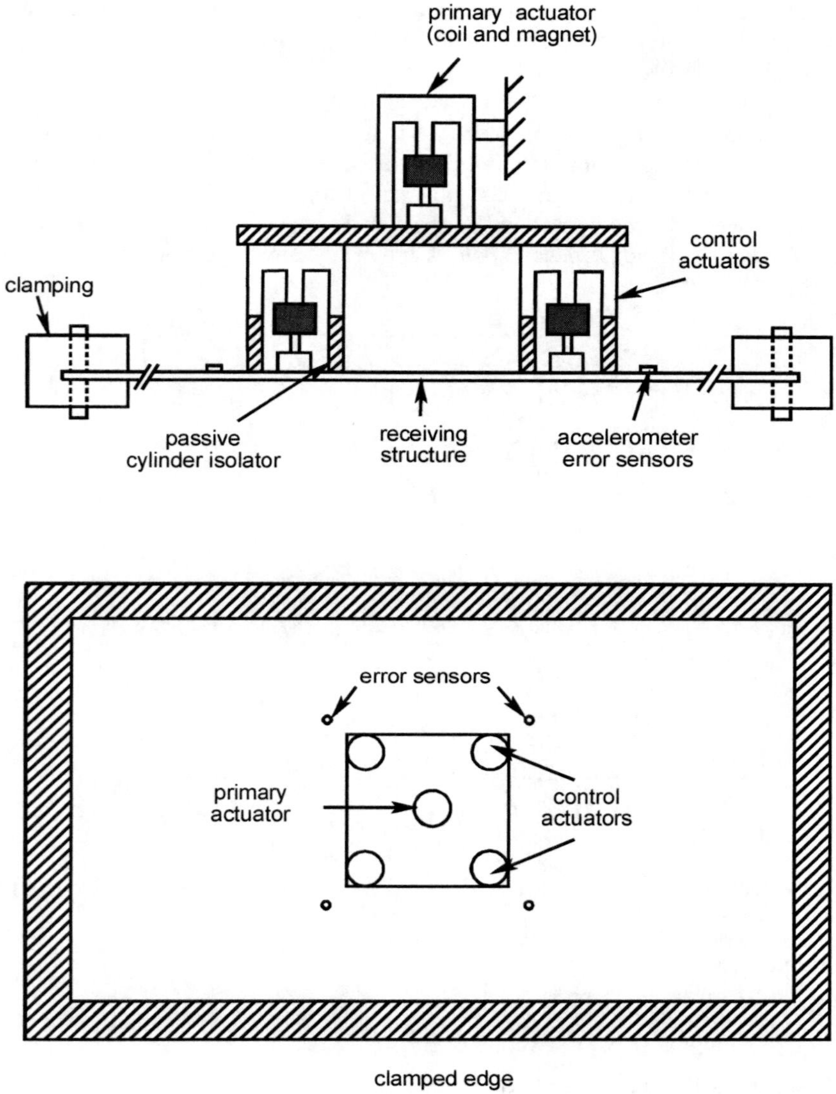

Figure 12.59 Active isolation of a rigid machine from a flexible clamped panel.

which is a Hermitian quadratic form with a unique minimum (provided $\boldsymbol{R}^{\mathrm{H}}\boldsymbol{R}$ is positive). Finding the minimum of Equation (12.6.4) results in the following optimal control force vector (the length of which is equal to the number of control forces, which in this case is four):

$$\boldsymbol{f}_c = -\left(\boldsymbol{R}^{\mathrm{H}}\boldsymbol{R}\right)^{-1}\boldsymbol{R}^{\mathrm{H}}\boldsymbol{W}_p \tag{12.6.5}$$

which is of the well-known form (see Chapters 6 and 8):

$$\boldsymbol{f}_c = -\boldsymbol{a}^{-1}\boldsymbol{b} \tag{12.6.6}$$

In practice, these optimal control forces can be determined by using a feedforward controller containing an adaptive filter, the weights of which are updated using a gradient descent algorithm as discussed in Chapter 6.

The reduction in vibration levels (averaged over 81 locations on the panel) for the system illustrated in Figure 12.59 are shown in Figure 12.60. Experimental data were obtained using a controller that minimised the acceleration levels at eight locations on the panel, and the theoretical curves were obtained using finite element analysis. The large difference between the theoretical and experimental results at low frequencies is possibly a result of limitations in the actuator responses and the dynamic range of the control system.

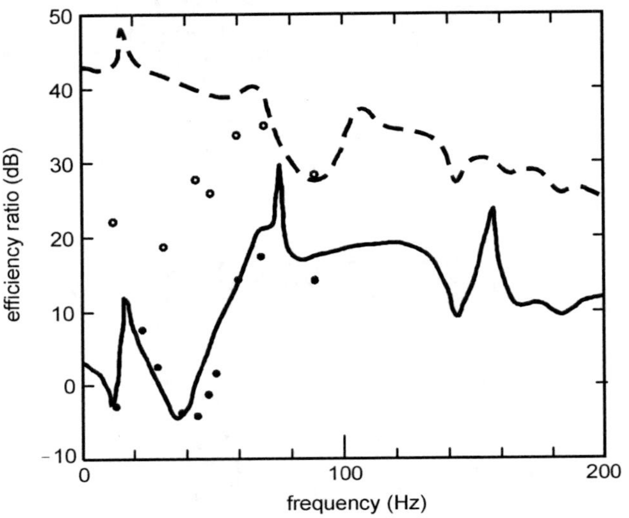

Figure 12.60 Reduction in average panel acceleration levels for the four isolator system shown in Figure 12.59.
——— finite element analysis prediction for passive isolators only;
– – – finite element analysis prediction for active plus passive isolation;
• • • • measured passive isolation;
∘ ∘ ∘ ∘ measured active plus passive isolation.

12.6.2 Generalised Excitation Forces

Unfortunately, the ideal situation discussed in the previous section in which only vertical forces are transmitted to the flexible support structure is difficult to realise in practice, although in some cases replacing the passive isolators with an air bag system to prevent transmission of horizontal forces and moments may be feasible (Jenkins et al., 1990).

In this section, a complete analysis of the transmission of vibratory power from a rigid machine to a flexible, simply supported rectangular panel which takes into account all possible force and moment components is undertaken. This is important, as in most practical systems, power transmission through an isolator can be very complicated, because equipment supported on isolators has many degrees of freedom. Thus, at the point of connection of each mount to the support structure, various forces and moments exist that excite various types of waves in the support structure. In fact, the vibration of the machine and the support structure is coupled through the mounts and the modal response of the coupled system is directly associated with the magnitude of the power injected by each of the exciting forces and moments. In support of the preceding argument, White (1990) showed that the power transmission due to moment excitation induces more flexural wave power transmission into a flexible support panel than does excitation by normal forces at higher frequencies. Also, it is well known that longitudinal and shear waves, which do not radiate much noise or contribute significantly to feelable vibration levels on structures, can have large amounts of their energy converted to flexural waves at structural discontinuities; thus it is important to control the injection of power into these wave types as well.

Experimental work undertaken by Pinnington and White (1981), Pinnington (1987), and Pavic and Oreskovic (1977) indicates that vibratory power transmission through each isolator is a suitable cost function for minimising the overall vibration response of the flexible support panel.

In the analysis to be outlined here, all of the power transmission components into the flexible support panel associated with all the force and moment components from each elastic mount are taken into account. The total power transmission into the support panel from all of the mounts is analysed by identifying the vibration modes corresponding to the coupled suspension system. The machine is modelled as a six-degrees-of-freedom rigid body, which is supported by a flexible rectangular panel through multiple elastic mounts, as shown in Figure 12.61.

As shown in the figure, the panel is modelled with both simply supported and corner supported boundary conditions. The former edge condition is realised in practice by supporting the edge of the panel with a strip of thin shim spring steel attached to a rigid foundation, as illustrated in Figure 12.62(a). Alternatively, grooves could be machined on each of the two faces of the plate, close to the edge, and small diameter steel rods placed in the groove and clamped between two solid pieces of steel, as shown in Figure 12.62(b). The corner supported edge condition is realised in practice by supporting the panel at a point under each corner. This could be achieved by machining an indentation in the plate close to each corner and resting a small spherical steel ball in the indentation and clamping the other side of the balls between solid steel supports, similar to what was done for the rods shown in Figure 12.62(b). The type of panel edge support only affects the panel mode shape functions, as will become apparent in the analysis to follow.

12.6.2.1 Rigid Body Equation of Motion

To begin the analysis, it is assumed that the rigid body is excited by forces and moments that can be represented by a generalised force vector given by Equation (12.5.1). An equation describing the motion of the rigid body can be derived using the same technique as used to derive the equation of motion (Equation (12.5.2)) for the rigid body acted upon by a single isolator in Section 12.5. The only difference in this case is that multiple isolators are acting. Thus,

$$\boldsymbol{Z}_0 \boldsymbol{D}_0 = \boldsymbol{Q}_0 + \sum_{J=1}^{L_1} \boldsymbol{R}_J{}' \boldsymbol{Q}_J{}' \tag{12.6.7}$$

where the variables without the subscript J were defined in Section 12.5, and L_1 is the number of isolators. Note that the subscript J also needs to be added to the elements of R'_J.

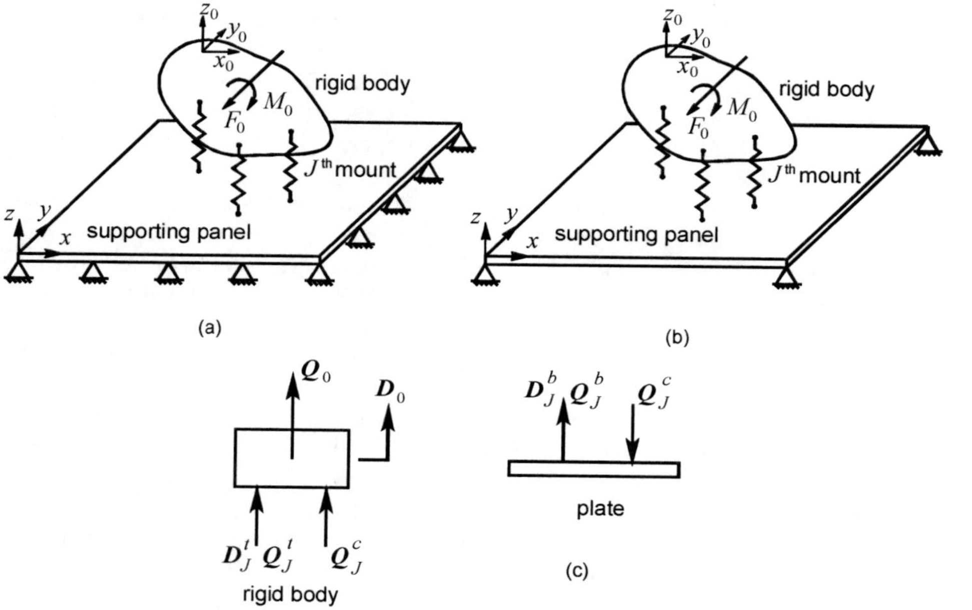

Figure 12.61 Rigid body isolated from a flexible panel by multiple isolation mounts: (a) simply supported panel; (b) corner supported, free edge panel; (c) sign conventions showing directions of positive forces and displacements.

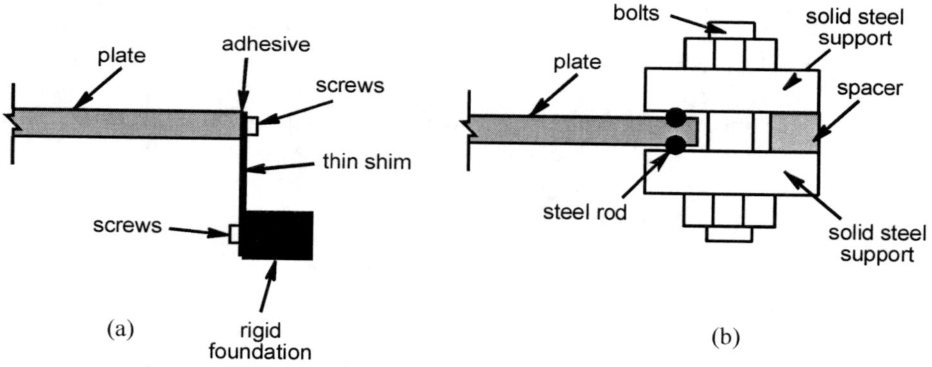

Figure 12.62 Two ways of realising a simply supported edge in practice.

12.6.2.2 Supporting Panel Equations of Motion

At the bottom of each elastic mount, the supporting panel is driven by forces and moments, which for the Jth mount can be written as:

$$\boldsymbol{Q}_J^b = -\boldsymbol{Q}_J^t = \left[F_{xJ}^b,\ F_{yJ}^b,\ F_{zJ}^b,\ M_{xJ}^b,\ M_{yJ}^b,\ M_{zJ}^b\right]^{\mathrm{T}} \tag{12.6.8a,b}$$

Here, the force and moment components in Q_J^b are assumed to be concentrated at one point, σ_J, on the panel, so that Dirac delta functions can be used to describe the external forces and moments per unit area. The effect of the two shear forces at each mount, on the panel vibration, can be modelled as two moments around the x- and y-axes. The drilling freedom of the thin panel is suppressed; therefore, the twisting moments M_{zJ}^b ($J = 1,...., L_1$) on the panel surface do not generate any flexural waves in the panel. By considering the force (in the z-direction) and moments (around the x- and y-axes) equilibria, by including the influence of the external forces and moments and by modelling the mass loading m_J^b of each mount on the supporting panel, at frequency ω as a concentrated inertial force $m_J^b\omega^2 w(\sigma,\omega)\delta(\sigma - \sigma_J)$, the panel displacement $w = w(\boldsymbol{\sigma},\omega)$ can be described by the following partial differential equation:

$$\rho h\frac{\partial^2 w}{\partial t^2} + \frac{Eh^3}{12(1-\nu^2)}\nabla^4 w = \sum_{J=1}^{L_1}\left[F_{zJ}^b\delta(\boldsymbol{\sigma}-\boldsymbol{\sigma}_J) + \frac{hF_{xJ}^b}{2}\frac{\partial\delta(\boldsymbol{\sigma}-\boldsymbol{\sigma}_J)}{\partial x} + \frac{hF_{yJ}^b}{2}\frac{\partial\delta(\boldsymbol{\sigma}-\boldsymbol{\sigma}_J)}{\partial x} - M_{xJ}^b\frac{\partial\delta(\boldsymbol{\sigma}-\boldsymbol{\sigma}_J)}{\partial y} + M_{yJ}^b\frac{\partial\delta(\boldsymbol{\sigma}-\boldsymbol{\sigma}_J)}{\partial x} + m_J^b\omega^2 w\,\delta(\boldsymbol{\sigma}-\boldsymbol{\sigma}_J)\right] \tag{12.6.9}$$

where ρ, E, ν and h are the density, Young's modulus of elasticity, Poisson's ratio and thickness respectively of the panel. The quantity $\boldsymbol{\sigma}_J = (x_J, y_J)$ is the location vector of the Jth mount on the supporting panel surface, and $\delta(\boldsymbol{\sigma} - \boldsymbol{\sigma}_J)$ is the Dirac delta function.

Using modal analysis, the panel displacement $\boldsymbol{w}(\boldsymbol{\sigma}, \omega)$ can be expressed in terms of a mode shape matrix $\boldsymbol{\psi}_p(\boldsymbol{\sigma}) = [\psi_1(\boldsymbol{\sigma}), \psi_2(\boldsymbol{\sigma}), \cdots, \psi_N(\boldsymbol{\sigma})]^T$ and a modal amplitude coefficient matrix $\boldsymbol{w}_p = [w_1, w_2, \cdots, w_N]^T$, as:

$$w = w(\boldsymbol{\sigma},\omega) = \boldsymbol{\psi}_p^T(\boldsymbol{\sigma})\boldsymbol{w}_p \tag{12.6.10a,b}$$

where $\psi_i(\boldsymbol{\sigma})$ is the ith panel mode shape function, which for a simply supported rectangular panel of dimensions L_x and L_y, is (Wallace 1972):

$$\psi_i(\boldsymbol{\sigma}) = \psi_{m,n}(x,y) = \sin\frac{m\pi x}{L_x}\sin\frac{n\pi y}{L_y} \tag{12.6.11}$$

where m and n are integers corresponding to the ith mode order.

For a corner supported plate (that is, one which is point supported at each of its four corners), the mode shape functions may be expressed as (Reed, 1965):

$$\psi_i(\boldsymbol{\sigma}) = \sum_{m=0}\sum_{n=1}\left[a_{mn}\cos\frac{m\pi x}{L_x}\sin\frac{n\pi y}{L_y} + b_{mn}\cos\frac{m\pi y}{L_y}\sin\frac{n\pi x}{L_x}\right] \tag{12.6.12}$$

where in this case the coefficients a_{mn} and b_{mn} are different for each mode and dependent on the ratio L_x/L_y.

Substituting Equation (12.6.10) into Equation (12.6.9) and using the orthogonal property of the mode shape functions and Dirac delta functions, the coefficient w_i for the ith panel mode becomes:

$$m_i(\omega_i^2 + j\eta_i\omega_i^2 - \omega^2)w_i = \sum_{J=1}^{L_1}\left[\frac{h}{2}\psi_{ix}(\boldsymbol{\sigma_J}),\ -\frac{h}{2}\psi_{iy}(\boldsymbol{\sigma_J}),\ \psi_i(\boldsymbol{\sigma_J}),\ -\psi_{iy}(\boldsymbol{\sigma_J}),\ -\psi_{ix}(\boldsymbol{\sigma_J}),\ 0\right]\boldsymbol{Q_J^t}$$

$$+\sum_{I'=1}^{P} C_{i',i}w_{i'} \qquad (i = 1, \cdots, N) \tag{12.6.13}$$

where the modal masses m_i are defined as:

$$m_i = \rho h \int_{A_f} \psi_i^2(\boldsymbol{\sigma})\,\mathrm{d}\boldsymbol{\sigma} \tag{12.6.14a,b}$$

For a simply rectangular plate, the right side of Equation (12.6.14b) is $\rho L_x L_y h/4$ and the angular resonance frequency ω_i of the ith panel mode (with modal indices (m, n)) is given by:

$$\omega_i = h\left[\left(\frac{m\pi}{L_x}\right)^2 + \left(\frac{n\pi}{L_y}\right)^2\right]\left[\frac{E}{12\rho(1-\nu^2)}\right]^{1/2} \tag{12.6.15}$$

For a corner supported plate, the resonance frequencies must be calculated using the Rayleigh–Ritz method as described by Reed (1965).

In the preceding equations, ρ, L_x, L_y and h are respectively the density, length dimensions and thickness of a rectangular plate of area A_f, and η_i is the loss factor of the ith panel mode, defined by Equation (12.5.14). $C_{i',i}(i', I = 1,...,N)$ is the concentrated mass contribution given by:

$$C_{i',i} = \sum_{J=1}^{L_1} m_J^b\,\omega^2\,\psi_{i'}(\boldsymbol{\sigma_J})\,\psi_i(\boldsymbol{\sigma_J}) \tag{12.6.16}$$

When i in Equation (12.6.13) increments from 1 to N, N simultaneous equations can be obtained for the coefficients $w_1, w_2, \cdots, w_N$, and the matrix equation for $\boldsymbol{w_p}$ is obtained as:

$$\boldsymbol{Z_p w_p} = \sum_{J=1}^{L_1} \boldsymbol{R_J}^b \boldsymbol{Q_J}^b \tag{12.6.17}$$

where $\boldsymbol{Z_p}$ is the uncoupled panel characteristic matrix including the influence of the concentrated masses of the mounts and is given by:

$$\boldsymbol{Z_p} = \begin{bmatrix} \chi_1 - C_{1,1} & -C_{1,2} & -C_{1,3} & \cdots & -C_{1,N} \\ -C_{2,1} & \chi_2 - C_{2,2} & -C_{2,3} & \cdots & -C_{2,N} \\ \vdots & \vdots & \vdots & \vdots & \vdots \\ -C_{N,1} & -C_{N,2} & -C_{N,3} & \cdots & \chi_N - C_{N,N} \end{bmatrix} \tag{12.6.18}$$

where χ_i ($i = 1, \cdots, N$) is defined as:

$$\chi_i = m_i(\omega_i^2 + \mathrm{j}\eta_i\omega_i^2 - \omega^2) \tag{12.6.19}$$

The quantity R_J^b is the force coupling matrix (on the supporting panel) for the Jth mount and is defined as follows:

$$\boldsymbol{R_J^b} = \begin{bmatrix} -\frac{h}{2}\psi_{1x}(\boldsymbol{\sigma_J}) & -\frac{h}{2}\psi_{1y}(\boldsymbol{\sigma_J}) & \psi_1(\boldsymbol{\sigma_J}) & \psi_{1y}(\boldsymbol{\sigma_J}) & -\psi_{1x}(\boldsymbol{\sigma_J}) & 0 \\ -\frac{h}{2}\psi_{2x}(\boldsymbol{\sigma_J}) & -\frac{h}{2}\psi_{2y}(\boldsymbol{\sigma_J}) & \psi_2(\boldsymbol{\sigma_J}) & \psi_{2y}(\boldsymbol{\sigma_J}) & -\psi_{2x}(\boldsymbol{\sigma_J}) & 0 \\ \vdots & \vdots & \vdots & \vdots & \vdots & \vdots \\ -\frac{h}{2}\psi_{Nx}(\boldsymbol{\sigma_J}) & -\frac{h}{2}\psi_{Ny}(\boldsymbol{\sigma_J}) & \psi_N(\boldsymbol{\sigma_J}) & \psi_{Ny}(\boldsymbol{\sigma_J}) & -\psi_{Nx}(\boldsymbol{\sigma_J}) & 0 \end{bmatrix} \tag{12.6.20}$$

For the simply supported rectangular panel and at $\boldsymbol{\sigma_J} = (x_{bJ}, y_{bJ})$, the functions $\psi_{ix}(\boldsymbol{\sigma_J})$ and $\psi_{iy}(\boldsymbol{\sigma_J})$ ($i = 1, \cdots, N$) are respectively:

$$\psi_{ix}(\boldsymbol{\sigma_J}) = \frac{\partial\psi_i(\boldsymbol{\sigma_J})}{\partial x} = \frac{m\pi}{L_x}\cos\frac{m\pi x_{bJ}}{L_x}\sin\frac{n\pi y_{bJ}}{L_y} \tag{12.6.21a,b}$$

$$\psi_{iy}(\boldsymbol{\sigma_J}) = \frac{\partial\psi_i(\boldsymbol{\sigma_J})}{\partial y} = \frac{n\pi}{L_y}\sin\frac{m\pi x_{bJ}}{L_x}\cos\frac{n\pi y_{bJ}}{L_y} \tag{12.6.22a,b}$$

For the corner supported panel, the functions are:

$$\psi_{ix}(\boldsymbol{\sigma_J}) = \frac{\pi}{L_x}\sum_{m=0}^{\infty}\sum_{n=1}^{\infty}\left[-ma_{mn}\sin\frac{m\pi x_{bJ}}{L_x}\sin\frac{n\pi y_{bJ}}{L_y} + nb_{mn}\cos\frac{m\pi y_{bJ}}{L_y}\cos\frac{n\pi x_{bJ}}{L_x}\right] \tag{12.6.23}$$

$$\psi_{iy}(\boldsymbol{\sigma_J}) = \frac{\pi}{L_y}\sum_{m=0}^{\infty}\sum_{n=1}^{\infty}\left[na_{mn}\cos\frac{m\pi x_{bJ}}{L_x}\cos\frac{n\pi y_{bJ}}{L_y} - mb_{mn}\sin\frac{m\pi y_{bJ}}{L_y}\sin\frac{n\pi x_{bJ}}{L_x}\right] \tag{12.6.24}$$

12.6.2.3 System Equations of Motion

Equations (12.6.7) and (12.6.17) are coupled through the elastic force vectors Q_J^b and Q_J^t,($J = 1, ..., L_1$) at the mounts. These force vectors are related to the relative displacements between the top and bottom ends of the mounts, as described by Equation (12.5.22) with the subscript J added to all quantities. The local displacements of the top and bottom surfaces of each mount can be related to the displacement vector $\boldsymbol{D}_0$ of the rigid body and $\boldsymbol{w_p}$ of the panel, because the local top displacements of the Jth mount are related to the rigid body displacement vectors $\boldsymbol{s_0}$ and $\boldsymbol{\theta_0}$ by:

$$\left[\xi^t_{xJ}, \xi^t_{yJ}, w^t_J\right]^{\mathrm{T}} = \boldsymbol{s}_0 + \boldsymbol{\theta}_0 \times \boldsymbol{r}_{0J} \qquad (12.6.25)$$

$$\left[\theta^t_{xJ}, \theta^t_{yJ}, \theta^t_{zJ}\right]^{\mathrm{T}} = \boldsymbol{\theta}_0 \qquad (12.6.26)$$

where $\boldsymbol{s}_0 = [x^0, y^0, z^0]^{\mathrm{T}}$ and $\boldsymbol{\theta}_0 = [\boldsymbol{\theta}_x^0, \boldsymbol{\theta}_y^0, \boldsymbol{\theta}_z^0]^{\mathrm{T}}$, as indicated by Equation (12.5.4).

The corresponding matrix form for displacements at the top of the mount is:

$$\boldsymbol{D}^t_J = (\boldsymbol{R}^t_J)^{\mathrm{T}} \boldsymbol{D}_0 \qquad (12.6.27)$$

where $\boldsymbol{R}^t$ is defined in Equation (12.5.4).

The local bottom displacements $\boldsymbol{D}_J^{\mathrm{b}}$ of the Jth mount are related to the panel displacement vector $w(\boldsymbol{\sigma}_J)$ at $\boldsymbol{\sigma}_J$ by:

$$\boldsymbol{D}^b_J = \left[\xi^b_{xJ}, \xi^b_{yJ}, w^b_J, \theta^b_{xJ}, \theta^b_{yJ}, \theta^b_{zJ}\right]^{\mathrm{T}} = \left[-\frac{h}{2}\frac{\partial w}{\partial x}, -\frac{h}{2}\frac{\partial w}{\partial y}, w, \frac{\partial w}{\partial y}, -\frac{\partial w}{\partial x}, 0\right]^{\mathrm{T}} \qquad (12.6.28a,b)$$

Substituting Equation (12.6.10) into Equation (12.6.28) allows the following matrix relation for displacements at the bottom of the mount to be obtained:

$$\boldsymbol{D}^b_J = (\boldsymbol{R}^b_J)^{\mathrm{T}} \boldsymbol{w}_p \qquad (12.6.29)$$

where $\boldsymbol{R}_J^{\mathrm{b}}$ is defined in Equation (12.6.20).

Consideration of the difference in displacement between the top and bottom of the isolators allows the following expression to be written:

$$\boldsymbol{Q}^t_J = -\boldsymbol{Q}^b_J = \boldsymbol{K}_J(\boldsymbol{D}^b_J - \boldsymbol{D}^t_J) \qquad (12.6.30)$$

where

$$\boldsymbol{K}_J = \begin{bmatrix} k_{J1}(1+\mathrm{j}\eta_{J1}) & & & \\ & k_{J2}(1+\mathrm{j}\eta_{J2}) & & \\ & & \ddots & \\ & & & k_{J6}(1+\mathrm{j}\eta_{J6}) \end{bmatrix} \qquad (12.6.31)$$

Substituting Equations (12.5.19) (with the subscript J added so that it applies for each mount), (12.6.8), (12.6.27) and (12.6.29) into Equations (12.6.7) and (12.6.17) gives a matrix equation describing the response of the coupled system as follows:

$$\begin{bmatrix} \boldsymbol{A}_{11} & \boldsymbol{A}_{12} \\ \boldsymbol{A}_{21} & \boldsymbol{A}_{22} \end{bmatrix} \begin{bmatrix} \boldsymbol{D}_0 \\ \boldsymbol{w}_p \end{bmatrix} = \begin{bmatrix} \boldsymbol{Q}_0 \\ \boldsymbol{0} \end{bmatrix} \qquad (12.6.32)$$

where **0** is a ($P \times 1$) zero vector and

$$\boldsymbol{A}_{11} = \boldsymbol{Z}_0 + \sum_{J=1}^{L_1} \boldsymbol{R}_J^t \boldsymbol{K}_J (\boldsymbol{R}_J^t)^{\mathrm{T}} \tag{12.6.33}$$

$$\boldsymbol{A}_{12} = -\sum_{j=1}^{L_1} \boldsymbol{R}_J^t \boldsymbol{K}_J (\boldsymbol{R}_J^b)^{\mathrm{T}} \tag{12.6.34}$$

$$\boldsymbol{A}_{21} = -\sum_{j=1}^{L_1} \boldsymbol{R}_J^b \boldsymbol{K}_J (\boldsymbol{R}_J^t)^{\mathrm{T}} \tag{12.6.35}$$

$$\boldsymbol{A}_{22} = \boldsymbol{Z}_p + \sum_{J=1}^{L_1} \boldsymbol{R}_J^b \boldsymbol{K}_J (\boldsymbol{R}_J^b)^{\mathrm{T}} \tag{12.6.36}$$

If $\boldsymbol{Q}_0 = 0$, Equation (12.6.32) describes the free vibration behaviour of the coupled system, the mode shapes and system resonance frequencies of which can be determined by solving this free vibration problem. For a given external forcing function acting on the rigid body $\boldsymbol{Q}_0$, the displacement response at any location can be calculated from Equation (12.6.32).

A passive isolator can be made active by introducing a force actuator connected in parallel with it. When used with a suitable feedforward control system, the actuator exerts a control force on the rigid body and the support point on the panel simultaneously. For active isolators acting at the same locations as the passive isolators, the right side of Equation (12.6.32) is replaced by a combined force vector, and the equation of motion becomes:

$$\begin{bmatrix} \boldsymbol{A}_{11} & \boldsymbol{A}_{12} \\ \boldsymbol{A}_{21} & \boldsymbol{A}_{22} \end{bmatrix} \begin{bmatrix} \boldsymbol{D}_0 \\ \boldsymbol{w}_p \end{bmatrix} = \begin{bmatrix} \boldsymbol{Q}_0 + \sum_{J=1}^{L_1} \boldsymbol{R}_J^t \boldsymbol{Q}_J^c \\ -\sum_{J=1}^{L_1} \boldsymbol{R}_J^b \boldsymbol{Q}_J^c \end{bmatrix} \tag{12.6.37}$$

where $\boldsymbol{Q}_J^c$ are the control force vectors generated by each actuator acting on the support panel.

12.6.2.4 Optimal Control Forces and Minimum Power Transmission

As before, the cost function chosen for minimisation is the time-averaged power transmission into the supporting panel through the isolators, and can be expressed as follows:

$$\begin{aligned} P_0 &= \mathrm{Re}\left\{-\frac{1}{2}\mathrm{j}\omega\sum_{J=1}^{L_1}(\boldsymbol{Q}_J^b - \boldsymbol{Q}_J^c)^{\mathrm{H}}\boldsymbol{D}_J^b\right\} = \frac{\omega}{2}\mathrm{Im}\left\{\sum_{J=1}^{L_1}(\boldsymbol{Q}_J^b - \boldsymbol{Q}_J^c)^{\mathrm{H}}\boldsymbol{D}_J^b\right\} \\ &= -\frac{\omega}{2}\mathrm{Im}\left\{\sum_{J=1}^{L_1}(\boldsymbol{D}_J^b)^{\mathrm{H}}(\boldsymbol{Q}_J^b - \boldsymbol{Q}_J^c)\right\} \end{aligned} \tag{12.6.38a–c}$$

Sign conventions are the same as those used for the beam problem of Section 12.5.

The time-averaged power transmission from the rigid body into the isolators can be described as:

$$P_t = \mathrm{Re}\left\{-\frac{1}{2}\mathrm{j}\omega\sum_{J=1}^{L_1}(\boldsymbol{Q}_J^t + \boldsymbol{Q}_J^c)^{\mathrm{H}}\boldsymbol{D}_J^t\right\} \tag{12.6.39}$$

and the input power due to the external driving force is given by:

$$P_i = \mathrm{Re}\left\{-\frac{1}{2}\mathrm{j}\omega\boldsymbol{Q}_0^{\mathrm{T}}\boldsymbol{D}_0\right\} \tag{12.6.40}$$

It is possible to express the total output power P_0 by using an explicit quadratic function. Thus, the term $\sum_{J=1}^{L_1}(\boldsymbol{D}_J^b)^{\mathrm{H}}(\boldsymbol{Q}_J^b - \boldsymbol{Q}_J^c)$ in Equation (12.6.38) can be expressed as:

$$\sum_{J=1}^{L_1}(\boldsymbol{D}_J^b)^{\mathrm{H}}(\boldsymbol{Q}_J^b - \boldsymbol{Q}_J^c) = (\boldsymbol{Q}^c)^{\mathrm{H}}\boldsymbol{a}\boldsymbol{Q}^c + (\boldsymbol{Q}^c)^{\mathrm{H}}\boldsymbol{b}_1 + \boldsymbol{b}_2\boldsymbol{Q}^c + \boldsymbol{c} \tag{12.6.41}$$

where $\boldsymbol{Q}^c = [\boldsymbol{Q}_1^c, \boldsymbol{Q}_2^c, \ldots, \boldsymbol{Q}_1^c]^{\mathrm{T}}$.

The form of the matrices $\boldsymbol{a}$, $\boldsymbol{b}_1$, $\boldsymbol{b}_2$ and $\boldsymbol{c}$ is dependent on the number of isolators used and to make the solution of this problem tractable, only four isolators between the rigid body and the support panel will be considered here; that is, $L_1 = 4$. In this case, the matrices $\boldsymbol{a}$, $\boldsymbol{b}_1$, $\boldsymbol{b}_2$, and $\boldsymbol{c}$ may be defined as follows:

$$\sum_{J=1}^{4}(\boldsymbol{D}_J^b)^{\mathrm{H}}(\boldsymbol{Q}_J^b - \boldsymbol{Q}_J^c) = (\boldsymbol{Q}^c)^{\mathrm{H}}\boldsymbol{a}\boldsymbol{Q}^c + (\boldsymbol{Q}^c)^{\mathrm{H}}\boldsymbol{b}_1 + \boldsymbol{b}_2\boldsymbol{Q}^c + \boldsymbol{c} \tag{12.6.42}$$

where

$$\boldsymbol{a} = \boldsymbol{G}_1\boldsymbol{K}\boldsymbol{G}_2 - \boldsymbol{G}_1 \tag{12.6.43}$$

$$\boldsymbol{b}_1 = \boldsymbol{G}_1\boldsymbol{K}\boldsymbol{G}_3 \tag{12.6.44}$$

$$\boldsymbol{b}_2 = \boldsymbol{G}_4\boldsymbol{K}\boldsymbol{G}_2 - \boldsymbol{G}_4 \tag{12.6.45}$$

$$\boldsymbol{c} = \boldsymbol{G}_4\boldsymbol{K}\boldsymbol{G}_3 \tag{12.6.46}$$

and

$$\boldsymbol{G}_1 = \begin{bmatrix} \boldsymbol{H}_1^{\mathrm{H}}\boldsymbol{R}_1^b & \boldsymbol{H}_1^{\mathrm{H}}\boldsymbol{R}_2^b & \boldsymbol{H}_1^{\mathrm{H}}\boldsymbol{R}_3^b & \boldsymbol{H}_1^{\mathrm{H}}\boldsymbol{R}_4^b \\ \boldsymbol{H}_2^{\mathrm{H}}\boldsymbol{R}_1^b & \boldsymbol{H}_2^{\mathrm{H}}\boldsymbol{R}_2^b & \boldsymbol{H}_2^{\mathrm{H}}\boldsymbol{R}_3^b & \boldsymbol{H}_2^{\mathrm{H}}\boldsymbol{R}_4^b \\ \boldsymbol{H}_3^{\mathrm{H}}\boldsymbol{R}_1^b & \boldsymbol{H}_3^{\mathrm{H}}\boldsymbol{R}_2^b & \boldsymbol{H}_3^{\mathrm{H}}\boldsymbol{R}_3^b & \boldsymbol{H}_3^{\mathrm{H}}\boldsymbol{R}_4^b \\ \boldsymbol{H}_4^{\mathrm{H}}\boldsymbol{R}_1^b & \boldsymbol{H}_4^{\mathrm{H}}\boldsymbol{R}_2^b & \boldsymbol{H}_4^{\mathrm{H}}\boldsymbol{R}_3^b & \boldsymbol{H}_4^{\mathrm{H}}\boldsymbol{R}_4^b \end{bmatrix} \tag{12.6.47}$$

$$G_2 = \begin{bmatrix} (R_1^t)^T H_5 - (R_1^b)^T H_1 & (R_1^t)^T H_6 - (R_1^b)^T H_2 & (R_1^t)^T H_7 - (R_1^b)^T H_3 & (R_1^t)^T H_8 - (R_1^b)^T H_4 \\ (R_2^t)^T H_5 - (R_2^b)^T H_1 & (R_2^t)^T H_6 - (R_2^b)^T H_2 & (R_2^t)^T H_7 - (R_2^b)^T H_3 & (R_2^t)^T H_8 - (R_2^b)^T H_4 \\ (R_3^t)^T H_5 - (R_3^b)^T H_1 & (R_3^t)^T H_6 - (R_3^b)^T H_2 & (R_3^t)^T H_7 - (R_3^b)^T H_3 & (R_3^t)^T H_8 - (R_3^b)^T H_4 \\ (R_4^t)^T H_5 - (R_4^b)^T H_1 & (R_4^t)^T H_6 - (R_4^b)^T H_2 & (R_4^t)^T H_7 - (R_4^b)^T H_3 & (R_4^t)^T H_8 - (R_4^b)^T H_4 \end{bmatrix} \tag{12.6.48}$$

$$G_3 = \begin{bmatrix} (R_1^t)^T B_{11} - (R_1^b)^T B_{21} \\ (R_2^t)^T B_{11} - (R_2^b)^T B_{21} \\ (R_3^t)^T B_{11} - (R_3^b)^T B_{21} \\ (R_4^t)^T B_{11} - (R_4^b)^T B_{21} \end{bmatrix} Q_0 \tag{12.6.49}$$

$$G_4 = Q_0^T \begin{bmatrix} B_{21}^H R_1^b & B_{21}^H R_2^b & B_{21}^H R_3^b & B_{21}^H R_4^b \end{bmatrix} \tag{12.6.50}$$

$$K = \begin{bmatrix} K_1 & & & \\ & K_2 & & \\ & & K_3 & \\ & & & K_4 \end{bmatrix} \tag{12.6.51}$$

where matrices H_1 to H_8 are defined as follows:

$$H_1 = B_{21} R_1^t - B_{22} R_1^b \tag{12.6.52}$$

$$H_2 = B_{21} R_2^t - B_{22} R_2^b \tag{12.6.53}$$

$$H_3 = B_{21} R_3^t - B_{22} R_3^b \tag{12.6.54}$$

$$H_4 = B_{21} R_4^t - B_{22} R_4^b \tag{12.6.55}$$

$$H_5 = B_{11} R_1^t - B_{12} R_1^b \tag{12.6.56}$$

$$H_6 = B_{11} R_2^t - B_{12} R_2^b \tag{12.6.57}$$

$$H_7 = B_{11} R_3^t - B_{12} R_3^b \tag{12.6.58}$$

$$H_8 = B_{11} R_4^t - B_{12} R_4^b \tag{12.6.59}$$

and the matrices $\boldsymbol{B}_{11}$, $\boldsymbol{B}_{12}$, $\boldsymbol{B}_{21}$, $\boldsymbol{B}_{22}$ in the above expressions are submatrix elements of the inverse of the system matrix $\boldsymbol{A}$, and are defined in Equations (12.5.35) to (12.5.39).

The control force vector $\boldsymbol{Q}^c$ of dimension (24 × 1) in Equation (12.6.42) is a combined vector of control forces as follows:

$$\boldsymbol{Q}^c = \left[\boldsymbol{Q}_1^c \;\; \boldsymbol{Q}_2^c \;\; \boldsymbol{Q}_3^c \;\; \boldsymbol{Q}_4^c\right]^{\mathrm{T}} \tag{12.6.60}$$

Thus, the total time-averaged output power P_0 can be expressed as the following real quadratic function:

$$P_0 = \mathrm{Re}\left\{-\frac{1}{2}\mathrm{j}\omega\sum_{J=1}^{4}(\boldsymbol{Q}_J^b - \boldsymbol{Q}_J^c)^{\mathrm{H}}\boldsymbol{D}_J^b\right\} = -\frac{\omega}{2}\mathrm{Im}\left\{\sum_{J=1}^{4}(\boldsymbol{D}_J^b)^{\mathrm{H}}(\boldsymbol{Q}_J^b - \boldsymbol{Q}_J^c)\right\}$$

$$= -\frac{\omega}{2}\left\{(\boldsymbol{q}^c)^{\mathrm{T}}\begin{bmatrix}\boldsymbol{a}^i & \boldsymbol{a}^r \\ -\boldsymbol{a}^r & \boldsymbol{a}^i\end{bmatrix}\boldsymbol{q}^c + (\boldsymbol{q}^c)^{\mathrm{T}}\begin{bmatrix}\boldsymbol{b}_1^i \\ -\boldsymbol{b}_1^r\end{bmatrix} + \begin{bmatrix}\boldsymbol{b}_2^i & \boldsymbol{b}_2^r\end{bmatrix}\boldsymbol{q}^c + c^i\right\} \tag{12.6.61a–c}$$

Following the same procedure as outlined in Section 12.5, Equations (12.5.49) to (12.5.57) (noting that the control force now consists of four vectors as shown in Equation (12.6.60), an expression for the optimal control forces can be obtained, which is identical to Equation (12.5.54), and the maximum achievable reduction in power transmission into the plate as a result of active control is given by:

$$(\Delta P)_{\max} = \frac{\omega}{2}\beta^{\mathrm{T}}\alpha^{-1}\beta \tag{12.6.62}$$

From the control force vector $\boldsymbol{q}^c_{\mathrm{opt}}$, which results in maximum power reduction, the amplitudes and phases of a set of optimal active control forces $(\boldsymbol{Q}^c{}_1)_{\mathrm{opt}}$, $(\boldsymbol{Q}^c{}_2)_{\mathrm{opt}}$, $(\boldsymbol{Q}^c{}_3)_{\mathrm{opt}}$ and $(\boldsymbol{Q}^c{}_4)_{\mathrm{opt}}$ can be obtained.

For a two isolator system, the preceding analysis can also be used, except that the matrices need to be reduced in size to exclude the contributions from the terms represented by $J = 3, 4$.

The minimum power transmission for multiple mounts is theoretically zero if the control forces act along the same axes as the excitation forces. The preceding analysis, however, allows evaluation of cases where the controller is non-ideal or where the control force along a particular axis is limited or unavailable.

12.6.3 Rigid Mass as the Intermediate Structure

The use of an intermediate rigid mass between a vibrating rigid body and a flexible support structure as a means of improving the vibration isolation of a passive system was mentioned in Section 12.2 where feedback control of the motion of the rigid mass was considered. For periodic excitation, the vibration of the rigid mass could be controlled using a feedforward system, the idea being that control of the rigid mass vibration will minimise the vibration transmitted to the support structure.

Jenkins et al. (1990) reported on the use of four isolators involving an intermediate rigid mass for the isolation of vibration of a diesel engine from a ribbed steel plate. The lower elastic mount beneath the rigid intermediate mass was an air bag spring which had little shear stiffness, resulting in no measurable transmission of rotational or horizontal translational forces. Thus, the active system was only required to control the transmission of vertical forces. A schematic representation of the actuator used is shown in Figure 12.63.

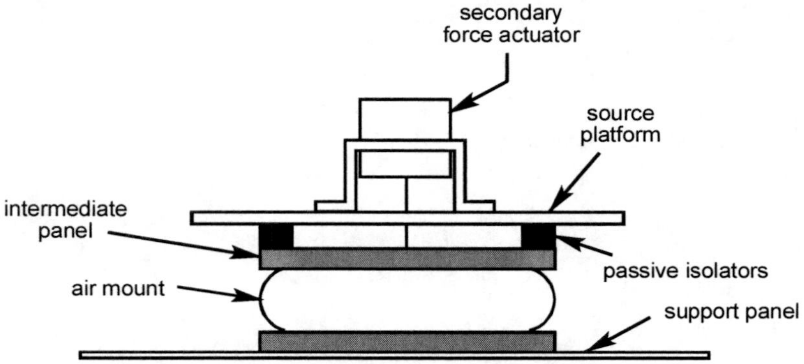

Figure 12.63 Active isolator with an air bag suspension (after Jenkins et al., 1990).

Good isolation (12 dB) was achieved by Jenkins et al. (1990) for the sixth harmonic of the diesel engine rotational speed. However, results obtained for the other harmonics were disappointing as the electrodynamic shaker used did not have a high enough force capability. An electrodynamic shaker of sufficient size would together have weighed half as much as the diesel engine. Thus, it has been suggested that the upper resilient mount should be redesigned to reduce vertical stiffness while retaining the shear stiffness that is necessary to maintain alignment of the shaker and its attachment point on the intermediate mass. This would result in a smaller force requirement of the shaker and potentially better performance in regard to isolation of the more important low order harmonics of the engine rotational speed.

12.7 FEEDFORWARD CONTROL: MULTIPLE ISOLATORS BETWEEN A RIGID BODY AND A FLEXIBLE CYLINDER

The particular case that will be analysed here is illustrated in Figure 12.64. More details of the analysis and some numerical results are given by Howard et al. (1996), Howard and Hansen, 2000, and Howard (1999), although some small errors in the equations have been corrected here.

The analysis presented here is based on the assumption that only radial vibrations are significant. Howard (1999) extended this simplified model to include all axes of vibration. His comparison of the two models shows that the simplified theory, which only considers radial vibration modes, is however, sufficiently accurate and the additional precision obtained by including all modes is insignificant.

As in the previous section, the cost function to be minimised will be the total vibratory power transmission into the flexible cylinder. The active isolators act on the rigid body and react on the support cylinder, and each acts in parallel with a passive element consisting of a stiffness and a viscous damper.

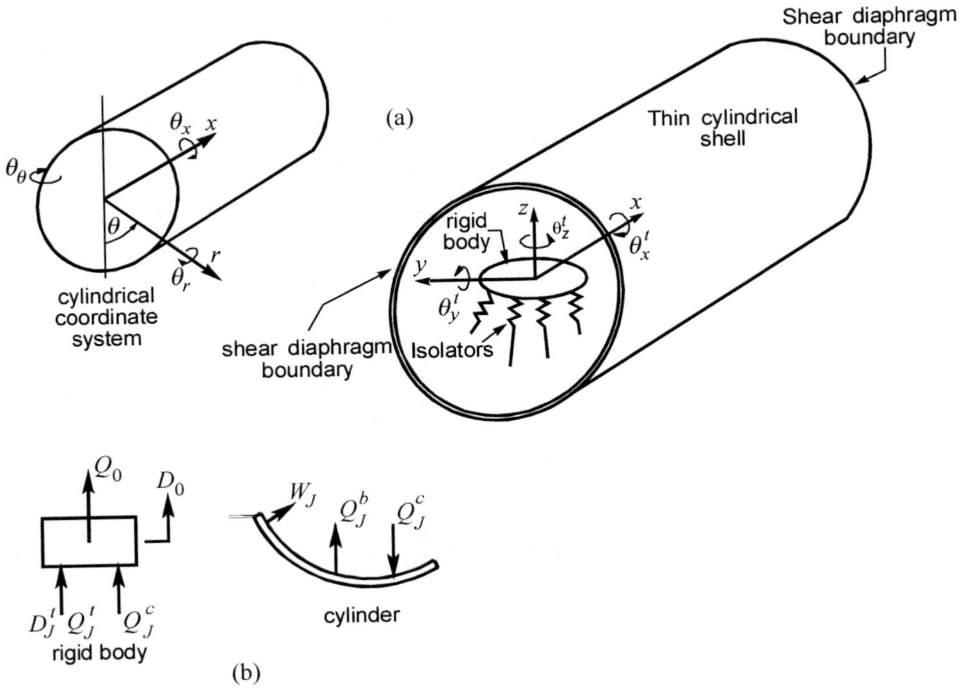

Figure 12.64 Rigid body isolated from a flexible cylinder using multiple mounts: (a) theoretical model; (b) sign conventions showing directions of positive forces and displacements.

12.7.1 Rigid Body Equation of Motion

The same equation of motion as used previously for the rigid body with a flexible panel support, Equation (12.6.7), also describes the rigid body motion in this case.

12.7.2 Supporting Thin Cylinder Equations of Motion

The supporting cylinder is driven by L_1 force vectors, defined as $\boldsymbol{Q}_J^b = [F_{xJ}^b, F_{yJ}^b, F_{zJ}^b, M_{xJ}^b, M_{yJ}^b, M_{zJ}^b]^{\mathrm{T}}$ in the Cartesian coordinate system or $\boldsymbol{Q}_J = [F_{xJ}, F_{\theta J}, F_{rJ}, M_{xJ}, M_{\theta J}, M_{rJ}]^{\mathrm{T}}$ in the cylindrical coordinate system as shown in Figure 12.64, where $J = 1, \cdots, L_1$. The force vectors acting on the top of each isolator are related to those acting on the bottom as follows:

$$\boldsymbol{Q}_J^b = -\boldsymbol{Q}_J^t \tag{12.7.1}$$

and the forces expressed in the cylindrical coordinate system are related to those expressed in Cartesian coordinate system as follows:

$$\boldsymbol{Q}_J = \boldsymbol{T}_J \boldsymbol{Q}_J^b \tag{12.7.2}$$

where $\boldsymbol{T}_J$ is a coordinate transformation matrix between the Cartesian and cylindrical coordinates of the support point $\boldsymbol{\sigma}_J$ on the internal surface of the cylinder at $r = R - h/2$, and is defined as follows:

$$\boldsymbol{T}_J = \begin{bmatrix} \boldsymbol{T}_{0J} & \boldsymbol{0} \\ \boldsymbol{0} & \boldsymbol{T}_{0J} \end{bmatrix} \tag{12.7.3}$$

where $\boldsymbol{0}$ is a 3 × 3 zero-order matrix and

$$\boldsymbol{T}_{0J} = \begin{bmatrix} 1 & 0 & 0 \\ 0 & \cos\theta_J & -\sin\theta_J \\ 0 & \sin\theta_J & \cos\theta_J \end{bmatrix} \tag{12.7.4}$$

It can be shown that $(\boldsymbol{T}_J)^{-1} = (\boldsymbol{T}_J)^{\mathrm{T}}$. The force and moment components in $\boldsymbol{Q}_J^b$ and $\boldsymbol{Q}_J$ are also assumed to be concentrated point actions at support points $\boldsymbol{\sigma}_J$ on the thin shell, so that Dirac delta functions and their partial derivatives can be used to describe the external force distribution on the inner surface of the cylinder. Compared with the radius R, the thickness h of the cylinder is sufficiently small for the vibration of the cylinder to be primarily radial, with the axial x and tangential θ displacements being small enough to allow the corresponding inertia terms in the axial and tangential directions in the equation of motion of the cylindrical shell to be neglected. Forces acting in the axial and tangential directions excite vibrational displacements in these directions, which in turn give rise to some radial motion as a result of coupling between the different directions of motion. However, the radial vibration produced in this way may be considered small compared to that which is produced directly by moments and forces acting in the radial direction. The assumption of small indirect radial vibration is important because it simplifies the analysis enormously, allowing the RHS of the first two of the following three equations of motion to be set equal to zero. Nevertheless, it is possible that including the effect of indirect radial vibration could lead to more accurate results. Note that the axial and tangential forces produced on the inside of the cylinder at the base of the mounts will produce some direct radial displacement due to the induced moment about the centre of the cylinder thickness. This is taken into account in the following analysis.

The bending displacement, $w\,(\theta,s{:}t)$, of the shell and resulting tangential displacements, $\xi_s\,(s,\ \theta;\ t)$, in the x-direction and $\xi_\theta\,(s,\ \theta;\ t)$ in the θ direction (where $s = x/R$) can be then described by the following equations (Donnell–Mushtari shell theory):

$$\frac{\partial^2 \xi_s}{\partial s^2} + \frac{(1-\nu)}{2}\frac{\partial^2 \xi_s}{\partial \theta^2} + \frac{(1+\nu)}{2}\frac{\partial^2 \xi_\theta}{\partial s \partial \theta} + \nu\frac{\partial w}{\partial s} = 0 \tag{12.7.5}$$

$$\frac{(1+\nu)}{2}\frac{\partial^2 \xi_s}{\partial s \partial \theta} + \frac{(1-\nu)}{2}\frac{\partial^2 \xi_\theta}{\partial s^2} + \frac{\partial^2 \xi_\theta}{\partial \theta^2} + \frac{\partial w}{\partial \theta} = 0 \tag{12.7.6}$$

$$\nu\frac{\partial \xi_s}{\partial s} + \frac{\partial \xi_\theta}{\partial \theta} + w + \kappa\nabla^4 w + \frac{\rho(1-\nu^2)R^2}{E}\frac{\partial^2 w}{\partial t^2} = \frac{(1-\nu^2)}{Eh}\left[q_1(x,\theta) + q_2(x,\theta)\right]\mathrm{e}^{\mathrm{j}\omega t} \tag{12.7.7}$$

where the gradient operator is defined as:

$$\nabla^4 = \nabla^2\nabla^2 = \left\{\frac{\partial^2}{\partial s^2} + \frac{\partial^2}{\partial \theta^2}\right\}^2 \tag{12.7.8}$$

The quantities $\xi_s(s, \theta; t)$, $\xi_\theta(s, \theta; t)$, and w are the orthogonal components of the displacement in the axial x, circumferential θ and radial w directions and the force distribution functions q_1 and q_2 on the right side of Equation (12.7.7) are:

$$q_1 = \sum_{J=1}^{L_1} m_J^b \omega^2 w(\sigma_J)\, \delta(\sigma - \sigma_J) \tag{12.7.9}$$

$$q_2 = p(s,\theta) \tag{12.7.10}$$

where L_1 is the number of shell modes considered and where ρ, E and v are respectively the density, Young's modulus and Poisson's ratio of the shell material, R and h are respectively the radius and thickness of the cylinder, κ is a dimensionless shell thickness parameter defined as $\kappa = h^2/(12R^2)$, $\boldsymbol{m}_J^b$ is a concentrated mass modelling the base at the bottom of the Jth mount, and the forcing function $p(x, \theta)$ can be expressed as follows:

$$\begin{aligned} p(s,\theta) &= \sum_{J=1}^{L_1} \left[-\frac{h}{2}\frac{\partial\delta(\boldsymbol{\sigma}-\boldsymbol{\sigma}_J)}{\partial x},\ -\frac{h}{2R}\frac{\partial\delta(\boldsymbol{\sigma}-\boldsymbol{\sigma}_J)}{\partial \theta},\ \delta(\boldsymbol{\sigma}-\boldsymbol{\sigma}_J),\right. \\ &\qquad \left. -\frac{1}{R}\frac{\partial\delta(\boldsymbol{\sigma}-\boldsymbol{\sigma}_J)}{\partial \theta},\ \frac{\partial\delta(\boldsymbol{\sigma}-\boldsymbol{\sigma}_J)}{\partial x},\ 0\right] \boldsymbol{Q}_J \\ &= \sum_{J=1}^{L_1} \left[-\frac{h}{2}\frac{\partial\delta(\boldsymbol{\sigma}-\boldsymbol{\sigma}_J)}{R\,\partial s},\ -\frac{h}{2R}\frac{\partial\delta(\boldsymbol{\sigma}-\boldsymbol{\sigma}_J)}{\partial \theta},\ \delta(\boldsymbol{\sigma}-\boldsymbol{\sigma}_J),\right. \\ &\qquad \left. -\frac{1}{R}\frac{\partial\delta(\boldsymbol{\sigma}-\boldsymbol{\sigma}_J)}{\partial \theta},\ \frac{\partial\delta(\boldsymbol{\sigma}-\boldsymbol{\sigma}_J)}{R\,\partial s},\ 0\right] \boldsymbol{Q}_J \end{aligned} \tag{12.7.11a,b}$$

The cylinder response is the summation of its orthogonal harmonic solutions: the 'odd' modes, which are described by a sine function in the circumferential direction, and the 'even' modes, which are described by a cosine function. For a simply supported circular cylindrical shell, the following harmonic solutions can be employed:

$$\xi_s(s, \theta; t) = \xi_s(s, \theta)\mathrm{e}^{\mathrm{j}\omega t},\ \xi_\theta(s, \theta; t) = \xi_\theta(s, \theta)\mathrm{e}^{\mathrm{j}\omega t},\ w(s, \theta; t) = w(s, \theta)\mathrm{e}^{\mathrm{j}\omega t} \tag{12.7.12a–c}$$

and

$$\xi_s(s, \theta) = \sum_{m=1,n=0}^{\infty} a_{mn} \cos\lambda s \sin n\theta + a'_{mn} \cos\lambda s \cos n\theta \tag{12.7.13}$$

$$\xi_\theta(s, \theta) = \sum_{m=1,n=0}^{\infty} b_{mn} \sin\lambda s \cos n\theta + b'_{mn} \sin\lambda s \sin n\theta \tag{12.7.14}$$

$$w(s,\theta) = \sum_{m=1,n=0}^{\infty} c_{mn}\sin\lambda s\sin n\theta + c'_{mn}\sin\lambda s\cos n\theta \tag{12.7.15}$$

where $\lambda = m\pi R/L_0$, $a_{mn}, \ldots, c_{mn}$, are odd mode amplitude coefficients for mode (m, n), $a'_{mn}, \ldots, c'_{mn}$ are even mode amplitude coefficients for mode (m, n), and L_0 is the length of the cylinder. Substituting Equation (12.7.10) into Equations (12.7.5) and (12.7.6) gives:

$$a_{mn} = \lambda\alpha_{mn}c_{mn} = \frac{\lambda(\nu\lambda^2 - n^2)}{(\lambda^2 + n^2)^2}c_{mn} \tag{12.7.16a,b}$$

where $\lambda = m\pi R/L_0$ and $m,n = 1, 2, \cdots$; $\alpha_{mn} = (\nu\lambda^2 - n^2)(\lambda^2 + n^2)^{-2}$ and $\beta_{mn} = [(2+\nu)\lambda^2 + n^2]\times(\lambda^2 + n^2)^{-2}$.

For convenience, the modes will be identified by a single index k rather than the double index mn, and arranged in ascending order of resonance frequency. Thus, c_{mn} will be represented as c_k from now on. Substituting Equations (12.7.11) to (12.7.16) into Equation (12.7.7), multiplying each side by $\psi_k(\sigma_J)$ and integrating over the cylinder length L_0 (orthogonal property of the mode shape functions), and using the results $\int_0^{L_0/R} \sin\lambda_1 s\sin\lambda_2 s\, \mathrm{d}s = \frac{L_0}{2R}$ if $\lambda_1 = \lambda_2$ and 0 if $\lambda_1 \neq \lambda_2$ and $\int_0^{2\pi} \sin n_1\theta\sin n_2\theta\, \mathrm{d}\theta = \pi$ if $n_1 = n_2$ and 0 if $n_1 \neq n_2$, and adding loss (damping) factors to every mode gives the following equation for the radial modal amplitude coefficients c_k, $(k = 1,2,\cdots)$ for all shell modes:

$$\frac{m_s}{4}(\omega_k^2 + j\eta_k\omega_k^2 - \omega^2)c_k =$$

$$\sum_{J=1}^{L_1}\left\{\left[\frac{h}{2R}\psi_{ks}(\boldsymbol{\sigma_J}),\ \frac{h}{2R}\psi_{k\theta}(\boldsymbol{\sigma_J}),\ \psi_k(\boldsymbol{\sigma_J}),\ \frac{1}{2R}\psi_{k\theta}(\boldsymbol{\sigma_J}),\ -\frac{1}{R}\psi_{ks}(\boldsymbol{\sigma_J}),\ 0\right]\boldsymbol{Q_J} + \sum_{i=1}^{\infty}C_{ik}c_i\right\} \tag{12.7.17}$$

where c_i and c_k are the coefficients for the ith shell mode and kth shell mode respectively The quantity $m_s = 2\pi\rho RhL_0$ is the mass of the cylindrical shell, ω_k and η_k are respectively, the shell mode resonance angular frequencies and loss factors arranged in ascending order, and C_{ik} is the concentrated mass contribution given by:

$$C_{ik} = \sum_{J=1}^{L_1} m_J^b\omega^2\psi_i(\boldsymbol{\sigma_J})\psi_k(\boldsymbol{\sigma_J}) \tag{12.7.18}$$

The three functions $\psi_k(\boldsymbol{\sigma_J})$, $\psi_{ks}(\boldsymbol{\sigma_J})$ and $\psi_{k\theta}(\boldsymbol{\sigma_J})$ are dimensionless and are evaluated at the point, $\boldsymbol{\sigma_J}(s_J, \theta_J)$, on the shell surface and defined as follows:

$$\psi_k(\boldsymbol{\sigma_J}) = \sin\lambda s_J\sin n\theta_J \tag{12.7.19}$$

$$\psi_{ks}(\boldsymbol{\sigma_J}) = \lambda\cos\lambda s_J\sin n\theta_J \tag{12.7.20}$$

$$\psi_{k\theta}(\boldsymbol{\sigma_J}) = n\sin\lambda s_J\cos n\theta_J \tag{12.7.21}$$

where n is the circumferential order of the kth mode.

The shell mode resonance angular frequencies, ω_k, (k = 1, 2, ···) can be calculated by using the following relation:

$$\omega_k^2 = \left[\frac{(1 - \nu^2)\lambda^4}{(\lambda^2 + n^2)^2} + \kappa(\lambda^2 + n^2)^2 \right] \frac{E}{\rho(1 - \nu^2)R^2} \tag{12.7.22}$$

When only the first N modes are taken into account, Equation (12.7.17) can be written in the following matrix form:

$$\boldsymbol{Z}_s \boldsymbol{c}^s = \sum_{J=1}^{L_1} \boldsymbol{R}_J^b \boldsymbol{Q}_J \tag{12.7.23}$$

where $\boldsymbol{Z}_s$ is the uncoupled shell characteristic matrix, including the influence of the concentrated masses of the isolating mounts:

$$\boldsymbol{Z}_s = \begin{bmatrix} \Omega_1 & & & & & \\ & \ddots & & & & \\ & & \Omega_N & & & \\ & & & \Omega_1' & & \\ & & & & \ddots & \\ & & & & & \Omega_N' \end{bmatrix} - \sum_{J=1}^{L_1} m_J^b \omega^2 \begin{bmatrix} [\boldsymbol{\Gamma}_J]^{\mathrm{T}} \boldsymbol{\Gamma}_J & [\boldsymbol{\Gamma}_J']^{\mathrm{T}} \boldsymbol{\Gamma}_J \\ [\boldsymbol{\Gamma}_J]^{\mathrm{T}} \boldsymbol{\Gamma}_J' & [\boldsymbol{\Gamma}_J']^{\mathrm{T}} \boldsymbol{\Gamma}_J' \end{bmatrix} \tag{12.7.24}$$

where Ω_k and Ω_k' (k = 1, ···, N) are defined as:

$$\Omega_k = \Omega_k' = \frac{1}{4} m_s (\omega_k^2 + \mathrm{j}\eta_k \omega_k^2 - \omega^2) \tag{12.7.25}$$

$\boldsymbol{\Gamma}_J$ and $\boldsymbol{\Gamma}_J'$ are defined as:

$$\boldsymbol{\Gamma}_J = [\sin\lambda_1 s_J \sin n_1 \theta_J, \; \dots \; , \sin\lambda_N s_J \sin n_N \theta_J] \tag{12.7.26a}$$

and

$$\boldsymbol{\Gamma}_J' = [\sin\lambda_1 s_J \cos n_1 \theta_J, \; \dots \; , \sin\lambda_P s_J \cos n_P \theta_J] \tag{12.7.26b}$$

and $\boldsymbol{c}^s = [c_1, c_2, \dots, c_P, c_1', c_2', \dots, c_P']^{\mathrm{T}}$.

The quantity $\boldsymbol{R}_J^b$ is a force coupling matrix corresponding to force vector $\boldsymbol{Q}_J$ acting on the Jth support point on the cylindrical shell and is defined as follows:

$$\boldsymbol{R}_J^b = \begin{bmatrix} \frac{h}{2R}\psi_{1s}(\boldsymbol{\sigma}_J) & \frac{h}{2R}\psi_{1\theta}(\boldsymbol{\sigma}_J) & \psi_1(\boldsymbol{\sigma}_J) & \frac{1}{R}\psi_{1\theta}(\boldsymbol{\sigma}_J) & -\frac{1}{R}\psi_{1s}(\boldsymbol{\sigma}_J) & 0 \\ \frac{h}{2R}\psi_{2s}(\boldsymbol{\sigma}_J) & \frac{h}{2R}\psi_{2\theta}(\boldsymbol{\sigma}_J) & \psi_2(\boldsymbol{\sigma}_J) & \frac{1}{R}\psi_{2\theta}(\boldsymbol{\sigma}_J) & -\frac{1}{R}\psi_{2s}(\boldsymbol{\sigma}_J) & 0 \\ \vdots & \vdots & \vdots & \vdots & \vdots & \vdots \\ \frac{h}{2R}\psi_{Ns}(\boldsymbol{\sigma}_J) & \frac{h}{2R}\psi_{N\theta}(\boldsymbol{\sigma}_J) & \psi_N(\boldsymbol{\sigma}_J) & \frac{1}{R}\psi_{N\theta}(\boldsymbol{\sigma}_J) & -\frac{1}{R}\psi_{Ns}(\boldsymbol{\sigma}_J) & 0 \\ \frac{h}{2R}\psi'_{1s}(\boldsymbol{\sigma}_J) & \frac{h}{2R}\psi'_{1\theta}(\boldsymbol{\sigma}_J) & \psi'_1(\boldsymbol{\sigma}_J) & \frac{1}{R}\psi'_{1\theta}(\boldsymbol{\sigma}_J) & -\frac{1}{R}\psi'_{1s}(\boldsymbol{\sigma}_J) & 0 \\ \frac{h}{2R}\psi'_{2s}(\boldsymbol{\sigma}_J) & \frac{h}{2R}\psi'_{2\theta}(\boldsymbol{\sigma}_J) & \psi'_2(\boldsymbol{\sigma}_J) & \frac{1}{R}\psi'_{2\theta}(\boldsymbol{\sigma}_J) & -\frac{1}{R}\psi'_{2s}(\boldsymbol{\sigma}_J) & 0 \\ \vdots & \vdots & \vdots & \vdots & \vdots & \vdots \\ \frac{h}{2R}\psi'_{Ns}(\boldsymbol{\sigma}_J) & \frac{h}{2R}\psi'_{N\theta}(\boldsymbol{\sigma}_J) & \psi'_N(\boldsymbol{\sigma}_J) & \frac{1}{R}\psi'_{N\theta}(\boldsymbol{\sigma}_J) & -\frac{1}{R}\psi'_{Ns}(\boldsymbol{\sigma}_J) & 0 \end{bmatrix} \qquad (12.7.27)$$

In the cylindrical coordinate system, the rotational displacements of the elastic elements at a point on the shell surface can be calculated by using the following equations:

$$\theta_x = \frac{1}{R}\frac{\partial w}{\partial \theta}, \qquad \theta_\theta = -\frac{\partial w}{\partial x} \qquad (12.7.28a,b)$$

Note that θ_r is zero. Using Equations (12.7.12) to (12.7.16) and (12.7.27), and the preceding definition of $\boldsymbol{c}^s$, the following matrix expression can be obtained:

$$\boldsymbol{W}_J = [R_J^b]^{\mathrm{T}}\boldsymbol{c}^s \qquad (J = 1, \cdots, L_1) \qquad (12.7.29)$$

where $\boldsymbol{W}_J = [\xi_s, \xi_\theta, w, \theta_s, \theta_\theta, \theta_w]^{\mathrm{T}}$ is the six-dimensional displacement vector of the *J*th support point $\boldsymbol{\sigma}_J$ in the cylindrical coordinate system.

12.7.3 System Equation of Motion

Adapting Equation (12.5.19) to multiple mounts, the relationship between the top and bottom displacements for each mount is:

$$\boldsymbol{Q}_J^t = -\boldsymbol{Q}_J^b = \boldsymbol{K}_J(\boldsymbol{D}_J^b - \boldsymbol{D}_J^t) \qquad (12.7.30a,b)$$

where

$$\boldsymbol{D}_J^t = (\boldsymbol{R}_J^t)^{\mathrm{T}}\boldsymbol{D}_0 \qquad (12.7.31)$$

Sign conventions are the same as those used for the beam problem in Section 12.5. The quantity $\boldsymbol{D}_0$ is defined by Equation (12.5.4), $\boldsymbol{R}'_J$ is defined by Equation (12.5.5), and $\boldsymbol{K}_J$ is defined by Equation (12.5.6) with a subscript *J* added.

The displacement vector D^b_J of the shell support point σ_J in the Cartesian coordinate system can be expressed in terms of W_J in the cylindrical coordinate system by using the following relationship:

$$W_J = T_J D_J^b \tag{12.7.32}$$

Synthesising Equations (12.6.7), (12.6.8), (12.6.26), (12.7.1), (12.7.2), (12.7.23), (12.7.29) and (12.7.32) gives the following equation of motion for the coupled system:

$$\begin{bmatrix} A_{11} & A_{12} \\ A_{21} & A_{22} \end{bmatrix} \begin{bmatrix} D_0 \\ c^s \end{bmatrix} = \begin{bmatrix} Q_0 \\ 0 \end{bmatrix} \tag{12.7.33}$$

where the element matrices $A_{11}, \cdots, A_{22}$ are given by the following expressions:

$$A_{11} = Z_0 + \sum_{J=1}^{L_1} R_J^t K_J (R_J^t)^{\mathrm{T}} \tag{12.7.34}$$

$$A_{12} = -\sum_{j=1}^{L_1} R_J^t K_J T_J^{\mathrm{T}} [R_J^b]^{\mathrm{T}} \tag{12.7.35}$$

$$A_{21} = -\sum_{j=1}^{L_1} R_J^b T_J K_J (R_J^t)^{\mathrm{T}} \tag{12.7.36}$$

$$A_{22} = Z_s + \sum_{J=1}^{L_1} R_J^b T_J K_J T_J^{\mathrm{T}} [R_J^b]^{\mathrm{T}} \tag{12.7.37}$$

The resonance frequencies and mode shapes of the coupled system can be obtained by solving the eigenvalue problem of the coefficient matrix containing the elements A_{ij} $(i, j = 1,4)$ when $Q_0 = 0$.

The passive isolators can be made active by using force actuators connected in parallel with each of them. When used with a suitable feedforward control system, these actuators exert control forces on the rigid body and the shell support points simultaneously, to change the system response to some optimal condition. In this case, the right-hand side of Equation (12.7.33) is replaced by a combined force vector, and the equation becomes:

$$\begin{bmatrix} A_{11} & A_{12} \\ A_{21} & A_{22} \end{bmatrix} \begin{bmatrix} D_0 \\ c^s \end{bmatrix} = \begin{bmatrix} Q_0 + \sum_{J=1}^{L_1} R_J^t Q_J^c \\ -\sum_{J=1}^{L_1} R_J^b T_J Q_J^c \end{bmatrix} \tag{12.7.38}$$

where Q_J^c is the control force vector acting on the rigid body from the Jth actuator connected in parallel with the Jth mount attaching the rigid body to the flexible shell, and is defined for the Jth actuator in terms of forces and moments in the Cartesian coordinate system as

$\boldsymbol{Q}_J^c = [F_x^c, F_y^c, F_z^c, M_x^c, M_y^c, M_z^c]_J^{\mathrm{T}}$. It is defined as positive when acting in the direction of negative cylinder displacement at the bottom of the isolator, which results in a control force acting in the direction of positive rigid body displacement at the top of the isolator.

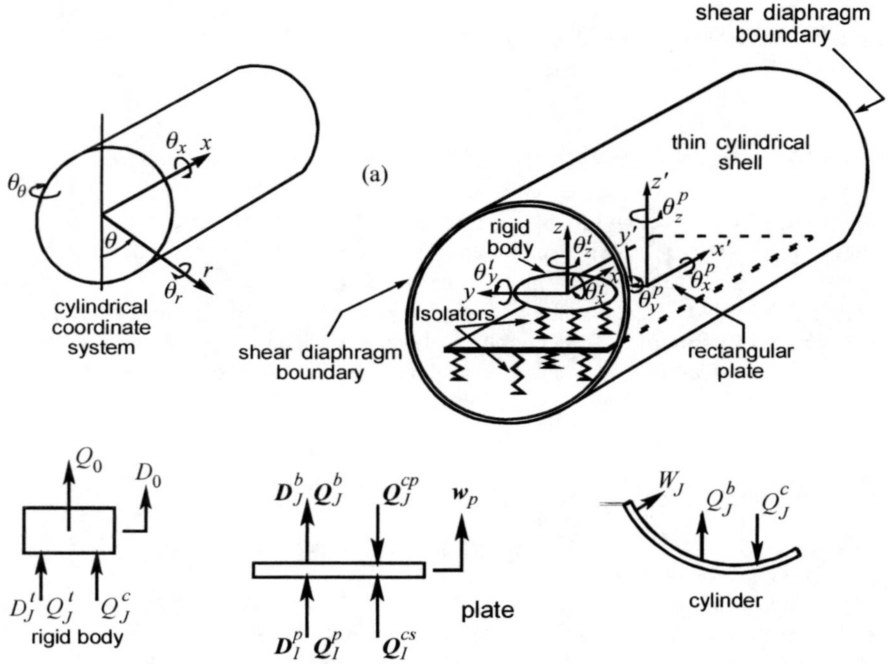

Figure 12.65 Rigid body isolated through an intermediate flexible panel from a flexible cylinder using multiple mounts: (a) theoretical model; (b) sign conventions showing directions of positive forces and displacements.

12.7.4 Minimisation of Power Transmission into the Support Cylinder

The total power transmission P_0 into the cylinder through the isolating mounts can be calculated as follows:

$$P_0 = -\mathrm{Re}\left\{\frac{1}{2}\mathrm{j}\omega\sum_{J=1}^{L_1}(\boldsymbol{Q}_J - \boldsymbol{T}_J\boldsymbol{Q}_J^c)^{\mathrm{H}}\boldsymbol{W}_J\right\} = -\frac{\omega}{2}\mathrm{Im}\left\{\sum_{J=1}^{L_1}\boldsymbol{W}_J^{\mathrm{H}}(\boldsymbol{Q}_J - \boldsymbol{T}_J\boldsymbol{Q}_J^c)\right\} \quad (12.7.39a,b)$$

This is the quantity which may be used as the active control cost function to be minimised. It can be seen that the output power transmission, P_0, depends upon the parameters of the passive system as well as on the control forces. For a given system, it is possible to express the output power by using an explicit quadratic function as follows:

$$\sum_{J=1}^{L_1}\boldsymbol{W}_J^{\mathrm{H}}(\boldsymbol{Q}_J - \boldsymbol{T}_J\boldsymbol{Q}_J^c) = (\boldsymbol{Q}^c)^{\mathrm{H}}\boldsymbol{a}\boldsymbol{Q}^c + (\boldsymbol{Q}^c)^{\mathrm{H}}\boldsymbol{b}_1 + \boldsymbol{b}_2\boldsymbol{Q}^c + c \quad (12.7.40)$$

The matrices $\boldsymbol{a}$, $\boldsymbol{b}_1$, $\boldsymbol{b}_2$ and the constant c are dependent upon the number of isolators used. For a four-isolator system, the analysis is as outlined in Section 12.6.2.4, Equations (12.6.40) to (12.6.58), except that the quantities $\boldsymbol{R}_1^b$, $\boldsymbol{R}_2^b$, $\boldsymbol{R}_3^b$, $\boldsymbol{R}_4^b$ and $\boldsymbol{D}_J^b$ are replaced by $\boldsymbol{R}_1^b T_1$, $\boldsymbol{R}_2^b T_2$, $\boldsymbol{R}_3^b T_3$, $\boldsymbol{R}_4^b T_4$ and W_J respectively.

The comments made in Section 12.6.2.5 and 12.6.2.6 regarding zero power transmission and the case of a control actuator with restoring force in a single axis are equally valid for the isolators considered here.

12.8 FEEDFORWARD CONTROL: SUMMARY

As mentioned at the beginning of Section 12.4, feedforward control is appropriate in situations where periodic machinery vibration is to be isolated from rigid or flexible support structures. The analyses outlined in Sections 12.5 to 12.8 can be used to estimate the maximum achievable reduction in vibratory power transmission from a machine into a support structure assuming the availability of an ideal feedforward electronic controller. Aspects to be considered in the design of such a controller are discussed in Chapters 6 and 13.

REFERENCES

Abdel–Rohman, M. (1987). Feasibility of active control of tall buildings against wind. *Journal of Structural Engineering*, **113**, 349–362.

Akatsu, Y., Fukushima, N., Tukahashi, K. Satoh, M. and Kawarazaki, Y. (1990). Active suspension employing an electrohydraulic pressure control system. In *Proceedings of the 18th Fisita Congress: The Promise of New Technology in the Automotive Industry*. SAE, Warrendale, PA, 949–959.

Alanoly, J. and Sankar, S. (1987). A new concept in semi-active vibration isolation. *Journal of Mechanics, Transmissions and Automation in Design*, **109**, 242–247.

Arzanpour, S. and Golnaraghi, M.F. (2008). A novel semi-active magnetorheological bushing design for variable displacement engines. *Journal of Intelligent Material Systems and Structures*, **19**, 989–1003.

Asano, S., Takahashi, J., Saito, T. and Matsumoto, H. (1991). Development of acceleration sensor and acceleration evaluation system for super low range frequency. In *Proceedings of the SAE International Congress on Sensors and Actuators*. SAE Publication, 242, 37–49.

Bender, E.K. (1968). Optimum linear preview control with application to vehicle suspension. *Journal of Basic Engineering*, **90**, 213–221.

Bies, D.A. (1968). *Feasibility Study of a Hybrid Vibration Isolation System*. Bolt Beranek and Newman Inc., Report 1620.

Bies, D.A. and Hansen, C.H. (1995). *Engineering Noise Control*, 2nd ed., Unwin Hyman, London, Ch. 10.

Blackwood, G.H. and von Flotow, A.H. (1993). Active control for vibration isolation despite resonant structural dynamics: a trade study of sensors, actuators and configurations. In *Proceedings of the Second Conference on Recent Advances in Active Control of Sound and Vibration*, Ed. R.A. Burdisso. Technomic Publishing Co., Blacksburg, VA, 482–494.

Bluethmann, B., Herrera, E., Hulse, A., Figuered, J., Junkin, L., Markee, M. and Ambrose, R.O. (2010). An active suspension system for lunar crew mobility. In *Proceedings of the 2010 IEEE Aerospace Conference*, Big Sky, MT.

Bonello, P., Brennan, M.J. and Elliott, S.J. (2005). Vibration control using an adaptive tuned vibration absorber with a variable curvature stiffness element. *Smart Materials and Structures*, **14**, 1055–1065.

Brennan, M.J. (2006). Some recent developments in adaptive tuned vibration absorbers/neutralisers. *Shock and Vibration*, **13**, 531–543.

Brennan, M.J., Ananthaganeshan, K.A. and Elliott, S.J. (2007). Instabilities due to instrumentation phase-lead and phase-lag in the feedback control of a simple vibrating system. *Journal of Sound and Vibration*, **304**, 466–478.

Brennan, M.J., Elliott, S.J. and Huang, X. (2006). A demonstration of active vibration isolation using decentralized velocity feedback control. *Smart Materials and Structures*, **15**, N19–N22.

Cao, D., Song, X. and Ahmadian, M. (2011). Editors' perspectives: road vehicle suspension design, dynamics, and control. *Vehicle System Dynamics*, **49**, 3–28.

Cao, J., Liu, H., Li, P. and Brown, D.J. (2008). State of the art in vehicle active suspension adaptive control systems based on intelligent methodologies. *IEEE Transactions on Intelligent Transportation Systems*, **9**, 392–405.

Chen, H. and Guo, K.-H. (2005). Constrained $\mathcal{H}^{\infty}$ control of active suspensions: an LMI approach. *IEEE Transactions on Control Systems Technology*, **13**, 412–421.

Cheok, K.C. and Huang, N.J. (1989). Lyapunov stability analysis for self-learning neural model with application to semi-active suspension control system. In *Proceedings of the IEEE International Symposium on Intelligent Control*, pp. 326–331.

Cheok, K.C., Hu, H.X. and Loh, N.K. (1989). Discrete-time frequency-shaping parametric control with application to active seat suspension control. *IEEE Transactions Industrial Electronics*, **36**, 383–390.

Choi, S.-B., Choi, Y.-T., Cheong, C.-C. and Jeon, Y.-S. (1999). Performance evaluation of a mixed mode ER engine mount via hardware-in-the-loop simulation. *Journal of Intelligent Material Systems and Structures*, **10**, 671–677.

Choi, S.-B. and Song, H.-J. (2002). Vibration control of a passenger vehicle utilizing a semi-active ER engine mount. *International Journal of Vehicle Mechanics and Mobility*, **37**, 193–216.

Choi, S.-B., Hong, S.-R., Sung, K.-G. and Sohn, J.-W. (2008). Optimal control of structural vibrations using a mixed-mode magnetorheological fluid mount. *International Journal of Mechanical Sciences*, **50**, 559–568.

Chopra, I. (2000). Status of application of smart structures technology to rotorcraft systems. *Journal of the American Helicopter Society*, **45**, 228–252.

Corl, J. (2009). *A comparison of helicopter active rotor gust rejection and vibration alleviation methods*. Master of Science thesis, Department of Mechanical Engineering, Pennsylvania State University, University Park, PA.

Craighead, I.A. (1988). An active suspension system for an ambulance stretcher. In *Proceedings of the International Conference on Advanced Suspensions*. Institution of Mechanical Engineers, London.

Crolla, D.A. (1988). Theoretical comparisons of various active suspension systems in terms of performance and power requirements. In *Proceedings of the Institution of Mechanical Engineers International Conference on Advanced Suspensions*. Mechanical Engineering Publications, Edmonds, WA.

Crolla, D.A. and Abdel–Hady, M.B.A. (1991). Active suspension control; performance comparisons using control laws applied to a full vehicle model. *Vehicle System Dynamics*, **20**, 107–120.

Darsivan, F.J., Martono, W. and Faris, W.F. (2009). Active engine mounting control algorithm using neural network. *Shock and Vibration*, **16**, 417–437.

Davies P.O.A.L. (1965). *Introduction to Dynamic Analysis and Automatic Control*. John Wiley & Sons, New York.

de Benito, C.D. and Eckert, S.J. (1990). Control of an active suspension system subject to random component failures. *Transactions of ASME, Journal of Dynamic Systems, Measurement and Control*, **112**, 94–99.

Den Hartog J.P. (1956). *Mechanical Vibrations*, 4th ed., McGraw-Hill, New York.

Decker, H., Schramm, W. and Kallenback, R. (1988). A practical approach towards advanced semi-active suspension systems. In *Proceedings of the International Conference on Advanced Suspensions*. Institution of Mechanical Engineers, London, 93–100.

Dong, X.-M., Yu, M., Liao, C.-R and Chen, W.M. (2010). Comparative research on semi-active control strategies for magneto-rheological suspension. *Nonlinear Dynamics*, **59**, 433–453.

Du Val, R.W., Gregory, C.Z. Jr. and Gupta, N.K. (1984). Design and evaluation of a state-feedback vibration controller. *Journal of the American Helicopter Society,* **29**, 30–37.

Elliott, S.J., Benassi, L., Brennan, M.J., Gardonio, P. and Huang, X. (2003). Mobility analysis of active isolation systems. *Journal of Sound and Vibration,* **271**, 297–321.

Elliott, S.J., Serrand, M. and Gardonio, P. (2001). Feedback stability limits for active isolation systems with reactive and inertial actuators. *Journal of Vibration and Acoustics,* **123**, 250–261.

El–Madany, M.M. and Qarmoush, A.O. (2011). Dynamic analysis of a slow-active suspension system based on a full car model. *Journal of Vibration and Control,* **17**, 39–53.

El–Madany, M.M. (1990). Ride performance potential of active fast load levelling systems. *Vehicle System Dynamics,* **19**, 19–47.

El–Madany, M.M. and Abduljabbar, Z. (1989). On the statistical performance of active and semi-active car suspension systems. *Computers and Structures,* **33**, 785–790.

El–Madany, M.M. and El–Razaz, Z.S. (1988). Performance of actively suspended cabs in highway trucks: evaluation and optimization. *Journal of Sound and Vibration,* **126**, 423–435.

El–Madany, M.M., El–Tamimi, A. and Al–Swailam, S.I. (1987). On some aspects of active control of road vehicles. *Modelling Simulation and Control,* **10**, 13–23.

Federspiel–Labrosse, G.M. (1954). Contribution a l'etude et au perfectionement de la suspension des vehicles. *Journal de la Society Ing. Auto.*, December, 427–436.

Fenn, R.C., Downer, J.R., Gondhalekar, V. and Johnson B.G. (1990). An active magnetic suspension for space-based microgravity isolation. In *ASME Publication NCA*, Vol. 8, 49–56.

Foag, W. (1988). A practical control concept for passenger car active suspensions with preview. In *Proceedings of the International Conference on Advanced Suspensions*. Institution of Mechanical Engineers, London, 43–50.

Fialho, I.J. (2000). $\mathcal{H}^\infty$ control design for the active rack isolation system. In *Proceedings of the American Control Conference*, Chicago, IL, 2082–2086.

Fialho, I. and Balas, G.J. (2002). Road adaptive active suspension design using linear parameter-varying gain-scheduling. *IEEE Transactions on Control Systems Technology,* **10**, 43–54.

Friedmann, P.P. and Millott, T.A. (1995). Vibration reduction in rotorcraft using active control: a comparison of various approaches. *Journal of Guidance, Control, and Dynamics,* **18**, 664–673.

Fuller, C.R. Elliott, S.J. and Nelson, P.A. (1996). *Active Control of Vibration*, Academic Press, London.

Gao, B., Darling, J., Tilley, D.G., Williams, R.A., Bean, A. and Donahue, J. (2006). Control of a hydropneumatic active suspension based on a non-linear quarter car model. *Proceedings of the Institution of Mechanical Engineers. Part I. Journal of Systems and Control Engineering,* **220**, 15–31.

Gevarter, W.B. (1970). Basic relations for the control of flexible vehicles. *AIAA Journal,* **8**, 666–672.

Gimenez, J.G., Busturia, J.M., Abete, J.M. and Vinyolas, J. (1988). Theoretical-experimental modelling of active suspensions for railway vehicles. *ASME Applied Mechanics Division*, AMD, **96**, 149–156.

Goodall, R.M. and Kortum, W. (1990). Active suspensions for railway vehicles: an affordable luxury or an inevitable consequence? In *Proceedings of the 11th Triennial World Congress of the International Federation of Automatic Control,* **5**, Pergamon Press, New York.

Goodall, R.M. and Mei, T.X. (2006). Active suspensions. Chapter 11 in *Handbook of Railway Vehicle Dynamics* (Simon Iwnicki), 327–358, Taylor & Francis Group, New York.

Gopala Rao, L.V.V. and Narayanan, S. (2008). Preview control of random response of a half car vehicle model traversing rough road. *Journal of Sound and Vibration,* **310**, 352–365.

Graf, P.L. and Shoureshi, R. (1987). *Modelling and Implementation of Semi-active Hydraulic Engine Mounts*. ASME paper 87-WA/DSC-28.

Gupta, N.K. and Du Val, R.W. (1982). A new approach for active control of rotorcraft vibration. *Journal of Guidance and Control,* **5**, 143–150.

Grodsinsky, C.M. and Whorton, M.S. (2000). Survey of active vibration isolation systems for microgravity applications. *Journal of Spacecraft and Rockets,* **37**, 586–596.

Guglielmino, E., Sireteanu, T., Stammers, C.W., Gheorghe, G. and Giuclea, M. (2008). *Semi-active Suspension Control: Improved Vehicle Ride and Road Friendliness,* Springer, London.

Hall, S.R. and Wereby, N.M. (1989). Linear control issues in the higher control of helicopter vibrations. In *Proceedings of the 45th Annual Forum of the American Helicopter Society*, Boston, MA.

Hall, S.R. and Wereley, N.M. (1992). Performance of higher harmonic control algorithms for helicopter vibration reduction. *Journal of Guidance,* **16**, 793–797.

Hattori, K., Kizu, R. Yokoyo, Y. and Ohno, H. (1990). Linear pressure control valve for active suspension. In *Proceedings of the ASME International Computers in Engineering Conference*, 583–588.

Hennecke, D. and Zieglmeier, F.J. (1988). Frequency dependent variable suspension damping: theoretical background and practical success. In *Proceedings of the International Conference on Advanced Suspensions*. Institution of Mechanical Engineers, London, 101–105.

Hine, P.J. and Pearce, P.T. (1988). A practical intelligent damping system. In *Proceedings of the International Conference on Advanced Suspensions*. Institution of Mechanical Engineers, London, 141–148.

Howard, C.Q. (1999). *Active isolation of machinery vibration from flexible structures*. PhD thesis, School of Mechanical Engineering, The University of Adelaide, Adelaide, Australia.

Howard, C.Q., Hansen, C.H. and Pan, J.-Q. (1997). Power transmission from a vibrating body to a circular cylindrical shell through passive and active isolators. *Journal of the Acoustical Society of America*, **101**, 1479–1491.

Howard, C.Q., and Hansen, C.H. (2000). Erratum: Power transmission from a vibrating body to a circular cylindrical shell through passive and active isolators. *Journal of the Acoustical Society of America*, **108**, 2443–2444.

Hrovat, D. (1997). Survey of advanced suspension developments and related optimal control applications. *Automatica*, **33**, 1781–1817.

Hrovat, D. (1991). Optimal suspension performance for 2-D vehicle models. *Journal of Sound and Vibration*, **146**, 93–110.

Huang, X., Elliott, S.J. and Brennan, M.J. (2003). Active isolation of a flexible structure from base vibration. *Journal of Sound and Vibration*, **263**, 357–376.

Iwata, Y and Nakano, M. (1976). Optimum preview control of vehicle air suspensions. *Bulletin of the Japanese Society of Mechanical Engineers*, **19**, 1485–1489.

Jalili, N. (2002). A comparative study and analysis of semi-active vibration control systems. *ASME Journal of Vibration and Acoustics*, **124**, 593–605.

Jenkins, M.D., Nelson, P.A. and Elliott, S.J. (1988). A finite element model for the prediction of the performance of an active vibration isolation system. In *Proceedings of Inter-Noise '88*. Institute of Noise Control Engineering, 1057–1060.

Jenkins, M.D., Nelson, P.A. and Elliott, S.J. (1990). Active isolation of periodic machinery vibration from resonant substructures in *Active Noise and Vibration Control-1990*. ASME Publication NCA, Vol. 8.

Jenkins, M.D., Nelson, P.A. Pinnington, R.J. and Elliott, S.J. (1993). Active isolation of periodic machinery vibrations. *Journal of Sound and Vibration,* **166**, 117–140.

Jones, D.I., Ownes, A.R. and Owen, R.G. (1990). A microgravity facility for in-orbit experiments. In *ASME Publication NCA*, Vol. 8, 67–74.

Junker, H. and Seewald, A. (1984). Theoretical investigation of an active suspension system for wheeled tractors. In *Proceedings of the 8th International Conference of the International Society for Terrain Vehicle Systems: The Performance of Off-Road Vehicles and Machines*, 185–214.

Kaplow, C.E. and Velman, J.R. (1980). Active local vibration isolation applied to a flexible space telescope. *Journal of Guidance and Control*, **3**, 227–233.

Karnopp, D. (1989). Analytical results for optimum actively damped suspensions under random excitation. *Transactions of the ASME, Journal of Vibration, Acoustics Stress and Reliability in Design*, **111**, 278–282.

Karnopp, D., Crosby, M.J. and Harwood, R.A. (1974). Vibration control using semi-active force generators. *Transactions of the ASME, Journal of Engineering for Industry*, **96**, 619–626.

Kela, L. and Vähäoja, P. (2009). Recent studies of adaptive tuned vibration absorbers/neutralizers. *Applied Mechanics Review*, **62**, 060801(1–9).

Kessler, C. (2011a). Active rotor control for helicopters: motivation and survey on higher harmonic control. *CEAS Aeronautical Journal*, **1**, 3–22.

Kessler, C. (2011b). Active rotor control for helicopters: individual blade control and swashplateless rotor designs. *CEAS Aeronautical Journal*, **1**, 23–54.

Kim, J.-S., Wang, K.W. and Smith, E.C. (2007).Development of a resonant trailing-edge flap actuation system for helicopter rotor vibration control. *Smart Materials and Structures*, **16**, 2275–2285.

Kim, S.-M., Elliott, S.J. and Brennan, M.J. (2001). Decentralized control for multichannel active vibration isolation. *IEEE Transactions on Control Systems Technology*, **9**, 93–100.

Kimica, S. (1965). *Servo-Controlled Pneumatic Isolators: Their Properties and Applications*. ASME publication 65-WA/MD-12.

King, S.P. (1988). Minimisation of helicopter vibration through blade design. *Aeronautical Journal*, **92**, 247–263.

Koo, J.-H., Ahmadian, M., Setareh, M. and Murray, T. (2004). In search of suitable control methods for semi-active tuned vibration absorbers. *Journal of Vibration and Control*, **10**, 163–174.

Korkmaz, S. (2011). A review of active structural control: challenges for engineering informatics. *Computers and Structures*, **89**, 2113–2132.

Lee, B.-H. and Lee, C.-W. (2009). Model based feedforward control of electromagnetic type active control engine-mount system. *Journal of Sound and Vibration*, **323**, 574–593.

Levitt, J.A. and Zorka, N.G. (1991). Influence of tire damping in quarter car active suspension models. *Journal of Dynamic Systems, Measurement and Control: Transactions of the ASME*, **113**, 134–137.

Litkouhi, B. and Boustany, N.M. (1988). On board sensor failure detection of an active suspension system using the generalized likelihood ratio approach. In *Proceedings of the IEEE 27th Conference on Decision and Control*.

Lin, C.-C., Chang, C.-C. and Wang, J.-F. (2010). Active control of irregular buildings considering soil-structure interaction effects. *Soil Dynamics and Earthquake Engineering*, **30**, 98–109.

Lin, J.-S. and Kanellakopoulos, I. (1997). Nonlinear design of active suspensions. *IEEE Control Systems*, **17**, 45–59.

Liu, Y., Matsuhisa, H. and Utsuno, H. (2008). Semi-active vibration isolation system with variable stiffness and damping control. *Journal of Sound and Vibration*, **313**, 16–28.

Lund, R.A. (1980). Active damping of large structures in wind. In *Structural Control*, Ed. H.H.E. Leipholtz. North Holland Publishing, Amsterdam.

Malek, K.M. and Hedrick, J.K. (1985). Decoupled active suspension design for improved automotive ride quality/handling performance. *Vehicle System Dynamics*, **14**, 78–81.

Mansour, H., Arzanpour, S., Golnaraghi, M.F. and Parameswaran, A.M. (2011a). Semi-active engine mount design using auxiliary magneto-rheological fluid compliance chamber. *Vehicle System Dynamics: International Journal of Vehicle Mechanics and Mobility*, **49**, 449–462.

Mansour, H., Arzanpour, S. and Golnaraghi, M.F. (2011b). Active decoupler hydraulic engine mount design with application to variable displacement engine. *Journal of Vibration and Control*, **17**, 1498–1508.

Marzbanrad, J., Ahmadi, G., Hojjat, Y. and Zohoor, H. (2002). Optimal active control of vehicle suspension systems including time delay and preview for rough roads. *Journal of Vibration and Control*, **8**, 967–991.

Marzbanrad, J., Ahmadi, G., Zohoor, H. and Hojjat, Y. (2004). Stochastic optimal preview control of a vehicle suspension. *Journal of Sound and Vibration*, **275**, 973–990.

McHugh, F.J. and Shaw, J. (1978). Helicopter vibration reduction with higher harmonic blade pitch. *Journal of the American Helicopter Society*, **23**, 26–35.

Mehendale, C.S., Fialho, I.J. and Grigoriadis, K.M. (2009). A linear parameter-varying framework for adaptive active microgravity isolation. *Journal of Vibration and Control*, **15**, 773–800.

Meller, T. and Frühauf, F. (1988). Variable damping: philosophy and experiences of a preferred system. In *Proceedings of the Institution of Mechanical Engineers International Conference on Advanced Suspensions*. Mechanical Engineering Publications, Edmonds, WA.

Metered, H., Bonello, P. and Oyadiji, S.O. (2010). The experimental identification of magnetorheological dampers and evaluation of their controllers. *Mechanical Systems and Signal Processing*, **24**, 976–994.

Miller L.R. (1988a). Tuning passive, semi-active and fully active suspension systems. In *Proceedings of the IEEE 27th Conference on Decision and Control*. Austin, TX, 2047–2053.

Miller, L.R. (1988b). The effect of hardware limitations on an on/off semi-active suspension. In *Proceedings of the International Conference on Advanced Suspensions*. Institution of Mechanical Engineers, London, 199–206.

Moore, J.H. (1988). Linear variable inductance position transducer for suspension systems. In *Proceedings of the International Conference on Advanced Suspensions*. Institution of Mechanical Engineers, London, 75–82.

Morselli, R. and Zanasi, R. (2008). Control of port Hamiltonian systems by dissipative devices and its application to improve the semi-active suspension behaviour. *Mechatronics*, **18**, 364–369.

Narayanan, S. and Raju G.V. (1990). Stochastic control of non-stationary response of a single-degree-of-freedom vehicle model. *Journal of Sound and Vibration*, **141**, 449–463.

Nelson, P.A. (1991). Active vibration isolation. In *Active Control of Noise and Vibration Course Notes*, Ed. C.H. Hansen. University of Adelaide, Adelaide, South Australia.

Nelson, P.A., Jenkins, M.S. and Elliott, S.J. (1987). Active isolation of periodic vibrations. In *Proceedings of Noise Con '87*. Institute of Noise Control Engineering, 425–430.

Nguyen, K. and Chopra, I. (1990). Application of higher harmonic control to rotors operating at high speed and thrust. *Journal of the American Helicopter Society*, **35**, 78–89.

Okamoto, I., Koyanagi, S., Higaki, H., Terada, K., Sebata, M. and Takai, H. (1987). Active suspension system for railroad passenger cars. In *Proceedings of the Joint ASME/IEEE/AAR Railroad Conference*. IEEE Service Centre, Piscataway, NJ, 141–146.

Olsson, C. (2006). Active automotive engine vibration isolation using feedback control. *Journal of Sound and Vibration*, **294**, 162–176.

Pan, J.-Q. and Hansen, C.H. (1993). Active control of power flow from a vibrating rigid body to a flexible panel through two active isolators. *Journal of the Acoustical Society of America*, **93**, 1947–1953.

Pan, J.-Q. and Hansen, C.H. (1994). Power flow from a vibration source through an intermediate flexible panel to a flexible cylinder. *ASME Journal of Vibration and Acoustics*, **116**, 496–505.

Pan, J.-Q., Hansen, C.H. and Pan, J. (1993). Active isolation of a vibration source from a thin beam using a single active mount. *Journal of the Acoustical Society of America*, **94**, 1425–1434.

Pan, J., Pan, J.-Q. and Hansen, C.H. (1992). Total power flow from a vibrating rigid body to a thin panel through multiple elastic mounts. *Journal of the Acoustical Society of America*, **92**, 895–907.

Parker, G.A. and Lau, K.S. (1988). A novel valve for semi-active vehicle suspension systems. In *Proceedings of the International Conference on Advanced Suspensions*. Institution of Mechanical Engineers, London, 69–74.

Patt, D., Liu, L., Chandrasekar, J., Bernstein, D.S. and Friedmann, P.P. (2005). Higher-harmonic-control algorithm for helicopter vibration reduction revisited. *Journal of Guidance, Control, and Dynamics*, **28**, 918–930.

Pavic, G. and Oreskovic, G. (1977). Energy flow through elastic mountings. In *Proceedings of the 9th International Congress on Acoustics*, Paper G3.

Petersen, N.R. (1980). Design of large scale tuned mass dampers. In *Structural Control*, Ed. H.H.E. Leipholtz. North Holland Publishing, Amsterdam.

Pinnington, R.J. (1987). Vibrational power transmission to a seating of a vibration isolated motor. *Journal of Sound and Vibration*, **118**, 515.

Pinnington, R.J. and White, R.G. (1981). Power flow through machine isolators to resonant and non-resonant beams. *Journal of Sound and Vibration*, **75**, 179–197.

Pollard, N.G., and Simons, N.J.A. (1984). Passenger comfort: the role of active suspensions. *Railway Engineering*, **1**, 17–31.

Prasad, J.V.R. (1991). Active vibration control using fixed order dynamic compensation with frequency shaped cost functionals. *IEEE Control Systems Magazine*, **11**, 71–78.

Preumont, A., Francois, A., Bossens, F. and Abu–Hanieh, A. (2002). Force feedback versus acceleration feedback in active vibration isolation. *Journal of Sound and Vibration*, **257**, 605–613.

Quinlan, D.C. (1992). Automotive active engine mount systems. *AVL Conference on Engine and Environment* '92.

Rakheja, S. and Sankar, S. (1985). *Vibration and a Shock Isolation Performance of a Semi-Active On-Off Damper*. ASME paper 85-DET-15.

Reed, R.E. (1965). *Comparison of Methods in Calculating Resonance Frequencies for Corner Supported Rectangular Plates*. NASA TN D-3030.

Rockwell, T.H. and Lawther, J.M. (1964). Theoretical and experimental results on active vibration dampers. *Journal of the Acoustical Society of America*, **36**, 1507–1515.

Ross, C.F., Scott, J.F. and Sutcliffe, S.G. (1988). Active control of vibration. International Patent Application No. PCT/GB87/00902.

Sam, Y.M., Osman, J.H.S and Ghani, M.R.A. (2004). A class of proportional-integral sliding mode control with application to active suspension system. *Systems and Control Letters*, **51**, 217–223.

Samali, B., Yang, J.N. and Liu, S.C. (1985). Active control of seismic excited buildings. *Journal of Structural Engineering*, **111**, 2165–2180.

Sasaki, M., Kamiya, J. and Shimogo, T. (1976). Optimal preview control of vehicle suspension. *Bulletin of the Japanese Society of Mechanical Engineers*, **19**, 265–273.

Satoh, M. and Osada, K. (1991). Long life potentiometric position sensor. Its material and application. In *Proceedings of the SAE International Congress on Sensors and Actuators*. SAE Publication P-242, 1–12.

Savaresi, S.M., Poussot–Vassal, C., Spelta, C., Dugard, L. and Sename, O. (2010). *Semi-Active Suspension Control Design for Vehicles*, Butterworth–Heinemann, Oxford.

Savaresi, S.M., Silani, E. and Bittanti, S. (2005). Acceleration-driven-damper (ADD): an optimal control algorithm for comfort-oriented semi-active suspensions. *ASME Journal of Dynamic Systems, Measurement, and Control*, **127**, 218–229.

Savaresi, S.M., Silani, E., Bittanti, S. and Porciani, N. (2003). On performance evaluation methods and control strategies for semi-active suspension systems. In *Proceedings of 42nd IEEE Conference on Decision and Control*, **3**, 2264–2269.

Savaresi, S.M. and Spelta, C. (2007). Mixed sky-hook and ADD: approaching the filtering limits of a semi-active suspension. *ASME Journal of Dynamic Systems, Measurement, and Control*, **129**, 382–392.

Scribner, K.B., Sievers, L.A. and von Flotow, A.H. (1990). *Active Narrow Band Vibration Isolation of Machinery Noise from Resonant Substructures*. ASME publication NCA-8, 101–111.

Serrand, M. and Elliott, S.J. (2000). Multichannel feedback control for the isolation of base-excited vibration. *Journal of Sound and Vibration*, **234**, 681–704.

Shannan, J.E. and Vanderploeg, M.J. (1989). Vehicle handling model with active suspensions. *Journal of Mechanisms, Transmissions and Automation in Design*, **111**, 375–381.

Sharp, R.S. and Crolla, D.A. (1987). Road vehicle suspension system design: a review. *Vehicle System Dynamics*, **16**, 167–192.

Sharp, R.S. and Hassan, S.A. (1986). The relative performance capabilities of passive, active and semi-active car suspension systems. In *Proceedings of the Institution of Mechanical Engineers*, **200**, part D3, 219–228.

Sharp, R.S. and Hassan, S.A. (1987a). Performance and design considerations for dissipative semi-active suspension systems for automobiles. In *Proceedings of the Institution of Mechanical Engineers*, **201,** Part D2, 149–153.

Sharp, R.S. and Hassan, S.A. (1987b). On the performance capabilities of active automobile suspension systems of limited bandwidth. *Vehicle System Dynamics*, **16**, 213–225

Shaw, J. (1967). *A Feasibility Study of Helicopter Vibration Reduction by Self-Optimising Higher Harmonic Blade Pitch Control*. MS thesis, Department of Aeronautics and Astronautics, MIT, Cambridge, MA.

Shaw, J. and Albion, N. (1980). Active control of rotor blade pitch for vibration reduction: a wind tunnel demonstration. *Vertica*, **4**, 3–11.

Shaw, J. and Albion, N. (1981). Active control of the helicopter rotor for vibration reduction. *Journal of the American Helicopter Society*, **26**, 32–39.

Shaw, J., Albion, N., Hanker, E. Jr. and Teal, R. (1989). Higher harmonic control: wind tunnel demonstration of fully effective vibratory hub force suppression. *Journal of the American Helicopter Society*, **34**, 14–25.

Shen, Y., Golnaraghi, M.F. and Heppler, G.R. (2006). Semi-active vibration control schemes for suspension systems using magnetorheological dampers. *Journal of Vibration and Control*, **12**, 3–24.

Shiozaki, M., Kamiya, S., Kuroyanagi, M. and Matsui, K. (1991). *High Speed Control of Damping Force Using Piezoelectric Elements*. SAE, Warrendale, PA, 149–154.

Sommerfeldt, S.D. and Tichy, J. (1990). Adaptive control of a two-stage vibration isolation mount. *Journal of the Acoustical Society of America*, **88**, 938–944.

Soong, T.T. and Spencer, Jr., B.F. (2002). Supplemental energy dissipation: state-of-the-art and state-of-the-practice. *Engineering Structures*, **24**, 243–259.

Spencer, Jr., B.F. and Nagarajaiah, S. (2003). State of the art of structural control. *Journal of Structural Engineering*, **129**, 845–856.

Sproston, J.L. and Stanway, R. (1992). Electrorheological fluids in vibration isolation. In *Proceedings of Actuator '92, the 3rd International Conference on New Actuators.* VDI/VDE-Technogiezentrum Informationstechnik GmbH, Bremen, Germany, 116–117.

Sproston, J.L., Stanway, R., Williams, E.W. and Rigby, S. (2004). The electrorheological automotive engine mount. *Journal of Electrostatics*, **32**, 253–259.

Stampleman, D.S. and von Flotow, A.H. (1990). Microgravity isolation mounts based upon piezoelectric film. In *ASME Publication NCA*, Vol. 8, 57–66.

Stanway, R., Sproston, J. and Firoozian, R. (1989). Identification of the damping law of an electro-rheological fluid: a sequential filtering approach. *Journal of Dynamic Systems Measurement and Control*, **111**, 91–96.

Stayner, R.M. (1988). Suspensions for agricultural vehicles. In *Proceedings of the International Conference on Advanced Suspensions*. Institution of Mechanical Engineers, London, 133–140.

Stein, G,J.. (1997). A driver's seat with active suspension of electropneumatic type. *Journal of Vibration and Acoustics*, **119**, 230–235.

Sun, J.Q., Jolly, M.R., and Norris, M.A. (1995). Passive, adaptive and active tuned vibration absorbers – a survey. *ASME Journal of Vibration and Acoustics*, **117**, 234–242.

Sunwoo, M., Cheok, K.C., and Huang, N.J. (1991). Application of model reference adaptive control to active suspension systems. *IEEE Transactions, Industrial Electronics*, **38**, 217–222.

Sussman, N.E. (1974). Statistical ground excitation models for high speed vehicle dynamic analysis. *High Speed Ground Transportation Journal*, **8**, 145–154.

Tanaka, N. and Kikushima, Y. (1988). Rigid support active vibration isolation. Journal of *Sound and Vibration*, **125**, 539–559.

Tani, K. Usui, S. Horiuchi, E., Shirai, N. and Hirobe, S. (1990). Computer controlled active suspension for a wheeled terrain robot. *International Journal of Computer Applications in Technology*, **3**, 100–104.

Teel, A.R., Zaccarian L. and Marcinkowski J.J. (2006). An anti-windup strategy for active vibration isolation systems, *Control Engineering Practice*, **14**, 17–27.

Thompson, A.G. (1976). An active suspension with optimal linear state feedback. *Vehicle System Dynamics* **5**, 187–203.

Thompson, A.G. and Pearce, C.E.M. (1998). Physically realisable feedback controls for a fully active preview suspension applied to a half car model. *Vehicle System Dynamics* **30**, 17–35.

Thompson, A.G. and Pearce, C.E.M. (1979). *An Optimal Suspension for an Automobile on a Random Road*. SAE paper 790478.

Thompson, A.G., Davis B.R. and Pearce C.E.M. (1980). *An Optimal Linear Active Suspension with Finite Road Preview*. Society of Automotive Engineers, paper 800520.

Todd, K.B. (1990). Handling performance of road vehicles with different active suspensions. *ASME Applied Mechanics Division Transportation Systems,* AMD **108**, 19-26.

Tomizuka, M. (1975). Optimal continuous finite preview problems. *IEEE Transactions on Automatic Control*, **AC 20**, 362–365.

Tomizuka, M. (1976). Optimal linear preview control with application to vehicle suspension: revisited. *ASME Journal of Dynamic Systems, Measurement and Control*, September, 309–315.

Tseng, T. and Hrovat, D. (1990). Some characteristics of optimal vehicle suspensions based on quarter car models. In *Proceedings of the IEEE 29th Conference on Decision and Control*, 2232–2237.

Ulsoy, A.G., Hrovat, D. (1990). Stability robustness of LQG active suspensions. In *Proceedings of the 1990 American Control Conference*, 1347–1356.

Valasek, M., Novak, M., Sika, Z. and Vaculin, O. (1997). Extended groundhook – new concept of semi-active control of truck's suspension. *Vehicle System Dynamics* **27**, 289–303.

Wallace, C.F. (1972). Radiation resistance of a rectangular plate. *Journal of the Acoustical Society of America*, **51**, 946–952.

Wang, A.-P. And Lin, Y.-H. (2007). Vibration control of a tall building subjected to earthquake excitation. *Journal of Sound and Vibration*, **299**, 757–773.

Wang, D.H. and Liao, W.H. (2009a). Semi-active suspension systems for railway vehicles using magnetorheological dampers. Part I. System integration and modelling. *Vehicle System Dynamics*, **47**, 1305–1325.

Wang, D.H. and Liao, W.H. (2009b). Semi-active suspension systems for railway vehicles using magnetorheological dampers. Part II. Simulation and analysis. *Vehicle System Dynamics*, **47**, 1439–1471.

Wendel, G.R. and Stecklein, G.L. (1991). A regenerative active suspension system. *Vehicle Dynamics and Electronic Controlled Suspensions*. SAE Special Publication Number 861, 129–135.

White, A.D. and Cooper, D.G. (1984). An adaptive controller for multivariable active noise control. *Applied Acoustics*, **17**, 99–109.

Whorton, M.S. (2005). Robust control for microgravity vibration isolation. *Journal of Spacecraft and Rockets*, **42**, 152–160.

Williams, R.A. (1985). Active suspensions: classical or optimal? *Vehicle System Dynamics*, **14**, 127–132.

Wilson, D.A., Sharp, R.S. and Hassan, S.A. (1986). Application of linear optimal control theory to the design of automobile suspensions. *Vehicle System Dynamics*, **15**, 103–118.

Yagiz, N., Hacioglu, Y. and Taskin, Y. (2008). Fuzzy sliding-mode control of active suspensions. *IEEE Transactions on Industrial Electronics*, **55**, 3883–3890.

Yagiz, N., Yuksek, I. and Sivrioglu, S. (2000). Robust control of active suspensions for a full vehicle model using sliding mode control. *JSME International Journal. Series C, Mechanical Systems, Machine Elements and Manufacturing*, **43**, 253–258.

Yan, B., Brennan, M.J., Elliott, S.J. and Ferguson, N.S. (2010). Active vibration isolation of a system with a distributed parameter isolator using absolute velocity feedback control. *Journal of Sound and Vibration*, **329**, 1601–1614.

Yao, G., Yap, F.F., Chen, G., Li, W.H. and Yeo, S.H. (2002). MR damper and its application for semi-active control of vehicle suspension system. *Mechatronics*, **12**, 963–973.

Yoshimura, T., Ananthanarayana, N. and Deepak, D. (1986). An active vertical suspension for track/vehicle systems. *Journal of Sound and Vibration*, **106**, 217–225.

Yoshimura, T. and Ananthanarayana, N. (1987). An active lateral suspension to a track/vehicle system using stochastic optimal control. *Journal of Sound and Vibration*, **115**, 473–482.

Yoshimura, T. and Ananthanarayana, N. (1991). Stochastic optimal control of vehicle suspension with preview on an irregular surface. *International Journal of Systems Science*, **22**, 1599–1611.

Yoshimura, T., Kume, A., Kurimoto, M. and Hino, J. (2001).Construction of an active suspension system of a quarter car model using the concept of sliding mode control. *Journal of Sound and Vibration*, **239**, 187–199.

Youn I. and Ha , A. (1995).Semi-active suspensions with adaptive capability. *Journal of Sound and Vibration*, **180**, 475–492.

Yue, C., Butsuen, T. and Hedrick, J.K. (1989). Alternative control laws for automotive active suspensions. *Transactions of ASME, Journal of Dynamic Systems, Measurement and Control*, **111**, 286–291.

Yu, M., Dong, X.M., Choi, S.B. and Liao, C.R. (2009). Human simulated intelligent control of vehicle suspension system with MR dampers. *Journal of Sound and Vibration*, **319**, 753–767.

Yu, Y., Naganathan, N.G. and Dukipati, R.V. (2001). A literature review of automotive vehicle engine mounting systems. *Mechanism and Machine Theory*, **36**, 123–142.

Zhou, D., Hansen, C.H., Li, J. and Change, W. (2010). Review of coupled vibration problems in EMS maglev vehicles. *International Journal of Acoustics and Vibration*, **15**, 10–23.

Zhou, D., Hansen, C.H. and Li, J. (2011a). Suppression of maglev vehicle-girder self-excited vibration using a virtual tuned mass damper. *Journal of Sound and Vibration*, **330**, 883–901.

Zhou, D., Li, J. and Hansen, C.H. (2011b).Application of least mean square algorithm to suppression of maglev track-induced self-excited vibration. *Journal of Sound and Vibration*, **330**, 5791–5811.

Zhou, D., Li, J. and Hansen, C.H. (2011c). Suppression of maglev track-induced self-excited vibration using an adaptive cancellation algorithm. *Applied Mechanics and Materials*, **44–47**, 586–590.

CHAPTER 13

Control System Implementation

When implementing active control systems to reduce structural vibration, to reduce sound transmission in ducts, to reduce vibration transmission through isolation systems or to reduce noise radiated by vibrating surfaces, there are many variables associated with the physical system arrangement that need to be optimised to achieve the maximum possible system performance. There are also a number of hardware and software aspects associated with the electronic controller that must be considered and accounted for when a laboratory system is adapted for operation in an industrial or commercial environment. There needs to be some sort of hierarchy in the approach to optimising the many variables, which include type, locations and numbers of control sources: quality of the reference signal for feedforward systems, type, locations and numbers of error sensors; software structure and algorithm considerations; and hardware architecture considerations (Hansen et al., 2007).

Control system software variables include convergence step size, number of filter taps for control and cancellation path filters, gain settings for both error signals in and control signals out, convergence coefficient and leakage coefficient size, control filter type, cancellation path model filter type and control algorithm. Control system hardware variables include type and accuracy of the A/D converter and general control system architecture. This chapter will introduce a number of issues associated with the implementation of active noise and vibration systems that have been investigated in the book, especially some of the issues associated with implementation of some of these algorithms in an electronic (digital) controller. This consideration will cover what is physically required in terms of hardware and performance-related issues in hardware selection. The discussion will be largely qualitative, as there are no fixed rules for optimising hardware arrangements. The discussion will also be brief, concentrating on a few issues which are commonly encountered in active control system implementation. A detailed discussion of hardware issues is beyond the scope of this book, and would probably be out of date by the time it is read.

13.1 HIERARCHY OF ACTIVE CONTROL SYSTEM IMPLEMENTATION

As mentioned above, the design of a practical active noise or vibration control system involves many choices and it is often difficult to make the correct choices unless the design of a system is approached in a systematic manner. This is why it is extremely difficult for a general purpose active noise control system to be optimal for any practical application because each application will be characterised by a different optimum set of choices (Hansen 1997; Hansen et al., 2007). Figure 13.1 shows the hierarchy of active noise and vibration control system design, where the order in which choices should be made and the different aspects that should be considered in the optimisation process are outlined. For industrially based controllers, an additional consideration may be added at the bottom of the figure and that is the effect of an industrial environment on the overall performance of the control system. Things that need to be taken into account include the existence of transient events which could affect the stability of the controller, the ability for continuous monitoring on-site in a control room as well as at a remote location, power failures, transducer failures and electronic component failures.

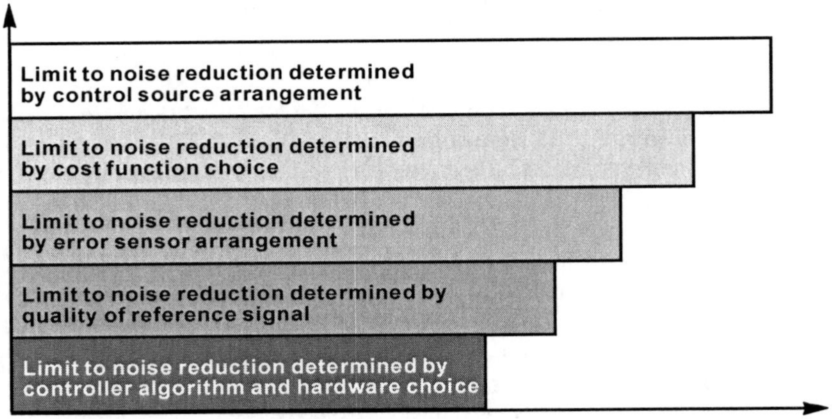

Figure 13.1 Hierarchy of active noise and vibration control system design.

The first parameter that must be optimised is the control source arrangement (number and location), as this will determine the maximum amount of cost function control achievable with an ideal error sensor arrangement and an ideal electronic controller. The control sources may be optimally arranged by using a genetic algorithm to determine the arrangement that minimises the specified cost function, which could be any one of a number of quantities including radiated sound power, global average sound pressure, energy density, pressure at one or more locations or structural vibration measures such as space averaged surface velocity.

The next parameter to be optimised is the practical choice of cost function that is to be minimised by the controller. This cost function is not necessarily the same as the cost function used to determine the optimum control source arrangement. For example, it may be the goal for the active noise or vibration control system to minimise radiated sound power and this would be the cost function used in an analytical model and genetic algorithm to determine the optimum control source arrangement. However, it is often not possible to sense an ideal cost function such as sound power and an optimal alternative cost function has to be substituted.

As another example, the cost function to be minimised may be the sound pressure at a person's ear location and this ear location may also be continually moving. In this case, it is not practical to put a microphone in the person's ear and minimisation of the sound pressure at a virtual location may be the optimal cost function. This would allow fixed microphones to be used to minimise the sound field at a moving location. Generally, the final choice of cost function will depend on many things, including the inconvenience or otherwise of implementing the required sensors and the susceptibility of the sensors to extraneous noise that is not part of the primary noise to be controlled.

The next parameter to be optimised is the error sensor arrangement (number and locations), which determines how close it is possible to get with an ideal controller to the maximum achievable control set by the control source arrangement. For example, if microphones are used to measure a cost function based on minimising global average sound pressure levels, it can be shown that the optimal locations for them are where there is the greatest difference between the primary sound field and the theoretically optimally controlled field. This procedure, which involves the use of multiple regression, is discussed at length in Chapter 8.

Next is the optimisation of the reference signal, which is needed for feedforward but not feedback systems. This generally means that if the controller is to reduce sound pressure levels or radiated sound power, the reference signal should be obtained by non-acoustic means if possible (such as with a tachometer); and if a microphone is used, care must be taken to ensure that the reference signal is not influenced by flow noise or by the control signal. These considerations are discussed in detail in Chapter 14.

The one remaining aspect not yet considered is the design of the electronic controller and this includes the software and hardware. For gradient descent algorithms to be effective in optimising the control filter weights (for FIR or IIR filters) and to minimise extraneous unwanted noise being introduced into the system through the control sources, it is necessary that the reference signal be well correlated with the error signals and with the sound to be minimised.

As mentioned in Chapter 1, an active noise and vibration control system can be divided into two major subsystems: the 'physical' system and the electronic control system. The physical system consists of the control sources and the error sensors, providing the structural/acoustic interface for the active control system. The electronic control system drives the physical system in such a way that the unwanted primary source noise and/or vibration field is attenuated. The design of the physical system, comprising the arrangement and type of control sources and error sensors, limits the maximum noise or vibration control that can be achieved by an ideal active controller, while the control electronics and algorithm limit the ability of the active control system to reach this maximum achievable result. This can also be found in Figure 13.1, where the first four limitations (down from the top of the figure) are decided by the physical system, and the last limitation is determined by the electronic control system. The design requirements for the electronic and physical subsystems are different from and relatively independent of each other, and the design of these subsystems varies from application to application.

Development of a typical active noise and vibration control system can be divided into the following four stages, which may need be applied recursively to obtain a good design (Elliott, 2001):

1. Analysis of the physical system by using the analytical models of simplified arrangements to determine the fundamental physical limitations of the active noise and vibration control system in the given application.
2. Calculation of the optimum performance by using different control strategies under ideal control conditions, using measured data taken from the physical system.
3. Simulation of different control strategies by using data recorded from the physical system under a variety of operating conditions.
4. Implementation of a real-time controller and testing of the system under all conditions to ensure that its behaviour is as predicted.

13.2 ANALOGUE CIRCUIT CONTROLLERS

Although it is difficult to design an analogue circuit controller for complicated or multiple channel active noise and vibration control systems, such controllers do have some advantages such as simple design, low cost and low system delay for feedback control. In fact, most currently available commercial active noise control headsets or earplugs use analogue circuit controllers (Pawelczyk, 2002), and analogue circuit controllers can also be used for active noise control systems in ducts to provide reasonable performance at low-cost.

13.2.1 Feedforward Controller

As discussed in Section 6.3, a block diagram of a single-channel feedforward control system is shown in Figure 13.2, where all quantities have been restated in the frequency domain.

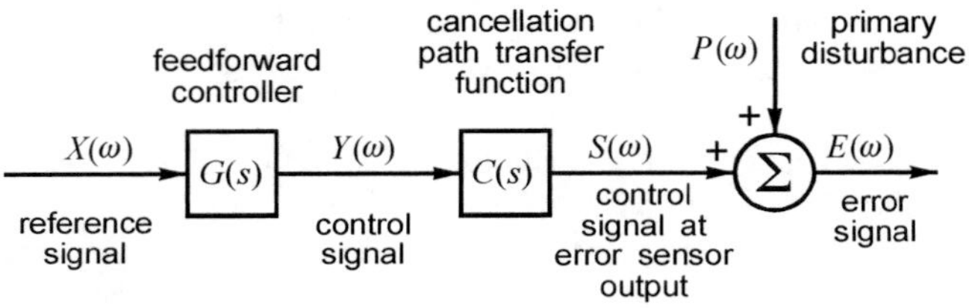

Figure 13.2 Equivalent block diagram of a duct feedforward control system in Figure 6.5.

The reference signal $X(\omega)$, provided to the controller is modified by the controller transfer function $G(s)$ to produce a control signal $Y(\omega)$:

$$Y(\omega) = G(s)X(\omega) \tag{13.2.1}$$

The control signal $S(\omega)$ at the error sensor is obtained by propagating the above signal through the 'physical' system, being modified by the cancellation path transfer function $C(s)$ in such a way that:

$$S(\omega) = C(s)Y(\omega) \tag{13.2.2}$$

The error signal $E(\omega)$ is the superposition of the primary and control source components:

$$E(\omega) = P(\omega) + S(\omega) = P(\omega) + C(s)Y(\omega) \tag{13.2.3}$$

and the optimal (although not necessarily causal) controller transfer function can be obtained as:

$$G(s) = -\frac{P(\omega)}{F(\omega)} \tag{13.2.4}$$

where the filtered reference signal $F(\omega)$ is the reference signal modified by the frequency response of the cancellation path transfer function:

$$F(\omega) = C(s)X(\omega) \tag{13.2.5}$$

For single-channel active control systems with a simple acoustical or vibration path, the above optimal controller might be approximated by using an analogue circuit. For example, for active noise control in a headset or in a long duct with little reflection from the end, assuming both the primary path (the acoustical or vibration path from the reference signal to the error signal) and the cancellation path (from the control source to the error signal) can be approximated by pure delays, the optimal controller given by Equation (13.2.4) is also a delay, which can be approximated by using a simple first-order RC circuit shown in Figure 13.3 in the low-frequency range. The frequency response of the circuit is:

$$G(\omega) = \frac{1}{1 + j\omega R_1 C} \tag{13.2.6}$$

and its gain, phase and delay are shown in the figure. It can be seen from the figure that the gain of this circuit is unity and that it has almost constant phase shift and delay at low frequencies; thus it can be used to approximate a pure delay in the low-frequency range.

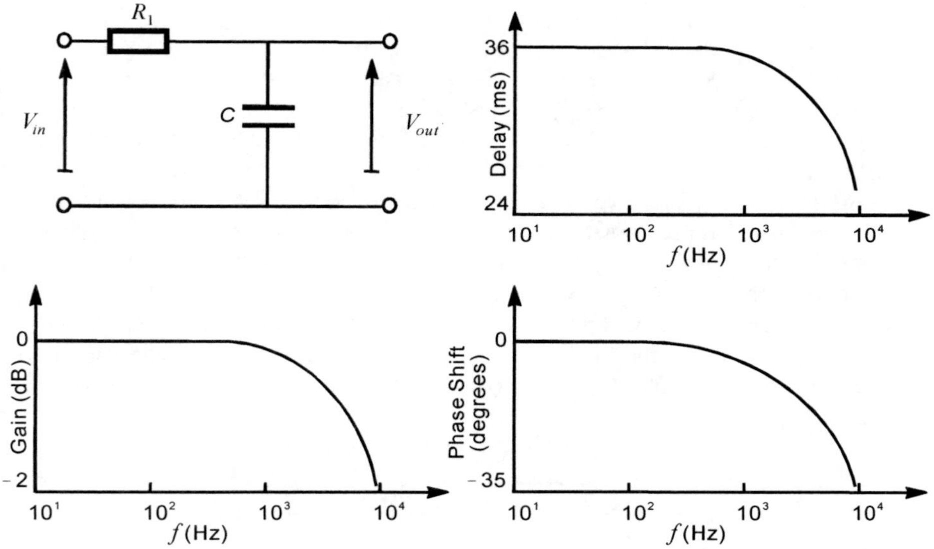

Figure 13.3 Circuit diagram and frequency response of a first-order RC circuit for feedforward active control, with $R_1C = 0.00001$.

Note, there are many kinds of analogue circuits that can be used to design an active noise controller by many different design methods and the practical primary path and the cancellation path are often not a pure delay, so usually higher-order and more complicated analogue circuits are used. Some commercial systems use active noise cancellation integrated circuits (ICs), which combine a pre-amplifier for the error sensor and a power amplifier for the control source, with all the necessary circuitry to enable active noise control to be implemented with a very low power consumption. Further information on active noise cancellation ICs can be found in technical notes available from manufacturers, such as Austriamicrosystems AG.

13.2.2 Feedback Controller

As mentioned in Section 7.2.1, the feedback structure is often used in active noise and vibration control systems when the reference signal is not available. It has been found that the noise attenuation at different frequency ranges in a feedback active control system is not independent of each other due to the mechanisms associated with feedback control. The disturbance attenuation over a finite range of frequencies is often accompanied by the disturbance amplification at other frequencies, and this phenomenon is called the waterbed effect (Doyle et al., 1992).

The aim of designing a feedback controller is to obtain good noise attenuation in the specified frequency bands with sufficient stability, at the cost of minimum noise amplification in the other uncontrolled bands. The $\mathcal{H}^\infty$ loop-shaping technique has been used to shape the open-loop frequency response, and the $\mathcal{H}^\infty$ optimisation technique has been used to synthesise the robust controller (Whidborne et al., 1994; Bai and Lee, 1997). By employing $\mathcal{H}_2$ performance criterion with $\mathcal{H}_2$ and $\mathcal{H}^\infty$ constraints, an $\mathcal{H}_2/\mathcal{H}^\infty$ feedback controller has been designed by utilising the sequential quadratic programming (Rafaely and Elliott, 1994). Through using the measured frequency response of the plant, the controller design problem can be formulated as a constrained optimisation problem, where the $\mathcal{H}_2$ performance objective can be minimised with various frequency dependent constraints by using Recursive Quadratic Programming (Yu and Hu, 2001).

The problem with some methods mentioned above is that the need to translate the performance requirements from practice into various design criteria and constraints for the optimisation problem. Although the intention of the constraints is to ensure optimal performance, sometimes they may cause sub-optimal performance due to the necessary trading off of various parameter settings. Extensive experience and repeated attempts are often required to obtain reasonable parameters. Additionally, it has been found that some numerical algorithms for solving the optimisation problem cannot be guaranteed to find a global optimisation solution.

A block diagram of a feedback active noise and vibration control system is shown in Figure 13.4, where $H(s)$ is the Laplace-domain transfer function of the feedback controller, $G(s)$ is the transfer function of the plant, $d(t)$ is the disturbance signal and $e(t)$ is the residual error signal.

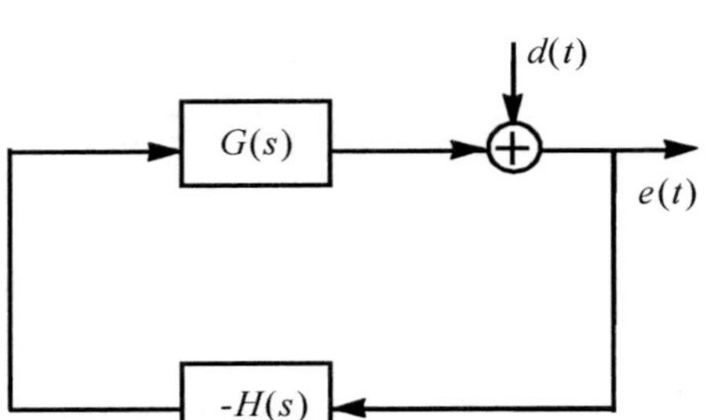

Figure 13.4 A block diagram of a feedback active noise and vibration control system.

The Laplace transform of the error signal can be written as:

$$E(s) = D(s) - G(s)H(s)E(s) \tag{13.2.7}$$

so the transfer function from the disturbance to the error signal is equal to:

$$E(s)/D(s) = \frac{1}{1 + G(s)H(s)} \tag{13.2.8}$$

This transfer function is also known as the sensitivity function of the system. Since the objective of this control system is to achieve a good reduction of the disturbance, the sensitivity function needs to be as small as possible. Substituting $s = \mathrm{j}\omega$ in Equation (13.2.8) allows the frequency response of the sensitivity function to be written as:

$$S(\mathrm{j}\omega) = \frac{1}{1 + G(\mathrm{j}\omega)H(\mathrm{j}\omega)} \tag{13.2.9}$$

where $G(\mathrm{j}\omega)$ is the frequency response of the plant, $H(\mathrm{j}\omega)$ is the frequency response of the controller, and $G(\mathrm{j}\omega)H(\mathrm{j}\omega)$ is the open-loop frequency response of the control system.

From the definition of the sensitivity function, it is clear that the control system provides a disturbance attenuation for $|S(\mathrm{j}\omega)| < 1$ and a disturbance amplification for $|S(\mathrm{j}\omega)| > 1$. The noise reduction NR is defined as $NR(\omega) = -20\log10|S(\mathrm{j}\omega)|$, and can be calculated according to the following equation:

$$NR(\omega) = 20\log_{10}|1 + G(\mathrm{j}\omega)H(\mathrm{j}\omega)| \tag{13.2.10}$$

If the open-loop gain is large from ω_1 to ω_2, so that:

$$|1 + G(\mathrm{j}\omega)H(\mathrm{j}\omega)| > 1 \; ; \qquad \omega \in [\omega_1, \omega_2] \tag{13.2.11}$$

then,

$$|S(\mathrm{j}\omega)| < 1, \qquad \omega \in [\omega_1, \omega_2] \tag{13.2.12}$$

and the control system can achieve a good noise reduction in this frequency range. However, the stability of the control system cannot be always maintained if the open-loop gain is too large, which thus limits the smallest value of the sensitivity function, and hence the greatest possible attenuation of the disturbance.

It is well known that the magnitude of the sensitivity function in the frequency domain follows the Bode's sensitivity integral theorem (Bode, 1945). Under the condition that the number of poles of the open-loop transfer function exceeds the number of zeroes by at least 2 (that is, a phase lag of $-180°$ will occur at high frequencies) and the plant and controller are stable, the following equation holds:

$$\int_0^{+\infty} 20\log_{10}|S(\mathrm{j}\omega)|\,\mathrm{d}\omega = 0 \tag{13.2.13}$$

If the logarithm Bode diagram of $|S(\mathrm{j}\omega)|$ is drawn, the area enclosed by the 0 dB line for the region $|S(\mathrm{j}\omega)| > 1$ should be the same as that for the region $|S(\mathrm{j}\omega)| < 1$. Therefore, the disturbance attenuation ($|S(\mathrm{j}\omega)| < 1$) at a finite range of frequencies is always accompanied by a disturbance amplification ($|S(\mathrm{j}\omega)| > 1$) at other frequencies. The above conditions are usually fulfilled in practice for systems with a high-order low-pass process. Based on the above analysis, it is feasible that the peak value of the undesired disturbance amplification due to the waterbed effect can be reduced to a minimum by averaging the disturbance amplification into the whole disturbance amplification frequency band.

Figure 13.5 shows a simple first-order RC circuit with its gain, phase and delay. The circuit can be used in feedback active control, and its frequency response is:

$$H(j\omega) = \frac{1 + j\omega R_2 C}{1 + j\omega (R_1 + R_2) C} \tag{13.2.14}$$

It can be seen from the figure that the gain of this circuit is almost unity at low frequencies, and then it is attenuated significantly as the frequency increases, while the phase shift (lag) of the circuit increases with frequency at low frequencies but becomes small after a certain middle frequency and is almost zero at high frequencies. The circuit makes little difference to the response of the original feedback system at low frequencies, but at mid frequencies, it provides an additional 20 dB or so of gain attenuation at the cost of about 60° additional phase lag at these frequencies. At higher frequencies, the additional phase lag reduces to zero but the additional gain attenuation remains at about 20 dB. These phase lag properties can increase the gain and phase margins of the feedback system.

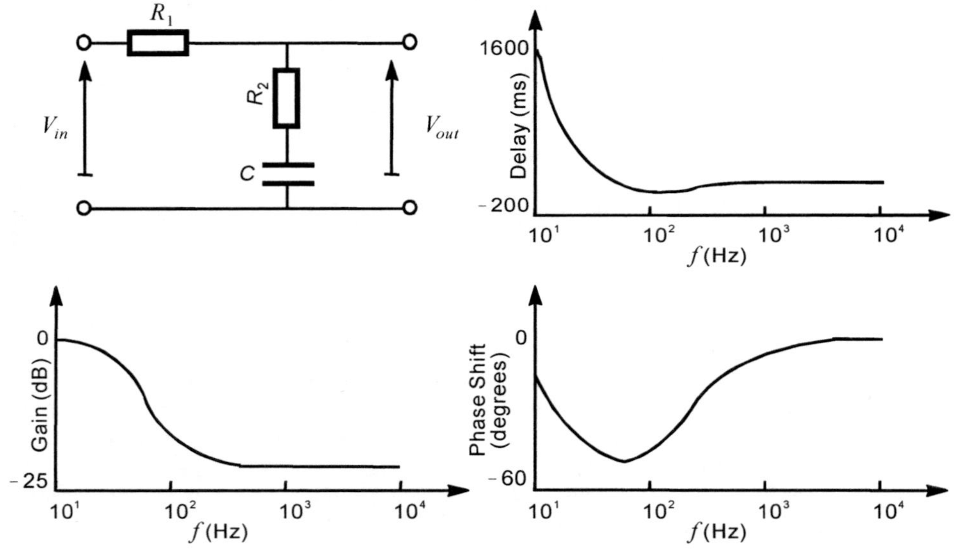

Figure 13.5 Circuit diagram and frequency response of a first-order RC circuit for feedback active control, with $(R_1+R_2)\,C = 0.005$ and $R_1/R_2 =/ 10$.

13.3 DIGITAL CONTROLLERS

The usage of digital controllers in active noise and vibration control systems has increased significantly over the years and is becoming the preferred solution for active noise and vibration control problems. The reasons for this are similar to those underlying the replacement of analogue signal processing by digital signal processing in many applications.

The main advantages of a digital controller are its flexibility for various algorithms, its adaptability with adaptive control algorithms and its accuracy with 16 bit or 32 bit processing. Although the cost of a digital system is still not low, the cost of the digital signal processing chips is decreasing continuously while their performance is also rapidly increasing.

The structure of a digital active controller can be divided into feedforward and feedback types, and the discussion in this section will be largely directed at the implementation of feedforward control systems, as it is these systems that have been of principal interest in this book. The hardware requirements for feedforward and feedback implementations are, for the most part, the same. What will be lacking here is a discussion of strictly feedback implementation issues, such as integrator windup. For a discussion of strictly feedback control issues, the reader is directed towards any good digital control text, such as Middleton and Goodwin (1990) and Franklin et al. (1990).

Illustrated in Figure 13.6 is a block diagram of the main components of a digital active control system. As was outlined in Chapter 5, digital implementation of control systems involves the inclusion of a number of components in addition to the standard textbook control arrangement. These additions include: an anti-aliasing filter, sample-and-hold circuitry, and an analogue-to-digital converter (ADC) on the input to the controller; and a digital-to-analogue converter (DAC), sample-and-hold circuitry, and a reconstruction filter on the output of the controller, and the addition of a clock to synchronise events such as sampling.

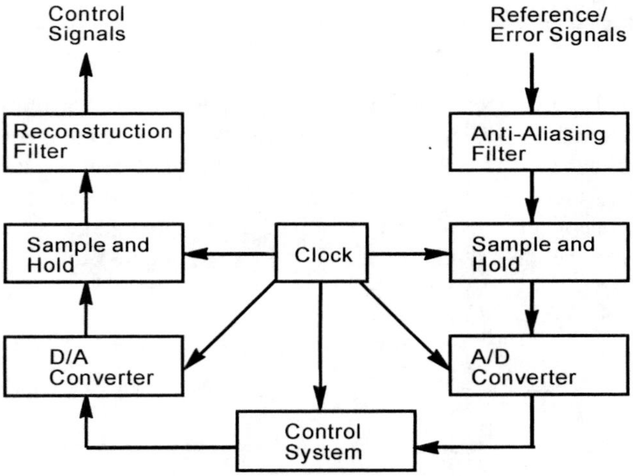

Figure 13.6 Block diagram of the main components of a digital control system.

Referring to Figure 13.7, this arrangement can be separated into two parts: the analogue/digital interface, and the main controller 'body' on which the control algorithms run. This body is typically a micro-processor. Discussion in this chapter will also be separated along the lines of Figure 13.7: design of the analogue-to-digital interface, and implementation of active control software on a micro-processor.

13.3.1 Analogue/Digital Interface

Recall from Chapter 5 that the ADCs and DACs provide an interface between the real (continuous) world and the world of a digital system. ADCs take some physical variable, usually an electrical voltage, and convert it to a stream of numbers, which are sent to controller for use in the control algorithm. These numbers usually arrive at increments of some fixed time period, or sampling period. The numbers arriving from the ADC are usually representative of the value of the signal at the start of the sampling period, as the data input to the ADC is normally sampled and then held constant during the conversion process to enable an accurate conversion, as will be discussed shortly. Commonly, the sampling period is implicitly referred to by a sampling rate, which is the number of samples taken in one second.

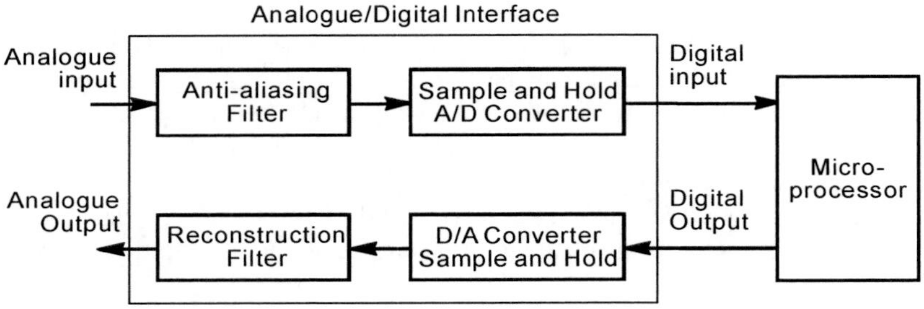

Figure 13.7 Digital control system separated into two parts: an analogue/digital interface, and the main controller body.

The digital signal coming from the ADC is quantised in level. This simply means that the stream of numbers sent to the digital control system has some finite number of digits, hence finite accuracy. This accuracy is normally quantified by the number of bits used by the ADC to represent the measured signal. For example, a 16-bit ADC converter will represent a sampled physical system variable as a set of 16 bits, each with a value of 0 or 1. It follows that the accuracy of the digital representation of the analogue value is limited by the 'quantum size', given by:

$$\text{quantum size} = \frac{\text{full scale range}}{2^n} \tag{13.3.1}$$

where n is the number of bits. For example, if the full scale range of the ADC is ±10 volts, the accuracy of the 16 bit digital representation is limited to better than (20 volts)/(2^{16}) = 0.305 millivolts. The difference between the actual analogue value and its digital representation is referred to as the quantisation error. The dynamic range of the ADC is also determined by the number of bits used to digitally represent the analogue value, and is usually expressed in decibels, or dB. For example, a 16 bit ADC has a dynamic range of (20 log (2^{16})), or 96.3 dB.

One point that should be mentioned here is that the 'ideal' dynamic range of a converter, as defined above, is often significantly less than the effective dynamic range that is achieved in implementation. This is partly because it is unlikely that the high end of the voltage range is consistently used, rendering some of the more significant bits useless. Measurement noise also often corrupts the signal, effectively rendering one or more of the less significant bits ineffective.

The DAC works in the opposite fashion to the ADC, in the sense that it provides a continuous output signal in response to an input stream of numbers. This continuous output is achieved using the sample-and-hold circuit, normally incorporated 'on-chip'. This circuit is designed to progressively extrapolate the output signal between successive samples in some prescribed manner. The most commonly used hold circuit is the zero order hold, which simply holds the output voltage constant between successive samples. With the zero order hold circuit, the output of the DAC/sample-and-hold circuitry is continuous in time, but quantised in level. To smooth out this pattern, a reconstruction filter is normally placed at the output of the DAC/sample-and-hold circuitry. This filter is low-pass, which has the effect of removing the high-frequency 'corners' from the stepped signal.

There are a wide range of variables associated with the design of the analogue/digital interface. Discussion here will concentrate on the most general parameters, such as analogue filter requirements for ADCs and DACs and sample rate selection. The discussion will specifically avoid package-specific issues such as how the data are made available to the micro-processor (parallel versus serial chips). Further discussion of these issues can be found in application notes available from most component manufacturers, such as Analog Devices or Texas Instruments.

13.3.1.1 Sample Rate Selection

The first issue of interest here is selection of a suitable sampling rate. An absolute limit on the lower value of the sampling rate can be derived by studying the effect of sampling on the (continuous) system transfer function. To do this, it is useful to consider the simplified ADC/DAC arrangement shown in Figure 13.8, comprising two parts: the sampler and the hold. The signal coming from the sampler can be viewed as a set of impulses, or an impulse train, as follows:

$$x^*(t) = \sum_{k=-\infty}^{\infty} x(t)\,\delta(t-kT) \tag{13.3.2}$$

where T is the sampling period, taken here to be fixed. The Laplace transform of this is:

$$\mathcal{L}\{x^*(t)\} = x^*(s) = \sum_{k=-\infty}^{\infty} x(kT)\mathrm{e}^{-skT} \tag{13.3.3}$$

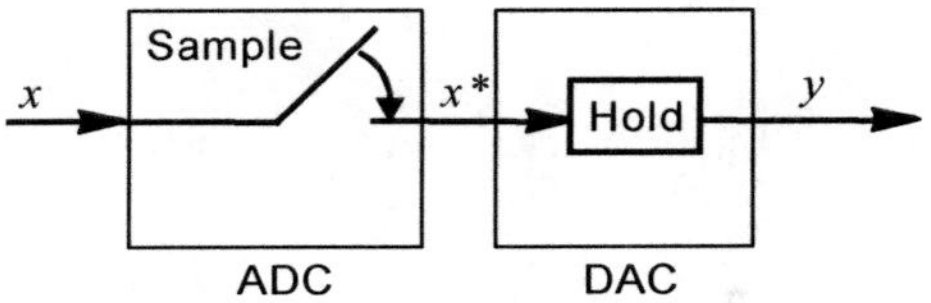

Figure 13.8 Simple ADC/DAC arrangement viewed as two parts: a sampler and a hold.

For the zero-order hold considered here, the output at time t is equal to:

$$x_n(t) = x(kT)\ ; \qquad kT \le t < kT+T \tag{13.3.4}$$

This can be expressed in terms of a unit step function as:

$$x_n(t) = x(kT)(1(t) - 1(t-T)) \tag{13.3.5}$$

which enables the Laplace transform to be calculated:

$$\mathcal{L}\{x_n(t)\} = x_n(s) = \mathcal{L}\{x(kT)\}\frac{1 - e^{-sT}}{s} \tag{13.3.6}$$

Combining the sample of Equation (13.3.3) and the hold of Equation (13.3.6), the continuous time transfer function of the sample-and-hold is:

$$y(s) = \sum_{k=-\infty}^{\infty} x(kT)e^{-skT}\frac{1 - e^{-sT}}{s} \tag{13.3.7}$$

Therefore, the digital control system can be modelled as in block diagram form as shown in Figure 13.9.

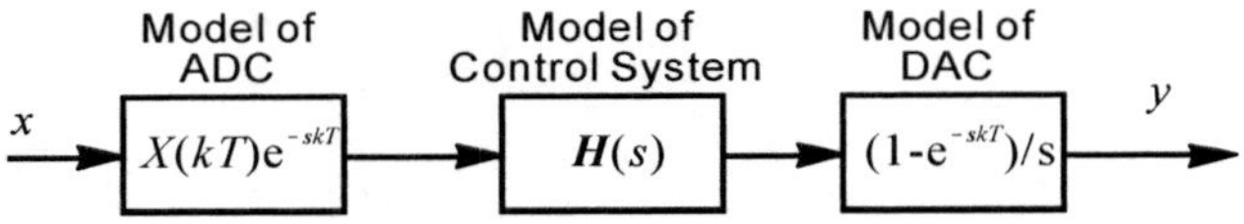

Figure 13.9 Block diagram of the simple ADC/DAC arrangement.

Consider now the frequency domain representation of Equation (13.3.2). The impulse train can be re-expressed as a Fourier series:

$$\sum_{k=-\infty}^{\infty} \delta(t - kT) = \sum_{n=-\infty}^{\infty} h_n e^{j\left(\frac{2\pi n}{T}\right)t} \tag{13.3.8}$$

where the coefficients h_n are found by integrating over a single sampling period:

$$h_n = \frac{1}{T}\int_{-T/2}^{T/2} \delta(t)e^{-j\omega_s n t}\,dt \tag{13.3.9}$$

Here, ω_s is the sampling frequency in rad s^{-1}, defined by:

$$\omega_s = \frac{2\pi}{T} \tag{13.3.10}$$

Evaluation of this integral results in:

$$h_n = \frac{1}{T} \tag{13.3.11}$$

Therefore, the impulse train can be expressed in frequency domain format as:

$$\sum_{k=-\infty}^{\infty} \delta(t - kT) = \frac{1}{T}\sum_{n=-\infty}^{\infty} x(s - jn\omega_s) \tag{13.3.12}$$

Substituting this result back into Equation (13.3.2), and taking the Laplace transform, it is found that:

$$x^*(s) = \frac{1}{T}\sum_{n=-\infty}^{\infty} x(s - jn\omega_s) \tag{13.3.13}$$

The significance of Equation (13.3.13) is that it states that images of the true value of the sampled spectrum repeat themselves at infinite numbers of intervals of ω_s. This phenomenon is termed 'aliasing'.

Practically, the phenomenon of aliasing means that it is impossible to tell the difference between two (or more) sinusoids based upon the sampled signal. This effect is illustrated in Figure 13.10, where two sinusoids have exactly the same sampled values, and can therefore not be distinguished from one another based upon the sampled data. Aliasing can have a significant detrimental effect upon control system performance if there are substantial levels of 'high' frequency data, $\omega > \omega_s/2$, which are allowed to alias onto the 'low' frequency data, $\omega < \omega_s/2$. To combat this problem, anti-aliasing filters placed in front of the input to the sample-and-hold/ADC arrangement. These are analogue low-pass filters, which remove frequency components greater than half the sampling frequency, $\omega > \omega_s/2$, from the input spectrum.

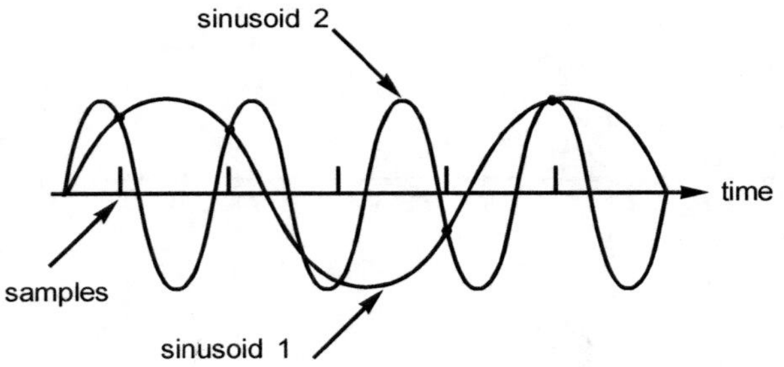

Figure 13.10 A demonstration of aliasing, where two sinusoids have the same sampled values.

From the preceding discussion, it can be surmised that the absolute lower value of sampling rate for a given problem is twice the highest frequency of interest. However, actually implementing a system with this sampling rate is not advisable. First, while it is theoretically possible to reconstruct a harmonic signal sampled at twice its frequency, the filter required to do so is of infinite length, and not BIBO stable (see the discussion of Shannon's reconstruction theory in Middleton and Goodwin (1990)). Second, there is no margin for error in the upper frequency limit; any slight change in upper frequency results in aliasing.

So what is a good sampling rate? It was noted in Chapter 6 that for tonal excitation, synchronising the sampling to be at four times the excitation frequency has some beneficial effects from the standpoint of adaptive algorithm performance (the possibility of a uniform error surface, or equal eigenvalues, which leads to the 'best' compromise between algorithm speed and stability). In general, however, it cannot be expected to be able to synchronise the sample rate of the controller with the unwanted disturbance, even if it is harmonic.

To paint a qualitative picture of a 'good' sampling rate, consider the problem of sampling a step response of a system. If, for ease of computation, it is assumed that the system has a bandwidth of 1 Hz, then the continuous signal will be defined by:

$$y(t) = 1 - e^{-2\pi t} \tag{13.3.14}$$

Referring to Figure 13.11, if the step response is sampled at 2 Hz, the step is indistinguishable to the viewer. Sampled at 5 Hz, the characteristics begin to appear. At 10 Hz, the step is apparent. In fact, if the samples are connected with straight lines, the reconstruction of the step is in error by less than 4% (Middleton and Goodwin, 1990). Intuitively, it seems that the filtering exercise, which is analogous the reconstruction of a signal, becomes 'easier' as the sampling rate increases.

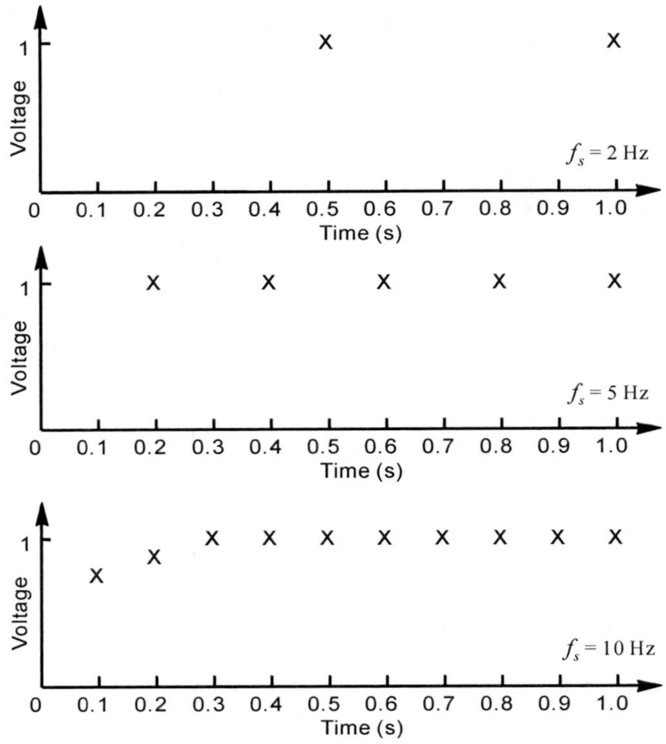

Figure 13.11 Step response of a system with a 1 Hz bandwidth, sampled at 2 Hz, 5 Hz and 10 Hz.

There is, however, a limit to this process. At high sampling rates, tens or even hundreds of times the disturbance frequency, there are numerical problems. In Chapter 6 it was shown that as the level of over-sampling increases, the disparity in the eigenvalues of the auto-correlation matrix of the input samples increases. These eigenvalues, to a large extent, govern the convergence behaviour of the algorithm. A large disparity in eigenvalues generally translates into slow algorithm convergence and reduced algorithm stability. Fast sampling also enhances the numerical inaccuracies associated with finite word length (integer) calculations (Middleton and Goodwin, 1990).

Based on the above discussion, it can be surmised that the optimum sampling rate is a compromise between fast and slow; both of these extremes lead to problems with adaptive algorithm convergence and, especially with fast sampling, stability. The 'optimum' sampling rate compromise is often cited as ten times the frequency of interest. In practice, this sampling rate provides for rapid convergence of the adaptive algorithm and reasonable levels of stability. In implementing active control systems it is often found that for a given sampling rate ω_s, the system will work reasonably well from frequencies approaching $\omega_s/100$ to frequencies up to $\omega_s/4$. On the low end of the scale, adaptation of the controller with frequencies below $\omega_s/100$ is often (extremely) slow, and not a particularly stable operation. While this is sometimes improved by increasing the length of the digital (control) filter, the only real solution is a reduction in sampling rate. On the high end of the scale, the adaptive algorithm appears ineffective with excitation frequencies above $\omega_s/4$.

13.3.1.2 Converter Type and Group Delay Considerations

A second important variable in the design of analogue/digital interface is the converter type, a selection which is influenced by group delay considerations. For the discussion here, group delay is a measure of the time it takes for a signal to pass from the input of the analogue/digital interface to the output. For analogue-to-digital conversion, what is of interest is the group delay through the anti-aliasing filters and the ADC. For digital-to-analogue conversion, what is of interest is the group delay through the DAC and the reconstruction filters.

Before discussing factors related to group delay, it is necessary to understand why group delay is important. Group delay has a significant influence upon (at least) three variables in active control system design: physical size, level of control and stability. Physical size is most obviously influenced by group delay in a causal feedforward active control system, such as a system designed to control (broadband) sound propagation in an air handling duct. If an upstream microphone is being used to supply a reference signal to the active control system, the group delay will limit the proximity of the reference sensor and control source, and hence the physical size of the system. If the group delay through each converter and filter is a few milliseconds, then for sound travelling at 343 ms^{-1} to be controlled, the active control system must be at least 2 m in length (once sensed, the sound will travel 2 m by the time the controlling disturbance is introduced into the system). For compact systems, group delay minimisation is desirable.

If a feedback control system is used, the effect of group delay is slightly different. First, group delay through the analogue/digital interface will be responsible for a phase shift in the signal provided to the controller such that the phase of the signal provided to the control source at time t will not be the same as the signal at the sensor at time t. If at all significant (say, >10°), this phase shift must be taken into account at the time of controller design. Even if it is taken into account, group delay often reduces the 'average' level of control that is achievable. This is because feedback systems are often implemented to attenuate the reverberant response of a system. If it takes 'longer' to sense the disturbance, it takes longer to control it. Averaged over the operating cycle of the system, this results in reduced levels of attenuation with longer group delays.

Group delay can also influence the stability of adaptive algorithms used in feedforward active control systems. As there is a finite time delay between the derivation of a control signal and its 'appearance' in the error signal, there is a delay between a change in filter weights and its effect being measured. As was discussed in Chapter 6, when the weight adaptation is done at time intervals shorter than this delay, the stability of the adaptive algorithm is reduced. With long group delays this is often the case.

In active control system implementation, perhaps the two most common types of ADCs employed are successive-approximation converters and sigma-delta converters. Successive-approximation converters function by a sequence of comparisons between the measured signal and known voltage levels. These converters often require a separate sample-and-hold amplifier to 'capture' the signal for conversion, and external anti-aliasing filters. They range widely in price, but can be relatively expensive for highly accurate devices. The group delay through the converter (governed by the conversion time) can vary, but is often significantly less than the group delay through the anti-aliasing filter that precedes it.

Sigma-delta converters are functionally different from successive-approximation converters, and are often significantly cheaper. These converters sample internally at a rate significantly faster than the 'observed' sampling rate, such as 128 times per delivered sample. An internal conversion with an accuracy of only 1 bit is followed by a 'decimation' filtering process to derive the final sampled signal. Because of the high internal sampling rates, the anti-aliasing filter cut-off can be an extremely high frequency. This means that a single pole filter can be used, which is usually on-chip. The high sampling rate also means that the sample-and-hold amplifier is not required. The overall result is often a (significantly) cheaper converter with no additional (sample-and-hold and anti-aliasing filter) circuitry required.

Sigma-delta converters also have some negative features. The most significant of these is group delay. The filtering process inherent in the sigma-delta conversion technique requires time. At the time of writing this book, the time was typically 30 samples at the nominal (no internal) sampling rate. Therefore, if a sampling rate of 5000 Hz was used, the group delay on the input to the controller is in the order of 5 to 6 ms. This is significantly longer than the group delay that could be expected using a successive-approximation converter and an anti-aliasing filter with a cut-off frequency of 2000 Hz. For harmonic excitation, this is not a problem, as the controller can be non-causal. If an adaptive algorithm is used, stability will be reduced, but this can often be compensated for. However, for a causal system (required for random noise control), this may be unacceptable.

13.3.1.3 Input /Output Filtering

Analogue circuit low-pass filters are often used to precede the ADCs to prevent anti-aliasing, and the fall-off rate of the filter depends on the spectrum and the dynamic range of the signal as well as the type of signal. Usually, a higher fall-off rate filter is needed for a tonal signal than for a random signal, and also a higher fall-off rate filter is needed for the reference signal than for the error signal (Elliott, 2001). For example, for a tonal disturbance with two frequencies of f_t and $f_s - f_t$, where f_s is the sampling rate of an active control system and the objective is to control f_t, then the fall-off rate of the low-pass anti-aliasing filter must be sufficiently high to prevent the aliased version of $f_s - f_t$, which is coherent with the target signal at f_t, to affect the spectrum at f_t. However, for random noise, because the aliasing signal of $f_s - f_t$, is not coherent with the signal at f_t, it will not affect the steady-state noise control performance of the system if the signal is an error signal. If the random signal is used as the reference signal, although the aliasing signal of $f_s - f_t$ is not coherent with the signal at f_t, it will propagate from the control filter and control source to the control point, so this aliasing signal of $f_s - f_t$ should be as small as possible. Therefore, a higher fall-off rate is desired.

The function of the reconstruction filter is to 'smooth' the output signal of the DAC, which is a stepped signal following the zero order hold device. Although the zero order hold has some filtering effects, its main effect is to suppress the frequency components in the original sampled signal at frequencies around the sampling frequency. The other frequency components above the sampling frequency need to be attenuated by the reconstruction filter. As for the anti-aliasing filter, the required fall-off rate of the reconstruction filter also depends on the spectrum of the signal as well as the requirements of attenuation. A higher fall-off rate offers better performance at the cost of a longer filter length, a more complex filter and a longer delay. A summary of the specifications of various anti-aliasing and reconstruction filters required in active noise and vibration control has been tabled by Elliott (2001).

The group delay through the anti-aliasing filter and the reconstruction filter is influenced by a number of variables. The two most significant are the number of poles in the filters and the cut-off frequency. Increasing the number of poles, and decreasing the cut-off frequency, both increase the group delay in an approximately linear fashion. The following equation has been found to be a useful rule of thumb in the initial design of active noise and vibration control system (Elliott, 2001):

$$\tau_A = \left(1.5 + \frac{3n}{8} \right) T \tag{13.3.15}$$

where τ_A is the total group delay of the whole controller, T is the sampling period, n is the number of poles in both the anti-aliasing and the reconstruction filters, and the 1.5 constant is derived from the one sample inherent delay of digital processing and a 0.5 sample delay associated with the zero order hold.

Sometimes, additional analogue bandpass filters are inserted before the ADC and after the DAC to filter out unnecessary interference to the desired signals. Often these bandpass filters can substitute for the anti-aliasing and reconstruction filters. The bandpass filters can also be implemented digitally in software in the digital signal processing module, but in this case they cannot act as anti-aliasing filters. In either case, the bandpass filters can result in significant additional delay in the system.

13.3.2 Micro-Processor Selection

In a digital active control system implementation, the actual control algorithms, including the control filters, will run on a micro-processor of some type. It is no coincidence that interest in active noise control has paralleled advances in micro-processor technology; it is these advances that have made active noise and vibration control practically realisable. As micro-processor technology continues to improve, the applicability of active control will expand.

This section briefly describes the types of micro-processors that are most commonly used at this stage, and the differences between them. Specific brands will not be discussed, only broad categories. The three types of micro-processors of interest here are 'general' micro-processors, digital signal processors (DSPs) and micro-controllers. In the future it is quite possible that the differences that separate these types of micro-processors will begin to evaporate, and the chip 'classes' described here will begin to merge as the technology advances.

General micro-processors, for the purposes of the discussion here, refer to micro-processors that have large instruction sets and can do a wide range of tasks. Such a micro-processor can be viewed as a jack of all trades but master of none. There are several classes of chip that fall into this category. Perhaps the most common is a complex instruction set computer, or CISC chip. These are general purpose microcomputers that have assembly language instruction sets with complex instructions that take several micro-processor clock cycles to implement. These are typically the micro-processors found in personal computers.

A more advanced version of this type of chip is the reduced instruction set computer, or RISC chip. RISC chips have only a simple instruction set, where each instruction can be implemented in one cycle of the micro-processor clock. While they are typically faster than their CISC cousins, RISC chips tend to be more expensive and more complex. Their compilers are also more complex, as all complex routines must be broken down into a series of simple instructions.

In active control implementations, general purpose micro-processors often have the disadvantage that they are relatively slow to do the most important task which is to multiply. It is not uncommon for a general purpose micro-processor to take tens of clock cycles to do a single multiplication. When implementing digital filters, this can be a serious problem, greatly limiting the length of filters that can be implemented. General purpose micro-processors do, in general, have greater flexibility than the other two types of micro-processors being discussed here. This alone, however, is usually not enough to make them the most attractive option.

Digital signal processors (DSPs) are micro-processors with architectures that are optimised for digital signal processing and numeric computation. They can typically perform a multiply and accumulate (add) operation in a single micro-processor clock cycle. This makes them extremely attractive for active control system implementation, which relies heavily on multiplications and additions (the functions of a digital filter). These micro-processors tend not to support 'other' operations, such as division, as well as general purpose micro-processors, but they are often still the micro-processor of choice.

Micro-controllers are essentially complete computer systems on one chip. They come with a wide variety of on-board options, including ADCs and DACs. They are also very inexpensive, commonly finding use in mass-produced products such as white goods. They are, however, lacking in some of the functions important in active control. They generally have less accuracy than the other devices discussed here, being, for example, 8-bit devices instead of 16- or 32-bit devices (which the other devices typically are). The ADCs and DACs are also typically 8 bits, whereas for the other devices, 16-, 22-, or 24-bit ADCs and DACs are common. Micro-controllers are also slow to multiply, as they lack the specialised architecture of the DSP. Their main asset is low-cost, although at the time of writing this book, DSPs in quantity were rapidly approaching the cost of micro-controllers.

Finally, there is the choice between fixed point and floating point processors. Basically, fixed point code is implemented using integer numbers, while floating point code is implemented using real numbers. Micro-processors, especially DSPs, can be purchased that explicitly support fixed point and floating point calculations. In general, fixed point processors are significantly less expensive than floating point processors. However, fixed point calculations are more prone to a range of numerical problems than floating point calculations. As most fixed point devices have floating point 'libraries', which allow them to emulate the functions of fixed point processors, it is not uncommon to implement a controller on a fixed point chip with fixed point real-time filtering code and floating point weight adaptation. The floating-point processors have the advantages of easy and flexible programming as well as good arithmetic properties, so they are extremely attractive for experimental and development active noise and vibration control systems (Elliott, 2001). Although the cost of floating-point processors is becoming less, in many cost sensitive applications, it is still desirable to use cheaper, fixed-point processors, especially for large-volume mass-produced products. It is believed that the performance achieved with a fixed-point processor can be very close to the maximum possible performance, providing that the signals are properly scaled (Elliott, 2001).

13.3.3 Software Considerations

Once a micro-processor and analogue/digital interface has been chosen, some thought must be given to the layout of the software. The issues of interest here are division of the control code, language used and the choice of various control architectures and algorithms.

In a typical adaptive feedforward active control system, the controller software can be divided into two parts: that which must be done in real-time (at each sample period), and that which can be done at a rate slower than real-time (over several sample periods). Typically, calculation of the control filter output(s) must be done in real-time, while all other operations can be done 'off-line'. Real-time code is typically 'interrupt driven'. With this, when a data sample is ready, a signal is sent to the chip which causes it to stop what it is doing so it can perform a specified operation (in this instance, calculation of the filter output). Splitting controller code into an interrupt-driven real-time part and an off-line background part often leads to the most efficient use of the micro-processor capabilities, especially for large implementations.

With most micro-processors, a number of programming languages can be used to generate controller code. These usually include a low-level assembly language and a variety of high-level languages, such as C. While it is generally possible to program tasks in any of the available languages, the speed of execution of the compiled code can vary significantly. In general, code written using the micro-processor assembly language executes faster (in other words, is more efficient) than code written using a higher level language. This means that it is advantageous to write the real-time portion of the control code in assembly language, to give it maximum speed. The off-line code could then be written in a higher level language, where speed is not as important as the speed in with which a program can be written and debugged (higher level languages tend to be far simpler to program than assembly language).

There is a great temptation to program active noise and vibration control systems in a high-level language, particularly C. This is exacerbated by the proliferation of PC-based DSP cards, which typically come with high-level language-based development tools. High level languages might work for small systems running on powerful DSPs, and in research settings where price is less important than flexibility. However, for commercial products, C-coded controllers are a poor choice. A routine programmed in C will typically take five to ten times as many clock cycles to execute as an assembler routine with the same functionality. Given the multiplier effect outlined previously, this is not tolerable. Even some DSP manufacturers, in particular Texas Instruments, are marketing their newest chips as being fast enough to tolerate the inefficiencies associated with high-level language programming. Still, it seems a pity to effectively run a DSP at 15 to 20% of its maximum throughput (Snyder, 1999).

In the implementation of an adaptive digital filter, there are two sources of finite precision error the quantisation error that occurs in the analogue-to-digital signal conversion, and the truncation error that occurs when multiplying two numbers in a finite precision environment. It may be tempting to ignore these errors in the implementation of the adaptive algorithm, as they would appear on the surface to be random in sign, and of an order less than the least significant bit of the system. However, such assumptions can lead to disastrous results. To fix the problem, a simple but effective method is to use the leakage adaptive algorithm that was introduced in Chapter 6.

13.3.4 Controller Architectures

In deciding on the appropriate architecture for an active noise and vibration controller, one must choose from a number of possible fundamental options (Hansen, 2004). Briefly, there are three practical fundamental options, which are:

1. Use currently available DSP boards and I/O boards that can be inserted into a PC and programmed in C and run via the Windows operating system. Thus, if more powerful boards become available, they can simply replace existing boards with negligible, if any, impact on the software requirements. Unfortunately, the Windows operating system suffers from a large overhead in terms of memory, and a system programmed in this way will invariably suffer due to processing speed limitations. This type of system is also too expensive and insufficiently robust for industrial applications.
2. Put a general-purpose DSP board in a custom enclosure with power supply and custom I/O boards, programmed using DSP programming tools. Unfortunately, the continued supply of these boards is often problematic, their specifications are unreliable and they have many components that are not needed for active noise control. There are also considerable problems interfacing general purpose DSP boards with I/O boards from another manufacturer.
3. Construct a multi-processor DSP board from scratch with sufficient power and memory to meet the most demanding active noise and vibration control applications. Include it in a modular system that allows the use of multiple DSP boards and multiple I/O modules that can be tailored to the number of channels needed and the processing power and memory required.

The third option offers more flexibility in terms of producing a system optimised for active noise and vibration control. Figure 13.12 shows an example architecture for this type of controller, where the central digital signal processing module is the centre of the active controller and carries out the core tasks such as the adaptive control filter update and cancellation path modelling.

A number of different control algorithms could be implemented on the platform, including time and frequency domain algorithms optimised for both tonal and broadband sound field control. The feedforward, feedback and hybrid structures should all be available, and advanced users would also be able to program their own control algorithms. Cancellation path identification would be possible, both on-line and off-line. The hardware usually consists of a high-end, powerful, floating point DSP processor or an array of DSP processors (Hansen, 2004; Qiu et al., 2004; Berkhoff, 2010; Chen et al., 2012) and would be configured to minimise digital delays and to provide a large capacity for filter taps for both control filters and cancellation path estimation.

The pre-processing and general management module carries out the tasks of pre-processing of the signals, monitoring of each channel status and management of communication with the digital-analogue interface module and the human-controller interface module. The pre-processing includes multi-rate filtering, sub-band signal generation and full-band signal synthesis, or time and frequency domain transformation.

Sometimes even real-time control signal generation can be carried out in this module to minimise the delay. The hardware usually adopts low-cost DSP chips, one for each analogue I/O channel (Hansen, 2004), or high-speed Field Programmable Gate Array (FPGA) (Berkhoff, 2010), or an ARM micro-processor (Chen et al., 2012). The hardware provides ample processing power for all the I/O management tasks and multi-rate filtering as well as transducer failure and signal overload management.

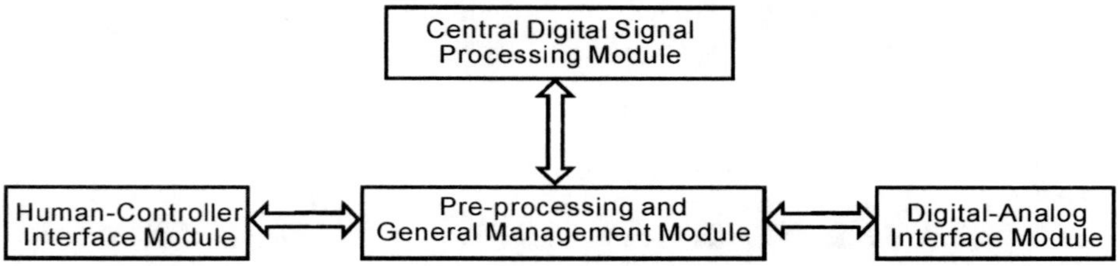

Figure 13.12 An architecture of general active noise and vibration controller.

The digital-analogue interface module usually consists of AD/DA converters, anti-aliasing filters and reconstruction filters. Note, pre-amplifiers for sensors, and power amplifiers for control sources are usually not included in the controller because different kinds of transducers require different amplifiers, and they are all also normally commercially available. The old-generation controllers usually use AD/DA converters with a sampling frequency of 10 to 20 times the centre frequency of the control band, while current controllers tend to use much higher sampling frequencies, such as 100 to 300 kHz, to achieve benefits of a much shorter delay and flexibility for a wide range of active noise and vibration control projects. The pre-processing and general management module manages the required down-sampling and up-sampling filtering so that the central digital signal processing module processes the signal at a much reduced sampling rate of 10 to 20 times the control frequency (Hansen, 2004; Berkhoff, 2010; Chen et al., 2012).

Special attention should be paid to selecting AD and DA converters for minimising their delays. The old-generation controllers usually use sigma-delta type of converters because they do not need separate anti-aliasing and reconstruction filters, they are low-cost and they have high precision. However, the problem with this type of converter for active noise and vibration control is their long delay, so current controllers tend to use specially designed AD and DA converters for active control or extra short delay successive-approximation AD and DA converters (Chen et al., 2009; Chen et al., 2012). For example, AD73322 from Analog Devices is a specially designed AD and DA converter, which has very short group delay, theoretically 25 μs in AD and 50 μs in DA at a 64 kHz sampling rate.

The human-controller interface module provides an interface for users so that they can customise the graphic user interface, set-up parameters of the algorithms and control system, and monitor the performance of the control system. There are various ways to implement this; for example, by communication with a PC via RS 232 protocol, via a PCI bus, via Ethernet protocol, or directly by using an embedded touch-screen so that no external PC is required (Chen et al., 2012). All these functions can run on the pre-processing and general management module.

With the architecture mentioned in Figure 13.12, it is also possible to have different configurations for different applications. For example, the controller can be configured as a multiple single-channel controller. Although the performance of the multiple single-channel controller is not as good as the fully coupled multi-channel controller, they may find application in industrial situations as a result of their simplicity, low-cost, tracking ability and adaptation speed (Hansen, 2004). The controller can also be configured as a combination of high-authority control (HAC) and low-authority control (LAC), which was shown to lead to improvements with respect to parametric uncertainties and un-modelled dynamics (Berkhoff, 2010). In their particular application, the HAC (adaptive fully coupled multi-channel control) is implemented on the central digital signal processing module, while the LAC (decentralised control) is implemented on the pre-processing and general management module (a high-speed Field Programmable Gate Array).

13.3.5 Procedures for Implementing Digital Active Controllers

The first step in implementing an active noise and vibration system is to select a structure for the system. As mentioned in Section 7.2, there are two fundamentally different structures of controller: feedback and feedforward. Whenever a high quality reference signal is available, feedforward control should be implemented due to its superior noise reduction performance and system robustness.

If feedback control is to be used, the sampling frequency of the system should be as high as possible to reduce the system delay. The control filter can be an FIR or IIR type (see Chapter 6), and it can be fixed or adaptive. Various methods that are used for designing analogue circuit feedback controllers can also be used to design the fixed digital feedback controllers. To implement adaptive feedback controllers, the internal model of the cancellation path can be used to transform a feedback control problem into a feedforward control problem, so that various algorithms introduced in Chapter 6 can be adopted (Elliott, 2001).

For feedforward control, the first issue is to check the coherence between the reference signal and the disturbance, which can be used to give a simple measure of the maximum reduction that would be achieved by the active noise and vibration control system (Kuo and Morgan, 1996). Then, depending on the properties of the physical paths and resources of the digital signal platform, FIR or IIR type filters are selected and the length of the filters for control and cancellation path modelling is determined.

After that, various algorithms described in Chapter 6 can be selected to update the control filter. The selection criteria depend on the noise properties, primary and cancellation path properties, as well as the resources and limitation of the digital processing platform. For tonal noise control, simple algorithms and hardware should be sufficient, while for multi-channel broadband control, more complex algorithms such as those incorporation multi-rate signal processing techniques, such as the frequency domain and sub-band algorithms introduced in Chapter 6, may need to be implemented to reduce the computational burden and to improve the control performance.

Sometimes, a combined feedback/feedforward structure, with either analogue or digital feedback can be adopted to improve the performance of an active noise and vibration control system. Modification of the effective plant response by the feedback controller benefits the adaptive feedforward control because it can reduce the variability of the effective plant response and reduce the amplitude of any resonances in the physical plant, thus improving the convergence speed of the adaptive feedforward system. In other words, feedforward and feedback ANC systems can work together in a complementary way in controlling broadband and tonal disturbances (Elliott, 2001).

While programming and debugging the selected algorithm on the digital signal processing platform, the code should be optimised to increase its efficiency. For time critical and repeated tasks, special attention should be paid to the code and in some cases it may be necessary to use assembly language to optimise its performance. For other background tasks and user interface, a high level language, such as C, is generally preferred. The completed active control system should be tested under various conditions before installing on site so that its various parameters, such as the convergence coefficient, the leakage coefficient, the update rate, etc. can be optimised.

Finally, some supervising software modules should also be programmed to handle the unexpected issues such as the failure of a channel or a sensor or an actuator etc. All hardware associated with the active control system, including the sensors, control sources, cables, circuits, etc. should be able to function properly on site where many environmental issues such as high temperature and strong electrical and magnetic field interference might exist. The system should also have ability to self-start after a power failure (Hansen, 2004).

13.4 AN EXAMPLE OF ACTIVE CONTROL SYSTEM IMPLEMENTATION

The implementation of successful active noise and vibration control systems is best illustrated by an example of an actual installation that includes many of the capabilities listed in the previous section. The chosen example is a system used to control sound propagation in the exhaust stack of a spray dryer unit used in a dairy factory for making powdered milk. Perhaps the most complicating aspects of this example were the frequency at which control was desired was above the higher-order mode cut-on frequency, and the sound field in the duct was never steady so that the error sensor outputs varied rapidly and continually in amplitude and phase (Hansen, 2004).

The aim of the active noise control system was to reduce the amplitude of tonal noise, at the fan blade passing frequency, which emanated from the top of the spray dryer exhaust stack and radiated into the surrounding community. The diameter of the exhaust stack is 1.6 m, the temperature range of exhaust air is from 60 to 90°C and the blade passing frequency ranges from 170 to 190 Hz. Calculations showed that for this stack, two higher-order modes propagate, thus greatly complicating the control system required. Not only do higher-order modes exist, but temperature variations in the stack resulted in rapidly shifting nodal planes, which resulted in quite rapid and large sound pressure fluctuations at any given location in the duct. Of course, a conservative approach would have been to weld axial splitter plates in the duct so that only plane waves could propagate between the control source and error sensors. For these to be effective, the welds would have to have been air-tight. As the duct was stainless steel and vertical, it was considered too difficult to implement this option, which is why a multi-mode control system was developed.

One of the more expensive and time consuming aspects of the installation was the signal cabling. Extensive lengths of signal cabling were needed to connect the reference and error sensors to the controller and to connect the controller to the loudspeakers, all of which were physically separated by many tens of metres. Care had to be taken in joining wires to minimise the possibility of corrosion of the joints and the introduction of extraneous electrical noise. In all cases, water-proof junction boxes were used. Due to the sensitive nature of the DSP chips and other electronic components, the controller had to be located in a temperature controlled room and space was found along side other process control instrumentation.

Interesting practical aspects associated with each element of the control system are described in the following sections, where the main components, such as the reference sensor, the error sensors, the control sound sources, the controller and their associated technical challenges, are discussed.

13.4.1 Reference Sensor

The reference sensor is used to derive a signal that is used by the controller to generate the driving signal for the control loudspeakers. For the spray dryers, the reference signal was derived using a Hall effect tachometer, which produced an electronic pulse each time a gear tooth passed its field of view. The gear wheel was manufactured especially for the project and was mounted on the fan drive shaft. The reference signal was generated digitally by passing the tachometer signal into an Analog Devices Sharc EZ-kit 21061 DSP board. The time between successive tachometer pulses was measured by passing the signal into the digital interrupt port, which interrupts the central processor and allows it to count clock cycles between successive pulses. As the CPU runs at 40 MHz, the resolution is 25 nano-seconds. The DSP was programmed to generate a sine wave with the same period (or any required multiple thereof) and output it with a sample rate of 16 kHz through a low-pass filter into the control system reference signal input. The reference signal sine wave thus remains synchronised with the fan blade passing frequency at all times.

13.4.2 Error Sensors

The large turbulent pressure fluctuations in the duct propagate at the speed of the air flow and not the speed of sound, and thus their presence complicates the error sensor design, which ideally should only respond to acoustic pressure fluctuations if undesirable control signals are to be avoided. Initially, microphones were placed in the end of a 2 m long porous tube with the intention of amplifying the acoustic pressure fluctuations and attenuating the turbulent pressure fluctuations.

This approach had two main problems. The turbulent pressure fluctuation rejection was insufficient, and anchoring the tubes in the vertical duct as well as protecting them during cleaning operations was problematical. They also represented a possible site for bacterial growth. The final design was a flush wall-mounted sensor, consisting of an electret microphone inserted in a tube with its own pre-amplifier, connected to an enclosure filled with a block of acoustic foam protected by a metallic foil to filter out some of the turbulence related pressure fluctuations. Small disks of Vyon, a porous plastic, were inserted in the tube in front of the microphone to attenuate the sound level incident on the microphones so as to avoid mechanical saturation of the microphone membrane, because of the high acoustic levels present inside the duct. A continuous air flow is maintained over the microphones to keep them cool. The aluminum foil facing on the acoustic foam insert prevents contamination of the microphone system by milk powder.

With this design, background pseudo sound was reduced to more than 13 dB below the targeted acoustic error signal. Twelve error sensors were installed on the exhaust stack, flush with the duct wall. They are organised in three rings of four, five and three sensors respectively. The three rings are respectively at a distance of 0.20 m, 0.70 m and 1.85m below the exhaust duct outlet plane.

13.4.3 Sound Sources

Due to the high temperature and sometimes wet environment, it was necessary to provide protection for the loudspeaker cones and cooling for the coil. At first, a membrane of Viton, a rubber-like material, was tried. Even though this was limp and relatively lightweight with a transmission loss close to zero at 200 Hz, it had a dramatic effect on the loudspeaker output, reducing the maximum achievable sound level by 20 dB, probably as a result of the Viton material acoustically loading the loudspeaker and changing its radiation impedance. As the loudspeakers have to generate sound levels in the duct of up to 138 dB, this amount of loss was unacceptable. The final design consisted of a 600 W paper cone, low-frequency loudspeaker, coated with a silicon compound to protect it against humidity, and then a reflective coating to protect it against dust particles and heat build-up. A later version of this speaker arrangement used a planar membrane of thin Mylar sheet over the face of the speaker, which only reduced the low-frequency speaker output by 3 dB.

To further discourage the build-up of milk powder dust on the speaker diaphragm, it would have been useful to introduce a small air jet close to where build-up is possible. The enclosure at the back of the speaker is connected to the front face of the speaker with a small tube to ensure pressure equalisation of the front and back surfaces of the speaker cone. Without this equalisation, the speaker cone would suffer a DC shift and fail to function correctly. The heat generated by the speaker coils in operation is dissipated using pressurised air injected in the vented speaker backing enclosures.

Each speaker enclosure is mounted flush with the duct wall and one to two speakers are mounted at each of four different axial locations, covering a total of seven different locations. The top speakers are mounted approximately 3500 mm below the duct exit plane. During operation, the 600 W speakers are driven at about 5 W, which minimises harmonic distortion and maximises speaker life expectancy.

13.4.4 Controller

A feedforward controller using a periodic block filtered-x LMS algorithm mentioned in Section 6.9.6.3 (with the control filter updates carried out every fifteenth sample) was chosen for the task. The periodic block filtered-x LMS algorithm is less computationally intensive than the standard filtered-x LMS algorithm as it allows control filter updates to be done after a block of error sensor data has been collected. There are twelve input channels for the error signals and six output channels for the control signals. The reference signal is derived from the tachometer on the fan shaft so there is no acoustic feedback from the control sources to the reference sensor.

The cost function is the average sound pressure from all twelve microphones and even though the sound pressure recorded by individual microphones changes quite quickly, the average changes much more slowly, and this allows the control system to converge to an optimum. Note that the cost function also includes a quantity proportional to the control signal outputs, which is equivalent to including leakage. Other cost functions such as weighting the microphone signals differently, so that more effort would be directed towards controlling the larger signals, were tried but not found to be any more effective than the average mentioned above. The optimum size of the convergence coefficient was found by trial and error as a compromise between convergence speed and instability risk, and was adjusted on site during system set-up. Needless to say, the acceptable stability risk was zero so a relatively small convergence coefficient was used.

It is well known that to ensure stability of the control system, it is necessary to have an accurate estimate of the cancellation path impulse response or transfer function. In many laboratory experiments, this is done off-line prior to starting up the controller, using random or pseudo random noise injected into each control source in turn. For any industrial application, this is the preferred way to start up the system. However, once the system begins to operate, it is generally not possible to shut it down to check the cancellation path transfer function and an on-line procedure is needed. Various schemes were tried on the spray dryer system, and it was eventually decided to use low level random noise, 30 dB below the level of the blade passing tone, introduced into each control speaker in sequence. Using such low level noise, which was not detectable by ear, required very long averaging times to determine the cancellation path transfer function but that has not proven to be a problem.

The controller hardware used for the project consists of two DSP boards based on the Analog Devices ADSP21062 processor, which is suitable for executing both system-level software (such as the interface to the PC) and signal processing software, such as system modelling and control filter adaptation. In the current system, one DSP board is used for each purpose, with one acting as a slave to the other, which is the master. The interface between the modules is implemented using the processor's high-speed link ports. Not shown in the above layout is the Analog Devices EZ-Kit 21061, which was used to process the reference signal input pulse train to produce a sine wave at the blade passing frequency, which was used as the reference signal input to the system.

Some of the problems that were overcome during commissioning include excessive noise on the tachometer signal due to electricians laying signal cables next to power cables; electromagnetic pickup in DSP board communication lines due to radiation from motor speed controllers and other process equipment; and power failures resulting in a requirement for operator intervention during restart. Specific features of the control system include automatic identification and alarm set state for any failed speaker or microphone channels; system shutdown and temporary rest if total noise levels exceeded primary noise levels; system reset if algorithm instability is detected; continuous low-level system identification; and an ability to self-start after a power failure.

13.4.5 Control System Performance

The performance of the system was measured in the community rather than at the error sensors in the duct (where the performance was much better). Noise reductions in the community ranged from 9 to 14 dB at the blade passing frequency, with the reduction being such that the tone was no longer noticeable.

REFERENCES

Bai, M. and Lee, D. (1997). Implementation of an active headset by using the H∞ robust control theory. *Journal of the Acoustical Society of America*, **102**, 2184–2190.

Berkhoff, A.P. (2010). Rapidly converging multichannel controllers for broadband noise and vibrations. In *Proceedings of the 39th International Congress & Exhibition on Noise Control Engineering* (*Inter-Noise 2010*).

Bode, H.W. (1945). *Network Analysis and Feedback Amplifier Design*. D. Van Nostrand Company, Inc., Princeton, NJ.

Chen, K., Lu, J. and Qiu, X. (2009). Multi-channel broadband active noise control hardware system design. In *Proceedings of Active 09*, 1–6.

Chen, K., Lu, J. and Qiu, X. (2012). The design of a multi-channel active noise controller with ultra low latency. In *Proceedings of ACOUSTICS 2012 HONG KONG Conference and Exhibition.*

Doyle, J.C., Francis, B.A. and Tannenbaum, A.R. (1992). *Feedback Control Theory.* Macmillan, London, UK.

Elliott, S. J. (2001). *Signal Processing for Active Control.* Academic Press.

Franklin, G.F., Powell, J.D. and Workman, M.L. (1990). *Digital Control of Dynamic Systems.* Addison-Wesley, Reading, MA.

Hansen, C.H. (1997). Active noise control – from laboratory to industrial implementation. In *Proceedings of Noise-Con '97*, 3– 38.

Hansen, C.H. (2004). Current and future industrial applications of active noise control. In *Proceedings of Active 04*, 1–18.

Hansen, C.H., Qiu, X., Petersen, C., Howard, C. and Singh, S. (2007). Optimization of active and semi-active noise and vibration control systems. In *Proceedings of 14th International Congress on Sound and Vibration*, 1–18.

Kuo, S.M. and Morgan D.R. (1996). *Active Noise Control Systems: Algorithms and DSP Implementations.* John Wiley & Son Inc., New York.

Middleton, R.H. and Goodwin, G.C. (1990). *Digital Control and Estimation: A Unified Approach.* Prentice Hall, Englewood Cliffs, NJ.

Pawelczyk, M. (2002). Analogue active noise control. *Applied Acoustics*, **63**, 1193–1213.

Qiu, X., Li N. and Chen G. (2004). Multiprocessor DSP systems for active control. In *Proceedings of 18th International Congress on Acoustics*, 1277–1280.

Rafaely, B. and Elliott, S.J. (1999). Active control of sound in a headrest: design and implementation, *IEEE Transactions on Control Systems Technology*, **7**, 79–84.

Snyder, S.D. (1999). Microprocessors for active control: bigger is not always enough. In *Proceedings of Active 99*, 1–18.

Whidborne, J.F., Postlethwaite, I. and Gu, D.W. (1994). Robust controller design using loop-shaping and the method of inequalities. *IEEE Transactions on Control Systems Technology*, **2**, 455–461.

Yu, S.H. and Hu, J.S. (2001). Controller design for active noise cancellation headphones using experimental raw data, *IEEE/ASME Transactions on Mechatronics*, **6**, 483–490.

CHAPTER 14

Sound Sources and Sound Sensors

The active noise control systems discussed in previous chapters require the use of transducers to either generate or measure sound. To choose the most appropriate transducer for a particular application, and to predict the maximum controllability of a particular physical system, it is useful to have a basic understanding of the physical principles governing the operation of the various transducers, so these are discussed in the following paragraphs.

14.1 LOUDSPEAKERS

A loudspeaker is an electro-acoustic actuator that radiates acoustical energy in response to an electrical input signal. Moving coil, electrostatic, optical and parametric array are four loudspeaker types that can be used for active noise control. Traditional moving coil loudspeakers are by far the most widely used transducer type in active noise control systems because of their low-cost, ease of manufacture and high performance level. Electrostatic loudspeakers have the advantage of low harmonic distortion and highly directional radiation properties while optical loudspeakers are insensitive to electrical, magnetic and radioactive fields. Parametric array loudspeakers produce a spatially focussed beam of low-frequency sound and can be used in active noise control to concentrate the control effort at the desired location of attenuation.

14.1.1 Traditional Moving-Coil Loudspeakers

Figure 14.1 shows the structure of a typical moving-coil loudspeaker. The loudspeaker contains a lightweight rigid diaphragm or cone usually manufactured from paper, plastic or metal. The cone is attached to the voice coil and supported by a flexible spider that allows axial but not radial movement. The voice coil is a lightweight tube wrapped with a specified length of wire. It is suspended within an annular gap in a permanent magnet and drives the centre of the cone to produce sound waves when a current is applied. The perimeter of the cone is supported by a flexible surround (or outer suspension) that adds additional mechanical damping to the cone. The end of the voice coil is covered by a dust cap to protect the magnet chamber from environmental contaminants.

When selecting a loudspeaker for an active noise control system, the important parameter for specification is cone volume displacement capability rather than power handling capacity. The electrical power needed to generate the required cone displacement depends on the volume of the enclosure backing the loudspeaker. At low frequencies (in the range generally of interest for active control), the effective stiffness associated with the loudspeaker will be inversely proportional to the enclosure volume. Thus, the electrical input power to the speaker will also be approximately inversely proportional to the square of the effective stiffness, as is shown by the following analysis.

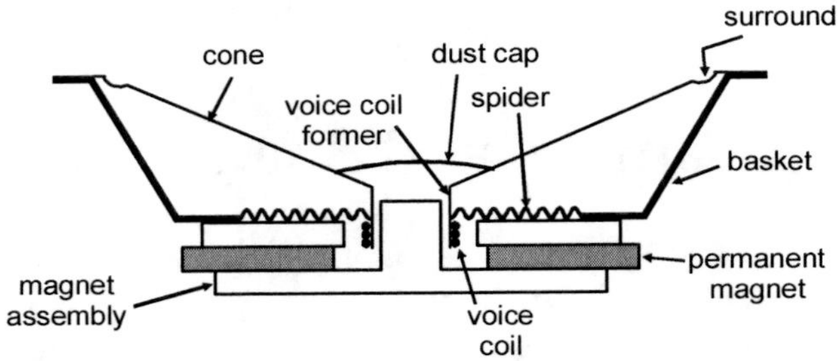

Figure 14.1 Schematic diagram of a moving-coil loudspeaker.

At low frequencies where the speaker cone can be assumed to behave like a rigid piston, the thrust T applied by the voice coil is:

$$T = BLi \tag{14.1.1}$$

where B is the magnetic flux density, L is the length of the conductor and i is the current flowing through it. This thrust is balanced by the acoustic load and the mechanical impedance of the loudspeaker suspension and cone. Equating the two gives:

$$BLI = Z_m v / S \tag{14.1.2}$$

where v is the cone volume velocity, S is the effective cone area and Z_m is the total mechanical impedance, which is the sum of that due to the acoustic load (Z_{ma}) and that due to the loudspeaker (Z_{me}). At frequency ω, this latter quantity is given by:

$$Z_{me} = \mathrm{j}\omega m + C_d + K/\mathrm{j}\omega \tag{14.1.3}$$

where m is the effective mass of the cone and armature, C_d is the suspension damping and k is the suspension stiffness (including the effect of the enclosure backing the loudspeaker). Note that at low frequencies and for small enclosures, the enclosure stiffness (which is inversely proportional to the volume) will generally be much larger than the suspension stiffness and thus will dominate the term K:

$$K = \frac{pA_c\omega}{u} \tag{14.1.4}$$

where A_c is the area of the speaker cone, and the impedance associated with the enclosure volume is given by Equation 9.37d in Bies and Hansen (2009):

$$\frac{p}{uA_c} = -\mathrm{j}\frac{\rho c^2}{V\omega} \tag{14.1.5}$$

where V is the enclosure volume. Thus,

$$K = -\mathrm{j}\frac{\rho c^2 A_c^2}{V} \tag{14.1.6}$$

The terminal voltage on the voice coil is given by:

$$E = Z_E i + BLu \tag{14.1.7}$$

where Z_E is the blocked electrical impedance given by:

$$Z_E = \mathrm{j}\omega L + R + C_c/\mathrm{j}\omega \tag{14.1.8}$$

where L, R and C_c are respectively the inductance, resistance and capacitance of the voice coil. The quantity u is the voice coil velocity which is equal to v/A if it is assumed that the cone is moving as a piston. Note that BLu is the back EMF due to motion of the coil. Shepherd et al. (1986) reported that using this back EMF as a feedback signal to the coil driving signal improves the smoothness of the speaker frequency response and makes the speaker behave more like a constant volume velocity source (infinite internal impedance).

The terminal voltage can be related to the cone movement by combining Equations (14.1.2) and (14.1.7) to give:

$$E = u\left[\frac{Z_E Z_m}{BL} + BL\right] \tag{14.1.9}$$

The electrical power consumed by the voice coil is given by:

$$W = \frac{1}{2}i^2 R + \frac{1}{2}u^2 \mathrm{Re}\{Z_m\} \tag{14.1.10}$$

which can be written in terms of the volume velocity of the cone (of area S) using Equation (14.1.2) to give:

$$W = \frac{v^2}{2S^2}\left[\frac{|Z_m|^2 R}{(BL)^2} + \mathrm{Re}\{Z_m\}\right] \tag{14.1.11}$$

The first term in brackets in Equation (14.1.11) represents the electrical losses and the second term the mechanical losses.

For small backing enclosures and low frequencies, the stiffness term Z_m will dominate, and it can be seen from Equation (14.1.11) that the power will then be approximately proportional to this quantity squared. It is also of interest to note that the required electrical power to the speaker is also proportional to the square of the volume velocity. The required control source volume velocity to achieve optimal control in a particular physical system is determined using the physical system analyses discussed elsewhere in this book.

In some cases it will be found that multiple loudspeakers are needed to obtain the required cone volume velocity. For sound propagating in a rectangular section duct, the required cone volume velocity can be shown to be related to the propagating sound pressure contribution of each loudspeaker, which for plane waves can be written as (Shepherd et al., 1986):

$$v = \frac{2pS(\omega d/2c_0)\sin(\omega\tau)}{\rho_0 c_0 S_p \sin(\omega d/2c_0)} \tag{14.1.12}$$

where p and v represent respectively the acoustic pressure amplitude and cone volume velocity amplitude at frequency ω, d is the side length of the square piston representing the speaker cone, c_0 is the speed of sound, S is the duct cross-sectional area and S_p is the speaker cone area. For a circular cone, d may be replaced by $\pi d/4$, where d is the diameter of the cone, to give an approximate result. If p in Equation (14.1.12) is an r.m.s. quantity, then v will be also.

In Equation (14.1.12), the term $\sin(\omega\tau)$ is included to account for a speaker source consisting of two speakers to make it directional (see Section 7.2 for a full discussion). The quantity τ is the time that the signal into one speaker is delayed before being input to the other. If only one speaker is used, the term $\sin(\omega\tau)$ is set equal to unity. If the loudspeaker is small compared to a wavelength of sound, the term $(\omega d/2c_0)/\sin(\omega d/2c_0)$ from Equation (14.1.12) is approximately unity and the required speaker response is much easier to realise in practice, thus minimising signal distortion. Thus, use of a number of smaller speakers is always preferred over one large loudspeaker if it is practical. For example, in a duct system it would be preferable to use a number of smaller speakers placed in the duct walls at the same axial location along the duct, rather than a single large loudspeaker. For random noise, the peak cone velocity requirements for active control are likely to be four or five times the estimated r.m.s. velocity requirement (Shepherd et al., 1986). Note also that loudspeakers should not be driven at maximum power or maximum cone deflection, as this results in substantial shortening of the speaker life. In fact, for active noise control systems, speakers should be designed so that the speaker produces no more than 10% to 20% of its rated power, thus minimising distortion and maximising speaker life.

Shepherd et al. (1986) and Ford (1984) show that the speaker electrical power requirement can vary over three orders of magnitude for the same acoustic power output. They also show that the electrical power requirements are minimised at the frequency corresponding to the mechanical resonance of the loudspeaker. Thus, it is suggested that the loudspeaker be designed for each application so that its mechanical resonance lies in the centre of the frequency range of interest. Unfortunately, this is not usually a practical alternative and doing this will require the use of motional feedback of the speaker cone to smooth the resulting frequency response. Most large loudspeakers have a fundamental mechanical resonance in free air of between 25 and 50 Hz. This can be varied to suit a particular application by either adding mass to the cone (to reduce it) or by increasing the effective suspension stiffness by adding a backing enclosure to the speaker (to increase the resonance frequency).

In cases where the required volume velocity exceeds the capability of available speakers (usually 200 to 400 W of electrical power), five alternative options are available. As a first alternative, the number of speakers at each control source location can be increased (for example, by placing them around the perimeter of a duct cross-section at the desired axial location).

As a second alternative, the loudspeaker coil and magnet could be replaced by a low-inertia, high-speed d.c. servo-motor and belt-drive system. The belt is used to convert the motor rotary motion to linear motion. Because of its low inertia and ability to rotate in either direction, the motor is capable of responding to audio signals up to about 130 Hz. Peak-to-peak deflections achievable with this arrangement are about 30 mm, compared to 10 mm for a typical bass speaker (Leventhall, 1988). Speakers are commercially available which can produce 140 dB at 1 m, between 28 Hz and 125 Hz with less than 2% harmonic distortion and 400 W of electrical power (Servo-Drive BassTech 7).

As a third alternative, the loudspeaker could be replaced with an air modulated source, which is a device that directs large volumes of air through slots that are adjusted in size using a vibration generator or servo-motor. Other techniques involve the continuous modulation of the size of an orifice through which high-pressure air is passing by using a plug or valve attached to a vibration generator or servo-motor (Glendinning et al., 1988). Because of the low acoustic efficiency of these types of sources, they need to be used in conjunction with a suitable horn, otherwise the compressed air requirements are very large; for example, it was found that two industrial air compressors typically used on building sites were needed to provide sufficient sound to actively control noise in an automotive exhaust system.

A fourth alternative is to use a horn to improve the coupling efficiency of the speaker in the frequency range of interest. The design of horns is discussed in the next section.

A fifth alternative is to use a well designed, tuned (and probably ported) backing enclosure optimised for the frequency range of interest. A ported loudspeaker enclosure contains a tuned aperture or duct that extends inwards from the enclosure wall. If designed correctly, ported loudspeaker systems can have lower distortion, increased efficiency and lower cut-off frequencies than their sealed counterparts. The design of low-frequency, closed-boxed and ported loudspeaker systems has been analysed by Benson (1969), Thiele (1971), and Small (1972, 1973). Their design technique is known as the Thiele-Small analysis and is based on electric filter design theory to determine the optimum system parameters. Keele (1981) extended the Thiele-Small theory to develop a simplified design methodology for low-frequency, closed-box and ported loudspeaker systems. Rutt (1985) has also developed a low-frequency loudspeaker design technique using root-locus analysis. Software is commercially available for the design of optimised loudspeaker systems (for example Leap), using as input, speaker electrical and mechanical specifications and the characteristics of the space into which the speaker will be radiating. If the speaker is to be used in a duct, one of dimensions of the radiating space is set very large and the other two are the duct cross-sectional dimensions.

Loudspeaker enclosures generally need drain holes to prevent water build-up caused by cyclic heating and cooling of enclosures, especially those located outdoors. The drain holes must be sufficiently small not to affect the stiffness of the enclosure at the frequencies of interest or alternatively must be taken into account in the design of the enclosure. Note that drain holes are also essential to allow mean pressure equalisation between the rear and front faces of the speaker cone. If the mean air pressure on one side of the cone is larger than on the other side, the cone will be forced against one of its stops and the resulting sound production will be very distorted.

A loudspeaker backed by a small enclosure can be approximated as a constant volume velocity source for the purpose of physical system analysis in the low-frequency range generally of interest for active noise control (for example below 400 Hz). Larger enclosures could result in the suspension resonance lying in the frequency range of interest; and unless cone motional feedback is used (Shepherd et al., 1986), the speaker cannot be considered a constant volume velocity source at frequencies near this resonance. However, at higher frequencies or for speakers with no backing enclosure, the constant acoustic pressure source (see Bies and Hansen, 2009, Chapter 6) offers a closer approximation to actual behaviour.

Special precautions must be taken in the loudspeaker design to take account of the operating environment. For example, in applications involving high humidity and/or high temperature, water resistant, high temperature adhesives should be used on the loudspeaker cone/coil connection. Special care is also needed with the speaker cone design; either the paper cone must be sprayed with a protective coating or substituted with some inert plastic, depending on the nature of the environment to be encountered. In cases of extreme heat, the loudspeaker cone must be protected with a heat shield, which may require the addition of cooling air flowing past the shield. Alternatively, the loudspeaker could be removed from extreme conditions by placing it at the end of a short tube, although the space for this may not always be available.

Loudspeaker non-linearities can degrade the performance of an active noise control system as signal distortion components are not cancelled by the primary noise. Moving-coil loudspeakers behave in a sufficiently linear manner if the excursion amplitude of the voice coil is small. At low frequencies, a small excursion amplitude requires a large diaphragm area to generate the required volume velocity, which is often impractical. Alternatively, digital controllers have been developed for active non-linear loudspeaker control. Klippel (1999) developed a digital controller to actively compensate for loudspeaker distortion using the back-EMF from the electrical signals at the loudspeaker terminals as the control input. The back-EMF defines the relationship between the velocity of the voice coil and the electrical signals at the loudspeaker terminals. The digital controller generates a pre-distorted electrical drive signal to counteract harmonic distortion in the loudspeaker output. Toso and Johnson (2009) developed an active control system based on the FXLMS algorithm to reduce harmonic distortion in an infrasonic loudspeaker driven by a DC motor. In this case, non-linearities are produced by the DC motor due to the commutation of the brushes on the rotor sectors.

14.1.2 Electrostatic Loudspeakers

Electrostatic loudspeakers commonly consist of a thin, moveable diaphragm suspended in the air gap between two stationary, perforated electrodes as shown in Figure 14.2. A high d.c. polarising voltage is applied to the diaphragm and ground potential is applied to the electrodes. With no input signal, the diaphragm is suspended at an equal distance to the two electrodes. When an input signal is applied, the two electrodes are driven in anti-phase so that an equal voltage of opposite polarity appears at each electrode. As like charges repel and opposite charges attract, the diaphragm moves towards one electrode and away from the other. The diaphragm movement drives the air on either side of it, thus producing sound waves.

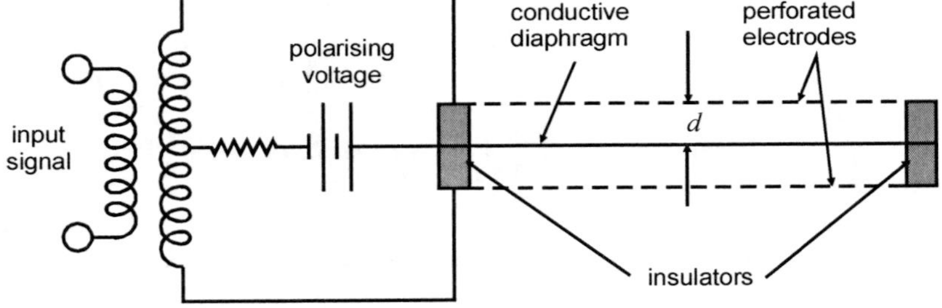

Figure 14.2 Section view of an electrostatic loudspeaker.

Despite a few unsuccessful attempts to manufacture electrostatic loudspeakers in the 1920s and 1930s, the first commercial wide range electrostatic loudspeaker was developed by the Acoustical Manufacturing Company (now Quad) in 1957. Their successful electrostatic loudspeaker design was based on the pioneering work of Walker (1955) and Williamson (1957) and used the principle of constant-charge operation (Hunt, 1954), whereby a constant polarising charge is applied to the surface of the diaphragm to ensure low non-linearity distortion. For constant-charge operation, the diaphragm's conductive coating is selected to have sufficiently high surface resistivity and/or a large-value external charging resistor is placed in series with the diaphragm. This ensures that the charge associated with the capacitance of the space between the two electrodes does not migrate along the surface during the period corresponding to the lowest frequency of interest.

The general equation governing electrostatic loudspeaker constant-charge operation is:

$$Q = CE \tag{14.1.13}$$

where Q is the charge between the diaphragm and the electrodes, E is the applied voltage and C is the total capacitance associated with the two electrodes and the space in between. Due to the large value resistor placed in series with the diaphragm, the charge Q will remain constant once the polarising charge process has reached equilibrium. Equation (14.1.13) indicates that for the charge to remain constant, the capacitance and voltage must have an inverse relationship.

The capacitance between the diaphragm and the electrodes in an electrostatic loudspeaker is given by:

$$C = 2\varepsilon_0 \frac{A}{d} \tag{14.1.14}$$

where d is the distance between the diaphragm and one electrode, $\varepsilon_0 = 8.854 \times 10^{-12}$ Farads/m and A is the area of the diaphragm.

When the electrodes are supplied with a signal voltage, the force exerted on the charged diaphragm suspended in the electric field is:

$$F = \frac{\varepsilon_0 A E_{pol} E_{sig}}{d^2} \tag{14.1.15}$$

where E_{pol} is the polarising voltage and E_{sig} is the signal voltage. Equation (14.1.15) states that the magnitude of the force on the diaphragm is directly proportional to the applied signal voltage and inversely proportional to the square of the distance between the electrodes.

The appeal of electrostatic loudspeakers for active noise control is in their low harmonic distortion and highly directional radiation properties. Harmonic distortion can degrade the performance of an active noise control system controlling tonal noise, as even though the amplitude of the fundamental tone is reduced, the amplitude of higher-order harmonics is increased. The diaphragm of an electrostatic loudspeaker is driven equally at all points on its surface, resulting in very low levels of harmonic distortion. Generally, the levels of harmonic distortion of an electrostatic loudspeaker are one or two orders of magnitude below those of a conventional cone loudspeaker.

In cases where local control at selected locations as a result of local cancellation is the target, highly directional sound sources are often desirable so as to concentrate the control effort at the desired locations of attenuation. This has the added benefit of reducing control spillover at locations outside the zone of quiet. Electrostatic loudspeakers have the advantage of much higher directivity than traditional cone loudspeakers as their radiating area is typically much larger. The motion of the electrostatic loudspeaker diaphragm radiates equal acoustical power in opposite directions, meaning that electrostatic loudspeakers are highly directional dipole radiators.

Electrostatic loudspeakers are generally large in area compared to their thickness, but they can also consist of curved or splayed panels. These configurations allow for improved control of the sound distribution pattern through appropriate selection of the radius of curvature. For example, the Quad ESL-63 is an electrostatic loudspeaker composed of concentric annular segments that produce a hemispherical radiation pattern.

The electrostatic loudspeaker output at low frequencies is limited by the maximum displacement of the diaphragm and this is determined by the spacing between the fixed electrodes and the diaphragm, and the damping in the suspension. The dipole nature of the system means the roll-off of 6 dB/oct will result in a less sharp reduction of loudspeaker acoustical output as the frequency becomes lower. A larger diaphragm area can be used to produce a lower useful frequency limit but this will, however, result in irregular high-frequency directional properties. To overcome this problem, electrostatic loudspeakers can consist of several panels with diaphragms of different sizes, where the tension and size of each diaphragm is optimised for a certain frequency range.

In terms of acoustic pressure output, Walker's equation (Baxandall, 2001) can be used to calculate the performance of an electrostatic loudspeaker over a large frequency range. The acoustic pressure output p at a distance r in the far-field on the axis of the loudspeaker is:

$$p = i \frac{E_{pol}}{2\pi c_0 r d} \tag{14.1.16}$$

where i is the input signal current and c_0 is the speed of sound. Equation (14.1.16) shows that there is a simple relationship between the input signal current and the sound pressure on-axis at large distances.

Electrostatic loudspeakers have the advantage of a thin, lightweight panel design compared to traditional moving-coil loudspeakers that can be heavy and require large installation volumes. Warwick Audio Technologies have recently developed a highly directional electrostatic loudspeaker, named the Flat Flexible Loudspeaker (FFL), whose shape is customisable depending on the application. The FFL is constructed from light-weight, ultra–thin, flexible materials and has the advantage of easy installation and the ability for large scale manufacturing at a low cost. The FFL has a flat (±3 dB) frequency response from 200 Hz to 20 kHz and sound pressure levels of a maximum of about 80 dB are achievable, which means that this type of loudspeaker may have only limited application to industrial active noise control systems.

14.1.2.1 Electrostrictive Loudspeakers

An electrostrictive polymer film (EPF) loudspeaker has been proposed by Heydt et al. (1998, 2000) as an alternative to electrostatic loudspeakers for use in active noise control systems. EPF loudspeakers consist of a thin polymer film sandwiched between two electrodes. The polymer film is compressed by electrostatic forces when voltage is applied to the electrodes. The force per unit area F that acts on the film is (Heydt et al., 1998):

$$F = \varepsilon_r \varepsilon_0 F_s^2 = \frac{\varepsilon_r \varepsilon_0 E^2}{h^2} \tag{14.1.17}$$

where F_s is the electric field strength, ε_r is the relative dielectric constant of the polymer, ε_0 is the dielectric constant of free space, h is the film thickness and E is the applied signal voltage. The thickness strain s of the polymer film resulting from compression is equal to the force per unit area of the film divided by the elastic modulus of the polymer, Y, according to (Heydt et al., 1998):

$$s = \frac{\varepsilon_r \varepsilon_0 F_s^2}{Y} = \frac{\varepsilon_r \varepsilon_0 E^2}{h^2 Y} \tag{14.1.18}$$

The polymer film and the two electrodes sit on top of a dense grid of holes. The enclosed volume below the grid is kept at a slightly negative air pressure. When the voltage signal is applied, the film thickness decreases. As the polymer is incompressible, the film area increases, and the negative enclosure pressure causes the film to deform or bubble downwards through the holes. Sound is produced by the volume acceleration of air resulting from bubble oscillations.

A typical EPF loudspeaker has an active area of 5 cm × 5 cm, a silicone film thickness of 83 μm and can produce on-axis sound pressures levels of over 80 dB at 1 m at frequencies above about 1 kHz.

EPF loudspeakers are suitable for use in active noise control systems that target relatively high-frequency noise as they are compact, lightweight and constructed from inexpensive materials. Both electrostatic and EPF loudspeakers are highly directional and generate high sound pressure by diaphragm displacement but a larger force per unit area is produced by an EPF actuator. The polymer film of an EPF actuator can be driven at higher electric field strengths than are possible with electrostatic loudspeakers. Additionally, EPF loudspeakers can operate at lower voltages than electrostatic loudspeakers as the same thickness strain can be generated at lower voltages by reducing the film thickness.

Harmonic distortion is unavoidable in EPF loudspeakers as the thickness strain of the film varies with the square of the applied voltage as shown in Equation (14.1.18). Heydt et al. (2000) showed that harmonic distortion can be reduced by shaping the input waveform to counteract harmonic distortion.

14.1.3 Optical Loudspeakers

An optical loudspeaker is a light powered electro-acoustic transducer that is insensitive to electrical, magnetic and radioactive fields. Kahana et al. (2004) developed the optical loudspeaker, referred to as the OptiSpeaker, for use in an active headset operating in an magnetic resonance imaging (MRI) environment. MRI devices generate noise levels exceeding 130 dB(A), which can be harmful to patients and those working in the control room. In addition, they produce strong electromagnetic and radio frequency fields, thus preventing conventional loudspeakers from being used in the active headset due to interference with the MRI.

Figure 14.3 shows the operational principle of the OptiSpeaker. The OptiSpeaker consists of three transduction stages:

1. The electro-optical conversion of analogue voltage to modulated light.
2. The opto-electrical conversion of light to analogue voltage.

3. The electro-mechanical conversion of analogue voltage to membrane displacement producing sound waves.

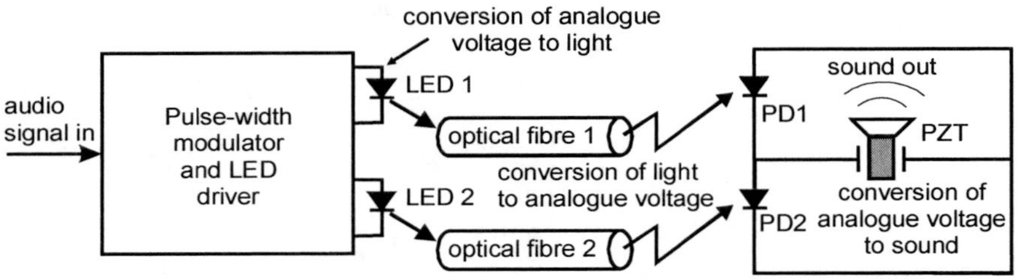

Figure 14.3 Principle of operation of the OptiSpeaker developed by Kahana et al. (2004).

In their experiments, Kahana et al. (2004) implemented stage one in the control room and stages two and three in the MRI device room. In stage one, light pulses are produced by two power LEDs driven by a pulse width modulator. These light pulses are transmitted through optical fibres and detected by photo-diodes in stage two. The voltage signal generated by the photo-diodes drives a piezoelectric actuator producing sound waves in stage 3.

The loudspeaker driver is constructed from a piezoelectric bimorph actuator consisting of two round disks of piezoelectric ceramic material bonded to a centre support vane. When an electric field is applied, one ceramic layer expands laterally while the other contracts. In this way, the ceramic layers bend and deflect according to the waveform of the applied voltage. The maximum displacement at the centre of the piezoelectric bimorph actuator, y_0, can be estimated according to (Kahana et al., 2004) as:

$$y_0 = s_{31}\left(\frac{d}{T}\right)^2\left(1 + \frac{t}{T}\right) E_{PZT} B \tag{14.1.19}$$

where s_{31} is the piezoelectric strain constant, d is the diameter of the piezoelectric disks, T is overall bimorph thickness, t is the combined thickness of the epoxy and centre vane, E_{PZT} is the voltage applied to the piezoelectric driver and B is the empirical weighting factor which has a value between 0.95 and 1.15 depending on the material type, clamping method, etc.

The maximum sound pressure level SPL, in dB, of the OptiSpeaker in the active headset is (Kahana et al., 2004):

$$SPL = 191 + 20\log_{10}\left[1 - \left(1 - \frac{\frac{\pi}{8} d^4 s_{31}\left(1 + \frac{t}{T}\right) E_{PZT} B}{T^2\left(V_E + \frac{\pi}{4} d^2 \delta\right)}\right)^{\gamma}\right] \tag{14.1.20}$$

where V_E is the average human ear-canal volume, δ is the distance between the bimorph and the headset housing, and $\gamma = 1.4$.

The bandwidth of the OptiSpeaker is determined by its electrical properties. The high-frequency limit of the OptiSpeaker in the active headset is given by:

$$f_{\max} = \frac{WR}{\pi C E_{PZTp}} \tag{14.1.21}$$

where W is the light power, R is the response of the photo-diodes, C is the capacity of the piezoelectric speaker and E_{PZTp} is the peak-to-peak drive voltage. Equation (14.1.21) shows that the high-frequency limit of the loudspeaker is determined by the maximum current drawn by the photo-diodes, I_{max} (where $I_{max} = WR$), and the properties of the piezoelectric driver including material and diameter.

For the control of MRI noise, Kahana et al. (2004) required that the sound pressure level output of the OptiSpeaker in the active headset be 100 to 110 dB over a frequency range of 100 to 3500 Hz. They used Equations (14.1.20) and (14.1.21) to optimise the loudspeaker and headset design to achieve their desired acoustic performance. A similar analysis can be used to design the headset for any desired acoustic performance.

14.1.4 Parametric Array

Conventional ANC systems use one or more loudspeakers as control sources. These speakers generally exhibit poor directionality characteristics within the audible frequency range and thus their incorporation in an ANC system can result in the sound pressure at locations away from the desired control point increasing significantly. As discussed above, it is possible to use electrostatic and electrostrictive loudspeakers to achieve better directionality. However, even better directional characteristics can be achieved using a parametric array, as described in this section.

It was known as far back as the nineteenth century that as a sound wave propagates it also distorts (Ingard and Pridmore-Brown, 1956). Non-linear interaction between two carrier waves of angular frequency ω_1 and ω_2 respectively causes demodulation to occur, producing sound waves at the sum $\omega_1 + \omega_2$ and difference $\omega_1 - \omega_2$ frequencies.

From the 1930s until the early 1960s, both theoretical and experimental analyses of the production of these extraneous frequencies were undertaken (Ingard and Pridmore-Brown, 1956; Thuras et al., 1935; Westervelt, 1957a,b; Bellin and Beyer, 1960; and Dean, 1962). Westervelt (1962) and Berktay (1965) presented a theoretical model of what has come to be termed the *parametric array* (also known as a parametric array loudspeaker (PAL) or an audio spotlight). The parametric array emits highly-directional, high-amplitude ultrasonic waves that demodulate into directional audible sound due to non-linear interaction with the medium through which they propagate. The secondary sound beams produced through this demodulation process, including the low-frequency audible difference signal, will have a similar directivity to that of the primary ultrasonic carrier beam. A parametric array is thus an effective means of producing highly directional low-frequency sound.

Through consideration of a geometric model, the far-field amplitude of the low-frequency component of the demodulated secondary sound field (subscript s) was derived by Berktay (1965) to be:

$$p_s(r,\theta) = \frac{P_1 P_2 (\omega_1 - \omega_2)^2 S e^{-\alpha_s r}}{4\pi \rho_0 c_0^4 r \sqrt{(\alpha_1 + \alpha_2 - \alpha_s)^2 + \left(2(k_1 - k_2)\sin^2(\theta/2)\right)^2}} \tag{14.1.22}$$

where P_1 and P_2 are sound pressure amplitudes of the two carrier waves, S is the area of the beam, r is the distance from the source, k is the wavenumber, θ is the angle of observation from the axis of the array, and α_i is the absorption coefficient of the ith carrier wave. This expression was derived assuming $r > \omega_s/\omega_s^2 c_0$ and $l \gg 1/(\alpha_1 + \alpha_2)$, where l is the effective source length. As this expression was derived under a far-field, small angle, assumption, it would not be practical for active noise control (Brooks et al., 2005). For this reason, Kidner et al. (2006) derived a numerical solution to the KZK (Khokhlov–Zabolotskaya–Kuznetsov) equation, which does not rely on the far-field assumption; however, Tanaka and Tanaka (2010) suggested that a closed-form solution would be preferable for the sound pressure results to be used in the development of an active noise control system. Thus, they derived the following closed-form expression:

$$p_s(r,\theta,t) = \frac{P_0^2(\gamma+1)m\,\omega_s^2 S}{8\pi\rho_0 c_0^4}\int_l^0 \frac{\mathrm{e}^{-\alpha_s r_q}\,\mathrm{e}^{-2\alpha_p x}\,\Phi}{r_q}\,\mathrm{d}x \tag{14.1.23}$$

where x is the axial distance from the source, t is time, γ is the specific heat ratio, m is the modulation index, $r_q = (r^2 + x^2 - 2rx\cos\theta)^{0.5}$ *and*

$$\Phi = \sin\left(\omega_s\left(t - \frac{r_q}{c_0}\right) - k_s x\right) - m\cos\left(2\omega_s\left(t - \frac{r_q}{c_0}\right) - k_s x\right) \tag{14.1.24}$$

Equation (14.1.23) has been shown to agree well with experimental data (Tanaka and Tanaka, 2010) in both the near and far-fields and thus was used by them for designing an optimal control law for minimising sound pressure at a target point.

In order to achieve sound pressure control at a point away from the parametric array's main axis, the entire array could be rotated mechanically, although this would not be ideal for a moving control target. Tanaka and Tanaka (2010) therefore used a phased array scheme to steer the parametric array in any given direction, and validated their ability to achieve off-axis control both numerically and experimentally.

The parametric array has also shown promise for global active noise applications. A parametric array with beam focussing has been used in a locally global control scheme to create a virtual sound source collocated with a target primary source, enabling global control (Tanaka and Tanaka, 2011).

14.2 HORNS

Horns are essentially tubes whose cross-sectional area increases from a small throat at one end to a large mouth at the other end. Horns are used as a means of improving the electro-acoustic efficiency of a loudspeaker over a restricted frequency range.

Common shapes for horns include conical, exponential and catenoidal. Conical horns have the simplest shape, with straight sides and a cross-sectional area defined by (Holland, 2001):

$$S(x) = S_0\,(x/x_0)^2 \tag{14.2.1}$$

where $S(x)$ is the cross-sectional area at axial location x from the throat of the horn, S_0 is the cross-sectional area of the throat and x_0 is the distance from the virtual apex of the horn to the throat. While conical horns do not have a characteristic frequency 'cut-off' which defines the low-frequency limit of the horn, they do exhibit poor sound transmission at low frequencies.

An exponential horn has a cross-sectional area of (Holland, 2001):

$$S(x) = S_0 \, e^{\frac{2x}{h}} \tag{14.2.2}$$

where h is a constant derived from the cut-off frequency of the horn, f_0, determined by:

$$f_0 = \frac{c_0}{2\pi h} \tag{14.2.3}$$

The catenoidal horn cross-sectional shape is (Morse, 1948):

$$S(x) = S_0 \cosh^2 (x/h) \tag{14.2.4}$$

where h again defines the cut-off frequency of the horn according to Equation (14.2.3).

Use of an exponential or conical horn in a conventional installation will result in a very uneven frequency response, both in amplitude and phase, thus making control over a broad frequency range more difficult (Ford, 1984). This is because the horn impedance controls the acoustic response in regions of the horn resonance, and the driver resistance controls the acoustic response in regions of anti-resonance. Ford concludes that the use of horns is likely to introduce more practical difficulties than any improvement in efficiency could justify.

Nevertheless, the authors have found that horn drivers fitted with a catenoidal horn can provide large volume velocities over a relatively wide frequency range with a small horn exit diameter. Catenoidal horns have an advantage over exponential or conical shapes because they have a much higher efficiency, especially just above the cut-off frequency of the horn.

The efficiency of any horn is reduced by reflections from the mouth back towards the throat which occur when the mouth diameter is small compared to a wavelength of sound. At low frequencies, Morse (1948) shows that the impedance presented by the throat of the horn to the driver is given by (assuming positive time dependence):

$$\frac{Z_0}{\rho_0 c_0} = \frac{\tau/2(a\omega/c_0)^2 + \mathrm{j}\tan(\omega\tau L_e/c_0)}{\tau + (\mathrm{j}\tau^2/2)(a\omega/c_0)^2 \tan(\omega\tau L_e/c_0)} \tag{14.2.5}$$

where ω is the driving frequency, c_0 is the speed of sound in free space, L_e is the effective length of the horn ($= L + 8a/3\pi$), L is the actual length of the horn, a is the radius of the horn mouth and τ is defined by:

$$\tau^2 = 1 - (\omega_0/\omega)^2 \tag{14.2.6}$$

where $\omega_0 = 2\pi f_0$.

The power radiated from the horn (assuming negligible losses in the horn and assuming a constant volume velocity driver) is given by:

$$W = \frac{1}{2}\bar{u}_0^2 S_0 \,\mathrm{Re}\{Z_0\} \tag{14.2.7}$$

where $\bar{u}_0$ is the velocity amplitude of the driver.

The derivation of Equation (14.2.5) is based on the assumption that the horn mouth is surrounded by an infinitely large baffle and radiates into free space. Radiation into a duct is a considerably more complex problem, and the resulting expression would depend on the impedance presented by the duct (or the location of the horn along the duct axis). Nevertheless, some general conclusions can be made as follows.

1. The acoustic volume velocity generated at the mouth of the horn will be strongly frequency dependent, as it depends on the impedance presented by the duct to the mouth as well as the horn transfer impedance from the throat to the mouth.
2. Although the horn driver at the throat may behave like a constant volume velocity source, the mouth of the horn will behave more like a constant sound pressure source. Thus, optimum locations for horns will be different to those for enclosure backed loudspeakers for the same physical system to be actively controlled.
3. At certain frequencies, it will be possible to achieve much larger volume velocities for a certain diameter source than would be possible using a loudspeaker. Thus, horn sources may prove advantageous in some situations requiring large reductions in tonal noise.

14.3 OMNI-DIRECTIONAL MICROPHONES

In modern digitally based active noise control systems, the frequency response of the microphone used is not very critical, as any lack of flatness in amplitude or phase is taken into account in the system identification algorithms. For this reason, it is common to find relatively inexpensive microphones in active control systems. Three of the most common types of microphone used in active noise control systems are the pre-polarised condenser or electret microphone, the piezoelectric microphone and the optical microphone. The electret microphone is sensitive to dust and moisture on its diaphragm, but it is capable of reliable operation at elevated temperatures and is relatively insensitive to vibration. The piezoelectric microphone is less sensitive to dust and moisture but can be damaged by exposure to elevated temperatures and, in general, it tends to be quite microphonic; that is, a piezoelectric microphone may respond about equally well to vibration and sound. Optical microphones have the advantage of being insensitive to electrical, magnetic and radioactive fields with intensity-modulated optical transducers being robust and highly resistant to changes in heat and moisture.

14.3.1 Condenser Microphone

A condenser microphone consists of a diaphragm which serves as one electrode of a condenser (or capacitor), and a polarised rigid backing plate, parallel to the diaphragm and separated from it by a very narrow air gap, which serves as the other electrode. A schematic diagram of the condenser microphone is shown in Figure 14.4. The condenser is polarised by means of a bound charge, so that small variations in the air gap due to sound pressure-induced displacement of the diaphragm result in corresponding variations in the voltage on the condenser.

The bound charge on the backing plate may be provided either by means of an externally supplied bias voltage of the order of 200 V, or by use of an electret which forms either part of the diaphragm or the backing plate. Condenser microphones that use an electret instead of applying a bias voltage are known as electret microphones. It is possible to purchase very inexpensive electret microphones for less than US$10; however, a manual selection from a large batch is necessary to select those with sufficient sensitivity, stability and frequency bandwidth. Details of the electret construction and its use are discussed in the literature (Fredericksen et al., 1979). For the purpose of the present discussion, however, the details of the latter construction are unimportant. The essential features of a condenser microphone and a sufficient representation of its electrical circuit for the present purpose are provided in Figure 14.4.

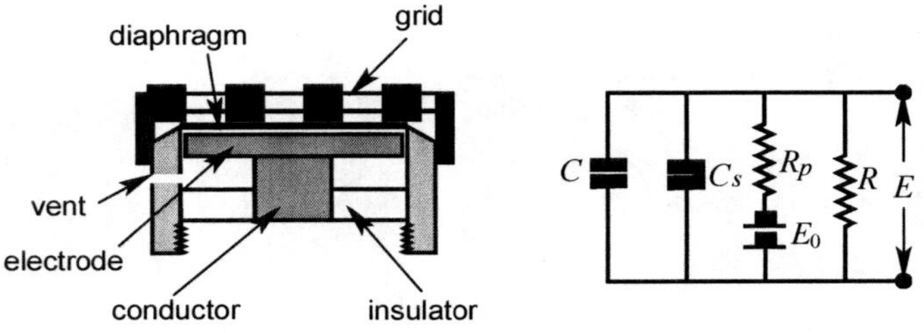

(a) (b)

Figure 14.4 Condenser microphone schematic and electrical circuit.

Referring to Figure 14.4, the bound charge Q may be supplied by a d.c. power supply of voltage E_0 through a very large resistor R_p. Alternatively, the branch containing the d.c. supply and resistor R_p may be thought of as a schematic representation of the electret. The microphone response voltage is detected across the load resistor R. A good signal can be obtained at the input to a high internal impedance detector, even though the motion of the diaphragm is only a small fraction of the wavelength of light.

It is of interest to derive the equation relating the microphone output voltage to the diaphragm displacement. By definition, the capacitance of a device is its charge divided by the voltage across it. From Figure 14.4, for the diaphragm at rest with a d.c. bias voltage of E_0:

$$\frac{Q}{C + C_s} = E_0 \tag{14.3.1}$$

where C is the microphone capacitance and the quantity C_s is the stray capacitance in the electrical circuit. The microphone capacitance is inversely proportional to the spacing at rest, h, between the diaphragm and the backing electrode. If the microphone diaphragm moves a distance x inward (negative displacement, positive sound pressure) so that the spacing becomes $h - x$, the microphone capacitance will increase from C to $C + \delta C$, and the voltage across the capacitors will decrease in response to the change in capacitance by an amount E to $E_0 - E$. Thus,

$$E = -\frac{Q}{C + \delta C + C_s} + E_0 \tag{14.3.2}$$

The microphone capacitance is inversely proportional to the spacing between the diaphragm and the backing electrode; thus,

$$\frac{C + \delta C}{C} = \frac{h}{h - x} \tag{14.3.3}$$

Equation (14.3.3) may be rewritten as:

$$\delta C = C\left(\frac{1}{1 - x/h} - 1\right) \tag{14.3.4}$$

Substitution of Equation (14.3.4) into (14.3.2) and use of Equation (14.3.1) gives the following relation:

$$E = -\frac{Q}{C}\left[\frac{1 - x/h}{1 + (C_s/C)(1 - x/h)} - \frac{1}{1 + C_s/C}\right] \tag{14.3.5}$$

Equation (14.3.5) may be rewritten in the following form:

$$E = -\frac{Q}{C}\left[\frac{1 - x/h}{1 + C_s/C}\left(1 + \frac{(x/h)(C_s/C)}{1 + C_s/C} + \dots\right) - \frac{1}{1 + C_s/C}\right] \tag{14.3.6}$$

By design, $C_s/C << 1$ and $x/h << 1$; thus, in a well designed microphone, the higher-order terms in Equation (14.3.6) may be omitted and by defining an empirical constant $K_1 = 1/(Ch)$, Equation (14.3.6) takes the following approximate form:

$$E \approx \frac{K_1 Qx}{1 + C_s/C} - \frac{K_1 Qx^2}{h(1 + C_s/C)^2} \tag{14.3.7}$$

As can be seen by the preceding equations, the constant K_1 depends upon the spacing, at rest, h, between the microphone diaphragm and the backing electrode and the capacitance of the device, C, and must be determined by calibration. For good linear response, the capacitance ratio C_s/C must be kept as small as possible and, similarly, the microphone displacement relative to the condenser spacing x/h should be very small. In a well designed microphone, the second term in Equation (14.3.7) is thus negligible.

The diaphragm of the condenser microphone is extremely light and is typically 5 to 50 μm thick. Having a small, low-mass rigid diaphragm means the microphone is relatively insensitive to vibration. The first resonance frequency of the diaphragm is determined by its mass and stiffness, and sets the upper limit of the microphone operating frequency range. The diaphragm can be constructed from a metal such as titanium or nickel or made from a polymer with a thin metallic layer. Condenser microphones with a metal diaphragm and other metallic parts are generally the most stable and resistant to changes in the environment and temperature.

14.3.2 Piezoelectric Microphone

A piezoelectric microphone, also known as a crystal microphone, is a transducer based on the piezoelectric properties. A sketch of the essential features of a typical piezoelectric microphone and a schematic representation of its electrical circuit are shown in Figure 14.5. In this case, sound incident upon the diaphragm tends to stress or unstress the piezoelectric element which, in response, induces a bound charge across its capacitance. The effect of the variable charge is like that of a voltage generator, as shown in the circuit in Figure 14.5.

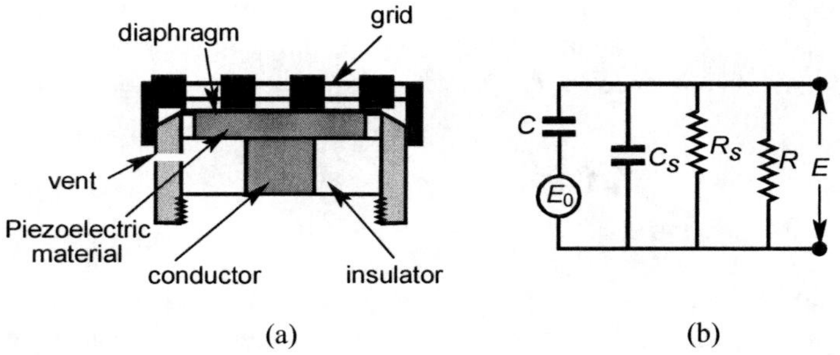

Figure 14.5 Piezoelectric microphone schematic and electrical circuit.

For a piezoelectric microphone, the equivalent circuit of which is shown in Figure 14.5, the microphone output voltage is given by:

$$E = \frac{Q}{C + C_s} = \frac{Q/C}{1 + C_s/C} \tag{14.3.8a,b}$$

As $E_0 = Q/C$:

$$E = \frac{E_0}{1 + C_s/C} \tag{14.3.9}$$

The voltage E_0 generated by the piezo crystal is proportional to its displacement. Thus,

$$E_0 = K_2 x \tag{14.3.10}$$

Substituting Equation (14.3.10) into (14.3.9) gives:

$$E = \frac{K_2 x}{(1 + C_s/C)} \tag{14.3.11}$$

Equation (14.3.11) is essentially the same as the linear term of Equation (14.3.7). In either case, an acoustic wave of sound pressure p incident upon the surface of the microphone diaphragm of area S will induce a mean displacement x determined by the compliance K_3 of the diaphragm. Thus, the relation between acoustic pressure and mean displacement is:

$$x = -K_3 S p \tag{14.3.12}$$

Substitution of Equation (14.3.12) into either (14.3.7) or (14.3.11) results in a relation, formally the same in either case, between the induced microphone voltage and the incident acoustic pressure. Thus, for either microphone:

$$E = KSp/(1 + C_s/C) \tag{14.3.13}$$

In Equation (14.3.13), the constant K must be determined by calibration.

Piezoelectric microphones tend to be narrowband devices, with an operating range typically between 80 Hz and 7 kHz. The piezoelectric materials used in modern-day microphones such as lead-zirconium titanate and barium titanate are robust and fairly resistant to dust and moisture. Piezoelectric microphones do not require a phantom power supply or amplification and, as such, they are relatively inexpensive devices. Contact transducers that measure vibration commonly employ piezoelectric elements so it is important to note that piezoelectric microphones tend to respond equally well to both vibration and sound.

14.3.3 Optical Microphones

Optical microphones are light-modulating acoustic sensors. Incident sound waves modulate light that is guided by an optical fibre or an optical waveguide to a sensing device. Optical microphones are categorised based on their light-modulating property, which can be intensity, phase, polarisation or wavelength. All modulation must, however, be reduced to intensity as this is the only quantity measurable by a photo-detector or a photo-multiplier. Intensity-modulated transducers are the simplest devices and can be operated with LEDs. They have the added advantage of being resistant to changes in heat and moisture. A thorough description of different optical microphone transduction techniques is given by Bilaniuk (1997).

The conversion of sound signals to light signals occurs in optical microphones without intermediate electronics. Optical microphones are therefore insensitive to electrical, magnetic and radioactive fields. Because of this unique property, optical microphones have been used in active noise control systems for magnetic resonance imaging (MRI) scanners (Kahana et al., 2004; Chambers et al., 2007; Kida et al., 2009). MRI scanners generate strong magnetic fields and noise exceeding 130 dB. Conventional transducers interfere with MRI imaging and their signals can in turn be contaminated in the audio frequency range.

Phone-Or Ltd. has recently developed a range of simple, cost effective optical microphones using the intensity modulation principle and Micro-Electro-Mechanical-Systems (MEMS) technology (Kots and Paritsky, 1999; Paritsky and Kots, 1999; Kahana et al., 2003). In Phone-Or's Fibre Optical Microphone (FOM), light emitted by an LED is beamed through an optical fibre onto a MEMS membrane. A gold-coated spot located in the centre of the membrane acts as a reflective surface for the light. Incident sound waves cause the membrane to vibrate, changing the intensity of the light reflected off the spot in the centre of the membrane. The reflected light is directed through a second optical fibre to a photo-detector where its intensity is measured and converted into an electrical signal. The MEMS membrane has a high mechanical sensitivity, a high level of linearity and high stability at a range of operating temperatures and humidity conditions. The FOM has a bandwidth typically from 1 Hz to 10 kHz, a sensitivity of −20 dB re 1 V Pa^{-1}, a dynamic range of greater than 85 dB, and a high signal-to-noise ratio, typically in the order of 70 dB.

14.3.4 Microphone Sensitivity

Microphone sensitivity is a measure of the electrical output of a microphone for a given acoustical sound pressure level. It is customary to express the sensitivity of microphones in decibels relative to a reference level. Following accepted practice, the reference voltage E_{ref} is selected to be 1 V and the reference sound pressure p_0 is 1 Pa. The sensitivity S_m of a microphone is written as:

$$S_m = 20\log_{10}[Ep_0/(E_{\text{ref}}p)] \tag{14.3.14}$$

Equation (14.3.14) may, in turn, be rewritten in terms of sound pressure level L_p:

$$S_m = 20 \log_{10} E - L_p + 94 \qquad \text{(dB)} \tag{14.3.15}$$

where E is the voltage generated by an incident sound of sound pressure level, L_p.

Typical condenser microphone sensitivities range between −25 and −60 dB re 1 V Pa^{-1}. Piezoelectric microphone sensitivities lie at the lower (more negative) end of this range. As an example, if the incident sound pressure level is 74 dB, then a microphone of −30 dB re 1 V Pa^{-1} sensitivity will produce a voltage which is down from 1 V by 50 dB. The voltage thus produced is $E = 10^{-50/20} = 3.15$ mV.

Reference to Equation (14.3.11) shows that the output voltage from a microphone is directly proportional to the area of the diaphragm. Thus, the smaller a microphone of a given type, the smaller will be its sensitivity. As the example shows, the output voltage may be rather small, especially if very low sound pressure levels are to be measured, and the magnitude of the gain which is possible in practice is limited by the internal noise of the amplification devices. These considerations call for a microphone with a large diaphragm which will produce a corresponding relatively large output voltage.

In applications where very low level error signals must be detected, microphones with a low self-noise floor must be used and these are more expensive. For example, electrets with a reasonable frequency response from 20 Hz to 6 kHz and with a self-noise floor of 50 dB are very low-cost and can be purchased for around US$10. However, electrets with a flat frequency response from 20 Hz to 20 kHz and a self-noise floor of −20dB can cost up to 50 times as much. Piezoelectric microphones are generally less expensive and less sensitive than electrets of the same diameter and are often adequate for use in active noise control systems.

14.4 DIRECTIONAL MICROPHONES

In cases where the direction of propagation of the acoustic disturbance is known, such as in a duct, it is desirable for the reference sensor to be directional so that it is insensitive to sound emanating from the control sources and only sensitive to sound arriving from the direction of the disturbance. In practice, the desired result can be achieved in any one or combination of three ways: using a directional control source, using a directional microphone or using an IIR filter to include the effect of the acoustic feedback signal in the electronic control system design. In this section, only directional microphones are discussed. However, it should be pointed out that in many cases it is not possible to obtain a high-performance, stable, active noise control system in a duct by just using an IIR filter, and the reference sensor often needs some directional properties as well, so that acoustic feedback from the control speakers to the reference microphone is minimised.

14.4.1 Tube Microphones

A very simple directional microphone consists of a long tube with a small axial slit and with a microphone mounted at one end, as first suggested by Tamm and Kurtz in 1954. The principle of operation of this device relies on the pressure fluctuations generated by the external acoustic field travelling down the tube towards the microphone at the speed of sound. If the acoustic wave generating the pressure fluctuations in the tube is also travelling from the far end of the tube towards the microphone, then the pressure fluctuations generated inside the tube will all be in-phase on arrival at the microphone and will reinforce one another. However, if the external acoustic wave is incident on the tube at some angle to the tube axis, the pressure fluctuations generated by it inside the tube will not be in-phase on arrival at the microphone, as the trace velocity of the external acoustic wave outside of the tube along the tube axis will be less than the velocity of the pressure fluctuations in the tube. Thus, the sensitivity of the device will decrease as the angle of the incident external sound field with respect to the tube axis increases from 0° to 180°, with 0° corresponding to a sound wave incident first on the end of the tube opposite the microphone. Thus, if this type of microphone is mounted in a duct, it should always point towards the sound source, regardless of the direction of any flow which is present.

The type of microphone just described is often referred to as a 'shotgun' microphone for obvious reasons. If the end opposite the one containing the microphone can be made totally absorptive, then the frequency response (or sensitivity as a function of frequency) of the device will be relatively flat. The best way to achieve an anechoic termination to a probe tube is to attach a long flexible, coiled tube to each end (Davy and Dunn, 1993). In this case the microphone is attached to the probe tube by way of a 'T-piece', which for the low-frequency range generally of interest is insignificantly different to having it at one end.

If an anechoic termination is not achieved, the result is a wavy frequency response (±1.5 dB for a 100% reflecting end, tube diameter 13 mm and tube length 500 mm). This frequency-response characteristic can be made smoother by adding a hollow tube (with open ends), filled with porous acoustic material, to the inside of the tube. The hollow tube diameter would be about half the probe tube diameter, as shown in Figure 14.6. Also, increasing the tube length and reducing its diameter will improve the flatness of frequency response (Munjal and Eriksson 1989).

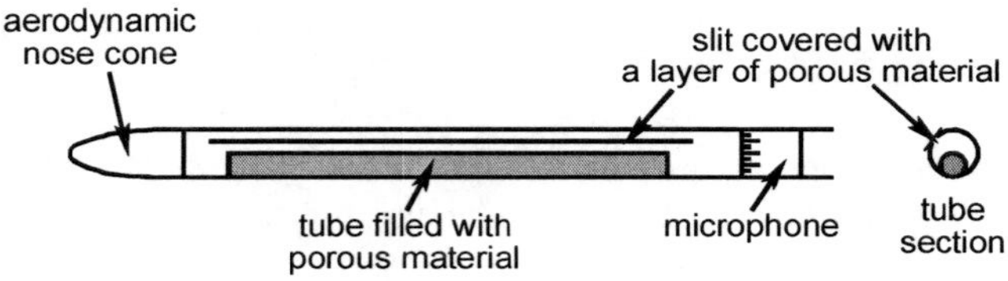

Figure 14.6 Directional microphone with internal absorption tube.

In practice, it is necessary to cover the slit in the probe tube with a porous cloth to reduce the damping of the pressure fluctuations travelling through the inside of the tube. If this is not done, the pressure fluctuations generated at the far end of the tube away from the microphone will differ in amplitude to those produced close to the microphone, thus reducing the effect of the phase cancelling mechanism at the microphone location, which minimises the contribution of acoustic signals resulting from waves not parallel to the tube axis. Thus, addition of the porous cloth enhances the directional properties of the device. However, if the cloth is too impervious, the overall sensitivity of the device will be substantially reduced. A good compromise, in practice, is to use a cloth with a flow resistance of about $2\rho c$ MKS rayls, where ρc is the characteristic impedance of the fluid in the duct. This results in a device with a sensitivity slightly higher (1 or 2 dB) than the microphone with no tube (Neise, 1975). Compare this to a reduction in sensitivity of 15 dB if a flow resistance of $20\rho c$ is used. The preceding values correspond to a slit width to tube circumference ratio of 0.025 and sound field frequency ω (rad s^{-1}) characterised by $\omega L/c > 2$ (Munjal and Eriksson, 1989), where L is the tube length (m) and c is the speed of sound in the fluid external to the tube (m s^{-1}).

The directional properties of the microphone and tube become more marked for longer tubes and higher frequencies. As a guide, the sensitivity difference between a sound wave arriving on-axis from the most sensitive direction and one arriving on-axis from the opposite direction is 0 dB for $\omega L/c \le 1$, 20 dB for $\omega L/c = 10$, and 40 dB for $\omega L/c = 100$. That is, the difference in dB varies in proportion to the log of the increase in the product of frequency and length, provided $\omega L/c \ge 1$. This assumes that the end of the tube opposite the microphone is 100% absorptive. If this is not so, the difference between the 0° and 180° response is much less, reducing to a constant 5 to 10 dB in the limit of a perfectly reflecting end for $\omega L/c > 2$.

As mentioned earlier, if the tube has a reflecting end opposite the microphone, the frequency response of the device will be wavy. This waviness increases if the flow resistance of the cloth over the slit increases, so this is another reason for keeping the flow resistance to a minimum.

14.4.2 Microphone Arrays

Microphone arrays have also been used in the past to achieve directional sensing. The principle of operation is similar to a tube containing a microphone at one end, except that the array is made up of a number of discrete sensing points arranged in a line, rather than the continuous sensor represented by the tube. The signal from each microphone is phase shifted to account for the time taken for an acoustic signal to travel directly from one to the other, resulting in suppression of acoustic waves arriving off-axis, or from the wrong on-axis direction. The array is effective below frequencies for which the microphone separation distance is less than half a wavelength. Increasing the number of microphones increases the effective directivity, but very good results can be obtained with just two microphones. At very low frequencies, microphone arrays are much more effective than probe tubes; and in a duct containing flow and low-frequency noise, a combination of two or more microphones in a probe tube is likely to be necessary (at least for the reference sensor) to attack the dual problem of turbulence noise reduction and lack of microphone directivity. A disadvantage of the discrete microphones is the variability as a function of signal frequency required in the phase delay applied to the signal from the microphone closest to the primary noise source. This means that a fixed phase delay will result in a probe with good directivity properties over a narrow frequency range. This may be adequate for some problems and if not, the working frequency range can be increased by using more microphones or by using circuitry capable of providing a time delay independent of frequency. The authors have achieved a directivity of better than 20 dB between 30 and 40 Hz using just two microphones spaced 1 m apart and a constant phase delay. The analysis of two and three speaker arrays (which is similar to that required for microphone arrays) was discussed in Section 7.2.2.3. Essentially, for a fixed time delay of τ_0 (which must be equal to the acoustic delay experienced by a signal travelling between the two microphones), the frequencies (in Hz) of maximum array sensitivity (twice that of a single microphone) are given by $f = 0.5n/\tau_0$, where n is an odd integer (see Figure 7.8). At frequencies below $0.17\tau_0$, the microphone sensitivity is less than that of a single microphone. As the number of (equally spaced) microphones in the array increases, the frequency below which the sensitivity is less than that of a single microphone decreases ($0.07\tau_0$ for five microphones – see Figure 7.10).

The preceding discussion is essentially applicable to one-dimensional sound fields such as in a duct. For three-dimensional sound fields, the use of microphone arrays is a little more complex and the ability to arrange microphones in an array that has a specified directivity is known as 'beamforming'.

Beamforming measures the amplitude and phase of the sound pressure over a planar or spherical or linear array of many microphones and this is used to maximise the total summed output of the array for sound coming from a specified direction, while minimising the response due to sound comin`g from different directions. By inserting an adjustable delay in the electronic signal path from each microphone, it is possible to 'steer' the array so that the direction of maximum response can be varied. In this way, the relative intensity of sound coming from different directions can be determined and the data analysed to produce a map of the relative importance of different parts of a sound source to the total far-field sound pressure level. This principle underpins the operation of a device known as an acoustic camera. However, it can be applied to the derivation of a directional reference or error signal for an active noise control system in a three-dimensional space.

Typical frequency ranges and recommended distances between the array and the radiating noise source for various commercially available array types are listed in Table 14.1. The dynamic range of a beamforming measurement varies from about 6 dB for ring arrays to up to 15 dB for spiral arrays. However, spiral arrays have the disadvantage of poor depth of field, so that it is more difficult to focus the array on the sound source, especially if the sound source is non-planar. Poorly focussed arrays used on more than one sound source existing at different distances from the array but in a similar direction can result in sources cancelling one another so that they disappear from the beamforming image altogether.

Table 14.1 Beamforming array properties.

Array Name	Array Size	Number of Mics.	Distance from Source (m)	Frequency Range (Hz)	Backward Attenuation[a]
Star	3×2 m arms	36	3–300	100–7,000	-21 dB
Ring	0.75 m dia	48	0.7–3	400–20,000	0 dB
Ring	0.35 m dia	32	0.35–1.5	400–20,000	0 dB
Ring	1.4 m dia	72	2.5–20	250–20,000	0 dB
Cube	0.35 m across	32	0.3–1.5	1,000–10,000	-20 dB
Sphere	0.35 m dia	48	0.3–1.5	1,000–10,000	-20 dB

[a]This is how much a wave is attenuated if it arrives from behind the array.

When a spherical array is used inside an irregular enclosure such as a car passenger compartment, it is necessary to use a CAD model of the interior of the enclosure and accurately position the beamforming array within it. Then the focus plane of the array can be adjusted in software for each direction to which the array is steered.

Beamformers have the disadvantage of poor spatial resolution of noise source locations, especially at low frequencies. For a beamforming array of largest dimension D and distance from the source, L, the resolution (or accuracy with which a source can be located) is given by:

$$Res = 1.22\frac{L}{D}\lambda \tag{14.4.1}$$

Beamforming array design is also important as there is a trade-off between depth of focus of the array and its dynamic range. The spiral array has the greatest dynamic range (up to 15 dB) but a very small depth of focus, whereas the ring array only has a dynamic range of 6 dB but a large depth of focus, allowing the array to focus on noise sources at differing distances and not requiring such precision in the estimate of the distance of the noise source from the array. The dynamic range is greatest for broadband noise sources and least for low-frequency and tonal sources.

14.4.2.1 Summary of the Underlying Beamforming Theory

Beamforming theory is generally quite complex, so only a brief summary will be presented here. For more details, the reader is referred to Christensen and Hald (2004) and Johnson and Dudgeon (1993). There are two types of beamforming: infinite-focus distance and finite-focus distance. For the former, plane waves are assumed and for the latter, spherical waves are assumed to originate from the focal point of the array.

In essence, infinite-focus beamforming in the context of interest here is the process of summing the signals from an array of microphones and applying different delays to the signals from each microphone so that sound coming from a particular direction causes a maximum summed microphone response, and sound coming from other directions causes no response. Of course in practice, sound from any direction will still cause some response but the principle of operation is that these responses will be well below the main response due to sound coming from the direction of interest.

Consider a planar array made up of L microphones at locations (x_ℓ, y_ℓ, $\ell = 1, ..., L$) in the x-y plane. If the measured sound pressure signals p_ℓ are individually delayed and then summed, the output of the array is:

$$p(\boldsymbol{n}, t) = \sum_{\ell=1}^{L} w_\ell p_\ell(t - \Delta_\ell(\boldsymbol{n})) \tag{14.4.2}$$

where w_ℓ is the weighting coefficient applied to sound pressure signal p_ℓ, and its function is to reduce the importance of the signals coming from the array edges, which in turn reduces the amplitudes of side lobes in the array response. Side lobes are peaks in the array response in directions other than the design direction and serve to reduce the dynamic range of the beamformer. The quantity $\boldsymbol{n}$ in Equation (14.4.2) is the unit vector in the direction of maximum sensitivity of the array, and the time delays Δ_ℓ are chosen to maximise the array sensitivity in direction $\boldsymbol{n}$. This is done by delaying the signals associated with a plane wave arriving from direction $\boldsymbol{n}$ so that they are aligned in time before being summed. The time delay Δ_ℓ is the dot product of the unit vector $\boldsymbol{n}$ and the vector $\boldsymbol{r}_\ell = (x_\ell, y_\ell)$ divided by the speed of sound c. That is,

$$\Delta_\ell = \frac{\boldsymbol{n} \cdot \boldsymbol{r}_\ell}{c} \tag{14.4.3}$$

If the analysis is done in the frequency domain, the beamformer output at angular frequency ω is:

$$P(\boldsymbol{n}, \omega) = \sum_{\ell=1}^{L} w_\ell P_\ell(\omega) \mathrm{e}^{-\mathrm{j}\omega\Delta_\ell(\boldsymbol{n})} = \sum_{\ell=1}^{I} w_\ell P_\ell(\omega) \mathrm{e}^{-\mathrm{j}\boldsymbol{k} \cdot \boldsymbol{r}_\ell} \tag{14.4.4a,b}$$

where $\boldsymbol{k} = -k\boldsymbol{n}$ is the wavenumber vector of a plane wave incident from the direction $\boldsymbol{n}$, which is the direction in which the array is focussed. More detailed analysis of various aspects affecting beamformer performance are discussed by Christensen and Hald (2004) and Johnson and Dudgeon (1993).

Finite-focus beamforming using a spherical wave assumption to locate the direction of a source and its strength at a particular distance from the array (array focal point) follows a similar but slightly more complex analysis than outlined above for infinite-focus beamforming. For the array to focus on a point source at a finite distance, the various microphone delays should align in time, the signals of a spherical wave radiated from the focus point. Equation (14.4.4a) still applies but the delay Δ_ℓ is defined as:

$$\Delta_\ell = \frac{|\boldsymbol{r}| - |\boldsymbol{r} - \boldsymbol{r}_i|}{c} \tag{14.4.5}$$

where $\boldsymbol{r}$ is the vector location of the source from an origin point in the same plane as the array, $\boldsymbol{r}_\ell$ is the vector location of microphone ℓ in the array with respect to the same origin, and $|\boldsymbol{r} - \boldsymbol{r}_\ell|$ is the scalar distance of microphone ℓ from the source location.

14.4.3 Gradient Microphones

As the sound pressure gradient is greatest in the direction normal to the wavefront of an acoustic disturbance, measurement of this quantity will provide a signal related to the direction of propagation of the wave. For a plane progressive wave of amplitude A, the first and second-order sound pressure gradients along the x-axis are respectively:

$$\frac{\partial p}{\partial x} = -\mathrm{j}k_x A\mathrm{e}^{-\mathrm{j}k_n x}\mathrm{e}^{-\mathrm{j}\omega t} \tag{14.4.6}$$

$$\frac{\partial^2 p}{\partial x^2} = -k_x^2 A\mathrm{e}^{-\mathrm{j}k_n x}\mathrm{e}^{-\mathrm{j}\omega t} \tag{14.4.7}$$

where

$$k_x = \frac{\omega \cos\theta}{c_0} \tag{14.4.8}$$

Note the dependence of the sensitivity on frequency ω for the first-order gradient and ω^2 for the second-order gradient. The first-order gradient microphone sensitivity is also dependent on cos θ, where θ is the angle between the direction of propagation of the wave and the x-axis (which is normal to the surface of a gradient microphone). Clearly, a second-order gradient device will have a $\cos^2\theta$ directivity. In the near-field of a sound source, the sound pressure gradients will be larger than those given by Equations (14.4.6) and (14.4.7) due to the decaying near-field. However, this is in addition to the $\cos\theta$ or $\cos^2\theta$ dependence just discussed and will rarely be of importance in active noise control systems.

A first-order gradient microphone may be realised by exposing both sides of an electret diaphragm (see section 14.3) to the sound field. A second-order gradient microphone can be realised by placing two first-order gradient microphones in small baffles spaced a short distance apart (Sessler et al., 1989) as shown in Figure 14.7.

The baffles in the second-order gradient serve to delay the signal between each side of each of the two first-order gradient microphones. The signal from the first-order microphone on the left is delayed by $-\tau$ and the one on the right is delayed by $+\tau$. In practice, this is achieved by delaying the output of the right-hand gradient microphone by 2τ and the left one by zero. Note that the outputs of the two first-order gradient microphones are poled in anti-phase as shown in the figure.

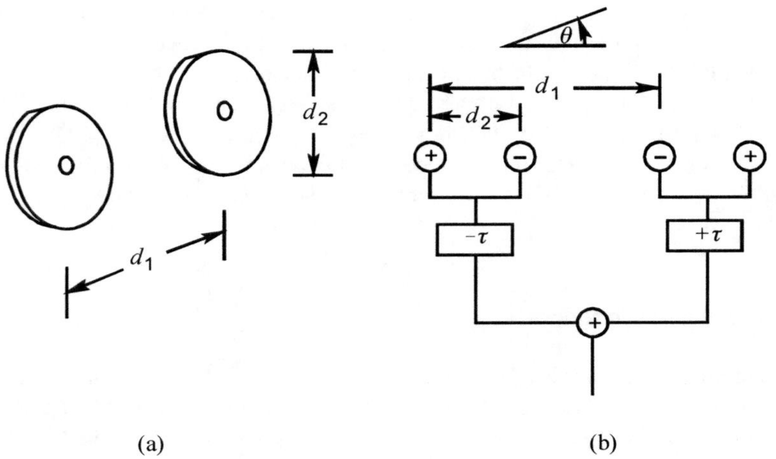

Figure 14.7 Second-order gradient microphone arrangement: (a) physical arrangement; (b) schematic representation.

The sensitivity of the second-order gradient microphone shown in Figure 14.7 at low frequencies is given by Sessler et al. 1989 as:

$$M_s = M_{os} k^2 d_1 d_2 \left[\frac{2c_0 \tau}{d_1} + \cos\theta \right] \cos\theta \tag{14.4.9}$$

where M_{os} is the sensitivity of each of the first-order gradient microphones, k is the wavenumber at the frequency of interest and the other quantities are defined in the figure.

Although gradient microphones are probably unsuitable for use as active control system sensors in ducts containing flow, they may be very useful as directional sensors for systems used to control sound radiation into free or enclosed spaces. A disadvantage of the simple first-order gradient microphone is that it is bi-directional; that is, it cannot distinguish the difference between waves arriving from in front of it or behind it. On the other hand, the second-order gradient microphone just described is uni-directional and has a very strong null corresponding to sound waves arriving from the $\theta = 180°$ direction.

A further disadvantage of gradient microphones in general is their poor signal-to-noise ratio at low frequencies, although this may not be a limitation in situations where very high level sound fields are being controlled.

Knowles produces a range of inexpensive commercially available electret pressure gradient microphones. The Knowles NR series gradient microphones have high sensitivity, high background noise rejection, a very small sensor size and a flat frequency response up to 3 kHz. These gradient microphones have been used previously in an active noise control system that minimises both the sound pressure and the pressure gradient along all three orthogonal axes with four control sources in a three-dimensional enclosure (Moreau et al., 2010).

14.5 TURBULENCE FILTERING SENSORS

When active noise control systems are applied to duct systems containing flow, the performance is often compromised by contamination of the reference and error microphone signals with signals generated by turbulent pressure fluctuations that travel at the flow speed rather than the speed of sound. The turbulent pressure fluctuations contribute very little to the noise radiated from the end of the duct; thus, it is essential that their influence is removed from both the reference and error microphones used in an active noise control system. If this is not done, the performance of the active control system in reducing sound propagation will be severely impaired.

14.5.1 Probe Tube Microphones

The most common and simplest way of reducing the influence of turbulent pressure fluctuations is to use a probe tube microphone, which consists of a microphone placed at one end of a long tube of diameter similar to the microphone diameter. The walls of the tube are either porous or contain holes or an axial slit. For best results, the microphone must be located at the end of the tube furthest from the sound source (the tube is oriented so that its longitudinal axis is parallel to the direction of sound propagation – see Figure 14.8).

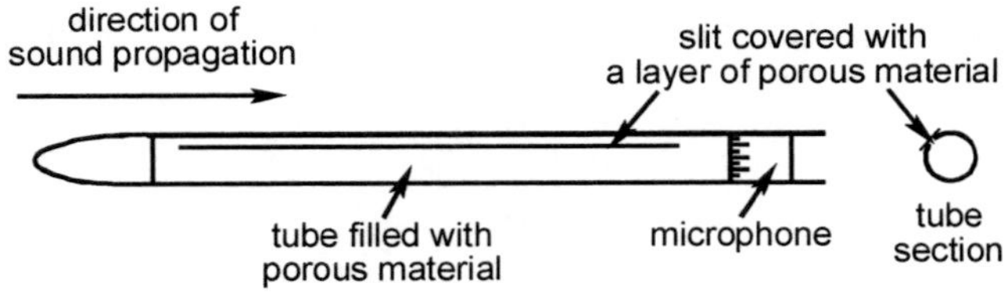

Figure 14.8 Probe tube microphone in a duct.

The principle of operation of the probe tube microphone in minimising the effect of the turbulent pressure fluctuations on the microphone signal relies on the difference in speed of propagation between acoustic waves and turbulent pressure fluctuations. Any fluctuating pressure field (whether acoustic or turbulent flow) acting on the outside of the tube will generate pressure disturbances inside the tube which will propagate with the speed of sound to either end of the tube. To simplify the explanation, it is assumed that the tube end opposite the microphone completely absorbs acoustic waves incident on it from inside the tube. Then the pressure at the microphone is only determined by the waves directly propagating toward it from the tube wall.

Pressure fluctuations along the inside of the tube excited by an external acoustic field will propagate towards the microphone at the same speed as the acoustic wave in the external duct. Thus, the pressure fluctuations excited inside the tube at various locations along the length will all arrive in-phase at the microphone. On the other hand, pressure fluctuations excited inside the tube by an external turbulent pressure field will not arrive in-phase at the microphone, as the turbulent pressure field external to the tube propagates at the speed of the flow. Thus, the pressure fluctuations generated on the inside of the tube by the turbulent flow will cancel out one another at the microphone, due to their random differences in relative phase. As a further illustration of the principle of operation, Shepherd et al. (1989) show mathematically that to an observer moving at the speed of sound, the turbulent pressure fluctuations in a duct average to zero while the acoustic component appears as a steady signal. It is also clear that the faster the convection speed of the turbulent pressure fluctuations, the longer the probe tube will need to be to adequately suppress their influence at the microphone. Of course, to completely suppress the turbulent pressure contribution at the microphone, the probe tube would need to be infinitely long. On the other hand, if the flow speed is too great, there will be a significant difference between the speed of sound in the duct and the speed of sound in the probe tube. This will result in a reduction in the sensitivity of the probe tube microphone to the acoustic signal. This is discussed in more detail in Section 14.5.1.2.

For reasons outlined in Section 14.4, the probe tube microphone also has a significant directional response, being much more sensitive to sound incident from one end than from the other end. Thus, in an active noise control system it serves the dual purpose of minimising the effect of turbulent pressure fluctuations and also the acoustic feedback arriving at the reference microphone from the control source (in a duct).

The sensitivity of the microphone in the probe tube to acoustic pressure fluctuations and to turbulent pressure fluctuations is dependent upon a number of parameters, each of which were investigated by Neise (1975) and which will be discussed in the following sections. As pointed out by Munjal and Eriksson (1989), the analysis used by Neise only considers the effect of the external sound field in the duct on the internal field in the probe tube and neglects to consider the effect of the sound field in the probe tube (generated by the external field) on the sound field in the duct. This has the effect of introducing errors in both the acoustic and turbulent pressure sensitivity of the microphone, which become significant at frequencies ω (rad s^{-1}) characterised by $\omega L/c < 0.5$, where L is the probe tube length and c is the speed of sound in the fluid external to the probe tube. Specific values quoted in the sections to follow apply to the probe tube discussed by Neise (1975) and Munjal and Eriksson (1989). This tube was 13 mm in diameter and had a slit-width to tube-circumference ratio of 0.025. The tube length dependence is included in the parameters quoted.

14.5.1.1 Effect of Slit Flow Resistance on Acoustic Sensitivity

To optimise the performance of the probe tube microphone, it is necessary to cover the slit with acoustically porous cloth, which serves to minimise the damping effect of the slit on acoustic waves propagating in the tube. Without the cloth cover, the acoustic waves generated in the tube at the far end away from the microphone would be significantly reduced in amplitude on reaching the microphone, and cancellation at the microphone of the various pressure fluctuations generated by the turbulent pressure field would not be as effective. However, if the cloth is insufficiently porous, the sensitivity of the microphone to the external acoustic field will be substantially reduced. Although the turbulent pressure sensitivity is also reduced by a similar amount, it is clearly undesirable to reduce the acoustic sensitivity of the microphone unnecessarily as the microphone signal will then be more susceptible to contamination by electronic noise.

A convenient compromise is to use fibreglass cloth having a flow resistance of approximately $2\rho_0 c_0$ MKS rayls. This results in an acoustic sensitivity of the probe tube plus microphone of approximately 2 dB over that of the microphone with no tube, for frequencies above $\omega L/c_0 = 5$ and -3 dB at frequencies below $\omega L/c_0 = 0.5$. In between, the sensitivity may be reasonably approximated as a linear function of $\omega L/c_0$.

For a cloth having a flow resistance of $2\rho c_0$, the sensitivity of the microphone plus tube is -15 dB at a frequency corresponding to $\omega L/U = 10$ (where U is the mean flow speed in the duct) and this decreases by 10 dB each time $\omega L/U$ increases by a factor of 10). For example, the sensitivity corresponding to $\omega L/U = 100$ is -25 dB. At lower frequencies, the sensitivity probably approaches 0 dB close to $\omega L/U = 1$, but this is uncertain, as the only analysis available is that by Neise (1975), which suffers from the one-way acoustic interaction limitation mentioned earlier.

Provided the acoustic sensitivity of the probe tube and microphone were independent of flow speed in the duct, the amplification of the acoustic signal over the turbulent pressure fluctuation signal would be given by the difference in the acoustic and turbulent pressure sensitivities, which can be calculated as just outlined for a specified flow speed and frequency. Although the acoustic sensitivity does depend on flow speed at higher values of $\omega L/c_0$, it will be shown in the next section that this occurs outside the frequency range and flow speed range of general concern in active noise control systems.

The previous discussion has been based on the assumption that the end of the tube opposite the microphone is 100% absorbing. In practice, achieving a condition even close to this is difficult and expensive, so a more realistic condition is to assume that both ends of the probe tube are completely reflecting. The effect of the reflecting end on the acoustic sensitivity corresponding to a slit covered by a cloth of flow resistance equal to $2\rho c$ is to cause it to vary between 0 and +3 dB as a function of frequency for frequencies above $\omega L/c = 2$ and to be equal to zero at lower frequencies. The effect of the reflecting end on the turbulent pressure sensitivity is to increase the average sensitivity by about 2 dB and introduce a small fluctuation (±1.5 dB) as a function of frequency (Neise, 1975). As the effect of the reflecting end on the difference between the acoustic and turbulent pressure sensitivity is small (of the order of 2 to 3 dB), it is generally not considered worthwhile to attempt to make the probe tube end opposite the microphone absorptive. The ±1.5 dB waviness in the frequency response could be reduced by adding into the probe tube a hollow smaller diameter tube open at the ends filled with porous material as illustrated in Section 14.4.1 and described by Neise (1975), but this is usually unnecessary for digital active noise control systems.

14.5.1.2 Effect of Flow Speed on Acoustic Sensitivity

The effect of the mean flow speed in the duct on the turbulent pressure sensitivity of the probe tube microphone was discussed in the previous section. Here, the effect of flow speed U on the acoustic sensitivity will be discussed. Neise showed that for the tube he investigated, the flow speed begins to have an effect on the acoustic sensitivity at frequencies corresponding to $\frac{\omega L}{c_0} > \frac{2}{M}$ where M is the Mach number of the mean flow. It appears that the sensitivity is reduced by about 20 dB for each factor of 10 increase in the product ωL, although the actual sensitivity curve is a wavy line as a function of frequency, varying ±3 dB about the mean. The reason for this high-frequency roll-off in microphone acoustic sensitivity is that as the flow speed increases, the phase speed of the acoustic wave in the duct becomes more different to that in the probe tube (in which there is no mean flow). Thus, the same cancellation mechanism begins to occur as for the turbulent pressure fluctuations. Fortunately, at reasonable flow speeds and probe tube lengths, this high-frequency roll-off occurs beyond the highest frequency usually of interest for active control. In general, the difference between the acoustic pressure and turbulent pressure sensitivities of the microphone increases by 3 to 4 dB for each halving of the flow speed (Neise, 1975), at least in the frequency range below the frequency at which the acoustic sensitivity begins to roll off. Note that flow speeds characterised by $M = 0.05$ to 0.07 are typical of commercial air conditioning and industrial air handling systems.

14.5.1.3 Effect of Probe Tube Orientation

It is clear from the discussion in Sections 14.4 and 14.5 so far, that the probe tube should be oriented so that the microphone end is furthest from the acoustic source generating the sound field to be measured. The direction of mean flow is not important. Orientation of the probe tube the opposite way round results in a much smaller difference between the acoustic and turbulent pressure sensitivities, resulting in a poorer turbulent pressure fluctuation filter.

14.5.1.4 Effect of Probe Tube Diameter

Munjal and Eriksson (1989) show that increasing the probe tube diameter (for a fixed duct cross-sectional area) for a tube with completely reflecting ends results in reduced acoustic sensitivity and increased waviness in the acoustic sensitivity frequency response. They conclude that the maximum permissible probe tube diameter should be less than 0.1 of the effective duct diameter. Note that the waviness in the frequency response increases from ±1.5 dB for a diameter ratio of 0.085 to ± 6 dB for a ratio of 0.34 (Munjal and Eriksson 1989).

14.5.1.5 Effect of a Reflective Duct Termination

If the probe tube is being used to sample the acoustic field in a duct in which waves are reflected from the exit, then the waviness in the acoustic pressure sensitivity frequency-response curve can be as much as ±10 dB. This could make the electronic controller design more difficult as stability margins could decrease.

14.5.1.6 Probe Tube Design Guidelines

Guidelines for designing probe tubes for active noise control systems are listed below.

1. Probe tube diameter should be less than 0.1 of the effective duct diameter.

2. Sound absorbing material in the probe tube or at the end opposite the microphone is unnecessary.
3. The flow resistance of the porous material covering the slit should be approximately $2\rho_0 c_0$.
4. The probe tube should be oriented so that its microphone is furthest from the sound source generating the sound field to be measured, irrespective of the direction of mean flow.
5. The ratio of slit width to probe tube circumference should be about 0.025.
6. The probe tube length should be at least 500 mm and longer if possible.
7. The cut-off frequency below which turbulent pressure fluctuations will not be attenuated is given by (Shepherd et al., 1989) as:

$$f_c = \frac{c_0 M}{L(1-M)} \quad \text{(Hz)} \tag{14.5.1}$$

where M is the flow Mach number, c_0 the speed of sound and L the probe tube length. At a flow speed of Mach number $M = 0.05$, the efficiency of a 13 mm probe tube, 500 mm long, designed using the preceding guidelines varies from about 10 dB at 50 Hz to about 25 dB at 500 Hz. (Here, efficiency refers to the relative sensitivity to acoustic and turbulent pressure fields.) Above about 500 Hz, the efficiency does not increase significantly.

Shepherd et al. (1989) point out that the effect of the probe tube on the microphone phase response as a function of frequency constitutes a complication for active noise control systems. However, probe tube microphones are the preferred means of sampling the sound field in the majority of commercial systems used to active control noise in ducts containing air flow. Shepherd et al. also show that in some cases, active control using a microphone with a standard, commercially available nose cone only allows the acoustic field to be attenuated to the level of the turbulent pressure fluctuations which may be a negligible amount in some cases. The probe tube microphone design described above can result in a further 10 to 25 dB reduction in acoustic levels and thus it is considered an essential part of many active control installations in air ducts.

14.5.1.7 Other Probe Tube Designs

Other probe tube designs which have been used with some success include a tube with small holes covering its surface (Noiseux and Horwath, 1970; Bolleter et al., 1970), a porous tube made of glass fibre (Nakamura et al., 1971) and a porous rigid ceramic, plastic or sintered metal tube (Hoops and Eriksson, 1991) with an average cell dimension less than 100 μm. The porous tube is now the preferred device in commercial active control systems for noise in ducts because of its low cost, simplicity of manufacture and simplicity of installation. It is difficult to judge from the literature, the relative performance of the various designs in terms of turbulence rejection capability. However, it appears that the tube with holes in its surface (and a cloth covering) does not perform as well as the tube with a slit (and cloth covering) or the porous ceramic tube.

14.5.2 Microphone Arrays

Another method of suppressing the turbulent pressure signal is to use a microphone array consisting of a number of discrete microphones arranged in a line directed at the noise source. If the signal from each microphone is electronically delayed by the acoustic propagation time from it to the furthest microphone from the source and the signals all combined, then the slower propagating turbulent pressure signal will be suppressed, the amount of suppression being dependent on the number of microphones used. This system is rather like a discrete version of the probe tube, although it is much less practical, more expensive and no more effective. If the discrete microphones are separated sufficiently so that the turbulent pressure fluctuations at each microphone are uncorrelated, then Shepherd et al. (1989) show that the suppression of the turbulent pressure signal will be 10 log N dB, where N is the number of microphones.

The discrete microphone array also has a cut-off frequency below which there will be no attenuation of the turbulent pressure fluctuations. This is given by:

$$f_c = \frac{c_0 M}{N L_s (1 - M)} \quad \text{(Hz)} \tag{14.5.2}$$

where c_0 is the speed of sound, M is the flow Mach number, N is the number of microphones in the array and L_s the separation between adjacent microphones. Note the similarity of this equation and Equation (14.5.1).

14.5.3 Use of Two Microphones and a Recursive Linear Optimal Filter

A more sophisticated means of filtering out the turbulent pressure contribution from a single microphone signal was proposed by Bouc and Felix (1987). They used a micro-processor to construct a Kalman filter (see Chapter 5) to provide an optimal estimate of the acoustic and turbulent pressure fluctuations. This allowed single frequency acoustic signals to be extracted successfully. The same authors also addressed the problem of extracting a broadband acoustic signal (or suppressing the turbulent pressure contribution in a broadband acoustic signal). This was done by using a linear auto-regressive moving average (ARMA) stochastic model, using an optimisation technique to estimate the parameters, and using the output signals from two microphones separated by one duct diameter and at the same axial location along the duct. Although they demonstrated a successful application of their technique, it has not received wide acceptance, probably due to the complexity of its implementation.

14.5.4 Microphone Boxes

In duct systems, turbulence-induced noise can be minimised by placing the error and reference microphones in outer boxes connected to the duct. The microphone is placed at the downstream end of the box and a slit connects the inside of the box to the duct interior. The boxes are often filled with a porous material for added turbulence rejection. The operation principle of microphones boxes is similar to that of probe tubes (see Section 14.5.1). As for probe tubes, the cut-off frequency below which turbulent pressure fluctuations will not be attenuated is:

$$f_c = \frac{c_0 M}{L_s(1-M)} \quad \text{(Hz)} \tag{14.5.3}$$

where L_s is the length of the slit. According to Equation (14.5.1), a longer slit length will result in increased turbulence rejection.

Microphone boxes have a number of advantages over conventional probe tubes and in particular, they protect the microphone from contaminants and moisture within the duct, increasing microphone life. In addition, the microphone box configuration increases microphone accessibility and ease of maintenance. Microphone boxes have been shown to out-perform probe tubes in terms of the level of turbulence rejection (Kuo and Morgan, 1996). However, placing probe tubes inside microphone boxes can be used to suppress turbulent pressure signals to a much greater level than microphone boxes or probe tubes alone (Larsson et al., 2005).

The construction of microphone boxes from commercially available T-duct pieces has been investigated by Larsson et al. (2009). Both a thin membrane and a metallic netting were used as the transition piece between the horizontal and vertical duct sections. The transition material allows the acoustic sound to propagate to the microphone while preventing air flow from influencing the microphone measurement. Larsson et al. (2009) found that microphones were less affected by turbulent pressure fluctuations when in T-ducts compared to in conventional microphone boxes with a slit.

A similar arrangement for minimising turbulence noise contamination of a sound measurement in a flow duct was presented by Hansen (2009), whereby a microphone is mounted in a small side branch with protective Mylar and acoustic foam inserts as illustrated in Figure 14.9.

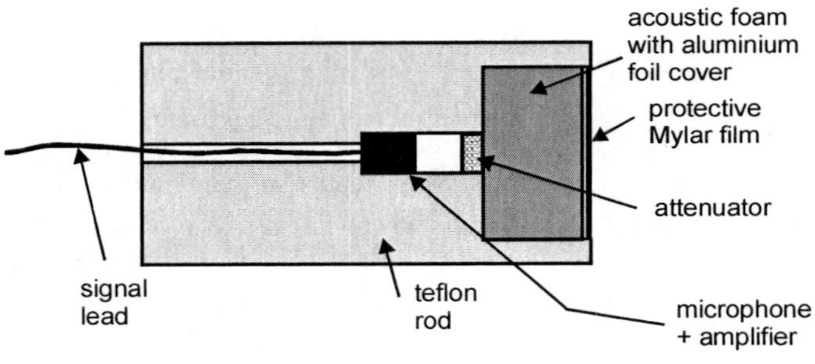

Figure 14.9 Arrangement for a side branch mounted microphone in a flow duct.

The microphone holder is positioned so that the protective Mylar film is flush with the inside wall of the duct. It is important that the front and rear faces of the microphone diaphragm are connected by a narrow duct or tube for ambient pressure equalisation, which will ensure that the microphone diaphragm will not be displaced and thus produce a non-linear dynamic response as a result of a mean pressure difference between its two sides. If the duct noise levels are too high for the microphone to handle, a small noise attenuating insert made of material such as Vyon (a porous plastic) may be used as illustrated in Figure 14.9. This arrangement is also suitable for use in contaminated flows. For flows that are very hot, it may be necessary to provide cooling air flow over the microphone and then exhaust it into the duct through another opening in the Teflon rod.

14.6 VIRTUAL SENSING ALGORITHMS FOR ACTIVE NOISE CONTROL

Traditional local active noise control systems create a zone of quiet at the physical error sensor by minimising the sound pressure measured at the sensor location with a control source. While significant attenuation may be achieved at the sensor location, the zone of quiet is generally small. This requires that the physical sensor be placed at the location of desired attenuation, which is often inconvenient. Virtual acoustic sensors have been developed to shift the zone of quiet away from the physical sensor to a remote location referred to as the virtual location. Using a physical sensor signal at some other location, a virtual sensing algorithm is used to estimate the sound pressure at the virtual location. Instead of minimising the physical sensor signal, the estimated sound pressure is minimised with the active noise control system to generate a zone of quiet centred on the virtual location. A number of virtual sensing algorithms have been developed to estimate the sound pressure at a fixed virtual location, including the virtual microphone arrangement (Elliott and David, 1992), the remote microphone technique (Roure and Albarrazin, 1999), the forward difference prediction technique (Cazzolato, 1999), the adaptive LMS virtual microphone technique (Cazzolato, 2002), the Kalman filtering virtual sensing method (Petersen et al., 2007), and the Stochastically Optimal Tonal Diffuse Field (SOTDF) virtual sensing technique (Moreau et al., 2007).

It is likely that the desired location of attenuation is not spatially fixed. This occurs, for example, when the desired location of attenuation is at the ear of a seated observer who moves their head. As a result, a number of moving virtual sensing algorithms have also been developed to generate a virtual microphone capable of tracking a moving virtual location. These include the remote moving microphone technique (Petersen et al., 2006), the adaptive LMS moving virtual microphone technique (Petersen et al., 2007), the Kalman filtering moving virtual sensing method (Petersen, 2007) and the Stochastically Optimal Tonal Diffuse Field (SOTDF) moving virtual sensing method (Moreau et al., 2009a).

Virtual sensing configurations are used in feedforward active noise control systems that employ the FXLMS algorithm, so the notation that has been used in Section 6.6 is also used here.

14.6.1 Virtual Sensing Problem Formulation

When describing the virtual sensing algorithms in the following sections, physical microphone quantities are denoted without a subscript while quantities measured at the virtual microphones are denoted with the subscript v. It is assumed here that there are N_e physical microphones, N_v spatially fixed virtual microphones and N_c control sources. The vector of the total sound pressures, $\boldsymbol{e}(k)$, at the N_e physical microphones is defined as:

$$\boldsymbol{e}(k) = \left[e_1(k) \;\; e_2(k) \;\; \dots \;\; e_{Ne}(k)\right]^{\mathrm{T}} \tag{14.6.1}$$

The total disturbance $\boldsymbol{e}(k)$ measured at the N_e physical microphones is the sum of the sound field contributions produced by the primary and control sound sources at the physical microphone locations, and may be written as:

$$\boldsymbol{e}(k) = \boldsymbol{p}(k) + \boldsymbol{s}(k) = \boldsymbol{p}(k) + \boldsymbol{H}\boldsymbol{y}(k) \tag{14.6.2a,b}$$

where $\boldsymbol{p}(k)$ is a vector of the primary sound pressures at the N_e physical microphones, $\boldsymbol{s}(k)$ is a vector of the control source sound pressures at the N_e physical microphones, $\boldsymbol{H}$ is a matrix of size $N_e \times N_c$ whose elements are the cancellation path transfer functions between the control source inputs and the physical microphone outputs, $\boldsymbol{y}(k)$ is a vector of the control source strengths and k is the time step.

Similarly, the vector of the total sound pressures, $\boldsymbol{e}_v(k)$, at the N_v spatially fixed virtual locations is defined as:

$$\boldsymbol{e}_v(k) = \begin{bmatrix} e_{v1}(k) & e_{v2}(k) & \dots & e_{vN_v}(k) \end{bmatrix}^{\mathrm{T}} \tag{14.6.3}$$

The total sound pressures $\boldsymbol{e}_v(k)$ at the N_v virtual microphones is the sum of the sound fields produced by the primary and control sources at the N_v virtual locations and may be written as:

$$\boldsymbol{e}_v(k) = \boldsymbol{p}_v(k) + \boldsymbol{s}_v(k) = \boldsymbol{p}_v(k) + \boldsymbol{H}_v \boldsymbol{y}(k) \tag{14.6.4a,b}$$

where $\boldsymbol{p}_v(k)$ is the vector of the primary sound pressures at the N_v virtual locations, $\boldsymbol{s}_v(k)$ is the vector of control source sound pressures at the N_v virtual locations and $\boldsymbol{H}_v$ is a matrix of size $N_v \times N_c$ whose elements are the cancellation path transfer functions between the control source inputs and the virtual locations.

Using the physical error signals $\boldsymbol{e}(k)$, a virtual sensing algorithm is used to estimate the sound pressures at the spatially fixed virtual locations $\hat{\boldsymbol{e}}_v(k)$, where the ^ indicates an estimated quantity. Instead of minimising the physical error signals, the estimated sound pressures are minimised with the active noise control system to generate zones of quiet centred at the virtual locations.

14.6.2 Spatially Fixed Virtual Sensing Algorithms

Spatially fixed virtual sensing algorithms are used to obtain estimates of the error signals at a number of spatially fixed virtual locations using the error signals from one or more remotely located physical error sensors, the control signal and knowledge of the system. These virtual sensing algorithms are then combined with an active noise control algorithm to generate zones of quiet centred at the fixed virtual locations.

14.6.2.1 Virtual Microphone Arrangement

The virtual microphone arrangement, proposed by Elliott and David (1992), was the first virtual sensing algorithm suggested for local active noise control. As suggested by Petersen (2007), the virtual microphone arrangement is most easily implemented when the number of physical microphones N_e equals the number of virtual microphones N_v. For this case, the microphones are located in N_v pairs, each consisting of a single physical and virtual microphone. The primary disturbance is assumed to be equal at the physical and virtual microphones in each pair so that:

$$\boldsymbol{p}_v(k) = \boldsymbol{p}(k) \tag{14.6.5}$$

This assumption holds provided that the primary sound field is essentially the same at the locations of the physical and virtual microphones in each pair.

A block diagram of the virtual microphone arrangement is shown in Figure 14.10. This virtual sensing algorithm first requires a preliminary identification stage in which the matrices $\boldsymbol{H}$ and $\boldsymbol{H}_v$ of cancellation path transfer functions are measured. These cancellation path transfer functions are modelled as matrices of FIR or IIR filters. Once the preliminary identification stage is complete, the microphones placed at the virtual locations are removed. As shown in Figure 14.10, estimates $\hat{\boldsymbol{e}}_v(k)$ of the total error signals at the virtual locations are now calculated as:

$$\hat{\boldsymbol{e}}_v(k) = \boldsymbol{e}(k) - (\boldsymbol{H} - \boldsymbol{H}_v)\boldsymbol{y}(k) \tag{14.6.6}$$

The performance of the virtual microphone arrangement has been thoroughly investigated in both tonal and broadband sound fields by a number of authors (Diaz et al., 2006; Holmberg et al., 2002; Horihata et al., 1997; Garcia-Bonito and Elliott, 1995; Garcia-Bonito et al., 1997; Matuoka et al., 1996; Pawelczyk, 2006; Rafaely et al., 1999). The performance of the virtual microphone arrangement has also been theoretically analysed in a pure tone diffuse sound field (Moreau, et al., 2007). At low frequencies, the virtual microphone arrangement generates a zone of quiet that is comparable in size to that achieved by directly minimising the measured sound pressure using a physical microphone located at the virtual location (Garcia-Bonito et al., 1996, 1997). At higher frequencies (above about 500 Hz), the 10 dB zone of quiet generated with the virtual microphone arrangement is substantially reduced compared to that obtained using a physical microphone at the virtual location. This is due to the assumption of equal primary sound pressure at the physical and virtual locations becoming invalid as the wavelength decreases.

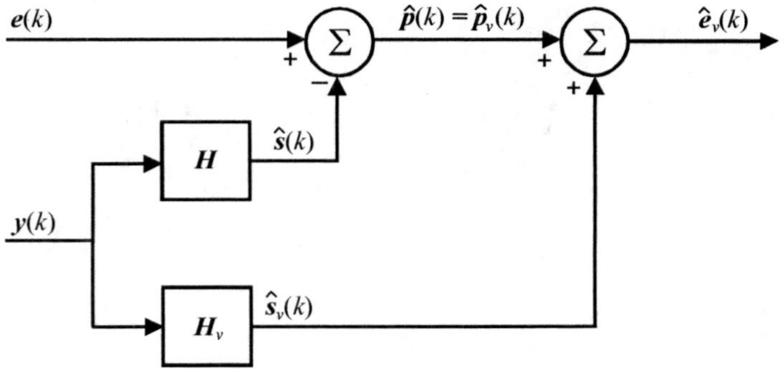

Figure 14.10 The virtual microphone arrangement.

14.6.2.2 Remote Microphone Technique

The remote microphone technique developed by Roure and Albarrazin (1999) is an extension of the virtual microphone arrangement (Elliott and David, 1992), and it uses an additional matrix of filters to compute estimates of the primary disturbances at the virtual sensors from a measurement of the primary disturbances at the physical sensors.

A block diagram of the remote microphone arrangement is shown in Figure 14.11. Like the virtual microphone arrangement, the remote microphone technique requires a preliminary identification stage in which the cancellation path transfer matrices, $\boldsymbol{H}$ and $\boldsymbol{H}_v$, are modelled as matrices of FIR or IIR filters. The $N_v \times N_e$ size matrix, $\boldsymbol{M}$, of primary transfer functions between the virtual locations and the physical locations is also estimated as a matrix of FIR or IIR filters during this preliminary identification stage. The cancellation path transfer function matrix $\boldsymbol{H}$ is identified using the control sources and the physical microphones while microphones temporarily placed at the virtual locations are used to obtain matrices $\boldsymbol{M}$ and $\boldsymbol{H}_v$.

As shown in Figure 14.11, estimates of the primary contribution at the physical error sensors, $\hat{\boldsymbol{p}}(k)$, are first calculated using:

$$\hat{\boldsymbol{p}}(k) = \boldsymbol{e}(k) - \hat{\boldsymbol{s}}(k) = \boldsymbol{e}(k) - \boldsymbol{H}\boldsymbol{y}(k) \tag{14.6.7a,b}$$

Next, estimates of the primary disturbances $\hat{\boldsymbol{p}}_v(k)$ at the virtual locations are found to be:

$$\hat{\boldsymbol{p}}_v(k) = \boldsymbol{M}\hat{\boldsymbol{p}}(k) \tag{14.6.8}$$

Finally, estimates of the total virtual error signals $\hat{\boldsymbol{e}}_v(k)$ are calculated as:

$$\hat{\boldsymbol{e}}_v(k) = \hat{\boldsymbol{p}}_v(k) - \hat{\boldsymbol{s}}_v(k) = \boldsymbol{M}\hat{\boldsymbol{p}}(k) + \boldsymbol{H}_v\boldsymbol{y}(k) \tag{14.6.9a,b}$$

Radcliffe and Gogate (1993) demonstrated that theoretically, a perfect estimate of the tonal disturbance at the virtual location can be achieved with this virtual sensing algorithm provided accurate models of the tonal transfer functions are obtained in the preliminary identification stage. In a three-dimensional model of a car cabin, the tonal control achieved at a number of virtual microphones generated with the remote microphone technique was equivalent to that achieved by directly minimising the measured signals from microphones at the virtual locations.

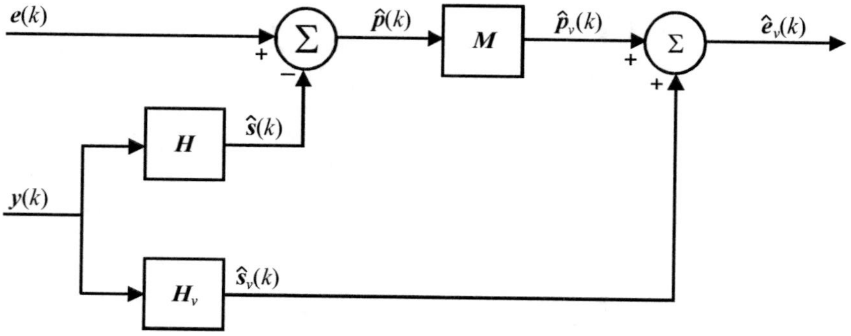

Figure 14.11 The remote microphone technique.

14.6.2.3 Forward Difference Prediction Technique

The forward difference prediction technique, as proposed by Cazzolato (1999), fits a polynomial to the signals from a number of physical microphones in an array. The sound pressure at the virtual location is estimated by extrapolating this polynomial to the virtual location. This virtual sensing algorithm is suitable for use in low-frequency sound fields, when the virtual distance and the spacing between the physical microphones are much less than a wavelength. At low frequencies, the spatial rate of change of the sound pressure between the microphones is small and extrapolation can therefore be applied.

The sound pressure at a virtual location x estimated by a first-order finite difference method, using $N_e = 2$ physical microphones, separated by a distance of $2h$, is given by:

$$\hat{e}_v(k) = e_2(k) + \frac{e_2(k) - e_1(k)}{2h}x \tag{14.6.10}$$

The sound pressure at a virtual location x estimated by a second-order finite difference method, using $N_e = 3$ physical microphones, each separated by a distance of h, is:

$$\hat{e}_v(k) = \frac{x(x + h)}{2h^2}e_1(k) + \frac{x(x + 2h)}{h^2}e_2(k) + \frac{(x + 2h)(x + h)}{2h^2}e_3(k) \tag{14.6.11}$$

The forward difference prediction technique has several advantages over other virtual sensing algorithms. First, the assumption of equal primary sound pressure at the physical and virtual locations is not necessary, but also preliminary identification is not required, nor are FIR filters or similar required to model the complex transfer functions between the sensors and the sources. Furthermore, this is a fixed gain prediction technique which is robust to physical system changes that may alter the complex transfer functions between the error sensors and the control sources.

Using either linear or quadratic prediction techniques, forward difference virtual sensors have been shown to out-perform physical microphones in terms of the level of attenuation achieved at a virtual location (removed from the physical microphones) in both a duct and in a free-field (Kestell et al., 2000; Kestell et al., 2001; Munn et al., 2001a; Munn et al., 2001b; Munn et al., 2003). While the second-order estimate is theoretically more accurate than the first-order estimate, real-time feedforward experiments in a narrow duct demonstrated that quadratic prediction techniques are adversely affected by short wavelength extraneous noise. It was also shown by Petersen (2007) that the estimation problem is ill-conditioned for the three-sensor arrangement, explaining the difference between numerical and experimental results.

In an attempt to improve the prediction accuracy of the forward difference algorithm, higher-order forward difference prediction virtual sensors, which act to spatially filter out the extraneous noise, were developed (Munn et al., 2002). Additional physical microphones were added to the array resulting in a greater number of microphones than system order. The microphone weights for this over-constrained system were then calculated using a least-squares approximation.

The sound pressure at a virtual location, x, estimated by a first-order finite difference estimate using $N_e = 3$ physical microphones, each separated by a distance of h, is:

$$\hat{e}_v(k) = \frac{(-3x - h)}{6h}e_1(k) + \frac{1}{3}e_2(k) + \frac{(3x + 5h)}{6h}e_3(k) \tag{14.6.12}$$

The sound pressure at a virtual location x estimated by a first-order finite difference estimate using $N_e = 5$ physical microphones, separated by a distance of $h/2$, is given by:

$$\hat{e}_v(k) = \frac{(-2x + 3h)}{5h}e_1(k) + \frac{(-x + 2h)}{5h}e_2(k) + \frac{1}{5}e_3(k) + \frac{x}{5h}e_4(k) + \frac{(2x - h)}{5h}e_5(k)$$

(14.6.13)

The sound pressure at a virtual location x estimated by a second-order finite difference estimate using $N_e = 5$ physical microphones, separated by a distance of $h/2$, is:

$$\hat{e}_v(k) = \frac{(20x^2 - 54xh + 31h^2)}{35h^2}e_1(k) + \frac{(-10x^2 + 3xh + 9h^2)}{35h^2}e_2(k)$$
$$+ \frac{(-20x^2 - 40xh - 31h^2)}{35h^2}e_3(k) + \frac{(-10x^2 - 27xh - 5h^2)}{35h^2}e_4(k) \quad (14.6.14)$$
$$+ \frac{(20x^2 - 26xh + 3h^2)}{35h^2}e_5(k)$$

In experiments, the accuracy of these higher-order forward difference prediction virtual sensors was found to be adversely affected by sensitivity, phase mismatches and relative position errors between microphone elements in the array (Munn, 2004). Such phase mismatches and position errors are unavoidable when a large number of physical microphones are used. It has also been demonstrated by Petersen (2007) that the estimation problem is ill-conditioned for higher-order forward difference extrapolations.

14.6.2.4 Adaptive LMS Virtual Microphone Technique

The adaptive LMS virtual microphone technique developed by Cazzolato (2002) employs the adaptive LMS algorithm (see Section 6.6) to adapt the weights of physical microphones in an array so that the weighted summation of these signals minimises the mean square difference between the predicted sound pressure at the virtual location and that measured by a microphone temporarily placed at the virtual location.

For the case of $N_v = 1$ virtual microphones, an estimate of the total disturbance $\hat{e}_v(k)$ at the virtual microphone location is calculated as the sum of the weighted physical sensor signals at N_e physical sensors in an array and is given by:

$$\hat{e}_v(k) = \Sigma w_i e_i(k) = \boldsymbol{w}^{\mathrm{T}}\boldsymbol{e}(k) \quad (14.6.15a,b)$$

where $\boldsymbol{w}$ is a vector containing the N_e physical error sensor weights:

$$\boldsymbol{w} = \begin{bmatrix} w_1 & w_2 & \cdots & w_{Ne} \end{bmatrix}^{\mathrm{T}} \quad (14.6.16)$$

The weights, $\boldsymbol{w}$, are calculated in a preliminary identification stage by switching the primary source off and exciting the control source with band-limited white noise. A modified version of the adaptive LMS algorithm is used to adapt the microphone weights. This algorithm can be used to find the optimal solution for the weights that minimises the mean square difference between the predicted sensor quantity $\hat{s}_v(k)$, and $s_v(k)$ measured by a physical sensor placed temporarily at the virtual location. A block diagram of the adaptive LMS virtual microphone used to estimate the physical error sensor weights is shown in Figure 14.12. As only a single temporal tap is used, the real valued weights correspond to a pure gain and are calculated using:

$$\boldsymbol{w}(k+1) = \boldsymbol{w}(k) + 2\mu\boldsymbol{s}(k)\varepsilon(k) \tag{14.6.17}$$

where μ is the convergence coefficient, $\boldsymbol{s}(k)$ is the vector of contributions to the physical microphone location from all of the control sources and $\varepsilon(k)$ is the error term. This error term, $\varepsilon(k)$, is defined as the difference between the actual control source contribution at the virtual sensor location and the estimated control source contribution, given by:

$$\varepsilon(k) = s_v(k) - \hat{s}_v(k) \tag{14.6.18}$$

where the estimated control source contribution at the virtual location is given by:

$$\hat{s}_v(k) = \boldsymbol{w}^{\mathrm{T}}\boldsymbol{s}(k) \tag{14.6.19}$$

Once the weights have converged, they are fixed and the temporary microphone is removed from the virtual location.

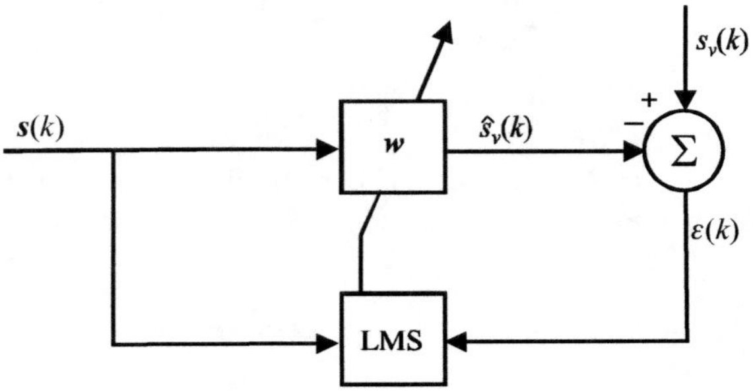

Figure 14.12 The adaptive LMS algorithm used to calculate the physical sensor weights.

The adaptive virtual sensors are unaffected by sensitivity mismatches and relative position errors that adversely affect the forward difference prediction technique. In addition, these adaptive sensors have been found to predict the sound pressure at a virtual location more accurately than the equivalent forward difference prediction virtual sensor (Cazzolato, 2002).

Petersen (2007) investigated the performance of the adaptive LMS virtual microphone technique in a broadband sound field with a frequency range of 50 to 500 Hz, in a long narrow duct. For an array of N_v = 2, 3 and 5 physical sensors, the overall estimation performance decreased with increasing distance between the physical sensor array and the virtual location, for all three configurations of physical sensors. The best estimation performance is theoretically achieved with an array of five physical sensors; however, this configuration was found to be ill-conditioned in experiments, and a similar estimation performance was achieved with all three physical sensor configurations.

Despite being calculated by exciting the control source only, the weights $\boldsymbol{w}$ in Equation (14.6.17), are applied to both the primary and control contributions. It has thus been assumed that the weights are optimal for the estimation of both contributions. This, however, may not always be true, especially in the near-field of the control source where the spatial properties of the primary and control sound fields are very different. As a result, Petersen (2007) suggested that the optimal weights for the estimation of both the primary and control source contributions should be found separately, with the adaptive LMS virtual microphone technique being implemented as shown in Figure 14.13.

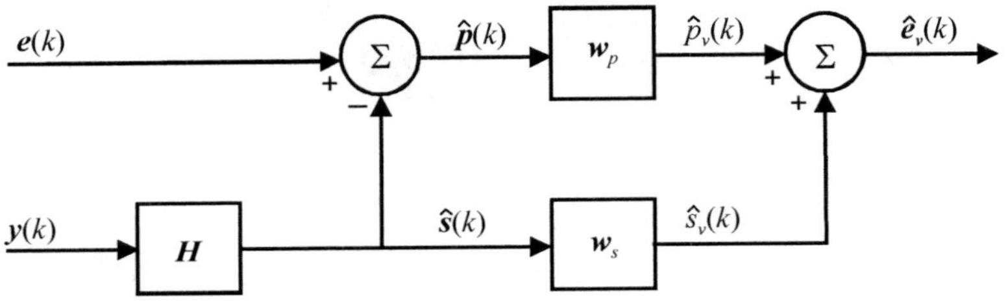

Figure 14.13 Adaptive LMS virtual microphone technique.

As shown in Figure 14.13, the virtual sensing algorithm separates the physical error signals into their primary and control components using the vector of the physical cancellation path transfer functions, $\boldsymbol{H}$. This vector of FIR or IIR filters is estimated in the preliminary identification stage. The primary component of the physical error signals is calculated as:

$$\hat{\boldsymbol{p}}(k) = \boldsymbol{e}(k) - \hat{\boldsymbol{s}}(k) = \boldsymbol{e}(k) - \boldsymbol{H}y(k) \qquad (14.6.20\text{a,b})$$

Once the primary and control signal weights have been estimated separately using Equation (14.6.17), the sound pressure at the virtual location is estimated, as shown in Figure 14.13, using:

$$\hat{e}_v(k) = \hat{p}_v(k) - \hat{s}_v(k) = \boldsymbol{w}_p^{\mathrm{T}}\hat{\boldsymbol{p}}(k) - \boldsymbol{w}_s^{\mathrm{T}}\hat{\boldsymbol{s}}(k) \qquad (14.6.21\text{a,b})$$

where $\boldsymbol{w}_p$ and $\boldsymbol{w}_s$ are vectors containing the N_e optimal physical primary and control signal weights, and $\hat{\boldsymbol{p}}(k)$ and $\hat{\boldsymbol{s}}(k)$ are vectors containing estimates of the primary and control source contributions at the N_e physical sensor locations.

14.6.2.5 Kalman Filtering Virtual Sensing Method

The Kalman filtering virtual sensing method (Petersen et al., 2008) uses Kalman filtering theory to obtain estimates of the error signals at the virtual locations. In this virtual sensing method, the active noise control system is first modelled as a state-space system (see Chapter 5) whose outputs are the physical and virtual error signals. The inputs are the primary and control source input signals. A Kalman filter is formulated to compute estimates of the plant states and subsequently estimate the virtual error signals using the physical error signals.

Following standard Kalman filter formulation, the active noise control system plant is described by the following state-space model:

$$\begin{aligned} z(k+1) &= \boldsymbol{A}z(k) + \boldsymbol{B}_s y(k) + \boldsymbol{B}_u \boldsymbol{u}(k) \\ \boldsymbol{e}(k) &= \boldsymbol{C}z(k) + \boldsymbol{D}_s y(k) + \boldsymbol{D}_u \boldsymbol{u}(k) + \boldsymbol{v}_p(k) \\ \boldsymbol{e}_v(k) &= \boldsymbol{C}_v z(k) + \boldsymbol{D}_{vs} y(k) + \boldsymbol{D}_{vu} \boldsymbol{u}(k) + \boldsymbol{v}_v(k) \end{aligned} \qquad (14.6.22\text{a–c})$$

where $z(k)$ are the N plant states, $\boldsymbol{v}_p(k)$ are the physical measurement noise signals, $\boldsymbol{v}_v(k)$ are the virtual measurement noise signals and $\boldsymbol{u}(k)$ are the N_p primary disturbance signals. In the state-space model, $\boldsymbol{A}$ is the state matrix of size $N \times N$ in discrete form, $\boldsymbol{B}_s$ is the discrete control signal input matrix of size $N \times N_c$, $\boldsymbol{B}_u$ is the discrete primary signal input matrix of size $N \times N_p$, $\boldsymbol{C}$ is the discrete physical output matrix of size $N_e \times N$, $\boldsymbol{C}_v$ is the discrete virtual output matrix of size $N_v \times N$, $\boldsymbol{D}_s$ and $\boldsymbol{D}_u$ are the discrete feedforward matrices for the physical microphones of size $N_e \times N_c$ and $N_e \times N_p$ respectively, and $\boldsymbol{D}_{vs}$ and $\boldsymbol{D}_{vu}$ are the discrete feedforward matrices for the virtual microphones of size $N_v \times N_c$ and $N_v \times N_p$ respectively. Inclusion of the measurement noise signals, $\boldsymbol{v}_p(k)$ and $\boldsymbol{v}_v(k)$, in the state-space model account for measurement noise on the microphones at the physical and virtual locations during the preliminary identification stage. Once the preliminary identification stage is complete, the microphones temporarily positioned at the virtual locations are removed.

Implementation of the Kalman filtering virtual sensing method is shown in the block diagram in Figure 14.14. In this figure, $\boldsymbol{G}$ is the generalised plant of the acoustic system, $\hat{\boldsymbol{G}}$ is an estimate of the generalised plant given by the state-space model in Equation (14.6.22) and $\boldsymbol{K}$ are the Kalman filter gains.

The covariance properties of the stochastic signals $\boldsymbol{u}(k)$, $\boldsymbol{v}_p(k)$ and $\boldsymbol{v}_v(k)$ are required when using Kalman filtering theory to estimate the error signals at the virtual locations. These covariance properties and the state-space model of the active noise control system plant are estimated during a preliminary identification stage with microphones temporarily positioned at the virtual locations. The primary disturbance signals $\boldsymbol{u}(k)$, the physical measurement noise signals $\boldsymbol{v}_p(k)$, and the virtual measurement noise signals $\boldsymbol{v}_v(k)$ are all assumed to be zero mean white stationary random processes with the following covariance properties:

$$\mathrm{E}\left\{\begin{bmatrix}\boldsymbol{u}(k)\\ \boldsymbol{v}_p(k)\\ \boldsymbol{v}_v(k)\end{bmatrix}\begin{bmatrix}\boldsymbol{u}(k)\\ \boldsymbol{v}_p(k)\\ \boldsymbol{v}_v(k)\\ 1\end{bmatrix}^{\mathrm{T}}\right\} = \begin{bmatrix}\mathbf{I} & \boldsymbol{S}_{pu}^{\mathrm{T}} & \boldsymbol{S}_{vu}^{\mathrm{T}} & \mathbf{0}\\ \boldsymbol{S}_{pu} & \boldsymbol{R}_p & \boldsymbol{R}_{pv} & \mathbf{0}\\ \boldsymbol{S}_{vu} & \boldsymbol{R}_{pv}^{\mathrm{T}} & \boldsymbol{R}_{vu} & \mathbf{0}\end{bmatrix}\delta_{nk} \tag{14.6.23}$$

where E{ } denotes the expectation operator, $\mathbf{I}$ is the identity matrix and δ_{nk} is the Kronecker delta function.

The term $\boldsymbol{B}_u\boldsymbol{u}(k)$ in Equation (14.6.22) can be interpreted as process noise, and the combined influence of the measurement noise signals and disturbance signals can be interpreted as an auxiliary measurement noise signal $\boldsymbol{v}(k)$, where

$$\boldsymbol{v}(k) = \begin{bmatrix}\boldsymbol{D}_u\boldsymbol{u}(k) + \boldsymbol{v}_p(k)\\ \boldsymbol{D}_{vu}\boldsymbol{u}(k) + \boldsymbol{v}_v(k)\end{bmatrix} \tag{14.6.24}$$

Using these definitions, the following covariance matrix can be defined as:

$$\mathrm{E}\left\{\begin{bmatrix}\{\boldsymbol{B}_u\boldsymbol{u}(k)\}\\ \boldsymbol{v}(k)\end{bmatrix}\begin{bmatrix}\{\boldsymbol{B}_u\boldsymbol{u}(k)\}\\ \boldsymbol{v}(k)\end{bmatrix}^{\mathrm{T}}\right\} = \begin{bmatrix}\overline{\boldsymbol{Q}}_u & \overline{\boldsymbol{S}}_u^{\mathrm{T}}\\ \overline{\boldsymbol{S}}_u & \overline{\boldsymbol{R}}_u\end{bmatrix}\delta_{nk} \tag{14.6.25}$$

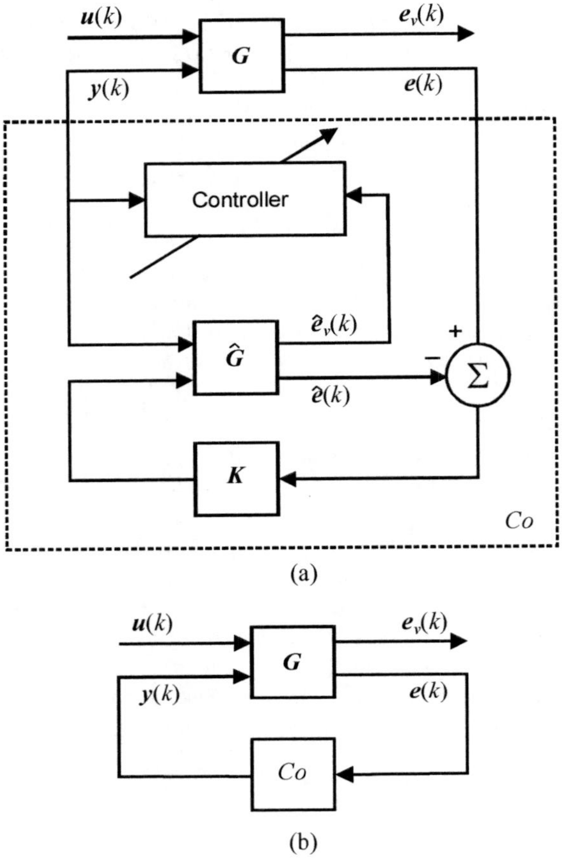

Figure 14.14 Block diagram of (a) implementation of the Kalman filtering virtual sensing method and (b) the generalised control configuration with two sets of inputs and two sets of outputs.

The covariance matrix $\overline{\boldsymbol{Q}}_u$ of the process noise is given by:

$$\overline{\boldsymbol{Q}}_u = \boldsymbol{B}_u \boldsymbol{B}_u^{\mathrm{T}} \tag{14.6.26}$$

The covariance matrix $\overline{\boldsymbol{R}}_u$ of the auxiliary measurement noise $\boldsymbol{v}(k)$ is given by:

$$\overline{\boldsymbol{R}}_u = \begin{bmatrix} \overline{\boldsymbol{R}}_p & \overline{\boldsymbol{R}}_{pv} \\ \overline{\boldsymbol{R}}_{pv}^T & \overline{\boldsymbol{R}}_v \end{bmatrix} = \begin{bmatrix} \boldsymbol{R}_p + \boldsymbol{S}_{pu}^{\mathrm{T}}\boldsymbol{D}_u + \boldsymbol{D}_u\boldsymbol{S}_{pu} + \boldsymbol{D}_u\boldsymbol{D}_u^{\mathrm{T}} & \boldsymbol{R}_{pv}^{\mathrm{T}} + \boldsymbol{S}_{pu}^{\mathrm{T}}\boldsymbol{D}_{vu} + \boldsymbol{D}_u\boldsymbol{S}_{vu} + \boldsymbol{D}_u\boldsymbol{D}_{vu}^{\mathrm{T}} \\ \boldsymbol{R}_{pv}^{\mathrm{T}} + \boldsymbol{S}_{vu}^{T}\boldsymbol{D}_u^{\mathrm{T}} + \boldsymbol{D}_{vu}\boldsymbol{S}_{pu} + \boldsymbol{D}_{vu}\boldsymbol{D}_u^{\mathrm{T}} & \boldsymbol{R}_v + \boldsymbol{S}_{vu}^{\mathrm{T}}\boldsymbol{D}_{vu} + \boldsymbol{D}_{vu}\boldsymbol{S}_{vu} + \boldsymbol{D}_{vu}\boldsymbol{D}_{vu}^{\mathrm{T}} \end{bmatrix}$$

(14.6.27a,b)

The covariance matrix $\overline{\boldsymbol{S}}_u$ between the process noise and the auxiliary measurement noise $\boldsymbol{v}(k)$ is given by:

$$\overline{\boldsymbol{S}}_u = \begin{bmatrix} \overline{\boldsymbol{S}}_{pu} \\ \overline{\boldsymbol{S}}_{vu} \end{bmatrix} = \begin{bmatrix} \boldsymbol{D}_u\boldsymbol{B}_u^T + \boldsymbol{S}_{pu}\boldsymbol{B}_u^T \\ \boldsymbol{D}_{vu}\boldsymbol{B}_u^T + \boldsymbol{S}_{vu}\boldsymbol{B}_u^T \end{bmatrix} \tag{14.6.28a,b}$$

The virtual sensing algorithm in state-space form, that estimates the virtual error signals $\hat{e}_v(k)$, given measurements of the physical error signals $e(i)$, up to $i = k$, is as follows:

$$\begin{bmatrix} \hat{z}(k+1) \\ \hat{e}_v(k) \end{bmatrix} = \begin{bmatrix} A - K_{ps}C & B_U - K_{ps}D_s & K_{ps} \\ C_v - M_{vs}C & D_{vs} - M_{vs}D_s & M_{vs} \end{bmatrix} \begin{bmatrix} \hat{z}(k-1) \\ s(k) \\ e(k) \end{bmatrix} \tag{14.6.29}$$

where K_{ps} is referred to as the Kalman gain matrix and M_{vs} is referred to as the virtual innovation gain matrix. The Kalman gain matrix and the virtual innovation gain matrix are:

$$K_{ps} = \left(AP_{ps}C^{\mathrm{T}} + \overline{S}_{pu}\right)R_{p\varepsilon}^{-1} \tag{14.6.30}$$

$$M_{vs} = \left(C_v P_{ps} C^{\mathrm{T}} + \overline{R}_{pv}^{-1}\right) R_{p\varepsilon}^{-1} \tag{14.6.31}$$

with $P_{ps} = P_{ps}^{\mathrm{T}}$, the unique solution to the discrete algebraic Riccati equation is given by:

$$P_{ps} = AP_{ps}A^{\mathrm{T}} - \left(AP_{ps}C^{\mathrm{T}} + \overline{S}_{pu}\right)\left(CP_{ps}C^{\mathrm{T}} + \overline{R}_p\right)^{-1}\left(AP_{ps}C^{\mathrm{T}} + \overline{S}_{pu}\right)^{\mathrm{T}} + \overline{Q}_u \tag{14.6.32}$$

where $R_{p\varepsilon}$ is the covariance matrix of the innovation signals $\varepsilon(k) = e(k) - \hat{e}(k-1)$ given by:

$$R_{p\varepsilon} = \left[\varepsilon(k)\varepsilon(k)^{\mathrm{T}}\right] = CP_{ps}C^{\mathrm{T}} + \overline{R}_p \tag{14.6.33a,b}$$

To implement the Kalman filtering virtual sensing method, the state-space matrices A, B_u, C, C_v, D_{ps} and D_{vs} of the state-space model in Equation (14.6.22) and the covariance matrices $\overline{Q}_u$, $\overline{S}_{pu}$, $\overline{R}_p$ and $\overline{R}_{pv}$ need to be estimated, in practice, in a preliminary system identification stage using sub-space identification techniques (Haverkamp, 2001). Together, the state-space model in Equation (14.6.22) and covariance matrices describe the behaviour of the active noise control system and the covariance properties of the input signals. Subspace identification techniques estimate a model of the active noise control system in an innovations form. Therefore, the Kalman filtering virtual sensing method needs to be reformulated for practical implementation with an innovations model of the active noise control system. The steps to practical implementation of the Kalman filtering virtual sensing method using an innovations model of the active noise control system are as follows (Petersen, 2007):

1. Temporarily locate physical sensors at the spatially fixed virtual locations and measure an input-output data-set:

$$\left\{ s(k), \begin{bmatrix} e(k) \\ e_v(k) \end{bmatrix} \right\}_{k=1}^{N_c} \tag{14.6.34}$$

2. Use sub-space identification techniques (Haverkamp, 2001) to estimate an innovations model of the physical and virtual error signals:

$$\begin{aligned}
\hat{z}(k+1) &= \hat{A}\hat{z}(k-1) + \hat{B}_s\, y(k) + \hat{K}_s\left[\varepsilon(k)^{\mathrm{T}}\varepsilon_v(k)^{\mathrm{T}}\right]^{\mathrm{T}} \\
e(k) &= \hat{C}\hat{z}(k-1) + \hat{D}_s\, y(k) + \varepsilon(k) \\
e_v(k) &= \hat{C}_v\hat{z}(k-1) + \hat{D}_{vs}\, y(k) + \varepsilon_v(k)
\end{aligned} \tag{14.6.35a,b,c}$$

and estimate the covariance matrix of the white innovation signals:

$$\hat{R}_\varepsilon = \begin{bmatrix} \overline{\hat{R}}_{p\varepsilon} & \overline{\hat{R}}_{pv\varepsilon} \\ \overline{\hat{R}}_{pv\varepsilon}^{\mathrm{T}} & \overline{\hat{R}}_{v\varepsilon} \end{bmatrix} \tag{14.6.36}$$

3. Implement the Kalman filtering virtual sensing method as:

$$\begin{bmatrix} \hat{z}(k+1) \\ \hat{e}_v(k) \end{bmatrix} = \begin{bmatrix} \hat{A}-\hat{K}_{ps}\hat{C} & \hat{B}_s-\hat{K}_{ps}\hat{D}_s & \hat{K}_{ps} \\ \hat{C}_v-\hat{M}_{vs}\hat{C} & \hat{D}_{vs}-\hat{M}_{vs}\hat{D}_s & \hat{M}_{vs} \end{bmatrix} \begin{bmatrix} \hat{z}(k-1) \\ s(k) \\ e(k) \end{bmatrix} \tag{14.6.37}$$

where the Kalman gain matrix $\hat{K}_{ps}$ and the virtual innovation gain matrix $\hat{M}_{vs}$ are calculated as follows:

$$\hat{K}_{ps} = \left(\hat{A}X_s\hat{C}^{\mathrm{T}} + \hat{K}_s \begin{bmatrix} \overline{\hat{R}}_{p\varepsilon} \\ \overline{\hat{R}}_{pv\varepsilon} \end{bmatrix} \right) \left(\hat{C}X_s\hat{C}^{\mathrm{T}} + \overline{\hat{R}}_{p\varepsilon}\right)^{-1} \tag{14.6.38}$$

with $X_s = X_s^T > 0$, the unique solution to the discrete algebraic Riccati equation given by:

$$\hat{M}_{vs} = \left(\hat{C}_v X_s \hat{C}^{\mathrm{T}} + \overline{\hat{R}}_{pv\varepsilon}^{\mathrm{T}}\right)\left(\hat{C}X_s\hat{C}^{\mathrm{T}} + \overline{\hat{R}}_{p\varepsilon}\right)^{-1} \tag{14.6.39}$$

$$X_s = \hat{A}X_s\hat{A}^{\mathrm{T}} - \hat{K}_{ps}\left(\hat{C}X_s\hat{C}^{T} + \overline{\hat{R}}_{p\varepsilon}\right)^{-1}\hat{K}_{ps}^{T} + \hat{K}_s\hat{R}_\varepsilon\hat{K}_s^{T} \tag{14.6.40}$$

14.6.2.6 Stochastically Optimal Tonal Diffuse Field (SOTDF) Virtual Sensing Method

The stochastically optimal tonal diffuse field virtual sensing method generates stochastically optimal virtual microphones and pressure gradient sensors specifically for use in pure tone diffuse sound fields (Moreau et al., 2007, 2009b). Like the forward difference extrapolation technique, this virtual sensing method does not require a preliminary identification stage nor models of the complex transfer functions between the error sensors and the sources.

In this section, the primary and control acoustic fields are considered diffuse and a different notation set will be adopted for convenience. The sound pressure at a position $\boldsymbol{x}$ in a single diffuse acoustic field is denoted $p_i(\boldsymbol{x})$. The quantity $g_i(\boldsymbol{x})$ denotes the x-axis component of the pressure gradient. In this section, the subscript i refers to a single diffuse acoustic field, whereas a lack of subscript indicates the total acoustic field produced by superposition of the primary and control diffuse acoustic fields.

For a displacement vector $\boldsymbol{r}$, the following functions are defined:

$$A(\boldsymbol{r}) = sinc(k|\boldsymbol{r}|) \tag{14.6.41}$$

$$B(\boldsymbol{r}) = \frac{\partial A(\boldsymbol{r})}{\partial r_x} = -k\left(\frac{sinc(k|\boldsymbol{r}|)-\cos(k|\boldsymbol{r}|)}{k|\boldsymbol{r}|}\right)\left(\frac{r_x}{|\boldsymbol{r}|}\right) \tag{14.6.42a,b}$$

$$C(\boldsymbol{r}) = \frac{\partial^2 A(\boldsymbol{r})}{\partial r_x^2} = -k^2\left[sinc(k|\boldsymbol{r}|)\left(\frac{r_x}{|\boldsymbol{r}|}\right)^2 + \left(\frac{sinc(k|\boldsymbol{r}|) - \cos(k|\boldsymbol{r}|)}{k|\boldsymbol{r}|}\right)\left(1 - 3\left(\frac{r_x}{|\boldsymbol{r}|}\right)^2\right)\right] \tag{14.6.43a,b}$$

The quantity $\boldsymbol{r} = r_x\boldsymbol{i} + r_y\boldsymbol{j} + r_z\boldsymbol{k}$ where $\boldsymbol{i}$, $\boldsymbol{j}$ and $\boldsymbol{k}$ are unit vectors in the x-, y- and z-directions respectively, and k is the wavenumber. The correlations between the sound pressures and pressure gradients at two different points $\boldsymbol{x}_j$ and $\boldsymbol{x}_k$, separated by $\boldsymbol{r}$ are given by (Elliott and Garcia-Bonito, 1995):

$$\langle p_i(\boldsymbol{x}_j)p_i^*(\boldsymbol{x}_k)\rangle = A(\boldsymbol{r})\langle|p_i|^2\rangle \tag{14.6.44}$$

$$\langle p_i(\boldsymbol{x}_j)g_i^*(\boldsymbol{x}_k)\rangle = -B(\boldsymbol{r})\langle|p_i|^2\rangle \tag{14.6.45}$$

$$\langle g_i(\boldsymbol{x}_j)p_i^*(\boldsymbol{x}_k)\rangle = B(\boldsymbol{r})\langle|p_i|^2\rangle \tag{14.6.46}$$

$$\langle g_i(\boldsymbol{x}_j)g_i^*(\boldsymbol{x}_k)\rangle = -C(\boldsymbol{r})\langle|p_i|^2\rangle \tag{14.6.47}$$

where $\langle\cdot\rangle$ denotes spatial averaging and * indicates complex conjugation. In the case that $\boldsymbol{x}_j$ and $\boldsymbol{x}_k$ are the same point, the limits of $A(\boldsymbol{r})$, $B(\boldsymbol{r})$ and $C(\boldsymbol{r})$ as $\boldsymbol{r} \to 0$ must be taken. If there are N_e sensors in the field, each measuring sound pressure or pressure gradient, then p is defined as an $N_e \times 1$ matrix, $\boldsymbol{p}$, whose elements are the relevant sound pressures and pressure gradients measured by the sensors. The sound pressure and pressure gradient at any point can be expressed as the weighted sum of the N_e components, each of which are perfectly correlated with a corresponding element of $\boldsymbol{p}$ and a component which is perfectly uncorrelated with each of the elements. Therefore, for each position $\boldsymbol{x}$, $p(\boldsymbol{x})$ and $g(\boldsymbol{x})$ can be written as:

$$p(\boldsymbol{x}) = \boldsymbol{H}_p(\boldsymbol{x})\boldsymbol{p} + p_u(\boldsymbol{x}) \tag{14.6.48}$$

$$g(\boldsymbol{x}) = \boldsymbol{H}_g(\boldsymbol{x})\boldsymbol{p} + g_u(\boldsymbol{x}) \tag{14.6.49}$$

where $\boldsymbol{H}_p(\boldsymbol{x})$ and $\boldsymbol{H}_g(\boldsymbol{x})$ are matrices of weights which are functions of the position $\boldsymbol{x}$ only, and $p_u(\boldsymbol{x})$ and $g_u(\boldsymbol{x})$ are perfectly uncorrelated with the elements of $\boldsymbol{p}$. It can be shown, by post-multiplying the expressions for $p(\boldsymbol{x})$ and $g(\boldsymbol{x})$ by $\boldsymbol{p}^{\mathrm{H}}$ and spatially averaging, that:

$$\boldsymbol{H}_p(\boldsymbol{x}) = \boldsymbol{L}_p(\boldsymbol{x})\boldsymbol{M}^{-1} \tag{14.6.50}$$

$$\boldsymbol{H}_g(\boldsymbol{x}) = \boldsymbol{L}_g(\boldsymbol{x})\boldsymbol{M}^{-1} \tag{14.6.51}$$

where

$$\boldsymbol{L}_p(\boldsymbol{x}) = \frac{\langle p_i(\boldsymbol{x})\,\boldsymbol{p}_i^{\mathrm{H}}\rangle}{\langle |p_i|^2\rangle} \tag{14.6.52}$$

$$\boldsymbol{M} = \frac{\langle \boldsymbol{p}_i\boldsymbol{p}_i^{\mathrm{H}}\rangle}{\langle |p_i|^2\rangle} \tag{14.6.53}$$

$$\boldsymbol{L}_g(\boldsymbol{x}) = \frac{\langle g_i(\boldsymbol{x})\boldsymbol{p}_i^{\mathrm{H}}\rangle}{\langle |p_i|^2\rangle} \tag{14.6.54}$$

The aim here is to estimate the sound pressure and pressure gradient at a virtual location. In order to do this, $p(\boldsymbol{x})$ and $g(\boldsymbol{x})$ must be estimated from the known quantities in $\boldsymbol{p}$. The sound pressure and pressure gradient at any point $\boldsymbol{x}$ are given by Equations (14.6.48) and (14.6.49). If only the measured quantities in $\boldsymbol{p}$ are known, then the best possible estimates of $p_u(\boldsymbol{x})$ and $g_u(\boldsymbol{x})$ are zero, since they are perfectly uncorrelated with the measured signals. Therefore, the best estimates of sound pressure and pressure gradient at any point $\boldsymbol{x}$ are given by:

$$\hat{p}(\boldsymbol{x}) = \boldsymbol{H}_p(\boldsymbol{x})\boldsymbol{p} \tag{14.6.55}$$

$$\hat{g}(\boldsymbol{x}) = \boldsymbol{H}_g(\boldsymbol{x})\boldsymbol{p} \tag{14.6.56}$$

Therefore, in a diffuse sound field, the sound pressure and pressure gradient at a virtual location can be estimated using Equations (14.6.55) and (14.6.56). This requires use of the matrix $\boldsymbol{p}$ whose elements are the relevant sound pressures and pressure gradients measured by the sensors, and calculation of the weight matrices $\boldsymbol{H}_p(\boldsymbol{x})$ and $\boldsymbol{H}_g(\boldsymbol{x})$ using matrice, $\boldsymbol{L}_p(\boldsymbol{x})$, $\boldsymbol{L}_g(\boldsymbol{x})$ and $\boldsymbol{M}$, defined in Equations (14.6.52) to (14.6.54).

As the distance between the locations of the physical and virtual sensors increases, the estimates of the virtual quantities approach zero. This is because the virtual and measured quantities become uncorrelated as this distance increases. If none of the distances between the virtual location and the physical sensors are small, then the sound pressure and pressure gradient at the virtual location will be uncorrelated with the measured quantities, and the best estimate of the sound pressure and pressure gradient at the virtual location will be close to zero.

In a pure tone diffuse sound field, a perfect estimate of the sound pressure at the virtual location may be obtained with the deterministic remote microphone technique (Roure and Albarrazin, 1999) provided that an accurate measurement of the transfer functions occurs in the preliminary identification stage. Although greater control can be achieved with the remote microphone technique, the stochastically optimal tonal diffuse field virtual sensing technique is much simpler to implement because it is a fixed scalar weighting method requiring only sensor position information. Unlike the remote microphone technique, this virtual sensing method is independent of the source or sensor locations within the sound field. The weight functions only need to be updated if the geometric arrangement of physical and virtual locations change with respect to each other.

The performance of the stochastically optimal tonal diffuse field virtual sensing method in generating a zone of quiet, centred at a virtual sensor located a distance of 0.1λ from the physical sensor array, has been investigated theoretically and using experimentally measured data (Moreau et al., 2007, Moreau et al., 2009b). Control at a virtual microphone, using the measured sound pressure and pressure gradient at a point, achieved a maximum attenuation of 24 dB at the virtual location and generated a 10 dB zone of quiet with a diameter of $\lambda/10$. This is the same size zone of quiet as that achieved by Elliott et al. (1988), when minimising the measured sound pressure at the physical sensor location with a single control source. Similar control performance was obtained using two closely spaced physical microphones to estimate the sound pressure at the virtual location. Minimising the sound pressure and pressure gradient at a virtual location with two control sources, using the measured sound pressures and pressure gradients at two points, achieved a maximum attenuation of 45 dB and extended the zone of quiet to a diameter of $\lambda/2$. This is the same size zone of quiet as that achieved by Elliot and Garcia-Bonito (1995), when minimising the measured sound pressure and pressure gradient with two control sources. A similar control performance was obtained using physical microphones at four closely spaced points to estimate the sound pressure and pressure gradient at the virtual location.

14.6.3 Moving Virtual Sensing Algorithms

As the virtual location may not be spatially fixed, moving virtual sensing algorithms have been developed in recent years. These moving virtual sensing algorithms estimate the error signals at a number of virtual locations that move through the sound field.

14.6.3.1 Remote Moving Microphone Technique

The remote moving microphone technique (Petersen et al., 2006) uses the remote microphone technique (Roure and Albarrazin, 1999) to obtain estimates of the virtual error signals at the moving virtual locations. In this section it is assumed that there are N_c control sources, N_e physical sensors and N_v moving virtual sensors. The time-variant locations of the N_v moving virtual microphones are contained in matrix $\boldsymbol{x}_v(k)$, of size $3 \times N_v$, defined as:

$$\boldsymbol{x}_v(k) = \begin{bmatrix} \boldsymbol{x}_{v1}(k) & \boldsymbol{x}_{v2}(k) & \dots & \boldsymbol{x}_{vN_v}(k) \end{bmatrix} \tag{14.6.57}$$

where each of the moving virtual locations, $\boldsymbol{x}_{vn_v}(k)$ are defined by three spatial co-ordinates with respect to a reference frame and are given by:

$$\boldsymbol{x}_{vn_v}(k) = \begin{bmatrix} x_{vn_v}(k) & y_{vn_v}(k) & z_{vn_v}(k) \end{bmatrix}^{\mathrm{T}} \tag{14.6.58}$$

It is assumed here that the N_v moving virtual locations $\boldsymbol{x}_v(k)$ are known at every time step. In practice, the moving virtual locations could be measured using a 3D head tracking system based on camera vision or on ultrasonic position sensing (Petersen, 2007).

The remote moving microphone technique is used to compute estimates of the virtual error signals $\hat{e}_v(k)$ at the moving virtual locations $x_v(k)$. A block diagram of the remote moving virtual sensing algorithm is given in Figure 14.15.

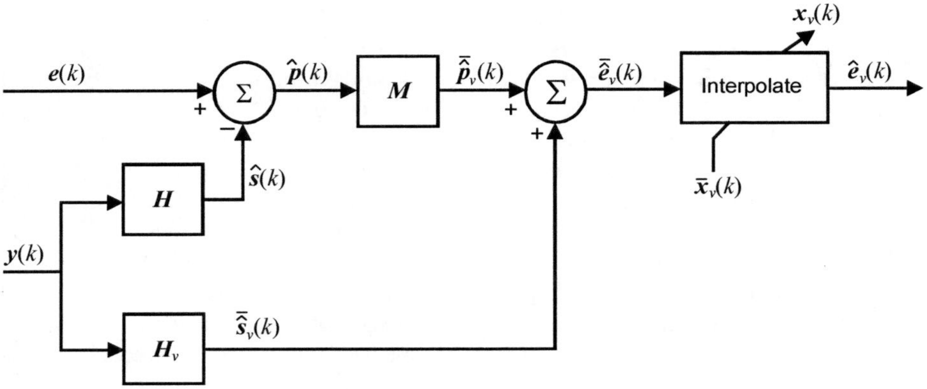

Figure 14.15 The remote moving microphone technique.

In this moving virtual sensing algorithm, the remote microphone technique is first used to obtain estimates of the virtual error signals $\bar{\hat{e}}_v(k)$ at $\bar{N}_v$ spatially fixed virtual microphone locations $\bar{x}_v$. It is assumed here that the moving virtual locations $x_v(k)$ are confined to a three-dimensional region and that the spatially fixed virtual microphone locations $\bar{x}_v$ are therefore located within this region. The vector of the $\bar{N}_v$ spatially fixed virtual microphone locations is given by:

$$\bar{x}_v = \left[\bar{x}_{v1} \;\; \bar{x}_{v2} \;\; \cdots \;\; \bar{x}_{v\bar{N}_v}\right] \tag{14.6.59}$$

where each of the spatially fixed virtual locations $\bar{x}_{v\bar{n}_v}$ are defined by three spatial co-ordinates with respect to a reference frame and are given by:

$$\bar{x}_{v\bar{n}_v} = \left[\bar{x}_{v\bar{n}_v} \;\; \bar{y}_{v\bar{n}_v} \;\; \bar{z}_{v\bar{n}_v}\right]^{\mathrm{T}} \tag{14.6.60}$$

The virtual error signals $\bar{\hat{e}}_v(k)$ at the spatially fixed virtual locations $\bar{x}_v$ are calculated using the remote microphone technique as described in Section 14.6.2.2. The remote microphone technique requires a preliminary identification stage in which the cancellation path transfer matrices $\boldsymbol{H}$, of size $N_e \times N_c$, and $\boldsymbol{H}_v$ of size $\bar{N}_v \times N_c$, are modelled as matrices of FIR or IIR filters. The $\bar{N}_v \times N_e$ size matrix, $\boldsymbol{M}$, of primary transfer functions between the spatially fixed virtual sensor locations and the physical sensor locations is also estimated as a matrix of FIR or IIR filters during this preliminary identification stage.

Estimates of the primary disturbances, $\hat{p}(k)$, at the N_e physical error sensors are first calculated using:

$$\hat{p}(k) = e(k) - \hat{s}(k) = e(k) - \boldsymbol{H}y(k) \tag{14.6.61a,b}$$

Next, estimates of the primary disturbances $\bar{\hat{p}}_v(k)$ at the spatially fixed virtual locations $\bar{x}_v$ are found to be:

$$\bar{\hat{p}}_v(k) = \boldsymbol{M}\hat{p}(k) \tag{14.6.62}$$

Finally, estimates of the total virtual error signals, $\bar{\hat{e}}_v(k)$, at the spatially fixed virtual locations $\bar{x}_v$ are calculated as:

$$\bar{\hat{e}}_v(k) = \bar{\hat{p}}_v(k) + \bar{\hat{s}}_v(k) = \boldsymbol{M}\boldsymbol{e}(k) + \left(\boldsymbol{H}_v - \boldsymbol{M}\boldsymbol{H}\right)\boldsymbol{y}(k) \qquad (14.6.63a,b)$$

As shown in Figure 14.15, estimates $\hat{e}_v(k)$ of the virtual error signals at the moving virtual locations $x_v(k)$ are now obtained by spatially interpolating the virtual error signals $\bar{\hat{e}}_v(k)$ at the spatially fixed virtual locations $\bar{x}_v$.

The performance of the remote moving microphone technique has been experimentally investigated in an acoustic duct, at an acoustic resonance (Petersen et al., 2006). In the acoustic duct, the virtual microphone moved sinusoidally between a virtual distance of $x_v = 0.02$ m and 0.12 m with a period of 10 s. Minimising the moving virtual error signal using a feedforward control approach achieved greater than 34 dB of attenuation at the moving virtual location. This is 20 dB of attenuation greater than that achieved by minimising the error signal at a fixed physical microphone at $x_v = 0$ or a fixed virtual microphone at $x_v = 0.02$ m. Moreau et al. (2008) then extended the remote moving virtual microphone technique to generate a virtual microphone capable of tracking the ear of a rotating artificial head inside a three-dimensional cavity. For head rotations of $\pm 45^\circ$ with a period of 10 s, between 30 dB and 40 dB of attenuation was experimentally achieved at the ear of the rotating artificial head at an acoustic resonance.

14.6.3.2 Adaptive LMS Moving Virtual Microphone Technique

The adaptive LMS moving virtual microphone technique (Petersen et al., 2007) uses the adaptive LMS virtual microphone technique (Cazzolato, 2002) to obtain estimates of the virtual error signals at the moving virtual locations. The adaptive LMS moving virtual microphone technique is used to compute estimates of the virtual error signals, $\hat{e}_v(k)$, at the moving virtual locations $x_v(k)$. A block diagram of the adaptive LMS moving virtual microphone technique is shown in Figure 14.16.

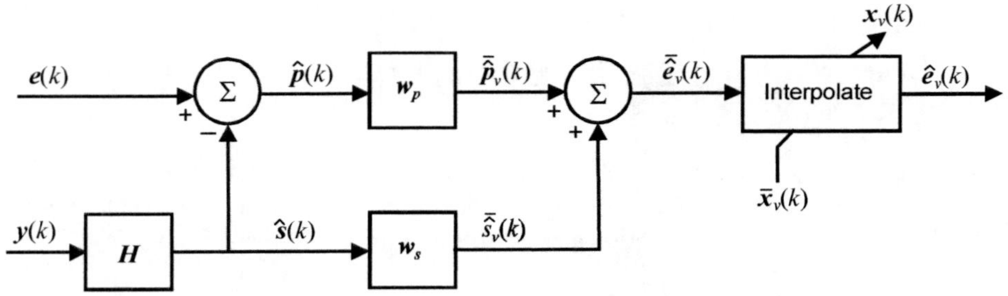

Figure 14.16 The adaptive LMS moving virtual microphone technique.

In this moving virtual sensing algorithm, the adaptive LMS virtual microphone technique, as described in Section 14.6.2.4, is first used to obtain estimates of the virtual error signals, $\bar{\hat{e}}_v(k)$, at the spatially fixed virtual locations $\bar{x}_v$. As shown in Figure 14.16, the primary component of the physical error signals is first calculated using the matrix of physical cancellation path transfer functions, $\boldsymbol{H}$, and is given as:

$$\hat{p}(k) = e(k) - \hat{s}(k) = e(k) - \boldsymbol{H}y(k) \qquad (14.6.64a,b)$$

Matrices of the primary and control signal weights, $\overline{w}_p$ and $\overline{w}_s$, of size $N_e \times \overline{N}_v$, at the $\overline{N}_v$ spatially fixed virtual locations $\overline{x}_v$ are then estimated separately using Equation (14.6.17). Estimates of the virtual error signals, $\hat{\overline{e}}_v(k)$, at the spatially fixed virtual locations $\overline{x}_v$ can then be calculated as:

$$\hat{\overline{e}}_v(k) = \hat{\overline{p}}_v(k) - \hat{\overline{s}}_v(k) = \overline{w}_p^{\mathrm{T}} \hat{p}(k) - \overline{w}_s^{\mathrm{T}} \hat{s}(k) \tag{14.6.65a,b}$$

As shown in Figure 14.16, estimates $\hat{e}_v(k)$ of the virtual error signals at the moving virtual locations $x_v(k)$ are now obtained by spatially interpolating the virtual error signals $\hat{\overline{e}}_v(k)$ at the spatially fixed virtual locations $\overline{x}_v$.

The performance of the adaptive LMS moving virtual microphone technique has also been experimentally investigated in an acoustic duct at an acoustic resonance (Petersen et al., 2007). Again, the virtual microphone was moved sinusoidally between a virtual distance of x_v = 0.02 m and 0.12 m with a period of 10 s. Experimental results demonstrated that minimising the moving virtual error signal using a feedforward control approach achieves an additional 18 dB of attenuation at the moving virtual location compared to minimising the error signal at a fixed physical microphone at x_v = 0 or a fixed virtual microphone at x_v = 0.02 m.

14.6.3.3 Kalman Filtering Moving Virtual Sensing Method

The Kalman filtering moving virtual sensing method (Petersen, 2007) uses Kalman filtering theory to obtain estimates of the virtual error signals at the moving virtual locations. The Kalman filtering virtual microphone method, as described in Section 14.6.2.5, is first used to obtain estimates of the virtual error signals, $\hat{\overline{e}}_v(k)$, at the spatially fixed virtual locations $\overline{x}_v$. A state-space realisation of the Kalman filtering virtual sensing algorithm, which estimates the virtual error signals $\hat{\overline{e}}_v(k)$, given measurements of the physical error signals $e(i)$ up to $i = k$, is as follows (Petersen, 2007):

$$\begin{bmatrix} \hat{z}(k+1) \\ \hat{\overline{e}}_v(k) \end{bmatrix} = \begin{bmatrix} A - K_{ps}C_p & B_s - K_{ps}D_{ps} & K_{ps} \\ \overline{C}_v - \overline{M}_{vs}C_p & \overline{D}_{vs} - \overline{M}_{vs}D_{ps} & \overline{M}_{vs} \end{bmatrix} \begin{bmatrix} \hat{z}(k-1) \\ s(k) \\ e(k) \end{bmatrix} \tag{14.6.66}$$

where $\overline{C}_v$ and $\overline{D}_{vs}$ are the state-space matrices of the virtual cancellation path matrix H_v at the spatially fixed virtual locations $\overline{x}_v$. The Kalman gain matrix K_{ps} can be found using Equation (14.6.30), and the innovation gain matrix $\overline{M}_{vs}$ of size $\overline{N}_v \times N_e$, is given by:

$$\overline{M}_{vs} = \left(\overline{C}_v P_{ps} C_p^T + \overline{R}_{pv}^{-1} \right) R_{p\varepsilon}^{-1} \tag{14.6.67}$$

with $P_{ps} = P_{ps}^T$; the unique solution to the discrete algebraic Riccati equation is given in Equation (14.6.32). The covariance matrix $\overline{R}_{pv}$ between the auxiliary measurement noises on the physical sensors and virtual sensors, spatially fixed at $\overline{x}_v$, is defined in Equation (14.6.27).

Estimates $\hat{e}_v(k)$ of the virtual error signals at the moving virtual locations $x_v(k)$ are now obtained by spatially interpolating the virtual error signals $\hat{\overline{e}}_v(k)$ at the spatially fixed virtual locations $\overline{x}_v$.

The performance of the Kalman filtering moving virtual sensing method has also been experimentally investigated in an acoustic duct at an acoustic resonance (Petersen, 2007). Again, the virtual microphone moved sinusoidally between a virtual distance of $\boldsymbol{x}_v = 0.02$ m and 0.12 m with a period of 10 s. Experimental results demonstrated that minimising the moving virtual error signal using a feedforward control approach achieves an additional 14 dB of attenuation at the moving virtual location compared to minimising the error signal at a fixed physical microphone at $\boldsymbol{x}_v = 0$ or a fixed virtual microphone at $\boldsymbol{x}_v = 0.02$ m. While this moving virtual sensing algorithm achieves significant attenuation at the moving virtual location, it is limited to use in systems of relatively low order such as an acoustic duct system.

14.6.3.4 Stochastically Optimal Tonal Diffuse Field (SOTDF) Moving Virtual Sensing Method

The SOTDF moving virtual sensing method (Moreau et al., 2009a) generates a stochastically optimal moving virtual microphone that tracks a three-dimensional trajectory in a three-dimensional sound field. The SOTDF moving virtual sensing method employs the SOTDF virtual sensing method (Moreau et al., 2009b) to obtain an estimate of the moving virtual error signal. The SOTDF virtual sensing method has been presented in Section 14.6.2.6.

As stated in Section 14.6.2.6, the SOTDF virtual sensing method calculates a stochastically optimal estimate of the virtual error signal at a spatially fixed virtual location using diffuse field theory. In the derivation of this algorithm, the primary acoustic field is considered diffuse, and the sound field contributions due to each of the control sources are modelled as uncorrelated single diffuse acoustic fields. It is assumed that there are N_e sensors in the field, each measuring the sound pressure or pressure gradient and $\boldsymbol{p}$ is defined as an $N_e \times 1$ matrix whose elements are the relevant sound pressures and pressure gradients measured by the sensors.

Equation (14.6.55) states that the best estimate of the sound pressure at a fixed point $\boldsymbol{x}$ is given by:

$$\hat{p}(\boldsymbol{x}) = \boldsymbol{H}_p(\boldsymbol{x})\boldsymbol{p} \tag{14.6.68}$$

where the weight matrix $\boldsymbol{H}_p(\boldsymbol{x})$ is defined in Equation (14.6.49).

It follows that the best estimate of the sound pressure at a single moving virtual location $\boldsymbol{x}_{v1}(k)$ is given by:

$$\hat{p}(\boldsymbol{x}_{v1}(k)) = \boldsymbol{H}_p(\boldsymbol{x}_{v1}(k))\boldsymbol{p} \tag{14.6.69}$$

Moreau et al. (2009a) used the SOTDF moving virtual sensing method to generate a virtual microphone capable of tracking the ear of a rotating artificial head inside a three-dimensional cavity. Three physical microphones were arranged on the corners of an isosceles triangle around the head. For head rotations of $\pm 45^\circ$ with a period of 10 s, between 20 and 28 dB of attenuation was experimentally achieved at the ear of the rotating artificial head at an acoustic resonance.

It is worth noting that while greater control can be achieved at the moving virtual location with the deterministic remote moving microphone technique, the SOTDF moving virtual sensing method is much simpler to implement, as it is a fixed weighting technique requiring only sensor position information. This also means that, unlike the remote moving microphone technique, the SOTDF moving virtual sensing method is robust to changes in the sound field that may alter the transfer functions between the error sensors and the control sources.

REFERENCES

Baxandall, P.J. (2001). Electrostatic Loudspeakers. In *Loudspeaker and Headphone Handbook.* Butterworths, London.

Bellin, J.L.S. and Beyer, R.T. (1960). Scattering of sound by sound. *Journal of the Acoustical Society of America*, **32**, 339–341.

Benson, J.E. (1969). *Theory and Design of Loudspeaker Enclosures*. Butterworth–Heinemann Ltd., Indianapolis.

Berktay, H. O. (1965). Possible exploitation of nonlinear acoustics in underwater transmitting applications. *Journal of Sound and Vibration*, **2**, 435–461.

Bies, D.A. and Hansen, C.H. (2009). *Engineering Noise Control*, 4th ed., Spon Press, London.

Bilaniuk, N. (1997). Optical microphone transduction techniques. *Applied Acoustics* **50(1)** 35–63.

Bolleter, U., Crocker, M.J. and Baade, P. (1970). Tubular microphone windscreen for in-duct fan sound power measurements. In *Proceedings of the 80th Meeting of the Acoustical Society of America*.

Bouc, R. and Felix, D. (1987). A real-time procedure for acoustic turbulence filtering in ducts. *Journal of Sound and Vibration*, **118**, 1–10.

Brooks, L.A., Zander, A.C. and Hansen, C.H. (2005). Investigation into the feasibility of using a parametric array control source in an active noise control system. In *Proceedings of Acoustics 2005*, Busselton, Western Australia, 39–45.

Cazzolato, B.S. (1999). *Sensing Systems for Active Control of Sound Transmission into Cavities*. PhD thesis, School of Mechanical Engineering, The University of Adelaide, SA, 5005.

Cazzolato, B.S. (2002). An adaptive LMS virtual microphone. In *Proceedings of Active 02*, Southampton, UK, 105–116,

Cazzolato, B.S., Halim, D., Petersen, D., Kahana, Y. and Kots, A. (2005). An optical 3D intensity and energy density probe. In *Proceedings of Acoustics 2005*, Busselton, Western Australia.

Chambers, J., Bullock, D., Kahana, Y., Kots, A. and Palmer, A. (2007). Developments in active noise control sound systems for magnetic resonance imaging. *Applied Acoustics*, **68**, 281–295.

Christensen, J.J. and Hald, J. (2004). *Beamforming*. Technical Review 1–2004. Brüel and Kjær, Copenhagen.

Dean, L.W. (1962). Interactions between sound waves. *Journal of the Acoustical Society of America*, **34**, 1039-1044.

Davy, J.L. and Dunn, I.P. (1993). The development of a flush mounted microphone turbulence screen for use in a power station chimney flue. *Noise Control Engineering Journal*, **41**, 313–322.

Diaz, J., Egana, J. and Vinolas, J. (2006). A local active noise control system based on a virtual-microphone technique for railway sleeping vehicle applications. *Mechanical Systems and Signal Processing* **20**, 2259–2276.

Elliott, S. and David, A. (1992). A virtual microphone arrangement for local active sound control. In *Proceedings of the 1st International Conference on Motion and Vibration Control*, 1027–1031.

Elliott, S. and Garcia-Bonito, J. (1995). Active cancellation of pressure and pressure gradient in a diffuse sound field. *Journal of Sound and Vibration*, **186**, 696–704.

Elliott, S., Joseph, P., Bullmore, A. and Nelson, P. (1988). Active cancellation at a point in a pure tone diffuse sound field. *Journal of Sound and Vibration*, **120**, 183–189.

Flanagan, J.L., Berkley, D.A., Elko, G.W., West, J.E. and Sondhi, M.M. (1991). Autodirective microphone systems. *Acustica*, **73**, 58–71.

Ford, R.D. (1984). Power requirements for active noise control in ducts. *Journal of Sound and Vibration*, **92**, 411–417.

Frederiksen, E., Eirby, N. and Mathiasen, H. (1979). *Technical Review*, No. 4, Brüel and Kjær, Copenhagen.

Garcia-Bonito, J. and Elliott, S. (1995). Strategies for local active control in diffuse sound fields. In *Proceedings of Active 95*, 561–572, Newport Beach, CA.

Garcia-Bonito, J., Elliott, S. and Boucher, C. (1996). A virtual microphone arrangement in a practical active headrest. In *Proceedings of Inter-Noise 96*, 1115–1120.

Garcia-Bonito, J., Elliott, S. and Boucher, C. (1997). Generation of zones of quiet using a virtual microphone arrangement. *Journal of the Acoustical Society of America*, **101**, 3498–3516.

Glendinning, A.G., Elliott, S.J. and Nelson, P.A. (1988). *A High Intensity Acoustic Source for Active Attenuation of Exhaust Noise*. ISVR Technical Report No. 156.

Hansen, C.H. (2009). Adventures in active noise control. In *Proceedings of EuroNoise 2009*, Edinburgh, 26–28 October.

Haverkamp, B. (2001). *State Space Identification: Theory and Practice*. PhD thesis, System and Control Engineering Group, Delft University of Technology.

Heydt, R., Kornbluh, R., Pelrine, R. and Mason, V. (1998) Design and performance of an electrostrictive-polymer-film acoustic actuator. *Journal of Sound and Vibration*, **215**, 297–311.

Heydt, R., Pelrine, R., Joseph, J., Eckerle, J. and Kornbluh, R. (2000). Acoustical performance of an electrostrictive polymer film loudspeaker. *Journal of the Acoustical Society of America*, **107**, 833–839.

Holland, K.R. (2001). Principles of sound radiation. In *Loudspeaker and Headphone Handbook.* Butterworths, London

Holmberg, U., Ramner, N. and Slovak, R. (2002). Low complexity robust control of a headrest system based on virtual microphones and the internal model principle. In *Proceedings of Active 02*, Southampton, UK, 1243–1250.

Hoops, R.H. and Eriksson, L.J. (1991). Rigid Foraminous Microphone Probe for Acoustic Measurement in Turbulent Flow. US Patent 4 903 249.

Horihata, S., Matsuoka, S., Kitagawa, H. and Ishimitsu, S. (1997). Active noise control by means of virtual error microphone system. In *Proceedings of Inter-Noise 97*, Budapest, Hungary, 529–532.

Hunt, F.V. (1954). *Electroacoustics*. John Wiley & Sons, New York.

Ingard, U. and Pridmore-Brown, D.C. (1956). Scattering of sound by sound. *Journal of the Acoustical Society of America*, **28**, 367–369.

Johnson, D.H. and Dudgeon, D.E. (1993). *Array Signal Processing: Concepts and Techniques.* Prentice Hall, Englewood Cliffs, NJ.

Kahana, Y., Paritsky, A., Kots, A. and Mican, S. (2003). Recent advances in optical microphone technology. In *Proceedings of Inter-Noise 2003*, Seogwipo, Korea.

Kahana, Y., Kots, A., Mican, S., Chambers, J. and Bullock, D. (2004). Optoacoustical ear defenders with active noise reduction in MRI communication systems. In *Proceedings of Active 04*, Williamsburg, VA.

Keele, Jr., D.B. (1981). Direct low-frequency driver synthesis from system specifications. In *Proceedings of the 69th Convention of the Audio Engineering Society*, Los Angeles, CA.

Kestell, C., Hansen, C.H. and Cazzolato, B.S. (2000). Active noise control with virtual sensors in a long narrow duct. *International Journal of Acoustics and Vibration*, **5**, 1–14.

Kestell, C., Hansen, C.H. and Cazzolato, B.S. (2001). Active noise control in a free field with virtual error sensors. *Journal of the Acoustical Society of America* **109**, 232–243.

Kida, M., Hirayama, R., Kajikawa, Y., Tani, T. and Kurumi, Y. (2009). Head-mounted active noise control system for MR noise. In *Proceedings of the IEEE Conference on Acoustics, Speech and Signal Processing,* 245–248.

Kidner, M.R.F., Zander, A.C. and Hansen, C.H. (2006). Active control of sound using a parametric array. In *Proceedings of Active 2006*, Adelaide, Australia.

Klippel, W. (1999). Active reduction of nonlinear loudspeaker distortion. In *Proceedings of Active 99*, Ottawa, Canada.

Kots, A. and Paritsky, A. (1999). Fibre optic microphone for harsh environment. In *Proceedings of the International Society for Optical Engineering.* Boston, MA.

Kuo, S.M. and Morgan, D.R. (1996). *Active Noise Control Systems*. John Wiley & Sons, New York.

Larsson, M., Johansson, S., Claesson, I. and Hakansson, L. (2009). A module-based active noise control system for ventilation systems. Part II. Performance evaluation. *International Journal of Acoustics and Vibration*, **14**, 196–206.

Larsson, M., Johansson, L., Hakansson, L. and Claesson, I. (2005). Microphone windscreens for turbulent noise suppression when applying active noise control to ducts. In *Proceedings of the 12th International Congress on Sound and Vibration*, Lisbon.

Leventhall, H.G. (1988). Problems of transducers for active attenuation of noise. In *Proceedings of Inter-Noise '88*. Institute of Noise Control Engineering, 1091–1094.

Matuoka, S., Kitagawa, H., Horihata, S., Ishimitu, S. and Tamura, F. (1996). Active noise control using a virtual error microphone system. *Transactions of the Japan Society of Mechanical Engineers*, **62**, 3459–3464.

Moreau, D.J., Ghan, J., Cazzolato, B.S. and Zander, A.C. (2007). Active noise control with a virtual acoustic sensor in a pure-tone diffuse sound field. In *Proceedings of the 14th International Congress on Sound and Vibration*, Cairns, Australia.

Moreau, D.J., Cazzolato, B.S. and Zander, A.C. (2008). Active noise control at a moving location in a modally dense three-dimensional sound field using virtual sensing. In *Proceedings of Acoustics 08*, Paris, France.

Moreau, D.J., Cazzolato, B.S. and Zander, A.C. (2009a). Active noise control at a moving virtual microphone using the SOTDF moving virtual sensing method. In *Proceedings of Acoustics 2009: Research to Consulting*, Adelaide, Australia.

Moreau, D.J., Ghan, J., Cazzolato, B.S. and Zander, A.C. (2009b). Active noise control in a pure tone diffuse sound field using virtual sensing. *Journal of the Acoustical Society of America*, **125**, 3742–3755.

Moreau, D.J, Cazzolato, B.S. and Zander, A.C. (2010). Active noise control at a virtual acoustic energy density sensor in a three-dimensional sound field. In *Proceedings of the 20th International Congress on Acoustics (ICA 2010)*, Sydney, Australia.

Morse, P.M. (1948) *Vibration and Sound*. McGraw-Hill, New York (reprinted by the Acoustical Society of America, 1981).

Morse, P.M. (1976). *Vibration and Sound*. American Institute of Physics, New York, 281–285.

Munjal, M.L. and Eriksson, L.J. (1989). An exact one-dimensional analysis of the acoustic sensitivity of the antiturbulence probe in a duct. *Journal of the Acoustical Society of America*, **85**, 582–587.

Munn, J. (2004). *Virtual Sensors for Active Noise Control*. PhD thesis, Department of Mechanical Engineering, The University of Adelaide.

Munn, J., Kestell, C., Cazzolato, B.S. and Hansen, C.H. (2001a). Real-time feedforward active control using virtual sensors in a long narrow duct. In *Proceedings of Acoustics 2001: Noise and Vibration Policy the Way Forward*, Canberra, Australia.

Munn, J., Kestell, C., Cazzolato, B.S. and Hansen, C.H. (2001b). Real-time feedforward active noise control using virtual error sensors. In *Proceedings of the 2001 International Congress and Exhibition on Noise Control Engineering*, The Hague, The Netherlands.

Munn, J., Cazzolato, B.S., Hansen, C.H. and Kestell, C. (2002). Higher order virtual sensing for remote active noise control. In *Proceedings of Active 02*, Southampton, UK.

Munn, J., Cazzolato, B.S., Kestell, C. and Hansen, C.H. (2003) Virtual error sensing for active noise control in a one-dimensional waveguide: performance prediction versus measurement (l). *Journal of the Acoustical Society of America*, **113**, 35–38.

Nakamura, A. Sugiyama, A., Tanaka, T. and Matsumoto, R. (1971). Experimental investigation for detection of sound pressure level by a microphone in an airstream. *Journal of the Acoustical Society of America*, **50**, 40–46.

Neise, W. (1975). Theoretical and experimental investigations of microphone probes for sound measurements in turbulent flow. *Journal of Sound and Vibration*, **39**, 371–400.

Noiseux, D.U. and Horwath, T.G. (1970). Design of a porous pipe microphone for the rejection of axial flow noise. In *Proceedings of the 79th Meeting of the Acoustical Society of America*.

Paritsky, A. and Kots, A. (1999). System for Attenuation of Noise. US Patent 5 969 838.

Pawelczyk, M. (2006). Polynomial approach to design of feedback virtual-microphone active noise control system. In *Proceedings of the 13th International Congress on Sound and Vibration*, Vienna, Austria.

Petersen, C. (2007). *Optimal Spatially Fixed and Moving Virtual Sensing Algorithms for Local Active Noise Control*. PhD thesis, School of Mechanical Engineering, The University of Adelaide.

Petersen, C., Cazzolato, B.S., Zander, A.C. and Hansen, C.H. (2006). Active noise control at a moving location using virtual sensing. In *Proceedings of the 13th International Congress on Sound and Vibration*, Vienna, Austria.

Petersen, C., Zander, A.Z., Cazzolato, B.S. and Hansen, C.H. (2007). A moving zone of quiet for narrowband noise in a one-dimensional duct using virtual sensing. *Journal of the Acoustical Society of America*, **121**, 1459–1470.

Petersen, C., Fraanje, R., Cazzolato, B.S., Zander, A.C. and Hansen, C.H. (2008). A Kalman filter approach to virtual sensing for active noise control. *Mechanical Systems and Signal Processing*, **22**, 490–508.

Radcliffe, C. and Gogate, S. (1993). A model based feedforward noise control algorithm for vehicle interiors. *Advanced Automotive Technologies ASME* **52(DSC)**, 299–304.

Rafaely, B., Elliott, S. and Garcia-Bonito, J. (1999). Broadband performance of an active headrest. *Journal of the Acoustical Society of America*, **106**, 787–793.

Roure, A. and Albarrazin, A. (1999). The remote microphone technique for active noise control. In *Proceedings of Active 99*, 1233–1244.

Rutt, T.E. (1985). Root-locus technique for vented-box loudspeaker design. *Journal of the Audio Engineering Society* **33**, 659–668.

Sessler, G.M., West, J.E. and Kubli, R.A. (1989). Uni-directional, second order gradient microphone. *Journal of the Acoustical Society of America*, **86**, 2063–2066.

Shepherd, I.C., Cabelli, A. and La Fontaine, R.F. (1986). Characteristics of loudspeakers operating in an active noise attenuator. *Journal of Sound and Vibration*, **110**, 471–481.

Shepherd, I.C., La Fontaine, R.F. and Cabelli, A. (1989). The influence of turbulent pressure fluctuations on an active attenuator in a flow duct. *Journal of Sound and Vibration*, **130**, 125–135.

Small, R.H. (1972). Direct-radiator loudspeaker system analysis. *Journal of the Audio Engineering Society*, **20**, 383–395.

Small, R.H. (1973). Vented-box loudspeaker systems. Part I. Small-signal analysis. *Journal of the Audio Engineering Society*, **21**, 363–372.

Tamm, K. and Kurtze, G. (1954). Ein neuartiges mikrofon groβer Richtungs-selektivität. *Acustica*, **4**, 469–470.

Tanaka, N. and Tanaka, M. (2010). Active noise control using a steerable parametric array loudspeaker. *Journal of the Acoustical Society of America*, **127**, 3526–3537.

Tanaka, N. and Tanaka, M. (2011). Mathematically trivial control of sound using a parametric beam focusing source. *Journal of the Acoustical Society of America*, **129**, 165–172.

Thiele, A.N. (1971). Loudspeakers in vented boxes: Part I. *Journal of the Audio Engineering Society*, **19**, 382–392.

Thuras, A.L., Jenkins, R.T., and O'Neil, H.T. (1935). Extraneous frequencies generated in air carrying intense sound waves. *Journal of the Acoustical Society of America*, **6**, 173–180.

Toso, A. and Johnson, M. (2009). Harmonic distortion control of a nonlinear loudspeaker. In *Proceedings of Active 2009*, Ottawa Canada.

Walker, P.J. (1955). Wide range electrostatic loudspeakers. *Wireless World*, **61**, 208–211.

Warwick Audio Technologies Ltd, www.warwickaudiotech.com/, site last viewed 27/02/2011.

Westervelt, P.J. (1962). Parametric acoustic array. *Journal of the Acoustical Society of America*, **35**, 535–537.

Westervelt, P.J. (1957a). Scattering of sound by sound. *Journal of the Acoustical Society of America*, **29**, 199–203.

Westervelt, P.J. (1957b). Scattering of sound by sound. *Journal of the Acoustical Society of America*, **29**, 934–935.

Williamson, D.T.N. (1957). The electrostatic loudspeaker. *Journal of the Institution of Electrical Engineers*, 460–463.

CHAPTER 15

Vibration Sensors and Vibration Sources

In this chapter, vibration sensors and actuators that have been used in the past in active noise and vibration control systems are discussed. Semi-active dampers, which are of major importance in vehicle suspension design, are strictly neither sensors nor actuators. As they are discussed in detail in Section 12.3, they will not be discussed further here. This discussion begins with a description of vibration sensors, followed by a description of various vibration actuators.

Prior to discussing the various sensor types, it is useful to point out the relationships between the acceleration, velocity and displacement of a vibrating object. For single frequencies or narrow bands of noise, the displacement d, velocity v, and acceleration a are related by the frequency ω (rad s^{-1}), as $d/\omega^2 = v/\omega = a$. In terms of phase angle, velocity leads displacement by 90° and acceleration leads velocity by 90°. For narrowband or broadband signals, velocity can also be derived from acceleration measurements using electronic integrating circuits. On the other hand, deriving velocity and acceleration signals by differentiating displacement signals is generally not practical due primarily to the limited dynamic range of displacement transducers and secondarily to the cost of differentiating electronics.

15.1 ACCELEROMETERS

Vibratory motion is often sensed by using an accelerometer attached to the vibrating surface. This type of transducer may be either piezoresistive or piezoelectric.

Piezoelectric accelerometers consist of a small mass attached to a piezoelectric crystal. The inertia force due to acceleration of the mass causes stress in the crystal, which in turn produces a voltage proportional to the stress (or acceleration of the mass). The mass may be mounted to produce either compressive/tensile stress or, alternatively, shear stress in the crystal. The latter arrangement allows a smaller (and lighter) accelerometer for the same sensitivity. Three piezoelectric accelerometer types are illustrated in Figures 15.1(a), (b) and (c), and a schematic of the delta shear type is shown in Figure 15.1(d).

Another type of piezoelectric accelerometer which is much less expensive is one made using piezoelectric polymer film (polyvinylidene fluoride, or PVDF) in place of the piezoelectric crystal. As it is not possible to use the shear arrangement shown in Figure 15.1(a) for PVDF film material, the resulting accelerometer sensitivity is usually about ten times less than that achieved with a piezo crystal mounted in a shear arrangement. However, the flexibility of the PVDF film allows it to be used in a beam-type arrangement (see Figure 15.2) for greater sensitivity, at the expense of a reduced upper limiting operating frequency.

PVDF film has also been used to make rotational acceleration transducers which are now commercially available. Both linear and rotational PVDF film accelerometers are likely to be preferred in future due to their high performance-to-cost ratio, especially when ordered in large quantities.

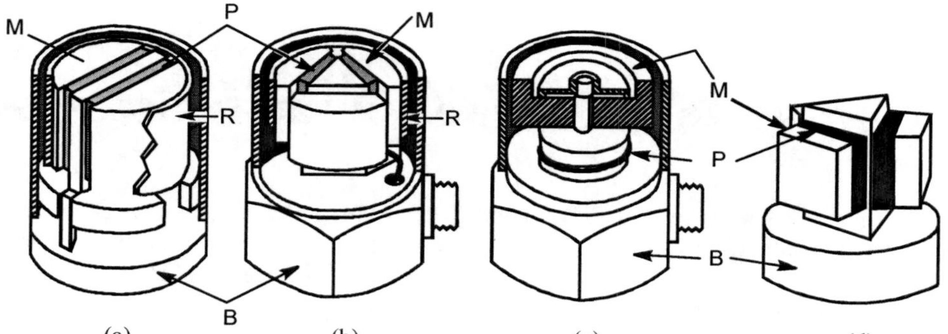

Figure 15.1 Piezoelectric accelerometer configurations: (a) planar shear type; (b) delta shear type; (c) compression type; (d) schematic of delta shear type. (Brüel) and Kjaer.)
M = seismic mass, P = piezoelectric element, R = clamping ring and B = base.

Piezoresistive accelerometers rely on the measurement of resistance change in a piezoresistive element subjected to stress. The element is generally mounted on a beam as in Figure 15.2. They are less common than piezoelectric crystal accelerometers and generally less sensitive by an order of magnitude for the same size and frequency response. They also require a stable d.c. power supply to excite the piezoresistive element (or elements). However, piezoresistive accelerometers are capable of measuring down to d.c. (or zero frequency), are easily calibrated and can be used effectively with low impedance voltage amplifiers.

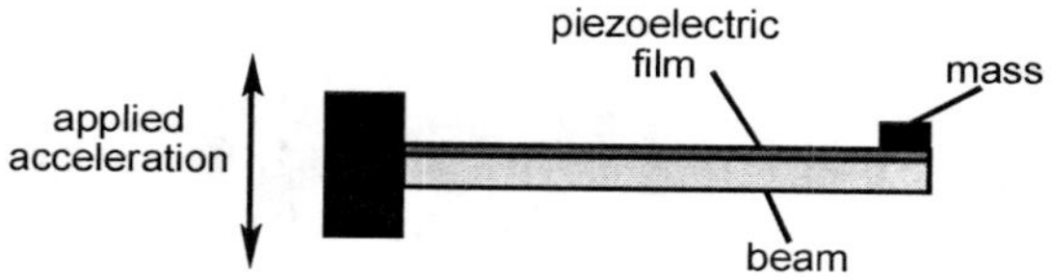

Figure 15.2 Beam-type accelerometer.

When choosing an accelerometer, some compromise must always be made between weight and sensitivity. Small accelerometers are more convenient to use, can measure higher frequencies and are less likely to affect the vibration characteristics of the structure by mass loading it. However, they have low sensitivity, which puts a lower limit on the acceleration amplitude that can be measured. Accelerometers range in weight from miniature 0.65 g for high level vibration amplitude (up to a frequency of 18 kHz) on light-weight structures, to 500 g for low level vibration measurement on heavy structures (up to a frequency of 1 kHz). Thus, prior to choosing an accelerometer, it is necessary to know approximately the range of vibration amplitudes and frequencies to be expected as well as detailed accelerometer characteristics, including the effect of various amplifier types. The latter information should be readily available from the accelerometer manufacturer.

Unlike a piezoresistive accelerometer, which may only be treated as a voltage source, a piezoelectric accelerometer may be treated as either a charge or voltage source. Thus, its sensitivity can be expressed in terms of charge or voltage per unit of acceleration (pC ms^{-2} or millivolts ms^{-2}). The piezoelectric element acts as a capacitor C_a in parallel with a very high internal leakage resistance R_a, which for practical purposes can be ignored. The element may be treated either as an ideal charge source in parallel with C_a and the cable capacitance C_c or as a voltage source V_a in series with C_a and loaded by C_c as shown in Figure 15.3.

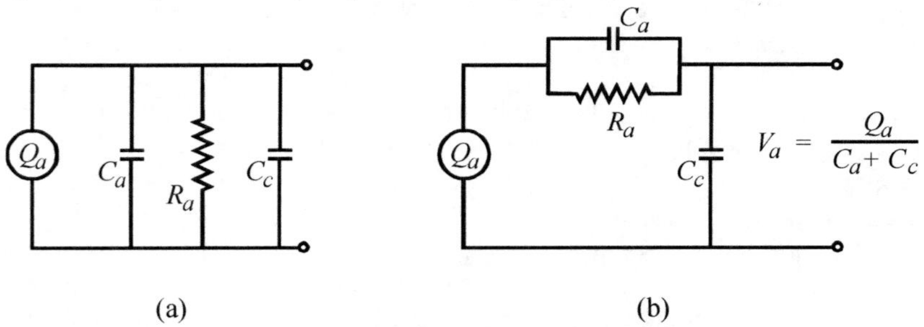

Figure 15.3 Equivalent circuits for a piezoelectric accelerometer.

Where voltage amplification is used, the sensitivity is dependent on the length of cable between the accelerometer and amplifier. Any motion of the connecting cable can also result in spurious acceleration signals. Because the voltage amplifier decreases the electrical time constant of the accelerometer, the amplifier must have a very high input impedance to measure low-frequency vibration and to not significantly load the accelerometer electrically, which would effectively reduce its sensitivity. Commercially available, high impedance voltage amplifiers allow accurate measurement down to about 20 Hz.

Alternatively, charge amplifiers (which unfortunately, are more expensive, although more commonly used due to the higher accuracy obtained for the measurements) are preferred as they have a very high input impedance and thus do not load the accelerometer output; they allow measurement of acceleration down to frequencies of 0.2 Hz; they are insensitive to cable lengths up to 500 m and they are relatively insensitive to cable movement. Many charge amplifiers also have the capability of integrating acceleration signals to produce signals proportional to velocity or displacement. Particularly at low frequencies, this facility should be used with care, as phase errors and high levels of electronic noise will be present, especially if double integration is used to obtain a displacement signal. Most accelerometers, except for miniature ones weighing less than 0.2 grams, have inbuilt charge pre-amplifiers, which avoids many of the cable induced noise problems associated with separate amplifiers.

The minimum vibration level that can be measured by an accelerometer is dependent upon its sensitivity and can be as low as 10^{-4} ms^{-2}. The maximum level is dependent upon size and can be as high as 10^{6} ms^{-2} for small shock accelerometers. Most commercially available accelerometers at least cover the range 10^{-2} to 5×10^{4} ms^{-2}. This range is then extended at one end or the other, depending upon the accelerometer type.

The transverse sensitivity of an accelerometer is its maximum sensitivity in a direction at right angles to its main axis. The maximum value is usually quoted on calibration charts and should be less than 5% of the axial sensitivity. Clearly, this can affect acceleration readings significantly if the transverse vibration amplitude at the measurement location is an order of magnitude larger than the axial amplitude. Note that the transverse sensitivity is not the same in all directions; it can vary from zero up to the maximum, depending on the direction of interest. Thus, it is possible to virtually eliminate the transverse vibration effect if the transverse vibration only occurs in one known direction.

Accelerometer base strain, due to the structure on which it is mounted undergoing strain variations, will generate vibration signals. These effects are reduced by using a shear type accelerometer and are virtually negligible for piezoresistive accelerometers.

Magnetic fields have a negligible effect on an accelerometer. The effect of intense electric fields can be minimised by using a differential pre-amplifier with two outputs from the same accelerometer such that voltages common to the two outputs are cancelled out. This arrangement is generally necessary when using accelerometers near large generators or alternators.

Accelerometers are generally lightly damped and have a single-degree-of-freedom resonance (characterised by the seismic mass and piezoelectric crystal stiffness) well above the operating frequency range. Care should be taken to ensure that high-frequency vibrations do not excite the accelerometer resonance and thus produce pre-amplifier or amplifier overloading or errors in measurements, especially when no frequency analysis is used. The effect of this resonance can be minimised by inserting a mechanical filter between the accelerometer and its mounting point. This results in loss of accuracy at lower frequencies, effectively shifting the ±3 dB error point down in frequency by a factor of five (see Figure 15.4). However, the transverse sensitivity at higher frequencies is also much reduced.

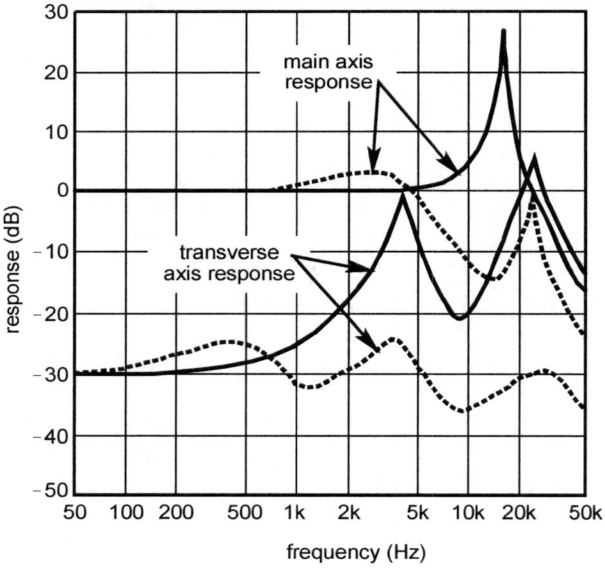

Figure 15.4 Typical frequency response of an accelerometer.

The frequency response of an accelerometer is regarded as essentially flat over the frequency range for which its electrical output is proportional to within ±5% of its mechanical input. The lower frequency limit has been discussed previously. The upper frequency limit is generally just less than one third of the resonance frequency (see Figure 15.5). The resonance frequency is dependent upon accelerometer size and may be as low as 1000 Hz or as high as 180 kHz. In general, accelerometers with higher resonance frequencies are smaller in size and less sensitive.

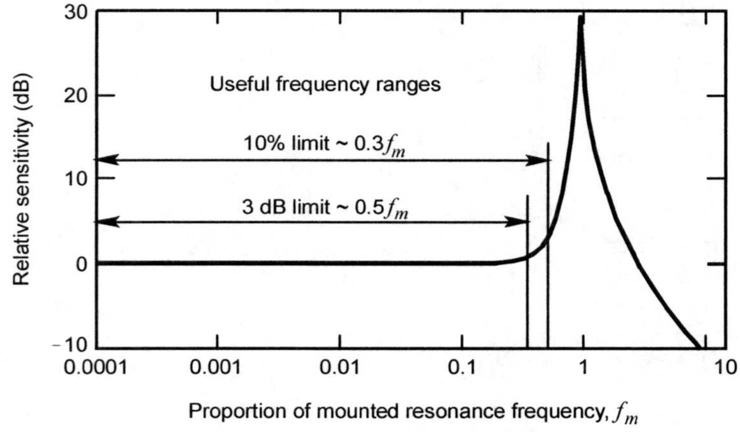

Figure 15.5 Useful frequency range of a piezoelectric accelerometer.

15.1.1 Accelerometer Mounting

There are numerous methods of attaching an accelerometer to a vibrating surface. Some are illustrated in Figure 15.6. The method used will determine the upper frequency for which measurements will be valid.

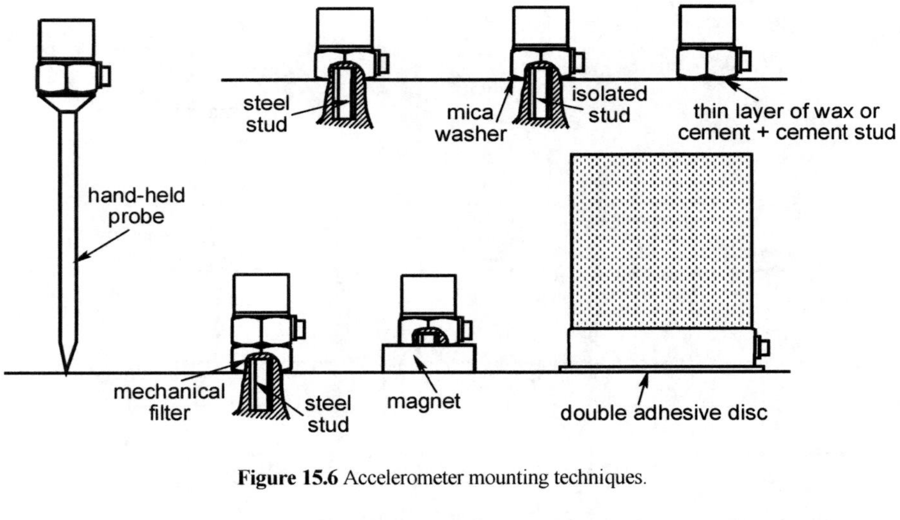

Figure 15.6 Accelerometer mounting techniques.

The effect of various mounting techniques on the frequency response of a Brüel and Kjaer general purpose accelerometer is shown in Figure 15.7. Mounting the accelerometer by bolting it to the surface to be measured with a steel stud is by far the best method and results in the resonance frequency quoted in the calibration chart. Best results are obtained if a thin layer of silicon grease is used between the accelerometer base and the structure, especially if the mounting surface is a little rough. If electrical isolation is desired, or if drilling holes in

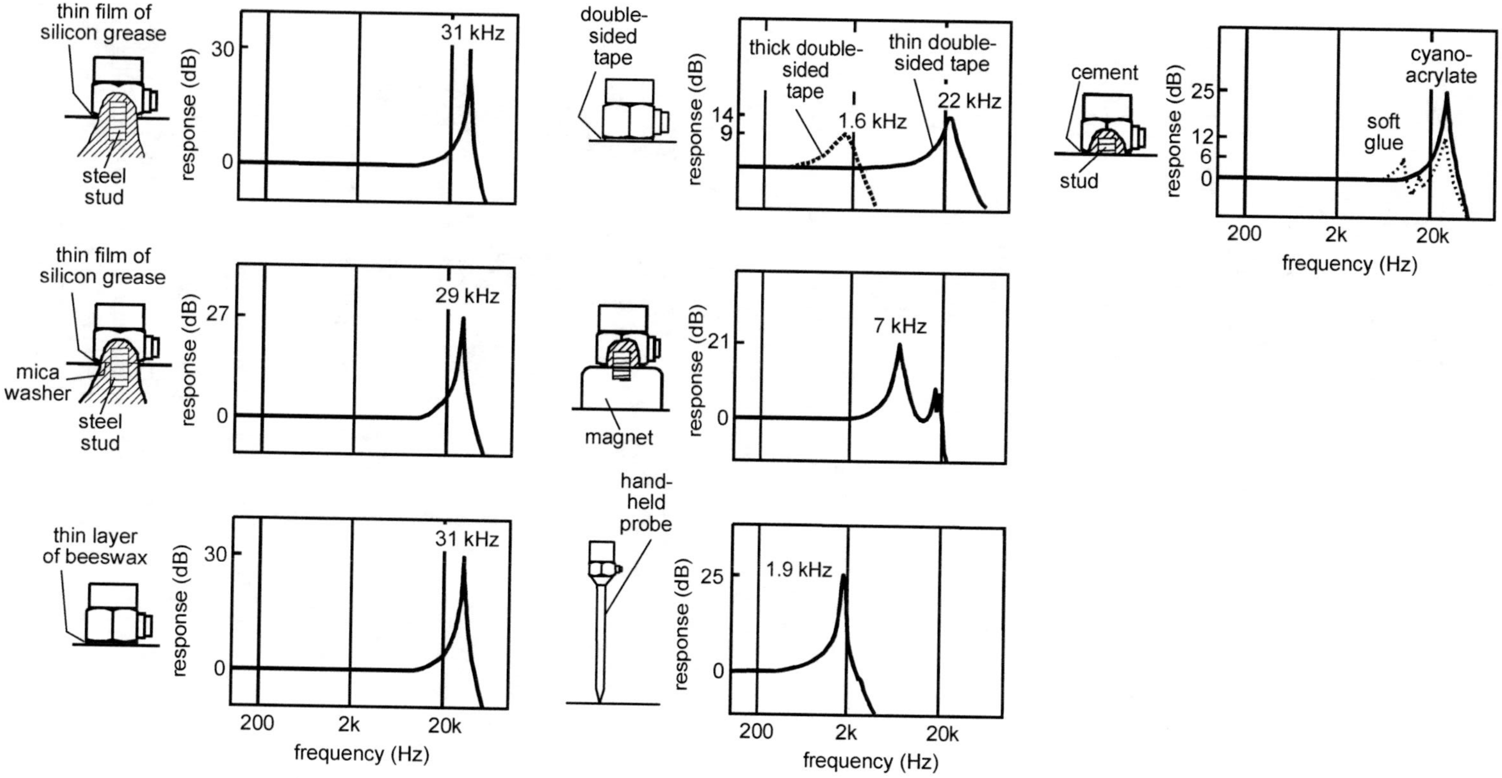

Figure 15.7 Effect of mounting technique on the frequency response of an accelerometer with a solid mounted resonance frequency of 32 kHz.

the surface is undesirable, a cementing stud may be used with little loss in performance. This stud is fixed to the surface using epoxy or cyanoacrylate adhesive. If electrical isolation is desired, a *thin* layer of epoxy can be spread on the surface and allowed to dry, and the stud then stuck to the dried layer. Alternatively, the accelerometer could be fixed to the surface directly using adhesive but this can damage it superficially. If this is done, it is important to remove the accelerometer using a twisting motion, which is one reason for the hexagonal base on some accelerometers. An alternative method of electrical isolation if holes are allowed in the test surface is to use an isolated stud with a mica washer.

Beeswax can be used for fixing the accelerometer to the vibrating surface with little loss in performance up to temperatures of 40°C provided only a thin layer is used. Thin, double-sided adhesive tape can also be used at the expense of reducing the useable upper frequency limit of the accelerometer by approximately 25%. Thick, double-sided tape (0.8 mm) reduces the upper frequency limit by up to 80%. The use of a permanent magnet to mount the accelerometer also affects its performance, reducing its mounted resonance frequency by approximately 50%. This mounting method is restricted to ferromagnetic materials.

For piezoelectric accelerometers with no integral pre-amplifier, the accelerometer cable should be fixed with tape to the vibrating surface to prevent the cable from excessive movement. This will minimise the effect of tribo-electric noise, which results from the cable screen being separated from the insulation around the inner core of the cable. This separation creates a varying electric field that results in a minute current flowing into the screen, which will be superimposed on the accelerometer signal as a noise signal. For this reason, low noise accelerometer cable should always be used.

15.1.2 Phase Response

Accelerometers generally have a flat, zero-degree phase response, as a function of frequency, between the mechanical input and electrical output signals at frequencies below about half the mounted resonance frequency as shown in Figure 15.8. If accelerometers are heavily damped to minimise the resonance effect, then the phase error between the mechanical input and electrical output will be significant, even at low frequencies.

15.1.3 Temperature Effects

Temperatures above 100°C can result in small reversible changes in accelerometer sensitivity of up to 12% at 200°C. If the accelerometer base temperature is kept down using a heat sink and mica washer with forced air cooling, then the sensitivity will change by less than 12% up to 400°C. Accelerometers cannot generally be used at temperatures in excess of 400°C.

15.1.4 Earth Loops

If the test object is connected to ground, the accelerometer must be electrically isolated from it or an earth loop may result, producing a high level hum in the resulting acceleration signal at the mains power supply frequency (50 or 60 Hz).

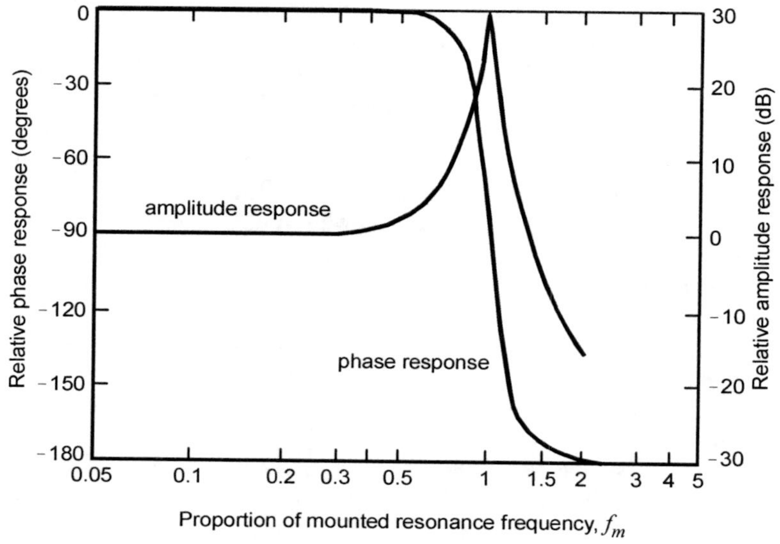

Figure 15.8 Typical piezoelectric accelerometer phase response.

15.1.5 Handling

Accelerometers should not be dropped on hard surfaces or subjected to temperatures in excess of 400°C. Otherwise, permanent damage (such as cracking of the brittle piezoelectric element), indicated by a significant change in accelerometer sensitivity, will result.

15.2 VELOCITY TRANSDUCERS

Velocity transducers are generally one of two types. The least common type is the non-contacting magnetic type consisting of a cylindrical permanent magnetic on which is wound an insulated coil. A voltage is produced by the varying reluctance between the transducer and the vibrating surface. This voltage is proportional to the surface velocity and the mean distance between the transducer and the surface. When non-ferrous vibrating surfaces are to be measured, a high permeability disc may be attached to the surface. This type of transducer is generally unsuitable for absolute measurements, but is very useful for relative vibration velocity measurements such as needed for vehicle active suspension systems. Another device suitable for vehicle active suspension systems consists of a magnet fixed to the vehicle body and a coil fixed to the suspension. Relative motion between the two produces a coil voltage proportional to the velocity (Wolfe and Jolly, 1990). This type of device is a subset of the most common general type of velocity transducer consisting of a moving coil surrounding a permanent magnet. Inductive EMF, which is proportional to the velocity of the coil with respect to the permanent magnet, is set-up in the coil when it is vibrated. In the 10 Hz to 1 kHz frequency range, for which the transducers are suitable, the permanent magnet remains virtually stationary and the resulting voltage is directly proportional to the velocity of the surface on which it is mounted. Outside this frequency range, the velocity transducer electrical output is not proportional to its velocity. This type of velocity transducer is designed to have a low natural frequency (below its useful frequency range); thus, it is generally quite heavy and can significantly mass-load light structures. Some care is needed in mounting but this is not as critical as for accelerometers, due to the relatively small upper frequency limit characterising the basic transducer.

Velocity transducers generally cover the dynamic range of 1 to 100 mm s^{-1}. Some allow measurements down to 0.1 mm s^{-1} while others extend them to 250 mm s^{-1}. Sensitivities are generally of the order of 20 mV mm s^{-1}. Low impedance, inexpensive voltage amplifiers are suitable for amplifying the signal. Temperatures during operation or storage should not exceed 120°C. Due to their limited dynamic range, they are not as useful as accelerometers.

Another type of velocity transducer is the laser Doppler velocimeter, which allows vibration velocity measurement without the need to fix anything (except reflective tape in the case of dark, dull surfaces) to the vibrating surface. One type of commercially available system is shown schematically in Figure 15.9. The laser light (frequency f) is split by the beam-splitter into two separate beams – the target beam and the reference beam.

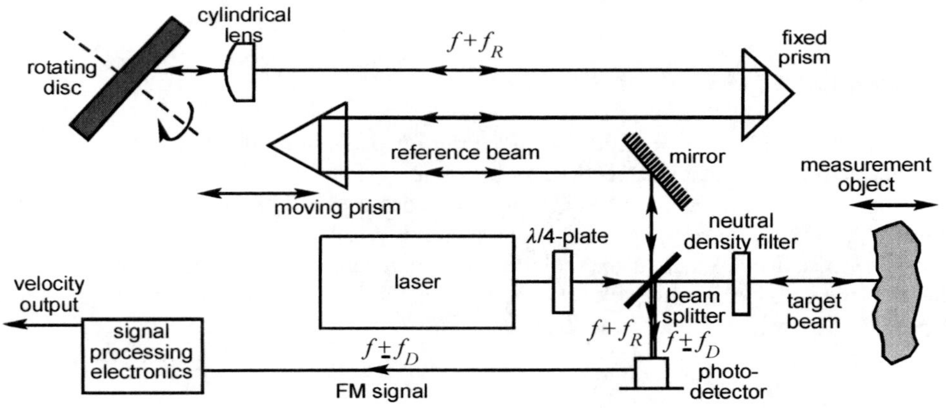

Figure 15.9 Laser Doppler velocimeter, schematic representation.

The target beam is directed on to the surface of a vibrating target, and is reflected back along its incident path. The returned light is frequency modulated by a variable Doppler shift due to the velocity of the target. The frequency range of the modulation is $f \pm f_D$, where f_D is the Doppler shift due to the maximum velocity of the target. This frequency modulation (FM) contains the required measurement but, since the frequencies are of the order of 10^{15} Hz, it is not practical to demodulate the signal directly. The returned target beam has first to be heterodyned with a reference beam so that the high-frequency component can be reduced. The reference beam is directed, by an optical system, onto a rotating reference disc set at an angle to the direction of incidence. The surface of the disc is covered with retroreflective tape, and its speed of rotation is controlled by a crystal oscillator. The frequency of the returned reference beam, ($f + f_R$), is increased by a constant Doppler shift due to the constant speed of the reference disc.

The two returned beams are recombined at the beam splitter and directed onto the surface of a photo-detector where they heterodyne. The photo-detector measures the intensity of the combined light, which has a beat frequency equal to the difference in frequency between the reference beam and the beam that has been reflected from the vibrating object. The output from the photo-detector is an FM signal with a frequency range $f_R \pm f_D$. It is necessary to frequency-shift the reference beam to prevent loss of signal as the target velocity passes through zero, which also allows the direction of the velocity to be determined as either positive or negative.

The output from the photo-detector is a frequency modulated signal centred on the reference frequency. Appropriate electronic signal processing is used to convert this to a voltage output proportional to the instantaneous velocity of the vibrating surface.

For a surface vibrating at many frequencies simultaneously, the beat frequency will contain all of these frequency components in the correct proportions, thus allowing broadband measurements to be made and then analysed in very narrow frequency bands.

Currently available laser vibrometer instrumentation has a dynamic range typically of 80 dB or more. Instruments can usually be adjusted using different processing modules so that the minimum and maximum measurable levels can be varied, while maintaining the same dynamic range. Instruments are available that can measure velocities up to 20 ms^{-1} and down to 1 μms^{-1} (although not with the same processing electronics) over a frequency range from DC up to 20 MHz.

15.3 DISPLACEMENT TRANSDUCERS

Although the dynamic range of displacement transducers is typically much smaller than it is for accelerometers, displacement transducers are often more practical at very low frequencies (0 to 10Hz), where vibration amplitudes are measured in terms of tenths or hundredths of a millimetre, and where corresponding accelerations are small. Also, in some active systems, displacement is the preferred control variable and attempts to derive this from an acceleration signal by double integration usually lead to excessive electronic noise, especially at very low frequencies (less than 10 Hz).

15.3.1 Proximity Probe

The most common type of displacement transducer is the proximity probe. Two types of proximity probe are available: the capacitance probe and the eddy current probe. The first relies on the measurement of the change in electrical capacitance between the vibrating machine surface and the stationary probe. The eddy current probe relies on the generation of a magnetic field at the probe tip by a high-frequency (>500 kHz) voltage applied to a coil of fine wire. This magnetic field induces eddy currents proportional to the size of the gap between the probe tip and machine surface. These currents oppose the high-frequency voltage and reduce its amplitude in proportion to the size of the gap. Typical gap ranges in which the amplitude is a linear function of gap size vary from between 1 mm and 4 mm for smaller diameter probes to between 2 mm and 20 mm for larger probes. The carrier amplitude signal is demodulated to give a low impedance voltage output proportional to gap size over the linear range of the transducer.

Eddy current probes are more common and easier to use than capacitative type pickups so further discussion will be restricted to the former type. When mounting an eddy current proximity probe to measure the vibration of a rotating shaft, the surface or shaft to be measured must be free of all irregularities such as scratches, corrosion, out-of-roundness, chain marks, etc. Any irregularity will cause a change in probe gap which is not a shaft position change, resulting in signal errors. This is called mechanical run-out.

The shaft material must be of uniform composition all the way round its surface so that the resistivity does not vary as the shaft rotates, resulting in an unwanted electronic noise signal. This is called electrical run-out noise. Care should also be taken with the use of plated shafts, as thin plating can allow the eddy currents to penetrate to the main shaft material, resulting in two different material resistivities being detected as well as the rough interface surface between shaft and surface treatment. As probes are generally matched to a particular shaft material, this can lead to calibration problems, as well as electronic noise problems due to the rough interface.

Proximity probes require a power supply and low impedance voltage amplifier, both of which are generally supplied in the same module. Note that the length of cable between the power supply and probe significantly affects the probe sensitivity.

The dynamic range of a proximity probe is typically 100:1, although some have a range of 150:1 and others 60:1. The resolution to which the probe can measure varies from 0.02 mm to 0.4 mm, depending upon the absolute range (2 mm to 25 mm). The limited dynamic range restricts its practical application to frequencies of less than 200 Hz.

15.3.2 Linear Variable Differential Transformer (LVDT)

This type of transducer consists of a single primary and two secondary coils wound around a cylindrical bobbin. A moveable nickel iron core is positioned inside the windings and it is the movement of this core, which is measured (see Figure 15.10).

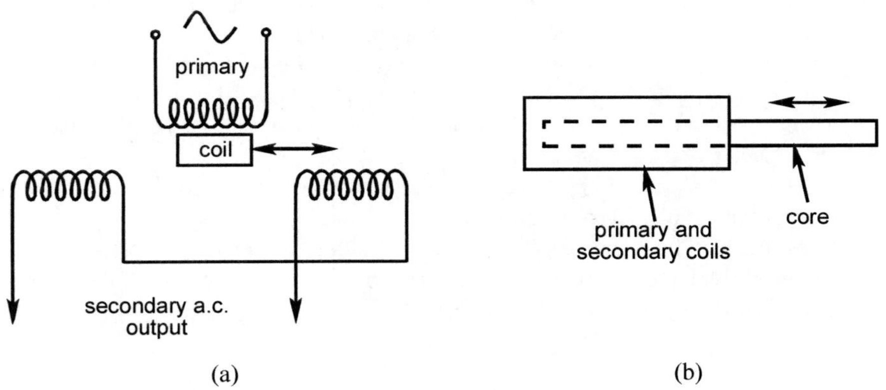

(a) (b)

Figure 15.10 (a) Schematic of an LVDT transducer; (b) LVDT transducer.

To operate the transducer, it is necessary to drive the primary coil with a sine wave, usually at a frequency of between 1 and 15 kHz and amplitude between 1 and 10 volts (r.m.s.). The output from the secondary coil is a sine wave that contains positional information in terms of its amplitude and phase. The output with the core at the centre of the stroke is zero rising to a maximum amplitude with the core at either end of its stroke. The output is in-phase with the primary drive with the core on one side of the centre and 180° out of phase when the core is on the other side of the centre.

Commercially available transducers are usually supplied with an in-built oscillator to generate the a.c. signal and appropriate demodulating electronics to convert the a.c. output into a d.c. level proportional to displacement. Alternatively, the LVDT, oscillator and demodulating electronics can be purchased separately, allowing greater flexibility. A typical dynamic range for this type of transducer is about 100:1, with maximum displacements measurable ranging from 1 mm to 100 mm, depending on the transducer selected. Note that because of the limited dynamic range, it is important to select the transducer with the appropriate upper limit for a particular task.

The frequency range characterising most commercially available LVDT transducers is d.c. to 100 Hz. For long stroke (±15 to ±100 mm) transducers, the diameter is typically between 12 and 20 mm, with a body length of about three times the maximum stroke (or displacement from centre). Short stroke transducers usually have a fixed length not less than 30 to 50 mm. More detail on the principles of operation of both of these devices are available in publications from commercial suppliers.

15.3.3 Linear Variable Inductance Transducers (LVIT)

This type of transducer is particularly suited to measuring relative displacements in vehicle suspension systems, and makes use of a cylindrical coil, the inductance of which is arranged to vary with movement of a spoiler over it. The coil is excited at approximately 100 kHz and the variation in its inductance is caused by the introduction of a highly conductive, non-ferrous spoiler into the resulting electromagnetic field. The spoiler takes the form of a co-axial rod sliding inside the coil or a co-axial cylinder sliding over the outside of the coil.

The most effective way of determining variations in inductance, and one which is relatively insensitive to variations in inductor coil resistance due to temperature effects or manufacturing irregularities, is to use a resonant circuit. The inductor is used as one component of a simple LC oscillator, the frequency of which changes with inductance changes, resulting in a frequency modulated carrier signal that can be converted to a voltage proportional to relative displacement between the coil and spoiler by using a commercially available frequency demodulation device.

Transducer sizes vary from diameters of a few millimetres to tens of millimetres, and lengths depend on the displacement amplitude to be measured.

Improvements in transducer performance can be made by using two coils and detecting the frequency differential between two resonant circuits, each of which contains one coil. This is described in more detail by Moore (1988).

15.4 STRAIN SENSORS

When a structure vibrates either flexurally or longitudinally, its surface is subjected to cyclic strains that can be detected with a strain sensor. Three types of strain sensor will be considered here: conventional resistive strain gauges, PVDF film and optical fibres. Piezoceramic crystals can also be used, but because of their brittle nature, relatively high cost and lower sensitivity, they cannot compare with PVDF film for this purpose and thus will not be considered here as sensors, although they are more suitable than PVDF film as actuators due to their higher force generating capability; thus, they are considered as actuators later on in this chapter.

15.4.1 Resistive Strain Gauges

Strain gauges are constructed of a thin, electrical-conducting wire, foil or semi-conductor sandwiched between two plastic sheets as shown in Figure 15.11. Larger areas at the ends of the grid facilitate connection of cables. The plastic sheets, which are bonded to the active element, simplify handling and protect the element from mechanical damage.

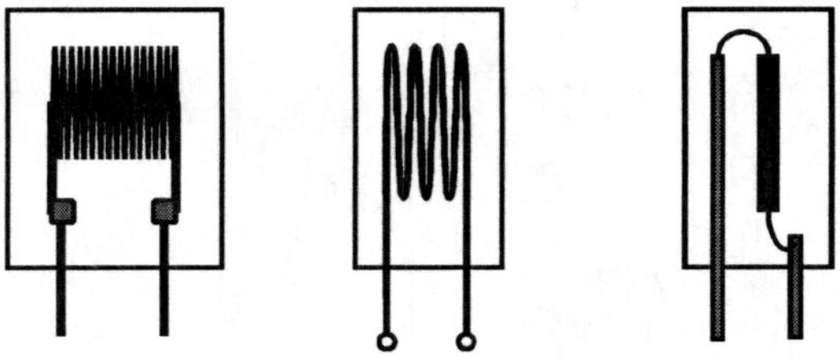

Figure 15.11 Types of strain gauge.

Foil strain gauges are produced by photo-etching a metallic foil (3 to 5 μm thick). The material used for the metallic element is usually an alloy of nickel and copper (constantan), the exact composition of which is dependent upon the material on which the strain gauge is to be used. Some effort is made to match the temperature coefficient of expansion of the strain gauge with the material on which it is to be used.

Wire strain gauges are produced using metallic wire, 15 to 25 μm in diameter. Foil gauges are easier to make if the gauge length is approximately 6 mm or less, whereas wire gauges are easier for lengths greater than approximately 6 mm. Wire gauges are also more suitable for high temperature applications, as more suitable materials are available and better mounting techniques can be used.

Semi-conductor strain gauges contain a semi-conductor element a few hundred micrometres wide and 20 to 30 μm thick. This type of gauge is the most expensive and the least preferred type for most applications.

The operation of metallic strain gauges is based on the principle that the electrical resistance of a metallic conductor changes if the conductor is subjected to an applied strain. The change in resistance of the strain gauge is partly due to the change in the geometry of the conductor and partly due to the change in the specific conductivity ρ_s of the conductor material due to changes in the material structure. The change in resistance ΔR as a fraction of the nominal resistance R_0 is given by:

$$\frac{\Delta R}{R_0} = \varepsilon\left(1 + 2\nu + \frac{1}{\rho_s}\frac{\partial \rho_s}{\partial \varepsilon}\right) \tag{15.4.1}$$

where υ is Poisson's ratio, ε is the strain imposed on the strain gauge and the first two terms in brackets represent the geometrical component. The quantity in brackets in Equation (15.4.1) is a constant over a wide range of values of ε for materials, such as constantan, which are used in strain gauges. Thus, Equation (15.4.1) can be written as:

$$\frac{\Delta R}{R_0} = K\varepsilon \tag{15.4.2}$$

where K is the gauge factor (usually in the vicinity of 2).

The operation of semi-conductor strain gauges is based on the piezoelectric effect; that is, mechanical stress applied to a semi-conductor will result in a change in resistance as a result of a change in electron mobility.

The type of doping material implanted in the semi-conductor will determine the conductivity (or gauge factor). It is also possible to obtain both negative or positive gauge factors.

The strain gauge relationship for semi-conductor gauges is:

$$\frac{\Delta R}{R_0} = \frac{T_0}{T} K_1 \varepsilon + \left(\frac{T_0}{T}\right)^2 K_2 \varepsilon^2 + \dots \tag{15.4.3}$$

It is this non-linearity and high unit cost that limit the use of this type of gauge.

As the resistance change ΔR is usually a very small value, it cannot be accurately measured directly. The most common means of determining ΔR is to insert the strain gauge as one leg in a Wheatstone bridge (see Figure 15.12). The bridge is balanced by varying R_1 to produce zero output, with the vibrating structure at rest. When the structure begins to vibrate, a voltage proportional to the strain (and thus structural vibration bending moment) will appear at the bridge output. Note that the bending moment is the double spatial derivative of the structural displacement.

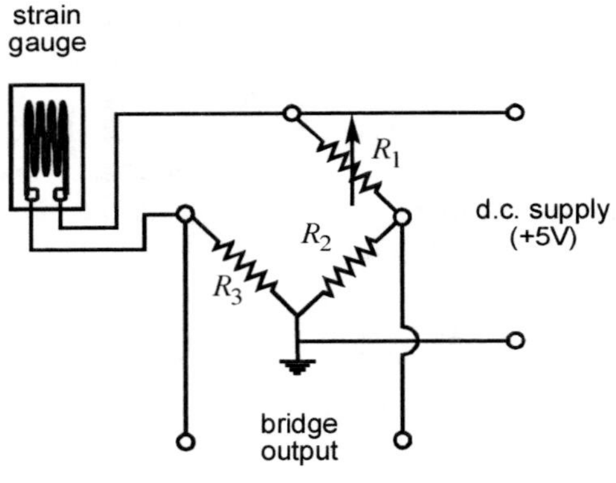

Figure 15.12 Wheatstone bridge circuit.

The bridge output voltage V_0 is related to the change in resistance ΔR of the strain gauge, the bridge d.c. supply voltage V_s and the structural strain as:

$$\frac{V_0}{V_s} = \frac{\Delta R}{R} = K\varepsilon \tag{15.4.4a,b}$$

As V_s is normally about 5 V and K is about 2 V, the output voltage V_0 is about ten times the strain level. Thus, 1 μ strain would produce 10 μV. Clearly, a high gain, low noise operational amplifier is needed to amplify the voltage to a useable level if strains as low as a few μ strain are to be measured. Also, the d.c. power supply to the bridge must be characterised by very low noise levels; often batteries rather than regulated mains power are necessary. The dynamic range of a strain gauge and its associated electrical system is from

2 to 5 μ strain to 10,000 μ strain over a frequency range limited only by the amplifying electronics. The frequency range of general interest for active noise and vibration control (0 to 500 Hz) is easily covered with standard circuitry.

In some cases, the sensitivity of the measurement system can be increased by replacing resistances R_3 or R_2 in the wheatstone bridge with a second strain gauge. If resistance R_2 is replaced with a strain gauge, the two strain gauges in the bridge must be tensioned and compressed out of phase with one another for the two effects to add together. This is achieved, for example, by using two strain gauges on opposite sides of a beam to detect flexural vibration. This configuration has the added advantage of minimising the d.c. drift in the output signal as a result of temperature changes occurring in the system being measured. If only one strain gauge is used (or if the second one replaces R_3 in the Wheatstone bridge circuit), then a temperature change in the system being measured will cause the structure to expand differently to the strain gauge (as the coefficients of thermal expansion of the test structure and the strain gauge are not exactly matched in practice), thus resulting in a strain gauge output not related to stress in the structure. However, this is rarely a problem when strain gauges are used just to sense vibration, as the temperature fluctuations generally occur well below the frequency range of interest.

When installing strain gauges on a surface for vibration measurement, many factors need to be considered such as adhesive selection, protection of the installed gauge from the environment and selection of a gauge size appropriate for the task. Larger gauges average the vibration over a larger surface area and are usually easier to install. Suitable adhesives include cyanoacrylate (super glue) provided the completed installation is covered with a water-proof compound. Other adhesives and protective compounds are available from strain gauge manufacturers.

15.4.2 PVDF Film

When bonded to a surface, polyvinylidine diflouride film (PVDF film) produces either a charge or a voltage proportional to the strain of the surface. The charge may be amplified by a high impedance charge amplifier to produce a voltage proportional to the strain. The circuits used to describe the electrical characteristics of PVDF film are similar to those discussed in Section 15.1 for accelerometers. The advantage offered by PVDF film when compared to strain gauges is its ability to act as a distributed sensor and to be shaped so that it senses particular vibration modes, or combinations of modes, if so desired. For example, if it is desired to control sound radiation, then the sensor could be shaped so that it sensed the surface vibration distribution that contributed most to the sound radiation. This is discussed in more detail in Chapter 8. When acting as a distributed sensor, PVDF film provides an output charge or voltage proportional to the surface strain integrated over the area covered by the sensor.

Before discussing the use of shaped sensors in more detail, the general properties of PVDF film will be outlined, so that some understanding may be gained of how it works and the meaning of the coefficients used to define its properties.

PVDF film is an extremely flexible polymer that can be polarised across its thickness (usually between 9 and 110 μm) by a strong electric field. It can act either as a sensor or an actuator. In this section, only its function as a sensor will be discussed; its function as an actuator will be discussed in Section 15.11. Once polarised, PVDF film will provide a charge or voltage when subjected to an applied tensile force or strain. Conversely, it will produce a strain when subjected to an applied voltage, but this must be well below its original polarisation voltage. The maximum operating voltage is 30 V μm^{-1} thickness and the breakdown voltage at which polarisation is lost is 100 V μm^{-1} thickness. The polarisation axis for PVDF film is usually normal to the surface, across the thickness (shown as the z-axis in Figure 15.13).

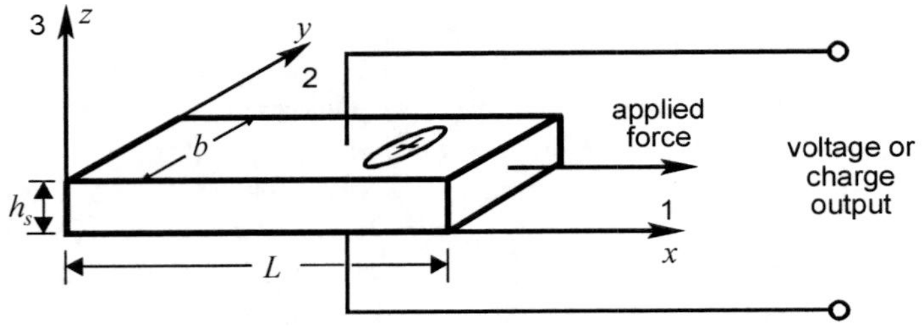

Figure 15.13 PVDF film.

For a compression or tension force F applied in the x-direction, the voltage V and charge Q outputs are given by:

$$\frac{Q}{Lb} = \frac{F}{h_s b} d_{31} = \sigma d_{31} \quad \mathrm{C\,m^{-2}} \tag{15.4.5a,b}$$

$$\frac{V}{h_s} = \frac{F}{h_s b} g_{31} = \sigma g_{31} \quad \mathrm{V\,m^{-2}} \tag{15.4.6a,b}$$

where σ is the stress in the PVDF film, and x, y, b, h_s and L are defined in Figure 15.13.

For a compression or tension force F applied in the y-direction, the voltage and charge outputs are found by substituting d_{32} and g_{32} for d_{31} and g_{31} in Equations (15.4.5) and (15.4.6). The d_{ij} coefficients are referred to as charge (or strain) coefficients and are constants that characterise a particular PVDF film. Similarly, the g_{ij} coefficients are referred to as voltage (or stress) coefficients.

The charge coefficients are the ratios of electric charge generated per unit area to force applied per unit area, and the voltage coefficients are the ratios of the electric field produced (volts/metre) to the force applied per unit area. The subscripts 1 and 2 refer to force applied in the x- and y-directions respectively, and the subscript 3 refers to the axis of polarisation, which is usually the z-axis. Typical values of the charge and voltage coefficients are:

$$g_{32} = g_{31} = 216 \times 10^{-3} \ \frac{\mathrm{V\,m^{-2}}}{\mathrm{N\,m^{-2}}}$$

$$d_{31} = 23 \times 10^{-12} \ \frac{\mathrm{C\,m^{-2}}}{\mathrm{N\,m^{-2}}}$$

$$d_{32} = 3 \times 10^{-12} \ \frac{\mathrm{C\,m^{-2}}}{\mathrm{N\,m^{-2}}}$$

The charge output due to an applied strain is given by:

$$\frac{Q}{Lb} = \varepsilon e_{31} \tag{15.4.7}$$

where ε is the applied strain and the coefficients e_{ij} are related to d_{ij} as (Lee and Moon, 1990a):

$$\begin{bmatrix} e_{31} \\ e_{32} \\ e_{36} \end{bmatrix} = \begin{bmatrix} E_p/(1-\nu_p^2) & \nu_p E_p/(1-\nu_p^2) & 0 \\ \nu_p E_p/(1-\nu_p^2) & E_p/(1-\nu_p^2) & 0 \\ 0 & 0 & E_p/2(1+\nu) \end{bmatrix} \begin{bmatrix} d_{31} \\ d_{32} \\ d_{36} \end{bmatrix} \tag{15.4.8}$$

The subscript 6 takes into account the possibility of the PVDF film principal axes not being coincident with the principal axes of the structure to which it is bonded. The quantity d_{36} is zero if the axes are coincident. E_p is the PVDF film elastic modulus (typically 2×10^9 Pa) and ν is Poisson's ratio (typically 0.35).

Typical thicknesses of PVDF film range from 9 μm to 110 μm. However, it can be seen from Equation (15.4.7) that if the film is attached to a vibrating surface, the charge produced is independent of the film thickness, and only dependent upon the strain induced by the vibrating surface and the area of film used. On the other hand, Equation (15.4.6) shows that the voltage produced is dependent on the thickness but not the area of film. However, the same disadvantages associated with using a piezoelectric accelerometer as a voltage source (discussed in Section 15.1) apply to the PVDF film as well.

As a guide to the sensitivity of PVDF film compared to a strain gauge for measuring dynamic strain, Equation (15.4.7) shows that a piece 10 mm × 10 mm will produce a charge of approximately 5.5 pC when subjected to 1 μ strain. A conventional charge amplifier noise floor is approximately 0.003 pC, so the strain detected can be as low as 10^{-9} strain. This compares very favourably with a strain gauge, which is limited to about 10^{-6} strain. The upper limit of the dynamic strain that can be measured is governed by the tensile and compressive strength of the PVDF film and is approximately 0.015, which is similar to a strain gauge.

It is of interest to examine how PVDF sensors may be shaped to respond to certain modes on a simply supported rectangular thin plate, of thickness h, illustrated in Figure 15.14.

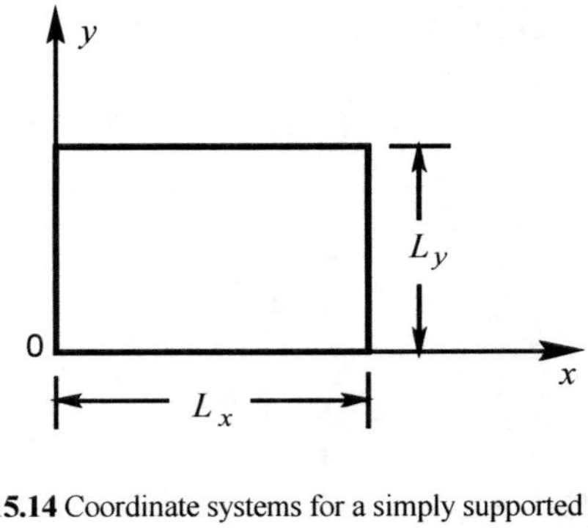

Figure 15.14 Coordinate systems for a simply supported thin plate.

Lee and Moon (1990b) give the equation describing the closed circuit charge signal $q(t)$ measured at the electrode of a piezoelectric film as:

$$q(t) = \frac{h+h_s}{2} \int_0^{L_x} \int_0^{L_y} S_f P_0 \left[e_{31} \frac{\partial^2 w}{\partial x^2} + e_{32} \frac{\partial^2 w}{\partial y^2} + 2e_{36} \frac{\partial^2 w}{\partial x \partial y} \right] \mathrm{d}y \mathrm{d}x \tag{15.4.9}$$

The quantities h and h_s are the thicknesses of the plate and piezo film respectively. The polarisation profile is P_0 which, for the case considered here, is either +1 or −1. The quantity S_f represents the shape function of the piezo-film, and w is the flexural displacement of the panel.

15.4.2.1 One-Dimensional Shaped Sensor

If the sensor is narrow and long ($x >> y$), then Equation (15.4.9) can be written as:

$$q_x(t) = \frac{h+h_s}{2} \int_0^{L_x} S_f P_0 e_{31} \frac{\partial^2 w}{\partial x^2} \mathrm{d}x \tag{15.4.10}$$

where $q(t)$ is the charge generated by a PVDF sensor that is relatively thin in the y-direction.

Neglecting time dependence, the flexural displacement amplitude of a simply supported panel for any specified frequency can be expressed as:

$$\bar{w}(x,y) = \sum_{m=1}^{\infty} \sum_{n=1}^{\infty} A_{mn} \sin\left(\frac{m\pi x}{L_x}\right) \sin\left(\frac{n\pi y}{L_y}\right) \tag{15.4.11}$$

where m and n represent the modal indices of the panel modes and A_{mn} represents the complex modal amplitude for mode (m, n) at a specified frequency of vibration. The quantities $\bar{w}(x, y)$ and A_{mn} are functions of the excitation frequency of the panel.

Substituting Equation (15.4.11) into Equation (15.4.10) gives for the charge amplitude:

$$\bar{q}_x = -\frac{(h+h_s)}{2} \int_0^{L_x} S_f P_0 e_{31} \sum_{m=1}^{\infty} \sum_{n=1}^{\infty} A_{mn} \left(\frac{m\pi}{L_x}\right)^2 \sin\left(\frac{m\pi x}{L_x}\right) \sin\left(\frac{n\pi y_s}{L_y}\right) \mathrm{d}x \tag{15.4.12}$$

The quantity y_s is the y-coordinate of the sensor extending in the x-direction.

Two different shapes for the PVDF film will now be considered. For the first, a narrow strip of constant width will be examined (see Figure 15.15(a)). For a strip such as this, $S_f = 1$ and P_0 = constant:

$$\bar{q}_x = -\frac{(h+h_s)}{2} e_{31} P_0 \sum_{m=1}^{\infty} \sum_{n=1}^{\infty} A_{mn} \left(\frac{m\pi}{L_x}\right)^2 \sin\frac{n\pi y_s}{L_y} \int_0^{L_x} \sin\left(\frac{m\pi x}{L_x}\right) \mathrm{d}x \tag{15.4.13}$$

Thus, if y_s is set equal to $L_y/2$; that is, if the strip is placed across the centre of the plate, it can be seen that all even (m even, n even) and odd even (m odd, n even) modes (that is, all modes with a nodal line along the sensor) will not contribute to the sensor output.

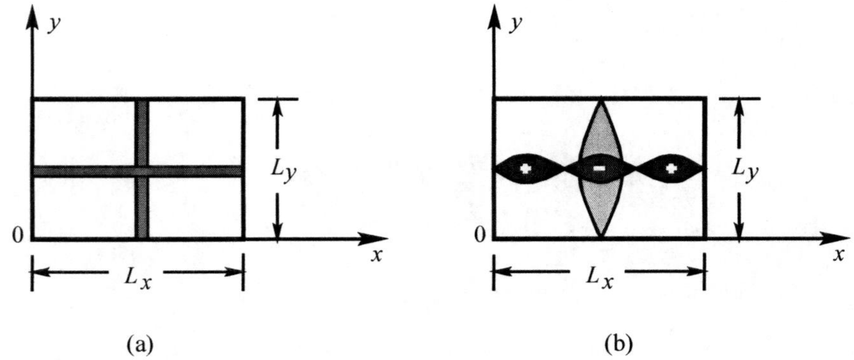

Figure 15.15 Strip sensors for detection of desired modes: (a) linear sensors; (b) shaped sensors $m' = 3$, $n' = 1$.

A sensor shape and pole connection will be considered, such that:

$$S_f P_0 = \mu_{m'} \sin\left(m'\pi \frac{x}{L_x} \right) \tag{15.4.14}$$

where $\mu_{m'}$ is the sensor scaling factor. That is, the sensor shape and polarity is made proportional to the modal strain distribution across the panel for modes with the coefficient m' in the. The quantity m' can also be considered the modal index of the sensor. A sensor with $m' = 3$ in the x-direction and another with $n' = 1$ in the y-direction are shown in Figure 15.15(b).

Substituting Equation (15.4.10) into Equation (15.4.8) and using the orthogonality property of the modes, the following is obtained:

$$\bar{q}_x = -\left(\frac{h + h_s}{2} \right) e_{31}\, \mu_{m'} \left(\frac{m\pi}{L_x} \right)^2 \frac{L_x}{2}\, \delta_{m'm} \sum_{n=1}^{\infty} A_{mn} \sin\left(\frac{n\pi y_s}{L_y} \right) \tag{15.4.15}$$

where

$$\delta_{m'm} = 1 \text{ if } m' = m$$
$$= 0 \text{ if } m' \neq m$$

It can be seen from Equation (15.4.15) that the sensor will not detect modes where $m \neq m'$. If the sensor is placed along the centre line of the panel so that $y_s = L_y/2$, it will only detect modes with $m = m'$ and with n odd.

A similar analysis procedure to that described above can be done for a sensor parallel to the y-axis with similar results.

15.4.2.2 Two-Dimensional Shaped Sensor for Simply Supported Boundary Conditions

For a two-dimensional sensor configuration with zero skew, Equation (15.4.9) can be written as:

$$q(t) = \left(\frac{h+h_s}{2}\right) \int_{x_1}^{x_2} \int_{y_1}^{y_2} S_f P_0 \left[e_{31} \frac{\partial^2 w}{\partial x^2} + e_{32} \frac{\partial^2 w}{\partial y^2} \right] \mathrm{d}x\,\mathrm{d}y \tag{15.4.16}$$

where

$$x_1 = (L_x - L_x')/2, \quad x_2 = x_1 + L_x'$$
$$y_1 = (L_y - L_y')/2, \quad y_2 = y_1 + L_y'$$

and L_x' and L_y' are the sensor dimensions (see Figure 15.16).

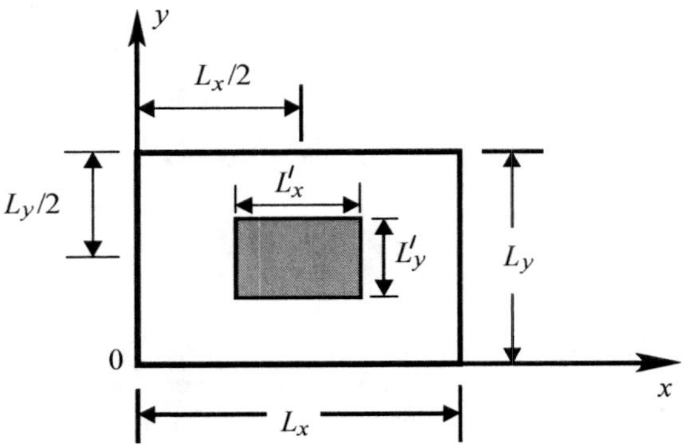

Figure 15.16 Two-dimensional sensor for detecting only odd/odd panel modes.

Following the same steps as for a one-dimensional sensor, Equation (15.4.16) can be used to derive the following expression for a two-dimensional rectangular sensor centred on a simply supported rectangular panel:

$$\bar{q}_{xy} = -\left(\frac{h+h_s}{2}\right) \int_{x_1}^{x_2} \int_{y_1}^{y_2} S_f P_0 \sum_{m=1}^{\infty} \sum_{n=1}^{\infty} A_{mn} \left[e_{31} \left(\frac{m\pi}{L_x}\right)^2 + e_{32} \left(\frac{n\pi}{L_y}\right)^2 \right] \sin\left(\frac{m\pi}{L_x}\right) \sin\left(\frac{n\pi y}{L_y}\right) \mathrm{d}y\,\mathrm{d}x \tag{15.4.17}$$

Setting the sensor polarity P_0 constant over the PVDF film, noting that for this shape, the shape factor, $S_f = 1$ and solving the integral gives $q_{xy} = 0$ for all modes except those for which m and n are both odd. In this case:

$$\bar{q}_{xy} = -\left(\frac{h+h_s}{2}\right) P_0 \sum_{m=1}^{\infty} \sum_{n=1}^{\infty} \left[A_{mn} \left(e_{31} \left(\frac{m\pi}{L_x}\right)^2 + e_{32} \left(\frac{n\pi}{L_y}\right)^2 \right) \times \frac{4 L_x L_y}{m n \pi^2} \cos\left(\frac{m\pi x_1}{L_x}\right) \cos\frac{n\pi y_1}{L_y} \right] \tag{15.4.18}$$

Note that for a lightly damped panel, A_{mn} may be considered real. Thus, a rectangular PVDF film sensor placed in the centre of a rectangular panel will only detect odd/odd modes.

Shaped sensors for the detection of acoustic power radiation are discussed in Chapter 8. See also Tanaka et al. (1996).

15.4.2.3 Two-Dimensional Shaped Sensor for Arbitrary Boundary Conditions

If a plate is excited into flexural vibration in only one dimension so that it is effectively a beam, the analysis outlined in Section 15.4.2.1 can be applied for arbitrary boundary conditions, with the sensor design being based on the curvature of the structural eigen functions (Clark and Burke, 1996).

For the case of two-dimensional structures, it is important to note that the analysis in Section 15.4.2.2 can only be applied to simply supported plates – it is not theoretically possible to apply it rigorously to plates or two-dimensional structures with other boundary conditions, although approximate modal sensors may still be realised (Clark and Burke, 1996). However, even small errors in the sensor shape can lead to large errors in the ability of the sensor to sense just one structural vibration mode or family of modes (Clark and Burke, 1996). Researchers have made a number of attempts to overcome this problem by using a number of approaches including variable thickness film (which is very difficult to implement in practice), replacing the variable thickness film with many small uniform thickness segments (Sun et al., 2002) and using multi-layered PVDF films (Kim et al., 2002). More recently, Donoso and Bellido (2009) presented a shape optimisation procedure that they claim is valid for arbitrary boundary conditions but only demonstrated it on a beam and simply supported plate. Suffice it to say that sensing vibration modes using shaped sensors is very difficult on any but the simplest of structures and even on those structures, small errors in the sensor shape can lead to large errors in the sensor's ability to exclude modes that are not intended to be sensed. Any errors are compounded by damping present in the plate structure as the analysis and derivation of the shape assumes that there is no structural damping at all.

15.4.3 Optical Fibres

The use of optical fibres as strain sensors was first discussed by Butter and Hocker (1978) and has since been applied as a sensor for active control of beam vibration by Cox and Lindner (1991). An optical fibre sensor consists of a small diameter single strand of glass that can be embedded in or attached to a dynamic structure producing negligible dynamic loading.

Optical fibre sensing is expensive, requiring a laser and complex signal processing electronics, but for some applications it is the only feasible choice. Such applications include structures subjected to high level electromagnetic fields, high temperatures and pressures, and/or harsh chemical environments.

Modal domain (MD) optical fibre sensors appear to be the most practical (Cox and Lindner, 1991). They were first investigated by Layton and Bucaro (1979) and utilise the interference phenomenon between two or more modes propagating in a multi-mode optical fibre to sense the axial strain in that fibre. This configuration, which does not require the traditional interferometer reference arm, simplifies sensor mounting and makes it more stable. Power distributions in the fibres were stabilised by the introduction of elliptic core fibres (Kim et al., 1987) which eliminated power coupling between spatially degenerate modes in the fibre as the fibre was strained. Another recent development was the attachment of insensitive lead-in and lead-out optical fibres to the sensing element.

An MD optical fibre sensor is characterised by an intensity pattern at its output, which is affected by the integrated strain over the length of the fibre. The effect of strain on the physical characteristics of the fibre is threefold: first, the axial length is changed; second, the Poisson effect will result in a change in the fibre diameter; and third, the photo-elastic effect results in the indices of refraction being dependent on strain. Thus, the functional dependence of the light intensity output on strain is complex. A typical optical fibre is constructed of an inner core and an outer cladding as shown in Figure 15.17. Usually the index of refraction n_1 of the inner core is similar to that of the outer cladding n_2 so that $(n_1 - n_2) << 1$. This assumption leads to a simplified description of the electromagnetic waves in the fibre and will be used in the analysis to follow.

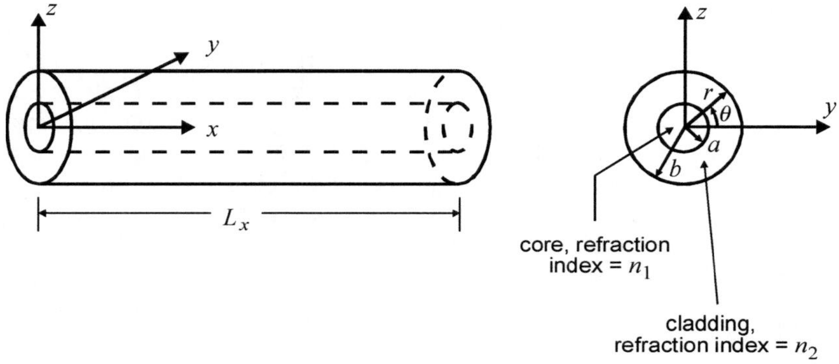

Figure 15.17 Optical fibre coordinate system.

To determine the nature of the functional dependence of the light output intensity on the fibre strain, it is of interest to examine the solutions to Maxwell's electromagnetic wave equation, which resembles the acoustic wave equation discussed at some length in Chapter 2. Although the analysis referred to here assumes a circular section fibre core, the results are applicable in general terms to the elliptical section core which is used in practice for reasons mentioned previously (Cox and Lindner, 1991). The electric field $\boldsymbol{E}$ for a propagating electromagnetic field is described by the following wave equation:

$$\nabla^2 \boldsymbol{E} - \frac{\partial^2}{\partial t^2}\boldsymbol{E} = 0 \tag{15.4.19}$$

For single frequency laser light propagating in one direction (the x or axial direction down the fibre, as in Figure 15.17), a solution to Equation (15.4.19) in cylindrical coordinates is:

$$\bar{\boldsymbol{E}}(r, \theta, t) = \boldsymbol{E}(r, \theta)e^{\mathrm{j}(\omega t - \beta x - \psi)} \tag{15.4.20}$$

where β is a propagation constant equivalent in meaning to an acoustic wavenumber, and ψ is a relative phase angle.

The electric field $E(r,\theta)$ can be decomposed into its vector components. Thus,

$$\boldsymbol{E}(r,\theta) = E_x(r,\theta)\boldsymbol{i} + E_y(r,\theta)\boldsymbol{j} + E_z(r,\theta)\boldsymbol{k} \tag{15.4.21}$$

where $\boldsymbol{i}, \boldsymbol{j}, \boldsymbol{k}$ are unit vectors in the x-, y- and z-directions.

If $\boldsymbol{E}(r,\theta)\bullet\boldsymbol{k}$ is a constant, then the wave is said to be linearly polarised (LP). Solutions are found for $\boldsymbol{E}(r,\theta)$ in both the central core and outer cladding of the fibre by applying the boundary conditions imposed by the shape of the core and the cladding.

As laser light is usually polarised in a direction corresponding to the z- or y-axis (linearly polarised), it is convenient to decompose the field in the y-z plane into two linearly independent solutions:

$$\boldsymbol{E}(r,\theta) = \left[E_{xy}\boldsymbol{i} + E_{yy}\boldsymbol{j}\right]\left[E_{xz}\boldsymbol{i} + E_{zz}\boldsymbol{k}\right] \tag{15.4.22}$$

The first subscript corresponds to the field and the second to the polarisation direction. Thus, the first two terms in Equation (15.4.22) correspond to light polarised in the y-direction, while the last two terms correspond to light polarised in the z-direction.

For laser light polarised in only one direction, it is sufficient to consider only the first solution of Equation (15.4.22). This is obtained in terms of β satisfying the boundary conditions in both the central core and the outer cladding of the fibre.

Solutions for the propagation constant β are found by solving the characteristic equation, which is determined by matching the tangential components of the electric field across the boundary between the core and cladding. Each value of β corresponds to a particular mode of electromagnetic propagation along the x-axis. Values of ψ corresponding to each mode are determined by the conditions of entry of the electromagnetic wave into the sensitive section. The lowest frequency at which a particular mode can carry power is referred to as the cut-off frequency for that mode. As the light is linearly polarised, the modes are referred to as LP_{ij} modes, where the subscript i refers to the electric field distribution in the y-z plane for fixed x while the subscript j refers to the distribution along the x-axis.

The LP_{01} mode has a zero cut-off frequency; it always carries power. For frequencies characterised by $0 < V < 2.405$, it is the only mode that carries power, and optical fibres designed to operate in this regime are called single mode fibres. Note V is the normalised frequency defined as:

$$V = \frac{\omega}{c} a\left(n_1^2 - n_2^2\right)^{1/2} \tag{15.4.23}$$

where ω is the radian frequency of the light, c is the speed of light, a is the radius of the inner core, and n_1, n_2 are respectively the indices of refraction of the inner core and outer cladding.

For frequencies defined by $2.405 < V < 3.832$, only the LP_{01} and LP_{11} modes carry power. Thus, for an optical fibre designed to operate in this frequency range, the intensity output is given by:

$$I_f(r,\theta,z) = |E_{01}(r,\theta,z) + E_{11}(r,\theta,z)|^2 \tag{15.4.24}$$

At any cross-section of the fibre optic sensor analysed above, the two propagating modes interfere to generate a well defined intensity pattern. As each mode is characterised by a different propagation constant β_{ij}, the interference pattern is a function of axial location x.

When the fibre is subjected to axial strain, the optical path length for each mode is changed by a different amount. This change is caused by changes in the axial length of the fibre, in the fibre diameter (as a result of the Poisson effect) and in the fibre index of refraction (as a result of the photo-elastic effect). As a result, the interference pattern at the fibre end face changes in a predictable way as a function of the applied strain. Thus, the sensor is essentially an interferometer with both legs contained in the optical fibre.

If it is assumed that a photo-detector can be configured to measure just a small part of the intensity at the output of the sensitive part of the sensor, then the output of the photo-detector can be shown to be (Cox and Lindner, 1991):

$$y_{PD} = K_e(I_0 + I_1 \cos\Gamma) \tag{15.4.25}$$

where K_e is the gain of the photo-detector and Γ is the phase of the intensity given by:

$$\Gamma = \Delta\beta L_x + \Delta\psi \tag{15.4.26}$$

where $\Delta\beta = \Delta\beta_{01} - \beta_{11}$ and $\Delta\Psi = \Psi_{01} - \Psi_{11}$, and L_x is the length of sensitive fibre. Note that the total intensity emerging from the end of the fibre is unaffected by the strain in the fibre; only the intensity distribution is affected. Thus, it is important that the photo-detector be configured to observe only part of the end face of the fibre. Means of doing this are discussed later.

Following the approach of Cox and Lindner (1991), the first-order effect of strain is on the phase of the intensity and is given by:

$$\Gamma(\varepsilon) = \Delta\beta(\varepsilon) L_x(\varepsilon) + \Delta\psi \tag{15.4.27}$$

By considering the phase change due to axial strain on each fibre segment dx, it can be shown that:

$$\Gamma(\varepsilon) = \Gamma_0 + \Delta\bar{\beta} \int_0^{L_x} \varepsilon(x,t)\,\mathrm{d}x \tag{15.4.28}$$

where $\Delta\bar{\beta}$ represents the amplitude of the time varying quantity $\Delta\beta$. Thus, the photo-detector output may be expressed as:

$$y_{PD}(t) = K_e I_1 \left[1 + \cos\left(\Gamma_0 + \Delta\hat{\beta} \int_0^{L_x} \varepsilon(x,t)\,\mathrm{d}x \right) \right] \tag{15.4.29}$$

where, for convenience, $I_1 = I_0$ has been assumed.

The resulting photo-detector output is a sinusoid with a d.c. offset. This d.c. offset can removed using a high-pass filter and Γ_0 can be set equal to $2\pi m + 3\pi/2$ (m an integer) by applying some pre-strain to the fibre. This pre-strain is necessary to ensure that the sensor will operate in its linear range. After these operations, the output is given by:

$$y_{MD}(t) = K_{MD} \sin\left[\Delta\hat{\beta} \int_0^{L_x} \varepsilon(x,t)\,\mathrm{d}x \right] \tag{15.4.30}$$

where K_{MD} is a constant. Note that the sinusoidal non-linearity can be ignored for small amplitudes of strain but must be taken into account when large amplitudes are to be measured.

As mentioned previously, when the modal domain (MD) sensor just described is implemented in practice, it is necessary to use an elliptical core to eliminate the problem of unpredictable coupling between four spatially degenerate LP_{11} modes which are all characterised by the same propagation constant β. It is also necessary to use a low-pass (as well as the high-pass) filter on the output to eliminate photo-detector noise. To ensure that $\Delta\psi$ remains constant, it is essential to use an insensitive 'lead-in' optical fibre that allows propagation of only the LP_{01} mode.

To measure light arriving from only part of the output face of the sensor it is necessary to splice it with an off-set (as shown in Figure 15.18) to an insensitive 'lead-out' fibre that only allows propagation of the LP_{01} mode. This has the effect of transforming the phase modulation (due to a time varying strain) at the output of the sensitive fibre to amplitude modulation of the light at the output of the lead-out fibre.

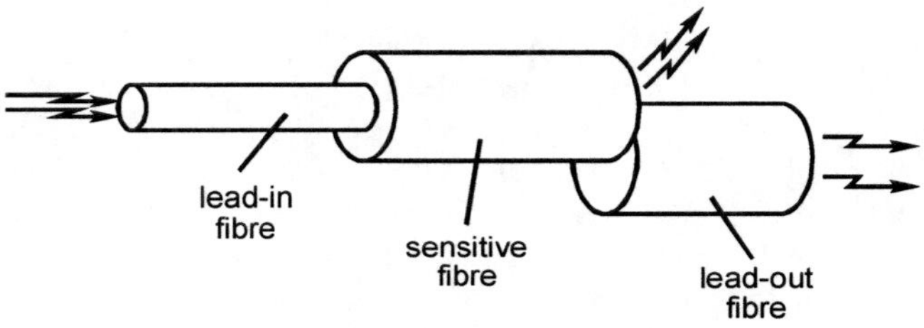

Figure 15.18 Optically spliced modal domain strain sensor.

In practice, the fusion of the lead-out fibre to the sensitive fibre is performed while the sensor is actually sensing cyclic strain so that the misalignment can be adjusted for optimal modulation.

15.5 HYDRAULIC ACTUATORS

Hydraulic actuators generally consist of a piston in a hydraulic cylinder, which has hydraulic fluid openings at each end. Introduction of high pressure fluid into one end of the cylinder causes the piston to move to the other end, expelling fluid from the opening at that end. Switching the supply of high pressure fluid to the other end of the cylinder causes the piston to move in the opposite direction. The most common method to alternate the hydraulic supply from one end of the cylinder to the other is to use a servo-valve, which consists of a moveable spool connected to a solenoid.

Applying voltage to the solenoid coil causes the spool to move, which in turn results in the opening and closing of valves that are responsible for directing hydraulic fluid to either end of the hydraulic cylinder (see Figure 15.19). Note that the servo-valve spool is usually maintained in its inactive position with springs at either end.

Servo-valves can be quite expensive, ranging from US$100 to US$3000 depending on the upper frequency at which they are required to operate, which can be 3 to 20 Hz for an active vehicle suspensions or 150 Hz for structural vibration control.

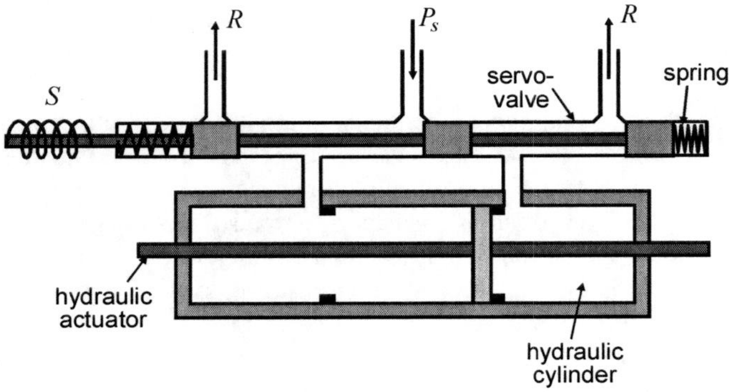

Figure 15.19 Schematic of a hydraulic actuator driven by a servo-valve shown with the actuator driven to the right.
R = hydraulic return line;
P_s = hydraulic supply pressure;
S = solenoid.

One advantage of hydraulic actuators is their large displacement and high force generating capability for a relatively small size. Disadvantages include the need for a hydraulic power supply, which can be inherently noisy, and non-linearities between the servo-valve input voltage and the hydraulic actuator force or displacement output. It is extremely important that the servo-valve be mounted as closely as possible to the hydraulic cylinder to minimise loss of performance, especially at high frequencies (above about 20 Hz). For best results at high frequencies, the servo-valve should be incorporated in the actuator assembly as was done by King (1988).

In an attempt to make the actuator motion correspond more closely to the solenoid drive current, some manufacturers have developed two-stage servo-valves, in which a torque motor is used to operate a valve, which in turn directs hydraulic fluid to one side or the other of the servo-valve spool.

Note that the hydraulic supply is generated by a hydraulic pump with an accumulator (or large pressurised reservoir) between the pump and the servo-valve to smooth out any pressure pulsations.

Hydraulic actuators have been used in the past in active vehicle suspensions (Crolla, 1988; Stayner, 1988) and active control of helicopter cabin vibration (King, 1988).

15.6 PNEUMATIC ACTUATORS

Pneumatic actuators are very similar in operation to hydraulic actuators, except that the hydraulic fluid is replaced by air. One advantage of active pneumatic actuators is that they can use the same air supply as passive air springs, which may be mounted in parallel with them. Pneumatic actuators may also be the preferred option in cases where an existing compressed air supply is available (such as in rail vehicle active suspension applications; see Buzan and Hedrick, 1983 and Cho and Hedrick, 1985).

The major disadvantage of pneumatic actuators is their relatively low bandwidth (less than 10 Hz) due to the compressibility of the air. Nevertheless, some success has been reported (Cho and Hedrick, 1985) in using pneumatic actuators in an active suspension for a rail vehicle.

15.7 PROOF MASS (OR INERTIAL) ACTUATOR

This type of actuator consists of a mass which is free to slide along a track (for horizontal excitation only) or on a suspension system for horizontal or vertical excitation or any other direction excitation as well. The mass is accelerated using an electromagnetic field (see Figures 15.20(a), (b) and (c)) so in one sense this type of actuator is similar to an electrodynamic shaker, which will be described in the next section. The difference is that in a proof mass actuator, the excitation force is a reaction to a relatively large vibrating mass, which is usually the permanent magnet. On the other hand, the excitation force from an electrodynamic shaker is generated by the electromagnet or armature, attached to a rigid ground or suspended on a flexible suspension (such as bicycle tubes), forcing the coil to vibrate. For this type of actuator, the coil is usually attached to the structure via a thin rod referred to as a 'stringer', whereas for the proof mass actuator, the coil is attached directly to the structure.

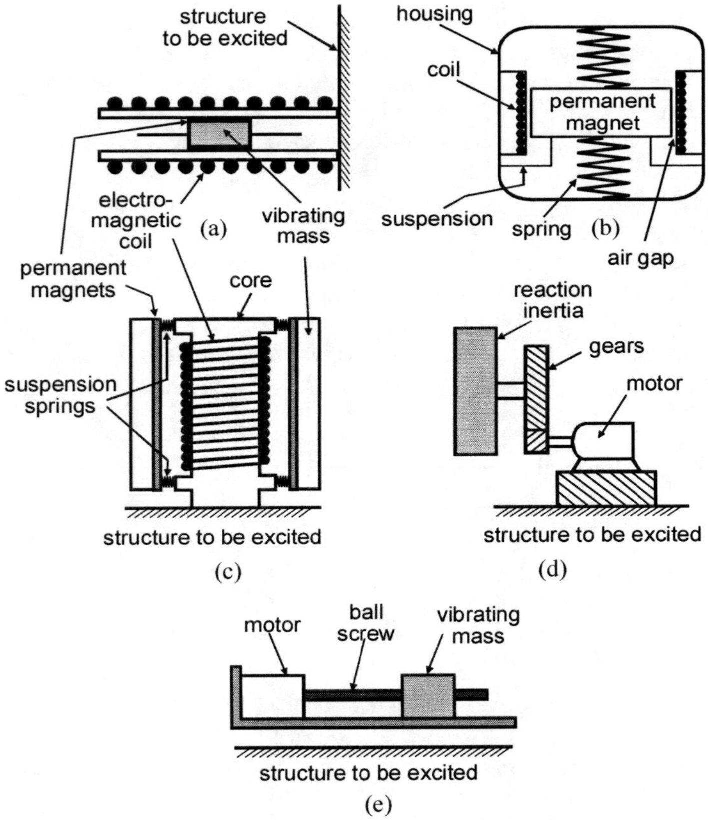

Figure 15.20 Proof mass actuators: (a) linear, electromagnetic drive for horizontal excitation; (b) and (c) linear, electromagnetic drive for excitation in any direction; (d) rotational; (e) linear, ball screw drive.

One problem that sometimes occurs with linear proof mass actuators is the limited motion of the reaction mass, which becomes more serious when low-frequency vibration modes in a structure are to be controlled. However, the problem is much less for the ball screw type actuator as high forces can be produced on the reaction mass without producing large motions. In cases where it is not possible to prevent the proof mass from encountering the limits of its travel, Politansky and Pilkey (1989) showed that a sub-optimal feedback control system yielded good control results.

The phase shift in the actuator frequency response at the fundamental resonance frequency can cause instability if the actuators are used in high-gain velocity feedback control systems. To achieve stability of the controlled system, it is important that the resonance frequency of the actuator is below the fundamental frequency of the structure whose vibration is to be controlled (Benassi and Elliott, 2005). Internal velocity feedback has been used in the past (Inman, 1990) to minimise the instability problem by adding damping to the resonance peak. In this arrangement, the relative velocity between the permanent magnets and coil is used as the feedback signal for the vibration control system. This relative motion can be measured using a secondary coil wound on the inside of the main driving coil (Paulitsch et al., 2004, 2005). More recently, Benassi and Elliott (2004, 2005) used displacement feedback and a PID controller to control the vibration of a structure on which the actuator was mounted. The displacement used for the feedback loop was the relative displacement between the coil and vibrating mass, and it was measured using strain gauges attached to a thin metal foil suspension that separated the coil from the vibrating mass.

The proportional term (corresponding to direct displacement feedback) of the PID controller was used to adjust the resonance frequency of the actuator. The integral term was used to provide self-levelling and thus prevent the sagging problem associated with soft suspensions, and also prevent the mass from hitting the stops when the entire unit was subjected to an external acceleration such as in an aircraft manouevre. The derivative term (corresponding to velocity feedback) was used to provide the system with sufficient damping to achieve a good stability margin (gain margin more than 6 dB).

For the purpose of the following analysis, the proof mass actuators shown in Figures 15.20(a–c) (with velocity feedback included) may be represented in block diagram form as shown in Figure 15.21.

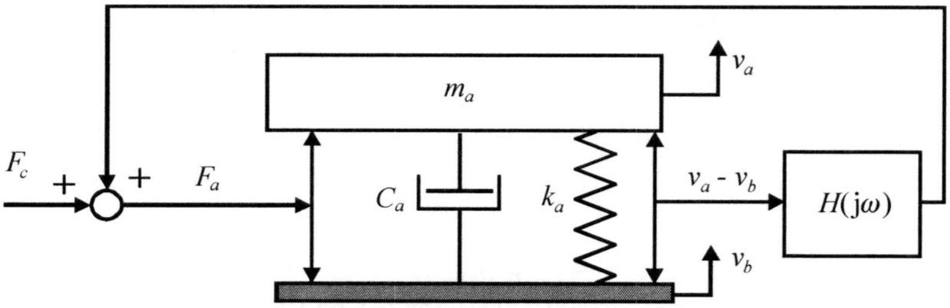

Figure 15.21 Representation of a proof mass actuator with velocity feedback. F_c is the actuator force in the absence of feedback, v_a is the velocity of the actuator mass, v_b is the velocity of the structure on which the actuator is mounted, C_a is the actuator damping, k_a is the stiffness of the actuator suspension and $H(j\omega)$ is the transfer function of the velocity feedback term.

The motion of the mass in Figure 15.21 may be expressed as:

$$j\omega m_a v_a + C_a(v_a - v_b) + k_a(v_a - v_b)/j\omega = F_a \tag{15.7.1}$$

Addition of a simple, frequency dependent gain for the term, $H(\mathrm{j}\omega)$, results in an addition of damping to the actuator (and thus the structure being controlled), which assists in maintaining system stability. The maximum gain that can be used is given by (Ananthaganeshan et al., 2002):

$$\frac{K_v}{C_a} = \frac{1 - 16\tau^2}{8\tau\zeta_a} \tag{15.7.2}$$

where

$$\tau = \frac{2\pi T}{\omega_a} \tag{15.7.3}$$

and T is the time delay (seconds) through the control system, the actuator and the relative velocity sensor. The resonance frequency of the actuator system is given by:

$$\omega_a = \sqrt{k_a / m_a} \tag{15.7.4}$$

and the critical damping ratio ζ_a is given by:

$$\zeta_a = \frac{C_a}{2\sqrt{k_a m_a}} \tag{15.7.5}$$

Note that when the gain reaches the amount indicated in Equation (15.7.2), it must be made constant and no longer allowed to increase with frequency. Thus, it is best if the maximum frequency of interest for control is well below this frequency (referred to as the critical frequency) and given by:

$$\omega = \frac{\omega_a}{4\tau} \tag{15.7.6}$$

The open-loop transfer function for the actuator mounted to a rigid structure is given by:

$$G(\mathrm{j}\omega)H(\mathrm{j}\omega) = \frac{1}{-\omega^2 m_a + \mathrm{j}\omega C_a + k_a}(\mathrm{j}\omega K_v) \tag{15.7.7}$$

The term in brackets represents the velocity feedback, and the denominator represents the actuator dynamics. Note that this equation is only valid below the frequency defined by Equation (15.7.6).

If the proof mass actuator is mounted on a rigid structure, the force transmitted to that structure is given by:

$$F_t = C_a(v_a - v_b) + k_a(v_a - v_b)/\mathrm{j}\omega - F_a \tag{15.7.8}$$

where the positive direction is upwards as shown in Figure 15.21. However, if the structure on which the actuator is mounted is flexible, the dynamics of the structure must be included in the analysis to determine the force that will be transmitted to the structure (see, for example, Paulitsch et al., 2004).

15.8 ELECTRODYNAMIC AND ELECTROMAGNETIC ACTUATORS

Electrodynamic actuators or shakers consist of a moveable core (or armature) to which is fixed a cylindrical electrical coil. The core and coil move back and forth in a magnetic field which is generated by a permanent magnet (smaller shakers) or by a d.c. current flowing through a second coil fixed to the stationary portion of the shaker (stator), as shown in Figure 15.22. The moveable core is supported on mounts which are very flexible in the axial direction in which it is desired that the core move and rigid in the radial direction to prevent the core contacting the outer armature. To maximise lightness and stiffness, the core is usually constructed of high strength aluminium alloy and is usually hollow. Mechanical stops are also usually provided to prevent the core from being over-driven. In some cases, the driving coil is air cooled.

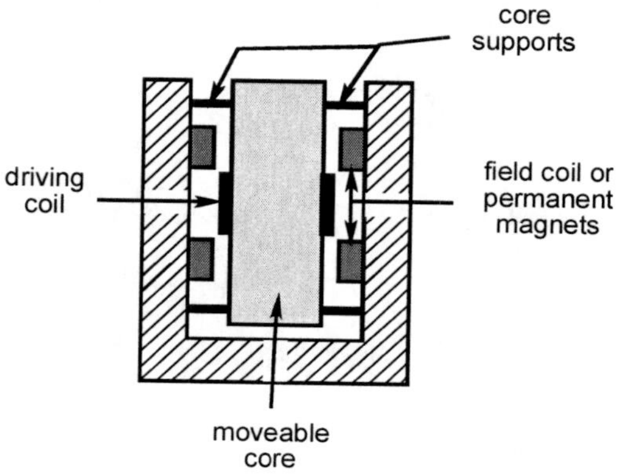

Figure 15.22 Electrodynamic shaker schematic diagram.

When a sinusoidal voltage is applied to the drive coil, the polarity and strength of the drive coil magnetic field changes in phase with the applied voltage and so then does the force of attraction between the field coil (or permanent magnets) and the driving coil, resulting in axial movement of the core (or armature).

Electromagnetic actuators are similar in construction to electrodynamic shakers except that the inner core as well as the armature is fixed (see Figure 15.23). It is constructed by surrounding an electrical coil with a permanent magnet. Supplying a sinusoidal voltage to the coil results in the production of a sinusoidally varying magnetic field that can be used to shake ferrous-magnet structures, or structures made from other materials if a thin piece of shim steel is bonded to them. Clearly, the electromagnetic actuator must be mounted on a rigid fixture, preferably not connected to the structure to be excited.

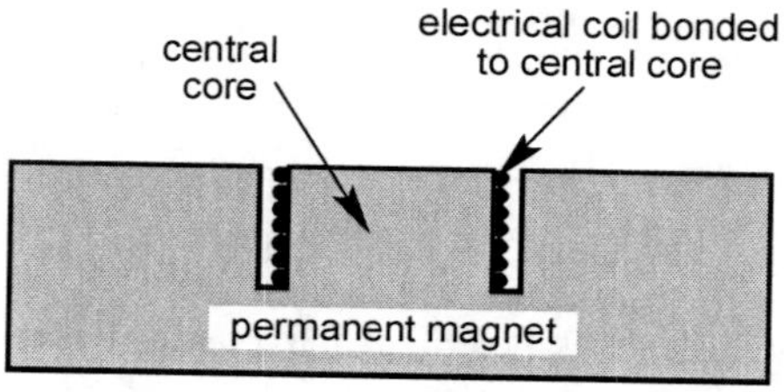

Figure 15.23 Electromagnetic shaker.

If the permanent magnet were not used, the attraction force between the coil and the structure being shaken would not vary sinsoidally but would vary as shown in Figure 15.24(b), resulting in most of the excitation energy being at twice the coil driving frequency.

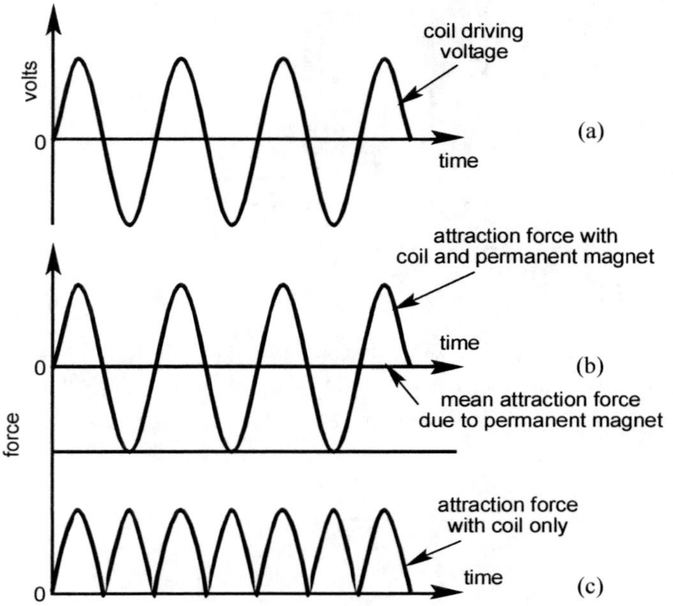

Figure 15.24 Electromagnetic attraction force versus coil driving voltage: (a) sinusoidal coil driving voltage; (b) corresponding attraction force for transducer with permanent magnet; (c) corresponding attraction force for transducer without permanent magnet.

Electromagnetic drivers such as those just described are easily and cheaply constructed by bonding the coil from a loudspeaker driver into the core of the permanent magnet where it is normally located in its at rest condition.

15.9 MAGNETOSTRICTIVE ACTUATORS

The phenomenon of giant magnetostriction in rare-earth Fe_2 compounds was first reported by Clark and Delson in 1972. The most effective alloy is referred to as terfenol, with the composition $Tb_{0.3}\,Dy_{0.7}\,Fe_{1.93}$. The subscripts refer to the atomic proportions of each element (terbium, dysprosium and iron). Multiplying these by the atomic weights gives the weight proportions (18%, 42% and 40% respectively). The notable properties of this material are its high strain ability (25 times that of nickel, the only other commonly used magnetostrictive material and 10 times that of piezoelectric ceramics – see next section) and high energy densities (100 times that of piezoceramics). Thus, terfenol-D can produce high force levels and high strains relative to other expanding materials. The maximum possible theoretical strain is 2440 μ strain, and 1400 μ strain is achievable in practice. The properties of terfenol are discussed in detail in the literature by Hiller et al. (1989) and Moffet et al. (1991a,b). A typical commercially available terfenol actuator is illustrated in Figure 15.25 (Edge Technologies, 1990).

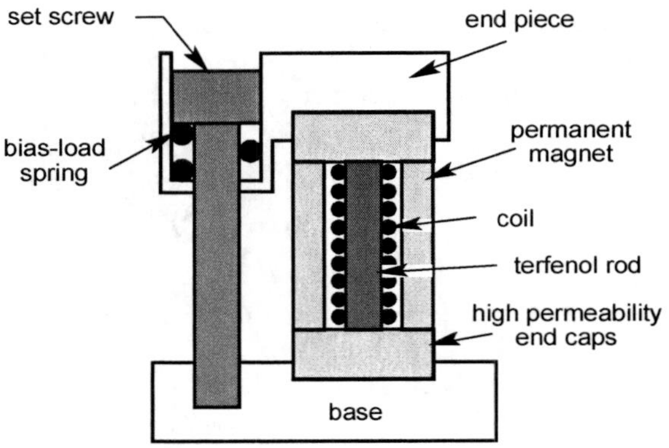

Figure 15.25 Cylindrical magnetostrictive actuator cross-section.

The alternating magnetic field is applied to the terfenol rod by a coil that can produce a strain of 1000 μstrain in the rod for an applied magnetic field, *H*, of 700 Oe. This field strength is a function of the current flowing in the coil and can be calculated using:

$$H = \frac{f 2\pi N I}{d_i(\alpha - 1)} \sqrt{\frac{(\alpha^2 - 1)}{2\pi\beta}} \tag{15.9.1}$$

$$f = 0.2 \sqrt{\frac{2\pi\beta}{(\alpha^2 - 1)}} \log_e \left[\frac{\alpha + \sqrt{\alpha^2 + \beta^2}}{1 + \sqrt{1 + \beta^2}} \right] \tag{15.9.2}$$

where $\alpha = d_0/d_i$, I is the coil current (amps), N is the number of turns in coil, d_0 is the coil outer diameter (m), d_i is the coil inner diameter (m), $\beta = L/d_i$ and L is the coil length (m). The geometry factor f gives some indication of the efficiency of the particular coil design, and has a maximum theoretical value of 0.179.

Thus, for an actuator containing a 50 mm long terfenol rod, the maximum possible displacement will be $50 \times 10^{-3} \times 1400 \times 10^{-6}$ m = 70 μm. In practice, actuators are generally limited to 1000 μ strain (or 50 μm for a 50 mm long terfenol rod) to minimise non-linearity.

Examination of the actuator illustrated in Figure 15.25 shows the use of high permeability iron end caps to help reduce fringing of the magnetic flux lines at the rod ends. Flux density is also improved by the coil extending past the ends of the terfenol rod.

15.9.1 Magnetic Bias

As application of a magnetic field of any orientation will cause the terfenol rod to expand, a d.c. magnetic bias is necessary if the terfenol rod expansion and contraction is to follow an a.c. input signal into the coil. Thus, a d.c. magnetic bias, supplied either with a d.c. voltage signal applied to the coil or with a permanent magnet, is required so that the terfenol rod expands and contracts about a mean strain (usually about 500 μ strain). Use of d.c. bias current is characterised by problems associated with drive amplifier design and transducer overheating; thus, it is more common to use a permanent magnet bias. The cylindrical permanent magnet bias shown in Figure 15.25 also acts as the flux return path for the exciter coil.

15.9.2 Mechanical Bias or Prestress

To optimise performance (maximise the strain ability of the actuator), it is necessary to apply a longitudinal pre-stress of between 7 and 10 MPa. The pre-stress ensures that the magnetically biassed actuator will contract as much as it expands when a sinusoidal voltage is applied to the drive coil. It is essential that the pre-stress arrangement also be highly compliant to avoid damping the drive element. The mechanical bias also helps to mate components and allows transfer of forces with low loss.

15.9.3 Frequency Response, Displacement and Force

Terfenol actuators are usually characterised by a resonance frequency in the few kilohertz range and generally have a reasonably flat frequency response at frequencies below about half the resonance frequency. The actual resonance frequency depends on the mass driven and the stiffness of the terfenol rod. Longer, thinner drives and larger masses result in lower resonance frequencies.

The force output for a clamped actuator may be calculated using $F = EA\Delta L/L$, where E is the elastic modulus (3.5×10^{10} N m^{-2}), A is the cross-sectional area of the terfenol rod (m^2) and $\Delta L/L$ is the strain. Thus, a 6 mm diameter rod operating in a magnetic field capable of producing ±500 μ strain about the bias value of +500 can produce a force of ±500 N. Specifications for one such actuator consisting of a 50 mm long × 6 mm terfenol rod are listed in Table 15.1.

15.9.4 Disadvantages of Terfenol Actuators

Terfenol is very brittle and must be carefully handled. Its tensile strength is low (100 MPa) although its compressive strength is reasonably high (780 MPa). Another important disadvantage is the low displacement capability, which would be a problem in low-frequency vibration control (below 100 Hz). One way around this limitation is to use displacement expansion devices which rely on the lever principle to magnify the displacement at the expense of reduced force output. The main disadvantage associated with magnetostrictive actuators, however, is the hysteresis inherent in terfenol, which results in a non-linear actuation force for a linear voltage input. This results in excitation frequencies being generated that are not related to the voltage input frequencies. This requires the use of non-linear controllers such as those based on neural networks.

15.9.5 Advantages of Terfenol Actuators

The main advantage of terfenol is its high force capability for a relatively low cost. Electrodynamic shakers capable of the same force capability cost and weigh between 25 and

100 times as much and require a much larger driving amplifier. Thus, another advantage is small size and light weight, which makes the actuator ideal for use in situations where no reaction mass is necessary. Actuator arrangements for active vibration control with no need for a reaction mass are discussed in Section 15.12.1.

Table 15.1 Typical terfenol actuator specifications.

Actuator	
Useable displacement	±25 μm
Current input for normal operation	±1 amp
Force capability±500 N	
Overall length 75 mm	
Overall diameter	60 mm
Terfenol length 50 mm	
Terfenol diameter	6 mm
Overall weight 0.6 kg	
Coil	
Rating 300 Oe amp^{-1},	1181 amp turns
Coil length	53 mm
Coil inner diameter	6.7 mm
Coil outer diameter	16.5 mm
Wire gauge	26 AWG

Note: See Edge Technologies manual for more information (Butler, 1988).

15.10 SHAPE MEMORY ALLOY ACTUATORS

A shape memory alloy is a material which, when plastically deformed in its low temperature (or martensitic) condition, and the external stresses removed, will regain its original (memory) shape when heated. The process of regaining the original shape is associated with a phase change of the solid alloy from a martensite to an austenite crystalline structure.

The most commonly used shape memory alloy in active vibration control systems is nitinol (an alloy of approximately equi-atomic composition of nickel and tin). Room temperature plastic strains of typically 6 to 8% can be completely recovered by heating the material beyond its transition temperature (which ranges from 45°C to 55°C). Restraining the material from regaining its memory shape can yield stresses of up to 500 MPa (for an 8% plastic strain and a temperature of 180°C – see Rogers, 1990). On being transformed from its martensite to its austenite phase, the elastic modulus of nitonol increases by a factor of three (from 25 to 75 GPa) and its yield strength increases by a factor of eight (from 80 to 600 MPa).

The attractiveness of nitinol over other shape memory alloys for active control is its high resistivity which makes resistive heating feasible. This means that nitinol can be conveniently heated by passing on electrical current through it.

Nitinol is commercially available in the form of thin wire or film, which can be used for the active control of structural vibration by either embedding the material in the structure or fixing it to the surface of the structure. Before being embedded or fixed to the surface of the structure, the nitinol is plastically elongated. When used in composite structures, the nitinol fibres must be constrained so that they cannot contract to their normal length during the high temperature curing of the composite.

Active control of structural vibration using nitinol can be achieved either directly or indirectly. When nitinol wire or film is heated, it tries to contract to its normal size. If it is bonded to the surface of a structure, it will impose a bending moment on the structure as it tries to contract. If the structure were a vibrating thin panel or a beam, then the vibration could be controlled by using a shape memory alloy element on each side of the panel or beam. Applying current to the element on one side will cause a controlling moment acting in one direction and applying current to the element on the other side will cause a controlling moment in the opposite direction. Thus, alternating the supply of current from one side to the other will induce an alternating moment in the structure, which can be used to control the structural vibration. This is referred to as direct control and application of this method to the control of the vibration of a cantilevered beam has been demonstrated successfully by Baz et al. (1990). The upper frequency limit for control is about 10 Hz, due to the difficulty in cooling the nitinol between cycles. If the nitinol is embedded in the structure (on each side of the neutral axis), the upper frequency limit for control would be much lower due to the inherent cooling problems.

A second active control technique involves the use of shape memory alloy elements embedded in a structure, which when activated, change the dynamic characteristics of the structure by changing its effective stiffness. For this technique, shape memory fibres are usually embedded on the neutral axis of a composite structure. Applying a current to the fibres causes them to apply a strain to the structure which can increase its resonance frequencies. This technique has been demonstrated for a beam (Rogers, 1990) and a plate (Rogers et al., 1991, Baz and Chen, 1995). In both cases, it was found that heating the nitinol resistively to about 120°C changed the resonance frequencies of the first few modes by up to 200%. Lower temperatures resulted in smaller changes. It follows that this technique can be used to optimise the sound radiation characteristics of a structure for a particular excitation. If the excitation frequency spectrum slowly changes, the structural stiffness could be readjusted by adjusting the current flow through the nitinol elements to maintain optimal sound radiation characteristics (Saunders et al., 1991). This has obvious applications to composite submarine hulls, aircraft fuselages and other aerospace structures.

15.11 PIEZOELECTRIC (ELECTROSTRICTIVE) ACTUATORS

This type of actuator may be divided into two distinct categories: the thin plate or film type and the stack (shaker) type. The thin type may be further categorised into piezoceramic plates (PZT) or piezoelectric film (PVDF). The thin type is usually bonded to the structure it is to excite and generates excitation by imparting a bending moment to the structure. The stack type is used in a similar way to an electrodynamic shaker or magnetostrictive actuator, generally applying to the structure a force distributed over a small area. The thin type and stack type actuators will be considered separately in the following discussion. Both types of actuator generate forces and displacements when subjected to an electric field – hence their alternative description, electrostrictive.

15.11.1 Thin Actuators

A detailed review of work on thin piezo actuators was done by Smits et al. (1991) and the reader is referred to this for a detailed survey of the uses of this material.

The behaviour of piezoceramic plate and PVDF film actuators is similar (although their properties differ significantly); thus, the two types will be considered together. Note that these actuator types can also act as sensors, and PVDF film acting in this capacity was discussed in detail in Section 15.4.

For thin actuators, the extension as a result of an applied voltage is in a direction normal to the direction of polarisation as shown in Figure 15.26. For an applied voltage V, the free extension ΔL *(m)* is given in terms of strain by:

$$\frac{\Delta L}{L} = \frac{d_{31} V}{h_a} = \Lambda \tag{15.11.1a,b}$$

where h_a is the piezoelectric actuator thickness and d_{31} is its charge or strain constant, representing the strain produced by an applied electric field.

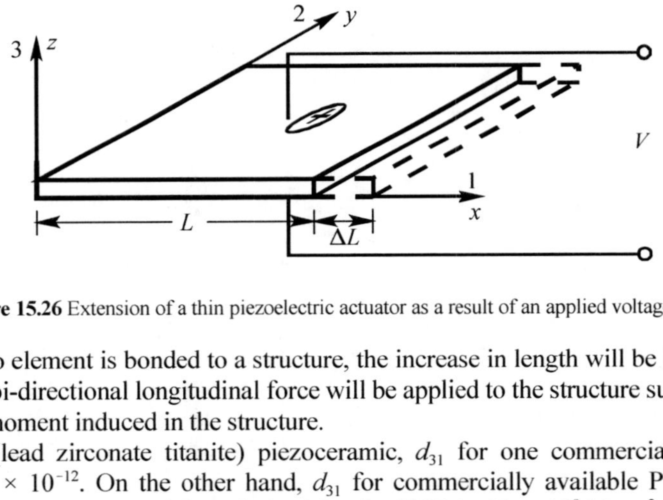

Figure 15.26 Extension of a thin piezoelectric actuator as a result of an applied voltage.

If the piezo element is bonded to a structure, the increase in length will be less than this and a resulting bi-directional longitudinal force will be applied to the structure surface, which will result in a moment induced in the structure.

For PZT (lead zirconate titanite) piezoceramic, d_{31} for one commercially available material is 166×10^{-12}. On the other hand, d_{31} for commercially available PVDF film is 23×10^{-12}. Also, Young's modulus of elasticity for PZT is 63×10^9 N m^{-2}, whereas for PVDF film it is only 2×10^9 N m^{-2}. It will be shown in the following section that the strain induced in a simple structural element such as a beam (which is not too thin) is proportional to the product of d_{31} and E for a given applied voltage. Thus, it is clear that for the same thickness material and same applied voltage, PZT will produce about 150 times the strain in a beam than will PVDF film. Note, however, that the maximum operating voltage for PVDF film is about 25 times that for PZT (the latter being about 1 V r.m.s. μm^{-1} material thickness).

A PZT layer or PVDF film bonded to a structure such as a beam and subjected to an applied voltage will effectively generate a moment approximately equivalent to a line moment at each free edge, as shown in Figure 15.27(a). If a second actuator is bonded to the opposite side of the beam and subjected to an applied voltage equal and opposite to that applied to the first actuator, then the beam will bend about its neutral axis. The moment distribution generated in the beam by the two assumed line moments is shown as the dashed line in Figure 15.27(b). The actual distribution is more like that shown by the solid line (Crawley and de Luis, 1987). However, the effect of this deviation from the assumed ideal case is insignificant from the viewpoint of calculating the beam response.

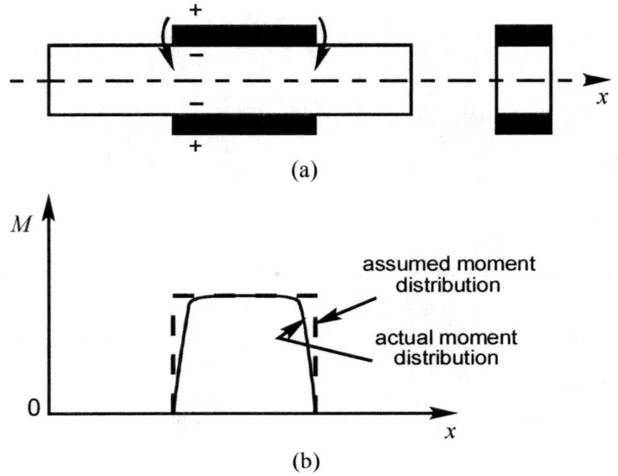

Figure 15.27 Beam bending moment distribution due to a PZT layer.

Although PZT actuators can be bonded directly on to structures using epoxy adhesive, their brittleness makes this a task requiring considerable skill, especially on curved surfaces. Usually, a thin layer of epoxy is first bonded to the surface under a layer of glass which has release agent sprayed on it. When dry, the glass is removed, leaving a smooth surface onto which the PZT is bonded. Some researchers prefer to order the PZT from manufacturers already bonded to a thin brass shim. This shim is then bonded (with a lot less trouble) directly to the structure to be controlled.

Thin PZT actuators for active control were first used by Forward and Swigert in 1981 for the control of vibration of a tall mast. In 1987, Crawley and de Luis presented a detailed analysis for the case of a beam excited by two PZT crystals bonded to it on opposite sides and driven 180° out of phase. Unfortunately, this work suffers from the incorrect assumption of uniform strain across the thickness of the actuators. However, many of the results obtained remain very useful. In 1988, Baz and Poh developed a finite element model of a beam excited by a singe PZT crystal bonded to one side. In 1991, four important papers were published which explained how to accurately model PZT excitation of beams (Crawley and Anderson 1991; Clark et al., 1991) and plates (Dimitriadis et al., 1991; Kim and Jones, 1991). Unfortunately, the assumption of the slope of the stress distribution in the actuator being the same as that in the beam, which was made by Clark et al. (1991) and Dimitriadis et al. (1991) is not strictly correct, as the elastic modulus of the actuator material is not the same as that of the beam. The implications of this will be made clear in the next section.

15.11.1.1 One-Dimensional Actuator Model: Effective Moment

It is useful to derive the bending moment generated in a one-dimensional structure such as a beam by a pair of piezoelectric actuators placed on opposite sides of it. Once derived, the bending moment can be used with the equations of Chapter 10 to calculate the response of a beam with any desired end conditions.

The configuration being analysed is shown in Figure 15.28, which also shows the assumed strain distribution and corresponding stress distribution for a positive beam displacement in the z-direction, when the actuators are energised and produce pure positive bending in the beam.

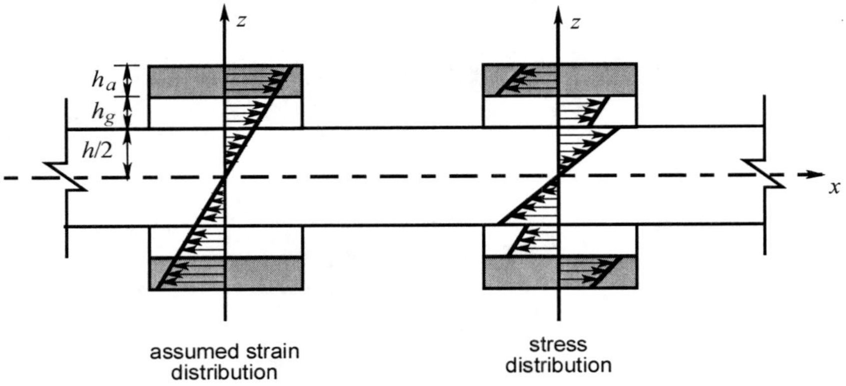

Figure 15.28 One-dimensional actuator model.

The linear strain distribution implies that the strains are continuous across each layer interface and the centres of the radii of curvature for each layer are concurrent. That is, the glue layer is not deformed during bending such that its thickness at the edges varies differently to its thickness in the centre.

Note that a glue layer has been included, although if this is very thin, it has a negligible effect on the results (Kim and Jones, 1991). Note also that the stress in the top actuator is compressive even though the stress in the glue layer and beam beneath it is tensile. This is because the strain in the actuator is less than it would be if it were not bonded to the beam (due to the applied electric field – see Equation (15.11.1)).

The bending moment generated in the beam by the actuators per unit length on a plane perpendicular to the x-axis is given by:

$$M_x = \int_{-h/2}^{h/2} \sigma_x z \, \mathrm{d}z \tag{15.11.2}$$

where h is the thickness of the beam, z is the distance from the centre of the beam to where σ_x is acting, and σ_x is the axial normal stress acting along the x-axis, varying from zero at the neutral axis to a maximum at the top or bottom of the beam. The moment M_x exists in the beam over the full length of the actuator and may be simulated by two external line moments acting in opposite directions (see Figure 15.27(a)) applied to the beam at the free edges of the actuators, where each line moment is equal in magnitude to bM_x (b is the width of the beam).

The stress σ_x may be written in terms of the stress in the beam at the interface between the bonding (or glue) layer and the beam (σ_{ib}) as follows:

$$\sigma_x = \frac{2\sigma_{ib} z}{h} \tag{15.11.3}$$

To obtain an expression for σ_{ib} in terms of known quantities, the following moment equilibrium condition is used:

$$\int_0^{\frac{h}{2}} \sigma_x z \, \mathrm{d}z + \int_{\frac{h}{2}}^{\frac{h}{2}+h_g} \sigma_g z \, \mathrm{d}z + \int_{\frac{h}{2}+h_g}^{\frac{h}{2}+h_g+h_a} \sigma_a z \, \mathrm{d}z = 0 \quad (15.11.4)$$

Before solving Equation (15.11.4) for σ_{ib}, it is necessary to express σ_x, σ_g and σ_a in terms of σ_{ib}.

The necessary equivalence for σ_x is given by Equation (15.11.3). The expressions relating σ_g and σ_a to σ_{ib} can be derived from the assumption of uniform strain slope across the beam, glue and actuator, as shown in Figure 15.28. That is,

$$\varepsilon(z) = \varepsilon_b = \varepsilon_g = \varepsilon_a = \mu z \quad (15.11.5a–d)$$

where μ is the strain slope, which can be found by developing an expression for ε_b in terms of the beam stress σ_x and the beam modulus of elasticity E_b. For a beam whose thickness is similar to its width, the plane stress assumption can be used, which gives:

$$\varepsilon_b = \frac{\sigma_x}{E_b} \quad (15.11.6)$$

Substituting Equation (15.11.3) into Equation (15.11.6) and using Equation (15.11.5) gives

$$\mu = \frac{2\sigma_{ib}}{E_b h} \quad (15.11.7)$$

Clearly, the stress in the actuator is a related to the free strain Λ due to the electric field (see Equation (15.11.1)), Young's modulus of elasticity for the actuator and the actuator strain ε_a as a result of bending, as follows:

$$\sigma_a = E_a(\varepsilon_a - \Lambda) \quad (15.11.8)$$

As Λ is always larger than ε_a, the stress in the actuator will be of opposite sign to that in the beam.

Substituting Equations (15.11.5) and (15.11.7) into Equation (15.11.8) gives:

$$\sigma_a = E_a\left(\frac{2\sigma_{ib}z}{E_b h} - \Lambda\right) \quad (15.11.9)$$

The stress in the bonding layer may be written as:

$$\sigma_g = E_g \varepsilon_g = 2\frac{E_g}{E_b}\sigma_{ib}\frac{z}{h} \quad (15.11.10a,b)$$

Substituting Equations (15.11.3), (15.11.9) and (15.11.10) into Equation (15.11.4) and integrating gives:

$$\sigma_{ib} = \frac{3hE_a h_a (h + 2h_g + h_a)\Lambda}{2\left(\frac{h^3}{4} + \frac{E_g}{E_b}h_g\left(\frac{3}{2}h^2 + 3hh_g + 2h_g^2\right) + \frac{E_a}{E_b}h_a\left(3\frac{h^2}{2} + 6h_g^2 + 2h_a^2 + 6hh_g + 3hh_a + \right.\right.} \tag{15.11.11}$$

Substituting Equation (15.11.11) into Equation (15.11.13), then into Equation (15.11.2) and integrating gives:

$$M_x = \frac{h^3 E_a h_a (h + 2h_g + h_a)\Lambda}{\frac{h^3}{4} + \frac{E_g}{E_b}h_g\left(\frac{3}{2}h^2 + 3hh_g + 2h_g^2\right) + \frac{E_a}{E_b}h_a\left(\frac{3h^2}{2} + 6h_g^2 + 2h_a^2 + 6hh_g + 3hh_a + 6h_g h_a\right)} \tag{15.11.12}$$

If the bonding layer is very thin, its effect can be neglected (Kim and Jones, 1991) and the following is obtained:

$$M_x = \frac{\rho_a(2 + \rho_a)}{4\left(1 + \frac{E_a}{E_b}\rho_a\left(3 + \rho_a^2 + 3\rho_a\right)\right)} h^2 E_a \Lambda \tag{15.11.13}$$

$$\text{where} \quad \rho_a = \frac{2h_a}{h} \tag{15.11.14}$$

The preceding formulation is for a beam completely covered with the actuator. In cases where the actuators only extend partially along the beam, free edges will exist where the equilibrium condition requires that the normal stress at the actuator boundaries be zero, effectively invalidating the relationships just derived. However, it is generally accepted (Kim and Jones, 1991; Dimitriadis et al., 1991) that the actuator stress field for a distributed actuator is unaffected by the free edge except within about four actuator thicknesses of the boundary. This has the effect of slightly altering the moment distribution in the beam (to the solid curve shown in Figure 15.27(b)), which has a negligible effect on the resulting beam response.

Using the results of Chapter 2.3, the classical beam equation of motion for excitation by a piezoelectric actuator pair is:

$$\frac{\partial^4 w}{\partial x^4} + \frac{\rho S}{E_b I_b}\frac{\partial^2 w}{\partial t^2} = -\frac{bM_x}{E_b I_b}\left[\delta'(x - x_1) - \delta'(x - x_2)\right] \tag{15.11.15}$$

where the prime denotes differentiation with respect to x, x_1 and x_2 are the axial coordinates of the two edges of the piezo actuator and M_L of Equation (2.3.50) is equal to bM_x. The quantity I_b is the second moment of area of the beam cross-section ($bh^3/12$), and E_b is the elastic modulus for the beam material. Note that the effects of transverse shear in the beam and rotary inertia are neglected in this equation, which makes it valid only for frequencies where the wavelength is much larger than the beam cross-sectional dimensions. If this is not so, then the more complex equation of motion derived in Section 2.3 must be used.

If the actuators do not cover the full width of the beam, the right-hand side of Equation (15.11.15) must be multiplied by b_a / b, where b_a is the actuator width and b is the beam width.

Note that the preceding analysis only applies to a beam whose thickness is similar to its width. For thin beams, the analysis and corresponding expression for the moment M_y acting on a thin plate, as derived in the next section, must be used. Furthermore, the elastic modulus E_b for the beam in Equation (15.11.15) must be replaced by $E_b/(1 - \nu^2)$.

15.11.1.2 Two-Dimensional Actuator Analysis

The analysis for a two-dimensional actuator is similar to that just done for the one-dimensional case with the main difference being the assumption of plane strain (due to the thinness of the plate with respect to its dimensions).

For the plate, the following stress strain relationships can be written (Timoshenko and Woinowsky-Krieger, 1959):

$$\sigma_{xp} = \frac{E_p}{1 - \nu_p^2} \left(\varepsilon_{xp} + \nu_p \varepsilon_{yp} \right) \tag{15.11.16}$$

$$\sigma_{yp} = \frac{E_p}{1 - \nu_p^2} \left(\varepsilon_{yp} + \nu_p \varepsilon_{xp} \right) \tag{15.11.17}$$

For an actuator which is bonded to a homogeneous plate and which gives equal free strains in the x- and y-directions (that is, $d_{31} = d_{32}$), $\varepsilon_{xp} = \varepsilon_{yp}$ (Dimitriadis et al., 1991); thus, the stresses in the x- and y-directions will also be equal. If this assumption is not made (and it may not be valid for some PVDF film), the analysis is still possible but the algebra becomes more complex.

Using the above mentioned equality, the subscripts x and y (denoting the x and y directions across the plate – see Figure 15.29) may be dropped, and Equations (15.11.16) and (15.11.17) become:

$$\sigma_p = \frac{E_p}{1 - \nu_p} \varepsilon_p \tag{15.11.18}$$

In terms of the stress σ_{ip} in the plate at the interface between the plate and the glue layer, the stress in the beam may be written as:

$$\sigma_p = \frac{\sigma_{ib} z}{h} \tag{15.11.19}$$

The slope μ of the strain curve is now given by:

$$\mu = \frac{2\sigma_{ip}(1 - \nu_p)}{E_p h} \tag{15.11.20}$$

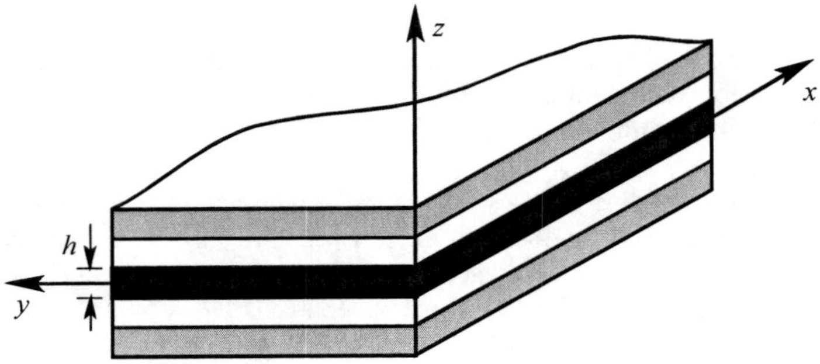

Figure 15.29 Plate model for two-dimensional actuator analysis.

The stress distribution in the bonding layer may be written as:

$$\sigma_g = \frac{E_g \varepsilon_g}{(1-v_g)} = \frac{E_g z \mu}{(1-v_g)} = \frac{2E_g(1-v_p)}{E_p(1-v_g)} \sigma_{ip} \frac{z}{h} \tag{15.11.21a–c}$$

Modifying Equation (15.11.8) to account for the Poisson effect, the following is obtained for the actuator stress:

$$\sigma_a = \frac{E_a}{1-v_a}(\varepsilon_a - \Lambda) \tag{15.11.22}$$

Substituting Equations (15.11.5) and (15.11.20) into Equation (15.11.22) gives:

$$\sigma_a = \frac{2E_a(1-v_p)\sigma_{ip}}{E_p(1-v_a)h} - \frac{E_a \Lambda}{(1-v_a)} \tag{15.11.23}$$

Substituting Equations (15.11.19), (15.11.21) and (15.11.23) into Equation (15.11.4) (where σ_x has been replaced by σ_p) gives:

$$\sigma_{ip} = \frac{3\gamma h h_a (h + 2h_g + h_a)\Lambda}{2\left(\frac{h^3}{4} + \alpha h_g\left(\frac{3}{2}h^2 + 3hh_g + 2h_g^2\right) + \beta h_a\left(\frac{3}{2}h^2 + 6h_g^2 + 2h_a^2 + 6hh_g + 3hh_a + 6h_g h_a\right)\right)} \tag{15.11.24}$$

where

$$\alpha = \frac{(1-v_p)E_g}{(1-v_g)E_p} \tag{15.11.25}$$

$$\beta = \frac{(1-v_p)E_a}{(1-v_a)E_p} \tag{15.11.26}$$

$$\gamma = \frac{E_a}{1 - \nu_a} \tag{15.11.27}$$

Values of E_a, ν_a and d_{31} for typical PZT and PVDF materials are given in Table 15.2.

Table 15.2 Properties of some commercially available thin piezo-electric actuators.

	PZT			
Material	G-1195	G-1512	G-1278	PVDF Film
E_a (GPa)	63	63	60	2
ν_a	0.35	0.35	0.35	0.35
d_{31} (m V^{-1} or C N^{-1})	1.66×10^{-10}	2.3×10^{-10}	2.7×10^{-10}	0.23×10^{-10}
d_{33} (m V^{-1})	-3.6×10^{-10}	-4.8×10^{-10}	-5.85×10^{-10}	-0.33×10^{-10}

The bending moments in the plate per unit length about planes perpendicular to the x- and y-axes are equal and are given by:

$$M_x = M_y = \int_{-h}^{h} \sigma_p z \, dz = \frac{1}{h} \int_{-h}^{h} \sigma_{ip} z^2 dz = \frac{2}{3} h^2 \sigma_{ip} \tag{15.11.28a–d}$$

and the equivalent external line moments are $M_{xL} = M_y(y_2 - y_1)$ and $M_{yL} = M_x(x_2 - x_1)$. Substituting Equation (15.11.24) into Equation (15.11.28d) gives the required expression for M_x and M_y.

For the case of zero glue layer thickness (which can usually be assumed), the following is obtained:

$$M_x = M_y = \frac{\rho_a(2 + \rho_a)}{4\left(1 + \beta\rho_a\left(3 + \rho_a^2 + 3\rho_a\right)\right)} h^2 \gamma \Lambda \tag{15.11.29a,b}$$

where ρ_a is defined by Equation (15.11.14).

Equations (15.11.24) to (15.11.29) can be used to calculate the applied moment as a function of the ratio of actuator thickness to plate thickness. When this is done, it is found that for a constant applied electric field strength (volts per mm of actuator thickness), there is an optimum actuator thickness for producing the maximum applied moment (Kim and Jones, 1991). It can be shown that for commercially available piezoceramics, the optimum actuator thickness is about half of the plate thickness for a steel substructure and about one quarter of the plate thickness for an aluminium substructure. A similar analysis to find the optimum actuator thickness for a beam could be done using Equation (15.11.12).

The moments generated in the plate by the piezoactuator can be approximated by external line moments acting on the plate at the edges of the piezoelectric layer. Using the results of Section 2.3, Equation (2.3.105), the plate equation of motion (classical plate theory) for excitation by a piezoceramic actuator bonded to the plate surface as shown in Figure 15.30 can be written as:

$$D\nabla^4 w + \rho h\ddot{w} = M_x\Big[-\big(\delta'(x-x_1) - \delta'(x-x_2)\big)\big(u(y-y_1)-u(y-y_2)\big) + \big(\delta^{\dagger}(y-y_1) - \delta^{\dagger}(y-y_2)\big)\big(u(x-x_1) - u(x-x_2)\big)\Big] \quad (15.11.30)$$

where M_x is defined by Equation (15.11.29), $u(\,)$ is the unit step function (introduced as the piezoelectric layer does not extend to the plate edges), $\delta(\,)$ is the Dirac delta function, and D is the flexural rigidity of the plate given by:

$$D = \frac{Eh^3}{12(1-\nu^2)} \quad (15.11.31)$$

The prime denotes differentiation with respect to x, the symbol † denotes differentiation with respect to y, the symbol • denotes differentiation with respect to time, ρh is the mass per unit area of the plate, and h is the plate half-thickness. It is assumed that the piezoelectric actuators do not add significantly to the mass or stiffness of the plate.

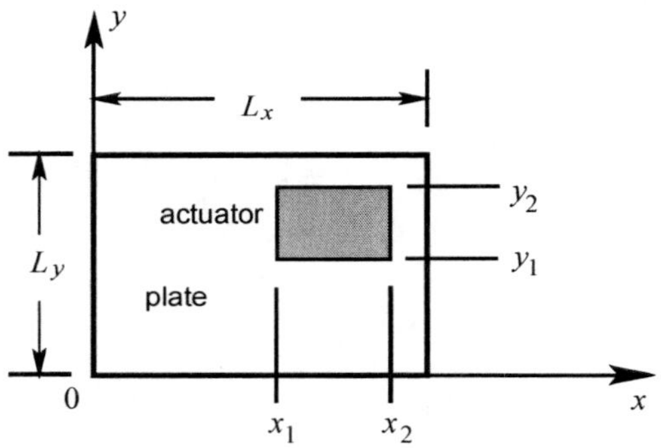

Figure 15.30 Rectangular plate excited by a rectangular actuator (note that a second identical actuator is located in the same position on the other side of the plate).

The plate forcing function represented by the right-hand side of Equation (15.11.30) is a result of external line moments acting around the edges of the actuators which act to generate the uniformly distributed moment in the plate.

It is of interest to examine the solution, w, to Equation (15.11.30) for a simply supported rectangular plate for a harmonically varying piezo-actuator exciting voltage of frequency ω. Using the results of Chapter 3, the following can be written for the plate displacement $w(x,y,t)$ at any location (x,y) and time t in terms of its vibration modes (m,n):

$$w(x,y,t) = e^{j\omega t} \sum_{m=1}^{\infty} \sum_{n=1}^{\infty} A_{mn}\,\psi_{mn}(x,y) \quad (15.11.32)$$

where A_{mn} is the amplitude of mode m,n and $\psi_{mn}(x,y)$ is its mode shape at location x,y.

Using the results of Section 2.3, the following equation describes the mode shape of a simply supported panel:

$$\psi_{mn}(x,y) = \sin\frac{m\pi x}{L_x}\sin\frac{n\pi y}{L_y} \tag{15.11.33}$$

The unknown modal amplitudes A_{mn} are found by substituting Equation (15.11.33) into Equation (15.11.32) and the result into Equation (15.11.30). Both sides of Equation (15.11.30) are then multiplied by $\psi_{m'n'}(x, y)$ and the result integrated over the surface of the plate. When this is done, the left side of the equation becomes:

$$\int_0^{L_y}\int_0^{L_x} \sin(\gamma_{m'}x)\sin(\gamma_{n'}y) \sum_{m=1}^{\infty}\sum_{n=1}^{\infty} A_{mn}\sin(\gamma_m x)\sin(\gamma_n y)\left[D(\gamma_m^2+\gamma_n^2) - \rho h\omega^2\right]\mathrm{d}x\,\mathrm{d}y \tag{15.11.34}$$

where $\gamma_m = m\pi/L_x$ and $\gamma_n = n\pi/L_y$.

As the modes are orthogonal, the product of any two different mode shape functions when integrated over the surface of the plate is zero. Thus, with $m' = m$, $n' = n$ and $S = L_xL_y$ being the area of the plate, Equation (15.11.34) becomes:

$$\frac{S}{2}A_{mn}\left[D(\gamma_m^2+\gamma_n^2) - \rho h\omega^2\right] \tag{15.11.35}$$

The plate resonance frequencies ω_{mn} are defined as the values of ω required to make Equation (15.11.35) equal to zero. Thus,

$$\omega_{mn}^2 = \frac{D}{\rho h}\left(\gamma_m^2+\gamma_n^2\right) \tag{15.11.36}$$

and Equation (15.11.35) can be written as:

$$\frac{\rho h S}{2}A_{mn}\left[\omega_{mn}^2 - \omega^2\right] \tag{15.11.37}$$

Multiplying the right side of Equation (15.11.30) by $\psi_{m'n'}$ and integrating over the area of the plate gives:

$$M_x\int_0^{L_x}\int_0^{L_y} \sin(\gamma_{m'}x)\sin(\gamma_{n'}y)$$

$$\times\left[\frac{\partial}{\partial x}\left(\delta(x-x_1) - \delta(x-x_2)\right)\left(u(y-y_1) - u(y-y_2)\right)\right. \tag{15.11.38}$$

$$\left. + \frac{\partial}{\partial y}\left(\delta(y-y_1) - \delta(y-y_2)\right)\left(u(x-x_1) - u(x-x_2)\right)\right]\mathrm{d}y\,\mathrm{d}x$$

The above expression can be evaluated to give:

$$\frac{-2M_x(\gamma_m^2+\gamma_n^2)}{\gamma_m\gamma_n}\left(\cos\gamma_m x_1 - \cos\gamma_m x_2\right)\left(\cos\gamma_n y_1 - \cos\gamma_n y_2\right) \tag{15.11.39}$$

where $m' = m$ and $n' = n$.

An expression for the modal amplitude A_{mn} is obtained by equating expressions (15.11.35) and (15.11.39) to give:

$$A_{mn} = \frac{4M_x(\gamma_m^2 + \gamma_n^2)}{\rho h S(\omega_{mn}^2 - \omega^2)\,\gamma_m\,\gamma_n}\left(\cos\gamma_m x_2 - \cos\gamma_m x_1\right)\left(\cos\gamma_n y_1 - \cos\gamma_n y_2\right) \qquad (15.11.40)$$

where M_x is defined by Equation (15.11.28).

An analysis of piezoelectric actuators attached to a cylinder is more complex than what has just been discussed for a plate. However, interested readers can consult Tani et al. (1995), who undertook the analysis for a cylinder.

Recently, piezoelectric crystal actuators have been developed that have an inter-digital electrode pattern (IDE actuators) that allows the actuators to be poled along its length rather than the standard electrode configuration of an electrode on the top and bottom surfaces, which allows poling through the actuator thickness. IDE actuators exhibit a 50% increase in strain over actuators with standard electrodes for the same applied voltage.

When controlling sound transmission through windows using an active control system, the actuators are often loudspeakers mounted in the window architrave. A possible alternative that has been demonstrated is to cover the entire window with a PVDF film with transparent electrodes. PVDF film is naturally transparent but considerable effort is required to achieve transparent electrodes. One example using transparent single-walled carbon nanotube electrodes was reported by Yu et al. (2006).

15.11.2 Thick Actuators

These actuators are generally thick compared to their other dimensions and act by expanding across their thickness in the direction of polarisation. The voltage is applied to the polarised surfaces as for the thin actuators, but in this case, the expansion of the thickness rather than expansion in the other dimensions is used to perform useful work. Stack actuators are made by bonding many thin actuators together as shown in Figure 15.31.

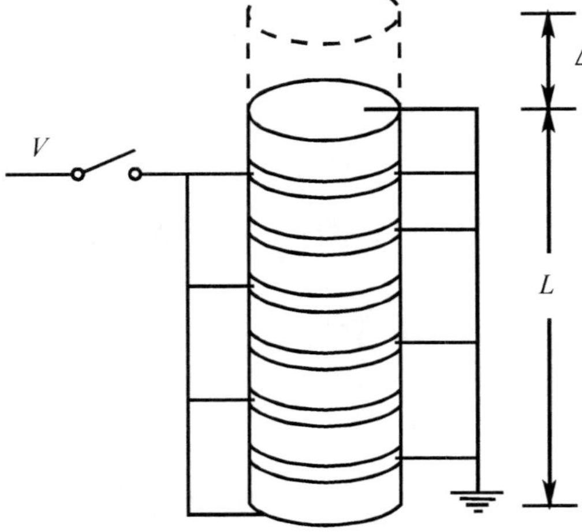

Figure 15.31 Stack piezoelectric actuator.

Commercially available stack actuators usually consist of up to 100 thin actuators (0.3 to 1 mm thick) stacked together, with diameters ranging from 15 to 50 mm. When unloaded, the free expansion of this type of actuator is given by:

$$\Delta L_0 = d_{33}\, n\, V \tag{15.11.41}$$

where n is the number of actuators in the stack, V is the applied voltage and d_{33} is the charge or strain constant for expansion in the thickness direction.

If the stacked actuator is to be used as a shaker, it should be pre-loaded with a compressive force greater than the maximum required dynamic force, as the actuators work much better at pushing than pulling.

The extension of a stack actuator when driving a stiffness k_s is given by:

$$\Delta L = \Delta L_0 \frac{k_a}{k_a + k_s} \tag{15.11.42}$$

where k_a and k_s represent the stiffnesses of the actuator and structure being driven respectively.

The effective force that the actuator can generate is given by:

$$F = k_a \Delta L_0 \left[1 - \frac{k_a}{k_a + k_s} \right] \tag{15.11.43}$$

The resonance frequency of an unloaded actuator is given by:

$$f_0 = \frac{1}{2\pi} \sqrt{\frac{2k_a}{m}} \tag{15.11.44}$$

where m is the mass of the actuator and ΔL_0 is its extension.

When the actuator is used to drive an additional mass M or a structure with a modal mass M, the effective resonance frequency f_0' is given by:

$$f_0' = f_0 \sqrt{\frac{m}{m + 2M}} \tag{15.11.45}$$

The upper limiting frequency of operation is given by:

$$f_{\max} = \frac{i_{\max}}{2\, c\, U_0} \tag{15.11.46}$$

The quantity C is the total capacitance of the actuator (farads), i_{max} (amps) is the available current from the driving amplifier and U_0 is the nominal operating voltage (usually 1000 V).

One problem with piezoelectric actuators (both the thin sheet type and stack type), which is also experienced by magnetostrictive actuators, is their hysteresis property, which means that the expansion as a function of a given applied electric field (or voltage) is dependent upon whether the field level is rising or falling. It is also a non-linear function of the applied electric field in either direction. This results in non-linear excitation of a structure attached to one of these actuator types. Thus, if the exciting voltage used to drive the actuator is sinusoidal, the resulting structure motion will contain additional frequencies. In practice, a strong signal at the first harmonic of the exciting frequency is usually found. This poses obvious problems when these actuators are used in active control systems with linear controllers. Clearly, to be effective, these actuators need to be driven with non-linear controllers, an example of which could contain a tapped delay line whose weights are adjusted using a neural network with a non-linear layer.

15.12 SMART STRUCTURES

Smart structures are those that contain sensors and/or actuators as an integral part. Passive smart structures are those that contain only sensing elements which allow their state at any particular time to be determined. Such sensors could include piezoelectric film (PVDF) or optical fibres. Active smart structures contain in-built actuators as well as sensors that enable them to respond automatically to correct some undesirable state detected by the sensors.

Actuators that have been used in laboratory experiments in the past include piezoelectric film or piezoceramic crystal or shape memory alloy. Magnetostrictive terfenol may also be useful for this purpose, but is more likely to be bonded to the external surface of structures rather than embedded in them.

Structures that particularly lend themselves to the integration of sensors and actuators are carbon fibre and glass fibre composite structures, which are made by laminating the glass or carbon cloth with a suitable epoxy resin. However, other non-composite structures can be made into smart structures by bonding actuators and sensors to their surface.

15.12.1 Novel Actuator Configurations

One problem with applying piezoceramic stacks, magnetostrictive rods, hydraulic or electrodynamic actuators to a structure in the traditional way is the need for a reaction mass and support for the actuator. The need for a reaction mass makes the application of active vibration control using these actuators impractical in many situations. However, the need for this reaction mass can be eliminated by using a more imaginative actuator configuration that applies control bending moments rather than control forces to the structure. For stiffened structures such as aircraft fuselages or submarine hulls, implementation of the actuator configurations is even more convenient.

Two possible actuator configurations, which need no reaction mass, are illustrated in Figures 15.32(a) and (b). The first figure illustrates the arrangement where stiffeners already exist on the panel, and the second figure shows an alternative arrangement which might be used if no prior stiffeners exist. The configurations shown for panel vibration control can easily be extended to cylindrical structures such as aircraft fuselages and submarine hulls.

15.12.2 Shunted Piezoelectric Dampers

Thin, piezoelectric plates bonded to a structure deform in bending when the structure vibrates and the strain of deformation results in the production of an electric charge. Thus,

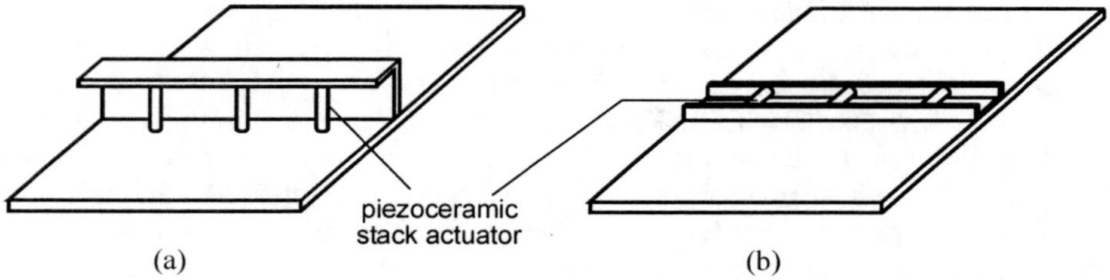

Figure 15.32 Actuator configurations for applying a bending moment to a panel, with no need for a reaction mass.

connecting the electrodes of a piezoelectric element to an electrical circuit containing a resistor only or a resistor and inductor can be used to damp structural vibration, as illustrated in Figure 15.33. The concept of using piezoelectric elements as structural dampers was first introduced by Forward (1979), but the first complete analysis of the use of these elements as dampers as well as vibration absorbers was published by Hagood and von Flotow in 1991.

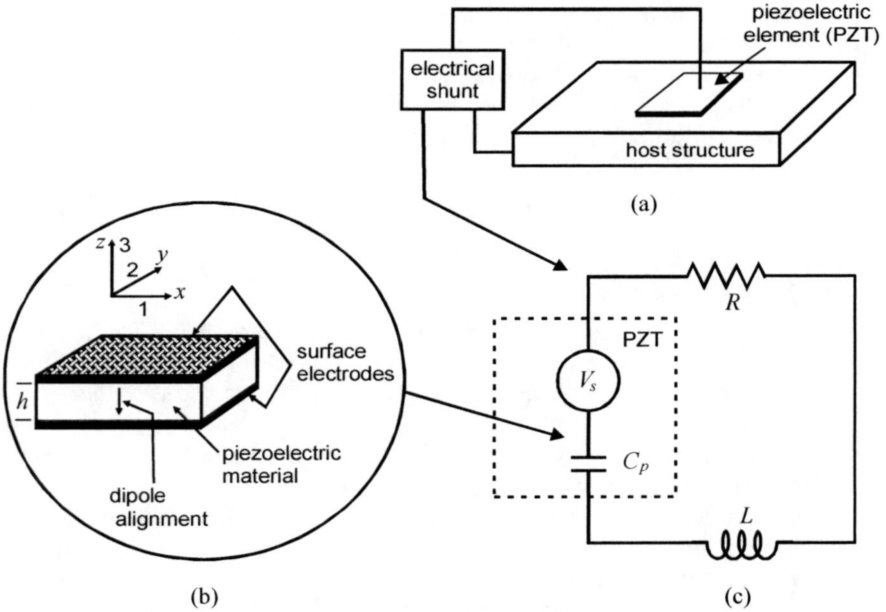

Figure 15.33 Basic form of the piezoelectric shunt vibration absorber with poling (or polarisation) in the '3' or *z*-direction: (a) piezoelectric element mounted on the substructure; (b) piezoelectric element details; (c) circuit schematic.

The combination of the capacitance that characterises the electrical properties of the piezoelectric element and the added inductance results in a resonant circuit that can be tuned, by varying the inductance, to the frequency to be controlled, which is usually a structural resonance. The resistor controls the bandwidth over which some vibration reduction will be achieved and it also controls the amount of reduction that will be achieved at the tuned frequency. These properties are linked in that the wider the frequency bandwidth over which some reduction is obtained, the smaller will be the reduction at the tuned frequency.

The circuit that contains only a resistance acts like a damping element and suppresses vibration at structural resonances, with the effectiveness increasing as the piezoelectric element position becomes closer to a modal antinode.

On the other hand, the circuit that contains both an inductor and resistor acts in a similar way to a mechanical vibration absorber and thus can be used to suppress vibrations at frequencies that do not correspond to structural resonances as well as at frequencies that do correspond to structural resonances. The former action is similar to that of a vibration neutraliser, while the latter action is similar to a vibration absorber (see Bies and Hansen, 2009, Chapter 10).

Following the same approach as that used by Hagood and von Flotow (1991) for a single-degree-of-freedom system, the displacement of a structure, containing a resonant shunted piezoelectric element bonded to it, can be modelled as:

$$\frac{X(\mathrm{j}\omega)}{F(\mathrm{j}\omega)} = \frac{1}{-M\omega^2 + \mathrm{j}\omega Z + K} \tag{15.12.1}$$

where M is the modal mass of the structure (for the mode to be damped), Z is the frequency dependent mechanical impedance of the resonant shunted piezoelectric element (which is dependent on the electrical impedance of the shunt) and K is the modal stiffness. The mechanical impedance may be normalised for convenience by the frequency dependent mechanical impedance, $Z_{11}^{D}(\mathrm{j}\omega)$, of the open circuit piezoelectric element for strain in the 1 or x-direction, which is given by:

$$Z_{11}^{D}(\mathrm{j}\omega) = \frac{K_{11}^{E}}{\mathrm{j}\omega(1 - k_{31}^{2})} \tag{15.12.2}$$

where the superscript D implies an open circuit for the piezoelectric element and the superscript E implies a short circuit measurement (constant electric field). The quantity K_{11}^{E} is the stiffness of the shorted piezoelectric element and k_{31} is the electromechanical coupling coefficient (for a piezoelectric element polarised in the z-direction and deforming in the x-direction), the square of which represents the fraction of mechanical energy due to expansion of the piezoelectric element in the x- or 1-direction that is converted into electrical energy in the z- or 3-direction. It is related to the permittivity $\varepsilon_3^{\mathrm{T}}$ (dielectric displacement per unit electric field) at constant stress in the 3- or z-direction, the piezoelectric constant d_{31} (induced strain in the 1- or x-direction due to an electric field applied in the 3- or z-direction) and the compliance s_{11}^{E} (strain produced per unit of applied stress in the x-direction – reciprocal of Young's modulus) of the piezoelectric element at constant electric field (piezoelectric electrodes shorted) as follows:

$$k_{31} = \frac{d_{31}}{\sqrt{s_{11}^{E}\varepsilon_3^{T}}} \tag{15.12.3}$$

The superscript T refers to the quantity measured under constant (zero) stress, which means that it is not attached to the structure.

The electrical impedance Z_{Es} of the shunt circuit at frequency ω is given by (see Figure 15.33(b)):

$$Z_{Es} = \mathrm{j}\omega L + R \tag{15.12.4}$$

The overall electrical impedance of the shunted circuit is the impedance of the shunt, given by Equation (15.12.4), in parallel with the impedance C_p^T of the piezoelectric element, where the superscript T refers to the capacitance being measured at constant stress:

$$Z_E = \frac{\mathrm{j}\omega L + R}{-\omega^2 L C_p^T + \mathrm{j}\omega R C_p^T + 1} \qquad (15.12.5)$$

The capacitance of the piezoelectric element can be expressed in terms of its area, S, perpendicular to the 3-direction, its thickness h in the 3-direction and its permittivity ε_3^T which are measurable properties, as:

$$C_p^T = \frac{S\varepsilon_3^T}{h} \qquad (15.12.6)$$

The electrical impedance of Equation (15.12.5) can be normalised by the open circuit electrical impedance of the piezoelectric element as follows:

$$\bar{Z}_E = \frac{Z_E}{1/\mathrm{j}\omega C_p^T} = \frac{-LC_p^T\omega^2 + \mathrm{j}\omega R C_p^T}{-\omega^2 L C_p^T + \mathrm{j}\omega R C_p^T + 1} \qquad (15.12.7)$$

The normalised mechanical impedance $\bar{Z}$ is related to the normalised electrical impedance by (Hagood and von Flotow, 1991):

$$\bar{Z} = \frac{1 - k_{31}^2}{1 - k_{31}^2 \bar{Z}_E} \qquad (15.12.8)$$

Substituting Equation (15.12.7) into Equation (15.12.8) gives:

$$\bar{Z} = \frac{-LC_p^S\omega^2 + \mathrm{j}\omega R C_p^S + 1 - k_{31}^2}{-\omega^2 L C_p^S + \mathrm{j}\omega R C_p^S + 1} \qquad (15.12.9)$$

where the superscript S implies refers to the piezoelectric element capacitance being measured at constant strain and this is related to the capacitance at constant stress by:

$$C_p^S = C_p^T(1 - k_{13}^2) \qquad (15.12.10)$$

The mechanical impedance, Z, of Equation (15.12.1) is obtained by multiplying the normalised impedance of Equation (15.12.9) by the open circuit impedance of the piezoelectric element given by Equation (15.12.2).

Equation (15.12.9) for the normalised mechanical impedance at frequency ω can be rewritten as:

$$\bar{Z} = 1 - k_{31}^2\left[\frac{\delta^2}{\gamma^2 + \delta^2 r\gamma + \delta^2}\right] \qquad (15.12.11)$$

where

$$\omega_e = \frac{1}{\sqrt{LC_p^S}} \; ; \qquad \delta = \frac{\omega_e}{\omega_n} \; ;$$
$$r = RC_p^S \omega_n \; ; \qquad \gamma = \frac{\mathrm{j}\omega}{\omega_n} \tag{15.12.12a–d}$$

and ω_n is the resonance frequency of the structure with the piezoelectric element attached and shorted.

Defining the generalised electromechanical coupling coefficient as:

$$K_{31}^2 = \frac{K_{11}}{K_{tot}} \frac{k_{31}^2}{(1 - k_{31}^2)} \tag{15.12.13}$$

Equation (15.12.1) can be written as:

$$\frac{X(\mathrm{j}\omega)}{F(\mathrm{j}\omega)/K_{tot}} = \frac{\gamma^2 + \delta^2 r\gamma + \delta^2}{(\gamma^2 + 1)(\gamma^2 + \delta^2 r\gamma + \delta^2) + K_{31}^2(\gamma^2 + \delta^2 r\gamma)} \tag{15.12.14}$$

where K_{tot} is the sum of the structural modal stiffness K and the modal stiffness K_{11} of the shorted piezoelectric element. The shorted stiffness is used because the open circuit piezoelectric element stores energy as electrical energy and returns it to the structure as mechanical energy (Hollkamp, 1994). The preceding equation is the one used to evaluate the extent of vibration reduction that is achieved with the shunt damper (see Bies and Hansen, 2009, Chapter 10 for a discussion of this process).

The optimal tuning ratio (Hagood and von Flotow, 1991) is given by:

$$\delta_{opt} = \sqrt{1 + K_{13}^2} \tag{15.12.15}$$

and the optimal value of r is:

$$r_{opt} = \sqrt{2} \frac{K_{31}}{1 + K_{31}^2} \tag{15.12.16}$$

The preceding equations were derived on the assumption that the structural damping was zero. If the structure can be characterised by a critical viscous damping ratio of ζ, then Equation (15.12.14) becomes (Hollkamp (1994):

$$\frac{X(\mathrm{j}\omega)}{F(\mathrm{j}\omega)/K_{tot}} = \frac{\gamma^2 + \delta^2 r\gamma + \delta^2}{(\gamma^2 + 2\zeta\gamma + 1)(\gamma^2 + \delta^2 r\gamma + \delta^2) + K_{31}^2(\gamma^2 + \delta^2 r\gamma)} \tag{15.12.17}$$

However, the optimum values in Equations (15.12.15) and (15.12.16) are approximately correct for damped structures, provided that the damping is light (Hollkamp, 1994).

For damping more than one mode with the same piezoelectric element, Moheimani (2003) presented the circuits shown in Figure 15.34, with the one on the right being much easier to implement, as the two inductors in each branch can be combined into a single element. The RL part of each leg in the circuit is used to provide the damping, while the additional LC circuit in each leg acts as a current flowing circuit. The RL part in each leg in combination with the capacitance of the piezoelectric element is tuned to a particular mode in the structure being controlled.

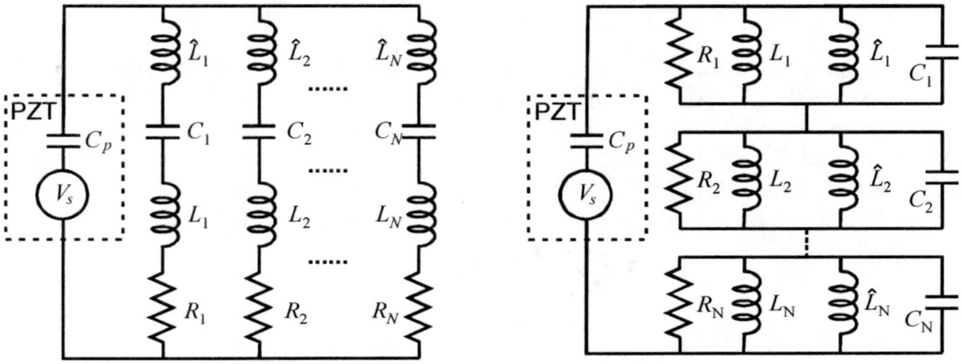

Figure 15.34 Shunt resonant circuits for damping N modes simultaneously (after Moheimani, 2003).

In practice, implementation of shunted dampers at low frequencies is often made difficult by the common requirement for an inductors to be quite large in size and the requirement for the number of operational amplifiers in a circuit of twice the number of modes to be controlled. In addition, it is not possible to optimise the individual shunts one at a time – rather they all must be optimised simultaneously using such techniques as genetic algorithms (Steffen and Inman, 1999; Steffen et al., 2000).

Moheimani (2003) proposed the use of active impedances for the shunts and used feedback control design methods to optimise the impedance (Fleming and Moheimani, 2005). Optimal tuning of multi-modal shunts was also discussed by Morgan and Wang (2006).

If only a resistor is used in the shunting circuit and the inductor in Figure 15.33(b) is discarded, the piezoelectric element can act as a viscous damper, with the extent of damping being frequency dependent and the energy being dissipated by the resistor. An analysis may be undertaken following the same procedure as for the resonant shunt, except that now Equation (15.12.4) only has the resistance term on the right-hand side. The results of this analysis are (Hagood and von Flotow, 1991):

$$\frac{X(\mathrm{j}\omega)}{F(\mathrm{j}\omega)/K_{tot}} = \frac{\gamma r + 1}{\gamma^2(1 + r\gamma) + 1 + r\gamma(1 + K_{31}^2)} \tag{15.12.18}$$

The maximum loss factor is:

$$\eta_{\max} = \frac{k_{31}^2}{2\sqrt{1 - k_{31}^2}} \tag{15.12.19}$$

at a frequency given by:

$$\omega = \frac{1}{RC_p^S}\sqrt{1 - k_{31}^2} \tag{15.12.20}$$

which gives the optimum resistance for damping at frequency ω as:

$$R_{opt} = \frac{1}{\omega C_p^S}\sqrt{1 - k_{31}^2} \tag{15.12.21}$$

The preceding analysis is applicable to piezoelectric elements that are polarised in their thickness direction. However, there are materials now available for which the direction of polarisation is along the x-axis and for these, the piezoelectric effect is up to 50% greater. Benjeddou and Ranger (2006) presented an analysis for a resistive shunt circuit including a piezoelectric element that had its poling direction along the x-axis that is similar to that presented above for the poling direction parallel to the z-axis, through the material thickness. The result is:

$$\frac{X(j\omega)}{F(j\omega)/K_{tot}} = \frac{\gamma r + 1}{\gamma^2(1 + r\gamma) + 1 + r\gamma(1 + K_{15}^2)} \tag{15.12.22}$$

where

$$K_{15}^2 = \frac{K_{55}\, k_{15}^2}{(K + K_{55})\,(1 - k_{15}^2)} \tag{15.12.23}$$

and K_{55} is the shear stiffness of the shorted piezoelectric element, given by:

$$K_{55} = G_{55}\frac{S}{h} \tag{15.12.24}$$

where G_{55} is the shear modulus of the shorted piezoelectric element.

15.12.3 Piezoelectric Sensoriactuators

Previous sections have discussed piezoelectric elements as either sensors or actuators. In this section, the use of the same element as a sensor and actuator simultaneously is discussed. The underlying motivation to use the same element as a sensor and actuator arises from the phenomenon that in feedback active vibration control, the system stability can be maximised if the sensor and actuator are collocated.

As a piezoelectric element develops a direct charge due to an applied voltage and also develops a charge due to its mechanical deformation as a response to the vibration of the structure to which it is attached. The challenge is to extract the charge due to the mechanical deformation from the total charge signal using a compensation technique. The problem is made difficult because the actuation voltage is usually much greater than the sensing voltage, (usually 20 to 25 dB greater), so it is a challenge to isolate the sensing voltage with the required accuracy for a vibration control system.

The most common way used to separate the actuation and sensor signals from the piezoelectric element output is to use a bridge circuit such as used with strain gauges for measuring strain in a structure. In using this compensation method, the greatest challenge is to maintain the balance of the bridge circuit under changing environmental conditions such as changes in the temperature of the piezoelectric element, which causes changes in its capacitance. Compounding this problem is the difficulty in accurately determining the capacitance of the piezoelectric element in its 'blocked' state. The 'blocked' state is the condition of the piezoelectric element attached to the structure but with no voltage applied to it. However, a small voltage has to be applied in order to measure its capacitance and this causes a small strain in the piezoelectric element which changes its capacitance. Thus, the measured capacitance will always be a little larger than the actual capacitance.

Central to the bridge circuit is the use of a reference capacitor, which is required to match as closely as possible the capacitance of the piezoelectric element used for sensing and actuation. Due to the problems mentioned above, it is not possible to exactly match the piezoelectric capacitance at all times and this can cause stability problems with the vibration control system that is using the sensoriactuator, which, in many cases, negates the advantage of exact collocation that is offered by this arrangement. Just a 1% variation in the piezoelectric element capacitance, which is easily achieved by a 6°C change in temperature, can destabilise a previously stable control system (Wang and Wang, 2010).

Dosch et al. (1992) presented two bridge circuits: one for obtaining a voltage output proportional to the strain and one to obtain a voltage output proportional to the *rate* of strain (see Figures 15.35(a) and (b)). In addition to the other limitations of the bridge compensator mentioned above, the strain measurement system shown in Figure 15.35(a) suffers from DC off-set drift due to there only being capacitors in the circuit. Thus, future publications using various refinements to the basic bridge compensation circuit are all based on the circuit in Figure 15.35(b).

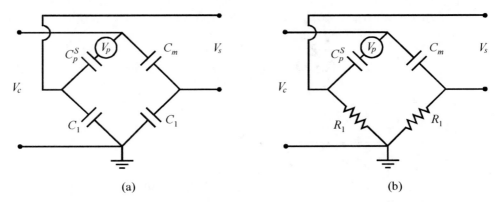

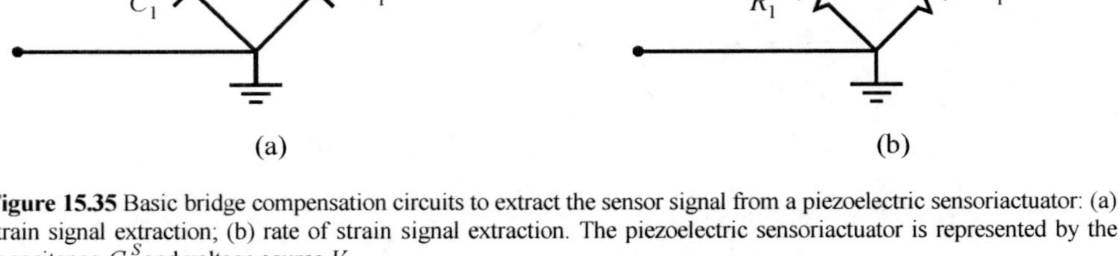

Figure 15.35 Basic bridge compensation circuits to extract the sensor signal from a piezoelectric sensoriactuator: (a) strain signal extraction; (b) rate of strain signal extraction. The piezoelectric sensoriactuator is represented by the capacitance C_p^S and voltage source V_p.

The sensor output voltage at frequency ω obtained using Figure 15.35(a), for $C_p^S = C_m$, is given by:

$$V_s(\mathrm{j}\omega) = V_1(\mathrm{j}\omega) - V_2(\mathrm{j}\omega) = \frac{C_p^S}{C_p^S + C_1} V_p(\mathrm{j}\omega) \tag{15.12.25}$$

where the superscript S refers to the piezoelectric element 'blocked' capacitance measured at constant (zero) strain. Similarly, the sensor output voltage obtained using Figure 15.35(b) for $C_p^S = C_m$ is given by:

$$V_s(\mathrm{j}\omega) = V_1(\mathrm{j}\omega) - V_2(\mathrm{j}\omega) = \frac{R_1 \mathrm{j}\omega C_p^S}{1 + R_1 \mathrm{j}\omega C_p^S} V_p(\mathrm{j}\omega) \tag{15.12.26}$$

From Equation (15.12.26), it can be seen that for $R_1 \omega C_p^S \ll 1$, $V_s = \dot{V}_p$, which indicates that the sensor output voltage is proportional to the rate of strain. Zhang et al. (2004) suggested making the right-hand resistor in Figure 15.35(b) a potentiometer to allow tuning and balancing of the bridge circuit prior to excitation of the structure.

Later versions of the strain rate sensor used capacitors and resistors in two or all four legs of the bridge to obtain a more stable circuit that also had less DC drift problems, and also use operational amplifiers to condition the signal output as shown in Figure 15.36. The signal conditioning is used to convert the high impedance input to a low impedance output, with gain dependent on the choice of resistors R_2, which are not necessarily all the same. The signal conditioning also removes common mode noise, resulting in a better capability to isolate the low level sensor voltage.

The signal conditioning part on the right-hand side of Figure 15.36 follows Sodano et al. (2004) and placing a resistor and capacitor in all four legs of the bridge follows Jones et al. (1994). For the circuit shown in Figure 15.36 and for a balanced bridge with R_1 equal to the variable resistance, the sensor output voltage at frequency ω is given by:

$$V_s(\mathrm{j}\omega) = V_1(\mathrm{j}\omega) - V_2(\mathrm{j}\omega) = \frac{R_1 R_0 \mathrm{j}\omega C_p^S}{(R_0 + R_1) + R_1 R_0 \mathrm{j}\omega (C_p^S + C_1)} V_p(\mathrm{j}\omega) \tag{15.12.27}$$

The circuit used by Sodano et al. (2004) did not include the resistance R_0 shown in Figure 15.36 and in this case the equation for the sensor voltage is given by (Simmers et al., 2004) as:

$$V_s(\mathrm{j}\omega) = V_1(\mathrm{j}\omega) - V_2(\mathrm{j}\omega) = \frac{R_1 \mathrm{j}\omega C_p^S}{1 + R_1 \mathrm{j}\omega (C_p^S + C_1)} V_p(\mathrm{j}\omega) \tag{15.12.28}$$

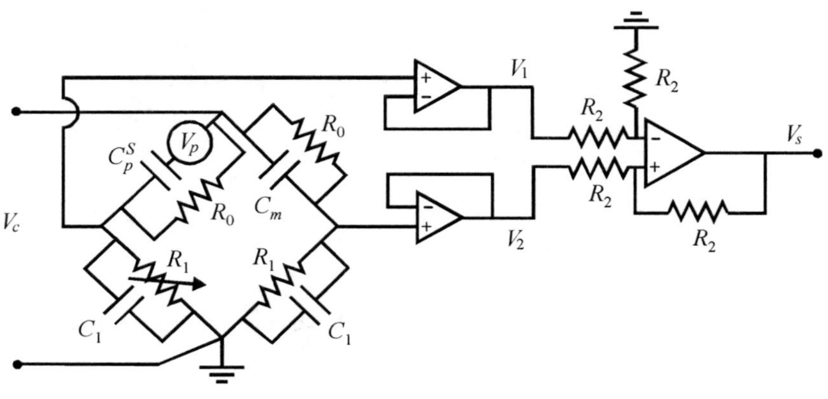

Figure 15.36 Alternative, more stable bridge compensator circuit for deriving a sensoriactuator voltage signal proportional to the rate of strain.

Simmers et al. (2004) suggested a way of making the stability of the circuit shown in Figure 15.36 less sensitive to capacitance variations in the piezoelectric element by adding two additional capacitances, C_1. One of these is either in series or in parallel to the piezoelectric element and the other is in series or parallel with the matching capacitor C_m respectively, with the series arrangement performing slightly better. The values of the added capacitances were similar to that of the piezoelectric element. This had the effect of halving the effect of capacitance variations of the piezoelectric element on the mismatch of capacitance in the two legs of the bridge circuit containing the piezoelectric element and the matching capacitor respectively. This meant that twice the temperature variation was possible before the piezoelectric element capacitance changed sufficiently that the system became unstable. The cost of increased stability was reduced control performance or less noise reduction.

Although the circuit shown in Figure 15.36 was a considerable improvement over those shown in Figure 15.35, the drift of the piezoelectric element capacitance as the temperature changed meant that the bridge had to be regularly tuned manually by adjusting the potentiometer. For this reason, Vipperman and Clark (1996) developed an automatic tuning circuit that kept the bridge balanced with no manual input (see Figure 15.37). They based their automatic tuning circuit on a slightly different (to Figure 15.36) manual tuning circuit which is illustrated in Figure 15.38. Note that in all figures, C_p represents the piezoelectric element and V_p is the voltage it produces as a result of the strain caused by the actuator voltage V_c.

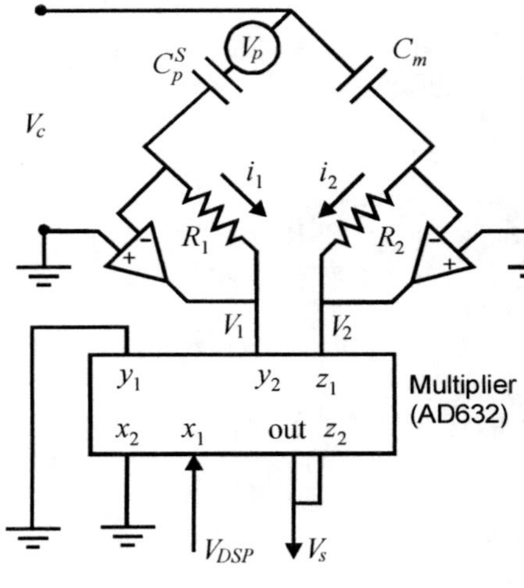

Figure 15.37 Sensoriactuator bridge circuit with automatic tuning (after Vipperman and Clark, 1996).

For the circuit illustrated in Figure 15.37, the sensor output voltage is given by:

$$\begin{aligned} V_s(\mathrm{j}\omega) &= \frac{V_{DSP}(\mathrm{j}\omega)[-V_2(\mathrm{j}\omega)]}{10} + V_1(\mathrm{j}\omega) \\ &= \frac{V_{DSP}(\mathrm{j}\omega)[-\mathrm{j}\omega R_2 C_m V_c(\mathrm{j}\omega)]}{10} - \mathrm{j}\omega R_1 C_p^S V_c(\mathrm{j}\omega) - R_1 i_{mech}(\mathrm{j}\omega) \end{aligned} \tag{15.12.29}$$

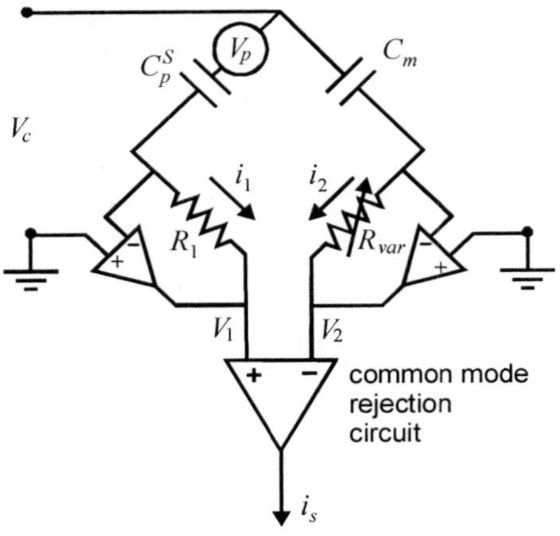

Figure 15.38 Sensoriactuator bridge circuit with manual tuning (after Vipperman and Clark, 1996).

Clearly, the optimal value for V_{DSP} is that which results in $V_s(j\omega) = -R_1 i_{mech}(j\omega)$ and this requires that:

$$V_{DSP} = \frac{10R_1 C_p^S}{R_2 C_m} \tag{15.12.30}$$

An LMS algorithm adapting a single-tap FIR filter is used to minimise $V_s(j\omega)$. Vipperman and Clark (1996) show that the update value for V_{DSP} is given by:

$$V_{DSP} \equiv w(k+1) \approx w(k) + 2\mu V_s(k) V_2(k)/10 \tag{15.12.31}$$

The quantity $w(k)$ is the single coefficient FIR filter at time step k which is equivalent to the voltage V_{DSP}. The quantity μ is the step size parameter (going from step k to step $k+1$), which controls the system stability and also the convergence of V_s to its optimum value (which is its minimum value), and thus the convergence of V_{DSP} to its optimum value given by Equation (15.12.30).

The difference between the scheme implemented in Equation (15.12.31) and the traditional LMS implementation is that here it is implemented in analogue form.

The complete circuit for the digital adaptive filter is illustrated in Figure 15.39. This circuit was used by Vipperman and Clark (1996) to adapt the filter coefficient. Low level white noise was used as the training signal to achieve this, and it is possible to do this at the same time as the sensoriactuator is used to control the response of the structure to which it is attached.

The scheme described above was modified by Fannin (1997), who added an additional, variable resistor R_c between C_m and the operational amplifier, which can be adjusted to compensate for any phase errors that result from imperfect matching of the matching capacitor to the piezoelectric element capacitance.

More recently, in an effort to overcome the limitations of the bridge circuit approaches discussed above, Wang and Wang (2010) proposed a switching technique, whereby the piezoelectric element acted as an actuator for a short time, followed by acting as a sensor for the same amount of time and repeating this pattern indefinitely as illustrated in Figure 15.40.

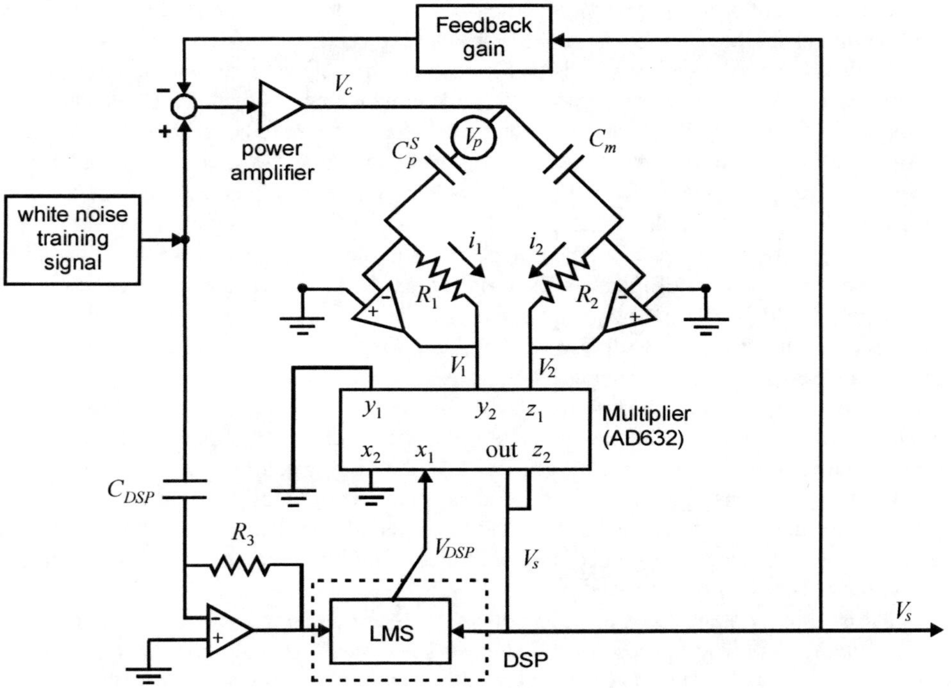

Figure 15.39 Schematic of the analogue digital adaptive filter used to estimate the dynamic capacitance.

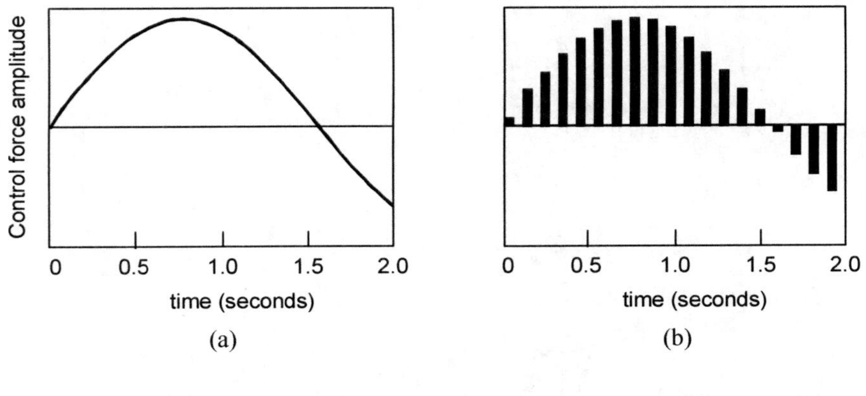

Figure 15.40 Illustration of the smooth and chopped actuation signals. The black bars represent the actuation on and the sensing circuit off while the white spaces in between represent the actuation off and the sensing circuit on (after Wang and Wang, 2010).

The short bursts of time were such that the sum of one off period and one on period was less than the reciprocal of the frequency to be controlled. This approach has its own difficulties, and the interested reader can consult the above reference to learn more.

Due to the problems with bridge balancing and signal extraction difficulties, the use of piezoelectric element sensoriactuators is not at all widespread. Most researchers achieve sensor/actuator collocation by bonding the sensor onto the top of the actuator and then bonding the other side of the actuator to the structure to be sensed and controlled.

A disadvantage associated with using piezoelectric elements as sensors is that they provide a sensor signal proportional to displacement, which means that direct feedback of the sensor signal for vibration control of the structure adds stiffness rather than adding damping, which would be achieved with velocity feedback. Of course, the sensor signal can be integrated to provide a signal proportional to velocity but the presence of an integrator in the circuit adds a delay that could be detrimental to the stability of the system.

A disadvantage of using piezoelectric elements as actuators for low-frequency damping is that they are not very efficient in exciting structures at low frequencies, resulting in a requirement for high controller gains with the associated instability problems that this brings. For these reasons, Paulitsch et al. (2004, 2006) and Okada et al. (1995) have introduced the use of a self-sensing electrodynamic shaker.

The electrodynamic shaker self-sensing circuit uses as the sensing voltage, V_s, a proportion of the input the induced voltage in the shaker driving coil generated by the back EMF resulting from the excitation of the shaker coil by the structure to which it is attached. Equations for this sensing voltage are dependent on the point mobility of the structure at the point of attachment of the control shaker and the transfer mobility between the points of the structure that experience the primary and control excitations respectively and an expression for excitation of a simply supported plate has been derived by Paulitsch et al. (2006).

The sensing voltage V_s is also dependent on the shunt impedance applied across the shaker driving coil. The purpose of the shunt impedance is to compensate for the internal electrical impedance of the shaker, which results in optimal damping to reduce the structural vibration at the point of attachment of the shaker. Paulitsch et al. (2006) investigated four different compensation circuits as shown in Figure 15.41.

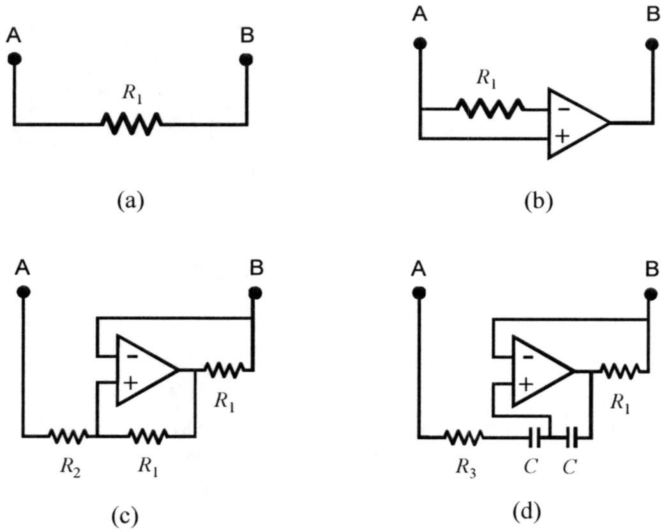

Figure 15.41 Various compensation circuits applied across the driving coil of an electrodynamic shaker to maximise the damping at the point of application of the shaker force (after Paulitsch et al., 2006): (a) shunted resistor; (b) positive current feedback; (c) induced voltage feedback; (d) induced voltage feedback with inductance compensation.

For the circuit shown in Figure 15.41(a), they obtained a 5 dB reduction for the first vibration mode of the plate and 2 dB for the second and third modes. For the circuits shown in parts (b) and (c), they obtained a 12 dB reduction for the first mode vibration amplitude and 8 dB for the second and third modes, but with high-frequency spillover occurring above 200 Hz. For the circuit shown in part (d), they obtained a 9 dB reduction for the first mode and at least 5 dB for all modes up to the eleventh one, which had a resonance frequency of 400 Hz. However, for case (d), high-frequency spillover occurred above about 500 Hz.

In 2004, Paulitsch et al. reported on a self-sensing inertial actuator design that used as a sensor, a second coil wound concentrically and inside with the main driving coil. The second coil results in an increase in the effective bandwidth of control.

15.12.4 Energy Harvesting from Vibration

One of the problems with Steiglitz–McBride active noise and vibration control systems is the extent of wiring necessary, especially if the number of channels is large, as may be the case in some smart structure configurations. One way to ameliorate this is to use radio transmission of the sensor signals and a local power source to power the sensor and transmitter. Unfortunately, mains supplied power sources require wiring, which can sometimes be extensive, and battery power sources require regular battery replacement. Thus, there is considerable motivation to be able to use structural vibration to generate sufficient energy to power the sensors and transmitters directly, or to recharge batteries that power the sensors and transmitters. With this motivation, the amount of research undertaken in this area in the past 5 to 8 years has been enormous to the point that a book has been published that is devoted entirely to this subject (Priya and Inman, 2009).

Almost all forms of energy harvesting rely on the use of a piezoelectric element, which produces a current when deformed as a result of vibration of the structure to which it is attached. The efficiency associated with converting the deformation mechanical energy to electrical energy has been found to range between 30% and 50%. The energy can be used to drive a device directly, stored in a rechargeable battery of stored in a capacitor. In a small number of applications, electromagnetic energy harvesters have been used and these also make use of structural vibration as the power source.

There are a number of commercially available energy harvesters based on piezoelectric elements and many of these are listed in Priya and Inman (2009, Chapter 1). Most products rely on bending vibration of the piezoelectric element to produce the current to drive a device and these usually consist of piezoelectric elements mounted on cantilevered beams (see Figure 15.42a–c) or a piezoelectric element in a cantilevered bimorph configuration where the piezoelectric elements cover both sides of a thin steel or aluminium beam, with one element on each side. Typical vibration amplitudes of the tip of the cantilever are of the order of 1 to 3 mm. It is important that the resonance frequency of the cantilever beam matches as closely as possible the frequency of maximum structural vibration. For this reason, it is often necessary to tune the resonance frequency of the cantilever by using small masses attached to the free end.

The power generated by commercial devices ranges from tens of microwatts to about 10 mW, depending on the vibration frequency and amplitude. Typical displacements of the tip of the cantilever beam range from 0.5 to 3 mm. Typical cantilever lengths range from 25 to 75 mm and widths range from 4 to 30 mm.

Alternatively, some piezoelectric energy harvesters rely on compression of a much thicker piezoelectric element via a mass mounted on top of the piezoelectric element (see

Figure 15.42d). These devices are mounted on parts of a structure where vibration levels are sufficient to generate the required current. The voltage from the harvester is supplied across the load resistance R_L, and a rechargeable battery or capacitor can be charged by replacing the load resistance with either one. The efficiency of power extraction from the piezoelectric element has been investigated by a number of researchers and some novel circuit designs have resulted. Some of these have their own components that need power to operate, so in some cases the minimum vibration level required before useful power extraction can occur is higher than in other cases.

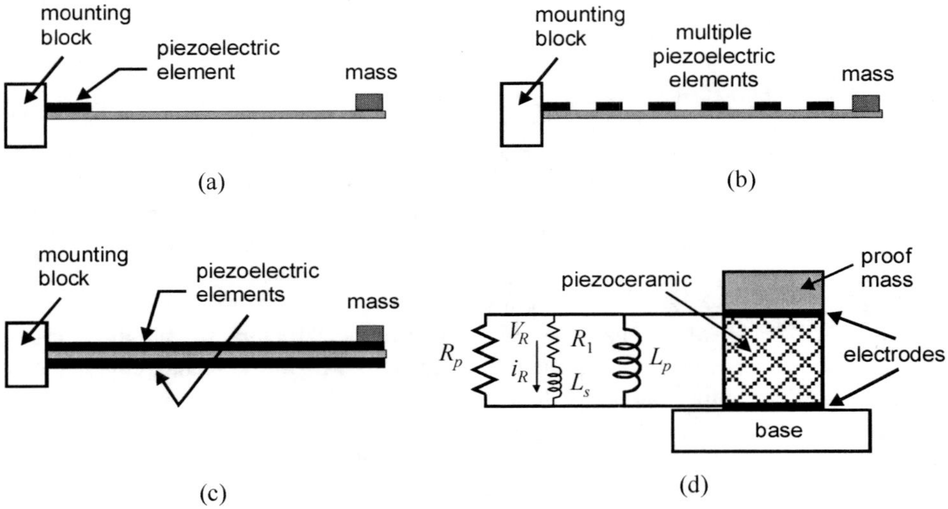

Figure 15.42 Piezoelectric element configurations for energy harvesting from a vibrating structure: (a) piezoelectric elements attached to a cantilevered beam near the support; (b) multiple piezoelectric elements attached to a cantilever beam; (c) bimorph with a piezoelectric element covering each side of a cantilever beam; (d) piezoelectric element stack harvester (including the harvester circuit).

Some of the circuits used for providing a supply current and voltage are illustrated in Figure 15.43.

All circuits include a rectifier, consisting of diodes and a smoothing capacitor. This is necessary, as the voltage or current generated by the piezoelectric element, due to vibration of the structure on which it is mounted, is AC and the voltage needed to power small electronic devices or to recharge batteries is DC.

The circuit shown in Figure 15.43(a) is the most basic circuit and only suitable for a demonstration using a load resistor. The voltage across the load resistor is smoothed by the capacitor C_R. This circuit was analysed by Lefeuvre et al. (2006, 2007), who modelled the piezoelectric element as a current source in parallel with a capacitor and controlled by the mechanical velocity of the structure at the point of attachment. They began with the equation that relates the force F_P developed by the piezoelectric element as a result of generated voltage V and structural displacement $\bar{w}$, the equation that relates the generated charge Q to the displacement of the structure and the generated voltage, and the equation that relates the generated current to the structural velocity and the rate of change of the generated voltage as follows (Lefeuvre et al. (2006):

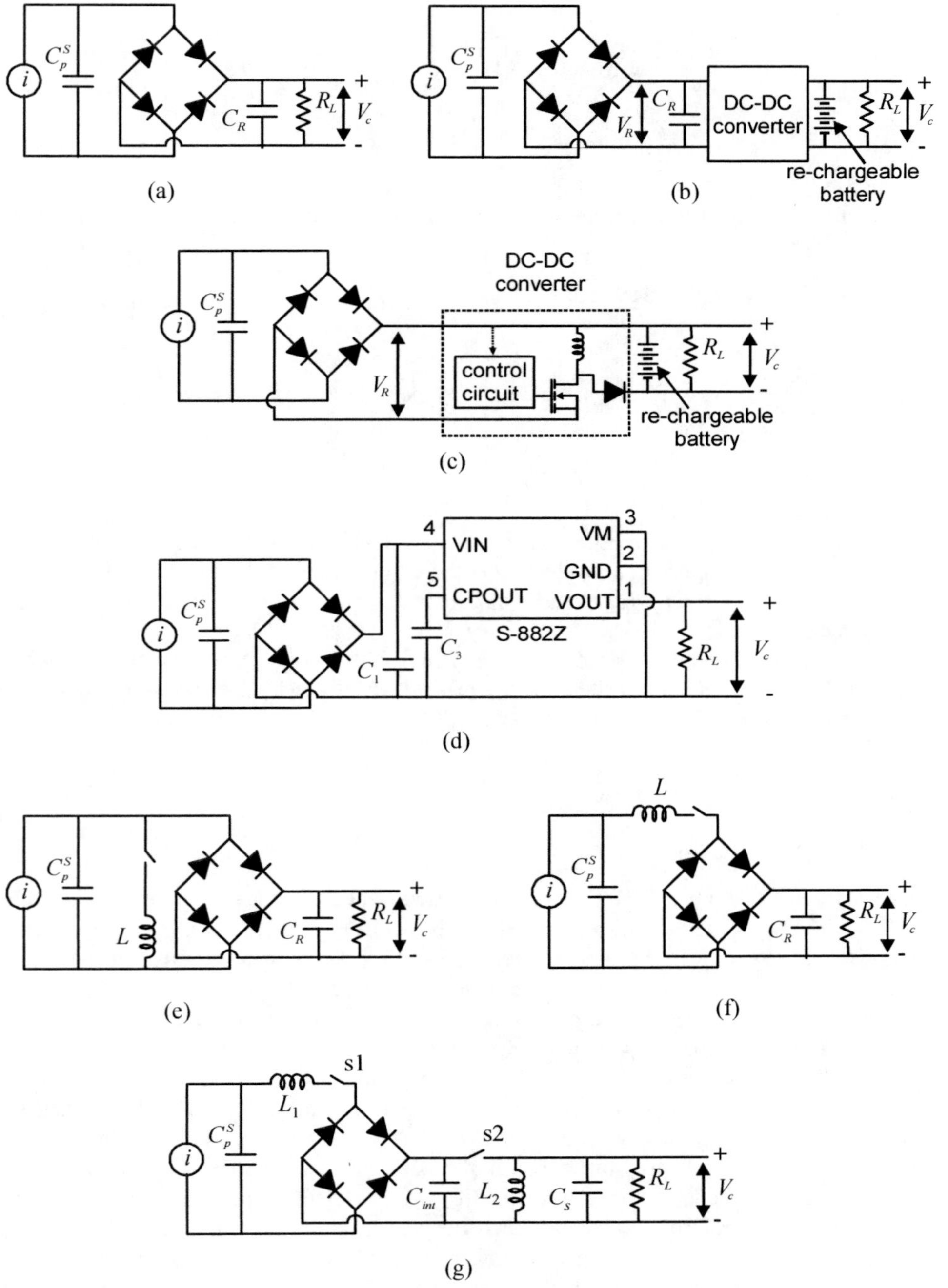

Figure 15.43 Circuits for energy harvesting from piezoelectric elements in compression or bending: (a) basic circuit; (b) simplified impedance matching circuit; (c) synchronous circuit to maximise power extraction (SECE, after Lefeuvre et al., 2005, 2007); (d) commercial implementation of an impedance matching and synchronous switching circuit (adapted from Volture data sheet, Mide Technology Corporation, 2010); (e) SSHI parallel circuit (after Lefeuvre et al., 2006); (f) SSHI series circuit (after Lefeuvre et al., 2006); (g) ESSH circuit (after Qiu, 2009).

$$f_P = K_{31}\bar{w} + \alpha V$$

$$Q = \alpha\bar{w} - C_p^S V \qquad (15.12.32a–c)$$

$$i = \alpha\frac{\partial\bar{w}}{\partial t} - C_p^S\frac{\mathrm{d}V}{\mathrm{d}t}$$

They then showed that the maximum possible power P_{max} at the optimum load impedance R_{Lopt} is given by:

$$P_{\max} = \frac{\alpha^2\omega\bar{w}^2}{2\pi C_p^S} \; ; \text{ and } \; R_{Lopt} = \frac{\pi}{2C_p^S\omega} \qquad (15.12.33)$$

where α is the force factor for the piezoelectric element (Newtons/Volt), given by:

$$\alpha = \frac{e_{31}S}{h} \qquad (15.12.34)$$

where S is the surface area of the piezoceramic element, h is its thickness, K_{11} is the short circuit stiffness of the piezoelectric element in the x- (or 1-) direction and C_p^S is the blocked capacitance of the piezoelectric element at constant (zero) strain (element attached to the non-excited structure).

If the resistive load is replaced by a rechargeable battery or storage capacitor, a constant load voltage V_R is considered instead of the load R_L. In this case, the maximum power and corresponding load voltage are (Lefeuvre et al., 2007):

$$P_{\max} = \frac{2\omega C_p^S V_{Ropt}^2}{\pi} \; ; \text{ and } \; V_{Ropt} = \frac{\alpha\bar{w}}{2C_p^S} \qquad (15.12.35)$$

The circuit in Figure 15.43(b) is a generalised impedance matching circuit showing a DC-DC converter. This idea was extended, as shown in Figure 15.43(c), by Lefeuvre et al. (2005, 2007) to include synchronous switching, which involves using a transistor to turn the DC-DC converter on when the rectified voltage V_R reaches an extremum (maximum or minimum) and to turn the DC-DC converter off when the rectified voltage V_R reaches zero. Thus, the charge developed by the piezoelectric element is removed entirely and transmitted to the load twice during each vibration cycle (for tonal vibration). The inductor value is chosen such that the time needed to extract the electric charge is much shorter than the vibration period. Lefeuvre et al. (2007) showed that the power extraction capability of the circuit in Figure 15.43(c) was 4 times that for the standard circuit with no switching.

A variation of the circuit in Figure 15.43(c) was implemented commercially in the form shown in Figure 15.43(d) which has been adapted from the Voltare data sheet (Mide Technology Corporation, 2010). This circuit is activated when the voltage produced at V_{IN} by the piezoelectric element via the small bypass capacitor C_1 (10 μF) exceeds the minimum start-up voltage of 1.8 V for the S-882Z DC-DC converter. When the voltage across the C_3 (100 μF) capacitor reaches 2.4 V, the DC-DC converter output is enabled, allowing the voltage to be applied to the load. When the voltage across the C_3 capacitor drops to 1.8 V, the load is disconnected by the S-882Z device, allowing the capacitor to be charged once again.

This circuit is designed primarily to power small sensors. A more complex commercial circuit that can be used for isolating the end application's impedance from the piezoelectric element and providing good impedance matching of the circuit to the piezoelectric element. The circuit is ideal for powering intermittently operating sensors or re-chargeable batteries and may be found in the Voltare data sheet (2010).

The SSHI parallel circuit shown in Figure 15.43(e) (Shu, 2009; Qiu et al., 2009) consists of the circuit shown in Figure 15.43(a) with the addition of an inductor and switch (connected together in series) in parallel with the piezoelectric element. The switch is turned on when the voltage on the piezoelectric element electrodes reaches an extremum (maximum or minimum). This results in an oscillation electrical circuit being established between the inductor and piezoelectric element capacitance and the voltage across the piezoelectric element is inverted. Following the voltage inversion process, which, by proper selection of the inductor value, takes a much shorter time than the period of mechanical vibration, the switch is opened. Immediately after the inversion process, the voltage across the piezoelectric element increases due to the induced vibration strain energy. The result of the voltage inversion process is to produce a voltage on the piezoelectric element that is closer to a square wave than a sine wave, with the result that the extremum voltages are experienced for longer times, resulting in more efficient energy harvesting. Energy losses in the inductor reduce the time for which the voltage is at an extremum, so it is important that a high quality inductor with low losses is used.

The SSHI series circuit shown in Figure 15.43(f) (Taylor et al., 2001; Lefeuvre et al., 2006) consists of the circuit shown in Figure 15.43(a) with the addition of an inductor L and switch in series with the piezoelectric element. The switching control is identical to that for the parallel circuit and the performance of both circuits in terms of power extraction capability is similar. The reader is referred to Lefeuvre et al. (2006) for details of typical component values and harvested power as a function of the electro-mechanical coupling coefficient, which is defined as the square root of (the electrical energy stored divided by the mechanical energy supplied).

The ESSH circuit shown in Figure 15.43(g) (Qiu et al., 2009) is similar to the series SSHI circuit but includes an additional capacitor, inductor and switch. When the piezoceramic voltage reaches either extremum, the switch s1 is closed and energy is transferred to the intermediate capacitor C_{int} through the inductor, L_1 and the rectifier. When the energy transfer is complete, switch s1 is opened. When the voltage across C_{int} increases above a preset level, switch s2 is closed and the charge is then transferred from C_{int} to the inductor L_2. When the voltage across C_{int} decreases below a preset level, switch, s2 is opened, the energy stored in L_2 is transferred to the smoothing capacitor C_R through the diode D5 and the process begins again.

Energy harvesters constructed using cantilever beams are usually directed at obtaining power from structures whose vibration amplitude is large at a single frequency and the resonance frequency of the harvester is tuned to match the excitation frequency. However, in many cases, the available excitation is broadband and it is preferable for the energy harvester to be able to use the available excitation frequency range. Although the cantilever beam energy harvesters will extract energy from the random noise at their resonance frequencies, only the first few resonance frequencies will contribute to the total power extraction (Lefeuvre et al., 2007). An alternative harvester configuration, illustrated in Figure 15.42(d), can be used that will make use of a wide bandwidth of random noise (Adhikari et al., 2009). The average normalised power for excitation of the harvester base by white noise of uniform acceleration power spectral density as a function of frequency is given by:

$$\tilde{P} = \frac{|V|^2}{R_1 [PSD]_{base\ acceleration}} = \frac{m \alpha \kappa^2 \pi}{(1 + 2\alpha\zeta)(\alpha\kappa^2 + 2\zeta)} \tag{15.12.36}$$

where

$$\alpha = \omega_n C_p^S R_1 ; \quad \beta = \omega_n L C_p^S = 1 \text{ at optimum}$$

$$\kappa = \frac{\theta^2}{K_{33} C_p^S} ; \quad \zeta = \frac{c_d}{2 m \omega_n} \tag{15.12.37a–d}$$

and K_{33} is the effective stiffness of the energy harvester in its axial direction, m its effective mass, c_d its damping and ω_n is its natural frequency, given by $\omega_n = \sqrt{K_{33}/m}$. The inductance L and resistance R_1 are defined in Figure 15.42(d). The quantity θ is the electromechanical coupling coefficient relating the voltage V generated by the harvester to the mechanical compression strain of the piezoelectric element, C_p^S is the capacitance of the piezoelectric element at constant (zero) strain. The quantity $[\tilde{P}SD]_{base\ acceleration}$ is the power spectral density of the acceleration of the base of the energy harvester. Equation (15.12.36) is only valid for the case of $\beta = 1$, which is its optimal value as for this value, Equation (15.12.37b) shows that the electrical and mechanical natural frequencies of the harvester are equal. For fixed ζ and κ, the value of α that corresponds to the maximum generated power is $\alpha = 1/\kappa$.

In addition to piezoelectric elements, it is also possible to use electromagnetic devices as energy harvesters (Sodano and Inman, 2005; Beeby and O'Donnell, 2009). In many applications, especially those for which the excitation frequency is a few tens of hertz or lower, electromagnetic energy harvesters are capable of producing more energy more efficiently than their piezoelectric counterparts. However, for most applications, piezoelectric elements will be more effective at charging batteries and are capable of providing good levels of current in random vibration environments as well as at resonance. In one reported test (Sodano and Inman, 2005), battery charging times at resonance were 70% of those achieved with a random noise signal of the same energy content.

An electromagnetic energy harvester is usually constructed so that the magnets move much more than the much lighter coil. This is achieved by attaching the magnets to the end of a flexible cantilever beam and fixing the coil to a solid base as illustrated schematically in Figure 15.44.

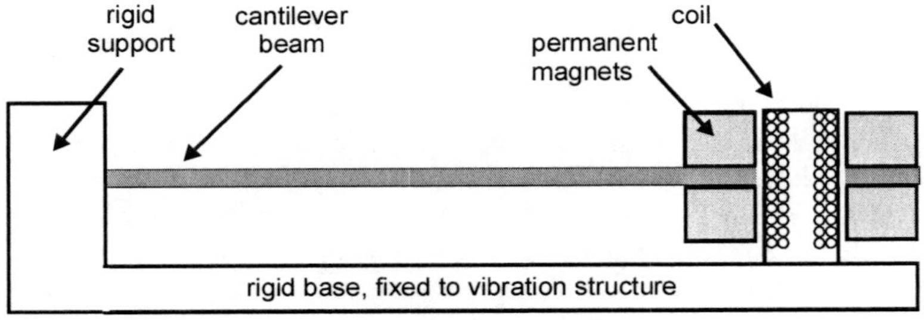

Figure 15.44 Electromagnetic energy harvester.

Electromagnetic energy harvesters are commercially available. One such device (see www.ferrosi.com) has a very high Q so that its power generating capability drops rapidly as the excitation frequency moves from the resonance frequency defined by the magnet mass on the cantilevered beam. That device can generate power levels of 10 mW for a vibration acceleration level of 0.1 g at 60 Hz. However, as the excitation frequency moves from the tuning frequency of 60 Hz, the power produced is reduced by 50% for a 1.5 Hz frequency shift. Another device (available from www.perpetuum.co.uk) can generate over 0.6 mW of power over a bandwidth of 2 Hz with a peak of 1 mW for a vibration level of 2.5×10^{-2} g. At 0.25 g, the device produces over 2 mW of power over a 20 Hz bandwidth with a peak of 12 mW. At 1 g, the device produces over 20 mW over a bandwidth of 20 Hz with a peak of 50 mW. The usual rating is a maximum of 4 mA at 5 V. The reason that the bandwidth increases as the acceleration level increases is that the electromagnetic damping increases as the relative motion between the coil and magnets increases. Devices are available for tuning frequencies of 50 Hz, 60 Hz, 100 Hz and 120 Hz, although custom designs for other frequencies are possible. Yet another device, developed by Hadas et al. (2007), occupies 45 cm^3 volume, resonates at 34.5 Hz and produces 3.5 mW from a 0.3 g acceleration.

Miniature energy harvesters are also available (www.lumedynetechnologies.com) and their manufacturer claims to be able to obtain useful power from vibration levels as low as 10^{-3} g.

The average mechanical power generated by an electromagnetic energy harvester with magnets of mass M and a relative vibration amplitude between the magnets and coil of $\bar{w}$ at frequency ω is given by:

$$P_{av} = \frac{M\bar{w}^2\omega^3}{2} \tag{15.12.38}$$

The maximum electrical power is extracted when the electromagnetic damping is made equal to the mechanical damping, c_d (N-s/m), by choosing an appropriate load resistance. The maximum possible electrical power is then,

$$P_{\max} = \frac{(M\bar{w}^2\omega^2)^2}{8c_d} \tag{15.12.39}$$

The load resistance required to maximise the generated electrical power for a coil of N turns and resistance R_c is then (Beeby and O'Donnell (2009):

$$R_L = \frac{N^2(\mathrm{d}\varphi/\mathrm{d}x)^2}{c_d} - R_c \tag{15.12.40}$$

where $\mathrm{d}\varphi/\mathrm{d}x$ is the magnetic flux linkage of the electrical circuit.

In a practical situation, it is desirable to maximise load power rather than generated power, and the expression for the load resistance that achieves this is (Beeby and O'Donnell (2009):

$$R_L = \frac{N^2(\mathrm{d}\varphi/\mathrm{d}x)^2}{c_d} + R_c \tag{15.12.41}$$

In this case, if the mechanical damping is much greater than the electrical damping (represented by the flux linkage gradient term), the optimum load resistance is actually the coil resistance.

15.13 ELECTRO-RHEOLOGICAL FLUIDS

Electro-rheological (ER) fluids are used in semi-active damping systems such as seismic oscillation dampers in building structures, motor vehicle suspensions and motor vehicle engine mounts. The advantage over passive dampers is that the amount of damping force applied can be varied as a function of time to minimise the vibration transmission and response of the system being damped. In vibration isolation systems, the introduction of passive damping results in a trade-off having to be made between the reduction in vibration transmission at the isolation system resonance frequency and the increase in vibration transmission at higher frequencies. The use of a semi-active damping system such as electro-rheological fluid eliminates the need to make this trade-off, and a reduction in vibration transmission at both the resonance frequency and at higher frequencies is possible. The use of semi-active damping systems for vehicle suspensions and engine mounts is discussed in Section 12.3. A cross-sectional view of a typical electro-rheological fluid damper configuration is shown in Figure 15.45 (Choi and Han, 2007), which is basically in the form of a valve. The ER valve comprises two concentric cylindrical electrodes which form an annular channel through which the ER fluid passes. The inner electrode usually has the high voltage applied to it while the outer cylinder is earthed. This generates an electric field perpendicular to the direction of the flow. Increasing the electric field increases the viscosity and hence the available damping force.

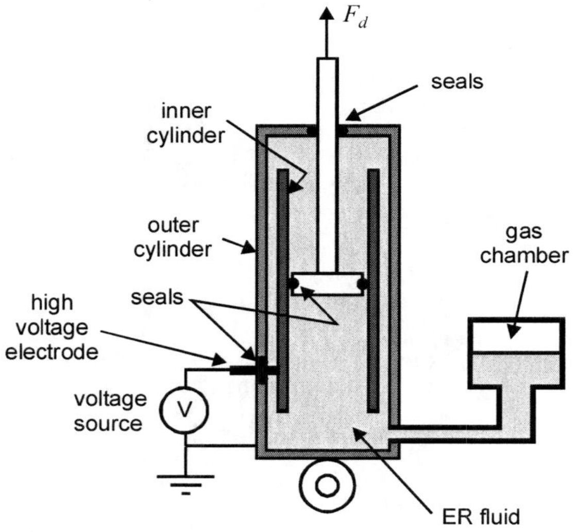

Figure 15.45 Typical electro-rheological fluid damper configuration (after Choi and Han, 2007).

The accumulator shown in Figure 15.45 is necessary to account for the rod being only on one side of the piston, causing the volume available to the electro-rheological fluid to change as the piston moves back and forth.

The dynamic viscosity η of an electro-rheological fluid is given by:

$$\eta = \eta_0 - C_v E^2 \tag{15.13.1}$$

where η_0 is the dynamic viscosity of the fluid/particle mix with no electric field applied, C_v is a constant dependent on the fluid and E is the electric field strength (kV/mm). The magnitude of the damping force that can be supplied by an electro-rheological damper is proportional to the shear stress τ that the fluid can withstand and this is given by (Choi et al., 2007):

$$\tau = \eta\dot{\gamma} + \tau_y \; ; \qquad \tau_y = \alpha E^{\beta} \tag{15.13.2a,b}$$

where α and β are constants dependent on the fluid, with β varying between 1.2 and 2.5 (Symans and constantinou, 1999) and $\dot{\gamma}$ is the shear strain rate (s^{-1}). More recently, Kugi et al. (2005) proposed the following relationship for the dynamic yield shear stress τ_y:

$$\tau_y = \alpha_2 E^2 + \alpha_3 E^3 \tag{15.13.3}$$

Even more recently, Zhao and Chen (2010) proposed the following relation:

$$\tau_y = \alpha_0 + \alpha_1 E + \alpha_2 E^2 \tag{15.13.4}$$

For laminar flow of the electro-rheological fluid in the electrode gap, the resistance to flow is given by (Choi and Han, 2007):

$$R_e = \frac{12\eta L}{bh^3} \tag{15.13.5}$$

where L is the length of the electrodes, h is the gap between the electrodes and b is the width, which for cylindrical electrodes, such as those shown in Figure 15.45, is given by, $b = \pi D$, where D is the diameter of the midpoint of the gap between the two electrodes.

The pressure drop due to the increase in yield stress is given by:

$$P_{ER} = \frac{2L\tau_y}{h} = \frac{2L\alpha E^{\beta}}{h} \tag{15.13.6a,b}$$

and the corresponding damping force at frequency, ω, is:

$$\begin{aligned} F_d &= \frac{S_r^2}{C_g} x_p + (S_p - S_r)^2 R_e \dot{x}_p + (S_p - S_r) P_{ER} \cdot sgn(\dot{x}_p) \\ &= k_e x_p + C_e \dot{x}_p + F_{ER} \end{aligned} \tag{15.13.7a,b}$$

where x_p and $\dot{x}_p$ are the frequency dependent piston excitation displacement and velocity respectively, S_p is the area of the piston, S_r is the cross-sectional area of the rod that is attached to the piston, C_g is the compliance of the gas in the accumulator and *sgn*( ) is the signum function and in this case is the sign of x_p. The first term in Equations (15.13.7a and b) would be zero for the case of a rod mounted on both sides of the piston, a configuration that is quite common. The last term in Equation (15.13.7a) is the damping force due to the electro-rheological effect and is rewritten in Equation (15.13.7b) as F_{ER}. Perhaps one term that is missing is the static friction force F_{st} which should be added to the LHS of the preceding equation. It is often relatively small, which is why it is ignored in some analyses. It can be included by adding the term $F_{st} \cdot sgn(\dot{x}_p)$ to the LHS of Equation (15.13.7a and b) in a similar way as done by Milecki (2001) in his analysis of the damping force of a magneto-rheological damper.

Using Equations (15.13.6b) and (15.13.7), the electric field required to produce the damping force F_{ER} can be written as:

$$E = \left[\frac{F_{ER}}{(S_p - S_r)} \frac{h}{2L\alpha} \right]^{1/\beta} \tag{15.13.8a,b}$$

The optimal damping force is determined by modelling the complete system to be controlled using optimal feedback analysis and control as described in Chapter 5. For an example of the application of the electro-rheological damper feedback control analysis to a seat, see Choi and Han (2005); for application to a vehicle suspension, see Choi et al. (2001); for application to a two-stage vibration isolation system, see Zhao and Chen (2010); and for application to seismic protection of building structures, see Ribakov and Gluck (1999).

The force generated by the damper must be measured and it should be noted that an electro-rheological damper can only generate a damping force in a direction opposite to the motion of the piston. In practice, the control law must be simplified to prevent 'chatter' and once the optimal control force is determined, the following control law is implemented. If the measured damping force is less than the optimal control force and the two forces have the same sign, the voltage applied to the electrodes is increased until electric field saturation is reached or until the damping force is equal to the optimal damping force. If the measured and optimal control forces are equal, then the existing voltage applied to the electrodes is maintained. For all other cases, the voltage applied to the electrodes is zero (see for example Dyke et al., 1996).

The most common type of electro-rheological fluid being investigated commercially is the class of dielectric oils (such as silicone oil) doped with semi-conductor particle suspensions with a typical concentration of particles representing a weight of 30% of the fluid. The particles can be ferroelectric (that is, they are characterised by an ability to exhibit spontaneous electric polarisation, which can be reversed by the application of an external electric field). Alternatively, the particles can be made from a conducting material coated with an insulator. The particles are typically less than 50 μm in diameter. On application of an electric field of sufficient strength, the particles form chains, which link across the electrodes, resulting in an apparent change in viscosity (or resistance to flow) by a factor of up to 10^5; that is, the fluid changes from a liquid consistency to that of a gel. Although it is possible to switch the fluids from their inactive to active state in 3 to 5 ms, it takes somewhat longer (up to 20 ms) to reverse the process, which limits to some extent the upper frequency limit of isolators and dampers constructed from these materials. Other problems are associated with the relatively high voltage requirements (2 to 10 kV) and separation of the fluid/solid particles when the fluid is in its inactive state for any extended period of time. This latter problem has largely been overcome with recently developed materials, which use either nano-size particles or match the density of the fluid with the density of the particles. One commercially available electro-rheological fluid consists of polymer particles suspended in a silicone oil of matched density 1.46 kg/m^3 (Smart Technology Limited data sheet). It has an operating temperature range of −20°C to 150°C and a viscosity of 110 mPa s at 30°C. Its dynamic viscosity is defined by:

$$\eta = \eta_0 - C_v E^2 \tag{15.13.9}$$

and its yield strength, τ_y is defined by:

$$\tau_y = \alpha E^2 = \alpha(V/h)^2 \tag{15.13.10}$$

where μ_0 is the fluid viscosity when no electric field is applied and in this case it is equal to 100 mPa s, C_v is a constant equal to 17.2, E is the electric field strength in kV/mm, V is the electric field voltage in kV (maximum ranges from 2 up to 5 kV) and h is the gap between electrodes (usually of the order of 1 mm). The constant $\alpha = 0.19$ kPa·mm^2/V^2 at 30°C and for any other temperature, it is given approximately by:

$$\alpha \approx 0.11 + 0.02T \tag{15.13.11}$$

where T is the temperature in °C.

Georgiades and Oyadiji (2003) showed that the introduction of electrodes made with square teeth (see Figure 15.46) doubled the damping force for the same electric field. However, these electrodes would be quite a bit more expensive to manufacture and assemble.

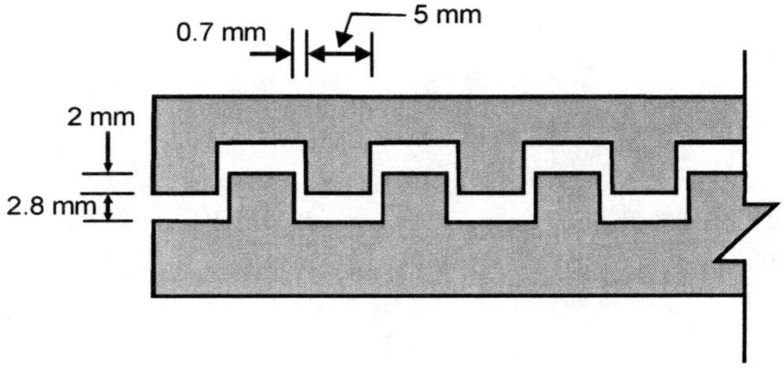

Figure 15.46 Electrodes with square teeth to enhance the electro-rheological effect (after Georgiades and Oyadiji, 2003).

In 2003, Wen et al. reported the discovery of a 'giant' electro-rheological effect exhibited by a suspension of nano-size barium titanyl oxide particles coated with urea , in silicone oil. When an electric field of 5.2 kV/mm was applied, the yield stress of the fluid reached 130 kPa, which is much greater than the normal 5 kPa exhibited by traditional electro-rheological fluids. They also noted that there was no observable sedimentation after the mixture had been left to stand for 2 weeks.

One disadvantage of electro-rheological dampers compared to magneto-rheological dampers is their relatively low force generation capability, of the order of 150 N for a damper with a 34 mm diameter piston, an electrode gap of 0.95 mm and an electrode voltage of 4 kV (Choi and Han, 2007). While this may be acceptable for vehicle suspensions and engine mounts, it is inadequate for damping large structures such as buildings subject to seismic loading. One way of greatly increasing the force capability of electro-rheological dampers is to use bypass ducts as shown in Figure 15.47 (Symans and Constantinou, 1999; Zhao and Chen, 2010) for a single duct. McMahon and Makris (1997) reported on the development of a 445 kN capacity electro-rheological damper that used multiple bypass ducts. The advantages associated with using multiple bypass ducts include less heating of the fluid and larger orifice sizes, resulting in a reduction of non-linear effects that are difficult to model and control.

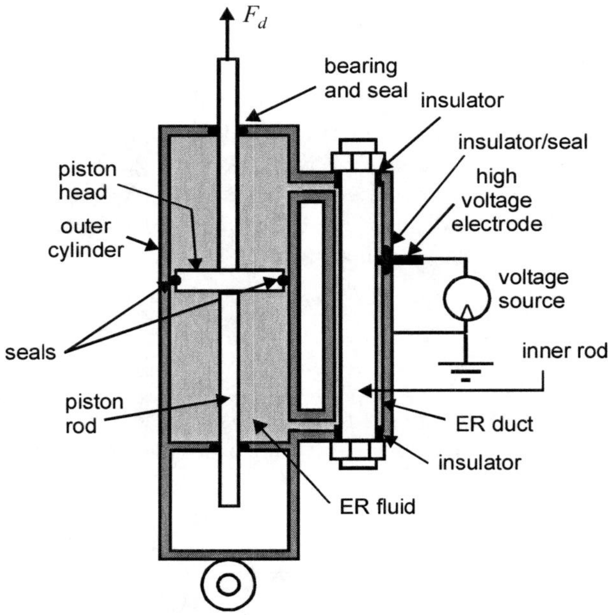

Figure 15.47 Electro-rheological damper showing a bypass configuration to greatly increase the damping force capability (after Symans and Constantinou, 1999; Zhao and Chen, 2010).

15.14 MAGNETO-RHEOLOGICAL FLUIDS

Magneto-rheological fluids are similar to electro-rheological fluids in that they are suspensions of particles in a carrier medium. The difference is that electro-rheological fluids become viscous when subjected to an electric field, whereas magneto-rheological fluids become viscous when subjected to a magnetic field. In addition, magneto-rheological fluids are stronger, more stable and generation of the necessary magnetic field requires much lower operating voltages and low power (10 to 20 watts typically), resulting in them being much easier to use. Typically, magneto-rheological fluids exhibit maximum yield strengths of 50 to 100 kPa for applied magnetic fields of 150 to 250 kAm^{-1}, compared to 5 to 10 kPa for electro-rheological fluids (Symans and Constantinou, 1999). The volume fraction of particles for magneto-rheological fluids ranges from 15% to 50% and the operating temperature ranges from −20 °C to 150 °C. A typical magneto-rheological fluid consists of iron particles of 3 to 10 μm diameter, suspended in a liquid such as mineral oil, synthetic oil or glycol. Additives are introduced to minimise sedimentation, which can occur when the device is left to stand unused for long periods of time. As with electro-rheological fluids, settling problems are also being reduced by using carrier fluids with similar densities to the particles being carried and by using particles that are much smaller in size (nano-particles). Unlike electro-rheological fluids, magneto-rheological fluids are insensitive to temperature and moisture contamination.

When not in the presence of a magnetic field, magneto-rheological fluids have a viscosity similar to engine oil but when subjected to a magnetic field, the viscosity changes to something like 'cold peanut butter' in 2 to 25 ms (Milecki, 2001; Truong and Ahn, 2010).

One disadvantage of magneto-rheological dampers is that they cannot be made as small as electro-rheological dampers because they require a bulky coil to generate the magnetic field rather than just a thin electrode. Perhaps the main disadvantage of magneto-rheological fluid dampers is that eventually there will be a residual magnetism that has to be removed by a secondary circuit when the dampers are switched off.

Truong and Ahn (2010) provided some technical data on a commercially available MR damper. The fluid had a density of about 3000 kgm^{-3}, an 81% by weight of solid particles, a viscosity of 92 mPa s at 40°C. The maximum current was 2A and the maximum voltage was 12 V. The response time to a 0 to 1A step input was less than 25 ms and the maximum damping force available was 4.4 kN.

Two examples of a magneto-rheological fluid damper are illustrated in Figure 15.48 (Unsal et al., 2008; Dyke et al., 1996). In one, the coil is mounted on the inside of the cylinder containing the magneto-rheological fluid and in the other, the coil is mounted on the outside. Although the latter configuration results in a larger diameter device, it has the great advantage that the coil does not heat up the fluid during operation and the coil is more accessible for maintenance. Neither design (a) nor (b) requires an accumulator as the shaft appears on both sides of the piston so the volume available for the MR fluid does not change as the damper operates. For the case of a shaft on only one side of the piston, an accumulator must be used as shown in Figure 15.45 for an electro-rheological damper.

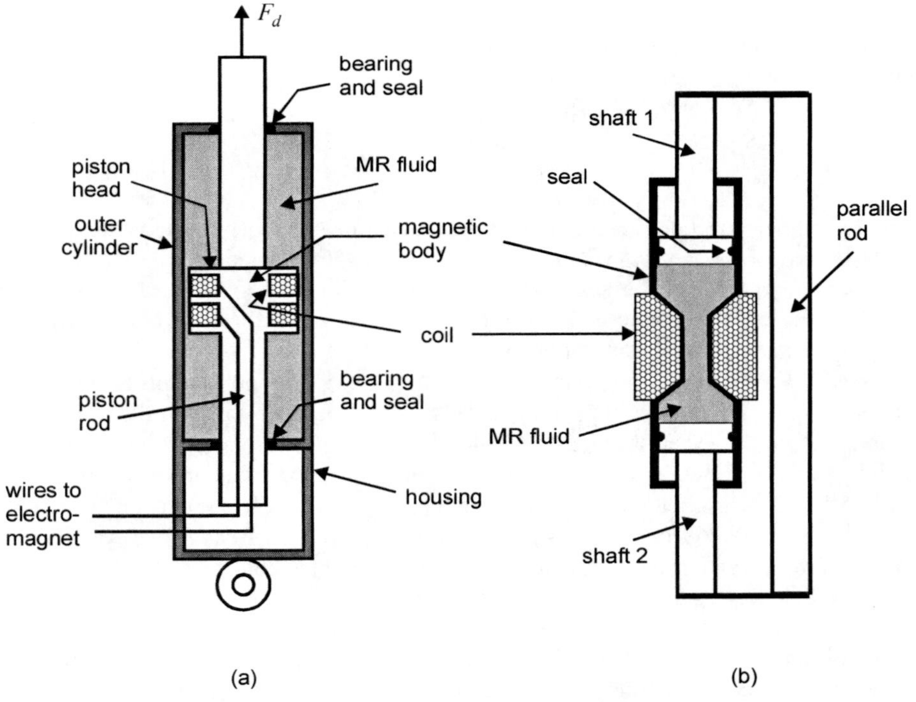

Figure 15.48 Two alternative configurations of magneto-rheological damper: (a) internal coil (after Unsal et al., 2008); (b) external coil (after Unsal et al., 2008).

In 2001, Milecki provided some conceptual designs for rotary dampers as illustrated in Figure 15.49. As magneto-rheological (MR) dampers (and ER dampers) rely on pushing fluid through a narrow passage, they have some damping effect even when no voltage is applied to the magnetising coil.

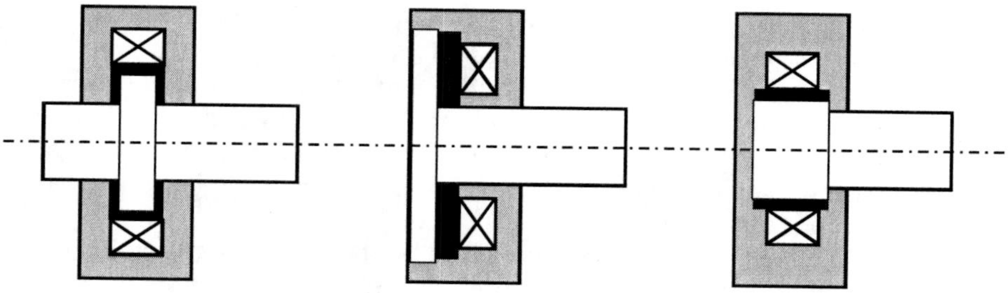

Figure 15.49 Conceptual designs for rotary magneto-rheological dampers (after Milecki, 2001).

The equation describing the damping force at frequency ω is similar to Equation (15.13.7) and is obtained from Milecki (2001) with the addition of a term to account for the effect of the accumulator

$$F_d = \frac{S_r^2}{C_g} x_p + F_{st} \cdot sgn(\dot{x}_p) + (S_p - S_r)^2 R_e \dot{x}_p + (S_p - S_r) P_{MR} \cdot sgn(\dot{x}_p) \tag{15.14.1}$$

where P_{MR} is the pressure drop due to the increase in yield stress and the other variables are defined in Section 15.13 and

$$(S_p - S_r) P_{MR} = \frac{b_H k_H}{(\mathrm{j}\omega T_m + 1)(\mathrm{j}\omega T_e + 1)} E \tag{15.14.2}$$

where b_H is the gain coefficient of the force generated by the magnetic field, k_H is a coefficient proportional to the coil resistance, T_m is the magnetic time constant, which describes the dynamics of the viscosity changes of the fluid when subjected to a magnetic field, T_e is the electrical time constant (coil inductance divided by the coil resistance at frequency ω), and E is the voltage applied to the coil.

As with ER dampers, the optimal voltage E to be applied to the coil to generate the required damping force is determined by undertaking a full feedback analysis of the system. Examples of such analyses have been provided by Milecki (2001) for a servo-motor coupled to a sliding mass, by Dyke et al. (1996) for controlling the seismic response of a building (see Figure 15.50), by Unsal et al. (2008) for a six-degrees-of-freedom support platform and by Dominguez et al. (2006) for controlling a truss structure. In 2010, Truong and Ahn presented an analysis for a self-sensing MR damper that can be used as both sensor and actuator in a semi-active damping system.

Once the optimal damping force has been determined (and this will vary rapidly as a function of time), the control law is implemented. As for the ER dampers, the control law is modified in practice to the following (Dyke et al., 1996) in order to prevent 'chatter'. If the measured damping force is less than the optimal control force and the two forces have the same sign, the voltage applied to the electrodes is increased until electric field saturation is reached or until the damping force is equal to the optimal damping force. If the measured and optimal control forces are equal, then the existing voltage applied to the electrodes is maintained. For all other cases, the voltage applied to the electrodes is zero.

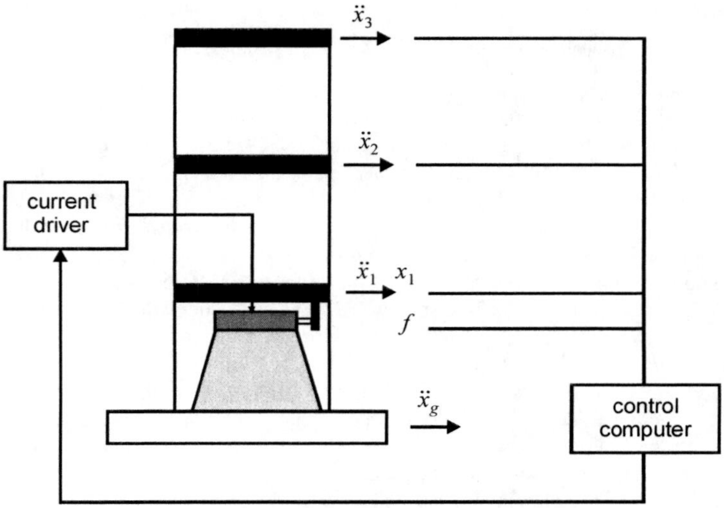

Figure 15.50 An example of the implementation of an MR damper to control the seismic response of a building (after Dyke et al., 1996).

REFERENCES

Adhikari, S., Friswell, M.I. and D J Inman, D.J. (2009). Piezoelectric energy harvesting from broadband random vibrations. *Smart Materials and Structures*, **18**, 1–7.

Ananthaganeshan, K.A., Brennan, M.J. and Elliott, S.J. (2002). High and low frequency instabilities in feedback control of a single-degree-of-freedom system. In *Proceedings of Active 2002*, Southampton, July, 2002, 1317–1326.

Austin, S.A. (1993). The vibration damping effect of an electrorheological fluid. *Journal of Vibration and Acoustics*, **115**, 136–140.

Baz, A.M. and Chen, T.-H. (1995). Active control of the lateral buckling of nitinol-reinforced composite beams. In *Proceedings of SPIE, Active Materials and Smart Structures*, edited by Anderson, G.L. and Lagoudas, **2427**, 30-48.

Baz, A.M. and Poh, S. (1988). Performance of an active control system with piezoelectric actuators. *Journal of Sound and Vibration*, **126**, 327–343.

Baz, A., Imam, K. and McCoy, J. (1990). Active vibration control of flexible beams using shape memory actuators. *Journal of Sound and Vibration*, **140**, 437–456.

Beeby, S.P. and O'Donnell, T. (2009). Electromagnetic energy harvesting, in S. Priya and D.J. Inman (Eds.), *Energy Harvesting Technologies*, Springer Science+Business Media, New York.

Benassi, L. and Elliott, S.J. (2004). Active vibration isolation using an inertial actuator with local displacement feedback control. *Journal of Sound and Vibration*, **278**, 705–724.

Benassi, L. and Elliott, S.J. (2005). Global control of a vibrating plate using a feedback-controlled inertial actuator. *Journal of Sound and Vibration*, **283**, 69–90.

Benjeddou, A. and Ranger, J.-A. (2006). Use of shunted shear-mode piezoceramics for structural vibration passive damping. *Computers and Structures*, **84**, 1415–1425.

Bies, D.A. and Hansen, C.H. (2009). *Engineering Noise Control*, 4th ed. Spon Press.

Butter, C.D. and Hocker, G.B. (1978). Fibre optics strain gauge. *Applied Optics*, **17**, 2867–2869.

Butler, J.L. (1988). *Application Manual for the Design of Etrema Terfenol-D Magnetostrictive Transducers*. Etrema Products Inc., 306 South 16th St, Ames, IA.

Buzan, F.T. and Hedrick, J.K. (1983). Lateral active pneumatic suspensions for rail vehicle control. In *Proceedings of the 1983 American Control Conference*. San Francisco, 263–269.

Cazzolato, B.S. and Hansen, C.H. (1999). Structural radiation mode sensing for active control of sound radiation into enclosed spaces. *Journal of the Acoustical Society of America*, **106**, 3732–3735.

Cho, D. and Hedrick, J.K. (1985). Pneumatic actuators for vehicle active suspension applications. *ASME Journal of Dynamic Systems Measurement and Control*, **107**, 67–72.

Choi, S.-B., Lee, H.K. and Chang, E.G. (2001). Field test results of a semi-active ER suspension system associated with skyhook controller. *Mechatronics*, **11**, 345–353.

Choi, S.-B. and Han, Y.-M. (2007). Vibration control of electrorheological seat suspension with human-body model using sliding mode control. *Journal of Sound and Vibration*, **303**, 391–404.

Clark, R.L. and Burke, S.E. (1996). Practical limitations in achieving shaped modal sensors with induced strain materials. *Journal of Vibration and Acoustics*, **118**, 668–675.

Clark, W.W., Robertshaw, H.H. and Warrington, T.J. (1990). A comparison of actuators for vibration control of the planar vibrations of a flexible cantilevered beam. *Journal of Intelligent Material Systems and Structures*, **1**, 289–308.

Clark, R.L., Fuller, C.R. and Wicks, A. (1991). Characterisation of multiple piezoelectric actuators for structural excitation. *Journal of the Acoustical Society of America*, **90**, 346–357.

Columbia Research Laboratories. *Theory and Application of Linear Variable Differential Transformers*. Columbia Technical Publications, Woodlyn, PA.

Cox, D.E. and Lindner, D.K. (1991). Active control for vibration suppression in a flexible beam using a modal domain optical fibre sensor. *ASME Journal of Vibration and Acoustics*, **113**, 369–382.

Crawley, E.F. and de Luis, J. (1987). Use of piezoelectric actuators as elements of intelligent structures. *AIAA Journal*, **25**, 1373–1385.

Crawley, E.F. and Anderson, E.H. (1990). Detailed models of piezoceramic actuation of beams. *Journal of Intelligent Material Systems and Structures*, **1**, 4–25.

Crolla, D.A. (1988). Theoretical comparisons of various active suspension systems in terms of performance and power requirements. In *Proceedings of the Institution of Mechanical Engineers International Conference on Advanced Suspensions*. London, 1–9.

Dimitriadis, E.K., Fuller, C.R. and Rogers, C.A. (1991). Piezoelectric actuators for distributed vibration excitation of thin plates. *ASME Journal of Vibration and Acoustics*, **113**, 100–107.

Dominguez, A., Sedaghati, R. and Stiharu, I. (2006). Semi-active vibration control of adaptive structures using magnetorheological dampers. *AIAA Journal*, **44**, 1563–1571.

Donoso, A. and Bellido, J.C. (2009). Systemmatic design of shaped piezoelectric modal sensors/actuators for rectangular plates by optimizing the polarization profile. *Structural and Multidisciplinary Optimization*, **38**, 347–356.

Dosch, J.J., Inman, D.J. and Garcia, E. (1992). A self-sensing piezoelectric actuator for collocated control. *Journal of Intelligent Material Systems and Structures*, **3**, 166–185.

Dyke, S.J., Spencer Jr., B.F., Sain, M.K. and Carlson, J.D. (1996). Modeling and control of magnetorheological dampers for seismic response reduction. *Smart Materials and Structures*, **5**, 565–575.

Edge Technologies (1990). RA101 *Actuator Handbook*.

Fannin, C.A. (1997). *Design of an Analog Adaptive Piezoelectric Sensoriactuator*. PhD thesis, Virginia Polytechnic Institute and State University, Blacksburg, VA.

Fein, O.M. and Gaul, M. (2004). On the application of shunted piezoelectric material to enhance structural damping of a plate. *Journal of Intelligent Material Systems and Structures*, **15**, 737–743.

Fleming, A.J. and Moheimani, S.O.R. (2005). Control orientated synthesis of high-performance piezoelectric shunt impedances for structural vibration control. *IEEE Transactions on Control Systems Technology*, **13**, 98–112.

Forward, R.L. (1979). Electronic damping of vibrations in optical structures. *Journal of Applied Optics* **18**, 690–697.

Forward, R.L. and Swigert, C.J. (1981). Electronic damping of orthogonal bending modes in a cylindrical mast. *Journal of Spacecraft and Rockets*, **18**, 5–17.

Georgiades, G. and Oyadiji, S.O. (1996). Effects of electrode geometry on the performance of electrorheological fluid valves. *Journal of Intelligent Material Systems and Structures*, **14**, 105–111.

Hagood, N.W. and Von Flotow, A. (1991). Damping of structural vibrations with piezoelectric materials and passive electrical networks. *Journal of Sound and Vibration*, **146**, 243–268.

Hiller, M.W, Bryant, M.D. and Umega, J. (1989). Attenuation and transformation of vibration through active control of magnetostrictive terfenol. *Journal of Sound and Vibration*, **134**, 507–519.

Hollkamp, J. (1994). Multimodal passive vibration suppression with piezoelectric materials and resonant shunts. *Journal of Intelligent Material Systems and Structures*, **5**, 49–57.

Jones, L., Garcia, E. and Waites, H. (1994). Self-sensing control as applied to a PZT stack actuator used as a micropositioner. *Smart Materials and Structures*, **3**, 147–156.

Kim, B.Y., Blake, J.H., Huong, S.Y. and Shaw, H.J. (1987). Use of highly elliptical core fibres in two-mode fibre devices. *Optical Letters*, **12**, 739.

Kim, J., Ryou, J.-K. and Kim, S.J. (2002). Optimal gain distribution for two-dimensional modal transducer and its implementation using multi-layered PVDF films. *Journal of Sound and Vibration*, **251**, 395–408.

Kim, S.J. and Jones, J.D. (1991). Optimal design of piezoactuators for active noise and vibration control. *AIAA Journal*, **29**, 2047–2053.

King, S.P. (1988). The minimisation of helicopter vibration through blade design and active control. *Aeronautical Journal*, **92**, 247–263.

Kugi, A., Holzmann, K. and Kemmetmüller, W. (2006). Active and semi-active control of electrorheological fluid devices. In *Proceedings of the IUTAM Symposium on Vibration Control of Nonlinear Mechanisms and Structures*, 203–212.

Layton, M.R. and Bucaro, J.A. (1979). Optical fibre acoustic sensor utilising mode-mode interference. *Applied Optics*, **18**, 666.

Lee, C.K. and Moon, F.C. (1990a). Modal sensors/actuators. *Transactions of ASME: Journal of Applied Mechanics*, **57**, 434–441.

Lee, C.K. and Moon F.C. (1990b). Laminated piezoelectric plates for torsion and bending sensors and actuators. *Journal of the Acoustical Society of America*, **85**, 2432–2439.

Lefeuvre, E., Badel, A., Richard, C. and Guyomar, D. (2005). Piezoelectric energy harvesting device optimization by synchronous electric charge extraction. *Journal of Intelligent Material Systems and Structures*, **16**, 865–876.

Lefeuvre, E., Badel, A., Richard, C., Petit, L. and Guyomar, D. (2006). A comparison between several vibration-powered piezoelectric generators for standalone systems. *Sensors and Actuators A*, **126**, 405–416.

Lefeuvre, E., Badel, A., Richard, C. and Guyomar, D. (2007). Energy harvesting using piezoelectric materials: case of random vibrations. *Journal of Electroceramics*, **19**, 349–355.

McMahon, S. and Makris N. (1997). Large-scale ER-damper for seismic protection of bridges. In *Proceedings of Structures Congress XV*, Portland, OR, pp. 1451–1455.

MIDE Technology Corporation (2010). Volture Piezoelectric Energy Harvesters Data Sheet.

Milecki, A. (2001). Investigation and control of magneto-rheological dampers. *International Journal of machine Tools and Manufacture*, **41**, 379–391.

Moffett, M.B., Clark, A.E., Wun–Fogle, M., Linberg, J., Teter, J.P. and McLaughlin, E.A. (1991a). Characterisation of terfenol-D for magnetostrictive transducers. *Journal of the Acoustical Society of America*, **89**, 1448–1455.

Moffett, M.B., Powers, J.M. and Clark, A.E. (1991b). Comparison of terfenol-D and PZT-4 power limitations. *Journal of the Acoustical Society of America*, **90**, 1184–1185.

Moheimani, S.O.R. (2003). A survey of recent innovations in vibration damping and control using shunted piezoelectric transducers. *IEEE Transactions on Control Systems Technology*, **11**, 482–494.

Moore, J.H. (1988). Linear variable inductance position transducer for suspension systems. In *Proceedings of the Institution of Mechanical Engineers International Conference on Advanced Suspensions*. London, 75–82.

Morgan, R.A. and Wang, K.W. (2002). Active-passive piezoelectric absorbers for systems under multiple non-stationary harmonic excitations. *Journal of Sound and Vibration*, **255**, 685–700.

Morishita, S. and Mitsui, J. (1992). Controllable squeeze film damper (an application of electrorheological fluid). *Journal of Vibration and Acoustics*, **114**, 354–357.

Okada, Y., Matsuda, K. and Hashitani, H. (1995). Self-sensing active vibration control using the moving-coil-type actuator. *Journal of Vibration and Acoustics*, **117**, 411–415.

Paulitsch, C., Gardonio, P. and Elliott, S.J. (2006). Active vibration control using an inertial actuator with internal damping. *Journal of the Acoustical Society of America*, **119**, 2131–2140.

Paulitsch, C., Gardonio, P., Elliott, S.J., Sas, P. and Boonen, R. (2004). Design of a lightweight electrodynamic inertial actuator with integrated velocity sensor for active vibration control of a thin lightly-damped panel. In *Proceedings of ISMA 2004*, 239–253.

Politansky, H. and Pilkey, W.D. (1989). Suboptimal feedback vibration control of a beam with a proof-mass actuator. *Journal of Guidance and Control*, **12**, 691–697.

Priya, S. and Inman, D.J., Eds. (2009). *Energy Harvesting technologies*. Springer, New York.

Qiu, J.H., Li, H.L. and Shen, H. (2009). Energy harvesting and vibration control using piezoelectric elements and a nonlinear approach. In *Proceedings of the IEEE International Symposium on the Applications of Ferroelectrics*, Xian, China, August 23–27, pp. 1–8.

Renno, J.M., Daqaq, M.F. and Inman, D.J. (2009). On the optimal energy harvesting from a vibration source using a piezoelectric stack, in S. Priya and D.J. Inman (Eds.), *Energy Harvesting Technologies*, Springer Science+Business Media, New York.

Ribakov, Y. and Gluck, J. (1999). Active control of mdof structures with supplemental electrorheological fluid dampers. *Earthquake Engineering and Structural Dynamics*, **28**, 143–156.

Rogers, C.A. (1990). Active vibration and structural acoustic control of shape memory alloy hybrid composites: experimental results. *Journal of the Acoustical Society of America*, **88**, 2803–2811.

Rogers, C.A., Liang, C. and Fuller, C.R. (1991). Modelling of shape memory alloy hybrid composites for structural acoustic control. *Journal of the Acoustical Society of America*, **89**, 210–220.

Saunders, W.R., Robertshaw, H.H. and Rogers, C.A. (1991). Structural acoustic control of a shape memory alloy composite beam. *Journal of Intelligent Material Systems and Structures*, **2**, 508–527.

Serridge, M. and Torben, R.L. (1987). *Piezoelectric Accelerometer and Vibration Preamplifier Handbook*. Brüel and Kjaer, Copenhagen.

Shu, Y.C. (2009). Performance evaluation of vibration-based piezoelectric energy scavengers, in S. Priya and D.J. Inman (Eds.), *Energy Harvesting Technologies*, Springer Science+Business Media, New York.

Simmers, G.E. Jr., Hodgkins, J.R., Mascarenas, D.D., Park, G. and Sohn, H. (2004). Improved piezoelectric self-sensing actuation. *Journal of Intelligent Material Systems and Structures*, **15**, 941–953.

Smits, J.G., Dalke, S.I. and Cooney, T.K. (1991). The constituent equations of piezoelectric bimorphs. *Sensors and Actuators A*, **28**, 41–61.

Sodano, H.A., Park, G. and Inman, D.J. (2004). An investigation into the performance of macro-fiber composites for sensing and structural vibration applications. *Mechanical Systems and Signal Processing*, **18**, 683–697.

Sodano, H.A. and Inman, D.J. (2005). Comparison of piezoelectric energy harvesting devices for recharging batteries. *Journal of Intelligent Material Systems and Structures*, **16**, 799–807.

Sproston, J.L. and Stanway, R. (1992). Electrorheological fluids in vibration isolation. In *Proceedings of Actuator '92, the 3rd International Conference on New Actuators*. VDI/VDE–Technogiezentrum Informationstechnik GmbH, Bremen, Germany, 116–117.

Stayner, R.M. (1988). Suspensions for agricultural vehicles. In *Proceedings of the Institution of Mechanical Engineers International Conference on Advanced Suspensions*. London, 133–137.

Steffen, V. Jr. and Inman, D.J. (1999). Optimal design of piezoelectric materials for vibration damping in mechanical Systems. *Journal of Intelligent Material Systems and Structures*, **10**, 945–955.

Steffen, V. Jr., Rade, D.A. and Inman, D.J. (2000). Using passive techniques for vibration damping in mechanical systems. *Journal of the Brazilian Society of Mechanical Sciences*, **22**, 411–421.

Sun, D., Tong, L. and Wang, D. (2002). Modal actuator/sensor by modulating thickness of piezoelectric layers for smart plates. *AIAA Journal*, **40**, 1676–1679.

Symans, M.D. and Constantinou, M.C. (1999). Semi-active control systems for seismic protection of structures: a state-of-the-art review. *Engineering Structures*, **21**, 469–487.

Tanaka, N., Snyder, S.D. and Hansen, C.H. (1996). Distributed parameter modal filtering using smart sensors. *Journal of Vibration and Acoustics*, **118**, 630–640.

Tani, J., Qiu, J. and Miura, H. (1995). Vibration control of a cylindrical shell using piezoelectric actuators. *Journal of Intelligent Material Systems and Structures*, **6**, 380–388.

Taylor, G.W., Burns, J.R., Kammann, S.M., Powers, W.B. and Welsh, T.R. (2001). The energy harvesting eel: a small subsurface ocean/river power generator, *IEEE Journal of Oceanic Engineering*, **26**, 539–547.

Timoshenko, S.P. and Woinowsky–Krieger, S. (1959). *Theory of plates and shells*, 2nd ed. McGraw-Hill, New York, Chapters 1, 2 and 4.

Truong, D.Q. and Ahn, K.K. (2010). Identification and application of black-box model for a self-sensing damping system using a magneto-rheological fluid damper. *Sensors and Actuators A: Physical*, **161**, 305–321.

Unsal, M., Niezrecki, C. and Crane, C.D. III (2008). Multi-axis semi-active vibration control using magnetorheological technology. *Journal of Intelligent Material Systems and Structures*, **19**, 1463–1470.

Vippermann, J.S. and Clark, R.L. (1996). Implementation of an adaptive piezoelectric sensoriactuator. *AIAA Journal*, **34**, 2102–2109.

Wallace, C.E. (1972). Radiation resistance of a rectangular panel. *Journal of the Acoustical Society of America*, **51**, 946–952.

Wang, A.B. and Wang, B.R. (2010). Piezoelectric self-sensing actuator for vibration suppression based on time-sharing method. In *Proceedings of the 8th IEEE International Conference on Control and Automation,* Xiamen, China, June 9–11, pp. 1403–1408.

Weiss, K.D., Coulter, J.P. and Carlson, J.D. (1992). Electrorheological materials and their usage in intelligent material systems and structures. Part 1. Mechanisms, formulations and properties. In *Recent Advances in Sensory and Adaptive Materials and their Applications.* Techonomic Publishing Co., Lancaster, PA.

Wen, W., Huangi, X., Yang, S., Lu, K. and Sheng, P. (2003). The giant electrorheological effect in suspensions of nanoparticles. *Nature Materials*, **2**, 727–730.

Wolfe, P.T. and Jolly, M.R. (1990). Velocity Transducer for Vehicle Suspension System. US Patent 4 979 573.

Yu, X., Rajamani, R., Stelson, K.A. and Cui, T. (2006). Active control of sound transmission through windows with carbon nanotube based actuators and moving noise source identification. In *Proceedings of the 2006 American Control Conference*, Minneapolis, MN, June 14–16, pp. 1227–1232.

Zhang, W., Qiu, J. and Tani, J. (2004). Robust vibration control of a plate using self-sensing actuators and piezoelectric patches. *Journal of Intelligent Material Systems and Structures*, **15**, 923–931.

Zhao, C. and Chen, D. (2010). A two-stage floating raft isolation system featuring electrorheological damper with semi-active static output feedback variable structure control. *Journal of Intelligent Material Systems and Structures*, **21**, 387–399.

APPENDIX A

Brief Review of Some Results of Linear Algebra

The purpose of this appendix is to provide a brief review of many of the particular results of linear algebra used in this book. For more extensive treatments, the reader should consult any of the standard textbooks on linear algebra, such as Bellman (1970), Noble (1969) and Noble and Daniel (1977). There are also a number of specialised software packages which deal explicitly with linear algebra manipulations. These packages are useful for solving the matrix equations encountered in active noise and vibration control research.

A.1 MATRICES AND VECTORS

An $(m \times n)$ matrix is a collection of mn numbers (complex or real), a_{ij} $(i=1,2,...,m, j=1,2,...,n)$, written in an array of m row and n columns:

$$\boldsymbol{A} = \begin{bmatrix} a_{11} & a_{12} & \cdots & a_{1n} \\ a_{21} & a_{22} & \cdots & a_{2n} \\ \vdots & \vdots & \vdots & \vdots \\ a_{m1} & a_{m2} & \cdots & a_{mn} \end{bmatrix} \tag{A1}$$

The term a_{ij} appears in the ith row and jth column of the array. If the number of rows is equal to the number of columns, the matrix is said to be square.

An m vector, also referred to as an $(m \times 1)$ vector or column m vector, is a matrix with 1 column and m rows:

$$\boldsymbol{a} = \begin{bmatrix} a_1 \\ a_2 \\ \vdots \\ a_m \end{bmatrix} \tag{A2}$$

Throughout this book, matrices are denoted by bold capital letters, such as $\boldsymbol{A}$, while vectors are denoted by bold lower case letters, such as $\boldsymbol{a}$.

A.2 ADDITION, SUBTRACTION AND MULTIPLICATION BY A SCALAR

If two matrices have the same number of rows and columns, they can be added or subtracted. When adding matrices, the individual corresponding terms are added. For example, if

$$\boldsymbol{A} = \begin{bmatrix} a_{11} & a_{12} & \cdots & a_{1n} \\ a_{21} & a_{22} & \cdots & a_{2n} \\ \vdots & \vdots & \vdots & \vdots \\ a_{m1} & a_{m2} & \cdots & a_{mn} \end{bmatrix} \tag{A3}$$

and

$$\boldsymbol{B} = \begin{bmatrix} b_{11} & b_{12} & \cdots & b_{1n} \\ b_{21} & b_{22} & \cdots & b_{2n} \\ \vdots & \vdots & \vdots & \vdots \\ b_{m1} & b_{m2} & \cdots & b_{mn} \end{bmatrix} \tag{A4}$$

then,

$$\boldsymbol{A} + \boldsymbol{B} = \begin{bmatrix} a_{11}+b_{11} & a_{12}+b_{12} & \cdots & a_{1n}+b_{1n} \\ a_{21}+b_{21} & a_{22}+b_{22} & \cdots & a_{2n}+b_{2n} \\ \vdots & \vdots & \vdots & \vdots \\ a_{m1}+b_{m1} & a_{m2}+b_{m2} & \cdots & a_{mn}+b_{mn} \end{bmatrix} \tag{A5}$$

When subtracting matrices, the individual terms are subtracted. Note that matrix addition is commutative, as:

$$\boldsymbol{A} + \boldsymbol{B} = \boldsymbol{B} + \boldsymbol{A} \tag{A6}$$

It is also associative, as:

$$(\boldsymbol{A} + \boldsymbol{B}) + \boldsymbol{C} = \boldsymbol{A} + (\boldsymbol{B} + \boldsymbol{C}) \tag{A7}$$

Matrices can also be multiplied by a scalar. Here, the individual terms are each multiplied by a scalar. For example, if k is a scalar:

$$k\boldsymbol{A} = \begin{bmatrix} ka_{11} & ka_{12} & \cdots & ka_{1n} \\ ka_{21} & ka_{22} & \cdots & ka_{2n} \\ \vdots & \vdots & \vdots & \vdots \\ ka_{m1} & ka_{m2} & \cdots & ka_{mn} \end{bmatrix} \tag{A8}$$

A.3 MULTIPLICATION OF MATRICES

Two matrices $\boldsymbol{A}$ and $\boldsymbol{B}$ can be multiplied together to form the product $\boldsymbol{AB}$ if the number of columns in $\boldsymbol{A}$ is equal to the number of rows in $\boldsymbol{B}$. If, for example, $\boldsymbol{A}$ is an $(m \times p)$ matrix and

B is a $(p \times n)$ matrix, then the product **AB** is defined by:

$$\boldsymbol{C} = \boldsymbol{AB} \tag{A9}$$

where **C** is an $(m \times n)$ matrix, the terms of which are defined by:

$$c_{ij} = \sum_{k=1}^{p} a_{ik} b_{kj} \tag{A10}$$

Matrix multiplication is associative, with the product of three (or more) matrices defined by:

$$\boldsymbol{ABC} = (\boldsymbol{AB})\boldsymbol{C} = \boldsymbol{A}(\boldsymbol{BC}) \tag{A11}$$

Matrix multiplication is also distributive, where

$$\boldsymbol{A}(\boldsymbol{B} + \boldsymbol{C}) = \boldsymbol{AB} + \boldsymbol{AC} \tag{A12}$$

However, matrix multiplication is not commutative, as, in general:

$$\boldsymbol{AB} \neq \boldsymbol{BA} \tag{A13}$$

In fact, while the product **AB** may be formed, it may not be possible to form the product **BA**.

The identity matrix **I** is defined as the $(p \times p)$ matrix with all principal diagonal elements equal to 1, and all other terms equal to zero:

$$\boldsymbol{I} = \begin{bmatrix} 1 & 0 & \cdots & 0 \\ 0 & 1 & \cdots & 0 \\ & & \ddots & \\ 0 & 0 & \cdots & 1 \end{bmatrix} \tag{A14}$$

For any $(m \times p)$ matrix **A**, the identity matrix has the property:

$$\boldsymbol{AI} = \boldsymbol{A} \tag{A15}$$

Similarly, if the identity matrix is $(m \times m)$:

$$\boldsymbol{IA} = \boldsymbol{A} \tag{A16}$$

A.4 TRANSPOSITION

If a matrix is transposed, the rows and columns are interchanged. For example, the transpose of the $(m \times n)$ matrix **A**, denoted by $\boldsymbol{A}^{\mathrm{T}}$, is defined as the $(n \times m)$ matrix **B**:

$$\boldsymbol{A}^{\mathrm{T}} = \boldsymbol{B} \tag{A17}$$

where

$$b_{ij} = a_{ji} \tag{A18}$$

The transpose of a matrix product is defined by:

$$(\boldsymbol{AB})^{\mathrm{T}} = \boldsymbol{B}^{\mathrm{T}}\boldsymbol{A}^{\mathrm{T}} \tag{A19}$$

This result can be extended to products of more than two matrices, such as:

$$(\boldsymbol{ABC})^{\mathrm{T}} = \boldsymbol{C}^{\mathrm{T}}\boldsymbol{B}^{\mathrm{T}}\boldsymbol{A}^{\mathrm{T}} \tag{A20}$$

If:

$$\boldsymbol{A} = \boldsymbol{A}^{\mathrm{T}} \tag{A21}$$

then the matrix $\boldsymbol{A}$ is said to be symmetric.

The Hermitian transpose of a matrix is defined as the complex conjugate of the transposed matrix (when taking the complex conjugate of a matrix, each term in the matrix is conjugated). Therefore, the Hermitian transpose of the $(m \times n)$ matrix $\boldsymbol{A}$, denoted by $\boldsymbol{A}^{\mathrm{H}}$, is defined as the $(n \times m)$ matrix $\boldsymbol{B}$:

$$\boldsymbol{A}^{\mathrm{H}} = \boldsymbol{B} \tag{A22}$$

where

$$b_{ij} = a_{ji}^{*} \tag{A23}$$

If $\boldsymbol{A} = \boldsymbol{A}^{\mathrm{H}}$, then $\boldsymbol{A}$ is said to be a Hermitian matrix.

A.5 DETERMINANTS

The determinant of the (2×2) matrix $\boldsymbol{A}$, denoted $|\boldsymbol{A}|$, is defined as:

$$|\boldsymbol{A}| = \begin{vmatrix} a_{11} & a_{12} \\ a_{21} & a_{22} \end{vmatrix} = a_{11}a_{22} - a_{12}a_{21} \tag{A24}$$

The minor $\boldsymbol{M}_{ij}$ of the element a_{ij} of the square matrix $\boldsymbol{A}$ is the determinant of the matrix formed by deleting the ith row and jth column from $\boldsymbol{A}$. For example, if $\boldsymbol{A}$ is a (3×3) matrix:

$$\boldsymbol{A} = \begin{bmatrix} a_{11} & a_{12} & a_{13} \\ a_{21} & a_{22} & a_{23} \\ a_{31} & a_{32} & a_{33} \end{bmatrix} \tag{A25}$$

the minor $\boldsymbol{M}_{11}$ is found by taking the determinant of $\boldsymbol{A}$ with the first column and first row of numbers deleted:

$$\boldsymbol{M}_{11} = \begin{vmatrix} a_{22} & a_{23} \\ a_{32} & a_{33} \end{vmatrix} \tag{A26}$$

The cofactor $\boldsymbol{C}_{ij}$ of the element $\boldsymbol{a}_{ij}$ of the matrix $\boldsymbol{A}$ is defined by:

$$\boldsymbol{C}_{ij} = (-1)^{i+j}\boldsymbol{M}_{ij} \tag{A27}$$

The determinant of a square matrix of arbitrary size is equal to the sum of the products of the elements and their cofactors along any column or row. For example, the determinant of the (3 × 3) matrix $\boldsymbol{A}$ above can be found by adding the products of the elements and their cofactors along the first row:

$$|\boldsymbol{A}| = a_{11}\boldsymbol{C}_{11} + a_{12}\boldsymbol{C}_{12} + a_{13}\boldsymbol{C}_{13} \tag{A28}$$

Therefore, the determinant of a large square matrix can be broken up into a problem of calculating the determinants of a number of smaller square matrices.

If two matrices $\boldsymbol{A}$ and $\boldsymbol{B}$ are square, then,

$$|\boldsymbol{AB}| = |\boldsymbol{A}|\,|\boldsymbol{B}| \tag{A29}$$

A matrix is said to be singular if its determinant is equal to zero.

A.6 MATRIX INVERSES

The inverse $\boldsymbol{A}^{-1}$ of the matrix $\boldsymbol{A}$ is defined by:

$$\boldsymbol{A}\boldsymbol{A}^{-1} = \boldsymbol{A}^{-1}\boldsymbol{A} = \boldsymbol{I} \tag{A30}$$

The matrix $\boldsymbol{A}$ must be square and, as will be seen, be non-singular for the inverse to be defined.

The inverse of a matrix $\boldsymbol{A}$ can be derived by first calculating the adjoint $\hat{\boldsymbol{A}}$ of the matrix. The adjoint $\hat{\boldsymbol{A}}$ is defined as the transpose of the matrix of cofactors of $\boldsymbol{A}$:

$$\hat{\boldsymbol{A}} = \begin{bmatrix} \boldsymbol{C}_{11} & \boldsymbol{C}_{12} & \cdots & \boldsymbol{C}_{1m} \\ \boldsymbol{C}_{21} & \boldsymbol{C}_{22} & \cdots & \boldsymbol{C}_{2m} \\ \vdots & \vdots & \vdots & \vdots \\ \boldsymbol{C}_{m1} & \boldsymbol{C}_{m2} & \cdots & \boldsymbol{C}_{mm} \end{bmatrix} \tag{A31}$$

The inverse $\boldsymbol{A}^{-1}$ of the matrix $\boldsymbol{A}$ is equal to the adjoint of $\boldsymbol{A}$ multiplied by the reciprocal of the determinant of $\boldsymbol{A}$:

$$\boldsymbol{A}^{-1} = \frac{1}{|\boldsymbol{A}|}\hat{\boldsymbol{A}} \tag{A32}$$

Note that if the matrix $\boldsymbol{A}$ is singular, such that the determinant is zero, the inverse is not defined.

While the definition given in Equation (A32) is correct, using it to calculate a matrix inverse is inefficient for all but the smallest matrices (as the order of operations increases with the size m of the matrix by $m!$). There are a number of algorithms which require on the order of m^3 operations to compute the inverse of an arbitrary square matrix (outlined in many of the standard texts and in numerical methods books such as Press et al. (1986)). For matrices of special form, such as Toeplitz matrices, which are symmetric matrices in which the elements along any diagonal are equal (often encountered in adaptive signal processing work), the order of operations can be further reduced (to m^2 operations for Toeplitz matrices).

If the matrix $\boldsymbol{A}$ is singular, it is possible to define the Moore–Penrose pseudo-inverse $\boldsymbol{A}'$ such that $\boldsymbol{A}'\boldsymbol{A}$ acts as the identity matrix on as large a set of vectors as possible. $\boldsymbol{A}'$ has the properties:

$$(\boldsymbol{A}')' = \boldsymbol{A}, \quad \boldsymbol{A}'\boldsymbol{A}\boldsymbol{A}' = \boldsymbol{A}', \quad \boldsymbol{A}\boldsymbol{A}'\boldsymbol{A} = \boldsymbol{A} \tag{A33}$$

If $\boldsymbol{A}$ is non-singular, then $\boldsymbol{A}^{-1} = \boldsymbol{A}'$.

A.7 RANK OF A MATRIX

The rank of the $(m \times n)$ matrix $\boldsymbol{A}$ is the maximum number of linearly independent rows of $\boldsymbol{A}$ and the maximum number of linearly independent columns of $\boldsymbol{A}$. Alternatively, the rank of $\boldsymbol{A}$ is a positive integer r such that some $(r \times r)$ submatrix of $\boldsymbol{A}$, formed by deleting $(m-r)$ rows and $(n-r)$ columns, is non-singular, whereas no $((r+1) \times (r+1))$ submatrix is non-singular. If rank $\boldsymbol{A}$ is equal to the number of columns or the number of rows of $\boldsymbol{A}$, then $\boldsymbol{A}$ is said to have full rank.

A.8 POSITIVE AND NON-NEGATIVE DEFINITE MATRICES

A matrix $\boldsymbol{A}$ is said to be positive definite if the quantity $\boldsymbol{x}^{\mathrm{H}}\boldsymbol{A}\boldsymbol{x}$ is positive for all non-zero vectors $\boldsymbol{x}$; if the quantity is simply non-negative, then $\boldsymbol{A}$ is said to be non-negative definite.

For $\boldsymbol{A}$ to be positive definite, all of the leading minors must be positive; that is,

$$a_{11} > 0, \quad \begin{vmatrix} a_{11} & a_{12} \\ a_{21} & a_{22} \end{vmatrix} > 0, \quad \begin{vmatrix} a_{11} & a_{12} & a_{13} \\ a_{21} & a_{22} & a_{23} \\ a_{31} & a_{32} & a_{33} \end{vmatrix} > 0, \; \ldots \text{ etc} \tag{A34}$$

For $\boldsymbol{A}$ to be non-negative definite, all of the leading minors must be non-negative.

A.9 EIGENVALUES AND EIGENVECTORS

Let $\boldsymbol{A}$ be a (square) $(n \times n)$ matrix. The polynomial $|\lambda\boldsymbol{I}-\boldsymbol{A}|$ is referred to as the characteristic equation of $\boldsymbol{A}$. The solutions to the characteristic equation are the eigenvalues of $\boldsymbol{A}$. If λ_i is an eigenvalue of $\boldsymbol{A}$, then there exists at least one vector $\boldsymbol{q}_i$ which satisfies the relationship:

$$Aq_i = \lambda_i q_i \tag{A35}$$

The vector q_i is an eigenvector of A. If the eigenvalue λ_i is not repeated, then the eigenvector q_i is unique. If an eigenvector λ_i is real, then the entries in the associated eigenvector q_i are real; if λ_i is complex, then so too are the entries in q_i.

The eigenvalues of a Hermitian matrix are all real, and if the matrix is also positive definite, the eigenvalues are also all positive. If a matrix is symmetric, then the eigenvalues are also all real. Further, it is true that:

$$|A| = \prod_{i=1}^{n} \lambda_i \tag{A36}$$

If A is singular, then there is at least one eigenvalue equal to zero.

A.10 ORTHOGONALITY

If a square matrix A has the property $A^H A = AA^H = I$, then the matrix A is said to be orthogonal. The eigenvalues of A then have a magnitude of unity. If q_i is an eigenvector associated with λ_i, and q_j is an eigenvector associated with λ_j, and if $\lambda_i \neq \lambda_j$ and $q_i^H q_j = 0$, then the vectors q_i and q_j are said to be orthogonal.

The eigenvectors of a Hermitian matrix are all orthogonal. Further, it is common to normalise the eigenvectors such that $q_i^H q_i = 1$, in which case the eigenvectors are said to be orthonormal. A set of orthonormal eigenvectors can be expressed as columns of a unitary matrix Q:

$$Q = (q_1, q_2, \cdots, q_n) \tag{A37}$$

which means that:

$$Q^H Q = QQ^H = I \tag{A38}$$

The set of equations which define the eigenvectors, expressed for a single eigenvector in (A35), can now be written in matrix form as:

$$AQ = Q\Lambda \tag{A39}$$

where Λ is the diagonal matrix of eigenvalues:

$$\Lambda = \begin{bmatrix} \lambda_1 & 0 & \cdots & 0 \\ 0 & \lambda_2 & \cdots & 0 \\ & & \ddots & \\ 0 & 0 & \cdots & \lambda_n \end{bmatrix} \tag{A40}$$

Post-multiplying both sides of Equation (A39) by Q^H yields:

$$A = Q\Lambda Q^H \tag{A41}$$

or

$$Q^{H}AQ = \Lambda \tag{A42}$$

Equations (A.41) and (A.42) define the orthonormal decomposition of A, where A is re-expressed in terms of its eigenvectors and eigenvalues.

A.11 VECTOR NORMS

The norm of a vector A, expressed as $\|x\|$, is the length or size of the vector x. The most common norm is the Euclidean norm, defined for the vector $x = (x_1, x_2, \ldots, x_n)$ as:

$$\|x\| = \left(\sum_{i=1}^{n} x_i^2 \right)^{1/2} \tag{A43}$$

Three properties of vector norms are:

1. $\|x\| \geq 0$ for all x, where the norm is equal to zero only if $x = \mathbf{0}$.
2. $\|ax\| = |a| \, \|x\|$ for any scalar a and all x.
3. $\|x+y\| \leq \|x\| + \|y\|$ for all x and y.

REFERENCES

Press, W.H., Flannery, B.P., Tenkolsky, S.A. and Vettering, W.T. (1986) *Numerical Recipes: The Art of Scientific Computing*. Cambridge University Press, Cambridge.

Index